Lecture Notes in Computer Science 3644

Commenced Publication in 1973
Founding and Former Series Editors:
Gerhard Goos, Juris Hartmanis, and Jan van Leeuwen

Editorial Board

David Hutchison
  Lancaster University, UK
Takeo Kanade
  Carnegie Mellon University, Pittsburgh, PA, USA
Josef Kittler
  University of Surrey, Guildford, UK
Jon M. Kleinberg
  Cornell University, Ithaca, NY, USA
Friedemann Mattern
  ETH Zurich, Switzerland
John C. Mitchell
  Stanford University, CA, USA
Moni Naor
  Weizmann Institute of Science, Rehovot, Israel
Oscar Nierstrasz
  University of Bern, Switzerland
C. Pandu Rangan
  Indian Institute of Technology, Madras, India
Bernhard Steffen
  University of Dortmund, Germany
Madhu Sudan
  Massachusetts Institute of Technology, MA, USA
Demetri Terzopoulos
  New York University, NY, USA
Doug Tygar
  University of California, Berkeley, CA, USA
Moshe Y. Vardi
  Rice University, Houston, TX, USA
Gerhard Weikum
  Max-Planck Institute of Computer Science, Saarbruecken, Germany

Lecture Notes in Computer Science

Commenced Publication in 1973
Founding and Former Series Editors:
Gerhard Goos, Juris Hartmanis, and Jan van Leeuwen

Editorial Board

David Hutchison
Lancaster University, UK

Takeo Kanade
Carnegie Mellon University, Pittsburgh, PA, USA

Josef Kittler
University of Surrey, Guildford, UK

Jon M. Kleinberg
Cornell University, Ithaca, NY, USA

Friedemann Mattern
ETH Zurich, Switzerland

John C. Mitchell
Stanford University, CA, USA

Moni Naor
Weizmann Institute of Science, Rehovot, Israel

Oscar Nierstrasz
University of Bern, Switzerland

C. Pandu Rangan
Indian Institute of Technology, Madras, India

Bernhard Steffen
University of Dortmund, Germany

Madhu Sudan
Massachusetts Institute of Technology, MA, USA

Demetri Terzopoulos
New York University, NY, USA

Doug Tygar
University of California, Berkeley, CA, USA

Moshe Y. Vardi
Rice University, Houston, TX, USA

Gerhard Weikum
Max-Planck Institute of Computer Science, Saarbruecken, Germany

De-Shuang Huang   Xiao-Ping Zhang
Guang-Bin Huang (Eds.)

Advances in Intelligent Computing

International Conference on Intelligent Computing, ICIC 2005
Hefei, China, August 23-26, 2005
Proceedings, Part I

Volume Editors

De-Shuang Huang
Chinese Academy of Sciences, Institute of Intelligent Machines
P.O. Box 1130, Hefei, Anhui, 230031, China
E-mail: dshuang@iim.ac.cn

Xiao-Ping Zhang
Ryerson University
Department of Electrical and Computer Engineering
350 Victoria Street, Toronto, Ontario, Canada, M5B 2K3
E-mail: xzhang@ee.ryerson.ca

Guang-Bin Huang
Nanyang Technological University
School of Electrical and Electronic Engineering
Nanyang Avenue, Singapore 639798
E-mail: egbhuang@ntu.edu.sg

Library of Congress Control Number: 2005930888

CR Subject Classification (1998): F.1, F.2, I.2, G.2, I.4, I.5, J.3, J.4, J.1

ISSN   0302-9743
ISBN-10   3-540-28226-2 Springer Berlin Heidelberg New York
ISBN-13   978-3-540-28226-6 Springer Berlin Heidelberg New York

This work is subject to copyright. All rights are reserved, whether the whole or part of the material is concerned, specifically the rights of translation, reprinting, re-use of illustrations, recitation, broadcasting, reproduction on microfilms or in any other way, and storage in data banks. Duplication of this publication or parts thereof is permitted only under the provisions of the German Copyright Law of September 9, 1965, in its current version, and permission for use must always be obtained from Springer. Violations are liable to prosecution under the German Copyright Law.

Springer is a part of Springer Science+Business Media

springeronline.com

© Springer-Verlag Berlin Heidelberg 2005
Printed in Germany

Typesetting: Camera-ready by author, data conversion by Scientific Publishing Services, Chennai, India
Printed on acid-free paper   SPIN: 11538059   06/3142   5 4 3 2 1 0

Preface

The International Conference on Intelligent Computing (ICIC) was set up as an annual forum dedicated to emerging and challenging topics in the various aspects of advances in computational intelligence fields, such as artificial intelligence, machine learning, bioinformatics, and computational biology, etc. The goal of this conference was to bring together researchers from academia and industry as well as practitioners to share ideas, problems and solutions related to the multifaceted aspects of intelligent computing.

This book constitutes the proceedings of the International Conference on Intelligent Computing (ICIC 2005), held in Hefei, Anhui, China, during August 23–26, 2005. ICIC 2005 received over 2000 submissions from authors in 39 countries and regions. Based on rigorous peer reviews, the Program Committee selected 563 high-quality papers for presentation at ICIC 2005; of these, 215 papers were published in this book organized into 9 categories, and the other 348 papers were published in five international journals.

The organizers of ICIC 2005 made great efforts to ensure the success of this conference. We here thank the members of the ICIC 2005 Advisory Committee for their guidance and advice, the members of the Program Committee and the referees for reviewing the papers, and the members of the Publication Committee for checking and compiling the papers. We would also like to thank the publisher, Springer, for their support in publishing the proceedings in the Lecture Notes in Computer Science Series. Particularly, we would like to thank all the authors for contributing their papers. Without their high-quality papers, the success of the conference would not have been possible. Finally, we are especially grateful to the IEEE Computational Intelligence Society and the IEEE Hong Kong Computational Intelligence Chapter as well as their National Science Foundation of China for their sponsorship.

15 June 2005

De-Shuang Huang
Institute of Intelligent Machines, Chinese Academy of Sciences, China

Xiao-Ping Zhang
Ryerson University, Canada

Guang-Bin Huang
Nanyang Technological University, Singapore

ICIC 2005 Organization

ICIC 2005 was organized and sponsored by the Institute of Intelligent Machines in cooperation with the University of Sciences & Technology, Hefei, and technically cosponsored by the IEEE Computational Intelligence Society and the Hong Kong Computational Intelligence Chapter.

Committees

General Chairs De-Shuang Huang, China
Jiming Liu, Hong Kong, China
Seong-Whan Lee, Korea

International Advisory Committee

Songde Ma, China
Marios M. Polycarpou, USA
Shoujue Wang, China
Yuanyan Tang, Hong Kong, China
Guanrong Chen, Hong Kong, China
Guoliang Chen, China
DeLiang Wang, USA

Horace H.S. Ip, HK
Paul Werbos, USA
Qingshi Zhu, China
Yunyu Shi, China
Yue Wang, China
Youshou Wu, China
Nanning Zheng, China
Fuchu He, China
Okyay Knynak, Turkey

Roger H. Lang, USA
Zheng Bao, China
Yanda Li, China
Ruwei Dai, China
Deyi Li, China
Yixin Zhong, China
Erke Mao, China

Program Committee Chairs Yiu-Ming Cheung, Hong Kong, China
Xiao-Ping Zhang, Canada

Organizing Committee Chairs Tao Mei, China
Yunjian Ge, China

Publication Chair Hujun Yin, UK
Publicity Chair Anders I. Morch, Norway
Registration Chair Haiyan Liu, China
International Liaison Chair Prashan Premaratne, Australia
Exhibition Chair Guang-Zheng Zhang, China
Finance Chair Hongwei Liu, China

Program Committee

Yuanyan Tang (Hong Kong, China), Jonathan H. Manton (Australia), Yong Xu (UK), Kang Li (UK), Virginie F. Ruiz (UK), Jing Zhang (China), Roberto Tagliaferri (Italy), Stanislaw Osowski (Poland), John Q. Gan (UK), Guang-Bin Huang (Singapore), Donald C. Wunsch (USA), Seiichi Ozawa (Japan), Zheru Chi (Hong Kong, China), Yanwei Chen (Japan), Masaharu Adachi (Japan), Huaguang Zhang (China), Choh Man Teng (USA), Simon X. Yang (Canada), Jufu Feng (China), Chenggang Zhang (China), Ling Guan (Canada), Ping Guo (China), Xianhua Dai (China), Jun Hu (China), Jinde Cao (China), Hye-Ran Byun (Korea), Key-Sun Choi (Korea), Dewen Hu (China), Sulin Pang (China), Shitong Wang (China), Yunping Zhu (China), Xuegong Zhang (China), Jie Tian (China), Daqi Zhu (China), Nikola Kasabov (New Zealand), Hai Huyen Dam (Australia), S. Purushothaman (India), Zhu Liu (USA), Kazunori Sugahara (Japan), Chuanlin Zhang (China), Hailin Liu (China), Hwan-Gye Cho (Korea), Hang Joon Kim (Korea), Xiaoguang Zhao (China), Zhi-Cheng Ji (China), Zhigang Zeng (China), Yongjun Ma (China), Ping Ao (USA), Yaoqi Zhou (USA), Ivica Kopriva (USA) Derong Liu (USA), Clement Leung (Australia), Abdelmalek Zidouri (Saudi Arabia), Jiwu Huang (China), Jun Zhang (China), Gary Geunbae Lee (Korea), Jae-Ho Lee (Korea), Byoung-Tak Zhang (Korea), Wen-Bo Zhao (China), Dong Hwa Kim (Korea), Yi Shen (China), C.H. Zhang (Japan), Zhongsheng Wang (China), YuMin Zhang (China), HanLin He (China), QiHong Chen (China), Y. Shi (Japan), Zhen Liu (Japan), K. Uchimura (Japan), L. Yun (Japan), ChangSheng Xu (Singapore), Yong Dong Wu (Singapore), Bin Zhu (China), LiYan Zhang (China), Dianhui Wang (Australia), Kezhi Mao (Singapore), Saeed Hashemi (Canada), Weiqi Chen (China), Bjonar Tessem (Norway), Xiyuan Chen (China), Christian Ritz (Australia), Bin Tang (Canada), Mehdi Shafiei (Canada), Jiangtao Xi (Australia), Andrea Soltoggio (UK), Maximino Salazar-Lechuga (UK), Benjamin S. Aribisala (UK), Xiaoli Li (UK), Jin Li (UK), Jinping Li (China), Sancho Salcedo-Sanz (Spain), Hisashi Handa (Japan)

Organizing Committee

Xianda Wu, Jinhuai Liu, Zengfu Wang, Huanqing Feng, Zonghai Chen, Gang Wu, Shuang Cong, Bing-Yu Sun, Hai-Tao Fang, Xing-Ming Zhao, Zhan-Li Sun, Ji-Xiang Du, Hong-Qiang Wang, Fei Han, Chun-Hou Zheng, Li Shang, Zhong-Hua Quan, Bing Wang, Peng Chen, Jun Zhang, Zhong-Qiu Zhao, Wei Jia

Secretary   Hai-Mei Zhang, China

Table of Contents – Part I

Perceptual and Pattern Recognition

A Novel Approach to Ocular Image Enhancement with Diffusion and Parallel AOS Algorithm
Lanfeng Yan, Janjun Ma, Wei Wang, Qing Liu, Qiuyong Zhou 1

An Iterative Hybrid Method for Image Interpolation
Yan Tian, Caifang Zhang, Fuyuan Peng, Sheng Zheng 10

Research on Reliability Evaluation of Series Systems with Optimization Algorithm
Weijin Jiang, Yusheng Xu, Yuhui Xu 20

Image Registration Based on Pseudo-Polar FFT and Analytical Fourier-Mellin Transform
Xiaoxin Guo, Zhiwen Xu, Yinan Lu, Zhanhui Liu, Yunjie Pang 30

Text Detection in Images Based on Color Texture Features
Chunmei Liu, Chunheng Wang, Ruwei Dai 40

Aligning and Segmenting Signatures at Their Crucial Points Through DTW
Zhong-Hua Quan, Hong-wei Ji 49

A SAR Image Despeckling Method Based on Dual Tree Complex Wavelet Transform
Xi-li Wang, Li-cheng Jiao 59

Rotation Registration of Medical Images Based on Image Symmetry
Xuan Yang, Jihong Pei, Weixin Xie 68

Target Tracking Under Occlusion by Combining Integral-Intensity-Matching with Multi-block-voting
Faliang Chang, Li Ma, Yizheng Qiao 77

Recognition of Leaf Images Based on Shape Features Using a Hypersphere Classifier
Xiao-Feng Wang, Ji-Xiang Du, Guo-Jun Zhang 87

A Robust Registration and Detection Method for Color Seal Verification
  Liang Cai, Li Mei 97

Enhanced Performance Metrics for Blind Image Restoration
  Prashan Premaratne, Farzad Safaei 107

Neighborhood Preserving Projections (NPP): A Novel Linear
Dimension Reduction Method
  Yanwei Pang, Lei Zhang, Zhengkai Liu, Nenghai Yu, Houqiang Li .. 117

An Encoded Mini-grid Structured Light Pattern for Dynamic Scenes
  Qingcang Yu, Xiaojun Jia, Jian Tao, Yun Zhao 126

Linear Predicted Hexagonal Search Algorithm with Moments
  Yunsong Wu, Graham Megson..................................... 136

A Moving Detection Algorithm Based on Space-Time Background
Difference
  Mei Xiao, Lei Zhang, Chong-Zhao Han 146

Bit-Precision Method for Low Complex Lossless Image Coding
  Jong Woo Won, Hyun Soo Ahn, Wook Joong Kim, Euee S. Jang ... 155

Texture Feature-Based Image Classification Using Wavelet Package
Transform
  Yue Zhang, Xing-Jian He, Jun-Hua Han.......................... 165

Authorization Based on Palmprint
  Xiao-yong Wei, Dan Xu, Guo-wu Yuan........................... 174

A Novel Dimension Conversion for the Quantization of SEW in
Wideband WI Speech Coding
  Kyung Jin Byun, Ik Soo Eo, He Bum Jeong, Minsoo Hahn 184

Adaptive Preprocessing Scheme Using Rank for Lossless Indexed Image
Compression
  Kang-Soo You, Tae-Yoon Park, Euee S. Jang, Hoon-Sung Kwak 194

An Effective Approach to Chin Contour Extraction
  Junyan Wang, Guangda Su, Xinggang Lin 203

Robust Face Recognition Across Lighting Variations Using Synthesized
Exemplars
  Sang-Woong Lee, Song-Hyang Moon, Seong-Whan Lee 213

Low-Dimensional Facial Image Representation Using FLD and MDS
  Jongmoo Choi, Juneho Yi 223

Object Tracking with Probabilistic Hausdorff Distance Matching
  Sang-Cheol Park, Seong-Whan Lee 233

Power Op-Amp Based Active Filter Design with Self Adjustable Gain
Control by Neural Networks for EMI Noise Problem
  *Kayhan Gulez, Mehmet Uzunoglu, Omer Caglar Onar,
  Bulent Vural* 243

Leaf Recognition Based on the Combination of Wavelet Transform and
Gaussian Interpolation
  Xiao Gu, Ji-Xiang Du, Xiao-Feng Wang 253

Fast Training of SVM via Morphological Clustering for Color Image
Segmentation
  Yi Fang, Chen Pan, Li Liu, Lei Fang 263

Similarity Measurement for Off-Line Signature Verification
  Xinge You, Bin Fang, Zhenyu He, Yuanyan Tang 272

Shape Matching and Recognition Base on Genetic Algorithm and
Application to Plant Species Identification
  Ji-Xiang Du, Xiao-Feng Wang, Xiao Gu 282

Detection of Hiding in the LSB of DCT Coefficients
  Mingqiao Wu, Zhongliang Zhu, Shiyao Jin 291

Tracking People in Video Camera Images Using Neural Networks
  Yongtae Do 301

How Face Pose Influence the Performance of SVM-Based Face and
Fingerprint Authentication System
  Chunhong Jiang, Guangda Su 310

A VQ-Based Blind Super-Resolution Algorithm
  Jianping Qiao, Ju Liu, Guoxia Sun 320

Sequential Stratified Sampling Belief Propagation for Multiple Targets
Tracking
  Jianru Xue, Nanning Zheng, Xiaopin Zhong 330

Retrieving Digital Artifacts from Digital Libraries Semantically
  Jun Ma, YingNan Yi, Tian Tian, Yuejun Li 340

An Extended System Method for Consistent Fundamental Matrix
Estimation
  Huixiang Zhong, Yueping Feng, Yunjie Pang 350

Informatics Theories and Applications

Derivations of Error Bound on Recording Traffic Time Series with
Long-Range Dependence
  Ming Li 360

A Matrix Algorithm for Mining Association Rules
  Yubo Yuan, Tingzhu Huang 370

A Hybrid Algorithm Based on PSO and Simulated Annealing and Its
Applications for Partner Selection in Virtual Enterprise
  *Fuqing Zhao, Qiuyu Zhang, Dongmei Yu, Xuhui Chen,
  Yahong Yang* 380

A Sequential Niching Technique for Particle Swarm Optimization
  Jun Zhang, Jing-Ru Zhang, Kang Li 390

An Optimization Model for Outlier Detection in Categorical Data
  Zengyou He, Shengchun Deng, Xiaofei Xu 400

Development and Test of an Artificial-Immune-Abnormal-Trading-
Detection System for Financial Markets
  Vincent C.S. Lee, Xingjian Yang 410

Adaptive Parameter Selection of Quantum-Behaved Particle Swarm
Optimization on Global Level
  Wenbo Xu, Jun Sun 420

New Method for Intrusion Features Mining in IDS
  Wu Liu, Jian-Ping Wu, Hai-Xin Duan, Xing Li 429

The Precision Improvement in Document Retrieval Using Ontology
Based Relevance Feedback
  Soo-Yeon Lim, Won-Joo Lee 438

Adaptive Filtering Based on the Wavelet Transform for FOG on the
Moving Base
  Xiyuan Chen 447

Text Similarity Computing Based on Standard Deviation
  Tao Liu, Jun Guo 456

Evolving Insight into High-Dimensional Data
  Yiqing Tu, Gang Li, Honghua Dai 465

SVM Based Automatic User Profile Construction for Personalized Search
  Rui Song, Enhong Chen, Min Zhao 475

Taxonomy Building and Machine Learning Based Automatic Classification for Knowledge-Oriented Chinese Questions
  Yunhua Hu, Qinghua Zheng, Huixian Bai, Xia Sun, Haifeng Dang .. 485

Learning TAN from Incomplete Data
  Fengzhan Tian, Zhihai Wang, Jian Yu, Houkuan Huang 495

The General Method of Improving Smooth Degree of Data Series
  Qiumei Chen, Wenzhan Dai 505

A Fault Tolerant Distributed Routing Algorithm Based on Combinatorial Ant Systems
  Jose Aguilar, Miguel Labrador 514

Improvement of HITS for Topic-Specific Web Crawler
  Xiaojun Zong, Yi Shen, Xiaoxin Liao 524

Geometrical Profile Optimization of Elliptical Flexure Hinge Using a Modified Particle Swarm Algorithm
  Guimin Chen, Jianyuan Jia, Qi Han 533

Classification of Chromosome Sequences with Entropy Kernel and LKPLS Algorithm
  Zhenqiu Liu, Dechang Chen 543

Computational Neuroscience and Bioscience

Adaptive Data Association for Multi-target Tracking Using Relaxation
  Yang-Weon Lee 552

Demonstration of DNA-Based Semantic Model by Using Parallel Overlap Assembly
  Yusei Tsuboi, Zuwairie Ibrahim, Osamu Ono 562

Multi-objective Particle Swarm Optimization Based on Minimal
Particle Angle
  Dun-Wei Gong, Yong Zhang, Jian-Hua Zhang 571

A Quantum Neural Networks Data Fusion Algorithm and Its
Application for Fault Diagnosis
  Daqi Zhu, ErKui Chen, Yongqing Yang 581

Statistical Feature Selection for Mandarin Speech Emotion Recognition
  Bo Xie, Ling Chen, Gen-Cai Chen, Chun Chen 591

Reconstruction of 3D Human Body Pose Based on Top-Down Learning
  Hee-Deok Yang, Sung-Kee Park, Seong-Whan Lee 601

2D and 3D Full-Body Gesture Database for Analyzing Daily Human
Gestures
  Bon-Woo Hwang, Sungmin Kim, Seong-Whan Lee 611

Sound Classification and Function Approximation Using Spiking Neural
Networks
  Hesham H. Amin, Robert H. Fujii 621

Improved DTW Algorithm for Online Signature Verification Based on
Writing Forces
  Ping Fang, ZhongCheng Wu, Fei Shen, YunJian Ge, Bing Fang 631

3D Reconstruction Based on Invariant Properties of 2D Lines in
Projective Space
  Bo-Ra Seok, Yong-Ho Hwang, Hyun-Ki Hong 641

ANN Hybrid Ensemble Learning Strategy in 3D Object Recognition
and Pose Estimation Based on Similarity
  Rui Nian, Guangrong Ji, Wencang Zhao, Chen Feng 650

Super-Resolution Reconstruction from Fluorescein Angiogram
Sequences
  Xiaoxin Guo, Zhiwen Xu, Yinan Lu, Zhanhui Liu, Yunjie Pang ... 661

Signature Verification Using Wavelet Transform and Support Vector
Machine
  Hong-Wei Ji, Zhong-Hua Quan 671

Visual Hand Tracking Using Nonparametric Sequential Belief
Propagation
  Wei Liang, Yunde Jia, Cheng Ge 679

Locating Vessel Centerlines in Retinal Images Using Wavelet Transform:
A Multilevel Approach
  Xinge You, Bin Fang, Yuan Yan Tang, Zhenyu He, Jian Huang 688

Models and Methods

Stability Analysis on a Neutral Neural Network Model
  Yumin Zhang, Lei Guo, Chunbo Feng 697

Radar Emitter Signal Recognition Based on Feature Selection and
Support Vector Machines
  Gexiang Zhang, Zhexin Cao, Yajun Gu, Weidong Jin, Laizhao Hu .. 707

Methods of Decreasing the Number of Support Vectors via k-Mean
Clustering
  Xiao-Lei Xia, Michael R. Lyu, Tat-Ming Lok, Guang-Bin Huang.... 717

Dynamic Principal Component Analysis Using Subspace Model
Identification
  Pingkang Li, Richard J. Treasure, Uwe Kruger 727

Associating Neural Networks with Partially Known Relationships for
Nonlinear Regressions
  Bao-Gang Hu, Han-Bing Qu, Yong Wang 737

SVM Regression and Its Application to Image Compression
  Runhai Jiao, Yuancheng Li, Qingyuan Wang, Bo Li 747

Reconstruction of Superquadric 3D Models by Parallel Particle Swarm
Optimization Algorithm with Island Model
  Fang Huang, Xiao-Ping Fan 757

Precision Control of Magnetostrictive Actuator Using Dynamic
Recurrent Neural Network with Hysteron
  *Shuying Cao, Jiaju Zheng, Wenmei Huang, Ling Weng,
  Bowen Wang, Qingxin Yang* 767

Orthogonal Forward Selection for Constructing the Radial Basis
Function Network with Tunable Nodes
  Sheng Chen, Xia Hong, Chris J. Harris 777

A Recurrent Neural Network for Extreme Eigenvalue Problem
  Fuye Feng, Quanju Zhang, Hailin Liu 787

Chaos Synchronization for a 4-Scroll Chaotic System via Nonlinear Control
Haigeng Luo, Jigui Jian, Xiaoxin Liao 797

Global Exponential Stability of a Class of Generalized Neural Networks with Variable Coefficients and Distributed Delays
Huaguang Zhang, Gang Wang 807

Designing the Ontology of XML Documents Semi-automatically
Mi Sug Gu, Jeong Hee Hwang, Keun Ho Ryu 818

A Logic Analysis Model About Complex Systems' Stability: Enlightenment from Nature
Naiqin Feng, Yuhui Qiu, Fang Wang, Yingshan Zhang, Shiqun Yin 828

Occluded 3D Object Recognition Using Partial Shape and Octree Model
Young Jae Lee, Young Tae Park 839

Possibility Theoretic Clustering
Shitong Wang, Fu-lai Chung, Min Xu, Dewen Hu, Lin Qing 849

A New Approach to Predict N, P, K and OM Content in a Loamy Mixed Soil by Using Near Infrared Reflectance Spectroscopy
Yong He, Haiyan Song, Annia García Pereira, Antihus Hernández Gómez 859

Soft Sensor Modeling Based on DICA-SVR
Ai-jun Chen, Zhi-huan Song, Ping Li 868

Borderline-SMOTE: A New Over-Sampling Method in Imbalanced Data Sets Learning
Hui Han, Wen-Yuan Wang, Bing-Huan Mao 878

Real-Time Gesture Recognition Using 3D Motion History Model
Ho-Kuen Shin, Sang-Woong Lee, Seong-Whan Lee 888

Learning Systems

Effective Directory Services and Classification Systems for Korean Language Education Internet Portal Sites
Su-Jin Cho, Seongsoo Lee 899

Information-Theoretic Selection of Classifiers for Building Multiple Classifier Systems
   Hee-Joong Kang, MoonWon Choo 909

Minimal RBF Networks by Gaussian Mixture Model
   Sung Mahn Ahn, Sung Baik 919

Learning the Bias of a Classifier in a GA-Based Inductive Learning Environment
   Yeongjoon Kim, Chuleui Hong 928

GA-Based Resource-Constrained Project Scheduling with the Objective of Minimizing Activities' Cost
   Zhenyuan Liu, Hongwei Wang 937

Single-Machine Partial Rescheduling with Bi-criterion Based on Genetic Algorithm
   Bing Wang, Xiaoping Lai, Lifeng Xi 947

An Entropy-Based Multi-population Genetic Algorithm and Its Application
   Chun-lian Li, Yu sun, Yan-shen Guo, Feng-ming Chu, Zong-ru Guo 957

Reinforcement Learning Based on Multi-agent in RoboCup
   Wei Zhang, Jiangeng Li, Xiaogang Ruan 967

Comparison of Stochastic and Approximation Algorithms for One-Dimensional Cutting Problems
   Zafer Bingul, Cuneyt Oysu 976

Search Space Filling and Shrinking Based to Solve Constraint Optimization Problems
   Yi Hong, Qingsheng Ren, Jin Zeng, Ying Zhang 986

A Reinforcement Learning Approach for Host-Based Intrusion Detection Using Sequences of System Calls
   Xin Xu, Tao Xie 995

Evolving Agent Societies Through Imitation Controlled by Artificial Emotions
   Willi Richert, Bernd Kleinjohann, Lisa Kleinjohann 1004

Study of Improved Hierarchy Genetic Algorithm Based on Adaptive
Niches
  *Qiao-Ling Ji, Wei-Min Qi, Wei-You Cai, Yuan-Chu Cheng,
  Feng Pan* 1014

Associativity, Auto-reversibility and Question-Answering on Q'tron
Neural Networks
  Tai-Wen Yue, Mei-Ching Chen 1023

Associative Classification in Text Categorization
  Jian Chen, Jian Yin, Jun Zhang, Jin Huang 1035

A Fast Input Selection Algorithm for Neural Modeling of Nonlinear
Dynamic Systems
  Kang Li, Jian Xun Peng 1045

Delay-Dependent Stability Analysis for a Class of Delayed Neural
Networks
  Ru-Liang Wang, Yong-Qing Liu 1055

The Dynamic Cache Algorithm of Proxy for Streaming Media
  Zhiwen Xu, Xiaoxin Guo, Zhengxuan Wang, Yunjie Pang 1065

Fusion of the Textural Feature and Palm-Lines for Palmprint
Authentication
  Xiangqian Wu, Fengmiao Zhang, Kuanquan Wang, David Zhang ... 1075

Nonlinear Prediction by Reinforcement Learning
  Takashi Kuremoto, Masanao Obayashi, Kunikazu Kobayashi 1085

Author Index 1095

Table of Contents – Part II

Genomics and Proteomics

Protein Secondary Structure Prediction Using Sequence Profile and
Conserved Domain Profile
  Seon-Kyung Woo, Chang-Beom Park, Seong-Whan Lee 1

Correlating Genes and Functions to Human Disease by Systematic
Differential Analysis of Expression Profiles
  Weiqiang Wang, Yanhong Zhou, Ran Bi 11

On the Evolvement of Extended Continuous Event Graphs
  Duan Zhang, Huaping Dai, Youxian Sun, Qingqing Kong 21

Combined Literature Mining and Gene Expression Analysis for
Modeling Neuro-endocrine-immune Interactions
  Lijiang Wu, Shao Li 31

Mean Shift and Morphology Based Segmentation Scheme for DNA
Microarray Images
  Shuanhu Wu, Chuangcun Wang, Hong Yan 41

The Cluster Distribution of Regulatory Motifs of Transcription in Yeast
Introns
  Jun Hu, Jing Zhang 51

E-Coli Promoter Recognition Using Neural Networks with Feature
Selection
  Paul C. Conilione, Dianhui Wang 61

Improve Capability of DNA Automaton: DNA Automaton with Three
Internal States and Tape Head Move in Two Directions
  Xiaolong Shi, Xin Li, Zheng Zhang, Jin Xu 71

A DNA Based Evolutionary Algorithm for the Minimal Set Cover
Problem
  Wenbin Liu, Xiangou Zhu, Guandong Xu, Qiang Zhang, Lin Gao ... 80

DNA Computing Model of Graph Isomorphism Based on Three
Dimensional DNA Graph Structures
  Zhixiang Yin, Jianzhong Cui, Jing Yang, Guangwu Liu 90

A DNA-Based Genetic Algorithm Implementation for Graph Coloring
Problem
  *Xiaoming Liu, Jianwei Yin, Jung-Sing Jwo, Zhilin Feng,
  Jinxiang Dong* 99

Adaptation and Decision Making

Studies on the Minimum Initial Marking of a Class of Hybrid Timed
Petri Nets
  Huaping Dai 109

A Fuzzy Neural Network System Based on Generalized Class Cover
and Particle Swarm Optimization
  *Yanxin Huang, Yan Wang, Wengang Zhou, Zhezhou Yu,
  Chunguang Zhou* 119

Simulation-Based Optimization of Singularly Perturbed Markov
Reward Processes with States Aggregation
  Dali Zhang, Hongsheng Xi, Baoqun Yin 129

Semi-active Control for Eccentric Structures with MR Damper by
Hybrid Intelligent Algorithm
  Hong-Nan Li, Zhiguo Chang, Jun Li 139

Control Chaos in Brushless DC Motor via Piecewise Quadratic State
Feedback
  Hai Peng Ren, Guanrong Chen 149

Support Vector Machine Adaptive Control of Nonlinear Systems
  Zonghai Sun, Liangzhi Gan, Youxian Sun 159

Improved GLR Parsing Algorithm
  Miao Li, ZhiGuo Wei, Jian Zhang, ZeLin Hu 169

Locomotion Control of Distributed Self-reconfigurable Robot Based on
Cellular Automata
  Qiu-xuan Wu, Ya-hui Wang, Guang-yi Cao, Yan-qiong Fei 179

Improvements to the Conventional Layer-by-Layer BP Algorithm
  *Xu-Qin Li, Fei Han, Tat-Ming Lok, Michael R. Lyu,
  Guang-Bin Huang* 189

An Intelligent Assistant for Public Transport Management
  Martin Molina 199

FPBN: A New Formalism for Evaluating Hybrid Bayesian Networks
Using Fuzzy Sets and Partial Least-Squares
  Xing-Chen Heng, Zheng Qin 209

π-Net ADL: An Architecture Description Language for Multi-agent
Systems
  Zhenhua Yu, Yuanli Cai, Ruifeng Wang, Jiuqiang Han 218

Automatic Construction of Bayesian Networks for Conversational Agent
  Sungsoo Lim, Sung-Bae Cho 228

Stochastic Lotka-Volterra Competitive Systems with Variable Delay
  Yi Shen, Guoying Zhao, Minghui Jiang, Xuerong Mao 238

Human Face Recognition Using Modified Hausdorff ARTMAP
  Arit Thammano, Songpol Ruensuk 248

A Dynamic Decision Approach for Long-Term Vendor Selection Based
on AHP and BSC
  Ziping Chiang 257

A Fuzzy-Expert-System-Based Structure for Active Queue Management
  Jin Wu, Karim Djemame 266

Parameter Identification Procedure in Groundwater Hydrology with
Artificial Neural Network
  Shouju Li, Yingxi Liu 276

Design of Intelligent Predictive Controller for Electrically Heated Micro
Heat Exchanger
  *Farzad Habibipour Roudsari, Mahdi Jalili-Kharaajoo,
  Mohammad Khajepour* 286

A Group Based Insert Manner for Storing Enormous Data Rapidly in
Intelligent Transportation System
  Young Jin Jung, Keun Ho Ryu 296

A RDF-Based Context Filtering System in Pervasive Environment
  Xin Lin, Shanping Li, Jian Xu, Wei Shi 306

Applications and Hardware

Mobile Agent Based Wireless Sensor Network for Intelligent
Maintenance
  Xue Wang, Aiguo Jiang, Sheng Wang 316

The Application of TSM Control to Integrated Guidance/Autopilot
Design for Missiles
  Jinyong Yu, Daquan Tang, Wen-jin Gu, Qingjiu Xu 326

The Robust Control for Unmanned Blimps Using Sliding Mode Control
Techniques
  Guoqing Xia, Benkun Yang 336

Real-Time Implementation of High-Performance Spectral Estimation
and Display on a Portable Doppler Device
  Yufeng Zhang, Jianhua Chen, Xinling Shi, Zhenyu Guo 346

Application of Wavelet Transform in Improving Resolution of
Two-Dimensional Infrared Correlation Spectroscopy
  Daqi Zhan, Suqin Sun 356

Intelligent PID Controller Tuning of AVR System Using GA and PSO
  Dong Hwa Kim, Jin Ill Park 366

Design and Implementation of Survivable Network Systems
  Chao Wang, Jianfeng Ma, Jianming Zhu 376

Optimal Placement of Active Members for Truss Structure Using
Genetic Algorithm
  Shaoze Yan, Kai Zheng, Qiang Zhao, Lin Zhang 386

Performance Comparison of SCTP and TCP over Linux Platform
  Jong-Shik Ha, Sang-Tae Kim, Seok J. Koh 396

Multiresolution Fusion Estimation of Dynamic Multiscale System
Subject to Nonlinear Measurement Equation
  Peiling Cui, Quan Pan, Guizeng Wang, Jianfeng Cui 405

On a Face Detection with an Adaptive Template Matching and an
Efficient Cascaded Object Detection
  Jin Ok Kim, Jun Yeong Jang, Chin Hyun Chung 414

Joint Limit Analysis and Elbow Movement Minimization for Redundant
Manipulators Using Closed Form Method
  Hadi Moradi, Sukhan Lee 423

Detecting Anomalous Network Traffic with Combined Fuzzy-Based
Approaches
  Hai-Tao He, Xiao-Nan Luo, Bao-Lu Liu 433

Support Vector Machines (SVM) for Color Image Segmentation with
Applications to Mobile Robot Localization Problems
  An-Min Zou, Zeng-Guang Hou, Min Tan 443

Fuzzy Logic Based Feedback Scheduler for Embedded Control Systems
  *Feng Xia, Xingfa Shen, Liping Liu, Zhi Wang,
  Youxian Sun* 453

The Application of FCMAC in Cable Gravity Compensation
  Xu-Mei Lin, Tao Mei, Hui-Jing Wang, Yan-Sheng Yao 463

Oscillation and Strong Oscillation for Impulsive Neutral Parabolic
Differential Systems with Delays
  Yu-Tian Zhang, Qi Luo 472

Blind Estimation of Fast Time-Varying Multi-antenna Channels Based
on Sequential Monte Carlo Method
  Mingyan Jiang, Dongfeng Yuan 482

Locating Human Eyes Using Edge and Intensity Information
  Jiatao Song, Zheru Chi, Zhengyou Wang, Wei Wang 492

A Nonlinear Adaptive Predictive Control Algorithm Based on OFS
Model
  Haitao Zhang, Zonghai Chen, Ming Li, Wei Xiang, Ting Qin 502

Profiling Multiple Domains of User Interests and Using Them for
Personalized Web Support
  Hyung Joon Kook 512

Analysis of SCTP Handover by Movement Patterns
  Dong Phil Kim, Seok Joo Koh, Sang Wook Kim 521

A Fuzzy Time Series Prediction Method Using the Evolutionary
Algorithm
  Hwan Il Kang 530

Modeling for Security Verification of a Cryptographic Protocol with
MAC Payload
  Huanbao Wang, Yousheng Zhang, Yuan Li 538

Bilingual Semantic Network Construction
  Jianyong Duan, Yi Hu, Ruzhan Lu, Yan Tian, Hui Liu 548

Other Applications

Approaching the Upper Limit of Lifetime for Data Gathering Sensor Networks
Haibin Yu, Peng Zeng, Wei Liang 558

A Scalable Energy Efficient Medium Access Control Protocol for Wireless Sensor Networks
Ruizhong Lin, Zhi Wang, Yanjun Li, Youxian Sun 568

Connectivity and RSSI Based Localization Scheme for Wireless Sensor Networks
Xingfa Shen, Zhi Wang, Peng Jiang, Ruizhong Lin, Youxian Sun ... 578

A Self-adaptive Energy-Aware Data Gathering Mechanism for Wireless Sensor Networks
Li-Min Sun, Ting-Xin Yan, Yan-Zhong Bi, Hong-Song Zhu 588

An Adaptive Energy-Efficient and Low-Delay MAC Protocol for Wireless Sensor Networks
Seongcheol Kim 598

Sensor Management of Multi-sensor Information Fusion Applied in Automatic Control System
Yue-Song Lin, An-ke Xue 607

A Survey of the Theory of Min-Max Systems
Yiping Cheng 616

Performance Bounds for a Class of Workflow Diagrams
Qianchuan Zhao 626

A Hybrid Quantum-Inspired Genetic Algorithm for Flow Shop Scheduling
Ling Wang, Hao Wu, Fang Tang, Da-Zhong Zheng 636

Stability and Stabilization of Impulsive Hybrid Dynamical Systems
Guangming Xie, Tianguang Chu, Long Wang 645

Fault Tolerant Supervisory for Discrete Event Systems Based on Event Observer
Fei Xue, Da-Zhong Zheng 655

A Study on the Effect of Interference on Time Hopping Binary PPM Impulse Radio System
YangSun Lee, HeauJo Kang, MalRey Lee, Tai-hoon Kim 665

A Study on the Enhanced Detection Method Considering the Channel
Response in OFDM Based WLAN
   *Hyoung-Goo Jeon, Hyun Lee, Won-Chul Choi, Hyun-Seo Oh,
Kyoung-Rok Cho* 675

Block Error Performance Improvement of DS/CDMA System with
Hybrid Techniques in Nakagami Fading Channel
   Heau Jo Kang, Mal Rey Lee 685

Performance Evaluation of Convolutional Turbo Codes in AWGN and
ITU-R Channels
   Seong Chul Cho, Jin Up Kim, Jae Sang Cha, Kyoung Rok Cho 695

Adaptive Modulation Based Power Line Communication System
   Jong-Joo Lee, Jae-Sang Cha, Myong-Chul Shin, Hak-Man Kim 704

A Novel Interference-Cancelled Home Network PLC System Based on
the Binary ZCD-CDMA
   Jae-Sang Cha, Myong-Chul Shin, Jong-Joo Lee 713

Securing Biometric Templates for Reliable Identity Authentication
   Muhammad Khurram Khan, Jiashu Zhang 723

Intelligent Tracking Persons Through Non-overlapping Cameras
   Kyoung-Mi Lee 733

A New LUT Watermarking Scheme with Near Minimum Distortion
Based on the Statistical Modeling in the Wavelet Domain
   Kan Li, Xiao-Ping Zhang 742

A Secure Image-Based Authentication Scheme for Mobile Devices
   Zhi Li, Qibin Sun, Yong Lian, D.D. Giusto 751

A Print-Scan Resistable Digital Seal System
   Yan Wang, Ruizhen Liu 761

ThresPassport – A Distributed Single Sign-On Service
   Tierui Chen, Bin B. Zhu, Shipeng Li, Xueqi Cheng 771

Functional Architecture of Mobile Gateway and Home Server for
Virtual Home Services
   Hyuncheol Kim, Seongjin Ahn 781

Validation of Real-Time Traffic Information Based on Personal
Communication Service Network
   Young-Jun Moon, Sangkeon Lee, Sangwoon Lee 791

A PKI Based Digital Rights Management System for Safe Playback
Jae-Pyo Park, Hong-jin Kim, Keun-Wang Lee, Keun-Soo Lee 801

Scheduling Method for a Real Time Data Service in the Wireless ATM Networks
Seung-Hyun Min, Kwang-Ho Chun, Myoung-Jun Kim 811

Road Change Detection Algorithms in Remote Sensing Environment
Hong-Gyoo Sohn, Gi-Hong Kim, Joon Heo 821

Safe RFID System Modeling Using Shared Key Pool in Ubiquitous Environments
Jinmook Kim, Hwangbin Ryou 831

Robust 3D Arm Tracking from Monocular Videos
Feng Guo, Gang Qian 841

Segmentation and Tracking of Neural Stem Cell
Chunming Tang, Ewert Bengtsson 851

License Plate Tracking from Monocular Camera View by Condensation Algorithm
İlhan Kubilay Yalçın, Muhittin Gökmen 860

A Multi-view Approach to Object Tracking in a Cluttered Scene Using Memory
Hang-Bong Kang, Sang-Hyun Cho 870

A Robust 3D Feature-Based People Stereo Tracking Algorithm
Guang Tian, Feihu Qi, Yong Fang, Masatoshi Kimachi, Yue Wu, Takashi Iketani, Xin Mao, Panjun Chen 880

Self-tuning Fuzzy Control for Shunt Active Power Filter
Jian Wu, Dian-guo Xu, Na He 890

Optimal Production Policy for a Volume-Flexibility Supply-Chain System
Ding-zhong Feng, Li-bin Zhang 900

Flame Image of Pint-Sized Power Plant's Boiler Denoising Using Wavelet-Domain HMT Models
Chunguang Ji, Ru Zhang, Shitao Wen, and Shiyong Li 910

Fast and Robust Portrait Segmentation Using QEA and Histogram Peak Distribution Methods
Heng Liu, David Zhang, Jingqi Yan, Zushu Li 920

Suppressing Chaos in Machine System with Impacts Using Period Pulse
Linze Wang, Wenli Zhao, Zhenrui Peng 929

The Cognitive Behaviors of a Spiking-Neuron Based Classical
Conditioning Model
Guoyu Zuo, Beibei Yang, Xiaogang Ruan 939

Probabilistic Tangent Subspace Method for M-QAM Signal
Equalization in Time-Varying Multipath Channels
Jing Yang, Yunpeng Xu, Hongxing Zou 949

Face Recognition Based on Generalized Canonical Correlation Analysis
Quan-Sen Sun, Pheng-Ann Heng, Zhong Jin, De-Shen Xia 958

Clustering Algorithm Based on Genetic Algorithm in Mobile Ad Hoc
Network
Yanlei Shang, Shiduan Cheng 968

A Pair-Ant Colony Algorithm for CDMA Multiuser Detector
Yao-Hua Xu, Yan-Jun Hu, Yuan-Yuan Zhang 978

An Enhanced Massively Multi-agent System for Discovering HIV
Population Dynamics
Shiwu Zhang, Jie Yang, Yuehua Wu, Jiming Liu 988

An Efficient Feature Extraction Method for the Middle-Age Character
Recognition
*Shahpour Alirezaee, Hasan Aghaeinia, Karim Faez,
Alireza Shayesteh Fard* 998

Author Index 1007

A Novel Approach to Ocular Image Enhancement with Diffusion and Parallel AOS Algorithm

Lanfeng Yan[1], Janjun Ma[1], Wei Wang[2], Qing Liu[1], and Qiuyong Zhou[1]

[1] People's Hospital of Gansu Province, China, 730000
[2] School of Information Science and Engineering, Lanzhou University
Yanlanfenggslz@163.com

Abstract. This paper suggests a new diffusion method, which based on modified coherence diffusion for the enhancement of ocular fundus images (OFI) and parallel AOS scheme is applied to speed algorithm, which is faster than usual approach and shows good performance. A structure tensor integrating the second-order directional differential information is applied to analyze weak edges, narrow peak, and vessels structures of OFI in diffusion. The structure tensor and the classical one as complementary descriptor are used to build the diffusion tensor. The several experiment results are provided and suggest that it is a robust method to prepare image for intelligent diagnosis and instruction for treatment of ocular diseases. The modified diffusion for the enhancement of OFI can preserve important oriented patterns, including strong edges and weak structures.

1 Introduction

The aim of medical images enhancement is to save the helpful information when images are denoised to prepare for segmentation and registration before image fusion, then it can be available for disease diagnoses, treatment and evaluate after operation, so it is very important for theory and appliance. The enhancement of medical images is the base for registration and fusion, vessels segmentation, feature extraction, registration, classification and fusion. This provides a foundation for further research. Medical image enhancement is an important preprocessing step that removes noise while preserving semantically important structures such as edges. This may give great help to simplify subsequent image analysis like segmentation and understanding and diagnosis. However, general image enhancement methods often remove much important information while removing noise.

The aim of ocular fundus images enhancement and vessels extracting is to get more special information for disease diagnoses, treatment and evaluation after therapy, so it is very important for theory and appliance. General image enhancement methods often remove more important information while removing noise. Image enhancement is an important step to removes noise for simplifying image segmentation and extraction. Nonlinear-based diffusion for image enhancement has attracted much attention in a purely data-driven way that is flexible to deal with the rich image structures [1]. Most of Nonlinear based image diffusion use a scalar diffusivity thus the diffusion flux is along gradient direction, which may blur edges. Moreover, such isotropic diffusion cannot preserve the oriented structures precisely.

With the increasing appearance of oriented structures in many computer vision and image processing problems, for instance eye fundus vessels, many attempts on oriented patterns enhancement have been made [2]. Partial differential equations based coherence diffusion has proved to be a very useful method [3,4].

This study suggests a new structure tensor integrating the second-order directional differential information, which can be used to analyze narrow peaks. The structure tensor as complementary descriptor plays a roles at different diffusion stage controlled by a switch parameter. This diffusion is called as switch diffusion. The switch diffusion is controlled by the diffusion tensor constructed from the switch structure tensor, which can preserve strong edges and weak edges. Additive operator splitting (AOS) scheme showed a nice performance in implement.

2 Principle of Diffusion

Diffusion is a physical process that equilibrates concentration differences without creating or destroying mass. Fick's Diffusion Law expresses the equilibration property [1]:

$$j = -D \bullet \nabla u \tag{1}$$

The observation that diffusion does only transport mass without creating or destroying mass is expressed by the continuity equation:

$$\frac{\partial u}{\partial t} = -div(j) \tag{2}$$

Thus results in diffusion equation:

$$\frac{\partial u}{\partial t} = div(D \bullet \nabla u) \tag{3}$$

Where D is diffusion tensor, j is mass flux. When D is a scalar, flux j is parallel to ∇u, which is isotropic diffusion. While D is a positive definite symmetric matrix, j and ∇u are not parallel, which is anisotropic diffusion. In the latter case, it is desirable to rotate the flux towards the orientation of interesting features that steers coherence-enhancing diffusion.

Precise coherence analysis is crucial to the diffusion behavior. Cottete and Germain used the tensor product of gradient as the original structure tensor [2]: $J_0 = \nabla u_\sigma \otimes \nabla u_\sigma^T$, where $u_\sigma = G_\sigma * u$ is the slightly smooth version image by convolving u with gauss kernel G_σ. The eigenvalues of J_0 are $\mu_1 = |\nabla u_\sigma|^2, \mu_2 = 0$, which provides coherent measurement $k = (\mu_1 - \mu_2)^2$ and the eigenvector $e_2 = \frac{\nabla u_\sigma^\perp}{|\nabla u_\sigma|}$ provides the coherent direction. However, the simple structure tensor fails in analyzing corners or parallel structures. To solve this problem,

Weickert proposed the structure tensor [4,5]: $J_\rho = G_\rho * (\nabla u_\sigma \otimes \nabla u_\sigma^T)$. The eigenvalues of J_ρ measure the variation of the gray values within a window size of order ρ. This structure tensor is useful to analyze strong edges, corners and T-junctions. However, it is a linear smoothing of J_0 and uses only the local average of the first-order differential information. Many image features such as narrow peaks, ridge-like edges could not be accurately described by J_ρ for the gradient is close to zero on these structures. Recently, Brox and Weickert proposed a nonlinear structure tensor by diffusing J_0 under the image gradient field ∇u, which may bring much computational cost [6,7].

3 Switch Diffusion

Narrow peaks and ridge-like structure are important features in OFI images. Classical structure tensors use only the local average of the gradient for estimating coherence. Therefore, it may not provide precise estimation on which the gradient is weak. Second-order directional derivative can cope with these structures well. We denote the intensity image by $u(x, y)$. The directional derivative of u at point (x, y) in the direction $\alpha = (\cos\theta, \sin\theta)^T$ is denoted by $u'_\alpha(x, y)$. It is defined as:

$$u'_\alpha(x,y) \stackrel{\Delta}{=} \lim_{h \to 0} \frac{u(x + h\cos\theta, y + h\sin\theta) - u(x, y)}{h}$$

$$= \frac{\partial u}{\partial x}\cos\theta + \frac{\partial u}{\partial y}\sin\theta \tag{4}$$

$$= <\nabla u, \alpha>$$

The second-order directional derivative of u along α is denoted by $u''_\alpha(x, y)$, and it follows that:

$$u''_\alpha(x, y) = \frac{\partial}{\partial \alpha}(\frac{\partial u}{\partial \alpha}) = <u'_\alpha, \alpha> = <<\nabla u, \alpha>, \alpha>$$

$$= \frac{\partial^2 u}{\partial x^2}\cos^2\theta + 2\frac{\partial^2 u}{\partial x \partial y}\sin\theta\cos\theta + \frac{\partial^2 u}{\partial y^2}\sin^2\theta$$

$$= (\cos\theta \quad \sin\theta) \begin{pmatrix} \frac{\partial^2 u}{\partial x^2} & \frac{\partial^2 u}{\partial x \partial y} \\ \frac{\partial^2 u}{\partial x \partial y} & \frac{\partial^2 u}{\partial y^2} \end{pmatrix} \begin{pmatrix} \cos\theta \\ \sin\theta \end{pmatrix} \tag{5}$$

$$= \alpha^T H \alpha$$

Where H is Hessian matrix u. Let μ_1, μ_2, $(\mu_1 \geq \mu_2)$ denote the eigenvalues of H and e_1, e_2 the corresponding eigenvectors. From Rayleigh's quotient, we can derive that:

$$\mu_2 \leq \alpha^T H \alpha = u_\alpha'' \leq \mu_1 \qquad (6)$$

So the eigenvalues of Hessian matrix are exactly the two extreme of u_α'' and the corresponding eigenvectors are the directions along which the second directional derivative reaches its extreme. Hessian matrix can describe the second-order structure of the local intensity variations along the eigenvectors, which can cope with narrow peaks and ridge-like structures. When there are dark (bright) narrow long structures in the image, second-order directional derivative u_α'' reaches its maximum (minimum) along eigenvectors $e_1(e_2)$. So the coherent direction should be along $e_2(e_1)$ respectively. The new structure tensor is defined by $J_H = H_\sigma$. Note that H_σ is the Hessian Matrix of u_σ. The proposed structure tensor is a desirable descriptor for analyzing narrow peaks or ridge-like structures.

The classical structure tensor J_ρ is useful to analyze strong edges, corners and T-junctions while fails in detecting weak edges precisely. However, the new proposed structure tensor J_H can capture weak edges as narrow peaks and ridge-like structures while fails in detecting strong edges. Therefore, the two structure tensors can be complementary to each other and provide reliable coherence estimation on different structures. A switch parameter T to control the roles of the two structure tensors is introduced. When $|\nabla u_\sigma| \geq T$, J_ρ is available, whereas $|\nabla u_\sigma| < T$, J_H is more reliable. The switch structure tensor is given by:

$$J = \begin{cases} J_\rho & if \quad |\nabla u_\sigma| \geq T \\ J_H & else \end{cases} \qquad (7)$$

A desirable diffusion tensor D should encourage coherent diffusion. A natural way to construct D is from structure tensor J such that D have the same eigenvectors as J and its eigenvalues prefers the diffusion along the coherent direction than across to it. Let μ_1, μ_2, $(\mu_1 \geq \mu_2)$ denote the eigenvalues of J and e_1, e_2 the corresponding eigenvectors. The coherent direction estimated from J is denoted by $e_{(co)}$ and the orthogonal direction by $e_{(co)}^\perp$. Note that the coherent direction estimated from J_ρ is always along e_2, while from J_H it can be divided into the following two cases: when there are dark curvilinear structure in the bright background, the coherent direction is along e_2; otherwise there are bright curvilinear structure in the dark background, the coherent direction is along e_1. Not only the eigenvectors, but also the eigenvalues can provide useful structure information: When $\mu_1 \gg \mu_2$, it corresponds to anisotropic oriented structure. When $\mu_1 \approx \mu_2$, it corresponds to isotropic structures. To encourage coherent diffusion, the eigenvalues of D can be chosen as [3,4]:

$$\lambda_{(co)} = \begin{cases} c, & \text{if } \mu_1 = \mu_2 \\ c + (1-c)\exp(-\dfrac{\beta}{(\mu_1 - \mu_2)^2}), & \text{else} \end{cases} \quad (8)$$

$$\lambda_{(co)}^{\perp} = c$$

Where $\lambda_{(co)}$ and $\lambda_{(co)}^{\perp}$ are the diffusivity along the direction $e_{(co)}$ and $e_{(co)}^{\perp}$ respectively. $\beta > 0$ serves as a threshold parameter. $\lambda_{(co)}$ is an increasing function with respect to the coherence measurement $(\mu_1 - \mu_2)^2$. When $(\mu_1 - \mu_2)^2 \gg \beta$, $\lambda_{(co)} \approx 1$; otherwise, it leads to $\lambda_{(co)} \approx c$. $c \in (0,1)$ is small positive parameter that guarantees D is positive definite.

Therefore, D can be obtained by $D = P \begin{pmatrix} \lambda_{(co)} \\ & \lambda_{(co)}^{\perp} \end{pmatrix} P^T$, where $P = (e_{(co)}, e_{(co)}^{\perp})$ is the eigenvector matrix. Let D_ρ, D_H denote the diffusion tensor constructed from J_ρ and J_H respectively. Therefore, the diffusion tensor is given by:

$$D = \begin{cases} D_\rho & \text{if } |\nabla u_\sigma| \geq T \\ D_H & \text{else} \end{cases} \quad (9)$$

When $|\nabla u_\sigma| \geq T$, the coherent direction of D is along $\nabla u_\sigma^{\perp}$; Otherwise, it is along $e_{(co)}^{(H)}$ that is the coherent direction estimated from J_H. With equation (9), we can obtain the following diffusion equation with the reflecting boundary condition and the original f as the initial condition.

$$\begin{cases} \dfrac{\partial u}{\partial t} = \mathrm{div}(D \bullet \nabla u) = \begin{cases} \mathrm{div}(D_\rho \bullet \nabla u) & \text{if } |\nabla u_\sigma| \geq T \\ \mathrm{div}(D_H \bullet \nabla u) & \text{else} \end{cases} & \text{on } \Omega \\ <D \bullet \nabla u, n>\big|_{\partial\Omega} = 0 & \text{on } \partial\Omega \\ u(\cdot, 0) = f(\cdot) & \text{on } \Omega \end{cases} \quad (10)$$

Equation (10) states that the diffusion tensor can steer a switch diffusion controlled by the parameter T, where two complementary diffusion processes can preserve the edge, corner as well as narrow peaks, ridge-like structures while smoothing the interior of the image.

4 Parallel AOS Algorithm

The implementation of the diffusion equation (10) can be implemented by the widely used explicit discretization or the implicit scheme. However, explicit scheme needs very small time steps that lead to poor efficiency and the implicit scheme brings much computational cost. Weickert in [3,6] proposed parallel AOS scheme based on semi-implicit discretization, resulting in a very fast and efficient algorithm. As the diffusion tensor D is a positive definite symmetric matrix, Let $D = \begin{pmatrix} d_{11} & d_{12} \\ d_{12} & d_{22} \end{pmatrix}$.

$$div(D \bullet \nabla u) = \sum_{i,j=1}^{m} \partial_{x_i}(d_{ij}\partial_{x_j}u) \qquad (11)$$

Where m is the dimension of the image. In our case, $m = 2$.

The discretization of equation (10) is given by the finite difference scheme:

$$\frac{U^{k+1} - U^k}{\Delta t} = \sum_{i,j=1}^{m} L_{ij}^k U^k \qquad (12)$$

The upper index denotes the time level and L_{ij} is a central difference approximation to the operator $\partial_{x_i}(d_{ij}\partial_{x_j}u)$. Using semi-implicit discretization, equation (12) can be rewritten as:

$$\frac{U^{k+1} - U^k}{\Delta t} = \sum_{l=1}^{m} L_{ll}^k U^{k+1} + \sum_{i=1}^{m}\sum_{j \neq i} L_{ij}^k U^k \Rightarrow U^{k+1} = (I - \Delta t \sum_{l=1}^{m} L_{ll}^k)^{-1} \cdot (I + \Delta t \sum_{i=1}^{m}\sum_{j \neq i} L_{ij}^k) U^k \qquad (13)$$

The semi-implicit scheme makes it more stable and efficient than explicit discretization. The AOS scheme can be given by:

$$U^{k+1} = \frac{1}{m}\sum_{l=1}^{m}(I - m\Delta t L_{ll}^k)^{-1} \cdot (I + \Delta t \sum_{i=1}^{m}\sum_{j \neq i} L_{ij}^k) U^k \qquad (14)$$

Equation (14) has the same first-order Taylor expansion in Δt as that of (13). The central difference approximation of L_{ll} ($l = 1,2$) guarantees it is diagonally dominant tridiagonal. Therefore, the diffusion equation (14) converts to solving a diagonally dominant tridiagonal system of linear equation, which can be fast solved by Thomas algorithm [7]. The AOS algorithm is ten times faster than the usual numerical method.

5 Experiments Results

Selection of parameters is important for the diffusion process. In the following experiments, we choose $c = 0.001$, and β the 90% quantile of the histogram for $(\mu_1 - \mu_2)^2$. Selection of parameter T is crucial to switch the two diffusion processes. If T is chosen too small, some narrow long structures can not be captured precisely; However if T is selected too large, some gradient information may lose and computational cost increases as more pixels are involved in second-order differential computing. Experiments show that setting T to be 5%~10% quantile of the histogram for $|\nabla u_\sigma|^2$ is sufficient for wide categories of images. The noise scale σ and integration scale ρ is given for each cases. Our switch diffusion for image coherence enhancement, the diffusion equation is give by (7). Isotropic diffusion by P-M equation for image enhancement, which corresponds to the case that D is a scalar diffusivity: $D = g(|\nabla u_\sigma|^2) = \dfrac{1}{1+|\nabla u_\sigma|^2/\lambda^2}$. Where λ is the contrast parameter that can be chosen as the 90% quantile of the histogram for $|\nabla f|$.

The experimental results are shown in fig1. It is an eye image selected from fundus fluorescence angiography (FFA). Fig 1(a) is the original images. Fig 1(b) is the results by the proposed switch diffusion. Fig 1(c) is the results by nonlinear diffusion. Fig 1(d) is the results by gradient magnitude method. From fig 1(b), we can see that the vessel branches are preserved precisely while most of the noise is removed. F=fig 1(c) shows that some thin narrow vessel branches cannot be preserved well and the noise is magnified somewhere. Fig 1(d) badly blurs the edges and could not close interrupted vessel branches. Experiments show that the switch diffusion is preferred along the coherent direction and the image features are preserved while the noise is removed very well. To nonlinear diffusion, it cannot close the interrupted line structures and also blurs the edge. However, gradient magnitude method deforms some narrow peak and ridge-like oriented structures and the noise of background is greatly magnified. The proposed switch diffusion is robust and reliable for enhancing and extracting oriented structures of vessels. It is useful to prepare the vessels images for segmentation and automatic analysis.

6 Conclusions and Discussion

A novel approach of ocular fundus image enhancement with an improved coherence diffusion or called switch diffusion was proposed. A new structure tensor integrating second-order directional differential information was propose as a complement of classical structure tensor, which can capture narrow, weak peak and ridge-like structures precisely from bad ocular fundus images. The two structure tensor play roles in different diffusion stage controlled by a redirecting parameter or called switch parameter, which can provide precise coherence estimation on different structures. Switch diffusion was controlled by the diffusion tensor constructed from the switch structure tensor, which can preserve many important edges, corners, T-junctions as

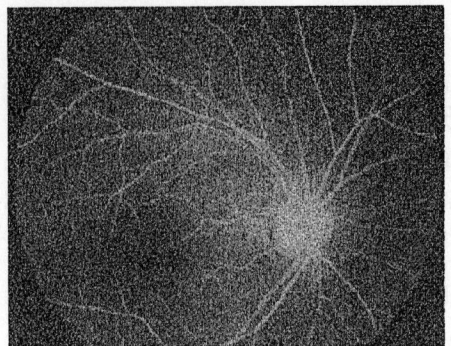

Fig. 1. (a) Original Image with Noise

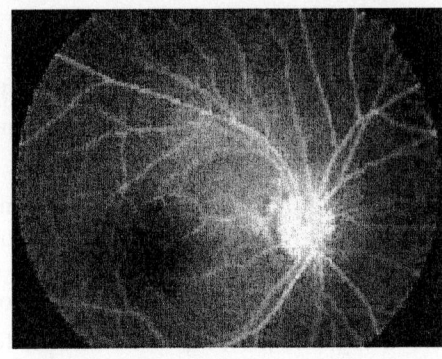

Fig. 1. (b) $\sigma = 0.18$, Iteration Time=9 Switch Diffusion

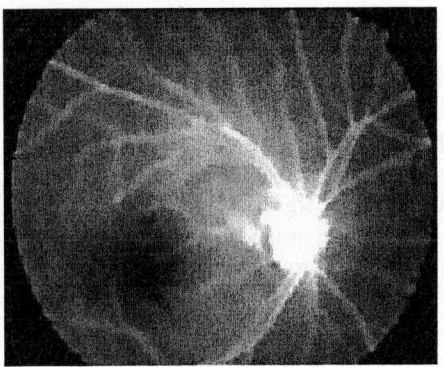

Fig. 1. (c) $\sigma = 0.18$, Iteration Time=9 Nonlinear Diffusion

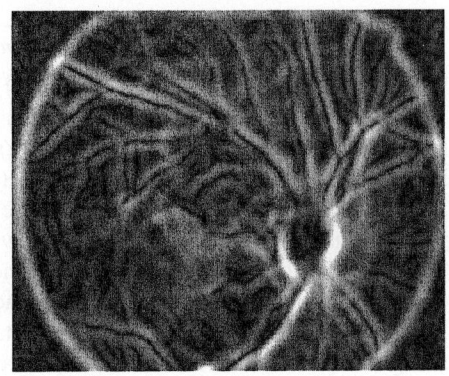

Fig. 1. (d) $\sigma = 0.18$, Iteration Time=9 Gradient Magnitude Method

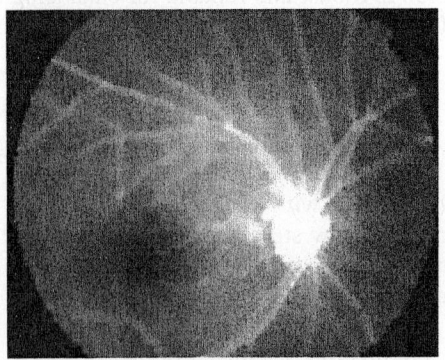

Fig. 2. (a) $\sigma = 0.18$, Iteration Time=26 Nonlinear Diffusion

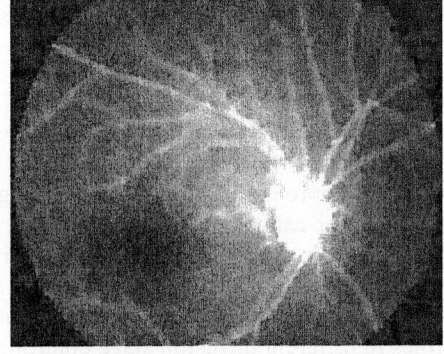

Fig. 2. (b) $\sigma = 0.18$, Iteration Time=26 Switch Diffusion

Acknowledgement

The authors would like to thank for the help from Hongmei Zhang PhD and Gang yu PhD of Xi'an Jiaotong University.

References

1. Diewald, U., Preusser, T., Rumpf, M., Strzodka, R.: Diffusion models and their accelerated solution in image and surface processing, Acta Math. Univ. Comenianae. 1(2001) 15-31
2. Joachim, W.: A review of nonlinear diffusion filtering. Scale-Space Theory in Computer Vision. 1252(1997) 3-28
3. Joachim, W.: Coherence-enhancing diffusion of colour images. Image and Vision Computing. 17(1999) 201-212
4. Bakalexis, S.A., Boutalis, Y.S., Mertzios, B.G.: Edge detection and image segmentation based on nonlinear anisotropic diffusion.Digital Signal Processing. 2(2002) 1203 -1206
5. Pietro, P., Jitendra, P.: Scale-Space and Edge Detection Using Anisotropic Diffusion. IEEE Transactions on Machine Intelligence. 12(1990) 629-639
6. Thomas, B., Joachim W. Nonlinear matrix diffusion for optic flow estimation. LNCS 2449(2002) 446-453
7. Joachim, W., Zuiderveld, K.J., ter Haar Romeny, B.M., Niessen, W.J.: Parallel implementations of AOS schemes: A fast way of Nonlinear diffusion filtering, Proc. 1997 IEEE International Conference on Image Processing, 3(1997) 396-399

An Iterative Hybrid Method for Image Interpolation

Yan Tian[1], Caifang Zhang[1], Fuyuan Peng[1], and Sheng Zheng[2]

[1] Electronic and Information Engineering Department,
Huazhong University of Science and Technology,
1037 Luoyu Road, Wuhan 430074, P.R. China
tianyan2000@126.com, wind_cf@163.com, pfuyuan@163.com
[2] Institute for Pattern Recognition and Artificial Intelligent,
Huazhong University of Science and Technology,
1037 Luoyu Road, Wuhan 430074, P.R. China
zhengsheng6511@sohu.com

Abstract. An iterative hybrid interpolation method is proposed in this study, which is an integration of the bilinear and the bi-cubic interpolation methods and implemented by an iterative scheme. First, the implement procedure of the iterative hybrid interpolation method is described. This covers (a) a low resolution image is interpolated by using the bilinear and the bi-cubic interpolators respectively; (b) a hybrid interpolated result is computed according to the weighted sum of both bilinear interpolation result and bi-cubic interpolation result and (c) the final interpolation result is obtained by repeating the similar steps for the successive two hybrid interpolation results by a recursive manner. Second, a further discussion on the method -- the relation between hybrid parameter and details of an image is provided from the theoretical point of view, at the same time, an approach used for the determining of the parameter is proposed based on the analysis of error parameter curve. Third, the effectiveness of the proposed method is verified based on the experimental study.

1 Introduction

Remote sensing images are one of the major data resources for capturing information on earth surface. In order to obtain detailed information, high-resolution satellite images are provided recently, such as IKONOS and QuikBird images. However, for further enhancing the images for more detailed information about earth surface, the high-resolution images need to be further processed by the technologies, such as image fusion, image enhancement and image interpolations.

Beside the application in remote sensing, image interpolation technique also plays an important role in other applications of image processing, such as digital zooming in CCD, restoration for compression data and photographic printing. Up to now, there are many interpolation methods have been developed for various purposes [1]. Using an example sensor model, Price and Hayes proposed an optimal pre-filter for image interpolation [2]. By applying edge-directed interpolation and edge sharpening operations, an interpolation technique based scheme for image expansion is introduced by Wang and Ward [3]. For robotic position compensation, Bai and Wang introduced a dynamic on-line fuzzy error interpolation technique [4]. For image enlargement, which aims at obtaining a better view of details in the image, a smart interpolation based on anisotropic diffusion was proposed by Battiato and Gallo [5]. Considering the level of activity in local regions of the image, Hadhoud et al. presented an adaptive warped distance method for image interpolation [6]. However, for the sake of implementation and computational speed, conventional techniques including nearest neighbour, linear, cubic and spline interpolation have been frequently utilized [2].

As we know, it is possible to obtain optimizing result to apply the sinc kernel interpolating function to an image. However, the sinc kernel decays too slowly and it will produce assignable error if we cut off it. Therefore, various or interpolation methods were proposed to solve this problem. Among them, the bilinear and bi-cubic methods are two classical interpolation techniques with their own advantages. From the viewpoint of filter theory, the bilinear and the bi-cubic methods have good response ability for low frequency and high-frequency components contained in an image, respectively. Therefore, in this study we will develop a new iterative hybrid interpolation method, which can combine the advantages of these two methods and implemented by a recursive scheme.

The remainder is arranged as follows. In section2, the novel iterative hybrid method is described in detail. A theoretical analysis for discussing the relation between hybrid parameter and frequency component is given also in this section. At the same time, applying the method based on error parameter analysis, an approach used to estimate the hybrid parameter is presented. To illustrate the effectiveness of the proposed method, several simulations based on the standard test images are studied in section3. Some conclusions, which drawn from this article and future works need to be further researched for this issue, are pointed out in the last section.

2 The Iterative Hybrid Interpolation Method

In this section, a novel interpolation method based on the combining of the bilinear and bi-cubic is described at first, and then the relationship between the hybrid parameter and the high and low frequency components contained in the original image is discussed, at last, a method to determine the hybrid parameter is given based on the analysis of error parameter curve.

2.1 The Proposed Method

An interpolation problem can be regarded as a filter problem [7]. In a filter design, we often try to make the frequency response of the designed filter to approximate frequency of an image as closely as possible. If an image contains much lower frequency information, it is reasonable to apply the bilinear interpolation than applying the bi-cubic interpolation. On the contrary, it is suitable to use bi-cubic and than the bilinear. For a general image, it contains images with both low frequency and high frequency. Therefore, it is natural to generate a high-resolution image by an iterative hybrid interpolation method, which possesses the advantages of both the bilinear and the bi-cubic methods.

We have mentioned that the target of the iterative hybrid interpolation method (proposed in this paper) is to generate a high-resolution image with both the merits of both the bilinear and the bi-cubic interpolation methods. For this purpose, in the implementation operation, the steps of the proposed method can be designed as follows:

Step1: we first interpolate a low-resolution image f_0 by applying the bilinear interpolators and obtain the linear interpolated result f_1.

Step2: the bi-cubic method is applied and the nonlinear interpolated result f_2 can thus be obtained.

Step3: for these two interpolated results, we endow the hybrid parameters ρ and $1-\rho$ ($0 \leq \rho \leq 1$), respectively. And then a hybrid result f_3 is obtained by summing up these two weighted results.

Step4: the kth interpolation result is obtained by above hybrid method for (k-1)th result and (k-2)th result by applying a recursive scheme $f_k = \rho f_{k-1} + (1-\rho) f_{k-2}$.

For step4, there are two ways to determine the ending condition of the iterative algorithm. One way is given by the root mean square error (RMSE) of f_k and f_{k-1}, that is if $RMSE(f_k(i,j), f_{k-1}(i,j)) \leq \varepsilon$ (ε is a threshold value given in advance), the iterative algorithm is finished,

The root mean square error (RMSE) is given by

$$RMSE(f_k, f_{k-1}) = \sqrt{\frac{\sum_{i=1}^{n}\sum_{j=1}^{m}(f_k(i,j) - f_{k-1}(i,j))^2}{n \times m}} \quad (1)$$

here $m \times n$ is the size of $f_k(i,j)$ and $f_{k-1}(i,j)$, $f_i(i,j)$ is the grey level of $f_k(i,j)$ on the coordinate (i,j).

Another way to determine the ending condition for the iterative algorithm is to set up certain steps for the iterative procedure (10,100, or 1000 steps, and so on). For sake of simplicity, in this paper, the second meaner will be utilized.

Notice that the first and second steps of the above iterative method are implemented based on the bilinear and the bicubic interpolation results, while the later steps are obtained by taking two successive hybrid interpolation results. Each of these steps is a hybrid interpolative scheme. The flow chart of the hybrid interpolation method based on the bilinear and the bi-cubic interpolation methods is illustrated in the Fig.1 below.

The iterative hybrid interpolation method, described in previous paragraphs, is implemented by successive hybrid interpolation algorithms; each of these hybrid algorithms is followed by a judgment condition. The whole procedure can be simply shown in figure 2.

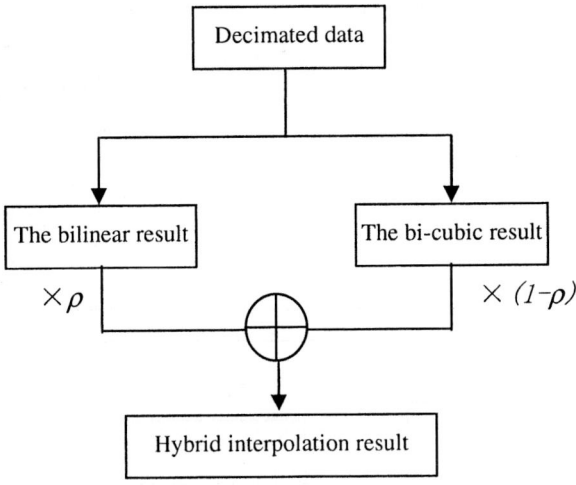

Fig. 1. The general approach of the hybrid interpolation method

We know the bilinear interpolation expression is

$$f_1(x, y) = a_1 x + a_2 y + a_3 xy + a_4 \qquad (2)$$

The implementation of this method is to utilize four corner points of grid and interpolate linearly along the boundaries of the grid. The 10-term bi-cubic interpolation, which is another widely used technique for image zooming, is given by

$$\begin{aligned}f_2(x, y) = &b_1 + b_2 x + b_3 y + b_4 x^2 + b_5 xy + b_6 y^2 \\ &+ b_7 x^3 + b_8 x^2 y + b_9 xy^2 + b_{10} y^3\end{aligned} \qquad (3)$$

The bi-cubic method is the lowest order polynomial interpolation technique in the 2-dimension case, which keeps the smoothness of the function and its first derivatives across grid boundaries.

The hybrid interpolation method based on the bi-linear and by bicubic is then proposed and given by

$$I = \rho A + (1-\rho) B, \ (0 \le \rho \le 1) \qquad (4)$$

here I, A and B represent the hybrid interpolator, bilinear interpolator and bi-cubic interpolator respectively. The constant ρ is called the hybrid parameter. Clearly, the bi-cubic and bi-linear interpolation methods are actually the special case of the hybrid interpolation method when hybrid parameter $\rho = 0$ and $\rho = 1$, respectively. It can be seen that the hybrid interpolation method is similar to the bi-cubic method in terms of time cost.

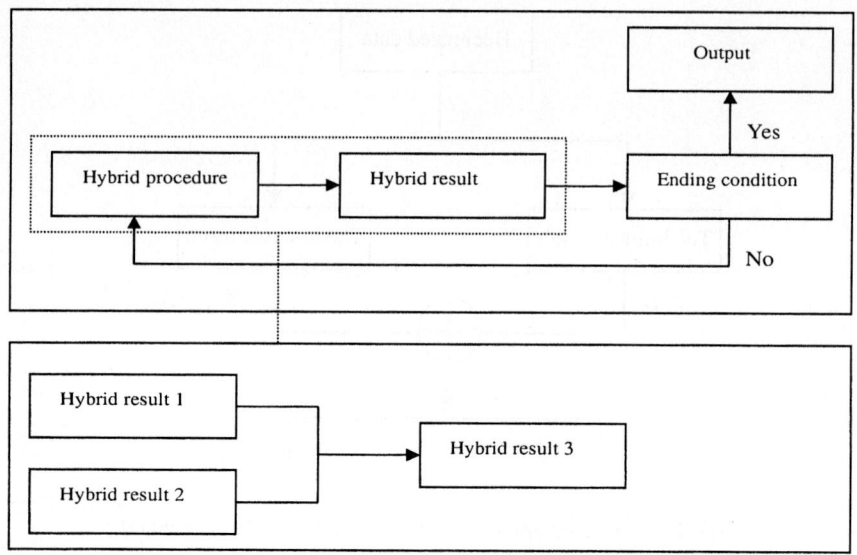

Fig. 2. The flow chart of the iterative hybrid interpolation method

2.2 Discussion on the Hybrid Parameter

Obviously, the hybrid parameter ρ plays an important role in this algorithm. Now we analyse this issue from the theoretical point of view.

For sake of discussion, we let f and f_0 denote the original data and sampled data, f_1, f_2 and f^ρ represent the bilinear, bi-cubic and hybrid interpolation results of f_0, respectively. Under this denotations, the mean square error (MSE) is defined by

$$MSE = \frac{\sum_{i=1}^{n}\sum_{j=1}^{m}(f_{i,j} - f_{i,j}^{\rho})^2}{n \times m} \tag{5}$$

here $f_{i,j}$ and $f_{i,j}^{\rho}$ are the elements of f and f^{α}. Now, let $\Delta f = | f - f^{\alpha} |$, we consider the error of f^{ρ} as follows:

$$\| \Delta f \|^2 = < f - f^\rho, f - f^\rho > \\ = < f, f > + < f^\rho, f^\rho > - 2 < f, f^\rho > . \tag{6}$$

Here, $< \bullet, \bullet >$ means the inner product of vectors. Hence, (5) is equal with (6) except a constant. Thus

$$\frac{\partial \| \Delta f \|^2}{\partial \rho} = \frac{\partial < f^\rho, f^\rho >}{\partial \rho} - 2\frac{\partial < f, f^\rho >}{\partial \rho} \tag{7}$$

Notice that $f^\rho = \rho A f_0 + (1-\rho) B f_0$, we have

$$\begin{aligned}<f^\rho, f^\rho> &=<\rho A f_0+(1-\rho)Bf_0, \rho Af_0+(1-\rho)Bf_0>\\ &=\rho^2 \|Af_0\|^2+(1-\rho)^2\|Bf_0\|^2+2\rho(1-\rho)<Af_0,Bf_0>,\end{aligned} \quad (8)$$

$$\begin{aligned}<f,f_\rho> &=<f, \rho Af_0+(1-\rho)Bf_0>\\ &=\rho<f,Af_0>+(1-\rho)<f,Bf_0>.\end{aligned} \quad (9)$$

So

$$\frac{\partial <f^\rho, f^\rho>}{\partial \rho} = 2\rho \|Af_0\|^2 - 2(1-\rho)\|Bf_0\|^2 \\ +2<Af_0,Bf_0>-4\rho<Af_0,Bf_0> \quad (10)$$

$$\frac{\partial <f,f^\rho>}{\partial \rho} = <f,Af_0>-<f,Bf_0>=<f,Af_0-Bf_0> \quad (11)$$

Thus

$$\frac{\partial \|\Delta f\|^2}{\partial \rho} = 2\rho\|Af_0\|^2-2(1-\rho)\|Bf_0\|^2+2<Af_0,Bf_0>\\ -2\rho<Af_0,Bf_0>-2<f,Af_0-Bf_0> \quad (12)$$

Let $\frac{\partial \|\Delta f\|^2}{\partial \rho}=0$, by Eq.(12), we have

$$\rho(\|Af_0\|^2+\|Bf_0\|^2-2<Af_0,Bf_0>)\\ =\|Bf_0\|^2+<f,Af_0-Bf_0>-<Af_0,Bf_0> \quad (13)$$

Then

$$\rho = \frac{<Bf_0-Af_0, Bf_0-f>}{\|Af_0-Bf_0\|^2} \quad (14)$$

From the above equation (14) we can find that the smaller the difference between Bf_0 and f, the smaller the parameter ρ. This again confirms the conclusion that the hybrid parameter α stands for the proportion of the low frequency contained in the original data.

From above discussion, we see that the hybrid interpolation method is very simple and very easily to implement. However, it takes both the high and low frequency components of an image into the consideration, and is thus a method with higher accuracy.

2.3 An Estimation Method for Hybrid Parameter

For the proposed method, a very important problem is how to select a suitable hybrid parameter. Motivated by the method based on the analysis of error parameter curve that was first proposed for the estimation of blurring function [8], we present an alternative method to estimate the hybrid parameter.

For sake of convenience, we denote the hybrid interpolation result and the original low-resolution image as f_0 and f_ρ, respectively. The estimation method used for hybrid parameter is designed as follows:

Step 1: given an estimative interval for the hybrid parameter $[a,b]$, that is to suppose $\rho \in [a,b]$. The initial iterative value, iterative step and the time of the iterative are denoted by ρ_0, $\Delta\rho$ and K, respectively.

Step 2: for $i = 1 : K$
$$\rho = \rho_0 + (i-1)\Delta\rho;$$
the hybrid interpolated result $\tilde{f}_\rho$ is calculated by the parameter ρ;

decimated $\tilde{f}_\rho$ by ratio 2:1, the obtained result is denoted as f_ρ;

calculate the error $E = \| f_0 - f_\rho \|^2$
end

Step 3: draw the error parameter ($E - \rho$) curve, and achieve an approximation of the hybrid parameter.

Remark: as claimed as step 3, a reasonable estimation of the hybrid parameter is obtained by analysis the $E - \rho$ curve. The method is to select a point from the flat region of the curve from the right side to the left side of the coordinate system.

3 Experimental Study and Analysis

In this section, two experimentations are conducted to illustrate the effectiveness of the iterative hybrid method. The simulation scheme is designed as the following:

Step1: To select a high-resolution image as the original image;
Step2: To re-sample the original image by certain level of ratio to generate a decimated image;
Step3: To interpolate the decimated image by the bilinear, bi-cubic and the iterative hybrid interpolation methods respectively; and
Step4: To assess the results obtained by these methods either qualitatively or quantitatively.

The first data set used for the iterative hybrid algorithm assessment is the woman image. Fig. 3 (a) is the decimated image with the decimation ratio 2:1. Fig.3 (b), (c) and (d) are the results obtained by the bi-cubic, bilinear and the iterative hybrid methods, respectively. For the iterative hybrid method, the parameter $\rho = 0.312$ and the final result is obtained by 3 steps. From Fig.3, we can find that there is not obvious difference between the three methods by a cursory visual examination. However, if we take a close observation, it can be found that the bilinear has the worst visual effect; the result provided by the iterative hybrid method is the best, while the result from bi-cubic interpolated result is in the middle in terms of visual effect.

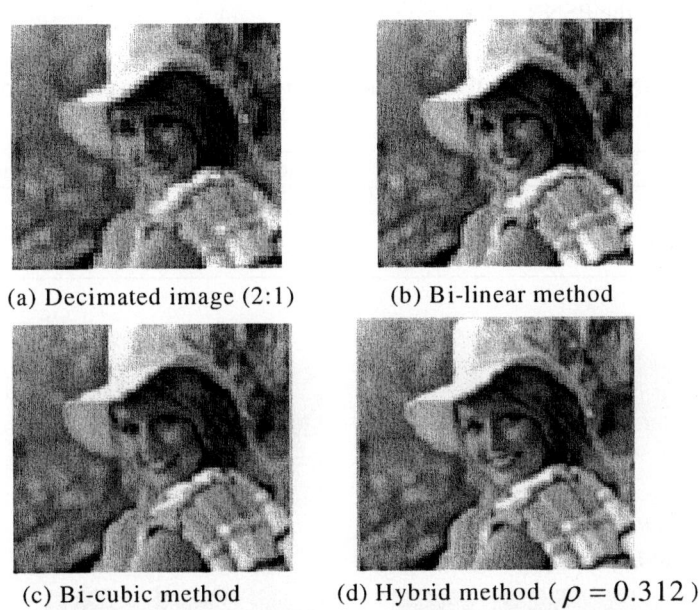

(a) Decimated image (2:1) (b) Bi-linear method
(c) Bi-cubic method (d) Hybrid method ($\rho = 0.312$)

Fig. 3. A comparison of the interpolation results based on three different interpolation methods

An objective assessment method is provided based on the Figure of Merit (FOM) -- peak-signal-noise-ratio (PSNR). The PSNR of these three methods is shown in Table 1. We see that PSNR of the bilinear, bi-cubic and the iterative hybrid methods are 28.5043 (db), 29.1147 (db) and 30.5504 (db) respectively. The difference between that of the iterative hybrid method and the bilinear method are 2.0461 (db). The result obtained by iterative hybrid method is greater than the result obtained by the bi-cubic by 1.4357 (db).

The second data is a remote sensing with clouds, the decimated image (decimation ration is set by 4:1) is shown in Fig.4 (a). The interpolation results obtained by the bilinear, bi-cubic and iterative hybrid interpolation methods are shown in Fig.4 (b), (c) and (d) respectively. In this case, the hybrid parameter ρ is 0.223, and the number of the iterative step is set to 4.

From Fig.4 (d), we see that the result obtained by the bilinear method also has a relatively poorer visual effect. However, there are slightly differences between the results obtained by the bi-cubic and iterative hybrid methods. The result obtained by the iterative hybrid method is smoother than that obtained by the bi-cubic for the whole scene of the image.

From Table 1, we can see that the PSNRs of the results obtained by the bilinear, bi-cubic and iterative hybrid interpolation methods are 26.5909 (db), 26.0036 (db) and 27.6711(db) respectively. Obviously, the PSNR of the result obtained by the iterative hybrid method is the largest among the results from the three methods. Comparing with the bilinear and the bi-cubic methods, the iterative hybrid interpolation is the best in terms of the PSNR value.

Table 1. The PSNRs of three different methods ($\rho = 0.312$ for Lena image, $\rho = 0.223$ for the cloud image)

	Bilinear	Bi-cubic	Iterative hybrid
Lena	28.5043	29.1147	30.5504
Cloud	26.5909	26.0036	27.6711

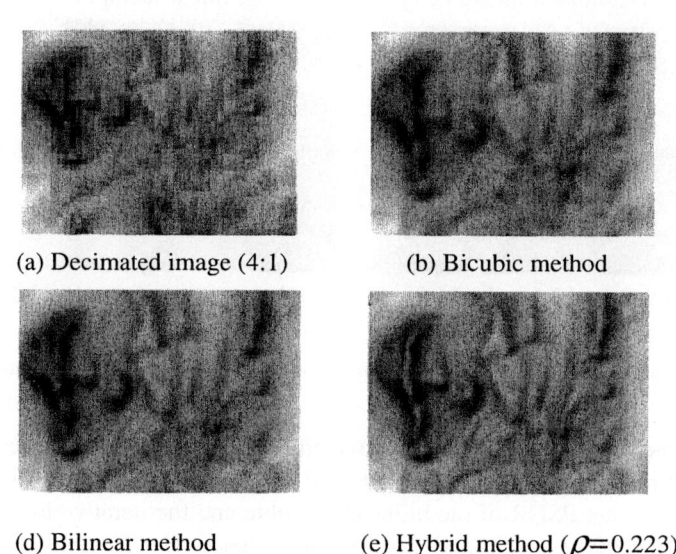

(a) Decimated image (4:1) (b) Bicubic method

(d) Bilinear method (e) Hybrid method (ρ=0.223)

Fig. 4. A comparison of the interpolation results of the remote sensing image with cloud based on three different interpolation methods

4 Conclusions

In this paper, an iterative hybrid method was proposed for interpolating images, which is an integration of the bilinear and bi-cubic interpolation methods. This model was developed based on an understanding that both low and high frequency components are normally contained in an image. A hybrid parameter adjusts the model for fitting the ratio of low and high frequency components in the hybrid model.

The value of the hybrid parameter is related to the level of complexity of the image. This method integrates the advantages of the bilinear and bi-cubic methods. The experiments showed that the iterative hybrid method has advantages than either the bilinear or bi-cubic methods in terms of PSNR. Future development of this research will be on adaptive and anisotropic determinations of the value of the hybrid parameter.

References

1. Robinson, J.A.: Efficient General Purpose Image Compression with Binary Tree Predictive Coding. IEEE Transactions on Image Processing, Vol.6, (1997)601-608
2. Price, J.R., Hayes, M.H.: Optimal Pre-filtering for Improved Image Interpolation. Conference Record of the Thirty-Second Asilomar Conference on Signals.Systems & Computers, 1-4 Nov. 1998, Vol.2, III, (1998)959-963
3. Wang, Q., Ward, R.: A New Edge-directed Image Expansion Scheme. Proceedings of the 2001 International Conference on Image Processing, 7-10 Oct. 2001, Vol.3, (2001)899 - 902
4. Bai, Y., Wang, D.: On The Comparison of Interpolation Techniques for Robotic Position. IEEE International Conference on Compensation, Systems, Man and Cybernetics, 5-8 Oct., 2003, Vol. 4, (2003)3384 -3389
5. Hadhoud, M.M., Dessouky, M.I., et al.: Adaptive Image Interpolation Based on Local Activity Levels. Proceedings of the 20th National Radio Science Conference, 18-20, March, 2003, (2003) C4_1 -C4_8
6. Shenoy, R.G., Parks, T.W.: An Optimal Recovery Approach to Interpolation. IEEE Transactions on Signal Processing, Vol.40, (1992)1987-1996
7. Zou, M.Y.: Deconvolution and Signal Recovery. Defense industry press, Beijing (2001)

Research on Reliability Evaluation of Series Systems with Optimization Algorithm

Weijin Jiang[1], Yusheng Xu[2], and Yuhui Xu[1]

[1] Department of computer, Hunan University of Industry,
Zhuzhou 412008, P.R. China
jwjnudt@163.com
[2] College of Mechanical Engineering and Applied Electronics,
Beijing University of Technology,
Beijing 100022, P.R. China
yshxu520@163.com

Abstract. The failure probability of a system can be expressed as an integral of the joint probability density function within the failure domain defined by the limit state functions of the system. Generally, it is very difficult to solve this integral directly. The evaluation of system reliability has been the active research area during the recent decades. Some methods were developed to solve system reliability analysis, such as Monte Carlo method, importance sampling method, bounding techniques and Probability Network Evaluation Technique (PNET). This paper presents the implementation of several optimization algorithms, modified Method of Feasible Direction (MFD), Sequential Linear Programming (SLP) and Sequential Quadratic programming (SQP), in order to demonstrate the convergence abilities and robust nature of the optimization technique when applied to series system reliability analysis. Examples taken from the published references were calculated and the results were compared with the answers of various other methods and the exact solution. Results indicate the optimization technique has a wide range of application with good convergence ability and robustness, and handle problems under generalized conditions or cases.

1 Introduction

The evaluation of system reliability analysis has been the active research area during the recent decades. Generally, the failure probability of a system can be expressed as

$$p_f = \int \cdots \int_D f_x(X) dX \qquad (1)$$

Where X is a vector of the random variables, f_x is the joint probability density function and D is the failure domain defined by the limit state functions of the system. It is very difficult to solve the integral Eq.(1) directly, and numerical simulation methods (such as Monte Carlo method, importance sampling method) and bounding techniques, probabilistic Network Evaluation Technique (PNET) were always employed [1-5].

The present paper focuses on the optimization techniques to evaluate system reliability. First, the multi-limit state functions of the series system are transformed into an equivalent limit state function, and to calculate system reliability is to calculate the minimum distance from the origin to the single equivalent limit state function in standard normal space according to the concept of the First Order Second Moment (FOSM). Three optimization algorithms are employed to solve this optimization problem, which are modified of Feasible Direction (MFD), Sequential Linear programming (SLP) and Sequential Quadratic programming (SQP) [6-8]. To demonstrate the convergence abilities and robust nature of the optimization technique for system reliability analysis, the Monte Carlo method, the importance sampling method and HL-RF algorithm are also used, and their results are compared as well.

2 Equivalent Limit State Function of Series System

There are many types systems, all of which can be categorized into two fundamental systems, i. e. series system and parallel system. Series systems can be thought of by a "weakest link" analogy. If one component fails, the entire system reaches failure. Parallel systems are also known as "redundant system" If one element fails, the system does not necessarily reach failure. In the present paper, series systems are studied [9].

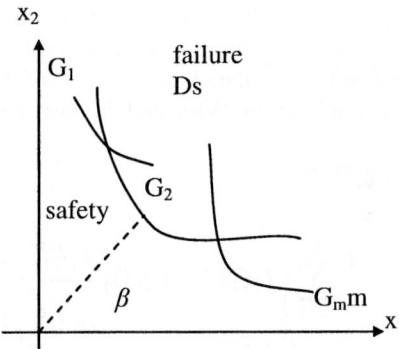

Fig. 1. Series system

For a series system, the failure region is the union of the failure regions of each limit state function, and the failure probability is expressed as Fig.1

$$P_f = P(\bigcup_{i=1}^{m} G_i \leq 0) = \int \cdots \int_{D_s} f_x(X) dX \qquad (2)$$

The equivalent limit state function of the series system can defined as

$$G_{series} = \min(G_1, G_2, \cdots, G_m) \qquad (3)$$

Thus, Eq.(2) can be written as

$$P_f = P(G_{series} \leq 0) = \int \cdots \int_{Gseries \leq 0} x(X)dX \qquad (4)$$

3 Structural Reliability Analysis

Using the equivalent limit state function defined in Eq.3, the reliability analysis methods of the component case can also be used to solve the system reliability problem. Many methods have been developed and implemented to solve the limit state functions for component failure probability[10-12].

3.1 Monte Carlo Method

$$P_f \approx J_1 = \frac{1}{N}\sum_{j=1}^{N} I[G(\hat{X}_j) \leq 0], \qquad (5)$$

Where $I[\ ]$ is an indicator function, equals to 1 if $[\ ]$ is true and equals to 0 if $[\ ]$ is false, $\hat{X}_j$ is the *jth* sample of the vector of random variable, N is the total number of samples. A very large number of samples are needed for the Monte Carlo method to the problem with low failure probability, which means the Monte Carlo method is time consuming. A strength of this method is its broad problem applicability. As long as the limit state function can be calculated, continuity and the ability to derivate the function is not a problem in application. With such higher can be relatively accurate.

3.2 Importance Sampling Method

$$P_f \approx J_2 = \frac{1}{N}\sum_{j=1}^{N}\left\{I[G(\hat{X}_j) \leq 0]\frac{f_x(\hat{X}_j)}{h_x(\hat{X}_j)}\right\}, \qquad (6)$$

Where $h_v()$ is called the importance sampling probability density function. The proper selection of $h_v()$ will generate more samples in the failure region, and thus can reduce the number samples needed.

3.3 First Order Second Moment Method

The First Order Second Moment method (FOSM) is a technique used to transform the integral into a multi-normal joint probability density function. This is done by utilizing a linear representation of a limit state function by way of its first two moments, the mean and standard deviation. Hasofer and Lind defined reliability index as the shortest distance from the origin to the limit state surface, which can be expressed as

$$\beta = \min \left(\sum_{i=1}^{n} y_i^2 \right)^{1/2} = \min(y^T \bullet y)^{1/2}, \qquad (7)$$

subject to $G(y)=0$, where $G(y)$ is the limit state function in standardized normal space, y_i is the independent random variable of standard normal distribution on the limit state surface.

For the limit state function with general random variables of non-normal distribution, the random variables can be transformed into the independent equivalent normal random variables by normal tail transformation, Rosenblatt transformation or Nataf transformation.

The relation between reliability index and failure probability is

$$Pf = \phi(-\beta), \qquad (8)$$

where ϕ represents the standard normal distribution. Eq.8 is accurate only for the specific case of random variables with normal distribution and linear limit state function.

It should be notes that the FOSM method is in fact a type of optimization, as shown in Eq.7. Various algorithms can be employed to solve such an optimization problem. The famous HL-RF algorithm is an efficient choice for many problems, in which the reliability index is found by an iterative process after a linear approximation of the limit state function, defined at design point (expressed by Eq.7)[2,3]

$$\beta = -\sum_{i=1}^{n} y_i^{\#} (\partial G / \partial y_i) / [\sum_{i=1}^{n} (\partial G / \partial y_i)^2]^{1/2} \qquad (9)$$

4 Optimization Algorithms

A general optimization problem can be expressed as:

$$\begin{aligned} &\min_{x} \quad W(X) \\ &\text{s.t.} \quad G_j(X) \le 0 \quad j=1,\ldots,m \\ &\quad X^L \le X \le X^U \end{aligned} \qquad (10)$$

Three optimization algorithms available in DOT program (Design Optimization Tools) are used in the present paper, which are the Modified method of Feasible Direction (MFD), Sequential Linear Programming (SLP) and sequential Quadratic programming (SQP) [4].

The feasible direction method is in the class of direct search algorithms, which can be stated as:

$$X^{k+1} = X^k + a^k d^k, \qquad (11)$$

where X^k and X^{k+1} are the kth and $(k+1)$th point, respectively, in design space, d^k is the search direction and a^k is the distance of travel between these two design points. The critical parts of the optimization task are finding a usable search direction and travel distance. Here we use the Fletcher-Reeves conjugate direction method, the search direction as:

$$dk = -\nabla W(Xk-1) + \frac{\|W(X^{k-1})\|^2}{\|W(X^{k-2})\|^2} d^{k-1} . \qquad (12)$$

Having determined a usable-feasible search direction, the problem now becomes one-dimensional search that minimizes $W\left(X^{k-1} + a^k d^k\right)$ which can be solved by many available algorithms.

4.1 Sequential Linear Programming (SLP)

The basic concept of SLP is quite simple. First, create a Taylor Series approximation to the objective and functions

$$\tilde{W}(X) = W(X^{k-1}) + \nabla W(X^{k-1})^T (X^k - X^{k-1}) , \qquad (13)$$

$$\tilde{g}_j(X) = g_j(X^{k-1}) + \nabla g_j(X^{k-1})^T (X^k - X^{k-1}) . \qquad (14)$$

Then, use this approximation for optimization, instead of the original nonlinear functions. During the optimization process, define move limits on the design variables. Typically, during one cycle, the design variables will be allowed to change by 20%~40%, but this is adjusted during later cycles.

4.2 Squential Quadratic Programming (SQP)

The basic concept is very similar to that of SLP. First, create a Taylor Series approximation of a quadratic approximate objective function and linearized constraints, with which a direction finding problem is formed as follows:

$$\min\ Q(d) = W^0 + \nabla W^T d + \frac{1}{2} d^T B d \qquad (15)$$
$$\text{s.t.}\quad g_j^0 + \nabla g_j^T d < 0 \qquad j = 1,...,m$$

This sub-problem is solved using MFD. The matrix B in Eq.15 is a positive define matrix, which is initially the identity matrix. On subsequent iterations, B is updated to approach the Hessian of the Lagrangian function.

$$\Phi = W(X) + \sum_{j=1}^{m} u_j \max[0, g_j(X)] , \qquad (16)$$

Where

$$X = X^{k-1} + ad \qquad (17)$$

After the one-dimensional search is completed, the matrix B is updated using the BFGS formula [4].

5 Numerical Examples

The design Optimization Tools (DOT) program and the reliability part of programs for Reliability Analysis and Design of Structural Systems (PRADSR) are employed in the present paper. PRADSR is a program that solves for the probability of failure, coefficient of variance, and reliability index by the Monte Carlo method, the FOSM method, and importance sampling.

All the examples evaluated in the present paper are taken form the published papers. The first two examples are component reliability problems to show the general applicability, convergence ability and robust nature of the optimization technique. Example 3~5 are reliability analysis for series systems. The results for each optimization method will be represented by SQP algorithm since all three optimization algorithms produce nearly identical results. The number of samples for Monte Carlo method and importance sampling method are 100000 and 500, respectively. Error is represented by relative error with respect to the accurate results or the results in the reference where exact values are not known.

(1) Example 1

The span and height of this frame Fig.2 are 20 ft and 15 ft[8]. Its significant potential failure modes are

$$g_1 = M_1 + 3M_2 + 2M_3 - 15S_1 - 10S_2$$
$$g_2 = 2M_1 + 2M_2 - 15S_1$$
$$g_3 = M_1 + M_2 + 4M_3 - 15S_1 - 10S_2$$
$$g_4 = 2M_1 + M_2 + M_3 - 15S_1$$
$$g_5 = M_1 + M_2 + 2M_3 - 15S_1$$
$$g_6 = M_1 + 2M_2 + M_3 - 15S_1$$

All the random variables have lognormal distribution with the statistical parameters shown in table 1. The equivalent limit state function of this frame is G_{series}=min($g_1, g_2, g_3, g_4, g_5, g_6$).

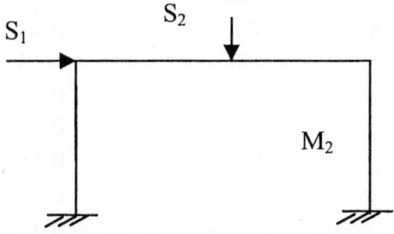

Fig. 2. One bay and one story frame

Table 1. Random variables of example 1

Variables	Mean	COV
M_1, M_2	500	0.15
M_3	667	0.15
S_1	50	0.30
S_2	100	0.15

Table 2. Results of example 1

	β_{min}	Monte	Importance	FOSM	Optim
β	3.252	3.280	3.220	3.247	3.252
Error%		0.861	0.984	0.153	0

(2) Example 2

The span and height of this frame Fig.3 are 20 ft and 15 ft[8]. and its significant potential failure modes are such as example 1.

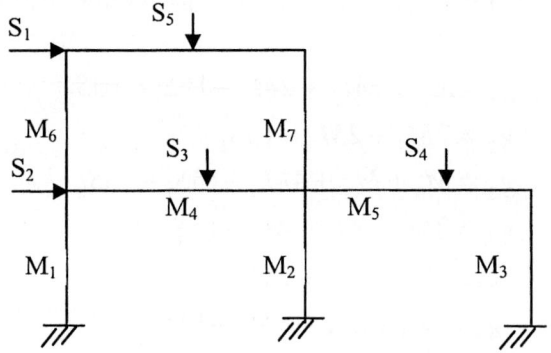

Fig. 3. Two-bay and two-story frame

Table 3. Random variables of example 2

Variables	Mean	COV
M_1, M_2, M_3	70	0.15
M_6, M_7	70	0.15
M_4	150	0.15
M_5	120	0.15
M_8	90	0.15
S_1	5	0.25
S_2	10	0.25
S_3	26.5	0.15
S_4	18	0.25
S_5	14	0.25

Table 4. Results of example 2

	β_{min}	Monte	Importance	FOSM	Optimize
β	3.100	2.900	3.199	3.205	3.100
Error%		6.452	3.194	3.387	0

(3) **Example 3**

$$g = x_1^5 + x_2^k - 18, \quad x_1, x_2 \sim N(10,5)$$

Table 5. Results of example 3 [7]

k		Ref.[7]	Monte	Importance	FOSM	Optimize
1	β	1.7	1.714	1.726	1.310	1.696
	Error%		0.824	1.541	22.941	0.255
2	β	2.4	2.316	2.275	1.326	2.137
	Error%		8.210	6.290	38.037	0.142
3	β	2.31	2.507	2.493	1.487	2.305

(4) **Example 4**

$$g_1 = 0.1(u_1^2 + u_2^2 - 2u_1u_2) - (u_1 + u_2)/\sqrt{2} + 2.5$$
$$g_2 = -0.5(u_1^2 + u_2^2 - 2u_1u_2) - (u_1 + u_2)/\sqrt{2} + 3.0$$

u_1 and u_2 are standard normal random variables.

Table 6. Results of example 4 [6]

		Accurate	Monte	Importance	FOSM	Optimize
g_1	β	2.50	2.644	2.624	2.529	2.500
	Error%		5.764	4.956	1.160	0.005
g_2	β	1.658	1.329	1.310	1.520	1.659
	Error%		8.515	7.931	35.628	0.222

6 Conclusions

The results of the examples demonstrated the convergence abilities robust nature of the optimization technique when applied to structural reliability analysis. For both component and system reliability analysis, the optimization algorithms employed in the present paper (MFD, SLP, SQP) can produce satisfying solutions with negligible error. The FOSM method can also give satisfying results for some cases (such as example 4 and 5) and may produce high error by linear representation of highly nonlinear functions at the design point for other cases, such as the results of example 3 with the relative error of 49.12%. Monte Carlo method can always produce good

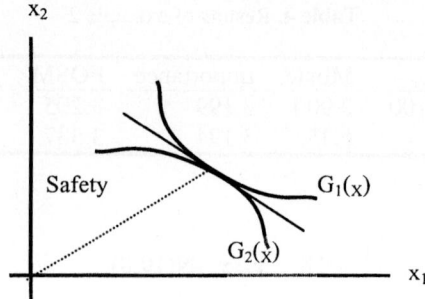

Fig. 4. FOSM Error

results as long as the number of the samples is enough, which results in expensive computational efforts. Importance sampling method can reduce the number of samples a lot by introducing the importance sampling probability density function, however, the design point should be determined at first.

In reliability analysis, two standard indices are used; the probability of failure and the reliability index. They can be transformed into each other by Eq.8, which is accurate only for the specific case of random variables with normal distribution and linear limit state functions. The Monte Carlo method solves the former while the FOSM and optimization methods solves for the later. As shown by the above examples, this transformation can introduce error when soling for one specific index. A graphical explanation is shown in Fig.4. The failure region, and thus the probability of failure, is greater for $G_2(x)$. The shortest distance, however, is the same for both functions. Therefore, the errors between the different methods of Monte Carlo, importance sampling, FOSM and optimization algorithms are caused by (1) the algorithms (such as FOSM method) and (2) the transformation between reliability index and failure probability by Eq.8 for nonlinear limit state function with random variables other than normal random varia0vles (such as Monte Carlo method and importance sampling).

Optimization technique is a good method for finding the reliability index of a structural reliability problem. As with all methods, application should fit the research or design problem. The use of other methods in conjunction with optimization method is recommended for reference and comparative use. Future research could focus on optimization technique when applied to parallel and combined systems considering especially on how to apply optimization technique to such systems.

Acknowledgments

The paper was supported by the National Natural Science Foundation of China (No. 60373062), and the Natural Science Foundation of Hunan Province of China (No. 04JJ3052).

References

1. Melchers, R. E.: Structural Reliability Analysis and Prediction [M]. John Wiley & Sons , (1999)
2. Hasofer, A. M., Lind, N. C.: An exact and invariant first order reliability format [J]. J Eng Mech Div, ASCE, 100 (EM1),(1974)111-121
3. Rackwitz R, Fiessler B: Structural reliability under combined random load sequences [J]. Computers and Structures, 9(1978) 489-494
4. DOT user manual: Vanderplaats Research and Development[M]. Inc. , Colorado Springs, CO, (1999)
5. Sorensen, J. D.: PRADSR: Reliability part of PRADSS{r}. Aalborg University, Denmark, (1994)
6. Borri,A., Speranzini, E.: Structural reliability analysis using a standard deterministic finite element code [J]. Structural Safety, , 19(4),(1997) 361-382
7. Xu, L., Cheng, G. D.: Discussion on moment methods for structural reliability analysis [J]. Structural Safety, 25(2003) 1993-1999
8. Zhao Y G, Ono T: System reliability evaluation of ductile frame structures [J]. Journal of Structure Engineering, 124(6), (1998)678-685
9. Weijin Jiang: Hybird Genetic algorithm research and its application in problem optimization. Proceedings of 2004 International Conference on Manchine Learning and Cybernetics, (2004)222-227
10. Weijin Jiang: Research on Optimize Prediction Model and Algorithm about Chaotic Time Series, Wuhan University Journal of Natural Sciences, 9(5), (2004) 735-740
11. Jiang Weijin. Modeling and Application of Complex Diagnosis Distributed Intelligence Based on MAS. Journal of Nanjing University (Natural Science), 40(4), (2004) 1-14
12. Jiang Weijin, Xu Yusheng, Sun Xingming. Research on Diagnosis Model Distributed Intelligence and Key Technique Based on MAS. Journal of Control Theory & Applications, 20(6), (2004) 231-236

Image Registration Based on Pseudo-Polar FFT and Analytical Fourier-Mellin Transform[*]

Xiaoxin Guo, Zhiwen Xu, Yinan Lu, Zhanhui Liu, and Yunjie Pang

College of Computer Science and Technology, Jilin University,
Key Laboratory of Symbol Computation and Knowledge Engineering of the Ministry of Education, Jilin University,
Qianjin Street 2699, Changchun, 130012, P.R. China
xiaoxin@mail.jl.cn

Abstract. This paper proposes a novel registration algorithm based on Pseudo-Polar Fast Fourier Transform (FFT) and Analytical Fourier-Mellin Transform (AFMT) for the alignment of images differing in translation, rotation angle, and uniform scale factor. The proposed algorithm employs the AFMT of the Fourier magnitude to determine all the geometric transformation parameters with its property of the invariance to translation and rotation. Besides, the proposed algorithm adopt a fast high accuracy conversion from Cartesian to polar coordinates based on the pseudo-polar FFT and the conversion from the pseudo-polar to the polar grid, which involves only 1D interpolations, and obtain a more significant improvement in accuracy than the conventional method using cross-correlation. Experiments show that the algorithm is accurate and robust regardless of white noise.

1 Introduction

Proper integration of useful data obtained from the separate images is often desired, since information gained from different images acquired in the track of the same events is usually of a complementary nature. A first step in this integration process is to find an optimal transformation between an image pair and to bring the contents involved into spatial alignment, a procedure referred to as registration. After registration, a fusion step is required for the integrated display of the data involved.

Many classical techniques for registering two images with misalignments due to the geometric transformation involve using the invariants (for example, the log-polar transformed Fourier magnitudes of the two images) and calculating the 2D cross-correlation function with respect to the invariants to determine the optimal rotation angle and scale factor [1]-[6]. Although the cross-correlation technique is reliable, efficient, and immune to white noise, but it causes resampling error during the conversion from Cartesian to polar coordinates. The disadvantage directly results in low peak correlations and low signal-to-noise ratio.

---

[*] This work was supported by the foundation of science and technology development of Jilin Province, China under Grant 20040531.

The optical research community first introduced the FMT for pattern recognition [7][8]. Several sets of rotation- and scale-invariant features based on the FMT modulus have been designed, but numerical estimation of the Mellin integral brings up crucial difficulties [9]. A solution for the convergence of the integral has been given by using the Analytical Fourier-Mellin Transform (AFMT), and a complete set of similarity-invariant features for planar gray-level images was proposed [10].

2 The Analytical Fourier-Mellin Transform and Approximation

Let $f(r, \theta)$ be the irradiance function representing a gray-level image defined over a compact set of $\mathcal{R}^2$. The origin of the polar coordinates is located on the image center in order to offset translation. The analytical Fourier-Mellin transform (AFMT) of f is given by [10]:

$$\forall (k,v) \in \mathcal{Z} \times \mathcal{R}, \quad M_{f_\sigma}(k,v) = \frac{1}{2\pi} \int_0^\infty \int_0^{2\pi} f(r,\theta) r^{\sigma-iv} e^{-ik\theta} d\theta \frac{dr}{r}, \quad (1)$$

with $\sigma > 0$. M_{f_σ} is assumed to be summable over $\mathcal{Z} \times \mathcal{R}$. The AFMT of an image f can be seen as the usual FMT of the distorted image $f_\sigma(r,\theta) = r^\sigma f(r,\theta)$ with $\theta > 0$.

Let g denote the rotation and size change of a gray-level image f through angle $\beta \in \mathcal{S}^+$ and scale factor $\alpha \in \mathcal{R}^*$, i.e. $g(r,\theta) = f(\alpha r, \theta + \beta)$. The AFMT of g is

$$\forall (k,v) \in \mathcal{Z} \times \mathcal{R}, \quad M_{g_\sigma}(k,v) = \frac{1}{2\pi} \int_0^\infty \int_0^{2\pi} f(\alpha r, \theta + \beta) r^{\sigma-iv} e^{-ik\theta} d\theta \frac{dr}{r}, \quad (2)$$

and performing a simple change of the variables r and θ gives the following relation:

$$\forall (k,v) \in \mathcal{Z} \times \mathcal{R}, \quad M_{g_\sigma}(k,v) = \alpha^{-\sigma+iv} e^{ik\beta} M_{f_\sigma}(k,v). \quad (3)$$

The relation in Eq. (3) makes the AFMT appropriate for extracting features that are invariant to scale and rotation changes. However, the usual modulus-based FMT descriptors are no longer invariant to scale because of the $\alpha^{-\sigma}$ term.

The digital AFMT approximation consists of resampling $f(i_1, i_2)$ in discrete polar coordinates and estimating the Fourier-Mellin integrals (1). The polar sampling grid is built from the intersection between M concentric circles with increasing radii of fixed spacing and N rays (or radial lines) originating from the image center. The angular and radial sampling steps are $\Delta\theta = 2\pi/N, \Delta r = R/M$, respectively, where R denotes the radius of the smallest disk required to contain the whole image.

Wherever the polar sampling point does not correspond to a grid location, the gray-level value is estimated by pseudo polar conversion using 1D interpolation with the higher accuracy than the other direct methods. Hence, the polar representation of an image is an $[M, N]$-matrix whose values correspond to

$$\hat{f}(\hat{r}_m, \hat{\theta}_n), \quad m \in [0, M-1], n \in [0, N-1], \tag{4}$$

where $r_m = m\Delta r$ and $\theta_n = n\Delta\theta$ are respectively the ray with index m and the circle with index n. Replacing integrals over circles and rays in Eq. (1), we get the approximation $\hat{M}_{f_\sigma}$ of f:

$$\hat{M}_{f_\sigma}(k, v) = \Delta r \Delta\theta \sum_{n=0}^{N-1} \sum_{m=0}^{M-1} \hat{f}(\hat{r}_m, \hat{\theta}_n) \exp(-ikn/N)(\hat{r}_m)^{\sigma-iv-1}, \tag{5}$$

where $\forall k \in [-K, K], \forall v \in [-V, V]$.

3 The Proposed Scheme

Consider for registration two functions denoted by f and g, representing two gray-level images defined over a compact set of $\mathcal{R}^2$, respectively, which are related by a four-parameter geometric transformation that maps each point in g to a corresponding point in f

$$g(x, y) = f(\alpha(x\cos\beta + y\sin\beta) - \Delta x, \alpha(-x\sin\beta + y\cos\beta) - \Delta y), \tag{6}$$

where Δx and Δy are translations, α is the uniform scale factor, and β is the rotation angle. According to the translation, reciprocal scaling and rotation properties, it may be readily shown that the magnitudes of the Fourier transform of these images are invariant to translation but retain the effects of rotation and scaling, as follows

$$|G(p, q)| = \frac{1}{\alpha^2} \left| F\left(\frac{p\cos\beta + q\sin\beta}{\alpha}, \frac{-p\sin\beta + q\cos\beta}{\alpha}\right) \right|, \tag{7}$$

where $F(p, q)$ and $G(p, q)$ are the Fourier transforms of $f(x, y)$ and $g(x, y)$, respectively.

As mentioned above, due to the $\alpha^{-\sigma}$ term, the modulus-based FMT descriptors are variant to scale, which contribute to the determination of scale factor. Note that in Eq. (7) a geometric transformation in the image domain corresponds to in the Fourier magnitude domain a combination of rotation angle α and uniform scale factor $1/\beta$, without translation, so the AFMT method can be readily used to find a set of appropriate transformation parameters when applying the AFMT in Fourier magnitudes rather than the original images.

3.1 The Conversion

The approximation process mentioned above can be implemented by an alternative process called polar fast Fourier transform (PFFT) [11][12]. We compute the polar-Fourier transform values based on a different grid for which fast algorithm exists, and then go to the polar coordinates via an interpolation stage. However, instead of using the Cartesian grid in the first stage, we use the pseudo-polar one. Since this grid is closer to the polar destination coordinates, there is a reason to believe that this approach will lead to better accuracy and thus lower oversampling requirements. However, in addition to the proximity of the pseudo-polar coordinates to the polar ones, the other very important benefit is the ability to perform the necessary interpolations via the pure 1D operation without loosing accuracy. To be brief, PFFT decomposes the problem into two steps: first, a Pseudo-Polar FFT is applied, in which a pseudo-polar sampling set is used, and second, a conversion from pseudo-polar to polar Fourier transform is performed.

Pseudo-Polar FFT. The pseudo-polar Fourier transform (PPFT) evaluates the 2D Fourier transform of an image on the pseudo-polar grid, which is an approximation to the polar grid. Formally, the pseudo-polar grid is given by two sets of samples: the Basically Vertical (BV) and the Basically Horizontal (BH) subsets, defined by (see Fig. 1(a))

$$BV_0 = \{(-\frac{2sl}{N}, s)\}, BH_0 = \{(s, -\frac{2sl}{N})\}, -N \leq s < N - 1, -\frac{N}{2} \leq l < \frac{N}{2} - 1. \tag{8}$$

As can be seen in Fig. 1(a), s serves as a "pseudo-radius" and l serves as a "pseudo-angle". If we ignore overlapped intersection between concentric squares and rays, the resolution of the pseudo-polar grid is N in the angular direction and $M = 2N$ in the radial direction. Using (r, θ) representation, the pseudo-polar grid is given by

$$BV_0(s, l) = (r_s^1, \theta_l^1), BH_0(s, l) = (r_s^2, \theta_l^2), \tag{9}$$

$$r_s^1 = s(4(l/N)^2 + 1)^{1/2}, r_s^2 = s(4(l/N)^2 + 1)^{1/2}, \tag{10}$$

$$\theta_l^1 = \pi/2 - \arctan(2l/N), \theta_l^2 = \arctan(2l/N), \tag{11}$$

where $s = -N, ..., N-1$ and $l = -N/2, ..., N/2 - 1$. The pseudo-polar Fourier transform is defined on the pseudo-polar grid BV and BH, given in Eq. (8). Formally, the pseudo-polar Fourier transform $\tilde{F}_{PP}^j (j = 1, 2)$ is a linear transformation, which is defined for $s = -N, ..., N-1$ and $l = -N/2, ..., N/2 - 1$, as

$$\tilde{F}_{PP}^1(l, s) = \tilde{F}^1(-\frac{2l}{N}s, s) = \sum_{i_1=-N/2}^{N/2-1} \sum_{i_2=-N/2}^{N/2-1} f(i_1, i_2) \exp(-\frac{\pi i}{N}(-\frac{2l}{N}si_1 + si_2)),$$
(12)

$$\tilde{F}_{PP}^2(l, s) = \tilde{F}^2(s, -\frac{2l}{N}s) = \sum_{i_1=-N/2}^{N/2-1} \sum_{i_2=-N/2}^{N/2-1} f(i_1, i_2) \exp(-\frac{\pi i}{N}(si_1 - \frac{2l}{N}si_2)),$$
(13)

where $f(i_1, i_2)$ is a discrete image of size $N \times N$, and $\tilde{F}^j (j = 1, 2)$ is the 2D Fourier transform of $f(i_1, i_2)$. As we can see in Fig. 1(a), for each fixed angle l,

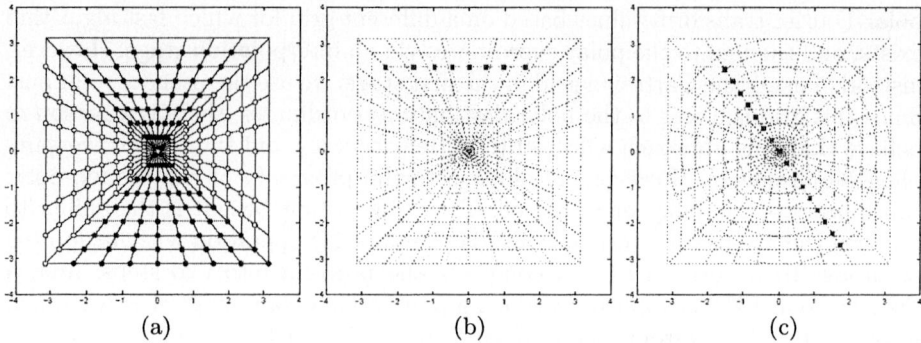

Fig. 1. (a) The pseudo-polar grid and its separation into BV and BH coordinates ($N = 8$); (b) First interpolation stage; (c) Second interpolation stage

the samples of the pseudo-polar grid are equally spaced in the radial direction. However, this spacing is different for different angles. Also, the grid is not equally spaced in the angular direction, but has equally spaced slopes.

According to Eq. (12) and (13), we can obtain two important properties of the pseudo-polar Fourier transform, i.e., it is invertible and that both the forward and inverse pseudo-polar Fourier transforms can be implemented using fast algorithms. Moreover, their implementations require only the application of 1D equispaced FFTs with complexity of $120N2\log(N)$ operations in terms of its separability property, i.e., in terms of the following formulation [12],

$$\tilde{F}_{PP}^1(l,s) = \sum_{i_1=-N/2}^{N/2-1} (\exp(\frac{2\pi i l s i_1}{N^2}) \sum_{i_2=-N/2}^{N/2-1} f(i_1,i_2)\exp(-\frac{\pi i s i_2}{N})), \quad (14)$$

$$\tilde{F}_{PP}^2(l,s) = \sum_{i_1=-N/2}^{N/2-1} (\exp(-\frac{\pi i s i_1}{N}) \sum_{i_2=-N/2}^{N/2-1} f(i_1,i_2)\exp(\frac{2\pi i l s i_2}{N^2})). \quad (15)$$

Grid Conversion: From Pseudo-Polar to Polar. Based on the pseudo-polar coordinate system, next we define the polar coordinate system and perform the grid conversion, with manipulations that lay out the necessary interpolation stages discussed later on. The polar basically vertical and basically horizontal frequency sampling points are obtained from the BV and BH subsets in the pseudo-polar grid as given in Eq. (8) by two operations:

Rotate the Rays: In order to obtain a uniform angle ray sampling as in the polar coordinate system, the rays must be rotated. This is done by replacing the term $2l/N$ in Eq. (8) with $\tan(\pi l/2N)$, being

$$BV_1 = \left\{\left(-s\cdot\tan\left(\frac{\pi l}{2N}\right), s\right)\right\}, -N \leq s < N-1, -\frac{N}{2} \leq l < \frac{N}{2}-1, \quad (16)$$

$$BH_1 = \left\{\left(s, -s\cdot\tan\left(\frac{\pi l}{2N}\right)\right)\right\}, -N \leq s < N-1, -\frac{N}{2} \leq l < \frac{N}{2}-1. \quad (17)$$

The result is a set of points organized on concentric squares as before, but the rays are spread differently with linearly growing angle instead of linearly growing slope. Rotating the rays amounts to 1D operation along horizontal lines for the BV points and vertical lines for the BH points. A set of N uniformly spread points along this line are to be replaced by a new set of N points along the same line in different locations (see Fig. 1(b)) owing to the uniform angle sampling of the new rays.

Circle the Squares: In order to obtain concentric circles as required in the polar coordinate system, we need to circle the squares. This is done by dividing each ray by a constant spacing, based on its angle, and therefore a function of the parameter l, being $R[l] = (1 + \tan^2(\pi l/2N))^{1/2}$. The resulting grid is given by

$$BV_2 = \left\{\left(-\frac{s}{R[l]}\tan\left(\frac{\pi l}{2N}\right), \frac{s}{R[l]}\right)\right\}, -N \leq s < N-1, -\frac{N}{2} \leq l < \frac{N}{2} - 1, \tag{18}$$

$$BH_2 = \left\{\left(\frac{s}{R[l]}, -\frac{s}{R[l]}\tan\left(\frac{\pi l}{2N}\right)\right)\right\}, -N \leq s < N-1, -\frac{N}{2} \leq l < \frac{N}{2} - 1. \tag{19}$$

Circling the squares also amounts to 1D operation along rays. A set of $2N$ uniformly spread points along this line are to be replaced by a new set of $2N$ points along the same line in different locations (see Fig. 1(c)). However, this time the destination points are also uniformly spread.

After the two-stage interpolation both with 1D operation, we get the polar Fourier spectrum of $f(i_1, i_2)$, including its magnitude spectrum $\hat{F}$, which will be substituted for the term $\hat{f}(\hat{r}_m, \hat{\theta}_n)$ in Eq. (5) to perform the AFMT for registration purpose.

The accuracy and error analysis of the algorithm in conversion process was studied both from the empirical and the theoretical view points in [12]. The approach is far more accurate than known state-of-the-art methods based on standard Cartesian grids. The analysis shows that the conversion scheme produces highly accurate transform values near the origin, where "typical" signals concentrate. According to the analysis, it is reasonable to expect that the following registration algorithm based on the pseudo polar grid approach can achieve satisfactory results with a low computation complexity.

3.2 The Algorithm

Eq. (3) shows that the AFMT converts a similarity transformation in the original domain into a complex multiplication in the Fourier-Mellin domain. Its logarithmic magnitude and phase representation in the Fourier-Mellin domain is written below

$$\ln A_{g_\sigma}(k, v) = -t\sigma + \ln A_{f_\sigma}(k, v), \tag{20}$$

$$\ln \Phi_{g_\sigma}(k, v) = vt + k\beta + \ln \Phi_{f_\sigma}(k, v), \tag{21}$$

where $A(k, v)$ and $\Phi(k, v)$ denote the magnitude and phase representation of the Fourier-Mellin domain, and $t = \ln \alpha$.

The proposed algorithm employs the AFMT of the Fourier magnitude, where the Fourier magnitude removes the effects of translation, whereas the AFMT eliminates the effects of rotation. Therefore, it is ready to determine the scaling factor, and further, to determine the rotation factor with the modulus of the Fourier-Mellin transform. Once the two parameters are estimated, one of the original images is appropriately scaled and rotated, and the images are cross-correlated to find the optimal translation parameters. For completeness, the algorithm is written as follows:

1. Calculate the corresponding pseudo polar Fourier magnitude $\tilde{F}_{PP}$ and $\tilde{G}_{PP}$ of an image pair f and g (Eq. (12) and (13)), respectively.
2. Perform conversion from the pseudo polar to the polar grid, and obtain the corresponding polar Fourier magnitude $\hat{F}$ and $\hat{G}$.
3. Calculate the AFMT of $\hat{F}$ and $\hat{G}$ by using the AFMT approximation given by Eq. (5), and obtain $\hat{M}_{\hat{F}_\sigma}(k,v)$ and $\hat{M}_{\hat{G}_\sigma}(k,v)$, respectively. Following the recommendation of Goh [13], σ is set to 0.5.
4. Using $\hat{M}_{\hat{F}_\sigma}(k,v)$ and $\hat{M}_{\hat{G}_\sigma}(k,v)$ in the form of Eq. (20) and (21), calculate

$$t = (\sigma(2K+1)(2V+1))^{-1} \sum_{v=-V}^{V} \sum_{k=-K}^{K} (\ln \hat{A}_{\hat{F}_\sigma}(k,v) - \ln \hat{A}_{\hat{G}_\sigma}(k,v)). \quad (22)$$

5. With the value t, calculate

$$\beta = ((2K+1)(2V+1))^{-1} \sum_{v=-V}^{V} \sum_{k=-K}^{K} \left(\ln \hat{\Phi}_{\hat{G}_\sigma}(k,v) - \ln \hat{\Phi}_{\hat{F}_\sigma}(k,v) - vt\right)/k, \quad (23)$$

$$\alpha = e^{-t}. \quad (24)$$

6. One of the image pair is appropriately scaled and rotated with the parameter α and β, and the images are cross-correlated to find the optimal translation parameters $(\Delta x, \Delta y)$.

Note that because of the use of Fourier magnitudes in place of the original images, we can readily derive the Eq. (24) according to the reciprocal scaling property.

3.3 Practical Considerations

In most applications, the range of rotations expected is limited. If the valid range of rotations is not known *a-priori*, then the effective size of the representation is one half of the size of the AFMT, because of its symmetry property.

In practice, since the images are finite, and rotating, scaling, or translating a finite image causes some of the pixel data to move out of the image frame or some new pixel data to enter the frame during transformations, the occlusion error must be considered. In order to reduce the error, the parameters K and V in Eq. (5) should be selected carefully according to image spectral content.

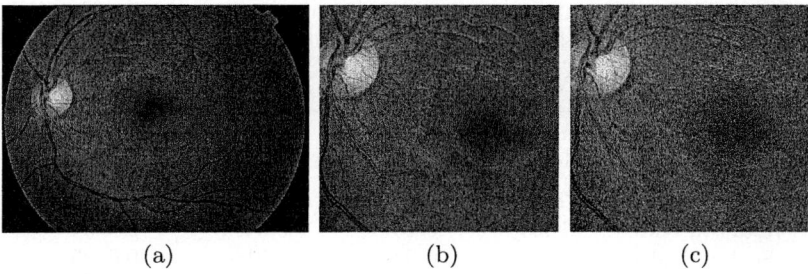

Fig. 2. (a) Digital fundus image of a normal subject (768 × 576 pixels); (b) Noiseless image with simulated misalignment; (c) Simulated Gaussian white noise added to image (b) ($\sigma^2 = 0.02$)

Moreover, considering the artifact introduced by the implicit tiling of finite images when the Fourier transform is computed for the rotation and scale phases of registration, it is desirable to perform a prefiltering prior to the Fourier transform. The filter H, for this purpose, is used to minimize the effects of the tiling. DeCastro and Morandi [14] recommend using a rotationally symmetric image frame to avoid seeing this artifact. For example, a mask shaped like a disk can be used as filter H, which zeros out pixels outside a certain radius. Here, we use a filter that blurs the borders of an image against the opposite borders. With this filter, very few pixels need be altered. In general, within a few pixels of each border there is no effect, so that a majority of pixels are unchanged. Like the round mask, this filter successfully removes the artifact.

4 Experiments

In this section, we present the experimental results for registration performance of the proposed method compared with the cross-correlation method in the case of simulated misalignment and simulated noise, respectively.

The accuracy of a registration algorithm may be evaluated by simulating a misalignment between two images. When the actual transformation between the two images is known, the error produced by the registration algorithm can be quantified.

Fig. 2(a) shows a digital fundus image that will be used in these simulations. The image is 768 × 576 pixels. 20 simulated misaligned images were generated by applying different geometric transformations, whose parameter values were chosen randomly from a uniform distribution, and then the 215 × 215 pixel region in the original and the transformed images were cropped and used for registration, thus avoiding the complications of 'edge-effects' associated with simulated transformations. Fig. 2(b) shows one of these simulations. Experimental results of the proposed method compared with cross-correlation method under the measures of root-mean-squared (RMS) errors and maximum absolute (MA) errors are given in Table 1(middle).

In order to test registration performance in presence of noise, the 20 simulated random transformations described above were regenerated with the addition of

Table 1. Experimental results with simulated misalignment and noise show the root-mean-squared (RMS) errors and maximum absolute (MA) errors of translation by pixels, rotation angle by degrees, and scale factor for 20 simulations

Errors	Simulated misalignment		Simulated noise	
	Proposed	Cross-correlation	Proposed	Cross-correlation
Translation RMS	0.2190	0.4447	0.6793	0.7621
Rotation RMS	0.1335	0.2311	0.5194	0.5954
Scaling RMS	0.0006	0.0045	0.0134	0.0185
Translation MA	0.5297	0.9501	1.8310	1.9214
Rotation MA	0.4835	0.6860	0.7289	0.9019
Scaling MA	0.0256	0.0791	0.0471	0.0886

noise. The noise was spatially uncorrelated and its amplitude was normally distributed with a zero mean. A different sequence of pseudo-random numbers was used for each image. Fig. 2(c) shows the image previously displayed in Fig. 2(b) with random noise ($\sigma^2 = 0.02$). Table 1(right) lists experimental results of the proposed method compared with cross-correlation method under the measures of RMS and MA errors.

As experimental results show, the proposed registration algorithm is able to find transformation parameters to a resolution better than the available discretization of the polar parameter space. In addition, the registration algorithm is resistant to white noise, and white noise only minimally affects the registration accuracy in the simulations.

5 Conclusions

In this paper, we propose a novel method for image registration based on analytical Fourier-Mellin transform. In implementation, we compute the polar-FT values based on the pseudo-polar grid for which fast algorithm exists, and then convert the pseudo-polar grid to the polar coordinates via an interpolation stage. In addition to the proximity of the pseudo-polar coordinates to the polar ones, the necessary interpolations can be performed only by 1D operations without loosing accuracy. After the conversion, we apply the AFMT in the polar Fourier magnitude. With the property of the invariance to translation and rotation, we can readily determine all the transformation parameters. Experimental results show that the proposed method outperforms the traditional cross-correlation method while white noise is present.

References

1. Anuta, P.E.: Spatial Registration of Multispectral and Multitemporal Digital Imagery Using Fast Fourier Transform Techniques. IEEE Trans. Geo. Elec., Vol. 8. (1970) 353–368
2. Casasent, D., Psaltis, D.: Position, Rotation, and Scale Invariant Optical Correlation. Applied Optics, Vol. 15. (1976) 1795–1799

3. Casasent, D., Psaltis, D.: Space-Bandwidth Product and Accuracy of the Optical Mellin Transform. Applied Optics, Vol. 16. (1977) 1472
4. Casasent, D., Psaltis, D.: Accuracy and Space Bandwidth in Space Variant Optical Correlators. Optics Comm., Vol. 23. (1977) 209–212
5. Casasent, D., Psaltis, D.: Deformation Invariant, Space-Variant Optical Pattern Recognition. In: Wolf, E. (ed.): Progress in Optics, North-Holland Publishing Co, Amsterdam (1978) 290–356
6. Aitmann, J., Reitbock, H.J.P.: A Fast Correlation Method for Scale- and Translation-Invariant Pattern Recognition. IEEE Trans. on Pattern Analysis and Machine Intelligence, Vol. 6. (1984) 46–57
7. Casasent, D., Psaltis, D.: Scale Invariant Optical Transform. Optical Engineering, Vol. 15, No. 3. (1976) 258–261
8. Yatagay, T., Choji, K., Saito, H.: Pattern Classification Using Optical Mellin Transform and Circular Photodiode Array. Optical Communication, Vol. 38, No. 3. (1981) 162–165
9. Zwicke, P.E., Kiss, Z.: A New Implementation of the Mellin Transform and Its Application to Radar Classification. IEEE trans. on Pattern Analysis and Machine Intelligence, Vol. 5, No. 2. (1983) 191–199
10. Ghorbel, F.: A Complete Invariant Description for Gray-Level Images by the Harmonic Analysis Approach. Pattern Recognition Letters, Vol. 15. (1994) 1043–1051
11. Averbuch, A., Shkolnisky, Y.: The 3D Discrete Radon Transform. Applied Computational Harmonic Analysis, Vol. 15, No. 1. (2003) 33–69
12. Averbuch, A., Shkolnisky, Y.: 3D Discrete X-Ray Transform. SIAM Conf. on Imaging Science 2004, Salt Lake City, Utah, USA (2004) 3–5
13. Goh, S.: The Mellin Transformation: Theory and Digital Filter Implementation. Ph.D. dissertation, Purdue University, West Lafayette, I.N. (1985)
14. DeCastro, E., Morandi, C.: Registration of Translated and Rotated Images Using Finite Fourier Transforms. IEEE Trans. on Pattern Analysis and Machine Intelligence, Vol. 9, No. 5. (1987) 700–703

Text Detection in Images Based on Color Texture Features

Chunmei Liu, Chunheng Wang, and Ruwei Dai

[1] Institute of Automation, Chinese Academy of Sciences, Beijing, China
{chunmei.liu, chunheng.wang, ruwei.dai}@ia.ac.cn

Abstract. In this paper, an algorithm is proposed for detecting texts in images and video frames. Firstly, it uses the variances and covariancs on the wavelet coefficients of different color channels as color textural features to characterize text and non-text areas. Secondly, the k-means algorithm is chosen to classify the image into text candidates and background. Finally, the detected text candidates undergo the empirical rules analysis to identify text areas and project profile analysis to refine their localization. Experimental results demonstrate that the proposed approach could efficiently be used as an automatic text detection system, which is robust for font-size, font-color, background complexity and language.

1 Introduction

Text detection is an arising research area, which plays an important role in system to index, browse and retrieve multimedia information. Texts contained in images and video frames have significant and detailed information about images, such as name and address in the name card, caption in the video, and so on. Nowadays commercial OCR systems only can handle the texts which are separated from the background and transformed to a binary image. When facing with complex background, they usually achieve poor performance. It increases the need for the automatic system of texts detection and extraction from the images and video frames.

In the past several years some research efforts have been concentrated on detecting texts and extracting texts in images. There has not been a good way to resolve the problem of text detection because it is hard to tackle the problems such as variation of font-size, font-color, language, spacing, distribution, the background complexity, influence of luminance, and so on. There are four main approaches for text localization in image. One is based on edge detection [3, 5, 9, 10, 16]. The second approach is connected-component-based analysis [4, 6, 16]. The third method is color analysis [2, 5, 6, 8]. The fourth is texture-based algorithm [1, 4, 7, 8, 9, 10, 11, 12]. These methods have different merit and shortcoming in reliability, accuracy and computation complexity. Usually the researchers combined methods mentioned above in order to abundantly absorb the information of text detection.

Texture information is very useful for image analysis. To some extent text has weak and irregular texture property, so it can be done as a special texture. In this aspect some works have been done and made some progress. llavata, J. et al. [1] apply the distribution of high-frequency wavelet coefficients to statistically characterize text and non-text areas and use the k-means algorithm to classify text areas in the image. Wu et al. [7] propose an algorithm based on the image gradient produced by nine second-order Gaussian derivatives. The pixels that have large gradient are considered as strokes of text blocks based on several empirical rules. Wenge Mao et al. [11] propose a method based on local energy analysis of pixels in images, calculated in a local region based on the wavelet transform coefficients of images. Gllavata, J. et al. [12] apply a wavelet transform to the image, and use the distribution of high-frequency wavelet coefficients to statistically characterize text and non-text areas. Chen et al. [13] propose a two-step text detection algorithm in complex background, which uses the edge information for initial text detection and employs the "distant map" as the feature and SVM as the classifier for learn-based text verification. In these above methods, the color information has been neglected to some extent. Texts in images often have the same color, and contrast clearly with the background. Color is also important information for text detection.

In contrast to other approaches, the proposed approach applies the color texture features in a local region based on the wavelet transform coefficients of images to characterize the text and non-text area. Then it uses unsupervised method to detect the text blocks from the background. It works as follows: Firstly, a wavelet transform is respectively applied to three color channels of the image. Secondly, the transformed image is scanned with a fixed size sliding window. On each window the color texture features based on color correlations of the channels are calculated. Thirdly, the k-means algorithm is used to classify the color texture features into two clusters: text candidates and background. Finally, the detected text candidates undergo the empirical rules analysis to identify text areas and project profile analysis to refine their localization. This approach is robust for font-size, font-color, background complexity and language to detect texts.

The paper is organized as follows. Section 2 presents the individual step of our approach to text detection. Section 3 discussed the experiment results. In the final section conclusions are provided.

2 Text Detection

In this section, the processing steps of the proposed approach are presented. Our aim is to build an automatic text detection system which is capable of handling still images with complex background, horizontally or vertically aligned texts, arbitrary font and color. The system needs to comply with the following assumptions: (a) input is a color image; (b) texts can not exceed a certain font size; (c) the strokes of a character have the same color. From Figure 1 we can see that the proposed approach is mainly performed by four steps: wavelet transform, color texture feature extraction, unsupervised text candidates detection, text refinement detection, which will be described in detail below.

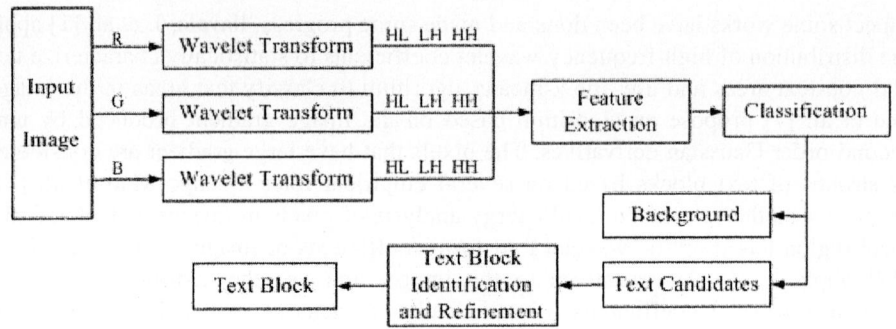

Fig. 1. Flow chart of the proposed approach

2.1 Wavelet Transform of the Image

Text is mainly composed of the strokes in horizontal, vertical, up-right, up-left direction. It has weak and irregular texture property, and can be done as a special texture. At the same time texts in images often have the same color, and contrast clearly with the background. According to these properties texture and color features are combined to detect text in images. We decompose the input image into three color channels C_i, where $i = 1, 2, 3$. On each channel a wavelet transform is applied to capture the texture property.

The main characteristic of wavelets transformation is to decompose a signal into sub-bands at various scales and frequencies, which is useful to detect edges with different orientations. In the 2-D case, when the wavelet transform is performed by a low filter and a high filter, four sub-bands are obtained after filtering: LL (low frequency), LH (vertical high frequency), HL (horizontal high frequency), and HH (high frequency). In the three high-frequency sub-bands (HL, LH, HH), edges in horizontal, respectively vertical or diagonal direction were detected. Since text areas are commonly characterized by having high contrast edges, high valued coefficients can be found in the high-frequency sub-bands.

2.2 Color Texture Feature Extraction

We employ the statistical features on the transformed image of each color channel to capture the color texture property. A sliding window of size $w \times h$ pixels is moved over the transformed image. For each window position and for each sub-band (HL, LH, HH), the features are computed. The features are mean, standard deviation, energy, entropy, inertia, local homogeneity and correlation. They are computed using the formula as followed:

$$f_1 = \frac{1}{w \times h} \sum_{i=1}^{w} \sum_{j=1}^{h} W(i, j), \qquad (1)$$

$$f_2 = \sqrt{\frac{1}{w \times h} \sum_{i=1}^{w} \sum_{j=1}^{h} [W(i, j) - \mu]^2}, \qquad (2)$$

$$f_3 = \sum_{i,j} W^2(i,j), \tag{3}$$

$$f_4 = \sum_{i,j} W(i,j) \cdot \log W(i,j), \tag{4}$$

$$f_5 = \sum_{i,j} (i-j)^2 W(i,j), \tag{5}$$

$$f_6 = \sum_{i,j} \frac{1}{1+(i-j)^2} W(i,j), \tag{6}$$

$$f_7 = \sum_{i,j} \frac{1}{1+(i-j)} W(i,j), \tag{7}$$

$$f_8 = \frac{\sum_{i,j}(i-\mu_x)(j-\mu_y)W(i,j)}{\sigma_x \sigma_y}, \tag{8}$$

Here, W is the sliding window, w is the width of the window, h is the height of the window, and (i,j) is the pixel position in the window. For each color channel C_i and for each sub-band B_j, where $i,j = 1, 2, 3$, the features f_1-f_8 form the feature vector $F_{C_i}^{B_j}$.

We compute the color texture feature vector which is based on CTF (Color Texture Feature) [18], using the formula as followed:

$$CTF^{B_j}(C_m, C_n) = Cov(F_{C_m}^{B_j}, F_{C_n}^{B_j}), \tag{9}$$

C_m, C_n, represent two color channels, where $m, n = 1, 2, 3$. In order to save computation costs, we select energy and local homogeneity features to form feature vector $F_{C_i}^{B_j}$. After the features computation, for the 1-level wavelet transformation the corresponding vectors consist of 36 features, which are energy and local homogeneity features, multiplied by 3 sub-bands and by addition between 3 color variances and 3 color covariances.

2.3 Unsupervised Text Candidates Detection

It could be assumed that the image is composed of two clusters: text and background when we detect the texts in the image. So we apply the k-means algorithm to classify the feature vectors of pixels into two clusters: text candidate and background. Here the k-means algorithm is selected because it can avoid the dependency on training data and data selection of supervised methods. Furthermore, before classification each component of the feature vector need be normalized in the range from 0 to 1. After this step, we get the initial text candidates which are binary. Seen from Figure 2-b, the result of the initial text candidates was shown after unsupervised text candidates detection.

 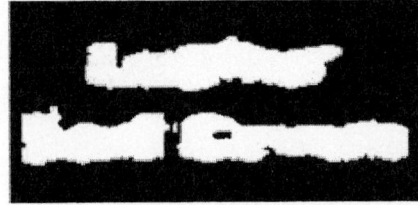

(a) Original image   (b) Initial text candidates

Fig. 2. Initial text candidates: (a) is original image, and (b) is the result of initial text candidates after classification

2.4 Text Refinement Detection

Identification and refinement of text candidates is mainly performed by five steps. Firstly, connected components of text candidates are formed by morphology operations "open" and "dilate" on the binary images of the initial text candidates, and too small isolated objects are discarded as background. Secondly, the transformed images of HL, LH, HH sub-bands are added together to get the edge map of the original image. Thirdly, we project the position of every connected component of text candidates on the edge map, and binarize every edge map using Otsu's thresholding. Fourthly, some empirical rules are performed to remove non-text blocks from the candidates. These rules are noted as followed:

- edge_area > t_1;
- edge_area / text_block_area > t_2;
- Text_block_width > t_3, Text_block_height > t_3;
- min(Text_block_width / Text_block_height, Text_block_height / Text_block _width) < t_4;
- 0 < Text_block_strokes'number / Text_block_area < t_5;

Here, t1, t2, t3, t4, t5, are the respective thresholding for above rules. According to experiments, they are appropriate as noted in Table 1.

Table 1. Parameters of empirical rules

t_1	t_2	t_3	t_4	t_5
20~120	0.2~0.3	8<=	0.8~0.9	0.01~0.05

If the text candidate doesn't satisfy the rules, it is considered that isn't text block. On the contrary, the candidate is text block if it satisfies the rules. Finally, the text candidates undergo a project profile analysis to refine the text localization, and label the result using the red rectangle to circle the text blocks. From Figure 3, the results of some samples are shown.

3 Experiment

In order to evaluate the proposed approach described here, a dataset of 100 images was obtained in magazine covers, www web images and real-life videos. We preprocess all the images into standard width (or height) of 128 pixel depending on the original image size to save computation costs. A bior3.5 wavelet was used with low-pass filter coefficients (-0.0138, 0.0414, 0.0525, -0.2679, -0.0718, 0.9667, -0.0718, -0.2679, 0.0525, 0.0414, -0.0138) and high-pass filter coefficients (-0.1768, 0.5303, -0.5303, 0.1768). The experiments with parameters w = 16, h = 8, t_1 = 100, t_2 = 0.25, t_3 = 8, t_4 = 0.85, t_5 = 0.015, which select energy and local homogeneity features, achieved a recall of 86.7% and a precision of 90.5% for the test sets. Precision and recall are defined as:

$$recall = \frac{a}{c}, \tag{10}$$

$$precision = \frac{a}{a+b}. \tag{11}$$

We checked the output for each image and measured the number of correctly detected text lines as well as the number of falsely detected text lines. Here a is the number of text lines classified as text, and b is the number of non-text lines predicted as text, and c is the total number of truly text lines in the test set.

Figure 3 shows the experimental results of various kinds of images selected from advertisements, video caption and news. In Figure 3-a, English texts on the book cover in natural image are successfully detected. Figure 3-c demonstrates that this approach is effective for the texts with the complicated background in the video frame. Figure 3-b,d shows that the texts with different font size including small size are correctly located, which demonstrates the proposed approach is robust for the different language and font size.

Considering the computation costs, we select several features to join the algorithm. For different kind of images, the features perform different performance. Here for the test sets we do the experiment to observe the detection results with different features. In Table 2, the results of the experiments are presented, where recall and precision are listed for different features.

In order to confirm the validity of the proposed approach, we conduct the experiment to detect the texts respectively in color space and in gray space. From the above diagram (Figure 4), we can see the better detection result in the color space, which illustrates the method using color texture features avails text detection, and validates the good performance of the proposed approach

(a) Scene image

(b) Magazine cover

(c) Video frame

(d) Magazine cover

Fig. 3. Experiment results of some samples

Table 2. Detection results by different features

Feature	Recall	Precision
mean	81.91%	85.85%
Standard deviation	84.62%	78.89%
energy	83.59%	88.87%
entropy	85.46%	76.90%
inertia	76.47%	70.67%
local homogeneity	78.86%	76.78%
correlation	81.25%	79.29%

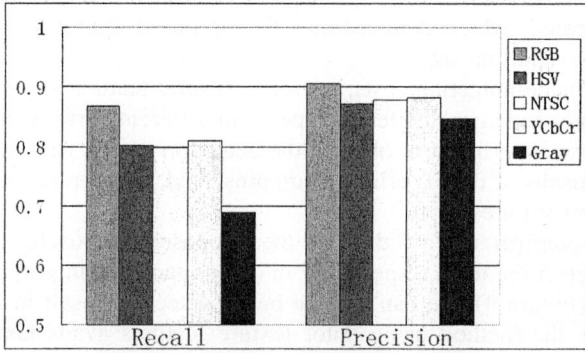

Fig. 4. Detection results in different color space and gray space

4 Conclusions

In this paper, we have proposed an algorithm based on the color texture features for detecting texts in images and video frames. According to the texture property of the texts and color contrast between texts and background in the images, the color texture features are applied to detect the texts in the images, which effectively fuse texture with color. Firstly, the wavelet transform is respectively applied to three color channels of the image. The color texture features based on color correlation of the "texture" channel are calculated in the local region. Then the classification is done by a k-means algorithm to detect the text candidates. Finally, the text candidates are refined by empirical rules. The experiment with various kinds of the images and video frames shows that the proposed method is effective on the distinction between regions and non-text regions. It is robust for font-size, font-color, background complexity and language. But the algorithm needs to be improved to save the computation cost.

We also apply this method on the images captured by the camera. And it is found that the proposed method has difficulties in detecting texts where there are strong illuminations changes and text distortion. These problems need to be tackled in the future research.

References

1. J. llavata, R. Ewerth, B. Freisleben: Text Detection in Images Based on Unsupervised Classification of High-frequency Wavelet Coefficients. Pattern Recognition, 2004. ICPR 2004. Proceedings of the 17th International Conference on, Vol. 1. (2004) 425–428
2. Kim, K.C., Byun, H.R., Song, Y.J., Choi, Y.W., Chi, S.Y., Kim, K.K., Chung, Y.K.: Scene text extraction in natural scene images using hierarchical feature combining and verificationn. Pattern Recognition, 2004. ICPR 2004. Proceedings of the 17th International Conference on , Vol. 2. (2004) 679–682
3. M. Cai, J.Q. Song, and M.R.Lyu: A new approach for video text detection. Image Processing. 2002. Proceedings. 2002 International Conference on, Vol. 1. (2002) I-117–I-120
4. V.Wu, R.Manamatha, E.Riseman: Finding text in images. 20th Int. ACM Conf. Research and Development in Information Retrieval. (1997) 3–12
5. Jiang Wu, Shao-Lin Qu, Qing Zhuo, Wen-Yuan Wang: Automatic text detection in complex color image. Machine Learning and Cybernetics, 2002. Proceedings. 2002 International Conference on, Vol. 3. (2002) 1167–1171
6. K.Jain, and B.Yu.: Automatic text location in images and video frames. Pattern recognition, Vol. 31 (1998) 2055–2076
7. V.Wu, R.Manamatha, E.Riseman: Textfinder: an automatic system to detect and recognized text in images. IEEE Trans. On PAMI, Vol. 20. (1999) 1224–1229
8. Yu Zhong, Karu, K., and Jain, A.K.: Locating text in complex color images. Document Analysis and Recognition, 1995., Proceedings of the Third International Conference on, Vol. 1. (1995) 146–149
9. Agnihotri, L., Dimitrova, N.: Text detection for video analysis. Content-Based Access of Image and Video Libraries, 1999. (CBAIVL '99) Proceedings, IEEE Workshop on. (1999) 109–113

10. Qixiang Ye, Wen Gao, Weiqiang Wang, Wei Zeng: A robust text detection algorithm in images and video frames. Information, Communications and Signal Processing, 2003 and the Fourth Pacific Rim Conference on Multimedia, Proceedings of the 2003 Joint Conference of the Fourth International Conference on, Vol. 2. (2003) 802–806
11. Wenge Mao, Fu-lai Chung, Lam, K.K.M., Wan-chi Sun: Hybrid Chinese/English Text Detection in Images and Video Frames. Pattern Recognition, 2002. Proceedings. 16th International Conference on, Vol. 3. (2002) 1015–1018
12. Gllavata, J., Ewerth, R., Freisleben, B.: Text Detection in Images Based on Unsupervised Classification of High-Frequency Wavelet Coefficients. Pattern Recognition, 2004. ICPR 2004. Proceedings of the 17th International Conference on, Vol. 1. (2004) 425–428
13. D.T. Chen, H.Bourlard, J-P., Thiran.: Text Identification in complex background using SVM. Int.Conf. on CVPR. (2001)
14. H.Li, D.Doermann, O.Kia.: Automatic text detection and tracking in digital video. IEEE Trans on Image Processing, Vol. 9. (2000) 147–156
15. R. Lienhart, A. Wernicke.: Localizing and Segmenting Text in Images and Videos. In IEEE Transactions on Circuits and Systems for Video Technology, Vol. 12. (2002) 256–258
16. L. Agnihotri, N. Dimitrova.: Text Detection for Video Analysis. In Proc. Int'l Conference on Multimedia Computing and Systems, Florence. (1999) 109–113
17. Y.Zhong, H.J. Zhang, A.K. Jain.: Automatic caption localization in compressed video. IEEE trans on Pattern Analysis and Machine Intelligence, Vol. 22. (2000) 385–392
18. Iakovidis.D.K, Maroulis.D.E, Karkanis.S.A, Flaounas. I.N: Color texture recognition in video sequences using wavelet covariance features and support vector machines. Euromicro Conference, 2003. Proceedings. (2003) 199–204
19. Yuanyan Tang, Ling Wang.: Wavelet analysis and character recognition. Science press, Beijing, China (2003)
20. Datong Chena, Jean-Marc Odobeza, Jean-Philippe Thiran.: A localization/verification scheme for finding text in images and video frames based on contrast independent features and machine learning methods, Signal Processing: Image Communication 19, (2004) 205–217
21. Xilin Chen; Jie Yang; Jing Zhang; Waibel, A.: Automatic detection and recognition of signs from natural scenes. Image Processing, IEEE Transactions, Vol. 13. Issue.1. (2004) 87–99
22. Xiangrong Chen, Yuille, A.L.: Detecting and reading text in natural scenes. Computer Vision and Pattern Recognition, 2004. CVPR 2004. Proceedings of the 2004 IEEE Computer Society Conference, Vol. 2. (2004) II366–II373
23. Xi Zhu, Xinggang Lin.: Automatic date imprint extraction from natural images. Information, Communications and Signal Processing, 2003 and the Fourth Pacific Rim Conference on Multimedia. Proceedings of the 2003 Joint Conference of the Fourth International Conference, Vol. 1. (2003) 518-522
24. Rainer Lienhart.: Video OCR: A Survey and Practitioner's Guide. In Video Mining, Kluwer Academic Publisher. (2003) 155–184
25. Bong-Kee Sin, Seon-Kyu Kim, Beom-Joon Cho.: Locating characters in scene images using frequency features. Pattern Recognition, 2002. Proceedings. 16th International Conference, Vol. 3. (2002) 489–492

Aligning and Segmenting Signatures at Their Crucial Points Through DTW*

Zhong-Hua Quan[1,2] and Hong-wei Ji[1]

[1] Hefei Institute of Intelligent Machines, Chinese Academy of Science
[2] Department of Automation, University of Science and Technology of China
{quanzhonghua, hwji}@iim.ac.cn

Abstract. This paper presents a novel approach that uses the dynamic time warping (DTW) to match the crucial points of signatures. Firstly, the signatures are aligned through the DTW and the crucial points of signatures are matched according to the mapping between the signatures. Then the signatures are segmented at these matched crucial points and the comparisons are accomplished between these segments. Experimental results show that such a strategy is quite promising.

1 Introduction

Signature verification is one of the oldest means of identity validation both for the author of a document or the initiator of a transaction. And as a result of the growing automation, many approaches were proposed over the past decades to perform the automated authentication through one's signature. Literatures [1] and [2] gave a review of the earlier works. Here we focus on on-line signature verification, which means that signatures are collected using a special instrument such as a digital tablet. For on-line signature verification, according to different kinds of features dealt with, it can be roughly classified into two groups. In the first group, signatures are represented by a numbers of global parameters which are computed from the signals. In the second group, signatures are represented by the complete signals collected when they are produced. Such complete signals are regarded as a function of time, $F(t)$. The functions reserve the complete information of signing process and features of the local shape. It is believed that the approach based on functions may lead to better results [1].

In general, signature is considered as a ballistic movement, that is, a motion controlled without instantaneous position feedback, from a motor program [1]. Some of the psychomotor reality can be discovered by studying the various steps a typical imitator must do to copy any signatures. According to [4], one of the first steps a forger has to do is look into a signature to extract its perceptually important points with which the signatures can be rebuilt. It can be deduced that if a signature is divided into a group of segments by such crucial points, the segments are probably corresponding to the psychomotor process. So segmentation that involves partitioning a signature into segments and extracting pertinent local information may be helpful in signature verification. Literatures [3,4,5,6,7] all proposed their method of segmentation. Literatures [4] and [3]

---

* This work was supported by the NSF of China (Nos.60472111 and 60405002).

proposed a method to segment signatures on their perceptually important points and establish the correspondence of segments, but the method cannot be used in comparison between the functional features of signatures since it requires re-sampling on the length of curves. Literatures [5, 6, 7] are all involved in extracting and segmenting signatures at crucial points (e.g., the corner of strokes where signatures change their direction, end points etc.), but there are some mismatching of crucial points that would result in deterioration in system performance. Literature [8] proposed a method to improve the segmentation of signatures through dynamic time warping (DTW). This work can improve the association between segments of signatures, since the DTW is an efficient method to achieve a perfect alignment between series those have non-linear distortion [9]. But in this work, the signatures were segmented on a uniform spatial interval, which violated the principle for segmentation proposed in [4]. So this paper presents a new method that employs DTW to match the crucial points of signatures, then segments the signatures at the matched crucial points. For these aligned segments, a method of linear time warping is employed to accomplish point-to-point comparison.

This paper is organized as follows. In Section 2 a new method of extracting and matching crucial points is presented. Section 3 discusses and explains comparison of signatures and decision making. Section 4 and section 5 give the related experimental results and conclusions respectively.

2 Segmenting Signatures at Crucial Points

2.1 Extracting Crucial Points

A signature is often represented by a time series of points, and each point may take two-dimensional coordinates, pressure and other information. A signature may has a form as follows:

$$P = (p_1, p_2, \ldots, p_i, \ldots, p_n), \quad p_i = (x_i, y_i, t_i). \tag{1}$$

here each point is assumed to be sampled in equal time interval.

The crucial points involved in this paper include end points of strokes and geometric extrema. The end points can be detected by pen tip leaving the surface of tablet. Geometric extrema are the points where signatures change their direction horizontally, vertically or both, which can be detected by finding the zero crossing of deviation of x, y sequences. Each detected extrema is labelled as one of 16 types according to the reason why it is detected [5]. The types of extrema are displayed in Fig. 1.

As presented in [5], each extrema should be detected as a horizontal maxima(or minima), a vertical maxima(or minima) or their combination. However, because of the instability of handwriting, not all of the points where the deviation of x (or y) crosses zero are correctly an extrema. Therefore a processing step aiming at reducing the number of candidate extrama is necessary.

Assume $e^x = e_1^x, e_2^x, \ldots, e_p^x, \ldots$ to represent a sequence of extrema detected by zero crossing of deviation of x. And each element e_p^x is described by its two dimensional coordinate $(e_p^x(x), e_p^x(y))$, time index $(e_p^x(t))$ and type $(e_p^x(c))$. We define a membership function of e_p^x for each candidate extrema as follows:

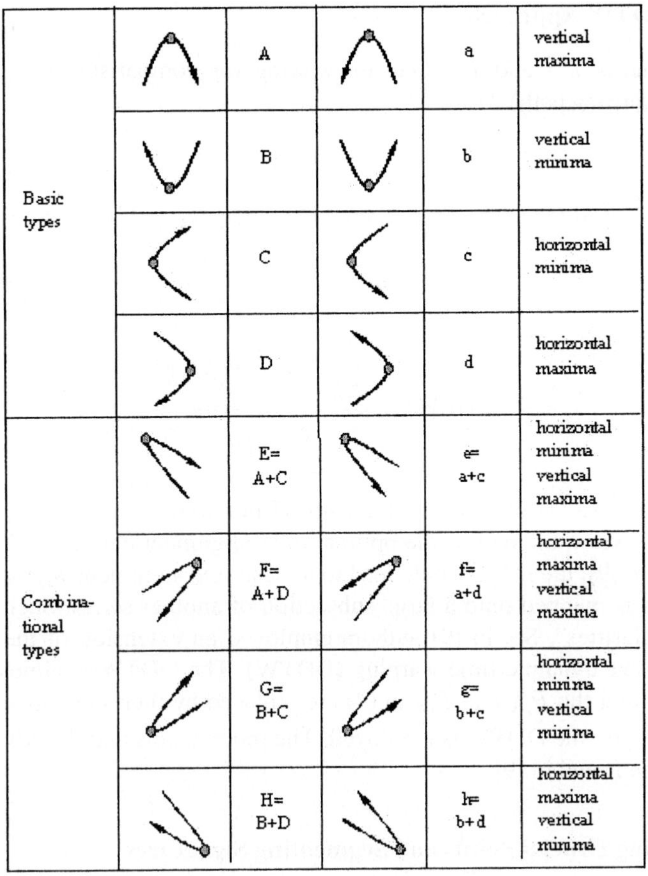

Basic types	⋀	A	⋀	a	vertical maxima
	⋁	B	⋁	b	vertical minima
	＜	C	＜	c	horizontal minima
	＞	D	＞	d	horizontal maxima
Combinational types	↙	E= A+C	↙	e= a+c	horizontal minima vertical maxima
	↗	F= A+D	↗	f= a+d	horizontal maxima vertical maxima
	↙	G= B+C	↙	g= b+c	horizontal minima vertical minima
	↘	H= B+D	↘	h= b+d	horizontal maxima vertical minima

Fig. 1. Types of extrema

$$\mu(e_p^x) = \frac{min(|x_{e_p^x(t)} - x_{e_p^x(t)-l}|, |x_{e_p^x(t)} - x_{e_p^x(t)+r}|)}{width}. \tag{2}$$

where $l = (e_p^x(t) - e_{p-1}^x(t))/2$ and $r = (e_{p+1}^x(t) - e_p^x(t))/2$ define a neighbor area of e_p^x, and $width = \max_{i=(1,\ldots,n)}(x_i) - \min_{i=(1,\ldots,n)}(x_i)$. Then the candidate whose $\mu(e_p^x)$ is less than a threshold is deleted.

The same operation is repeated on the sequence $e^y = e_1^y, e_2^y, \ldots, e_q^y, \ldots$ detected by zero crossing of deviation of y. The sequence e^x, e^y and those end points are integrated into one sequence e at last. Considering that sometimes a combinational type may appear as successive two basic types, if $|e_p^x(t) - e_q^y(t)| \leq T_0$ then e_p^x and e_q^y are combined into one extrema in e that has a combinational type. The other candidates are inserted into the sequence e according to their time index. Here each element e_i is described by its two dimensional coordinate $(e_i(x), e_i(y))$, time index $e_i(t)$ and type $e_i(c)$.

2.2 The DDTW Approach

For two patterns P^R and P^T that are waiting for comparison, we define a time-alignment (warping path) M as:

$$M = ((1,1), \ldots, (t_k^R, t_k^T), \ldots). \quad (3)$$

where t_k^R and t_k^T refer to the time index of the mapped points in the two signatures. Assume:

$$d(M_k) = ||P_{t_k^R}^R - P_{t_k^T}^T||. \quad (4)$$

$$D(M) = \frac{\sum_{k=1}^{K} w(k) * d(M_k)}{\sum_{k=1}^{K} w(k)}. \quad (5)$$

The goal of the DTW is to find the path M that minimizes $D(M)$ [11]. From the DTW point of view, the path is the optimal time-alignment between P^R and P^T. But as presented in [9], the DTW would lead to undesirable alignment where a single point on one series is mapped onto a large subsection of another series. Such a behavior is called "singularities". So, in [9] authors employed an extension of the DTW that is called derivative dynamic time warping (DDTW). The DDTW is almost the same as DTW except that the $P_{t_k^R}^R$ and $P_{t_k^T}^T$ in (4) are replaced by their derivative, respectively. And in this paper, the DDTW is employed. The more details and the advantages of the DDTW can be found in [9].

2.3 Matching Crucial Points and Segmenting Signatures

In this paper, we use the DDTW to establish the correspondence of crucial points. Assume that we have got the extrema sequences e^R and e^T of the two signatures and the mapping function M between P^R and P^T. Now the distance between two crucial points e_i^R and e_j^T can be defined as:

$$D(e_i^R, e_j^T) = \frac{\min |e_i^R(t) - e_i^R(t')| + \min |e_j^T(t) - e_j^T(t')|}{2 \times S(e_i^R(c), e_j^T(c))}. \quad (6)$$

where $(e_i^R(t), e_i^R(t')) \in M$, $(e_j^T(t'), e_j^T(t)) \in M$,
and $S(e_i^R(c), e_j^T(c))$ is defined as:

$$S(e_i^R(c), e_j^T(c)) = \begin{cases} 1, & e_i^R(c) \approx e_j^T(c) \\ \epsilon, & e_i^R(c) \neq e_j^T(c). \end{cases} \quad (7)$$

where $\epsilon \geq 0$ is a small enough constant. The meaning of the equation 6 can be described by the Fig. 2.

A recursive method is employed here to work out the alignment of the two sequence (e^R and e^T), and the arithmetic is displayed as follows.

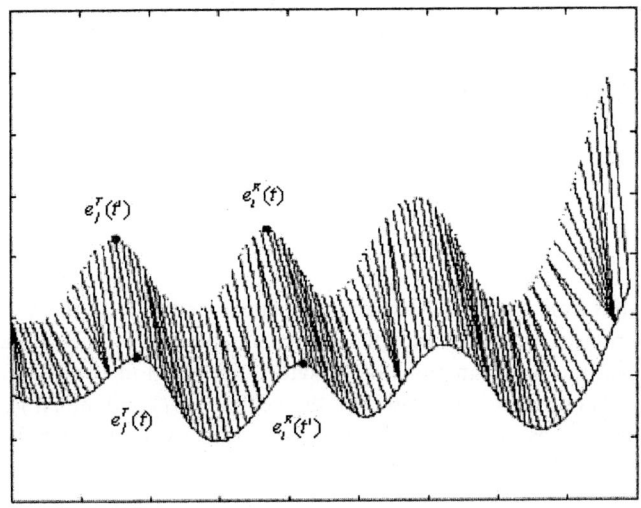

Fig. 2. The distance between a pair of crucial points e_i^R and e_j^T

Begin
    Initial : $row = 1, col = 1$, map[length of e^R];
    while $row \leq length\ of\ e^T$ and $col \leq length\ of\ e^R$
        if $D(e_{jj}^R, e_{ii}^T) = \min\limits_{\substack{i \in (row, row+3) \\ j \in (col, col+3)}} D(e_j^R, e_i^T)$
        and $D(e_{jj}^R, e_{ii}^T) \leq threshold$,
        map[jj]=ii,$row = ii + 1, col = jj + 1$;
        else $row = row + 1, col = col + 1$;
    return map
End

The "map" returned records the mapping between the sequence of crucial points. Therefore the signatures can be segmented at the crucial points which is matched in "map", and the segments are matched according to their end points.

3 Comparison of Signatures and Classification

3.1 Comparison of Signatures

When segmenting signatures according to the matching of crucial points, the signatures compared (P^T and P^R) are composed of a series of segments $S_1^T, S_2^T, \ldots, S_K^T$ and $S_1^R, S_2^R, \ldots, S_K^R$ respectively. The S_i^T is corresponded to S_i^R, and the K equals to pairs of matched crucial points. Then the linear time warping can be applied on the aligned segments. Since the time functions of signatures are discrete and the intervals are equal, the linear time warping can be accomplished by re-sampling on segments of testing pattern. Let $n_{S_i^T}$ and $m_{S_i^R}$ be the points number of segments S_i^T and S_i^R respectively, the time interval involved to re-sample the S_i^T can be computed from the following equation.

$$\Delta t'_i = \frac{m_{S_i^R} * \Delta t}{n_{S_i^T}}. \tag{8}$$

where the Δt is the original time interval.

After re-sampling, the testing patten P'^T and the reference pattern P^R contains the same number of points. The distance between them can be computed as follows:

$$D(F) = \frac{\sum_{i=1}^m F_i^R - F'^T_i}{m}. \tag{9}$$

where F^R is the time function of reference pattern and F'^T is the time function of testing patten after re-sampling, m is the points number of reference pattern.

In this paper, the time functions of signatures involved in verification include the two dimensional coordinate $x(t)$ and $y(t)$, the velocity $v(t)$ and the tangent angle $\theta(t)$. They can be denoted as $x(i)$, $y(i)$, $v(i)$ and $\theta(i)$ respectively because the intervals are equal. The $\theta(i)$ and the $v(i)$ can be defined as follows:

$$\theta(i) = \arctan(y(i+1) - y(i), x(i+1) - x(i)) \times 0.5 \\ + \arctan(y(i+2) - y(i), x(i+2) - x(i)) \times 0.5 \tag{10}$$

$$v(i) = \sqrt{(y(i+1) - y(i))^2 + (x(i+1) - x(i))^2}. \tag{11}$$

where $0 \leq i \leq m$.

These time sequences of testing pattern should be re-sampled at first, then the distance between the testing pattern and the reference pattern can be computed according to 9. Let now $D(x)$, $D(y)$, $D(\theta)$ and $D(v)$ be the distance computed from these time sequences. And $D(t)$ is the cumulated time warping of testing pattern that can be computed as:

$$D(t) = \frac{\sum_{j=1}^K \sum_{i=1}^{m_S^j} |\Delta t - \Delta t'| \times i}{m}. \tag{12}$$

3.2 Classification

The final stage in the verification procedure is the actual classification. In this paper, we use the Mahalanobis decision making.

Let $D = (D(x), D(y), D(\theta), D(v), D(t))$, $\overline{D}$ is mean of D over an initial set of original signatures, and

$$\Sigma = \begin{bmatrix} \sigma_{D(x)}^2 & 0 & 0 & 0 & 0 \\ 0 & \sigma_{D(y)}^2 & 0 & 0 & 0 \\ 0 & 0 & \sigma_{D(\theta)}^2 & 0 & 0 \\ 0 & 0 & 0 & \sigma_{D(v)}^2 & 0 \\ 0 & 0 & 0 & 0 & \sigma_{D(t)}^2 \end{bmatrix}. \tag{13}$$

where $\sigma_{D(x)}$, $\sigma_{D(y)}$, $\sigma_{D(\theta)}$, $\sigma_{D(v)}$ and $\sigma_{D(t)}$ are standard deviation of $D(x)$, $D(y)$, $D(\theta)$, $D(v)$ and $D(t)$ respectively.

Then the combined measure for matching reference and testing pattern is given by a simplified Mahalanobis distance as:

$$D_{mahal} = (D - \overline{D})' \Sigma^{-1} (D - \overline{D}). \tag{14}$$

We call D_{mahal} the simplified Mahalanobis discriminant function. In the next section, we will use D_{mahal} to accomplish the actual classification.

4 Experimental Results

The data used for experiment in this paper is collected by ourselves among the students in our lab. This data involves 20 volunteers, each of which provided 10 signatures. Thus there are total 200 signatures.

4.1 Matching of Crucial Points

The performance of the matching of crucial points has an important influence to the segmenting and the time warping of signatures. Some examples for mismatching crucial points are displayed in Fig. 3. In Fig. 3, the crucial points correctly matched are marked as asterisk, and the mismatched crucial points are marked as circle. In Fig. 3b some crucial points are mismatched, and it can be found that the mismatching of crucial points can distort the time warping of the signatures.

To evaluate the performance of the approach matching the crucial points, we chose 60 samples from the data set (3 samples from each signer), and matched the crucial

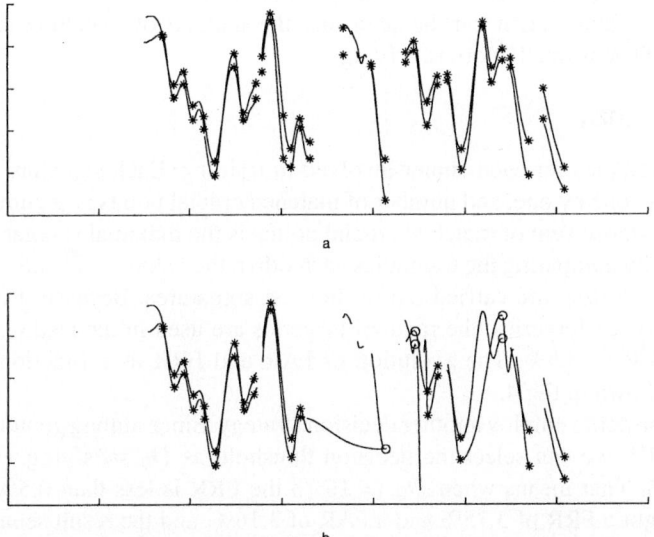

Fig. 3. Some examples for mismatching crucial points

Table 1. The performance comparison of matching crucial points by the approach in [5] and the approach using DTW

	The method in [5]	The method in this paper
C	91.6	99.2

points sequence of the signatures from the single signer in manual work. Then the result is referred to as the expected value. Now we can define the accuracy of matching as:

$$C = \frac{\sum_{i=1}^{m} \Gamma'(i)}{\sum_{i=1}^{m} \Gamma(i)} \times 100\%. \tag{15}$$

where $\Gamma(i)$ and $\Gamma'(i)$ are referred to as:

$$\Gamma(i) = \begin{cases} 1 \; ; \; if \; p(i) \neq 0. \\ 0 \; ; \; else \end{cases} \tag{16}$$

$$\Gamma'(i) = \begin{cases} 1 \; ; \; if \; p(i) \neq 0 \; and \; p(i) = p'(i). \\ 0 \; ; \; else \end{cases} \tag{17}$$

where the $p(i)$ is the mapping function between the sequences of crucial points worked out by the DTW. And the $p'(i)$ is referred to as the desirable mapping function.

For comparison, we implemented both the approach of matching crucial points in [5] and that proposed in this paper. Figure 3 is an example for comparison, in this figure the (a) is accomplished by DTW, and the (b) is accomplished by the method in [5]. The corresponding correct rate are displayed in the following table.

From the Table 4.1, it can be seen that the method of matching crucial points through the DTW is much more satisfying.

4.2 Classification

There are 6 samples for each signer involved in training. Each signature is compared with the others one by one, and number of matched crucial points is accumulated. Then the signature whose sum of matched crucial points is the maximal is regarded as the templet. And by comparing the 6 samples each other, the values of $\overline{D}$ and Σ are worked out . Then the testing are carried out on the rest signatures. Because it is difficult to acquire the skilled forgeries, the random forgeries are used in the testing. As a result, we got an EER of 3.8%. The evolution of FAR and FRR as a function of decision threshold is shown in Fig.4.

Here we prefer to employ another decision strategy. Since among genuine signatures $D_{mahal} \sim \mathcal{X}^2$, we can select the decision threshold as D_0 satisfying $P(D_{mahal} \geq D_0) = 0.005$. That means when $D_0 = 16.75$ the FRR is less than 0.5%. On testing samples, we got a FRR of 3.75% and a FAR of 3.16%, and the result seems consistent with Fig. 4. Considered that random forgeries are involved in experiments, the FAR looks a little high, but the false accepted signatures are focus in two signers. Some samples for building templet of the two signers are displayed in the Fig. 5.

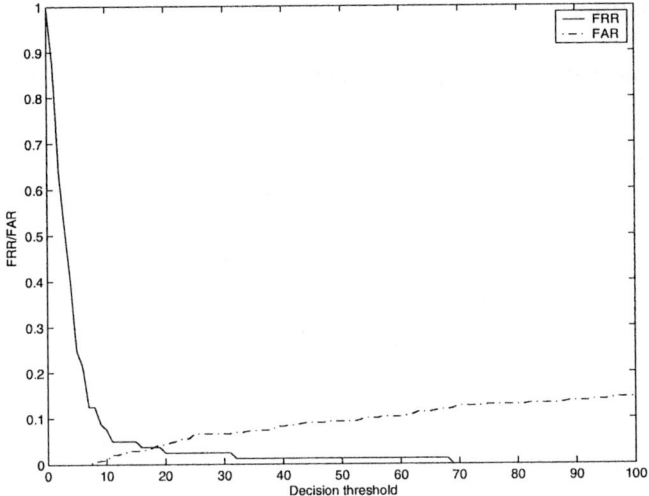

Fig. 4. Curves of FAR and FRR varying with decision threshold

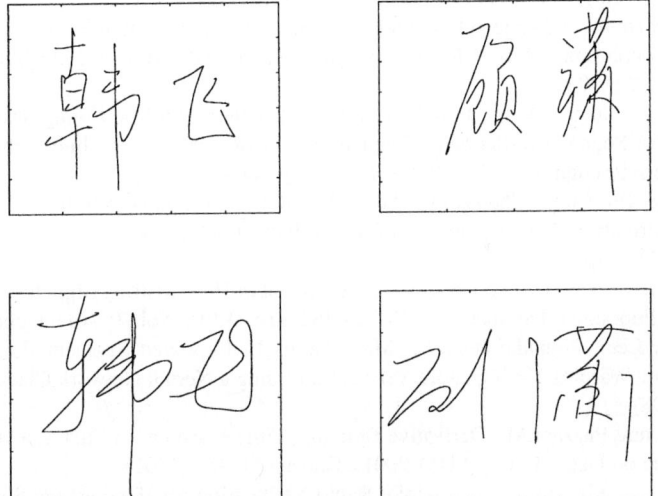

Fig. 5. Samples of the goats

From the Fig. 5 it can be found that these signatures of the two signers are varying between two different shapes. In signature verification, such signers whose signatures are not stable enough are usually refused in enrollment in order to reduce the risk. If this strategy is employed here, the FAR would reduce to 0 while the FRR is 4.16%.

These results show that signature verification is not available for some people whose signatures are not stable enough. It is believed that it is necessary to exclude such signers when they register for the sake of guaranteeing the safety of the system.

5 Conclusions

This paper presented a method that use the DTW to align and segment signatures at their crucial points, and then the linear time warping is employed on aligned segments to deal with the unequal duration and non-linear time distortion of signatures. Experiments on the random forgeries showed that such a method is competent. Nevertheless, the consistency of the segments haven't be taken in consideration although the approach is a segment-based method. And the Mahalanobis decision is a statistical method that needs lots of samples, but it is almost impossible to be satisfied for signature verification. In the future, we would like to polish the solution for these problems.

References

1. Plamondon,R., and Lorette,G.: Automatic Signature Verification and Writer Identification-The state of the Art. Pattern Recognition, Vol.22, No.7 (July 1989)
2. Leclerc,F., Plamondon,R.: Automatic Signature Verification: the state of the art-1989-1993, Int. J. Pattern Recognition Artif. Intel., Vol.8, No.3 (1994)
3. Yue,K.W., Wijesoma,W.S.: Improved Segmentation and Segment Association for on-line Signature Verification. Systems, Man, and Cybernetics, 2000 IEEE International Conference on, Vol. 4 (2000)
4. Jean-Jules Brault and Rejean Plamondon: Segmenting Handwritten Signatures at Their Perceptually Important Points. IEEE Transactions on Pattern Analysis and Machine Intelligence, Vol.15, No.9 (1993)
5. Jaeyeon Lee, Ho-Sub Yoon, Jung Soh, Byung Tae Chun, Yun Koo Chung: Using Geometric Extrema for Segment-to-Segment Characteristics Comparison in Online Signature Verification. Pattern Recognition, Vol. 37, Iss. 1 (January 2004)
6. Guo Hong, Jin Xianji: The Extract Algorithm of Special Points in Signature Based on Dynamic Information, J. of Wuhan Uni. of Sci. & Tech., (Natural Science Edition), Vol.24, No.2 (June 2001)
7. Zhang Kui, Jin Xianji, Pei Xiandeng: An Approach of Handwritting Signatures Comparision Based on Functional Parameter, MINI-MICRO SYSTEMS, Vol.20, No.6 (June 1999)
8. Wan-Suck Lee, Mohankrishnan,N., Mark Paulik,J.: Improved Segmentation through Dynamic Time Warping for Signature Verification Using a Nerual Network Classifier. ICIP'98, Vol.2 (1998)
9. Keogh,E., and Pazzani,M.: Derivative Dynamic Time Warping. In First SIAM International Conference on Data Mining (SDM'2001), Chicago, USA. (2001)
10. Zhao Guimin, Xia Limin, Chen Aibin: Rapid Verification for Handwriting Signature, Computer Engineering, Vol.29, No.7 (May 2003)
11. Ronny Martens, Luc Claesen: On-Line Signature Verification by Dynamic Time-Warping. IEEE Proceedings of ICPR'96 (1996)

A SAR Image Despeckling Method Based on Dual Tree Complex Wavelet Transform[1]

Xi-li Wang[1] and Li-cheng Jiao[2]

[1] School of Computer Science, Shannxi Normal University, 710062, Xi'an, China
wangxili@snnu.edu.cn
[2] Institute of Intelligent Information Processing, Xidian University, 710071, Xi'an, China
lchjiao@mail.xidian.edu.cn

Abstract. Based on the dual tree complex wavelet transform and edge detection, a SAR image despeckling algorithm is proposed. It can be used to remove white Gauss additive noise (WGAN) too. The DT-CWT has the properties of shift invariance and more directions. Edges are effectively extracted based on this complex transform and adjacent scales coefficients multiplication. According to the statistical property of the edge and non edge wavelet coefficients, Laplacian and Gaussian distribution are used to describe them respectively. Bayesian MAP estimator is used to estimate the noiseless wavelet coefficient values. Analysis and experiments illustrate the effectiveness of the proposed algorithm.

1 Introduction

The wavelet transform has become an important tool for removing noise from corrupted image. This is due to its energy compaction property. Wavelet thresholding [1] is often used for such task. According to wavelet transform energy compaction property, small coefficients are more likely due to noise, and large ones due to important features (such as edges). So noise can be filtered by thresholding. The threshold acts an important role in such methods. A threshold adaptively selected in each subband is better than a uniform one [2].

Recently, statistical models for images and their wavelet transform coefficients are developed. Under these models, Bayesian estimation techniques such as maximum a posteriori (MAP) estimator, maximum likelihood (ML) estimator can be used. Now the denoising task is to estimate clean coefficients using an a priori probability distribution of the coefficients. Sometimes coefficients are regarded as independent random variables described by some probability distribution function, such as Gaussian, Laplacian, and generalized Gaussian distribution. Some threshold shrinkage function can be gained using models assumed by Gaussian, Laplacian distribution. When coefficients are regarded as dependent random variables, more complicated models and better results can be obtained [3], in the mean time, the computing complexity increases greatly.

---

[1] Supported by the National Science Foundation of China under Grant No.60133010.

Speckles in synthetic aperture radar (SAR) images are multiplicative noises. Arsenault et al. [4] demonstrate speckles can be regarded as white Gauss additive noise (WGAN) after transforming image to logarithm domain. So we change speckles to Gauss noises.

Images are usually denoised in orthogonal wavelet domain. Although such transforms have no redundancy, they lack of shift invariance and directional selectivity. Some shift invariant denoising methods are proposed to get better results [5,6,7]. Since shift invariance, redundant representation outperforms the orthogonal basis; we use the dual tree complex wavelet transform (DT-CWT)[8] that has such properties in this paper.

If edge information can be obtained, we can preserve such main image features better when filtrate noises. In order to detect edges in noisy images effectively, we design an edge detection method by adjacent scale coefficients multiplication based on DT-CWT.

Based on DT-CWT and edge information, we propose a new wavelet domain image denoising approach. In section 2 the DT-CWT is described. In section 3 and 4, the edge detection and the denoising algorithms are described in detail. Then denoising results are given. Finally is conclusion.

2 Dual Tree Complex Wavelet Transform

DT-CWT is proposed by professor N.G.Kingsbury [8]. It extends discrete wavelet transform (DWT) via separable filters to complex transform. Compared with other complex transforms, it has the advantage of perfect reconstruction. Compared with the orthogonal wavelet transforms, it has the advantages of shift invariance and more directional selectivity.

Let $x(s)$, $s = (s_1, s_2) \in R^2$ represents an image, the DT-CWT decompose it using dilations and translations of a complex scaling function $\phi_{j_0,k}$ and six complex wavelet functions $\psi_{j,k}$:

$$x(s) = \sum_{k \in Z^2} u_{j_0,k} \phi_{j_0,k}(s) + \sum_{b \in B} \sum_{j \geq j_0} \sum_{k \in Z^2} c_{j,k}^b \psi_{j,k}^b(s) \qquad (1)$$

Where $B = \{\pm 15^0, \pm 45^0, \pm 75^0\}$, $\phi_{j_0,k} = \phi_{j_0,k}^r + j\phi_{j_0,k}^i$, $\psi_{j,k} = \psi_{j,k}^r + j\psi_{j,k}^i$, and $\phi_{j_0,k}^r$、 $\phi_{j_0,k}^i$ ($\psi_{j,k}^r$、 $\psi_{j,k}^i$) are themselves real scaling (wavelet) functions. Thus DT-CWT is combinations of two real wavelet transforms. When implement, two decomposition trees based on two real filters give the real and imaginary parts of the complex coefficients. DT-CWT uses real but not complex filters to generate complex coefficients. Down sampling by 2 is eliminated to approximate shift invariance with the real DWT at each level. This is equivalent to two parallel full-decimated trees, bringing about redundancy of 4:1 for 2-D signals. There are six subbands of DT-CWT at each level that provide better directionality than DWT.

3 Edge Detection Based on DT-CWT and Scale Multiplication

Signal features and noises have large wavelet coefficients. Large coefficients caused by sharp edges can transfer with increasing scale, those caused by noises decay fast with increasing scale. Most coefficients other than feature and noise have very small values. This shows coefficients at the same direction and position have some correlation. Xu et al [9] use the direct spatial multiplication of wavelet coefficients at several adjacent scales to detect the location of edges. According to the above analysis, this multiplication can increasing feature coefficients times, decreasing noisy coefficients times. So features are enhanced while noises are suppressed. Then improving the accuracy of locating edges is more easily. But feature coefficients cannot persistent well along scales using DWT since it is not shift invariance, whereas large feature values persistent well along scales by DT-CWT. So multiplying coefficients at adjacent scales is more effective for locating edges using DT-CWT.

For a certain scale m, we multiply the coefficients of this scale n and its father coefficient in the coarse neighbor scale $p(n)$:

$$Cor(m,n) = w(m,n) \cdot w(m+1, p(n)) \qquad (2)$$

$n = 1, 2, \cdots N$. N is the number of pixels at this scale. We also square the coefficients of this scale:

$$S(m,n) = w^2(m,n) \qquad (3)$$

From above we know that for a pixel describing an edge, its value of Cor and S are both large. For a pixel describing a noise, both values decrease with increasing m, and Cor are much smaller than S. Both values are small for other pixels. Then we give a simple rule to distinguish edge and non edge points. If the three conditions: ① $Cor(m,n)$ is large, ② $Cor(m,n) \geq S(m,n) \cdot c$, ③ the parent of n is an edge point, are satisfied simultaneously, then n is an edge point, else it is not an edge point. For the coarsest scale, the first two conditions must be satisfied. c is a constant. Maybe Cor is slightly smaller than S for a weak edge, so c is added in order to extract such weak edge. Obviously, its value should be among $(0.5, 1)$.

The coefficients of DT-CWT are complex $c_i = u_i + jv_i$. The magnitude of the complex coefficient is used: $|c_i| = \sqrt{u_i^2 + v_i^2}$. This is a more reliable measure than either the real or the imaginary part. Since if there has a slightly signal shift, this do not affect the magnitude but do affect the real or the imaginary parts.

Figure 1 shows the image features are enhanced while noises are suppressed by coefficients multiplication between scales in complex wavelet domain. (a) shows a noisy image (white Gaussian noise with standard deviation 20 is added). (b) shows the S value of a subband. (c) shows the Cor value of the same subband (bright colour represent large values).

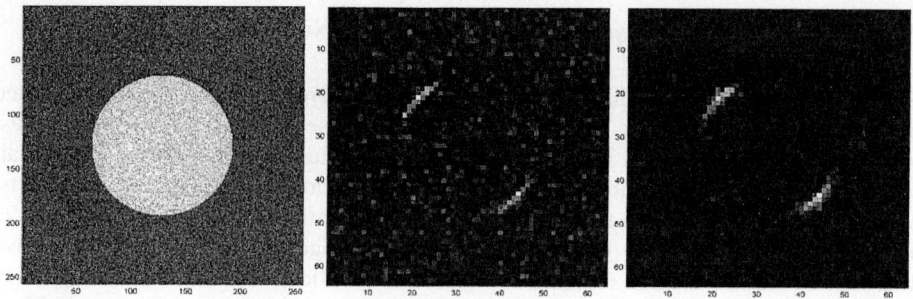

Fig. 1. (a) Noisy image (b) S value of a subband (c) Cor value of the same subband

4 Denoising Algorithm

Assume the original image f is corrupted by white Gaussian noise n, $f = g + n$, g is the observed image. In wavelet domain, the model is $y = x + \varepsilon$, x, y, ε representing the wavelet coefficient of clean, noisy image and noises respectively. The aim is to estimate x from the noisy observation y. We use Bayesian MAP estimator for this purpose:

$$\hat{x} = \arg\max_{x} P_{x|y}(x \mid y) = \arg\max_{x}(P_{y|x}(y \mid x) \cdot P_{x}(x)) = \arg\max_{x}(P_{n}(y - x) \cdot P_{x}(x)) \quad (4)$$

If $P_x(x)$ is assumed to be a zero mean Gaussian density with variance σ^2, we obtain the estimator:

$$\hat{x} = \frac{\sigma^2}{\sigma^2 + \sigma_n^2} \cdot y \quad (5)$$

If $P_x(x)$ is assumed to be Laplacian, then the estimator is:

$$\hat{x} = sign(y)(\mid y \mid - \frac{\sqrt{2}\sigma_n^2}{\sigma_x})_+ \quad (6)$$

Where $(a)_+$ means if $a < 0, (a)_+ = 0$, otherwise $(a)_+ = a$.

Equation (6) is the classical soft shrinkage function. Although the two distributions cannot describe wavelet coefficients very accurately, they are often applied because the MAP estimator is simple to compute and they are convenient assumption for $P_x(x)$.

We estimate σ_y for every noisy coefficient spatial adaptively. Now most edge points are distinguished from non edge, we can estimate them respectively to get more reliable value by formula (7) and (8).

$$\sigma_{y_i}^2 = \frac{1}{2}(y_i^2 + y_{ip}^2) \qquad (7)$$

$$\sigma_{y_i}^2 = \frac{1}{N_w^2}\sum_{k=1}^{N^2} y_k^2 \qquad (8)$$

y_{ip} is parent of y_i. A window of $N_w \times N_w$ centred at i is used in (8) to compute the variance. Instead of using all coefficients inside the window, we can only use those non edge points, and exclude the edge points.

The variance of the noiseless coefficients can be calculate by:

$$\sigma_{x_i}^2 = \sigma_{y_i}^2 - \sigma_n^2 \qquad (9)$$

Study on the coefficients of wavelet subband, we find that a Laplacian density is more fitted to the log histogram of the edge coefficients. A Gaussian density is more fitted to the log histogram of non edge points. Figure 2 is an example. Then the coefficients belong to edge are modelled as i.i.d. Laplacian distribution, those belong to non edge are modelled as i.i.d. Gaussian distribution.

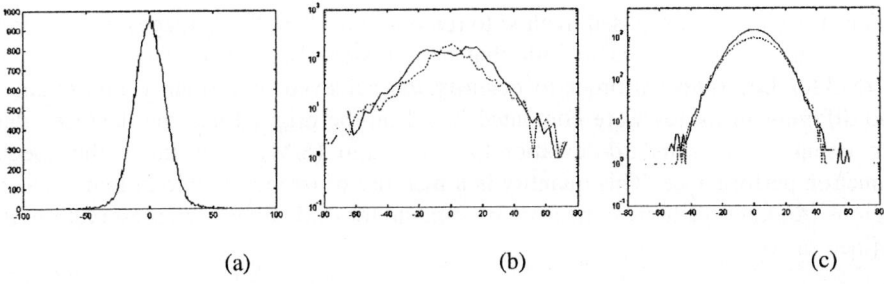

Fig. 2. (a) Coefficient histogram of second level, +45 degree of Lena image. (b) Log histogram of edge coefficients of the same subband (*solid line*), and a fitted Laplacian r.v.'s log histogram (*dotted line*). (c) Log histogram of non edge coefficients (*solid line*), and a fitted Gaussian r.v.'s log histogram (*dotted line*).

The denoising algorithm can be summarized as follows:

1. Compute the image logarithmically
2. Decompose image to DT-CWT domain
3. Extract edge image for every high subband at every scale, using our edge detection algorithm
4. Compute variance of the edge and non edge coefficients
5. Modify coefficients belong to edge and non edge respectively

6. Reconstruct logarithmical image by the modified coefficients
7. Obtain the denoised image by exponential computation

If the noise is WGAN, the first and last step can be cancelled.

Edge detection and the denoising procedure are top-down. Starting from the coarsest scale, algorithm goes from parents to children subbands. After completion of a subband, we obtain both edge image and modified coefficient.

5 Experimental Results

First, we use 256 greyscale images Barbara and Lena as test images. i.i.d. Gaussian noise at different levels are added to the images. Five levels of decomposition are used for denoising. In fact, image is decomposed with six levels in order to extract edge. The results are compared with the results of soft threshold based on DWT. The thresholds are estimated spatially adaptive. The wavelet we use is Daubechies wavelet (db8). The performance is tested using the PSNR measure.

The results are listed in table 1. The results of part of Lena and Barbara image using DWT soft threshold and our method are given in figure 3 (noise standard deviation $\sigma = 20$). As seen from the PSNR values and the denoised images, the noise is reduced effectively, most sharp edges are preserved, and ringing artifacts are little.

Then, we use the proposed method to remove speckle in SAR images. We compare the results of our approach with that of classical Lee filter. The window size is 3×3 for Lee filter. In order to quantify the achieved performance improvement, two different measures were computed based on the original and the denoised data. We compute the standard-deviation-to-mean ratio (S/M) to quantify the speckle reduction performance. This quantity is a measure of image speckle in homogeneous regions. Another qualitative measure is compute for evaluating edge preservation. It is defined as [10]:

$$EP = \frac{\Gamma(g'-\overline{g}', \hat{g}'-\overline{\hat{g}}')}{\sqrt{\Gamma(g'-\overline{g}', g'-\overline{g}') \cdot \Gamma(\hat{g}'-\overline{\hat{g}}', \hat{g}'-\overline{\hat{g}}')}} \quad (10)$$

Where g', $\hat{g}'$ are the high pass filtered versions of the original image g and the despeckled image $\hat{g}$ respectively. The over line operator represents the mean value. And $\Gamma(s_1, s_2) = \sum_{i=1}^{N \times N} s_{1_i} \cdot s_{2_i}$. The EP value should be close to unity for an optimal effect of edge preservation.

The obtained values of S/M and EP for the methods applied to two SAR images are given in Table 2. The original and result images are in figure 3. Our method gives better results in terms of the S/M and EP measure, which indicates that the technique

Table 1. Results for the test images with several noise levels

Image	Noise level	Soft threshold (DWT)	Proposed
Lena	$\sigma = 10$	34.05	35.33
	$\sigma = 20$	30.46	32.35
	$\sigma = 30$	28.28	30.50
	$\sigma = 50$	25.48	28.00
Barbara	$\sigma = 10$	31.55	33.01
	$\sigma = 20$	27.23	28.92
	$\sigma = 30$	24.93	26.78
	$\sigma = 50$	22.20	24.16

Table 2. Results for SAR images

	Image 1		Image 2	
	S/M	EP	S/M	EP
Original image	0.2417	-	0.2544	-
Lee filter	0.1091	0.0111	0.1297	0.1198
Proposed method	0.0585	0.3404	0.0651	0.2633

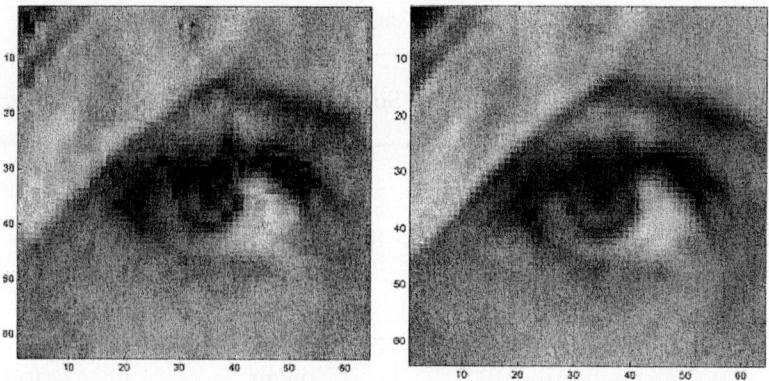

(a) Denoising image by DWT soft threshold (b) The proposed algorithm result

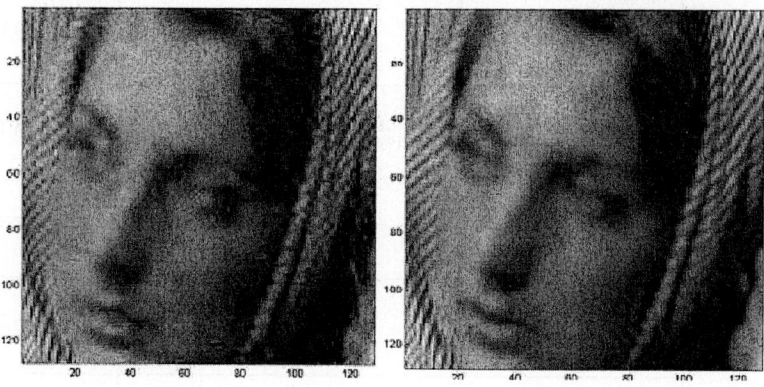

(c) Denoising image by DWT soft threshold (d) The proposed algorithm result

Fig. 3. Part of Lena and Barbara image denoising results

exhibits a clearly better performance in terms of both speckle reduction and edge preservation.

6 Conclusions

In this paper, we propose an effective and low complexity method to remove speckle noise for SAR images. The DT-CWT with shift invariance is used. Edges are effectively extracted based on this complex transform and adjacent scales coefficients multiplication. According to the statistical property of the edge and non edge wavelet coefficients, Laplacian and Gaussian distribution are used to describe them respectively. MAP estimator is used to estimate the noiseless coefficient values. Experiments for standard test images and SAR images show that, the proposed algorithm performs better in both speckle reduction and edge preservation.

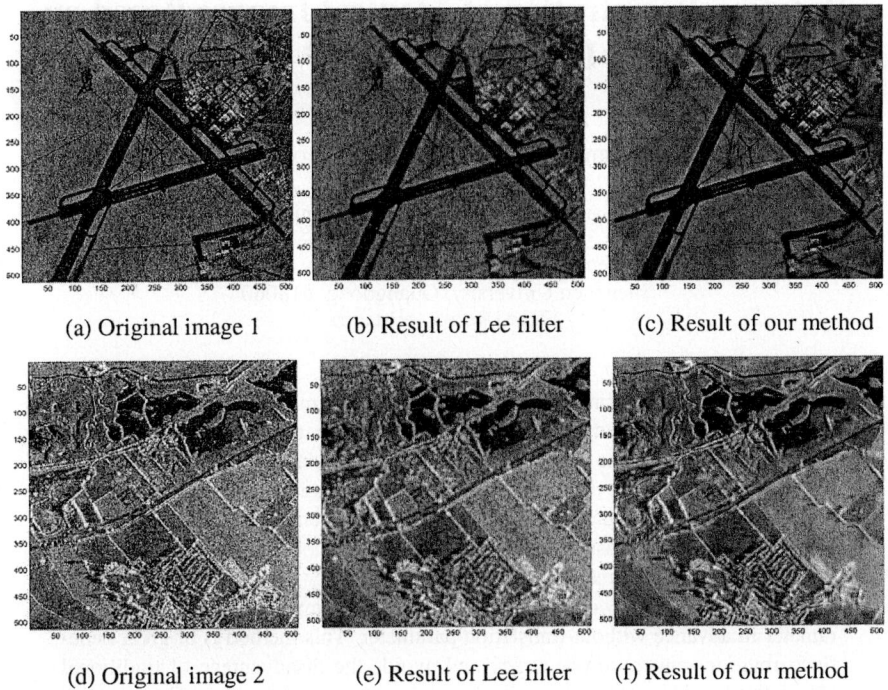

(a) Original image 1 (b) Result of Lee filter (c) Result of our method

(d) Original image 2 (e) Result of Lee filter (f) Result of our method

Fig. 4. SAR images and the results

References

1. Donoho, D. L., Johnstone, I. M.: Ideal Spatial Adaptation via Wavelet Shrinkage. Biometrika. vol.81. (1994) 425-455
2. Chang, S., Yu, B., Vetterli, M.: Adaptive Wavelet Thresholding for Image Denoising and Compression. IEEE Trans. Image Processing. vol.9. (2000) 1532-1546
3. Crouse, M. S., Nowak, R. D., Baraniuk, R. G.: Wavelet-based Statistical Signal Processing Using Hidden Markov Models. IEEE Trans. Signal Processing. vol.46. (1998) 886-902
4. Arsenault, H. H., April, G.: Properties of Speckle Integrated with a Finite Aperture and Logarithmically Transformed. J. Opt. Soc. Amer. vol.66. (1976) 1160–1163
5. Li, X., Orchard, M. T.: Spatially Adaptive Image Denoising under Overcomplete Expansion. Proc. ICIP. (2000)
6. Sendur, L., Selesnick, I. W.: Bivariate Shrinkage Functions for Wavelet-based Denoising Exploiting Interscale Dependency. IEEE Trans. Signal Processing. vol.50. (2002) 2744-2756
7. Portilla, J., Strela, V., Wainwright, M. J., Simoncelli, E. P.: Image Denoising Using Scale Mixtures of Gaussians in the Wavelet Domain. IEEE Trans. Image Processing. vol.12. (2003) 1338-1351
8. Kingsbury, N. G.: Complex Wavelets for Shift Invariant Analysis and Filtering of Signals. Journal of Applied and Computational Harmonic Analysis. (2001) 234-253
9. Xu, Y., Weaver, J. B., Healy, D. M.: Wavelet Transform Domain Filters: a Spatially Selective Noise Filtration Technique. IEEE Trans. Image Processing. vol.3. (1994) 747-757
10. Sattar, F., Floreby, L., Salomonsson, G., Lövström, B.: Image Enhancement Based on a Nonlinear Multiscale Method. IEEE Trans. Image Processing, vol.6. (1997) 888-895

Rotation Registration of Medical Images Based on Image Symmetry*

Xuan Yang[1,2], Jihong Pei[2], and Weixin Xie[1,2]

[1] College of Information and Engineering, Shenzhen University, Guangdong, 518060
xyang0520@263.net
[2] ATR National Defense Technology Key labortory of Shenzhen University,
Shenzhen University, Guangdong, 518060
jhpei@szu.edu.cn

Abstract. Mutual Information has been used as a similarity metric in medical images registration. But local extrema impede the registration optimization process and rule out the registration accuracy, especially for rotation registration. In this paper, a novel approach to rotate registration based on image symmetry measure is presented. Image symmetry measure is defined to measure the symmetry about the possible axis. The symmetry measure is at its maximum when the possible symmetry axis is the real symmetry axis. The angle between the symmetry axes of two images can be used to estimate rotate registration parameter in advance without translation parameter. This method is of great benefit to rotation registration accuracy and avoids the disadvantage of traditional MI method searching in the multi-dimensional parameter space. Experiments show that our method is feasible and effective to rotation registration of medical images, which have obvious symmetry characteristics.

1 Introduction

Multimodal medical image produces different information, which complements each other. In clinical applications those images are frequently fused together to improve the diagnostic accuracy. Registration of multimodal medical images is an important first step in successful fusion of those images. Image registration is a procedure that determines the best match between multimodal images of the same object field. The mutual information (MI) between two images can be regarded as a statistical tool to measure the degree to which an image can be predicted from the other. It has been used to be a similarity measure for images registration problems [1, 4-9]. However, the local maxima of mutual information make it difficult to register images, because the search algorithm will converge to the local maximum easily. Ji[2] and Tsao [3] analyzed both sampling and interpolation effects of mutual information. Several pre-processing methods are discussed for reducing the interpolation effects also. Likar and Pernus [7] proposed a combination of prior and floating information on the joint probability and random re-sampling of one image to improve the registration. Pluim [4] combined both standard mutual information and gradient information to yield a better registration.

---

* Supported by Guangdong Province Nature Science Foundation (No. 31789).

In MI method, transformation parameters are obtained by searching in multi-dimensional parameter space using optimization algorithm. The local maxima of one-dimensional parameter would affect the searching results in other dimensions space. For rotation registration, rotation parameter would be affected by translation parameter easily. That means, when searching translation parameters run into local maxima, it is very difficult to search the rotation parameter accurately. In this paper, a registration method based on image symmetry used to solve rotation parameter is proposed. The symmetry characteristics of medical images are used to define the symmetry measure, which could be used to determinate the symmetry axis of images. The angle between symmetry axes of two images is used to determinate the rotation parameter. Our method could obtain more accurate rotation transformation parameter when rigid transformation parameters include translation, scale and rotation all together. Moreover, rotation parameter could be solved independently without translation and scale parameters, which could avoid the local maxima of these rigid transformation parameters. Experiments show that our method is feasible to solve rotation parameter for symmetry medical images. Rotation registration accuracy is better than that of MI method for rigid image registration.

2 Symmetry Measure

2.1 Mutual Information

Mutual information is a basic concept from information theory, measuring the statistical dependence between two random variables or the amount of information that one variable contains about the other. The mutual information I of two images X and Y is evaluated as [5,6]:

$$I(X,Y) = \sum_{x}\sum_{y} P_{XY}(x,y) \log \frac{P_{XY}(x,y)}{P_X(x)P_Y(y)} \quad (1)$$

where $P_X(x)$, $P_Y(y)$, and $P_{XY}(x,y)$ are the marginal and joint probability mass functions. Mutual information will be at its maximum when the images matched. However, the mutual information function could contain local maxima, which impede the registration optimization process and rule out the subpixel accuracy.

2.2 Symmetry Measure

In generally, many medical images are symmetry [10,11], which could be used to determinate the symmetry axes of images. Furthermore, the rotation degree could be represented using the symmetry axes. Symmetry measure is defined to estimate the symmetry axis of the image at first.

Suppose image f is mirror-symmetry about the symmetry axis passing through the center point (cx, cy). Base line is defined as the vertical line pass (cx, cy) in the image. The possible symmetry axis l_θ is the line passing through (cx, cy) and makes an angle of θ with base line, illustrated as fig 1.

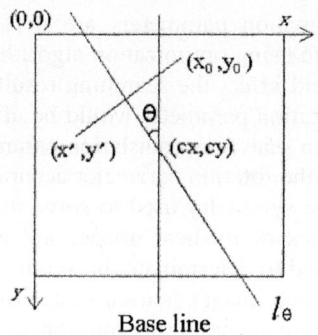

Fig. 1. The possible symmetry axis

Point set P and symmetry transform point set $ST(P)$ about the possible symmetry axis l_θ are defined as,

$$P = \{(x, y) \mid Ax + By + C > 0\}. \tag{2}$$

$$ST(P) = \left\{ (x', y') \,\middle|\, x' = \left\lfloor \frac{Ax_0 + By_0 + 2C + BKx_0 + By_0}{-A + KB} \right\rfloor, \; y' = \lfloor y_0 - K(x' - x_0) \rfloor \right\} \tag{3}$$

$$\forall (x_0, y_0) \in P$$

where $A = tg(\pi/2 + \theta)$, $B = 1$, $C = -cy - tg(\pi/2 + \theta)cx$, $K = tg\theta$, $\lfloor \bullet \rfloor$ is below truncation function. P is the pixel set, which located on one side of the possible symmetry axis l_θ. $ST(P)$ is the pixel set, which is the mirror symmetry of P about possible symmetry axis l_θ. For each pixel $(x_0, y_0) \in P$, the symmetry pixel about l_θ is (x', y'). In other words, the possible symmetry axis l_θ is the vertical bisector line between (x_0, y_0) and (x', y').

Denote by Ω the 2D point space, the mapping d is defined as, $d: \Omega \times \Omega \to R$

$$d(P, Q) = 1 - \frac{\sum_i \|p_i - q_i\|}{\sum_i \|p_i\| + \sum_i \|q_i\|}. \tag{4}$$

where $P, Q \in \Omega$, $\|\bullet\|$ is norm. The metric defines a distance function between two shapes in Ω.

The symmetry measure SM_θ of f about the possible symmetry line l_θ is defined as,

$$SM_\theta = d(f(P), f(ST(P))) \tag{5}$$

where $f(P) = \{f(x, y) \mid (x, y) \in P\}$, $f(x, y)$ is the intensity of (x, y).

SM_θ represents the mirror-symmetry degree of the image about the possible symmetry axis l_θ. The smaller SM_θ is, the larger the angle between the possible symmetry axis and symmetry axis is. SM_θ is at its maximum when the possible symmetry axis is the real symmetry axis. The angle of the symmetry axis in image f is defined as,

$$\theta_m = \max_\theta \{SM_\theta(f)\} \ . \tag{6}$$

The symmetry measure SM_θ is defined based on the intensity distribution. When the image is symmetry, the symmetry measure could be used to estimate the symmetry axis. Accordingly, the rotation parameter could be computed for images using the angle between the symmetry axes of two images. Suppose f_1 is the reference image, f_2 is the float image. Let the rotation angle be positive along clockwise. The rotation transformation parameter θ^* between f_1 and f_2 is,

$$\theta^* = \theta_m(f_1) - \theta_m(f_2) \ . \tag{7}$$

3 Symmetry Center Point (cx, cy)

From above analysis we can see that, it is necessary to locate the symmetry center point (cx, cy) at first. We segment an image using a threshold. Suppose (x_1, y_1) and (x_2, y_2) be the left-up corner and right-down corner of the rectangle that encloses the objects in an image. The symmetry center point could be

$$cx = \frac{x_1 + x_2}{2} \ \Box \ cy = \frac{y_1 + y_2}{2} \ . \tag{8}$$

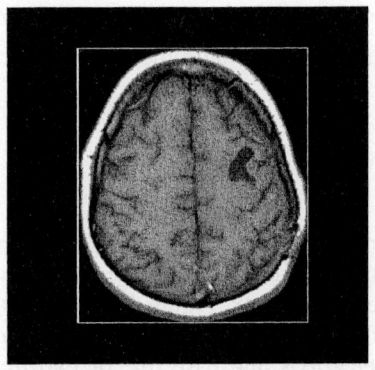
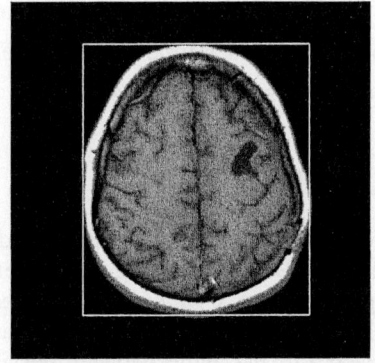

Fig. 2. Rectangles enclosing the object in an image using different threshold (the threshold of the left image is 50, the right image is 100)

Because the backgrounds of most medical images are dark, it is not difficult to segment objects from medical images. In generally, segmentation results of medical images are not sensitive to threshold because the difference between the object and background is obvious. So, for different threshold, the rectangle enclosing objects would not change greatly and the symmetry center point would not vary greatly also. An example is illustrated in figure 2. Moreover, in order to reduce the error of the symmetry center point, several axes paralleled to the possible symmetry axis are selected to compute SM_θ. The SM_θ with maximum value is selected as the symmetry measure about the possible symmetry axis l_θ, as illustrated in figure 3.

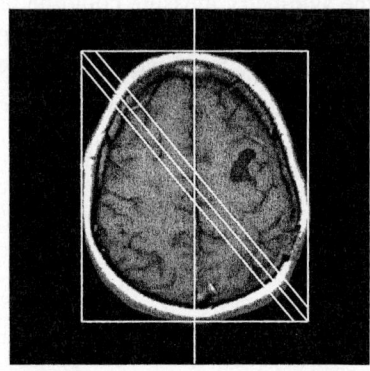

Fig. 3. Several axes paralleled to the possible symmetry axis. The possible symmetry axis is the middle one among the three parallel lines.

From above analysis, we can see that rotation registration is independent to translation and scale registration. That means rotation registration could be done at first without translation and scale parameters, which is benefit to image registration because the local maxima of other transformation parameters would not affect rotation parameter greatly.

In order to obtain the maximum of SM_θ, global search algorithm is used. It would not take too much time to search the maximum of SM_θ because the search algorithm is done in one-dimension space.

4 Experiments

We applied our method to multimodality symmetry medical images. Figure 4 shows T1 and PD images being experimented with. The size of two images is 210×180. Figure 5 shows CBF and MR images with size of 128×128. Figure 6 shows MR and PET image with size 122×122. (c) and (d) of figure 4,5,6 are responses of symmetry measure to the angle between possible symmetry axis and base line of two images alternatively. (e) of figure 4,5,6 are responses of mutual information to rotation angle. The abscissa is the angle, which is positive along clockwise.

All these medical images are registered already and the rotation parameters should be 0°. Let the left image be reference image and transform the right image. The rotation parameters obtained by our method are 0° in figure 4, 3° in figure 5 and 3° in figure 6 alternatively. The rotation parameters obtained by MI method are 0° in figure 4, 9° in figure 5 and -3° in figure 6 alternatively. The rotation registration accuracy of our method is better than that of MI method. From subjective view, it can be seen that the response of symmetry measure is smoother than that of mutual information and the maximum of symmetry measure is corresponding to the symmetry axis.

In order to illustrate the rotation registration accuracy, images are transformed with translation and rotation parameters. Our method and MI method are used to solve rotation transformation parameter alternatively. Powell's method is used to search transformation parameter in MI method. Powell's method repeatedly iterates the dimensions of the search space, performing one-dimensional optimizations for each dimension, until convergence is reached. However, this method could run into local maxima and reach local optimal results. For our method, only one-dimensional search of rotation parameter is done by global search algorithm. The maximum of SM could be searched and less time is taken.

Table 1,2,3 shows the rotation parameter obtained by two methods. It can be seen that the rotation registration accuracy of our method is better than that of MI method, because it is difficult to reach global maximum of MI when searching in

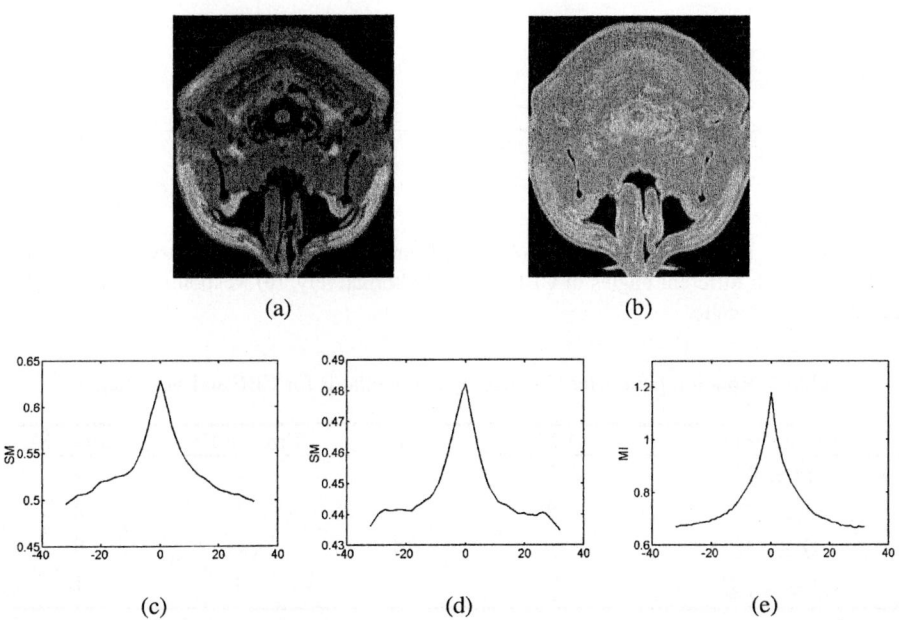

Fig. 4. (a) T1 image, (b) PD image, (c) and (d) are responses of Symmetry Measure about possible axis with different angles of T1 and PD alternatively. (e) Response of mutual information to rotation angle.

Table 1. Rotation parameter obtained by two methods for T1 and PD images

Rotation parameter	SM	MI	Error of SM	Error of MI
$\theta = 5°$	5	8	0	3
$\theta = -10°$	-10	-6	0	4
$\theta = 15°$	14	22	1	7
$\theta = -15°$	-15	-16	0	1

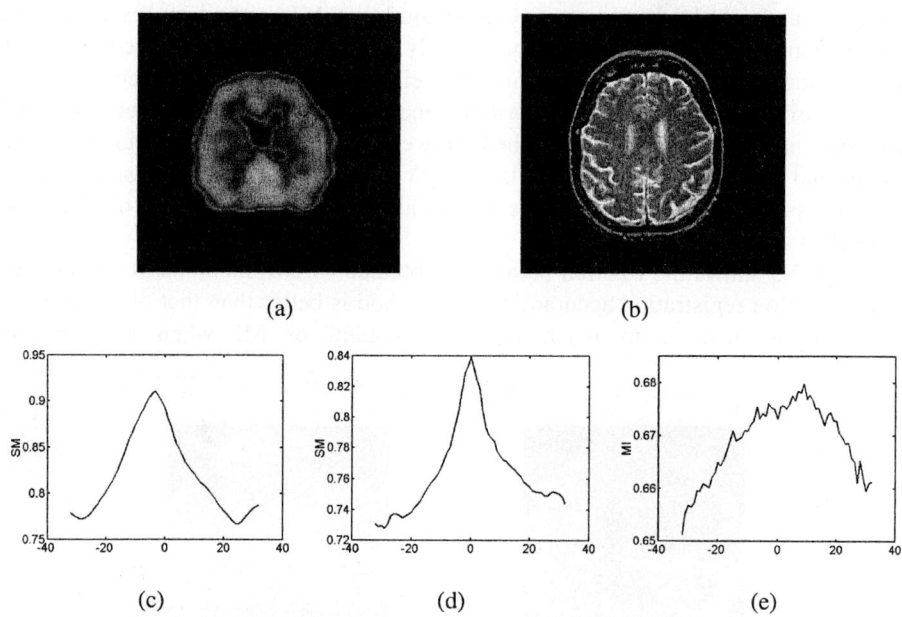

Fig. 5. (a) CBF image, (b) MR image, (c) and (d) are responses of Symmetry Measure about possible axis with different angles of CBF and MR alternatively. (e) Response of mutual information to rotation angle.

Table 2. Rotation parameter obtained by two methods for CBF and MR images

Rotation Parameter	EM	MI	Error of EM	Error of MI
$\theta = 8°$	9	5	1	3
$\theta = -7°$	-9	-5	2	2
$\theta = 18°$	19	18	1	0
$\theta = -12°$	-12	-11	0	1

multi-dimensional space. In additionally, When MI method does registration well, our method could do registration well also, as figure 4. That means the registration accuracy of our method is satisfied.

Table 3. Rotation parameter obtained by two methods for MR and PET images

Rotation Parameter	EM	MI	Error of EM	Error of MI
$\theta = 3°$	2	6	1	3
$\theta = -12°$	-13	-20	1	8
$\theta = 13°$	12	20	1	7
$\theta = -17°$	-19	-9	2	8

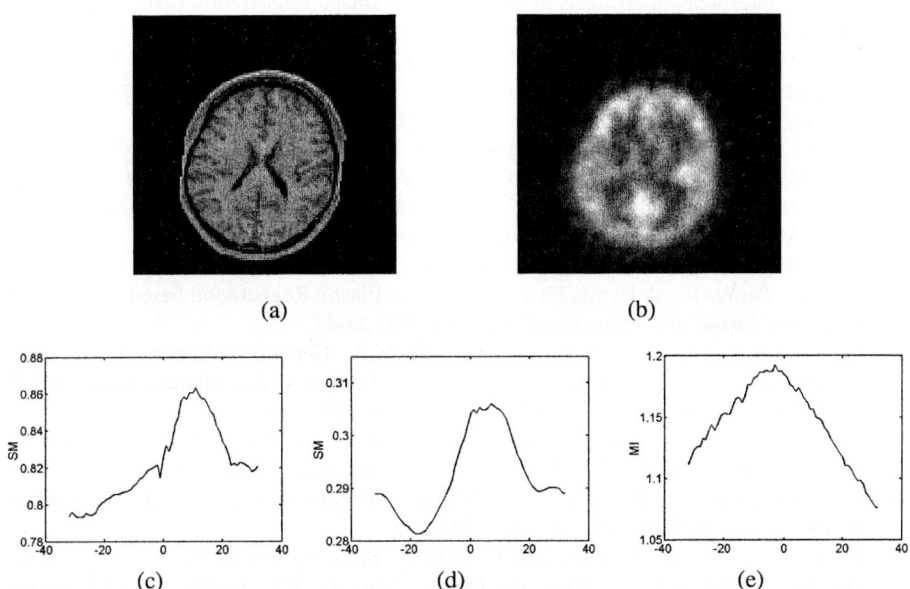

Fig. 6. (a) MR image, (b) PET image, (c) and (d) are responses of Symmetry Measure about possible axis with different angles of MR and PET alternatively. (e) Response of mutual information to rotation angle.

5 Conclusion

Mutual information has been developed into an accurate measure for multimodal medical image registration. But local extrema impede the registration optimization process and rule out the registration accuracy, especially for rotation registration. We proposed a novel rotation registration method based on image symmetry. Symmetry measure is defined to estimate the symmetry axis of images. When the symmetry measure is at its maximum, the possible symmetry axis is the real symmetry axis. The rotation transformation parameter is the angle between the symmetry axes of two images. Our method could be used to estimate rotation parameter independently. That means it is not necessary to estimate translation parameter in the meantime, which could reduce the influence of translation parameter to rotation parameter when searching in the multi-dimensional parameter space. Experiments show that the registration accuracy of our method is better than that of MI method for rigid image registration. When MI method does registration well, our method could do registration well also.

References

1. Josien Pluim,P.W., Antoine Maintz,J.B. and Max A. Viergever : Mutual-Information-Based Registration of Medical Images: A Survey. IEEE Trans Medical Imaging. 22(2003) 986-1004
2. Jim Xiuquan Ji, Hao Pan, Zhi-Pei Liang : Further Analysis of Interpolation Effects in Mutual Information-Based Image Registration. IEEE Trans. Medical Imaging. 22 (2003) 1131-1140
3. Tsao, J.: Interpolation Artifacts in Multimodality Image Registration based on Maximization of Mutual Information. IEEE Transactions on Medical Imaging. 22 (2003) 854 – 864
4. Pluim,J.P.W., Maintz,J.B.A. and Viergever,M.A. : Image Registration by Maximization of Combined Mutual Information and Gradient Information. IEEE Transactions on Medical Imaging. 19(2000) 809-814
5. Viola,P. and Wells III,W.M. : Alignment by Maximization of Mutual Information. In: Proceedings of the 5th International Conference on Computer Vision (1995) 16-23
6. Collignon,A., Maes,F., Delaere,D., Vandermeulen,D., Suetens,P. and Marchal,G. : Automated Multi-modality Image Registration based on Information Theory. In: Information Processing in Medical Imaing, The Netherlands: Kluwer (1995) 263-274
7. Likar,B., Pernus,F. : A Hierarchical Approach to Elastic Registration based on Mutual Information. Image and Vision Computing 19(2001) 33-44
8. Maintz, J.B.A., Meijering, H.W., & Viergever, M.A.: General Multimodal Elastic Registration based on Mutual Information. In K.M. Hanson (Ed.): Medical Imaging 1998, Vol.3338, Bellingham, WA: SPIE Press (1998) 144-154
9. Chen,H.M., Varshney,P.K. : A Pyramid Approach For Multimodality Image Registration Based On Mutual Information. In: ISIF'2000 (2000) Mod3-9~Mod3-15
10. Zabrodsky, H.; Peleg, S.; Avnir, D.: A Measure of Symmetry based on Shape Similarity. In: Proceedings IEEE CVPR '92 (1992) 703 –706
11. Tuzikov,A.V., Olivier Colliot, Isabelle Bloch : Brain Symmetry Plane Computation in MR Images Using Inertia Axes and Optimization. In: Proceedings. 16th International Conference on Pattern Recognition, Vol.1 (2002) 516 - 519

Target Tracking Under Occlusion by Combining Integral-Intensity-Matching with Multi-block-voting

Faliang Chang, Li Ma, and Yizheng Qiao

School of Control Science and Engineering, Shandong University,
73 Jingshi Road, 250061, Shandong Province, P. R. China
flchang@sdu.edu.cn, maryparis@163.com

Abstract. We propose a new method to solve the occlusion problem efficiently in rigid target tracking by combining integral-intensity-matching algorithm with multi-block-voting algorithm. If the target is occluded, means some blocks are occluded and tracked falsely. Then we don't let the occluded blocks participate in voting and integral-intensity-matching calculation, and use the remainder unoccluded blocks which can represent target' attribute to track the target unceasingly. Experimental results show that the adopted two algorithms are complementary, and effective combination can achieve reliable tracking performance under heavy occlusion.

1 Introduction

Moving target tracking is a substantial task of computer vision. It has been widely used in robot vision navigation, medical diagnosing, video-surveillance, etc. Occlusion is a serious and difficult problem which can cause loss of target in the tracking process.

To solve the occlusion problem, many scholars have proposed kinds of efficient methods: (1) Tracking based on target features' matching between successive frames [1], [2], [3], [4], [5], and the features can be the target intensity, binary image, edge points or corner points, etc. The Generalized Hough Transform algorithm, in which the unoccluded corner points vote the target position [6]. (2) Methods based on Multi-templates. Literature [7] uses matching error to determine occlusion and tracks target by integral intensity correlation matching. Literature [8] adopts the sub-templates matching based on maximum a posteriori probability and determines which target the sub-template belongs to, and it can solve the multi-moving-target occlusion problem. (3) Using dynamic Bayesian network which can model the occlusion process explicitly [9], [10]. (4) Particle filter based on color histogram can solve the partial occlusion problem [11]. (5) Tracking algorithm under occlusion based on mesh model [12], [13].

This paper solves occlusion problem based on intensity matching, with combined integral-intensity-matching algorithm and multi-block-voting algorithm. One contribution of our work is combining the above two related and complementary algorithms to track the target separately because each of the two adopted algorithms has its advantage and disadvantage but combining them can get good performance. Another contribution of our work is using the adaptive dividing method to get blocks with obvious characteristic and use blocks to measure the occlusion region. Not only can we avoid the difficulty of detection the target feature points like corner points in [6], but also can overcome the disadvantage of the fixed divided blocks which may match falsely if the characteristic of the blocks is unobvious [7], [8], [14].

Our method begins with adaptive dividing the target into blocks with each has obvious characteristic, then combining integral-intensity-matching algorithm with multi-block-voting algorithm to track the target. Under occlusion, the blocks which have been occluded don't participate in voting and integral matching calculation, and we use the remainder unoccluded blocks to track the target unceasingly.

The remainder of this paper is organized as follows: Section 2 introduces the proposed algorithm in detail. Including: 2.1 Adaptive target dividing. 2.2 Kalman prediction based on the current statistical model 2.3 Tracking algorithm based on combining integral-intensity-matching with multi-block-voting. 2.4 Target tracking algorithm under occlusion. 2.5 The abruptly varying illumination handling. Section 3 shows experimental results and makes analysis. Section 4 presents our conclusions.

2 Target Tracking Algorithm by Combining Integral-Intensity- Matching with Multi-block-voting UnderOcclusion

2.1 Adaptive Target Dividing

The purpose of adaptive dividing is to obtain fine characteristic blocks which can be tracked accurately. The basic idea is: Dividing the target into blocks of size 8*8 pixels and doing the merging and splitting operation based on the intensity and structure attribute. The intensity attribute is defined as intensity variance of each block which reflects differences of the gray levels and the amount of edge points. The structure attribute is defined as position of the block and its adjacent region attribute, and each block should have low similarity with its adjacent region.

Adaptive dividing depends on the following principles:

1. Calculating intensity variance of each block and if the value is less than a certain threshold, merging the block with one of its 4-adjacent blocks which has the largest variance. The choosing of threshold should depend on experiments but because of less and less unoccluded blocks left in the occlusion process, the threshold can't be high in order to get enough unoccluded blocks to vote even in the heavy occlusion.
2. Calculating the similarity of each block with its adjacent region. Because if the similarity is more than a threshold, the false matching probably occurs because its adjacent region in the next frame is within the searching window probably. So the block should be merged with another block in its 4-adjacent region which has the lowest similarity value.

    The calculation of block's similarity is: Moving the block to some new positions within its small adjacent region, and calculating the intensity correlation value between the block and the corresponding region of the new position respectively. Then we choose the highest value as the similarity of the block.

3. If some blocks have high variance and low similarity, we can split them into two blocks which satisfy the principle of (1) and (2) respectively.
4. After merging and splitting, the size of block should be more than or equal to 8*8.
5. For the blocks which are adjacent to the rectangular tracking window, using a lower threshold for the variance principle introduced in (1) and they are not subject to principle (2). Because they probably contain background information which changes all the time and can't represent the attribute of target. Along with the tracking process they will be determined to be occluded, having no voting and integral-intensity-matching right.

Fig.1 shows the experimental result of the target dividing before occlusion. We can see the blocks which are adjacent to the rectangular tracking window change little. Other blocks have possessed obvious characteristic.

Assuming the deformation of the target is small because the process of occlusion is temporary and the assuming accords with the reality mostly expect for the case of maneuvering like sudden turning, we detect the unoccluded blocks every 20 frames and adjust the dividing appropriately (through merging and splitting) to ensure the reliability of tracking and decrease the calculating time.

Fig. 1. Adaptive target dividing before occlusion. The tracking region is divided into 39 blocks and the blocks belong to the target have own obvious characteristic.

2.2 Kalman Prediction Based on the Current Statistical Model

We can use Kalman filter theory to predict the moving of target accurately (containing the location, velocity and acceleration of the target). Then we match the target and each block in searching window centered on the location of Kalman prediction in the next frame. The small searching window is very efficient in most cases and it decreases the calculation greatly. In the special cases of abrupt varying illumination and target being fully occluded, we adopt the predictive location of Kalman as the location of the target to track directly.

The basic idea of current statistical model is: if the target is maneuvering now with a certain acceleration, the acceleration in next frame is finite and only within the adjacent range of the "current acceleration" [15].

The Kalman filter equations are given by:

$$X_k = \Phi_{k,k-1} X_{k-1} + U_{k,k-1} \bar{a} + \Gamma_{k,k-1} W_{k-1}. \tag{1}$$

$$Z_k = H_k X_k + V_k. \tag{2}$$

where X_k and X_{k-1} are the state vectors at time k and $k-1$, Z_k is the observation vector at time k, $\bar{a}$ is the "current acceleration" that equals to the acceleration in one step prediction state vector $X_{k,k-1}$ and the $U_{k,k-1}$ is corresponding matrix, W and V are noises on the state and observation respectively. $\Phi_{k,k-1}$ is the state transition matrix and H_k is the observation matrix, $\Gamma_{k,k-1}$ is the noise input matrix. The Kalman theory gives the equations for optimal prediction $\hat{X}_{k,k-1}$ given the statistics of state and observation noises.

In the case of target tracking, choosing the state vector $x_k = [s_x, s_y, v_x, v_y, a_x, a_y]^T$ which is composed of location, velocity and acceleration of the moving target in the x and y axis respectively. The observation vector is $y_k = [s_x, s_y]^T$ which represents the location of the target.

2.3 Tracking Algorithm Based on Combining Integral-Intensity-Matching with Multi-block-voting

We can track the target accurately if the displacement vector can be calculated correctly between two successive frames. This paper derives two displacement vectors $(\Delta x_1, \Delta y_1)$ and $(\Delta x_2, \Delta y_2)$ from multi-block-voting algorithm and integral-intensity-matching algorithm respectively. If the two vectors are identical, the target moving vector is equal to them. Otherwise we should choose the better one based on the smoothness of target moving.

2.3.1 Multi-block-voting Algorithm

After adaptive dividing, each block searches for its optimal matching in the searching window (size of $w \times w$ and w is determined by the experiments) centered on the predictive location of Kalman in next frame. Every point in the window represents a possible moving vector (d_k, d_l), $k = 1, \ldots, w, l = 1, \ldots, w$, Then calculating the intensity matching error $S(d_k, d_l)$ to every displacement vector (d_k, d_l) between two successive frames. The intensity matching error, i.e.,

$$S_p(d_k, d_l) = \sum_{i=1}^{m}\sum_{j=1}^{n}\left|M_p(i,j) - I_p(i+d_k, j+d_l)\right|^2 .$$

$$p = 1, \ldots, N, k = 1, \ldots, w, l = 1, \ldots, w$$

(3)

where $M_p(i,j)$ is the gray value of block p in current frame, $I_p(i+d_k, j+d_l)$ is the gray value of the corresponding block p through displacement vector (d_k, d_l) in next frame. m and n represent the height and width of the block p respectively. N is the number of blocks via dividing.

We let (d_{kp}^*, d_{lp}^*) corresponds to the correct matching of block p, then its matching error corresponds to $\min(S_p(d_k, d_l))$ in the $w \times w$ matching errors. For each block, we calculate its minimum matching error respectively, then we can obtain N displacement vectors (d_{kp}^*, d_{lp}^*), $p=1,\ldots,N$. Voting is choosing the most frequent motion vector as the target displacement vector $(\Delta x_1, \Delta y_1)$.

2.3.2 Target Integral-Intensity-Matching Algorithm

Regarding the target as a whole, we calculate the intensity correlation matching error for every displacement vector (d_k, d_l), $k=1,\ldots,w, l=1,\ldots,w$. That is

$$sum(d_k, d_l) = \sum_{p=1}^{N} S_p(d_k, d_l). \tag{4}$$

We choose the displacement vector which corresponds to $\min(sum(d_k, d_l))$ as the target displacement vector $(\Delta x_2, \Delta y_2)$.

2.3.3 Combining the Two Above Algorithms

Tracking each block, i.e., searching for its optimal matching, we only use the information of each block, but have no use of its position in the target and structure information of its adjacent region. Even though adopting adaptive dividing, we only try our best to get blocks with distinct characteristic (depending on the target intensity attribute and the threshold which can not be high). Of course, we can ignore the few false matching of some blocks via voting, but it may vote inaccurately in the case of less and less unoccluded blocks left under heavy occlusion. It's the disadvantage of multi-block-voting algorithm. But we can use the integral-intensity-matching algorithm to overcome it. The second algorithm reckons the target as a whole and the position of each block is fixed, so it uses the whole information to supplement the incompleteness of block's local information. But this algorithm has disadvantages too. It will loss the target in the process of occlusion because it can't estimate the occluded region accurately, also it can be highly affected by abruptly varying of local intensity. However, these disadvantages can be handled by multi-block-voting algorithm, because we can determine the occluded region by the blocks and the local abruptly varying intensity only brings some blocks' false tracking and the other blocks which are in the majority can still vote correctly. So the two algorithms are complementary and the tracking is more reliable if combining them.

For the two results $(\Delta x_1, \Delta y_1)$ and $(\Delta x_2, \Delta y_2)$, if they are the same, the displacement vector of target is equal to them. If they are different, we choose the better one depending on the smoothness of target moving between successive frames (except for the maneuvering), that is

$$d_i = \left| \Delta \hat{x} - \Delta x_i \right| + \left| \Delta \hat{y} - \Delta y_i \right|, i = 1, 2. \tag{5}$$

Choosing $(\Delta x_i, \Delta y_i), i = 1 \text{ or } 2$ which corresponds to the $\min\{d_1, d_2\}$ as the target displacement vector $(\Delta x, \Delta y)$, where $\Delta \hat{x}$ and $\Delta \hat{y}$ correspond to the displacements of the predictive location of Kalman in x and y axis respectively.

2.4 The Tracking Algorithm Under Occlusion

2.4.1 Occlusion Handling

The target is occluded, means some blocks are occluded and tracked falsely. When the difference $|d_{kp}^* - \Delta x|$ or $|d_{lp}^* - \Delta y|$ is more than a certain threshold, it means block p has been occluded, and making a sign to it representing the occlusion. The choosing of threshold is easy and it should be lower such as 1 or 2.

The matching algorithm between current frame and last frame in occluding process is as follows: (We also combine the two algorithms as 2.3 section does.)

1. For each block, calculating the $\min(S(d_k, d_l))$ according to the formula (3) and deriving (d_k^*, d_l^*). The blocks which have occlusion sign have no right to vote and the remainder unoccluded blocks vote the displacement vector $(\Delta x_1, \Delta y_1)$. Although the occluded blocks don't participate in voting, we also calculate their (d_k^*, d_l^*) via matching. Because occluded blocks will be out of occlusion step by step. When their displacement vector (d_k^*, d_l^*) is equal to $(\Delta x, \Delta y)$, it means they have been out of occlusion and we should delete the occlusion sign of them.
2. For the integral-intensity-matching, we don't let the occluded blocks participate in calculation, that is:

$$\arg\min_{(d_k, d_l)} (sum(d_k, d_l)) = \sum_{p=1, p \neq q_i}^{N} S_p(d_k, d_l). \tag{6}$$

where $q_i, i = 1, \ldots, M$ represents occluded block and the number of occluded blocks is M. Displacement vector $(\Delta x_2, \Delta y_2)$ is equal to the result of the formula (6).

3. Combining the two results as the method introduced in 2.3.3. We can obtain the target displacement vector $(\Delta x, \Delta y)$.

2.4.2 Updating the Occluded Region

The first important task of tracking under occlusion is to determine the occluded blocks and they are different per frame in the process, so the occluded region should be updated real time.

The updating method of the occlusion region in current frame is as follows:

1. For the occluded blocks in the last frame, if their $|d_k^* - \Delta x|$ or $|d_l^* - \Delta y|$ is more than a certain threshold, means they are still occluded. Otherwise they have been out of occlusion and they can get correct tracking and deleting their occlusion sign.
2. For the unoccluded blocks in the last frame, if their $|d_k^* - \Delta x|$ or $|d_l^* - \Delta y|$ is more than the threshold, it means that they have entered into occlusion in current frame and we make occlusion sign for them.

2.5 The Abruptly Varying Illumination Handling

One disadvantage of intensity matching algorithm is the sensitivity of abruptly varying illumination. The tracking will be false if all of the target' gray values change greatly. We propose a method to solve the problem to gain a more reliable tracking performance. Firstly, determining whether the illumination changes or not, and if so, we examine the tracking result next. If it accords with the smoothness of moving, means the varying illumination has no serious effect on tracking. Otherwise it means false tracking. Then we regard the predictive location of Kalman as the location of the target to track directly.

The method of determining whether abruptly varying illumination happens or not is: Sampling uniformly per frame with the coordinate of $(8\times n, 8\times m)$ as the sample points, where $n=1,...,32, m=1,...,32$, (The image size is 256×256). Then we have 1024 sample points and the distribution of each point accords with Gauss distribution (excepting for the points in the target and the noise points). In order to adapt to the gradually varying illumination, we only reserve the statistical data of recent 100 frames and calculate the mean and variance of each point. If most of the points accord with their latest distribution(the mean and variance of the latest 100 frame) i.e., the difference between the gray value of each point in current frame and the mean value of the latest Gauss model is less than a certain threshold (The threshold is function of variance), then there is no abruptly varying illumination. Otherwise we determine that the illumination changes abruptly and build another model for each point in current frame. The subsequent frames adopt the same processing. After the abruptly varying illumination, if most of the points among a certain amount of frames accord with the second distribution model, we delete the first model and if there occurs another abruptly varying illumination, we build a new model for it again.

3 The Experimental Results and Analysis

The first experiment: Comparing the integral-intensity-matching algorithm with the proposed algorithm in this paper to indicate our algorithm can void the influence of abruptly intensity varying while the former can't, because the influence is handled by the combined multi-block-voting algorithm. The first image of Fig.2 shows that each algorithm gets the same accurate tracking before occlusion occurs. In the second image we can see false tracking of integral-intensity-matching algorithm because of a special local occlusion (i.e. local intensity abruptly varying). The third image shows accurate tracking of our algorithm because we combine the multi-block-voting method.

The second experiment: The experimental scenario is that one car is occluded by another still car belongs to the background. Slowed in Fig.3, the first image displays the tracking before occlusion. The target is divided into 39 blocks and some of the blocks have been reckoned as occluded which lie in the adjacency of rectangular tracking window and represent the background. The second image describes that the target is entering into occlusion. 20 blocks have been signed occlusion, and one block that has the occlusion sign isn't occluded actually (the reason is its similarity is not low enough and there exit noises). But it has no influence on voting and since the noise is stochastic, it will not affect the following track. The third image describes the most serious occlusion and 31 blocks have been occluded, but we can track the target accurately by using the only 8 unoccluded blocks. The fourth image shows the target is being out of the occlusion, with 17 blocks having the occlusion sign and most of them corresponding to the background. We choose the threshold which is equal to 1 for the determination of the occlusion and it accords with the reality. It can be seen that the tracking is very accurate in the occlusion process and we can estimate the different occlusion phases based on the number of occluded blocks.

The third experiment: Fig.4 shows the handling of abruptly varying illumination problem under occlusion. The first image shows the tracking under a dim luminance. In the second image the illumination changes suddenly, the algorithm based on intensity matching has false tracking. The third image shows the performance of adopted method aimed at solving the special problem. The processing is: we detect the case (the illumination changes abruptly) firstly and determine the tracking result doesn't accord with the smoothness of moving. So we make sure that the tracking result is false and adopt the predictive location of Kalman as the location of the target to track directly. The experiment indicates the proposed method can well solve the false tracking problem brought by the abruptly varying illumination efficiently.

Fig. 2. *Comparing our algorithm with the integral-intensity-matching algorithm.* The first image: Tracking before occlusion, and both of the two algorithms can track accurately. The second image: the target is occluded locally, and tracking is false using the integral-intensity-matching algorithm only. The third image: Accurate tracking using the proposed algorithm in this paper (overcoming the local abruptly varying intensity problem).

Fig. 3. *Tracking process using our proposed algorithm.* The first image: Tracking before occlusion. The second image: Tracking under occlusion. The third image: Tracking under the most serious occlusion. The fourth image: Tracking when the target is being out of occlusion.

Fig. 4. *The abruptly varying illumination handling.* The first image: The correct tracking under occlusion before the illumination changes (the luminance is very dim). The second image: The illumination abruptly changes and tracking is false before we adopt the proposed solving method. The third image: The correct tracking using our proposed method.

4 Conclusions

Occlusion brings serious problem in target tracking. The algorithm proposed in this paper can solve the problem efficiently and the experiments have showed its good performance. The algorithm has its own characteristics: (1) The algorithm bases on intensity matching and doesn't rely on the target segmentation and detection which is very difficult to operate accurately, so the algorithm has better accuracy and faster tracking. (2) Using adaptive dividing to overcome the disadvantage of fixed dividing which may bring false tracking because of the unobvious characteristic and the high similarity of some fixed blocks. (3) Combining integral-intensity-matching with multi-block-voting, and having the best use of information of the target to get reliable tracking. (4) Having no necessity to determine the beginning and the end of occlusion. (5) Using blocks to determine the occluded region and updating it per frame. Detection the occluded blocks based on the displacement vector error and the error's threshold is easy to choose.

In reality, most of the moving targets accord with the proposed algorithm but there also exist some exceptions: Small targets and targets which have single or very regular intensity distribution are not suitable for adaptive dividing. These problems and the tracking of non-rigid target under occlusion are our following work.

Acknowledgement. The paper is financial supported by the National Natural Science Foundation of China (No.Z60104009) and the Shandong Natural Science Foundation of China (No.Y2005G26).

References

1. Marecenaro,L., Ferrari,M., Marchesottl,L. and Regazzoni.C.S.: Multiple Object Tracking Under Heavy Occlusions by Using Kalman Filter Based on Shape Matching, Proceedings. International Conference on Image Processing, vol.3 (2002) 341-344
2. Natan Pterfreund: Robust Tracking of Position and Velocity with Kalman Snakes, IEEE Transactions on Pattern Analysis and Machine Intelligence ,vol.22,no.5 (1999) 564-569
3. Galvin,B., McCane,B., Novins,K.: Visual Snakes for Occlusion Analysis, IEEE Computer Society Conference on Computer Vision and Pattern Recognition (1999)

4. Loutast,E., Diamantarast,K., Pitast,I.: Occlusion Resistant Object Tracking, Proceedings. International Conference on Image Processing, vol.2 (2001) 65-68
5. Hieu T.Nguyen, Marcel Worring, Rein van den Boomgaard: Occlusion robust adaptive template tracking, Proceedings. Eighth IEEE International Conference on Computer Vision, Vol.1(2001) 678-683
6. Franco Oberti, Simona Calcagno, Michela Zara and Calo S.Regazzoni: Robust Tracking of Humans and Vehicles in cluttered Scenes with Occlusions, Proceedings. International Conference on Image Processing, vol.3 (2002) 629-632
7. Ken Ito, Shigeyuki Sakane: Robust View-based Visual Tracking with Detection of Occlusions, Proceedings. IEEE International Conference on Robotics and Automation, vol.2 (2001) 1207-1213
8. Shunsuke Kamijo, Yasuyuki Matsushita, Katsushi Ikeuchi, Masao Sakauchi: Occlusion Robust Tracking utilizing Spatio-Temporal Markov Random field Model, Proceedings.15th International Conference on Pattern Recognition, vol.1 (2000) 140-144
9. Ying Wu, Ting Yu, Guang Hua: Tracking Appearances with Occlusions, Proceedings of the IEEE Computer Society Conference on Computer Vision and Pattern Recognition, vol.1 (2003) 789-795
10. Min Hu, Weiming Hu, Tieniu Tan: Tracking People through Occlusion, Proceedings of the17th International Conference on Pattern Recognition, vol.2 (2004) 724-727
11. Nummiaro,K., Koller-Meier,E., Van Gool,L.: An adaptive Color-based Particle Filter, Image and Vision Computing, vol.21, issue 1 (2003) 99-1112
12. Yucel Altunbasak, A.Murat Tekalp: Occlusion-Adaptive, Content-Based Mesh Design and Forward Tracking, IEEE Transactions on Image Processing, vol.6, issue.9(1997) 1270-1280
13. Jian-Wei Zhao, Peng Wang, Cong-Qing Liu: An Object Tracking Algorithm based on occlusion mesh model, Proceedings. International Conference on Machine Learning and Cybernetics, vol.1 (2002) 288-292
14. Shrif Abd El-Azim: An Efficient Object Tracking Technique Using Block-Matching Algorithm, Nineteenth National Radio Science Conference (2002) 427-433
15. Zhou Hongren, Kumar K S P: A Current Statistical Model and Adaptive Algorithm for Estimation Maneuvering Targets, AIAA Journal, Guidance, Control and Dynamics,vol.7, no.5 (1984)

Recognition of Leaf Images Based on Shape Features Using a Hypersphere Classifier*

Xiao-Feng Wang[1,2], Ji-Xiang Du[1], and Guo-Jun Zhang[1]

[1] Institute of Intelligent Machines, Chinese Academy of Sciences, Hefei, 230031
[2] Department of Computer Science and Technology, Hefei University, Hefei, 230022
xfwang@iim.ac.cn

Abstract. Recognizing plant leaves has so far been an important and difficult task. This paper introduces a method of recognizing leaf images based on shape features using a hypersphere classifier. Firstly, we apply image segmentation to the leaf images. Then we extract eight geometric features including rectangularity, circularity, eccentricity, etc, and seven moment invariants for classification. Finally we propose using a moving center hypersphere classifier to address these shape features. As a result there are more than 20 classes of plant leaves successfully classified. The average correct recognition rate is up to 92.2 percent.

1 Introduction

Plant is one of the most important forms of life on earth. Plants maintain the balance of oxygen and carbon dioxide of earth's atmosphere. The relations between plants and human beings are also very close. In addition, plants are important means of livelihood and production of human beings. But in recent years people have been seriously destroying the natural environments, so that many plants constantly die and even die out every year. Fortunately, people now have realized that is a terrible mistake and are beginning to take steps to protect plants.

The first step of protecting plants is to automatically recognize or classify them, i.e., understand what they are and where they come from. But it is very difficult for ones to recognize a plant in hand correctly and immediately because there are so many kinds of plants unknown to us on earth. So we wish to use image processing and pattern recognition techniques to make up the deficiency of our recognition ability. This point can be performed just through computers and other image acquiring facilities. According to theory of plant taxonomy, it can be inferred that plant leaves are most useful and direct basis for distinguishing a plant from the others, and moreover, leaves can be very easily found and collected everywhere. By computing some efficient features of leaves and using a suitable pattern classifier it is possible for us to recognize different plants quickly.

Some recent work has focused on leaf feature extraction for recognition of plant. Im et al. [4] used a hierarchical polygon approximation representation of leaf shape to recognize the Acer family variety. Wang et al. [5] gave a method which combines different features based on centroid-contour distance curve, and

---

* This work was supported by the NSF of China (Nos.60472111 and 60405002).

adopted fuzzy integral for leaf image retrieval. Moreover, Saitoh et al. [6] required two images, a frontal flower image and a leaf image to recognize the plant.

In this paper we propose a method of recognizing leaf images based on shape features using a hypershphere classifier. 15 features are extracted from pre-processed leaf images, which include eight ratios of geometric features and seven Hu moment invariants. In addition, a moving center hypersphere (MCH) classifier is proposed for classifying a large number of leaves.

This paper is organized as follows: Section 2 introduces the leaf image segmentation and feature extraction methods. The moving center hypersphere classifier is described in Section 3. Section 4 gives some experimental results about the performances of the hypersphere classifier by some practical plant leaves. Finally, conclusions are in section 5.

2 Image Segmentation and Feature Extraction

2.1 Image Segmentation

The purpose of image segmentation is to separate leaf objects from background so that we can extract leaves' shape features exactly in the later procedures, and the output of image segmentation is a binary image in which the leaf objects are numerically displayed with 1 and the background is with 0. There are two kinds of background in the leaf images that we have collected, one is simple (as shown in Fig.1-a), another kind is complicated (as shown in Fig.1-d). In this paper we choose iterative threshold selection segmentation method to address leaf images with simple background. Marker-controlled watershed segmentation method is selected for those leaf images with complicated background.

(1) Iterative Threshold Selection Segmentation: It can be seen in the leaf images with simple background that the gray level of pixels within leaf objects is distinctly different from that of pixels within the background. In this case, there are some distinct peaks corresponding to leaf objects and background in the gray level histogram. Therefore, we could find a threshold to transform a gray leaf image into a binary leaf image. Here we use the iterative threshold selection segmentation method [7] to compute the threshold, the method is summarized as follows:

Step 1. Compute the maximum value G_{max} and the minimum value G_{min} of all gray level in gray leaf image. Then set threshold $T_k = (G_{max} + G_{min})/2, k = 0$.
Step 2. At iterative step k, segment image into objects and background using threshold T_k. Compute G_O and G_B as the mean gray level of objects and background.
Step 3. Set new threshold $T_{k+1} = (G_O + G_B)/2$.
Step 4. If $T_{k+1} = T_k$, output threshold $T = T_k$, iteration stop; otherwise,$k = k + 1$, return to step 2.

(2) Marker-Controlled Watershed Segmentation: In the leaf images with complicated background we could find that target leaves are touching or covering some background leaves. It's difficult to separate out the target leaves from the

background using the traditional thresholding methods. The watershed segmentation is a popular segmentation method coming from the field of mathematical morphology, which can separate touching objects in an image. So we consider using marker-controlled watershed segmentation method [8] to get the target leaf objects within the complicated background, this procedure can be summarized as follows:

Step 1. Transform a leaf image into a gray image and compute its gradient image.
Step 2. Mark the target leaf objects and background.
Step 3. Applying watershed segmentation to the gradient image.

After segmentation we can locate the leaf objects in binary images. Notice that there exist some variance on length and curvature of leafstalks. To keep the precision of shape features extraction these leafstalks should be removed. Therefore, we consider applying opening operation of mathematical morphology to binary images, which is defined as an erosion operation followed by a dilation operation using a same structuring element. By performing opening operation several times, we can successfully remove the leafstalks while preserving the main shape characteristics of leaf objects. The results of segmentation and removing leafstalks for leaf images with simple and complicated background are illustrated in Fig.1 (a) - Fig.1 (g).

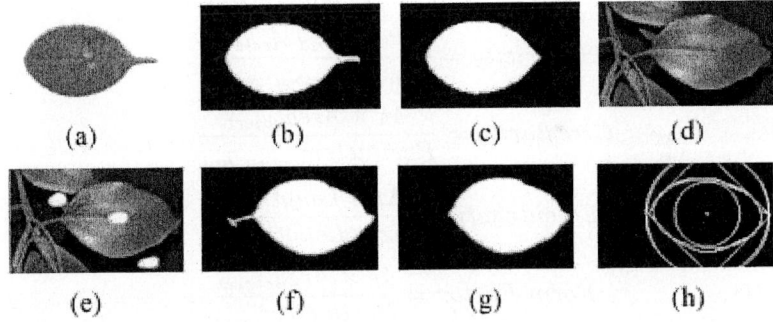

Fig. 1. (a) Leaf image with simple background. (b) Iterative threshold selection segmentation result. (c) Removing leafstalk result. (d) Leaf image with complicated background. (e) Markers of target leaf and background. (f) Watershed segmentation result. (g) Removing leafstalk result. (h) Some shape descriptors used to compute eight geometric features of Fig.1 (g).

2.2 Features Extraction

There are many kinds of image features such as shape features, color features and texture features, etc, which can be used for leaf images classification. According to theory of plant taxonomy, it can be inferred that the shape features are the most important and effective ones. Consequently, we consider using shape features for classification. Notice that there exist greater morphological differences in different kinds of leaves; even in the same kind of leaves there also exist some variance on scale. In this case, we should use ratios instead of those variable values such as area and perimeter, etc.

From the binary image we could get several shape descriptors that include bounding box (a rectangle that circumscribes a object), convex hull (the smallest convex shape that contains the object), inscribed circle (the largest possible circle that can be drawn interior to the object), circumscribed circle (a circle that passes through all vertices of a object and contains the entire object in its interior), centroid and contour. Fig.1 (h) illustrates these shape descriptors obtained from Fig.1 (g). Using these shape descriptors we could further compute eight ratios of geometric features, which include:

$$Aspect\ Ratio = \frac{length_{bounding\ box}}{width_{bounding\ box}}. \quad (1)$$

$$Rectangularity = \frac{Area_{object}}{Area_{bounding\ box}}. \quad (2)$$

$$Area\ Convexity = \frac{Area_{object}}{Area_{convex\ hull}}. \quad (3)$$

$$Perimeter\ Convexity = \frac{Perimeter_{object}}{Perimeter_{convex\ hull}}. \quad (4)$$

$$Sphericity = \frac{R_{inscribed\ circle}}{R_{circumscribed\ circle}}. \quad (5)$$

$$Circularity = \frac{4\pi \times Area_{object}}{Perimeter_{convex\ hull}^2}. \quad (6)$$

$$Eccentricity = \frac{Axis\ Length_{long}}{Axis\ Length_{short}}. \quad (7)$$

$$Form\ Factor = \frac{4\pi \times Area_{object}}{Perimeter_{object}^2}. \quad (8)$$

These ratios are naturally invariant to translation, rotation and scaling; it's very important and useful to leaf images classification. Besides the geometric features, the moments are also widely used as shape features for image processing and classification, which provide a more geometric and intuitive meaning than the geometric features. It was M K Hu [9] that first set out the mathematical foundation for two-dimensional moment invariants. Hu has also defined seven of these moment invariants computed from central moments through order three that are invariant under object translation, scaling and rotation. Accordingly, we consider using these Hu moment invariants as classification features in this paper. The values of them can be calculated from contours using Chen's improved moments [10] as follows: The Chen's improved geometrical moments of order ($p + q$) are defined as:

The Chen's improved geometrical moments of order $(p+q)$ are defined as :

$$M_{pq} = \int_C x^p y^q ds . \tag{9}$$

where $p, q = 0, 1, 2, \ldots$, $\int_C$ is the line integral along a closed contour C and $ds = \sqrt{(dx)^2 + (dy)^2}$.

For practical implementation M_{pq} could be computed in their discrete form:

$$M_{pq} = \sum_{(x,y) \in C} x^p y^q . \tag{10}$$

Then the contour central moments can be calculated as follows:

$$\mu_{pq} = \int_C (x - \bar{x})^p (y - \bar{y})^q ds . \tag{11}$$

$$\bar{x} = \frac{M_{10}}{M_{00}} \quad \bar{y} = \frac{M_{01}}{M_{00}} . \tag{12}$$

In the discrete case μ_{pq} above becomes:

$$\mu_{pq} = \sum_{(x,y) \in C} (x - \bar{x})^p (y - \bar{y})^q . \tag{13}$$

These new central moments are further normalized using the following formula:

$$\eta_{pq} = \frac{\mu_{pq}}{\mu_{00}^\gamma} . \tag{14}$$

where the normalization factor is $\gamma = p + q + 1$. The seven moment invariant values can then be calculated from the normalized central moments as follows:

$$\phi_1 = \eta_{20} + \eta_{02}$$
$$\phi_2 = (\eta_{20} - \eta_{02})^2 + 4\eta_{11}^2$$
$$\phi_3 = (\eta_{30} - 3\eta_{12})^2 + (\eta_{03} - 3\eta_{21})^2$$
$$\phi_4 = (\eta_{30} + \eta_{12})^2 + (\eta_{03} + \eta_{21})^2$$
$$\phi_5 = (3\eta_{30} - 3\eta_{12})(\eta_{30} + \eta_{12})[(\eta_{30} + \eta_{12})^2 - 3(\eta_{03} + \eta_{21})^2] + (3\eta_{21} - \eta_{03})(\eta_{21}$$
$$+ \eta_{03}) \times [3(\eta_{30} + \eta_{12})^2 - (\eta_{21} + \eta_{03})^2]$$
$$\phi_6 = (\eta_{20} - \eta_{02})[(\eta_{30} + \eta_{12})^2 - (\eta_{21} + \eta_{03})^2] + 4\eta_{11}(\eta_{30} + \eta_{12})(\eta_{21} + \eta_{03})$$
$$\phi_7 = (3\eta_{21} - \eta_{03})(\eta_{30} + \eta_{12})[(\eta_{30} + \eta_{12})^2 - 3(\eta_{21} + \eta_{03})^2] + (3\eta_{12} - \eta_{30})(\eta_{21}$$
$$+ \eta_{03}) \times [3(\eta_{30} + \eta_{12})^2 - (\eta_{21} + \eta_{03})^2] . \tag{15}$$

Table 1 shows the values of eight geometric features and seven Hu moment invariants of leaf in Fig.1 (g). From Table 1 it could be seen that values of geometric features and moment invariants are different greatly in order of magnitude, therefore, be-fore classification these features need to be normalized as follows:

$$F = \frac{F - F_{min}}{F_{max} - F_{min}} . \tag{16}$$

where F denote one feature, F_{max} is the maximum value of all features in the same class with F , F_{min} is the minimum one.

Table 1. Values of eight geometric features and seven Hu moment invariants corresponding to leaf in Fig.1 (g)

Aspect Ratio	1.098	ϕ_1	1.768187e-001
Rectangularity	0.751	ϕ_2	5.537947e-003
Area Convexity	0.977	ϕ_3	3.432757e-005
Perimeter Convexity	1.052	ϕ_4	2.816526e-006
Sphericity	0.540	ϕ_5	2.603669e-011
Circularity	0.921	ϕ_6	1.662384e-007
Eccentricity	1.566	ϕ_7	9.437735e-012
Form Factor	0.768		

3 Moving Center Hyperesphere Classifier

In this paper we regard one feature vector containing eight geometric features and seven moment invariants as a pattern. Considering that the number of patterns and the dimension of pattern space are both very large, if we use conventional classifier like K-NN or Neural Network, the corresponding classification process would be quite time-consuming and space-consuming. Therefore, we propose using a moving center hypersphere(MCH) classification method to perform the plant leaves classification, which fundamental idea is that we regard each class of patterns as a series of "hyper-spheres", while in conventional approaches these patterns from one class are all treated as a set of "points". The first step of this method is to compute the multidimentional median of the points of the considered class, and set the initial center as the closest point from that class to that median. Then we find the maximum radius that can encompass the points of the class. Through a certain iteration we remove the center of the hypersphere around in a way that would enlarge the hypersphere and have it encompass as many points as possible. This is performed by having the center "hop" from one data point to a neighboring point. Once we find the largest possible hypersphere, the points inside this hypersphere are removed, and the whole procedure is repeated for the remaining points of the class. We continue until all points of that class are covered by some hyperspheres. At that point, we tackle the points of the next class in a similar manner. Here, we take one class for example to summarize the whole iterating training procedure of the hypersphere classifier as follows:

Step 1. Put all the training data points into set S, set hypersphere index $k = 0$.
Step 2. Select the closest point to the median of points in S as the initial center of the hypersphere k.

Step 3. Find the nearest point to the center from all other classes, and denote the distance as $d1$.

Step 4. Find the farthest point of the same class inside the hypersphere k with radius $d1$ to the center. Let $d2$ denote the distance from the center to that farthest point.

Step 5. Set the radius of the hypersphere k as $(d1 + d2)/2$.

Step 6. Select the point in the most negative direction of the center to the nearest point of the other classes among the nearest m points in this class. The purpose is to move the center to the new point to enlarge the hypersphere. If point exists then set the point as new center of the hypersphere k, return to step 4; otherwise continue.

Step 7. Remove those points covered by the hypersphere k from the set S. If S is still not empty then $k = k+1$, return to Step 2; otherwise remove the redundant hyperspheres that are totally enclosed by larger hyperspheres of this class, training finishes.

After the training is finished, MCH system is required to be able to classify any given input data point. The perpendicular distances from the input data point to the outside surface of all hyperspheres (with that distance counting as negative if the point is inside the hypersphere) are selected as the classification criterion. Assume that there are totally hyperspheres after training, each of which has the radius of $r_i (i = 1, 2, \cdots, H)$, and let d_i denote the distance between the data point and the center of hypersphere h_i, then we can define the decision rule as follows:

$$I = arg\ min(d_i - r_i), i \in \{1, 2, \cdots, H\}\ . \tag{17}$$

where I means index for nearest neighbor hypersphere.

4 Experimental Results

To verify the MCH classifier we have taken 800 leaf samples corresponding to 20 classes of plants collected by ourselves such as ginkgo, seatung, maple, etc (as shown in Fig.2). Each class includes 40 leaf samples, of which 25 samples are selected randomly as training samples and the remaining is used for testing samples. In our implementation we used Visual C++ 6.0 on Windows XP operating system running on Intel PC 2.4GHZ 512MB RAM.

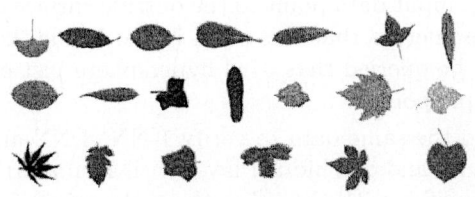

Fig. 2. Twenty classes of plant leaves used for recognition

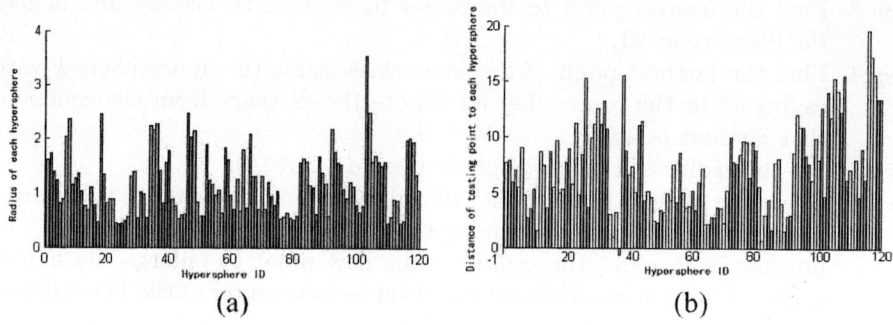

Fig. 3. (a) The Histogram for each hyperspheres' radius after training 500 samples corresponding to 20 classes. (b) The histograms for the distances of one Ginkgo testing sample to outside surface of each hypersphere.

Table 2. Class that each hypersphere belongs to

class ID	hypersphere ID	class ID	hypersphere ID	class ID	hypersphere ID
1	1 - 6	8	52 - 57	15	88 - 91
2	7 - 17	9	58 - 61	16	92 - 96
3	18 - 27	10	62 - 63	17	97 - 10
4	28 - 33	11	64 - 65	18	103 - 108
5	34 - 38	12	66 - 71	19	109 - 115
6	39 - 45	13	72 - 81	20	116 - 120
7	46 - 51	14	82 - 87		

* class ID explanation: 1 Sweet Osmanther 2 Seatung 3 Chinese Box 4 Photinia 5 Gingkgo 6 Tuliptree 7 London Planetree 8 Red Maple 9 Donglas Fir 10 SevenAngle 11 Beautiful Sweetgum 12 Panicled Goldraintree 13 China Floweringquince 14 Rose bush 15 Chrysanthemum 16 Sunflower 17 China Redbud 18 Bamboo 19 Plum 20 Willow.

Fig.3 (a) shows the histogram for each hyperspheres' radius after training 500 samples. Class that each hypersphere belongs to is listed in Table 2. It can be seen from Fig.3 (a) and Table 2 that there are total 120 hyperspheres obtained after training finished. The histogram for the distance of one ginkgo testing sample point to the outside surface of each hypersphere is shown in Fig.3 (b), in which the distance from the input data point to the outside surface of 37rd hypersphere is negative. It can be inferred that this point is just inside the 37rd hypersphere. From Table 2 it can be queried that 37rd hypersphere just encompasses part of ginkgo training sample points.

Here we also used the same data to verify 1-NN, 4-NN and BPNN (15 input nodes, 20 output nodes and one hidden layer with 12 nodes). Performance comparisons of MCH classifier with the other three classifiers are shown in Table 3 and Fig.4.

Table 3. Performance comparisons of four classifiers

	Training time(ms)	Classifying time(ms)	Storage vectors	Average correct rate(%)
1-NN	/	17.2	500	92.6
4-NN	/	42.7	500	92.3
BPNN	3720	7.6	/	92.4
H-S	36.2	9.8	120	92.2

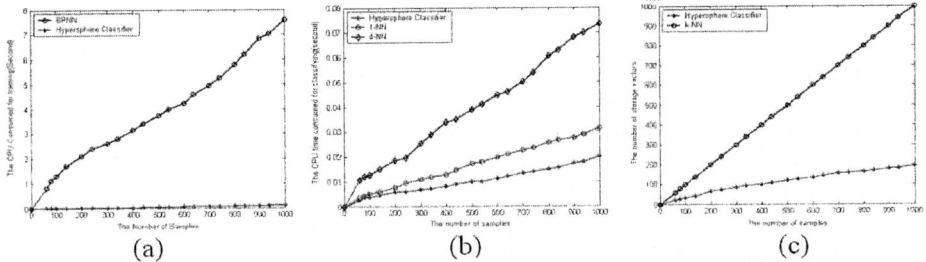

Fig. 4. (a) The CPU time consumed for training vs the number of samples (b) The CPU time consumed for classifying vs the number of samples (c) The numbers of stored vectors vs the number of samples

From Table 3 it can be seen that classifying time for the MCH classifier is shorter than the ones for 1-NN and 4-NN classifiers, and the storage vector number for the former is smaller than the ones for the latter two, and the correct recognition rate for the MCH classifier is similar to the ones for the latter two, and the MCH classifier needs to be trained while the latter two do not need at all. Here BPNN is also used to compare with MCH classifier. For the trained BPNN we only need to store its weight matrix so that storage vectors of BPNN can be ignored. However, there are two main drawbacks with BPNN, one is its slow convergence rate, and the other is that it sometimes falls into local minimum instead of global minimum. The slow convergence rate will pose a computational problem, and local minima will degrade the classification accuracy. In our experiment we found that if the number of training samples was greater than 5000, the convergence speed of BPNN would be very slow. As we know the leaf images classification is a large data set problem which has a serious speed issue and need to be made much more efficient. Fig.4 also shows that the larger the data set, the more improvement in speed and storage space which MCH classifier can make. So it can be concluded that the MCH classifier is a more preferred candidate for the leaf images classification than k-NN and BPNN because it can not only save the storage space but also reduce the time consumed without sacrificing the classification accuracy.

5 Conclusions

In this paper, we proposed using a moving center hypersphere(MCH) classifier to address shape features extracted from pre-processed images. Experimental results show that 20 classes of practical plant leaves are successfully recognized, and the average recognition rate is up to 92.2 percent. The performance of MCH classifier is satisfying while compared to nearest neighbor method and BPNN. Our future research works will include how to classify the leaves with deficiencies and combine adaptive neural networks with hypersphere technique to increase the correct recognition rate.

References

1. Milan Sonka, Vaclav Hlavac and Roger Boyle: Image Processing,Analysis and Machine Vision, Second Edition. Posts & Telecom Press, Beijing (2002)
2. Huang, D.S.: Systematic Theory of Neural Networks for Pattern Recognition. Publishing House of Electronic Industry of China, Beijing (1996)
3. Zhaoqi Bian and Xuegong Zhang: Pattern Recognition. TsingHua University Press, Beijing (1999)
4. Im, C., Nishida, H., Kunii, T.L.: Recognizing Plant Species by Leaf Shapes-a Case Study of the Acer Family. Proc. Pattern Recognition. **2** (1998) 1171–1173
5. Wang, Z., Chi, Z., Feng, D.: Fuzzy Integral for Leaf Image Retrieval. Proc. Fuzzy Systems. **1** (2002) 372–377
6. Saitoh, T., Kaneko, T.: Automatic Recognition of Wild Flowers. Proc. Pattern Recognition. **2** (2000) 507–510
7. Ridler, T.W. and Calvard, S.: Picture Thresholding Using An Iterative Selection Method. IEEE Transaction on System, Man and Cybernetics. **8, 8** (1978) 630–632
8. Vincent, L. and Soille, P.: Watersheds in Digital Spaces: An Efficient Algorithm Based on Immersion Simulations. IEEE Transactions on Pattern Analysis and Machine Intelligence. **13, 6** (1991) 583–598
9. Hu, M.K.: Visual Pattern Recognition by Moment Invariants. IRE Transaction Information Theory. **8, 2** (1962) 179–187
10. Chaur-Chin Chen: Improved Moment Invariants for Shape Discrimination. Pattern Rec-ognition. **26, 5** (1993) 683–686
11. Hart, P.E.: The Condensed Nearest Neighbor Rule. IEEE Transactions on Information Theory. **14, 3** (1967) 515–516
12. Huang, D.S.: Radial Basis Probabilistic Neural Networks: Model and Application. International Journal of Pattern Recognition and Artificial Intelli-gence. **13, 7** (1999) 1083–1101
13. Huang, D.S., Horace, H.S.Ip, Law Ken C.K. and Zheru Chi: Zeroing Polynomials Using Modified Constrained Neural Network Approach. IEEE Trans. On Neural Networks. **16, 3** (2005) 721–732
14. Huang, D.S., Horace, H.S.Ip and Zheru Chi: A Neural Root Finder of Polynomials Based on Root Moments. Neural Computation. **16, 8** (2004) 1721–1762
15. Huang, D.S.: Application of Generalized Radial Basis Function Networks to Recognition of Radar Targets. International Journal of Pattern Recognition and Artificial Intelligence. **13, 6** (1999) 945–962

A Robust Registration and Detection Method for Color Seal Verification

Liang Cai[1] and Li Mei[2]

[1] College of Computer Science and Technology, Zhejiang University,
310027 Hangzhou, Zhejiang, P.R. China
cai2000@21cn.com
[2] Hangzhou Sunyard System Co., Ltd., 310053 Hangzhou, Zhejiang, P.R. China
ml@sunyard.com

Abstract. As important premises of automatic seal verification system, candidate seal must be detected from processed image and done registration with the template seal. The paper gives such a robust method. After the candidate seal is detected from the processed image by contour skeleton analysis, FFT is performed for template and candidate seals. FFT Magnitude feature matrix describing global and invariant properties of the seal image is constructed by integrating the Fourier transformation over each of the regions of a wedge-ring-detector. Robust rotation angle is evaluated by minimizing the difference between two feature matrixes for the two seals. Then, relative translation can be evaluated by limited position enumerating in the space domain. Experiment results show that our method can deal with noise-corrupted images and complicate-background images. Seal detection and registration are fast and accurate, and the methods have been used in a real seal identification system successfully.

1 Introduction

Automatic color seal verification system in financial industry has a comprehensive application prospect. Though there are some bottleneck problems to be resolved in such systems, such as fast detection of candidate seal, robust registration between the candidate and template seals. In recent years, many papers about seal verification are presented, but seal detection task on a whole image is not involved. In [1][2], candidate seal is given as an input. Candidate seal locates in a specified region of the processed image in [3]. There are also many methods for seal registration between candidate and template seals. In [4], two-step approach (computing the center information of seal image and performing the correlation under rθ-domain) is used for registration. The center information depends strongly on the found of minimum circular region covering the entire seal imprint. However, under the real case of seal noise involved, this method will fail. It seems that many traditional techniques may also been applied to such a task, such as the moment-based method [5] and Fourier descriptor [6]. However, the moment invariants are only in reasonably close agreement for the geometrically modified versions of the same object. It is not suitable to be applied for the with seal noise. Similarly, the Fourier descriptor is usually to be applied to the shape analysis possessing continuous closed curve. In [7], contour analysis for finding the principal orientation of a seal image is developed for the registration task. But it will be driven to the last ditch when the peak for principal orientation is not prominent on the seal image with serious noise. Accordingly, in the following, we will first present an effective approach for candidate seal detection based on contour skeleton analysis, and then describe a robust two-step registration method. Relative rotation is restored in frequency domain firstly, and then translation is evaluated by limited position numerating in the space domain.

The rest of the paper is organized as follows. To simplify the processed image, a seal color extraction method is introduced in Section 2. Then a quick seal detection algorithm based on seal contour skeleton analysis is described in Section 3. In Section 4, we introduce the two-step approach for robust seal registration. Experiment results and conclusion are given in Section 5.

2 Seal Color Extraction

In practical seal verification system, the seal color is usually pure red or pure blue. To obtain resistance to noise and high recognition ratio, as well as to speed the seal detection procedure, we separate the pixels of seal color from others, that we call seal color extraction.

RGB space is subdivided into 8 subspaces by dichotomy. Each of the color subspace is a cube with size of 128. The colors on the 8 corner of the RGB space are 8 individual standard colors, which are black, red, green, yellow, blue, magenta, cyan and white. According to this order, We define the color index for these 8 standard colors are respectively 0, 1, 2, ..., 7. We can obtain the corresponding color index for an arbitrary point(r, g, b) in the RGB Space by equation (1), where "[]" represents integer Round operator.

$$\text{Index}(r, g, b) = [b/128] \times 4 + [g/128] \times 2 + [r/128] \ . \tag{1}$$

In the seal verification system, scanned images are in the presence of significant illumination variations. The most common method to compensate for illumination changes is to perform color normalization. We use equation (2) to do color normalization. Let I(i, j) be the gray value at pixel(i, j), M, VAR and M_0, VAR_0 are source

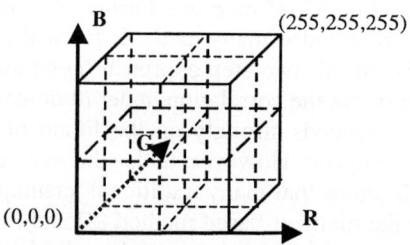

Fig. 1. 3D RGB space is subdivided into 8 subspaces by dichotomy

and destination gray mean, variance respectively. Then the normalized gray value N(i,j) for pixel(i, j) is

$$N(i,j) = \begin{cases} M_0 + \sqrt{\dfrac{VAR_0 + (I(i,j) - M)^2}{VAR}} & \text{if } I(i,j) > M, \\ M_0 - \sqrt{\dfrac{VAR_0 + (I(i,j) - M)^2}{VAR}} & \text{otherwise} \end{cases} \quad (2)$$

Where M_0, VAR_0 can be decided by experiential equation (3).

$$\begin{cases} M_0 = \min(255, \max(0, \ 140 + (M - 160)/10)) \\ VAR_0 = 4.0 + (VAR - 20.0)/150.0 \end{cases} \quad (3)$$

The normalized color(r, g, b) for source color (r, g, b) is as follows:

$$\begin{pmatrix} r \\ g \\ b \end{pmatrix} = \begin{pmatrix} \max(0, \min(255, r_0 + N(i,j) - I(i,j))) \\ \max(0, \min(255, g_0 + N(i,j) - I(i,j))) \\ \max(0, \min(255, b_0 + N(i,j) - I(i,j))) \end{pmatrix} \quad (4)$$

From the normalized image, seal color pixel can be extracted effectively by equation (1). For example, the seal color is red as shown in Fig. 2. The color index for standard red is 1. We Set standard red (255, 0, 0) for all pixels with color index 1, and standard white (255, 255, 255) for other pixels. The image at the right of Fig. 2 is the result after seal color extraction. Not only the processed image is simplified but also the stroke on the seal is very clean.

Fig. 2. The result before and after seal color (red) extraction

3 Candidate Seal Detection

To detect candidate seal on the processed images automatically, key information of the template seal should be analyzed at first. The key information of a template seal should include the center point (x_c, y_c), feature axes and their length.

3.1 Key Information Analysis for Template Seal

Template seal usually contains only one seal and the shape for template seal can be circular, rectangular or elliptic. We define feature axes for template seals accordingly,

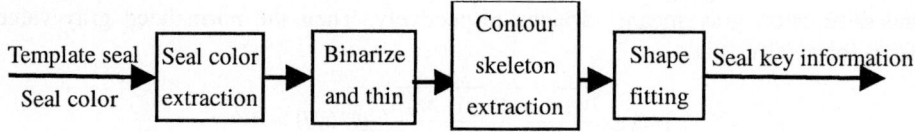

Fig. 3. The procedure for key information analysis of template seal

arbitrary a diameter for circular seal, major and minor axis for elliptic seal and two orthogonal midperpendiculars for rectangular seal.

Fig. 3 illustrates the procedure for key information analysis of the template seal. Followed by seal color extraction, we setting all the pixels with color index 1 as '1', and the other pixels as '0' to binarize the template seal image. Skeleton of the template seal is extracted by using thinning algorithm [8]. A set of 2D points on the outermost skeleton of the template seal is sampled. We can configure shape type by shape fitting. Firstly, Dave Eberly's circle fitting algorithm [9] runs on this set of points. The template seal is deemed to circular if the fitting error is small enough. Otherwise, least squares fitting program for ellipse [10] is further used. Again, the template seal is elliptic if the fitting error is small enough. Fitting error as well as the key information (center point, feature axes) is returned back by the fitting program. If the fitting error is more than a given threshold, then we take an ulterior step to detect two sets of parallel lines that are orthogonal. Key info for rectangular seal can be constructed from these two sets of parallel lines.

3.2 Candidate Seal Detection by Contour Skeleton Analysis

There are two sides that hold in play in detection procedure. One is to reduce the possible positions to be further judged as to speed the detection. On the other hand, we should avoid leaving out seal and inaccurate seal being detected.

In our method, different shapes of seals including rectangular, circular and elliptic are considered. Candidate seal is detected based on the contour localizer of the template seal. A contour localizer can be exclusively determined from the key information of the template seal. A contour localizer for elliptic template seal as an example is illustrated in Fig. 4.

A contour localizer is composed of three skeleton rings, outer, mid and inner rings. These rings are homocentric ellipses with different sizes. The mid ring is exactly the same size as that of the template seal's skeleton, and the outer and inner rings are the extended and shrunken version of the mid ring. When the contour localizer rotates or translates, three rings transforms in synchronization. If the localizer is completely overlaid on the candidate seal, the mid ring will hit many superposition points and the inner and outer rings will hit few points. The two endpoints of major axis of the mid ring are called anchor points. The quantity of translation and rotation for contour localizer are exclusively determined from the anchor points.

Algorithm for detecting candidate elliptic seals by using an elliptic contour localizer is as follows:

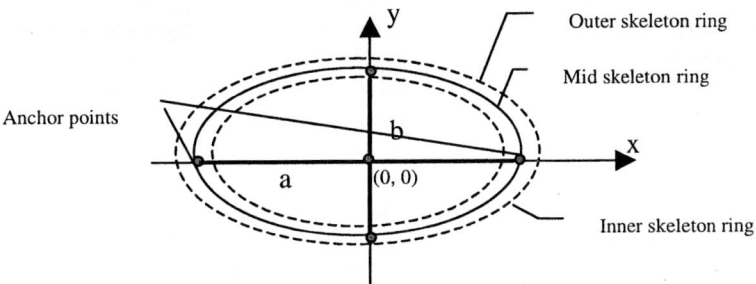

Fig. 4. Contour localizer for elliptic template seal

Step 1. After a downscale factor is determined according to the key information of the template seal, the processed image is downscaled.

Step 2. The seal color is extracted from the downscaled image. The pixels with seal color are set as '1' (black pixels), and the others are set as '0' (white pixels). Then the skeleton for the processed image is picked up [8].

Step 3. A standard elliptic contour localizer is constructed with key information of template seal and the downscale factor determined in step 1. 360 sampled points on the mid skeleton ring are computed beforehand. Let them be $(Cx_i, Cy_i)(i=1,..360)$. One point is sampled from the skeleton for every 1°. And 360 sample points on the outer and inner skeleton rings are also computed respectively.

Step 4. For arbitrary two black pixels $P_1(x_1, y_1)$, $P_2(x_2, y_2)$ on the thinned skeleton of the processed image, the Euclidean distance is computed. If the distance is close to the major length of the mid ring of the contour localizer, then a rotation version for contour localizer is computed by taking these two pixels as anchor points. 360 sample points $(Cx'_i, Cy'_i)(i=1,..360)$ on mid ring of the rotation version are computed by:

$$\begin{pmatrix} Cx'_i \\ Cy'_i \end{pmatrix} = \begin{pmatrix} \cos\alpha & \sin\alpha \\ -\sin\alpha & \cos\alpha \end{pmatrix} \begin{pmatrix} Cx_i \\ Cy_i \end{pmatrix} + \begin{pmatrix} \frac{x_1+x_2}{2} \\ \frac{y_1+y_2}{2} \end{pmatrix} \qquad (5)$$

is the inclining angle for mid ring. The rotated version of the localizer is overlaid on the processed image with P_1, P_2 overlapped with the two anchor points. If black-point number in the neighborhood of point (Cx'_i, Cy'_i) on the processed skeleton image are more than threshold, then we call the point (Cx'_i, Cy'_i) are hit. If the hit points for contour localizer are too less, then, Step 4 is repeated. Otherwise the hit points for outer and inner rings are configured. If the hit points for them are less than a predetermined threshold, then we suppose there is a candidate seal with P_1P_2 as anchor points. And we crop the corresponding region for this candidate seal.

Same detection method is for circular and rectangular seals. The only difference consist how to construct the corresponding contour localizer. To reduce detection time for circular seal, we choose the two endpoints of aclinic diameter as the anchor points. Similarly, two endpoints of the midperpendicular are chosen as anchor points for rectangular seal.

Seal detection can be done quickly because pixels overlapped on the anchor points of contour localizer are very limited and the rotation versions for contour localizer can be computed beforehand. Experiment results show that the average detection time spent on color image database with average size of 2000x1024 is less than 400ms. And the time for detect an elliptic seal is less than 200ms. Combined with the robust seal registration in Section 5, the false positive and false negative ratio for detection is very low.

4 Robust Seal Registration

Another bottleneck problem in seal verification system is how to do seal registration robustly. Inspired by a wedge-ring-detector (WRD), we construct feature magnitude matrix for two seals and restore their relative rotation in frequency domain firstly, and further relative translation is evaluated on the space domain.

4.1 Feature Magnitude Matrix is Constructed to Restore Relative Rotation

By observing the FFT magnitude image for template and candidate seals in Fig. 3 and 4(where magnitude is the square root of the square sum of real and imaginary components, and is normalized to [0, 255]), the origins locate in the center of the transformed images. Aiguille signal is generated by the recurrent noise in original image.

Fig. 5. Rectangular template seal, candidate seal detected, and their corresponding FFT magnitude image

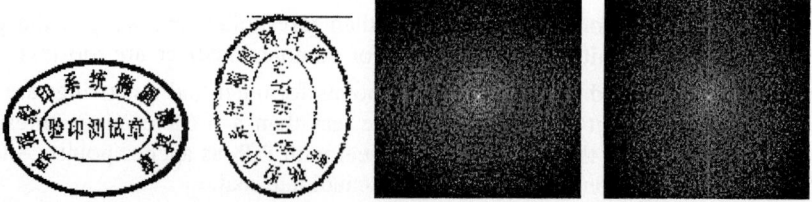

Fig. 6. Elliptic template seal, candidate seal detected, and their corresponding magnitude images

As known as Comparability theorem for 2D FFT, if f(x,y) is rotated by a angle q, then the spectrum of f(x, y) is also rotated by a same angle. Similarly, if gray distribution of the 2D seal image is treated as a function about the pixel's coordinate, and a certain rotation exists between different samples of the individual seal, then a same angle exists between their FFT magnitude spectrums.

We present an approach to restore relative rotation for two seals in the frequency domain, inspired by an opto-electronic device called the wedge-ring-detector (WRD) [11]. Fig. 6 illustrates a WRD with 2 rings and 16 wedges.

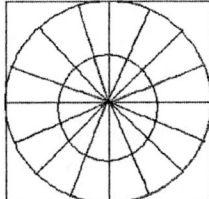

Fig. 7. Wedge-ring-detector

Due to the inherent periodic structure of a seal, stroke frequency and stroke orientation seem to be a good choice. It is well know that the Fourier transformation of an image can capture such global features. When browsing through the literature, we came across a device inspired by opto-electronics, called the wedge-ring-detector. The idea of the WRD is as follows:

1. Computing the discrete Fourier transform of the image.
2. Partition the Fourier transform into wedge-shaped and ring-shaped regions. These regions correspond to different ranges and orientations of spatial frequency.
3. Integrating the Fourier transform over each of these regions gives a figure describing the amount of spectral energy in the corresponding region.

Supposed sample steps for radial and latitude direction are r, Δ and the corresponding sample times are m and n times respectively, then m×n regions of wedge-shape or ring-shape is formed. By integrating (summing) the spectral energy for each region, Reg_{ij}, a magnitude feature matrix FM can be easily constructed.

$$FM = (F_1, F_2, ... F_m) = \begin{pmatrix} G_{11} & G_{21} & \cdots & G_{m1} \\ G_{12} & G_{22} & \cdots & G_{m2} \\ \vdots & \vdots & \vdots & \vdots \\ G_{1n} & G_{2n} & \cdots & G_{mn} \end{pmatrix}, \text{ where } G_{ij} = \sum_{(u,v) \in \text{Re } g_{ij}} |F(u,v)|$$

and F(u, v) is the Fourier transform for the seal image.

All features are (only to a certain extent) invariant under translation, since the WRD integrates over absolute values. The WRD thus provides us with a single matrix describing global and invariant properties of the seal image.

Let the corresponding magnitude feature matrixes for template and candidate seal are $FM_{template}$, $FM_{candidate}$ respectively, and $FM_{candidate}^k$ be the feature matrix after all elements on each line of $FM_{candidate}$ are moved down to nether k lines in turn. It comes down to seek a k_0, such that the difference of magnitude feature matrix $FM_{template}$ and $FM_{candidate}^{k0}$ for template and candidate seal reaches the minimum, that is to say, $k_0 = \underset{k=0,1,2,3,...n}{\operatorname{argmin}} |FM_{template} - FM_{candidate}^k|$, where $|\cdot|$ represents the norm operation, it can be sum of the absolute value of each element's difference. Here the two seals reaches the best registration for rotation, and the relative rotation angle is $\varDelta \times K_0$.

Rotation angle is fined in 3 levels from coarse to fine. First we set $\varDelta = 5°$, and get a coarse angle θ that reaches the best angle registration, then we set $\varDelta = 1°$, and find a finer angle θ_1 in range $[\theta-5, \theta+5]$ that also reaches the best angle registration. Lastly, we set $\varDelta = 0.1°$, and the best relative rotation angle is find in range $[\theta_1-1, \theta_1+1]$.

As we known, FFT should be performed with size of power of 2. Since real size for template and candidate seal is the not the case, two images are transformed at first. Let the size of them are $w_0 \times h_0$, $w_1 \times h_1$ respectively, and let maxLen = max(w_0, h_0, w_1, h_1). We configure a scale factor: scale= maxLen/128. Two seals are resampled using the same scale factor. So the max length of the resampled image is 128. Two scaled images are copied to a blank image with standard size of 128×128. The 128×128 blank image is originally filled with value '0'.

Since FFT magnitude spectrum is symmetry about the center, there are two versions for rotation parameters(α or α +180°). And the proper rotation angle can be accepted or rejected in the next step of translation parameters evaluation.

4.2 Translation Parameters Restored in Space Domain by Coarse-to-Fine Modulating

After the candidate seal is rotated, it is overlaid on the template seal. If there is a translation (x, y) between two seals' center, the point matching score is marked as $s1 = \dfrac{sp(x,y) \times 2}{mp + tp(x,y)}$, where $sp(x,y)$ is the total number for black points of superposition, mp and tp(x,y) are the number of black points on the template seal and valid region of the candidate seal. The part overlaid by the template seal is said of valid region of the candidate seal.

The procedure to find best translation parameters is the procedure to find the best point matching score. Two steps from coarse to fine modulating are used. In first step, we use a larger step, and in the second step, we use a small step of only one pixel. If the best point matching score is less than a predetermined threshold, then we abandon this false candidate seal. Fig. 7 shows the trichromatic maps for the true and false seal overlapped on the template seal with best point matching score.

Since the registration task is decomposed into two independent steps, rotation and translation parameters can be evaluated successively. The registration is done very quickly. And since the magnitude feature matrix depicted is a global concept, our method can deal with images with mass noise and complex background.

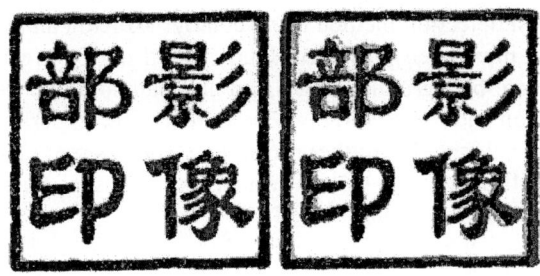

Fig. 8. Trichromatic maps for true and false seal overlapped on the template seal with the best point matching score

5 Experiment Results and Conclusion

10 different seals are used in our experiment. Seal 1 to 5 are rectangular seals (Type 1), and Seal 6, 7, 8 are circular seals (Type 2), and the last 2 are elliptic seals (Type 3). We impress two seals on each cheque. These two seals can either be of the same or the different type of shape. The seal can be located at anywhere of the cheque and with arbitrary rotation angle. The real seal color is red, and most of the cheque images take a look with red-dot-texture background. Different seal material is used, such as rubber, cattle horn.

To test the validity for our registration method, we also implement method called contour analysis in [7] for comparison. And the best point matching score is used to evaluate their validity. The higher match score, the more accurate the registration is. The match results are shown in Tab. 1. The average best point matching score are 82% and 74.9% for our method and method in [7], where two registration method run on the same image sets that are composed of the candidate seals detected in the last step and the corresponding template seals. Under test environment PIV 2.4G/ 512M/81G, the detect ratio and average time for 10 sets are also shown in Tab. 1.

Table 1. The detect ratio and average time for 10 sets

	Seal 1 Type 1	Seal 2 Type 1	Seal 3 Type 1	Seal 4 Type 1	Seal 5 Type 1	Seal 6 Type 2	Seal 7 Type 2	Seal 8 Type 2	Seal 9 Type 3	Seal 10 Type 3
Number for Actual seal	85	83	74	73	70	61	65	70	69	80
Number for seal detected	85	83	74	73	70	61	64	70	69	80
Average best point matching score (our method)	79%	83%	81%	84%	81%	88%	82%	82%	79%	81%
Average best point matching score (method in [7])	75%	74%	77%	76%	71%	80%	75%	76%	72%	75%
Average time for detection and registration(ms)	547	553	481	487	490	143	398	129	584	512
False negative ratio for detection	0%	0%	0%	0%	0%	0%	0%	1.53%	0%	0%
False positive ratio for detection	0%	0%	0%	0%	0%	0%	0%	0%	0%	0%

As a conclusion, a detection method using contour skeleton analysis and a two-step registration method have been presented. Our method not only can deal with noise-eroded and complicate-background images, but also can hold true for different kind of seal shapes. Detection and registration is done in real time. Experiments show that the registration result is valuable for further processing, such as seal identification system. In the future, we will try to deal with even more kind of seal shapes, and to test the effectiveness and robustness of our method.

References

1. Wen Gao, Shengfu Dong, Debin Zhao: Stroke Edge Matching Based Automatic Chinese Seal Imprint Verification. In: Proceeding of the Fourth International Conference for Young Computer Scientists (1995) 838–843
2. Song Yong, Liu Hong: A New Automatic Seal Image Retrieval Method Based on Curves Matching. In: ACTA Scientiarum Naturalium Universitatis Pekinensis 1 (2004) 85–90
3. Chen, Y.S.: Automatic Identification for a Chinese Seal Image. In: Pattern Recognition 11 (1996) 1807–1820
4. Yung-Sheng Chen: Computer Processing on the Identification of a Chinese Seal Image. In: Proceedings of the IEEE Third IAPR International Conference on Document Analysis and Recognition (1995) 422–425
5. Shapiro, L.G., Stochman, G.C.: Computer Vision. Prentice Hall, Inc., New Jersey (2001)
6. Gonzalez, R.C., Woods, R.E.: Digital Image Processing. Addison Wesley, New York (1992)
7. Yung-Sheng Chen: Registration of Seal Images Using Contour Analysis. In: Proceedings of the 13th Scandinavian Conference on Image Analysis (2003) 255–261
8. Ruwei Dai: Chinese Character Recognition System and Integrated Approach. Zhejiang Science and Technology Publishing House, Zhejiang, China (1998)
9. ftp://ftp.cs.unc.edu/pub/users/eberly/magic/circfit.c
10. http://www.astro.rug.nl/~gipsy/sub/ellipse.c
11. Schwaighofer, A.: Sorting it Out: Machine Learning and Fingerprints. In: Special Issue on Foundations of Information Processing of TELEMATIK 1 (2002) 18–20

Enhanced Performance Metrics for Blind Image Restoration

Prashan Premaratne and Farzad Safaei

School of Electrical, Computer & Telecommunications Engineering,
University of Wollongong, North Wollongong, NSW 2522, Australia
{prashan, farzad}@uow.edu.au

Abstract. Mean Squared Error (MSE) has been *the* performance metric in most performance appraisals up to date if not all. However, MSE is useful only if an original non degraded image is available in image restoration scenario. In blind image restoration, where no original image exists, MSE criterion can not be used. In this article we introduce a new concept of incorporating Human Visual System (HVS) into blind restoration of degraded images. Since the image quality is subjective in nature, human observers can differently interpret the same iterative restoration results. This research also attempts to address this problem by quantifying some of the evaluation criteria with significant improvement in the consistency of the judgment of the final result. We have modified some image fidelity metrics such as MSE, Correlation Value and Laplacian Correlation Value metrics to be used in iterative blind restoration of blurred images. A detailed discussion and some experimental results pertaining to these issues are presented in this article.

1 Introduction

Blind image restoration is the process where, original image has to be estimated without the explicit knowledge of the underlying degradation process [1]. This process is difficult since information about the original image or the underlying blurring process is not available in many practical applications such as in space exploration [2,3]. Hence, trying to retrieve unknown images, one has to incorporate human visual perception mechanism [4,5] in order to accurately restore the image. Interesting research has been carried out and reported incorporating the human perception in image processing and more research is needed to apply these concepts to blind image deconvolution scenario.

The dependence of the image restoration techniques on the human observer has been a critical factor in fast deconvolution algorithms. For instance, iterative image restorations such as Iterative Blind Deconvolution relies on human visual perception to terminate the process when an acceptable restoration is achieved [1]. This technique usually runs in excess of 5000 iterations for medium-sized (256x256 pixel) images. Even if the observer has to evaluate the result for every 10 iterations, in order to terminate the iteration process, it is a tedious process susceptible to error in judgment. Furthermore, since this measure is subjective

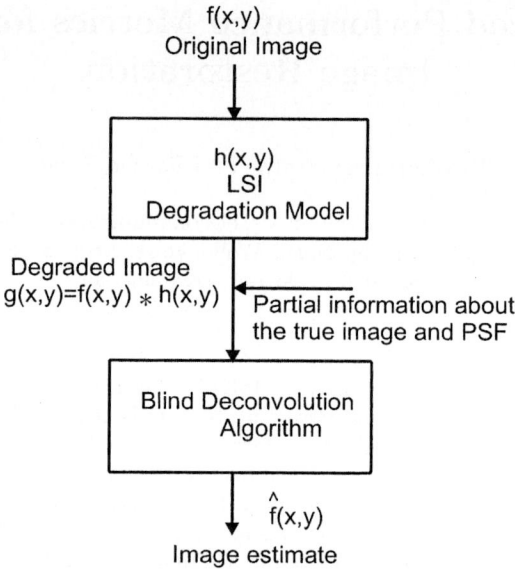

Fig. 1. Blind image restoration scenario

and a quantitative measure like Mean Square Error (MSE) estimation is not possible due to the nature of the blind deconvolution problem, different approaches to the problem are encouraged.

Even though, recent research has reported significant developments in artificial intelligence and neural network to fuzzy logic systems coming close to predicting complex relationships that only human mind is capable of performing, it will be far fetched to assume complete machine dominance in these analysis in near future.

In iterative image restoration, current restored image is compared against the previous version for every iteration. The human perception would determine the trend of the process that is whether the restoration tends to improve or not. These kind of subjective measures can be quantified for machine evaluation by capturing the human factor. Correlation of restored frames offers a possible solution to the above-mentioned scenario where correlation of each restored image is calculated against the previous frame and the values are plotted against the iteration number to see the trend in restoration. However, the human intervention cannot be completely disregarded, as the observer has to justify whether an image is really being restored.

Since blind image restoration of still images lack knowledge of the original image, it would be very challenging to restore it in the absence of a priori knowledge. However, assuming that modern technology is capable of receiving blurred images with limited blurring such that a human observer is capable of interpreting the content of the image, the proposed technique can be applicable to support the observer with quantifiable information. It may happen that different observers with different experiences might come to different subjective conclusions for the same information whereas the proposed measure will help them to come to similar and verifiable conclusions. Fig. 1 depicts the general blind deconvolution scenario.

Most attempts to find worthwhile image fidelity measures have been ad hoc in nature. A measure is postulated, perhaps on the basis of some physiological evidence, but more often upon analytical and computation expediency, and then evaluated. Another branch of thought on the problem is that an image fidelity measure that is to mimic human evaluation should operate in the same way as the human brain. Following this concept a preprocessing operation is performed on the image to be evaluated before invoking the fidelity measure. Blind deconvolution scenario approximates, as well as possible, the process that actually occurs in the initial stages of the human visual system.

2 The Human Visual System Model and Image Quality Assessment

The most widely used image quality measure in digital image compression research has been MSE. MSE estimates the mismatch between the original, unprocessed image and the processed image. However, it has been empirically determined that the MSE and its variants do not correlate well with subjective (human) quality assessments [6,7]. The reasons are not well understood, but one suspects that the MSE does not adequately track the types of degradations caused by digital image compression processing techniques, and that it does not adequately "mimic" what the human visual system does in assessing image quality. One possible reason for the poor performance of the MSE criteria is that it lacks the emphasis on structural information contained in any image. Human visual system emphasizes on structural information and the quality is assessed based on this information. Unfortunately, due to the simplicity of the MSE, the whole image is treated uniformly disregarding any emphasis on structural information.

Some researchers [8,9] have attempted to improve upon quality assessment by incorporating elaborate models of the visual process. Such models have been devised in an attempt to simulate the effects of many of the parameters affecting vision, such as orientation, field angle, and Mach bands, but their utility for practical problems is small due to their complexity, inherent unknowns, and need for some times detailed a priori knowledge of viewing condition parameter values. Incorporation of an elaborate visual system model into an image quality measure is not practical at present.

However, it has been found that several simplifying assumptions for the visual model can still lead to a quality measure that performs better than, for instance, the MSE, which does not incorporate a visual model [8,9]. If one assumes that the visual system is linear, at least for low contrast images, and is isotropic, and that the scenes viewed are monochrome and static, with observer-preferred length of time, then these assumptions lead to a single, straightforward function representing the visual system, which is amenable to incorporation in a quality measure. These assumptions are valid for certain classes of image observation, notably reconnaissance images being viewed for interpretation purposes.

It would be desirable to develop a tool for automatic quality assessment of monochrome imagery, the result of which would exhibit a high degree of correlation with the ratings assigned by trained observers. The HVS model is based upon two phenomena: the HVS sensitivity to background illumination level and to spatial frequencies. Sensitivity to background illumination level results from the fact that the eye is known to be more sensitive to small variations in dark surroundings than in light ones [10]. This warrants the incorporation of a suitable monotonically increasing convex intensity mapping function in the quality metric. The spatial frequency sensitivity is related to the fact that the eye contains channels sharply tuned to certain spatial frequencies, specifically those within the range of 3 to 9 cycles/degree (subtended viewing angle). This implies use of a spatial frequency sensitivity model in the form of a weighing function in the measure.

Another important factor in automatic quality assessment is that the human observer directs his attention to various subsections of a complex scene during cognition. If a subsection contains features of interest, e.g., man-made objects, the observer may place a higher emphasis on the rendered quality. Incorporation of this issue into the quality metric is complicated since it involves knowledge-based recognition and image understanding. This observation suggests that the quality metric should be a local operation, with its parameters being adjusted frequently according to the perceived local features of interest.

3 Monochrome Image Fidelity

There has been much effort toward the development and assessment of quantitative measures of monochrome image fidelity. A useful measure should correlate well with subjective testing for a broad class of imagery and be reasonably calculable. It is also highly desirable that the measure be analytic so that it can be used as an objective performance function in the optimization or parametric design of image processing systems.

Quantitative measure of monochrome image fidelity may be classed as univariate or bivariate. A univariate measure is a numerical rating assigned to a single image based upon measurements of the image field, and a bivariate measure is a numerical comparison between a pair of images.

In most of the research attempts to date [5,7] different image fidelity criteria has been introduced and their success has depended on the correlation of their criteria with the subjective assessment of the trained human observer. Over the years, many of these criteria have been abandoned due to low correlation and high computational complexity.

A classical measure of univariate image fidelity is the equivalent rectangular pass-band measure, defined as

$$Q = \sum_{u=0}^{N-1} \sum_{v=0}^{M-1} |\mathcal{F}(u,v)|^2 \ . \tag{1}$$

where $\mathcal{F}(u,v)$ is the two-dimensional Fourier transform of the image $f(x,y)$ with support size of the image is given by, $0 \leq x \leq N-1$, $0 \leq y \leq M-1$. The squaring operation of this measure gives a higher weighting to the low spatial frequency image components, which are generally of high magnitude. Again, however, the measure has not proved to be well correlated with subjective testing.

Attempts at the development of bivariate measures of image quality have met with somewhat more success. Consider a pair of image fields composed of some standard or ideal image $f(x,y)$ and an approximate or degraded version of the image denoted as $\hat{f}(x,y)$. One measure of the *closeness* of the two image fields is the cross-correlation function, defined as

$$\kappa = \sum_{x=0}^{N-1} \sum_{y=0}^{M-1} f(x,y) \hat{f}(x,y) \ . \tag{2}$$

Usually the cross-correlation function is normalized by the reference image energy so that the peak correlation is unity. The normalized cross-correlation measure is given by

$$\kappa = \frac{\sum_{x=0}^{N-1} \sum_{y=0}^{M-1} f(x,y) \hat{f}(x,y)}{\sum_{x=0}^{N-1} \sum_{y=0}^{M-1} [f(x,y)]^2} \ . \tag{3}$$

From Parseval's theorem, it is found that the normalized cross correlation can also be computed in terms of the Fourier transforms of the images according to the relation

$$\kappa = \frac{\sum_{u=0}^{N-1} \sum_{v=0}^{M-1} \mathcal{F}(u,v) \hat{\mathcal{F}}^*(u,v)}{\sum_{u=0}^{N-1} \sum_{v=0}^{M-1} [\mathcal{F}(u,v)]^2} \ . \tag{4}$$

Since edge rendition is of importance in image perception, Andrews [11] has proposed a Laplacian correlation measure for image evaluation defined by

$$\kappa = \frac{\sum_{u=0}^{N-1} \sum_{v=0}^{M-1} (u^2+v^2) \mathcal{F}(u,v) \hat{\mathcal{F}}^*(u,v)}{\sum_{u=0}^{N-1} \sum_{v=0}^{M-1} (u^2+v^2) [\mathcal{F}(u,v)]^2} \ . \tag{5}$$

Multiplication of the Fourier spectra $\mathcal{F}(u,v)$ by the quadratic frequency factor u^2+v^2 is related to the frequency, in analog format is equivalent to performing an edge sharpening Laplacian operation on the spatial domain fields $f(x,y)$

before correlation. Experiments by Andrews [11] on low-pass and high-pass linearly filtered images demonstrate that the basic correlation measure remains quite high even when an image is severely low-pass filtered and is of subjectively poor quality, but that the Laplacian correlation measure drops off rapidly as the degree of low-pass filtering increases. Conversely, however, it is possible to generate subjectively poor quality images with severe low spatial frequency distortion that yield a relatively large Laplacian correlation measure.

The most common error measure in image processing is the normalized MSE, given by

$$MSE(\hat{f}) = \frac{\sum_{\forall(x,y)} \left[a\hat{f}(x,y) - f(x,y)\right]^2}{\sum_{\forall(x,y)} f^2(x,y)}. \tag{6}$$

The MSE expression is generally preferred to the absolute error equation because the former is analytically tractable, while the latter is difficult to manipulate. For this reason there has been great effort to determine transformations of the image field that yield a mean-square error measure that correlates well with subjective results. The most basic transformation, of course, is the linear point transformation. The power low transformation has also received considerable attention, along with the logarithmic transformation. Combinations of the point spatial transformations mentioned above have also been considered.

4 Proposed Performance Metric

Considering the HVS model and the inherent difficulties with the blind deconvolution problem, where the original or true image is unavailable for usage in any of the measures above, we propose the following modifications.

To identify the trend in the iterative restoration process, we treat the kth iteration as the true image and the $(k+1)^{th}$ iteration as the restored image and the MSE, correlation value and Laplacian values are evaluated against the iteration number. This will result in following modified measures: From Eq. (3) with the mentioned modification, we arrive at

$$\kappa = \frac{\sum_{x=0}^{N-1} \sum_{y=0}^{M-1} f_k(x,y) f_{k+1}(x,y)}{\sum_{x=0}^{N-1} \sum_{y=0}^{M-1} [f_k(x,y)]^2}. \tag{7}$$

Similarly from Eq. (4)

$$\kappa = \frac{\sum_{u=0}^{N-1} \sum_{v=0}^{M-1} \mathcal{F}_k(u,v) \mathcal{F}_{k+1}^*(u,v)}{\sum_{u=0}^{N-1} \sum_{v=0}^{M-1} [\mathcal{F}_k(u,v)]^2}. \tag{8}$$

From Eq. (5)

$$\kappa = \frac{\sum_{u=0}^{N-1}\sum_{v=0}^{M-1}(u^2+v^2)\mathcal{F}_k(u,v)\mathcal{F}_{k+1}^*(u,v)}{\sum_{u=0}^{N-1}\sum_{v=0}^{M-1}(u^2+v^2)\left[\mathcal{F}_k(u,v)\right]^2} . \quad (9)$$

Finally from Eq. (6) the modification of MSE leads to a new MSE measure that no longer requires the original true (unblurred) image,

$$MSE(\hat{f}) = \frac{\sum_{\forall(x,y)}\left[af_{k+1}(x,y)-f_k(x,y)\right]^2}{\sum_{\forall(x,y)}f_k^2(x,y)} . \quad (10)$$

5 Simulation Results

We performed several simulations with the proposed image evaluation performance metrics and obtained some encouraging results that may be useful for future image evaluation techniques.

The 'cameraman' image cropped to the size of 225x225 pixels was used in our experiment. The Point Spread Function (PSF) used was of size 32x32 resulting in the blurred image of 256x256 (the convolution of the image and PSF). For the restoration of the blurred image using a blind deconvolution technique, we used a zero sheet separation algorithm citePremaratne. This algorithm does not assume the knowledge of the PSF and is evaluated with zero sheet separation techniques.

Fig. 2 depicts the blurred 'cameraman' image. In order to evaluate the viability of our proposed metrics in blind deconvolution scenario, we have used the modified equations (8), (9) and (10). The measures of MSE, Normalized

Fig. 2. Blurred image of 'cameraman'

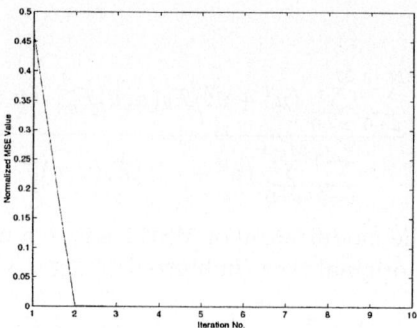

Fig. 3. MSE plotted against iteration number

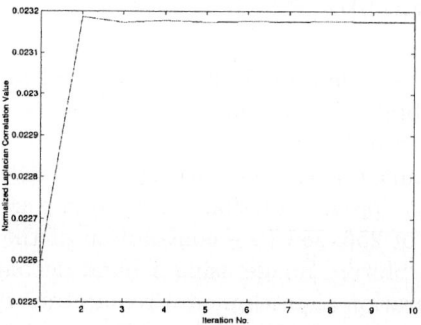

Fig. 4. Normalized Laplacian correlation value of iterative restoration

Correlation Value and Normalized Laplacian Correlation Value are calculated for different iterations in the restoration process. Figures 3 and 4 show two of the above measures. The restored images using 2 and 5 iterations are shown in Figures 5 and 6 respectively.

6 Discussion and Conclusions

Fig. 2 depicts a severe blur resulting from a 32x32 sized random PSF. During the iterative process to recover the true image, MSE resulting from each iterative step is calculated and plotted in Fig. 3. This value should be interpreted with cautious unlike in traditional MSE. The measured values suggest that the MSE converges to zero, which is the ideal case in any recovery process. However, since the MSE is measured relative to the previous restored frame, a zero MSE value is interpreted as the ideal convergence as expected by the human observer. The conventional MSE, which is used in almost all image restoration algorithms, is the sole metric for performance analysis despite its inherent limitations. Its limitations are evident from Fig. 4, which depicts the normalized Laplacian

Fig. 5. Recovered image after 2 iterations

Fig. 6. Recovered Image after 5 iterations

correlation measure of the iterative process. Even though, the MSE suggests that the restoration has converged after 2 iterations, correlation value suggests that this is attained only after 3 iterations. Although, the images shown in Fig. 5 and 6 can not be distinguished from the naked human eye, Fig. 6 is the truly restored and converged image in this iterative process, as the human visual system would interpret.

Our research was aimed at exploring better image fidelity metrics for blind image restoration scenario. This is encouraged by the limitations in the widely used MSE criteria as the sole performance metric in most of the image restoration technique. We believe that our effort to incorporate elements of HVS model to develop better metrics was successful and will be researched further.

References

1. Ayers, G.R. and Dainty,J.C.: Iterative Blind Deconvolution Method and its Applications. In: Optics Letters, Vol. 13, no. 7, (1988) 547–549
2. Biggs, D.S.C. and Andrews, M.: Asymmetric Iterative Blind Deconvolution of Multiframe Images. In: Proc. SPIE, vol. 33. (1998) 3461–3472
3. Premaratne, P. and Ko, C.C.: Blind Image Restoration via Separation of Pointzeros. In: IEE Proc. Vision,Image and Signal Processing, Vol. 148, no. 1, (2001) 65–69
4. Stockham, T.G.: Image Processing in the Context of a Visusal Model. In: Proc. IEEE, Vol. 60, no. 7, (1972) 828–842
5. Saghri,J.A., Cheatham, P.S. and Habibi, A.: Image Quality Measure Bbased on a Human Visual System Model. In: Opt. Eng., Vol. 28, no. 7, (1989) 813–818
6. Mannos, J.L. and Sakrison, D.J.: The Effects of a Visual Fidelity Criterion on the Encoding of Images. In: IEEE Trans. Inform. Theory, Vol. IT-20, (1974) 525–536
7. Lucas, F.X.J. and Budrikis, Z.L.: Picture Quality Prediction Based on a Visual Model. In: IEEE Trans. Commun., Vol. COM-30, (1982) 1679–1692
8. Overington,I.: Toward a Complete Model of Photopic Visual Threshold Performance. In: Opt. Eng., Vol. 21, (1982) 2–13
9. Lambrecht, C. J. V. D. B.: A Working Spatio-temporal Model of the Human Visual System for Image Restoration and Quality Assessment Applications. In: Proc. ICASSP-96, (1996).
10. Lowery, E.M., and DePalma, J.J: Sine Wave Response of the Visual Systems, In: J. Opt. Soc. Am., Vol. 51, (1961) 740–746
11. Andrews, H.C.: Computer Techniques in Image Processing. Academic Press, New York (1970)

Neighborhood Preserving Projections (NPP): A Novel Linear Dimension Reduction Method

Yanwei Pang[1], Lei Zhang[2], Zhengkai Liu[1], Nenghai Yu[1], and Houqiang Li[1]

[1] Information Processing Center, University of Science and Technology of China, Hefei 230027, China
{pyw, zhengkai, ynh, lihq}@ustc.edu.cn
[2] Microsoft Research Asia, Beijing 100080, China
leizhang@microsoft.com

Abstract. Dimension reduction is a crucial step for pattern recognition and information retrieval tasks to overcome the curse of dimensionality. In this paper a novel unsupervised linear dimension reduction method, *Neighborhood Preserving Projections* (NPP), is proposed. In contrast to traditional linear dimension reduction method, such as principal component analysis (PCA), the proposed method has good neighborhood-preserving property. The main idea of NPP is to approximate the classical locally linear embedding (i.e. LLE) by introducing a linear transform matrix. The transform matrix is obtained by optimizing a certain objective function. Preliminary experimental results on known manifold data show the effectiveness of the proposed method.

1 Introduction

To deal with tasks such as pattern recognition and information retrieval, one is often confronted with the curse of dimensionality [1]. The dimensionality problem arises from the fact that there are usually few samples compared to the sample dimension. Due to the curse of dimensionality, a robust classifier is hard to be built and the computational cost is prohibitive. Dimension reduction is such a technique that attempts to overcome the curse of the dimensionality and to extract relevant features. For example, although the original dimensionality of the space of all images of the same subject may be quite large, its intrinsic dimensionality is usually very small [2].

Many dimension reduction methods have been proposed and can be categorized into linear (e.g. PCA, MDS and LDA) and non-linear (e.g. LLE, ISOMAP, Laplacian Eigenmap, KPCA and KDA) methods. The differences between these methods lie in their different motivations and objective functions. Principal component analysis (PCA) [3] may be the most frequently used dimension reduction method. PCA seeks a subspace that best represents the data in a least-squares sense. Multidimensional scaling (MDS) [3] finds an embedding that preserves the interpoint distances, and is equivalent to PCA when those distances are Euclidean. Linear discriminant analysis (LDA), a supervised learning algorithm, selects a transformation matrix in such a way that the ratio of the between-class scatter and the within-class scatter is maximized [4]. By nonlinearly mapping the input space to a high-dimensional feature space, PCA and LDA can be evolved into KPCA (kernel PCA) [5] and KDA (kernel discriminant analysis) [6]. Though, compared to their linear forms PCA and LDA, KPCA and KLDA can deal with nonlinear problem to some extent, it is difficult to determine the optimal kernels.

Recently, several nonlinear manifold-embedding-based approaches were proposed such as locally linear embedding (LLE) [7], isometric feature mapping (Isomap) [8] and Laplacian Eigenmaps [9]. They all utilize local neighborhood relation to learn the global structure of nonlinear manifolds. But they have quite different motivations and derivations. Limitations of such approaches include their demanding for sufficiently dense sampling and heavy computational burden. Moreover, the original LLE, Isomap and Laplcacian Eigenmaps can not directly deal with the out-of-sample problem [10] . Out-of-sample problem states that only the low dimensional embedding map of training samples can be computed but the samples out of the training set (i.e. testing samples) cannot be calculated directly, analytically or even cannot be calculated at all.

Soon after the aforementioned nonlinear manifold embedding approaches were developed, much endeavor is made to improve and extend them. More recently, locality preserving projections (LPP) [11] was proposed based on Laplacian Eigenmaps. When applied to face recognition, this method is called Laplacianfaces [12]. LPP is a linear dimension reduction method which is derived by finding the optimal linear approximations to the eigenfunctions of the Laplace Beltrami operator on the manifold. Besides its capacity to resolve the out-of-sample problem, LPP shares the locality preserving property. The locality preserving property makes LPP distinct from conventional PCA, MDS and LDA. Motivated by LPP, in this paper, we propose novel dimension reduction method which we call *Neighborhood Preserving Projections* (NPP). While LPP is derived from Laplacian Eigenmaps, ours is derived from LLE. Since the proposed method is a linear form of the original nonlinear LLE, NPP inherits LLE's neighborhood property naturally.

The rest of this paper is organized as follows: Section 2 gives an overview of the proposed method, NPP. Section 3 provides a brief description of LLE. In section 4, the motivation and justification of NPP is presented. Preliminary experimental results are shown in Section 5. Finally, conclusions are drawn in section 6.

2 Overview of the Proposed Method: NPP

2.1 Dimension Reduction Problem

Given N points $\mathbf{X}=[\mathbf{x}_1,\mathbf{x}_2,\ldots,\mathbf{x}_N]$ in D dimensional space, dimension reduction is conducted such that these points are mapped to be new points $\mathbf{Y}=[\mathbf{y}_1,\mathbf{y}_2,\ldots,\mathbf{y}_N]$ in d dimensional space where $d \ll D$. Dimension reduction can be performed either in linear way or in non-linear way. Original LLE is a non-linear dimension reduction technique while our proposed method NPP is a linear one. For linear method, a linear transformation matrix is determined so that

$$\mathbf{y}_i = \mathbf{A}^T \mathbf{x}_i. \qquad (1)$$

The transformation matrix is not computed in an arbitrary way, it is obtained, instead, according to a certain objective function. It is the objective function that makes our proposed linear dimension reduction algorithm, NPP, distinct itself from other algorithms. Before presenting a detailed derivation of NPP algorithm, we will give an overview of it in next subsection.

2.2 Overview

The first two steps of NPP algorithm are the same as those of LLE. Our main contribution lies in third step. The details will be given in section 3 and section 4.

Step 1. Assign neighbors to each data point $\mathbf{x}_i$ (for example by using the K nearest neighbors)

Step 2. Compute the weights W_{ij} that best linearly reconstruct $\mathbf{x}_i$ from its neighbors, solving the constrained least-squares problem in equation (3).

Step 3. Compute the linear transform matrix $\mathbf{A}$ by solving the generalized eigenvalue problem:

$$\mathbf{L}\mathbf{A}^T = \lambda \mathbf{C}\mathbf{A}^T. \qquad (2)$$

Where

$$\mathbf{L} = \mathbf{X}\mathbf{M}\mathbf{X}^T$$
$$\mathbf{C} = \mathbf{X}\mathbf{X}^T$$
$$\mathbf{M} = (\mathbf{I} - \mathbf{W})(\mathbf{I} - \mathbf{W})^T.$$

Note that we will explain step 3 in detail in section 4.

Step 4. Dimension reduction is performed simply by

$$\mathbf{Y} = \mathbf{A}^T \mathbf{X}.$$

Because the proposed method is closely related to LLE algorithm, we will give a breif introduction of LLE before the detailed derivation of NPP.

3 Locally Linear Embedding (LLE)

To begin, suppose the data consist of N real-valued vectors $\mathbf{x}_i$, each of dimensionality D, sampled from a smooth underlying manifold. Provided the manifold is well-sampled, it is expected that each data point and its neighbors lie on or close to a locally linear patch of the manifold. We characterize the local geometry of these patches by linear coefficients W_{ij} that reconstruct each data point $\mathbf{x}_i$ from its K neighbors $\mathbf{x}_j$. Choose W_{ij} to minimize a cost function of squared reconstruction errors:

$$J_1(\mathbf{W}) = \sum_{i=1}^{N} \|\mathbf{x}_i - \sum_{j=1}^{K} W_{ij}\mathbf{x}_j\|^2. \qquad (3)$$

The reconstruction error can be minimized analytically using a Lagrange multiplier to enforce the constraint that (see [13] for details).

A basic idea behind LLE is that the same weights W_{ij} that reconstruct the ith data in D dimensions should also reconstruct its embedded manifold coordinates in d dimensions. Hence, each high-dimensional data $\mathbf{x}_i$ can be mapped to a low-dimensional vector $\mathbf{y}_i$ by minimizing the embedding cost function:

$$J_2(\mathbf{Y}) = \sum_{i=1}^{N} \|\mathbf{y}_i - \sum_{j=1}^{K} W_{ij}\mathbf{y}_j\|^2. \tag{4}$$

$$= \|\mathbf{Y}(\mathbf{I} - \mathbf{W})\|^2$$

$$= trace(\mathbf{Y}(\mathbf{I} - \mathbf{W})(\mathbf{I} - \mathbf{W})^T \mathbf{Y}^T)$$

$$= trace(\mathbf{Y}\mathbf{M}\mathbf{Y}^T).$$

where

$$\mathbf{M} = (\mathbf{I} - \mathbf{W})(\mathbf{I} - \mathbf{W})^T. \tag{5}$$
$$\mathbf{W} = \begin{bmatrix} \mathbf{w}_1 & \mathbf{w}_2 & \cdots & \mathbf{w}_N \end{bmatrix}.$$

$\mathbf{I}$ represents an identity matrix.

To make the optimization problem well posed, two constrains can be imposed to remove the translational and rotational degree of freedom:

$$\sum_{i=1}^{N} \mathbf{y}_i = 0 \quad or \quad \mathbf{Y}\mathbf{1} = 0. \tag{6}$$

$$\frac{1}{N-1}\sum_{i=1}^{N} \mathbf{y}_i \mathbf{y}_i^T = \mathbf{I} \quad or \quad \frac{1}{N-1}\mathbf{Y}\mathbf{Y}^T = \mathbf{I}. \tag{7}$$

where $\mathbf{1}$ stands for a summing vector: $\mathbf{1} = [1,1,\ldots,1]^T$

The constrained minimization can then be done using the method of Lagrange multipliers:

$$L(\mathbf{Y}) = \mathbf{Y}\mathbf{M}\mathbf{Y}^T + \lambda((N-1)\mathbf{I} - \mathbf{Y}\mathbf{Y}^T). \tag{8}$$

Setting the gradients with respect to $\mathbf{Y}$ to zero

$$\frac{\partial L}{\partial \mathbf{Y}} = 0 \Rightarrow \quad 2\mathbf{M}\mathbf{Y}^T - 2\lambda\mathbf{Y}^T = 0. \tag{9}$$

leads to a symmetric eigenvalue problem:

$$\mathbf{M}\mathbf{Y}^T = \lambda \mathbf{Y}^T. \tag{10}$$

We can impose the first constraint above (for zero mean) by discarding the eigenvectors associated with eigenvalue 0 (free translation), and keeping the eigenvectors, $\mathbf{u}_i$, associated with the bottom d nonzero eigenvalues. These produce the d rows of the d-by-N output matrix $\mathbf{Y}$ [15]:

$$\mathbf{Y} = \begin{bmatrix} \mathbf{y}_1 \, \mathbf{y}_2 \, \cdots \, \mathbf{y}_N \end{bmatrix}_{d \times N} = \begin{bmatrix} \mathbf{u}_1 \\ \mathbf{u}_2 \\ \vdots \\ \mathbf{u}_d \end{bmatrix}_{d \times N}. \tag{11}$$

4 The Proposed Method (NPP)

4.1 Motivation

Though LLE possesses some favorable properties [13], its computational cost is expensive than most linear dimension reduction methods. Moreover, it cannot map a new testing point directly, which is referred to as out-of-sample problem as stated in section 1. This problem arises from the fact that the embedding of $\mathbf{y}_i$ is obtained in a way that does not explicitly involve the input point $\mathbf{x}_i$. The cost function J_2 in equation (4) depends merely on the weights W_{ij}. To establish a bridge across this gap, we plug equation (1) into the cost function J_2 and the resultant cost function is optimized. The process of NPP has been presented in section 2. In the next subsection its justification will be given. Because the first two steps of NPP are the same as LLE, only justification related to step 3 is presented.

4.2 Justification

Here we rewrite equation (1)

$$\mathbf{y}_i = \mathbf{A}^T \mathbf{x}_i \quad or \quad \mathbf{Y} = \mathbf{A}^T \mathbf{X}. \tag{12}$$

where

$$\mathbf{A} = [\mathbf{a}_0, \mathbf{a}_1, \cdots, \mathbf{a}_d]$$

We plug equation (12) into the cost function J_2:

$$J_2(\mathbf{Y}) = \sum_{i=1}^{N} \|\mathbf{y}_i - \sum_{j=1}^{K} W_{ij} \mathbf{y}_j\|. \tag{13}$$

$$= trace(\mathbf{Y}\mathbf{M}\mathbf{Y}^T)$$

$$= trace((\mathbf{A}^T\mathbf{X})\mathbf{M}(\mathbf{A}^T\mathbf{X})^T)$$

$$= trace(\mathbf{A}^T(\mathbf{X}\mathbf{M}\mathbf{X}^T)\mathbf{A}).$$

The two constrains of equation (6) and (7) now becomes:

$$\mathbf{Y}\mathbf{1} = 0 \Rightarrow (\mathbf{A}^T\mathbf{X})\mathbf{1} = 0. \tag{14}$$

$$\frac{1}{N-1}\mathbf{Y}\mathbf{Y}^T = \mathbf{I} \Rightarrow \frac{1}{N-1}\mathbf{A}^T\mathbf{X}(\mathbf{A}^T\mathbf{X})^T = \frac{1}{N-1}\mathbf{A}^T(\mathbf{X}\mathbf{X}^T)\mathbf{A} = \mathbf{I}. \tag{15}$$

The constrained minimization can then be done using the method of Lagrange multipliers:

$$\mathcal{L}(\mathbf{A}) = \mathbf{A}^T(\mathbf{X}\mathbf{M}\mathbf{X}^T)\mathbf{A} + \lambda((N-1)\mathbf{I} - \mathbf{A}^T\mathbf{X}\mathbf{X}^T\mathbf{A}). \tag{16}$$

Setting the gradients with respect to $\mathbf{A}$ to zero we have

$$\frac{\partial \mathcal{L}}{\partial \mathbf{A}} = 0 \Rightarrow 2(\mathbf{X}\mathbf{M}\mathbf{X}^T)\mathbf{A}^T - 2\lambda\mathbf{X}\mathbf{X}^T\mathbf{A}^T.$$

By defining

$$\mathbf{L} = \mathbf{X}\mathbf{M}\mathbf{X}^T. \tag{17}$$

$$\mathbf{C} = \mathbf{X}\mathbf{X}^T. \tag{18}$$

we can rewrite equation (17) in the form of a generalized eigenvalue problem:

$$\mathbf{L}\mathbf{A}^T = \lambda\mathbf{C}\mathbf{A}^T. \tag{19}$$

If $\mathbf{C}$ is invertible, equation (20) can be transformed to a standard eigenvlaue problem:

$$(\mathbf{C}^{-1}\mathbf{L})\mathbf{A}^T = \lambda\mathbf{A}^T. \tag{20}$$

Once $\mathbf{A}$ is obtained by solving equation (20) or (21), $\mathbf{X}$ can be mapped to a low dimensional space by

$$\mathbf{Y} = \mathbf{A}^T\mathbf{X}.$$

The constraint (14) can be imposed on by subtracting the mean vector of training set from a training vector or testing vector:

$$\mathbf{y}_i = \mathbf{A}^T(\mathbf{x} - \bar{\mathbf{x}}). \tag{21}$$

where

$$\bar{\mathbf{x}} = \sum_{i=1}^{N} \mathbf{x}_i \tag{22}$$

5 Experimental Results

To demonstrate the effectiveness of the proposed method, NPP, experiments were conducted on data of the famous "swiss roll" and "s-curve" to compare with PCA.

The data set of 2000 points which are randomly chosen from the "swiss roll" (Fig.1 (a)) and "s-curve" (Fig.2 (a)) are shown in Fig. 1(b) and Fig. 2(b) respectively, which are used as training data. PCA seeks a direction onto which projected data has the maximum variance. Therefore, by PCA, data is projected onto a plane perpendicular to the paper plane and parallel to the vertical margin of the paper for our "swiss roll" and "s-curve" experiments. Examining Fig.1(d) and Fig.2 (d), one can find that projected points by PCA are blended. For example, in Fig.1(d) red points overlap largely with blue points and green points. In Fig.2(d) blue points overlap largely with yellow points.

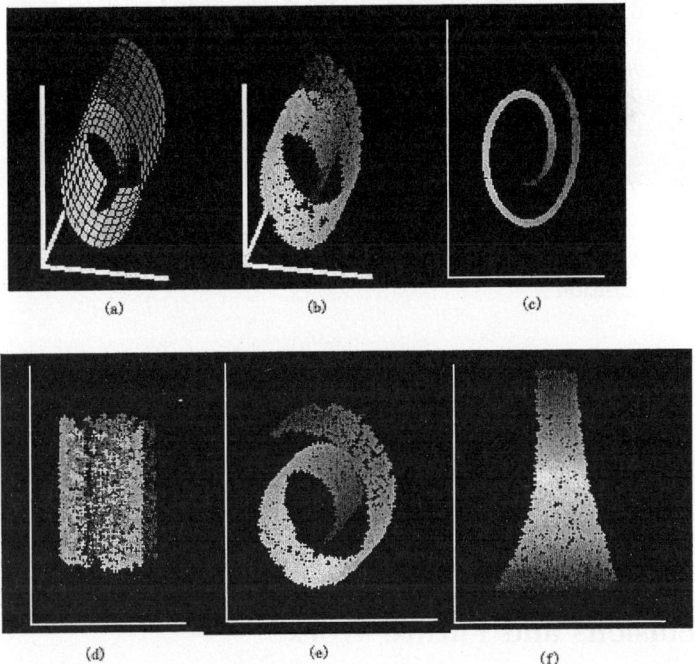

Fig. 1. (a) 3-D "swiss roll"; (b) 2000 points sampled from (a); (c) NPP representation; (d) PCA representationo

In contrast to PCA, our proposed NPP is able to search a direction projected onto which neighborhood relations are preserved along the curve of the manifold as possible. Therefore, by NPP, data are projected onto a plane parallel to the paper plane. Consequently, the projected data is show in fig 1(c) and fig 2(c).

Fig. 1(e) and fig. 2(e) show the results of LPP. From fig. 1(e), it is observed that LPP performs better than PCA. However, in fig. 1(e) blue points nearly connect to red points which is unfavorable. Fig. 2(e) is the result of LPP on "S-

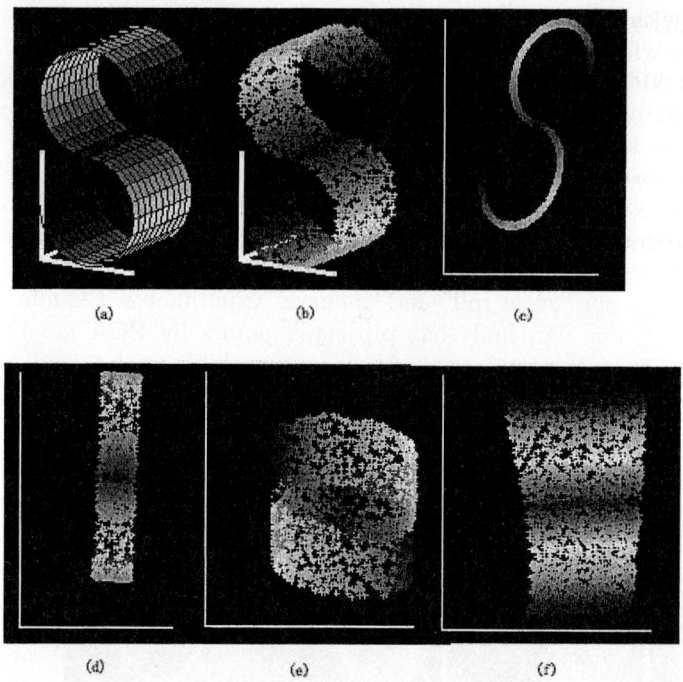

Fig. 2. (a) 3-D "s-curve"; (b) 2000 points sampled from (a); (c) NPP representation; (d) PCA representation

curve" data. We find that red points overlap with both blue and yellow points. Therefore, it is concluded that NPP outperforms LPP.

From fig. 1(c) and fig. 2(c), we can see that NPP can not always unfold the manifold as LLE (Fig. 1(f) and Fig. 2(f))can. Furthermore, many neighbors are collapsed into a single point in the low dimensional space.The reason is that NPP is a linear transform instead of nonlinear one like LLE. Nevertheless, the NPP has favorable properties against other linear transform methods such as PCA.

6 Conclusions and Future Work

By introducing a linear transform matrix into LLE algorithm, a novel unsupervised linear dimension reduction method , Neighborhood Preserving Projections (NPP),has been proposed in this paper. The linear transform matrix is obtained by optimizing a certain objective function which is similar to that of LLE. Hence, NPP inherits LLE's neighborhood property naturally. In contrast to traditional linear dimension reduction method, such as principal component analysis (PCA), the proposed method has good neighborhood-preserving property along the direction of the manifold.

Note that equation (20) is similar to equation (2) in [14] in some sense where another linear dimension reduction method, LPP was proposed. We will compare NPP with LPP both in theory and in applications. Moreover, additional experiments will be conducted on real data.

Though NPP as well as LPP, LDA and PCA, because of their linear nature, might not outperform nonlinear LLE, Isomap and Laplacian Eigenmaps, NPP is a novel and useful linear dimension reduction method.

As future work, we will perform NPP in a large high-dimensional space by introducing a kernel [16-17]. It is believed that kernel NPP ,which is a nonlinear dimension reduction method, can outperform NPP.

References

1. Jain,A.K., Duin, R.P.W., Mao,J.: Statistical Pattern Recognition: A Review. IEEE Transactions on Pattern Analysis and Machine Intelligence, Vol. 22, No. 1, (2000) 4-37
2. Belkin,M., Niyogi,P.: Laplacian Eigenmaps for Dimensionality Reduction and Data Representation. Neural Computation, Vol. 15, No. 6, (2003)1373-1396
3. R. O. Duda,R. O., Hart, P. E., Stork,D.G.: Pattern Classification. Wiley-Interscience, 2000
4. N.B. Peter,N.B., Joao,P.H., David,J.K.: Eigenfaces vs. Fisherfaces: Recognition Using Class Specific Linear Projection. IEEE Transactions on Pattern Analysis and Machine Intelligence, Vol. 19, No. 7, (1997)711-720
5. Scholkopf,B., Smola,A., Muller,K.R.: Nonlinear Component Analysis as a Kernel Eigenvalue Problem. Neural Computation, Vol 10, No. 5, (1998) 1299-1319
6. Baudat,G., Anouar,F.: Generalized Discriminant Analysis Using a Kernel Approach. Neural Computation, Vol. 12, (2000) 2385-2404
7. Roweis,S., Saul,L.: Nonlinear Dimensionality Reduction by Locally Linear Embedding. Science, Vol. 290, No. 5500 , (2000) 2323-2326
8. Joshua,B., Tenenbaum, Langford, J.: A Global Geometric Framework for Nonlinear Dimensionality Reduction. Science, Vol. 290, No. 5500, (2000) 2319-2323
9. Belkin,M., Niyogi,P.: Laplacian Eigenmaps for Dimensionality Reduction and Data Representation. Neural Computation, Vol. 5, No. 6, (2003)1373-1396
10. Bengio,Y., Paiement,J., Vincent,P., Dellallaeu,O., Roux,N.L, Quimet,M.: Out-of-sample Extensions for LLE, Isomap, MDS, Eigenmaps, and Spectral Clustering. Neural Information Processing Systems, 2003
11. He. X., Yan,S., Hu,Y., Zhang,H.: Learning a Locality Preserving Subspace for Visual Recognition. In Proc. IEEE International Conference on Computer Vison, 2003
12. He,X., Yan. S, Hu,Y., Niyogi,P., Zhang,H.J.: Face Recognition Using Laplacianfaces. IEEE Transactions on Pattern Analysis and Machine Intelligence, Vol. 27, No. 3, (2005)328-340
13. Saul,L.K., Roweis,S.T.: Think Globally, Fit Locally: Unsupervised Learning of Low Dimensional Manifolds. Journal of Machine Learning Research, Vol. 4, (2003) 119-155
14. He,X., Niyogi,P.: Locality Preserving Projection. Technical Report TR-2002-09, Department of Computer Science, the University of Chicago
15. Gering,D.: Linear and Nonlinear Data Dimensionality Reduction. Technical Report, the Massachusettes Institute of Technology (2002)
16. John, S.T., Nello, C.: Kernel Methods for Pattern Analysis, Cambridge University Press, 2004
17. Ham, J., Lee, D.D., Mika, S., and Scholkopf, B.: A Kernel View of the Dimensionality Reduction of Manifold. Proc. Int. Conf. Machine Learning, (2004) 369-376

An Encoded Mini-grid Structured Light Pattern for Dynamic Scenes*

Qingcang Yu[1,2,3], Xiaojun Jia[2], Jian Tao[4], and Yun Zhao[2]

[1] Zhejiang University, Hangzhou, Zhejiang, China 310029
qcyu@zist.edu.cn
[2] Zhejiang Sci-Tech University, Hangzhou, Zhejiang, China 310018
{qcyu, zhaoyun}@zist.edu.cn
[3] Zhejiang Provincial Key Lab of Modern Textile Machinery,
Hangzhou, Zhejiang, China 310018
[4] Zhejiang University of Science and Technology, Hangzhou, Zhejiang, China 310012
taojian@vip.sina.com

Abstract. This paper presents a structured light pattern for moving objects sensing in dynamic scenes. The proposed binary pattern can be projected by laser illumination, which aims at eliminate the affect of ambient sunlight, so as to widen application fields of depth sending approach based on structure light. Without the help of color information, the binary can provide great number of code words to make all sub-pattern own a globe unique code to make it suitable for moving objects sensing at one-shot. The propose patter offers more measurement spots than traditional patterns based on M-array, so as to acquire higher resolution. In this paper, the proposed pattern and codification are firstly presented. A new algorithm for fractured contour searching and searching strategies are discussed. An algorithm based on angle variation for contour character identification is given. Code points mapping and mapping regulations are also presented.

1 Introduction

Stereovision is a well-known approach for depth acquisition. For those stereovision approaches, which perform by utilizing two cameras, pixels matching between two images grabbed respectively by two cameras, should be addressed. Depth sensing based on structured light (SL)[1][2] is another research field of stereovision. It avoids the headache job of pixels matching in two images, so as to greatly reduces the difficulties in depth sensing.

A proper illumination pattern is critical for stereovision approaches based on structured light technique. A number of approaches have been investigated. Codification of these approaches can be summarized into five aspects as follow. (1) One axis or both axes. (2) Periodical or absolute. (3) Spatial, spacetime or direct. (4) Color or monochrome. (5) The shape, such as stripes, dot matrices, grids, or other special forms.

---

* This work is supported by Zhejiang NSF, China, Grant # M503237 to Yu Qingcang.

Practical SL patterns always integrate all aspect of above considerations. We don't intend to cite more references here since Jordi Pages et al.[3] and Mouaddib et al. [4] had given well summarizations on previous approaches based on structured light. In this paper, we aimed our SL pattern at dynamic scenes. Additional restrictions should be introduced due to such scenes. A dynamic scene seems unlikely to adopt spacetime codifications, and the pattern should be coded in both axes with globe unique code words. Furthermore, in order to employ laser illumination to eliminate the affect of ambient sunlight, a binary projection should be adopted and it becomes a headache problem that how to generate adequate code words utilizing monochrome light.

Perfect maps (PM) and M-array patterns are the most suitable methods for dynamic scenes. They are spatial encoded in both axes and all sub-matrices with size of n*m only appear once in whole pattern since the code word of a point is identified by its neighbors, i.e. within a window. And they are also capable of both color and monochrome illumination. Most perfect maps and M-arrays can be generated using the method based on the extension of pseudorandom sequence of Arrays (PRSA) [5] to the bi-dimension case. Ozturk et al. [6] proposed a method to generation of perfect map codes for an active stereo imaging system. Morita et al. [7] proposed an approach to reconstruct 3-D surfaces of objects by projecting a M-array pattern. The M-array pattern was generated by placing an aluminum plate with 32*27 holes before a projector. Kiyasu et al. [8] proposed a method to measure the 3-D shape of specular polyhedrons by utilizing M-array. The projected M-array is a 18*18 binary pattern with window size of 4*2. Vuylsteke and Oosterlinck[9] realized it's hard to generate adequate code words with binary modulation. A novel shape-modulation based on pseudonoise sequences was proposed in their work. A 64*63 grid points with different texture and orientation was generated and used as projecting pattern. Morano et al. [10] proposed a color pattern based on pseudorandom codes. The utilization of colors reduces the size of window. But it should be made sure the color-coding is feasible. Spoelder et al.[11] evaluated the accuracy and the robustness on color-coded pseudorandom binary arrays (PRBA's). Good results were obtained by using a 65*63 PRBA pattern with window size of 6*2. Petriu et al.[12] also proposed a grid light pattern of 15-by-15 with multi-valued color codification. But the colors used in the proposed pattern were not fixed. This approach generally suited for static sense since it assumed the objects kept stationary during the experiment.

But if we extend M-array to a large pattern for wide range sensing. For example, if we bespread a 1024*768 projector plate with code points which occupy 6*6 pixels and spaced by 1 pixel. There will be 146*109 code points on the plate and 15914 code words are required for global unique codification. 14 code bits and a minimum window size of 4*4 are required to generate such a large number of code words while utilizing monochrome illumination. With the increasing of window size, it's getting more complex to seek 'neighbors' in heavily distorted images on discontinuous surfaces.

In this paper, we proposed a mini-grid pattern. Related code points are seal by a loop, so as to simplify pattern recognition process. Proposed pattern and codification are firstly presented. A "seed" algorithm for fractured contour searching and searching strategies are discussed. An algorithm based on angle variation for contour character identification is given. Code points mapping and mapping regulations are presented at the end.

2 Encoded Mini-grid Pattern

Several mini-grid sub-patterns were investigated. But the pattern as shown in figure 1a was adopted in this paper. Grid lines in sub-patterns have width of two pixels. We named these squares that between grid lines code squares. The blank squares with size of 3*3 pixels makes it still be viewed in case of surrounding grid lines are widened by one pixel due to the image deformation caused by target surface. The codification is implemented by setting code squares with black or just leave then blank. Horns are used as corner markers which will be described in later section.

We don't except too many black code squares presented in a sub-pattern for they are regarded as undeterminable territory. It would be a large undeterminable area if two black code squares were connected together. The codification is absolute since each sub-pattern can be assigned with a global unique code.

Figure 1a shows partial image we grabbed by a digital camera with resolution of 2048*1536 while we projected sub-patterns onto a human truck with a projector with resolution of 800*600.

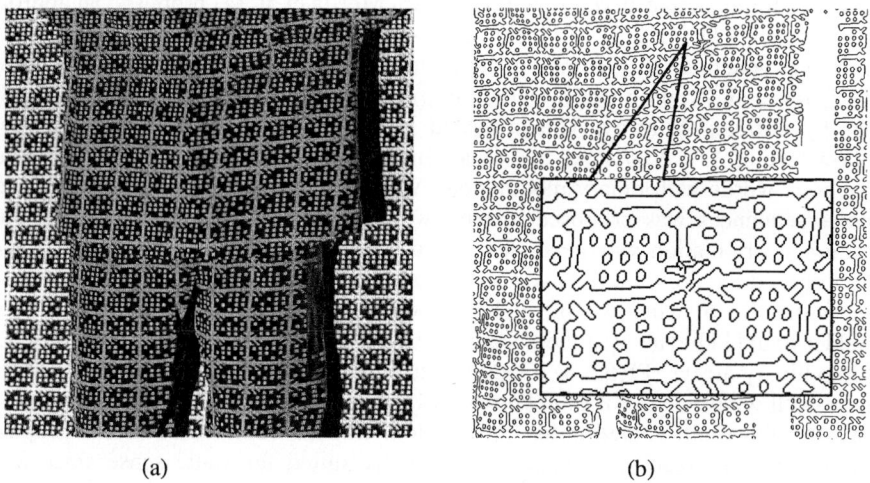

(a)   (b)

Fig. 1. (a) Partial image grabbed by a digital camera while sub-patterns were projected onto a human truck. (b) Partial image after Canny algorithm was applied on the original image.

3 Sub-pattern Retrieve

It's hard to define a fixed threshold for edge detection of sub-pattern due to saturation of the target. Canny edge detection algorithm is employed in this approach for pattern division. Figure 1b shows partial image after Canny algorithm was applied on original image as shown in figure 1a. Most sub-patterns can be well retrieved except those on discontinuous surface or polluted. Blank code squares are now changed to circles and enwrapped in sub-pattern contours. Contour extraction is the key issue for pattern division. But we found, as shown in figure 1b, some contours of sub-patterns are interconnected and some are fractured.

3.1 Espial Scope and Connectivity

Two regulations were proposed to over come the two problems mentioned above. First is that broadened espial scopes should be adopted. And second, the route with shortest length should be taken as searching goal.

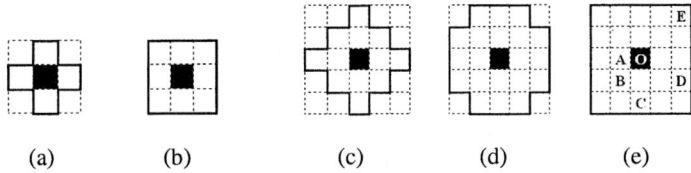

(a)   (b)   (c)   (d)   (e)

Fig. 2. Espial scopes.(a) 4-connectivity. (b) 8-connectivity. (c) 12-connectivity. (d) 20-connectivity. (e) 24-connectivity.

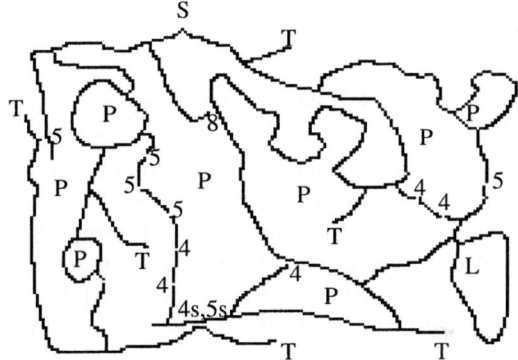

Fig. 3. Artificial route map for contour searching algorithm. Symbol 4, 5 and 8 stands for a jump connection of type 4, 5 and 8 respectively. Symbol 4s, 5s means there exists several jump connections of type 4 and 5 respectively. Symbol T indicates a terminal branch. Symbol L means a closed loop. Symbol P stands for parallel branches.

Traditional connectivity detection based on 4 or 8 directions, as shown in figure 2a and b, would surely fail to correct contour fragmentations. Broadened espial scopes were proposed as shown in figure 2c, d and e. As shown in figure 2e, pixel pairs O-A and O-B are regard without doubt as connected neighbors. To make fractured contour to be closed, we might have to treat pixel pair O-C as connected neighbors too. We defined pixel pair O-C have a jump connectivity of type 4 since the square distance between pixel O and C is 4. Similarly, we defined pixel pair O-D have a jump connectivity of type 5, pixel pair O-E have a jump connectivity of type 8. We created a test map as shown in figure 3 with all type of connections and tricks such as terminal branches, closed loops, parallel branches. Though the sub-pattern contour in figure 1b doesn't seem so complicated, but the program should be flexible to handle unforeseeable cases.

3.2 'Seed' Algorithm for Contour Searching

The widened espial scope is the way to overcome the gap in contours. But it also causes troubles in contour searching since it greatly increases alternations in path searching. Taking point M in figure 4 as an example, M is a usual point in 8-connection detection, but it become an intersection in 24-connection detection since it can jumps to points Q, S, T and O. Actually, each point would become an intersection for it has at least two jump connections in widened espial scope. Backdating progress would become complicated or even failed.

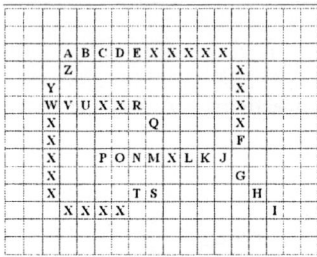

Fig. 4. A sample contour map

A new algorithm, called 'seeds' algorithm, was proposed in this paper to simplify contour searching. Each point is traveled only once. 'seeds' algorithm takes two steps of planting seeds and growing seeds alternatively and discussed briefly as follow.

Planting seeds is performed while searching progress reaches an intersection, e.g. intersection *F*. Three points *G*, *J* and *K* can be viewed by point *F*, and so we plants three seeds *FG(4)*, *FJ(2)* and *FK(5)* to seeds list. The number in parentheses indicates the priority of the seed which was given by the connection type discussed above.

Then we choose the first seed with minimum priority, that is seed *FJ(2)* in this case, and grow it. Growing a seed repeats espial searching and erasing pixels to blank until it reaches an intersection and turn to planting seeds. Seed growing process forms a continuous chain segment. In above example, the growing of seed *FJ(2)* immediately meets an intersection and two new seeds *JK(1)* and *JG(2)* are planted.

About 'dead seed'. In above example, seed *JK(1)* will be grow first for it has a minimum priority. Seed *JG(2)* will be grown prior to seed *FG(4)* for it has a relatively smaller priority. While seed *JG(2)*grows, it travels from point *G* to *I* and erase pixel *G*, *H* and *I* simultaneously. While seed *FG(4)* at last begin to grow, it will find the start point *G* is blank since other seed has already traveled and erased this point. It's no longer necessary to grow seed *FG(4)* and We named it 'dead seed'. Based on these operations, all pixels were traveled only once.

3.3 Optimization Strategy

As shown in figure 3, from start point S, we can find several closed routes. Each route has different route length and contains different number and different type of jump connections. The optimal route depends on how we treat these jump connections, i.e. optimization strategy. Different desires can be easily achieved in 'seed algorithm'. As we discussed in above section, we treat the jump connection as a certain amount of route length. Optimizing Strategy is implemented by setting different types of jump connection with different length weights. Figure 5 shows different results of the map shown in figure 3 while we gave different weight setting. In this paper, we set all the length weights of jump connection at 5000. This setting regards jump connections of type 4, 5 and 8 as the same, and aims at a minimum jumping time.

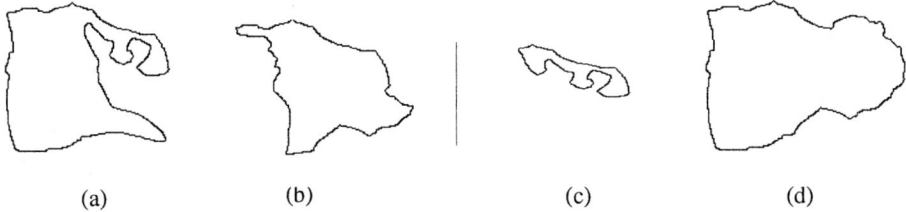

Fig. 5. Contour searching results on different searching strategies. (a)w4=2000, w5=2000, w8=8000. (b) w4=5,w5=5,w8=8000. (c) w4=5,w5=5,w8=5. (d) w4=2000,w5=5,w8=8000. w4,w5 and w8 stand for weights of jump connection of type 4,5 and 8 respectively.

4 Pattern Recognition and Decoding

We can make fractured pattern contour thoroughly closed after retrieved. And then we scanned inside the contour to search for code squares, which had changed to code circles after Canny edge detection algorithm applied. 'Seed' algorithm was still employed here to eliminate the interconnection of code circles.

We need to know bits information of retrieved pattern to implement pattern decoding, comparing the retrieved pattern with the standard coding map. It would be a rather arbitrary way if we treat the retrieved pattern as a linear rectangle and then map it to standard pattern model to retrieve the code word. Steps are taken in this section to perform a fine mapping.

4.1 Key Points Identification

Horns are settled in sub-pattern acting as key points and we realized that these points, such as point 1, 3, 5, 7 shown in figure 6a, have large angle variations.

Instead of evaluating angle variation by its direct conjunct neighbors, we defined the angle variation at one point, point p in figure 6b for an example, as follow.

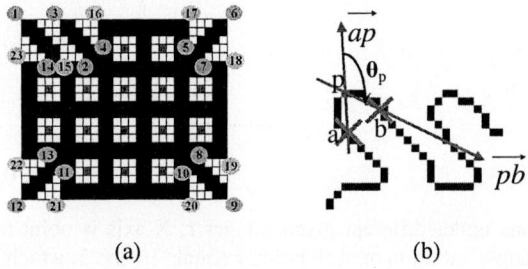

Fig. 6. (a)Key points definition (b)Graphical representation of angle variation

Giving an integer r, and assume Ip is the index of point p in contour sequence. Two points a and b can by created at coordinates of

$$x_a = \frac{1}{r}\sum_{i=Ip-r}^{Ip-1} x_i \qquad y_a = \frac{1}{r}\sum_{i=Ip-r}^{Ip-1} y_i . \qquad (1)$$

$$x_b = \frac{1}{r}\sum_{i=Ip+1}^{Ip+r} x_i \qquad y_b = \frac{1}{r}\sum_{i=Ip+1}^{Ip+r} y_i . \qquad (2)$$

respectively, where x_i and y_i are the coordinates of the point with index of i in contour sequence.

Defines vectors $\vec{ap}$ and $\vec{pb}$ as

$$\vec{ap} = complex(x_p, y_p) - complex(x_a, y_a) . \qquad (3)$$

$$\vec{pb} = complex(x_b, y_b) - complex(x_p, y_p) . \qquad (4)$$

Angle variation at point p can be computed as

$$\theta_p = angle(\vec{pb}) - angle(\vec{ap}) . \qquad (5)$$

So, if we always arrange contour sequence at one direction, say clockwise, the angle variations at key points (*KP* for short)1, 3, 6, 9 and 12 will have positive peak values, while the angle variations at points 2, 4, 5, 7, 8, 10, 11, 13 and 14 get negative peak values as shown in figure 7a.

Angle variations related closely with the given integer r. A small value of r is capable of view detail information of the chain and a large value of r is helpful to perceive general shape of the chain. We can find in figure 7a, there would be two negative peaks between two successive positive peaks. The only exception happens to *KP1*. A small value of r is subsequently adopted to obtain detailed angle variation as shown in figure 7b. *KP15* might conceal even under a very small r value. Key points identification results by applying angle variation method are displayed in figure 8. We only use *KP1* to *KP14* as reference points in next stage since they have high precision.

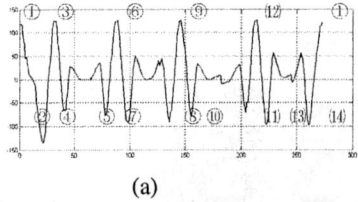
(a)

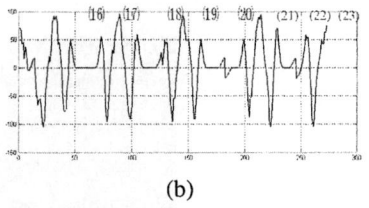
(b)

Fig. 7. Angle variations under different given integer r. X-axis is point index in contour sequence. Y-axis is the angle variation of each point. r equals 10 and 5, which is about one twenty fifth and one fifty second of chain length respectively in (a) and (b).

Fig. 8. Sample results of key points identification

4.2 Border Code Circles Regulations

It seems unlikely to find a mapping regulation which can cope properly with all possible deformations even under the assistances of the reference points. A basic conception offers great help. That is the relative position between code circles keeps unchanged however the deformation is.

Border code circles (*BCC*)should be introduced here. First, we defined segment between *KP16* and *KP17* as border *B1*, and so on. Assume r is radius of the *CCk* and *D* is the minimum distance from *CCk* to *B1*. If $D < 5r$, then *CCk* is defined as a *BCC* since it's unlikely to have extra code circles between *CCk* and B_i. Say *CCk* belongs to border B_i at I_k and note as $k \in (B_i, I_k)$.

We also computed the distances from the center of *CCk* to all *KP*. If *KPn* has a minimum distance value, we note as $p_k = n$. We present some basic rules, only for border circles, in brief as follow.

- If $k_1 \in (B_1, I_{k1}), k_2 \in (B_1, I_{k2})$. Then

  $k_1 = 8, k_2 = 12 \quad (I_{k1} < I_{k2})$

  $k_1 = 12, k_2 = 8 \quad (I_{k1} > I_{k2})$

- If $k_1 \in (B_1, I_{k1})$. Then

  $k_1 = 8 \quad (p_{k1} = 4 \quad or \quad p_{k1} = 16)$

  $k_1 = 12 \quad (p_{k1} = 5 \quad or \quad p_{k1} = 17)$

The principle of these regulations is quite effective and can also be applied to other borders.

5 Experimental Results

Figure 9 displays the experimental result while applying above recognition algorithm on partial image of human trunk. The result shows the algorithm works correctly for integrated sub-patterns.

6 Conclusions

We presented a novel structured light pattern for moving objects sensing in dynamic scenes at one-shot. The main features of the proposed pattern can be summarized as: (1) Globe unique code word for each sub-pattern made it suitable for dynamic scenes. (2) It's a binary pattern, ready to be projected by laser illumination to eliminate the affect of ambient sunlight, so as to widen application fields of depth sending approach based on structure light. (3) It offers more measurement spots than traditional patterns based on M-array, so it can acquire higher resolution. Experimental result shows that the proposed pattern works well on slightly deformed surface.

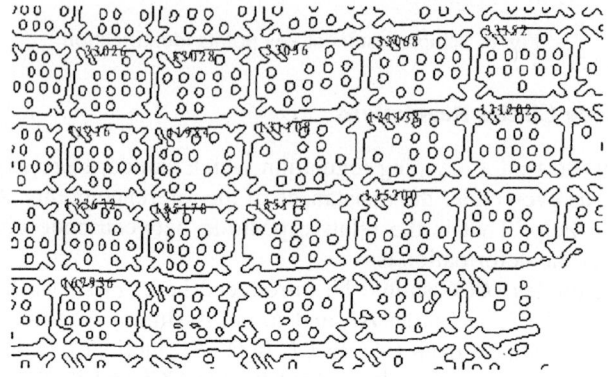

Fig. 9. Experimental decoding result on partial image

References

1. Gledhill, D., Gui Yun Tian, Taylor, D. and Clarke, D.: 3D Reconstruction of a Region of Interest Using Structured Light and Stereo Panoramic Images. Information Visualisation, 2004 Proceedings. Eighth International Conference on. (2004)1007–1012
2. Baglio, S. and Foti, E.: Non-invasive Measurements to Analyze Sandy Bed Evolution under Sea Waves Action. Instrumentation and Measurement, IEEE Transactions on. Vol.52, No.3 (2003)762–770
3. Pages, J., Salvi, J., Garcia, R., Matabosch, C.: Overview of Coded Light Projection Techniques for Automatic 3D Profiling. Robotics and Automation, 2003. Proceedings. ICRA '03. IEEE International Conference on. Vol.1 (2003)133–138
4. Mouaddib, E., Batlle, J., Salvi, J.: Recent Progress in Structured Light in Order to Solve the Correspondence Problem in Stereovision. Robotics and Automation, 1997. Proceedings., 1997 IEEE International Conference on. Vol.1 (1997)130–136
5. Ali, A., Halijak, C.A.: The Pseudorandom Sequence of Arrays. System Theory, 1989. Proceedings., Twenty-First Southeastern Symposium on. (1989)138–140
6. Ozturk, C., Nissanov, J., Dubin, S.: Generation of Perfect Map Codes for an Active Stereo Imaging System. Bioengineering Conference, 1996., Proceedings of the 1996 IEEE Twenty-Second Annual Northeast. (1996)76–77
7. Morita, H., Yajima, K., Sakata, S.: Reconstruction Of Surfaces Of 3-D Objects By M-array Pattern Projection Method. Computer Vision., 1988. Second International Conference on. (1988)468–473
8. Kiyasu, S., Hoshino, H., Yano, K., Fujimura, S.: Measurement of the 3-D Shape of Specular Polyhedrons Using an M-array Coded Light Source. Instrumentation and Measurement, IEEE Transactions on. Vol.44, No.3 (1995)775–778

9. Vuylsteke, P., Oosterlinck, A.: Range Image Acquisition with a Single Binary-encoded Light Pattern. Pattern Analysis and Machine Intelligence, IEEE Transactions on. Vol.12, No.2 (1990)148–164
10. Morano, R., Ozturk, C., Conn, R., Dubin, S., Zietz, S., Nissano, J.: Structured Light Using Pseudorandom Codes. Pattern Analysis and Machine Intelligence, IEEE Transactions on. Vol.20, No. 3 (1998)322–327
11. Spoelder, J., Vos, F., Petriu, E., Croen, F.: A Study of the Robustness of Pseudorandom Binary-array-based Surface Characterization. Instrumentation and Measurement, IEEE Transactions on. Vol.47, No.4 (1998)833–838
12. Petriu, E.M., Sakr, Z., Spoelder, J.W., Moica, A.: Object Recognition Using Pseudo-Random Color Encoded Structured Light. Instrumentation and Measurement Technology Conference, 2000. IMTC 2000. Proceedings of the 17th IEEE. (2000)1237–1241

Linear Predicted Hexagonal Search Algorithm with Moments

Yunsong Wu and Graham Megson

School of System Engineering, Reading University, Reading, UK, RG6 6AA
{sir02yw, g.m.megson}@rdg.ac.uk

Abstract. A novel Linear Hashtable Method Predicted Hexagonal Search (LHMPHS) method for block based motion compensation is proposed. Fast block matching algorithms use the origin as the initial search center, which often does not track motion very well. To improve the accuracy of the fast BMA's, we employ a predicted starting search point, which reflects the motion trend of the current block. The predicted search centre is found closer to the global minimum. Thus the center-biased BMA's can be used to find the motion vector more efficiently. The performance of the algorithm is evaluated by using standard video sequences, considers the three important metrics: The results show that the proposed algorithm enhances the accuracy of current hexagonal algorithms and is better than Full Search, Logarithmic Search etc.

1 Introduction

In this paper, we propose a Linear Hashtable Method Predicted Hexagonal Search (LHMPHS) consists of two parts, a Linear Hashtable Motion Estimation Algorithm (LHMEA) and Hexagonal Search (HEXBS) to predict motion vectors for intercoding. The objective of our motion estimation scheme is to achieve good quality video with very low computational complexity. There are a large number of motion prediction algorithms in the literature. The most popular motion compensation method so far has been the block-based motion estimation, which uses a block-matching algorithm (BMA) to find the best matched block from a reference frame. This approach is adopted in various video coding standards such as ITU-T H.261 [1] and MPEG-1/2 [2], [3]. Usually block with the least Mean Square Error (MSE) is considered as a match, and the difference of their positions describes the motion vector of the block in the current frame to be saved in the corresponding position on the motion map. However, motion estimation is computationally intensive and can consume up to 80% of the computational power of an encoder if the full search is used. Consequently it is highly desired to speed up the process without introducing too much distortion. Many computationally efficient variants of block matching are known, such as Two Level Search (TS), Two Dimensional Logarithmic Search (DLS) and Subsample Search (SS) 1, the Three-Step search (TSS), Four-Step Search (4SS) 2, Block-Based Gradient Descent Search (BBGDS) 3, and Diamond Search (DS) 4, 5 algorithms. A very interesting method called HEXBS has been proposed by Ce Zhu, Xiao Lin, and Lap-Pui Chau 6, together with some variant, such as the Enhanced

Hexagonal method 7, and Hexagonal method with Fast Inner Search8. However, most of these algorithms have the problem of being easily trapped in a non-optimum solution. Additionally most of these fast hierarchical BMA's use the origin of the searching window as the initial search center and have not exploited the motion correlation of the blocks within an image. To improve the fast BMA's accuracy, LHMEA and spatially related blocks can be used to predict an initial search center that reflects the current block's motion trend, and then the final motion vector can be found by the center-biased BMA's such as the N3SS, 4SS, and BBGDS. Because a proper predicted initial center makes the global optimal minimum closer to the predicted search center, the center-biased BMA's increase the chance of finding the global minimum with lower numbers of search points.

Our method attempts to predict the motion vectors using a linear algorithm incorporating a hashtable [9]. We employ LHMEA to quickly search all Macroblocks (MB) in picture. Motion Vectors (MV) generated from LHMEA are then used as predictors for HEXBS motion estimation, which only searches a small number of the MBs. Because LHMEA is based on a simple linear algorithm, the computation time is relatively small. Completed with HEXBS, is one of best motion estimation methods to date. The method proposed in this paper achieves the best results so far among all the algorithms investigated. A further test with moments invariants show how powerful the hashtable can be. The test with Moments shows that the more information hashtable has, the better it performs.

The rest of the paper is organized as follows. Section I begins with a brief introduction to HEXBS. The proposed LHMEA, the method and LHMPHS are discussed in Section II. Experiments conducted based on the proposed algorithm are then presented. Section III concludes the paper with some remarks and discussions about the proposed scheme.

1.1 Hexagonal Algorithm

HEXBS is an improved method based on the DS (Diamond Search) which has shown the significant improvement over other fast algorithms because fewer search points are evaluated. Compared to DS which uses a diamond search pattern, the HEXBS adopts a hexagonal search pattern. The motion estimation process normally comprises of two steps. First a low-resolution coarse search identifies a small area where the best motion vector is expected to lie, using a large hexagon search pattern. The coarse search continues uses gradient scheme until the center point of the hexagon has the current smallest distortion. Then a fine-resolution inner search is conducted to select the best motion vector in the located small region.

Most fast algorithms focus on speeding up the coarse search by taking various smart ways to reduce the number of search points in identifying a small area for inner search. There are two main directions to improve the coarse search:

1. usage of predictors [8, 10]
2. early termination 10

The Motion Vector Field Adaptive Search Technique (MVFAST) 11 based on DS improved significantly the preexisting HEXBS both in image quality and speed up by considering a small set of predictors as possible motion vector predictor candidates.

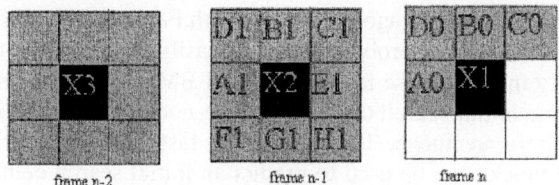

Fig. 1. Bocks correlated with the current one

The best motion vector predictor candidate is used as the center of a search. In general, the blocks correlated with the current block, which are likely to undergo the same motion, can be divided into three categories:

(1) Spatially correlated blocks (A0, B0, C0, D0),
(2) Neighboring blocks in the previous frame (A1, B1, C1, D1, E1, F1, G1, H1)
(3) Co-located blocks in the previous two frames (X2 and X3), which provide the Acceleration motion vector (MV).

Except for coarse search improvement, Inner search improvement includes:

1. 4 points 8 (search points:2,5,7,4)
2. 8 points 10 (search points:1-8)
3. Inner group search 10. (It divides 6 end points of hexagon pattern into 6 groups. Only search points near to the group with smallest MSE)

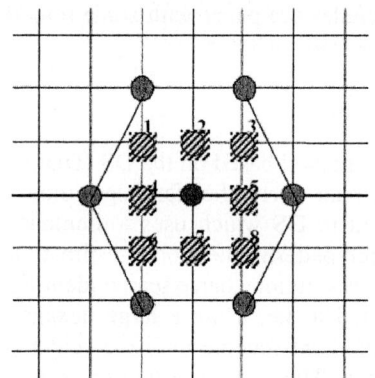

Fig. 2. Hexagon Inner Search Method

2 Linear Hashtable Method Predicted Hexagonal Search (LHMPHS)

Most of current Hexagonal search algorithms using predictive methods focus on relations between the current frame and previous frames. The method employed here constructs a predictor from the current frame information by using spatially related MB or pixel information. A vector hashtable lookup algorithm is employed to form an exhaustive search: it considers every macroblock in the search window while also using global information in the reference block. A computation performed on each block to set up a hashtable, which by definition is a dictionary in which keys are mapped to array positions by a hash function. The key is the one to help you open the right door of macroblock. Since it considers the whole frame it is less likely to be directed into local minima when finding the best match.

The way to constructing a fast hashtable algorithm is to design a joint hash function which is computationally simple. This implies using as few coefficients as possible to represent a block and so minimize the pre-processing stops required to construct the hashtable. The algorithms we present here have two projections, one of them is a scalar denoting the sum of all pixels in the macroblock. It is also the DC coefficient of macroblock. Another is the linear function of Y=Ax+B (y is luminance, x is the location in the table.) Each of these projections is mathematically related to the error metric. Under certain conditions, the value of the projection indicates whether the candidate macroblock will do better than the best-so-far match. The major algorithm we discuss here is linear algorithm.

2.1 Linear Hashtable Motion Estimation Algorithm (LHMEA)

The Linear Algorithm is interesting because it is the easiest and fastest way to calculate the hash function-using mostly additions. It is also very easy to perform in parallel. The basic method is to use polynomial approximation to get such result y=mx+c; y is luminance value of all pixels, x is the location of pixel in macroblocks. Sequentially the frame is scanned from left to right, from top to bottom and the equations (1) and (2) calculated from each MB. Coefficients m and c are then used as the key for hashtable linked to where MB located

$$m = \frac{N*\sum_{i=0}^{N}(x_i*y_i) - \sum_{i=0}^{N}x_i * \sum_{i=0}^{N}y_i}{N*\sum_{i=0}^{N}x_i^2 - \sum_{i=0}^{N}x_i * \sum_{i=0}^{N}x_i} \quad (1)$$

$$c = \frac{\sum_{i=0}^{N}y_i*\sum_{i=0}^{N}x_i^2 - \sum_{i=0}^{N}x_i * \sum_{i=0}^{N}x_i * y_i}{N*\sum_{i=0}^{N}x_i^2 - \sum_{i=0}^{N}x_i * \sum_{i=0}^{N}x_i} \quad (2)$$

When searching for a block (1) and (2) are calculated and used to hash to a set of coordinate blocks which can be used as predictors for a more powerful local search. In previous research methods, when people try to find a block that best matches a predefined block in the current frame, matching was performed by SAD (calculating difference between current block and reference block). In the current existing methods, the MB moves inside a search window centered on the position of the current block in the current frame. If coefficients are powerful enough to hold enough information of the MB, motion estimators should be accurate. So LHMEA increases accuracy, reduces computation time. Observe that since the x values are constructed for each MB much of (1) and (2) can be precomputed.

2.2 Linear Hashtable Method Predicted Hexagonal Search

After motion estimators are generated by LHMEA, they will be used as predictors for hexagonal algorithm. These predictors are different from all previous predictors. They are based on full search and current frame only. Because the predictors generated are accurate, they also reduce the time for a hexagonal search method to locate the best MB.

The original Hexagonal Search moves step by step, at a maximum two pixels per step. In our proposed method, The LHMEA motion vectors are used to move the hexagonal search directly to the area near to the MB where distortion is smallest.

This saves significant computation time on low-resolution coarse search.

In the Figure below, we compare our proposed scheme with a number of fast search methods. All the hexagonal search methods used 6-side-based fast inner search 10 and early termination criteria [10]. All the data here refers to P frames only. The LHMPHS can achieve nearly the same image quality as FS while only costs 10% time. The LHMPHS is better than HS on compression rate when time and PSNR are the same. HSM is the best algorithm.

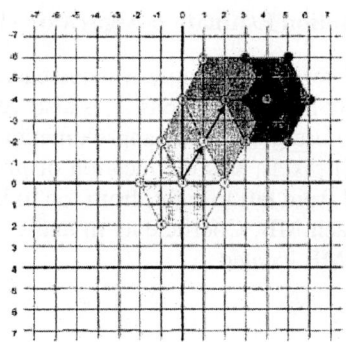

Fig. 3. Original Hexagonal Coarse Search[6]

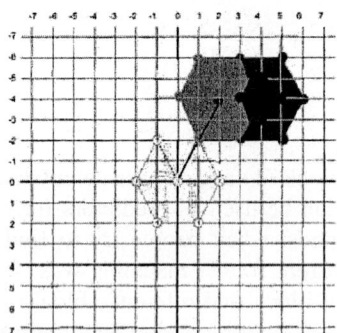

Fig. 4. Proposed Hexagonal Coarse Search

If we can find better coefficients in the hashtable to represent MB, the hashtable will have a better result. The prediction scheme in the Predictive HEXBS yields 10.5%-23% improvement in speed; The prediction scheme results in more contributions for the large motion video sequences; The prediction scheme produces minor contribution for the medium motion or especially low motion video sequences. On the other hand, the prediction provides improvement for both speed and MSE in fast motion sequences and only marginal improvement for slow motion sequences. The reason is that they are certain center biased algorithms. It was based on the fact that for most sequences motion vectors were concentrated in a small area around the center of the search. This suggests that, instead of initially examining the (0,0) position, we could achieve better results if the LHMEA predictor is examined first and given higher priority with the use of early termination threshold.

Flower garden data stream is also used as test source. As in the figure below, LHMPHS is better than HS, while worse than HSM on compression rate and PSNR.

Table 1. Comparison of compression rate, time and PSNR between FS, LS, SS, TLS, HEXBS, LHMPHS, HSM (based on 100 frames of Table Tennis)

Search Method	EXHAUSTIVE	LOGARITHMIC	SUBSAMPLE	TWOLEVEL	Hex No Predictor	Pred_Hashtable	Pred_Median
Inner Search					Inner Group	Inner Group	Inner Group
Early Termination					Hex_near	Hex_near	Hex_near
Compression Time(s)	11	1	4	3	1	1	1
(fps)	2.3726	24.770	6.428	7.317	19.4245	14.2857	17.5325
Compression Rate	48	42	48	48	37	39	42
PSNR	21.3	21.1	21.3	21.3	21.2	21.2	21.2

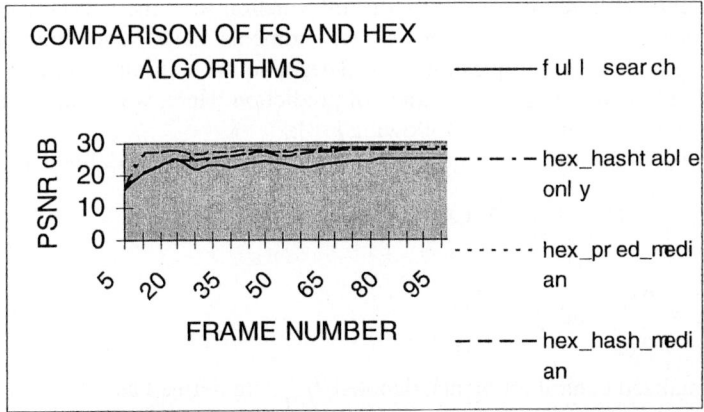

Fig. 5. Comparison of PSNR between FS (full search), HS (hex_hashtable only), HMPHS (hex_pred_median), and HSM (hex_hash_median) (based on 5-100 frames of Flower Garden)

In the pictures below I choose a frame from Flower Garden MPEG clips generated by FS and LHMPHS. Macroblock types are analyzed and MVs are displayed in the pictures. The pictures show FS is better than LHMPHS in macroblock type decision.

(a)   (b)

Fig. 6. Motion vectors and MB analysis in a Flower Garden frame using FS(a) and LHMPHS(b)

2.3 Moments Invariants

Hashtable generation is made on the number of entries in the hashtable. Representative coefficients from the reference frame are computed (using linear hashtable training algorithms) and stored in hashtable. Clearly, the better the hash key is, the less quantization error during encoding and the look-up will be faster. Therefore, the minimum bound on the size of hashtable should be at least equal to the maximum number of motion vectors that any subsequent predicted frame will require. Thus video fidelity becomes a function of the size of the hashtable. In addition to the coefficients from the Linear Algorithm, we put moments invariant into the hashtable as a test. The set of moments considered are invariants to translation, rotation, and scale change. Clearly the moments represent a lot more information than the coefficients m and c. The experimental result below shows that moments have some improvement on hashtable method. The key question is to determine the minimum number of coefficients needed to maintain good accuracy of prediction. Here, we consider moments of two-dimensional functions are as following [14]:

For a 2-D continuous function f(x,y) the central moments are defined as

$$\mu_{pq} = \sum_x \sum_y (x-\bar{x})^p (y-\bar{y})^q f(x,y) . \qquad (3)$$

where $\bar{x} = \dfrac{m_{10}}{m_{00}}$ and $\bar{y} = \dfrac{m_{01}}{m_{00}}$

The normalized central moments, denoted η_{pq}, are defined as

$$\eta_{pq} = \frac{\mu_{pq}}{\mu_{00}^\gamma} \quad \text{where} \quad \gamma = \frac{p+q}{2} + 1 \quad \text{for p+q=2,3,....}$$

A set of seven invariant moments can be derived from the second and third moments.

$$\phi_1 = \eta_{20} + \eta_{02}$$
$$\phi_2 = (\eta_{20} - \eta_{02})^2 + 4\eta_{11}^2$$
$$\phi_3 = (\eta_{30} - 3\eta_{12})^2 + (3\eta_{21} - \eta_{03})^2$$
$$\phi_4 = (\eta_{30} + \eta_{12})^2 + (\eta_{21} + \eta_{03})^2$$
$$\phi_5 = (\eta_{30} - 3\eta_{12})(\eta_{30} + \eta_{12})[(\eta_{30} + \eta_{12})^2 - 3(\eta_{21} + \eta_{03})^2]$$
$$+ (3\eta_{21} - \eta_{03})(\eta_{21} + \eta_{03})[3(\eta_{30} + \eta_{12})^2 - (\eta_{21} + \eta_{03})^2]$$
$$\phi_6 = (\eta_{20} - \eta_{02})[(\eta_{30} + \eta_{12})^2 - (\eta_{21} + \eta_{03})^2]$$
$$+ 4\eta_{11}(\eta_{30} + \eta_{12})(\eta_{21} + \eta_{03})$$
$$\phi_7 = (3\eta_{21} - \eta_{03})(\eta_{30} + \eta_{12})[(\eta_{30} + \eta_{12})^2 - 3(\eta_{21} + \eta_{03})^2]$$
$$+ (3\eta_{12} - \eta_{30})(\eta_{21} + \eta_{03})[3(\eta_{30} + \eta_{12})^2 - (\eta_{21} + \eta_{03})^2]$$

Table 2. Comparison of compression rate, time and PSNR among LHMPHS with different number of moments in hashtable (based on 150 frames of Table Tennis)

Data Steam	Table Tennis	Table Tennis	Table Tennis
Moment Test	No moments	With 2 Moments	With 7 Moments
compression time(s)	16	32	43
compression rate	94	94	95
Average P frame PSNR	40.2	40.6	40.6

Table 3. Comparison of compression rate, time and PSNR among LHMPHS with different number of moments in hashtable (based on 150 frames of Flower Garden)

Data Steam	Flower Garden	Flower Garden	Flower Garden
Moment Test	No moments	With 2 Moments	With 7 Moments
compression time(s)	17	44	45
compression rate	38	38	38
Average P frame PSNR	41.5	41.5	41.5

Table 2 and 3 shows experimental results using three different algorithms: LHMPHS without moments; LHMPHS with 2 moments (ϕ_1 and ϕ_2) in hashtable; and LHMPHS with 7 moments in the hashtable. The experiments result demonstrates that invariant moments improve compression rate and PSNR at the cost of compression time. There is not apparent improvement in Flower Garden and Football. Moments work better on smaller motion video stream. If we can find better coefficients in hashtable, the experimental result can be further improved.

3 Summary

In the paper we proposed a new algorithm called Linear Hashtable Motion Estimation Algorithm (LHMEA) in video compression. It uses linear algorithm to set up hashtable. The algorithm searches in hashtable to find motion estimator instead of by Full Search. Then the motion estimator it generated will be sent to Hexagonal algorithm, which is one of best motion estimation algorithm, as predictor. In this way, it is improved both in quality and speed of motion estimation. Moments invariants are used to prove the more information hashtable has, the better it is. The key point in the method is to find suitable coefficients to represent whole MB. The more information the coefficients in hashtable hold about pictures, the better result LHMPHS will get. This also leaves space for future development. Contributions from this paper are (1) first time hashtable concept is used in video compression. It uses several variables to represent whole MB and gives a direction for future research; (2) linear algorithm is used in video compression to improve speed and also leave space for future parallel coding; (3) the LHMPHS is proposed. MVs produced by the LHMEA will be used as predictors for the HEXBS. This makes up for the drawback of the coarse search in the HEXBS. This can also be used and leave space for research of nearly all kinds of similar fast algorithms for example Diamond Search etc. (4) Invariant moments are added into hashtable to check how many coefficients work best for hashtable. We also want prove that the more information hashtable has the better result the table will have. (5) Spatially related MB information is used not only in coarse search but also inner fine search.

References

1. Ze-Nian li: Lecture Note of Computer Vision on personal website (2000)
2. Po, L. M., and Ma, W. C.: A Novel Four-step Search Algorithm for Fast Block Motion Estimation. IEEE and Systems for Video Technology, vol. 6, pp. 313–317, (June 1996)
3. Liu, L. K., and Feig, E.: A Block-based Gradient Descent Search Algorithm for Block Motion Estimation in Video Coding. IEEE Trans. Circuits Syst. Video Technol., vol. 6, pp. 419–423, (Aug. 1996)
4. Zhu, S., and Ma, K.-K.: A New Diamond Search Algorithm for Fast Blockmatching Motion Estimation.IEEE Trans. Image Processing, vol. 9, pp. 287–290, (Feb. 2000)
5. Tham, J. Y., Ranganath, S., Ranganath, M., and Kassim, A. A.: A Novel Unrestricted Center-biased Diamond Search Algorithm for Block Motion Estimation. IEEE Trans. Circuits and Systems for Video Technology, vol. 8, pp. 369–377, (Aug. 1998)

6. Ce Zhu, Xiao Lin, and Lap-Pui Chau: Hexagon-Based Search Pattern for Fast Block Motion Estimation. IEEE Trans on Circuits and Systems for Video Technology, Vol. 12, No. 5, (May 2002)
7. Zhu, C., Lin, X. and Chau, L.P.: An Enhanced Hexagonal Search Algorithm for Block Motion Estimation. IEEE International Symposium on Circuits and Systems, ISCAS2003, Bangkok, Thailand, (May 2003)
8. Ce Zhu, Senior Member, IEEE, Xiao Lin, Lappui Chau, and Lai-Man Po: Enhanced Hexagonal Search for Fast Block Motion Estimation. IEEE Trans on Circuits and Systems for Video Technology, Vol. 14, No. 10, (Oct 2004)
9. Graham Megson &Alavi, F.N.: Patent 0111627.6 -- for SALGEN Systems Ltd
10. Paolo De Pascalis, Luca Pezzoni, Gian Antonio Mian and Daniele Bagni: Fast Motion Estimation With Size-Based Predictors Selection Hexagon Search In H.264/AVC encoding. EUSIPCO (2004)
11. Alexis M. Tourapis, Oscar C. Au, Ming L. Liou: Predictive Motion Vector Field Adaptive Search Technique (PMVFAST) Enhancing Block Based Motion Estimation. proceedings of Visual Communications and Image Processing, San Jose, CA, January (2001)
12. Rafael C. Gonzalez: Digital Image Processing. second edition, Prentice Hall (2002)
13. Jie Wei and Ze-Nian Li: An Efficient Two-Pass MAP-MRF Algorithm for Motion Estimation Based on Mean Field Theory. IEEE Transactions on Circuits and Systems for Video Technology, vol. 9, No. 6, (Sep. 1999)

A Moving Detection Algorithm Based on Space-Time Background Difference

Mei Xiao[1], Lei Zhang[2], and Chong-Zhao Han[1]

[1] School of Electronics and Information Engineering, Xi'an Jiaotong University,
710049 Xi'an, China
xiaomeijx@163.com, czhan@mail.xjtu.edu.cn
[2] Taibai Campus, Chang'an University, 710064 Xi'an, China
zhanglei@chd.edu.cn

Abstract. Based on the assumption that background figures have been extracted form the input image, we propose a method that can effectively detection the moving objects from image sequence in this paper. The background difference, background difference based neighborhood pixels and frame difference information are fused to get the seed points of real moving object, only the blobs in moving detection based on background difference that intersect with seed pixels are selected as the final moving segmentation result, then we can obtain the true moving foreground. Simulation results show that the algorithm can avoid the false detection due to the wrong in background model or background update and can handle situation where the background of the scene contains small motions, and motion detection and segmentation can be performed correctly.

1 Introduction

Identifying moving objects from a video sequence is a fundamental and critical task in video surveillance, traffic monitoring and analysis. Several approaches are known to separate foreground from background. A common approach to identifying the moving objects is background subtraction, where each video frame is compared against a reference or background model. Pixels in the current frame that deviate significantly from the background are considered to be moving objects. These "foreground" pixels are further processed for object localization and tracking. Since background subtraction is often the first step in many computer vision applications, it is important that the extracted foreground pixels accurately correspond to the moving objects of interest.

A large number of background subtraction methods for fixed cameras have been proposed in recent years. We classify background modeling techniques into statistical model[1][2][3][4][5][6] and approach based on background assumption [7][8][9][10] et al. On the assumption that background figures have been extracted form the input image, we focus only on algorithm of moving detection based on background subtraction in this paper. Some of the commonly-used moving detection techniques based on background subtraction are described below.

Foreground detection compares the input video frame with the background model, and identifies candidate foreground pixels from the input frame. Except for the

MoG[6] and the NPM[10] model, most of the techniques use a single image as their background models. The most commonly used approach for foreground detection is to check whether the input pixel is significantly different from the corresponding background estimate:

$$F_t(x,y) = \begin{cases} 1, & |I_t(x,y) - B_t(x,y)| > th \\ 0, & otherwise \end{cases}$$ (1)

Where, $I_t(x,y)$ represents the input value of the point (x,y) of the tth frame; $B_t(x,y)$ is current background image, here $B_t(x,y)$ denotes a single background image; $F_t(x,y)$ is moving object image, which is a binary mask, where the pixels with value 1, indicate the moving foreground, th is a suitable noise threshold.

Another popular foreground detection scheme [11] is to threshold based on the normalized statistics:

$$F_t(x,y) = \begin{cases} 1, & \dfrac{|I_t(x,y) - B_t(x,y) - u_d|}{\sigma_d} > th \\ 0, & otherwise \end{cases}$$ (2)

Where, μ_d and σ_d are the mean and the standard deviation of $|I_t(x,y) - B_t(x,y)|$ for all spatial locations (x,y).

Ideally, the threshold should be a function of the spatial location (x,y). For example, the threshold should be smaller for regions with low contrast. One possible modification is proposed by Fuentes and Velastin [12]. They use the relative difference rather than absolute difference to emphasize the contrast in dark areas such as shadow:

$$F_t(x,y) = \begin{cases} 1, & \dfrac{|I_t(x,y) - B_t(x,y)|}{B_t(x,y)} > th \\ 0, & otherwise \end{cases}$$ (3)

Nevertheless, this technique cannot be used to enhance contrast in bright images such as an outdoor scene under heavy fog.

A suppression the false detections has been proposed in [10] after the first stage of the background subtraction. Firstly, using this probability estimate, the pixel is considered a foreground pixel if $P_r(x_t) < th$ □we can obtain a foreground image after deal with the all pixel in current image. The probability estimate function $P_r(x_t)$ is:

$$P_r(x_t) = \frac{1}{N} \sum_{i=1}^{N} \prod_{j=1}^{d} \frac{1}{\sqrt{2\pi\sigma_j^2}} e^{-\frac{1}{2}\frac{(x_{t,j} - x_{i,j})^2}{\sigma_j^2}}$$ (4)

Where, $x_1, x_2, ..., x_N$ is a recent sample of intensity values for a pixel; x_t is pixel value at time t; σ_j is kernel bandwidths for the j color channel, and it assume independence between the different color channels; th is the global threshold.

Then, define the component displacement probability P_C:

$$P_C = \prod_{x \in C} P_A(x) . \qquad (5)$$

Where, $P_A(x_t) = \max_{y \in A(x)} Pr(x_t | B_y)$, B_y is the background sample for pixel.

For a connected component corresponding to a real target, the probability that this component has displaced from the background will be very small. So, a detected pixel x will be considered to be a part of the background only if $(P_A(x) > th_1) \wedge (P_C(x) > th_2)$.

A modified difference method based on the information of the neighborhood of each pixel is proposed in [13]. Given a couple of images with sx columns and sy rows, the operator that defines the Neighborhood-based difference ξ can be seen in Esq. (6):

$$\xi(I_1, I_2) = \sum_{i=-1}^{1} \sum_{j=-1}^{1} |I_1(x+i, y+j) - I_2(x+i, y+j)|, \qquad (6)$$
$$\forall x \in \{2, ..., sx-1\} and \forall y \in \{2, ..., sy-1\}.$$

Let I_t be the current image, and BKG_t be the background image, get the two result image F_b and F_t:

$$F_{b_t} = \begin{cases} 1, & \xi(I_t, B_t) > th_1 \\ 0, & otherwise \end{cases} . \qquad (7)$$

$$F_{f_t} = \begin{cases} 1, & \xi(I_t, I_{t-1}) > th_2 \\ 0, & otherwise \end{cases} . \qquad (8)$$

Where, th_1 and th_2 are suitable noise threshold.

Then, label the connected group of F_b and F_f, and finding intersections between both labeled pixel groups. Finally, only the connected areas in F_b which contain of the intersections are selected as the moving foreground objects. The method can avoid the error detection when foreground stop suddenly, however, it can not segment accurately the edge of the moving targets and the choose of noise threshold is very difficult.

Most schemes determine the foreground threshold experimentally. Another approach to introduce spatial variability is to use two thresholds with hysteresis[14]. The basic idea is to first identify "strong" foreground pixels whose absolute differences with the background estimates exceeded a large threshold. Then, foreground regions are grown from strong foreground pixels by including neighboring pixels with absolute differences larger than a smaller threshold.

Based on the assumption that background figures have been extracted form the input image, a new moving object detection algorithm based on space-time background difference has been proposed in this paper.

2 Moving Detection Based on Space-Time Background Difference

In spite of the background subtraction introduced as above is simply, but those approach show poor results in most real image sequences due to three main problems.

- Selection of threshold value. Noise in the image, gray level similarity between the background and the foreground objects made the selection of the threshold differently.
- Small movements in the scene background. Most of background algorithm is not represented the neighborhood pixels information, so when small movements in the scene background will be considered as a moving one. For instance, if a tree branch in scene moves to occupy a new pixel, then it will be detected as a foreground object because the background is not model the movement of the part of the scene.
- The false detection occurred at the process of background update or background reconstruction. It will take a long time to adapt the change of the scene for background model, usually, a lot of error detection would get because the alteration of background that are not represented in the background model. For example, if a new object comes into a scene and remains static for a long time.

A large number of background subtraction methods adopt morphological filters to remove noisy pixels and to fill the moving regions poorly segmented, moving detection result can be partially solved in some environment. However, if foreground is small(Fig.2), or part of foreground is divided into several small parts by occluded, this kind of false detection is difficulty to eliminate using morphology or noise filtering because these operations might also affect small moving targets. In this paper, we hope to avoid the above problems, a novel foreground detection algorithm based on Space-time information has been proposed. The method fuse background difference information, neighborhood-based background difference information and frames difference information to get the seed pixels of true moving object, the connected blobs in background difference that intersect with seed pixels are the final real moving objects. The steps of detection algorithm are described as follow:

Firstly, define three differences to detect motion:

- Moving segment image base background difference $Mb_t(x,y)$:

$$Mb_t(x, y) = \begin{cases} 1, \min_k \left| I_t(x, y) - B_t^k(x, y) \right| > \sigma_b \\ 0, \quad\quad\quad otherwise \end{cases} . \tag{9}$$

Where, $B_t^k(x,y)$ is k^{th} ($k = \{1,2,...q\}$) background image of pixel (x,y) at t frame, we assume there a q background image; $I_t(x,y)$ is current input value of pixel (x,y) at t frame; Mb is moving detection result image based on background difference, Mb will have 1's in pixels with motion, and 0's in pixels of static zones; σ_b is a suitable noise threshold.

- Moving detection image accounting for neighborhood-based background difference $Mn_t(x,y)$:

$$Mn_t(x,y) = \begin{cases} 1, & \min_k(\min_{i,j}(|I_t(x,y) - B_t^k(x+i, y+j)|)) > \sigma_n, \\ & and \quad i, j \in \{-1, 0, 1\} \\ 0, & otherwise \end{cases} \quad (10)$$

Where, Mn is moving detection result image of current input frame accounting for neighborhood-based background difference, 1 is motion object, 0 is static zeros, σ_n is a suitable noise threshold. In our implementation, 3×3 square neighborhood is used; however, we also can choose 5×5 square or diameter 5 circular as neighborhood area.

- Moving segmentation based on frame difference $Mf_t(x,y)$:

$$Mf_t(x,y) = \begin{cases} 1, & |I_t(x,y) - I_{t-s}(x,y)| > \sigma_f \\ 0, & otherwise \end{cases} \quad (11)$$

Where, s is the time interval, usually, from 1-3 are used; σ_f is a suitable noise threshold. Mf is moving detection result image accounting for frames information, 1 is motion object, 0 is static zeros.

Mb is composed of integrated moving, false moving and static zones. The static zones correspond to pixels with the same gray-level between consecutive frames but different in relation to the background image, for example a person who stops in the scene. False moving detection usually correspond to scene background areas that are not represented in the background model, for example, small movements in the scene are classified into moving zones. The result of Mb is show in Fig. 1(d), where the white pixels indicate the foreground.

Mn is consist of not integrated moving and static region, which exclude a lot of error regions of small movements in scene. On the one hand, moving object in Mn is not integrated but inane because the background information of neighborhood pixels have been take into accounted ; on the other hand, Mn contains most of the static zones in Mb. The result of Mn is show in Fig. 1(e), white pixels are foreground object.

Mf is composed of not integrated moving and false moving zones. Since it uses only a single previous frame, frame differencing may not be able to identify the interiors pixels of a large, uniformly-colored moving object. The false moving zones are mainly error detection due to pixel moving in scene. The result of Mf is show in Fig.1(f), white pixels mean foreground regions.

In a word, the detection result of *Mb*, *Mn* and *Mf* is complementary each other, we aim to obtain good detection by using the technique of information fusion.

Secondly, Seed of regions/pixels of moving object;

Seed regions/pixels of moving object $Os_t(x,y)$:

$$Os_t(x,y) = \begin{cases} 1, & Mb_t(x,y)=1, \text{ and } Mn_t(x,y)=1, \\ & \text{and } Mf_t(x,y)=1 \\ 0, & \text{otherwise} \end{cases} \quad (12)$$

Os is the intersection blobs of *Mb*, *Mf* and *Mn*, which is shown in Fig.1(g). Compared with Fig.1(d), those false initial moving points are removed and only the correct moving seed points, corresponding to the object of true target regions, are preserved.

Finally, Moving objects segment results.

Label connected components in *Mb* with 8-connected labeling, we select only the blobs of *Mb* that intersect with blobs of *Os* as the final moving segmentation result M_t.

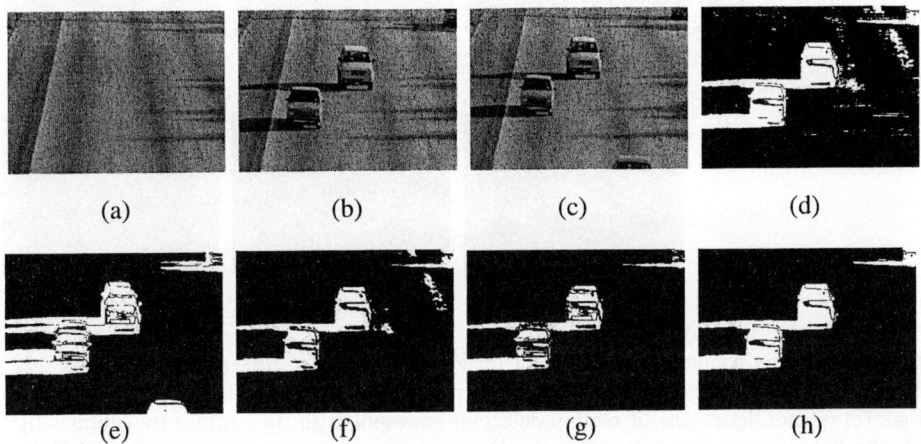

Fig. 1. Test image of detection method. (a) is background image based on mean method ; (b) is 125th frame; (c) is 124th frame; (d) is *Mb*; (e) is *Mn*; (f) is *Mf*; (g) is *Os*; (h) is *M*.

Note: There are three threshold parameters are mentioned in moving detection algorithm, the threshold's influence on detection result greatly decreases because of the multiple different information fusion, three thresholds select as follow $\sigma_b = \sigma_n = \sigma_f$.

3 Simulation Results and Comparisons

The system was tested on a variety of indoor and outdoor sequences. The computer parameter is pentium IV,1.0GHZ(256MBRAM), test software is matlab6.5, image intensity is 256 grade (8 bit). The same thresholds were used for all the sequences.

The values of important thresholds used were $\sigma_b = \sigma_n = \sigma_f = 25$, $s = 2$. The video shows the pure detection result without any morphological operations, noise filtering and any tracking information of targets.

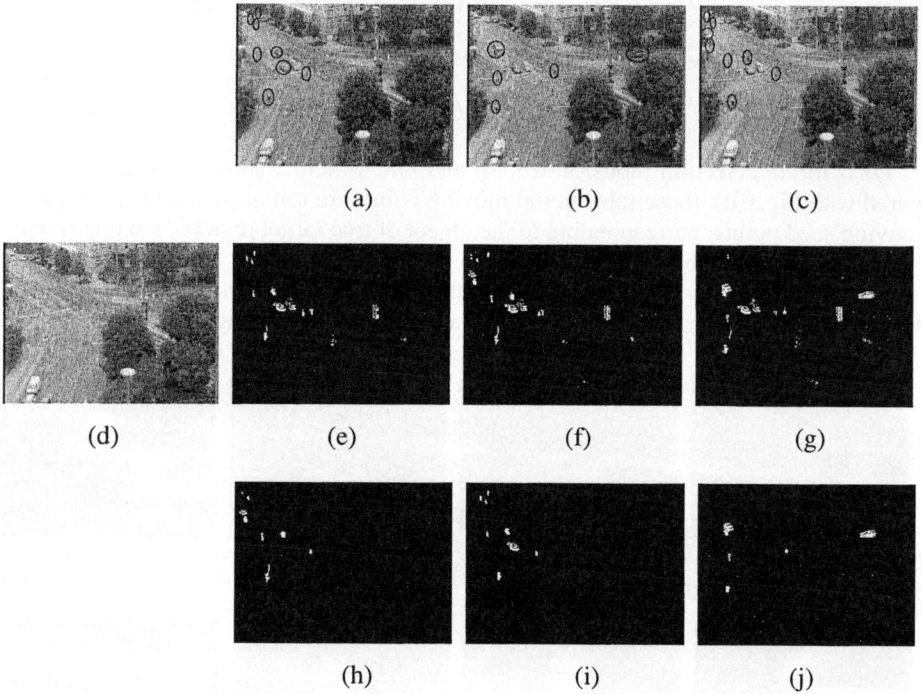

Fig. 2. Detection results for small object image sequence. Fig. 2(a)-(c) is 101st frame, 122nd frame and 177th frame, separately; fig.2(d) is the background image based on mean method; fig.2 (e)-(g) are the results of motion detection corresponds to fig.2 (a)-(c) by subtract fig.2 (d); fig.2 (h)-(j) are the results of motion detection corresponds to fig.2 (a)-(c) by using our moving detection algorithm.

Fig.2 is traffic sequence, 100 frames are used to reconstruct the background, the vehicles move at first, and come to a stop when the revolving light turns red. From the detection result, we can see the stop vehicle is considered as foreground successively because the stop vehicles are not represented in the background model; on the other hand, the revolving light is consider as foreground for ever because the multi-model (the similar multi-distribution scene such as revolving light are waving trees, camouflage, rippling water and sparkling monitor) can not be model by the mean. The false segmentation described as below can not eliminate by filtering. On the contrary, our motion segmentation method can eliminate the false detection regions due to the wrong background model, and motion detection and segmentation performed correctly. In addition, the size of the smallest object is 12 through computing, the result shows the detection results of small objects. The real moving object was masked with red ellipse.

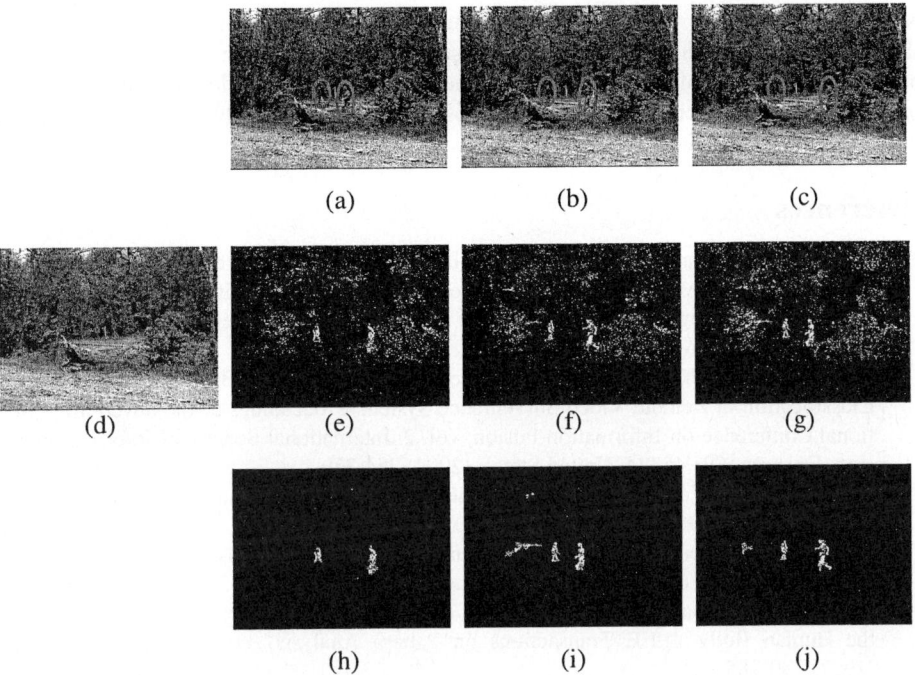

Fig. 3. Detection result for dynamic scene. Fig.3(a)-(c) is 342nd frame, 363rd frame and 381st frame, separately; fig.3(d) is the background image based on median method; fig.3 (e)-(g) are the results of motion detection corresponds to fig.3 (a)-(c) by subtract fig.3 (d); fig.3 (h)-(j) are the results of motion detection corresponds to fig.3 (a)-(c) by using our algorithm. The real moving object was masked with amethyst ellipse.

Fig.3 is jump image sequence, 100 frames are used to reconstruct the background, the person move in scene, and the leaf waving in the wind.

The background of the scene contains small motions, here the small motions mean waving trees, can not characterize by median, so there are a lot of error in segmentation result. The false detection can result in large errors in object localization and cause serious problems for algorithms that use segmentation results as their basic measurements. From the results of our method, most of the error areas are eliminated. Fig.3 shows the robustness of our detection algorithm for dynamic scene.

However, the algorithm proposed in the paper can not eliminate the false detection without an end or a limit. From fig.3 (i) and (j), there also are litter error segmentation due to the motion of background. As a result, a good background method is necessity in order to obtain correct result.

4 Conclusions

In this paper, a moving object detection algorithm based on space-time background subtraction is presented. Based on the assumption that background figures have been obtained form the input image, this method doesn't need any prior knowledge of background and foreground. Meanwhile, the threshold parameter can be chosen in a wide

wide range, which is needed to be determined in the algorithm. Simulation results show that the algorithm can avoid the false detection due to the wrong background model or the update of background, then the algorithm can handle situation where the background of the scene contains small motions, and motion detection and segmentation can be performed correctly.

References

1. Gloyer, B., Aghajan, H.K., Siu, K.Y., Kailath, T.: Video-based Freeway Monitoring System Using Recursive Vehicle Tracking. Proceedings of SPIE-The International Society for Optical Engineering, vol. 2421. Society of Photo-Optical Instrumentation Engineers, Bellingham, WA, USA (1995) 173-180
2. Hou, Z.Q., Han, C.Z.: A Background Reconstruction Algorithm Based on Pixel Intensity Classification in Remote Video Surveillance System. Proceedings of the Seventh International Conference on Information Fusion, vol. 2. International Society of Information Fusion, Fairborn, OH 45324, United States, (2004) 754-759
3. Long, W., Yang, Y.: Stationary Background Generation: An Alternative to The Difference of Two Images. Pattern Recognition. (1990) 1351-1359
4. Kornprobst, P., Deriche, R., Aubert, G.: Image Sequence Analysis via Partial Difference Equations. Journal of Mathematical Imaging and Vision. (1999) 5-26
5. Wren, C.R., Azarbayejani, A., Darrell, T., Pentland, A.P.: Pfinder: Real-time Tracking of the Human Body. IEEE Transactions on Pattern Analysis and Machine Intelligence. (1997) 780-785
6. Stauffer, C., Grimson, W.E.L.: Adaptive Background Mixture Models for Real-time Tracking. Proceedings of the IEEE Computer Society Conference on Computer Vision and Pattern Recognition, IEEE, Los Alamitos, CA, USA (1999) 246-252
7. Javed, O., Shafique, K., Shah M.: A Hierarchical Approach to Robust Background Subtraction Using Color and Gradient Information. Proceedings of the workshop on Motion and Video Computing. IEEE computer society, Los Alamitos, California (2002) 22-27
8. Lee, D.S., Hull, J.J., Erol, B.: A Bayesian Framework for Gaussian Mixture Background Modeling. Proceedings of 2003 International Conference on Image Processing, vol. 2. Institute of Electrical and Electronics Engineers Computer Society (2003) 973-979
9. Zivkovic, Z., Van, D.H.F.: Recursive Unsupervised Learning of Finite Mixture Models. IEEE Transactions on pattern analysis and machine intelligence. (2004) 651-656
10. Elgammal, A., Harwood, D., Davis, L.: Non-parametric Model for Background Subtraction. In 6th European Conference on Computer Vision: Dublin Ireland (2000)
11. Cheung, S.C.S., Kamath, C.: Robust Techniques for Background Subtraction in Urban Traffic Video. Proceedings of SPIE-The International Society for Optical Engineering, vol. 5308. International Society for Optical Engineering, Bellingham, WA 98227-0010, United States (2004) 881-892
12. Fuentes, L., Velastin, S.: From Tracking to Advanced Surveillance. Proceedings of IEEE International Conference on Image Processing, vol. 2. Institute of Electrical and Electronics Engineers Computer Society (2003) 121-125
13. Herrero, E., Orrite, C., Senar, J.: Detected Motion Classification with a Double-background and a Neighborhood-based Difference. Pattern Recognition Letters. (2003) 2079-2092
14. Cucchiara, R., Piccardi, M., Prati, A.: Detecting Moving Objects, Ghosts, and Shadows In Video Streams. IEEE Transactions on Pattern Analysis and Machine Intelligence. (2003) 1337-1342

Bit-Precision Method for Low Complex Lossless Image Coding

Jong Woo Won[1], Hyun Soo Ahn[1], Wook Joong Kim[2], and Euee S. Jang[1]

[1] Jong Woo Won Computer division, College of Information and Communications,
Hanyang University, 17 Hangdang-dong, Seongdong-gu, Seoul, 133-791, Korea
{jwwon@dmlab.hanyang.ac.kr, esjang@hanyang.ac.kr}

[2] Wook Joong Kim, Senior Engineer Broadcasting Media Research Department,
Radio & Broadcasting Research Laboratory, ETRI,
161 Gajeong-dong, Yuseong-gu, Daejeon, 305-350, Korea
{wjk@etri.re.kr}

Abstract. In this paper, we proposed a novel entropy coding called bit-precision method. Huffman coding and arithmetic coding are among the most popular methods for entropy-coding the symbols after quantization in image coding. Arithmetic coding outperforms Huffman coding in compression efficiency, while Huffman coding is less complex than arithmetic coding. Usually, one has to sacrifice either compression efficiency or computational complexity by choosing Huffman coding or arithmetic coding. We proposed a new entropy coding method that simply defines the bit precision of given symbols, which leads to a comparable compression efficiency to arithmetic coding and to the lower computation complexity than Huffman coding. The proposed method was tested for lossless image coding and simulation results verified that the proposed method produces the better compression efficiency than (single model) arithmetic coding and the substantially lower computational complexity than Huffman coding.

1 Introduction

Ever-increasing processing power and communication bandwidths are broadening the boundary of computing and communications devices from wired to wireless. As the computing power and communication bandwidths vary from environment to environment, the need of data compression with good compression efficiency and low computational complexity is more apparent than ever.

Data in data compression, in most cases, are the audio-visual data. In order to maximize the compression efficiency, a typical compression tool uses three steps to encode the data: prediction (or transform), quantization, and entropy coding. Prediction (or transform) is to reduce the input data range to represent the data in a more compact manner. While the prediction stage is a big portion of data compression, it is rather media specific. In other words, depending on media type, one has to use different prediction methods. Quantization is a process that also reduces the input data range in its precision from N bits per sample to M

bits per sample(where $N > M$). Quantization is a lossy process: the original data is distorted after quantization. After the quantization process, there comes entropy coding.

Entropy coding is to allocate the optimal bits depending on the information of the given symbol. Unlike from prediction and quantization, the entropy coding process guarantees the lossless compression. For this reason, any type of media compression tool has entropy coding as a core compression module. The compression performance of any entropy coding method is compared with the entropy of the given data as Shannon defined[1].

There are numerous entropy coding methods in the literature[2], [3]. The most popular methods are Huffman coding and arithmetic coding. Huffman coding, developed in 1950es, is one of the earliest entropy coding methods that are still in use. Huffman coding yields compression ratio close to the entropy of data by producing prefix codes for the given symbols. The Huffman decoder needs just a lookup table with prefix codes and their corresponding symbols. The decoding process using Huffman coding is as simple as (linear) search operation.

Arithmetic coding gained its popularity thanks to its improved compression efficiency over Huffman coding[4]. Arithmetic coding can assign real number of bits close to the information size unlike the integer number bit assignment in Huffman coding. Adaptive design that does not require the transmission of probability table is also another merit of arithmetic coding. The improved compression efficiency of arithmetic coding is, however, possible at the cost of increased computational complexity. Although many efficient designs are published by the researchers, arithmetic coding is still more expensive than Huffman coding in computational complexity. The details on various entropy coding methods including Huffman coding and arithmetic coding can be found in [6].

Overall, the battle between Huffman coding and arithmetic coding is not over. Many media standards do use the one or the other or both. The recent audiovisual standard, MPEG-4, uses both Huffman coding and arithmetic coding. This fact tells us that we ultimately need an entropy coding method that satisfies low complexity and high compression performance at the same time.

In this paper, we proposed the bit precision method, which is an extension of bit-precision based data representation. This technique is very simple to implement and, yet, can yield good compression comparable to arithmetic coding.

This paper is organized as follows. In section 2, the proposed bit-precision method is described in details. In order to evaluate the proposed entropy coding method, we have devised a lossless image coding. The lossless image coding with Huffman coding, arithmetic coding, and the proposed method is explained in the section 3. The experimental results are provided to evaluate the compression efficiency and computational complexity of the proposed method in section 4. Finally, we summarize our findings in conclusion.

2 Bit-Precision Based Entropy Coding

2.1 Bit-Precision Based Representation

Shannon defined the entropy (H) as the minimum number of bits to represent given symbols as

$$H = -\sum p_i \log_2 p_i .\qquad(1)$$

where p_i is the probability of i-th symbol and $\sum_i^n P_i = 1$. In bit-precision based representation, each symbol is represented by the multiples of bit precision value (bpv) bits as shown in Table 1. The last code word for the given bpv bits is the escape code for the upper value of symbol. For example, symbol value 7 can be represented as '111 000' when $bpv = 3$. When the bit-precision value (bpv) is one, the resulting code is the same as unary representation.

Table 1. Symbols and bit-precision based representation

Symbol	Bit precision value (bpv) = 1	bpv = 2	bpv = 3	bpv = 4
0	0	00	000	0000
1	10	01	001	0001
2	110	10	010	0010
3	1110	11 00	011	0011
4	11110	11 01	100	0100
5	111110	11 10	101	0101
6	1111110	11 11 00	110	0110
7	11111110	11 11 01	111 000	0111
...	...	...	...	...

From Table 1, one can see that the bit-precision based code is also a specific form of prefix codes like Huffman coding. The unique feature of the bit-precision based code is that it does not consider any probability distribution of the input symbols, if the input symbols are ordered based on their frequencies. From Table 1, the upper input symbols in order are assigned with less bits regardless of the size of bpv. For the best compression results, the input symbols have to be reordered as follows:

$$\begin{array}{c}s_i, s_j \in S, 0 \leq i < j \leq n-1 \\ p(s_i) \geq p(s_j) .\end{array}\qquad(2)$$

where s_i and s_j are the input symbols and $p(s_i)$ and $p(s_j)$ are the probabilities of the corresponding input symbols. Although we use the probability information of the given input symbols, it should be noted that there is no need of keeping probability information at the decoder side. This is similar to the case of Huffman coding.

In bit-precision based representation, the only information that the decoder has to know prior to decoding is *bpv*. The value of *bpv* can be embedded in the compressed bit stream as header information. In Huffman coding, the decoder has to have a lookup table that converts variable length codes to output symbols like Table 1. Therefore, bit-precision method is better suited for adaptive design of entropy coding, which is explained in the next section.

2.2 Block-Based Bit-Precision Based Representation

The entropy of a given data is calculated as global entropy with an assumption that the local entropy (i.e. entropy of some portion of the data) will be similar to the global entropy. In many cases, this assumption is not true. Although the global entropy approximates the local entropy, one can achieve the better compression ratio if the local entropy can be exploited.

Adaptive design of entropy coding is to exploit the local entropy. In order to design Huffman coding adaptively, one needs to transmit the code lookup table whenever the codes are changed. Therefore, there is no great gain in adaptive Huffman coding over fixed Huffman coding, considering the increased bit overhead and complexity. Arithmetic coding is better suited for adaptive design, since local statistics can be updated to the existing probability table with minimal overhead. However, the computation complexity to update the entire probability table is not marginal.

Using the bit-precision based representation, the decoder only keeps track of *bpv* information from block to block. In each block of data, the encoder will choose the best performing *bpv*. Hence, the decoder will simply use the *bpv* information to compute the proper output symbol from the given *bpv*.

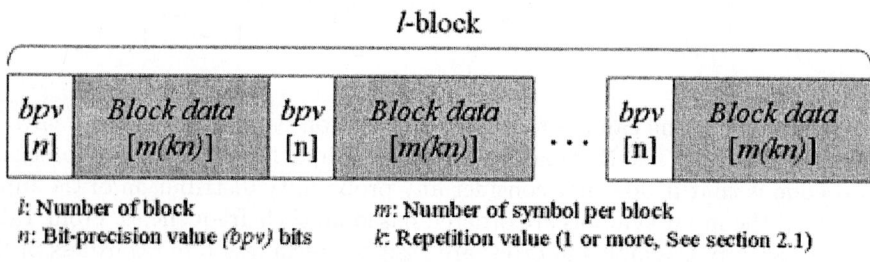

Fig. 1. Bitstream structure of block-based bit-precision

Fig. 1 shows the bitstream structure of block-based bit-precision based representation. Each block which is portion of entire bitstream consist of *bpv* and *block data*.

2.3 Complexity Analysis

High computational complexity of arithmetic coding over Huffman coding is very well known. Therefore, we provide the complexity analysis between Huffman coding and the proposed bit-precision codes. Normally, Huffman decoding process can be considered with $O(n)$ complexity, since the basic operation in Huffman decoding is the (linear) search of lookup table and its average time complexity will be n/2.

Bit-precision decoder computes the output symbol and the basic operation per symbol is a single computation as follows:

$$n = mb + r, 0 \leq r < b, 0 \leq m \leq \lfloor \tfrac{n}{b} \rfloor, b = 2^{bpv} - 1 \ . \qquad (3)$$

where n is the output symbol, m is the number of the largest code (b) given bpv, and r is the remainder value between 0 and b-1. As shown in the equation, the computation is $O(1)$. Therefore, we can conclude that the complexity of the proposed bit-precision based method is much lower than that of Huffman coding.

2.4 Encoding Considerations

The optimal compression of bit-precision method is determined by the value of bpv at the encoding stage. One way to find the best bpv is to compute the bit rate with every case of bpv. For instance, if we use three bits for bpv, the possible value of bpv is from 0 to 7, where we can check the bit rates with 8 different cases.

One may argue that the encoder complexity is greater than the decoder complexity. Nevertheless, the encoder complexity is still low and it does not affect the decoder complexity. It may be worthwhile to further research how to decrease the encoder complexity in the future.

3 Test Model: Lossless Image Compression

In order to check the performance of the proposed bit-precision based representation, we have tested the proposed method with Huffman coding and arithmetic coding in lossless image coding, where entropy coding performance is of great importance.

3.1 A Simple Image Coding Model

A lossless image coding consists of two processes: prediction and entropy coding (or residual errors). For prediction, we use simple prediction rule, which is used in JPEG lossless mode, as follows[5].

$$\hat{X}_{i,j} = \tfrac{A+B}{2}, A = X_{i-1,j}, B = X_{i,j-1} \\ e_{i,j} = X_{i,j} - \hat{X}_{i,j} \ . \qquad (4)$$

where $X_{i,j}$, $\hat{X}_{i,j}$, and $e_{i,j}$ are the original pixel value, the prediction value, and the residual error at the i-th column and j-th row, respectively.

3.2 Entropy Coding of Residual Errors

The residual error is to be entropy-coded as shown in Fig. 2. For evaluation of compression performance and computational complexity, we have applied the entropy coding methods that are listed in Table 2. In the case of Huffman coding, only one lookup table is used. In arithmetic coding, we chose two arithmetic coding methods: the one with a probability model and the other with multiple models. These arithmetic coding methods are adaptive, which means that the corresponding probability table is updated on-line. Multiple models in arithmetic coding are used when block-based coding is carried out. In bit-precision method, probability model is not used as described in section 2.

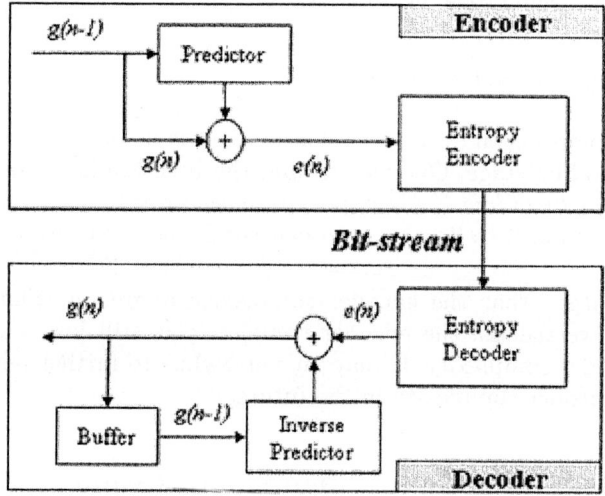

Fig. 2. Encoder and decoder structure of simple lossless image coding model

Table 2. Different entropy coding methods and their configurations

Methods	Description
Huffman coding (HC)	- Single probability model
	- Fixed probability model
Arithmetic coding (SAC)	- Single probability model
	- Adaptive probability model
Arithmetic coding (MAC)	- Multiple probability models
	- Adaptive probability model
	- Block-based coding only
Bit-precision codes (BC)	- No probability model

4 Experimental Results

4.1 Test Conditions

Experiments are performed on five standard 512x512 grey-scale images to test different entropy coding methods. We measured the compression efficiency and the decoding time to measure the computational complexity. In evaluating the computational complexity, the run-time is averaged over five times on the test PC which based on Windows XP with Intel Centrino1.5GHz CPU and 512MB DDR-RAM.

Since no software optimization on any entropy coding was performed, the decoding time results do not necessarily represent the absolute computational complexity. Rather, we measured the decoding time to analyze the relative time differences among the different entropy coding methods.

4.2 Compression Efficiency in Block-Based Coding Mode

Fig. 3 shows the compression performances of different entropy coding methods in lossless image coding on Lena image while changing the block size. The block size varies from 2x2 to 512x512 (size of image). In each block, the entropy coding can have its own probability model. In the case of Huffman coding, the size of lookup table becomes very large if the block size gets lower than 64x64, such that it is meaningless to use Huffman coding with a very small block size. Single model arithmetic coding method (SAC) uses one single model regardless of the block size, which makes its compression ratio look flat. Multiple model

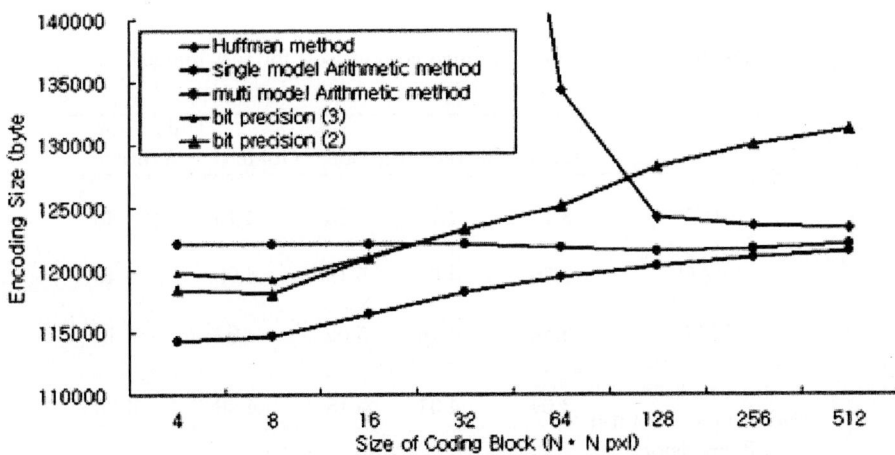

Fig. 3. Compression efficiency comparison on Lena error image. Bit-precision (3) indicates when the size of $bpv = 3$ (where a possible bpv values are from 0 to 7) and Bit-precision (2) when the size of $bpv = 2$.

arithmetic coding (MAC) uses one of eight different probability tables to encode each block. As the block size gets smaller, the compression efficiency gets higher. MAC shows the best performance in compression efficiency. Bit-precision based representation (BC) shows the similar tendency to MAC, in that the compression gain becomes higher with the smaller block size. Overall, BC shows its performance as the second best, which is even better than that of SAC.

4.3 Computational Complexity Analysis from Decoding Time

We measured the decoding times of different entropy coding methods and the results are shown in Table 3. The proposed BC showed lowest complexity in decoding time. It was about 30 times faster than MAC. The overall complexity with different block size was shown to be stable in any entropy coding method. And the tendency of Table 3 and Fig. 3 were also observed in other test images.

Table 3. Decoding time performance with variable block size (test image: Lena)

	4	8	16	32	64	128	256	512
SAC	91.53	91.83	91.03	91.23	91.03	89.93	90.43	89.83
MAC	191.77	192.57	174.64	171.23	194.77	196.08	179.35	187.06
bit precision (size of bpv = 3 bits)	6.79	6.33	6.12	6.01	5.8	6.01	5.61	6.21
bit precision (size of bpv = 2 bits)	6.68	6.31	6.05	5.99	5.78	5.99	5.58	6.13

Table 4. Coding performance comparison for test images

	Lena	Barbara	Baboon	Goldhill	Unit : bpp Average
HC	3.71	4.73	5.33	4.15	4.48
SAC	3.76	4.59	5.19	4.06	4.40
MAC	3.49	4.16	5.02	3.92	4.15
bit precision (size of bpv = 3 bits)	3.64	4.31	5.18	4.09	4.30
bit precision (size of bpv = 2 bits)	3.60	4.53	5.18	4.08	4.35

Table 5. Decoding time performance

	Lena	Barbara	Baboon	Goldhill	Unit : ms Average
HC	18.42	22.13	24.63	19.14	21.08
SAC	89.83	106.35	116.07	96.24	102.12
MAC	187.06	214.31	214.41	212.56	207.09
bit precision (size of bpv = 3 bits)	6.21	6.01	5.81	6.41	6.11
bit precision (size of bpv = 2 bits)	6.13	6.51	5.71	6.43	6.20

4.4 Best Performance Comparison

From the above results, we measured the best performance in compression efficiency and computational complexity in each entropy coding method. The test results on five test images are shown in Tables 4 and 5. BC showed its compression performance better than HC and SAC, worse than MAC. BC was shown to be the lowest computational complexity among the test methods. BC was about 3 times faster than Huffman coding.

The best results in compression efficiency were possible with MAC, but the computational complexity of MAC is more than 30 times. Huffman coding was neither faster in decoding time nor more efficient in compression than BC. BC clearly showed its feasibility as fast and simple entropy coding with competitive compression efficiency.

5 Conclusions

Bit-precision based representation showed its feasibility as a competitive entropy coding method when fast decoding capability with good compression efficiency is demanded. We tried this method in lossless image coding, but it may be worthwhile to extend to the other fields from general compression to specific media compression (such as video).

Acknowledgements

This work was supported by the Hanyang University Research Fund.

References

1. Shannon, C.E.: Prediction and Entropy of Printed English. Bell System Technical Journal **30(10)** (January) 54–63
2. Majid Rabbani, Paul W. Jones: Digital Image Compression Techniques. Donald C.O'Shea. Series Editor. Georgia Inc. Vol. TT7. (1991)

3. David Salomon. Data Compression. Springer-Verlag New York. Inc. Second Edition. (1998)
4. Alistair Moffat, Radford M.Neal, Ian H, Witten. Arithmetic Coding Revisited. ACM Transactions on Information Systems (TOIS). Vol. 16. Issue 3. (July 1998) 256–294
5. Wallace, G.K.: The JPEG Still Picture Compression Standard. IEEE Transactions. Vol 38. Issue 1. (Feb. 1992) xviii–xxxiv
6. Alistair Moffat, Andrew Turpin: Compression and Coding Algorithms. Kluew Academic Publishers. (1999)

Texture Feature-Based Image Classification Using Wavelet Package Transform[1]

Yue Zhang[+], Xing-Jian He, and Jun-Hua Han

Institute of Intelligent Machines, Chinese Academy of Sciences, P.O.Box 1130,
Hefei, Anhui 230031, China
{yzhang, xjhe, jhhan}@iim.ac.cn

Abstract. In this paper, a new method based on wavelet package transform is proposed for classification of texture images. It has been demonstrated that a large amount of texture information of texture images is located in middle-high frequency parts of image, a corresponding method called wavelet package transform, not only decomposing image from the low frequency parts, but also from the middle-high frequency parts, is presented to segment texture images into a few texture domains used for image classification. Some experimental results are obtained to indicate that our method for image classification is superior to the co-occurrence matrix technique obviously.

1 Introduction

Generally speaking, the first step of texture analysis is mainly to segment an image into some homogeneous sub-images. And then we can use many properties to determine the region homogeneity such as the texture, the color of a region and the gray-level intensity. When there are many texture images in practice, texture is the main information that can describe the texture images exactly. Most natural surfaces exhibit texture. Texture analysis plays an important role in many machine vision tasks , such as surface inspection , scene classification , surface orientation and shape determination , and its tasks are mainly to cover these fields such as classification , segmentation , and synthesis[1], [2], [3], [4]. Texture is characterized by the spatial distribution of gray levels in a neighborhood. Although the texture has been widely applied to image analysis, no specific and universe definition is proposed for texture ,A large number of techniques that is used to analyze texture image have been proposed in the past [5], [6], [7], [8], [9], texture analysis was based on the co-occurrence matrix proposed by Haralick et al.[10], [11], [12] Then, Gaussian Markov random fields (GMRF) and the $2-D$ autoregressive model(AR) were proposed to characterize textures[13], [14], but the traditional statistical approaches to texture analysis are restricted to the texture on a single scale. Recently a method based on wavelet transform have received a lot of attention, mainly because it can provides a precise and unifying frame work for the analysis and characterization of a signal at

---

[1] This work was supported by the National Science Foundation of China (Nos.60472111 and 60405002).
[+] The corresponding author.

different scales [15]. In this paper, we mainly focus on a particular approach to texture feature extraction which is referred to as wavelet package transform[16] based on wavelet transform, which is applied on a set of texture images and statistical features such as the average energy which is extracted from the approximation and detail regions of the decomposed images, at different scales. The reason for the popularity of wavelet packet transform ,compared with the tradition methods for extracting image texture feature, is mostly due to not only decomposing image from the low-frequency parts, but also from the high-frequency parts. After the texture images are properly segmented for two level, selecting a classifier is necessary for image classification, many techniques on classifier are discussed in some literatures [17], [18], [19], [20], [21], [22].In this paper, we classify the texture images by distance classification. A comparison in experimental section is made to show the efficiency and effectiveness of our approach with respect to the co-occurrence matrix technique.

2 Wavelet Package Transform

2.1 Wavelet Basis

The wavelet transform, also called as mathematical microscope, was developed in the mid and later 1980s. The basic ideal of the wavelet transform is to represent any arbitrary function as a superposition of wavelets .Any such superposition decomposes the given function into different scale levels where each level is further decomposed with a resolution adapted to that level . Currently, the wavelet transform has been applied to many fields such as image compression, image segmentation, and so on.

Suppose that a function $\psi(x) \in L^2(R)$, whose Fourier transform is $\hat{\psi}(\omega)$, satisfies the following admissibility condition:

$$C_\psi = \int \frac{|\hat{\psi}(\omega)|^2}{|\omega|} d\omega < \infty . \tag{1}$$

So $\psi(x)$ is called as mother wavelet, and consequently the wavelet sequences are generated from the single function $\psi(x)$ by translations and dilations as follows:

$$\psi_{a,b}(x) = |a|^{\frac{-1}{2}} \psi(\frac{x-b}{a}) . \tag{2}$$

where a is the scale factor, b is the translated factor. So the continuous wavelet transform of a $1-D$ signal $f(x)$ is defined as:

$$(W_a f)(b) = \int f(x) \psi_{a,b}(x) dx . \tag{3}$$

The mother wavelet $\psi(x)$ has to satisfy the admissibility criterion to ensure that it is a localized zero-mean function.

If a and b are discretized, assuming a to be denoted as 2 and b as 1, the continuous wavelet will become the binary wavelet:

$$\psi_{j,k}(x) = 2^{\frac{j}{2}}\psi(2^j - k) . \qquad (4)$$

Given $f(x) \in L^2(R)$, an expression of $f(x)$ can be got by multiresolution analysis:

$$f(x) = \sum_{j,k} d_{j,k}\psi_{j,k} . \qquad (5)$$

where $d_{j,k}$ is equivalent to $<f, \psi_{j,k}>$, j is the scale and k is the translation.

If there are two subspaces denoted by V_J and W_J, $V_{J+1} = V_J + W_J$, $V_J \perp W_J$, there will have:

$$L^2(R) = \cdots W_{-1} + W_0 + W_1 + \cdots + W_{N-1} + V_{N-1} . \qquad (6)$$

Given $\varphi \in V_0$, V_J consists of the expression $\psi_{j,k}(x) = 2^{\frac{j}{2}}\psi(2^j - k)$ spanned by φ. Accordingly, if ψ belongs to W_0, and W_J is spanned by ψ, then the double scale equation is obtained as follows:

$$\varphi(t) = \sum_{n \in z} h_n \varphi(2t - n) . \qquad (7)$$

$$\psi(t) = \sum_{n \in z} g_n \varphi(2t - n) . \qquad (8)$$

where $\varphi(t)$ is a scale function, $\psi(t)$ is a wavelet function. Their Fourier transform are expressed as:

$$\varphi(2\omega) = H(\omega)\varphi(\omega) . \qquad (9)$$

$$\psi(\omega) = G(\frac{\omega}{2})\varphi(\frac{\omega}{2}) . \qquad (10)$$

where H and G are called perfect reconstruction quadrature mirror filters (QMFs) if they satisfy the orthogonality conditions

$$HG^* = GH^* = 0 . \qquad (11)$$

$$H^*H + G^*G = I . \qquad (12)$$

where H^* and G^* are the adjoint operators of H and G, respectively, and I is the identity operator.

The decomposition and reconstruction formula by MALLAT [23] can be obtained using double scale function as follows:

$$c_n^j = 2^{-1/2} \sum_k c_k^{j+1} h_{k-2n} \ . \qquad (13)$$

$$d_n^j = 2^{-1/2} \sum_k c_k^{j+1} g_{k-2n} \ . \qquad (14)$$

$$c_n^{j+1} = 2^{-1/2} (\sum_k c_k^j h_{n-2k} + \sum_k d_k^j g_{n-2k}) \ . \qquad (15)$$

where c_n^j is a low-pass coefficient, and d_n^j is a high-pass coefficient.

Generally speaking, a low filter and a high filter (H and G) are used for segmenting image. In a classical wavelet decomposition, the image is split into an approximation and details images. The approximation is then split itself into a second level of approximation and details. With the decomposition and reconstruction formula mentioned in equation (11),(12),(13), the image is usually segmented into a so-called approximation image and into so-called detail images. The transformed coefficients in approximation and detail sub-images are the essential features, which are as useful for image classification.

2.2 Wavelet Package Decomposition

As pointed out above, much of information needed for extracting image texture features is to locate in the middle-high frequency parts [24], thus wavelet package transform based on wavelet transform, is introduced to further decompose the texture image with a large number of information in the middle-high frequency parts. Generally speaking, wavelet package transform can not only decompose image from the low frequency parts, but also do it from the middle-high frequency parts, The wavelet package decomposition, is a generalization of the classical wavelet decomposition that offers a richer signal analysis. In that case, the details as well as the approximations can be split. The tree-structured wavelet package transform [25] is displayed in Fig 1, where S denotes the signal, D denotes the detail and A the approximation.

In order to intuitively observe the decomposition based on wavelet package transform at two levels, an example is set to illustrate the distribution information in the sub-image, as shown in Fig 2.

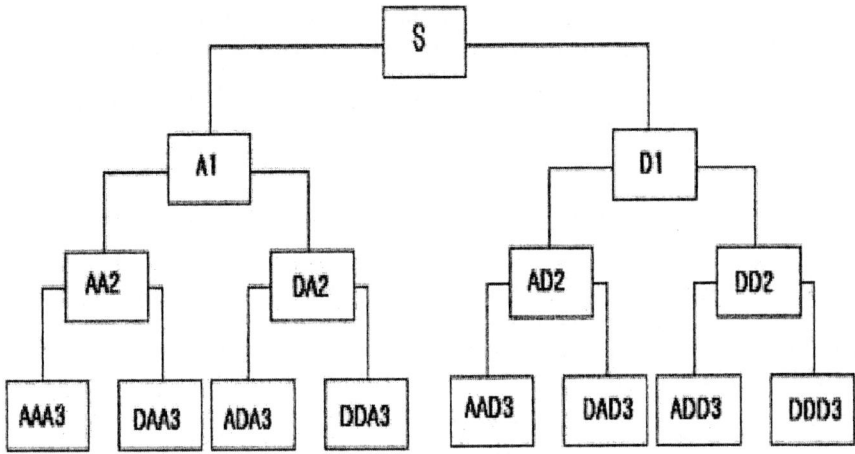

Fig. 1. The tree-structured wavelets transform

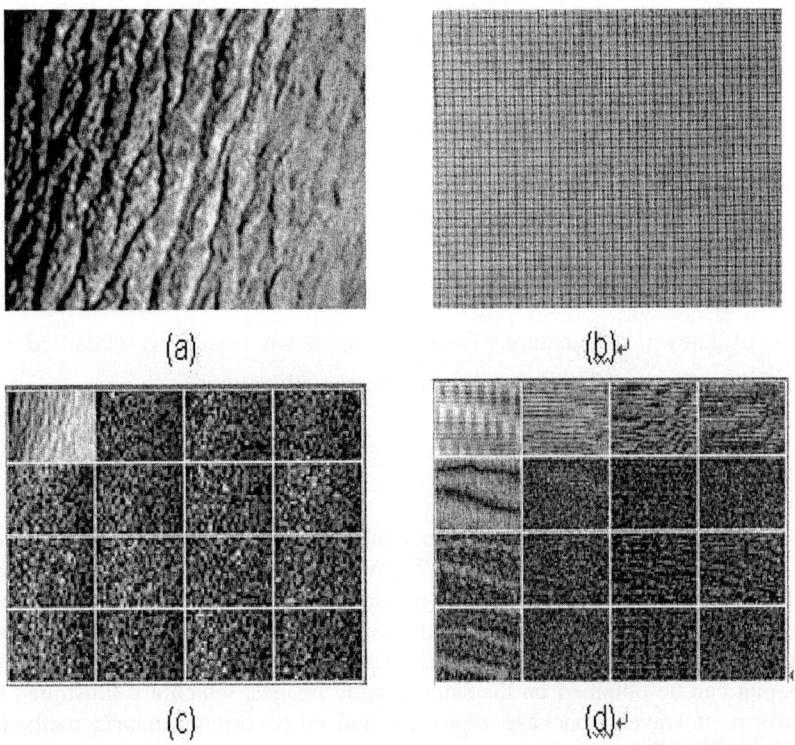

Fig. 2. Image decomposition (a) (b) The original images (c) (d) The decomposed images

As seen in Fig 2, the texture of the image is regularly segmented into the sub-image domains in which the approximation and the details are to locate, and all texture information which is used to extract texture features is displayed from the horizontal, vertical and diagonal directions. We can also observe much information in the middle-high parts.

3 Feature Extraction and Image Classification

Firstly the known texture images are decomposed using our proposed method, wavelet package transform. Then, the average energy of approximation and detail sub-image of two level decomposed images are calculated as features using the formulas given as follows:

$$E = \frac{1}{N \times N} \sum_{i=1}^{N} \sum_{J=1}^{N} |f(x, y)| . \tag{16}$$

where N denotes the size of sub-image, $f(x, y)$ denotes the value of pixel of image.

Secondly for texture classification, the unknown texture image is decomposed using wavelet package transform and a similar set of average energy features are extracted and compared with the corresponding feature values which are assumed to be known in advance using a distance vector formula, given in

$$D(j) = \sum_{i=1}^{N} abs[f_i(x) - f_i(j)] . \tag{17}$$

where $f_i(x)$ represents the features of unknown texture, while $f_i(j)$ represents the features of known j^{th} texture. Then, the unknown texture is classified as j^{th} texture, if the distance $D(j)$ is minimum among all textures.

4 Experimental Results

In this section, we use our method to conduct several experiments and illustrate the visual results obtained on MATLAB 7.0.We select 64 images obtained from the website (http://astronomy.swin.edu.au/~pbourke/texture/cloth/) for classification. The original color images are converted to the same size gray-scale images It is assumed that the sorts of image is known in advance, as shown in Fig 3 .All experimental results that can be obtained on the same sample images, which are illustrated by the comparison of wavelet package transform and co-occurrence matrix method , are shown in Table 1,2.

From the Table 1, it is seen that, using the proposed method based on wavelet package transform, the accuracy of classification for Stone ,Door, Brick, Bark, Noise, Ripple and Leaf are 90.9%, 90.9%, 83.3%, 85.7%, 60.0%,87.5% and 72.7%, respectively. Nevertheless, with co-occurrence matrix method, the corresponding accuracies with 63.6%, 63.6%, 66.7%, 70.4%, 50.0%, 62.5% and 54.4% in the Table 2 are obviously lower than that in the Table 1. In other words, the approach for image classification based on wavelet package transform is more efficient than the method using co-occurrence matrix. Although the effect of our approach for classifying the thin texture image such as leaf and noise is inferior to the thick texture image, it is still obvious relatively to the method based on co-occurrence matrix.

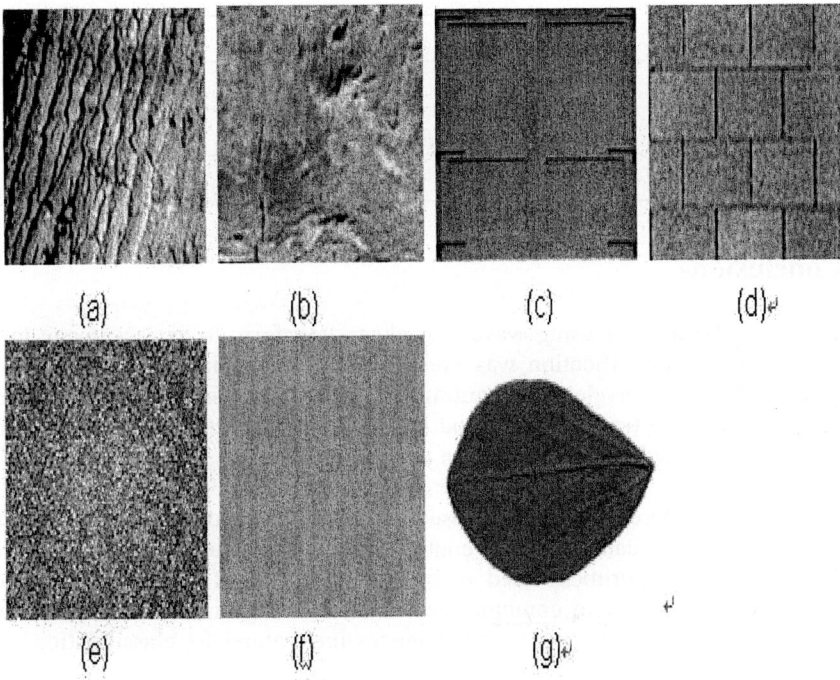

Fig. 3. Sorts of images known (a) Bark (b) Stone (c) Door (d) Brick (e) Noise (f) Ripple (g) Leaf

Table 1. Experimental result based on wavelet package transform

Sorts of images	Test samples	Misclassification samples	Accuracy (%)
Stone	11	1	90.9
Door	11	1	90.9
Brick	6	1	83.3
Bark	7	1	85.7
Noise	10	4	60.0
Ripple	8	1	87.5
Leaf	11	3	72.7

Table 2. Experimental result rased on co-occurrence matrix

Sorts of images	Test samples	Misclassification samples	Accuracy (%)
Stone	11	4	63.6
Door	11	4	63.6
Brick	6	2	66.7
Bark	7	2	70.4
Noise	10	5	50.0
Ripple	8	3	62.5
Leaf	11	5	54.4

5 Conclusions

In this paper, the method using wavelet package transform to extract image texture features for image classification was presented. As shown in our experiment, it is more efficient to use wavelet package transform to segment image texture for image classification than the traditional method based on co-occurrence matrix. The reason is that the wavelet package transform can not only decompose the image from the low frequency parts, but also from the high frequency parts. That is, all kinds of textures are divided into different sub-images used for extracting texture features. Although the proposed method can reach the content result, the accuracy with lower value obtained from the experiment need to be further improved. Therefore, the future works will focus on how to combine the wavelet package transform together with other methods for efficiently extracting image texture features for classification.

References

1. Livens, S.: Image Analysis for Material Characterization. PhD Thesis. University of Antwerp. Antwerp. Belgium. (1998)
2. Randen, T.: Filter and Filter Bank Design for Image Texture Recognition. PhD Thesis. NTNU. Stavanger. Norway. (1997)
3. Van de Wouwer, G.: Wavelets for Multiscale Texture Analysis. PhD Thesis. University of Antwerp. Antwerp. Belgium. (1998)
4. Gonzalez, R.C., Woods, R.E.: Digital Image Processing. Addison-Wesley. Reading. MA. (1992)
5. Haralick, R.M.: Statistical and Structural Approaches to Texture. Proc. IEEE 67. (1979) 768 - 804
6. Pitas I.: Digital Image Processing Algorithms and Applications. New York. Wiley. (2000)
7. Ahuja, N., Rosenfeld, A.: Mosaic Models for Textures. IEEE Trans. Anal. Mach. Intell. 3(1). 1-11
8. Huang, D.S.: Systematic Theory of Neural Networks for Pattern Recognition. Publishing House of Electronic Industry of China. Beijing. (1996)

9. Cohen, F.S., Fan, Z., Patel, M.A.: Classification of Rotation and Scaled Textured Images Using Gaussian Markov Random Field Models. IEEE Trans. Pattern Anal. 13 (2) (1991) 192 - 202
10. Haralick, R.M., Shanmugam, K., Dinstein, I.: Textural Features for Image Classification IEEE Trans. Systems Man Cybernet. SMC-3 (1973) 610 - 621
11. Haralick, R.M., Shapiro, L.S.: Computer Vision. Vol.1. Addision Wesley. Reading. MA. (1992)
12. Ohanian, P.P., Dubes, R.C.: Performance Evaluation for Four Classes of Textural Features. Pattern Recognition. 25(8) (1992) 819 - 833
13. Kashyap, R.L., Chellappa, R.: Estimation and Choice of Neighbors in Spatial-Interaction Models of Images. IEEE Trans. Inf. Theory TT-29(1) (1983) 60 - 72
14. Kashyap, R.L., Khotanzao, A.: A Model-Based Method for Rotation Invariant Texture Classification. IEEE Trans. Pattern Analysis Mach. Intell. PAMI-8(4) (July 1986) 472 - 481
15. Unser, M.: Texture Classification and Segmentation Using Wavelet Frames. IEEE Trans. Image Process. 4 (11) (1995) 1549–1560
16. Coifman, R.R., Meyer, Y., Wickerhauser, M.V.: Progress in Wavelet Analysis and Applications. Editions Frontieres. France. (1993)
17. Wu, W.-R., Wei, S.-C.:Rotation and Gray Scale Transform Invariant Texture Classification Using Spiral Resampling, Subband Decomposition and Hidden Markov Model. IEEE Trans. Image Process. 5 (10) (1996)1423 - 1433
18. Pontil, M. and Verri, A.: Supporl Vector Machines for 3D Object Recognition, IEEE Trans. Parrem Analysis and Machine Intelligence, vol. 20, no.6 (June 1998) 637-646
19. Huang, D.S., Ma, S.D.: Linear and Nonlinear FeedForward Neural Network Classifiers: A Comprehensive Understanding. Journal of Intelligent Systems. Vol.9. No.1. (1999) 1 - 38
20. Huang, D.S.: Radial Basis Probabilistic Neural Networks:Model and Application , International Journal of Pattern Recognition and Artificial Intelligence, 13(7),(1999)1083-1101
21. Ma, Y. J.: Texture Image Classification Based on Support Vector Machine and Distance Classification. Proceeding of 4th World Congress on Intelligence and Automation
22. Hwang, W J., Wen, K.W.: Fast kNN Classification Algorithm Based on Partial Distance Search. Electron. Lett. Vol. 34. No. 21. (1998) 2062 - 2063
23. Mallat, S.: A Theory for Multiresolution Signal Decomposition: the Wavelet Representation. IEEE Trans Patt Anal Machine Intell. 11. (1989) 674 – 693
24. Tiahorng Chang, Kuo, C.-C.Jay.: Texture Analysis and Classification with Tree-Structured Wavelet Transform. IEEE Transactions on Image Processing. Vol.2. No.4. (Oct. 1993) 429 – 441
25. Soman, A.K., Vaidyanathan, P.P.: On Orthonormal Wavelets and Paraunitary Filter Banks. IEEE Trans. Signal Processing, Vol. 41. No 3. (1993) 1170-1182

Authorization Based on Palmprint

Xiao-yong Wei[1,2], Dan Xu[1], and Guo-wu Yuan[1]

[1] Department of Computer Science, Information School, Yunnan University,
650091 Kunming, Yunnan, China
qsc99@hotmail.com
[2] Department of Computer Science, City University of Hong Kong,
Kowloon, Hong Kong
xiaoyong@cityu.edu.hk

Abstract. In this paper, palmprint features were classified into Local Features and Global Features. Based on this definition, we discussed the advantage and weakness of each kind of features and presented a new palmprint identification algorithm using combination features. In this algorithm, a new method for capturing the key points of hand geometry was proposed. Then we described our new method of palmprint feature extracting. This method considered both the global feature and local detail of a palmprint texture and proposed a new kind of palmprint feature. The experimental results demonstrated the effectiveness and accuracy of these proposed methods.

1 Introduction

User verification systems that use traditional scheme (such as password, ID cards, etc.) are vulnerable to the wiles of an impostor. So automatic human identification has become an important issue in today's information and network based society. The techniques for automatically identifying an individual based on his unique physical or behavioral characteristics are called biometrics [7].

Due to its stability and uniqueness, palmprint can be considered as one of the reliable means distinguishing a man from his fellows, and can be easily integrated with the existing bimetrics system to provide enhanced level of confidence.

Recent years, most of research in biometrics has been focused on palmprint identification, and many approaches have been developed. According to the operating space in which palmprint features be extracted, those approaches can be categorized into spatial domain approaches and frequency domain approaches. According to the type of palmprint features, those approaches can be categorized into structural features-based approaches and statistics features-based approaches.

A spatial domain approach has been firstly represented in 1998 [5] by Zhang and Shu. Because the matching of lines is more easily than that of curves, this approach viewed palmprint features as lines-constituted texture. As a result, palmprint matching can be transformed into lines matching based on that view. In 2002, Nicolae and Anil proposed anther spatial domain approach [6], which extracted some feature points from the palmprint and calculated the orientations of the lines with which these points associated. Both points and orientations were used as the features of an individual palmprint in this approach. Spatial domain approaches are easy to be understood and to be implemented. But it is sensitive to the non-linear transformation of soft palm skin.

A representative method of frequency domain approaches was known as the method based on Fourier Transform (W.Li et al., 2002) in [8]. This method took advantages of the relativity between a palm image's frequency domain representation and spatial domain representation. It translated the spatial domain image into its frequency domain representation, and extracted the statistics features in later representation. It performed well but is hypersensitive to the asymmetric illumination environment.

An approach which can be assigned to statistics features based approaches was presented in 2003 by Ajay Kumar et. [2]. It divided the palmprint image into a set of n smaller sub-regions, and then calculated the mean and standard deviation of each sub-region as the palmprint feature.

For enhancing the accuracy and reducing executive times of features matching, approach based on hierarchical structure was presented in [9, 10]. Other approaches including K-L Transform based in [11], Wavelet Transform based in [12], Gabor Filter based in [13,14] and Neural Network based in [15], were also be developed.

As you seen, different approaches have employed different features. We can divide these features into two classes of Global Features and Local Features. Global Features can be defined as the features that represent the trait of a region. For instance, the statistics feature in [2] and the frequency domain representation in [8] belonge to this class. Local Features is defined as the features that represent the trait of certain detail. Feature points and line orientation in [6] are of this class. Obviously, based on the definition above, some improvements will be possible by using the combination of these two kinds of features.

Therefore, in this paper we are interested in using combination features to do recognition. In Section 2 we introduce the image preprocessing used in this paper. Our new palmprint features extracting is described in depth in Section 3. Section 4 will introduce our features matching method. The experimental results are described in both Section 5 and Section 6 while Section 7 contains the discussion and conclusions.

2 Image Preprocessing

2.1 Direction Alignment Based on Morphology

To rotate all the hands into same direction, Ajay used an ellipse fitting based method to do this [2]. Because the binarized shape of the hand can be approximated by an ellipse, the orientation of the binarized hand image is approximated by the major axis of the ellipse and required angle of rotation is the difference between normal and the orientation of image. But the result of this method is sensitive to the captured wrist length and degrees of fingers opening. Han also presented a wavelet based method of alignment in [3], which search the corners along the outline of handgeometry and use these corners to do the alignment. However, the threshold value according to which corner points is classified is hardly to automatically choose. Furthermore, this approach is time consuming. Therefore we designed a new hand image alignment method based on morphological operations as follows.

By observing the hand image, we can find that the interval space between fingers except thumb forms three headstand triangle regions. There are three corner points (we called key points) exactly locate at the peaks of the triangles. Because their relative location is comparatively steady, these key points can be used as the reference point for the alignment operations. In our alignment method, there are three main steps. Step1: By using an array SE (known as structuring element) to dilate the binarized source hand image, a webbing will be added to the outline of source image, and some part of the triangle regions also will be filled. The result is shown in Fig 1(b). Step2: The same SE in Step1 will be used again to erode the dilated hand image to cut off some webbings added in Step1. As shown in Fig 1(c), because some parts of triangle regions have jointed with the fingers, the jointed parts will not be cut off. Step3: Subtracting source hand image from the result image obtained by step2 will result in some parts of triangle regions (Fig 1d). It is obviously easy to get the key points now. The whole processing can be formulated as fellow.

$$T\arg et = (G \oplus SE) \ominus SE - G \ . \tag{1}$$

where G is the binarized source hand image and SE is structuring element.

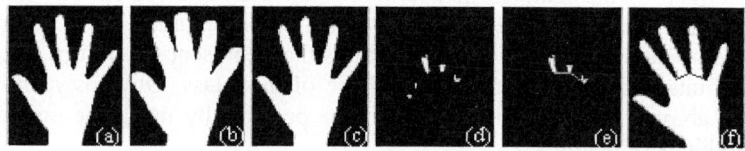

Fig. 1. Get the triangle regions from binarized source hand image, (a) Binarized hand image, (b) Dilated, (c) Eroded, (d) Triangle regions, (e) Connect the key points, (f) Rotated

After getting the triangle regions, the pecks of each triangle regions are exactly the key points. Then we can compute the relative location of the key points. Finally, as shown in Fig 1(f), we rotate the source image to normal orientation according to the relative location of key points. This method of alignment is simpler to be understood and faster than the methods in [2] and [3].

2.2 ROI Generation

To compute the region of interest (ROI), we can use key points obtained form 2.1 as reference points. By connecting the leftmost key point and the rightmost corner point, we can get a line as horizontal coordinate axis. Then draw a vertical line from the middle key point to horizontal coordinate axis, the cross point will be used as origin. After established above coordinate system, we can extract the rectangle region as ROI. A perfect ROI shall avoid the creasy region at thumb root and callus region at other fingers' root when choosing it. The result was shown as Fig 2.

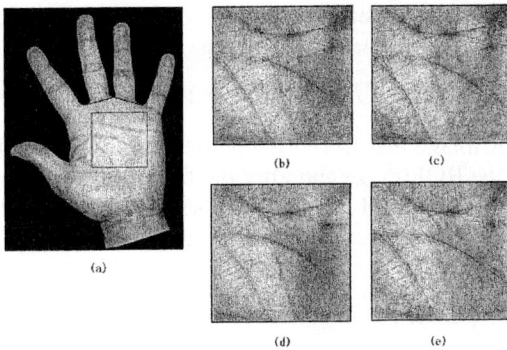

Fig. 2. ROI Generation, (a) Key points connected line and ROI rectangle,(b)(c)(d)(e) Captured ROIs from same class, ClassID=23 and SampleID=237,238,239,240

As shown in Fig 2, it is obviously that the ROIs from same class (person) are accurately resembled each other in relative location of source image. The ROI also effectively avoid the creasy region at thumb root and callus region at other fingers' root.

3 Feature Extraction

In many papers, the position and orientation of palmprint lines are usually used as the features of palmprint (e.g., in [5][6]). Line matching methods are employed in these papers to do the verification. But skin of palm is too soft to endure extrusion. So nonlinear transform of palmprint lines are inevitable and line matching methods are all hypersensitive to it. Instead of accurately characterize these palm lines, another method divides the palmprint image into a set of n smaller sub-regions, then calculates the mean and standard deviation of each sub-region as the palmprint feature(as in [2]). However, it is known that standard deviation of an image is directly associated with gray distribution of the image. Consider with different sub-regions, which have different texture, only if they have approximate image histogram, their standard deviation would be same. Using only standard deviation as feature and ignoring relationship of each line's position will result in classes unreasonably distributing in eigenspace. Obviously, according to the definition of Section 1, the position and orientation of palmprint lines belong to Local Features as well as the standard deviation of a sub-region belongs to Global Features. Local Features, the traits of detail, is easier to be observed by human vision. Consequently, the matching of them is easy to be understood and to be designed. However, because of the reasons method above (skin extrusion, lighting etc.), a matching with high accuracy degree is usually impossible. Global Features, the trait of a region, which with less dependence of accurate features matching, ignore the relative position of detail and in other word can't represent the arrangement of different details. Therefore, we designed a new method which considered both palm lines (Local Features) and their relative position (Global Features) to improve the performance of matching. It can be summarized as follows.

Firstly, we use four line detectors (Masks) to detect lines oriented at 0°, 45°, 90° and 135° in ROI image. It can be formulated as:

$$G_i = G * Mask_i \quad i=0, 1, 2, 3. \tag{2}$$

where G is the ROI image and $Mask_i$ is line detector of different direction. This operation can filtrate the ROI image and after the filtration the lines with different orientation but blending in same ROI image can be separated into four images G_i. Then we transform the four G_i with projection transform defined as follow:

$$R_\theta(x') = \int_{-\infty}^{\infty} G_i(x', y')dy' \quad i=0, 1, 2, 3 \text{ while } \theta = 0, 45, 90, 135$$

$$\text{where } \begin{pmatrix} x' \\ y' \end{pmatrix} = \begin{pmatrix} \cos\theta & -\sin\theta \\ \sin\theta & \cos\theta \end{pmatrix} \begin{pmatrix} x \\ y \end{pmatrix}. \tag{3}$$

Fig 3 can illustrate this projection transform.

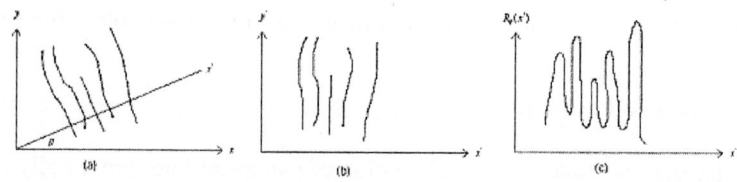

Fig. 3. Our projection transform, (a) Source image Gi(x, y), (b) Iimage transformed to coordinate system (x', y'), (c) Iimage projected to x'

Our projection transform compute the image's linear integral at certain orientation for each x'. Considering the digital image, it is equal to compute the gray summation of certain orientation for each rotated column. As shown in Fig3(c), the coordinate of x' hold good representation of position of lines, and value of $R_\theta(x')$ can also represent the length of certain line at x'. After this transformation, G_i will be transformed to $R_0(x')$, $R_{45}(x')$, $R_{90}(x')$ and $R_{135}(x')$. They are four one-dimensional vectors of different length, which can be stored as Vpi, i=0, 1, 2, 3.

For having a visualized impression of the classify capacity these vectors attributed to, Fig 4 (a) has plotted Vp0, which from 20 samples of 2 classes (10 per class), in same figure. It is easily to see that 20 curves congregated into 2 bunches (10 per bunch). This is exactly corresponding to the fact that 20 samples are from 2 classes. Obviously, Vpi have powerful capacity of classification. However, we must notice that Vpi contains overfull details, which are disadvantage factors to the processing of feature matching. To solve that problem, we use DWT (known as Discrete Wavelet Transform) to do twice decomposition with Vpi. In the processing, wavelets Daubechies-9 is employed. Four transformed vectors obtained from this operation can be stored into DWTVpi as the features of a palmprint. We also plotted DWTVp0 in Fig 4(b) as a comparison. As you seen, transformed vectors DWTVpi have lesser details but better classification capacity than Vpi.

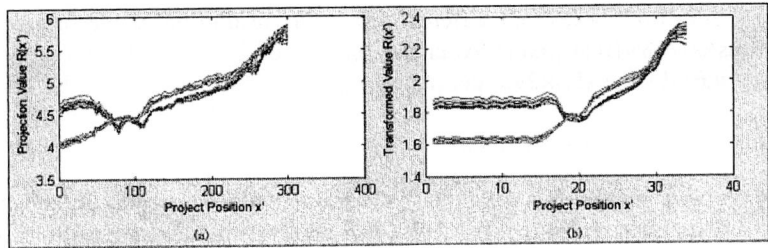

Fig. 4. Classification capacity visualization, (a) Vp0, (b) DWTVp0

4 Feature Matching and Information Fusion

Euclidian Square Norm(distance score) is a simple and powerful method to measure the dissimilarity of two vectors (as in [1]). However, it dose not consider the inner distribution density of a class. In our approach, we supposed that each class distributed in eigenspace as super sphere. By using appropriate training set, we can get the diameter (as D_k) and centroid(as template vector VC_k) of a super sphere of class k. To get the similarity of a test vector data(as VT) and claimed class k, we firstly compute the Euclidian distance between test vector and the claimed class' centroid(VC_k), then use computed Euclidian distance divide the 10* D_k, the result of division will indicate the dissimilarity of test vector data and certain class k. This processing can be formulated as:

$$S_k = 1 - DS_k = 1 - \|VT - VC_k\| / 10 D_k . \qquad (4)$$

where DS_k denote the dissimilarity of test vector data and claimed class k and S_k denoted the similarity of test vector data and claimed class k. In the processing of computing, negative value will occur when the distance between VT and VCk. This situation means VT and VCk are utterly dissimilar. We can simply set the negative Sk to zero. This formulation has considered both distance and inner distribution density. Apply it to palmprint feature vectors (DWTVpi), we can get 4 dissimilarities of palmprint(as ∂_i). To get a final similarity score between test data and claimed class, we must do some information fusion processing. In fusion of 4 palmprint dissimilarities, considering the weightiness of different orientation(e.g. in left hand, most of palm lines tend to 135°, so DWTVp3 has higher weightiness than other orientations), each orientation feature vector will be given a weigh value w_i. A final similarity score will be compute as:

$$\partial = \sum_{i=0}^{3} w_i \partial_i , \quad \sum_{i=0}^{3} k_i = 1 . \qquad (5)$$

5 Experimental Results

In our experiment, the scanner reformed by ourselves [4] is employed. To establish a experiment images database, we have collected 516 images of palmprints from 51 individuals with both sexes and different ages. The resolution of the original palmprint images is 768×1024 pixels. After use method of 2.1 to do direction alignment, we also get a ROI (size 290×290) per image.

5.1 Results of Features Extraction

By using method described in Section 3, we get Vpi of each palmprint and then use wavelet Coiflets5 transform them to DWTVpi with 5 times decomposition. Because the data is too complex, we plotted them into Fig 5 to analyze the accuracy of acquiring palmprint features.

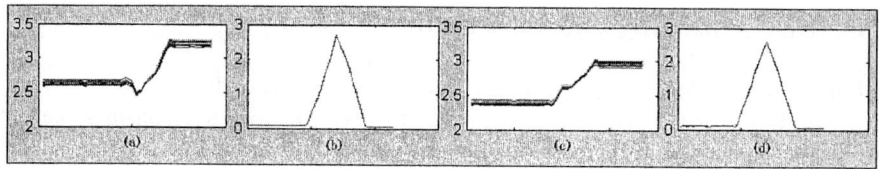

Fig. 5. Feature extraction of palmprint: (From 10 samples of same class) (a) DWTVp0, (b) DWTVp1,(c) DWTVp2, (d) DWTVp3

As shown in Fig 5, each curves overlapped well into 4 bunch of curves. It indicated that the stability of extracting each features can be perfectly controlled.

5.2 Test Classification Capacity of Different Wavelet

To get a perfect palmprint features after DWT, different resolution with different wavelet and different decomposition times will generate different result. We define max accuracy rate of palmprint verification based on certain wavelet and certain decomposition times to evaluate the classification capacity of a certain solution. In our experiment, 21 wavelets of Daubechies 1~9(D1~D9), Symlets 2~8(S2~S8) and Coiflets 1~5(C1~C5) were tested. Part of results were shown in Table 1 (WN=Wave Name, DT= Decomposition Times, MA=Max Accuracy Rate). Obviously, Coiflets 5 with 5 decompositions will be a better solution.

6 Verification and Identification

To observe and evaluate the performing efficiency of our approach, three palmprint recognition systems were implemented in our experiment. They included (1) recognition system based on the approaches of this paper, (2) recognition system based on the approaches in [5](Fourier Transform based approach, as a delegation of Global Features based system) and (3) recognition system based on the approaches in [8](lines matching based approach, as a delegation of Local Features based system). The experimental results are shown as follows.

Table 1. Test of classification capacity of different wavelet

WN	DT	MAR	WN	DT	MAR	WN	DT	MAR
C5	1	93.05	D1	1	93.19	S3	5	97.98
C5	2	95.44	D1	2	95.39	S3	6	97.93
C5	3	97.18	D1	3	97.09	S3	7	96.95
C5	4	97.13	D1	4	97.37	S3	8	95.82
C5	*5*	*98.36*	D9	3	97.51	S4	3	97.42
C5	6	98.07	D9	4	97.46	S4	4	97.37
C5	7	97.13	D9	5	97.98	S4	5	98.12
C5	8	96.10	D9	6	98.07	S4	6	98.07
C5	9	95.02	D9	7	96.48	C3	2	95.49
C5	10	95.35	D9	8	95.11	C3	3	97.5

6.1 Verification of One to One

To get the FRR (False Reject Rate) and FAR (False Accept Rate) of each system, 133386(516×515/2+516) times verification will be executed on each of them. We plotted their ROC (Receiver Operation Characteristic) curves in Fig 6a.

From the figure, it is seemed that our approaches are performed better than the method in both [5] and [8]. To get a more accurate comparison, Table 2(T=Threshold, E=Error rate, A=Accuracy rate) shows detailed digital comparative data stand on three different views: Equal Error Rate view, Total Minimum Error view and Max Accuracy Rate view (defined in 5.2).

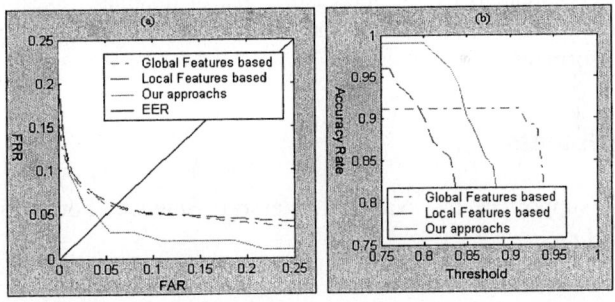

Fig. 6. ROC and Accuracy, (a) ROC of each system, (b) Accuracy of each verification systems

6.2 Identification of One to Many

We also used system (1)(2)(3) to do 40320(516×51) times identification test. The threshold of similarity from 0 to 1, which can determine acceptation or rejection, had be tested in each system and record the accuracy of classification at each threshold. The result was shown in Fig 6b. The accurate rate of (1) is 99.12%, (2) is 96.74%, (3) is 92.18%. It indicated that combination features had more advantages than Global Features and Local Features.

Table 2. Compare verification systems on different views

System	EER			Total Minimum Error				MAR
	T	E	A	T	FAR	FRR	A	
(1)	0.760	0.022	98.13	0.810	0.053	0.004	98.36	98.36
(2)	0.840	0.024	96.47	0.900	0.000	0.043	96.24	96.86
(3)	0.780	0.044	96.03	0.780	0.028	0.053	95.98	96.14

7 Conclusion and Future Work

In this paper, we have presented an approach for biometrics identification. In our approach, a method direction alignment based on morphology was firstly executed. This alignment method was simpler and faster than other methods. Then we take advantage of the key points captured in the alignment processing to get the ROI. With the ROI, a new method of palmprint feature extraction has been proposed. The results palmpritn features have powerful capacity of classification. In features matching, we presented a new method of similarity measurement, which has considered the inner distribution density of each class. Next, based on these methods, a verification and an identification experiment were performed. Experimental results indicated that our approach have perfect capability and our system performed best in three systems based on Global Features, Local Features and Combination Features.

However, these results may be biased by the small size of the data set which we have used. If more subjects are added, one should expect some overlap due to similar palms of different persons. To investigate this matter further, additional data will be collected. But accompanied by the increasing size of database, execution speed will be paid more attention. Some methods based on decision tree should be attempted to optimize the performance.

Acknowledgements

This work is supported by National Natural Science Foundation of China (NSFC.60162001).

References

1. Arun Ross, Anil Jain: Information Fusion in Biometrics. Pattern Recognition Letters. 24 (2003) 2115 - 2125
2. Ajay Kumar, David, C.M Wong, Helen, C.Shen, Anil Jain: Personal Verification Using Palmprint and Hand Geometry Biometric. AVBPA (2003) 668 - 678
3. Chin-Chuan Han, Hsu-Liang Cheng, Chih-lung Lin, Kuo-Chin Fan: Personal Authentication Using Palmprint Features. Pattern Recognition. 36 (2003) 371 - 381
4. Wei Xiao-yong, Xu Dan, Yuan Guo-wu: Research of Preprocessing in Palmprint Identification. The 13th National Conference on Multimedia Technology (NCMT). Ningbo. China. (Oct. 2004) 155 - 164

5. Zhang, D., Shu, W.: Two Novel Characteristics in Palmprint Verification: Datum Invariance and Line Feature Matching. Pattern Recognition. Vol. 32. No. 4. (1999) 691 - 702
6. Duta, N., Jain, A.K., Mardia, Kanti V.: Matching of Palmprint. Pattern Recognition Letters. Vol. 23. No. 4. (Feb. 2002) 477 - 485
7. Jain, A., Bolle, R., Pankanti, S., (eds.): Biometrics: Personal Identification in Networked Society. Boston: Kluwer Academic. (1999)
8. Li, W., Zhang, D., Xu, Z.: Palmprint Identification by Fourier Transform. International Journal Pattern. Recognition and Artificial Intelligence. Vol. 16. No. 4. (2002) 417 - 432
9. Shu, W., Rong, G., Bian, Z.Q., Zhang, D.: Automatic Palmprint Verification. International Journal of Image and Graphics. Vol. 1. No. 1. (2001)
10. You, J., Li, W., Zhang, D.: Hierarchical Palmprint Identification via Multiple Feature Extraction. Pattern Recognition. Vol. 35. (2002) 847 - 859
11. Lu, G.M., Zhang, D., Wang, K.Q.: Palmprint Recognition Using Eigenpalms Features. Pattern Recognition Letters. Vol. 24. (2003) 1463 - 1467
12. Kumar, A., Shen, H.C.: Recognition of Palmprints Using Wavelet-Based Features. Proc. Intl. Conf. Sys. Cybern. SCI-2002. Orlando. Florida. (Jul. 2002)
13. Kong, W.K., Zhang, D.: Palmprint Texture Analysis Based on Low-Resolution Images for Personal Authentication. Proc. ICPR-2002. Quebec City. Canada
14. Kong, W.K., Zhang, D., Li, W.X.: Palmprint Feature Extraction Using 2-D Gabor Filters. Pattern Recognition. Vol. 36. (2003) 2339 - 2347
15. Han, C.C., Cheng, H.L., Lin, C.L., Fan, K.C.: Personal Authentication Using Palm-Print Features. Pattern Recognition. Vol. 36. (2003) 371 - 381

A Novel Dimension Conversion for the Quantization of SEW in Wideband WI Speech Coding

Kyung Jin Byun[1], Ik Soo Eo[1], He Bum Jeong[1], and Minsoo Hahn[2]

[1] Electronics and Telecommunications Research Institute,
161 Gajeong-Dong Yuseong-Gu, Daejeon, Korea, 305-350
kjbyun@etri.re.kr
[2] Information and Communications University (ICU),
103-6 Moonji-Dong Yuseong-Gu, Daejeon, Korea, 305-714
mshahn@icu.ac.kr

Abstract. The waveform interpolation is one of the speech coding algorithms with high quality at low bit rates. In the WI coding, the vector quantization of SEW requires a variable dimension quantization technique since the dimension of the SEW amplitude spectrum varies depending on the pitch period. However, since the variable dimension vector makes a difficulty to employ conventional vector quantization techniques directly, some dimension conversion techniques are usually utilized for the quantization of the variable dimension vectors. In this paper, we propose a new dimension conversion method for the SEW quantization in order to reduce the cost of codebook storage space with a small conversion error in the wideband WI speech coding. This dimension conversion method would be more useful for the wideband speech because wideband speech requires larger codebook memory for the variable dimension vector quantization compared to narrowband speech.

1 Introduction

In recent years, various speech coding algorithms have been widely used in many applications such as mobile communication systems and digital storage systems for the speech signal to be represented in lower bit rates while maintaining its quality. On the one hand, the Code Excited Linear Predictive (CELP) algorithm has been known as the one of the best coding algorithms for bit rate between 8 kbps and 16 kbps. Especially, the Algebraic CELP (ACELP) algorithm has been widely adopted due to its outstanding performance for many standard speech coders such as G.729, Enhanced Variable Rate Coding (EVRC), and Adaptive Multi-Rate (AMR) coder. However, the speech quality of these CELP coders degrades rapidly at rate below 4 kbps.

On the other hand, waveform interpolation (WI) coder classified into parametric one is able to produce good quality speech even at the below 4 kbps rates. Most speech coders including standard coders mentioned above operate on narrow bandwidth limited to 200 - 3400 Hz. Similarly, most researches in the WI coder have also been focused on narrow band speech so far. However, as mobile systems are evolving from speech-dominated services to multimedia ones, the advent of the wideband coder becomes highly desirable because it is able to provide higher quality speech.

The wideband speech coder extends the audio bandwidth to 50 - 7000 Hz in order to achieve the high quality both in the sense of speech intelligibility and naturalness.

Waveform interpolation is one of the speech coding algorithms with high perceptual quality at low bit rates [1]. Most literatures for the WI coding have been concentrated on the narrow band speech coding, but recent researches in [2] and [3] show the potential of applying a WI algorithm to wideband speech. Moreover, the WI coding has an advantage in the view of embedded coding techniques. An embedded coding structure usually consists of a base-layer and an enhancement-layer. Therefore, an embedded coding structure usually requires an additional functional unit to implement the enhancement-layer [4]. However, since the WI coding uses an open-loop encoding method, the spectral and the excitation information are not tightly correlated. Therefore, the bit-rates can be easily controlled by changing the number of assigned bits and the transmission interval of excitation signals that consist of a slowly evolving waveform and a rapidly evolving waveform [5]. Consequently, by using WI coding scheme, a high-quality embedded system working at low bit-rates can be easily implemented without any additional functional unit.

In the WI coding, there are four parameters to be transmitted. They are the Linear Prediction (LP) parameter, the pitch value, the power and the characteristic waveform (CW). The CW parameter is decomposed into a slowly evolving waveform (SEW) and a rapidly evolving waveform (REW). Since the SEW and REW have very distinctive requirements, they should be quantized separately to enhance the coding efficiency. Among these parameters, the SEW is perceptually important and has a strong influence on the quality of the reconstructed speech [1]. In addition, the vector quantization of SEW requires a variable dimension quantization technique since the dimension of the SEW amplitude spectrum varies depending on the pitch period. However, this variable dimension vector has a difficulty to adopt conventional vector quantization techniques directly. One of the useful quantization methods for the variable dimension vectors is to utilize some dimension conversion techniques. By exploiting the dimension conversion technique, the quantization of SEW magnitude can be performed using a conventional vector quantization. Several dimension conversion vector quantization (DCVQ) techniques have been reported in the literature [6]-[9].

In this paper, we describe an investigation into the dimension conversion for the SEW quantization in order to reduce the cost of storage space of the codebook with a small conversion error in WI coding of wideband speech. Firstly, in section 2, we describe the overview of WI speech coding algorithm. In section 3, we present the efficient method of dimension conversion for the SEW magnitude vector quantization with a small size of codebook memory. Then, the experimental results are discussed in section 4. Finally, conclusions are made in section 5.

2 Overview of WI Speech Coder

The waveform interpolation coding has been extensively and steadily developed since it was first introduced by Kleijn [1]. The encoder block diagram of the WI speech coder is shown in figure 1.

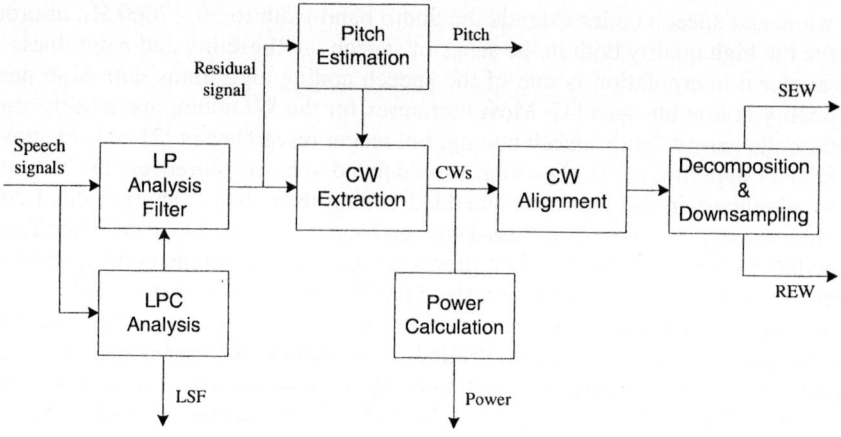

Fig. 1. Encoder block diagram of WI speech coder

The WI coder firstly performs LP analysis once per frame for the input speech. The LP parameter set is converted into the LSF for the efficient quantization and generally vector quantized using various quantization techniques. The pitch estimation is performed in the linear prediction residual domain. In the WI paradigm, the accuracy of this pitch estimator is very crucial to the performance of the coder. There are a variety of pitch estimation techniques available. Some of them are based on locating the dominant peak in each pitch cycle whereas others are based on finding the delay which gives the maximum autocorrelation or prediction gain for a frame of samples [10], [11]. After the pitch is estimated, WI coder extracts pitch-cycle waveforms which are known as CWs from the residual signal at a constant rate. These CWs are used to form a two-dimensional waveform which evolves on a pitch synchronous nature. The CWs are usually represented using the Discrete Time Fourier Series (DTFS) as follows:

$$u(n,\phi) = \sum_{k=1}^{\lfloor P(n)/2 \rfloor} [A_k(n)\cos(k\phi) + B_k(n)\sin(k\phi)] \quad 0 \leq \phi(\bullet) < 2\pi \quad (1)$$

where $\phi = \phi(m) = 2\pi n / P(n)$, A_k and B_k are the DTFS coefficients, and $P(n)$ is the pitch value. The extraction procedure performed in the residual domain provides a DTFS description for every extracted CW. Since these CWs are generally not in synchronized phase, i.e., the main features in the CWs are not time-aligned, the smoothness of the surface in the time direction should be maximized. This can be accomplished by aligning the extracted CW with the previously extracted CW by introducing a circular time shift to the current one. Since the DTFS description of the CW enables to regard the CW as a single cycle of a periodic signal, the circular time shift is indeed equivalent to adding a linear phase to the DTFS coefficients. The CWs are then normalized by their power, which is quantized separately. The main motivation of this normalization is to separate the power and the shape in CWs so that they can be quantized separately to achieve higher coding efficiency.

This two-dimensional surface then is decomposed into two independent components, i.e., SEWs and REWs, via low pass filtering prior to quantization. The SEW and the REW are down sampled and quantized separately. The SEW component represents mostly the periodic (voiced) component while the REW corresponds mainly to the noise-like (unvoiced) component of the speech signal. Because these components have different perceptual properties, they can be exploited to increase coding efficiency in the compression. In other words, the SEW component requires only a low update rate but has to be described with a reasonably high accuracy whereas the REW component requires a higher transmission rate but even a rough description is perceptually accurate. This property of the CW suggests that low pass filtering of the CW surface leads to a slowly evolving waveform. The rapidly evolving part of the signal can be obtained by simply subtracting the corresponding SEW from the CW as following equation:

$$u_{REW}(n,\phi) = u_{CW}(n,\phi) - u_{SEW}(n,\phi). \qquad (2)$$

In the decoder side, the received parameters are the LP coefficients, the pitch value, the power of the CW, the SEW and REW magnitude spectrum. The decoder can obtain a continuous CW surface by interpolating the successive SEW and REW and then recombining them. After performing the power de-normalization and subsequent realignment, the two-dimensional CW surfaces are converted back into the one-dimensional residual signal using a CW and a pitch value at every sample point obtained by linear interpolation. This conversion process also requires the phase track estimated from the pitch value at each sample point. The reconstructed one-dimensional residual signal is used to excite the linear predictive synthesis filter to obtain the final output speech signal.

3 Dimension Conversion of the SEW Magnitude Spectrum

Since the characteristic waveforms are extracted from LP residual signal at pitch-cycle rate, their length varies according to the pitch period. After the CWs are extracted and aligned, their powers are then normalized to separate the power and the shape in CWs. The CWs are then decomposed into two components of the SEW and the REW. Consequently, the dimension of the SEW amplitude spectrum varies depending on the pitch period. This variable dimension vector causes a problem in quantization because it gives a great difficulty to employ conventional vector quantization techniques directly.

3.1 Dimension Conversion

Theoretically, the optimal solution for the vector quantization of variable dimension vectors is to adopt a separate codebook for each different vector dimension. However, this approach is quite impractical because it requires enormous memory space due to a number of separate codebooks. One of the most practical quantization methods for the variable dimension vectors is to utilize some dimension conversion technique as shown in figure 2.

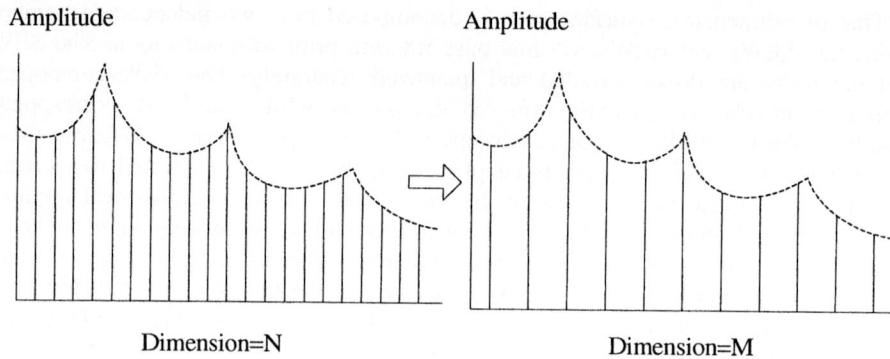

Fig. 2. Illustration of Dimension conversion

At the encoder side, the variable dimension vector is converted into a fixed dimension vector and then the fixed-dimension vector can be quantized by conventional quantization method. These procedures are reversed at the decoder in which the fixed dimension vector is converted into the original dimension vector using the transmitted pitch information from the encoder.

3.2 A New Dimension Conversion of SEW Magnitude

In this section, we describe a new dimension conversion method for the SEW magnitude spectrum quantization. Each of the CWs extracted from the LP residual represents a pitch-cycle and thus their length varies depending on the pitch period P(t). When the waveforms are converted into frequency domain for efficient quantization, the most compact representation contains frequency-domain samples at multiples of the pitch frequency. This results in spectral vectors with the variable dimension

$$M(t) = \left\lfloor \frac{P(t)}{2} \right\rfloor. \tag{3}$$

For the narrowband speech, the pitch period P can vary from 20 (2.5 msec) to 148 (18.5 msec) and thus possible value of M, the number of harmonics, covers the range from 10 to 74. However, for the wideband speech, M is doubled, i.e., the pitch period can vary from 40 to 296 resulting in 20 to 148 harmonics. As a result, the dimension of codebook becomes twice larger than that of narrowband speech. Thus, the storage space of the codebook becomes a more important issue for the wideband speech. In our implementation, for the wideband speech, the pitch period is allowed to vary from 40 to 256 resulting in 20 to 128 harmonics.

We designed three dimension-conversion schemes for the quantization as given in table 1. The first one, we call it 1_CB, has a single codebook with maximum fixed dimension. The second one, 2_CB, has two codebooks with two different fixed dimensions. The last one proposed in this paper, 1_CB_2, has a codebook with a fixed dimension but a different resolution according to frequency range. For the first scheme, all vectors are converted to a certain dimension of N with a single codebook.

For the second scheme, all vectors of dimension smaller than or equal to N are converted into dimension N, all vectors of dimension N+1 to 128 are converted into dimension 128 with two codebooks of dimension N and dimension 128, respectively. For the last scheme, the components of a vector in the lower sub-band less than 1000 Hz are converted into the maximum dimension of 16, and the other components in the higher sub-band are converted to a certain dimension of N-16. The second scheme is a kind of the multi-codebook vector quantization (MCVQ). It is clear that if we would have continued to separate the dimension vector, it would lead to the optimal solution whereas require huge memory space for a great number of codebooks.

Table 1. Three dimension conversion schemes and their dimensions

Schemes	Variable dimension		Fixed dimension	
1_CB	20 ~ 128		N	
2_CB	$P \leq 2N$:	20 ~ N	N	
	$P > 2N$:	N+1 ~ 128	128	
1_CB_2	Low band	High band	Low band	High band
	3 ~ 16	17 ~ 112	16	N-16

Figure 3 shows the spectral distortion between the original SEW and the dimension-converted SEW spectrum of the first scheme for various dimension. This figure gives us the insight that the larger, the dimension, the better, the codebook quality. In other words, the conversion error can be reduced if the variable dimension vector would be converted into a larger fixed dimension vector. However, a larger dimension vector requires more codebook memory. In addition, it has been known that the low frequency band of the SEW magnitude is more perceptually significant than the high frequency band [1]. Our experimental results given in table 2 also show that the spectral distortion between the original and the converted vector at below 1000 Hz is almost equal to that of higher frequency region (1000 to 8000 Hz). The dimension conversion error in this table was measured by using the spectral distortion measure (SD) as shown in the next section.

Table 2. Dimension conversion error of 1_CB scheme for each sub-band

Dimension	Lower sub-band SD (dB)	Higher sub-band SD (dB)	Overall SD (dB)
50	1.160	1.209	1.233
60	1.064	1.091	1.119
70	0.891	0.948	0.968
80	0.790	0.852	0.872
90	0.711	0.796	0.811
100	0.687	0.731	0.749

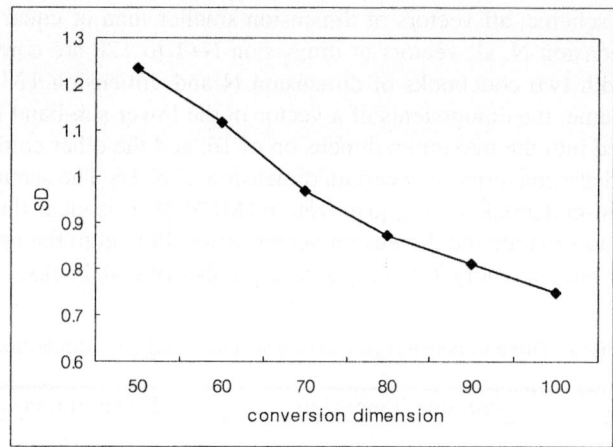

Fig. 3. Spectral distortion of the first scheme with dimension conversion

Figure 4 illustrates the SEW magnitude spectrums. Generally, the energy in the lower frequency band less than 1000 Hz is much larger than that of the higher frequency band as shown in figure 4. In this figure, SEW dimension is 36 and the subband region less than 1000 Hz includes only about five harmonics of the SEW. The converted-SEW_80 in this figure means original SEW which has dimension of 36 is converted into the dimension of 80 and converted back to the original dimension vector. The converted-SEW_128 can also be understood in the same context. As we can see in this figure, the converted-SEW_80 has a larger conversion error than that of converted-SEW_128, and it is also found that the conversion error in lower subband region is much more dominant compared to that of the higher sub-band region.

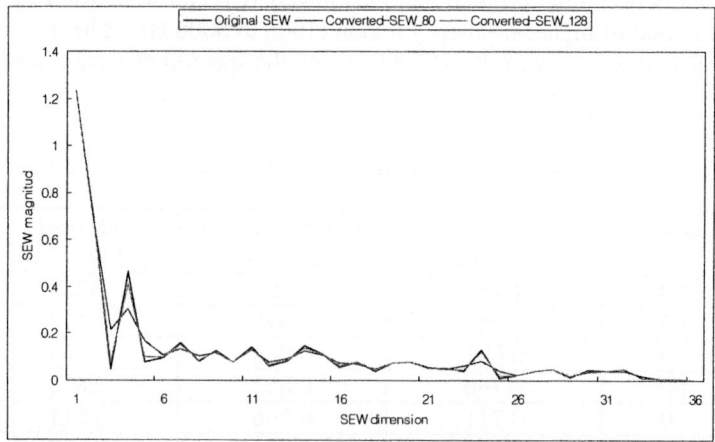

Fig. 4. Comparison between the original and the converted SEW magnitude spectrum

Based on the above mentioned reason, we propose a new dimension conversion scheme having a single codebook with a small conversion error and a small codebook memory. In the proposed conversion method, the variable dimension vector in the low frequency section is converted into the maximum dimension to get better quality and the variable dimension vector in the high frequency section is converted into a fixed dimension less than 128 to reduce codebook memory. The SEW magnitude spectrum is usually split into some sub-bands. The sub-bands are quantized separately and the base-band is quantized using more bits than other higher sub-bands. Such bit-allocation scheme is to accommodate the better resolving capability of the human ear for lower frequencies. In our implementation, the SEW magnitude spectrum is split into three non-overlapping sub-band; the low frequency section is 0 to 1000 Hz and the high frequency section are 1000 to 4000 Hz and 4000 to 8000 Hz. The bit-allocation for these three sub-bands is 8, 6 and 5 bits.

However, for the dimension conversion the band is split into two sub-bands as mentioned above. Thus, the number of the dimension vectors in the low frequency and the high frequency band can vary from 3 to 16 and from 17 to 112, respectively. Consequently, when the variable dimension vectors are converted into the fixed dimension vectors, all vectors involved within 1000 Hz are converted into the dimension of 16 and all vectors in the high frequency section (1000 to 8000 Hz) are converted into the dimension of N-16. Overall spectral distortion according to the variation of N and the amount of codebook memories will be shown in the next section as an example of dimension conversion schemes of this paper.

4 Experimental Results

For the objective comparison, we adopted the spectral distortion measure SD between the original spectral vector and the converted spectral one defined as follows:

$$SD = \sqrt{\frac{1}{L-1}\sum_{k=1}^{L-1}\left(20\log_{10}S(k) - 20\log_{10}\hat{S}(k)\right)^2} \qquad (4)$$

where SD in dB units, and L-1 is the number of spectral samples in the interested range. Figure 5 illustrate the performance comparison between the proposed and the conventional dimension conversion method.

In figure 5, 1_CB, 2_CB, and 1_CB_2 are the conversion schemes as explained in section 3.2. As shown in figure 5, although the proposed conversion method, 1_CB_2, uses a single codebook, the spectral distortion of the conversion error is smaller than that of the second scheme with two codebooks. It can be also easily expected from the explanation in section 3 that the conversion error for the second scheme, 2_CB, is smaller than that of the first scheme as shown in figure 5.

Moreover, because our method uses the conversion dimension which is less than the maximum dimension of 128 for the high frequency region, the memory requirement of the codebook can be reduced. Especially, this advantage is much more important in the wideband WI speech coding, because the codebook memory required for the SEW quantization in the wideband speech becomes twice larger than that of the narrow-band speech. Table 3 shows that the codebook memory required for an implementation of three different conversion schemes. As shown in this table, 1_CB_2 with fixed dimension of 80 is able to reduce the codebook memory about 20 % compared to 1_CB with fixed dimension of 128 while the performance of the 1_CB_2 is slightly less than that of the 1_CB for the case of table 3.

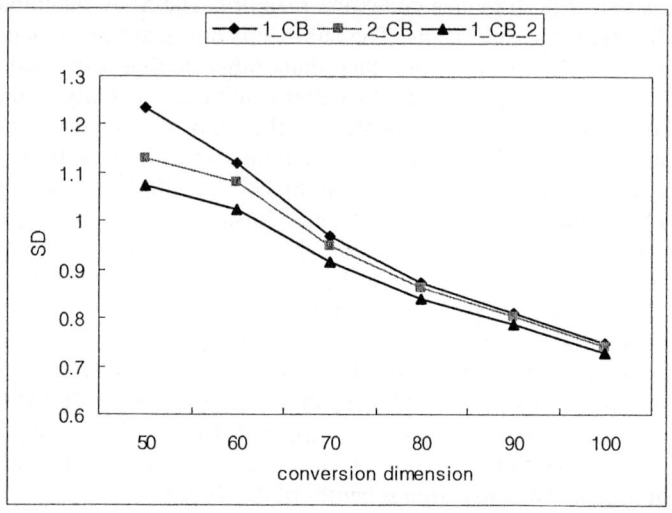

Fig. 5. SD comparison between proposed and convectional conversion method

Table 3. The amount of codebook memories for three different dimension conversion schemes

Schemes	Codebook memory for sub-bands			Total
1_CB	16x256	48x64	64x32	9,184 words
2_CB	10x256	30x64	40x32	14,944 words
	16x256	48x64	64x32	
1_CB_2	16x256	30x64	40x32	7,296 words

5 Conclusion

This paper presents a new dimension conversion method for the SEW magnitude quantization in wideband WI speech coding. Because the proposed conversion method uses a single codebook with different accuracies for the low and the high frequency region, it is possible to implement a vector quantization with a small size of codebook memory and a reasonably small conversion error. In the view of the memory space requirement, the proposed conversion method has an advantage of reducing the codebook memory about 20 % through an appropriate trade-off. Therefore, the proposed conversion method would be more useful for the wideband speech since the wideband speech requires larger codebook memory for the variable dimension SEW vector quantization compared to narrowband speech.

References

1. W. B. Kleijn, L., J. Haagen: Waveform interpolation for coding and synthesis. In Speech Coding and Synthesis, Elsevier Science B. V. (1995) 175–207
2. C. H. Ritz, I. S. Burnett, J. Lukasiak: Extending waveform interpolation to wideband speech coding. Proc. IEEE workshop on Speech Coding (2002) 32–34
3. C. H. Ritz, I. S. Burnett, J. Lukasiak: Low bit rate wideband WI speech coding. Proc. IEEE International Conference on Acoustics, Speech, and Signal Processing (2003) 804-807
4. K. J. Byun, I. S. Eo, H. B. Jeong, M. Hahn: An Embedded ACELP Speech Coding Based on the AMR-WB Codec. ETRI Journal, vol. 27, no. 2 (2005) 231-234
5. Hong-Goo Kang and D. Sen: Embedded WI coding between 2.0 and 4.8 kbit/s. Proc. IEEE workshop on Speech Coding (1999) 87–89
6. P. Lupini and V. Cuperman: Vector quantization of harmonic magnitudes for low rate speech coders. Proc. IEEE Globecom Conference (1994) 858-862
7. M. Nishiguchi, J. Mutsumoto, R. Wakatsuki, S. Ono: Vector quantized MBE with simplified V/UV decision at 3.0 kbps. Proc. IEEE International Conference on Acoustics, Speech, and Signal Processing (1993) 151-154
8. M. Nishiguchi, A. Inoue, Y. Maeda, J. Matsumoto: Parametric speech coding-HVXC at 2.0-4.0 kbps. Proc. IEEE workshop on Speech Coding (1999) 84–86
9. E. Shlomot, V. Cuperman, A. Gersho: Hybrid coding: combined harmonic and waveform coding of speech at 4 kb/s. IEEE Trans. Speech, and Audio Processing, vol. 9, no. 6 (2001) 632-646
10. K. J. Byun, S. Jeong, H. Kim and M. Hahn: Noise Whitening-Based Pitch Detection for Speech Highly Corrupted by Colored Noise. ETRI Journal, vol. 25, no. 1 (2003) 49-51
11. L. R. Rabiner, R. W. Shafer: Digital Processing of Speech Signals, Prentice-Hall (1978) 141-161

Adaptive Preprocessing Scheme Using Rank for Lossless Indexed Image Compression

Kang-Soo You[1], Tae-Yoon Park[2], Euee S. Jang[3], and Hoon-Sung Kwak[1]

[1] Dept. of Image Engineering,
Chonbuk National University, Jeonju 561-756, Korea
{mickey, hskwak}@chonbuk.ac.kr
[2] Dept. of Electrical & Computer Engineering,
Virginia Tech, Blacksburg VA 24061, USA
typark@vt.edu
[3] College of Information & Communication,
Hanyang University, Seoul 133-791, Korea
esjang@hanyang.ac.kr

Abstract. This paper proposes a brand-new preprocessing scheme using the ranking of co-occurrence count about indices in neighboring pixels. Original indices in an index image are substituted by their ranks. Arithmetic coding, then, is followed. Using this proposed algorithm, a better compression efficiency can be expected with higher data redundancy because the indices of the most pixels are concentrated to the relatively few rank numbers. Experimental results show that the proposed algorithm achieves a better compression performance up to 26–48% over GIF, arithmetic coding and Zeng's scheme.

1 Introduction

Highly compressed palette-based images are needed in many applications such as characters and logos produced by computer graphics and World Wide Web (WWW) online services. The index (or palette) image has been adopted by GIF (Graphic Interchange Format) which uses Lempel-Ziv algorithm as form of lossless compression[1,2].

It has been recognized that the index color image can be reindexed without any loss in order to compress palette-based images more efficiently[3,4]. It turns out that some reindexing scheme tends to produce more compressible index image than others. Palette reordering is a class of preprocessing methods aiming at finding a permutation of the color palette such that the resulting image of indices is more amenable for compression[5,6].

The particular scheme reindexing palette-based images was proposed by Zeng et al. [5] in 2000. It is based on the one-step look-ahead greedy approach, which aims at increasing the efficiency of lossless compression. The method reindexes one symbol at a time in a greedy approach. Each re-assignment is optimized based on the statistics collected from the initial index image and previously executed re-assignments. In general, the larger difference between the neighboring index values, the more bits it costs, so the goal of the method is to come up with a reindexing scheme that tends to reduce the overall difference of index values of adjacent pixels. The reindexed image tends to be smoother than the initial indexed image, thus is more amenable to most lossless compression algorithms[5,6].

In this paper, we propose a brand-new palette reordering method whose name is RIAC (*Rank Indexing with Arithmetic Coding*) to reindex palette-based image using the ranking of co-occurrence frequency for increasing the lossless compression efficiency of color index images. The proposed method is possible to compress index images efficiently, because the redundancy of the image data can be increased by ranks.

Although it is expected that the performance of compression will be improved for common sized images, the compression ratio is decreased for small sized images, because it needs to transmit additional data related to the ranked image. To make up for the disadvantage, we finally present another method enhancing the RIAC which is termed ARIAC (*Adaptive Rank Indexing with Arithmetic Coding*). The improved method can compress and decompress without additional data. Thus, it is expected to improve compression performance in relation to not only general sized images but also small sized images.

The remainder of this paper is organized as follows. In Section 2, we present the RIAC method in detail for the purpose of improving compression performance related to the palette-based images' reindexing using the way to replace index values with the ranks. In Section 3, we proposed the improved ARIAC method and described detailed steps about how to operate it. Section 4 provides the experimental results comparing the GIF, the plain arithmetic coding and previous reindexing method of Zeng *et al.* [?] in terms of compression improvement. Finally, in Section 5, we drew the conclusions.

2 Rank Indexing with Arithmetic Coding (RIAC) Method

Let us suppose we have an $n \times m$ indexed image and the set of index values $\mathbf{I} = \{I_1, I_2, \ldots, I_M\}$ corresponds to the set of color symbols denoted with $\mathbf{S} = \{S_1, S_2, \ldots, S_M\}$. Fig. 1 shows an example of an image with 4 colors. Let $\mathbf{p} = (p_1, p_2, \ldots, p_l), l = n \times m$, $\mathbf{p}$ denotes the sequence of indices (pixels), where p_0 is the index value of virtual pixel which is located in the left side of first pixel and p_i is the index value of the i-th pixel of the input image in row by row scanning order. $(1 \leq i \leq l)$

The algorithm is started by collecting the co-occurrence statistics of $\mathbf{p}$. The co-occurrence statistics are stored in a suitable matrix $\mathbf{C}$, named *Co-occurrence Count Matrix* (CCM). More formally the entry $c_{i,j}$ of $\mathbf{C}$ reports the number of successive occurrence of index values pair, (I_i, I_j), observed over $\mathbf{p}$. For the sample image in Fig. 1, the sequence of indices in row order is $\mathbf{p} = (0, 3, 2, 0,$

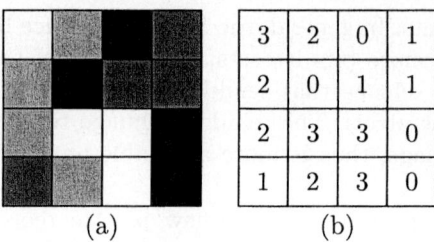

Fig. 1. An example of a 4 × 4 sample image with 4 colors; (a) shows original color image; (b) shows an index image using color palette

1, 2, 0, 1, 1, 2, 3, 3, 0, 1, 2, 3, 0), where the first element of the sequence, p_0 has value selected arbitrarily.

Fig. 2 (a) shows CCM (co-occurrence count matrix) when $p_0 = 0$, p_0 is equal to the first index value in the palette. After the construction of **C**, the algorithm ranks the co-occurrence counts of **C**. These values are stored in a suitable matrix **R**, named *Co-occurrence Count Rank Matrix* (CRM). The entry $r(i,j)$ of **R** represents the rank of $c(i,j)$ in decremental order over the co-occurrence counts sequence $c(i,j) = \{c(i,k)|0 \leq k \leq M\}$. Fig. 2 (b) shows CRM. Note that **R** from a **C** may vary according to the ranking rule. When two or more co-occurrences in a row of **C** have a same count value, the ranking rule gives a higher order to the co-occurrence (I_i, I_j) that i) has lower value of destination index value i, than others, ii) has closer distance diagonally from a chosen pixel's index value, and so on. Referencing **R**, the sequence of ranked images **p**′ gets generated by substituting each index value of the original image sequence **p** with the co-occurrence frequency rank value. The algorithm applies the mapping $p'_i \mapsto r_{p_{i-1}, p_i}$ to all image pixels. Fig. 2 (c) shows the ranked image of the sample input index image in Fig. 1.

The ranked image is compressed using the arithmetic coding. The reason that the arithmetic coding is chosen is because it shows a better compression rate than

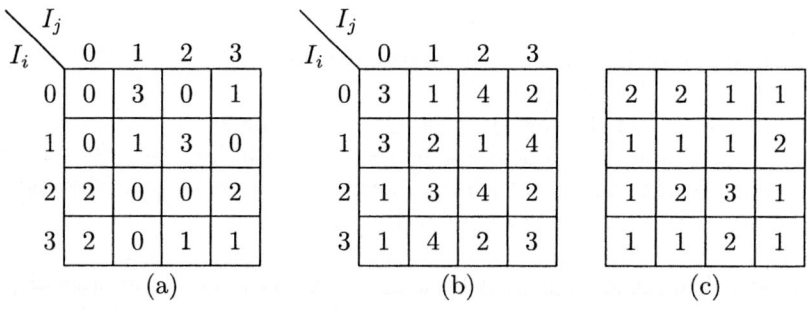

Fig. 2. The matrixes of the sample input image in Fig. 1.; (a) CCM (co-occurrence count matrix) **C**; (b) CRM (co-occurrence count rank matrix) **R**; (c) The ranked image **p**′

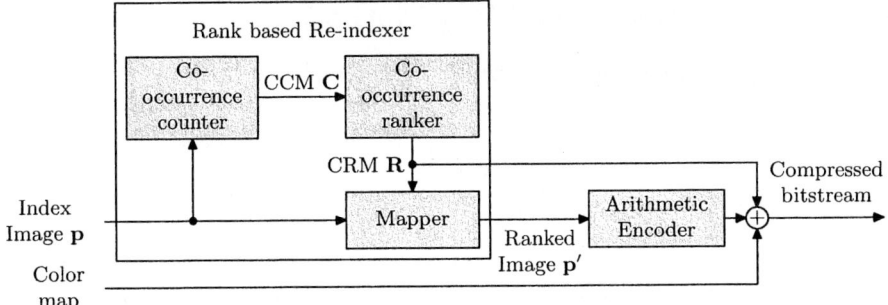

Fig. 3. The compression system architecture of RIAC scheme

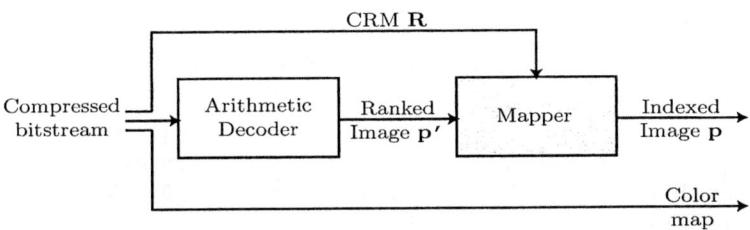

Fig. 4. The decompression system architecture of RIAC scheme

the Huffman coding when the symbols to compress have skewed distributions of little variance. The compression rates can be improved up to 10 15% by adapting global characteristics of the image as well as its local characteristics[2,7].

Fig. 3 and Fig. 4 show the architecture of the proposed RIAC method. Fig. 3 illustrates the encoder of the proposed method. An original image is transformed into indexed-color image, and CCM and CRM are generated to create the ranked image. Here, the number of compressed bits from arithmetic coding, the color map of original image and CRM information must be stored or transmitted together, because they are necessary for the recovery of the original image from the ranked image. The number of used color M can be different in every image, so the size of CRM depends on the number of index of original image.

The decoder can be realized by reversing the encoder process. Fig. 4 illustrates the decoder of the proposed method. Arithmetic decoding gets performed on the compressed bit sequence from the coder, which then is converted into a ranked image based on CRM and its original index image gets reconstructed. The reconstructed image with color map has the same resolution and size as the original image due to the lossless nature of a compression algorithm.

As mentioned earlier, the CRM-related information must be stored or transmitted when the index image is converted into the ranked image. This may be inefficient depending on the image size and the number of color, especially when the original image is small. In next section, we show how to convert into the ranked image with no need for transmission of CRM information.

3 Adaptive RIAC (ARIAC) Method

This section proposes an *adaptive rank-indexing with arithmetic coding* (ARIAC) method that re-indexes one symbol at a time in a greedy fashion. Each re-assignment is optimized based on the statistics collected from the part of initial index image and previously executed re-assignments.

Let $\mathbf{C}^i$ denotes i-th partial CCM, where superscript i represents that the algorithm collecting the co-occurrence statistics of a partial input index image. The entry $c^i_{j,k}$ of $\mathbf{C}^i$ reports the number of times the pair of successive index values, (I_j, I_k) observed over the ordered set $\mathbf{p}^i$ which denotes a sequence that includes the first $i+1$ elements of $\mathbf{p}$, $\mathbf{p}^i = (p_0, p_1, \ldots, p_i)$.

The algorithm starts generating the partial CCM, $\mathbf{C}^0$. After this step, all entries $c^0_{i,j}$ of $\mathbf{C}^0$ are initialized by 0, because $\mathbf{p}^0$ has only one element, 0. After the construction of $\mathbf{C}^0$, the proposed algorithm ranks the co-occurrence count of $\mathbf{C}^0$. These values are stored in a 0-th partial CRM, $\mathbf{R}^0$. The entry $r^0_{i,j}$ of $\mathbf{R}^0$ represents the rank of $c^0_{i,j}$ in descending order over the co-occurrence counts sequence $c^0_{i,j} = \{c^0_{i,k} | 0 \leq k \leq M\}$. Referencing $\mathbf{R}^0$, the algorithm sets the value of the first pixel p'_1 in ranked image $\mathbf{p}'$ as the co-occurrence frequency rank value of successive index (p_0, p_1). Fig. 5 (a) shows the $\mathbf{C}^0$. And Fig. 5 (b) and (c) show $\mathbf{R}^0$ and the ranked image respectively.

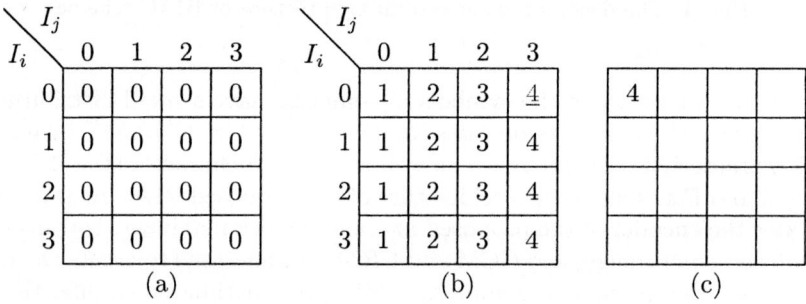

Fig. 5. Related matrixes for a generation of the first pixel in the ranked image from the index image in Fig. 1.; (a) co-occurrence count matrix $\mathbf{C}^0$ of $\mathbf{p}^0$; (b) co-occurrence count rank matrix $\mathbf{R}^0$ of (a); (C) the rank value of first pixel

To get p'_2, the algorithm makes $\mathbf{p}_1$ by putting p_1 to right side of $\mathbf{p}_0$ and makes $\mathbf{C}_1$ by updating the co-occurrence count matrix $\mathbf{C}_0$ referring $\mathbf{p}_1$. The algorithm iterates previous steps until the rank of all pixels in the ranked image are filled. Fig. 6 (a), (b) and (c) show $\mathbf{C}^1$, $\mathbf{R}^0$ and ranked image respectively. Also Fig. 7 shows final ranked image and related matrixes.

4 Performance Evaluation

In order to evaluate the compression efficiency of the proposed method, we have performed tests over seven synthetic graphic images and three natural color

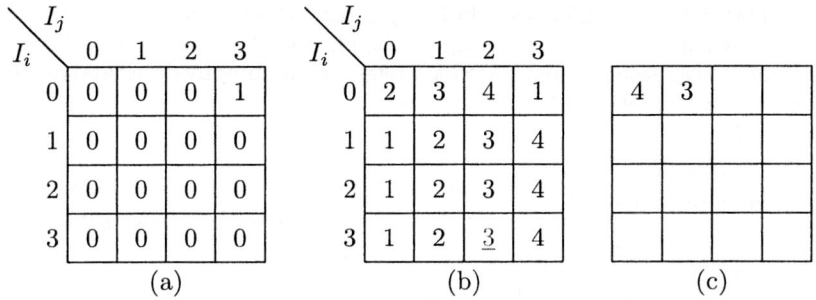

Fig. 6. Related matrixes for the generation of the second pixel in the ranked image from the index image in Fig. 1.; (a) co-occurrence count matrix $\mathbf{C}^1$ of $\mathbf{p}^1$; (b) co-occurrence count rank matrix $\mathbf{R}^1$ of (a); (C) the rank values of first two pixels

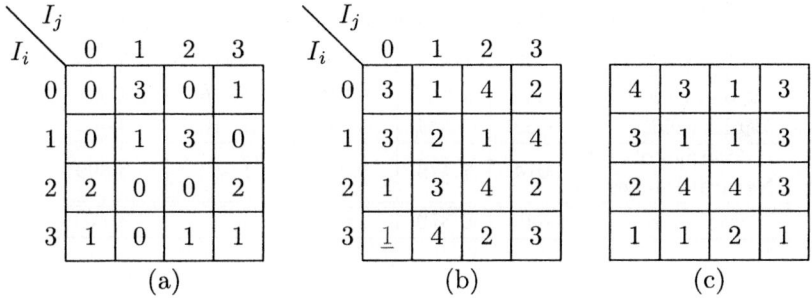

Fig. 7. Related matrixes for the generation of the last pixel in the ranked image from the index image in Fig. 1.; (a) co-occurrence count matrix $\mathbf{C}^{15}$ of $\mathbf{p}^{15}$; (b) co-occurrence count rank matrix $\mathbf{R}^{15}$ of (a); (c) the rank values of the pixels

images. All of these range in only a limited number of colors from 12 to 256. We have compared the performance of proposed schemes with the LZW algorithm of GIF, the plain arithmetic coding and the Zeng's algorithm.

Table 1 shows the simulation results, in terms of *bits per pixel* (bpp) in three different methods and the proposed RIAC and ARIAC techniques respectively. The bpp indicates 8/CR (*Compression Ratio*). The CR is obtained dividing original image size into compressed image size. The bpp values are calculated including the size of uncompressed color map in various schemes and specially added to the size of CRM in only RIAC algorithm. As Table 1 demonstrates, for synthetic graphic images, the proposed ARIAC algorithm reduces the bit rates than the other coding schemes, more specifically 15%, 54%, 16% and 12% for GIF, AC, Zeng and RIAC, respectively on the average. Table 1 also shows the results for three natural color images - "monarch", "girl" and "lena" - which are palletized using 256 colors. Similar results are observed that ARIAC reduces bit rates, 35%, 48%, 26% and 14%, respectively, than other coding algorithms.

Fig. 8 displays "party8", 526×286 with 12 colors. The original image is shown in Fig. 8 (a). And Fig. 8 (b) shows the ranked image obtained from the proposed algorithm on the original image. The ranked image appears to be much smoother

Table 1. Performance comparisons, in bits per pixel (bpp), between different lossless compression schemes. Whereas AC refers to the plain arithmetic cording scheme, Zeng refers to the Zeng's scheme, and (A)RIAC refers to the proposed (Adaptive) Ranked Indexing with Arithmetic Coding scheme in this paper.

Images (Num. of colors)	GIF	AC	Zeng	RIAC	ARIAC
party8 (12)	0.429	1.730	0.318	0.355	0.289
netscape (32)	2.121	3.572	1.791	2.245	2.020
benjerry (48)	1.254	2.425	1.154	1.695	1.210
ghouse (256)	4.999	7.157	4.841	4.759	4.157
clegg (256)	5.699	7.617	5.836	4.628	4.175
cwheel (256)	2.769	7.241	3.058	3.076	2.857
serrano (256)	2.897	7.208	3.393	2.637	2.379
Average of Synthetic Images	2.881	5.279	2.913	2.771	2.441
monarch (256)	4.948	7.436	4.325	3.710	3.233
Lena (256)	6.535	7.657	5.710	4.776	4.055
girl (256)	6.559	7.344	5.727	5.170	4.409
Average of Natural Images	6.014	7.479	5.254	4.552	3.899
Total Average	4.448	6.379	4.084	3.661	3.170

Fig. 8. The images of "party8"; (a) original index image; (b) ranked image by ARIAC scheme

than the original image. This is expected to greatly facilitate the subsequent lossless coding.

The histogram of the color indices distribution on the given index image is shown in Fig. 9 (a). And Fig. 9 (b) shows the histogram of the rank indices distribution on the ranked image. As can be seen Fig. 9 (b), the histogram is condensed in higher rank (ie. lower rank number), therefore this makes that we can expect the higher efficiency in the compression ratio.

Fig. 10 displays "lena", 512×512 with 256 colors. And Fig. 11 displays the histogram of the original index image and the ranked image with top 10 bins. As can be seen Fig. 10 and Fig. 11, the ranked image have more redundant ranks

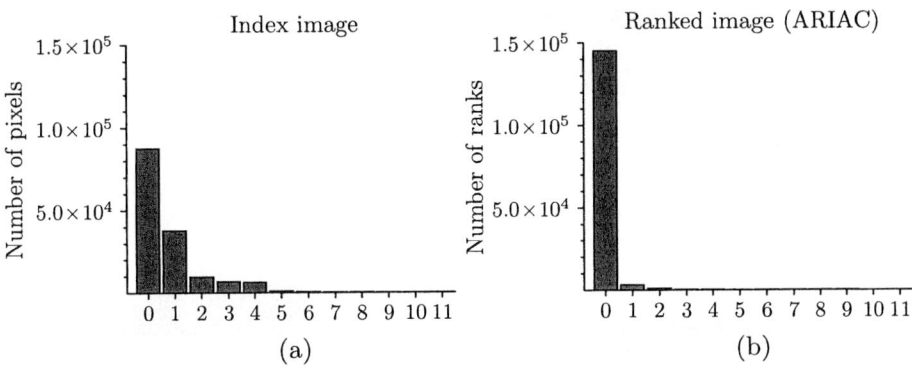

Fig. 9. Histogram comparison between original "party8" image (a) and ranked image using ARIAC scheme (b)

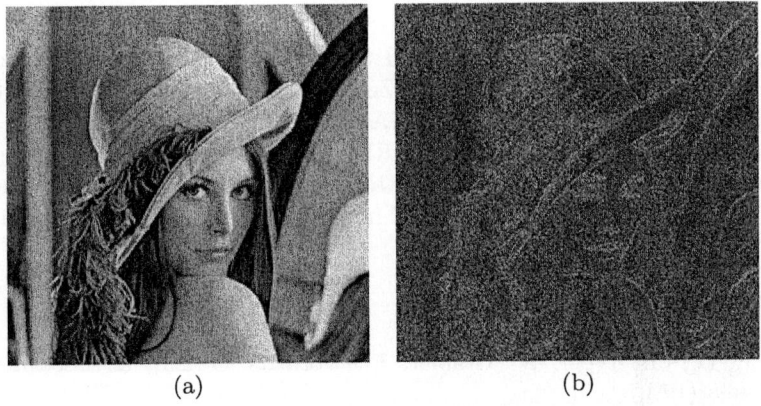

Fig. 10. The images of "lena"; (a) original index image; (b) ranked image by ARIAC scheme

and the histogram is condensed in higher rank. It also can reduces the bit rate than other lossless image compression scheme.

5 Conclusions

In this paper, we proposed a brand-new reindexing algorithms, RIAC and ARIAC, to achieve a better compression without any loss on palette based color images. The proposed algorithm calculates the index co-occurrence frequency between the colors of two neighboring pixels in index image, then using this, converts original index image into the ranked image with more redundancy of information. The ranked image can be compressed efficiently using the arithmetic encoder. In particular, ARIAC scheme dose not need to send CRM information including the index rank information, additionally.

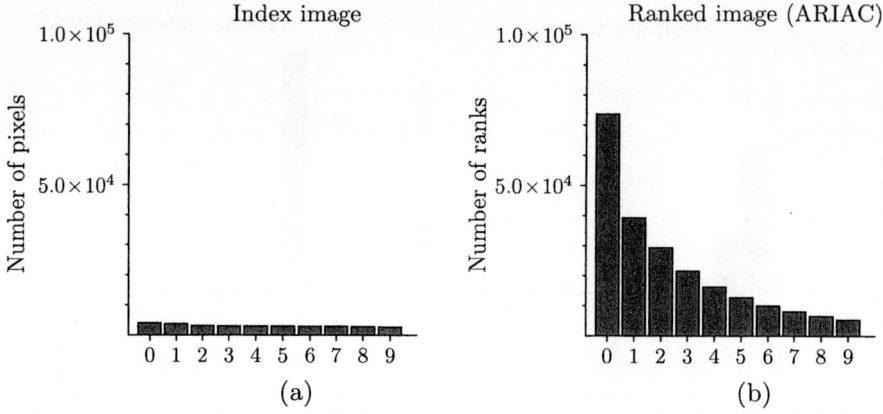

Fig. 11. The histograms of "lena"; (a) original index image; (b) ranked image by ARIAC scheme

From the simulations, it is verified that our proposed algorithm improved compression ratio over GIF, AC and Zeng's algorithm by 0.94 bpp, 3.06 bpp, 0.74 bpp, respectively. Especially ARIAC scheme reduces the bit rate by 13% approximately, when compared to the RIAC scheme. This proposed preprocessing method can be utilized and applied to storing of biomedical images requiring losslessness and to fast transmission of images over low-bandwidth channel.

References

1. Murray, D., van Ryper, W.: Graphics File Formats, O'Reilly & Associates, INC., California (1996)
2. Kangsoo, Y., Hangjeong, L., Eueesun, J., Hoonsung, K.: An Efficient Lossless Compression Algorithm using Arithmetic Coding for Indexed Color Image, The Korean Institute of Communication Sciences, Vol. 30, No. 1C (2005) 35–43
3. Zaccarin, A., Liu, B.: A novel approach for coding color quantized images. IEEE Tran. Image Proc., Vol. 2 No. 4 (1996) 442–453
4. Memon, N., Venkateswaran, A.: On ordering color maps for lossless predictive coding. IEEE Tran. Image Proc., Vol. 5 No. 11 (1996) 1522–1527
5. Zeng, W., Li, J., Lei, S.: An efficient color re-indexing scheme for palette-based compression. Proc. Of the 7th Conf. on Image Processing (2000) 476–479
6. Pinho Armando J., Neves Antonio J. R.: A note on Zeng's technique for color reindexing of palette-based images. IEEE Signal Processing Letters, Vol. 11, Issue 2 (2004) 232–234
7. Crane R.: A Simplified Approach to Image Processing, Prentice Hall, New Sersey (1997)

An Effective Approach to Chin Contour Extraction

Junyan Wang, Guangda Su, and Xinggang Lin

Dept of Electronic Engineering, Tsinghua University,
Beijing, China
wjy01@mails.tsinghua.edu.cn

Abstract. In front-view facial images, chin contour is a relative stable shape feature and can be widely used in face recognition. But it is hard to extract by conventional edge-detection methods due to the complexities of grayscale distribution in chin area. This paper presents an effective approach to chin contour extraction using the facial parts distributing rules and approved snake model. We first approximately localize a parabola as the initial contour according to prior statistical knowledge, then use approved active contour model to find the real chin contour through iteration. Experimental results show that by this algorithm we can extract the precise chin contour which preserves lots of details for face recognition.

1 Introduction

Face recognition is one of mankind's best abilities. Being an area of both theoretical and practical interest, automatic face recognition has recently attracted a lot of attention, leading to many significant achievements [1]. In recent years, facial profile is widely used for face alignment, face normalization and face classification [2] in face recognition. Chin contour is the main part of facial profile and contains most shape information of facial contour. So chin contour has attracted a few researchers' attention as a stable shape feature, and has been extracted in many different methods for face recognition.

The main method to analyze chin contour is building geometric model to approach the boundary by dynamic curve fitting. Conventional chin model is parabola: single parabola approximation [3] and dual half parabola approximation [4]. This model is satisfying in curve approximation, but it loses numerous details of chin contour (only the curvature of parabola is useful information). Yu Dongliang [5] uses canny operator to extract chin contour with a lot of details, but that method needs manual work to position the start and the end point which limits its application.

We often use edge-detect operators to extract the object boundary during image contour processing. But it doesn't work well when the image grads have no obvious differentiation around the object boundary. The figure of contour lines of the image (Fig.1(2)) and its Sobel edge (Fig.1(3)) illustrate the grey of boundary often blends into the background. The points with bigger grads form the boundary of hair, eyes and other contours of strong edge with different grey, which do not include the interesting points of chin contour. Simple edge-detect operator can not extract the interesting chin contour, so we have to adopt other methods.

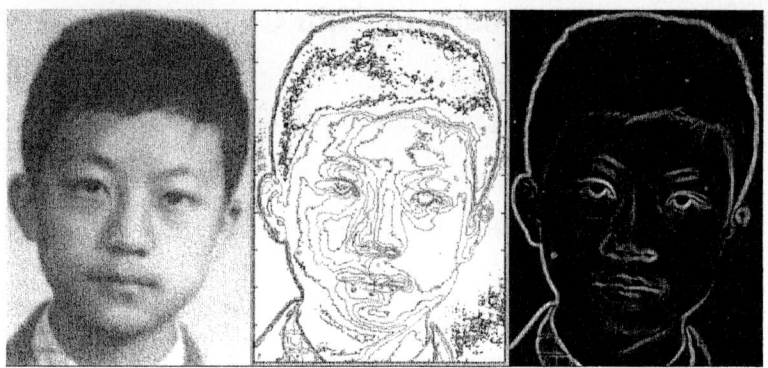

Fig. 1. 1) Source Image   (2) Contour Lines   (3) Sobel Image

Besides the methods mentioned above, active contour model (snake) [6][7] is often used for extracting closed contours. Snake presents another way to solve the problem of edge detection which can converge to the optimal profile in theory. So improved snake method may satisfy our application.

In this paper, an effective approach is proposed to extract chin contour for front-view gray scale Images. Firstly we locate the far eye's corners automatically and normalize the face image; secondly get the Y-coordinates of facial parts using X-directional integral projection; at last extract the chin contour by approved active contour model with a fit initial contour. This approved snake converges to the real chin contour after a few times of iteration and reduces the complex calculation of conventional snake. The experimental results show the efficiency and accuracy of this method.

The remainder of this paper is organized as the following. The method of facial parts position is described in Sec.2. The approved snake method, which is used for extracting the chin contour, is given in Sec.3. The last part is the experimental results and discussion.

2 Facial Part Position

2.1 Face Position and Normalization

Before extracting chin contour, we need locate the interesting area of chin. The facial parts distribution satisfies the laws in ratio. Some positions of facial organs, such as eyes and mouth, may help us to find the interesting area of chin. So we can position the organs which are easily detected first, and then according to these information to locate probable chin area.

Gu Hua [8] proposed an approach to locate the vital feature points on faces, including eye balls, eye's corners, nose and so on. Here we use the method to locate the far eye's corners. This approach locates face area by template matching. Then position the far eye's corners by using SUSAN operator. Fig2(1) shows the facial area location result. The rectangle in the figure is the rough facial area and the two circles are the far eye's corners.

Fig. 2. (1) Face area   (2) Adjusted Image   (3) Normalized Image

After getting the positions of far eye's corners, we can rotate the face image through affine transform to let the Y-coordinates of far eye's corners equal. Thus the mathematical model of chin contour may be simplified. Fig 2(2) shows the result of image rotating. In order to let the chin extraction be effective, the image is normalized by zooming in or zooming out and image cutting, and as the result the size of face image is 300*300 pixels with far eye's corners position (75, 60) and (225, 60) respectively. The result of normalization is shown in Fig 2(3).

2.2 Get Y-coordinate of the Lowest Point in Chin Contour

As shown in Fig 3, in the normalized face images, the facial parts locate in a certain order. From top to down, there are eyebrows, eyes, nose, mouth and chin. So we can get the Y-coordinates of the facial parts by X-directional integral projection.

Because of different illumination conditions, the projection curves of various images have minimum values of different amounts and different positions. These add the difficulty of detecting the positions of minimum values and getting the accurate

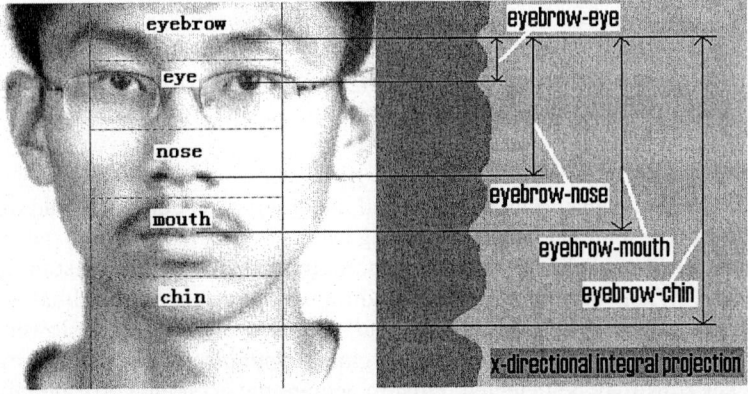

Fig. 3. Facial X-directional projection and distances between facial parts

corresponding relations between minimum values and facial parts. We propose a better position method, including two stages. The first is getting all the minimal values on the projected curve. The second stage is finding out the corresponding relation between minimal values and Y-coordinates of facial parts.

2.2.1 Find Minimal Values of X-directional Integral Projection Curve

Fig.3 shows the X-directional integral projection curve. The minimal values of the curve is not very obvious and not easily to be detected. So we transform the curve to detect the minimal values. First filter the curve with a mean filter (Fig.4(1)); and then calculate its first derivative (Fig.4(2)) and second derivative. The peak values and the lowest values of the second derivative correspond to the minimal and maximal values of the projection curve respectively. Because we only concern the minimal values, we maintain the peak values of the second derivative, shown in Fig.4(3). Now we only need to detect the peak values of second derivative. Contrasting with the detection of minimal values of integral projection curve, the peaks of the second derivative are more shallow and high, and are easier to be detected. The small rectangles in Fig.4(4) are the detection results. This method can detect small minimal values with high accuracy.

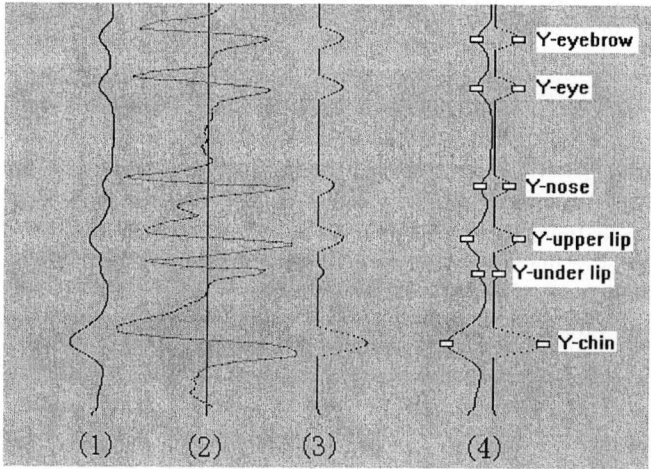

Fig. 4. (1) X-directional integral projection curve (2) first derivative (3) second derivative (4) minimal values and position result

2.2.2 Find the Y-coordinate of Chin

On studying face images and X-directional integral project curve, we find out there are several probable minimal values, including eyebrow, eye, lower eyelid, nose, the upper lip, the under lip and chin. But not all the minimal values always exist in the curve of every face. When the illumination conditions change, some minimal values may disappear. For example, under blazing light, the minimal values of the lower eyelid and the under lip may disappear. If having on glasses with black rims, the minimal value of the glasses rims may be detected. How to acquire the accurate corresponding relation between the minimal values and the Y-coordinates of facial parts is another difficulty after detecting the minimal values of the projection curve.

Based on the positions of far eye's corners, we could firstly consider that the minimal value closest to the Y-coordinate of far eye's corner was the Y-coordinate of eyes, and the upper minimal value was the Y-coordinate of eyebrows; and the lower minimal value might be the Y-coordinate of nose or glasses rims. According to this method and our prior knowledge of faces, we could get the Y-coordinates of other facial parts. Because of the difference of faces and face images, this detection may be error. So if we merely search the minimum of chin, we may get the error result easily. In order to fit for these variations and reduce error, we use the distributing rules of facial parts. So we calculate the mean values and variance of the ratios of the distances between the Y-coordinate facial parts of 600 images. Table 1 and table 2 show the statistical results.

Table 1. The mean values of the ratios of distances between facial parts to the distance between eyebrow and chin

ratio	mean value
eyebrow-eye / eyebrow-chin	0.1517
eyebrow-nose / eyebrow-chin	0.5010
eyebrow-upper lip / eyebrow-chin	0.6887
eyebrow-under lip / eyebrow-chin	0.8147
eyebrow-chin / eyebrow-chin	1.000

Table 2. The variances of the ratios of distances between facial parts to the distance between eyebrow and chin

facial part	eyes	nose	upper lip	under lip	chin
eyebrow	0.0176	0.0220	0.0191	0.0186	0.0146
eyes	0	0.0227	0.0186	0.0212	0.0160
nose	0.0227	0	0.0163	0.0216	0.0284
upper lip	0.0186	0.0163	0	0.0173	0.0273
under lip	0.0212	0.0216	0.0173	0	0.0298

According to these statistical results, we can find the best matching result of the minimal values to Y-coordinates of the facial parts. Here we use iteration to find out the best matching result with most matching points and least matching error. The result of the matching is shown in Fig.4(4). We test our method in our database with 637 face images and the result is shown in table 3. We can see that this method can position the Y-coordinates of facial parts with high accuracy.

Table 3. The result of acquiring the Y-coordinates of facial parts (637 images)

	eyebrow	nose	mouth	chin	total
correct	634	634	633	626	625
ratio	99.53%	99.53%	99.37%	98.27%	98.12%

3 Chin Contour Extraction

3.1 Snake

Active contour models are the energy-minimizing spline guided by internal constraint forces and influenced by external image forces that pull it toward features like lines and edges [6]. Let $v(s) = (x(s), y(s))$ be the parametric description of the snake ($s \in [0,1]$). The conventional energy function in the snake is defined as follows:

$$E(v) = E_s(v) + P(v) \quad (1)$$

where

$$E_s(v) = \int_0^1 (\varpi_1(s)|v_s|^2 + \varpi_2(s)|v_{ss}|^2) ds \quad (2)$$

$\varpi_1(s)$ and $\varpi_2(s)$ are parameters.

$$P(v) = \int_0^1 p(v(s)) ds \quad (3)$$

The goal is to find the snake to minimize equation (1).

3.2 Snake for Chin Contour Extraction

As the boundary extracted by snake method is only determined by its energy function in theory, how to define the optimal energy function is the most important but difficult problem. Because of the abundant freedom of boundary searching, using snake model attains the optimal position through repetitious iteration and complex calculation.

3.2.1 Energy Function

For our application, we rewrite equation (1) into the following form.

$$\sum_{i=1}^{N} E(i) = \sum_{i=1}^{N} (\alpha E_c(i) + \beta E_s(i) + \gamma E_i(i)) \quad (4)$$

When converges, equation (4) minimizes. N is the account of chin contour points to be extracted. $E_c(i)$ is the continuous energy of the *i*th point. $E_s(i)$ is the curvature energy of the *i*th point. $E_i(i)$ is the image energy of the *i*th point. α, β, γ are parameters.

$$E_c(i) = \frac{\left\| x(i) - x(i-1) \right\| + \left| y(i) - y(i-1) \right| - d_{mean}}{\left| E_{c\max} - d_{mean} \right|} \quad (5)$$

In equation (5), d_{mean} is the mean distance between the two adjoining points on chin contour. The distance between the adjoining points is more close to the mean distance, the energy is smaller. So this energy definition not only assures smoothness but also avoids cumulus.

$$E_s(i) = \frac{|2 \times x(i) - x(i-1) - x(i+1)| + |2 \times y(i) - y(i-1) - y(i+1)|}{E_{s\max}} \quad (6)$$

$g(i)$ in equation (6) is the Y-directional second derivative.

$$g(i) = \frac{\sum_{j=-l}^{l}(I(x(i), y(i) - 2l + j) + I(x(i), y(i) + 2l + j) - 2I(x(i), y(i) + j))}{2l+1} \quad (7)$$

$(x(i), y(i))$ is the coordinate of ith point. $I(x(i), y(i))$ is the gray level of ith point. $\pm l$ is the length of the scope of the points' moving on one step and in our experimental $l = 7$. As the shade exists on the neck, the points in chin contours have the characteristic of big second derivative and small grey level. And the second derivative defined in equation (7) is a positive number. According to the thought, we design the image energy function showing in equation (8).

$$E_i(i) = \begin{cases} \dfrac{\sum_{j=-1}^{1}\sum_{k=-1}^{1} I(x(i)+j, y(i)+k)/9}{g(i)} & \text{if} \quad g(i) > 0 \\ 10000 & \text{else} \end{cases} \quad (8)$$

If the initial points are given, calculate $E_{c\max}$, $E_{s\max}$, $E_{i\max}$ and d_{mean}, then find the point in the varying area which minimizes the energy function for each initial point. After doing this for each point one time, calculate $E_{c\max}$, $E_{s\max}$, $E_{i\max}$ and d_{mean} over again. Continue this circulation until the time that all the coordinates of the points don't change. Thus the points on the chin contour are acquired.

3.2.2 Initial Contour

In introduction, we describe that the shape of chin contour is similar to parabola, so we initiate the snake with one parabola as $y = ax^2 + bx + c$. Let the top left corner be the grid origin. We collect 100 chin contours and find $\bar{a} = 0.005$. (x_{chin}, y_{chin}) is acquired in 2.2. So the initial points on chin contour are $\{(x, y) | y = ax^2 + bx + c, 75 \le x \le 225\}$, and a, b, c follows the equation (9). The initial contour is shown in Fig.5(2).

$$\begin{cases} -\dfrac{b}{2a} = x_{chin} \\ \dfrac{4ac-b^2}{4a} = y_{chin} \\ a = 0.005 \end{cases} \qquad (9)$$

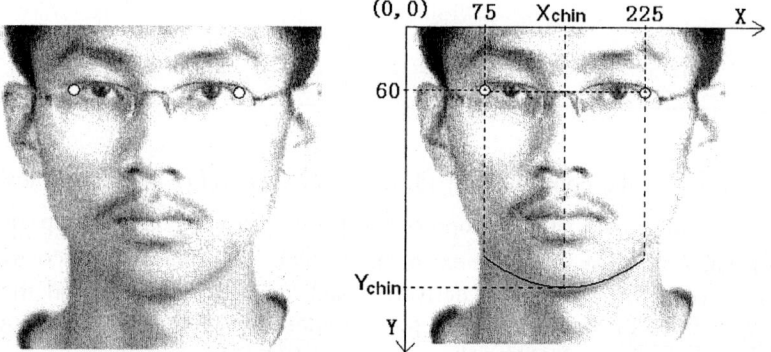

Fig. 5. (1) source image (2) initial contour

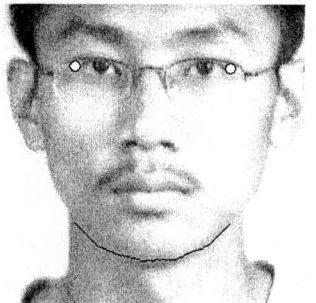 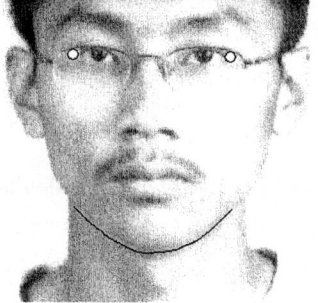

Fig. 6. (1) points after once iteration (2) points after convergence

3.2.3 Iteration

After getting initial points, the iteration is working under the rule of minimizing the energy function. During the iteration, the X-coordinates of chin dots are invariable, and the Y-coordinates are changing in the scope to find the best position. Here the scope is an experimental scope of [y-7, y+7] which assures finding the optimal point and avoiding divergence. Under these rules, the number of points in the snake is invariable. This method accords with the requirement of active contour model.

After once iteration, the snake is closer to the real chin contour as shown in Fig.6(1). At last the snake converges accurately to the real chin contour shown in Fig.6(2).

4   Experimental Results and Conclusion

Fig.7 shows the experimental results of the approach proposed in this paper. The black points are the points on chin contour. We can see that this approach is very efficient with high accuracy.

This approved snake can be used for extracting unclosed edge with feasible algorithm description, solving the problem that snake often is applied for close profile. The initial contour is appropriate, so the approved snake can converge to the real chin contour after a few times of iteration. This approach reduces the computation complexity of snake, so the speed of extraction is very high. The experimental results show that the chin contours extracted through this method have a lot of details for face recognition.

Although this method is approved for front-view face image, it is also applicable for faces with different poses. The result of this approach may be influenced by illumination conditions, so we can do illumination proportion in preprocessing. Our test set contains many gray level face images, but the method can be used for color images.

Fig. 7. Samples of chin contour extraction

References

1. Zhao W., Chellappa, R., Phillips, P.J.: Face Recognition: A Literature Survey. ACM computing surveys, v 35, n 4, (2003) 399-458
2. Wang Jun-Yan, Su Guang-Da: The Research of Chin Contour in Fronto-Parallel Images. International conference on machine learning and cybernetics, v5, (2003) 2814-2819
3. Xiaobo Li, Nicholas Roeder: Face Contour Extraction from Front-View Images. Pattern Recognition, 28(8), (1995) 1167-1179

4. Kampmann M: Estimation of the Chin and Cheek Contours for Precise Face Model Adaptation. Paper Proceedings. International Conference on Image Processing. IEEE Compute. Soc. Part vol.3, (1997).300-303
5. YU Dong-liang, SU Guang-da: Research of Chin Contour Extraction. Pattern Recognition and Artificial Intelligence, 15(1), (2002) 75-79
6. Kass, M., Witkin, A., and Terzopoulos, D.: Snakes: Active Contour Models, in IJCV, (1988), 321-331
7. HyunWook Park, Todd Schoepflin, and Yongmin Kim: Active Contour Model with Gradient Directional Information: Directional Snake, IEEE trans on circuits and systems for video technology, 11(2), (2001) 252-256
8. Gu Hua, Guangda Su, and Du Cheng: Automatic Localization of the Vital Feature Points on Human Faces, Journal of Optoelectronics Laser, 15(8), (2004) 975-979(in Chinese)

Robust Face Recognition Across Lighting Variations Using Synthesized Exemplars

Sang-Woong Lee, Song-Hyang Moon, and Seong-Whan Lee*

Department of Computer Science and Engineering, Korea University,
Anam-dong, Seongbuk-ku, Seoul 136-713, Korea
{shmoon, sangwlee, swlee}@image.korea.ac.kr

Abstract. In this paper, we propose a new face recognition method under arbitrary lighting conditions, given only a single registered image and training data under unknown illuminations. Our proposed method is based on the exemplars which are synthesized from photometric stereo images of training data and the linear combination of those exemplars are used to represent the new face. We make experiments for verifying our approach and compare it with two traditional approaches. As a result, higher recognition rates are reported in these experiments using the illumination subset of Max-Planck Institute Face Database.

1 Introduction

Early works in illumination invariant face recognition focused on image representations that are mostly insensitive to changes under various lighting [1]. Various images representations are compared by measuring distances on a controlled face database. Edge map, second derivatives and 2D Gabor filters are examples of the image representations used. However, these kind of approaches have some drawbacks. First, the different image representations can be only extracted once they overcome some degree of illumination variations. Second, features for the person's identity are weakened whereas the illumination-invariant features are extracted.

The different approaches, called the photometric-stereo method, are based on the low dimensionality of the image space [2]. The images of one object with a Lambertian surface, taken from a fixed viewpoint and varying illuminations lie in a linear subspace. We can classify the new probe image by checking to see if it lies in the linear span of the registered gallery images. These gallery images are composed of at least three images of the same person under different illuminations. Since it recognizes the new image by checking that it is spanned in a linear subspace of the multiple gallery images, it cannot handle the new illuminated images of a different person.

To avoid the necessity of multiple gallery images, the bilinear analysis approach is proposed [3]. It applies SVD(Singular Value Decomposition) to a variety of vision problems including identity and lighting. The main limitation of

---

* To whom all correspondence should be addressed.

these bilinear analysis methods is that prior knowledge of the images, like the lighting direction of training data are required.

Unlike the methods described above, Blanz and Vetter use 3D morphable models of a human head [5]. The 3D model is created using a database collected by Cyberware laser scans. Both geometry and texture are linearly spanned by the training ensemble. This approach enables us to handle illumination, pose and expression variations. But it requires the external 3D model and high computational cost.

For illumination-robust face recognition, we have to solve the following problem : *Given a single image of a face under the arbitrary illumination, how can the same face under the different illumination be recognized?* In this paper, we propose a new approach for solving this problem based on the synthesized exemplars. The illuminated-exemplars are synthesized from photometric stereo images of each object and the new probe image can be represented by a linear combination of these synthesized exemplars. The weight coefficients are estimated in this representation and can be used as the illumination invariant identity signature.

For face recognition, our proposed method has several distinct advantages over the previously proposed methods. First, the information regarding the lighting condition of training data is not required. We can synthesize the arbitrary illuminated-exemplars from the photometric stereo images of training data. Second, we can perform recognition with only one gallery image by using linear analysis of exemplars in the same class. Third, the coefficients of exemplars are the illumination invariant identity signature for face recognition, which results in high recognition rates.

2 Background

We begin with a brief review of the photometric stereo method with Lambertian lighting model and bilinear analysis of illuminated training images. We will explain what is the Lambertian reflectance and how it can be used in the photometric stereo images for face recognition [2]. We will also explain recognition methods using the bilinear analysis of the training images [3],[4].

2.1 Photometric Stereo

We assume the face has the Lambertian surface, the illuminated image I can represented by

$$I = \rho N^T L = T^T L \qquad (1)$$

where n is the surface normal and ρ is the albedo, a material dependant coefficient. The object-specific matrix, T includes albedo and surface normal information of object. We have n images, $(I_1, I_2, ..., I_n)$ of one object under varying illumination. These images, called photometric stereo images, were observed at a fixed pose and different lighting sources. Assuming that they are from the same object a with single viewpoint and various illuminations, the following can be expressed

$$\mathbf{I} = \begin{pmatrix} I_1 \\ I_2 \\ \vdots \\ I_n \end{pmatrix} = \begin{pmatrix} T^T L_1 \\ T^T L_2 \\ \vdots \\ T^T L_n \end{pmatrix} = T^T \begin{pmatrix} L_1 \\ L_2 \\ \vdots \\ L_n \end{pmatrix} = T^T \mathbf{L} \qquad (2)$$

where $\mathbf{I}$, the collection images $\{I_1, I_2, ..., I_n\}$ of the same object under different lighting condition, is the observation matrix. $\mathbf{L} = \{L_1, L_2, ..., L_n\}$ is the light source matrix. If the lighting parameters are known, we can extract the surface normal orientation for objects. To solve T, the least squares estimation of $\mathbf{I}$ using SVD. We can classify a new probe image by computing the minimum distance between the probe image and n-dimensional linear subspace. Photometric stereo method requires at least 3 gallery images for one object. Multiple gallery images are large restrictions for a real face recognition system.

2.2 Bilinear Models

Bilinear models offer a powerful framework for extracting the two-factor structure, identity and lighting. Bilinear analysis approaches had applied SVD to a variety of vision problems including identity and lighting [3],[4]. For bilinear analysis, training images of different objects under the same set of illuminations are required. Theses approaches also assume the Lambertian surface and the image space $T^T L$, where both T and L vary. Let $L_1, L_2, ..., L_n$ be a basis of linearly independent vectors, $\mathbf{L} = \sum_{j=1}^{n} \beta_j L_j$ for some coefficients $\boldsymbol{\beta} = (\beta_1, \beta_2, ..., \beta_n)$. Let $\{T_1, ..., T_m\}$ be a basis for spanning all the possible products between albedo and surface normal of the class of objects, thus $T = \sum_{i=1}^{m} \alpha_i T_i$ for some coefficients $\boldsymbol{\alpha} = (\alpha_1, ..., \alpha_m)$. Let $\mathbf{A} = \{A_1, ..., A_m\}$ be the matrix whose columns are the images of one object, i.e., $A_k = \alpha_k T_k \sum_{j=1}^{n} \beta_j L_j$. A_k are n images of k-th object and the column of A_k, A_{k_j} is the image of k-th object under j-th illumination. Therefore we can represent the new probe image H by linear combination of $\{A_1, ..., A_m\}$ with the bilinear coefficients, $\boldsymbol{\alpha}$ and $\boldsymbol{\beta}$.

$$H = \rho_H N^T L = T_H^T L = (\sum_{i=1}^{m} \alpha_k T_k)(\sum_{j=1}^{n} \beta_j L_j) = \boldsymbol{\alpha \beta} \mathbf{A} \qquad (3)$$

The bilinear problem in the $m + 3$ unknowns is finding $\boldsymbol{\alpha}$ and $\boldsymbol{\beta}$. Clearly, we solve these unknowns, we can generate the image space of object H from any desired illumination condition simply by keeping $\boldsymbol{\alpha}$ fixed and varying $\boldsymbol{\beta}$. But these approaches require the same set of illuminations per object, so that we have to know about the lighting condition of training data in advance.

3 Linear Analysis of the Synthesized Exemplars

We propose an illumination invariant face recognition method based on the synthesized exemplars. We synthesize the illuminated-exemplars from photometric stereo images and represent a new probe image by linear combination of those exemplars.

The procedure has two phases: training and testing. Images in the database are separated into two groups for either training or testing. In the training procedure, we construct the training data to consist of at least three illuminated images per object. However we do not know the lighting conditions of training data and the training data can be constructed using different objects and different sets of illuminations unlike bilinear analysis method.

In our experiments, we construct the train matrix as m people under n different illuminated images. This is followed by computing the orthogonal basis images by the PCA for inverting the observation matrix per person. The orthogonal basis images of one person are used to synthesize the exemplars. We can then reconstruct a novel illuminated image using these basis images of the same face. In the testing procedure, we synthesize the exemplars under the same illumination as the input image. The lighting conditions of these m synthesized exemplars and input images are same. The input image can be represented by the linear combination of the exemplars, the weight coefficients are used as those signature identities for face recognition. In the registration, those gallery images are already saved for the recognition, we find the facial image that has the nearest coefficient by computing the correlation.

3.1 Synthesis of the Exemplars

We assume that the face has a Lambertian surface and the light source, whose locations are not precisely known, emits light equally in all directions from a single point. Then, an image I is represented by $T^T L$ as shown Eq. 1 and the matrix $\mathbf{I}$ that made n images can be represented by $T^T \mathbf{L}$ as shown Eq. 2. The photometric stereo images are from the same object, we can assume that they have the same object-specific matrix T and different illumination vector $\mathbf{L}$. If the light source matrix $\mathbf{L}$ is non-singular($|\mathbf{L}| \neq 0$) and $\{L_1, L_2, ..., L_n\}$ are linearly independent, the matrix $\mathbf{L}$ is invertible and then T can be expressed by the product of matrix $\mathbf{I}$ and the pseudo-inverse of $\mathbf{L}$, $\mathbf{L}^+$.

$$T = \mathbf{I}\mathbf{L}^+ \tag{4}$$

The light source matrix $\mathbf{L}$ can be invertible when $\{L_1, L_2, ..., L_n\}$ are linearly independent of each other. To make the images independent from each other, we transform the photometric stereo images into the orthogonal basis images, $\{B_1, B_2, ..., B_{n-1}\}$, by principal component analysis (PCA). By applying PCA to photometric stereo images, we can express a new illuminated image of the same object using the orthogonal basis images by changing the coefficients $\boldsymbol{\alpha}$ and the orthogonal basis images can be obtained in off-line training. Our method for synthesizing the image, called '*exemplar*', proposes that we use the input image as a reference. Since photometric stereo images have the same object-specific matrix and the input image is used as a reference, the synthesized exemplar's lighting condition is similar to that of input image. The input image H can be represented using a linear combination of orthogonal basis images.

$$H = \bar{B} + \boldsymbol{\alpha}\,\mathbf{B} \tag{5}$$

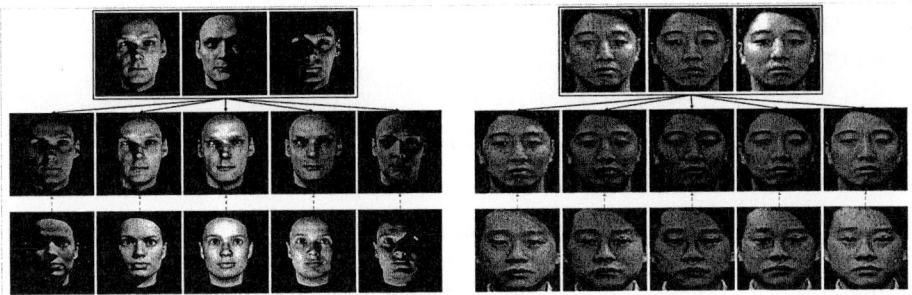

Fig. 1. Example of the synthesized exemplars

where $\bar{B}$ represents the mean of orthogonal basis images per object and $\alpha \in \Re^{n-1}$. We can find the coefficient α as follows. The columns of matrix are orthogonal to each other, the transpose is the inverse and we can now easily find the coefficient vector α^* by transpose instead inverse.

$$\alpha^* = \mathbf{B}^{-1}(H - \bar{B}) = \mathbf{B}^T(H - \bar{B}) \qquad (6)$$

In the photometric stereo images, we choose three images of random lighting directions, $\{\tilde{I}_1, \tilde{I}_2, \tilde{I}_3\}$ and we transform those images into the orthogonal coordinate system by PCA by eigenvectors $\{\tilde{B}_1, \tilde{B}_2\}$. Where $\bar{\tilde{B}}$ is the mean of $\{\tilde{B}_1, \tilde{B}_2\}$ and $\tilde{\alpha}^* = \{\tilde{\alpha}_1, \tilde{\alpha}_2\}$ is the coefficient for synthesizing the exemplar $\tilde{E}$, an exemplar using three images is as follows.

$$\tilde{E} = \bar{\tilde{B}} + \sum_{j=1}^{2} \tilde{\alpha}_j \tilde{B}_j = \bar{\tilde{B}} + \tilde{\alpha}^* \tilde{\mathbf{B}} \qquad (7)$$

Fig. 1 shows examples of the synthesized exemplars from the training data. We choose three images under random illumination of each person and those chosen images for each person are different set. The top row represents three different illuminated images of the same person from the training data. The middle row shows examples of the synthesized exemplars using the images from the top row. While bottom row shows examples of the different illuminated input images. Each synthesized exemplar image (middle row) references the illumination of input image found directly below it. As shown, the synthesized exemplars have very similar lighting conditions to that of the input image. One exemplar image is synthesized per object, so there are m exemplar images under the same lighting condition of the input image where the training data is collected by the images of m objects.

3.2 Linear Combination of Synthesized Exemplars

In the previous section, we described that how the exemplar is synthesized. Using both the photometric stereo images and input image as illumination reference,

m exemplars are synthesized per person. The exemplar $\tilde{E}_k$ of k-th person can be represent as

$$\tilde{E}_k = \bar{\bar{B}}_k + \sum_{j=1}^{2} \tilde{\alpha}_{k_j} \tilde{B}_{k_j} = \bar{\bar{B}}_k + \tilde{\boldsymbol{\alpha}}_k^* \tilde{\mathbf{B}}_k \qquad (8)$$

where $\bar{\bar{B}}$ is the mean of orthogonal basis images $\{\tilde{B}_1, \tilde{B}_2\}$ from three photometric stereo images, $\{\tilde{I}_{k_1}, \tilde{I}_{k_2}, \tilde{I}_{k_3}\}$. The column of $\mathbf{I}_k$, I_{k_i} is the image under i-th illumination of k-th person. The input image is represented well by the linear combination of the exemplars. At this time, the linear coefficients are estimated from the exemplars under the same illumination. That means, the coefficients depend on the m exemplars but not on the lighting conditions. Because the exemplars are for the object class only, the coefficients provide a signature identity that is invariant to illumination. The coefficient vector $\boldsymbol{f}$ is computed by the following equation.

$$H = \sum_{k=1}^{m} f_k \tilde{E}_k = \boldsymbol{f} \tilde{\mathbf{E}} \qquad (9)$$

where $\boldsymbol{f} = \{f_1, f_2, ..., f_m\}$ is the coefficient vector from the m exemplars and used for recognition. f_k is the weight coefficient for the k-th exemplar object. $\tilde{\mathbf{E}} = \{\tilde{E}_1, \tilde{E}_2, ..., \tilde{E}_m\}$ is the matrix of the synthesized exemplars. The problem is to choose $\boldsymbol{f}$ so as to minimize the cost function, $\mathcal{C}(\boldsymbol{f})$. We define the cost function as the sum of square errors which measures the difference between the input image and the linear sum of the exemplars. We can find the optimal coefficient $\boldsymbol{f}$, which minimizes the cost function, $\mathcal{C}(\boldsymbol{f})$.

$$\boldsymbol{f}^* = \arg\min_{\boldsymbol{f}} \mathcal{C}(\boldsymbol{f}) \qquad (10)$$

with the cost function,

$$\mathcal{C}(\boldsymbol{f}) = \sum_{i=1}^{d} (H(x_i) - \sum_{k=1}^{m} f_k \tilde{E}_k(x_i))^2 \qquad (11)$$

To represent the input image H using exemplars, we have to find $\boldsymbol{f}$ by the equation of $H = \tilde{\mathbf{E}} \boldsymbol{f}$, where $\tilde{\mathbf{E}} = \{\tilde{E}_1, \tilde{E}_2, ..., \tilde{E}_m\}$. The least square solution satisfies $\tilde{\mathbf{E}}^T H = \tilde{\mathbf{E}}^T \tilde{\mathbf{E}} \boldsymbol{f}$. If the columns of $\tilde{\mathbf{E}}$ are linearly independent, then $\tilde{\mathbf{E}}^T \tilde{\mathbf{E}}$ is non-singular and has an inverse.

$$\boldsymbol{f}^* = (\tilde{\mathbf{E}}^T \tilde{\mathbf{E}})^{-1} \tilde{\mathbf{E}}^T H \qquad (12)$$

We can express the input image H using the computed $\boldsymbol{f}^*$ on the assumption that the columns of matrix $\tilde{\mathbf{E}}$ are linearly independent. If they are not independent, the solution $\boldsymbol{f}^*$ will not be unique, in this case, the solution can be solved by the pseudo-inverse of $\tilde{\mathbf{E}}$, $\tilde{\mathbf{E}}^+$. But, that is unlikely to the happen for proposed

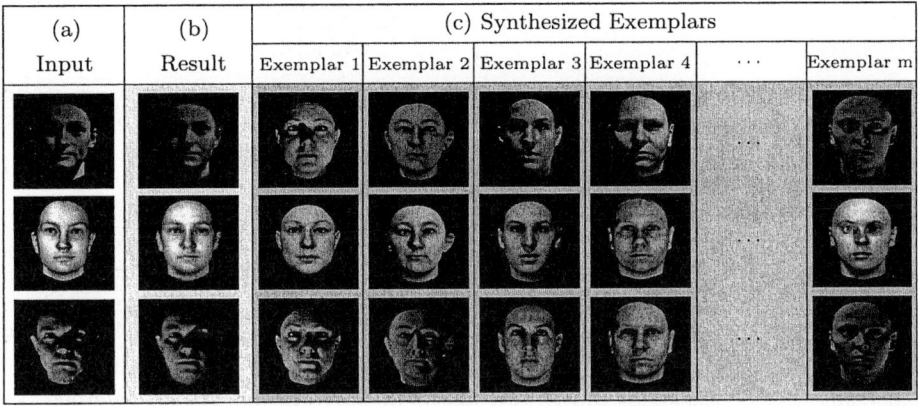

Fig. 2. Example of the reconstructed input images

method. The reconstructed image H^R of the input image H is represented as follows.

$$H^R = \sum_{k=1}^{m} f_k^* \tilde{E}_k = \boldsymbol{f}^* \tilde{\mathbf{E}} \tag{13}$$

By using Eq.(13), we can get the optimal weight coefficient vector to represent the input image. To verify the coefficients as the signature identity, we reconstruct the input image using the computed coefficients. Fig. 2 shows the example of reconstructed images using coefficients $\boldsymbol{f}^*$. In this figure, (a) shows the input image under the arbitrary lighting condition, (b) shows the reconstructed images using the linear combination of exemplars and (c) is the synthesized exemplars with the input image as illumination reference.

3.3 Recognition

In this section, we describe what kind of signature is used for recognizing the face. We use the linear coefficients of the synthesized exemplars for face recognition. When the gallery or probe image is taken, we synthesize exemplars from the photometric stereo images with each gallery or probe image. We analyze the input image, gallery and probe image, by the synthesized exemplars. We can then obtain the linear coefficients of both the gallery image and probe image, those coefficients are used the signatures for face recognition. Suppose that a gallery image G has its signature $\boldsymbol{f}_g^*$ and a probe image P has its signature $\boldsymbol{f}_p^*$.

$$\boldsymbol{f}_g^* = (\tilde{\mathbf{E}}_g^T \tilde{\mathbf{E}}_g)^{-1} \tilde{\mathbf{E}}_\mathbf{g}^T G, \quad \boldsymbol{f}_p^* = (\tilde{\mathbf{E}}_p^T \tilde{\mathbf{E}}_p)^{-1} \tilde{\mathbf{E}}_\mathbf{p}^T P, \tag{14}$$

where $\tilde{\mathbf{E}}_g$ and $\tilde{\mathbf{E}}_p$ are the matrices of synthesized exemplars using G and P as illumination reference image. The normalized correlation between a gallery and probe image is

$$corr(G, P) = \frac{Cov(\boldsymbol{f}_g^*, \boldsymbol{f}_p^*)}{sd(\boldsymbol{f}_g^*)sd(\boldsymbol{f}_p^*)} \tag{15}$$

where $sd(a)$ is the standard deviation of a and $Cov(a, b)$ means the covariance of a and b.

4 Experiments

We have conducted a number of experiments with our approach using the MPI (Max-Planck Institute) Face Database [5]. In these experiments, we compared the proposed method with 'Eigenface/WO3 [7]' and 'Bilinear analysis' [3] method. To solve the illumination problem, this method is applied without three principal components, the most influential factor in degradation of performance. We also implemented the bilinear analysis method for comparison.

4.1 Face Database

The MPI Face Database is used to demonstrate our proposed approach. We use 200 two-dimensional images of Caucasian faces that were rendered from a database of three-dimensional head models recorded with a laser scanner ($Cyberware^{TM}$) [6]. The images were rendered from a viewpoint 120cm in front of each face with ambient light only. For training, we use the images of 100 people. We use 25 face images in different illumination conditions, from $-60°$ to $+60°$ in the yaw axis and from $-60°$ to $+60°$ in the pitch axis, per person.

4.2 Experimental Results and Analysis

We present the recognition results when the images of training and testing sets are taken from the same database. We have conducted two experiments by changing the lighting directions of the gallery and probe set.

Gallery set of fixed direction and probe set of all lighting directions: Graph in Fig. 3 shows the position configuration of the lights and the recognition rates for the fixed gallery set of lighting conditions(100 images) with the probe sets of varying lighting conditions (100 images under each illumination). We use the gallery set under the first lighting condition, $L11$ and the probe sets under the other 24 lighting conditions, from $L12$ to $L55$ in the testing set. In this experiment, we obtain good recognition results although the illumination changes are rapidly. As shown Fig. 3, when the distance between the light sources of the gallery and probe sets are small, the recognition performance is high, conversely when the distance between the two are large, the recognition results are of lower quality, especially when using the eigenface/WO3 and the bilinear methods. The bilinear analysis method allows higher recognition rates than the eigenface/WO3 method, though neither method results in as high a performance as our proposed method.

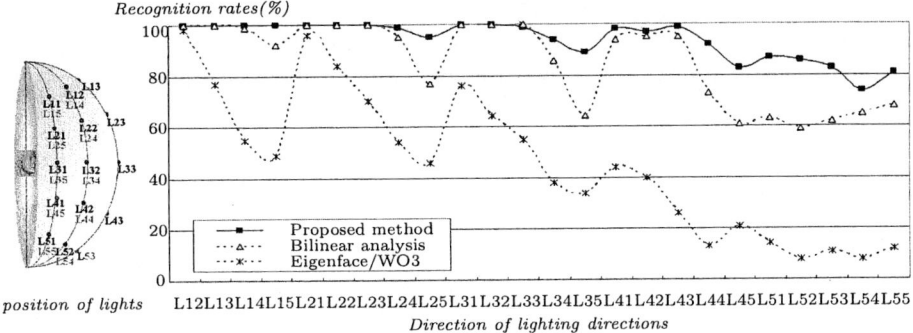

Fig. 3. Recognition rates for all the probe sets with a fixed gallery set

Gallery set and probe set of varying lighting directions: The next experiment is designed for the gallery and probe sets both under different lighting conditions. Table 1 represents the comparison results between our approach and bilinear analysis approach for the gallery and probe sets under the different directions of lighting, $\{L11, L22, L33, L44, L55\}$. P means the probe sets and G means the gallery sets. The right number is for bilinear analysis approach and the left one for our approach. The average rates obtained by bilinear analysis are 88.9%, while our approach outperforms it at an average of 95.1%.

Table 1. Recognition rates comparison

G \ P	L11	L22	L33	L44	L55	Avg.
L11	-	100/100	99/100	92/73	81/68	94.4/88.2
L22	100/100	-	100/100	100/79	86/51	97.2/86.0
L33	99/100	100/100	-	100/100	99/100	99.6/100.0
L44	85/77	99/87	100/100	-	100/100	96.8/92.8
L55	63/43	79/47	95/97	100/100	-	87.4/77.4
Avg.	89.4/84.0	95.6/86.8	98.8/99.4	98.4/90.4	93.2/71.0	95.1/88.9

5 Conclusions and Future Work

We have addressed a new approach for illumination invariant face recognition. The idea here is to synthesize exemplars using photometric stereo images and apply them to represent the new input image under the arbitrary illumination, while only one input image and one registered image per person are required for recognition. The weight coefficients are used as the signature identity, so that a new image can be represented as a linear combination of a small number of exemplars of training data. Experimental results on various face images have shown a good performance when compared with the previous approaches and our approach also shows a stable recognition performance even under the large changes of illumination. In the future, we need to make more experiments with the other face database. Furthermore, it can become particularly difficult when illumination is coupled with pose variation. Because there are the extreme lighting changes which are caused by pose variation, we are also trying to treat not only lighting changes but also pose changes.

Acknowledgments

This research was supported by the Intelligent Robotics Development Program, one of the 21st Century Frontier R&D Programs funded by the Ministry of Commerce, Industry and Energy of Korea and we would like to thank the Max-Planck-Institute for providing the MPI Face Database.

References

1. Adini, Y., Moses, Y., Ullman, S.: Face Recognition : the Problem of Compensating for Changes in Illumination Direction. IEEE Transactions on Pattern Analysis and Machine Intelligence, 19(7) (1997) 721–732
2. Basriand, R., Jacobs, D.: Photometric Stereo with General, Unknown Lighting. Proc. of IEEE Computer Society Conference on Computer Vision and Pattern Recognition, 2 (2001) 374–381
3. Freeman, W.T., Tenenbaum, J.B.: Learning bilinear models for two-factor problems in vision. Proc. of IEEE Computer Society Conference on Computer Vision and Pattern Recognition, 17 (1997) 554–560
4. Shashua, A., Raviv, T.R.: The Quotient Image : Class Based Re-rendering and Recognition with Varying Illuminations. IEEE Transactions on Pattern Analysis and Machine Intelligence, 23(2) (2001) 129–139
5. Blanz, V., Vetter, T.: Face recognition based on fitting a 3D morphable model. IEEE Transactions on Pattern Analysis and Machine Intelligence, 25(9) (2003) 1063–1074
6. Blanz, V., Romdhani, S., Vetter, T.: Face Identification across Different Poses and Illuminations with a 3D Morphable Model. Proc. of the 5th International Conference on Automatic Face and Gesture Recognition (2002) 202–207
7. Turk, M., Pentland, A.: Eigenfaces for recognition. Journal of Congnitive Neutoscience, 3 (1991) 72–86

Low-Dimensional Facial Image Representation Using FLD and MDS

Jongmoo Choi[1] and Juneho Yi[2]

[1] Intelligent Systems Research Center,
Sungkyunkwan University, Korea
[2] School of Information and Communication Engineering,
Sungkyunkwan University, Korea,
Biometrics Engineering Research Center
{jmchoi, jhyi}@ece.skku.ac.kr

Abstract. We present a technique for low-dimensional representation of facial images that achieve graceful degradation of recognition performance. We have observed that if data is well-clustered into classes, features extracted from a topologically continuous transformation of the data are appropriate for recognition when low-dimensional features are to be used. Based on this idea, our technique is composed of two consecutive transformations of the input data. The first transformation is concerned with best separation of the input data into classes and the second focuses on the transformation that the distance relationship between data points before and after the transformation is kept as closely as possible. We employ FLD (Linear Discriminant Analysis) for the first transformation, and classical MDS (Multi-Dimensional Scaling) for the second transformation. We also present a nonlinear extension of the MDS by 'kernel trick'. We have evaluated the recognition performance of our algorithms: FLD combined with MDS and FLD combined with kernel MDS. Experimental results using FERET facial image database show that the recognition performances degrade gracefully when low-dimensional features are used.

1 Introduction

This research presents a technique for low-dimensional data representation of which the recognition performance degrades gracefully. The technique reduces dimension of high-dimensional input data as much as possible, while preserving the information necessary for the pattern classification. The algorithms like PCA (Principal Components Analysis) [1][2], FLD (Fishers Linear Discriminant) [3][4] and ICA (Independent Components Analysis) [5][6] can be used for reduction of the dimension of the input data but are not appropriate for low-dimensional representation of high dimensional data because their recognition performance degrade significantly. For low-dimensional data representation, SOFM (Self-Organizing Feature Map) [7], PP (Projection Pursuit) [8] and MDS (Multi-Dimensional Scaling) [9] are proposed. These techniques suitable for data representation in low-dimensions, usually two or three dimensions. They try to represent the data points in a such way that the distances between points in low-dimensional space correspond to the dissimilarities between points in the original high dimensional space. However, these techniques do not yield high recognition rates mainly because they do not consider class specific information. Our idea is that these methods incorporated with class specific information can provide high recognition rates.

We have found that if data is well-clustered into classes, features extracted from a topologically continuous transformation of the data are appropriate for recognition when extremely low-dimensional features are to be used. Based on this idea, we first apply a transformation to the input data to achieve the most separation of classes, and then apply another transformation to maintain the topological continuity of the data that the first transformation produces. By Topological continuity [7], we mean that the distribution of data before and after dimensional reduction is similar in the sense that the distance relationship between data points is maintained.

To experimentally prove our claim, we have proposed a technique for extremely low-dimensional representation of data with graceful degradation of recognition performance. It is composed of two consecutive transformations of the input data. The first transformation is concerned with best separation of the input data into classes and the second focuses on the transformation in the sense that the distance relationship between data points is kept. The technique employs FLD and MDS for the transformations. MDS preserves the distance relationship before and after the data is transformed as closely as possible. This way, it is possible to represent data in low-dimensions without serious degradation of recognition performance. We have extended the classical (linear) MDS to a nonlinear version using 'kernel trick' [10]. The kernel MDS also preserves the distance relationship within nonlinear transformations.

The following section gives a brief overview of the feature extraction and dimensional reduction methods that have preciously been used for object recognition. In section 3, we describe the proposed FLD combined with MDS and the FLD combined with kernel MDS, respectively. (Let us call them 'FLD+MDS' and 'FLD+KMDS', respectively.) We report the experimental results on the recognition performance of FLD+MDS and FLD+KMDS in section 4.

2 Dimensional Reduction and Topological Continuity

There have been reported many algorithms for dimensional reduction and feature extraction. One group of dimensional reduction methods can be referred to as *topology-preserving mapping* and another group as *well-clustered mapping*. Among the former group are SOFM, MDS and GTM (Generative Topographic Mapping) [11] and these methods are used mainly for data visualization or data compression. FLD and Kernel FLD [12] are examples of the latter group and are mostly used for pattern classification [13].

We can achieve very low-dimensional data representation with graceful degradation of performance by using a *topology-preserving map* when the data is well clustered into classes. However, the typical facial image data in real environments do not have well-clustered distribution and it is not guaranteed to achieve high classification performance by a *topology-preserving map* although we can get a low-dimensional data set. Accordingly, we have to focus more on the discriminant power rather than dimensional reduction in the case.

3 Our Methods for Low-Dimensional Data Representation

3.1 Two-Stage Dimensional Reduction

We present two methods for extremely low-dimensional data representation by applying two different transformations in a row. The first stage is only concerned with best separation of classes. Once the data is rendered well-separated into classes by the first stage transformation, the second stage transformation only focuses on preservation of topological continuity before and after the transformation of the data. As previously described, the idea is based on the fact that if data is well-clustered into classes, features extracted from a topologically continuous transformation of the data are appropriate for recognition when extremely low-dimensional features are to be used.

3.2 Method I: FLD+MDS

FLD. Let $\mathbf{x}_k \in \mathbb{R}^N, k = 1, \cdots, M$ be a set of training data. FLD produces a linear discriminant function $\mathbf{f}(\mathbf{x}) = \mathbf{W}^T \mathbf{x}$ which maps the input data onto the classification space. We have employed FLD (Fisher's linear discriminant) as an instance of FLD techniques. FLD finds a matrix $\mathbf{W}$ that maximizes

$$J(\mathbf{W}) = \frac{|\mathbf{W}^T \mathbf{S}_b \mathbf{W}|}{|\mathbf{W}^T \mathbf{S}_w \mathbf{W}|} \quad (1)$$

where $\mathbf{S}_b$ and $\mathbf{S}_w$ are between- and within-class scatter matrices, respectively. $\mathbf{W}$ is computed by maximizing $J(\mathbf{W})$. That is, we find a subspace where, for the data projected onto the subspace, between-class variance is maximized while minimizing within-class variance. As a result of the first transformation, we obtain $\mathbf{z} = \mathbf{W}^T \mathbf{x}$.

After the stage of FLD, the next stage maps $\mathbf{z}$ onto a low-dimensional feature space $\mathbf{f} = \mathbf{G}(\mathbf{z})$ by MDS.

Classical MDS. Let us $\mathbf{x}_k \in \mathbb{R}^N, k = 1, \cdots, M$ be a set of observations and $\mathbf{D}$ be a dissimilarity matrix. Classical MDS is an algebraic method to find a set of points in low-dimensional space so that the dissimilarity are well-approximated by the interpoint distances.

In summary, the inner product matrix of raw data $\mathbf{B} = \mathbf{X}^T\mathbf{X}$ can be computed by $\mathbf{B} = -\frac{1}{2}\mathbf{HDH}$, where $\mathbf{X}$ is the data matrix $\mathbf{X} = [\mathbf{x}_1, \cdots, \mathbf{x}_M] \in \mathbb{R}^{N \times M}$ and $\mathbf{H}$ is a centering matrix $\mathbf{H} = \mathbf{I} - \frac{1}{M}\mathbf{1}^T\mathbf{1}$. $\mathbf{B}$ is real, symmetric and positive semi-definite. Let the eigendecomposition of $\mathbf{B}$ be $\mathbf{B} = \mathbf{V}\mathbf{\Lambda}\mathbf{V}^T$, where $\mathbf{\Lambda}$ is a diagonal matrix and $\mathbf{V}$ is a matrix whose columns are the eigenvectors of $\mathbf{B}$. The matrix $\hat{\mathbf{X}}$ for low-dimensional feature vectors can be obtained as $\hat{\mathbf{X}} = \mathbf{\Lambda}_k^{1/2}\mathbf{V}_k^T$ where $\mathbf{\Lambda}_k^{1/2}$ is a diagonal matrix of k largest eigenvalues and $\mathbf{V}_k$ is its corresponding eigenvectors matrix. Thus, we can compute a set of feature vectors, $\hat{\mathbf{X}}$, for a low-dimensional representation. See [14] for a detailed description.

Mapping onto an MDS Subspace via PCA. We could not map new input vectors to features by using the classical MDS because the map is not explicitly defined in the classical MDS. We used a method that achieves mapping onto an MDS subspace via PCA based on the relationship between MDS and PCA. Let $\mathbf{Y}_{\mathbf{MDS}}$ be a set of feature vectors in an MDS subspace and $\mathbf{Y}_{\mathbf{PCA}}$ be a set of feature vectors in a PCA subspace. Let $\mathbf{\Lambda}_{\mathbf{MDS}}$ denotes the digonal matrix of eigenvalues of inner product matrix $\mathbf{B}$. Then, the relationship between PCA and MDS is

$$\mathbf{Y}_{\mathbf{PCA}} = \mathbf{\Lambda}_{\mathbf{MDS}}^{1/2}\mathbf{Y}_{\mathbf{MDS}}. \qquad (2)$$

The derivation of equation (2) is described in the following. For centered data, the covariance matrix is $\mathbf{\Sigma} = \mathbf{E}\{\mathbf{X}\mathbf{X}^T\} = \frac{1}{M}\mathbf{X}\mathbf{X}^T$. PCA is concerned with the eigendecomposition of the covariance matrix as follows;

$$\mathbf{\Sigma}\mathbf{V}_{\mathbf{PCA}} = \frac{1}{M}\mathbf{X}\mathbf{X}^T\mathbf{V}_{\mathbf{PCA}} = \mathbf{V}_{\mathbf{PCA}}\mathbf{\Lambda}_{\mathbf{PCA}}. \qquad (3)$$

MDS is concerned with the eigendecomposition of the inner product matrix $\mathbf{B} = \mathbf{X}^T\mathbf{X}$ as follows;

$$\mathbf{B}\mathbf{V}_{\mathbf{MDS}} = \mathbf{X}^T\mathbf{X}\mathbf{V}_{\mathbf{MDS}} = \mathbf{V}_{\mathbf{MDS}}\mathbf{\Lambda}_{\mathbf{MDS}}. \qquad (4)$$

Using equations (3) and (4), we have

$$\mathbf{X}\mathbf{X}^T(\mathbf{X}\mathbf{V}_{\mathbf{MDS}}) = (\mathbf{X}\mathbf{V}_{\mathbf{MDS}})\mathbf{\Lambda}_{\mathbf{MDS}} \qquad (5)$$

and $\mathbf{V}_{\mathbf{PCA}} = \mathbf{X}\mathbf{V}_{\mathbf{MDS}}$, where $\mathbf{\Lambda}_{\mathbf{PCA}} \simeq \mathbf{\Lambda}_{\mathbf{MDS}}$. The feature vector set of PCA subspace is

$$\mathbf{Y}_{\mathbf{PCA}} = \mathbf{\Lambda}_{\mathbf{MDS}}^{\frac{1}{2}}\mathbf{Y}_{\mathbf{MDS}}. \qquad (6)$$

Note that, whereas the classical MDS computes inner product matrix $\mathbf{B}$ from the given dissimilarity matrix $\mathbf{D}$ without using input patterns $\mathbf{X}$, in this dimensional reduction problem for pattern recognition, we can obtain $\mathbf{B}$ directly from the input patterns $\mathbf{X}$. (Let us call that 'projective classical MDS'.) For the purpose of low-dimensional feature extraction, we need to compute projections onto FLD and MDS subspaces. Let $\mathbf{p}$ be an input pattern, then the feature vector in FLD+MDS space becomes

$$\mathbf{f}_{\mathbf{FLD+MDS}} = (\mathbf{\Lambda}_{\mathbf{PCA}}^{-1/2})\mathbf{W}_{\mathbf{PCA}}^T\mathbf{W}_{\mathbf{FLD}}^T\,\mathbf{p}. \qquad (7)$$

3.3 Method II: FLD+KMDS

The FLD+MDS method finds feature vectors by a linear transformation in order to extract low-dimensional representation. The linear method is very simple and fast because we need a matrix computation only. Moreover, linear model is rather robust against noise and most likely will not overfit. However, linear methods are often too limited for real-world data. In order to consider non-linear extension of the FLD+MDS method, we focus on the MDS because the transformation concerns dimensional reduction and it may contain some distortion due to low dimensional reduction. We have extended the classical (linear) MDS to a nonlinear version using 'kernel trick' [10] and developed a two-stages low-dimensional feature extraction based on the nonlinear extension.

Kernel PCA. Recently kernel based nonlinear analysis has more been investigated in pattern recognition. The kernel trick can efficiently construct nonlinear relations of the input data in an implicit feature space obtained by nonlinear kernel mapping. The method only depends on inner products in the feature space but does not need to compute the feature space explicitly. Kernel PCA combines the kernel trick with PCA to find nonlinear principal components in the feature space. The kernel trick is used firstly to project the input data into an implicit space called feature space by nonlinear kernel mapping, then PCA is empolyed to this feature space, thus a nonlinear principal components can be yielded in the input data[10][15].

Let $\mathbf{x}_k, k = 1, \cdots, M, \mathbf{x}_k \in \mathbf{R}^N$ be centered samples and Φ be a non-linear mapping to some feature space F, $\Phi : \mathbf{R}^N \to F$. In order to perform PCA in high-dimensional feature space, we have to compute eigenvectors and eigenvalues of the covariance matrix $\mathbf{C}$ in the feature space. The covariance matrix in feature space is defined as follows;

$$\mathbf{C} \equiv \frac{1}{M} \sum_{j=1}^{M} \Phi(\mathbf{x}_j) \Phi(\mathbf{x}_j)^T \tag{8}$$

where $\Phi(\mathbf{x}_j)^T$ denotes the transpose of $\Phi(\mathbf{x}_j)$. We cannot compute directly in the case when F is very high dimensional. The eigenvectors can be represented the span of the training data, and this leads to a dual eigenvalue problem for the solution [10]. To extract nonlinear features from a test point, we compute the dot product between $\Phi(\mathbf{x})$ and the nth normalized eigenvector in feature space,

$$< \mathbf{v}^k, \Phi(\mathbf{x}) > = \sum_{i=1}^{M} \alpha_i^k (\Phi(\mathbf{x}_i) \cdot (\Phi(\mathbf{x}))). \tag{9}$$

A Kernel MDS. We present a kernel MDS algorithm that combines the kernel trick with classical MDS. In this algorithm, we simply modified the kernel PCA algorithm in order to extend the classical MDS. The kernel MDS performs 'projective classical MDS' in a nonlinear feature space as follows. The relationship between classical MDS and PCA is

$$y_{MDS} = \mathbf{D}^{-1/2}\mathbf{V}_{PCA}^{T}\mathbf{x} \qquad (10)$$

where y_{MDS} and y_{PCA} are denote features projected onto MDS subspace and PCA subspace, respectively. $\mathbf{D}$ and $\mathbf{V}_{PCA}$ represent a squared eigenvalue matrix and eigenvector matrix of PCA, respectively. Based on the relationship, we can replace the eigenvectors, α_j, of kernel PCA by scaled eigenvectors $\beta_j = \frac{1}{\sqrt{\lambda_j}}\alpha_j$ using corresponding eigenvalues. We compute a feature vector in a nonlinear MDS subspace by a simple modification as follows;

$$< \mathbf{v}^k, \mathbf{\Phi}(\mathbf{x}) > = \sum_{i=1}^{M} \beta_i^k (\mathbf{\Phi}(\mathbf{x}_i) \cdot (\mathbf{\Phi}(\mathbf{x})). \qquad (11)$$

Feature Extraction Using FLD+KMDS. For the purpose of low-dimensional feature extraction, we need to compute projections onto FLD and MDS subspaces. Let $\mathbf{p}$ be an input pattern, then the feature vector in FLD+KMDS space becomes

$$\mathbf{f_{FLD+kMDS}} = < \mathbf{v}^k, \mathbf{\Phi}(\mathbf{W_{FLD}^T W_{PCA}^T p}) >$$
$$= \sum_{i=1}^{M} \beta_i^k (\mathbf{\Phi}(\mathbf{x}_i) \cdot \mathbf{\Phi}(\mathbf{W_{FLD}^T p})). \qquad (12)$$

The kernel MDS shares some advantages with kernel PCA. While most other MDS solutions need a iterative optimization of a stress function, the advantage of the proposed method is that it is computed from an eigenproblem. On the contrary, we need more computatinal time than the linear case.

4 Experimental Results

We have evaluated the recognition performance of the proposed FLD+MDS and FLD+KMDS methods as follows. We have compared the recognition performance of PCA [2], FLD [4] and the proposed FLD+MDS and FLD+KMDS methods using a part of FERET database [16].

4.1 FERET Database and Experimental Method

The FERET Database is a set of facial images collected by NIST from 1993 to 1997. For preprocessing, we closely cropped all images in the database which include internal facial structures such as the eyebrow, eyes, nose, mouth and chin. The cropped images do not contain the facial contours. Each face image is downsampled to 50x50 to reduce the computational complexity and histogram equalization is applied.

The whole set of images, U, used in the experiment, consists of three subsets named 'ba', 'bj' and 'bk'. Basically, the whole set U contains images of 200 persons and each person in the U has three different images within the 'ba', 'bj' and 'bk' sets. The 'ba' set is a subset of 'fa' which has images with normal frontal facial expression. The 'bj' set is a subset of 'fb'. The images of 'fb' have some other frontal facial expressions. The 'ba' and 'bj' set contain 200 images of 200 persons, respectively. The 'bk' set is equal to the 'fc' of which images were taken with different cameras and under different lighting conditions. The 'bk' set contains 194 images of 194 persons.

For the experiment, we have divided the whole set U into training set (T), gallery set (G) and probe set (P). In order to get an unbiased result of performance evaluation, no one within the training set (T) is included in the gallery and the probe sets. i.e. $T \cap \{G \cup P\} = \emptyset$. The experiment consists of two sub-experiments; The first experiment is concerned with evaluation regarding normal facial expression changes. We use the 'ba' set as the gallery and the 'bj' set as the probe. The second experiment is to evaluate the performance under illumination changes. We have assigned the 'ba' set to the gallery and the 'bk' set to the probe. In addition, we randomly selected 50% of the whole set in each sub-experiment in order to reduce the influence of a particular training set because a facial recognition algorithm based on statistical learning depends on the selection of training images. Thus, a training set contains 100 persons in each sub-experiment.

We have compared the recognition performance of our FLD+MDS with that of FLD. In each algorithm, we have computed a linear transformation matrix that contains a set of basis vectors for a subspace using the training set, and then have transformed the entire patterns in the gallery set into feature vectors. For the test, each input pattern in the probe set was transformed into its corresponding feature vector. We used a nearest neighbor classifier for recognition.

4.2 Results

FLD+MDS. As shown in Figure 1, FLD+MDS method performs better than the others in the case of low-dimensional representation. The experimental results show that low-dimensional data representation with graceful degradation of recognition performance can be achieved by using an inter-distance preserving transformation after the input data is rendered well clustered into classes. The recognition rate for a given number of features in these figures was obtained by averaging thirty experiments.

Figure 1 shows the recognition rates of FLD+MDS for three different distance measures, L1, L2 and cosine. We can see that there is no significant performance difference between the three distance measures.

FLD+KMDS. Figure 2 shows the recognition rate as a function of the number of dimensions for standard PCA, FLD and the proposed FLD+KMDS method, averaged across 30 test sets in 'babj' and 'babk'. We used the L2 distance metric for comparison. In this comparison, We used kernel MDS at degree d=2 of the polynomial kernel. From this results, it can be found that the proposed FLD+MDS method outperforms the others when represented in a low-dimensional space.

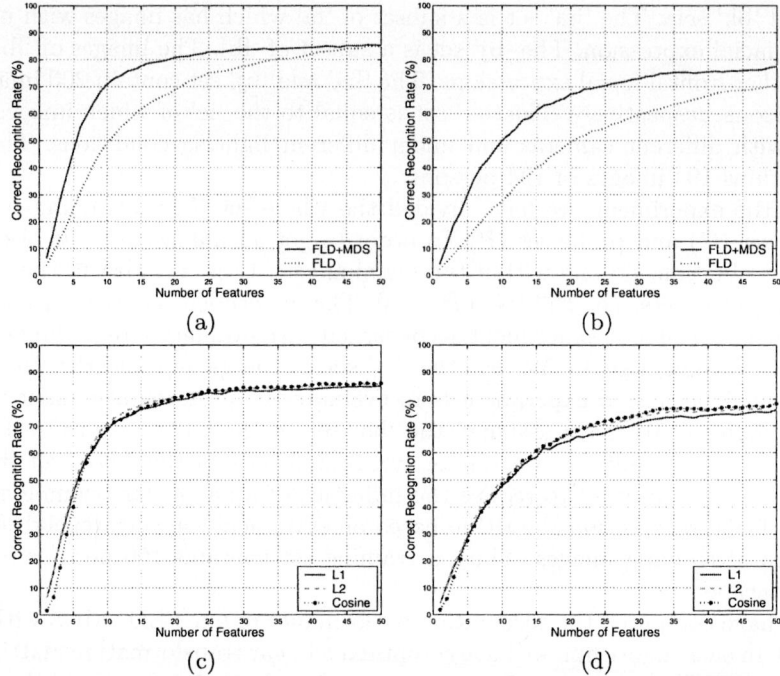

Fig. 1. Comparison of recognition rates: (a) and (b) represent recognition rates for 'ba'-'bj' set and 'ba'-'bk' set, respectively. (c) and (d) represent recognition rates for various distance measures in the case of 'ba'-'bj' set and 'ba'-'bk' set, respectively.

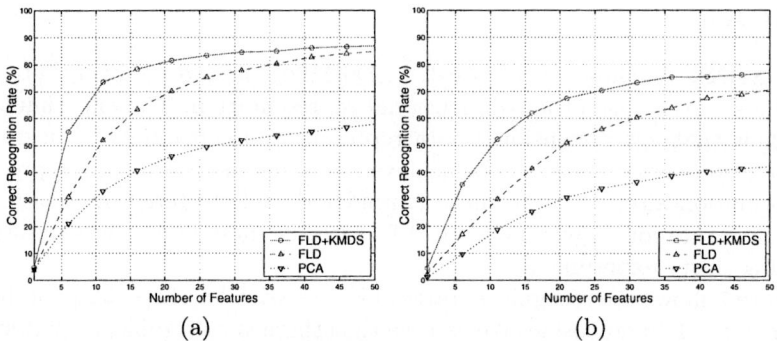

Fig. 2. Comparison of PCA, FLD, and FLD+KMDS algorithms: (a) and (b) represent recognition rates for 'ba'-'bj' set and 'ba'-'bk' set, respectively

The experiments in this FLD+KMDS are performed with the polynomial ($k(\mathbf{x}, \mathbf{x}') = \langle \mathbf{x}, \mathbf{x}' \rangle^2$), Gaussian RBF ($k(\mathbf{x}, \mathbf{x}') = \exp(-\|\mathbf{x} - \mathbf{x}'\|^2/(1.0 \times 10^7))$), and sigmoid kernels ($k(\mathbf{x}, \mathbf{x}') = \tanh(\langle \mathbf{x}, \mathbf{x}' \rangle/10^9 + 0.005)$), and the FLD+KMDS method is compared with standard PCA and FLD. Figure 3 shows the recognition rates for three kernel functions in 'babj' and 'babk', respectively. We can

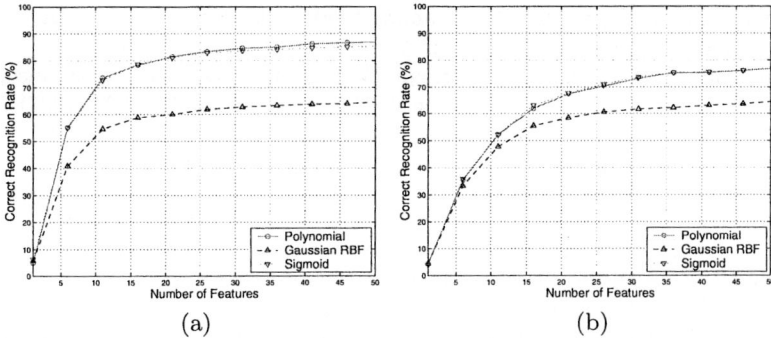

Fig. 3. Comparison of ploynomial (d=2), RBF, and Sigmoid kernel functions: (a) and (b) represent recognition rates for 'ba'-'bj' set and 'ba'-'bk' set, respectively

see that there is no significant performance difference between the sigmoid kernel and the polynomial kernel, and these two kernel functions outperform the Gaussian RBF kernel function.

5 Conclusion

This research presents a technique for low dimensional reduction of facial data that does not give significant degradation of the recognition rate. The FLD+MDS and FLD+KMDS methods outperform FLD method when represented in a low-dimensional space. These results experimentally prove that if data is tightly clustered and well separated into classes, a few features extracted from a topological continuous mapping of the data are appropriate low dimensional features for recognition without significant degradation of recognition performance.

Our methods are practically useful for face recognition, especially when facial feature data need to be stored in low capacity storing devices such as bar codes and smart cards. It is also readily applicable to real-time face recognition in the case of a large database. While representation of face images is the focus of this paper, the algorithm is sufficiently general to be applicable to a large variety of object recognition tasks.

Acknowledgement. This work was supported by the Korea Science Engineering Foundation (KOSEF) through the Biometrics Engineering Research Center (BERC) at Yonsei University and BK21.

References

1. Kirby, M., Sirovich, L.: Application of the Karhunen-Loeve Procedure for the Characterization of Human Faces. IEEE Trans. on PAMI. 12(1) (1990) 103-108
2. Turk, M., Pentland, A.: Eigenfaces for Recognition. Journal of Cognitive Neuroscience. 3(1) (1991) 71-86

3. Etemad, K., Chellappa, R.: Discriminant Analysis for Recognition of Human faces image. Journal of Optical Society of America. 14(8) (1997) 1724-1733
4. Belhumeur, P., Hespanha, J., Kriegman, D.: Eigenfaces vs. Fisherfaces: Recognition Using Class Specific Linear Projection. IEEE Trans. on PAMI. 19(7) (1997) 711-720
5. Bartlett, M.S., Martin, H., Sejnowski, T.J.: Independent Component Representations for Face Recognition. Proceedings of the SPIE. 3299 (1998) 528-539
6. Hyvärinen, A., Karhunen, J., Oja, E.: Independent Component Analysis, John Wiley & Sons, Inc. (2001)
7. Kohonen, T.: Self-Organizing Maps. Springer-Verlag (1995).
8. Friedman, J.K., Tukey, J.W.: A Projection Pursuit Algorithm for Exploratoty Data Analysis. IEEE Trans on computers. 23 (1974) 881-889
9. Duda, R. O., Hart, P.E., Stork, D.G.: Pattern Classification, John Wiley & Sons, Inc. (2001)
10. Schölkopf, B., Smola, A., Müller, K.: Nonlinear Component Analysis as a Kernel Eigenvalue Problem. Neural Computation, 10 (1998) 1299-1319, .
11. Bishop, C.M., Svensén, M.: GTM: The Generative Topographic Mapping. Neural Computation. 10(1) (1998) 215-234,
12. Mika, S., Rätsch, G., Weston, J., Schölkopf, B., Müller, K.R.: Fisher Discriminant Analysis with Kernels. IEEE Neural Networks for Signal Processing IX. (1999) 41-48
13. Carreira-Perpoñán, M.: A Reivew of Dimension Reduction Techniques. Technical Report CS-96-09. Dept. of Computer Science University of Sheffield. (1997)
14. Pcekalska, E., Paclík, P., Duin, R. P.W.: A Generalized Kernel Approach to Dissimilarity-based Classification. Journal of Machine Learning Research. 2 (2001) 175-211
15. Schölkopf, B., Smola, A.: Learning with Kernels. MIT Press. (2002)
16. Phillips, P.J., Moon, H.J., Rizvi, S.A., Rauss,. P.J.: The FERET Evaluation Methodology for Face-Recognition Algorithms. IEEE Trans. on PAMI. 22(10)(2000)1090–1104

Object Tracking with Probabilistic Hausdorff Distance Matching

Sang-Cheol Park and Seong-Whan Lee*

Department of Computer Science and Engineering, Korea University,
Anam-dong, Seongbuk-ku, Seoul 136-713, Korea
{scpark, swlee}@image.korea.ac.kr

Abstract. This paper proposes a new method of extracting and tracking a nonrigid object moving while allowing camera movement. For object extraction we first detect an object using watershed segmentation technique and then extract its contour points by approximating the boundary using the idea of feature point weighting. For object tracking we take the contour to estimate its motion in the next frame by the maximum likelihood method. The position of the object is estimated using a probabilistic Hausdorff measurement while the shape variation is modelled using a modified active contour model. The proposed method is highly tolerant to occlusion. Because the tracking result is stable unless an object is fully occluded during tracking, the proposed method can be applied to various applications.

1 Introduction

Thanks to the recent surge of interest in computer based video surveillance, video-based multimedia service and interactive broadcasting, the problem of tracking objects in video has attracted a lot of researchers around the world. Object tracking is so basic and in high demand that it is an indispensable component in many applications including robot vision, video surveillance, object based compression, etc. The problem of object tracking can be divided into two subproblems, extraction of a target object and tracking it over time. In general even those subtasks are not easy because of cluttered backgrounds and frequent occlusions. However, we can readily find a large number of studies on diverse methods using a variety of features such as color, edge, optical flow, and contour [1].

A target object under tracking involves motion, which we classify into three types; in the increasing difficulty of analysis, they are (1) object motion with a static camera, (2) camera motion over a static scene/object, and (3) simultaneous movement of camera and objects. To speak in terms of applications, the first type of motion has been heavily studied by the students in video surveillance, while the second type has been a prime target of analysis in video compression. The last type of hybrid motion, though more involved, has been an important topic in object based services, object based video compression, and interactive video.

---

* To whom all correspondence should be addressed.

This paper discusses an efficient method of extracting the contours of and tracking an interesting object in video from a non-static camera. First, object extraction steps estimate the rough contour of an object, and then locate and delineate the exact contour. Since the initial contour estimation starts from a given seed or a rough estimate, it is desired to make the initial error as small as possible. For this, we have developed a modified watershed algorithm based on Gaussian weighted edges. The resulting contour is then used to estimate the motion vectors and locate the objects in the succeeding frames. In the object tracking step, there are two subtasks, global motion estimation using the Hausdorff matching method and local deformation analysis using the modified active contour model.

The overall procedure is illustrated in Figure 1.

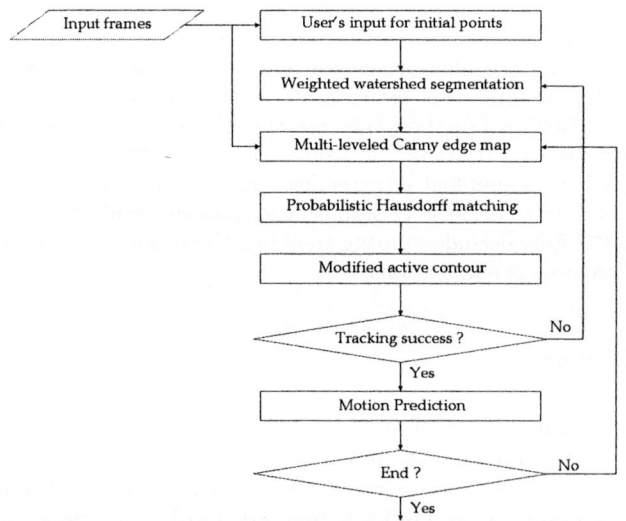

Fig. 1. The block diagram of the proposed method

2 Proposed Contour Extraction and Motion Tracking

2.1 Object Extraction with Watershed Algorithm

The watershed transform is a well-known method of choice for image segmentation in the field of mathematical morphology. The watershed transform can be classified as a region based segmentation approach. To date a number of algorithms have been developed to compute watershed transforms. They can be divided into two classes, one based on the specification of a recursive algorithm by Vincent and Soille[2], and the other based on distance functions by Meyer[4]. The details of the watershed algorithm of this paper are based on the works developed by Nguyen[3], and Vincent and Soille[2].

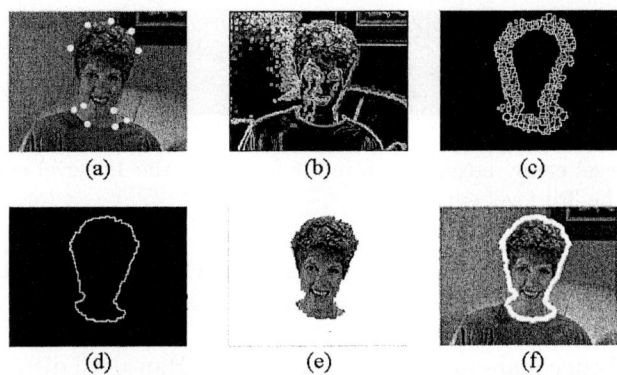

Fig. 2. Watershed segmentation process steps, (a) an input image and selected feature points (b) an edge image (c) initial small regions (d) merged region (e) segmented result (f) extracted object

Initial Watershed Segmentation. An important idea of the proposed method has a Gaussian weight to the initial feature points of an interesting object. This helps removing the need for selecting all detailed feature points of an object. Let $X = (X_1, ..., X_n)$ be an ordered sequence of feature points $X_i = [x_i, y_i]^t$, $i = 1, ..., n$. And let us denote the magnitude of the displacement from the origin to X be $m(X)$, and the Gaussian weight for the feature point X as $G(X)$. Then the weighted magnitude of the displacement vector to X for the watershed segmentation is given by

$$M_{new}(X_i) = G(X_i) * M(X_i)$$
$$G(X) = \frac{1}{\sqrt{2\pi}\sigma} e^{-\frac{x^2+y^2}{2\sigma^2}} \qquad (1)$$

where $*$ denotes 2D convolution operation. The watershed algorithm proceeds according to Equation 1 with a sequence of weighted feature points. The watershed segmentation algorithm is a region based technique. The result is a set of image regions dividing the input image.

Figure 2 illustrates the overall object extraction process, in which a numerous fragmental regions are produced and then merged into a single complete contour. In the region merging step, we simply merge the regions one by one until we're left with only two regions, foreground and background.

2.2 Contour Tracking of the Interesting Object

An object in an image is defined by a contour. A contour is sufficient for our task. The proposed tracking method takes the contour generated in preceding step to analyze the object contour's motion. The motion estimate is based on probabilistic Hausdorff distance measurement between the previous contour and all possible candidate regions.

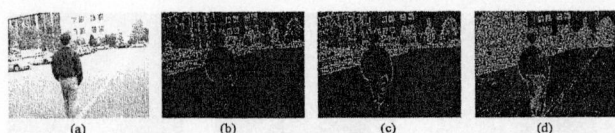

Fig. 3. Multi-level canny edge, (a) an input frame (b) the 1st level edge (c) the 2nd level edge (d) the 3rd level edge

Global Motion Tracking with Hausdorff Distance. For the candidate feature set in the next frame, we use the feature map which is calculated by multi-level canny edge detector.

These level edge maps are used to calculate the Hausdorff distance, then each level has the different weight. If we use three level edge maps, the real distance can be calculated by the following Equation 2

$$HD = H_{L1}(M,R) * \omega_1 + H_{L2}(M,R) * \omega_2 + H_{L3}(M,R) * \omega_3 \qquad (2)$$

where ω_i is the weight coefficient and H_{L1}, H_{L2} and H_{L3}, are the Hausdorff distance calculated from the first, second, and third level edge map respectively.

The Hausdorff matching score is the Hausdorff distance between two finite feature point sets[6]. The Hausdorff distance is used to measure the degree of similarity between the extracted contour and a search region in video frames. Let us assume that we are given by a sequence of reference feature points, $M = (m_1, m_2, ..., m_k)$ computed in the preceding frame, and let a new sequence of image feature points, $R = (r_1, r_2, ..., r_n)$ from the current frame. The Hausdorff distance between M and R can be written by

$$H(M,R) = max(h(M,R), h(R,M)) \qquad (3)$$

where

$$h(M,R) = \max_{m \in M} \min_{r \in R} \|m - r\| \qquad (4)$$

where $\|\;\|$ is the distance between two points, e.g., Euclidean norm. This distance is sensitive to noise and occlusion, we extend the function $h(M,R)$ so that the modified Hausdorff can measure the partial distance between a reference feature and an image feature. We use the likelihood function to find the position of the object.

First, to estimate the position of the model using the maximum likelihood, we define the set of distance measurements between reference feature points and input image feature points. Let $d(m,R)$ be the partial distance, we use some measurements of the partial distance. We consider the following distance measurements:

$$D_1(M,R) = \min_{m \in M} d(m,R) \qquad (5)$$

$$D_2(M,R) =^{75} K^{th}_{m \in M} d(m,R) \qquad (6)$$

$$D_3(M,R) = {}^{90}K^{th}_{m \in M} d(m,R) \qquad (7)$$

where the partial distance is defined as $d(m,R) = \min_{r \in R} \|m - r\|$ and ${}^x K^{th}_{m \in M}$ represents the Kth ranked distance that $K/m_k = x\%$[6].

$$D_4(M,R) = \min_{m \in M} \sum_W [I(m) - I(R)]^2 \qquad (8)$$

$$D_5(M,R) = \min_{m \in M} \sum_W |I(m) - I(R)| \qquad (9)$$

There are several distance measurements to estimate the partial distance, e.g., Euclidean distance, city-block distance, the distance transform, the sum of squared difference, optical-flow feature tracking method which can be used for the distance measurement defined as Equation 8.

Probabilistic Hausdorff Measurement. Let s be the position of an object in the current frame, then distance measurements are denoted $D_1(s), D_2(s),...,D_l(s)$. We use these distances to match maximum likelihood. In order to estimate the contour motion problem, we use the Olson's maximum-likelihood matching method[5]. Then the joint probability of the distance measurements $D_i(s)$ can be written as

$$p(D_1(s),...,D_m(s)|s) = \prod_{i=1}^{l} p(D_i(s)) \qquad (10)$$

where $p(D_i(s))$ is the probability density function of normalized distance of $D_i(s)$ at the position s, that is, $p(D_i(s)) = (1 - D_i(s))$, p must be maximized by Equation 10. The likelihood for p is formulated as the product of the prior probability of the position s and the probability in Equation 11:

$$L(s) = p(s) \prod_{i=1}^{m} p(D_i(s)). \qquad (11)$$

For convenience, we will take the logarithm of Equation 11 as follows Equation 12.

$$\ln L(t) = \ln p(t) + \sum_{i=1}^{m} \ln p(D_i(t)). \qquad (12)$$

We search s that maximizes this likelihood function, then this result uses to estimate the global motion for an interesting object.

Local Deformation with Active Contour Motion Model. The conventional active contour model consists of 2 terms, internal and external energy terms. The discrete energy function in the active contour model is defined as follows:

$$E(t) = \sum_t [E_{int}(t) + E_{ext}(t)] \qquad (13)$$

In this paper we propose a modified active contour motion model that adds an external motion term. The proposed energy function contains four terms: continuity, curvature, image force, and motion estimation confidence as follows:

$$E(t) = \sum_i [\alpha E_{cont}(v_i(t)) + \beta E_{curv}(v_i(t)) + \gamma E_{img}(v_i(t)) + \eta E_{match}(v_i(t))] \quad (14)$$

where contour energy functions, continuity energy, $E_{cont}(v(t))$, curvature energy, $E_{curv}(v(t))$, image force (edge/gradient), $E_{img}(v(t))$ and motion estimation confidence, $E_{match}(v(t))$. This last term in the above equation measures the motion variation between feature points in current frame and those in the previous frame. It can be defined as

$$E_{macth}(t) = \sum_i^n |\bar{v}(t) - v_i(t)|^2 \quad (15)$$

where $\bar{v}$ is the average motion vector at time t, v_i is the motion vector of the ith feature point in frame t.

The position of the contour must be predicted for the efficient tracking. To predict the position of the contour in the next frame, we calculate the following prediction energy, the predicted position of the contour at time $t+1$

$$E_{pre}(t+1) = E_{real}(t) + \sum_{v_i \in V} [\epsilon \cdot (E_{real}^{v_i}(t) - E_{real}^{v_i}(t-1)) + (1-\epsilon) \cdot (E_{real}^{v_i}(t) - E_{pre}(t))]$$

$$(16)$$

where $E_{pre}(t+1)$ is the predicted position of the contour at time $t+1$ and $E_{real}(t)$ is the real energy term of the moved contour at time t and ϵ is the matching rate between $E_{real}(t)$ and $E_{pre}(t)$. We proposed the probabilistic method to extract and to track an object using watershed and Hausdorff matching.

3 Experimental Results and Analysis

We have proposed the object tracking method using the modified Hausdorff distance matching and active contour algorithm. It has been tested over a set of VHS video clips recorded. The background includes streets, cars, indoor laboratory scenes. All the deformable targets are the human movements freely. The test images have been in monochrome MPEG-1 format with the dimension of 320 x 240. For comparison of the performance, we will refer to the method of Kass's active contour[7].

3.1 Simple Tracking

Figure 4 shows the three frames of a video sequence used in the work of Astrom[8]. We took the clip to compare the tracking performance with their method. The background of the video is relatively simple. The target objects include a cup,

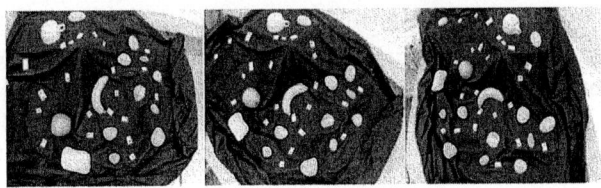

Fig. 4. Astrom and Kahl's test images

a pimento, a banana, an archetype and several polygonal objects. They are stationary, but the camera moves, apparently panning, tilting and zooming.

For the tracking performance $tr(t)$ at time t, we calculated the ratio of the number of pixels of the real area $np_r(t)$ to that of the segmented area $np_e(t)$. Also, for occlusion rate, $occ(t)$, at time, t, we used the pixel difference between pixels of the real area $np_r(t)$ and that of the occluded area $np_o(t)$.

$$tr(t) = \frac{np_e(t)}{np_r(t)} \qquad (17)$$

$$occ(t) = \frac{np_o(t)}{np_r(t)} \qquad (18)$$

Figure 5 shows the value of $tr(t)$ over the sequence of the video clip. The two straight lines downward show the performance degradation averaged over the entire sequences.

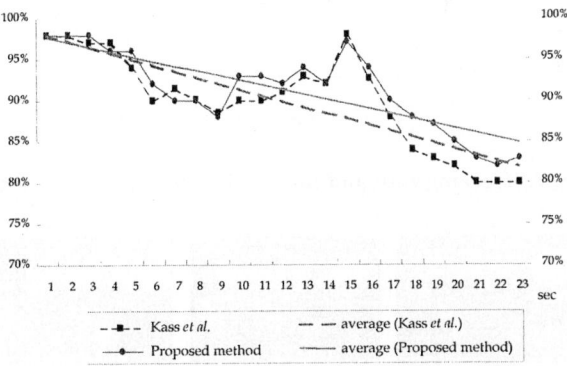

Fig. 5. The comparison to the tracking performance

3.2 Long Sequence Tracking Performance

Figure 6 shows selected samples from a video of 180 frames (numbered from 946 to 1126) where the human upper body is the target of tracking. The object boundary in the first frame was manually given by a user. It consists of a sequence of feature points appropriately defining the boundary. Note that the boundaries of subsequent tracking frames are highly accurate. The performance of tracking over the long sequences is summarized in Table 1 and Figure 6. The proposed method records 5 percent over the standard snake and template models. Moreover the accuracy degrades much slowly compared with those of snake and template.

Fig. 6. Human body tracking in a parking lot

Table 1. Experiment over a long sequence video

	average number of mis-tracked pixels	tracking rate
Snake model	255	88%
Adaptive template	287	86%
Proposed method	174	93%

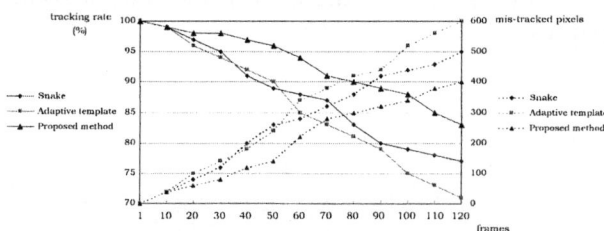

Fig. 7. Tracking rate and mis-tracked pixels for each frame

Fig. 8. The experimental result of a person tracking in a room

The object boundary was manually selected by a user in the first frame, several feature points of a object were simply selected by a user. Furthermore, a human body sequence captured in our lab was used to check the performance experiment in the different environment.

Figure 8 shows a successful result under the pose variation of the man. In order to compare the experimental results quantitatively, we measured the successful tracking rate and the mistaken pixels of tracking process from each frame in Figure 7.

4 Conclusions and Future Works

This paper proposed an effective contour tracking method for a deformable object which was described by a set of feature points. The proposed method works successfully in spite of non-static camera and nonlinear deformation over time with cluttered backgrounds. The extracting result has shown that the proposed method separates well target foreground objects from the background by analyzing the background texture.

Through a series of experimental results, we could confirm that the proposed method was quite stable and produced good results under object shape deformation and non-static camera motion. One problem with the implementation of the proposed system is that it sometimes fails if an object moves too fast. Another problem is that processing time is so slow, so our proposed method is not suited to realtime application. We believe that these problems can easily be overcome. The real issue of the paper is the performance and robustness of extracting and tracking deformable objects.

Acknowledgments

This research was supported by the Intelligent Robotics Development Program, one of the 21st Century Frontier R&D Programs funded by the Ministry of Commerce, Industry and Energy of Korea.

References

1. Leymarie, F., Levine, M.D.: Tracking Deformable Objects in the Plane Using an Active Contour Model. IEEE Trans. on Pattern Analysis and Machine Intelligence, Vol. 15, No. 6 (1996) 617–634
2. Vincent, L., Soille, P.: Watershed indigital spaces: an efficient algorithm based on immersion simulations. IEEE Trans. on Pattern Analysis and Machine Intelligence, Vol. 13 (1991) 583–598
3. Nguyen, H.T., Worring, M., Boongaard, R.: Watersnakes : Energy-Driven Watershed Segmentation. IEEE Trans. on Pattern Analysis and Machine Intelligence, Vol. 25, No. 3 (2003) 330–342
4. Meyer, F.: Topographic distance and watershed lines. Signal Processing, Vol. 38 (1994) 113–125

5. Olson, C.F.: Maximum-Likelihood Image Matching. IEEE Trans. on Pattern Analysis and Machine Intelligence, Vol. 24, No. 6 (2002) 853–857
6. Huttenlocher, D.P., Klanderman, G.A., Rucklidge, W.J.: Computing Images Using the Hausdorff Distance. IEEE Trans. on Pattern Analysis and Machine Intelligence, Vol. 15, No. 9 (1993) 850–863
7. Kass, M., Witkin, A., Terzopoulos, D.: Snakes: Active Contour Models. In Proc. of International Conference on Computer Vision, London, England, Vol. 1 (1987) 259–268
8. Astrom, K., Kahl, F.: Motion Estimation in Image Sequences Using the Deformation of Apparent Contours. IEEE Trans. on Pattern Analysis and Machine Intelligence, Vol. 21, No. 2 (1999) 114–127

Power Op-Amp Based Active Filter Design with Self Adjustable Gain Control by Neural Networks for EMI Noise Problem

Kayhan Gulez, Mehmet Uzunoglu, Omer Caglar Onar, and Bulent Vural,

Yildiz Technical University, Faculty of Electrical and Electronics,
Department of Electrical Engineering, 34349, Besiktas, Istanbul, Turkey
{gulez, uzunoglu, conar, bvural}@yildiz.edu.tr

Abstract. An induction motor control system fed by an AC/DC rectifier and a DC/AC inverter group is a nonlinear and EMI generating load causes harmonic distortions and EMI noise effects in power control systems. In this paper, a simulation model is designed for the control circuit and the harmonic effects of the nonlinear load are investigated using FFT analyses. Also, the EMI noise generated by the switching-mode power electronic devices measured using high frequency spectrum scopes. An LISN based active filter is used to damp the harmonic distortions and EMI noises in the simulation environment. Neural network based control system is used to tune the power op-amp gain of series active filter to obtain the voltage value stability at the equipment side as well.

1 Introduction

Recently, the broader use of power electronic based loads (rectifiers, inverters, motor control systems, etc.) has led to a growth of power pollution because of the nonlinear voltage-current characteristics of these loads. Current and voltage harmonics that are generated by the non-linear loads or by the switching devices in power electronics system may cause a serious damage to any electrical equipment [1]. Thus, load currents and voltages are nonsinusoidal and it is necessary to compensate voltage and current harmonics. Also, EMI noise caused by load characteristics and switching power electronic elements is an important problem for the network because of the conducted EMI emissions on the common mode line of the grid propagated by the switching elements. The compensation of these harmonics and EMI noise effects is recently being more and more important and causing widespread concern to the power system engineers have attracted special interests on active filtering.

The AF is classified into two types as series and shunt active filter. The series type active filter is installed series to the nonlinear loads or harmonic generating loads and works as the harmonic compensation voltage source. The shunt type active filter is usually installed parallel to the loads. It works as the current source and compensates the harmonic current of the load [2].

In the engineering environment, EMI filters are called as "Black Magic" because there has not been a well defined design method, the input and the output impedances in the related circuit are not constant over the band of interest and the filter insertion lost test method specifications often confuse or influence the design methods [1, 3]. In this paper, a three-phase active filter topology based on a basic one introduced and developed in [1], [3], [4], [5], [6] is reviewed and used.

In a series active filter application, usually the output voltage of the filter is lower than the input voltage of the filter because of the non-ideal cut-off characteristic of the filter. This non-ideal characteristic causes the filter to cut-off an amount of fundamental frequency component as leakage current. Also, the voltage drop on the filter and line impedances, occurs a lowered voltage at the load side. To provide the RMS voltage value stability at the equipment side, a self adjustable gain control mechanism is used in this study. In this paper, "Fast Back-propagation Algorithm" based artificial neural network architecture is taken as a control system to adjust the filter gain at fundamental frequency to maintain the voltage value stability as well. Artificial Neural Networks (ANN) is successfully used in a lot of areas such as control, DSP, state estimation and detection. Here, the results of a practical case simulation in MATLAB & Simulink are present using Simulink as a modeling environment and Neural Network Toolbox to create the neural network structure.

1.1 EMI Noise Effects

EMI noises have bad effects on power system elements. EMI noises caused by the switching-mode semiconductor devices or load characteristics can be divided into two parts: Conducted EMI Emissions and Radiated EMI Emissions. In this study, conducted EMI emissions are investigated in the simulation environment. Generally, EMI noises causes [7],

- Uncontrolled switching states on power electronic devices working near the EMI noise generating systems,
- Instability or oscillation problems in control systems,
- Parasitic effects on communication lines and data loss,
- Encoder feedback failures in motor control systems,
- Failures at programmable controller,
- Problems in remote controlled I/O systems,
- Wrong evaluations of sensing and measuring devices.

2 Simulation Network

A converter – inverter based induction motor control system is taken as a harmonic and EMI noise source in this study. A general block diagram of the system is given below in Fig. 1.

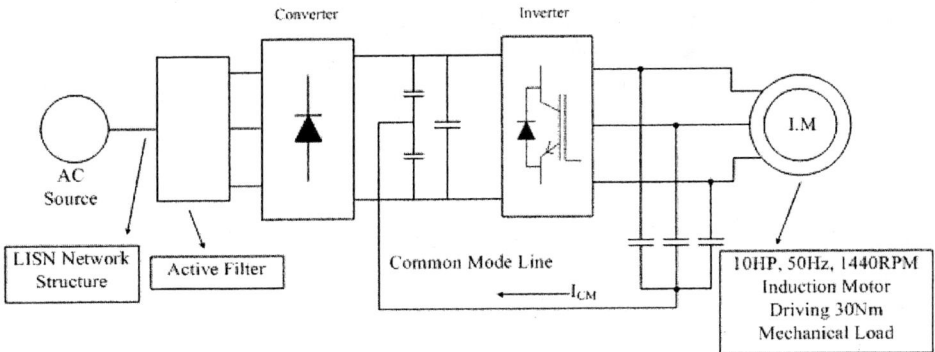

Fig. 1. Structure of the induction motor control system

The simulation network is modeled using MATLAB, Simulink and SimPowerSystems which provides an effective platform for dynamic system simulations [8, 9, and 10]. The Simulink model of the system is given in Fig. 2.

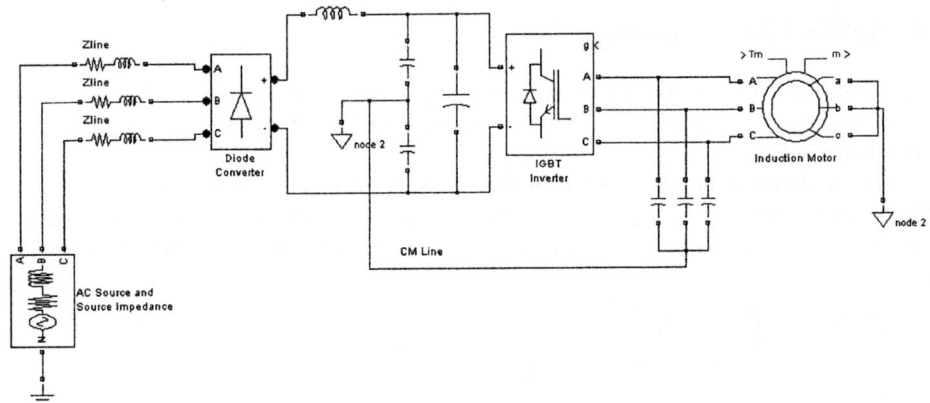

Fig. 2. Simulink model of the motor control system

3 EMI Measurement

A brief review of conducted EMI measurement is important before the filter design stages. The Line Impedance Stability Network (LISN) is based on military standards and is used by many electromagnetic interference and electromagnetic compatibility test institutions. The LISN, required in the measurement, contains 56µH inductors, 22.5µF capacitors and 50Ω resistors. The single phase schematic of LISN circuit is given in Fig. 3. of which topology is used by widespread EMI and EMC applications [3, 4, 5 and 6].

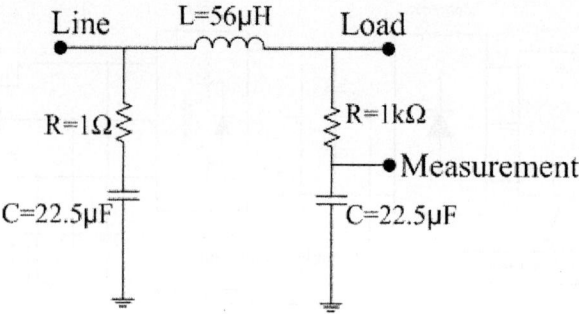

Fig. 3. Single phase schematic of LISN network

At EMI noise frequency, the inductors are essentially open circuit, the capacitors are essentially short circuit and the noise sees 50Ω measurement point. The noise voltage, measured from the 50Ω resistors contains both common-mode (CM) noise and differential-mode (DM) noise. Each mode of noise is dealt with by the respective section of an active filter.

4 Active Filter Topology

As mentioned before, the active filter topology used in the paper are based on and developed in [1], [3], [4], [5], [6]. The active filter topology is depicted as a basic schematic in Fig. 4.

The op-amp circuit is modeled in Simulink environment. In modeling procedure, the voltage controlled current source based equivalent circuit of an op-amp circuit taken as a sample. The Simulink model of the op-amp circuit is given below in Fig. 5.

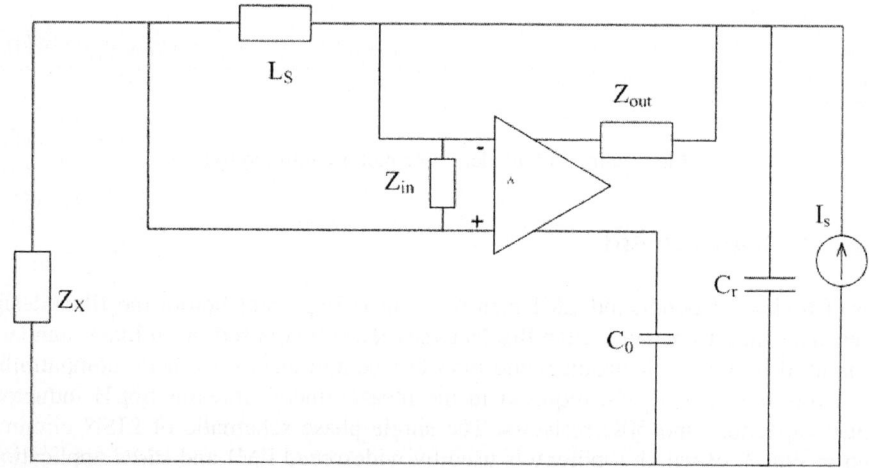

Fig. 4. Active filter topology

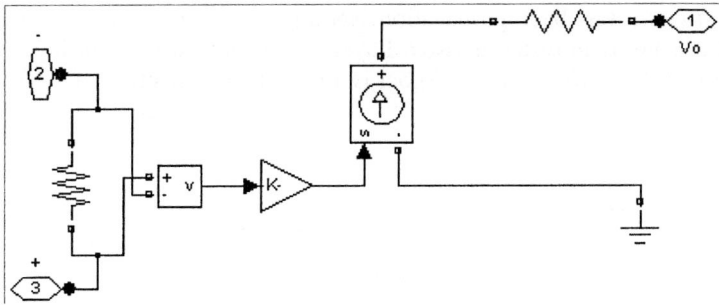

Fig. 5. The Simulink model of the equivalent op-amp circuit

The input and output impedance values and the internal gain of the op-amp are taken from a catalogue of a real power op-amp circuit especially used in industrial filtering applications.

In the active filter topology, the capacitor C_r absorbs most of the current source of power supply generating a ripple voltage which must be dropped across the active circuit with a minimal ripple current conducted through the utility. This function is realized and achieved by the amplifier with gain A, input and output impedances Z_{in} and Z_{out}, respectively. The other problem with the circuit is that the full line voltage is across the output of the active circuit. It requires that Z_{out} include a series blocking capacitor, C_O. Since this capacitor C_O appears in series with the output of the amplifier, its impedance is reduced by the magnitude of the amplifier's gain. The other important condition of active EMI filter is the compensation of feedback loop. The point is that the loop transmission is simply the product of the amplifier gain and the negative feed back [1, 2].

5 Artificial Neural Networks for Gain Control

A neural network can be trained to perform a particular function by adjusting the values of the connections (weights) between elements. Commonly neural networks are adjusted, or trained, so that a particular input leads to a specific target output. Such a situation is shown below. There, the network is adjusted, based on a comparison of the output and the target, until the network output matches the target. Typically many such input/target pairs are used, in this supervised learning, to train a network. The general diagram of such a training algorithm is given in Fig. 6 [11, 12].

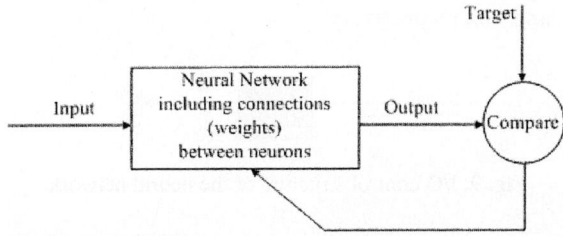

Fig. 6. General diagram of neural network training structure

In this study, a fast back-propagation NN algorithm with 2 hidden layers, which has 4 and 2 neurons in order, is used. Layers are tansig and output layer is purelin. Maximum epoch is 250 and wanted goal is 1e-3. The architecture of ANN controller is depicted in Fig. 7.

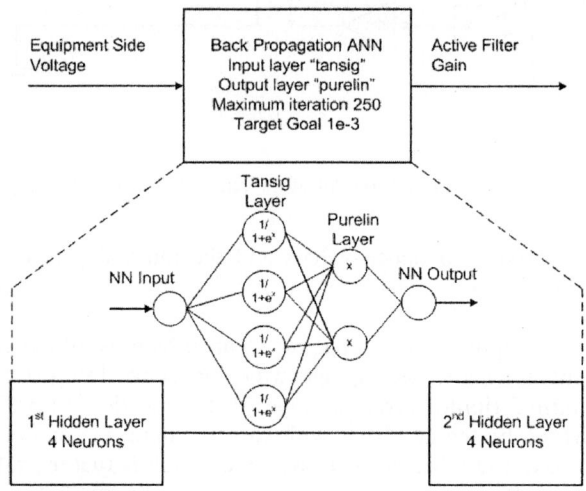

Fig. 7. Neural network structure control diagram

The neural network architecture including the layers is given in Fig. 8.

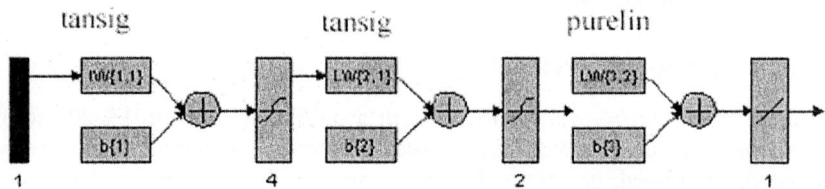

Fig. 8. Neural network layer and connection architecture

The Simulink diagram of the neural network structure is given in following figures. A general input/output structure of the neural network and weights connections are depicted in Fig. 9. and 10., respectively.

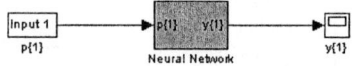

Fig. 9. I/O control structure of the neural network

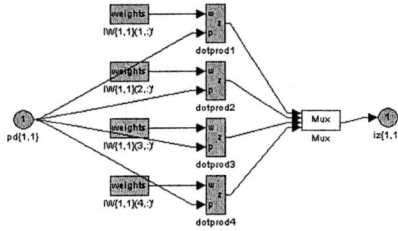

Fig. 10. Weights diagram

The gain value of the active filter is trained using the voltage values at the equipment side. To obtain the voltage value stability at the equipment side, the gain of the active filter is determined using the neural network algorithm given above.

Some values of phase-ground RMS voltage of the load side used in the training process is given in Table 1.

Table 1. Some values of training set of ANN

Voltage (V)	Gain
220	1
219	1.0046
218	1.0092
217	1.0138
216	1.0185
215	1.02223

The training performance is depicted in Fig. 11. including the goal and training performance of the neural network.

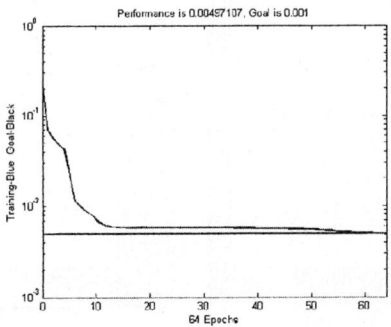

Fig. 11. Neural network training performance

6 Simulation Results

The simulation results are investigated in terms of two other criteria; the EMI noise suppress performance of the active filter and the neural networks based adaptive gain control performance of the filter, respectively.

6.1 Active Filter EMI Suppress Performance Analysis

In the induction motor control system, the common mode EMI noises are distributed to the other power system and control components by the common mode line of the inverter output phases. The common mode current waveform including high frequency EMI noises is given in Fig. 12 which obtained as a simulation result.

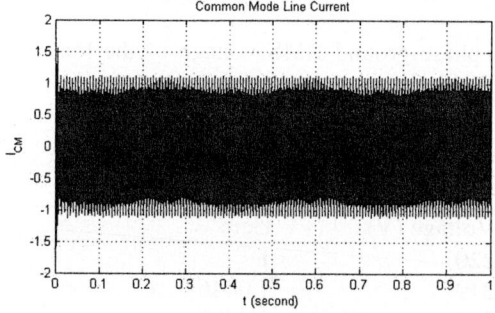

Fig. 12. Common mode line current of induction motor control system

The spectrum analyzer result showing noise condition before and after the active filter is given in Fig. 13.a and Fig. 13.b, respectively.

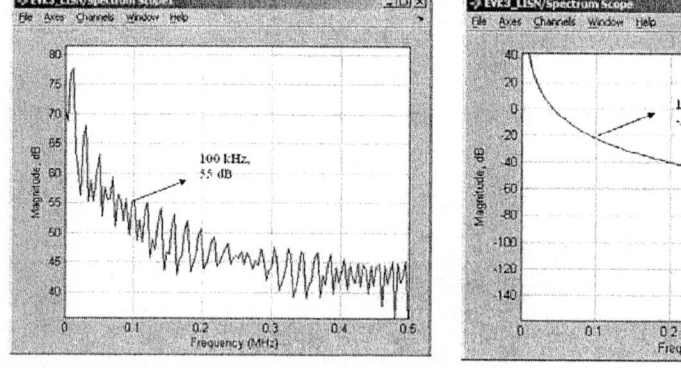

Fig. 13.a and **Fig. 13.b** Spectrum analyzer results and EMI suppress performance of the filter

6.2 Active Filter Gain Control Performance

In the neural network based active filter gain control performance, another induction motor control system with the same motor and mechanic load parameters is connected to equipment side of the power system. Because of the extra power demand and noise effects of the second induction motor control system, the filter loss and the voltage drop on the line and filter impedances are increased and the equipment side RMS voltage value is decreased. To connect the other motor and the control system, a three-phase switch is used closing at simulation time t=0.5 second. The simulation system used to analyze the gain control performance is given below in Fig. 14.

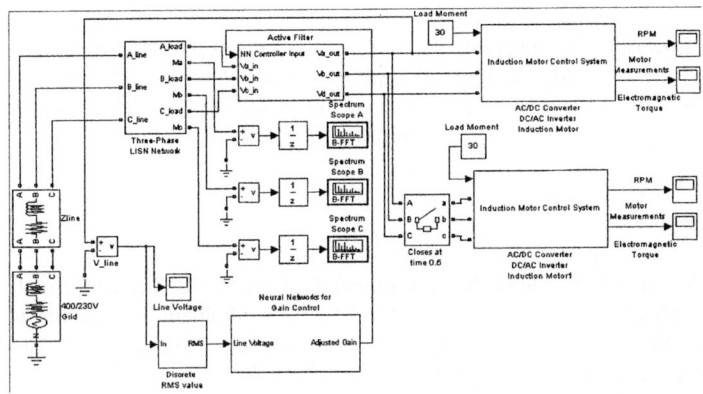

Fig. 14. Simulation model of the neural network based gain control

As given in Fig. 12., the line voltmeter and a RMS value calculator is used to measure the line-end voltage. The RMS value of the voltage is used as an input to the neural network based control system and the neural networks determine the gain of the active filter to maintain the voltage stability. The RMS value of the line end voltage of the power system is given with and without the NN based gain controller in Fig. 15.a and Fig. 15.b respectively.

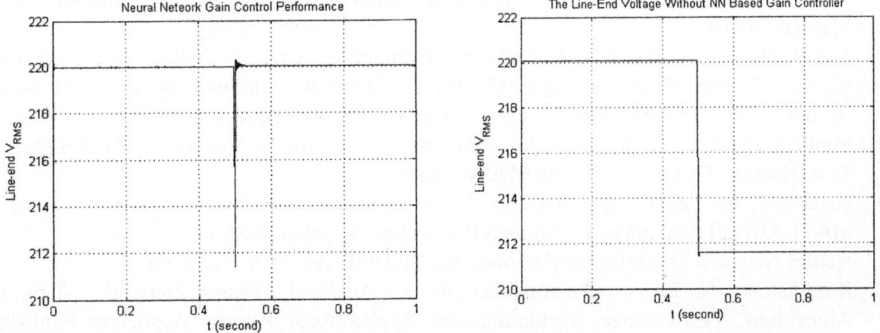

Fig. 15.a and **Fig. 15.b** Line end voltage with and without neural network controller

7 Conclusions

Current and voltage harmonics and EMI noise generated by the non-linear loads or by the switching devices in power electronics system may cause a serious damage to any electrical equipment. Also, EMI noise caused by switching power electronic elements is an important problem for the network. In this study, an active filter with self adjustable gain control by neural networks is designed for EMI noise problem in power control systems. The active filter gain is controlled by neural network algorithms to regulate the line end voltage which decreases under varying load conditions and filter loss. In the simulation results, not only the EMI noise effects suppressed but also the equipment side voltage is regulated.

Acknowledgment. The research carried out by the support of Scientific Research Coordinatorship of Yildiz Technical University for the research projects numbered 24-04-02-01 and 24-04-02-03. The authors are thankful for the support of Scientific Research Coordinatorship of Yildiz Technical University.

References

1. Gulez, K., Mutoh, N., Ogata, M., Harashima, F., Ohnishi, K.: Source Current Type Active Filter Application With Double LISN For EMI Noise Problem In Induction Motor Control Systems, SICE Annual Conference (2002) International Conference on Instrumentation, Control and Information Technology, Osaka, JAPAN
2. Peng, F., Z., Kohata, M., Akagi, H.: Compensation Characteristics of Shunt Active and Series Active Filter, T.IEE Japan, Vol. 110-D, No.1 (1993)
3. Ozenbough, R., L.: EMI Filter Design Second Edition Revised and Expanded, Mercel Dekker, Inc., New York, USA (2001)
4. Gulez, K., Hiroshi, Watanabe, Harashima, F.: Design of ANN (Artificial Neural Networks) Fast Backpropagation Algorithm Gain Scheduling Controller of Active Filtering, Vol. 1 TENCON (2000) 307-312
5. Farkas, T.: A Scientific Approach to EMI Reduction, M.S. Thesis, Massachusetts Institute of Technology, Cambridge, MA, August (1991)
6. Farkas, T.: Viability of Active EMI Filters for Utility Applications, IEEE Trans. On Power Electronics, Vol. 9 (1994) 328-336
7. MTE Corporation: EMI/RFI Filters, EMI Sources, Standards and Solutions Online Manual (2000)
8. MATLAB The Language of Technical Computing: Getting Started with MATLAB Version 7, The MathWorks Inc (MATLAB, Simulink, SimPowerSystems and Neural Network Toolbox are registred trademarks of The MathWorks Inc.)
9. SimPowerSystems For Use with Simulink: User's Guide Version 4, Hydro-Québec TransÉnergie Technologies, The MathWorks Inc.
10. Uzunoglu, M., Kızıl, A., Onar, O. C.: Her Yonu ile MATLAB 2e (Prificiency in MATLAB, 2e) (in Turkish), Turkmen Publication, Istanbul (2003)
11. Neural Network Toolbox: User's Guide Version 4.0, The MathWorks Inc.
12. Karayiannis, N. B., Venetsanopoulas, A. N.: Artificial Neural Networks –Learning Algorithms, Performance Evaluation and Applications, Kluwer Academic Publishers (1994) 161-195

Leaf Recognition Based on the Combination of Wavelet Transform and Gaussian Interpolation[1]

Xiao Gu[1], Ji-Xiang Du[1,2], and Xiao-Feng Wang[1]

[1] Institute of Intelligent Machines, Chinese Academy of Sciences, P.O.Box 1130,
Hefei, Anhui 230031, China
[2] Department of Automation, University of Science and Technology of China,
Hefei 230027, China
{xgu, du_jx, xfwang}@iim.ac.cn

Abstract: In this paper, a new approach for leaf recognition using the result of segmentation of leaf's skeleton based on the combination of wavelet transform (WT) and Gaussian interpolation is proposed. And then the classifiers, a nearest neighbor classifier (1-NN), a k-nearest neighbor classifier (k-NN) and a radial basis probabilistic neural network (RBPNN) are used, based on run-length features (RF) extracted from the skeleton to recognize the leaves. Finally, the effectiveness and efficiency of the proposed method is demonstrated by several experiments. The results show that the skeleton can be successfully and obviously extracted from the whole leaf, and the recognition rates of leaves based on their skeleton can be greatly improved.

1 Introduction

Plant recognition by computers automatically is a very important task for agriculture, forestry, pharmacological science, and etc. In addition, with the deterioration of environments☐more and more rare plant species are at the margin of extinction. Many of rare plants have died out. So the investigation of the plant recognition can contribute to environmental protection. At present, plant recognition usually adopts following classification methods, such as color histogram, Fourier transform (FT), morphologic anatomy, cell biology, etc. [1].

Plant recognition can be performed in terms of the plants' shape, flowers, and leaves, barks, seeds, and so on. Nevertheless, most of them cannot be analyzed easily because of their complex 3D structures if based on simple 2D images. Therefore, in this paper, we study plant recognition by its leaf, exactly by leaf's skeleton. So we can use leaves skeleton as their texture features to recognize them.

In this paper, a new method of segmenting leaf's skeleton is proposed. That is, the combination of wavelet transform and Gaussian interpolation is used to extract the leaf's contour and venation as well as skeleton. Usually, the wavelet transform is of the capability of mapping an image into a low-resolution image space and a series of detail image spaces. However, ones only keep an eye on the low-resolution images,

---

[1] This work was supported by the National Natural Science Foundation of China (Nos.60472111 and 60405002).

while the detail images are ignored. For the majority of images, their detail images indicate the noise or uselessness of the original ones. But the so-called noise of leaf's image is just the leaf's skeleton. So, we intentionally adopt the detail images to produce leaf's skeleton. And then, the extraction of texture features will be carried out based on the leaf's skeleton. After that, we derive run-length features (RF) from leaf's skeleton, which can reflect the directivity of texture in the image. Therefore, after computing these features of leaves, different species of plants can be classified by using three classifiers such as 1-NN, k-NN and radial basis probabilistic neural network (RBPNN) [2].

The rest of this paper is organized as follows: Section 2 reviews the WT, and Section 3 introduces how to use the combination of WT and Gaussian interpolation to segment the skeleton of leaf, and gives the corresponding algorithm. Section 4 reports some experimental results on leaf recognition. Finally, some concluding remarks are included in Section 5.

2 The Review of Wavelet Transform

The wavelet transform (WT), a linear integral transform that maps $L^2(\mathbb{R}) \rightarrow L^2(\mathbb{R}^2)$, has emerged over the last two decades as a powerful new theoretical framework for the analysis and decomposition of signals and images at multi-resolutions [3]. Moreover, due to its both locations in time/space and in frequency, this transform is to completely differ from Fourier transform [4,5].

2.1 Wavelet Transform in 1D

The wavelet transform is defined as decomposition of a signal $f(t)$ using a series of elemental functions called as wavelets and scaling factors, which are created by scaling and translating a kernel function $\psi(t)$ referred to as the mother wavelet:

$$\psi_{ab}(t) = \frac{1}{\sqrt{a}} \psi(\frac{t-b}{a}) \qquad \text{where} \quad a, b \in \mathbb{R}, a \neq 0 . \tag{1}$$

Thus, the continuous wavelet transform (CWT) is defined as the inner product:

$$W_f(a,b) = \int_{-\infty}^{\infty} \psi_{ab}(t) \bar{f}(t) dt = \langle \psi_{ab}, f \rangle . \tag{2}$$

And, the discrete wavelet representation (DWT) can be defined as:

$$W_f^d(j,k) = \int_{-\infty}^{\infty} \psi_{j,k}(x) \bar{f}(x) dx = \langle \psi_{j,k}, f \rangle \qquad j, k \in \mathbb{Z} . \tag{3}$$

Thus, the DWT of $f(x)$ can be written as:

$$W_\varphi(j_0,k) = \sum_x f(x)\varphi_{j_0,k}(x) \ . \tag{4}$$

$$W_\psi(j,k) = \sum_x f(x)\psi_{j,k}(x) \ . \tag{5}$$

where the wavelet and scaling functions are:

$$\psi_{j,k}(x) = 2^{-\frac{j}{2}}\psi(2^{-j}x - k) \ . \tag{6}$$

$$\varphi_{j,k}(x) = 2^{-\frac{j}{2}}\varphi(2^{-j}x - k) \ . \tag{7}$$

where j is the scale factor, and k is the shifting factor.

Therefore, 1D wavelet transform of a discrete signal is equal to passing the signal through a pair of low-pass and high-pass filters, followed by a down-sampling operator with factor 2 [6].

2.2 Wavelet Transform in 2D

Simply by applying two 1D transforms separately, 2D wavelet transform of an image $I = A_0 = f(x,y)$ of size $M \times N$ is then:

$$A_j = \sum_x \sum_y f(x,y)\varphi(x,y) \ . \tag{8}$$

$$D_{j1} = \sum_x \sum_y f(x,y)\psi^H(x,y) \ . \tag{9}$$

$$D_{j2} = \sum_x \sum_y f(x,y)\psi^V(x,y) \ . \tag{10}$$

$$D_{j1} = \sum_x \sum_y f(x,y)\psi^D(x,y) \ . \tag{11}$$

That is, four quarter-size output sub-images, A_j, D_{j1}, D_{j2}, and D_{j3}, are generated by convolving its rows and columns with h_l and h_h, and down-sampling its columns or rows, as shown in Figure 1.

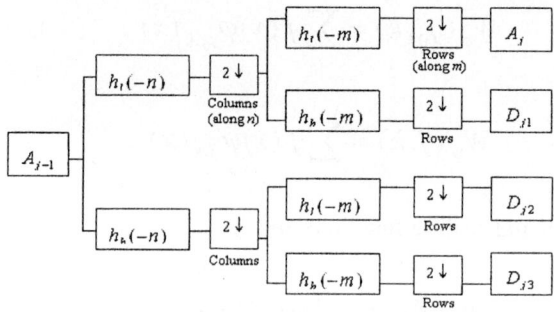

Fig. 1. Block diagram of 2-D discrete wavelet transform

Here, h_l and h_h are the low-pass and high-pass filters, respectively, j is the decomposition or reconstruction scale, $\downarrow 2,1$ ($\uparrow 2,1$) represents sub-sampling along the rows (columns) [7]. A_j, an approximation component containing its low-frequency information, is obtained by low-pass filtering, and it is therefore referred to as the low-resolution image at scale j. The detail images D_{ji} are obtained by high-pass filtering in specific direction that contains directional detail information in high-frequency at scale j [8]. The set of sub-images at several scales, $\{A_d, D_{ji}\}_{i=1, 2, 3, j=1...d}$, are known as the approximation and the detail images of the original image I, respectively [7].

3 Application of the Combination of Wavelet Transform and Gaussian Interpolation

In our study, WT is used to decompose leaf's images and produce low-resolution images and a series of detail images. Usually, the detail information distributed in three directions is the horizontal, vertical and diagonal details in corresponding three images. Generally, because of usefulness of the approximation of the image, these detail images are removed at the effort of blurring the image, removing noise, and detecting edges from image, etc.

However, the details where there exists leaf's contour and venation are the key of the whole image processing. Leaf's skeleton will be produced from these detail images by a synthesized manner. Here, a series of detail images will be obtained based on wavelet transformation. If a single image is decomposed for several times, then the number of detail images will be trebled. Thus, the total number of statistical features gotten from these detail images will be increased promptly. In addition, there also exists the redundant information in these features. Therefore, we will add Gaussian interpolation to the three detail images, D_{j1} D_{j2} and D_{j3}, and then reconstruct them at each scale after decomposition. At the same time, this operation will bring us the vivid leaf's skeleton at different scales. Thus, all the details distributed over three images will be focused on only a single image. It will shrink the

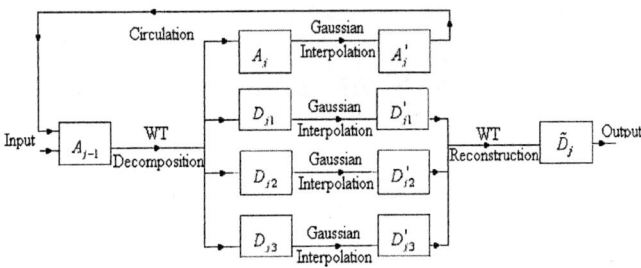

Fig. 2. Block diagram of application of WT to segmentation of leaf's skeleton

range for feature extraction subsequently. Consequently, the reconstructed detail images and the leaf's skeletons can be defined as $\tilde{D}_j$. All of these are clearly shown in Figure 2.

After getting the images $\{\tilde{D}_j\}$ of leaf's skeleton, we can extract texture features from them to recognize the leaves. Here we use the run-length features $(l,\ b,\ \theta)$, which is the sequence of l pixels with gray scale b at direction θ and $N(l,\ b,\ \theta)$ is generally defined to describe the number of run $(l,\ b,\ \theta)$ in the image. Let L denote the gray scale of image, and N_l denote the number of l. And the denominator in the following formulae will normalize these features.

The measure of short runs

$$SR = \frac{\sum_{b=1}^{L}\sum_{l=1}^{N_l}\frac{1}{l^2}N(l,\ b,\ \theta)}{\sum_{b=1}^{L}\sum_{l=1}^{N_l}N(l,\ b,\ \theta)}. \qquad (12)$$

The measure of long runs

$$LR = \frac{\sum_{b=1}^{L}\sum_{l=1}^{N_l}l^2 N(l,\ b,\ \theta)}{\sum_{b=1}^{L}\sum_{l=1}^{N_l}N(l,\ b,\ \theta)}. \qquad (13)$$

Distribution of gray scales

$$DG = \frac{\sum_{b=1}^{L}[\sum_{l=1}^{N_l}N(l,\ b,\ \theta)]^2}{\sum_{b=1}^{L}\sum_{l=1}^{N_l}(l,\ b,\ \theta)}. \qquad (14)$$

Distribution of lengths

$$DL = \frac{\sum_{l=1}^{N_l}[\sum_{b=1}^{L} N(l, b, \theta)]^2}{\sum_{b=1}^{L}\sum_{l=1}^{N_l}(l, b, \theta)}. \qquad (15)$$

Percentage of runs

$$PR = \frac{\sum_{b=1}^{L}\sum_{l=1}^{N_l} N(l, b, \theta)}{N_1 \times N_2}. \qquad (16)$$

Now, we can separate the algorithm into two main parts. Part 1 is the segmentation of leaf's skeleton by WT, and Part 2 is the extraction of features. The detailed steps of this algorithm can be stated as follows:

Part 1: Skeleton segmentation stage

Step 1. Input the original leaf's image I.

Step 2. If I is a color image, we can turn it into a gray scale image and denote it as $A_0 = A_0'$. Or just let $A_0 = A_0' = I$, and then turn to Step 3.

Step 3. Set the parameters of WT, like the levels of WT, the wavelet's name and its corresponding coefficients. Denote the variable for circulation as j, and let $j = 1$.

Step 4. Decompose A_{j-1}' to get four images (A_j, D_{j1}, D_{j2} and D_{j3}) by WT.

Step 5. Adding Gaussian interpolation to A_j, D_{j1}, D_{j2} and D_{j3}, to produce A_j', D_{j1}', D_{j2}' and D_{j3}'.

Step 6. Reconstruct D_{j1}', D_{j2}' and D_{j3}' to get the image of leaf's skeleton $\tilde{D}_j$.

Part 2: Feature extraction stage

Step 7. Extracting the run-length features (SR, LR, DG, DL and PR) from the image $\tilde{D}_j$. Meanwhile, save these features to a data file.

Step 8. If j equals to the parameter of levels of WT, turn to Step 8. Otherwise let $j = j+1$, and then turn to Step 4.

Step 9. End, and output the data file which saves the features.

4 Experiments and Results

All the following experiments are programmed by Microsoft Visual C++ 6.0, and run on Pentium 4 with the clock of 2.6 GHz and the RAM of 256M under Microsoft windows XP environment. Meanwhile, all of the results for the following Figures and Tables are obtained by the experiments more than 50 times, and then are averaged.

This database of the leaf images is built by us in our lab by the scanner and digital camera, including twenty species (as shown in Figure 3).

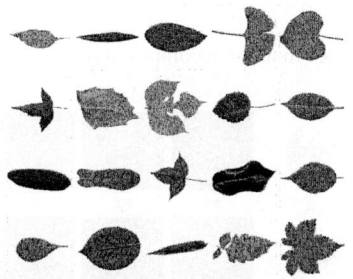

Fig. 3. Twenty species of leaf images

Before processing, the colored images will be turned into gray-scale images (as shown in Figure 4), ignoring the color information, since the majorities of leaves are green. In practice, the variety of the change of nutrition, water, atmosphere and season can cause the change of the color.

Fig. 4. The pre-processed images: Original image and Gray-scale image

After pre-processing procedure, the gray-scale images will be decomposed by WT of depth 3 using a biorthogonal spline wavelet of order 2 [9]. The decomposed images at each scale j, including approximate images A_j, horizontal detail images D_{j1}, vertical detail images D_{j2}, and diagonal detail images D_{j3}, $j=1,2,3$, are shown in Figure 5. After every wavelet is decomposed, by adding Gaussian interpolation, the results as images $\{A'_j, D'_{j1}, D'_{j2}, D'_{j3}\}_{j=1,2,3}$ can be seen in Figure 6. And Figure 7 clearly shows the reconstructed images $\{\tilde{D}_j\}_{j=1,2,3}$, from which we can see the leaf's skeleton obviously.

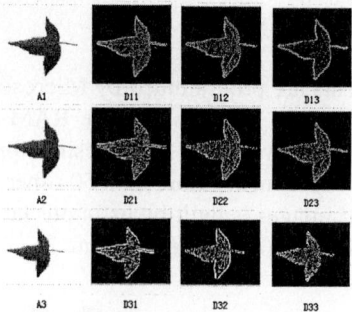

Fig. 5. Images that are decomposed by wavelet transform

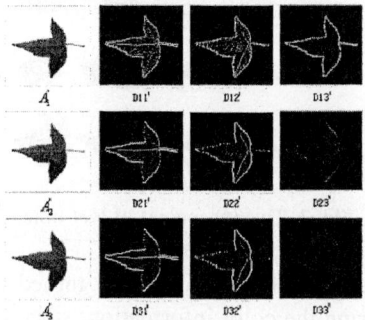

Fig. 6. Adding Gaussian interpolation to wavelet decomposition

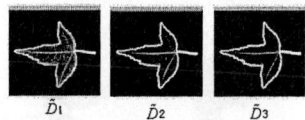

Fig. 7. The leaf's skeleton derived by reconstructing detail images after Gaussian interpolation at different scale j

After the segmentation of leaf's skeleton, three reconstructed detail images $\{\tilde{D}_j\}_{j=1,2,3}$ of a single image are generated. There are 60 ($5\times4\times3$) features of run-length for extraction. And assume that 1-NN, k-NN and RBPNN [10,11] with the recursive least square back propagation algorithm (RLSBPA) [12,13-17], are respectively used for recognition and comparison. The results are respectively shown in Figure 8 and Table 1.

It is shown in Figure 8 that, the correct rate of using original image for leaf recognition can only reach 60 percent or somewhat more than 75 percent, and the instance of only using WT can get an increase of 10 percent. Finally, using the method proposed in this paper, i.e., a combination of WT and Gaussian interpolation, the result is the best, which can be close to 95 percent. Moreover, we can carefully

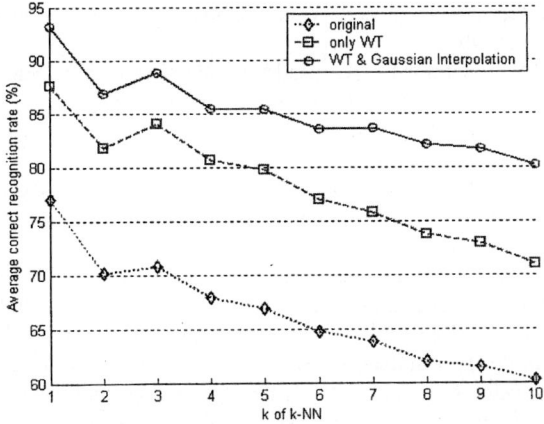

Fig. 8. k vs. the average correct recognition rates of using three different classification methods

Table 1. The performance comparison of different classifiers

Classifiers	1-NN	k-NN (k=5)	RBPNN
Average correct recognition rate (%)	93.1702	85.4681	91.1809

find that, with the increase of k, the downtrend of the correct recognition rate of using WT and Gaussian interpolation is slower than the former. It tells us that the stability of this method is the best. And Table 1 gives the distribution of the average correct rate by using our method.

5 Conclusions

This paper proposed a new wavelet decomposition and Gaussian interpolation method, which not only successfully separates the contour but also separates the venation, i.e., the skeleton of the whole leaf. Then, run-length features (RF) were used to perform the recognition of leaves by different classifiers such as 1-NN, k-NN and RBPNN. The experimental results found that our method can obviously segment the leaf's skeleton, and the correct recognition rate of using features extracted from the skeleton is more improved than other methods.

References

1. Du, J.X., Wang, X.F., Huang, D.S.: Automatic Plant Leaves Recognition System Based on Image Processing Techniques. Technical Report. Institute of Intelligent Machines. Chinese Academy of Sciences. (2004)
2. Huang, D.S, Ma, S.D.: Linear and Nonlinear Feed-Forward Neural Network Classifiers: A Comprehensive Understanding. Journal of Intelligent System 9(1) (1999) 1 - 38

3. Vetterli, M., Kovacevic, J.: Wavelets and Subband Coding. Prentice Hall. Englewood Cliffs. New Jersey. (1995)
4. Akansu, Ali N., Richard, A. Haddad: Multiresolution Signal Decomposition: Transforms, Subbands, Wavelets. Academic Press. Inc. (1992)
5. Vetterli, M., Herley, C.: Wavelets and Filter Banks: Theory and Design. IEEE Trans. on Signal Proc. Vol. 40. (1992) 2207-2231
6. Ahmadian, A. Mostafa: An Efficient Texture Classification Algorithm Using Gabor Wavelet. Proceedings of the 25th Annual International Conference of the IEEE EMBS. Cancum. Mexico. (2003)
7. Busch, A., Boles, W., Sridharan, S.: Logarithmic Quantisation of Wavelet Coefficients for Improved Texture Classification Performance. The International Conference on Acoustics, Speech, and Signal Processing. Proceedings. (ICASSP 2004)
8. Wouwer, G. Van de, Scheunders, P., Dyck, D. Van: Statistical Texture Characterization from Discrete Wavelet Representations. IEEE Transactions on Image Processing. (1999)
9. Unser, M., Aldroubi, A., Eden, M.: A Family of Polynomial Spline Wavelet Transforms. Signal Processing. Vol. 30. (1993) 141 - 162
10. Huang, D.S.: Radial Basis Probabilistic Neural Network: Model and Application. International Journal of Pattern Recognition and Artificial Intelligence. 13(7) (1999) 1083 - 1101
11. Zhao, W.B., Huang, D.S.: The Structure Optimization of Radial Basis Probabilistic Networks Based on Genetic Agorithm. The 2002 IEEE World Congress on Computational Intelligence. IJCNN02. (2002) 1086 - 1091
12. Huang, D.S.: Systematic Theory of Neural Networks for Pattern Recognition (in Chinese). Publishing House of Electronic Industry of China. Beijing. (1996) 49 - 51
13. Huang, D.S., Zhao, W.B.: Determining the Centers of Radial Basis Probabilities Neural Networks by Recursive Orthogonal Least Square Algorithms. Applied Mathematics and Computation. Vol. 162. No.1. (2005) 461 - 473
14. Huang, D.S., Ma, S.D.: A New Radial Basis Probabilistic Neural Network Model. The 3rd Int. Conf on Signal Processing (ICSP) Proceedings. Beijing. China. (Oct. 14-18 1996) 1449 - 1452
15. Huang, D.S.: The Local Minima Free Condition of Feedforward Neural Networks for Outer-Supervised Learning. IEEE Trans on Systems, Man and Cybernetics. Vol.28B. No.3. (1998) 477 - 480
16. Huang, D.S.: Application of Generalized Radial Basis Function Networks to Recognition of Radar Targets. International Journal of Pattern Recognition and Artificial Intelligence. Vol.13. No.6. (1999) 945 - 962
17. Huang, D.S.: The Bottleneck Behaviour in Linear Feedforward Neural Network Classifiers and Their Breakthrough. Journal of Computer Science and Technology. Vol.14. No.1. (1999) 34 - 43

Fast Training of SVM via Morphological Clustering for Color Image Segmentation

Yi Fang, Chen Pan, Li Liu, and Lei Fang

Key Laboratory of Biomedical Information Engineering of Education Ministry,
Xi'an Jiaotong University, Xi'an, 710049 China
fangyi@mail.xjtu.edu.cn

Abstract. A novel method of designing efficient SVM for fast color image segmentation is proposed in this paper. For application of large-scale image data, a new approach to initializing training set via pre-selecting useful training samples is adopted. By using a morphological unsupervised clustering technique, samples at the boundary of each cluster are selected for SVM training. With the proposed method, various experiments are carried out on the color blood cell images. Results show that the training set and time can be decreased considerably without lose of any segmentation accuracy.

1 Introduction

Image segmentation is a critical and complicated topic of image analysis [1-3], the precision of which directly influences the feasibility and reliability of the image analytical result. Over the years, many approaches have been proposed to overcome this difficulty [3-6]. Although these methods are capable of partitioning an image, more often than not, the segmented regions do not correspond well to the actual objects [2].

It is mentioned in [3] that image segmentation methods can be categorized as supervised or unsupervised learning/classification procedures in color space. Recently, with desirable properties such as independence on the feature dimensionality and good generalization performance, Support vector machine has become an increasingly popular tool for machine learning tasks involving classification, regression or novelty detection and has shown its great performance in many image processing applications such as face identification, texture segmentation and remote image classification [7-10]. Therefore, a method based on SVM is applied to the color image segmentation in this research.

As we know, however, SVM suffers from exceeded computation and long training time caused by large-scale image samples especially for color images. The prevailing efficient methods proposed to speed up SVM are subset selection techniques and pre-processing methods [11]. Generally speaking, a subset selection is a technique to speed up SVM training by dividing the original QP (quadratic programming) problem into small pieces, thus reducing the size of each QP problem [11]. There have been three efficient subset techniques such as decomposition methods [12], chunking methods [13] and SMO (sequential minimal optimization) [14] [15]. Although these methods do accelerate training, it is still time-consuming for necessity of learning all training data.

Besides, pre-processing methods are also applied to speed up SVM training usually. Hereby, some effective pre-processing methods have been developed. Zhan et al [16] proposed a data selection method via excluding the points that make the hyper-surface highly convoluted. Schohn et al [17] utilized active learning strategy, selecting samples closest to the classification hyper-plane to reconstruct training set. Li et al [18] trimmed the large-scale training set by training a random SVM using partial training set. However, it is still a waste of time for these pre-processing methods above to train SVM repeatedly while choosing data in a certain range.

It is well known that only training samples located near the hyper-plane position are required for SVM. The researches mentioned in [11][19-21] indicate that samples at the boundary of each cluster are critical data that decisively affect the decision boundary in higher feature space. Therefore, a novel method for fast training SVM based on morphological clustering is proposed. 2D-histogram is used to obtain a rapid but coarse clustering of the color image. Then, the boundary of each cluster can be found. The training samples at the boundary of each cluster could be pricked off quickly. Consequently, the number of training vectors is decreased dramatically, accelerating the SVM training and requiring less computation.

This paper follows four parts. Section 2 gives a brief introduction of SVM. Section 3 describes the technique of selecting boundary samples. Experiments and results are observed in section 4. Section 5 presents the discussion and conclusion.

2 Support Vector Machines

The paper only describes SVM briefly, a more detailed description can be found in [7]. The nonlinear SVM implements the following idea: it maps the input data nonlinearly into a higher dimension feature space where an optimal separable hyper-plane is constructed. Let the training set of size N be $\{(\mathbf{x}_i, y_i)\}_{i=1}^{N}$, where $\mathbf{x}_i \in R^d$ is the input vector and $y_i \in (-1, +1)$ is the corresponding desired response. The function $\varphi(\cdot)$ first maps $\mathbf{x}$ to $\mathbf{z} \in F$. Then the SVM constructs a hyper-plane $\mathbf{w}^T \mathbf{z} + b = 0$ in F where the data is linearly separable. It can be shown that the coefficients of the generalized optimal hyper-plane:

$$\mathbf{w} = \sum_{i=1}^{N} \alpha_i y_i z_i \tag{1}$$

where $\alpha = [\alpha_1 \ \alpha_2 \ \cdots \ \alpha_N]^T$ can be found via solving the following QP problem.

Maximize

$$W(\alpha) = 1^T - \frac{1}{2}\alpha^T \mathbf{H} \alpha \tag{2}$$

subject to

$$\alpha^T \mathbf{y} = \mathbf{0} \quad \alpha_i \geq 0 \tag{3}$$

where $H_{ij} = y_i y_j K(\mathbf{x}_i, \mathbf{x}_j) = y_i y_j \varphi(\mathbf{x}_i)^T \varphi(\mathbf{x}_j)$, and $\alpha = \begin{bmatrix} \alpha_1 & \alpha_2 & \cdots & \alpha_N \end{bmatrix}^T$, the kernel function is represented by $K(\cdot,\cdot)$.

In the case of linear non-separable, the SVM introduces non-negative slack variable $\xi_i \geq 0$ and the optimization problem below:

$$\min \frac{1}{2}\|\mathbf{w}\|^2 + C\sum_i^n \xi_i \tag{4}$$

subject to

$$y_i(\mathbf{w}^T \mathbf{x} + b) \geq 1 - \xi_i, i = 1, \cdots n. \tag{5}$$

where C is the misclassification penalty parameter controlling the tradeoff between the maximum margin and the minimum error.

The $\mathbf{x}_i$, whose corresponding α_i is nonzero, is so-called support vector, which lies on the decision boundary and determine the optimal hyper-plane.

3 Data Selection by Clustering

The data in the same cluster always have similar characteristic, so we can select some data in the same clusters rather than all of them to reduce the amount of training data. Various strategies can be employed in clustering color images, such as K-means, SOM, and etc. However, these methods suffer from time-consuming. Given the advantages [22][23] of 2D-histogram, an unsupervised morphological method for clustering 2D-histogram is adopted in this paper.

3.1 The Framework of Our Method

Fig. 1 shows the framework of the proposed method. First, a Gussisan filter is applied on the 3D-hitogram [24] (see Fig. 2. a) to suppress the noise caused by sparse distribution of the color data. Then a 2D-histogram is obtained from the projections of 3D-histogram. The 2D-histogram is clustered via unsupervised morphological clustering as discussed in section 3.2. Subsequently, boundary samples are picked out using a selection algorithm, which is described in section 3.3. Finally, learning with reduced training set, which is constructed by boundary samples, a classifier model can be produced.

3.2 Unsupervised Morphological Clustering

In the 2D-histogram, we assume that different classes of vectors surround different dominant peaks considered as the cluster centroids. For the 2D-historam can be regarded as a gray level image, a morphological clustering algorithm is adopted to cluster this gray image quickly. Although this clustering is coarse and cannot segment the image accurately (usually over-segmented), it doesn't influence the boundary samples selection. This method follows three steps as below:

Step 1: The 2D-histogram (see Fig. 2. b) is obtained from the projection of the 3D-histogram. High values in 2D-histogram are depicted by darker pixels.
Step 2: An erosion operation on 2D-histogram image reduces each bin to its main color (see Fig. 2. c), which is considered as a cluster centroid.
Step 3: Using these centroids as markers, watershed algorithm is implemented on the image and this provides the clustering of histogram (see Fig. 2.d).

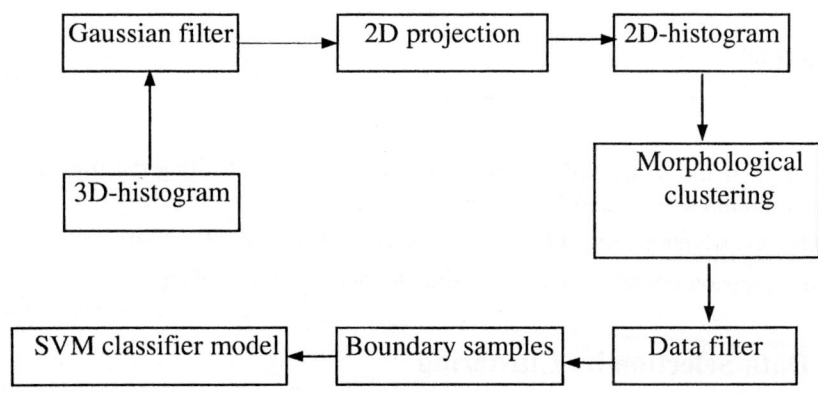

Fig. 1. Framework of the proposed method

3.3 Boundary Samples Selection Algorithm

The selection algorithm to choose boundary samples in the 3D-histogram for training is described as follows:

A. Transform Fig. 2. d to a binary image (see Fig. 3. a).
B. Two images (see Fig.3. (b, c)) result from eroding the Fig. 3.a. using structuring elements x and y (size of x < size of y) respectively. The purpose of eroding Fig.3 .a with structuring elements x is to avoid overlapping. We use two elements, x and y, of different sizes to erode Fig.3.a, so that the boundary can be produced by subtracting the two eroded images. Fig.3.b and Fig.3.c are the results of erosion with x and y respectively.
C. The boundary (see Fig. 3. b) of each cluster can be obtained via subtracting Fig. 3.c from Fig. 3.b.
D. In the 3D-histogram, color vectors relevant to the boundary extracted in step C are labeled to form the training set.

Fig.3.d shows an example of boundaries of four classes. There is no overlapping of each boundary in this case. The samples, which is the labeled pixels in step. **D**, are of a small number but critical, for they preserve the most valuable distribution information. And such samples are candidates of the support vectors.

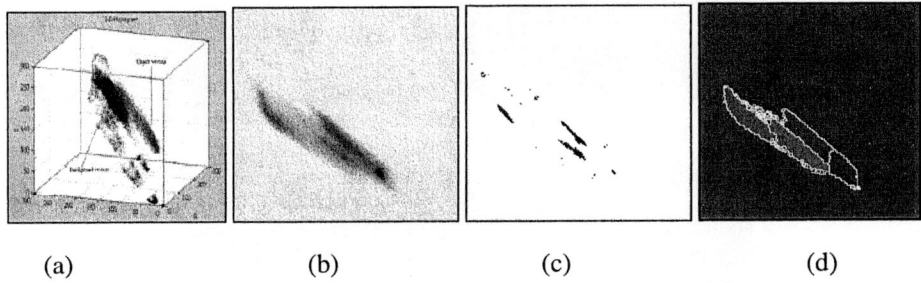

Fig. 2. (a) The 3D-histogram in the RGB color space; (b) The GB 2D-histogram; (c) The dominant colors; (d) The watershed clustering

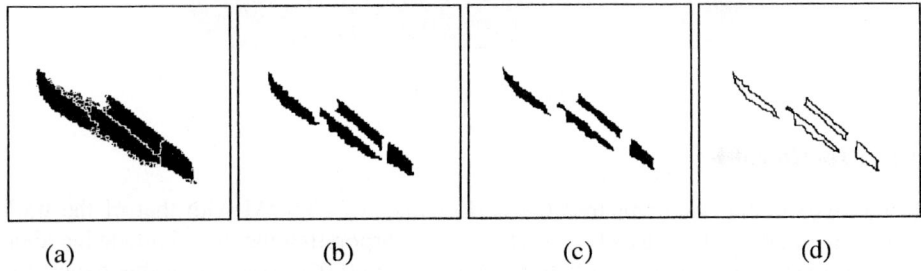

Fig. 3. (a) binary image; (b) eroded image (the structuring element is x); (c) eroded image (the structuring element is y); (d) boundaries of four classes

Note that the size of erosion structuring element is important. Oversized erosion elements result in getting the clusters center rather than boundaries, while overlapping of training data occurs if the size is too small. Thus, a proper size must be carefully selected in experiments.

4 Experiments and Results

In this paper, the proposed fast training method of SVM can be abbreviated to FTSVM. Two experiments were carried out with blood cell images (720*540). During the experiments, we implemented FTSVM on color blood cell images to compare its performance with that of a traditional training method, and to assess its generalization performance. The traditional training set is constructed by selecting samples that completely describe each class, whereas FTSVM only selects samples at the boundary of each class. In the experiments, the LIBSVM package was used to train SVM, and algorithms presented in the paper were implemented on Acer7500G-M computer with 256M RAM running Windows XP. The RBF kernel function and SMO algorithm were adopted in the experiments. A large amount of color blood cell images were used as the experimental database.

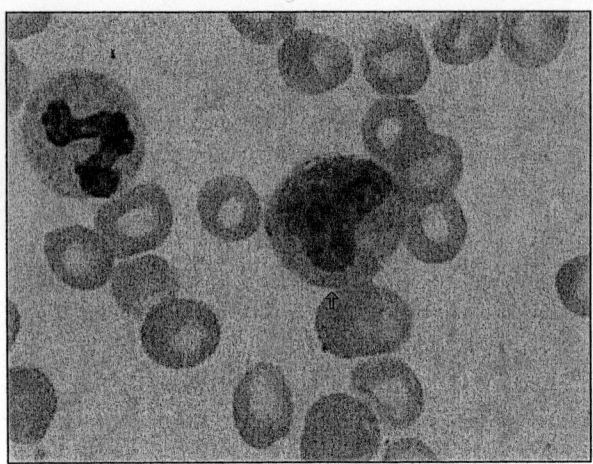

Fig. 4. Training image

4.1 Experiment I

In the experiment, we compared the performance of FTSVM with that of the traditional one on the database of Fig. 4. Here, by incorporating the prior knowledge about the test image, we can select a better 2D-histogram (GB-histogram in this paper) that includes valuable color information. The sizes of erosion operation structuring elements x and y (discussed in section 3.3) were set at 9-pixel connectivity and 16-pixel connectivity, and the kernel parameter q of RBF and the regulation parameter C (see formula (3)) were set at 50 and 100 respectively. Table 1 shows the comparative results.

Table 1. Comparative results

Method	Traditional method	FTSVM
Training set size	13061	1363
Number of Support vectors	542	15
Training time (s)	31.10	0.55
Classification accuracy	96.8%	95.9%

Results show that the size of training set can be reduced to 10% with the classification accuracy only reduced by 0.9% of the traditional method. And the training is almost real time (only 0.55s needed in this case). The number of support vectors is decreased greatly, from 542 to 15. Therefore, the FTSVM can be trained online and the segmentation algorithm can be implemented in real-time applications.

4.2 Experiment II

In order to assess the generalization performance of FTSVM, we select some other blood cell images (720*540). As we can see in Fig.5, the three test images display white blood cells at different growth level. And because of different sample sources, preparation protocols, staining techniques, microscope and camera parameters for the image acquisition, the color of cytoplasm varies. However, the white blood cells in all three images were extracted successfully in spite that the SVM classifier was trained by Fig. 4 only, and then directly applied to images in Fig. 5. The segmentation results show that even some new colors that did not appear in the training data can be classified correctly, which proves the high generalization performance of FTSVM.

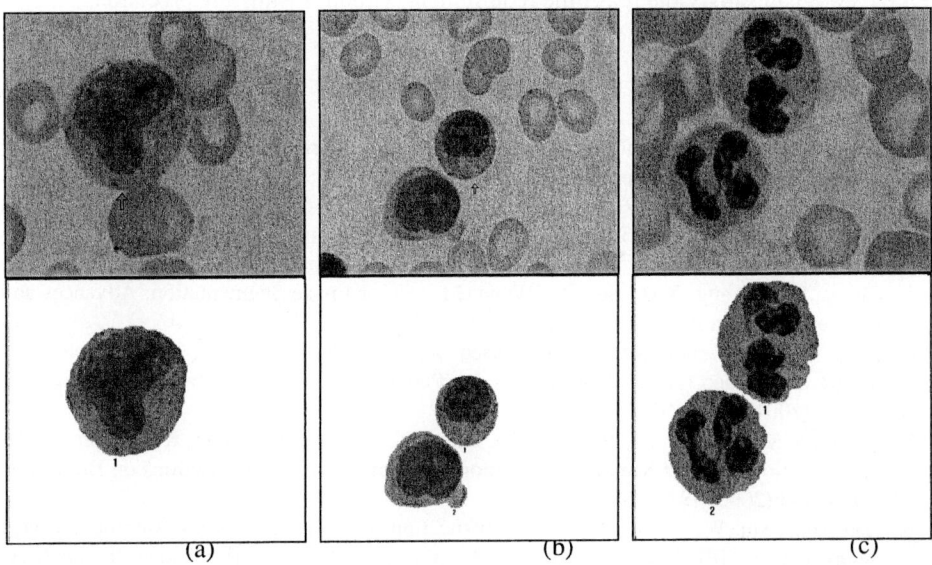

Fig. 5. Color image segmentation. (a) Tested image 1 and its segmentation; (b) Tested image 2 and its segmentation; (c) Tested image 3 and its segmentation.

5 Discussions and Conclusions

As we know, image segmentation is to extract the desired object from the background. In case of color blood cell images, color of cytoplasm varies because of different staining and illumination. Some conventional segmentation algorithms, based on constructing a color lookup table through mapping training data to color space, often fail. Obviously, due to great generalization performance, FTSVM can partly overcome this problem via finding a proper hyper-plane to classify color pixels.

Many researches indicate that the 1D-histogram and 3D-histogram suffer from the lack of local spatial information, and an effective way to solve this problem is to employ 2D-hitogram. Three different 2D-histograms can be derived from the projection of a 3D-histogram onto band-pair planes. By incorporating the prior knowledge about the test image, we can select a better one without comparing the all three.

Furthermore, despite a lack of stable theory supporting that the boundary samples in the input space could decisively affect boundary decision in higher feature space. Experiments in this research and others suggest the rationality of this hypothesis. The training set initialized by the proposed method furnishes the most useful support vectors that construct the optimal hyper-plane with maximized generalization ability.

In summary, a novel method of designing efficient SVM by pre-selecting support vectors is developed in this paper. By using a morphological unsupervised clustering technique, the process of selecting boundary samples is fast, simple and unsupervised, and the number of training vectors is reduced significantly as well. Experimental results show that FTSVM is able to tackle the problem of large-scale image data. Compared with conventional methods, the new image segmentation algorithm based on FTSVM can satisfy the real-time requirement without losing any precision.

References

1. Moghaddamzadeh, A., Bourbakis, N.: A Fuzzy Region Growing Approach for Segmentation of Color Images. Pattern Recognition, Vol. 30 (1997) 867–881
2. Zhu, S.C., Yuille, A.: Region Competition: Unifying Snakes, Region Growing, and Hayeshdl for Multi-band Image Segmentation. IEEE Trans. PAMI, Vol. 18(1996) 884–900
3. Cheng, H.D., Jiang, X.H., Sun, Y., Wang ,J.L.: Color Image Segmentation: Advances and Prospects. Pattern Recognition, Vol. 34 (2001) 2259–2281
4. Deng, Y., Manjunath, B.S.: Unsupervised Segmentation of Color-texture Regions inIimages and Video. IEEE Transactions on Pattern Analysis and Machine Intelligence, Vol.23 (2001)800 - 810
5. Naemura, M., Fukuda, A., Mizutani, Y., Izumi, Y., Tanaka, Y., Enami, K.: Morphological Segmentation of Sport Scenes Using Color Information. IEEE Transactions on Broadcasting, Vol.46(2000) 181 - 188
6. Gao, Hai., Siu, Wan-Chi., Hou, Chao-Huan.: Improved Techniques for Automatic Image Segmentation. IEEE Transactions on Circuits and Systems for Video Technology, Vol.11(2001)1273 - 1280
7. Vapnik, V.: Statistical Learning Ttheory. John Willey &Sons (1998)
8. Niyogi, P., Burges, C., Ramesh, P.: Distinctive Feature Detection Using Support Vector Machines. ICASSP'99, Phoenix, Arizona, USA, Vol. 1 (1999) 425–428
9. Scholkopf, B., Smola, A., Muller, K.R., Burges, C.J.C., Vapnik, V.: Support Vector Methods in Learning and Feature Extraction. Journal of Intelligent Information Processing Systems (1998) 3–9
10. Foody, G.M., Mathur, A.: A Relative Evaluation of Multiclass Image Classification by Support Vector Machines. IEEE Transactions on Geoscience and Remote Sensing, Vol.42 (2004)1335–1343

11. Sun, Sheng-Yu., Tseng, C.L., Chen, Y.H., Chuang, S.C., Fu, H.C.: Cluster-based Support Vector Machines in Text-independent Speaker Identification. IEEE International Joint Conference on Neural Networks Proceedings, Vol. 1 (2004) 734
12. Lin, Chih-Jen.: On the Convergence of the Decomposition Method for Support Vector Machines. IEEE Transactions on Neural Networks, Vol. 12 (2001) 1288–1298
13. Rychetsky, M., Ortmann, S., Ullmann, M., Glesner, M.: Accelerated Training of Support Vector Machines. IJCNN '99, International Joint Conference on Neural Networks, Vol. 2 (1999) 998–1003
14. Platt, J.: Fast Training of SVMs Using Sequential Minimal Optimization. Advances in Kernel Methods - Support Vector Learning, MIT press, Cambridge, MA (1999) 185–208
15. Takahashi, N., Nishi, T.: Rigorous Proof of Termination of SMO Algorithm for Support Vector Machines. IEEE Transactions on Neural Networks, Vol.16 (2005) 774–776
16. Zhan, Yiqiang., Shen, Dinggang.: Design Efficient Support Vector Machine for Fast Classification. Pattern Recognition, Vol. Vol. 38 (2005) 157–161
17. Schohn, G., Cohn, D.: Less is More: Active Learning with Support Vector Machines. Proceeding of the Seventeenth International Conference on Machines Learning (ICML2000), Standorf, CA, USA, Publisher: Morgen Kaugmann,(2000) 839–846
18. Li, Honglian., Wang, Chunhua., Yuan, Baozong.: A Learning Strategy of SVM Used to Large Training Set. Chinese Journal of Computers, Vol. 27 (2004) 715–719
19. Barros de Almeida, M., de Padua Braga, A., Braga, J.P.: SVM-KM: Speeding SVMs Learning with A Priori Cluster Selection and K-means. Neural Networks Proceedings, Sixth Brazilian Symposium (2000)162 –167
20. Foody, G. M., Mathur, A.: Toward Intelligent Training of Supervised Iimage Classifications: Directing Training Data Acquisition for SVM Classification. Remote Sensing of Environment, Vol. 93 (2004) 107–117
21. Xia, Jiantao., He, Mingyi., Wang, Yuying., Feng, Yan.: A Fast Training Algorithm for Support Vector Machine via Boundary Sample Selection. International Conference on Neural Networks and Signal Processing, Proceedings of the 2003, Vol. 1 (2003) 20–22
22. Yonekura, E.A., Facon, J.: 2-D Histogram-based Segmentation of Postal Envelopes. Computer Graphics and Image Processing, SIBGRAPI, XVI Brazilian Symposium (2003) 247 –253
23. Xue, H., Geraud, T., Duret-Lutz, A.: Multiband Segmentation Using Morphological Clustering and Fusion–application to Color Image Segmentation. International Conference on Image Processing, Vol. 1 (2003) 353–356
24. Vinod, V.V., Chaudhury, S., Mukherjee, J., Ghose, S.: A Connectionist Approach for Color Image Segmentation. IEEE International Conference on Systems, Man and Cybernetics, Vol. 1 (1992) 100–104

Similarity Measurement for Off-Line Signature Verification

Xinge You[1,3], Bin Fang[2,3], Zhenyu He[3], and Yuanyan Tang[1,3]

[1] Faculty of Mathematics and Computer Science,
Hubei University, 430062, China
xyou@hubu.edu.cn
[2] Center for Intelligent Computation of Information,
Chongqing University, Chongqing, China
[3] Department of Computer Science, Hong Kong Baptist University
{xyou, fangb, zyhe, yytang}@comp.hkbu.edu.hk

Abstract. Existing methods to deal with off-line signature verification usually adopt the feature representation based approaches which suffer from limited training samples. It is desired to employ straightforward means to measure similarity between 2-D static signature graphs. In this paper, we incorporate merits of both global and local alignment methods. Two signature patterns are globally registered using weak affine transformation and correspondences of feature points between two signature patterns are determined by applying an elastic local alignment algorithm. Similarity is measured as the mean square of sum Euclidean distances of all found corresponding feature points based on a match list. Experimental results showed that the computed similarity measurement was able to provide sufficient discriminatory information. Verification performance in terms of equal error rate was 18.6% with four training samples.

1 Introduction

An important issue in pattern recognition is the effect of insufficient samples available for training in classification accuracy. It is well known that when the ratio of the number of training samples to the number of the feature dimensionality is small, the estimates of the statistical model parameters are not accurate, and therefore the classification results may not be satisfactory. This problem is especially significant in off-line signature verification where usually only a few samples can be available for training such as 2-4 signatures when one open a bank account [1-9].

Most of the earlier work on off-line signature verification involves the extraction of features from the signatures image by various schemes. Y. Qi *et al.* [10] used local grid features and global geometric features to build multi-scale verification functions. Sabourin *et al.* [11] used an extended shadow code as a feature vector to incorporate both local and global information into the verification decision. B. Fang *et al.* [12] used positional variances of the 1 dimensional projection profiles of the signature patterns and the relative stroke positions of two dimensional patterns. Refer to Figure 1 for some examples of the genuine signatures and corresponding forgeries.

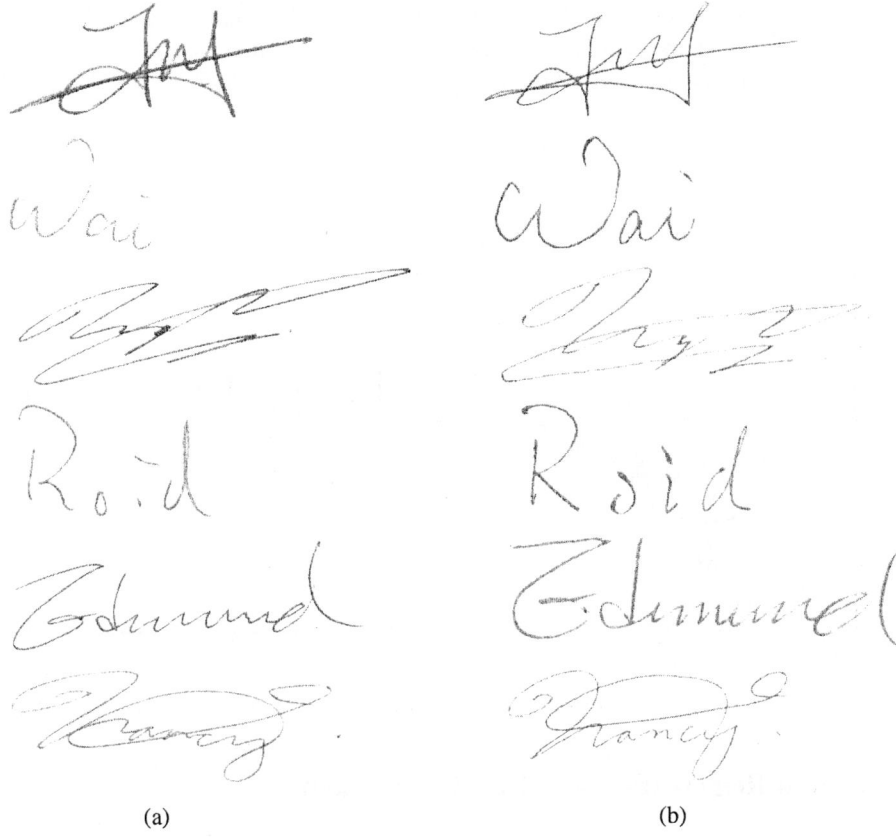

Fig. 1. (a) Samples of genuine signatures. (b) Samples of forgeries

Existing methods to deal with off-line signature verification usually adopt the feature based approaches which suffer from two important issues: first, how much these extracted features are discriminatory to represent the signature itself for the purpose of verification. While lots of different algorithms have been proposed till now, most of them are ad hoc and assumed built models are stable. Then secondly, in any cases, usually only a limited number of samples can be obtained to train an off-line signature verification system which further makes off-line signature verification a formidable task. Hence, it is desired to invent straightforward means to measure similarity between signatures by employing the 2-D signature graphs. The problem is how to define and compute similarity measurement between two signature patterns. Since no two signatures are identical, there are no parametric models no matter linear or non-linear that we can use to register or match two signature patterns. See Figure 2. If we apply some non-parametric schemes such as elastic matching algorithm to find out corresponding stroke or feature points between two signature patterns, since they are not globally aligned in terms of shift and rotation, the computed similarity quantity in certain from is just misleading to wrong measurement. Our idea to tackle this problem is to incorporate merits of both global and local alignment methods. In this paper, we propose algorithms to register two signatures using weak affine transformation and measure the similarity based on a match list of feature points produced by an elastic local alignment algorithm. Experimental results showed that the computed similarity measurement was able to provide sufficient discriminatory information. Verification performance in terms of equal error rate was 18.6% with four training samples.

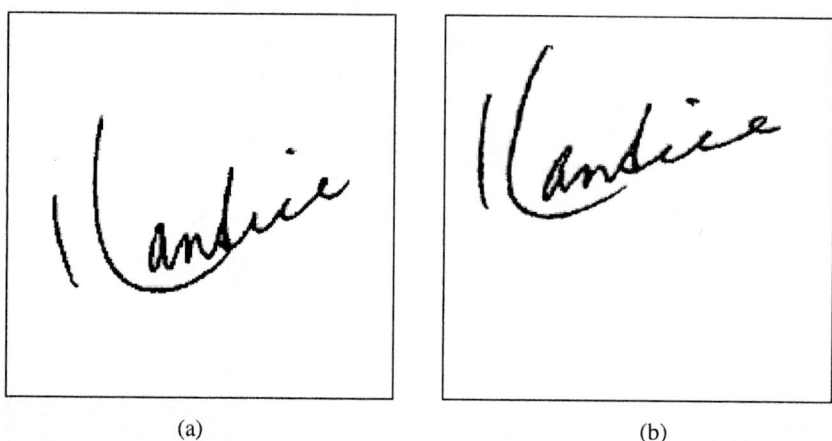

(a)   (b)

Fig. 2. (a) Original image of signature *Template*. (b) original image of signature *Input*

2 Global Registration and Local Alignment

In order to measure similarity between two signatures in terms of certain kind of distance computation such as Euclidean distance, it is important to accurately find out corresponding strokes or feature points of the signatures. Here, we employ a fast global registration algorithm based on the weak affine transform to align the signature patterns. It also functions as normalization in traditional pre-processing stage. An elastic local alignment algorithm is then applied to produce a match list of corresponding feature points between the signatures to facilitate the computation of similarity measure.

2.1 Fast Global Registration

Signature patterns are not just simply overlapped each other to find out corresponding feature points since shift and rotation have existed. Although no two signature from the same person are identical which means that no parametric registration model is applicable no matter linear or non-linear, considering the major two factors, we propose the use of a weak affine registration of translations and rotation to roughly align the two signature patterns in order to facilitate the elastic local alignment to find corresponding feature points for similarity measurement. The model can be mathematically expressed as follows:

$$\begin{bmatrix} x' \\ y' \end{bmatrix} = \begin{bmatrix} \cos\theta & -\sin\theta \\ \sin\theta & \cos\theta \end{bmatrix} \begin{bmatrix} x \\ y \end{bmatrix} + \begin{bmatrix} \Delta x \\ \Delta y \end{bmatrix} \quad (1)$$

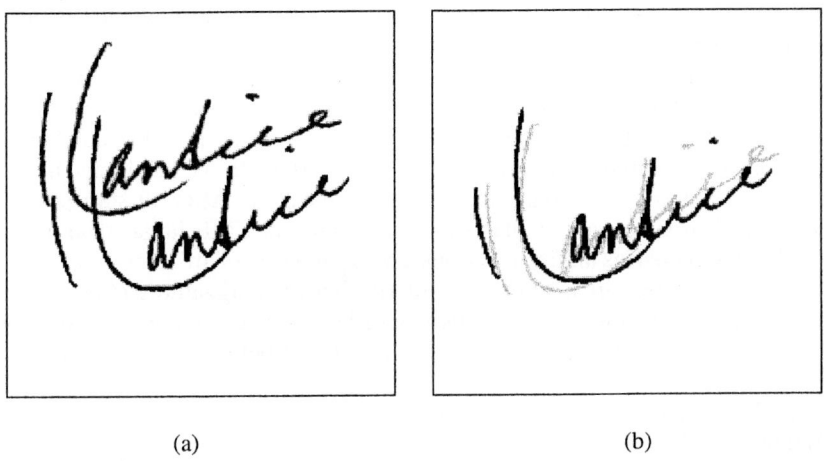

(a) (b)

Fig. 3. (a) Overlapped image of the signature *Template* and signature *Input* in Fig. 2. before matching. (b) Overlapped image of signature *Template* (black, Fig. 2(a)) and signature *Input* (grey, Fig. 2(b)) after global registration

In order to evaluate the goodness of fit between two signature patterns, a distance measure is computed in terms of the corresponding transformation. A search for the optimal transformation is to find the global minimum of the defined distance function. The search process typically starts with a number of initial positions in the parameter space by multi-resolution strategy.

The idea behind multi-resolution matching is to search for the local optimal transformation at a coarse resolution with a large number of initial positions. Only a few promising local optimal positions with acceptable centreline mapping errors of the resulting transformation are selected as initial positions before proceeding to the next level of finer resolution. The assumption is that at least one of them is a good approximation to the global optimal matching. The algorithm is detailed as follows.

One of the two signatures to be registered is called the *Template* and the other the *Input*. Thinning is performed for both the *Template* and the *Input* so that the resulting patterns consist of lines with one pixel width only. A sequential distance transformation (DT) is applied to create a distance map for the *Template* by propagation local distances [13]. The *Input* at different positions with respect to the corresponding transformations is superimposed on the *Template* distance map. A centreline mapping error (CME) to evaluate matching accuracy is defined as the average of feature point distance of the *Input* as follows:

$$CME = \frac{1}{N} \sum_{\substack{p(i,j) \in Input \\ p(i,j) \in Template}} DM_{Template}(p(i,j))^2 \quad (2)$$

where N is the total number of feature point $p(i,j)$ of the *Input* signature pattern and *DM* is the distance map created for the *Template* signature pattern. It is obvious that a perfect match between the *Template* and *Input* images will result in a minimum value of CME.

The search of minimum CME starts by using a number of combinations of initial model parameters. For each start point, the CME function are searched for neighboring positions in a sequential process by varying only one parameter at a time while keeping all the other parameters constant. If a smaller distance value is found, then the parameter value is updated and a new search of the possible neighbors with smaller distance continues. The algorithm stops after all its neighbors have been examined and there is no change in the distance measure. After all start points have been examined, transformations having local minima in CME larger than a prefixed threshold are selected as initial positions on the next level of finer resolution. The optimal position search of maximum similarity between signatures is operated from coarse resolution towards fine resolution with less and less number of start points. Refer to the pseudo-codes for the search method shown below.

```
Algorithm Search
Begin
FOR each initial position of the template
X' = X coordinate of initial position;
Y' = Y coordinate of initial position;
θ' = rotation angle of initial position;
X" = step-length of X translation parameter;
Y" = step-length of Y translation parameter;
θ" = step-length of rotation parameter θ;
dist_min = value of distance function at initial position;
LOOP
FOR x = X' – X", X' + X", step = X"
FOR y = Y' – Y", Y' + Y", step = Y"
FOR θ = θ' – θ", θ' + θ", step = θ"
        dist_current = value of distance function at position (x,y, θ);
        IF ( dist_current < dist_min ) THEN
                X' = x; Y' = y; θ' = θ;
                LOOP_FLAG = 0;
                EXIT FOR x;
                EXIT FOR y;
                EXIT FOR θ;
    ELSE LOOP_FLAG = 1;
    ENDIF
ENDFOR   // θ
ENDFOR   // y
ENDFOR   // x
IF (LOOP_FLAG = 1) EXIT LOOP
ENDLOOP
Optimal position = (X',Y',θ');
ENDFOR
```

The final optimal match is determined by the transformation which has the smallest centreline mapping error at level 0 (the finest resolution). Once the relative parameters for the global transformation model have been computed, the registration between two signatures is ready as illustrated in Fig. 3.

(a)   (b)

Fig. 4. (a) Overlapped image of the approximated skeletons of signature *Template* (black) and signature *Input* (grey) after global registration. (b) Overlapped images after elastic local alignment by deforming the *Template*.

2.2 Elastic Local Alignment

Since the two signature patterns have already been roughly aligned by global registration, the next work to do is to find out corresponding feature points by using a non-parametric elastic matching algorithm. Let *Template* and *Input* be the two globally registered signature patterns. Lines and curves are approximated by fitting a set of straight lines. Each resulting straight line is then divided into smaller segments of approximately equal lengths referred as an 'element' which is represented by its slope and the position vector of its midpoint. Both signature patterns are, in turn, represented by a set of elements. Hence, the matching problem is equal to matching two sets of elements. Note that the number of elements in the two patterns need not be equal.

The *Template* is elastically deformed in order to match the *Input* locally until the corresponding elements of both *Input* and *Template* meet, as illustrated in Fig. 4. The objective is to achieve local alignment while to maintain the regional structure as much as possible. We elaborately create an energy function whose original format can be found in [14] to guide the deformation process.

$$E_1 = -K_1^2 \sum_{i=1}^{N_I} \ln \sum_{j=1}^{N_T} \exp\left(-\left|\mathbf{T}_j - \mathbf{I}_i\right|^2 \Big/ 2K_1^2\right) f\left(\theta_{Tj,Ii}\right) \quad (3)$$

$$+ \sum_{j=1}^{N_T} \sum_{k=1}^{N_T} w_{jk} \left(d_{Tj,Tk} - d^0_{Tj,Tk}\right)^2$$

where
 N_T = number of *Template* elements,
 N_I = number of *Input* elements,
 T_j = position vector of the midpoint of the jth *Template* element,
 θ_{Tj} = direction of the jth *Template* element,
 I_i = position vector of the midpoint of the ith *Input* element,
 θ_{Ii} = direction of the ith *Input* element,
 $\theta_{Tj,Ii}$ = angle between *Template* element T_j and *Input* element I_i, restricted within 0-90°,
 $f(\theta_{Tj,Ii}) = \max(\cos \theta_{Tj,Ii}, 0.1)$,
 $d_{Tj,Tk}$ = current value of $|T_j - T_k|$,
 $d^0_{Tj,Tk}$ = initial value of $|T_j - T_k|$,

$$w_{jk} = \frac{\exp\left(-|T_j - T_k|^2 / 2K_2^2\right)}{\sum_{n=1}^{N_T} \exp\left(-|T_j - T_n|^2 / 2K_2^2\right)}$$

K_1 and K_2: size parameters of the Gaussian windows which establish neighbourhoods of influence, and are decreased monotonically in successive iterations.

The first term of the energy function is a measure of the overall distance between elements of the two patterns. As the size K_1 of the Gaussian window decreases monotonically in successive iterations, in order for the energy E_1 to attain a minimum, each I_i should have at least one T_j attracted to it.

The second term is a weighted sum of all relative displacements between each *Template* element and its neighbors within the Gaussian weighted neighborhood of size parameter K_2. Each *Template* element normally does not move towards its nearest *Input* element but tends to follow the weighted mean movement of its neighbors in order to minimize the distortions within the neighborhood. E_1 is minimized by a gradient descent procedure. The movement ΔT_j applied to T_j is equal to $-\partial E_1/\partial T_j$ and is given by

$$\Delta T_j = \sum_{i=1}^{N_I} u_{ij}(I_i - T_j) + 2\sum_{m=1}^{N_T} (w_{mj} + w_{jm})[(T_m - T_m^0) - (T_j - T_j^0)] \qquad (4)$$

where T_j^0 = initial value of T_j and

$$u_{ij} = \exp(-|I_i - T_j|^2 / 2K_1^2) f(\theta_{Ii,Tj}) / \sum_{n=1}^{N_T} \exp(-|I_i - T_n|^2 / 2K_1^2) f(\theta_{Ii,Tn})$$

3 Similarity Measurement

At the end of the iteration of the elastic local alignment, the corresponding elements of the two signature patterns should hopefully be the nearest to each other as shown in

Fig. 4(b). And a match list of feature points has been produced. Then it is trivial to compute the Euclidean distance between two matched feature points by referring to their original coordinate positions. We define the mean square of sum Euclidean distances of all found corresponding feature points as the measure quantity for similarity between two signatures (SQ) as given below:

$$SQ = \frac{1}{N}\sqrt{\sum_{i=1}^{N}\left(\|\mathbf{T}_i - \mathbf{S}_i^{'}\|\right)^2} \qquad (5)$$

where $\mathbf{T}_i$ is the position vector of one feature points of the *Template*, $\mathbf{S}_i^{'}$ is the position vector of the corresponding feature points of the *Input*, N is the total number of feature points in the *Template*. As we want to test the robust and effectiveness of our proposed algorithms, only one sample signature was used as *Template* in global and local alignment. Another 3 genuine samples were employed to determine the threshold which verifies the test signature whether it is genuine or not. Any test signatures with SQ value larger that the threshold will be rejected as a forgery.

4 Experimental Results

The database to test the proposed algorithms was collected from 55 authors who contributed 6 genuine signature samples and 12 forgers who produced 6 forgeries for each author. 4 out of the 6 genuine signatures were used as training samples where one was arbitrarily selected as the *Template* and the remaining three were used to determine the suitable threshold. Cross-validation strategy was adopted to compute performance in terms of EER which stands for Equal Error Rate when the FAR is equal to the FRR.

To globally register two signature patterns by using the weak affine model through multi-resolution approach, the depth of multi-resolution is set to 2, resulting in the size of the 2nd level images being 32×32. There are 54 initial positions at the lowest resolution, namely: 3×3 translation points, separated by 3 pixels, and 6 equidistant rotation angles.

In the practice to employ the elastic local alignment method to find out corresponding feature points between two signature patterns, each line or curve is approximated by fitting a sequence of short straight lines ('elements') of about 10 pixels long. The neighbourhood size parameters were set to 20 pixels.

On average, the EER was 18.6% which is comparable to other existing methods [2-5] for off-line signature verification. EER was computed by varying the threshold which is 1.0 to 2.5 standard deviation from the mean of the SQ values of the three training samples.

5 Conclusion

Feature based signature verification suffers from the lack of training samples which leads to unstable and inaccurate statistic model. In order to avoid building statistic model and measure the similarity between two signature patterns directly, we propose the use of a global registration algorithm to roughly align two signatures to facilitate computing similarity measurement between two signature patterns. After applying an elastic local alignment algorithm, we are able to successfully produce a match list which we can use to compute the similarity quantity (SQ) for verification. In the experiment, four sample signatures were employed as training samples to determine the threshold and results were promising. If we use more training samples and incorporate statistical approach, it is hopeful to achieve better performance which is our next work in the future.

Acknowledgments

This research was partially supported by a grant (60403011) from National Natural Science Foundation of China, and grants (2003ABA012) and (20045006071-17) from Science & Technology Department, Hubei Province and the People's Municipal Government of Wuhan respectively, China. This research was also supported by the grants (RGC and FRG) from Hong Kong Baptist University.

References

1. Plamondon, R., Srihari, S.N.: On-Line and Off-Line Handwriting Recognition: A Comprehensive Survey. IEEE Transactions on Pattern Analysis and Machine Intelligence. 22(1) (2000) 63-84
2. Ammar, M.: Progress in Verification of Skillfully Simulated Handwritten Signatures. Int. J. Pattern Recognition and Artificial Intelligence. 5(1) (1991) 337-351
3. Sabourin, R., Genest, G., Prêteux, F.: Off-line Signature Verification by Local Granulometric Size Distributions. IEEE Trans. Pattern Analysis and Machine Intelligence. 19(9) (1997) 976-988
4. Sabourin, R., Genest, G.: An Extended-shadow-code-based Approach for Off-line Signature Verification. Part I. Evaluation of The Bar Mask Definition. In Proc. Int. Conference on Pattern Recognition. (1994) 450-453
5. Raudys, S.J., Jain, A.K.: Small Sample Size Effects in Statistical Pattern Recognition. IEEE Tran. Pattern Recognition and Machine Intelligence. 13(3) (1991) 252-264
6. Fukunaga, K.: Introduction to Statistical Pattern Recognition. Second Edition. Academic Press, Boston (1990)
7. Murshed, N., Sabourin, R., Bortolozzi, F.: A Cognitive Approach to Off-line Signature Verification. Automatic Bankcheck Processing, World Scientific Publishing Co., Singapore. (1997)
8. O'Sullivan, F.: A Statistical Perspective on Ill-posed Inverse Problems. Statistical Science, (1986) 502-527
9. Fang, B., Tang, Y.Y.: Reduction of Feature Statistics Estimation Error for Small Training Sample Size in Off-line Signature Verification. First International Conference on Biometric Authentication. Lecture Notes in Computer Science, Vol. 3072. Springer-Verlag, Berlin Heidelberg New York (2004) 526-532
10. Qi, Yingyong., Hunt, B.R.: Signature Verification Using Global and Grid Features. Pattern Recognition. 27(12) (1994) 1621-1629

11. Sabourin, R., Genest, G., Prêteux, F.: Off-line Signature Verification by Local Granulometric Size Distributions. IEEE Trans. Pattern Analysis and Machine Intelligence. 19(9) (1997) 976-988
12. Fang, B., Leung, C.H., Tang, Y.Y., Tse, K.W., Kwok, P.C.K., Wong, Y.K.: Offline signature verification by the tracking of feature and stroke positions. Pattern Recognition. 36(1) (2003) 91–101
13. Borgefors, G.: Hierarchical Chamfer Matching: A Parametric Edge Matching Algorithm. IEEE Trans. Pattern Analysis Machine Intelligence. 10(6) (1988) 849-865
14. Leung, C.H., Suen, C.Y.: Matching of Complex Patterns by Energy Minimization. IEEE Transactions on Systems, Man and Cybernetics. Part B. 28(5) (1998) 712-720

Shape Matching and Recognition Base on Genetic Algorithm and Application to Plant Species Identification*

Ji-Xiang Du[1,2], Xiao-Feng Wang[1], and Xiao Gu[1]

[1] Institute of Intelligent Machines, Chinese Academy of Sciences, Hefei, Anhui, China
[2] Department of Automation, University of Science and Technology of China
{du_jx, xfwang, xgu}@iim.ac.cn

Abstract. In this paper an efficient shape matching and recognition approach based on genetic algorithm is proposed and successfully applied to plant special identification. Firstly, a Douglas-Peucker approximation algorithm is adopted to the original shape and a new shape representation is used to form the sequence of invariant attributes. Then a genetic algorithm for shape matching is proposed to do the shape recognition. Finally, the superiority of our proposed method over traditional approaches to plant species identification is demonstrated by experiment. The experimental result showed that our proposed genetic algorithm for leaf shape matching is much suitable for the recognition of not only intact but also blurred, partial, distorted and overlapped plant leaves due to its robustness.

1 Introduction

The shape feature is one of the most important features for characterizing an object, which is commonly used in object recognition, matching and registration. In addition, the shape matching and recognition is also an important part of machine intelligence that is useful for both decision-making and data processing. More importantly, the recognition based on shape feature is also a central problem in those fields such as pattern recognition, image technology and computer vision, etc., which have received considerable attention recent years. Face recognition, image preprocessing, computer vision, fingerprint identification, handwriting analysis, and medical diagnosis, etc., are some of the common application areas of shape recognition [1, 2, 17].

In this paper, we introduce the shape matching and recognition technique to a new application: plant species identification. Plant species identification is a process resulting in the assignment of each individual plant to a descending series of groups of related plants, as judged by common characteristics. It is important and essential to correctly and quickly recognize and identify the plant species in collecting and preserving genetic resources, discovery of new species, plant resource surveys and plant species database management, etc. Plant identification has had a very long history,

---

* This work was supported by the National Science Foundation of China (Nos.60472111 and 60405002).

from the dawn of human existence. However, so far, this time-consuming and troublesome task was mainly carried out by botanists. Currently, automatic plant recognition from color images is one of most difficult tasks in computer vision because of lacking of proper models or representations for plant; and there are a great number of biological variations that different species of plants can take on.

The shape of plant leaf is one of the most important features for characterizing various plants and plant leaves are approximately of two-dimensional nature. Therefore, the study of leaf image recognition will be an important and feasible step for plant identification. For this target, some research works have been done. Rui, et al. [3] proposed a modified Fourier descriptor (FD) to represent leaf shapes and recognize plant species. Abbasi, et al. [4] used a curvature scale space (CSS) image to represent leaf shapes for Chrysanthemum variety classification. Mokhtarian et al. [5] improved this method and applied it to leaf classification with self-intersection.

But most of the existing plant recognition methods are only focus on the intact plant leaves and not applicable for the non-intact leaves largely existing in practice, such as the deformed, partial, overlapped and blurred leaves. So to design a practical, accurate, and automatic plant species identification system based on plants leaf images, in our work, we propose a shape matching method based on genetic algorithm for plant leaf, which is robust and can handle not only the intact leaves but also partial or overlapped leaves.

This paper is organized as follows: In Section 2, a shape polygonal approximation and the corresponding invariant attributes sequence representation are described and discussed. In Section 3, the genetic algorithm for shape matching is presented in detail. The experimental results are reported in Section 4, and Section 5 concludes the whole paper and gives related conclusions.

2 Shape Polygonal Approximations and Invariant Attributes Sequence Representation

An extracted leaf contour often exhibits too many resolvable points, thus it should be not directly applied on shape matching and the shape representation should be compressed. In this paper we adopted the Douglas-Peucker approximation algorithm [6, 7], a pure geometrical algorithm, to get a smooth contour on a smaller number of vertices, which is a better method due to the simplicity and shorter computational time.

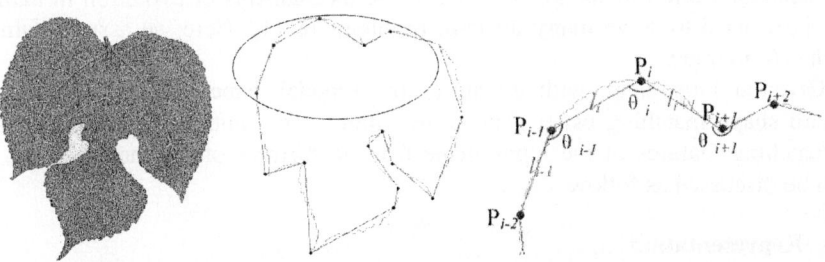

Fig. 1. A polygonal representation of contour

After performing the polygonal approximation of contour, a leaf shape can be represented as a sequence of vertices (as shown in Fig.1). The number of the points of this representation is largely smaller than the one of the original contours. But it is not invariant to rotation, scale and translation.

Further, to gain the invariant features, another representation method, referred to as sequence of invariant attributes, is used for representing the shape. Each shape can be represented as a sequence of features or attributes:

$$Q = \{Q_1, Q_2, Q_3, \cdots, Q_n\} \quad Q_i = (\rho_i, \theta_i, A_i, C_i) \tag{1}$$

where

$$\rho_i = \frac{l_i}{l_{i-1}} = \frac{|P_i P_{i-1}|}{|P_{i-1} P_{i-2}|} \tag{2}$$

$$\theta_i = \arccos \frac{|P_i P_{i-1}|^2 + |P_{i-1} P_{i-2}|^2 - |P_i P_{i-2}|^2}{2 |P_i P_{i-1}| \|P_{i-1} P_{i-2}|} \tag{3}$$

$$A_i = \frac{a_i}{a_{i-1}} = \frac{|P_i P_{i+1}| \sin(\theta_i)}{|P_{i-1} P_{i-2}| \sin(\theta_{i-1})} \tag{4}$$

$$C_i = \begin{cases} 1 & P_i \text{ is convex} \\ -1 & P_i \text{ is concave} \end{cases} \tag{5}$$

Each primitive Q_i consists of four attributes: the length ρ_i of the corresponding line segment; the relative angle θ_i; the area A_i with the preceding line segment P_{i-1} and the convexity C_i of the vertex P_i. All lengths ρ_i and areas A_i are locally normalized. So these four attributes are invariant to rotation, scale and translation.

3 Genetic Algorithm for Shape Matching

This section will describe and discuss genetic algorithm to partial and overlapped shape matching. The genetic algorithm is a model of machine learning, which derives its behavior from a metaphor of some of the mechanisms of evolution in nature and has been used to solve many difficult problems [8-11]. Here we assume familiarity with GA notation.

Given an input shape with n features and a model shape with m features, we can regard shape matching using genetic algorithm as the optimum problem of finding n matching features of the input shape from m features of the model shape, which will be discussed as follow.

3.1 Representation

Each individual in the population represents a candidate solution to the shape-matching problem. In our work, each individual can be regarded as mapping each of the input shape features to one of the model shape features. Thus each individual can be represented as a list in which each entry shows its corresponding feature of the matching model shape. And the notations are used as follow.

Each input shape can be represented as:

$$I = \{I_1, I_2, \cdots, I_n\}$$

where $I_k, (k=1,2,\cdots,n)$ is the kth feature of the input shape, consisting of four attributes $(\rho_k, \theta_k, A_k, C_k)$.

The model shapes are $M_1, M_2, \cdots, M_L$, each of which is represented as

$$M_j = \{M_{j1}, M_{j2}, \cdots, M_{jm_j}\} (j=1,2,\cdots,L)$$

where $M_{jk}, (k=1,2,\cdots,m_j)$ is the kth feature of the jth model shape, consisting of four attributes $(\rho_{jk}, \theta_{jk}, A_{jk}, C_{jk})$.

Each individual $D = \{D_1, D_2, \cdots, D_n\}$ can be regarded as a function $D = f(I)$ mapping input shape features to model shape features, i.e.,

$$D_k = f(I_k) = M_{ji}$$

where $1 \leq k \leq n, 1 \leq j \leq L, 1 \leq i \leq m_j$.

3.2 Fitness

The fitness evaluation is a mechanism used to determine the confidence level of the optimized solutions to the proposed problem. In our work, the fitness of an individual describes how well each feature of the input shape matches with the model shape feature to which it is matched. Fitness is calculated by testing the compatibility of the input shape features and the corresponding model features to which an individual maps them. The similarity measurement between the input shape feature I_k and the model feature $M_{ji} = D_k$ is defined as a distance function:

$$S(I_k, D_k) = f(S_C, S_\theta, S_\rho, S_A) \tag{6}$$

where $S_C, S_\theta, S_\rho, S_A$ are the similarity measurements for the four attributes of each feature pair between the input shape and the model shape and defined respectively as follow:

$$S_C = 0.5 * \left(1 + \frac{C(I_k)}{C(D_k)}\right) \tag{7}$$

$$S_\theta = 1 - \sin(0.5 * |\theta(I_k) - \theta(D_k)|) \tag{8}$$

$$S_\rho = 1 - \frac{|\rho(I_k) - \rho(D_k)|}{\max(I_k, D_k)} \tag{9}$$

$$S_A = 1 - \frac{|A(I_k) - A(D_k)|}{\max(I_k, D_k)} \tag{10}$$

Thus, the distance function $S(I_k, D_k)$ is defined as

$$S(I_k, D_k) = S_C \left[\frac{w_\theta S_\theta + w_\rho S_\theta S_\rho + w_A S_\theta S_\rho S_A}{w_\theta + w_\rho + w_A} \right] \tag{11}$$

where w_θ, w_ρ, w_A are the weight values for S_θ, S_ρ, S_A (in this paper, $w_\theta = w_\rho = w_A = 1$).
So fitness is calculated using the following formula:

$$Fitness = \frac{\sum_{k=1}^{n} S(I_k, D_k)}{n} \tag{12}$$

3.3 Selection Mechanism

The selection mechanism is responsible for selecting the parents from the population and forming the mating pool. The selection mechanism emulates the survival of the fittest mechanism in nature. It is expected that a fitter individual receives a higher number of offspring and thus has a higher chance of surviving on the subsequent evolution while the weaker individuals will eventually die. In this work we are using the roulette wheel selection (RWS), which is one of the most common, and easy-to implement selection mechanism. Basically it works as follows: each individual in the population is associated with a sector in a virtual wheel. According to the fitness value of the individual, the sector will have a larger area when the corresponding individual has a better fitness value while a lower fitness value will lead to a smaller sector. In additional we also used an elitist survival selection mechanism (ESS), i.e., the best fifth of all individuals in a generation is allowed to survive into the next generation.

3.4 Crossover and Mutation Operators

The purpose of the crossover operator is to produce new individuals that are distinctly different from their parents, yet retain some of their parents' characteristics. There are two commonly used crossover techniques, called single point crossover (SPC) and two-point crossover (TPC). In single point crossover, two parent individuals are interchanged at a randomly selected point thus creating two children. In two-point crossover, two crossover points are selected instead of just one crossover point. The part of the individuals between these two points is then swapped to generate two children. There are some other crossover techniques such as uniform crossover.

Some of the individuals in the new generation produced by crossover are mutated using the mutation operator. The most common form of mutation is to choose points on an individual and alter them with some predetermined probability. As mutation rates are very small in natural evolution, the probability with which the mutation operator is applied is set to a very low value.

Since we are representing an individual through a list, the single point and two-point crossover are applied to the individuals, producing two offspring. Mutation randomly chooses points on an individual and maps input shape features to the features of randomly chosen model shapes. All the individuals are mutated with 10% mutation rate.

4 Experimental Results and Discussions

The leaf image database used in our experiment was constructed by our lab, which consists of intact, blurred, partial, deformed, and overlapped leaf images of 25 plant species. There are 1800 images in our database (A subset was shown in Fig2). The experiment is designed to illustrate the superiority of our approach for plant species identification over traditional methods such as Fourier descriptors (FD) [3], Hu invariant moment (HM)[12], contour moment (CM) [13] and pair-wise geometrical histogram (PGH) [14] as well as geometrical features (GF) [15, 16]. Each method computes a distance for each pair of matched shapes.

Fig. 2. A subset of leaf image database

In the first experiment (Exp1), for each plant species, there are 20 intact leaf images selected from our leaves database as the model images for the GA method. In addition, there are at least 20 intact leaf images selected from the remaining leaves as the input samples. To recognize the class of one input leaf, it is compared to all model images, and the K nearest neighbors are selected. Each of the K nearest neighbors' votes in favor of its class, and the class that gets the most votes is regarded as the class of the tested leaf. In our experiment, we set $K = 4$. Then the performance comparison between the GA and the five competitors was implemented, and the results were shown in Table 1. From Table 1, it can be clearly seen that theses methods are efficient for the recognition of intact leaves, and specifically the GA and GF methods achieved the best performance. But, note that the GA algorithm will consume much more time.

Secondly, here we will test the capability of recognizing the partial plant leaf. The model samples are the same as the above, but for each contour of the input samples of every species used in the above experiment, a successive sub-contour is randomly extracted to form a partial shape, and the IR (Intact Rate) is defined as:

$$IR = \frac{P_{Sub\_C}}{P_C} * 100 \tag{13}$$

where P_{Sub_C} is the perimeter of the sub-contour.

Table 1. The performance comparison for different methods

Methods	Average Recognition Rate (%)	
	Exp1	Exp3
GA	93.6	83.2
GF	92.1	61.1
HM	83.4	62.6
CM	75.6	54.2
PGH	70.1	52.6
FD	89.2	60.7

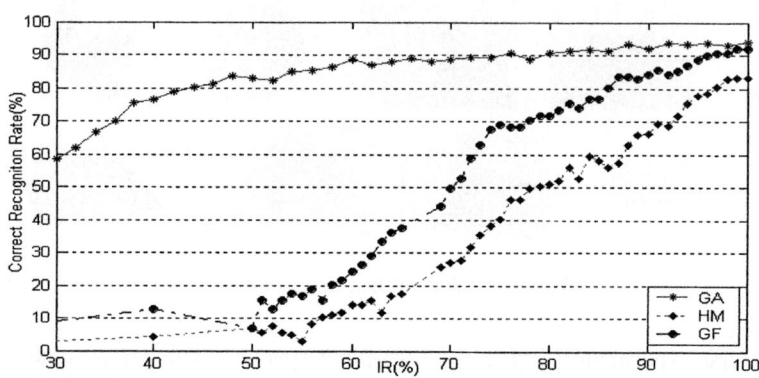

Fig. 3. The curves for the correct recognition rates *vs* IRs

This experiment was repeated for 10 times, and the averaged result is shown in Figure 3. For the GA method, it is shown that when the shape of leaf can keep the whole contour information over 40%, the average recognition rate is greater than

70%. But if using the GF and HM to recognize the partial leaves, the accuracy is very low. When IR is over 90, the correct recognition rate can be accepted. So, the GA method is a better choice for recognizing partial leaves than the GF or the HM method.

Finally, the model samples are also as same as the above and all the rest images are used as input samples. The results shown in Table 1 clearly demonstrated that our GA method achieved the best performance than other traditional methods.

In additional we compared the performance of different operations of GA (as shown in Table 2) using the data as same as Exp1. Table 2 showed that using the combined roulette wheel and elitist survival selection mechanism and two-point crossover operation will achieve a better recognition performance.

Table 2. The performance comparison for different operations

Selection	Crossover	Recognition Rate (%)
RWS	SPC	87.6
	TPC	90.8
RWS+ESS	SPC	91.3
	TPC	93.6

5 Conclusions

In this paper, an efficient plant identification method based on leaf images using genetic algorithm was proposed and performed. Firstly, the leaf shape was approximated to polygon through Douglas-Peucker algorithm, and the representation of invariant attribute sequence was used. Then a genetic algorithm was proposed to leaf shape matching. Our proposed approach was intrinsically invariant to rotation, translation and scale. The experimental result demonstrated that compared with other traditional methods, our proposed method is effective and efficient for recognition of both intact and non-intact leaf images, such as the deformed, partial and overlapped, due to its robustness. In the future work, those works of how to identify plant species using combined features (e.g., shape, texture and color) will be also included based on our proposed methods for practical applications.

References

1. Dengsheng, Z., Guojun, L.: Review of shape representation and description techniques. Pattern Recognition. Vol. 37. 1 (2004) 1-19
2. Loncaric, S.: A survey of shape analysis techniques. Pattern Recognition. Vol. 31. 8 (1998) 983-1001
3. Rui, Y., She, A.C., Huang, T.S.: Modified Fourier Descriptors for Shape Representation: A Practical Approach. In First International Workshop on Image Databases and Multimedia Search, Amsterdam, Netherlands (1996)

4. Abbasi, S., Mokhtarian, F., Kittler, J.: Reliable Classification of Chrysanthemum Leaves through Curvature Scale Space. ICSSTCV97 (1997) 284-295
5. Mokhtarian, F., Abbasi, S.: Matching Shapes with Self-intersection: Application to Leaf Classification. IEEE Trans on Image Processing Vol. 13 5 (2004) 653-661
6. Hershberger, J., Snoeyink, J.: Speeding Up the Douglas–Peucker Line Simplification Algorithm. Proceedings of the Fifth International Symposium on Spatial Data Handling, Vol. 1 (1992) 134 –143
7. Yingchao, R., Chongjun, Y., Zhanfu, Y., Pancheng, W.: Way to speed up buffer generalization by douglas-peucker algorithm. Geoscience and Remote Sensing Symposium, 2004. IGARSS '04. Proceedings. 2004 IEEE International 2916-2919
8. Ozcan, E., Mohan, C.K.: Shape Recognition Using Genetic Algorithms. Proc. IEEE Int. Conf. on Evol. Computation, Nagoya (Japan), (1996) 414-420
9. Abdelhadi, B., Benoudjit, A., Nait-Said, N.: Application of Genetic Algorithm With a Novel Adaptive Scheme for the Identification of Induction Machine Parameters. IEEE Transactions on Energy Conversion: Accepted for future publication (2005)
10. Chaiyaratana, N.; Zalzala, A.M.S.: Recent developments in evolutionary and genetic algorithms: theory and applications. Genetic Algorithms In Engineering Systems: Innovations And Applications, 1997. GALESIA 97. Second International Conference On (Conf. Publ. No. 446) 2-4 Sept. 1997 270 – 277
11. Lee, C.S.; Guo, S.M.; Hsu, C.Y.: Genetic-Based Fuzzy Image Filter and Its Application to Image Processing. IEEE Transactions on Systems, Man, and Cybernetics-Part B: Cybernetics: Accepted for future publication. (2005) 694 – 711
12. Chaur-Chin, Ch.: Improved moment invariants for shape discrimination, Pattern Recognition, Vol. 26. 5 (1993) 683-686
13. Gurta, L., Srinath, M.D.: Contour Sequence Moments for the Classification of Closed Planar Shapes. Pattern Recognition, Vol. 20. 3 (1987). 267-271
14. Iivarinen, J., Peura, M., Srel, J., and Visa, A.: Comparison of Combined Shape Descriptors for Irregular Objects. 8th British Machine Vision Conference (BMVC'97) 1997
15. Zhang, G.J., Wang, X.F., Huang, D.S., Chi, Z., Cheung, Y.M., Du, J.X., Wan, Y.Y.: A Hypersphere Method for Plant Leaves Classification. Proceedings of the 2004 International Symposium on Intelligent Multimedia, Video & Speech Processing (ISIMP 2004), Hong Kong, China, (2004). 165-168
16. Wan, Y.Y., Du, J.X., Huang, D.S., Chi, Z., Cheung, Y.M., Wang, X.F., and Zhang, G.J.: Bark texture feature extraction based on statistical texture analysis. Proceedings of the 2004 International Symposium on Intelligent Multimedia, Video & Speech Processing (ISIMP 2004), Hong Kong, China, (2004) 482-485
17. Huang, D.S.: Systematic Theory of Neural Networks for Pattern Recognition. Publishing House of Electronic Industry of China, Beijing (1996)

Detection of Hiding in the LSB of DCT Coefficients

Mingqiao Wu[1,2], Zhongliang Zhu[2], and Shiyao Jin[1]

[1] School of Computer, National University of Defense Technology,
410073 Changsha, China
wumingqiao1021@sohu.com
[2] Key Lab, Southwest Institute of Electron & Telecom Techniques,
610040 Chengdu, China

Abstract. In this paper, we provide a steganalysis method which can detect the hiding in the least significant bit of the DCT coefficients. The method is based on the thought that the DCT coefficients are correlative. So the LSB sequence of the DCT coefficients is not random as a pseudo-random sequence. The randomness the LSB sequence is measured by some statistical tests. We find, as the increase of the embedded secrets, the randomness of the LSB sequence increase. Using the statistical tests as the features, we train ε-support vector regression (ε-SVR) with train images to get the statistical mode of the estimation of the embed secrets. With the statistical mode, we can discriminate the stego-images from the clear ones. We test our method on Jsteg and OutGuess. The results of experiments show that our method can detect the hiding by Jsteg and OutGuess either.

1 Introduction

Steganography is the art of invisible communication. Its purpose is to hide the very presence of communication by embedding messages into innocuous-looking cover objects. Until now hundreds of steganographic techniques are available on Internet [1]. Some of them use JPEG images as covers. The JPEG images have the following characters. First, the JPEG format is currently the most common format for storing image data; Second, it is also supported by virtually all software applications that allow viewing and working with digital images. The stegangraphic techniques hiding data in JPEGs include Jsteg[2], F5 [3], OutGuess [4] and so on. In all programs, message bits are embedded by manipulating the quntized DCT coefficients. Jsteg and OutGuess embed message bits into the LSBs of quantized DCT coefficients. If steganographic techniques are misused by criminals for planning criminal activities, it will threaten the security of country. Many researchers have began to research the detection of steganography. Jsteg with sequential message embedding is detectable using the chi-square attack [5], Jsteg with random straddling is detectable by generlized chi-square attack [6]. But the chi-square attack can not detect the OutGuess as OutGuess preserves first-order statistics of image. In this paper, we present a method to detect the hiding in the LSB of DCT coefficients. We think the detection of stego-images is a classification problem, one class is the clear images, the other class is the stego-images. We want to use the development of machine learning to solve this problem. In section 2 we describe the basic thought of our steganalytic method and the extraction of features. In section 3 we test our method on the detection of Jsteg with random straddling. In section 4 we test our method on the detection of OutGuess. We conclude in section 5.

2 Steganalytic Method

In this paper we only study on gray image. We are illumined by Westfeld in [5]. It reveals that the LSBs of luminance values in digital images are not completely random, so replacement of LSBs could be detected by visual attacks. We believe the DCT coefficients in the same place of 8× 8 blocks are correlative. So if we sort the DCT coefficients in a suitable order, the LSB sequence is not complete random. As the embedding of the secret data, the randomness of the LSB sequence will increase.

2.1 LSB Sequence of DCT Coefficients

The reason that we think the DCT coefficients in the same place of 8× 8 blocks are correlative relies on two. First, for JPRG images, the DCT coefficients are get by making discrete cosine transformation to 8× 8 pixels blocks. The pixels in adjacent blocks are highly correlative. Second, research shows that the distribution of DCT coefficients is Gaussian distribution for DC coefficient and Laplacian distribution for AC coefficients [7]. Figure 1 shows the distribution of the DCT coefficients of Standard image "bridge".

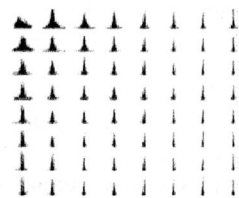

Fig. 1. (a) Standard image "bridge"   (b) Histogram of DCT coefficients of "bridge"

Let I describe the gray image of size $M \times N$, $dct_k(i, j)$ represent the DCT coefficient in 8× 8 block k. (i, j) is the coordinate of coefficient in the block. The blocks are ranged from left to right and up to bottom of image. For example, the DC coefficient of first block is expressed as $dct_1(1,1)$. We get the LSB sequence of DCT coefficients as follow:

$dct(i, j) = [];\quad DCT = [];$
$for\ k = 1:(M*N/64)$
$\quad if\ ((dct_k(i, j)!= 0)\ \&\ (dct_k(i, j)!= 1))$
$\quad\quad dct(i, j) = strcat(dct(i, j), dct_k(i, j))$
$\quad end$
end

```
for i = 1:8
    for j = 1:8
        DCT = strcat(DCT, dct(i, j))
    end
end
LSB_{DCT} = LSB(DCT)
```

In the above code, we concatenate the DCT coefficient in the same place in all blocks. $dct(i, j)$ is a string composed by the (i, j) DCT coefficients from block 1 to block $M*N/64$ which are neither 0 nor 1. DCT is the string of $dct(i, j)$. For Jsteg and OutGuess, DCT is the string of redundant bits which can be used to embedded secret data. $strcat(a,b)$ is a manipulation that concatenate b at the end of sting a. $LSB(DCT)$ is a manipulation to get the least significant bits of string DCT. So LSB_{DCT} is the LSB sequence of DCT coefficients, it is a binary sequence.

2.2 Feature Extraction

We think discriminating the stego-images from the clear ones is a two classes problem. We use statistical tests for the LSB sequence of DCT coefficients got in above section as the features for discrimination. With encryption the embedded secret data is almost pseudo-random. We think that with the embedding of the secret data, the randomness of the LSB sequence will also increase. The number of features is 12 and the features are divided into two parts. Let B denote the set $\{0,1\}$, $s^N = s_1, \cdots s_i, \cdots, s_N$, $s_i \in B$ denote the binary sequence with length N. Here is the LSB sequence of DCT coefficients. The first part of features includes following [8]:

1. maurer test:

The sequence s^N is portioned into adjacent non-overlapping blocks of length L. The total length of sample sequence s^N is $N = (Q+K)L$, where K is the number of steps of test and Q is the number of initialization steps. Let $b_n(s^N) = [s_{L(n-1)+1}, \cdots, s_{Ln}]$ for $1 \leq n \leq Q+K$ denote the n-th block of length L of the sequence s^N. Let the integer-valued quantity $A_n(s^N)$ be defined as taking on the value i if the block $b_n(s^N)$ has previously occurred and otherwise let $A_n(s^N) = n$. The maurer test function $f_{T_U}(s^N)$ is defined by:

$$f_{T_U}(s^N) = \frac{1}{K} \sum_{n=Q+1}^{Q+K} \log_2 A_n(s^N)$$

where for $Q+1 \le n \le Q+K$, $A_n(s^N)$ is defined by

2. frequency test

For a sequence $s^N = s_1, \cdots s_i, \cdots, s_N$, $s_i \in B$, the test function $f_{T_F}(s^N)$ is defined as:

$$f_{T_F}(s^N) = \frac{2}{\sqrt{N}} (\sum_{i=1}^{N} s_i - N/2)$$

3. serial test

The sample sequence s^N is cut into N/L consecutive blocks of length L (e.g., $L=8$), and the number $n_i(s^N)$ of occurrences of the binary representation of the integer i is determined for $0 \le i \le 2^L - 1$. f_{T_S} is defined as

$$f_{T_S}(s^N) = \frac{L2^L}{N} \sum_{i=0}^{2^L-1} (n_i(s^N) - \frac{N}{L2^L})^2$$

4. run test

In the run test with parameter L, the number $n_i^0(s^N)$ of 0-runs of length i and similarly the number $n_i^1(s^N)$ of 1-runs of length i in the sequence s^N are determined for $1 \le i \le L$ (e.g., $L=15$). $f_{T_R}(s^N)$ is defined as

$$f_{T_R}(s^N) = \sum_{b \in \{0,1\}} \sum_{i=1}^{L} \frac{(n_i^b(s^N) - N/2^{i+2})^2}{N/2^{i+2}}$$

5. autocorrelation test

An autocorrelation test $f_{T_A}(s^N, \tau)$ with delay τ for the sequence $s^N = s_1, \cdots s_i, \cdots, s_N$ is a frequency test for the sequence $s_1 \oplus s_{1+\tau}, s_2 \oplus s_{2+\tau}, \cdots, s_{N-\tau} \oplus s_N$, where $\oplus$ denotes addition modulo 2.

6. autocorrelation coefficient

First we compute the correlation with no normalization:

$$R_{s^N s^N}(m) = \sum_{i=1}^{N-m} s_{i+m} s_i$$

Then we get the normalized autocorrelation coefficient:

$$f_{xcoor}(s^N) = R_{s^N s^N}(1) / R_{s^N s^N}(0)$$

The second part of features is the changes of the statistical tests after embedding amount of secret data by . The embedding amount is 50% of redundant bits of image. Let s^N denotes the LSB sequence of original image DCT coefficients, $\hat{s}^N$ denotes the LSB sequence of the image after embedding. $abs(x)$ is the manipulation of get the absolute value of x. We get the following features:

7. change of maurer test

$$g_{T_U} = abs[f_{T_U}(\hat{s}^N) - f_{T_U}(s^N)]$$

8. change of frequency test

$$g_{T_F} = abs[f_{T_F}(\hat{s}^N) - f_{T_F}(s^N)]$$

9. change of serial test

$$g_{T_s} = abs[f_{T_s}(\hat{s}^N) - f_{T_s}(s^N)]$$

10. change of run test

$$g_{T_R} = abs[f_{T_R}(\hat{s}^N) - f_{T_R}(s^N)]$$

11. change of autocorrelation test

$$g_{T_A} = abs[f_{T_A}(\hat{s}^N, 1) - f_{T_A}(s^N, 1)]$$

12. change of autocorrelation coefficient

$$g_{xcoor} = abs[f_{xcoor}(\hat{s}^N) - f_{xcoor}(s^N)]$$

2.3 ε-Support Vector Regression (ε-SVR)

Using the 12 dimensions feature, we want to get the statistical model of estimation of embedding secret data. We are going to use the development of the machine learning-ε-Support Vector Regression (ε-SVR) to establish the statistical model. Given a set of data points, $\{(X_i, z_i), \cdots, (X_l, z_l)\}$, such that $X_i \in R^n$ is an input and $z_i \in R^1$ is a target output, the standard form of support vector regression is [9]:

$$\min_{W,b,\xi,\xi^*} \frac{1}{2}W^TW + C\sum_{i=1}^{l}\xi_i + C\sum_{i=1}^{l}\xi_i^*$$

subject to $W^T\phi(X_i) + b - z_i \leq \varepsilon + \xi_i,$

$z_i - W^T\phi(X_i) - b \leq \varepsilon + \xi_i^*,$

$\xi_i, \xi_i^* \geq 0, i = 1, \cdots, l.$

The dual is:

$$\min_{\alpha,\alpha^*} \frac{1}{2}(\alpha-\alpha^*)^T Q(\alpha-\alpha^*) + \varepsilon\sum_{i=1}^{l}(\alpha_i + \alpha_i^*) + \sum_{i=1}^{l}z_i(\alpha_i - \alpha_i^*)$$

subject to $\sum_{i=1}^{l}(\alpha_i - \alpha_i^*) = 0, 0 \leq \alpha_i, \alpha_i^* \leq C, i = 1, \cdots, l.$

where $Q_{ij} = K(X_i, X_j) \equiv \phi(X_i)^T \phi(X_j)$, $K(X_i, X_j)$ is the kernel function, ϕ is the function map vector X_i into higher dimensional space.

The approximate function is:

$$\sum_{i=1}^{l}(-\alpha_i + \alpha_i^*)K(X_i, X) + b$$

In our problem, input X_i is a 12 dimensions vector composed by 12 features described in section 2.2, output $z_i, 0 \leq z_i \leq 1$ is the estimated amount of embedded data for image i. Kernel function is radial basis function (RBF) :

$$K(X_i, X_j) = \exp(-\gamma \|X_i - X_j\|^2), \gamma > 0.$$

3 Tests on Jsteg

In this section we describe the experiment using our method to detect Jsteg with random straddling. We use 250 images as train images, and 100 test images. All the images are taken by digital camera then converted to gray images and cut to 512×512 size. For every train image and every test image, we embed 0,10%,20%,30%,40%,50%,60%,70%,80%,90%,100% secret data respectively using Jsteg with random straddling. After embedding, there are $250 \times 11 = 2750$ train images and $100 \times 11 = 1100$ test images. The experiment is done as following steps:

(1) Prepare data. For every image (train image or test images), we get it's 12 dimensions features. This step can be divided into two sub-steps. First one, we calculate the six statistical tests; Second one, after embedding 50% secret data with Jsteg, we calculate the six changes of statistical tests. Getting the 12 dimensions input data, we scaling each feature in the range [-1,+1].

(2) Search the best parameter for ε-SVR. Using the train data, we search the best parameter (C, γ, ε) for ε-SVR to make the mean squared error(mse) least.

(3) Using the train data and best parameter, we train the ε-SVR to get the statistical model of estimation of amount of secret data.

(4) Test the statistical mode with test data.

The plots of some features of train images are shown in Fig2. The best parameter searched for ε-SVR is $C=64, \gamma=1.0, \varepsilon=0.03125$, and $mse = 0.003236$. We use these parameters to train the ε-SVR and use it to predict the test data. In test, the Mean Squared Error =0.003347. The result of predict is plotted in Fig 3.

In figure 3, we can see that the the regression results of the clear images and the stego-images embedded 10% secret data overlap, so varying the detection threshold plays off the likelihood of false positive against missed detections (false negative results). In this paper, the false positive results are considered more serious the missed detections. We chose the median of the regression result of stego-images embedded 10% secret data as the threshold. Regression result that is larger than this threshold indicates a image with embedded data, inverse, it's a clear image. The classification results are shown in Table 1.

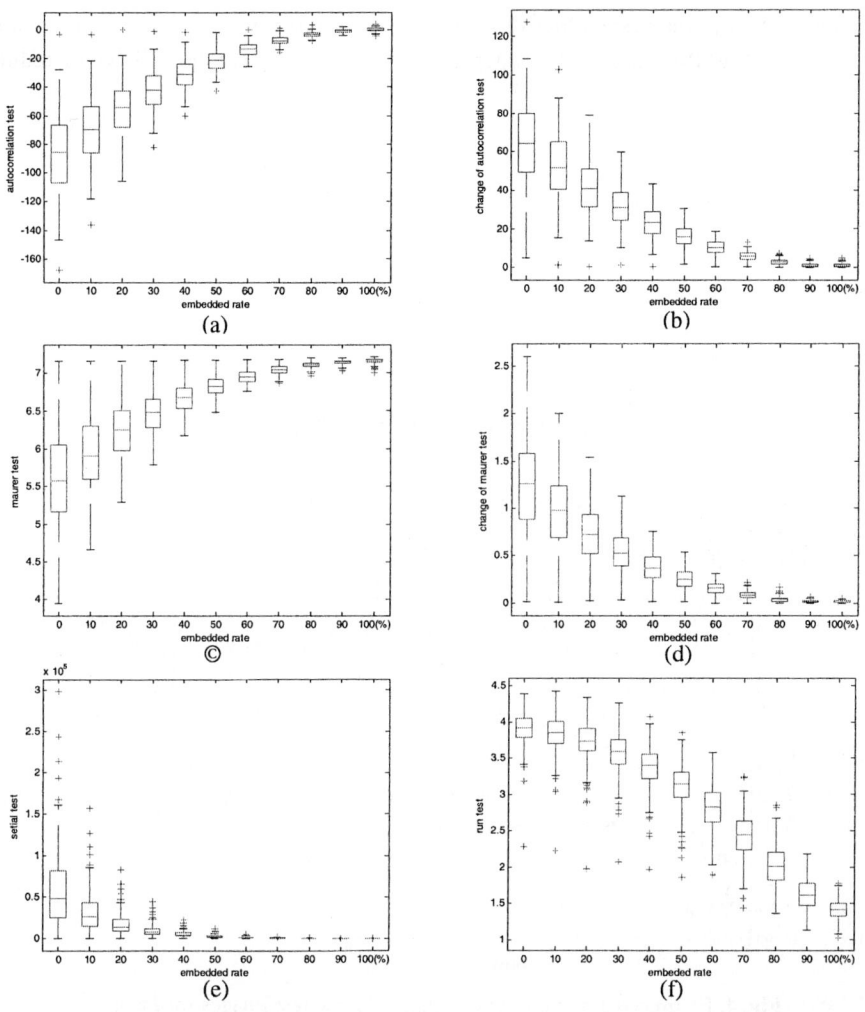

Fig. 2. Boxplot of some features of train images (a)Autocorrelation test (b)Change of autocorrelation test (c)Maurer test (d)Change of maurer test (e)Serial test (f)Run test

Table 1. Classification results (Threshold=0.0941)

Test images With embedded rate	0	10 %	20 %	30 %	40 %	50 %	60 %	70 %	80 %	90 %	100 %
Accuracy of classification	90 %	51 %	94 %	99 %	100 %	100 %	100 %	100 %	100 %	100 %	100 %

We can see from the results that this method can discriminate the steo-image using Jsteg random straddling especially when the embedded rate extends 20% of redundant bits.

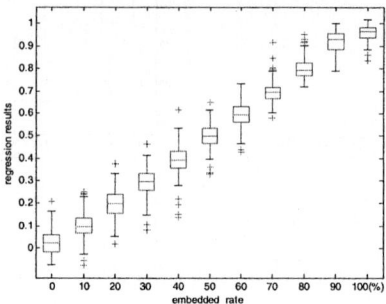

Fig. 3. Boxplot of the regression results of test images for Jsteg, for every embedded rate, there are 100 test images

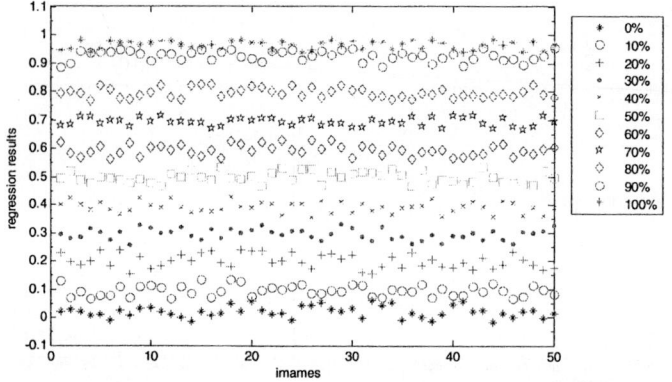

Fig. 4. Estimated amount of embedding of some test images for Jsteg

4 Tests on OutGuess

We test our method on OutGuess. The experiment is similar to experiment on Jsteg, except two modifications. One is: Because Outguess has the maximal capacity for embedding so we embed $0, 10\%, 20\%, 30\%, 40\%, 50\%, 60\%$ secret data in each of the images. After embedding, we get $250 \times 7 = 1750$ train images and $100 \times 7 = 700$ test images; Second, in the step (1) of experiment, we embed 50% secret data with OutGuess, then calculate the change of statistical tests. Fig 5 is the boxplot of regression results of test images for Outguess. The searched parameters are $C = 16, \gamma = 0.5, \varepsilon = 0.03125$, mse=0.002861. In test, the Mean Squared Error =0.002345.

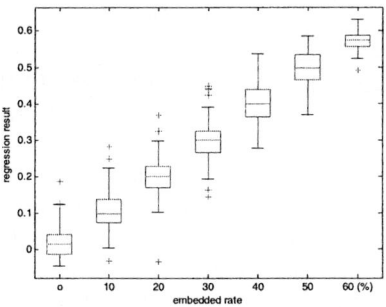

Fig. 5. Boxplot of the regression results of test images for OutGuess, for every embedded rate, there are 100 test images

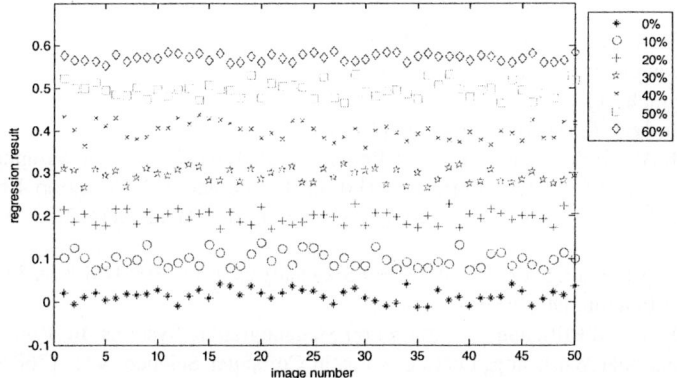

Fig. 6. Estimated amount of embedding of some test images for OutGuess

As in this paper the false positive results are considered more serious than missed detections, we set the threshold of discriminating stego-images from clear ones be the median of the regression of test images with 10% embedding. The classification results are shown in Table 2. We can see although the OutGuess preserve the first-order statistics of image histogram, it can not preserve the statistics of the LSB sequence so we can detect it. With the increase of embed data, the accuracy of detection increase rapidly.

Table 2. Classification results (Threshold=0.0972)

Test images With embedded rate	0	10 %	20 %	30%	40 %	50 %	60 %
Accuracy of classification	96 %	51 %	99 %	100 %	100 %	100 %	100 %

5 Conclusions

In this paper, we present a method to detect hiding in the LSB of DCT coefficients. We think as the DCT coefficients are correlative, the LSB sequence of DCT coefficients (extracted in suitable order) will not be so random as a random sequence. Embedding secret data in DCT coefficients, the randomness of LSB sequence will increase. We also believe discriminating stego-images from the clear ones is a classification problem. After extracting 12 features, we use ε-SVR to get the estimation of embedding amount in image and classify the image. Experiments show our method can detect hiding using Jsteg straddling and OutGuess. The next step of our work is to improve the detection accuracy and extend our method to color images.

References

1. Neilols F. Johnson. Steganography Tools and Software. http://www.jjtc.com/Steganography/toolmatrix.html
2. Steganography software for Windows, http://members.tripod.com/steganography/stego/software.html
3. Westfeld, A. High Capacity Despite Better Steganalysis (F5-A Stegnographic Algorithm). In : Moskowiz, I.S. (eds.): Information Hiding. 4th International Workshop. Lecture Notes in Computer Science, Vol.2137. Springer-Verlag, Berlin Heidelberg New York, 2001,pp. 289-302
4. Provos, N. Defending Against Statistical Steganalysis. Proc. 10th USENIX Security Symposium. Washington, DC, 2001
5. Westfeld, A. and Pfitmann, A. Attacks on Steganographic Systems. In: Pfitzmann A.(eds.): 3rd International Workshop. Lecture Notes in Computer Science, Vo;. 1768.Spring-Verlag, Berlin Heidelberg New York, 2000, pp. 61-76
6. Provos, N. and Honeyman, P. Detecting Steganographic Contend on the Internet. CITI Technical Report 01-11,2001
7. R.C. Reininger, J.D. Gibson: Distribution of the Two-Dimensional DCT Coefficients for Images. In: IEEE Transaction On Communicatins. Vol. Com-31,No.6.June 1983
8. Ueli M. Maurer. A Universal Statistical Test for Random Bit Generators. Journal of Cryptology, 5(2):89-105,1992
9. V. Vapnik. Statistical Learning Theory. Wiley, New York, NY, 1998

Tracking People in Video Camera Images Using Neural Networks

Yongtae Do

School of Electronic Engineering,
Daegu University, Kyungsan-City,
Kyungpook, 712-714,
South Korea
ytdo@daegu.ac.kr

Abstract. People are difficult targets to process in video surveillance and monitoring (VSAM) because of small size and non-rigid motion. In this paper, we address neural network application to people tracking for VSAM. A feedforward multilayer perceptron network (FMPN) is employed for the tracking in low-resolution image sequences using position, shape, and color cues. When multiple people are partly occluded by themselves, the foreground image patch of the people group detected is divided into individuals using another FMPN. This network incorporates three different techniques relying on a line connecting top pixels of the binary foreground image, the vertical projection of the binary foreground image, and pixel value variances of divided regions. The use of neural networks provides efficient tracking in real outdoor situations particularly where the detailed visual information of people is unavailable due mainly to low image resolution.

1 Introduction

Recently VSAM has received increasing attention in computer vision [1]. People and vehicles are two most important targets in VSAM. Although significant amount of research work has been carried out targeting vehicles, active study on people was only lately begun. One obvious reason is that there have been more commercial demands for automatic vehicle monitoring and traffic control. In addition, there are reasons related to technical difficulties. Specifically, people have higher degree of freedom and their shapes are varying in more flexible ways. There are also varieties of colors and texture patterns in human images. Their motion is less predictable and they can appear virtually at any place from any direction in a scene monitored. Identifying people in an image is also difficult especially when people are partly occluded by any object or by themselves.

Important aspects of people image processing for VSAM include detection, identification, tracking, shape analysis, and activity understanding. Fujiyoshi and Lipton [2] used a background subtraction technique for detecting moving objects in video image sequences, where a statistical background model is updated adaptively. For the identification, Lipton et al. [3] defined a simple measure of dispersedness with the perimeter and area of a target object in an image based on an observation that a person has more complex shape than a vehicle. Pfinder [4] is an example of tracking a person in a room and interpreting his or her behavior. Ghost^{3D} [5] is a system for estimating human postures using narrow-baseline stereo cameras. Brand and Kettnaker [6] presented an entropy minimization approach to clustering video sequences into events and creating classifiers to detect those events in the future.

In this paper, we address neural network application to people tracking. This work was motivated by three practical observations. First, we found that most existing computer vision techniques for tracking people rely on relatively high-resolution images. However, an outdoor VSAM camera is usually installed at a far distance from a scene to cover a large area and the target people in an image are likely in low-resolution as exampled in Fig. 1. Secondly, there is lack of techniques for identifying people occluded by themselves during motion. When two or more people merge by partial occlusion in an image, dividing the image patch of people into individuals is important not only to maintain tracking but also to understand the activities of people. Thirdly, when tracking non-rigid objects like people in low-resolution images, it is difficult to use a statistical motion model or build a reliable feature model. Although a human observer can understand and track a video sequence without any serious difficulty, it is hard to formulize. Under these circumstances, we expect artificial neural networks to learn the capability that a human observer implicitly possesses.

The system to be described in this paper is in a structure as shown in Fig. 2. Two separate FMPNs are employed, which are represented in dotted ellipses. One is for searching a person found in the previous image frame in current image frame. Section 2 describes the design of this network. The other is for checking partial occlusion in a foreground image patch and dividing it if necessary. Section 3 provides details of this network. Experimental results are given in Section 4. It is followed by a conclusion in Section 5.

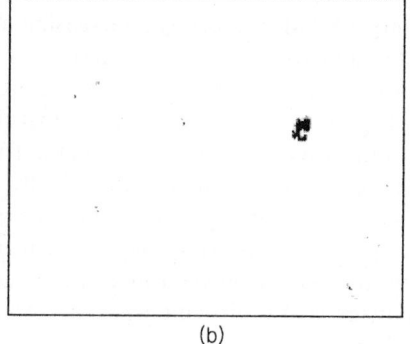

(a) (b)

Fig. 1. An example low resolution image of target people: (a) Original image, (b) Detected foreground

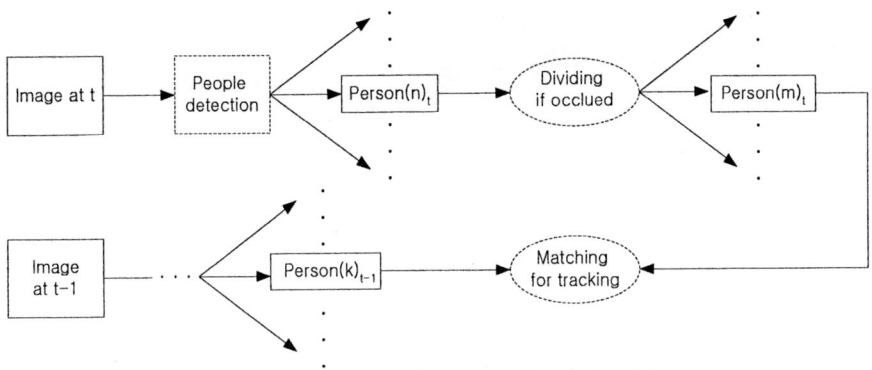

Fig. 2. Schematic block diagram of the people tracking system where solid boxes represent images, dotted boxes represent processes, and ellipses represent neural computations

2 Neural Network for Tracking

We define the tracking of people as the process of identifying each individual person in image sequences for recording or analysis. Most previous researches for people tracking in computer vision assume a single person [4] or use a statistical model [7]. In the former, the tracking problem becomes trivial but the assumption is not practical in most real outdoor scenes. In the latter, the performance of statistical estimation is not good if occlusions happen among multiple independently moving people in low-resolution images, or if image frame rate is low and irregular.

In this paper, an FMPN is employed for tracking people in low-resolution video image sequences of an irregular rate. We expect that a neural network can handle the tracking process effectively where conventional techniques cannot. Fig. 3 shows the network structure proposed. A person detected in the previous image frame is searched in current image frame using this network. The output s becomes closer to 1 as two image patches compared match more. Error back-propagation algorithm is used in network training.

The FMPN for tracking uses position, color, and shape cues. The adaptive background subtraction technique by Fujiyoshi and Lipton [2] is used for detecting moving people. Around people detected, minimum bounding boxes are drawn. One our observation is that the upper parts (head and torso) of a person have less variation than the lower parts (legs) in image sequences. This observation hints that processing with the information given from upper body parts is more reliable than that from the lower parts. The position is determined at the center coordinate of the top of a bounding box, which is an approximation of the head position. The network uses the positional difference between two image patches detected at t and $t-1$ as

$$\Delta i = |i_t - i_{t-1}|, \quad \Delta j = |j_t - j_{t-1}| \qquad (1)$$

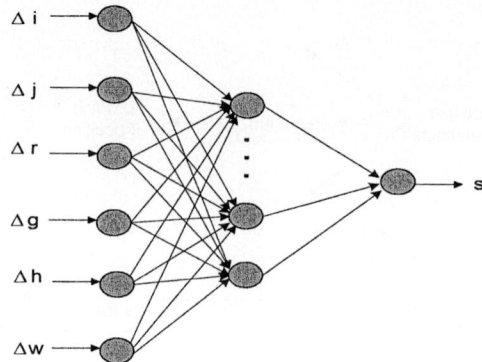

Fig. 3. Neural network structure for tracking

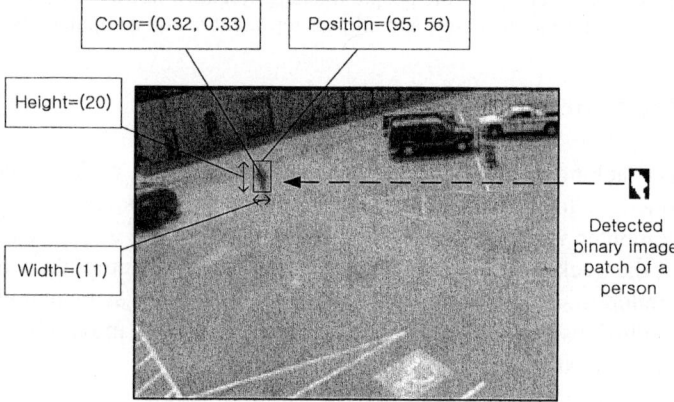

Fig. 4. An example of data extraction from an image for tracking

As the same way, the difference in the colors of torso parts is obtained for r and g values, which are normalized R and G color values respectively. The b value is dependent on other two values as $b = 1 - r - g$ and is not used for the network. Of small objects in image sequences, color information is not reliable as that of large objects. Thus, once matching is determined, each color value c is updated as

$$c_t = \alpha c_t + (1-\alpha)c_{t-1} \qquad (2)$$

where α is an adaptation factor. The network also uses the difference in the sizes of bounding boxes. In Fig. 3, those for height and width are represented by Δh and Δw respectively.

Fig. 4 shows an example of the process to detect a person and extract input data of the neural network. As described, the foreground image patch is detected first. Then, a minimum bounding box is drawn. The position, color, and size data are obtained from the box for the comparison with boxed image patches in previous image frame. The bottom line of the bounding box is determined at the 70% of the lowest point to overcome the instability of the lower body parts in video image sequences.

3 Neural Network for Dividing a People Group into Individuals

The tracking process discussed in the previous section meets a difficulty when two or more people are in close distance and an occlusion happens resulting in a mergence of foreground image patches. Sometimes one is behind of another completely and there is nothing to do but waiting the one reappears. In many other cases, only part of one's body is occluded by another person. The image patch of a group of people partly occluded by themselves is divided into individuals by analyzing three different features as described in following sub-sections.

3.1 Draping Line

For dividing an image patch of partially occluded people, we use a new technique, named 'draping line', assuming that people in a scene are in roughly upright postures. This assumption is acceptable in most practical situations if the camera angles are set properly and people are walking (or running) in normal way. In this case, the division can be limited to vertical direction only. The draping line is drawn by connecting the top pixels of columns of a foreground image as exampled in Fig. 5.

A real use of draping line is exampled in Fig. 6. For an image patch of Fig. 6(a), the foreground binary image is obtained as shown in Fig. 6(b). Then, a draping line is drawn like Fig. 6(c). At the local minimum (or minima if there are more than two people in an image patch) of the line, the image patch is vertically divided. Fig. 6(d) shows bounding boxes drawn after division. If necessary, the draping line can be smoothed against noise.

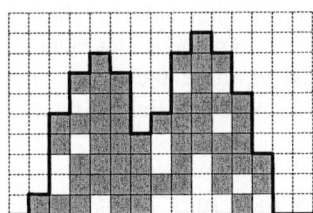

Fig. 5. Drawing a draping line for a foreground image patch

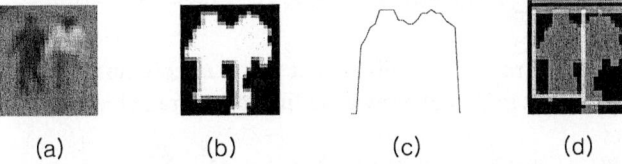

(a)   (b)   (c)   (d)

Fig. 6. Dividing an image patch of people into individuals by draping: (a) Original image, (b) Detected foreground, (c) Draping line, (d) Division at the local minimum of the draping line

3.2 Vertical Projection

In our experiments, the draping line based technique described in previous section showed good results in various situations. There are, however, cases where the technique is not effective as all pixels below the draping line in the image patch are ignored. Fig. 7 shows such an example. In this figure, using the vertical projection Pr

$$Pr(j) = \sum_{i=1}^{N} b(i,j), \quad 1 \leq j \leq M \tag{3}$$

of a binary image $\{b(i,j) \mid 1 \leq i \leq N, \ 1 \leq j \leq M\}$ as shown in Fig. 7(d) can be more effective than relying on draping.

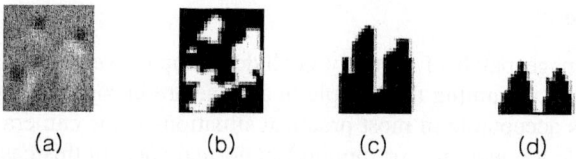

(a) (b) (c) (d)

Fig. 7. An example where vertical projection is more effective than draping: (a) Original image, (b) Foreground image, (c) Draping, (d) Vertical projection

3.3 Variances of Divided Regions

Ideally, if an image patch of two people in partial occlusion is optimally divided, the variance of pixel values inside each divided region will be low while the variance of total undivided region will be high. This is particularly true if two people wear clothes of prominently different colors. Often, however, people wear upper and lower clothes in different colors and it can raise the variance even in the region of the same person. Considering these facts, the variance of pixel values of each row, row_var, within a region, which is divided by a candidate dividing line, is first calculated like

$$row\_var(i) = \sum_{j=1}^{D}(p(i,j)b(i,j) - \overline{p}(i))^2 / \sum_{j=1}^{D} b(i,j) \tag{4}$$

$$\overline{p}(i) = \sum_{j=1}^{D} p(i,j)b(i,j) / \sum_{j=1}^{D} b(i,j), \quad \text{where } 1 \leq i \leq N \tag{5}$$

assuming that the two regions are divided at the $D'th$ column of $N \times M$ image. Then, all row_vars of the left region of a dividing line are averaged to get div_var_L

$$div\_var_L = \sum_{i=1}^{N} row\_var(i) \sum_{j=1}^{D} b(i,j) / \sum_{i=1}^{N} \sum_{j=1}^{D} b(i,j) \tag{6}$$

After calculating div_var_R of right region by the same way for $D+1 \leq j \leq M$, the average of the two variances, in_var, is obtained. The variance of total region, between_var, can be calculated by the same procedure but for $1 \leq j \leq M$.

3.4 Integration by a Neural Network

As three methods discussed so far check different aspects of image features, we can expect their integration to bring about better division. A neural network is employed to integrate the three methods proposed. Fig. 8 shows an FMPN structure designed for such a purpose. To make a smaller network, the *in_var* and *between_var* are obtained only for gray image (rather than for R, G, B color values separately). Centering at $j'th$ coordinate, where $3 \le j \le M-2$, arrays of five sequential elements of both draping line and projection graph are used for the input of network in addition to the values of *in_var* and *between_var*.

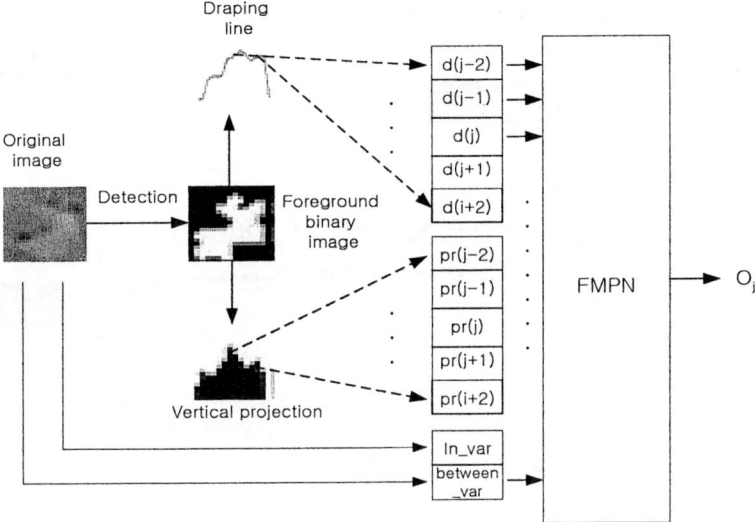

Fig. 8. A neural network for dividing an image patch of partially occluded people

4 Results

Experiments have been performed on real outdoor image sequences acquired with a fixed video camera. The image frame size was of 240 × 320 pixels. The size of the minimum bounding box of a person was varying depending on one's position and posture, but the smallest was in 15 × 8 pixels. Foreground image patches of moving people were extracted and stored on line. The time interval between image frames was rather irregular mainly depending on the number of people appear in the scene. The stored video files were analyzed later using an offline program written in Matlab.

Two different experiments were done. The first experiment was for testing the two neural networks designed. Fourteen target people from five video sequences taken at two different locations were used for the experiment. In all cases, the system described in this paper worked well. Fig. 9 shows a result of experimental tracking where the person carrying a black bag was tracked. The dividing network extracted the target person in all image frames except the last one where a complete occlusion happened. The tracking network successfully followed the target person after division except one on the last frame.

As the second experiment, the system designed was applied to counting the number of people in a scene. As stated in [8], counting pedestrians is often required for traffic control but currently done manually. In [8], Kalman filter and Learning Vector Quantization were used for high-resolution images. In our experiment, the image resolution was quite low and we used the dividing FMPN designed in Section 3. Fig. 10 shows the counting by division for the case of five walking people. They walked from close distance to far distance and correct division was possible only in the first two frames. The system can make correct counting if at least one correct division is possible and this result showed such a case.

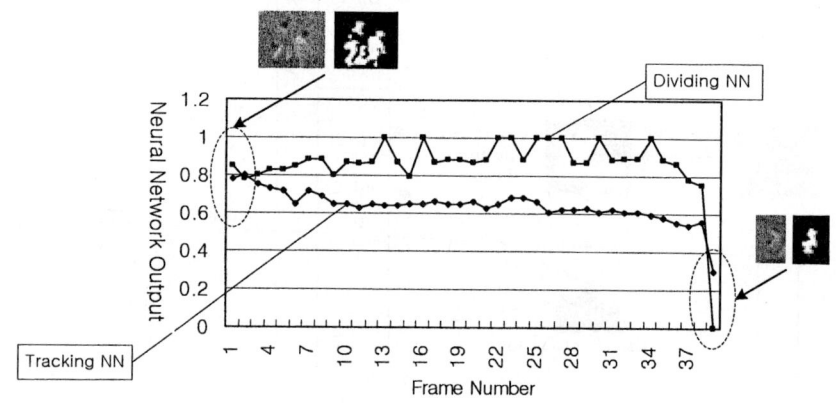

Fig. 9. An experimental result for tracking a person

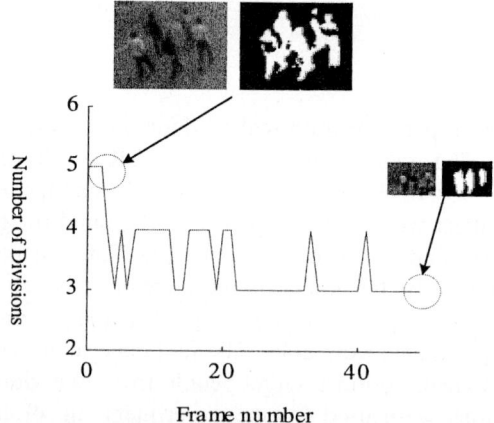

Fig. 10. Experimental result of counting the number of people in a scene

5 Conclusion

We have presented techniques to track moving people in low-resolution video images for outdoor surveillance and monitoring applications. Emphasis was given to two neural networks designed - one for dividing a group of partly occluded people into individuals, and the other for tracking each individual person after division. The dividing network uses the draping line and vertical projection of a binary foreground image in addition to the variance of the corresponding gray image. By the use of neural network, dividing could be effectively done in various situations only except the case of complete occlusion. Each divided individual could be tracked using the other neural network. The tracking network uses the position, color and shape information of a target person. Unlike most existing techniques, the system described in this paper can work effectively even for small targets in image sequences of irregular time interval without using explicit statistical or mathematical models. We could get good results when the system was tested in two different kinds of experiments; one for testing neural network performances in tracking people who were occluded by themselves during motion, and the other for counting the number of pedestrians in a scene.

Acknowledgement

This work was supported by the Korea Research Foundation Grant (KRF-2003-013-D00101).

References

1. Collins, R.T., Lipton, A.J., Kanade, T.: Introduction to the Special Section on Video Surveillance. IEEE Trans. Pattern Analysis and Machine Intelligence, Vol. 22, No. 8 (2000) 745-746
2. Fujiyoshi, H., Lipton, A.: Real-time Human Motion Analysis by Skeletonization. In: Proc. IEEE Workshop on Application of Computer Vision (1998) 15-21
3. Lipton, A.J., Fujiyoshi, H., Patil, R.S.: Moving Target Classification and Tracking from Real-Time Video. In: Proc. IEEE Workshop on Application of Computer Vision (1998) 8-14
4. Wren, C.R., Azarbayejani, A., Darrell,T., Pentland, A.P.: Pfinder: Real-Time Tracking of the Human Body. IEEE Trans. Pattern Analysis and Machine Intelligence, Vol. 19, No. 7 (1997) 780-785
5. Haritaoglu, I., Beymer, D., Flickner, M.: Ghost^{3D}: Detecting Body Posture and Parts Using Stereo. In: Proc. IEEE Workshop on Motion and Video Computing (2002) 175-180
6. Brand, M., Kettnaker, V.: Discovery and Segmentation of Activities in Video. IEEE Trans. Pattern Analysis and Machine Intelligence, Vol. 22, No. 8 (2000) 844-851
7. McKenna, S.J., Jabri, S., Duric, Z., Rosenfeld, A., Wechsler, H.: Tracking Groups of People, Computer Vision and Image Understanding, Vol. 80, No. 1 (2000) 42-56
8. Heikkila, J., Silven, O.: A Real-Time System for Monitoring of Cyclists and Pedestrians. In: Proc. IEEE Workshop on Visual Surveillance (1999) 74-81

How Face Pose Influence the Performance of SVM-Based Face and Fingerprint Authentication System

Chunhong Jiang and Guangda Su

Electronic Engineering Department, Tsinghua University, 100084, Beijing, China
jiangch@mail.tsinghua.edu.cn, sugd@ee.tsinghua.edu.cn

Abstract. How face pose (rotating from right to left) influence the fusion authentication accuracy in face and fingerprint identity authentication system? This paper firstly tries to answer this question. The maximum rotating degree that fusion system can bear is given out by experiment. Furthermore, theoretical analysis deals with how face pose influence the fusion performance is proposed in this paper. Experiment results show that faces with big rotated degree can not be helpful but harmful to fusion system. And the maximum rotated angle of face that fusion system can bear is 20 degree. On the other hand, theoretical analysis proved that the mathematical inherence of influence of face pose on fusion system is not only the reduction of variance but also the decrease of distance between the genuine and imposter classes.

1 Introduction

Recently biometric identity authentication (BIA) gains more and more research interesting in the world. BIA is a process of verifying an identity claim using a person's behavioral and physiological characteristics. BIA is becoming an important alternative to traditional authentication methods such as keys (something one has) or PIN numbers (something one knows), because it is essentially "who one is", i.e. by biometric information. Therefore, it is immune to misplacement, forgetfulness and beguilement.

The most prominent trend of BIA is to use of multiple biometric modalities (MBIA) to build the fusion authentication system [1~5], such as combining face and speech [1], face and fingerprint [2], face and iris [3]. These MBIA systems all selected face as a sub system because face has the good acceptability, collectable and performance. But as we know, face has the pose problem during acquisition. It is sure that face pose will affect the performance of fusion system. Whereas, how dose it influence? How much does it influence? And how many rotated degree can fusion system bear? Finally what is the essential inherence of the influence? Unfortunately these questions had not been involved in those references mentioned above. But, it is very important to application systems. Therefore, this paper tried to solve these problems and give the answer.

The remainder of this paper is organized as follows. The performance of face and fingerprint authentication system are described in section 2 and 3. The influence of face pose to fusion system is studied and analyzed in section 4. Finally we give out the main conclusions and future work in section 5.

2 Face Authentication

Face authentication involves face detection, feature extraction, feature matching process and decision making. In this paper, we use an automatic method for face detection and for eye and chin orientation [6], and adopted multimodal part face recognition method based on principal component analysis (MMP-PCA) to extract feature set [7]. The experiment face images are from the TH (Tsinghua University, China) Database. The TH database contains 270 subjects and 20 face images per subject with every other 5 degree turning from the front face to left (-) or right (+), and 10 fingerprint images from 2 different fingers with 5 images each. In our experiment, 186 subjects were selected for fusion of face and fingerprint authentication. We selected 13 face images and 5 fingerprint images for each subject. For face images, the first one, which is the one with zero turning degree, was selected as template, and the other 12 images as probes. Fig.1 shows the face and fingerprint images in TH database. Table 1 shows the training and testing protocols, the genuine and impostor match numbers. Protocol

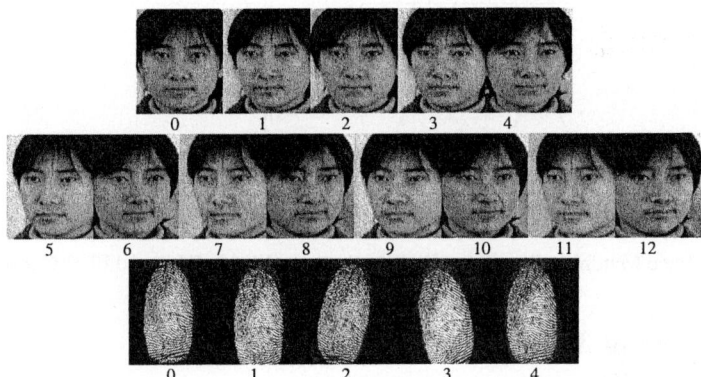

Fig. 1. Samples of Face and Fingerprint Images in the TH Database

Table 1. Authentication protocols. (SN--serial number; Degree--turning degree to right or left; Tr--for training; Te--for testing)

			Template	Probes							
		SN	0	1	2	3	4	5, 6	7, 8	9, 10	11, 12
		Degree	0	-5	+5	-10	+10	±15	±20	±25	±30
Face		Protocol 1 (P1)		Te1	Tr	Te1	Tr	Te2	Te3	Te4	Te5
		Protocol 2 (P2)		Tr	Te1	Tr	Te1	Te2	Te3	Te4	Te5
		Protocol 3 (P3)		Tr	Tr	Tr	Tr	Te2	Te3	Te4	Te5
		Number of genuine match		186	186	186	186	2×186	2×186	2×186	2×186
		Number of impostor match		185× 186	185× 186	185× 186	185× 186	2×185 ×186	2×185 ×186	2×185 ×186	2×185 ×186
Finger print	Set	Fusion with	Template	Probes							
	A	Tr, Te1	0	1		2		3		4	
	B	Te2,Te3	1	0		2		3		4	
	C	Te4,Te5	2	0		1		3		4	

(SN in fingerprint section)

1(*P1*), 2(*P2*) and 3(*P3*) are different from the training sets, +5 and +10 degree for *P1*, -5 and -10 degree for P2, ±5 and ±10 degree for *P3*. For training face images in *P1* and *P2*, we have 2 genuine match scores per subject, together 2×186 match scores constructing the training genuine distribution, and 2×185×186 impostor match scores constructing the training impostor distribution. Obviously every testing face sets has the same number. From table 1, we can see that each protocol of *P1* and *P2* has 1 training set and 5 testing sets (Te1~Te5). For *P3*, Te1 set of *P1* or *P2* was used for training, so the training set of P3 has 4×186 genuine match scores and 4×185×186 impostor match scores, and the testing sets (Te2~Te5) are same as sets in *P1* and *P2*.

Fig.2 shows face distributing of genuine and impostor match similarity (%) of Tr data set of P3 (Note that the following figures relate to face and fusion systems are all from this set except indicate). It is obvious that the genuine and impostor overlapped each other, and the decision errors are unavoidable. FAR and FRR curves of face authentication system are presented in Fig.3, EER is 0.044 when authentication threshold is 77%.

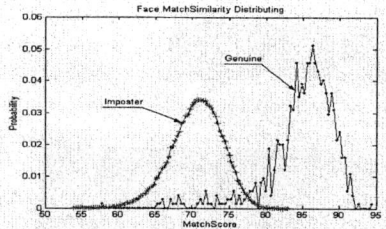 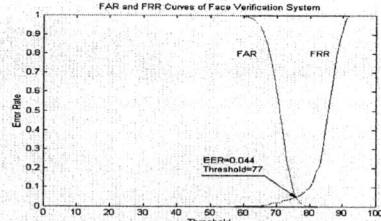

Fig. 2. Face Matching Similarity Distributing   **Fig. 3.** FAR and FRR Curves of Face

3 Fingerprint Authentication

In the Fingerprint authentication system, we use an automatic algorithm to locate the core point and extracted the local structure (direction, position relationship with the neighbor minutiaes) and global structure (position in the whole fingerprint) of all the minutiaes [8]. The matching algorithm used local and also global structures of every minutia. Fig.4(a) shows a sample in the TH fingerprint database, the core, the first orientation and minutiae points are presented on it and (b) shows the extracted ridge and minutiae points.

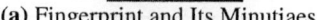

(a) Fingerprint and Its Minutiaes   (b) The Ridge and Minutiaes

Fig. 4. Sample in the TH Fingerprint Database

For fingerprint images, we selected 5 images from one finger. Table 1 shows the fingerprint protocol. One was selected to be template and the other four leaved to be probes. As to fusion with face, three data sets are built, i.e. A, B and C. Data in each set was used to generate 4×186 genuine match scores and 4×185×186 impostor match scores. Fig.5 shows fingerprint distributing of genuine and impostor match similarity (%) on data set A. FAR and FRR curves of fingerprint authentication system was presented in Fig.6. EER is 0.0107 when threshold is 18%. See Fig.6, FAR and FRR curves intersect and form a flatter vale, which predicate the range of threshold with respect to smaller FAR and FRR is larger, and the point of intersection is nearer to the x-axis, so the EER is smaller, both compared with face authentication in Fig.3. As a result, the authentication accuracy and robustness of fingerprint outperforms face authentication system obviously. In the next section, we will see fusion systems present a rather better performance than either of face and fingerprint system.

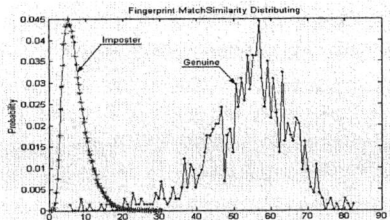 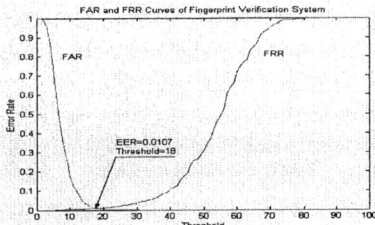

Fig. 5. Fingerprint Match Similarity Distributing   **Fig. 6.** FAR and FRR Curves of Fingerprint

4   Influence of Face Pose on Fusion Authentication Performance

As to fuse the face and fingerprint authentication systems, a confidence vector $X(x_1, x_2)$ represents the confidence output of multiple authentication systems was constructed, where x_1 and x_2 correspond to the similarity (score) obtained from the face and fingerprint authentication system respectively. Further more, for multi-biometric modalities more than 2, the problem turns to be N dimensional score vector $X(x_1, x_2, \cdots x_N)$ separated into 2 classes, genuine or impostor. In other words, the identity authentication problem is always a two-class problem in spite of any number of biometrics.

4.1   Fusion Results of SVM

Support vector machine (SVM) is based on the principle of structural risk minimization. It aims not only to classify correctly all the training samples, but also to maximize the margin from both classes. The optimal hyperplane classifier of a SVM is unique, so the generalization performance of SVM is better than other methods that possible lead to local minimum 9. In reference 10, we already compared SVM with other fusion methods, and some detail can be seen in it. The detailed principle of SVM can be seen in reference 10. In this paper, three kernel functions are used. They are:

Polynomials: $K(x,z) = (x^T z + 1)^d, d > 0$ (1)

Radial Basis Functions: $K(x,z) = \exp(-g\|x - z\|^2)$ (2)

Hyperbolic Tangent: $K(x,z) = \tanh(\beta x^T z + \gamma)$ (3)

Fig.7 shows the classification hyperplane of SVM-Pnm (d=2) on genuine and imposter. The fusion score distributing and FAR and FRR curves of SVM-Pnm are showed in fig.8 and 9. See fig.9, FAR and FRR curves intersect and form a very flat and broad vale; this means that for a large region of threshold value in which the FAR and FRR are both very small. Accordingly, not the accuracy but the robustness of the authentication system are both improved after fusion with SVM. ROC Curves of Face, Fingerprint and SVM System showed in fig.10, obviously the performance of SVM fusion systems are better than the fingerprint and face systems.

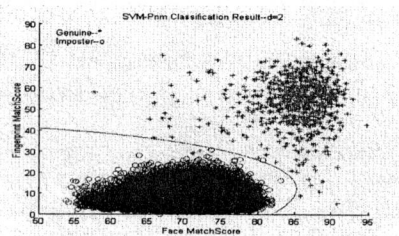

Fig. 7. Classification Hyperplane of SVM-Pnm

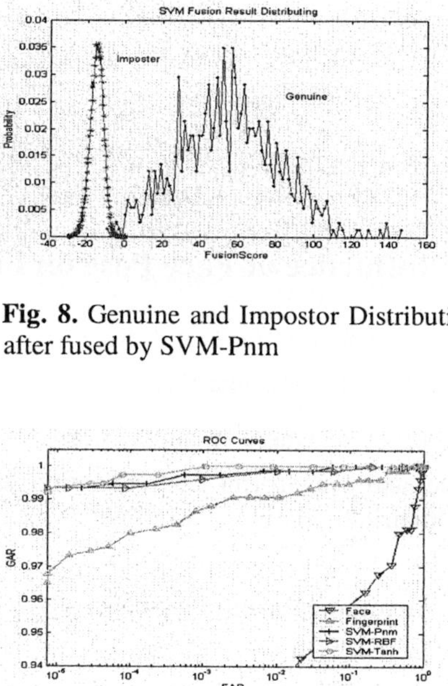

Fig. 8. Genuine and Impostor Distributing after fused by SVM-Pnm

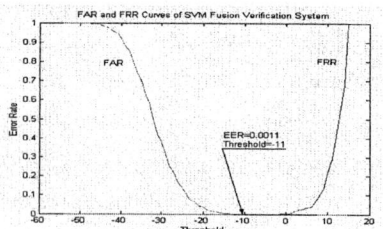
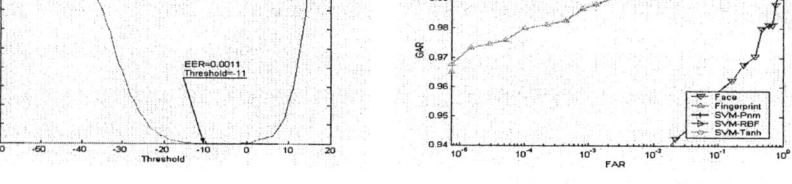

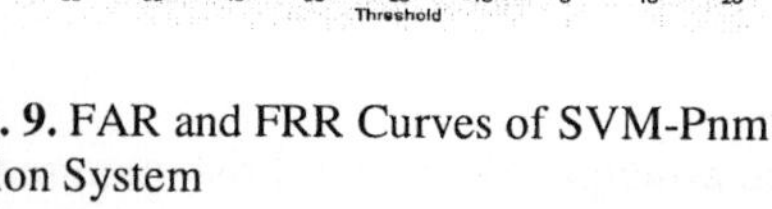

Fig. 9. FAR and FRR Curves of SVM-Pnm fusion System

Fig. 10. ROC Curves of Face, Fingerprint and SVM System

4.2 Influence of Face Pose on Authentication Performance

In practical authentication system, front face is not guarantied always, and face images acquired often have pose angle. It is not difficult to guess that with pose angle getting larger, face authentication accuracy is falling down. And how will this affect fusion performance? What is the theoretical evidence of this influence? Furthermore, how large angle can fusion authentication system bear? In other words, maintaining high authentication accuracy required, faces may rotate in what an allowable range? This paper will first investigate these problems, which are very important to practical system.

We select *P3* for example. See fig.11, compare (a) with (b), it is obvious that the genuine center is moving to the left because of face turning degree getting larger, whereas, the impostor center is comparatively stable, just a small shift. Fig.12 shows the mean and standard variance curves, noted that the class distances between genuine and imposter, see fig.12(a), become shorter and shorter with the face turning angle increasing, otherwise, the standard variances become smaller. Table 2 shows the mean (μ) and standard variance (σ) values of each face data set. For face genuine sets. The mean values become smaller from 87.0241 to 68.1473, and 70.9282 to 64.5138 for impostor sets, along with the rotated angle from -5 to +30 degree. While, the genuine standard variances change from 3.5310 to 3.1942 and the imposter standard variances change from 3.4203 to 2.327. Therefore the distances between genuine and imposter classes decrease from 16.0959 to 3.6299. From the σ values, we can see that the impostor data sets are much stable than the genuine. Table 3 shows the mean and standard variance values of different fingerprint data sets. From this table, we can see that distances between the genuine and imposter of fingerprint are much larger than face. Otherwise, the mean and standard variance of A, B and C data sets are very closer, so different fingerprint data sets are stable.

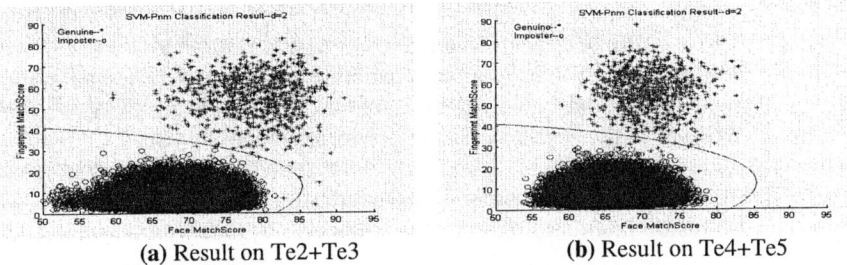

(a) Result on Te2+Te3   (b) Result on Te4+Te5

Fig. 11. SVM-Pnm Classification Results on Te2+Te3 and Te4+Te5

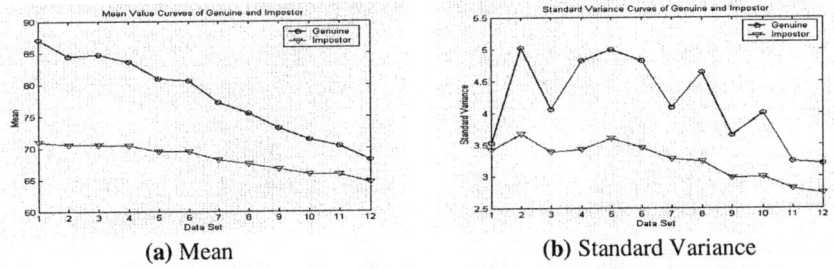

(a) Mean   (b) Standard Variance

Fig. 12. Mean and Standard Variance Curves of different face data sets

Table 2. Mean and standard variance of face data sets with different turning degree

μ & σ Degree	Genuine μ	Genuine σ	Impostor μ	Impostor σ
-5	87.0241	3.5310	70.9282	3.4203
+5	84.4497	5.0213	70.4596	3.6701
-10	84.7111	4.0597	70.5198	3.3837
+10	83.5352	4.8207	70.4046	3.4138
-15	80.9482	4.9914	69.4611	3.6018
+15	80.5642	4.8187	69.4207	3.4448
-20	77.1225	4.0876	68.1191	3.2706
+20	75.4831	4.6328	67.5016	3.2318
-25	73.1199	3.6429	66.6440	2.9652
+25	71.3474	3.9970	65.8601	2.9841
-30	70.2859	3.2360	65.7849	2.8058
+30	68.1473	3.1942	64.5138	2.7327

Table 3. Mean and standard variance of fingerprint data sets

μ & σ		Data sets A	B	C
Genuine	μ	53.72	54.94	54.96
Genuine	σ	11.32	11.20	11.44
Impostor	μ	7.270	7.310	7.360
Impostor	σ	3.280	3.300	3.330

The performance of face, fingerprint and SVM-Pnm authentication systems are showed in table 4. We select face authentication threshold as 82% and 30% for fingerprint, then FAR are both 7.265e-6, the same as SVM-Pnm. From table 4, we can see that SVM-Pnm' false rejected numbers are less than fingerprint when turning degree in the range of [5, 20], but more than fingerprint when turning degree not less than 25. Well, we can draw an experiential conclusion that if face turning degree is larger than 20, fusion face and fingerprint is not necessary. Under this condition, face

Table 4. Performance of face, fingerprint and SVM fusion systems. Note: for FAR and FRR column, the up value in bracket is the false accepted or false rejected number, under the number is the false rate

FAR, FRR Data sets	Face T=82% FAR	Face T=82% FRR	Fingerprint T=30% FAR	Fingerprint T=30% FRR	SVM Fusion T=0.0 FAR	SVM Fusion T=0.0 FRR
±5~±10	(1), 7.265e-6	(133), 0.1788	(1), 7.265e-6	(25), 0.0336	(1), 7.265e-6	(5), 0.0067
±15	(0), 0.0000	(192), 0.5161	(0), 0.0000	(7), 0.0188	(0), 0.0000	(1), 0.0027
±20	(0), 0.0000	(344), 0.9247	(0), 0.0000	(9), 0.0242	(0), 0.0000	(7), 0.0188
±25	(0), 0.0000	(367), 0.9866	(0), 0.0000	(11), 0.0296	(0), 0.0000	(14), 0.0376
±30	(0), 0.0000	(371), 0.9973	(0), 0.0000	(11), 0.0296	(0), 0.0000	(17), 0.0457

authentication is incapable of being helpful to fusion system, and even harmful to fusion system to some extent. Theoretical analysis of this influence will be presented in next sub section.

4.3 Theoretical Analysis

Suppose that the two classes, genuine and imposter follow Gaussian distribution and the probability density function $P(Z_i^{k=c})$ and $P(Z_i^{k=I})$ have mean u_i^k and variance $(\sigma_i^k)^2$, so the pdf is

$$P(Z) = \frac{1}{\sigma\sqrt{2\pi}} \exp(\frac{-(Z-u)^2}{2\sigma^2}) \quad (4)$$

In BA, FRR and FAR are both the function of decision threshold. Fig.13 is the sketch map of genuine and imposter distribution. Where, T is the decision threshold.

$$FRR(T) = \int_{-\infty}^{T} P(Z_i^{k=C} = Z)\,dz$$
$$= \int_{-\infty}^{T} \frac{1}{\sigma_i^C \sqrt{2\pi}} \exp[\frac{-(Z-u_i^C)^2}{2(\sigma_i^C)^2}]dz \quad (5)$$
$$= \frac{1}{2} + \frac{1}{2} erf(\frac{(T-u_i^C)}{\sigma_i^C \sqrt{2}})$$

$$FAR(T) = \int_{T}^{\infty} P(Z_i^{k=I} = Z)\,dz$$
$$= 1 - \int_{-\infty}^{T} P(Z_i^{k=I} = Z)\,dz \quad (6)$$
$$= 1 - [\frac{1}{2} + \frac{1}{2} erf(\frac{(T-u_i^I)}{\sigma_i^I \sqrt{2}})]$$
$$= \frac{1}{2} - \frac{1}{2} erf(\frac{(T-u_i^I)}{\sigma_i^I \sqrt{2}})$$

Where, $erf(x) = \frac{2}{\sqrt{\pi}} \int_0^x e^{-t^2} dt$ is the error function. Erf is a monotonous increase odd function, and its curve is showed in fig.14, and so $erf(-x) = -erf(x)$.

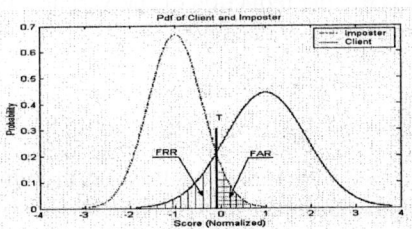 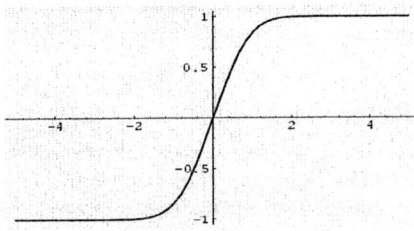

Fig. 13. Genuine and Imposter distribution   **Fig. 14.** Erf Curve

The minimal error happens when $FRR(T) = FAR(T) = EER(T)$, i.e. the Equal Error Rate. And substitute equation (5) and (6) into the upper formula, we have

$$erf(\frac{(T-u_i^C)}{\sigma_i^C \sqrt{2}}) + erf(\frac{(T-u_i^I)}{\sigma_i^I \sqrt{2}}) = 0 \qquad (7)$$

And using the odd function property, we can deduce that

$$T = \frac{u_i^I \cdot \sigma_i^C + u_i^C \cdot \sigma_i^I}{\sigma_i^I + \sigma_i^C} \qquad (8)$$

Substitute (8) into (7), then obtain

$$EER = \frac{1}{2} - \frac{1}{2} erf(\frac{1}{\sqrt{2}} \times \frac{u_i^C - u_i^I}{\sigma_i^C + \sigma_i^I}) \qquad (9)$$

Note that the ratio $R = u_i^C - u_i^I / \sigma_i^C + \sigma_i^I$ is similar to Fisher ratio. From equation (9), we can see that EER is inverse ratio with the between-class distance, see the numerator of R, and direct ratio with the total standard variance, see the denominator of R. Therefore the value of EER is dependent on not only the between-class distance but also the total standard variance (can be look at as the within class distance). To obtain smaller EER, one must enlarge the between-class distance and/or reduce the within class distance. Hence EER is inverse ratio with the ratio R, and ratio R can reflect system performance sufficiently. So the larger this ratio is, the better the genuine and imposter classes are discriminated, and the lower EER will be. Although the σ decreased, the EER become bigger in our experiments with face turning degree getting larger. This is because, the ratio R is smaller. So, reduced variance is not sufficient for judging whether EER become bigger or smaller. And the ratio R is just the key evidence.

5 Conclusions and Future Work

This paper firstly tries to interpret how face pose influence the fusion authentication accuracy in SVM-based face and fingerprint identity authentication system. Experiment results show that faces with big rotated degree can not be helpful but harmful to fusion system. And the maximum rotated angle of face that fusion system can bear is 20 degree. Furthermore, theoretical analysis proved that the mathematical inherence of influence of face pose on fusion system is not the reduction of the variance but the decrease of distance between the genuine and imposter classes.

References

1. Lin Hong and Anil Jain: Integrating Faces and Fingerprints for Personal Identification. IEEE Transactions on Pattern Analysis and Machine Intelligence. Vol.20(12).(1998) 1295-1307
2. Souheil Ben-Yacoub, et al.: Fusion of Face and Speech Data for Person Identity Authentication. IEEE Transactions on Neural Networks, Vol.10(5). (1999) 1065-1074

3. Wang Yunhong, Tan Tieniu: Combining Face and Iris Biometrics for Identity Verification. Proceedings of the Conference on Audio- and Video-Based Biometric Person Authentication, (2003) 805-813
4. Arun Ross, Anil Jain: Information Fusion in Biometrics. Pattern Recognition Letters, Vol.24. (2003) 2115-2125
5. Anil K. Jain, Arun Ross et al.: An Introduction to Biometric Recognition. IEEE Transactions on Circuits and Systems for Video Technology, Vol.14(1). (2004) 4-20
6. Gu Hua, Su Guangda, et al.: Automatic Extracting the Key Points of Human Face. Proceeding of the 4th Conference on Biometrics Recognition, Beijing, China, (2003)
7. Su Guangda, Zhang Cuiping et al.: MMP-PCA Face Recognition Method. Electronics Letters, Vol.38(25). (2002) 1654-1656
8. Huang Ruke: Research on the Multi-Hierarchical Algorithm for Fast Fingerprint Recognition. Bachelor thesis of Tsinghua University, (2002)
9. Sergios Theodoridis, Konstantinos Koutroumbas: Pattern Recognition, Elsevier Science, (2003)
10. Jiang Chunhong, Su Guangda: Information Fusion in Face and Fingerprint Identity Authentication System. Proceeding of ICMLC'04, Vol.1-7. (2004) 3529-3535

A VQ-Based Blind Super-Resolution Algorithm

Jianping Qiao[1,2], Ju Liu[1], and Guoxia Sun[1]

[1] School of Information Science and Engineering, Shandong University,
Jinan 250100, Shandong, China
[2] National Laboratory on Machine Perception, Beijing University,
Beijing 100871, Beijing, China
{jpqiao, juliu, sun_guoxia}@sdu.edu.cn

Abstract. In this paper, a novel method of blind Super-Resolution (SR) image restoration is presented. First, a learning based blur identification method is proposed to identify the blur parameter in which Sobel operator and Vector Quantization (VQ) are used for extracting feature vectors. Then a super-resolution image is reconstructed by a new hybrid MAP/POCS method where the data fidelity term is minimized by l_1 norm and regularization term is defined on the high frequency sub-bands offered by Stationary Wavelet Transform (SWT) to incorporate the smoothness of the discontinuity field. Simulation results demonstrate the effectiveness and robustness of our method.

1 Introduction

In many image applications such as remote sensing, military surveillance, medical diagnostics and HDTV, SR images are often required. However, the quality of image resolution depends on the physical characteristics of the imaging devices, and it is hard to improve the image resolution by replacing sensors because of the cost or hardware physical limits. Super-resolution image reconstruction is one promising technique to solve the problem which uses digital image processing techniques to obtain an SR image (or sequence) from several low resolution samples of the same scene. It has been one of the most active research areas in the field of image recovery.

The super-resolution idea was first addressed by Tsai and Huang [1], who used the aliasing effect to restore a high-resolution image from multiple low-resolution (LR) images. Many different methods were proposed then in this field, such as iterative back projection (IBP) [2], projection onto convex set (POCS) [3], Bayesian estimation [4], etc. However, only a few address the problem of blind SR. Meanwhile, SR restoration can be considered as a second-generation problem of image recovery, therefore some methods of image recovery can be extended to solve SR problem. In [5], a VQ-Based blur identification algorithm for image recovery was proposed by Nakagaki. But in this method if the variance of the band-pass filter was not chosen properly, it was hard to identify blur parameter.

In this paper, we propose a blind SR algorithm by improving the VQ-based approach. In our method, Sobel operator is introduced for extracting feature vectors and DCT is applied to reduce the dimensionality of the vector. After blur identification, we also propose a new algorithm for SR restoration by combining l_1 norm minimization and stationary wavelet transform. Simulations show that our method is robust to different types of images and different noise models. Also, it has fast convergence performance and low complexity.

The paper is organized as follows: The proposed SR blur identification algorithm is presented in section 2. Section 3 describes the SR restoration method. Simulations in section 4 show the effectiveness of our method. Section 5 concludes this paper.

2 Blur Identification

At present, there are two super-resolution models: the warping-blurring model and the blurring-warping model. The former model coincides with the imaging physics but it is usable only if the motion among SR images is known a priori, and the latter model is more appropriate when the motion has to be estimated [8]. In the following, the warping-blurring model is taken as an example to discuss the problem. The relationship between the low-resolution images and the high resolution image can be formulated as [4]:

$$\underline{Y}_k = D_k H_k F_k \underline{X} + \underline{E}_k \tag{1}$$

where $\underline{Y}_k$ is the $M_1 M_2 \times 1$ lexicographically ordered vector containing pixels from the k th LR frame, $\underline{X}$ is the $N_1 N_2 \times 1$ lexicographically ordered vector containing pixels from SR image, $\underline{E}_k$ is the system noise with the size $M_1 M_2 \times 1$, D_k is the decimation matrix of size $M_1 M_2 \times N_1 N_2$, H_k is the blurring matrix of size $N_1 N_2 \times N_1 N_2$, known as the Point Spread Function(PSF), F_k is a geometric warp matrix of size $N_1 N_2 \times N_1 N_2$, $1 \le k \le K$ and K is the number of low-resolution images.

In many practical situations, the blurring process is unknown or is known only within a set of parameters. Therefore it is necessary to incorporate the blur identification into the restoration procedure.

2.1 The VQ-Based Blur Identification Algorithm

Vector quantization is a technique used in the data compression field. In [5], it was applied to identify blur parameter of the degraded image where different vector represents different local characteristics of the image. The method consists of two stages: codebook design and blur identification. Details are shown as in Fig.1.

Assuming that the blurring process is known and blur function is parameterized by the parameter i. A number of VQ codebooks, each corresponding to a candidate blurring function are designed using band-pass filtered prototype images. A codebook with the minimum average distortion for a given degraded image is selected and the blur function used to create the codebook is identified as the unknown blurring function.

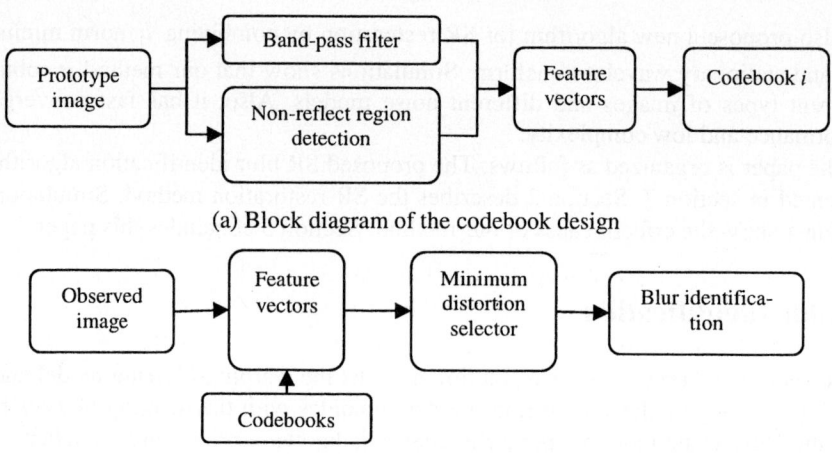

(a) Block diagram of the codebook design

(b) Block diagram of the blur identification

Fig. 1. Block diagram of the VQ-based blur identification approach

2.2 The Improved VQ-Based Blur Identification Algorithm

In [5], Nakagaki used LOG filter as the band-pass filter for the codebook design. But if the variance of the LOG filter is not chosen properly, it is hard to identify the blur parameter. Moreover, the parameter has to be changed when the type of the blurred image changes. In our method Sobel operator is used to detect the edge of the blurred image, then feature vectors are formed by some of the DCT coefficients of each edge detected image block. This process is shown as in Fig.2.

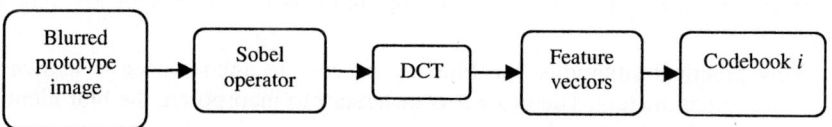

Fig. 2. Block diagram of the improved approach

Then the improved approach is extended to identify the blur function of the LR image. The blur identification algorithm for SR can be depicted as follows:

The first stage is codebook design:

1) Choose n candidates for each LR image and some prototype images (or training images) that belong to the same class with the LR image.

2) Blur and down-sample the prototype images according to the observed LR image.

3) Up-sample the LR image with bilinear or bicubic interpolation.

4) Detect the edge of the up-sampled image with Sobel operator and divide the edge detected image into blocks, the feature vectors are the low frequency DCT coefficients of each block.

5) The LBG algorithm is used for the creation of the codebook.

The second stage is blur identification:

Given a blurred low-resolution image, after the interpolation, edge detection and DCT in the same way as that in the first stage, the distance between the image and each codebook is calculated. A codebook with the minimum distance value is selected. The blurring function corresponding to the codebook is identified as that of the LR image.

The basic idea of the proposed method is that because the blurring function is considered as a low-pass filter and low-frequency regions contains little or no information about the PSF, information about PSF is not lost after edge detection. Meanwhile, in contrast to the LOG filter, Sobel operator enhances the robustness of the algorithm to different types of images. This is because what we need is the information that can distinguish different PSFs, the edge detected image by Sobel operator furthest preserves this information. In this way, the selection of the variance of the LOG filter is avoided.

3 Super-Resolution Reconstruction

Super-resolution reconstruction is an ill-posed problem. In [6], the wavelet representation of the image is utilized to construct a multi-scale Huber-Markov model. The model is incorporated into Bayesian MAP estimator to reconstruct the SR image. This method is effective to preserve the edges of the image, but it is very sensitive to the assumed model of data and noise. Sina Farsiu et al. [7] propose an approach using l_1 norm minimization to enhance the robustness to different data and noise models.

3.1 Robust Data Fusion

According to Bayesian theory, the ML estimation of the SR image is given by:

$$\hat{\underline{X}} = Arg\max_{\underline{X}}[p(\underline{Y}_k/\underline{X})] = Arg\min_{\underline{X}}\left[\sum_{k=1}^{K}\rho(\underline{Y}_k, D_k H_k F_k \underline{X})\right] \quad (2)$$

where ρ measures the "distance" between the model and measurements. In order to make the cost function robust to different data and noise models, we choose l_1 norm to minimize the measurement error [7]. Meanwhile, according to POCS idea, by incorporating constraining convex sets that represent a priori knowledge of the restored image into the reconstruction process, the optimization problem can be stated as:

$$\hat{\underline{X}} = Arg\min_{\underline{X}}\left[\sum_{k=1}^{K}\|D_k H_k F_k \underline{X} - \underline{Y}_k\|_1^1\right] \quad subject\ to\ \{\hat{\underline{X}} \in C_p, 1 \le p \le P\} \quad (3)$$

where C_p represents the additional constraints such as constraints on the output energy, phase, support and so on.

A unique and stable estimate $\underline{X}$ can be sought by introducing a priori information about the SR image into the reconstruction process:

$$\underline{\hat{X}} = \underset{\underline{X}}{Arg\ min}\left[\sum_{k=1}^{K}\|D_k H_k F_k \underline{X} - \underline{Y}_k\|_1^1 + \lambda\Phi(\underline{X})\right] \quad subject\ to\ \{\underline{\hat{X}} \in C_p, 1 \leq p \leq P\} \quad (4)$$

where λ is the regularization parameter and the function Φ has the following form:

$$\Phi(t) = \sqrt{1+t^2} - 1 \quad (5)$$

3.2 Stationary Wavelet Transform

Compared with the standard DWT decomposition scheme, the sub-sampling in SWT is dropped and results in a highly redundant wavelet representation. By sacrificing the orthogonality, SWT gets the shift invariance property which has been widely used in the image processing.

3.3 Super-Resolution Reconstruction Based on SWT Regularization

Super-resolution reconstruction is an image fusion technique. From this perspective, it is essential that the transformation should have the property of shift invariance if the image sequence or observed images can not be registered accurately. So in this method we exploit the stationary wavelet transform to extract the high-frequency information of the image in different directions which will be used as a priori information to be incorporated into the reconstruction process. Integrating the a priori information into (4), we define the following new optimization problem:

$$\underline{\hat{X}} = \underset{\underline{X}}{Arg\ min}\left[\sum_{k=1}^{K}\|D_k H_k F_k \underline{X} - \underline{Y}_k\|_1^1 + \lambda\sum_{m,n}\sum_{j=1}^{J}\sum_{i=1}^{3}\Phi(W_{j,i}\underline{X})\right] \quad (6)$$
$$subject\ to\ \{\underline{\hat{X}} \in C_p, 1 \leq p \leq P\}$$

where λ is the regularization parameter, $W_{j,i}$ is the 2-D stationary wavelet transform matrix, $j = 1,\ldots,J$ are the levels of the wavelet decomposition, $i = 1,2,3$ are the orientations of the wavelet transform corresponding to vertical, horizontal and diagonal orientations, respectively. Regularization parameter λ controls the tradeoff between fidelity to the data (the first term) and smoothness of the solution (the second term).

The steepest descent algorithm is used to solve the optimization problem:

$$\underline{\hat{X}}_{n+1} = \underline{\hat{X}}_n + \beta\left\{\sum_{k=1}^{K}F_k^T H_k^T D_k^T sign(D_k H_k F_k \underline{\hat{X}}_n - \underline{Y}_k) + \lambda\sum_{m,n}\sum_{j=1}^{J}\sum_{i=1}^{3}W_{j,i}^T R^{-1} W_{j,i}\underline{\hat{X}}_n\right\} \quad (7)$$

where β is step size, matrix R is a diagonal matrix which can be stated as follows:

$$R = diag\left[\sqrt{1+\left(W_{j,i}\hat{X}_n\right)^2}\right]_{n=1}^{N_1N_2} \tag{8}$$

Constraints $\{C_p\}_{p=1}^{P}$ are added to the new estimation in each iteration.

4 Simulations

In this section we present several experiments to demonstrate the effectiveness of our method. First we describe some of the experimental results obtained with the proposed blur identification algorithm in section 2, then present some of the experimental results obtained with the proposed SR restoration algorithm in section 3.

4.1 Blur Identification Algorithm

In this subsection, three types of images are used to demonstrate the effectiveness of the proposed approach. They are atmosphere images, medical images and nature images. We assume that the blurring process is out-of-focus blur and is parameterized by the radius r. Different types of noise were added into the simulated LR images, such

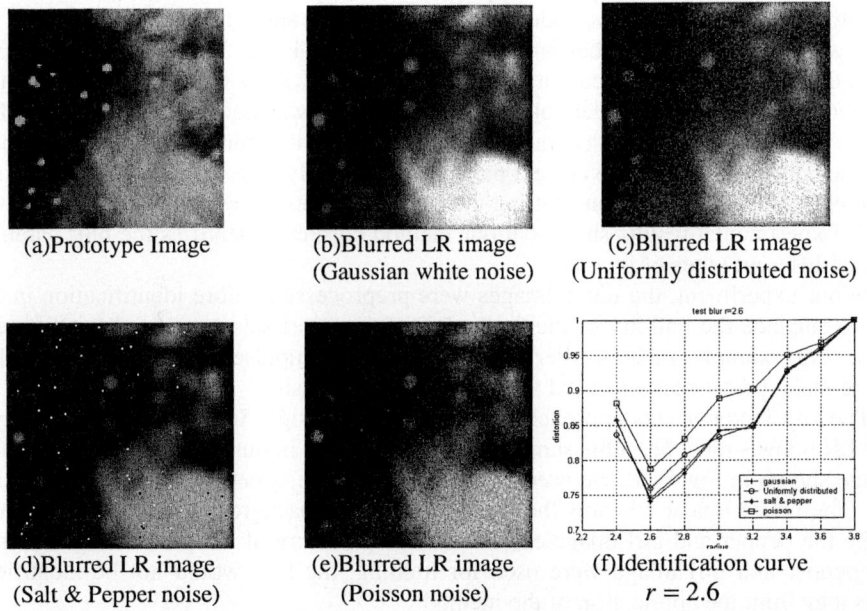

(a) Prototype Image (b) Blurred LR image (Gaussian white noise) (c) Blurred LR image (Uniformly distributed noise)

(d) Blurred LR image (Salt & Pepper noise) (e) Blurred LR image (Poisson noise) (f) Identification curve $r = 2.6$

Fig. 3. Atmosphere images

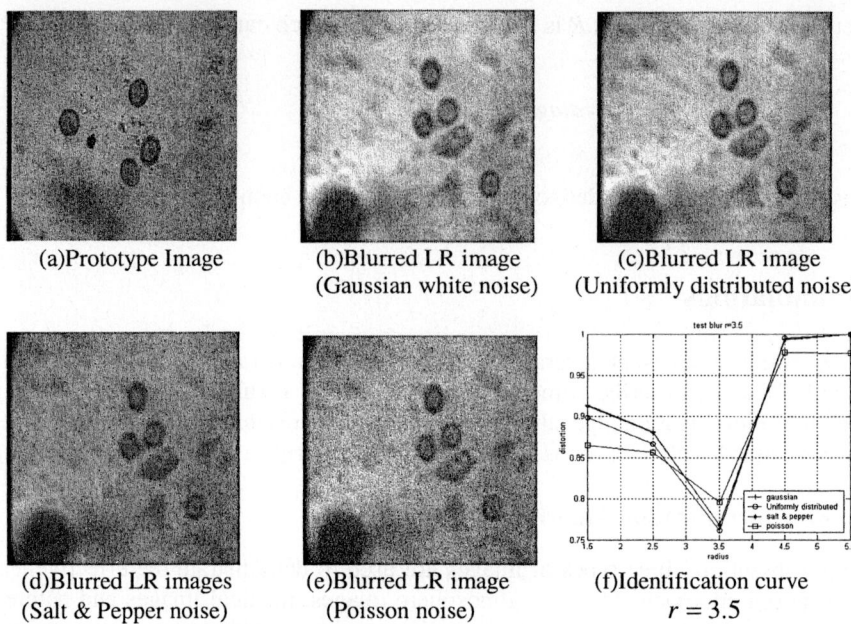

Fig. 4. Medical images

as Gaussian white noise, Salt and Pepper noise and Poisson noise, etc. Fig.3 (a) shows the prototype image used for codebook design, (b)~(e) show blurred LR images with different types of noise. The candidates were r ={2.4 2.6 2.8 3.0 3.2 3.4 3.6 3.8 4.0}.(d) shows the identification curve where X-axis shows the candidates of the blurred image used for codebook design, Y-axis shows the distortion between LR image and the codebooks. The value corresponding to the minimum point of the plot in X-axis is identified as the correct parameter. Similarly, Fig.4 shows the blur identification of medical images and setting the candidates were r = {2.0 2.5 3.0 3.5 4.0 4.5 5.0}. Experimental results show that our method correctly identifies the blur parameters of different types of images.

In our experiment, the noisy images were preprocessed before identification in order to enhance the validity of the method. Different methods were used for different noise. For example, median filter was utilized when impulse noise exists and multi-frame mean preprocess was used when poison noise exists.

The key factor in this approach is the codebook design. In detail, training images and LR images must be in the same class. For example, in our experiment, given a LR image shown in Fig.5 (b), we used an image of a house, a pepper and a girl for training. The house image contains the line feature in the background of the given image while the pepper and girl image contain the curve feature of the given image. If only the pepper and girl image were used for training, the blur would not be identified. This may limit the application of the method.

Moreover, the dimensionality of the vector was reduced by taking low frequency DCT coefficients. Here, we must emphasize that the dimensionality is relative to the type of the image. For the medical image shown in Fig.4, there are more flat regions and the dimensionality of the vector was reduced from 8×8=64 to 6, while for the nature image shown in Fig.5, there are more sharp edges and the dimensionality of the vector should be larger.

(a)Prototype images   (b)Blurred image

Fig. 5. Natural images

4.2 SR Restoration Algorithm

In this subsection, we show experimental results obtained by the proposed SR algorithm. Sixteen blurred, sub-sampled and noisy low-resolution images were generated by the high-resolution standard image BARBARA. The degradation includes affine motion (including rotation and translation), blurring with Gaussian kernel which is parameterized by the variance σ^2, and a 4:1 decimation ratio, and additive salt and pepper noise and white Gaussian noise. First of all, blur identification was performed. We still used an image of a house, a pepper and a girl for training. In fig.6 we can identify the parameter σ^2 is 3.5. Then the parameter was used for SR restoration. The first LR image without motion was selected as the reference image and db4 is used for SWT. The parameters in (7) were selected experimentally with $J=3$, $\lambda=0.5$, $\beta=8$. Fig.7(a)-(d) show the respective results of the four different methods: bilinear interpolation, method in [6], method in [7] and the method proposed in this paper. As l_2 norm is sensitive to noise models, salt and pepper noises are still visible in Fig.7(b). Fig.8(a) shows the MSE between the reconstructed image and the original image, (b)

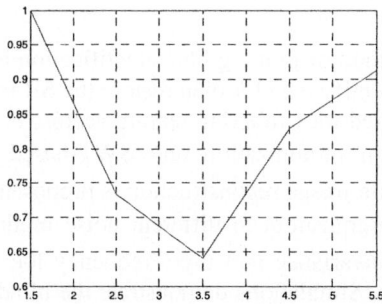

Fig. 6. Identification curve

(a)Bilinear interpolation (b) method in [6] (c) method in [7] (d) our method
(PSNR= 17.8612dB) (PSNR= 19.3054dB) (PSNR= 21.0906dB) (PSNR= 21.9328dB)

Fig. 7. Comparison of different methods

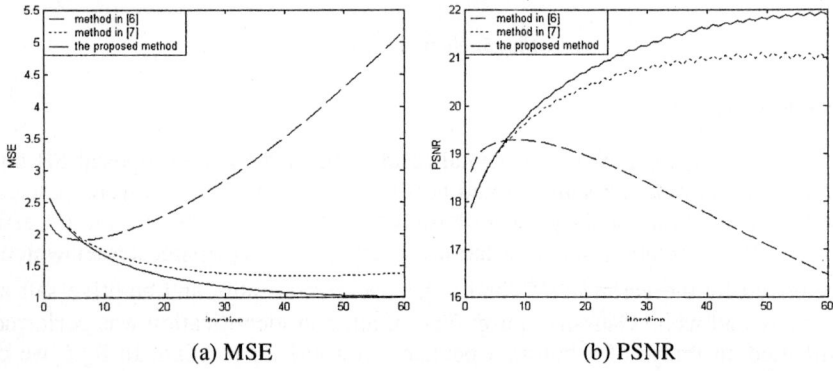

(a) MSE (b) PSNR

Fig. 8. Iteration curves (a) The MSE between the reconstructed image and the original image. (b) The PSNR between the reconstructed image and the original image

shows the PSNR between the reconstructed image and the original image. Evidently, the proposed method has better convergence performance and the image by our method has the highest PSNR.

5 Conclusion

In this paper, a novel method of jointing blur identification and SR image restoration was proposed. We first identify the blur then restore the SR image. Sobel operator and Vector quantization are used for extracting feature vectors in blur identification. Then by combining the l_1 norm minimization and SWT-based regularization, a novel method for super-resolution image reconstruction is proposed. It makes full use of the robustness of l_1 norm minimization to different noise models and errors in motion estimation and the shift-invariance and time-frequency localization property of stationary wavelet transform. Simulations demonstrate the blind SR method is robust to different types of noise and has faster convergence performance.

Acknowledgement. This Project was partially sponsored by the Excellent Young Scientist Award Foundation of Shandong Province (NO. 01BS04) and partially by the Open Foundation of National Laboratory on Machine Perception (0403) and The Project-sponsored by SRF for ROCS, SEM.

References

1. Tsai, R.Y., Huang, T.S.: Multiframe Image Restoration and Registration. Advances in Computer Vision and Image Processing, JAI Press Inc. 1 (1984) 317-339
2. Peleg, S., Keren, D.: Improving Image Resolution Using Subpixel Motion. pattern recognition letter 5(3) (1987) 223-226
3. Ozkan, M.K., Tekalp A.M., Sezan, M.I.: POCS-based Resolution of Space-varying Blurred Images. IEEE Trans. on Image Processing 3 (7) (1994) 450-454
4. Michael Elad, Arie Feuer: Restoration of A Single Super-resolution Image from Several Blurred, Noise, and Undersampled Measured Images. IEEE Trans. on Image Processing 6 (12) (1997) 1646-1658
5. Nakagaki, R., Katsaggelos, A.K.: A VQ-based Blind Image Restoration Algorithm. IEEE Trans. Image Processing 12 (9) (2003) 1044–1053.
6. Zhang Xinming, Shen Lansun: Super-resolution Restoration with Multi-scale Edge-preserving Regularization. Journal of software 14 (6) (2003) 1075-1081
7. Farsiu, S., Robinson, D., Elad, M., and Milanfar, P.: Fast and Robust Multi-Frame Super-resolution, IEEE Trans. On Image Processing 13(10) (2004) 1327- 1344
8. Zhaozhong Wang, Feihu Qi.: On Ambiguities in Super-Resolution Modeling. IEEE signal processing letters 11 (8) (2004) 678-681

Sequential Stratified Sampling Belief Propagation for Multiple Targets Tracking

Jianru Xue, Nanning Zheng, and Xiaopin Zhong

Institute of Artificial Intelligence and Robotics, Xi'an Jiaotong University,
No 28, Xianningxilu,Xi'an,Shaanxi Province, China (710049),
{jrxue, nnzheng, xpzhong}@aiar.xjtu.edu.cn

Abstract. In this paper, we model occlusion and appearance/disappearance in multi-target tracking in video by three coupled Markov random fields that model the following: a field for joint states of multi-target, one binary process for existence of individual target, and another binary process for occlusion of dual adjacent targets. By introducing two robust functions, we eliminate the two binary processes, and then apply a novel version of belief propagation called sequential stratified sampling belief propagation algorithm to obtain the maximum a posteriori (MAP) estimation in the resulted dynamic Markov network. By using stratified sampler, we incorporate bottom-up information provided by a learned detector (e.g. SVM classifier) and belief information for the messages updating. Other low-level visual cues (e.g. color and shape) can be easily incorporated in our multi-target tracking model to obtain better tracking results. Experimental results suggest that our methods are comparable to the state-of-the-art multiple targets tracking methods in several test cases.

1 Introduction

Data associations and state estimation are two core activities of classical multi-target tracking techniques [1]. However, the data association problem is NP-hard though some work has been done to generate k-best hypotheses in polynomial time [1][2]. If targets are distinctive from each other, they can be tracked independently by using multiple independent trackers with least confusion. However, its performance in tracking targets in video is heavily interfered with difficulties such as heavy cluttered background,the presence of a large, varying number of targets and their complex interactions.To adequately capture the uncertainties due to these factors, a probabilistic framework is required.

The Bayesian approaches for multiple target tracking solve association and estimation jointly in a maximum a posteriori(MAP) formulation [2]. Most of them fall into two categories. The first extends the state-space to include components for all targets of interest, e.g. [3][4][13], this allows the reduction of the multi-object case to less difficult single-object case, and will overcome the limitation of the single object state representation based particle filter which can't handle multi-modality. A variable number of objects can be accommodated by either dynamic changing the dimension of the state space, or by a corresponding set of indicator variables signifying whether an object is present or not.The second build multi-object trackers by multiple instantiations of single object tracker, e.g. [5][8][12]. Strategies with various levels of sophistication have been developed to interpret the output of the resulting trackers in case of occlusions and overlapping objects.

More recently, a boosted particle filter [7] and mixture particle filter [6] are proposed to track a varying number of hockey players. These two methods are actually single particle filter tracking framework to address the multiple target tracking problem with the help of mixture density model. Methods mentioned above alleviate the ambiguities due to interactions among multiple targets somewhat. However, how to model varying number of targets and their interactions still remains an open problem.

There are two major contributions in this paper. Firstly, we model targets' state and their interaction explicitly by using three MRFs and subsequently approximate it to a Markov network by introducing two robust functions. Secondly, we apply a novel version of belief propagation called sequential stratified sampling belief propagation algorithm to obtain the MAP estimation in the dynamic Markov network. With stratified sampler, we can incorporate bottom-up information from a learned detector (e.g. SVM classifier), mix backward-forward messages passing with information from belief node, and combine Other low-level visual cues (e.g. color and shape) easily into our model to improve algorithm performance.

The rest of paper is organized as follows: in Section 2, a novel model is proposed to explicitly represent existence, occlusions, and multi-target state in the Bayesian framework. In Section 3, stratified sampling belief propagation is applied to infer the target state. The basic model is then extended in Section 4 to integrate other cues such as temporal information and bottom-up information from a learned detector. The experimental results shown in Section 5 demonstrate that our model is effective and efficient. Finally, we summarize and suggest several promising lines of future work in Section 6.

2 Problem Formulation and Basic model

Assume the total number of targets is varying but bounded by M which is known. We denote the state of an individual target by x_i, $x_i \in S$, $i = 1, ..., M$, the joint state by $\mathbf{X} = \{x_1, ..., x_M\}$, $\mathbf{X} \in \mathbf{S}$ for M targets, the image observation of x_i by y_i, and the joint observation by $\mathbf{Y}$. Each individual target state at time t including information about the location, velocity, appearance and scale. In this paper, we use color histogram to represent the target appearance, and the size of bounding box within which the color histogram is computed to represent the scale. Given image observations $\mathbf{Y}_t$ at time t and $\mathbf{Y}_{1:t}$ till t, the tracking problem is to obtain the maximum a posterior probability of targets configuration $P(\mathbf{X}_t|\mathbf{Y}_{1:t})$. According to the Bayesian rule and the Markovian property, we have

$$P(\mathbf{X}_t|\mathbf{Y}_{1:t}) \propto P(\mathbf{Y}_t|\mathbf{X}_t) \int P(\mathbf{X}_t|\mathbf{X}_{t-1})P(\mathbf{X}_{t-1}|\mathbf{Y}_{1:t-1})d\mathbf{X}_{t-1} \qquad (1)$$

and $P(x_t^i|\mathbf{Y}_{1:t})$ can be obtain by marginalizing $P(\mathbf{X}_t|\mathbf{Y}_{1:t})$.

To estimate this probability, the configuration dynamics and the configuration likelihood need to be modeled. It is generally assumed that the objects are moving according to independent Markov dynamics. However, it is generally difficult to distinguish and segment these spatially adjacent targets from image observations, so the states of targets still couple through the observation when multiple targets move close or present occlusion.

To address this problem, we model multi-target configuration at time t by three coupled MRFs: $\mathbf{X}$ for the joint state of multi-target, each node represents a target, $\mathbf{D}$ for a binary process to indicate absence of each target, and $\mathbf{O}$ for a binary process located on the dual targets node to indicate their occlusion relationship. Using Bayes' rule, the joint posterior probability over $\mathbf{X}, \mathbf{D}$, and $\mathbf{O}$ given image observation $\mathbf{Y}$:

$$P(\mathbf{X},\mathbf{D},\mathbf{O}|\mathbf{Y}) = P(\mathbf{Y}|\mathbf{X},\mathbf{D},\mathbf{O})P(\mathbf{X},\mathbf{D},\mathbf{O})/P(\mathbf{Y}) \qquad (2)$$

For simplicity, we assumed that likelihood $p(\mathbf{Y}|\mathbf{X},\mathbf{D},\mathbf{O})$ is independent of $\mathbf{O}$, because the observation $\mathbf{Y}$ is target-based. Assuming that the observation noise follows an independent identical distribution (i.i.d.), we can define the likelihood $p(\mathbf{Y}|\mathbf{X},\mathbf{D})$ as

$$P(\mathbf{Y}|\mathbf{X},\mathbf{D}) \propto \prod_{i \notin \mathbf{D}} \exp(-L_i(x_i,\mathbf{Y})) \qquad (3)$$

where $L_i(x_i,\mathbf{Y})$ is matching cost function of the i th target, $i \in \{1,...,M\}$ labels all the targets, with state x given observation $\mathbf{Y}$, Our observation likelihood just considers the targets present, despite of occlusion.

As to the matching cost function, we can use either shape based observation model as in [15], color based observation model as in [11], or learned observation model [4], even sophisticated observation based on several visual cues integration. In the experiment in section 6, we use a color observation model [11] to track hockey players.

There is no simple statistical relationship between coupled fields $\mathbf{X}, \mathbf{O}$ and $\mathbf{D}$. In this paper, we ignore the statistical dependence between $\mathbf{D}$ and $\mathbf{X}, \mathbf{O}$. Assuming $\mathbf{X}, \mathbf{O}$ and $\mathbf{D}$ follow the Markov property, by specifying the first order neighborhood system $G(i)$ and $N(i) = \{j|d(x_i,x_j) < \delta, j \in G(i)\}$ of target i, where $d(x_i,x_j)$ is the distance between the two targets in the state space, δ is a threshold to determine the neighborhood, the prior (3) can be expanded as:

$$P(\mathbf{X},\mathbf{O},\mathbf{D}) = \prod_i \prod_{j \in N(i)} \exp(-\varphi_c(x_i,x_j,O_{i,j})) \prod_i \exp(-\eta_c(D_i)) \qquad (4)$$

where $\varphi_c(x_i,x_j,O_{i,j})$ is the joint clique potential function of sites x_i, x_j (neighbor of x_i) and $O_{i,j}$, $O_{i,j}$ is the binary variable between x_i and x_j, and $\eta_c(D_i)$ is the clique potential function of D_i. $\varphi_c(x_i,x_j,O_{i,j})$ and $\eta_c(D_i)$ are user-customized functions to enforce the contextual constraints for state estimation. To enforce spatial interactions between x_i, x_j, we define $\varphi_c(x_i,x_j,O_{i,j})$ as follows:

$$\varphi_c(x_i,x_j,O_{i,j}) = \varphi(x_i,x_j)(1-O_{i,j}) + \gamma(O_{i,j}) \qquad (5)$$

where $\varphi(x_i,x_j)$ penalizes the different assignments of neighboring sites when occlusion exists between them and $\gamma(O_{i,j})$ penalizes the occurrence of an occlusion between sites i and j. Typically, $\gamma(0) = 0$. By combining (3), (4), and (5), our basic model (2) becomes:

$$P(\mathbf{X},\mathbf{D},\mathbf{O}|\mathbf{Y}) \propto \prod_{i \notin \mathbf{D}} \exp(-L_i(x_i,\mathbf{Y})) \prod_i \exp(-\eta_c(D_i))$$
$$\times \prod_i \prod_{j \in N(i)} \exp(-\varphi(x_i,x_j)(1-O_{i,j}) + \gamma(O_{i,j})) \qquad (6)$$

3 Approximate Inference

3.1 Model Approximation

Maximization of the posterior (6) can be rewritten as

$$\max_{\mathbf{X},\mathbf{D},\mathbf{O}} P(\mathbf{X},\mathbf{D},\mathbf{O}|\mathbf{Y}) = \max_{\mathbf{X}} \{\max_{\mathbf{D}} \prod_i \exp(-(L_i(x_i,\mathbf{Y})(1-D_i) + \eta_c(D_i)D_i))$$

$$\times \max_{\mathbf{O}} \prod_i \prod_{j \in N(i)} \exp(-(\varphi(x_i,x_j)(1-O_{i,j}) + \gamma(O_{i,j}))\} \quad (7)$$

because the first two factors on the right hand side of (7) are independent of $\mathbf{O}$ and the last factor on the right hand of (7) is independent of $\mathbf{D}$.

Now we relax the binary process $O_{i,j}$ and D_i to analog process $O'_{i,j}$ and D'_i by allowing $0 \leq O'_{i,j} \leq 1$ and $0 \leq D'_i \leq 1$. According to [14], we obtain two robust estimators for two terms in the right hand of (7):

$$\psi_d(x_i) = \min_{D'_i}(L_i(x_i,\mathbf{Y})(1-D'_i) + \eta_c(D'_i)D'_i) \quad (8)$$

$$\psi_p(x_i,x_j) = \min_{O'_{i,j}}(\varphi(x_i,x_j)(1-O'_{i,j}) + \gamma(O'_{i,j})) \quad (9)$$

Then, we get the posterior probability over $\mathbf{X}$ defined by two robust functions.

$$P(\mathbf{X}|\mathbf{Y}) \propto \prod_i \exp(-\psi_d(x_i)) \prod_i \prod_{j \in N(i)} \exp(-\psi_p(x_i,x_j)) \quad (10)$$

Thus, we not only eliminate two analog processes via the outlier process but also convert the task of modeling the prior terms $\eta_c(D_i), \varphi(x_i,x_j), \gamma(O_{i,j})$ explicitly into defining two robust functions $\psi_d(x_i)$ and $\psi_p(x_i,x_j)$ that model occlusion and disappearance implicitly. In this paper, our robust functions are derived from the Total Variance (TV) model with the potential function $\rho(x) = |x|$ because of its discontinuity preserving property. We truncate this potential function as our robust function:

$$\psi_d(x_i) = -\ln((1-e_d)\exp(-\frac{|L_i(x_i,\mathbf{Y})|}{\sigma_d}) + e_d) \quad (11)$$

$$\psi_p(x_i,x_j) = -\ln((1-e_p)\exp(-\frac{|x_i - x_j|}{\sigma_p}) + e_p) \quad (12)$$

By varying parameters e and σ, we control the shape of the robust function and, therefore, the posterior probability.

3.2 Algorithm Approximation

Following, we describe below how the belief propagation algorithm is used to compute the MAP of the posterior distribution (10). Consider a Markov network $G = \{V,\varepsilon\}$, where V denotes node set and ε denotes edge set, is an undirected graph. Nodes $\{x_i, i \in V\}$ are hidden variables and nodes $\{y_i, i \in V\}$ are observed variables. By denoting $\mathbf{X} = \{x_i\}$ and $\mathbf{Y} = \{y_i\}$, the posterior $P(\mathbf{X}|\mathbf{Y})$ can be factorized as

$$P(\mathbf{X}|\mathbf{Y}) \propto \prod_i \rho_i(x_i,y_i) \prod_i \prod_{j \in N(i)} \rho_{i,j}(x_i,x_j) \quad (13)$$

where $\rho_{i,j}(x_i, x_j)$ is the compatibility function between nodes x_i and x_j, and $\rho_i(x_i, y_i)$ is the local evidence for node x_i. It can be observed that the form of our posterior (11) is the same form of (14), if we define

$$\rho_{i,j}(x_i, x_j) = \exp(-\psi_p(x_i, x_j)) \qquad (14)$$

$$\rho_i(x_i, y_i) = \exp(-\psi_d(x_i)) \qquad (15)$$

Thus inferring the joint state of multiple targets in our framework is defined as estimating belief in the graphical model.

To cope with the continuous state space of each target, the non-Gaussian conditionals between nodes, and the non-Gaussian likelihood, any algorithm for belief propagation with particle set at its heart, for examples, PAMAPS in [9] and NBP in [10], is a MC approximation to the integral in the message. Different from PAMAPS and NBP Sampling particles from Gaussian mixture to represent message, we use a totally importance sampling scheme. Since the efficiency of a Monte Carlo algorithm is strongly dependent on the sample positions, so if possible we would like to use all the available information when choosing these positions. We adopt a stratified sampler similar to PAMPAS but in its sequential version.

The stratified sampling propagation that consists of message updating and belief computation, detail is described in Table 1. Each message in BP is represented by a set of weighted particles, i.e., $m_{ji}(x_j) \sim \{s_j^{(n)}, w_j^{(i,n)}\}_{n=1}^N, i \in N(j)$, where $s_j^{(n)}$ and $w_j^{(i,n)}$ denote the sample and its weight of the message passing from x_i to x_j, respectively, then the message updating process is based on these set of weighted samples. The marginal posterior probability in each node is also represented by a set of weighted samples, i.e. $P(x_j|\mathbf{Y}) \sim \{s_j^{(n)}, \pi_j^{(n)}\}_{n=1}^N$. It should be noted that for graph with loops, such as our Markov network for multi-target tracking, the BP algorithm can not guarantee the global optimal solution[16]. Although we have not obtained the rigorous results on the convergence rate, we always observe the convergence in less than 5 iterations in our experiments.

4 Fusing Information from Temporal and Bottom-up Detector

To inferring $P(x_{i,t}|\mathbf{Y}_{1:t})$ in (1), we have to extend the BP algorithm discussed in section 3.2 to the dynamic Markov network shown in figure 2. Since

$$P(\mathbf{X}_t|\mathbf{X}_{t-1}) = \prod_i P(x_{i,t}|x_{i,t-1}), \quad i = 1, ..., M \qquad (16)$$

Given the inference results $P(x_{i,t-1}|\mathbf{Y}_{1:t-1})$ at previous time $t-1$, the message updating at time t is

$$m_{ij}^n(x_{j,t}) = \kappa \int \rho_{i,j}(x_{i,t}, x_{j,t})\rho_i(x_{i,t}, y_{i,t}) \\ \times \int P(x_{i,t}|x_{i,t-1})P(x_{i,t-1}|\mathbf{Y}_{1:t-1})dx_{i,t-1} \prod_{u \in N(i)\setminus i} m_{ui}^{n-1}(x_{i,t})dx_{i,t} \qquad (17)$$

Sequential Stratified Sampling Belief Propagation for Multiple Targets Tracking

Table 1. Stratified sampling message updating and belief computing algorithm

---

Generate $\{s_{j,t,k+1}^{(n)}, w_{j,t,k+1}^{(i,n)}\}_{n=1}^{N}$ and $\{s_{j,t,k+1}^{(n)}, \pi_{j,t,k+1}^{(n)}\}_{n=1}^{N}$ respectively from $\{s_{j,t,k}^{(n)}, w_{j,t,k}^{(i,n)}\}_{n=1}^{N}$ and $\{s_{j,t,k}^{(n)}, \pi_{j,t,k}^{(n)}\}_{n=1}^{N}$.

1. *Stratified Sampling from different proposal distributions,*
   (a) For $1 \leq n < \nu N$, and each $i \in N(j)$
   Draw Stratified Sampling according to α.
   If draw sample $s_{j,k+1}^{(n)}$ from $P(x_{j,t}|x_{j,0:t-1})$ set weight
   $$\tilde{w}_{j,t,k+1}^{(i,n)} = 1/(\frac{1}{N}\sum_{r=1}^{N} p(s_{j,t,k+1}^{(n)}|s_{j,t-1}^{(r)}))$$
   If draw sample $s_{j,k+1}^{(n)}$ from $Q_{svm}(x_{j,t}|x_{j,0:t-1},\mathbf{Y}_t)$ set weight
   $$\tilde{w}_{j,t,k+1}^{(i,n)} = 1/(\frac{1}{N}\sum_{r=1}^{N} Q_{svm}(s_{j,t,k+1}^{(n)}|s_{j,t-1}^{(r)},\mathbf{Y}_t))$$
   (b) for $\nu N \leq n \leq N$, and each $i \in N(j)$
   Draw sample $s_{j,k+1}^{(n)}$ from $\{s_{j,t,k}^{(n)}, \pi_{j,t,k}^{(n)}\}_{n=1}^{N}$ according to the weight $\pi_{j,k}^{(n)}$
   Set $\xi_{j,k+1}^{(i,n)} = 1/\pi_{j,k}^{(n)}$
   (c) for $\nu N \leq n \leq N$
   $$\tilde{w}_{j,k+1}^{(i,n)} = (1-v)\xi_{j,k+1}^{(i,n)}/(\sum_{l=vN}^{N}\xi_{j,k+1}^{(i,l)})$$

2. *Applying importance correction.* For $1 \leq n \leq N$ $w_{j,k+1}^{(i,n)} = \tilde{w}_{j,k+1}^{(i,n)} \cdot P_{ij}(s_{j,t,k+1}^{(n)})$
   $$P_{ij}(s_{j,k+1}^{(n)}) = \sum_{m=1}^{N}\{\pi_{i,t,k}^{(m)}\rho_i(y_{i,t,k}^{(m)}, s_{i,t,k}^{(m)})$$
   $$\times \prod_{l \in \Gamma(i)\setminus j} w_{i,t,k}^{(l,m)} \times [\sum_{r=1}^{N} p(s_{i,t,k}^{(m)}|s_{i,t-1}^{(r)})]\rho_{ij}(s_{i,t,k}^{(m)}, s_{j,t,k+1}^{(n)})\}$$

3. *Normalization:* Normalize $w_{j,t,k+1}^{(i,n)}, i \in N(j)$ and set
   $\pi_{j,t,k+1}^{(n)} = \rho_j(y_{j,t,k+1}^{(n)}, s_{j,t,k+1}^{(n)})\prod_{u \in N(j)} w_{j,k+1}^{(u,n)} \times \sum_r p(s_{j,t,k+1}^{(n)}|s_{j,t-1}^{(r)})$ and normalize it.
   Then We get $\{s_{j,t,k}^{(n)}, w_{j,t,k}^{(i,n)}\}_{n=1}^{N}$ and $\{s_{j,t,k}^{(n)}, \pi_{j,t,k}^{(n)}\}_{n=1}^{N}$.
4. $k \leftarrow k+1$, iterate 2.1→2.3 until convergence.

---

and belief of target i can be written as:

$$\hat{P}_{i,t}(x_{i,t}|\mathbf{Y}_{1:t}) = \alpha\rho_i(y_{i,t}|x_{i,t})\prod_{j \in N(i)} m_{ji}(x_{i,t})$$
$$\times \int P(x_{i,t}|x_{i,t-1})\hat{P}_{i,t-1}(x_{i,t-1}|\mathbf{Y}_{1:t-1})dx_{i,t-1} \quad (18)$$

From(18), We clearly see that at time instant t, the belief of i th target is determined by three factors: (1) the local evidence $\rho_i(x_i, y_i)$, (2) the prediction prior $\int P(x_{i,t}|x_{i,t-1})\hat{P}_{i,t-1}(x_{i,t-1}|\mathbf{Y}_{1:t-1})dx_{i,t-1}$ from previous time frame, and (3) the

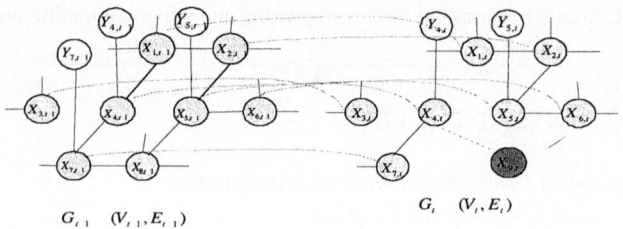

Fig. 1. Dynamic Markov network for multiple targets, the link between each paired hidden nodes indicates possible occlusion occurs. Graph G_{t-1} and G_t describe targets configuration at two consecutive time instants. Link and node in red indicate new occlusions with the new target addition.

Table 2. Sequential stratified sampling for belief propagation

---

Generate $\{s_{j,t}^{(n)}, \pi_{j,t}^{(n)}\}_{n=1}^N$ from $\{s_{j,t-1}^{(n)}, \pi_{j,t-1}^{(n)}\}_{n=1}^N$

1. *Initialization:*
   (a) *Re-sampling: for each* $j = 1,..,M$, *re-sampling* $\{s_{j,t-1}^{(n)}\}_{n=1}^N$ *according to the weights* $\pi_{j,t-1}^{(n)}$ *to get* $\{s_{j,t-1}^{(n)}, 1/N\}_{n=1}^N$.
   (b) *Prediction: for each* $j = 1,..,M$, *for each sample in* $\{s_{j,t-1}^{(n)}, 1/N\}_{n=1}^N$, *sampling from* $p(x_{j,t}|x_{j,t-1})$ *to get sample* $\{s_{j,t}^{(n)}\}_{n=1}^N$
   (c) *Belief and message Initialization: for each* $j = 1,..,M$, *assign weight* $w_{j,t,k}^{(i,n)} = 1/N$, $\pi_{j,t,k}^{(n)} = p_j(y_{j,t,k}^{(n)}|s_{j,t,k}^{(n)})$ *and normalize, where* $i \in N(j)$.
2. *Update message & compute belief as Tabel1*
3. *Inference result* $p(x_j|\mathbf{Y}) \sim \{s_j^{(n)}, \pi_j^{(n)}\}_{n=1}^M$, *where* $s_j^{(n)} = s_{j,k}^{(n)}$ *and* $\pi_j^{(i,n)} = \pi_{j,k+1}^{(i,n)}$.

---

incoming messages from neighborhood targets. We can still represent the messages and marginal posterior probabilities at each time instant as weighted samples, i.e. $m_{ji}(x_{j,t}) \sim \{s_{j,t}^{(n)}, w_{j,t}^{(i,n)}\}_{n=1}^N, i \in N(j)$ and $P(x_{j,t}|\mathbf{Y}) \sim \{s_{j,t}^{(n)}, \pi_{j,t}^{(n)}\}_{n=1}^N$.

To improve the reliability of the tracker, Our belief propagation framework is designed to incorporate bottom up information from a learned SVM classifier, any other detector can also be adopted in a same way. One expects detector to be noisy in that they will some times fail to detect targets or will find spurious target, even these noisy information provide valuable low-level cues. As described in, we use a similar mixture proposal as [7]: performing prediction.

$$P(\mathbf{X}_t|\mathbf{X}_{0:t-1}, \mathbf{Y}_{1:t}) = \alpha Q_{svm}(\mathbf{X}_t|\mathbf{X}_{t-1}, \mathbf{Y}_t) + (1-\alpha)p(\mathbf{X}_t|\mathbf{X}_{t-1}) \qquad (19)$$

where Q_{svm} is a Gaussian distribution center around the detected targets at time instant. The parameters $0 \leq \alpha \leq 1$ and $0 \leq v \leq 1$ can be set dynamically without affecting the convergence of belief propagation. By increasing α and v, we place more importance on the prediction by target motion model and messages than detector and belief.

Then following equations from (17-19), message propagation and belief computation is described in Table 1. the sequential Monte Carlo belief propagation is shown in Table 2.

5 Experiments

In this section, we test stratified sampling belief propagation tracking results on two video sequences. The first is a synthetic example, in which we didn't use stratified sampler, parameters in our algorithm were set as $\alpha = 0, v = 1$. The second is a real video sequence of hockey game, in which stratified sampler is used, the parameters are $\alpha = 0.8, v = 0.8$. We also compare our results with those obtained by multiple independent trackers, and the latest boosted particle filter[7].

5.1 Synthesized Video

In the synthesized video, there are five identical and moving balls in a noisy background. Each ball presents an independent constant velocity motion and is bounded by the image borders. The Synthesized sequence challenges many existing methods due to the frequent presence of occlusion.

We use different colored boxes to display the estimated positions. An index is also attached to each box to denote the identifier of each ball. We compare our results with those obtained by the multiple independent trackers(M.I.Tracker), which we implemented using Condensation [15]. Figure 3 shows some samples in handling occlusion, the results of M.I.Tracker are on the top and our results at the bottom. In both algorithms, 20 particles are used for each target. However, M.I.Tracker can't produce satisfactory results. The red lines in figure 3 that link different targets are the visual illustration of the structure of the Markov network in our algorithm. By observing the changing structure of the network over 768 frames, We find in our experiments that our approach can handle the occlusion at 98% of the sequence.

5.2 Multiple Hockey Players Tracking

Both our algorithm and M.I.Tracker have been tested on a real video sequence of hockey game. In our experiments, a SVM classifier is trained to detect hockey players. In order to train the detector, a total of 1300 figures of hockey players are used, we got 34 SVs. Figure 4 shows M.I.Tracker results on the top, and our algorithm tracking results on the bottom. As expected, our new method provides robust and stable results, while M.I.Tracker can't. We also apply our algorithm to the switching problem by combing with motion coherence and dynamic predictions. This can be easily validated by the subjective evaluation on the tracking sequence.

Finally, we also compare our algorithm with the latest multiple targets tracking algorithm, boosted particle filter [7]. Both the performance are almost the same, however, boosted particle filter use different sub-section color model for different players, while our algorithm use the same color model for all the players. It is obvious that in occlusion handling, boosted particle filter depends heavily on the discriminability of its likelihood function, while ours tracking algorithm can handle occlusion nicely due to the better modeling of interaction among multiple targets.

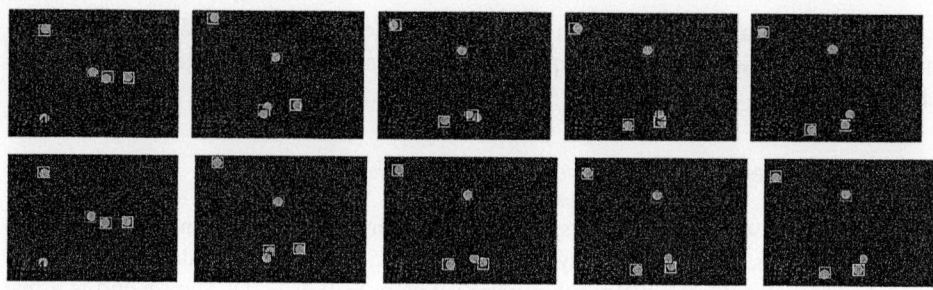

Fig. 2. Tracking balls by M.I.T and our algorithm. The result of MIT is Shown on the top with the result of our algorithm at the bottom. Frames 53 to 59 presents an occlusion event, identities by M.I.T becomes error while our algorithm produces right results.

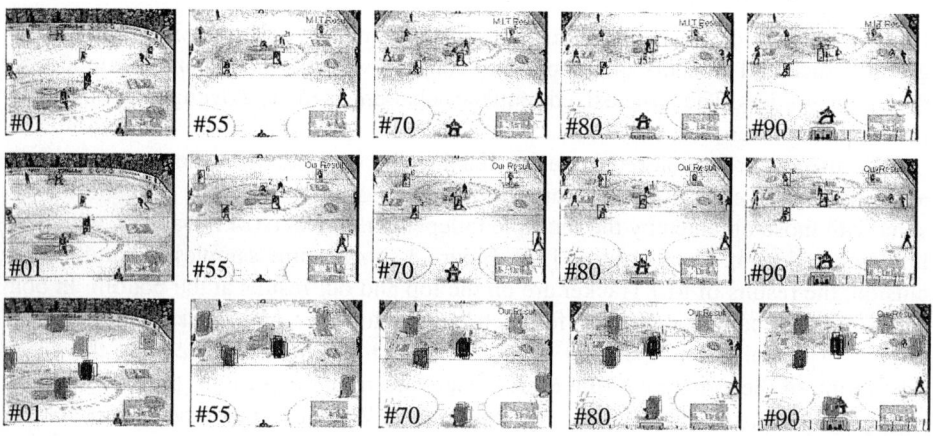

Fig. 3. Hockey players tracking result: The result of M.I.T is on the top, final result of our algorithm is in the middle, and the bottom shows the intermediate result of our algorithm. In both algorithms, 20 particles were used.

6 Discussion and Conclusion

This paper proposed to model multiple targets tracking problem in a dynamic Markov network which consists of three coupled Markov random fields that model the following: a field for joint states of multi-target, one binary process for existence of individual target, and a binary process for occlusion of every two adjacent targets. It was shown how a sequential stratified sampling belief propagation algorithm to obtain the maximum a posteriori (MAP) estimation in the dynamic Markov network. It was also shown that by using stratified sampler, how to incorporate bottom-up information from a learned detector (e.g. SVM classifier) and belief information for the messages updating. The new proposed method is able to track multiple targets and handle their interactions such as occlusion, sudden appearance and disappearance, as was illustrated on a synthetic and a real world tracking problem. Our future work includes the improvement of the current inference algorithm and the development of learning algorithm for the parameters of the model described in this paper.

Acknowledgement

This work was supported in part by National Foundation grants of China 60205001, 60405004, 60021302.

References

1. Bar-Shalom, Y., X.R., L.: Multitarget Multisensor Tracking:Principles and Techniques. YBS Publishing (1995)
2. I.J., C., Hingorani, S.L.: An Efficient Implementation of Reid's Multiple Hypotheses Tracking Algorithm and its Evaluation for the Purpose of Visual Tracking. IEEE T-PAMI (1996) 138-150
3. Hue, C., LeCadre, J.P., Perez, P.: Tracking Multiple Objects with Particle Filtering. IEEE Trans. on Aerospace and Electronic Systems 38(3)(2002) 791–812
4. Isard, M., MacCormick, J., Bramble, P.: A Bayesian Multiple-blob Tracker. Intl.Conf. on Computer Vision (2001) 34–41
5. Tweed, D., et.al.: Tracking Objects Using Subordinated Condensation. British Machine Vision Conf. (2002)
6. Jaco, V., Arnaud, D., Patrick, P.: Maintaining Multi-Modality through Mixture Tracking. Intl.Conf. on Computer Vision (2003)
7. Okuma, K., et.al.: A Boosted Particle filter: Multitarget Detection and Tracking. Eur.Conf. on Computer Vision (2004)
8. Yu, T., Wu, Y.: Collaborative Tracking of Multiple Targets. Intl' Conf. on Computer Vision and Pattern Recognition (2004)
9. Isard, M.: PAMPAS: Real-valued Graphical Models for Computer Vision, Intl' Conf.on Computer Vision and Pattern Recognition (2003) 613–620
10. Sudderth, E., Ihler, A., Freeman, W., Willsky, A.: Nonparametric Belief Propagation. Intl' Conf. on Computer Vision and Pattern Recognition (2003) 605–612
11. Comaniciu, D., Ramesh, V., Meer, P.: Real-time Tracking of Non-rigid Objects Using Mean Shift, Proc. Intl' Conf.on Computer Vision and Pattern Recognition (2000) 142-151
12. MacCormick, J.P., Blake, A.: A probabilistic Exclusion Principle for Tracking Multiple Objects. Intl.Conf. on Computer Vision (1999) 572–578
13. Tao, H., Sawhney, H.S., Kumar, R.: A Sampling Algorithm for Tracking Multiple Objects. Vision Algorithms 99 (1999)
14. Black, M.J., Rangarajan, A.: On the Unification of Line Processes, Outlier Rejection and Robust statistics with applications in Early vision. Int'l J. Computer Vision 19 (1) (1996) 57-91
15. Isard, M., Blake, A.: Condensation - conditional Density Propagation for Visual Tracking. Intl. J. of Computer Vision 29 (1) (1998) 5-28
16. Murphy, K., Weiss, Y., Jordan, M.: Loopy-belief Propagation for Approximate Inference: An Empirical Study. Proc. Fifteenth Conference on Uncertainty in Artificial Intelligence, Stockholm, Sweden (1999)

Retrieving Digital Artifacts from Digital Libraries Semantically

Jun Ma, YingNan Yi, Tian Tian, and Yuejun Li

School of Computer Science and Technology, Shandong Univ., Jinan 250061, China
majun@sdu.edu.cn

Abstract. The techniques for organizing and retrieving the artifacts from digital libraries (DLs) semantically are discussed, which include letting the taxonomies and semantic relations work in tandem to index the artifacts in DLs; integrating the techniques used in natural language processing and taxonomies to help users to start their retrieval processes; and ranking scientific papers on similarity in terms of contents or ranking the relevant papers on multi-factors. These techniques are verified through the design and implementation of a prototype of DLs for scientific paper management.

1 Introduction

Along with the repeat development of web technology it is an important research issue to find, store and share digital artifacts efficiently through the Internet/ Intranet. Many enterprises, research institutes and publishers have established their DLs in order to share the digital resources within organizations. Now the artifacts in most DLs are organized based on so-called taxonomies, which are established based on the principle of the traditional catalogs of libraries, e.g., the portal design of ACM DL [1], the classification of domains, e.g. the portal design of Yahoo!, or the mixture of domain classification, workflows in organizations and the attributes of the artifacts [4]. Users retrieve the artifacts form DLs by following the hierarchical catalogs or by inputting a keyword set to find the artifacts whose feature description can match the keyword set. However, there are shortcomings in such retrieval patterns. First, It is hard to collect all artifacts for a topic by following the taxonomies because the content of an artifact may cover several domains or classes of the taxonomy used in DLs. The keyword matching often misses a lot of artifacts because the artifacts in DLs may be represented in different keyword sets or users do not read them because the relevant artifacts are not at top positions in the output lists. Second, when a user finds an interesting artifact, DLs usually do not provide the artifacts that are relevant to the selected one at the same time. However, in knowledge sharing, the lack of these links often make users unable to understand the content of the selected one deeply Furthermore researchers are often unable to think in number of parallel ways by studying the ideas given in the relevant artifacts. Third, most DLs do not provide various rankings on the artifacts that are similar in term of contents. Users often miss some important artifacts because there are too many artifacts to be selected.

In recent years, Martin, etc. [5] addressed the metadata languages and knowledge representation in these languages as well as indexing artifacts using knowledge in theory. However, they did not discuss how to utilize these metadata to retrieve artifacts concretely. Ransom and Wu provided using knowledge anchors to reduce cognitive overhead, where a knowledge anchor is the point within the frame from which the user can trigger a link [7]. However, it is hard to maintain and establish such index systems. Liao, etc [3] and Kwan [2], addressed how to use enterprise, domain and information ontologies to organize DLs in theory. However, in many application domains recognizable ontologies have not been built. Shibata and Hori [7] show an automatic question-answer system by a 1-1 mapping function from the questions sets to the solution sets. However, knowledge is dynamic. Users often absorb knowledge in their own ways. It is difficult to establish a universal 1-1 mapping function in most application domains.

In this paper, we provide novel techniques to retrieve artifacts from DLs semantically and verify these techniques through a design and implement of a prototype of a special DL—Intelligent Paper Management Systems (IPMS), where taxonomies and the semantic relations work in tandem to index the papers in the DL. Users can trace the relevant papers by clicking the links to access these papers. Furthermore we integrate the techniques used in natural language processing and taxonomies to understand users' searching requirement and guide users in their information retrieval processes. In addition we study paper ranking on multi-factors as well as investigate their properties of these ranking algorithms throughout a series of experiments.

2 Inserting Semantic Relations into Taxonomy

We first established a taxonomy T for IMPS based on the classification of the research domains in computer science. The classification is consistent with that given by The Nature Scientific Foundation of China in 2005 for fund applications [6]. It is also consistent with the contents of the transactions and journals of ACM, IEEE and Elsevier at el. We collected about 600 research papers, which are in the PDF form, from the journals, magazines and proceedings of ACM, IEEE and Elsevier from 2001 to 2003 in the research areas of AI, IR, KM, Multimedia, CSCW et ac. These papers are used as our experiment artifacts. The metadata of these papers consist of the names of authors, titles, keyword set as well as the references. Based on the metadata these papers were classified automatically and the links to access these papers are inserted to the metadata of the leaf nodes of the taxonomy T.

The system architecture of IPMS is browser/ server, i.e., the indexing system of IPMS is installed on a server of the campus network of Shandong University. Users use the system via its portal.

Based on the metadata of papers, two kinds of semantic relations, denoted by $\leftrightarrow$ and $\rightarrow$, are inserted into the metadata of papers in order to let the taxonomy and semantic relations work in tandem. The semantic relations are defined as follows. Let A and B be two artifacts in IPMS, then define

1. $A \leftrightarrow B$ if A and B are two leaf nodes of T and refer to the same topic judged by similarity(A, B) > α, where α is a threshold value, 0<α≤1, and similarity is a function for similarity computation.
2. $A \to B$ if B is an artifact referring to the historical, background or relevant knowledge of A, which are mainly provided by the content of the paper as well as the references of the paper.

The → relations of a paper can be found by computing the intersection of the references of paper A and all papers in IPMS. The ↔ relation is calculated by

$$similarity(A,B) = \alpha * stru\_similarity(A,B,T) + \beta * unstru\_similarity(A,B).$$

where stru_similarity(A,B,T) is a function to compute the similarity between A and B based on their positions in T. T describes the semantic relations among the nodes of T according to domain knowledge. Unstru_similarity(A,B) is a function to computer the similarity of A and B based on traditional models, e.g. vector space or keyword set [4].

In our experiment, we choose

$$stru\_similarity(A,B,T) = ((\frac{2 \times depth(LCA(A,B,T))}{depth(A,T) + Depth(B,T)})^r$$

where depth(A,T) is the number of edges from the root of T to A, depth(root) = 0, and LCA(A,B,T) is the lowest common ancestor of A and B in T, $0 \le r \le 1$, and

$$unstru\_similarity(A,B) = \frac{|A \cap B|}{|A \cup B|})$$

z seems that height of the taxonomy T is 6; the average size of the keyword set of a paper is 4.5; and the values of stru_similarity(A,B,T) is usually greater than the second one, but the function unstru_similarity usually describes the similarity of two

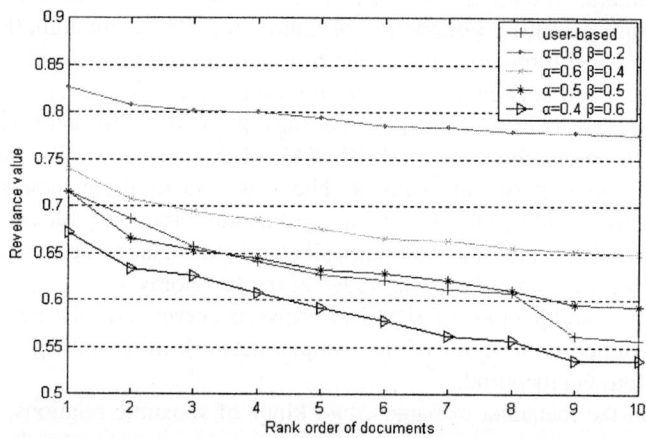

Fig. 1. The study on the choice of α and β

projects more exactly. Therefore the choice of α = 0.4 and β = 0.6 seems to be a better trade-off for IPMS. However, it is clear that the choice of α and β depends on the size of taxonomy and the average size of the keyword sets of the artifacts in a DL in practice. We also study the choice of r, the choice r will determine the distribution of the values of stru_similarity(A,B) in [0,1], the smaller r is, the better the values of the function will be distributed.

After these semantic relations are calculated, the semantic relations are inserted into the metadata of the nodes of T. Therefore the two kinds of semantic relations are inserted into the taxonomy T in this way. Clearly the index system is not longer a tree but a semantic network. The total process to build the indexing system of IPMS is described in Fig. 2.

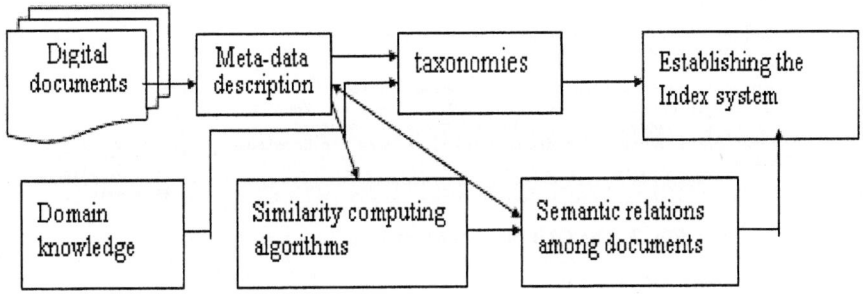

Fig. 2. The process to establish the index system

We will show that the established index system becomes the infrastructure of retrieving papers semantically from IPMS in the following sections.

3 Tracing the Relevant Papers

IPMS allows users to retrieve papers by inputting keywords. In addition, users can input their searching requirement in English or Chinese sentences. The principle of dealing with the natural languages will be discussed later. After a user find a paper A, the metadata of A, the hyperlink to access A and the hyperlinks to access the papers that have → or ↔ relations with A will be presented in the decreasing order on the similarity with A. Then the user can continues to retrieve the relevant papers by following the hyperlinks to access the relevant papers. The number of the relevant papers is controlled by a threshold value β, 0<β≤1. Only the papers those are similar to the selected paper with the similarity value greater than β will be shown to users. Clearly the greater β is, the less the relevant papers will be given. However, the choice of β depends on the number of papers on a topic in IPMS, users have to choose a reasonable β according to the number of papers shown in screen in practice. The GUI of the interface of IPMS is shown in Fig. 3.

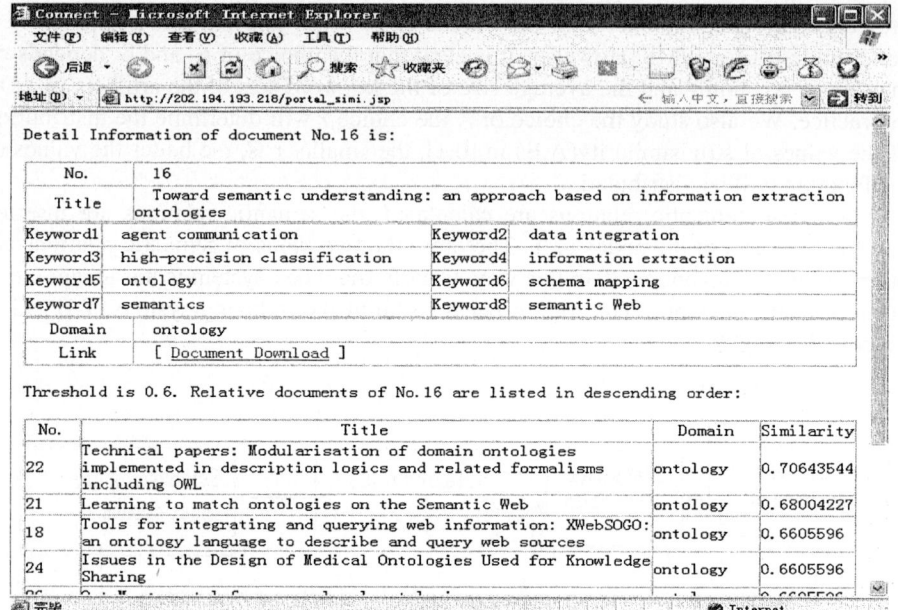

Fig. 3. The GUI of a paper retrieval process of IPMS

4 Starting a Retrieval Process Through a Conversation

In order to solve the problem that when a user has no knowledge on some aspect, s/he is often unable to provide suitable keywords for IR, we try to utilize the established taxonomy T, which describes the domain knowledge, to carry out a conversation between system users and IPMS. The purpose of the conversation is to guess users' retrieval purpose, and then help users to skip useless concept layers, finally locate a leaf node of T to access the papers they need.

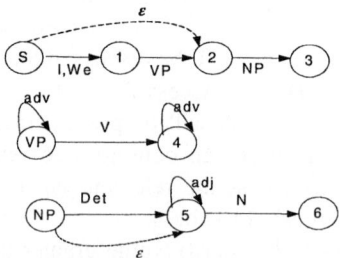

Fig. 4. The improved ATNs for parsing English sentence

The user interface is designed as follows. A user input a keyword set or a sentence in English or Chinese into the conversation window of IPMS. There is a thesaurus dictionary in IPMS. We propose an improved the ATN (augmented transition network), mentioned in [9], to parse the inputted English sentences on the assumption that all inputted sentences are simple. The principle of the improved ATN is shown in Fig 4. Its main function is to find the nouns as the keyword set.

As mentioned in section 2 the metadata of each node D of T has a keyword set that describe the concept corresponding to D. Further we put a question sentence in the metadata of D to ask a new question. A conversation is driven by matching the users' searching requirement with the keyword set in the metadata of the nodes of T. First IPMS generates a question, like "what kinds artifacts are you looking for", and then analyses the inputted keyword set or sentence. Parses the input and then gets a keyword set S as the users' requirement. Then IMPS chooses a sub-taxonomy T' from T as a search space. After then it matches the nodes of T' with S from the root of T' to the leaves of T'. It will stop at a node D of T', such as, D is of the biggest similarity value among all nodes of T'. After then IPMS generates a new question stored in node D to get users' input again. The process repeats until D is a leaf node of T, so users can access the papers s/he wants or returns with failure when the similarity value of node D and the keyword set S is too small. Fig. 5 demonstrates a segment of a conversation between a user and IPMS.

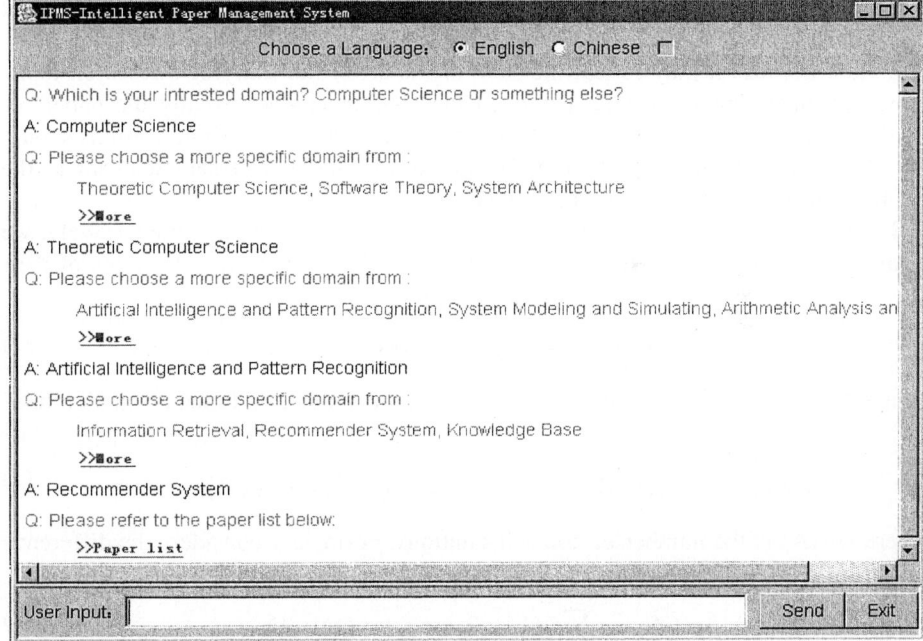

Fig. 5. A segment of a conversation between system and a user

The principle of dealing with Chinese sentences is almost the same as that mentioned for English sentences. However, since to segment a Chinese sentence to get a keyword set is different from English ones. We use a character-combination-based approach to find a keyword set in the inputted sentences. The principle is as follows: Let l be the length of the longest keyword used in the nodes of the taxonomy T. For a given sentence S, use all the combinations of adjacency characters in S with length k ($k \leq$ l) to match the keyword sets of the nodes of T'. Based on the known statistics that most Chinese words are in the length $l = 2, 3$ and the number of nodes of T is usually smaller, the method works well in our experiments.

5 Ranking Papers Based on Multi-factors

In section 3 we described how to collect the papers refer to the same topic and sort them on similarity. However, because there are too many papers that refer to same topic, researchers hope IPMS can rank these relevant papers on multi-factors. For the time being the importance of papers is often measured by their citation numbers, which can be found in SCI, EI databases or the web site CiteSeer [10]. However, in many cases the citation number of a paper A may not reflect the comprehensive value of A. For example, if A, B are two papers with the same number of citations but there are more self-citation in the statistic of the citation of B than that of the citation of A, whether A is better than B? Furthermore whether the number of application areas that a paper refers to should be taken into account also? The published time of a paper is important because it may reflect the state of the art of the study of a research topic. However, in the paper ranking based on the citation number usually the papers or books at the top positions are aged. Whether the impact factors of publications should be taken into account also because the quality of papers is closely relevant to the publications in which they were published.

In order to evaluate the ranking of scientific research papers comprehensively, we proposed four new formulae to rank papers based on their "value" in terms of their citation number, publications, the published time and so below.

$$Value(A) = N(A) \times DN(A). \qquad (1)$$

where N(A) is the citation number and DN(A) is the number of domains that A refers to.

$$Value(A) = (N(A) - UN(A) + UN(A)^2) \times DN(A). \qquad (2)$$

where UN(A) is the number of non-self-citations. Formula 2 considers the difference of self-citation and the citation by other people.

$$Value(A) = (\sum (C_i - P_y) \times f_i) \times DN(A). \qquad (3)$$

where P_y is the year A was published, C_i is the year of an author i who cited A, and f_i is the impact factors of the publication resource, $0 \leq f_i \leq 1.0$. Formula 3 takes the published time and impact factors of the publication where the paper is published into account.

$$Value(A) = \frac{(\sum(C_i - P_y) \times f_s) \times UN(A)^2 \times DN(A)}{P_t - P_y + 1}. \qquad (4)$$

where P_t is the year at present. Clearly all factors mentioned above are taken into account in formula (4).

Let us compare the new rankings on above four formulae with the ranking on pure citation number in decreased order, and name the latter *standard ranking*. In our experiment at each time we choose a topic and then collect the top 10 papers from the website citeseer in standard ranking as a sample set S, and then investigate the new ranking on S based on formula k, $1 \leq k \leq 4$. Without confusion let the integer i denote the top i paper in the standard ranking in the following discussion, $1 \leq i \leq 10$. Let us define several measure functions.

a) index(k,i) = j, j is the index of paper i in the ranking on the kth formula.
b) change_pos (i,k) = 1 if index(k,i) – $i \neq 0$; 0 otherwise.
c) Move(k) = $((\sum_{i=1}^{10} index(k,i) - i)/10$.
d) variance (k) = $\max_{1 \leq i \leq 10}$ |index(k,i) – i|.

Clearly Move(k) measures how many papers change their positions in the ranking on formula k at one times. Function variance (k) describes the span of maximum index change among the 10 papers in the new ranking compared with the standard ranking. Furthermore, since it is often more important to measure how many papers change their ranking order, then we define a new function move-ahead(i,k).

e) move-ahead(i,k) = 1 if $\exists j, j < i$, paper i is ahead paper j in the ranking based on formula k; 0 otherwise.
f) move-ahead(k) = $(\sum_{i=1}^{10} move\_ahead(i,k))/10$.

Move-ahead(k) is the mathematical expectation of number of papers that exchange their positions with at least one paper ahead of the papers in the standard ranking.

Then we utilize the standard ranking for a given topic and then collect the papers with more than 4 citations based on the citation information provided by the website citeseer [10]. We used 20 group samples in the experiment and define move(k,j), variance (k,j), move_ahead(k,j) are the values of the function move(k), variance(k) and move_ahead(k) for the jth selected sample set respectively. Define

g) $\overline{move(k)} = (\sum_{j=1}^{20} move(k,j))/20$;

h) $\overline{variance(k)} = (\sum_{j=1}^{20} variance(k,j))/20$; and

l) $\overline{move\_ahead(k)} = (\sum_{j=1}^{20} variance(k,j))/20$.

Table 1 gives the comparison of the five formulae based on functions given in g),h) and l)

Table 1. The comparison of the difference of the rankings on five formulae

%	$\overline{move}$	$\overline{variance}$	$\overline{move\_ahead}$
Formula 1	60%	3	28%
Formula 2	50%	2	31%
Formula 3	97%	6	71%
Formula 4	60%	2	29%

Figuer 6 shows a comparison of the rankings on five formulae on a given sample set graphically

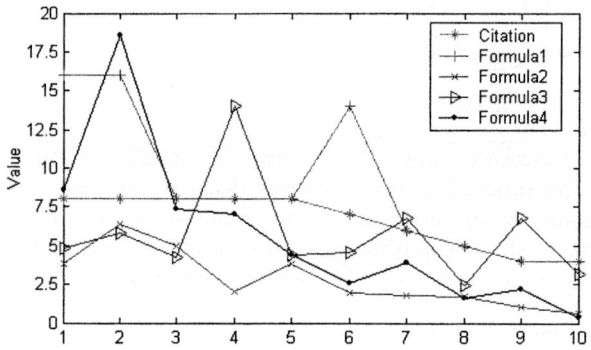

Fig. 6. A comparison on the rankings on five formulae for a given sample set

It seems that the ranking results on the five formulae are very different if we use function $\overline{move}$ to evaluate their difference. However, if we use the function $\overline{move\_ahead}$ and $\overline{variance}$ to measure the ranking difference of the five formulae, the papers that change their order on formulae 1,2 in these experiments are about 30%, and the span of the maximum index difference are very small. It shows the second measure is more suitable than the first one. In addition, it shows that the citation numbers of papers in fact dominate the rankings on formulae 1,2 although the two formulae take more factors into account. Clearly the ranking based on formula 3 is completely different from others. It is a new method to rank scientific papers, which emphasizes the levels of publications and published time. Although formula 4 takes all factors into account, the paper rankings on formula 4 are very similar with those on formulae 1 and 2. It is because the value of $UN(A)^2 \times DN(A)$ is too big so that it dominates the final ranking. It also shows the citation number of a paper reflect its importance.

6 Conclusions and Future Work

In this paper we demonstrate how to set up an index system for a special DL system, i.e. an Intelligent Paper Management System (IPMS). We insert two kinds of semantic relations into the nodes of the taxonomy used in IPMS in order to get a new kind of index system for DLs. Then we discuss how to retrieve papers from the IPMS semantically based on the established index system. We study how to integrate the techniques used in nature language processing and the domain knowledge described in taxonomy to drive a conversation to start a retrieval process. The technology is very suitable to the case that there are many branches in the classification of an application domain. In addition we explore the algorithms for ranking papers on multi-factors and study their properties. Although in our discussion the artifacts are scientific papers, however, these research results can be applied to develop of the DLs, where the metadata of the artifacts are used to organize the index system of the DLs. In fact, for the time being most index systems of DLs are developed based on the metadata of the artefacts stored in the DLs. It will be more popular in the future because metadata description is needed to wrap the audio and video files and XML, RDF and WSDL are widely adopted as metadata description languages in knowledge sharing.

Our future study will focus on organize artifacts of DLs based on multi-taxonomies, where several different taxonomies describe the same domain knowledge but from different viewpoints. We will study the efficient algorithms for similarity computation between two objects in the new index systems and the user interface design for retrieving the artifacts form DLs semantically in the new index systems .

References

1. ACM DL: http://portal.acm.org/dl.cfm
2. Kwan, M.M., Balasubramanian, P.: KnowledgeScope: Managing Knowledge in Context. Decision Support Systems. 5 (2003) 467-486
3. Liao, M., Abecker, A., Bernaridi, A., Hinkelmann, K., Sintek, M.: Ontologies for Knowledge Retrieval in Organizational Memories. In: Bomarius, F., Proc. of the Learning Software Organization Workshop. Kaiserslauten (1999) 19-26
4. Ma J. and Hemmje M.: Knowledge Management Support for Cooperative Research, 17th World Computer congress, Kluwer Academic Publisher, Montreal, Canada (2002) 280-284
5. Martin, P., Eklund, P.W.: Knowledge Retrieval and the World Wide Web. IEEE Intelligent Systems. 3 (2000) 18-25
6. NSF China: http://.nsfc.gov.cn
7. Ransom, S., Wu, X.: Using Knowledge Anchors to Reduce Cognitive Overhead. IEEE Computer. 11 (2002) 111-112
8. Shibata, H., Hori, K.: A System to Support Long-Term Creative Thinking in Daily Life and Its Evaluation In: Holt I. (ed.): Proceedings of the Fourth Conference on Creativity & Cognition. ACM Press, Loughborough (2002) 142-149
9. Rich, R., Knight, K.: Artificial Intelligence. 2nd edition. McGraw-Hill (1991) 377-428
10. Citeseer: http://citeseer.ist.psu.edu/

An Extended System Method for Consistent Fundamental Matrix Estimation

Huixiang Zhong, Yueping Feng, and Yunjie Pang

College of Computer Science and Technology, Jilin University,
Key Laboratory of Symbol Computation and Knowledge
Engineering of the Ministry of Education,
Jilin University, Changchun, 130012, P. R. China
{zhonghx, fengyp}@jlu.edu.cn, pyjcojlu@public.cc.jl.cn

Abstract. This paper is concerned with solution of the consistent fundamental matrix estimation in a quadratic measurement error model. First an extended system for determining the estimator is proposed, and an efficient implementation for solving the system-a continuation method is developed to fix on an interval in which a local minimum belongs. Then an optimization method using a quadratic interpolation is used to exactly locate the minimum. The proposed method avoids solving total eigenvalue problems. Thus the computational cost is significantly reduced. Synthetic and real images are used to verify and illustrate the effectiveness of the proposed approach.

1 Introduction

The fundamental matrix is the algebraic representation of epipolar geometry. It is the most powerful tool in the analysis of images pairs taken from uncalibrated cameras. The fundamental matrix encapsulates all the available information on the epipolar geometry. It is independent of the scene structure, and can be recovered from a set of image point matches without a priori knowledge of the parameters of the camera. Estimation of the fundamental matrix is the important step for many applications in computer vision [1,2], and is the focus of many researches [3,4,5].

The fundamental matrix is usually computed from the following epipolar constraints

$$v_i^T F u_i = 0 \quad \text{for} \quad i = 1, 2, \cdots N. \qquad (1)$$

where $u_i = [u_i(1), u_i(2), 1]^T \in \mathbb{R}^3$ and $v_i = [v_i(1), v_i(2), 1]^T \in \mathbb{R}^3$ represent the homogeneous pixel coordinates of matched points in the first and second image, respectively, and $F \in \mathbb{R}^{3\times 3}$ is the fundamental matrix. The rank of the fundamental matrix F is 2. A number of algorithms have been developed to estimate this matrix. Equation (1) can exactly be solved only in absence of noise, e.g. by using the eight-point algorithm [3]. For noisy images, the fundamental matrix is very sensitive to errors on the point locations. Because the first two components of the vectors u_i and v_i are located with errors, more matched points are needed and a measurement error model must be considered. We suppose that u_i and v_i can be written as

$$u_i = u_{0,i} + \tilde{u}_i \quad \text{and} \quad v_i = v_{0,i} + \tilde{v}_i \quad \text{for} \quad i = 1, 2, \cdots N. \tag{2}$$

and that there exists $F_0 \in \mathbb{R}^{3\times 3}$, such that

$$v_{0,i}^T F_0 u_{0,i} = 0 \quad \text{for} \quad i = 1, 2, \cdots N. \tag{3}$$

The matrix F_0 is the true fundamental matrix and rank$(F_0) = 2$. The vectors $u_{0,i}$ and $v_{0,i}$ are the true values of the measurements u_i and v_i, respectively, and $\tilde{u}_i$ and $\tilde{v}_i$ represent the measurement errors.

By interpreting the observations $a_i := u_i \otimes v_i$ as

$$a_i = u_{0,i} \otimes v_{0,i} + d_i \quad \text{for} \quad i = 1, 2, \cdots N. \tag{4}$$

where $d_1, \cdots, d_N$ are zero mean independently and identically distributed random vectors, Mühlich and Mester [1] proposed a total least-squares (TLS) estimator of F. The idea is to transform Eq.(1) in the form

$$(u_i \otimes v_i)^T f = 0 \quad \text{for} \quad i = 1, 2, \cdots N. \tag{5}$$

where $f = (F_{11}, F_{21}, F_{31}, F_{12}, F_{22}, F_{23}, F_{13}, F_{23}, F_{33})^T$.

The TLS estimator of F is found by solving

$$\min_f \|Af\|^2 = \min \sum_{i=1}^{N} r_i^2 \quad \text{s.t.} \quad f^T f = 1. \tag{6}$$

where $A := [a_1, a_2, \cdots, a_N]^T$, and $r_i := a_i^T f$ is the ith residual. As a_i involves the product of two spatial coordinates, the perturbations in the a_i^T rows are not Gaussian distributed. Therefore, the TLS solution is suboptimal, biased, and inconsistent [5]. For noisy data, some techniques have been tried for improving the accuracy of the eight-point algorithm in the presence of noise [1,3,4,5]. For large images, to reduce the condition number of $A^T A$, several scalings of the point coordinates have been proposed with good results [3]. One type of them is the statistical scaling of Hartley [3] which requires a centering and a scaling of the image feature points. This preprocessing has found a theoretical justification in [1] limited to the assumption of noise confined only in the second image. However, the assumption is not realistic. Leedan and Meer [4] further improved the method in [1] using a generalized TLS technique. But the improved estimation remains inconsistent and biased.

By using a quadratic measurement error model and taking more realistic assumptions, Kukush et al. [5] established a consistent fundamental matrix estimator. How to effectively calculate the estimator, however, has not been discussed. In this paper, first an extended system for determining the estimator is proposed, and an efficient implementation for solving the system-a continuation method [6] is developed to fix on an interval in which a local minimum belongs.

Next an optimization method using a quadratic interpolation is used to locate the minimum. The proposed method avoids solving total eigenvalue problems. Thus the computational cost is significantly reduced. Synthetic and real images are used to verify and illustrate the usefulness and effectiveness of the proposed approach.

2   The Consistent Fundamental Matrix Estimation

By using the following assumptions on the errors $\tilde{u}_i$ and $\tilde{v}_i$:

(i) The error vectors $\{\tilde{u}_i, \tilde{v}_i, i \geq 1\}$ are independent with $E[\tilde{u}_i] = E[\tilde{v}_i] = 0$, for $i \geq 1$,

(ii) $\mathrm{cov}(\tilde{u}_i) = \mathrm{cov}(\tilde{v}_i) = \sigma_0^2 T$, $i \geq 1$, with fixed $\sigma_0 > 0$, and $T = \mathrm{diag}(1,1,0)$, and using the quadratic measurement error model (2) and (3), Kukush et al. [5] established a consistent estimator for the fundamental matrix F.

The procedure for determining the estimator of the model is outlined as follows. For details, we refer readers to Kukush et al [5].
Given: N pairs of matched points $u_i \in \mathbb{R}^3$, $v_i \in \mathbb{R}^3$, $1 \leq i \leq N$ and upper bound d^2.
Stage 1: Computation of $\hat{F}_1$, $\|\hat{F}_1\|_F = 1$.
Compute $\hat{\sigma}^2 = \arg \min_{0 \leq \sigma^2 \leq d^2} |\lambda_{\min}(S_N(\sigma^2))|$ with

$$S_N(\sigma^2) := \sum_{i=1}^{N} (u_i u_i^T - \sigma^2 T) \otimes (v_i v_i^T - \sigma^2 T).$$

Compute the eigenvector $\hat{f}_1$ corresponding to $\lambda_{\min}(S_N(\hat{\sigma}^2))$.
Set

$$\hat{F}_1 = \begin{bmatrix} \hat{f}_1(1) & \hat{f}_1(4) & \hat{f}_1(7) \\ \hat{f}_1(2) & \hat{f}_1(5) & \hat{f}_1(8) \\ \hat{f}_1(3) & \hat{f}_1(6) & \hat{f}_1(9) \end{bmatrix}.$$

Stage 2: Computation of $\hat{F}$, rank $(\hat{F}) = 2$.

Compute the SVD of $\hat{F}_1$: $\hat{F}_1 = USV^T$ with $UU^T = I = VV^T$, $U \in \mathbb{R}^{3 \times 3}$, $V \in \mathbb{R}^{3 \times 3}$, $S = \mathrm{diag}(s_1, s_2, s_3)$ and $s_1 \geq s_2 \geq s_3$.
Set $\hat{F} = U\hat{S}V^T$ with $S = \mathrm{diag}(s_1, s_2, 0)$. *End*

In the computational procedure described above, the key problem is to find the estimator $\hat{\sigma}^2$ of the corresponding non-smooth optimization problem. In addition, due to the lack of homogeneity in image coordinates serving as input in the computation, the algorithm is very sensitive to noise. In next section, first data normalization is introduced, and then an efficient method for determining the estimator $\hat{\sigma}^2$ is presented.

3 Efficient Method of Solution

To reduce the sensitivity to noise in the specification of the matched points, data normalization should be used in fundamental matrix estimation. Following Hartley [3], the process of normalization consists of scaling and translating the data so that points u_i and v_i are transformed to $\hat{u}_i$ and $\hat{v}_i$ by using two transformation matrices T and T':

$$\hat{u}_i = T u_i, \quad \hat{v}_i = T' v_i. \tag{7}$$

where

$$T = \begin{bmatrix} k_u & 0 & -k_u u_{01} \\ 0 & k_u & -k_u u_{02} \\ 0 & 0 & 1 \end{bmatrix},$$

$$u_{0j} = \frac{1}{N}\sum_{i=1}^{N} u_i(j), \; j=1,2, \quad \sigma_u = \sqrt{\frac{1}{N}\sum_{i=1}^{N}\sum_{j=1}^{2}[(u_i(j)-u_{0j})^2]}, \quad k_u = \frac{\sqrt{2}}{\sigma_u}$$

and T' is defined in a way similar to T. Then the $\hat{F}$ matrix is estimated from the normalized matched points and, finally, it must be restored to obtain F using the equation

$$F = T^T \hat{F} T'. \tag{8}$$

The following discussions refer to the transformed data and omit the superscript for convenience.

Next, we consider the optimization problem of finding $\hat{\sigma}^2$. It should be noted that the optimization problem is not differentiable. To overcome this difficulty, we introduce a new eigenvalue problem: $[S_N(\sigma^2)]^2 \phi = \lambda \phi$, $\phi \neq 0$. Since $S_N(\sigma^2)$ is symmetric, all its eigenvalues are real and each eigenvalue of $[S_N(\sigma^2)]^2$ is square of the corresponding one of $S_N(\sigma^2)$. Thus we have

$$\hat{\sigma}^2 = \arg\min_{0 \leq \sigma^2 \leq d^2} |\lambda_{\min}(S_N(\sigma^2))| = \arg\min_{0 \leq \sigma^2 \leq d^2} \lambda_{\min}([S_N(\sigma^2)]^2).$$

Thus the original non-differentiable optimization problem becomes differentiable one.

For calculation of $\hat{\sigma}^2$, the following extended system for determining the curve of the smallest eigenvalue $\lambda_{\min}(\sigma^2)$ of $[S_N(\sigma^2)]^2$ versus σ^2 is first introduced:

$$F(y,\rho) := F(\phi, \lambda, \rho) := \begin{cases} [S_N(\rho)]^2 \phi - \lambda \phi \\ \phi^T \phi - 1 \end{cases} = 0. \tag{9}$$

where $\rho := \sigma^2$, $y := (\phi, \lambda) \in \mathbb{R}^9 \times \mathbb{R}$ are unknows and ρ is the continuation parameter. The predictor-corrector method in [6] will be used to solve (9). The starting solution $(\phi, \lambda, \rho) = (\phi_0, \lambda_0, \rho_0)$ to (9) can be achieved by calculating the smallest eigenvalue λ_0 and its corresponding eigenvector ϕ_0 of $[S_N(\rho_0)]^2$ with

$\rho_0 = 0$. Note that by using (9), we only need to compute the smallest eigenvalue and its corresponding eigenvector of $[S_N(\sigma^2)]^2$, without solving other eigenvalues and corresponding eigenvectors.

Assume that the solution $y^j = (\phi_j, \lambda_j)$ of (9) for $\rho = \rho_j$ has been obtained. The jth continuation step starts from a solution (y^j, ρ_j) of (9) and attempts to calculate the solution (y^{j+1}, ρ_{j+1}) for "next" $\rho = \rho_{j+1}$. With predictor-corrector methods, the step $j \to j+1$ is split into two steps:

$$(y^j, \rho_j) \xrightarrow{\text{predictor}} (\bar{y}^j, \rho_{j+1}) \xrightarrow{\text{corrector}} (y^{j+1}, \rho_{j+1}).$$

The predictor merely provides an initial guess for corrector iterations that home in on a solution of (9).

The tangent predictor is described as follows. Differentiating (9) with respect to ρ and evaluating the resulting expressions at $(\phi, \lambda, \rho) = (\phi_j, \lambda_j, \rho_j)$, we have

$$\begin{cases} ([S_N(\rho_j)]^2 - \lambda_j I_9)\left(\dfrac{d\phi}{d\rho}\right)_j - \phi_j \left(\dfrac{d\lambda}{d\rho}\right)_j = -2S_N(\rho_j)\left(\dfrac{\partial S_N(\rho)}{\partial \rho}\right)_j \phi_j, \\ 2\phi_j^T \left(\dfrac{d\phi}{d\rho}\right)_j = 0 \end{cases} \quad (10)$$

where

$$\left(\dfrac{\partial S_N(\rho)}{\partial \rho}\right)_j = \sum_{i=1}^N [(-T) \otimes (v_i v_i^T - \rho_j T) + (u_i u_i^T - \rho_j T) \otimes (-T)]. \quad (11)$$

The predictor point (initial approximation) for $(\phi_{j+1}, \lambda_{j+1})$ is

$$(\bar{\phi}_{j+1}, \bar{\lambda}_{j+1}) = (\phi_j, \lambda_j) + \Delta\rho \left(\left(\dfrac{d\phi}{d\rho}\right)_j, \left(\dfrac{d\lambda}{d\rho}\right)_j\right). \quad (12)$$

where $\Delta\rho = (\rho_{j+1} - \rho_j)$ is a step length. A constant step length is taken through this paper.

Now, we turn to corrector iteration. Consider one step of the Newton iteration, formulated for the vector $y = (\phi, \lambda)$

$$\begin{cases} ([S_N(\rho_{j+1})]^2 - \lambda^{(v)} I_9)\Delta\phi - \phi^{(v)} \Delta\lambda = -([S_N(\rho_{j+1})]^2 - \lambda^{(v)} I_9)\phi^{(v)}, \\ (\phi^{(v)})^T \Delta\phi = 0.5[1 - (\phi^{(v)})^T \phi^{(v)}], \end{cases} \quad (13)$$

$$(\phi^{(v+1)}, \lambda^{(v+1)}) = (\phi^{(v)}, \lambda^{(v)}) + (\Delta\phi, \Delta\lambda), \quad v = 0, 1, \cdots. \quad (14)$$

The iteration starts from the predictor $(\phi^{(0)}, \lambda^{(0)}) := (\bar{\phi}^{j+1}, \bar{\lambda}^{j+1})$. Equations (13) and (14) define a sequence of corrector iterations $(\phi^{(v)}, \lambda^{(v)})$ that is to converge to a solution $(\phi^{j+1}, \lambda^{j+1})$. The iteration will stop if

$$\|(\phi^{(v+1)} - \phi^{(v)}, \lambda^{(v+1)} - \lambda^{(v)})\| < \varepsilon$$

where ε is a specified error tolerance, which is defined as 10^{-8} in the present paper.

During the continuation, if for some integer m, the following inequality

$$\left(\frac{d\lambda}{d\rho}\right)_{m-1}\left(\frac{d\lambda}{d\rho}\right)_{m} < 0. \tag{15}$$

holds, then a minimum value of the function $\lambda_{\min}(\rho)$ is to be passed, i.e. the interval $(a,b) := (\rho_{m-1}, \rho_m)$ contains $\hat{\sigma}^2$. We are now solving a minimization problem of $\lambda_{\min}(\rho)$ for $\rho \in (\rho_{m-1}, \rho_m)$ in which the second derivative $\lambda''_{\min}(\rho)$ is expensive to evaluate. In the present implementation, we will propose an efficient method for solving the minimization problem, which avoids evaluating $\lambda''_{\min}(\rho)$ [7].

Given two points ρ^1 and ρ^2, values of function $\lambda_{\min}(\rho^1)$ and $\lambda_{\min}(\rho^2)$, and value of the first derivative $\lambda'_{\min}(\rho^1)$, we now construct a quadratic interpolation function $q(\rho) = a\rho^2 + b\rho + c$ such that

$$\begin{cases} q(\rho^1) = a(\rho^1)^2 + b\rho^1 + c = \lambda_{\min}(\rho^1) := \lambda_1, \\ q(\rho^2) = a(\rho^2)^2 + b\rho^2 + c = \lambda_{\min}(\rho^2) := \lambda_2, \\ q'(\rho^1) = 2a\rho^1 + b = \lambda'_{\min}(\rho^1) := \lambda'_1. \end{cases} \tag{16}$$

From (16), we get a, b and c, and the minimum value of the quadratic interpolation function $q(\rho)$ is achieved at

$$\tilde{\rho} = \rho^1 - \frac{(\rho^1 - \rho^2)\lambda'_1}{2\left[\lambda'_1 - \frac{\lambda_1 - \lambda_2}{\rho^1 - \rho^2}\right]}. \tag{17}$$

The expression above may be written as the following iterative formula:

$$\rho^{k+1} = \rho^{k-1} - \frac{(\rho^{k-1} - \rho^k)\lambda'_{k-1}}{2\left[\lambda'_{k-1} - \frac{\lambda_{k-1} - \lambda_k}{\rho^{k-1} - \rho^k}\right]}, \quad k = 1, 2, \cdots. \tag{18}$$

The initial inputs for the iteration are

$$\rho^0 = \rho_{m-1}, \quad \rho^1 = \rho_m.$$

The iteration will stop if

$$\left|\left(\frac{d\lambda}{d\rho}\right)_k\right| < Tol$$

where Tol is is a specified error tolerance, which is defined as 10^{-6} in this paper.

Note that, values of $\lambda_{\min}(\rho)$ and $\lambda'_{\min}(\rho)$ are not given by formulas; rather they are the outputs from computational procedures described in (10), (13) and (14), while $\lambda''_{\min}(\rho)$ is very difficult to calculate. Our iteration formula (18) requires only values of $\lambda_{\min}(\rho)$ and $\lambda'_{\min}(\rho)$.

The procedure of the proposed method is summarized as follows.

(1) Complete data normalization of the matched points by using (7).
(2) Compute the smallest eigenvalue λ_0 and its corresponding eigenvector ϕ_0 of $[S_N(\rho_0)]^2$ with $\rho_0 = 0$.

(3) Solve (9) by using the predictor-corrector method.
(4) During the continuation, using (15) to determine an internal containing $\hat{\sigma}^2$ where $\lambda_{\min}(\rho)$ arrives at its minimum value.
(5) Locate the value of $\hat{\sigma}^2$ by applying the iteration formula (18) and obtain the corresponding eigenvector $\hat{f}_1$ corresponding to $\lambda_{\min}(S_N(\hat{\sigma}^2))$.
(6) Set
$$\hat{F}_1 = \begin{bmatrix} \hat{f}_1(1) & \hat{f}_1(4) & \hat{f}_1(7) \\ \hat{f}_1(2) & \hat{f}_1(5) & \hat{f}_1(8) \\ \hat{f}_1(3) & \hat{f}_1(6) & \hat{f}_1(9) \end{bmatrix}.$$
(7) Form $\hat{F}$ from $\hat{F}_1$ by putting rank-2 constraint as described in Section 2.
(8) The fundamental matrix F for original image coordinates is obtained from (8).

4 Experimental Results

In this section, experimental results for the consistent estimation and the TLS one are presented. For the consistent estimation, the proposed implementation is used. The TLS estimation is obtained by using the data normalization and the best rank-two approximation of solution of the optimization problem (6) [3,4,5], seeing stage 2 in Section 2. To compare the quality of different estimation, epipoles defined as the null-spaces of the restricted, rank two matrix

$$Fe_1 = 0, \quad F^T e_2 = 0. \tag{19}$$

are used where e_1 and e_2 are the epipoles in the first and the second image, respectively.

4.1 Experiments with Synthetic Data

The experiments are based on a pair of synthetic images. The fundamental matrix F is constructed from two perspective projection matrices, and the true epipole coordinate in the second image is $e_2 = [500, 80]$. One hundred matched points are used. The coordinates of the points in each image are corrupted with zero-mean normal noise with standard deviation σ. Tests are performed for different values of σ. For each experimental condition, i.e. the value of σ and the employed estimation technique, 1000 trials were run. Table 1 shows the mean and standard deviation of epipole in the second image.

From Table 1, the following facts can be observed. The proposed method has very small bias for each level of noise; on the contrary, the strong bias and the large spread for the epipole estimation occur for the TLS method. With the rise of the standard deviation σ of noise, the differences between the results of the TLS method and those of the proposed method increase. The standard deviations from the proposed method are smaller than those from the TLS method for each value of σ. This indicates that the sensitivity of the proposed method to noise is weaker. The consistent estimations can always give very excellent results, but the errors of the LTS method increase with the rise of the standard deviation σ of noise and the corresponding results become unreliable. These results illustrate the importance of the consistent estimation and show good performance of the proposed implementation.

Table 1. Mean and standard deviation of epipole in the second image for different noise level σ

σ	The TLS method		The proposed method	
	Mean	Standard deviation	Mean	Standard deviation
0	[500, 80]	(0, 0)	[500, 80]	(0, 0)
0.1	[500.30, 80.02]	(3.61, 1.86)	[500.08, 80.11]	(3.60, 1.86)
0.2	[501.02, 79.44]	(6.93, 3.46)	[500.06, 79.82]	(6.91, 3.46)
0.4	[504.49, 78.99]	(15.02, 7.40)	[500.62, 80.52]	(14.76, 7.28)
0.6	[509.77, 76.61]	(23.23, 10.91)	[500.90, 80.11]	(22.35, 10.56)
0.8	[515.10, 73.95]	(32.66, 15.49)	[499.70, 80.06]	(31.17, 14.76)
1.0	[524.47, 69.56]	(39.40, 18.28)	[500.03, 79.25]	(35.82, 16.83)
1.2	[540.59, 66.83]	(62.73, 29.69)	[504.47, 80.94]	(54.64, 26.06)
1.5	[563.26, 52.88]	(95.26, 43.45)	[504.89, 76.17]	(76.26, 34.82)
2.0	[605.40, 37.69]	(150.24, 70.16)	[506.44, 76.53]	(107.72, 51.04)

4.2 Experiments with Real Images

The proposed method is tested on two different pairs of real images to evaluate its performance. In carrying out these tests, the proposed method as described above is compared with the TLS method for finding the epipole. In all cases, the matched points are obtained by using the method proposed in [8], and some outliers are detected and removed, based on least-median squares techniques [8]. The images are presented in Fig. 1 (a) and (b) to show the diversity of image types. There is a variation in the accuracy of the matched points for the different images, as will be indicated later.

Tables 2 and 3 show the experimental results of the two methods, with different numbers of points N, which are used to compute the fundamental matrix. Here, N ranges from 20 up to 90 percent of the total number of matched points to show the convergence of the proposed method. For each value of N, the algorithms are run 200 times using randomly selected sets of N matching points, and then mean and standard deviation of epipoles are calculated.

For the corridor scene, the matched points were known with extreme accuracy [3], whereas for the urban scene, the matches were less accurate. We can see that, with increase of N, the standard deviation decreases. In all cases, the proposed method performs better than the TLS method. In the cases of the corridor image the effect is not so great since the errors on the point locations are small. In the case of the urban scene with bigger errors on the point locations, the advantage

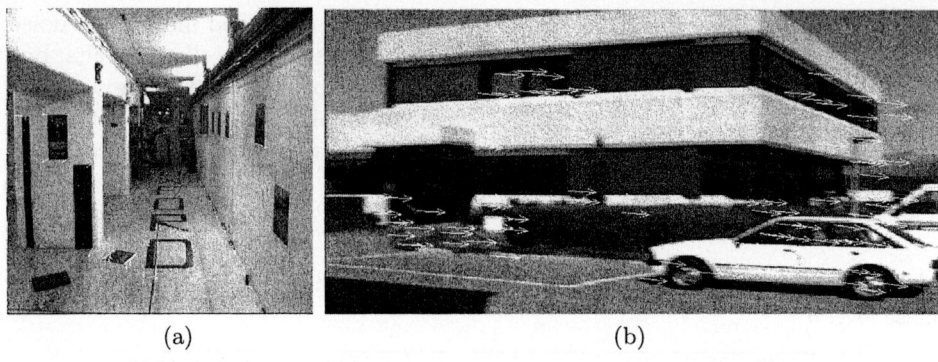

Fig. 1. Real images with matchings (a) Corridor scene (b) Urban scene

of the proposed method is dramatic. Experiment results of these images indicate again the importance of the consistent estimation.

Table 2. Mean and standard deviation of epipole of image (a) with different N

N	The proposed method		The TLS method	
	Mean	Standard deviation	Mean	Standard deviation
20	[269.29, 206.53]	(30.13, 39.96)	[271.90, 207.12]	(49.95, 56.15)
40	[274.69, 204.41]	(14.09, 16.80)	[279.22, 206.51]	(19.59, 21.79)
60	[275.52, 200.27]	(10.10, 10.84)	[279.07, 200.82]	(11.04, 12.13)
80	[278.80, 203.15]	(7.33, 7.59)	[282.53, 204.19]	(7.61, 7.95)
100	[281.33, 202.51]	(3.85, 4.36)	[284.98, 203.90]	(4.32, 4.87)

Table 3. Mean and standard deviation of epipole of image (b) with different N

N	The proposed method		The TLS method	
	Mean	Standard deviation	Mean	Standard deviation
20	[−170.07, 494.38]	(175.30, 186.92)	[−287.04, 585.05]	(200.70, 210.30)
40	[−183.16, 489.68]	(128.93, 113.52)	[−315.79, 589.78]	(165.05, 138.45)
60	[−216.51, 531.11]	(108.44, 98.58)	[−364.77, 649.74]	(132.50, 117.86)
80	[−206.07, 527.12]	(41.63, 41.24)	[−340.68, 638.79]	(51.65, 48.57)

5 Conclusion

In this paper, solution of consistent fundamental matrix estimation in a quadratic measurement error model has been studied. An extended system for determining the consistent estimator has been proposed, and an efficient implementation for solving the system-a continuation method has been developed. An optimization method using a quadratic interpolation has been used to exactly locate the minimum. The proposed method avoids solving total eigenvalue problems. Thus the computational cost is significantly reduced. Experiments with synthetic and real images have shown the effectiveness of the proposed method.

References

1. Mühlich, M., Mester, R.: The Role of Total Least Squares in Motion Analysis. In: Burkhardt, H. (ed.): Proceedings of the European Conference on Computer Vision (ECCV'98). Lecture Notes on Computer Science, Vol. 1407. Springer-Verlag, Berlin (1998) 305-321
2. Hartley, R.I., Zisserman, A.: Multiple View Geometry in Computer Vision. Cambridge University Press, London (2000)
3. Hartley, R.I.: In Defence of the Eight-point Algorithm. IEEE Trans. Pattern Anal. Mach. Intell. 19 (1997) 580-593
4. Leedan, Y., Meer, P.: Heteroscedastic Regression in Computer Vision: Problems with Bilinear Constraint. Int. J. Comput. Vision 37 (2000) 127-150
5. Kukush, A., Markovsky, I., Van Huffel, S.: Consistent Fundamental Matrix Estimation in A Quadratic Measurement Error Model Arising in Motion Analysis. Computational Statistics & Data Analysis 41 (2002) 3-18
6. Seydel, R.: Practical Bifurcation and Stability Analysis: From Equilibrium to Chaos, 2nd ed. World Publishing Corporation, Beijing (1999)
7. Yuan, Y.X., Sun, W.Y.: Optimization Theory and Methods. Science Press, Beijing (1999) (in Chinese)
8. Zhang, Z., Deriche, R., Faugeras, O., Luong, Q.-T.: A Robust Technique For Matching Two Uncalibrated Images through the Recovery of the Unknown Epipolar Geometry, Artificial Intelligence Journal 78 (1995) 87-119

Derivations of Error Bound on Recording Traffic Time Series with Long-Range Dependence

Ming Li

School of Information Science & Technology,
East China Normal University,
Shanghai 200026, P.R. China
ming_lihk@yahoo.com, mli@ee.ecnu.edu.cn

Abstract. Measurement of traffic time series plays a key role in the research of communication networks though theoretic research has a considerable advances. Differing from analytical analysis, quantities of interest are estimates experimentally analyzed from measured real life data. Hence, accuracy should be taken into account from a view of engineering. In practical terms, it is inappropriate to record data series that is either too short or over-long as too short record may not provide enough data to achieve a given degree of accuracy of an estimate while over-long record is usually improper for real-time applications. Consequently, error analysis based on record length has practical significance. This paper substantially extends our previous work [20,21] by detailing the derivations of error bound relating to record length and the Hurst parameter of a long-range dependent fractional Gaussian noise and by interpreting the effects of long-range dependence on record length. In addition, a theoretical evaluation of some widely used traces in the traffic research is also given.

1 Introduction

Measurement of traffic is required in many applications, such as performance analysis of communication systems, e.g. [1], traffic analysis, e.g. [2], [3], modeling and simulation, e.g. [4], [5], [6], real-time traffic data collection, e.g. [7], [8], [9] and so on. In this regard, record length of a traffic series should be predetermined according to a given degree of accuracy of an estimate (e.g. autocorrelation function (ACF)) before performing measurement. In practice, an appropriate record length is crucial to applications, especially real-time measurement because if the length of a measured series is too short, an estimate may not achieve a given accuracy. On the other hand, over-long record length will cost too much record time, storage space and computation time.

In the field of measurement, length requirements of a measured random sequence are traditionally for those with short-range dependence (SRD), see e.g. [10]. That is because random processes encountered in many fields of engineering are usually of SRD [11], [12], [13], [14], [15]. Intuitively, length requirements of long-range dependent (LRD) sequences should be distinctly different from those of SRD sequences because LRD processes evidently differ from SRD ones in nature [3], [16]. However, reports about record length requirements for LRD traffic measurement are rarely seen, to our best knowledge.

Our early work [17] gave a primary result and [18] presented a concluded result but derivations in detail have not been given there. As a supplementary to them, this paper provides the detailed derivations of error bound for recording LRD traffic. In the derivations, we take LRD fractional Gaussian noise (FGN) as a representative of LRD traffic. Due to the fact that FGN is an approximate model of real traffic and inequalities used in the derivations, the result in this paper may be conservative but it may yet be a reference guideline for record length of traffic in academic research and in practice.

In addition to the error bound, this paper also interprets how long-range dependence effects on record length, which will show that the length requirement of a measured LRD sequence drastically differs from that of SRD sequences.

The rest of paper is organized as follows. In Section 2, we give a theorem regarding error bound for requiring record length of measured LRD traffic and detailed derivations. Discussions are given in Section 3 and conclusions in Section 4.

2 Upper Bound of Standard Deviation and Its Derivation

Denote $x(t_i)$ ($i = 0, 1, 2, \ldots$) as a traffic time series, representing the number of bytes in a packet on a packet-by-packet basis at the time t_i. Then, $x(i)$ is a series, indicating the number of bytes in the ith packet. Mathematically, $x(i)$ is LRD if its ACF $r(k) \sim ck^{2H-2}$ for $c > 0$ and $H \in (0.5, 1)$ while $x(i)$ is called asymptotically self-similar if $x(ai)$ ($a > 0$) asymptotically has the same statistics as $x(i)$, where H is called the Hurst parameter and $\sim$ stands for the asymptotical equivalence under $k \to \infty$.

In mathematics, the true ACF of $x(i)$ is computed over infinite interval $r(k) = \lim_{L \to \infty} \frac{1}{L} \sum_{i=1}^{L} x(i) x(i+k)$ [10]. However, physically measured data sequences are finite. Let a positive integer L be the data block size of $x(i)$. Then, the ACF of $x(i)$ is estimated by $\frac{1}{L} \sum_{i=1}^{L} x(i) x(i+k)$. Obviously, L should be large enough for a given degree of accuracy of ACF estimation.

The ACF of FGN is given by $0.5\sigma^2[(k+1)^{2H} - 2k^{2H} + (k-1)^{2H}]$, where $\sigma^2 = \dfrac{\Gamma(1-2H)\cos(H\pi)}{H\pi}$ [16]. In the normalized case, the ACF of FGN is written by $0.5[(k+1)^{2H} - 2k^{2H} + (k-1)^{2H}]$. Suppose $r(\tau)$ is the true ACF of FGN and $R(\tau)$ is its estimate with L length. Let $M^2(R)$ be the mean square error in terms of R. Then, the following theorem represents $M^2(R)$ as a two-dimension (2-D) function of L and H, which establishes a reference guideline for requiring record length of traffic for a given degree of accuracy.

Theorem: Let $x(t)$ be FGN with $H \in (0.5, 1)$. Let $r(\tau)$ be the true ACF of $x(t)$. Let L be the block size of data. Let $R(\tau)$ be an estimate of $r(\tau)$ with L length. Let $\text{Var}[R(\tau)]$ be the variance of $R(\tau)$. Then,

$$\text{Var}[R(\tau)] \leq \frac{\sigma^4}{L(2H+1)}[(L+1)^{2H+1} - 2L^{2H+1} + (L-1)^{2H+1}]. \tag{1-a}$$

Proof: Mathematically, $r(\tau)$ is computed over infinite interval:

$$r(\tau) = E[x(t)x(t+\tau)] = \lim_{T \to \infty} \frac{1}{T} \int_0^T x(t)x(t+\tau)dt. \tag{P-1}$$

In practice, $r(\tau)$ can only be estimated with a finite sequence. Therefore,

$$r(\tau) \approx R(\tau) = \frac{1}{L} \int_{t_0}^{t_0+L} x(t)x(t+\tau)dt, \tag{P-2}$$

where t_0 is the start time. Usually, $\int_{t_0}^{t_0+L} x(t)x(t+\tau)dt \neq \int_{t_1}^{t_1+L} x(t)x(t+\tau)dt$ for $t_0 \neq t_1$ and R is a random variable.

As $M^2(R) = E[(r-R)^2]$, it is a function of L. The larger the L the smaller the $M^2(R)$. We aim at finding the quantitative relationship between $M^2(R)$ and L.

Let $b^2(R)$ be the bias of R. Then,

$$M^2(R) = \text{Var}(R) + b^2(R). \tag{P-3}$$

Since

$$E[R(\tau)] = \frac{1}{L} \int_{t_0}^{t_0+L} E[x(t)x(t+\tau)]dt = \frac{1}{L} \int_{t_0}^{t_0+L} r(\tau)dt = r(\tau), \tag{P-4}$$

$R(\tau)$ is the unbiased estimate of $r(\tau)$ and $M^2(R) = \text{Var}(R)$ accordingly. We need expressing $\text{Var}(R)$ by the following proposition to prove the theorem.

Proposition: Let $x(t)$ be a Gaussian process. Let $r(\tau)$ be the true ACF of $x(t)$. Let L be the block size of data. Let $R(\tau)$ be an estimate of $r(\tau)$ with L length. Let $\text{Var}[R(\tau)]$ be the variance of $R(\tau)$. Suppose $r(\tau)$ is monotonously decreasing and $r(\tau) \geq 0$. Then,

$$\text{Var}[R(\tau)] \leq \frac{2}{L} \int_0^L [r^2(t) + r(t+\tau)r(-t+\tau)]dt \leq \frac{4}{L} \int_0^L r^2(t)dt. \tag{P-5}$$

Proof: As $\text{Var}(R) = E\{[R - E(R)]^2\} = E(R^2) - E^2(R)$, according to (P-4), one has

$$\text{Var}[R(\tau)] = E(R^2) - r^2(\tau). \tag{P-6}$$

Expanding $E(R^2)$ yields

$$E(R^2) = E\left\{\left[\frac{1}{L}\int_{t_0}^{t_0+L} x(t)x(t+\tau)dt\right]^2\right\}$$

$$= E\left[\frac{1}{L^2}\int_{t_0}^{t_0+L} x(t_1)x(t_1+\tau)dt_1 \int_{t_0}^{t_0+L} x(t_2)x(t_2+\tau)dt_2\right]$$

$$= E\left[\frac{1}{L^2}\int_{t_0}^{t_0+L}\int_{t_0}^{t_0+L} x(t_1)x(t_2)x(t_1+\tau)x(t_2+\tau)dt_1 dt_2\right]$$

$$= \frac{1}{L^2}\int_{t_0}^{t_0+L}\int_{t_0}^{t_0+L} E[x(t_1)x(t_2)x(t_1+\tau)x(t_2+\tau)]dt_1 dt_2.$$

Thus,

$$\operatorname{Var}[R(\tau)] = \frac{1}{L^2}\int_{t_0}^{t_0+L}\int_{t_0}^{t_0+L} E[x(t_1)x(t_2)x(t_1+\tau)x(t_2+\tau)]dt_1 dt_2 - r^2(\tau). \tag{P-7}$$

Let

$$\begin{cases} X_1 = x(t_1) \\ X_2 = x(t_2) \\ X_3 = x(t_1+\tau) \\ X_4 = x(t_2+\tau). \end{cases} \tag{P-8}$$

Then, $E[x(t_1)x(t_2)x(t_1+\tau)x(t_2+\tau)] = E(X_1X_2X_3X_4)$. Since x is Gaussian, random variables X_1, X_2, X_3 and X_4 have a joint-normal distribution and $E(X_1X_2X_3X_4) = m_{12}m_{34} + m_{13}m_{24} + m_{14}m_{23}$, where

$$\begin{cases} m_{12} = E[x(t_1)x(t_2)] = r(t_2 - t_1) \\ m_{13} = E[x(t_1)x(t_1+\tau)] = r(\tau) \\ m_{14} = E[x(t_1)x(t_2+\tau)] = r(t_2 - t_1 + \tau) \\ m_{23} = E[x(t_2)x(t_1+\tau)] = r(t_1 - t_2 + \tau) \\ m_{24} = E[x(t_2)x(t_2+\tau)] = r(\tau) \\ m_{34} = E[x(t_1)x(t_2+\tau)] = r(t_2 - t_1). \end{cases} \tag{P-9}$$

Therefore,

$$\frac{1}{L^2}\int_{t_0}^{t_0+L}\int_{t_0}^{t_0+L} E[x(t_1)x(t_2)x(t_1+\tau)x(t_2+\tau)]dt_1 dt_2$$

$$= \frac{1}{L^2} \int_{t_0}^{t_0+L} \int_{t_0}^{t_0+L} E(X_1 X_2 X_3 X_4) dt_1 dt_2$$

$$= \frac{1}{L^2} \int_{t_0}^{t_0+L} \int_{t_0}^{t_0+L} (m_{12} m_{34} + m_{13} m_{24} + m_{14} m_{23}) dt_1 dt_2$$

$$= \frac{1}{L^2} \int_{t_0}^{t_0+L} \int_{t_0}^{t_0+L} [r^2(t_2 - t_1) + r^2(\tau) + r(t_2 - t_1 + \tau) r(t_1 - t_2 + \tau)] dt_1 dt_2$$

$$= \frac{1}{L^2} \int_{t_0}^{t_0+L} \int_{t_0}^{t_0+L} [r^2(t_2 - t_1) + r(t_2 - t_1 + \tau) r(t_1 - t_2 + \tau)] dt_1 dt_2 + r^2(\tau).$$

According to (P-7), the variance is expressed as

$$\mathrm{Var}[R(\tau)] = \frac{1}{L^2} \int_{t_0}^{t_0+L} \int_{t_0}^{t_0+L} [r^2(t_2 - t_1) + r(t_2 - t_1 + \tau) r(t_1 - t_2 + \tau)] dt_1 dt_2.$$

Replacing $(t_2 - t_1)$ with t in the above expression yields

$$\mathrm{Var}[R(\tau)] = \frac{1}{L^2} \int_{t_0}^{t_0+L} dt_1 \int_{t_0-t_1}^{t_0-t_1+L} [r^2(t) + r(t + \tau) r(-t + \tau)] dt.$$

Let $f(t) = r^2(t) + r(t + \tau) r(-t + \tau)$. Then,

$$\mathrm{Var}[R(\tau)] = \frac{1}{L^2} \int_{t_0}^{t_0+L} dt_1 \int_{t_0-t_1}^{t_0-t_1+L} [r^2(t) + r(t + \tau) r(-t + \tau)] dt$$

$$= \frac{1}{L^2} \int_{t_0}^{t_0+L} dt_1 \int_{t_0-t_1}^{t_0-t_1+L} f(t) dt. \quad \text{(P-10)}$$

Without losing the generality, let $t_0 = 0$. Then, the above becomes

$$\mathrm{Var}[R(\tau)] = \frac{1}{L^2} \int_0^L f(t) dt \int_0^{L-t} dt_1 + \frac{1}{L^2} \int_{-L}^0 f(t) dt \int_{-t}^L dt_1$$

$$= \frac{1}{L^2} \int_0^L (L - t) f(t) dt + \frac{1}{L^2} \int_{-L}^0 (L + t) f(t) dt.$$

Since ACFs are even functions, the above expression is written by

$$\text{Var}(R) = \frac{1}{L^2}\int_0^L (L-t)f(t)dt + \frac{1}{L^2}\int_{-L}^0 (L+t)f(t)dt = \frac{2}{L^2}\int_0^L (L-t)f(t)dt \qquad \text{(P-11)}$$

$$= \frac{2}{L^2}\int_0^L (L-t)[r^2(t) + r(t+\tau)r(-t+\tau)]dt.$$

By using inequality, (P-11) becomes

$$\text{Var}[R(\tau)] = \frac{2}{L}\int_0^L (1-t/L)[r^2(t) + r(t+\tau)r(-t+\tau)]dt$$

$$\leq \frac{2}{L}\int_0^L |1 - \frac{t}{L}| \| r^2(t) + r(t+\tau)r(-t+\tau) | dt \qquad \text{(P-12)}$$

$$\leq \frac{2}{L}\int_0^L | r^2(t) + r(t+\tau)r(-t+\tau) | dt \leq \frac{4}{L}\int_0^L r^2(t)dt.$$

Therefore, Proposition holds.

Now, replacing $r(t)$ with the ACF of FGN in (P-5) yields

$$\text{Var}[R(\tau)] \leq \frac{4}{L}\int_0^L r^2(t)dt = \frac{\sigma^4}{L}\int_0^L [(t+1)^{2H} - 2t^{2H} + (t-1)^{2H}]^2 dt$$

$$\leq \frac{\sigma^4}{L}\int_0^L [(t+1)^{2H} - 2t^{2H} + (t-1)^{2H}]dt$$

$$= \frac{\sigma^4}{L(2H+1)}[(L+1)^{2H+1} - 2L^{2H+1} + (L-1)^{2H+1}].$$

Thus, Theorem results. ◆

In the normalized case, $\text{Var}[R(\tau)]$ is given by

$$\text{Var}[R(\tau)] \leq \frac{1}{L(2H+1)}[(L+1)^{2H+1} - 2L^{2H+1} + (L-1)^{2H+1}]. \qquad \text{(1-b)}$$

Without losing the generality, we consider the normalized ACF here and below. Denote $s(L, H)$ as the bound of standard deviation in the normalized case. Then,

$$s(L, H) = \sqrt{\frac{1}{L(2H+1)}[(L+1)^{2H+1} - 2L^{2H+1} + (L-1)^{2H+1}]}. \qquad (2)$$

The above formula represents an upper bound of $\text{Var}[R(\tau)]$. Based on it, we illustrate $s(L, H)$ in terms of L by Fig. 1 for $H = 0.70, 0.75$, and 0.80.

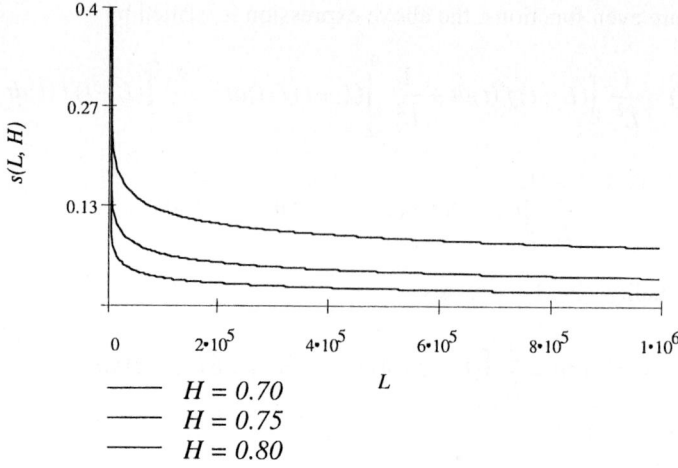

Fig. 1. Illustrating $s(L, H)$ for $H = 0.70, 0.75$, and 0.80

3 Discussion

From Fig. 1, we see that $s(L, H)$ is an increasing function of H for a given value of L and a decreasing function of L if H is fixed. Thus, a large L is required for a large value of H for a given accuracy of ACF estimation.

For facilitating the analysis, we draw up Table A (Appendix) to illustrate $s(L, H)$ for various values of Hs and Ls. In engineering, accuracy is usually considered from the perspective of order of magnitude. The following illustrates the relationship between H and L from the view of accuracy-order:

- The rows from the 2nd row ($L = 2^5$) to the 4th row ($L = 2^7$) in Table A cover two different orders of magnitude of s when H varies from 0.55 to 0.95. For instance, $s(L,H)|_{L=2^7, H=0.55} = 0.118$ while $s(L,H)|_{L=2^5, H=0.95} = 1.159$.
- The 5th row ($L = 2^8$) and the 6th row ($L = 2^9$) of Table A cover three different orders of magnitude of s for different Hs as we see that $s(2^8, 0.55) = 0.086$, $s(2^8, 0.75) = 0.306$, and $s(2^8, 0.95) = 1.045$.
- The rows from the 7th ($L = 2^{10}$) to the 12th ($L = 2^{15}$) in Table A cover two different orders of s when H varies from 0.55 to 0.95 as one has, for example, $s(2^{15}, 0.60) = 0.017$ but $s(2^{10}, 0.95) = 0.975$.
- The rows from the 13th ($L = 2^{16}$) to the 19th ($L = 2^{22}$) in Table A cover three different orders of s for different Hs, e.g., $s(2^{22}, 0.60) = 0.002$, $s(2^{20}, 0.70) = 0.018$ while $s(2^{16}, 0.95) = 0.792$.
- The rows from the 20th ($L = 2^{23}$) to the 23rd ($L = 2^{26}$) in Table A cover four different orders of s for different Hs since one has, e.g., $s(2^{26}, 0.55) = 3.767\text{--}e4$, $s(2^5, 0.70) = 0.007$, $s(2^{24}, 0.80) = 0.045$, and $s(2^{23}, 0.95) = 0.621$.

The above discussions show that the standard deviations vary in orders of magnitude when H changes from 0.55 to 0.95, implying a series with a larger value of H

requires a larger record length for a given accuracy of ACF estimation. This is the first point about record length in measurement suggested in this paper.

Now, we consider the record length requirement about SRD sequences by taking a look at a Poisson function $y(t)$, which presents a telegraph process. The ACF of $y(t)$ is given by $R_y(\tau) = e^{-2\lambda|\tau|}$, where λ is the average value of the number of changes of sign of $y(t)$ in $(t, t + \tau)$. Then, the bound of standard deviation, denoted as $sp(L, \lambda)$, is given by

$$sp(L, \lambda) = \sqrt{\frac{1}{L\lambda}(1 - e^{-4\lambda L})}. \qquad (3)$$

In fact, replacing $r(t)$ with $R_y(\tau) = e^{-2\lambda|\tau|}$ in (P-5) yields equation (3).

According to (3), we achieve $sp(2^5, 1) \approx 0.18$ even if $\lambda = 1$ and $L = 2^5$. Considering $s(2^9, 0.70) \approx 0.18$ and $2^9/2^5 = 16$, we see the record length required for a SRD sequence is noticeably shorter than LRD one for the same accuracy order. This comparison gives the second point for record length in measurement explained in this article.

Table 1. Six TCP traces in packet size

Dataset	Date	Duration	Packets
dec-pkt-1.TCP	08Mar95	10PM-11PM	3.3 million
dec-pkt-2.TCP	09Mar95	2AM-3AM	3.9 million
dec-pkt-3.TCP	09Mar95	10AM-11AM	4.3 million
dec-pkt-4.TCP	09Mar95	2PM-3PM	5.7 million
Lbl-pkt-4.TCP	21Jan94	2AM-3AM	1.3 million
Lbl-pkt-5.TCP	28Jan94	2AM-3AM	1.3 million

Notes:

- An exact value of $s(L, H)$ usually does not equal to the real accuracy of the correlation estimation of a measured LRD-traffic sequence because FGN is only an approximate model of real traffic. On the other hand, there are errors in data transmission, data storage, measurement, numerical computations, and data processing. In addition, there are many factors causing errors and uncertainties due to the natural shifts, such as various shifts occurring in devices, or some purposeful changes in communication systems. Therefore, the concrete accuracy value is not as pressing as accuracy-order for the considerations in measurement design. For that reason, we emphasize that the contribution of $s(L, H)$ lies in that it provides a relationship between s, L and H for a reference guideline in the design stage of measurement.
- Tables 1 lists the traces on WAN [19]. Now, we evaluate Lbl-pkt-4.TCP of 1.3×10^6 length, which is the shortest one in the table. For $H = 0.90$ (relatively strong LRD) and s being in the order of 0.1, we can select $L = 2^9$. Because Theorem provides a conservative guideline for requiring record length of traffic due to inequalities used in the derivations, we verify that those traces are long enough for accurate ACF estimation as well as general patterns of traffic.

4 Conclusions

We have derived a formula representing the standard deviation of ACF estimation of FGN as a 2-D function of the record length and the Hurst parameter. It may be conservative for real traffic but it may yet serve as a reference guideline for requiring record length in measurement. Based on the present formula, the noteworthy difference between measuring LRD and SRD sequences has been noticed.

References

1. Paxson, V.: Measurements and Analysis of End-to-End Internet Dynamics. Ph.D. Dissertation, Lawrence Berkeley National Laboratory, University of California (1997)
2. Csabai, I.: $1/f$ Noise in Computer Network Traffic. Journal of Physics A: Mathematical & General, 27 (1994) L417-L421
3. Willinger, W., Paxson, V., Riedi, R.H., Taqqu, M.S.: Long-Range Dependence and Data Network Traffic. In: Doukhan, P., Oppenheim, G., and Taqqu, M.S. (eds.): Long-Range Dependence: Theory and Applications, Birkhauser (2002)
4. Li, M., Zhao, W., et al.: Modeling Autocorrelation Functions of Self-Similar Teletraffic in Communication Networks based on Optimal Approximation in Hilbert Space. Applied Mathematical Modelling, 27 (2003) 155-168
5. Li, M., Jia, W., Zhao, W.: Correlation Form of Timestamp Increment Sequences of Self-Similar Traffic on Ethernet. Electronics Letters, 36 (2000) 1168-1169
6. Li, M., Chi, C.H.: A Correlation-based Computational Method for Simulating Long-Range Dependent Data. Journal of the Franklin Institute, 340 (2003) 503-514
7. Li, M.: An Approach to Reliably Identifying Signs of DDOS Flood Attacks based on LRD Traffic Pattern Recognition. Computer & Security, 23 (2004) 549-558
8. Li, M., Chi, C.H., Long, D.Y.: Fractional Gaussian Noise: a Tool of Characterizing Traffic for Detection Purpose. Springer LNCS, Vol. 3309 (2004) 94-103
9. Li. M., et al.: Probability Principle of a Reliable Approach to Detect Signs of DDOS Flood Attacks. Springer LNCS, Vol. 3320 (2004) 569-572
10. Bendat, J.S., Piersol, A.G.: Random Data: Analysis and Measurement Procedure. 3rd Edition, John Wiley & Sons (2000)
11. Gibson, J.D. (ed.): The Communications Handbook. IEEE Press (1997)
12. Li, M., et al.: H_2-Optimal Control of Random Loading for a Laboratory Fatigue Test. Journal of Testing and Evaluation, 26 (1998) 619-625
13. Li, M., et al.: An On-Line Correction Technique of Random Loading with a Real-Time Signal Processor for a Laboratory Fatigue Test. Journal of Testing and Evaluation, 28 (2000) 409-414
14. Li, M.: An Optimal Controller of an Irregular Wave Maker. Applied Mathematical Modelling, 29 (2005) 55-63
15. Li, M.: An Iteration Method of Adjusting Random Loading for a Laboratory Fatigue Test. International Journal of Fatigue, 27 (2005) 783-789
16. Beran, J.: Statistics for Long-Memory Processes. Chapman & Hall (1994)
17. Li, M., Jia, W., Zhao, W.: Length Requirements of Recorded Data of Network Traffic. IEEE ICII2001, Vol. 2 (2001) 45-49, China
18. Li, M.: Statistical Error Analysis on Recording LRD Traffic Time Series. Springer LNCS, Vol. 3222 (2004) 403-406
19. http://ita.ee.lbl.gov/

Appendix:

Table A. $s(L, H)$ for $L = 2^{5+m}$ ($m = 0, 1, ..., 21$), $H = 0.55 + 0.05n$ ($n = 0, 1, ..., 8$)

H	0.55	0.60	0.65	0.70	0.75	0.80	0.85	0.90	0.95
2^5	0.220	0.274	0.339	0.418	0.515	0.632	0.775	0.949	1.159
2^6	0.161	0.208	0.266	0.340	0.433	0.551	0.699	0.885	1.120
2^7	0.118	0.157	0.209	0.276	0.364	0.479	0.630	0.826	1.081
2^8	0.086	0.119	0.164	0.224	0.306	0.417	0.568	0.771	1.045
2^9	0.063	0.090	0.128	0.182	0.257	0.363	0.511	0.719	1.009
2^{10}	0.046	0.068	0.101	0.148	0.217	0.316	0.461	0.671	0.975
2^{11}	0.034	0.052	0.079	0.120	0.182	0.275	0.415	0.626	0.941
2^{12}	0.025	0.039	0.062	0.098	0.153	0.240	0.374	0.584	0.909
2^{13}	0.018	0.030	0.049	0.079	0.129	0.209	0.337	0.545	0.878
2^{14}	0.013	0.023	0.038	0.064	0.108	0.182	0.304	0.508	0.849
2^{15}	0.010	0.017	0.030	0.052	0.091	0.158	0.274	0.474	0.820
2^{16}	0.007	0.013	0.024	0.042	0.077	0.138	0.247	0.443	0.792
2^{17}	0.005	0.010	0.018	0.034	0.064	0.120	0.223	0.413	0.765
2^{18}	0.004	0.007	0.014	0.028	0.054	0.104	0.201	0.385	0.739
2^{19}	0.003	0.006	0.011	0.023	0.046	0.091	0.181	0.359	0.714
2^{20}	0.002	0.004	0.009	0.018	0.038	0.079	0.163	0.335	0.689
2^{21}	0.001	0.003	0.007	0.015	0.032	0.069	0.147	0.313	0.666
2^{22}	0.001	0.002	0.005	0.012	0.027	0.060	0.132	0.292	0.643
2^{23}	8.−e4	0.002	0.004	0.010	0.023	0.052	0.119	0.273	0.621
2^{24}	6−e4	0.001	0.003	0.008	0.019	0.045	0.108	0.255	0.601
2^{25}	5−e4	0.001	0.003	0.007	0.016	0.039	0.099	0.242	0.578
2^{26}	4−e4	8−e4	0.002	0.005	0.014	0.036	0.085	0.224	0.509

A Matrix Algorithm for Mining Association Rules*

Yubo Yuan and Tingzhu Huang

School of Applied Mathematics,
University of Electronic Science and Technology of China,
Chengdu, Sichuan, 610054, P.R. China
ybyuan@uestc.edu.cn

Abstract. Finding association rules is an important data mining problem and can be derived based on mining large frequent candidate sets. In this paper, a new algorithm for efficient generating large frequent candidate sets is proposed, which is called Matrix Algorithm. The algorithm generates a matrix which entries 1 or 0 by passing over the cruel database only once, and then the frequent candidate sets are obtained from the resulting matrix. Finally association rules are mined from the frequent candidate sets. Numerical experiments and comparison with the Apriori Algorithm are made on 4 randomly generated test problems with small, middle and large sizes. Experiments results confirm that the proposed algorithm is more effective than Apriori Algorithm.

Keywords: data mining; association rules; matrix; algorithm.

1 Introduction

Data mining is a generic term which covers research techniques and tools used to extract useful information from large databases. It has been recognized as a new area for database research.

Discovering association rules from databases is an important data mining problem[1]. Lots of researches on mining association rules focus on discovering the relationships among items in the transaction database. Many of them pay attention on designing efficient algorithms for mining association rules.

The problem of mining association rules from a transaction database can be decomposed into two subproblems. Let the support of an itemset Z be the ratio of the number of transactions containing itemset Z and the total number of transactions in the database. First, all itemsets whose supports are not less than the user-specified minimum support are identified. Each such itemset is referred to as a large itemset. Second, the association rules whose confidences are not less than the user-specified minimum confidence are generated from these large itemsets. For example, let Z be a large itemset. The confidence of a rule

---

* Supported by the Youth Key Foundations of Univ. of Electronic Science and Technology of China (Jx04042).

$X \Rightarrow Y$ is the ratio of the supports of itemset $X \cup Y$ and itemset X. All rules of the form $X \Rightarrow Y$ satisfying $X \cup Y = Z, X \cap Y = \emptyset$ and the minimum confidence constraint are generated. Once all large itemsets are discovered, the desired association rules can be obtained in a straightforward manner.

An algorithm for finding all association rules, referred to as the AIS algorithm, was first explored in [2]. The AIS algorithm requires to repeatedly scan the database. It uses the large itemsets discovered in the previous pass as the basis to generate new potentially large itemsets, called candidate itemsets, and counts their supports during the pass over the data. Specifically, after reading a transaction, it is determined which of the large itemsets found in the previous pass are contained in the transaction. New candidate itemsets are generated by extending these large itemsets with other items in the transaction. However, the performance study in [3] shows that AIS is not efficient since it generates too many candidate itemsets that are found not to be large itemsets finally.

In [3], Apriori Algorithm was proposed for efficiently mining association rules. Different from the AIS algorithm, the algorithm generate the candidate itemsets by using only the large itemsets found in the previous pass. For example, at the $(k-1)th$ iteration, all large itemsets containing $(k-1)$ items, called large $(k-1)$-itemsets, are generated. In the next iteration, the candidate itemsets containing k items are generated by joining large $(k-1)$-itemsets. The heuristic is that any subset of a large itemset must be large. By the heuristic, Apriori Algorithm can generate a much smaller number of candidate itemsets than AIS algorithm.

Another effective algorithm for the candidate set generation, called DHP, was proposed in [4]. By utilizing a hash technique, DHP can efficiently generate all the large itemsets and effectively reduce the size of the transaction database.The performance study in [4] shows that the number of candidate 2-itemsets generated by DHP is smaller than that by the a priori algorithm. Moreover, the transaction database size is trimmed at a much earlier stage of the iterations. As a result, the total execution time can be reduced significantly by DHP.

In [5], the PSI and PSI_{seq} algorithms were proposed for efficient large itemsets generation and large sequences generation, respectively. The two algorithms use prestored information to minimize the numbers of candidate itemsets and candidate sequences counted in each database scan.The prestored informations for PSI and PSI_{seq} include the itemsets and the sequences along with their support counts found in the last mining, respectively. Empirical results show that only using little storage space for the prestored information, the total computation time can be reduced effectively. In [6], an efficient graph-based algorithm to discover interesting association rules embedded in the transaction database and the customer database is presented.

The notion of mining multiple-level association rules was introduced in [7]. In many applications, association rules discovered at multiple concept levels are useful.Usually, the association relationship expressed at a lower concept level provides more specific information than that expressed at a higher concept level. The approach is to first find large items at the top-most level, and then progressively deepen the mining process into their descendants at lower levels. A similar idea of extracting generalized association rules using a taxonomy was presented in [8].

The issue of mining quantitative association rules in large relational databases was investigated in [9]. The attributes considered in a relation are quantitative or categorical.The values of the attribute are partitioned using an *equi−depth* approach (that is, each interval resulted from the partition contains roughly the same number of tuples), and then adjacent intervals are combined as necessary. A related problem is clustering association rules [10], which combines similar *adjacent* association rules to form a few general rules. Ref.[10] proposed a geometric-based algorithm to perform the clustering and applied the minimum description length principle as a means of evaluating clusters.

There are lots of papers mentioned on algorithms for mining association rules, such as [11-16].

In this paper, we will propose a new algorithm for finding association rules. The proposed algorithm only needs to pass over the database once. By this pass over the database with given transactions, we can get a matrix which only entries 0 and 1. Then, the frequent itemsets can be obtained from the operations of rows on this matrix, (see in section 3).

The rest of these paper is organized as follows: In section 2, we give a review of Apriori Algorithm for association rule mining and present the limitations of Apriori algorithm. In section 3, we present the new proposed algorithm for finding association rules, called Matrix Algorithm(MA) . In section 4, we discuss the concrete experiments using the new algorithm. Conclusions are given in section 5.

2 Apriori Algorithm

In this section, we first give a review of the Apriori algorithm[3] for finding association rules from large database. We mainly pay attention on the characteristic of multiple passes of this algorithm.

The Apriori algorithm is well-known([6]) and makes multiple passes over the database to find frequent itemsets. In the kth pass, the algorithm finds all frequent k-itemsets, consisting of k items, for example $\{i_{l_1}, i_{l_2}, \ldots, i_{l_k}\}$ is a frequent itemset if it's supporting number is beyond the minsupport.

Each pass consists of two phases: the candidate generation phase and the support counting phase. The candidate generation phase obtains C_k from L_{k-1}, where L_{k-1} is the set of all the frequent $(k-1)$−itemsets, C_k is the set of candidate k−itemsets. The support counting phase finds the frequent k−itemsets L_k from C_k. The algorithm terminates once either L_k or C_k is empty.

Given L_{k-1}, the set of all frequent $(k-1)$−itemsets, all candidate k−itemsets we is generated.

Candidate generation takes two steps as follows:

(a) Joint step: A superset C_k is created by linking L_{k-1} with itself, which can be done effectively when items and itemsets are lexicographically ordered.

(b) Prune step: Let c be a k-itemset, and $c \in C_k$. If c has some $(k-1)$–subset not in L_{k-1}, c will be deleted from C_k.

The Apriori algorithm generates all frequent itemsets after the support accounting and makes multiple passes over a given database.

The following example is used to show the characteristic of multiple passes of the Apriori algorithm.

Example 1. Let the item set $I = \{A, B, C, D, E, F\}$, the transactions be $t_1 = \{A, B, C\}, t_2 = \{A, C, D\}, t_3 = \{A, B, E, F\}, t_4 = \{A, B, C, D, F\}, t_5 = \{B, C, D, E\}$ and the minsupport=50%,(see Table 1).

Firstly, the 1-itemsets $\{A\}, \{B\}, \{C\}, \{D\}, \{E\}$ and $\{F\}$ are generated as candidates at the first pass over the dataset. It can be obtained from Table 1 that $\{A\}$.count = 4, $\{B\}$.count =4, $\{C\}$.count =4, $\{D\}$.count =3, $\{E\}$.count =2,and $\{F\}$.count=2. Because minsupp =50% and dbsize =5, $\{A\}, \{B\}, \{C\}$ and $\{D\}$ are frequent itemsets. The results are given in Table 2.

Secondly, the frequent 1-itemsets L_1 consists of $\{A\}, \{B\}, \{C\}$, and $\{D\}$. The second pass begins over the dataset to search for 2-itemset candidates. The candidate 2–itemset C_2 consists of $\{A, B\}, \{A, C\}, \{A, D\}, \{B, C\}, \{B, D\}$, and $\{C, D\}$, and, $\{A, B\}$.count = 3, $\{A, C\}$.count = 3 ,$\{A, D\}$.count = 2, $\{B, C\}$.count = 3 ,$\{B, D\}$.count = 2, $\{C, D\}$.count = 3 , then it follows that $\{A, B\}, \{A, C\}, \{B, C\}$ are frequent 2-itemsets.

Table 1. Transaction Database

Transaction ID	Transaction Set
t_1	$\{A, B, C\}$
t_2	$\{A, C, D\}$
t_3	$\{A, B, E, F\}$
t_4	$\{A, B, C, D, F\}$
t_5	$\{B, C, D, E\}$

Table 2. Frequent itemsets

Frequent itemsets	Itemsets	supp	>minsupport
Frequent 1-itemsets	$\{A\}$	4	yes
Frequent 1-itemsets	$\{B\}$	4	yes
Frequent 1-itemsets	$\{C\}$	4	yes
Frequent 1-itemsets	$\{D\}$	3	yes
Frequent 2-itemsets	$\{A, B\}$	3	yes
Frequent 2-itemsets	$\{A, C\}$	3	yes
Frequent 2-itemsets	$\{B, C\}$	3	yes
Frequent 2-itemsets	$\{C, D\}$	3	yes
Frequent 3-itemsets	$\{A, B, C\}$	2	no

The third pass begins over the dataset to search for 3- itemset candidates. In this pass, the linking operation ⋈ is used to generate the candidate sets. Frequent 3-itemset is $\{A, B, C\}$.

There is no frequent 4-itemset because the candidate set for generating the frequent 4-itemsets is empty. The algorithm is ended. Then the frequent itemsets are $L = \{\{A\}, \{B\}, \{C\}, \{D\}, \{A, B\}, \{A, C\}, \{B, C\}\}$.

For large database, for example databases with scale beyond 10^{12} bytes, the algorithm will take more time to have a pass over the database. The Apriori algorithm need k pass to generate frequent itemsets L_k. For a given large database, the Apriori algorithm used for identifying frequent itemsets, involves a search with little heuristic information in a space with an exponential amount of items and possible itemsets. These characteristics algorithm may be inefficient when the number of frequent itemsets is large.

In this paper, we mainly want to develop algorithm to overcome the first limitation. In next section, we will introduce a new algorithm to do this.

3 Generating Matrix and Matrix Algorithm

The Apriori algorithm needs to make multiple passes over the database. This makes the algorithm somewhat inefficient in practice. In this section we will give a new algorithm which only requires one pass over the database and generate an matrix, the rest work only need use inner product operator.

3.1 Generating Matrix

We first generate a matrix which only entries either 0 or 1 by passing over the database once. We call it generating matrix. Then, we can derive association rules from the matrix. The process of generating the matrix is as follows: We set the items in I as columns and the transactions as rows of the matrix.

Let the set of items be $I = \{i_1, i_2, \cdots, i_n\}$, and the set of the transactions be $D = \{t_1, t_2, \cdots, t_m\}$. Then the generating matrix $G = \{g_{ij}\}, (i = 1, 2, \cdots, m; j = 1, 2, \cdots, n)$ is an $m \times n$ matrix, where $g_{ij} = 0$ or 1 is determined by the following rule,

$$g_{ij} = \begin{cases} 1, \text{ if } i_j \in t_i, \\ 0, \text{ if } i_j \notin t_i. \end{cases}$$

Example 2. Let $I = \{A, B, C, D, E, F\}$, and the transactions be $t_1 = \{A, B, C\}$, $t_2 = \{A, C, D\}, t_3 = \{A, B, E, F\}, t_4 = \{A, B, C, D, F\}, t_5 = \{B, C, D, E\}$. Then the generating matrix is

$$G = \begin{pmatrix} 1 & 1 & 1 & 0 & 0 & 0 \\ 1 & 0 & 1 & 1 & 0 & 0 \\ 1 & 1 & 0 & 0 & 1 & 1 \\ 1 & 1 & 1 & 1 & 0 & 1 \\ 0 & 1 & 1 & 1 & 1 & 0 \end{pmatrix} \equiv \begin{pmatrix} g_1 \\ g_2 \\ g_3 \\ g_4 \\ g_5 \end{pmatrix}.$$

With this generating matrix, we can find frequent itemsets and generate association rules.

3.2 Matrix Algorithm

The candidate k-itemset is generated using the following process:

(a) The 1-itemset C_1 consists of the sets which are subsets of single item in I, that is, $C_1 = \{\{i_1\}, \{i_2\}, \cdots, \{i_n\}\}$.

In order to compute the support number for each set in C_1, we express every set in C_1 as a row vector in R^n, that is, we express $\{i_1\}$ as $S_1^1 = \{1, 0, \cdots, 0\}$, and $\{i_k\}$ as $S_k^1 = \{0, 0, \cdots, 1, \cdots, 0\}$, the kth element is 1 and others are 0. Then the support number of the set $\{i_k\}$ is calculated by :

$$supp(\{i_k\}) = \sum_{j=1}^{m} <g_j, S_k^1>, \qquad (1)$$

where $<\ ,\ >$ is the inner product of two row vectors and $g_j, j = 1, 2, ..., m$ are rows of the matrix G. For example, let $g_1 = \{1, 0, 1\}$. Then if $S_1^1 = \{1, 0, 0\}$, $<g_1, S_1^1> = 1$ and if $S_2^1 = \{0, 1, 0\}$, then $<g_1, S_2^1> = 0$. Then the set of all the frequent 1-itemsets, L_1, is generated from C_1. If the support number of $\{i_k\}$ is beyond the user-specified threshold $Minsupport$, that is,

$$supp(\{i_k\}) \geq Minsupport, \qquad (2)$$

then $\{i_k\} \in L_1$.

(b) The set of candidate 2-itemsets C_2 is the joint set of L_1 with itself, that is, $C_2 = L_1 \bowtie L_1$. Each subset in C_2 consists of two items and has the form $\{i_k, i_j\}, k < j$.

Similarly, we specify each set in C_2 a row vector in R^n. For example, for the set $\{i_k, i_j\}$, the specified vector is $S_{k,j}^2 = \{0, \cdots, 0, 1, 0, \cdots, 0, 1, 0, \cdots, 0\}$, where the kth and jth elements are 1 and others are 0. The support number of the set $\{i_k, i_j\}$ is :

$$supp(\{i_k, i_j\}) = \sum_{s=1}^{m} int[\frac{<g_s, S_{i,k}^2>}{2}], \qquad (3)$$

where $int[\cdot]$ is the integrating function that changes a real number to integer by discarding the number after decimal point. For example, $int[0.6] = 0$, and $int[2.3] = 2$.

The frequent itemset L_2 is generated from C_2 with the set whose support number is beyond the user specified threshold $Minsupport$, that is, if

$$supp(\{i_k, i_j\}) \geq Minsupport \qquad (4)$$

then $\{i_k, i_j\} \in L_2$.

After the frequent 2-itemsets L_2 is obtained, it can be used to generate C_3, for example, if $\{i_k, i_j\}, \{i_k, i_l\} \in L_2, k < j, k < l, j < l$, then $\{i_k, i_j\} \bowtie \{i_k, i_l\} = \{i_k, i_j, i_l\}$

(c) Repeat the above process with successively increasing number k until either C_k or L_k is empty, where each subset in C_k has the form $\{i_{l_1}, i_{l_2}, \cdots, i_{l_k}\}$ including k items, and is generated from the frequent $(k-1)$–itemsets L_{k-1}, and L_k is the frequent k–itemsets generated from C_k with the set whose support number is beyond the user specified threshold $Minsupport$. The support number can compute according to the following formula

$$supp(\{i_{l_1}, i_{l_2}, \cdots, i_{l_k}\}) = \sum_{j=1}^{m} int[\frac{<g_j, S^k_{l_1,l_2,\cdots,l_k}>}{k}]. \qquad (5)$$

where $S^k_{l_1,l_2,l_3,\cdots,l_k} = \{0, \cdots, 0, 1, 0, \cdots, 0, 1, 0, \cdots, 0\}$, and it's l_1th, l_2th, $\cdots$ and l_kth elements are 1 and others are 0.

At the end of procedure, we can get the all frequent itemsets by the following formula. Let the procedure is terminated after step k, then

$$L = \bigcup_{i=1}^{k-1} L_i. \qquad (6)$$

Throughout the process above only one pass over the database is used to generate the frequent itemsets, and hence the algorithm is called matrix algorithm.

We would like to say a few more words about this process from L_{k-1} generating C_k. The following three aspects are important when we implement experiments on the algorithm: (i)the generation of the generating matrix for given database; (ii)the joint operation $\bowtie$; (iii)generation of the frequent itemsets. At the end of the algorithm, we get the frequent itemset L. In the next subsection, we will derive the association rules from the frequent itemsets L.

3.3 Generating Association Rules

In this section, we derive all association rules

$$\{X \Rightarrow Y - X\}$$

with support s and confidence c, where X, Y are frequent itemsets and $X \subset Y$; $X, Y \in L$.

The $Minconfidence$ and the $Minsupport$ are specified by users or experts. The confidence of rule $X \Rightarrow Y - X$ is computed by the following statement,

$$conf(X \Rightarrow Y - X) = \frac{supp(Y)}{supp(X)}. \qquad (7)$$

The method of generating association rules is to select all X, Y from L, and $X \subset Y$, then compute $conf(X \Rightarrow Y - X)$. If $conf(X \Rightarrow Y - X)$ is beyond the $Minsupport$, $X \Rightarrow Y - X$ is an association rule.

4 Experiment Results

To evaluate the performance of the proposed Matrix algorithm, we performed extensive simulation experiments. Our goals are to study the amount of cpu times to generate frequent itemsets. We compare the performance of the proposed algorithm with the Apriori algorithm. In this section we present the experimental results with the times of finding frequent itemsets.

The experiments are implemented on 4 randomly generated databases with normal distribution and on a PC with 1.8G MHz Pentium IV and 256 MB SDRAM using MATLAB 6.1.

In Table 1, we give out the database scale. n is the number of items in I. m is the number of transactions in D.

In Table 4, is the comparison result between The Apriori Algorithm and our Proposed Matrix Algorithm with cpu times to generate frequent itemsets.

The following conclusion can be obtained from the results in Table 4.

For the small size problem, for example, size is smaller than one hundred, the difference between the Apriori Algorithm and the Matrix Algorithm is not obvious. But for large or huge size problems, for example, size is larger than one thousand, the difference is clear.

For generation of the 1-frequent itemsets, the Matrix Algorithm takes more time than the Apriori Algorithm. But, for generation of the k-frequent ($k \geq 2$) itemsets, the Matrix Algorithm does it more efficiently than the Apriori Algorithm does. We showed this result in Figure 1 about the database 2 with size 5000 and Figure 2 about the database 4 with size 500000.

For generation of association rules, there is no difference between the Matrix Algorithm and the Apriori Algorithm and hence they have the same times in generating association rules from frequent itemsets L, and hence we do not give these results.

Table 3. The database scale

Dataset	n	m
D1	20	500
D2	100	5000
D3	1000	50000
D4	10000	500000

Table 4. Experimental results

Dataset	Apriori	Matrix
D1	0.02sec	0.015sec
D2	4.13ecs	0.845sec
D3	76.4sec	8.24sec
D4	179.8sec	22.67sec

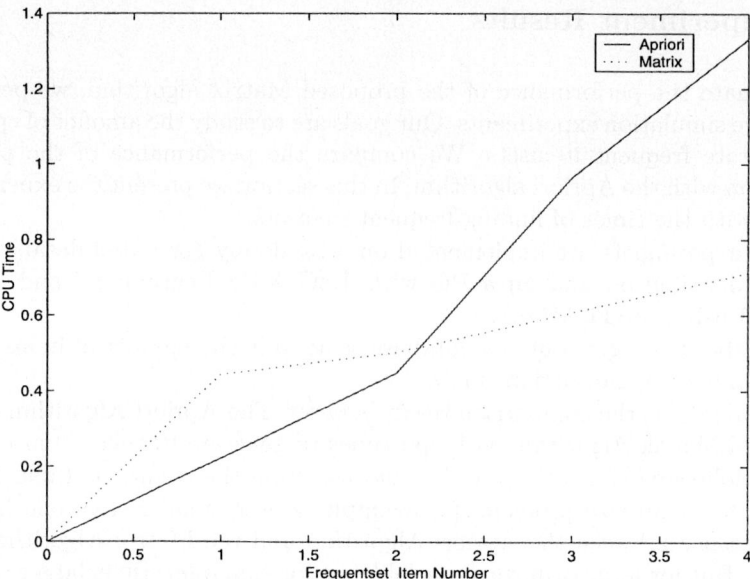

Fig. 1. Compare Apriori Algorithm and Matrix Algorithm with CPU time for D2

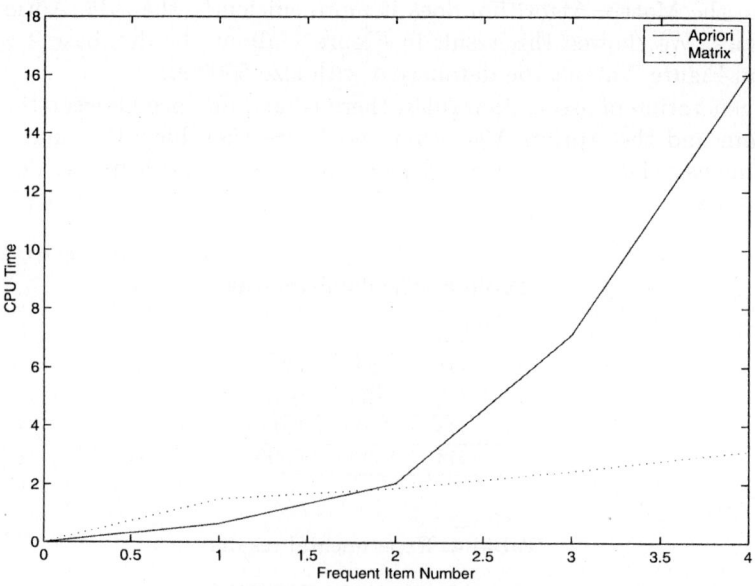

Fig. 2. Compare Apriori Algorithm and Matrix Algorithm with CPU time for D4

5 Conclusions

In this paper, a new algorithm is proposed for association rules finding from databases. The algorithm takes matrix which only entries either 1 or 0. It is generated from transaction databases. Then, the frequent itemsets are generated from the resulting matrix and the association rules can be derived from the resulting frequent itemsets. Comparing Numerical experiments results obtained by Matrix Algorithm with Apriori Algorithm, the proposed algorithm is efficient and robust.

References

1. Agrawal, R., Imielinski, T., Swami,A.: Database mining: A performance perspective. IEEE Trans. Knowledge and Data Eng., 5(6) (1993): 914-925.
2. Agrawal, R., Imielinski, T., Swami, A.: Mining association rules between sets of items in large databases. In: Proceedings of the ACM SIGMOD Conference on Management of Data, 1993: 207-216.
3. Agrawal, R., Srikant, R.: Fast algorithms for mining association rules. In: Proceedings of International Conference on Very Large Data Bases, 1994: 487-499.
4. Park, J.S., Chen, M.S., Yu, P.S.: Using a hash-based method with transaction trimming for mining association rules, IEEE Trans. on Knowledge Data Engrg. 9(5)(1997): 813-825.
5. Pauray S.M. Tsai, Chien-Ming Chen: Discovering knowledge from large databases using prestored information. Information Systems. 26(1)(2001): 3-16.
6. Pauray S.M. Tsai, Chien-Ming Chen: Mining interesting association rules from customer databases and transaction databases. Information Systems. 29(3)(2004):685-696.
7. Han, J., Fu, Y.: Discovery of multiple-level association rules from large databases. Proceedings of the VLDB Conference. 1995: 420-431.
8. Srikant, R., Agrawal, R.: Mining generalized association rules. Proceedings of the VLDB Conference. 1995: 407-419.
9. Srikant, R., Agrawal, R.: Mining quantitative association rules in large relational tables. Proceedings of the ACM SIGMOD. 1996:1-12.
10. Lent, B., Swami, A., Widom, J.:Clustering association rules. Proceedings of the IEEE International Conference on Data Engineering. 1997:220-231.
11. Agrawal, R., Shafer, J.: Parallel mining of association rules. IEEE Trans. on Knowledge and Data Engg., 8(6) (1996): 962-969.
12. Hipp, J., Gontzer, U., Nakhaeizadeh, G.: Algorithms for association rule mining – a general survey and comparison. SIGKDD Explorations, 2(1) (2000): 58-64.
13. Berzal, F., Cubero, J.C., Marrin, N., Serrano, J.M.: TBAR: An efficient method for association rules mining in relational databases. Data and Knowledge Engineering 37(2001):47-64.
14. John D. Holt, Soon M. Chung: Mining association rules using inverted hashing and pruning,Information Processing Letters, 83(2002): 211-220.
15. Ping-Yu Hsu, en-Liang Chen, Chun-Ching Ling: Algorithms for mining association rules in bagdatabases. Information Science. 166(1)(2004):31-47.
16. Shichao Zhang, Jingli Lu, Chengqi Zhang: A fuzzy logic based method to acquireuser threshold of minimum-support for mining association rules. Information Science. 164(1)(2004):1-16.

A Hybrid Algorithm Based on PSO and Simulated Annealing and Its Applications for Partner Selection in Virtual Enterprise

Fuqing Zhao[1], Qiuyu Zhang[1], Dongmei Yu[1], Xuhui Chen[1],
and Yahong Yang[2]

[1] School of Computer and Communication, Lanzhou University of Technology,
730050 Lanzhou, P.R. China
{zhaofq, zhangqy, yudm, xhchen}@mail2.lut.cn
[2] College of Civil Engineering, Lanzhou University of Techchnology,
730050 Lanzhou, P.R. China
yangyahong@mail2.lut.cn

Abstract. Partner selection is a very popular problem in the research of virtual organization and supply chain management, the key step in the formation of virtual enterprise is the decision making on partner selection. In this paper, a activity network based multi-objective partner selection model is put forward. Then a new heuristic algorithm based on particle swarm optimization(PSO) and simulated annealing(SA) is proposed to solve the multi-objective problem. PSO employs a collaborative population-based search, which is inspired by the social behavior of bird flocking. It combines local search(by self experience) and global search(by neighboring experience), possessing high search efficiency. SA employs certain probability to avoid becoming trapped in a local optimum and the search process can be controlled by the cooling schedule. The hybrid algorithm combines the high speed of PSO with the powerful ability to avoid being trapped in local minimum of SA. We compare the hybrid algorithm to both the standard PSO and SA models, the simulation results show that the proposed model and algorithm are effective.

1 Introduction

In order to adapt to globalization competition, new production model must be researched. The network alliance enterprise is a try and a research on how to adjust to the globalization competition for traditional enterprises[1]. As the VE environment continues to grow in size and complexity, the importance of managing such complexity increases. Once the dynamic alliance is to be established, how to select an appropriate partner becomes the key problem and has attracted much research attention recently[2–4]. For the partner selection problem, in addition to cost, due date and the precedence of sub-project, the risk of failure of the project is another important factor need to be considered. Therefore, an effective approach that can actually deal with the risk-based partner selection problem is a major concern in VEs. Qualitative analysis methods are commonly used to deal with the partner selection problem in many researches[5,6]. However, quantitative analysis methods for partner selection are still a challenge to VEs. Talluri and Baker [2] proposed a two-phase mathematical programming approach for partner selection by designing a VE where the factors of cost, time and distance were considered [4]. However, the precedence of sub-project and the risk factor, which are also important for partner selection, were not considered in their paper. In fact, the sub-projects contracted by partners compose an activity network with precedence [9,10]. From this point of view, the problem is considered as a partner selection problem embedded within project scheduling [11] and cannot easily be solved by general mathematical programming methods. Therefore, there is a need to formulate mathematical models and propose optimization methods for VEs to make decision on partner selection.

2 Problem Formulation of Partner Selection

The partner selection problem can be described as follows:

Assume an enterprise wins a bid for a large project consisting of several sub-projects. The enterprise is not able to complete the whole project using its own capacity. Therefore, it has to call tenderers for the sub-projects. It does so after it determines the upper bound of the payment for each sub-project first. The tenderers who can accept the payment condition will respond to the sub-project and propose the probability of success and time they need to finish the sub-project according to the resources they have. Then, the enterprise selects one tenderer for each sub-project to maximize the success of the project.

The project consists of n sub-projects. Because of the connected relationship between these sub-projects, they form an activity network. If sub-project j can only begin after the completion of sub-project i, we note the connected sub-project pair by $(i, j) \in H,$. Here H is the set of all connected sub-project pairs. Without loss of generality, the final sub-project is noted as sub-project n. Its completion time d_n is defined as the completion time of the project.

Each partner has different processing times and risk of failure, which is described by fail probability. For sub-project $i, i = 1, 2, \cdots, n$. There are m_i candidates responding to the tenderee invitation. For the candidate j of sub-project i, its fail probability is p_{ij} and its processing time is q_{ij} periods. The objective is to select the optimal combination of partner enterprises for all sub-projects in order to minimize the risk of the project including the risk of failure and the risk of tardiness. This is the same as the objective, which is to select the optimal combination of partner enterprises for all sub-projects to maximize the success of the project, including the success probability and finishing the project within the due date.

Defining the variables:

$$x_{ij}(t) \begin{cases} 1 & \text{job } i \text{ is contracted to candidate } j \text{ at period } t; \\ 0 & \text{otherwise.} \end{cases}$$

Then the problem can be modeled as follows:

$$\max_{w} F_s(w) = 1 - \prod_{i=1}^{n}\sum_{j=1}^{m_i}\sum_{t=1}^{d} x_{ij}(t)(1-p_{ij})c_{ij} + S_L(1-[[d_n(w)-D]^+]^-) \quad (1)$$

$$\max_{x} F_t = \sum_{i=1}^{n}\sum_{j=1}^{m_i}\sum_{t=1}^{d_n} x_{ij}(1-p_{ij})c_{ij} + \beta[d-D]^+ \quad (2)$$

$$\max_{i=1\cdots n}\{\sum_{j=1}^{m_n}\sum_{t=1}^{d_n}(t+q_{nj})x_{nj}(t)\} = d_n \quad (3)$$

$$(t+q_{ij})\sum_{j=1}^{m_i}\sum_{t=1}^{d_n} x_{ij}(t) \le t\sum_{j=1}^{m_k}\sum_{t=1}^{d_n} x_{kj}(t), \quad t=1,2,\ldots,d_n \quad \forall(i,k)\in H \quad (4)$$

$$x_{ij}(t) = 1 \text{ or } 0 \quad \forall i,j,t, \quad (5)$$

where $s_{ij} = 1 - p_{ij}$, is the probability of success for the candidate j of sub-project i; $S_L = \prod_{i=1}^{n} s_{im_i}$, $[w]^+$ stands for $\max\{0,w\}$; $[y]^-$ stands for $\min\{1,y\}$. D is the due date of the project.

The objective of the problem is a nonlinear one and it is easy to see that the formulated problem is not convex. Also the objective function is not continuous and differentiable, the model cannot be solved by general mathematical programming methods. Therefore, we attempt to introduce a hybrid algorithm which is based on PSO and SA for the problem.

3 The Hybrid Algorithm Based on PSO and SA

3.1 Fundamental Principle of PSO

PSO simulates a social behavior such as bird flocking to a promising position for certain objectives in a multidimensional space [12,13]. Like evolutionary algorithm, PSO conducts search using a population (called swarm) of individuals (called particles) that are updated from iteration to iteration. Each particle represents a candidate position (i.e., solution) to the problem at hand, resembling the chromosome of GA. A particle is treated as a point in an M-dimension space, and the status of a particle is characterized by its position and velocity [14]. Initialized with a swarm of random particles, PSO is achieved through particle flying along the trajectory that will be adjusted based on the best experience or position of the one particle (called local best) and the best experience or position ever found by all particles (called global best). The M-dimension position for the i th particle in the t th iteration can be denoted as $x_i(t) = \{x_{i1}(t), x_{i2}(t), \cdots, x_{iM}(t)\}$. Similarly, the velocity (i.e., distance change), also an M-dimension vector, for the i th particle in the t th iteration can be described as $v_i(t) = \{v_{i1}(t), v_{i2}(t), \cdots, v_{iM}(t)\}$. The particle-updating mechanism for particle flying (i.e.,search process) can be described as:

Initialize population
Do
  for i=1 to Population Size
  if $f(x_i) < f(p_i)$ then $p_i = x_i$
  $p_g = \min(p_i)$
    for d=1 to Dimension Space Size

$$v_{id} = \chi(wv_{id} + c_1 r_1 (p_{id} - x_{id}) + c_2 r_2 (p_{gd} - x_{id})) \qquad (6)$$

$$\text{if } (v_{id} > v_{max}) \text{ then } v_{id} = v_{d\,max} \qquad (7)$$

$$\text{if } (v_{id} < -v_{max}) \text{ then } v_{id} = -v_{d\,max} \qquad (8)$$

$$x_{id} = x_{id} + v_{id} \qquad (9)$$

$$\text{if } x_{id} > x_{max} \text{ then } x_{id} = x_{d\,max} \qquad (10)$$

$$\text{if } x_{id} < -x_{max} \text{ then } x_{id} = -x_{d\,max} \qquad (11)$$

    Next d
  Next i
Until termination condition is met
Where
p_i Pbest of agent i at iteration k;
p_g gbest of the whole group
χ compress factor
r_1, r_2 random numbers in (0,1)
c_1, c_2 weighting factor
ϖ inertia function, in this paper, the inertia weight is set to the following equation

$$\varpi = \varpi_{max} - \frac{\varpi_{max} - \varpi_{min}}{I_{max}} \times I.$$

    where ϖ_{max} Initial value of weighting coefficient
ϖ_{min} Final value of weighting coefficient
I_{max} Maximum number of iterations or generation
I current iteration or generation number
  Fig.1 shows the above concept of modification of searching points.

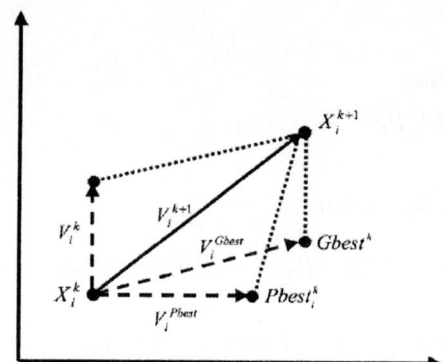

Fig. 1. The Mechanism of Particle swarm optimization

3.2 Hybrid PSO Algorithm

PSO algorithm is problem-independent, which means little specific knowledge relevant to a given problem is required. What we have to know is just the fitness evaluation for each solution. This advantage makes PSO more robust than many other search algorithms. However, as a stochastic search algorithm, PSO is prone to lack global search ability at the end of a run. PSO may fail to find the required optima in case when the problem to be solved is too complicated and complex. SA employs certain probability to avoid becoming trapped in a local optimal and the search process can be controlled by the cooling schedule of SA, we can control the search process and avoid individuals being trapped in local optimum more efficiently. Thus, a hybrid algorithm of PSO and SA, named HPSO, is presented as follows.

Begin
  STEP 1 Initialization
    Initialize swarm population, each particle's position and velocity;
    Evaluate each particle's fitness;
    Initialize gbest position with the lowest fitness particle in swarm;
    Initialize pbest position with a copy of particle itself;
    Initialize $\varpi_{max}, \varpi_{min}, c_1, c_2$, maximum generation, and generation=0.
    Determine T_0, T_{end}, B.
  STEP 2 Operation
    For PSO
    do {
      generate next swarm by Eq.(6) to Eq.(11);
      find new gbest and pbest;
      update gbest of swarm and pbest of the particle;
      generation++;
    }
    while (generation<maximum generation)
    For SA[15]

For gbest particle S of swarm
{ $T_k = T_0$
do {
   generate a neighbor solution S' from S;
   calculate fitness of S';
   Evaluate S' {
      $\Delta = f(S') - f(S)$;
      if ($\min[1, \exp(-\Delta / T_k)] > random[0,1]$)
         Accept S';
         Update the best solution found so far if possible;
      }
      $T_k = BT_{k-1}$;
   }
}
while ($T_k > T_{end}$)
STEP 3   Output optimization results.
END

From the chart, we can see that PSO provides initial solution for SA during the hybrid search process. Such hybrid algorithm can be converted to tradition SA by setting swarm size to one particle. HPSO implements easily and reserves the generality of PSO and SA. Moreover, such HPSO can be applied to many combinatorial optimization problems by simple modification.

4   Experiments Analysis

4.1   The Example

The example involves the real life problem of machine center , which bid for a project for the petroleum drilling machine . The project consists of 12 sub-projects, its due date is 48 months. The precedence relationship represented by the mode shown in Fig. 2.

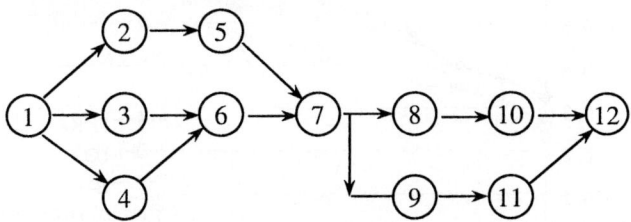

Fig. 2. The network of the example

The achieved solution is shown in Table 1. Its success probability is 0.621 and the completion time is 46 and is just in time. Compared with the solution obtained by the enumeration algorithm (Table 3), it can be seen that the solution is an optimal one and that it uses less time than the PSO and SA.

The generation process is shown in Fig. 3. From this figure, it can be found that the evolution process of the HPSO tends to be stable when the generation reaches more than 60.

Table 1. The job list and success probability and processing time

Job No.	Contractor	Succeed probability	Processing time	Begin time	Completion time
1	3	0.85	5	0	3
2	4	0.87	5	0	8
3	5	0.94	12	9	16
4	2	0.86	8	5	15
5	2	0.70	13	8	28
6	6	0.98	10	10	18
7	2	0.75	7	18	22
8	5	0.93	15	19	34
9	3	0.98	9	6	40
10	4	0.92	15	40	36
11	6	0.98	19	36	42
12	3	0.88	5	42	46

Table 2. The comparison of different algorithm

Best rate(100%)	HPSO result	PSO result	SA result
Best	0.621	0.621	0.621
Mean	0.575	0.432	0.405
Minimum	0.499	0.213	0.159
CPU time	28.00"	49.00"	1529.0"

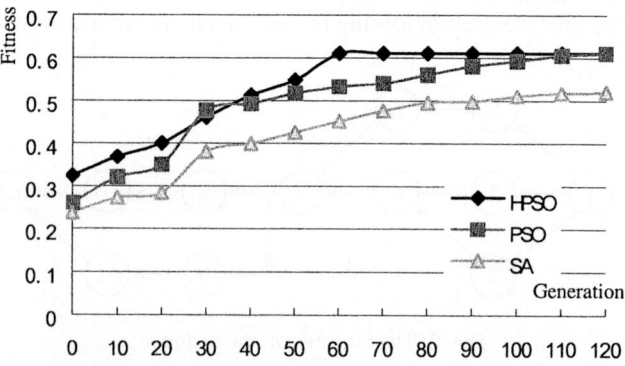

Fig. 3. The generation process of the different algorithm

Table 3. Comparison of the result of different problems with different sizes

N	Size	Alg.	CPU time	Best rate(%)	Best	Mean	Minimum
2	2.687×10^{30}	PSO	75"	87	0.067	0.036	0.002
		HPSO	32"	100	0.067	0.050	0.002
		SA	31456"	100	0.067	0.057	0.057
3	1.254×10^{36}	PSO	133"	68	0.253	0.161	0.007
		HPSO	63"	100	0.253	0.222	0.010
		SA	--				
4	4.333×10^{41}	PSO	202"	65	0.116	0.070	0.002
		HPSO	98"	100	0.117	0.092	0.002
		SA	--				
5	3.370×10^{46}	PSO	256"	46	0.064	0.037	0.0005
		HPSO	113"	100	0.066	0.048	0.001
		SA	--				

4.2 Performance Analysis

When considering the larger size problem, the PSO and SA usually hard to obtain the optimal solution in desired time. So the HPSO is introduced. To test the performance of the HPSO, we randomly produced some problems with different sizes. The result together with the comparison of the PSO without the embedded rule and SA are shown in Table 3. In the table, "Size" stands for the size of the solution space, "CPU time" for the computation time of each running of the CPU; "—" means that we cannot obtain a solution in an acceptable time. The "Best rate" is tested by 100 runs involving different random seeds.

From Table 3, we see that the problem size grows with the sub-project number extremely rapidly. The recommended HPSO can achieve the optimal solution with a higher probability and the computation time does not increase quickly with the size increase. The SA can guarantee the optimum solution when the problem size is small, but it will waste a huge amount of time and still cannot obtain this in an acceptable time when the problem size is larger. Although the PSO can solve the larger problem quickly, it cannot find the optimal solution generally. The comparison strongly suggests that the HPSO is able to efficiently improve the computational performance of complex combinatorial optimization problems.

6 Conclusions

Selecting partners is a key factor in the establishment phase of virtual enterprise. A VE is a dynamic alliance of member companies (tenderee and tenderers), which join to take advantage of a market opportunity. Partner selection is an inherent problem in VE. Minimizing risk in partner selection and ensuring the due date of the project are key to ensure the success of the VE. In this paper, the description of a partner selection problem is introduced. The HPSO is proposed to solve the complex combinatorial optimization problems. Compared with the PSO and SA, the HPSO has better synthetic performance in both the computation speed and optimality. The computation results suggest its potential to solve the practical partner selection and sub-project management problems.

Acknowledgements

This research is supported by Natural Science foundation of GANSU province(grant NO ZS032-B25-008) and (3ZS042-B25-005).

References

1. Davulcu, H., Kifer, M., Pokorny, L.R., Dawson. S.: Modeling and Analysis of Interactions in Virtual Enterprises. In: Proceedings of the Ninth International Workshop on Research Issues on Data Engineering: Information Technology for Virtual Enterprises. Sydney, Australia, 1999. p. 12–18
2. Talluri, S., Baker, R.C.: Quantitative Framework for Designing Efficient Business Process Alliances. In: Proceedings of 1996 International Conference on Engineering and Technology Management, Piscataway. 1996. p. 656–661
3. Gaonkar, Roshan S., Viswanadham, N.: Partner Selection and Synchronized Planning in Dynamic Manufacturing Networks. IEEE Transactions on Robotics and Automation . 2003,19(1):117-130
4. Chu, X.N., Tso, S.K., Zhang, W.J.et al.: Partnership Synthesis for Virtual Enterprises. International Journal of Advanced Manufacturing Technology 2002,19(5),384-391
5. Feng, D.Z.,Yamashiro, M.: A Pragmatic Approach for Optimal Selection of Plant-Specific Process Plans in a Virtual Enterprise. Production Planning and Control 2003,14(6):562-570
6. Brucker, P., Drexl, A., MohringR, Neumann K., Pesch, E.: Resource-Constrained Project Scheduling Notation, Classification, Models, and Methods. European Journal of Operational Research 1999;112:3–41
7. Hanne, Thomas: Global Multiobjective Optimization Using Evolutionary Algorithms. Journal of Heuristics 2000,6(3):347-360
8. Alcaraz, J., Maroto, C., Ruiz, R.: Solving the Multi-Mode Resource-Constrained Project Scheduling Problem with Genetic Algorithms. Journal of the Operational Research Society 2003,54(6):614-626
9. Xiaobing Liu, et al.: The Method and Realizing System of Selecting Enterprise Partners for Virtual Enterprises [J] . Industrial Engineering Journal,2001,4(2):10-13
10. Yongjun Ma, Shu Zhang: Selection Method for Design Partners in Network Extended Enterprises[J].Chinese Journal of Mechanical Engineering,2000,36(1):15-19
11. Wenjun Zheng, Xumei Zhang, et al.: Evaluation Architecture and Optimization Decision of Partner Choice in Virtual Enterprises[J]. Computer Integrated Manufacturing System —CIMS , 2000,6(5):63-67
12. Eberhart, R.C., Shi, Y.: Particle Swarm Optimization: Developments, Applications and Resources. Proceedings of the IEEE Conference on Evolutionary Computation, ICEC 2001, p 81-86

13. van den Bergh, Frans, Engelbrecht, Andries P.: A Cooperative Approach to Participle Swam Optimization. IEEE Transactions on Evolutionary Computation, 2004,8(3):225-239
14. Coello Coello, Carlos A., Pulido, Gregorio Toscano, Lechuga, Maximino Salazar: Handling Multiple Objectives with Particle Swarm Optimization. IEEE Transactions on Evolutionary Computation,2004 8(3):256-279
15. Kubotani Hiroyuki, Yoshimura Kazuyuki: Performance Avaluation of Acceptance Probability Functions for Multi-objective SA. Computers and Operations Research, 2003,30(3):427-442

A Sequential Niching Technique for Particle Swarm Optimization[*]

Jun Zhang[1,2], Jing-Ru Zhang[1], and Kang Li[3]

[1] Institute of Intelligent Machines, Chinese Academy of Sciences,
P.O.Box 1130, Hefei, Anhui 230031, China
[2] Department of Automation, University of Science and Technology of China,
Hefei Anhui, China
[3] School of Electrical & Electronic Engineering Queen's University Belfast
zhangjun@iim.ac.cn, jrzhang@iim.ac.cn, K.Li@qub.ac.uk

Abstract. This paper proposed a modified algorithm, sequential niching particle swarm optimization (SNPSO), for the attempt to get multiple maxima of multimodal function. Based on the sequential niching technique, our proposed SNPSO algorithm can divide a whole swarm into several sub-swarms, which can detect possible optimal solutions in multimodal problems sequentially. Moreover, for the purpose of determining sub-swarm's launch criteria, we adopted a new PSO space convergence rate (SCR), in which each sub-swarm can search possible local optimal solution recurrently until the iteration criteria is reached. Meanwhile, in order to encourage every sub-swarm flying to a new place in search space, the algorithm modified the raw fitness function of the new launched sub-swarm. Finally, the experimental results show that the SNPSO algorithm is more effective and efficient than the SNGA algorithm.

1 Introduction

The niching technique corresponding to the genetic algorithm (GA) and particle swarm optimization (PSO) is aimed at locating multiple optimal solutions in multimodal problems. The merit of the niching technique can be summarized below: first, with the multiple optima was being found, the chances of locating the global optimum may be improved. Second, a diverse set of high-quality solutions have been found will provide researcher innovative alternative solutions in some real world problems. There are many applications for this technique. Such application includes classification, multiobjective function optimization and artifical neural network (ANN) training. Contrasting to traditional ANN technique [1-4], niching technique can be used in all ANN training and has unique virtue. In fact, the goal of the original PSO algorithm is just to seek one solution for optimization problem effectively. However, the original algorithms generally can't locate multiple optimal solutions because of the particles of the swarm all sharing the social knowledge regarded as a global optimal solution. Most of researchers are only interested in maintaining the diversity of the particles to avoid

---

[*] This work was supported by the National Science Foundation of China (Nos.60472111 and 60405002).

converging to local optima. [5-9]. However, the investigations on the PSO niching technique have been neglected to a large extent. Comparatively, the literatures on the GA niching technique [10-17] are more than the ones on PSO niching technique [18,19].

The niching technique of GA can be divided into two sorts: the parallel and sequential niching method. The sequential technique has more advantages than the parallel one [13]. We shall in detail discuss this problem in Section 2. However, the research on PSO niching technique is just being in starting period, there only exists one parallel technique: Niching PSO algorithm [18], so far no one cares about how to apply sequential technique into PSO algorithm. In this paper, we shall firstly transfer the sequential technique to particle swarm optimization algorithm (SNPSO), and contrast the performance with the original sequential GA. This paper is organized as follow. In Section 2, we shall give a brief overview of the PSO algorithm and existing niching techniques. A sequential niching particle swarm optimization (SNPSO) algorithm is presented and discussed in Section 3, Section 4 gives the related experimental results. Finally, some conclusive remarks are included in section 5.

2 Particle Swarm Optimization and Niching Techniques

2.1 Particle Swarm Optimization

Particle swarm optimization (PSO) is a population-based stochastic search algorithm. The algorithm was firstly developed by Dr. Eberhart and Dr. Kennedy in 1995 [20], inspired by social behavior of bird flocking or fish schooling. Many individuals referred to as particles, are grouped into a swarm, which "flies" through multidimensional search space. Each particle in the swarm represents a candidate solution to the optimization problem, and it can evaluate their positions, referred to as fitness at every iteration. In addition, each particle can also keep its own experience in mind while those particles in a local neighborhood share the memories of their "best" positions. As a result, those memories can be used to adjust their own velocities and positions. The original PSO were modified by Shi and Eberhart [21] with the introduction of inertia weight. The equations for the manipulation of the swarm can be written as:

$$V_{id} = W*V_{id} + C1*rand1()*(P_{id} - X_{id}) + C2*rand2()*(P_{gd} - X_{id}). \quad (1)$$

$$X_{id} = X_{id} + V_{id}. \quad (2)$$

where i = 1,2,…N, W is called as inertia weight. C1 and C2 are positive constants, referred to as cognitive and social parameters, rand1 (*) and rand2 (*) are random numbers, respectively, uniformly distributed in [0..1]. A large inertia weight will encourage a global exploration, while a small one will promote local exploration, i.e., fine-tuning the current search area. The *ith* particle of the swarm is represented by the D dimensional vector X_{id}, and the best particle in the swarm denoted by the index g, the best previous position of the *ith* particle is recorded and represented as P_{id} and the velocity of the *ith* particle is as V_{id}.

An alternative version of PSO is to employ a parameter called as the constriction factor. It may be necessary to guarantee convergence of the PSO algorithm. Usually, the swarm can be manipulated according to the following equations:[22]

$$V_{id} = K * [V_{id} + C1 * rand1() * (P_{id} - X_{id}) + C2 * rand2() * (P_{gd} - X_{id})]. \quad (3)$$

where K is the constriction factor. The value of the constriction factor is obtained through the formula:

$$K = \frac{2}{\left|2 - \varphi - \sqrt{\varphi^2 - 4\varphi}\right|}. \quad (4)$$

where $\varphi = C_1 + C_2, \varphi > 4$.

2.2 Niching Techniques

Niching methods attempt to find multiple solutions to optimization problems. They have been studied extensively in the field of genetic algorithms (GA). Most of them are very interesting and fascinating. In contrast to niching technique of GA, the PSO niching method is so poor and boring. Usually, the niching technique can be divided into two sorts, the parallel and sequential niching methods. Parallel niching methods can identify and maintain several niches in a population simultaneously. Sequential niching methods can find multiple solutions by iteratively applying niching to a problem space, at the same time; mark a potential solution at each iteration to ensure that search efforts are not duplicated. Some GA niching methods used currently can be summarized below:

Fitness Sharing: Perhaps the most well known technique is Fitness Sharing. Goldberg [14] proposed to regard each niche as a finite resource, and share this resource among all individuals in the niche. Thus, an individual's fitness fi can be adapted to its shared fitness $f_i' = \frac{f_i}{\sum_j sh(d_{i,j})}$. The sharing function is defined as $sh(d) = 1 - (\frac{d}{\sigma_{share}})^\alpha$, if the distance d between individuals i and j is less than σ_{share}, otherwise is zero. The distance measure d can be genotypic or phenotypic, depending on the problem to be solved.

Deterministic Crowing(DC): De Jong [15] firstly proposed the idea of deterministic crowing (DC). The method can evolve a population by deriving offspring from parents and then let them compete against each other for a position in a next generation.

Restricted Tournament Selection : The method introduced in [16] is similar to DC, but it randomly adapts the selected individuals and then lets them compete against the most similar individuals from the population.

Sequential Niching Technique : This is a simple, fast algorithm that can locate multiple solutions by iterating simple GA [13]. To avoid converging to the same area of the search space, the algorithm is generally to modify the objective function's fitness through applying a fitness derating function. There are three main advantages with the algorithm. The first is its simplicity due to a simple add-on technique being required, the second is that the algorithm can work in small population, and the third is that the speed is faster with respect to other algorithms.

Although particle swarm optimization algorithm has been developed to locate multiple optimal solutions, it is not very popular. This algorithm can be listed below:

Niching Particle Swarm Optimizer : This algorithm [18] is a first PSO niching technique. The algorithm can train a main swarm using only cognitive model. When a particle fitness shows very little change over a small number of iterations, the algorithm then will create a sub-swarm around the particle in a small area so that the sub-swarm can be trained to locate multiple solutions. Whether the algorithm succeeds or not will depend on the proper initial distribution of particles throughout the search space.

3 A Sequential Niching Particle Swarm Optimization

The Sequential Niching Particle Swarm Optimization algorithm can successfully locate multiple solutions to multimodal function. The details of the algorithm will be presented in this section. Because of the algorithm using multi-sub-swarms to detect different solution sequentially, the stop training criterion for each sub-swarm is very important. Therefore, we propose a Space Convergence Rate (SCR) as the stop training criterion for each sub-swarm.

3.1 Space Convergence Rate (SCR)

Assume that we have an n-dimensional swarm consisting of m particles, each dimensional range of the swarm was confined in $[x_{min}, x_{max}]$. $X_i (x_{i1}, x_{i2}, \ldots x_{in})$ is the *ith* particle of a swarm, then the average position of the swarm is:

$$\overline{X}_{avg} = \frac{1}{m} \sum_{i=1}^{m} X_i . \tag{5}$$

The normalized Euclidean distance between the particle i and j can be denoted as:

$$\|X_i, X_j\| = \sqrt{\frac{1}{n} \sum_{k=1}^{n} (\frac{x_{ik} - x_{jk}}{x_{max} - x_{min}})^2} . \tag{6}$$

Then, we can get the definition of Space Convergence Rate (SCR):

$$SCR = \frac{1}{m} \sum_{i=1}^{m} \|X_i, \overline{X}_{avg}\| . \tag{7}$$

According to the above definition, SCR is always small than one. When all particles converge into one point, SCR will equal zero. So, the algorithm can use the SCR as convergence criterion of the particles in a swarm. When the SCR of a swarm reaches to a given value, it shows that the swarm has achieved the setting goal.

3.2 Sequential Niching PSO Algorithm (SNPSO)

SNPSO consists of several sub-swarms. The number of sub-swarms depends on the required number of optimal solutions (RNOS). In general, the RNOS is slightly greater than what we want to get the number of optimal solutions in multimodal problem. The

SCR of the sub-swarm will determine the search precision of the sub-swarm. If we want to get an approximate solution, we can choose a bigger SCR. On the contrary, if we want to get a more precise solution, then we must choose a smaller SCR. The first sub-swarm will be initialized like an ordinary PSO algorithm, and then the swarm will be trained to find a good solution in search space. When the SCR of the sub-swarm reaches to the given value, it means that the first sub-swarm has found an optimal solution. So, the algorithm will begin to launch a new sub-swarm for searching a new optimal solution. If the algorithm does not modify the fitness function, the new sub-swarm will possibly again "fly" to optimal solution found by first sub-swarm. Therefore, the algorithm must modify the fitness function according to the following principle.

The modified fitness function is adopted from Beasley's SNGA [13]. For an individual X, the fitness M(X) is computed from the raw fitness function F(X) multiplied by a number of single-peak derating functions. Initially we set $M_0(X)=F(X)$, and let $X_{1best}, X_{2best}....X_{nbest}$ be the best particle position of 1th, 2th...nth sub-swarm launched before. The modified fitness function can be then updated according to the following formulae:

$$M_{n+1}(X) = M_n(X) * G(X, X_{nbest}). \tag{8}$$

There are various forms single-peak derating functions. We only employed the power law equation to test our algorithm. In equation (9), r is the niching radius, $\|X, X_{best}\|$ as defined above, is a normalized Euclidean distance. α is the power factor, $\alpha > 1$ determining how concave the derating curve is, and $\alpha < 1$ then determining how convex the derating curve is, If $\alpha = 1$, the derating function will be a linear function.

$$G(X, X_{best}) = \begin{cases} (\frac{\|X, X_{best}\|}{r})^\alpha & if \quad \|X, X_{best}\| < r \\ 1 & otherwise \end{cases} \tag{9}$$

To determine the value of niching radius r, we use the same method as Deb [17].Thus, r is given as follows:

$$r = \frac{\sqrt{k}}{2 * \sqrt[k]{p}}. \tag{10}$$

In equation (10), k is the dimensional number of the search space; p is the number of maxima of the multimodal problem.

In this way, the algorithm can modify the fitness function according to the distance between current particle and the best position of the sub-swarm launched before. If the distance is smaller than the niching radius, the fitness function will become smaller; if the distance is larger than the niching radius, the fitness function is not changed. The modified fitness function can encourage the particles of the new launched sub-swarm explore the new area of the search space. And the particle of the new sub-swarm will "fly" to a new "best" position having not been found before.

3.3 Pseudo-Code for SNPSO

The parameters involved in the algorithm are the required number of optimal solutions (RNOS) and the search precision of each sub-swarm (i.e., SCR). The other parameter is niching radius r. Figure 1 shows the pseudo-code used in our experiments.

```
nCount=1;
nPrecision=given SCR value;
repeat
  create a new n-dimensional PSO population
  repeat
    Train the sub-swarm with modified fitness function
  until SCR<= nPrecision
  nCount=nCount+1;
until nCount>RNOS OR reach maximum iteration number
```

Fig. 1. Pseudo-code for SNPSO

4 Experimental Results

This section will give several experimental results of the SNPSO algorithm to find maxima of multimodal functions using five test functions.

4.1 Test Functions Employed

We consider five multimodal functions with different difficulty, which can be seen from Fig.2 to Fig.6. Here, we used 5 functions, which were firstly introduced by Goldberg and Richardson (1987) [14] and adopted by Beasley et al. (1993) [13], to evaluate our proposed approach. These functions can be described as follows:

$$F1(x) = \sin^6(5\pi x). \tag{11}$$

$$F2(x) = (e^{-2\log(2)\times(\frac{x-0.1}{0.8})^2}) \times \sin^6(5\pi x). \tag{12}$$

$$F3(x) = \sin^6(5\pi(x^{3/4} - 0.05)). \tag{13}$$

$$F4(x) = (e^{-2\log(2)\times(\frac{x-0.1}{0.8})^2}) \times \sin^6(5\pi(x^{3/4} - 0.05)). \tag{14}$$

$$F5(x, y) = 200 - (x^2 + y - 11)^2 - (x + y^2 - 7)^2. \tag{15}$$

Functions F1 and F3 both have 5 maxima with a function value of 1.0. In F1, maxima are evenly spaced, while in F3 maxima are unevenly spaced. Functions F2 and F4 are all oscillating functions; the local and global peaks exist at the same x-positions as in Functions F1 and F3. Assume that Functions F1 to F4 are investigated in the range of [0, 1]. For each of the functions, maxima locate at the following x positions:

Table 1. x position corresponding to the maxima

F1	0.1	0.3	0.5	0.7	0.9
F2	0.1	0.3	0.5	0.7	0.9
F3	0.08	0.246	0.45	0.681	0.934
F4	0.08	0.246	0.45	0.681	0.934

Function F5, the modified Himmelblau function, has 4 equal maxima with the value equal to 200, i.e., (-2.81, 3.13), (3.0, 2.0), (3.58, -1.85) and (-3.78,-3.28). The range of x and y is in between [-6, +6].

4.2 Experimental Setup and Results

For each of the 5 functions, 30 experiments were done with SNPSO algorithm. The inertia weight of every sub-swarm used in experiment is set to 0.729, C1 and C2 set to 1.49445, V_{max} set to the maximum range X_{max}. In addition, let the maximum iterative number be set to 10000. This setup is equivalent to the constriction factor method, where K=0.729, and $\varphi = 4.1$, C1 equals C2. Shi. and Eberhart [23] concluded that this parameter setup can ensure convergence and have better performance than others. The population size of each sub-swarm is set to 20. The SCR and required number of optimal solutions (RNOS) can all vary in the experiments. For F1 to F4, SCR is set to 0.001, RNOS set to 5; for F5, RNOS set to 4, The niching radius was 0.1 for F1-F4, and 0.35 for F5, which was determined by equation (8).

We usually use two criterions to measure the performance of niching technique: (1) *Accuracy:* How close the solution found are to the optimum solutions; (2) *Success rate:* The proportion of the experiments that can find all optimal solutions. Table 2 reports our experimental results. The fitness values in Table 2 represent mean fitness of the best particle in each sub-swarm, which inclueds suboptimal solutions for F2 and F4. Deviation represents the mean deviation in fitness value of all particles in all sub-swarms. In contrast to the original SNGA, our algorithm can get more accurate optimal solutions in fitness values. For F5, Beasley at al reported only 76% success rate, while our proposed algorithm was proved to be most effective.

Table 2. The performance comparison for five multimodal functions

Function	Fitness	Deviation	Success rate
F1	4.98e-06	6.91e-05	100%
F2	2.6e-02	5.8e-02	93%
F3	1.81e-05	2.3e-04	93%
F4	6.47e-04	1.32e-02	93%
F5	2.34e-06	1.41e-05	100%

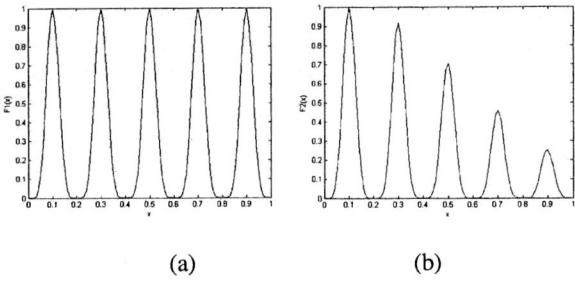

Fig. 2. (a) Function F1 with equal maxima (b) Function F2 with decreasing maxima

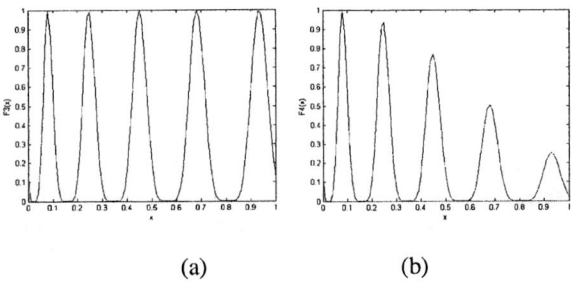

Fig. 3. (a) Function F3 with uneven maxima (b) Function F4 with uneven decreasing Maxima

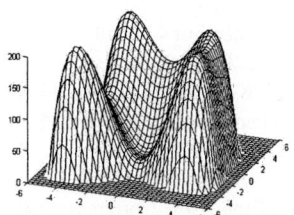

Fig. 4. F5 Modified Himmelblau's function

5 Conclusion and Future Work

This paper introduced how to transfer sequential niching technique into particle swarm optimization. The algorithm proposed a new PSO space convergence rate (SCR) as every sub-swarm launched criteria. The experimental results showed that the algorithm can successfully and efficiently locate multiple optimal solutions for multimodal optimization problems.

However, most niching techniques both in GA and PSO all face with a dilemma. On one side, the niching algorithm must sacrifice some global searching ability through setting the niching range or other means for multiple solutions. On the other hand; the algorithm must get a certain number of wanted results from a great number of existing solutions in multimodal problems. Still, how to solve the problem depends on whether some global or prophetic knowledge can be obtained in advance or not. Some niching techniques can solve the problem by giving prophetic knowledge such as the number of optimal solutions to get niching radius. Actually, in real world, some knowledge of the problem cannot be gotten in advance. In addition, the niching radius is a very important parameter for all niching technique. An accurate radius can easily promote the algorithm to converge faster. Therefore, the methods for estimating niching radius will be the one of further research tasks. From the nature ecological system, we observed that the niche is the results of dynamic equilibrium to every species. The next step is to simulate the process; every sub-population is like one species in natural that can expand their niches; also, they can merge and separate each other, when this sub-population get a dynamic equilibrium status, the real niching radius will be acquired.

References

1. Huang, D. S.: Systematic Theory of Neural Networks for Pattern Recognition. Publishing House of Electronic Industry of China, Beijing (1996)
2. Huang, D. S., Ma, S. D.: Linear and Nonlinear Feedforward Neural Network Classifiers: A Comprehensive Understanding. Journal of Intelligent Systems, Vol.9, No.1 (1999) 1-38
3. Huang, D. S.: The Local Minima Free Condition of Feedforward Neural Networks for Outer-supervised Learning. IEEE Trans on Systems, Man and Cybernetics, Vol.28B, No.3 (1998) 477-480
4. Huang, D. S.: The United Adaptive Learning Algorithm for the Link Weights and the Shape Parameters in RBFN for Pattern Recognition. International Journal of Pattern Recognition and Artificial Intelligence, Vol.11, No.6 (1997) 873-888
5. Kennedy, j.: Small Worlds and Mega-minds: Effects of Neighborhood Topology on Particle Swarm Performance. In Proceedings of the Congress on Evolutionary Computation, Washington DC, USA (1999) 1931-1938
6. Løvbjerg, M., Rasmussen, T. K., Krink, T.: Hybrid Particle Swarm Optimizer with Breeding and Subpopulations. Proceedings Genetic and Evolutionary Computation Conference, Morgan Kaufmann, San Francisco, CA, (2001)
7. Krink, T., Vestertroem, J.S., Riget, J.: Particle Swarm Optimisation with Spatial Particle Extension. In Proceedings of the IEEE Congress on Evolutionary Computation, Honolulu, Hawaii USA (2002)
8. Blackwell, T., and Bentley, P. J.: Don't Push Me! Collision-avoiding Swarms. IEEE Congress on Evolutionary Computation, Honolulu, Hawaii USA (2002)
9. Løvbjerg, and Krink, T.: Extending Particle Swarms with Self-organized Criticality. Proceedings of the Fourth Congress on Evolutionary Computation (2002)
10. Li, J-P., Balazs, M.E., Parks, G.T. and Clarkson, P.J: A Species Conserving Genetic Algorithm for Multimodal Function Optimization. Evolutionary Computation, 11 (1) (2003) 107-109
11. Streichert, F., Stein, G., Ulmer, H., Zell, A.: A Clustering Based Niching EA for Multimodal Search Spaces. Proceedings of the 6th International Conference on Artificial Evolution (2003) 293-304
12. Gan, J. and Warwick, K.: A Variable Radius Niche Technique for Speciation in Genetic Algorithms. In Proceedings of the Genetic and Evolutionary Computation Conference, Morgan-Kaufmann (2000) 96-103
13. Beasley, D., Bull, D.R., Martin, R.R.: A Sequential Niche Technique for Multimodal Function Optimization. Evolutionary Computation, 1(2), MIT Press (1993) 101-125

14. Goldberg, D.E., Richardson, J.: Genetic Algorithm with Sharing for Multimodal Function Optimization. In Proceedings of the Second International Conference on Genetic Algorithms (1987) 41-49
15. De Jong, K. A.: An Analysis of the Behavior of a Class of Genetic Adaptive systems. PhD Thesis, Dept. of Computer and Communication Sciences, University of Michigan (1975)
16. Harik, G.: Finding Multiple Solutions using Restricted Tournament Selection, In L. J. Eshelman (Ed.), Proceedings of the Sixth International Conference on Genetic Algorithms, San Francisco: Morgan Kaufmann (1995) 24-31
17. Deb, K.: Genetic Algorithms in Multimodal Function Optimization. Masters Thesis, TCGA Report No. 89002, The University of Alabama, Department of Engineering Mechanics
18. Brits, R., Engelbrecht, A. P., van den Bergh, F.: A Niching Particle Swarm Optimizer. Conference on Simulated Evolution and Learning, Singapore (2002)
19. Konstantinos, E. Parsopoulos and Michael, N. Vrahatis,:On the Computation of All Global Minimizers through Particle Swarm Optimization. IEEE Transactions on Evolutionary Computation, VOL. 8, NO. 3 (2004)
20. Kennedy, J., and Eberhart, R. C,: Particle Swarm Optimization. Proc. Of IEEE International Conference on Neural Networks (ICNN), Vol. IV, Perth, Australia (1995) 1942-1948
21. Shi, Y., and Eberhart, R. C.: A Modified Particle Swarm Optimizer. Proceedings of the 1998 IEEE International Conference on Evolutionary Computation, Piscataway, NJ: IEEE Press (1998) 69-73
22. Clerc, M.: The Swarm and the Queen: Towards a Deterministic and Adaptive Particle Swarm Optimization. Proc. Congress on Evolutionary Computation, Washington, DC Piscataway, NJ: IEEE Service Center (1999) 1951-1957
23. Eberhart, R. C., and Shi, Y.: Comparing Inertia Weights and Constriction Factors in Particle Swarm Optimization. Proc. Congress on Evolutionary Computation , San Diego, CA (2000) 84-88

An Optimization Model for Outlier Detection in Categorical Data

Zengyou He, Shengchun Deng, and Xiaofei Xu

Department of Computer Science and Engineering,
Harbin Institute of Technology, China
zengyouhe@yahoo.com, {dsc, xiaofei}@hit.edu.cn

Abstract. In this paper, we formally define the problem of outlier detection in categorical data as an optimization problem from a global viewpoint. Moreover, we present a local-search heuristic based algorithm for efficiently finding feasible solutions. Experimental results on real datasets and large synthetic datasets demonstrate the superiority of our model and algorithm.

1 Introduction

A well-quoted definition of outliers is firstly given by Hawkins [1]. Recently, more concrete meanings of outliers are defined [e.g., 3-37]. However, conventional approaches do not handle categorical data in a satisfactory manner, and most existing techniques lack for a solid theoretical foundation or assume underlying distributions that are not well suited for exploratory data mining applications. To fulfill this void, an optimization model is explored in this paper for mining outliers.

From a systematic viewpoint, a dataset that contains many outliers have a great amount of mess. In other words, removing outliers from the dataset will result in a dataset that is less "disordered". Based on this observation, the problem of outlier mining could be defined informally as an optimization problem as follows: finding a small subset of target dataset such that the degree of disorder of the resultant dataset after the removal of this subset is minimized.

In our optimization model, we first have to resolve the issue of what we mean by the "the degree of disorder of a dataset". In other words, we have to make our objective function clear. Entropy in information theory is a good choice for measuring the "the degree of disorder of a dataset". Hence, we will aim to minimize the expected entropy of the resultant dataset in our problem.

Consequently, we have to resolve the issue of what we mean by "a small subset of target dataset". Since it is very common in the real applications to report top-k outliers to end users, we set the size of this set to be k. That is, we aim to find k outliers from the original dataset, where k is the expected number of outliers in the data set.

So far, the optimization problem could be described in a more concise manner as follows: finding a subset of k objects such that the expected entropy of the resultant dataset after the removal of this subset is minimized.

In the above optimization problem, an exhaustive search through all possible solutions with k outliers for the one with the minimum objective value is costly since for n objects and k outliers there are $C(n, k)$ possible solutions. A variety of well known greedy search techniques, including simulated annealing and genetic algorithms, can be tried to find a reasonable solution. We have not investigated such approaches in detail since we expect the outlier-mining algorithm to be mostly applied large datasets, so computationally expensive approaches become unattractive. However, to get a feel for the quality-time tradeoffs involved, we devised and studied the greedy optimization scheme that uses local-search heuristic to efficiently find feasible solutions. Experimental results on real datasets and large synthetic datasets demonstrate the superiority of our model and algorithm.

2 Related Work

Previous researches on outlier detection broadly fall into the following categories.

Distribution based methods are previously conducted by the statistics community [1,5,6]. Recently, Yamanishi et al. [7,8] used a Gaussian mixture model to present the normal behaviors and discover outliers.

Depth-based is the second category for outlier mining in statistics [9,10].

Deviation-based techniques identify outliers by inspecting the characteristics of objects and consider an object that deviates these features as an outlier [11].

Distance based method was originally proposed by Knorr and Ng [12-15]. This notion is further extended in [16-18].

Density based This was proposed by Breunig et al. [19]. The density-based approach is further extended in [20-24].

Clustering-based outlier detection techniques regarded *small* clusters as outliers [25,27] or identified outliers by removing clusters from the original dataset [26].

Sub-Space based. Aggarwal and Yu [3] discussed a new projection based technique for outlier mining in high dimensional space. A frequent pattern based outlier detection method is proposed in [4]. Wei et al. [28] introduced a hypergraph model to detect outliers in categorical dataset.

Support vector based outlier mining was recently developed in [29-32].

Neutral network based. The replicator neutral network (*RNN*) is employed to detect outliers by Harkins et al. [33,34].

In addition, the class outlier detection problem is considered in [35-37].

3 Background and Problem Formulation

3.1 Entropy

Entropy is the measure of information and uncertainty of a random variable [2]. If X is a random variable, and $S(X)$ the set of values that X can take, and $p(x)$ the probability function of X, the entropy $E(X)$ is defined as shown in Equation (1).

$$E(X) = - \sum_{x \in S(X)} p(x) log(p(x)) \cdot \quad (1)$$

The entropy of a multivariable vector $\hat{x} = \{X_1,...,X_m\}$ can be computed as shown in Equation (2).

$$E(\hat{x}) = - \sum_{x_1 \in S(X_1)} \cdots \sum_{x_m \in S(X_m)} p(x_1,...,x_m) log(p(x_1,...,x_m)). \quad (2)$$

3.2 Problem Formulation

The problem we are trying to solve can be formulated as follows. Given a dataset D of n points $\hat{p}_1,..., \hat{p}_n$, where each point is a multidimensional vector of m categorical attributes, i.e., $\hat{p}_i = (p_i^1,...,p_i^m)$, and given a integer k, we would like to find a subset $O \subseteq D$ with size k, in such a way that we minimize the entropy of $D - O$. That is,

$$\min_{O \subseteq D} E(D - O) \quad \text{Subject to } |O| \models k. \quad (3)$$

In this problem, we need to compute the entropy of a set of records using Equation (2). To make computation more efficient, we make a simplification in the computation of entropy of a set of records. We assume the independences of the record, transforming Equation (2) into Equation (4). That is, the joint probability of combined attribute values becomes the product of the probabilities of each attribute, and hence the entropy can be computed as the sum of entropies of the attributes.

$$E(\hat{x}) = - \sum_{x_1 \in S(X_1)} \cdots \sum_{x_m \in S(X_m)} p(x_1,...,x_m) log(p(x_1,...,x_m)) = E(X_1) + E(X_2) + ... + E(X_n). \quad (4)$$

4 Local Search Algorithm

In this section, we present a local-search heuristic based algorithm, denoted by LSA, which is effective and efficient on identifying outliers.

4.1 Overview

The LSA algorithm takes the number of desired outliers (supposed to be k) as input and iteratively improves the value of object function. Initially, we randomly select k points and label them as outliers. In the iteration process, for each point labeled as non-outlier, its label is exchanged with each of the k outliers and the entropy objective is re-evaluated. If the entropy decreases, the point's non-outlier label is exchanged with the outlier label of the point that achieved the best new value and the algorithm proceeds to the next object. When all non-outlier points have been checked for possible improvements, a sweep is completed. If at least one label was changed in a sweep, we initiate a new sweep. The algorithm terminates when a full sweep does not change any labels, thereby indicating that a local optimum is reached.

4.2 Data Structure

Given a dataset D of n points $\hat{p}_1, \ldots, \hat{p}_n$, where each point is a multidimensional vector of m categorical attributes, we need m corresponding hash tables as our basic data structure. Each hash table has attribute values as keys and the frequencies of attribute values as referred values. Thus, in $O(1)$ expected time, we can determine the frequency of an attribute value in corresponding hash table.

4.3 The Algorithm

Fig.1 shows the LSA algorithm. The collection of records is stored in a file on the disk and we read each record t in sequence.

In the initialization phase of the LSA algorithm, we firstly select the first k records from the data set to construct initial set of outliers. Each consequent record is labeled as non-outlier and hash tables for attributes are also constructed and updated.

In iteration phase, we read each record t that is labeled as non-outlier, its label is exchanged with each of the k outliers and the changes on entropy value are evaluated. If the entropy decreases, the point's non-outlier label is exchanged with the outlier label of the point that achieved the best new value and the algorithm proceeds to the next object. After each swap, the hash tables are also updated. If no swap happened in one pass of all records, iteration phase terminates; otherwise, a new pass begins. Essentially, at each step we locally optimize the criterion. In this phase, the key step is computing the changed value of entropy. With the use of hashing technique, in $O(1)$ expected time, we can determine the frequency of an attribute value in corresponding hash table. Hence, we can determine the decreased entropy value in $O(m)$ expected time since the changed value is only dependent on the attribute values of two records to be swapped.

4.4 Time and Space Complexities

Worst-case analysis: The time and space complexities of the LSA algorithm depend on the size of dataset (n), the number of attributes (m), the size of every hash table, the number of outliers (k) and the iteration times (I).

To simplify the analysis, we will assume that every attribute has the same number of distinct attributes values, p. Then, in the worst case, in the initialization phase, the time complexity is $O(n*m*p)$. In the iteration phase, since the computation of value changed on entropy requires at most $O(m*p)$ and hence this phase has time complexity $O(n*k*m*p*I)$. Totally, the LSA algorithm has time complexity $O(n*k*m*p*I)$ in worst case.

The algorithm only needs to store m hash tables and the dataset in main memory, so the space complexity of our algorithm is $O((p+n)*m)$.

```
Algorithm LSA
Input:    D    // the categorical database
          k    // the number of desired outliers
Output:   k identified outliers
/* Phase 1-initialization */
01  Begin
02      foreach record t in D
03          counter++
04          if counter<=k then
05              label t as an outlier with flag "1"
06          else
07              update hash tables using t
08              label t as a non-outlier with flag "0"
/* Phase 2-Iteration */
09      Repeat
10          not_moved =true
11          while not end of the database do
12              read next record t which is labeled "0"   //non-outlier
13              foreach record o in current k outliers
14                  exchanging label of t with that of o and evaluating the change of entropy
15              if maximal decrease on entropy is achieved by record b then
16                  swap the labels of t and b
17                  update hash tables using t and b
18                  not_moved =false
19      Until not_moved
20  End
```

Fig. 1. The LSA Algorithm

Practical analysis: Categorical attributes usually have *small* domains. An important of implication of the compactness of categorical domains is that the parameter, p, can be regarded to be very small. And the use of hashing technique also reduces the impact of p, as discussed previously, we can determine the frequency of an attribute value in $O(1)$ expected time, So, in practice, the time complexity of LSA can be expected to be $O(n*k*m*I)$.

The above analysis shows that the time complexity of LSA is linear to the size of dataset, the number of attributes and the iteration times, which make this algorithm scalable.

5 Experimental Results

We ran our algorithm on real-life datasets obtained from the UCI Machine Learning Repository [38] to test its performance against other algorithms on identifying true outliers. In addition, some large synthetic datasets are used to demonstrate the scalability of our algorithm.

5.1 Experiment Design and Evaluation Method

We used two real life datasets (*lymphography* and *cancer*) to demonstrate the effectiveness of our algorithm against *FindFPOF* algorithm [4], *FindCBLOF* algorithm [27] and *KNN* algorithm [16]. In addition, on the *cancer* dataset, we add the results of *RNN* based outlier detection algorithm that are reported in [33,34] for comparison, although we didn't implement the *RNN* based outlier detection algorithm.

For all the experiments, the two parameters needed by *FindCBLOF* [27] algorithm are set to 90% and 5 separately as done in [27]. For the *KNN* algorithm [16], the results were obtained using the *5-nearest-neighbour*; For *FindFPOF* algorithm [4], the parameter *mini-support* for mining frequent patterns is fixed to 10%, and the maximal number of items in an itemset is set to 5. Since the LSA algorithm is parameter-free (besides the number of desired outliers), we don't need to set any parameters.

As pointed out by Aggarwal and Yu [3], one way to test how well the outlier detection algorithm worked is to run the method on the dataset and test the percentage of points which belong to the rare classes. If outlier detection works well, it is expected that the rare classes would be over-represented in the set of points found. These kinds of classes are also interesting from a practical perspective.

Since we know the true class of each object in the test dataset, we define objects in small classes as rare cases. The number of rare cases identified is utilized as the assessment basis for comparing our algorithm with other algorithms.

5.2 Results on Lymphography Data

The first dataset used is the Lymphography data set, which has 148 instances with 18 attributes. The data set contains a total of 4 classes. Classes 2 and 3 have the largest number of instances. The remained classes are regarded as rare class labels for they are small in size. The corresponding class distribution is illustrated in Table 1.

Table 1. Class distribution of lymphography data set

Case	Class codes	Percentage of instances
Commonly Occurring Classes	2, 3	95.9%
Rare Classes	1, 4	4.1%

Table 2 shows the results produced by different algorithms. Here, the *top ratio* is ratio of the number of records specified as *top-k* outliers to that of the records in the dataset. The *coverage* is ratio of the number of detected rare classes to that of the rare classes in the dataset. For example, we let LSA algorithm find the *top 7* outliers with the top ratio of 5%. By examining these 7 points, we found that 6 of them belonged to the rare classes.

In this experiment, the LSA algorithm performed the best for all cases and can find all the records in rare classes when the *top ratio* reached 5%. In contrast, for the *KNN* algorithm, it achieved this goal with the *top ratio* at 10%, which is the twice of that of our algorithm.

Table 2. Detected rare classes in lymphography dataset

Top Ratio (Number of Records)	Number of Rare Classes Included (Coverage)			
	LSA	FindFPOF	FindCBLOF	KNN
5% (7)	6(100%)	5(83%)	4 (67%)	4 (67%)
10%(15)	6(100%)	5(83%)	4 (67%)	6(100%)
11%(16)	6(100%)	6(100%)	4 (67%)	6(100%)
15%(22)	6(100%)	6 (100%)	4 (67%)	6(100%)
20%(30)	6(100%)	6 (100%)	6 (100%)	6(100%)

5.3 Results on Wisconsin Breast Cancer Data

The second dataset used is the Wisconsin breast cancer data set, which has 699 instances with 9 attributes, in this experiment, all attributes are considered as categorical. Each record is labeled as *benign* (458 or 65.5%) or *malignant* (241 or 34.5%). We follow the experimental technique of Harkins, et al. [33,34] by removing some of the *malignant* records to form a very unbalanced distribution; the resultant dataset had 39 (8%) *malignant* records and 444 (92%) *benign* records (The resultant dataset is available at: http://research.cmis.csiro.au/rohanb/outliers/breast-cancer/). The corresponding class distribution is illustrated in Table 3.

Table 3. Class distribution of wisconsin breast cancer data set

Case	Class codes	Percentage of instances
Commonly Occurring Classes	1	92%
Rare Classes	2	8%

For this dataset, we also consider the *RNN* based outlier detection algorithm [33]. The results of *RNN* based outlier detection algorithm on this dataset are reproduced from [33].

Table 4 shows the results produced by different algorithms. Clearly, among all of these algorithms, *RNN* performed the worst in most cases. In comparison to other algorithms, LSA always performed the best besides the case when top ratio is 4%. Hence, this experiment also demonstrates the superiority of LSA algorithm.

5.4 Scalability Tests

The purpose of this experiment was to test the scalability of the LSA algorithm when handling very large datasets. A synthesized categorical dataset created with the software developed by Dana Cristofor (The source codes are public available at:

http://www.cs.umb.edu/~dana/GAClust/index.html) is used. The data size (i.e., number of rows), the number of attributes and the number of classes are the major parameters in the synthesized categorical data generation, which were set to be 100,000, 10 and 10 separately. Moreover, we set the random generator seed to 5. We will refer to this synthesized dataset with name of DS1.

Table 4. Detected malignant records in wisconsin breast cancer dataset

Top Ratio (Number of Records)	Number of Rare Classes Included (Coverage)				
	LSA	FindFPOF	FindCBLOF	RNN	KNN
1%(4)	4 (10.26%)	3(7.69%)	4 (10.26%)	3 (7.69%)	4(10.26%)
2%(8)	8 (20.52%)	7 (17.95%)	7 (17.95%)	6 (15.38%)	8(20.52%)
4%(16)	15(38.46%)	14 (35.90%)	14 (35.90%)	11 (28.21%)	16(41%)
6%(24)	22(56.41%)	21 (53.85%)	21 (53.85%)	18 (46.15%)	20(51.28%)
8%(32)	29(74.36%)	28(71.79%)	27 (69.23%)	25 (64.10%)	27(69.23%)
10%(40)	33(84.62%)	31(79.49%)	32 (82.05%)	30 (76.92%)	32(82.05%)
12%(48)	38 (97.44%)	35 (89.74%)	35 (89.74%)	35 (89.74%)	37(94.87%)
14%(56)	39 (100%)	39 (100%)	38 (97.44%)	36 (92.31%)	39 (100%)
16%(64)	39 (100%)	39 (100%)	39 (100%)	36 (92.31%)	39 (100%)
18%(72)	39 (100%)	39 (100%)	39 (100%)	38 (97.44%)	39 (100%)
20%(80)	39 (100%)	39 (100%)	39 (100%)	38 (97.44%)	39 (100%)
25%(100)	39 (100%)	39 (100%)	39 (100%)	38 (97.44%)	39 (100%)
28%(112)	39 (100%)	39 (100%)	39 (100%)	39 (100%)	39 (100%)

We tested two types of scalability of the LSA algorithm on large dataset. The first one is the scalability against the number of objects for a given number of outliers and the second is the scalability against the number of outliers for a given number of objects. Our LSA algorithm was implemented in Java. All experiments were conducted on a Pentium4-2.4G machine with 512 M of RAM and running Windows 2000. Fig. 2 shows the results of using LSA to find 30 outliers from different number of objects. Fig. 3 shows the results of using LSA to find different number of outliers on DS1 dataset.

One important observation from these figures was that the run time of LSA algorithm tends to increase linearly as both the number of records and the number of outliers are increased, which verified our claim in Section 4.4.

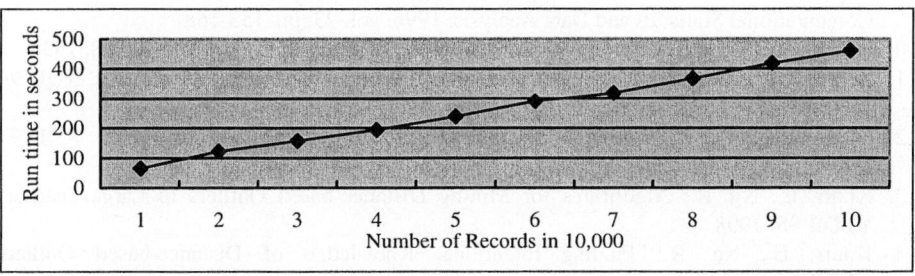

Fig. 2. Scalability of LSA to the number of objects when mining 30 outliers

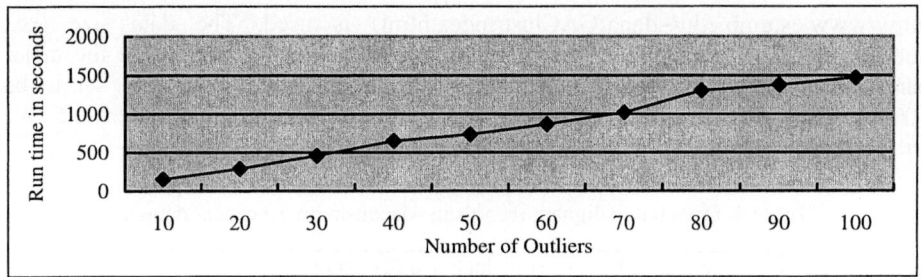

Fig. 3. Scalability of LSA to the number of outliers

6 Conclusions

The problem of outlier detection has traditionally been addressed using data mining methods. There are opportunities for optimization to improve these methods, and this paper focused on building an optimization model for outlier detection. Experimental results on real datasets and large synthetic datasets demonstrate the superiority of our new optimization-based method.

References

1. Hawkins, D.: Identification of Outliers. Chapman and Hall, Reading, London, 1980
2. Shannon, C.E.: A Mathematical Theory of Communication. Bell System Technical Journal, 1948, pp.379-423
3. Aggarwal, C., Yu, P.: Outlier Detection for High Dimensional Data. SIGMOD'01, 2001
4. He, Z., et al.: A Frequent Pattern Discovery Based Method for Outlier Detection. WAIM'04, 2004
5. Barnett, V., Lewis, T.: Outliers in Statistical Data. John Wiley and Sons, New York, 1994
6. Rousseeuw, P.: A. Leroy. Robust Regression and Outlier Detection. John Wiley and Sons, 1987
7. Yamanishi, K., Takeuchi, J., Williams G.: On-line Unsupervised Outlier Detection Using Finite Mixtures with Discounting Learning Algorithms. KDD'00, pp. 320-325, 2000
8. Yamanishi, K., Takeuchi, J.: Discovering Outlier Filtering Rules from Unlabeled Data-Combining a Supervised Learner with an Unsupervised Learner. KDD'01, 2001
9. Nuts, R., Rousseeuw, P.: Computing Depth Contours of Bivariate Point Clouds. Computational Statistics and Data Analysis. 1996, vol. 23, pp. 153-168
10. Johnson, T., et al.: Fast Computation of 2-dimensional Depth Contours. KDD'98, 1998
11. Arning, A., et al: A Linear Method for Deviation Detection in Large Databases. KDD'96, 1996
12. Knorr, E., Ng R.: A Unified Notion of Outliers: Properties and Computation. KDD'97, 1997
13. Knorr, E., Ng. R.: Algorithms for Mining Distance-based Outliers in Large Datasets. VLDB'98, 1998
14. Knorr, E., Ng. R.: Finding Intentional Knowledge of Distance-based Outliers. VLDB'99, 1999

15. Knorr, E., et al.: Distance-based Outliers: Algorithms and Applications. VLDB Journal, 2000
16. Ramaswamy, S., Rastogi, R., Kyuseok, S.: Efficient Algorithms for Mining Outliers from Large Data Sets. SIGMOD'00, pp. 93-104, 2000
17. Angiulli, F., Pizzuti, C.: Fast Outlier Detection in High Dimensional Spaces. PKDD'02, 2002
18. Bay, S. D., Schwabacher, M.: Mining Distance Based Outliers in Near Linear Time with Randomization and a Simple Pruning Rule. KDD'03, 2003
19. Breunig, M., et al.: LOF: Identifying Density-Based Local Outliers. SIGMOD'00, 2000
20. Tang J., et al.: Enhancing Effectiveness of Outlier Detections for Low Density Patterns. PAKDD'02, 2002
21. Chiu, A. L., Fu, A. W.: Enhancements on Local Outlier Detection. IDEAS'03, 2003
22. Jin, W., et al.: Mining top-n local Outliers in Large Databases. KDD'01, 2001
23. Papadimitriou, S., et al.: Fast Outlier Detection Using the Local Correlation Integral. ICDE'03, 2003
24. Hu, T., Sung, S. Y.: Detecting Pattern-based Outliers. Pattern Recognition Letters, 2003
25. Jiang, M. F., Tseng, S. S., Su, C. M.: Two-phase Clustering Process for Outliers Detection. Pattern Recognition Letters, 2001, 22(6-7): 691-700
26. Yu, D., Sheikholeslami, G., Zhang, A.: FindOut: Finding Out Outliers in Large Datasets. Knowledge and Information Systems, 2002, 4(4): 387-412
27. He, Z., et al.: Discovering Cluster Based Local Outliers. Pattern Recognition Letters, 2003
28. Wei, L., et al.: HOT: Hypergraph-Based Outlier Test for Categorical Data. PAKDD'03, 2003
29. Tax, D., Duin, R.: Support Vector Data Description. Pattern Recognition Letters, 1999
30. Schölkopf, B., et al.: Estimating the Support of a High Dimensional Distribution. Neural Computation, 2001, 13 (7): 1443-1472
31. Cao, L. J., Lee, H. P., Chong, W.K.: Modified Support Vector Novelty Detector Using Training Data with Outliers. Pattern Recognition Letters, 2003, 24 (14): 2479-2487
32. Petrovskiy, M.: A Hybrid Method for Patterns Mining and Outliers Detection in the Web Usage Log. AWIC'03, pp.318-328, 2003
33. Harkins, S., et al.: Outlier Detection Using Replicator Neural Networks. DaWaK'02, 2002
34. Willams, G. J., et al.: A Comparative Study of RNN for Outlier Detection in Data Mining. ICDM'02, pp. 709-712, 2002
35. He, Z., et al.: Outlier Detection Integrating Semantic Knowledge. WAIM'02, 2002
36. Papadimitriou, S., Faloutsos, C.: Cross-outlier Detection. SSTD'03, pp.199-213, 2003.
37. He,Z., Xu, X., Huang, J., Deng, S.: Mining Class Outlier: Concepts, Algorithms and Applications in CRM. Expert System with Applications, 2004
38. Merz, G., Murphy, P.: Uci Repository of Machine Learning Databases. http://www.ics.uci.edu/mlearn/MLRepository.html, 1996

Development and Test of an Artificial-Immune-Abnormal-Trading-Detection System for Financial Markets

Vincent C.S. Lee and Xingjian Yang

School of Business Systems, Faculty of Information Technology,
Monash University, Clayton Campus, Wellington Road, Victoria 3800, Australia
Tel: +61 3-99052360
vincent.lee@infotech.monash.edu.au
http://www.bsys.monash.edu.au/index.html

Abstract. In this paper, we implement a pilot study on the detection of abnormal financial asset trading activities using an artificial immune system. We develop a prototype *artificial immune abnormal-trading-detecting system* (AIAS) to scan the proxy data from the stock market and detect the abnormal trading such as insider trading and market manipulation, etc. among them. The rapid and real time detection capability of abnormal trading activities has been tested under simulated stock market as well as using real intraday price data of selected Australian stocks. Finally, three parameters used in the AIAS are tested so that the performance and robustness of the system are enhanced.

1 Introduction

Recent years have seen increased interests in applying biologically inspired systems, e.g. neural networks, evolutionary computation and DNA computation, natural immune system, etc. to solve real world complex problems. Current research challenges have focused on the exploitation of the key features of the natural immune system (NIS), such as Recognition, Feature Extraction, Diversity, Learning, Memory, Distributed Detection, Self-regulation and Adaptability, etc [1]. A NIS comprises a complex system of cells, molecules and organs that jointly represent an identification mechanism capable of perceiving and combating exogenous infectious microorganisms [1], which contain many antigens that are substances that can trigger immune responses, resulting in production of antibodies as part of the body's defence against infection and disease to neutralize related antigens [2].

Generally speaking, an artificial immune system (AIS) is a specific computational algorithm which takes its inspiration from the way how a NIS learns to respond to those exogenous invaders. It simulates the key features, such as adaptability, pattern recognition, learning, and memory acquisition of the NIS in order to deal with the problems [3] in computer security, anomaly detection, fault diagnosis, pattern recognition and a variety of other applications [4] in science and engineering, etc. As one of the main areas of the financial market, the stock market has an important concept – *noise*, which is defined as the fluctuations of price and volume that can confuse interpretation of market direction [5]. Accordingly, those investors undertaking trades which generate such confusions are termed as *noise traders*. Some noise traders are described as essential players of the stock market [6] whereas some (*insiders*) illegitimately take advantage of exclusive information which is still unavailable to the public to trade securities and disclose some information through a public signal consisting of a noisy transformation of his or her own private information [7]. Some *market manipulators* try to influence the price of a security in order to create false or misleading patterns of active trading to bring in more traders. Often, the newly brought in traders further cause significant or even disastrous deviation of security prices from the underlying values of the related assets resulting to the failure of correct interpretation of the market direction.

Currently, there are a number of security surveillance software packages available in the market, some of which have flaws. For example, applying the same benchmark for all securities have been proved improper because different types of securities represent various specific characteristics [8] and furthermore the metrics[1] are not evenly distributed over the day or week [9], e.g., the recurring intra-daily pattern in trade volume in Australian markets is high at the beginning and end of the trading day and lower in the mid trading day; Mondays generally have lower volume of trades than other days during the week [10]. Meanwhile, as for the metrics themselves, there is no consensus about which is the most appropriate one [8].

Hence, an *expansile* and *adaptive* abnormal-trading detecting system with the characteristics of good *self-learning* and *memory* capacities is proposed in this paper. Our proposed artificial abnormal detection system has the following advantages. Firstly, its *adaptivity* means that the system is able to learn the trading patterns of different stocks; and also that it is able to learn the different trading patterns of the same stock according to each economic period, trading period, and the locality of financial markets. Secondly, it is anticipated that better proxies will be continuously found; the *expansibility* of our proposed system allows those newly discovered proxies be added to the system. Thirdly, the system can be used as a tool to compare proxies so as to search for a more proper proxy for a specified stock in a certain period at a certain place. Finally, those proxies used in the system before they are out of date[2] will be eliminated or superseded. Thus with the continuous introduction of the new proxies into the system, the limited memory capacity of the system can be effectively utilized.

2 The Proposed Artificial Immune Abnormal-Trading-Detection System (AIAS)

The idea of the proposed Artificial Immune Abnormal-Trading-Detecting System (AIAS) is predicated on the process of natural immune system. In the AIAS, *antigens* are the format used to represent the proxy data of the stock market. Such an antigen is consisted of epitopes, namely, *No.*, *Type*, *Date*, *Weekday*, *Time* and *Value*, shown in Table 1.

---

[1] A *metric* is a proxy devised on the basis of the measurable attributes of the stock market, such as "traded volume" and "traded price change" etc.
[2] If a proxy is out of date, it is no longer capable of tracking and reflecting trace of the illegal trading.

Table 1. The Structure of an AIAS Antigen

No.	Type	Date	Weekday	Time	Value
Ag09	Volume	20/09/2004	Mon	15:10	200

On the other hand, *candidate antibodies* are the ones created on the basis of antigens. Those candidate antibodies comprise paratopes, namely, *No.*, *Type*, *Weekday Group*, *10-min-interval Group* and *99.5 Percentile Threshold* as shown in Table 2.

Table 2. The Structure of an AIAS Candidate Antibody

No.	Type	Weekday Group	10-min-interval Group	99.5 Percentile Threshold
Ab01	Change of Price	Wed-Thu	15:40-16:00	0.3

Candidate antibodies are generated on the basis of the antigens in the *Pools* and *Hatches* (see next two paragraphs) and they are the top values (a certain percentile, say above 99.5) of the grouped-up antigens (grouped into Weekday & 10-min-interval Combined Groups in accordance with antigens' Weekday Group and 10-min-interval Group epitopes). They are used just like benchmarks to compare with each incoming security data (i.e., each antigen) and are used to detect which of them is abnormal.

The AIAS has two main parts, namely, *Candidate Antibody Memory Repertoire* (or *Repertoire*), which corresponds to the peripheral lymphoid organ where antibodies meet antigens; and *Antibody Factory* (or *Factory*) which is corresponding to the central lymphoid organ of natural immune systems that are responsible for the production of antibodies. *Repertoire* is the environment (memory store) where candidate antibodies are stored and at where each incoming antigen into AIAS, based on the stock market data at the interval of 10 minutes [9], compares and matches with the candidate antibodies.

In *Factory*, there are numbers of places called *Antigen Archive Pools* (or *Pools*) that each is connected with its related *Antigen Archive Hatch Pools* (or *Hatches*). *Pools* are the places where the most recent 30-day antigen data [9] are archived. These antigen data in the *Pools* have sufficient resources to initialize and to update the candidate antibodies in the *Repertoire*, and the antigen data can help the AIAS learn the trading patterns of the participating security during a certain period at a certain location [9]. Linked to those *Pools*, there are corresponding *Hatches* of the same proxy type, which archive each incoming antigen from the stock market at a certain constant frequency. *Hatches* serve the function of a "buffer tank" and are used as a data source for iteration cycle in the adaptive learning of the AIAS.

The main function of the AIAS is first to extract and update the trading patterns (how a certain metric is distributed over all the time intervals) out of the proxy data in antigen format stored in the *Pools* and *Hatches* every day; then the candidate antibodies (benchmarks) are created on the basis of those trading patterns in order to compare with each invading antigen into the system so that the abnormality of the invading antigen is determined. Figure 1 shows the schematic block diagram of the proposed AIAS.

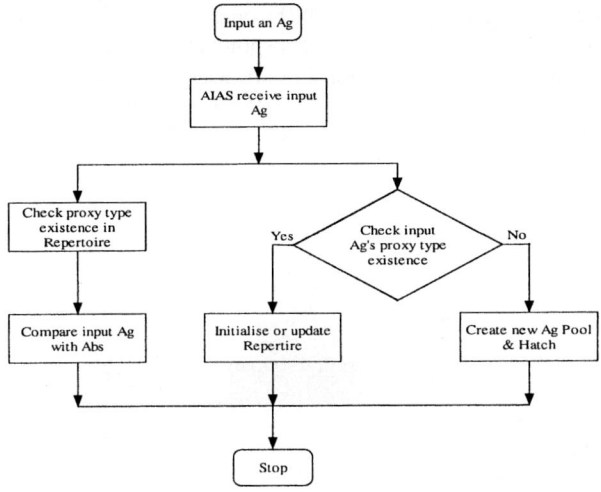

Fig. 1. The Schematic Block Diagram of AIAS

3 AIAS Testing

Two independent high frequency (10 minutes interval) datasets are used in the testing. The first dataset is derived by asset derivative formula and the second dataset is made up of the real historical stock data.

3.1 Test with Artificial High-Frequency Anomaly-Added Price-Change Data

a. Dataset

Price-change proxy data at the frequency of every 10 minutes are used to test the performance of the AIAS. The first dataset of this proxy used to test the AIAS is generated by *Random Walk* method [11].

$$S_{i+1} = S(1+\mu\delta t + \sigma\varphi\delta t^{1/2}) . \qquad (1)$$

where μ is the drift rate, δt is the timestep, σ is the volatility of the asset, φ is a random number from a Normal distribution.

Regarding this project, S_i is the initial asset price used to generate the 10-min-interval security price data of Qantas Airway Limited, the value of which is the close price ($3.71) of Qantas Airways Limited on 31/12/2004 quoted from the historical stock database of Finance.Yahoo.com. On the other hand, φ here is realised by

$$\left(\sum_{i=1}^{12} \text{RAND}() \right) . \qquad (2)$$

which is the sum of the 12 random numbers generated by the spreadsheet subtracted by 6. Additionally, μ is assigned with 0.15, σ 0.4, and the timestep 0.0003 (as for timesteps, each of them represents a weekday and 10-min-interval combination, for example, Timestep 0 corresponds to 16:00 on 31/12/2004, which the initial Qantas price belongs to; and Timestep 0.0003 corresponds to 10:00 on 04/01/2005).

After all the parameters of equation (1) are assigned, Qantas prices are simulated by the spreadsheet by invoking the Random Walk formula, which price changes have been shown to be normally distributed by SPSS as follows.

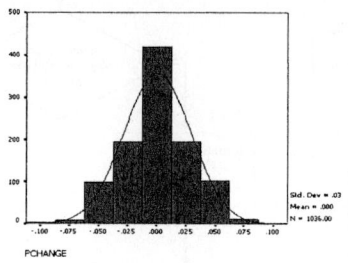

Fig. 2. The Histogram of Price_changes Data by SPSS

Then the price data are transferred into Price-change data, which are the absolute values of the differences between each 10-min-interval price and its previous one. But because the high-frequency Price-change dataset generated by Random Walk does not include the abnormal elements (such as insider trading, market manipulation, etc.), the real stock market has, therefore, some man-made purposive abnormal pieces of data (*anomalies*, which are generated values that are larger than those normal data they replace by a random amount) are aggregated with the high-frequency normal Price-change dataset generated before by replacing some random chosen data in it.

b. System Initiation and Validation

In order to initialise the Candidate Antibody Memory Repertoire, the immediate past 30-day proxy data are used as inputs into the AIAS. After the trading patterns are extracted, the result of the candidate antibodies of Price-change is shown in Table 3.

The thresholds of each antibody are supported by SPSS as shown in Figure 3.

Table 3. Candidate Antibodies of Price-change after Initialisation

	A	B	C	D	E
1	**Candidate Antibody Memory Repertoire**				
2	Listed by Proxy Types:				
3	TYPE - PC Candidate Antibodies				
4	No.	Type	Weekday Group	10-min-interval Group	99.5 Percentile Threshold
5	Ab 1	PC	Mon~Fri	10:00~10:10	0.10
6	Ab 2	PC	Mon~Fri	10:20~16:00	0.09

Statistics				Statistics		
PCCHANGE				PCCHANGE		
N	Valid	44		N	Valid	770
	Missing	0			Missing	0
Percentiles	99.5	.1000		Percentiles	99.5	.0914

Fig. 3. "99.5th Percentile" Calculated by SPSS for Each Initialised Antibody

Then the following date (31st) proxy data are input into the AIAS incorporated with two randomly man-made anomalies to check whether the system can detect those anomalies with the antibodies created before and also as a check whether the Repertoire can be updated on the basis of the new one day data. The results of the detection and update are shown in the two panels of Figure 4.

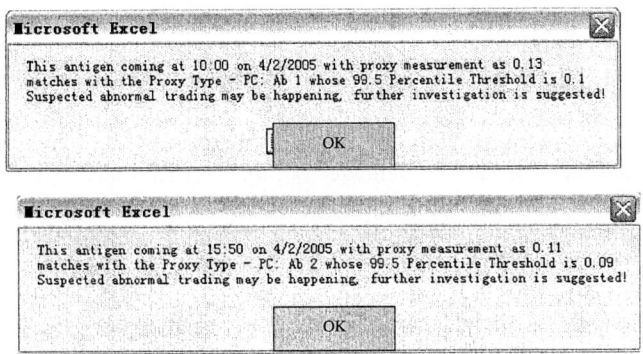

Fig. 4. Alerts Activated by Two Anomalies

Table 4. Candidate Antibodies of Price-change after Update

As shown in Figure 4, two man-made anomalies are detected by the AIAS and the related alerts are also launched to inform the users. Furthermore, the **Repertoire** is updated as well and the number of antibodies has increased to four compared with two given in Table 3. The result of the update is supported by SPSS as shown in Figure 5.

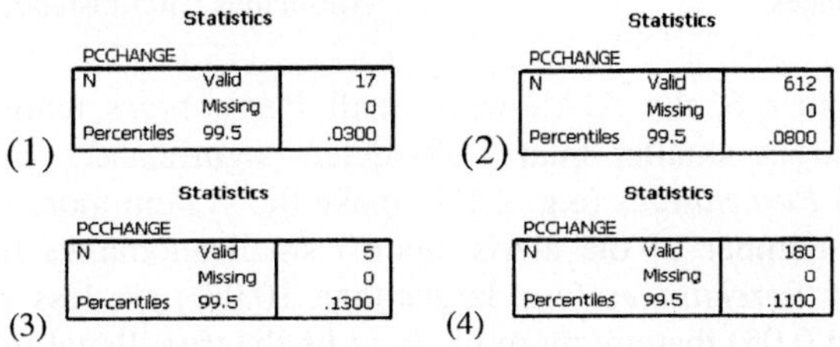

Fig. 5. "99.5th Percentile" Calculated by SPSS for Each Updated Antibody

3.2 Test with High-Frequency Real Stock Market Price-Change Data

According to *CORPORATE LAW ELECTRONIC BULLETIN*, Bulletin No. 52, December 2001 / January 2002, there was some insider trading (one on 24 of April, 2001 was proved) happening shortly before Qantas announced at the end of April 2001 that it would take over the operations of Impulse Airlines [12]. This test is to see whether the AIAS can detect some abnormal trading around the end of April 2001.

a. Parameter Comparisons:

In the proposed AIAS system, there are three parameters, namely, Percentage, Percentile and Multiplier, which directly influence the performance of the system.

The parameter *Percentage* is associated with the trading pattern extracting process, which is to group up those proxy data with similar measurements and of contiguous times into a certain trading pattern so that independent benchmark will be generated later according to the pattern. The "Percentage" controls the subtlety of how patterns differ from the others. If the "Percentage" is large, AIAS tends to ignore small proxy data differences and merge those close proxy data of contiguous times into one pattern. On the contrary, if it is small, the system tends to notice smaller differences and draw more patterns to distinguish them. To determine the appropriate value for assigning to this parameter, man-made anomalies (0.01, 0.02, 0.03, 0.04 and 0.05) are separately entered into the system with the *Percentage* equal to 2.5%, 5.0%, 7.5%, 10.0% and 12.5% respectively to replace all antigens through out a single week, e.g. the one-week (10:00~16:00 from Mon to Fri) antigens' measurements input into the AIAS all equal to 0.01 for the first time, then equal to 0.02 for the second and so on so forth. The performance of the system is shown in Figure 6.

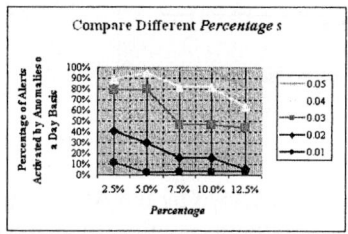

Fig. 6. The Performance of the AIAS with Different *Percentage*s under Various Anomalies Circumstances

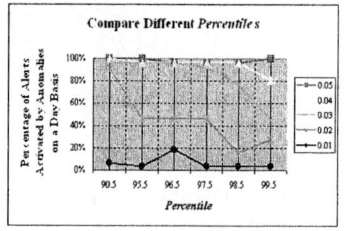

Fig. 7. The Performance of the AIAS with Different *Percentile*s under Various Anomalies Circumstances

As shown in Figure 6, the AIAS with small *Percentage*s tends to detect small anomalies (*Percentage*s smaller than 5.0% detect significantly more than others). However, too small *Percentage*s (e.g. 2.5%) make the system more sensitive to small anomalies and the number of the alerts against small anomalies increases. On the contrary, the higher *Percentage*s (e.g. larger than 10.0%) are less sensitive to large anomalies (0.04 and 0.05) that are more likely to be the true illegal trading signals. As for the anomaly 0.03 (the middle one), the AIAS with the *Percentage*s smaller than 7.5% seems over-sensitive. Therefore, relatively, *Percentage*s between 7.5% and 10.0% are more reasonable.

The parameter *Percentile* is associated to the likelihood of the AIAS to send an alert to the user. The higher this parameter is, the more alerts will be expected (more sensitive). The test results are shown in Figure 7. In Figure 7, the small parameters make the AIAS more sensitive to the small anomalies (e.g. 0.02), while the high ones make the system less sensitive to large anomalies (e.g. 0.04). This implies that middle *Percentile*s (95.5, 96.5 and 97.5) is better. Among the middle three, the 95.5 is best for detecting the 0.05 anomaly, hence the *Percentile*s 95.5 is used.

The parameter *Multiplier* is used by the AIAS to define which sort of abnormal antigens should be ignored by the system so that their measurements will not be referred to when antibodies are initialised or updated. The test, similar to the above two tests, which performance is shown in Figure 8.

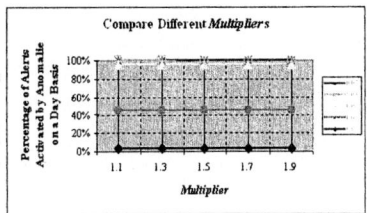

Fig. 8. The Performance of the AIAS with Different *Multiplier*s under Various Anomalies

As shown in Figure 8, the AIAS has almost same performance with the small *Multiplier*s (1.1 and 1.3) detects a little more 0.04 anomaly than the others. This small difference is caused by the extreme antigens, defined as not extreme by other *Multiplier*s, but are treated as extreme when the *Multiplier* equals 1.1 or 1.3 and discarded by the AIAS. To circumvent over-sensitivity of the system, 1.5 is used.

b. Dataset and System Testing:
The dataset in this test is the real 10-minute-interval traded-price data of Qantas Airways Limited (21/03/2001 ~ 24/04/2001). Among them, the data from 21/03/2001 to 19/04/2001 are used to initialise the candidate antibodies, the Repertoire is shown in Table 5.

Table 5. The Initialised Candidate Antibody Memory Repertoire

A	B	C	D	E
Candidate Antibody Memory Repertoire				
Listed by Proxy Types:				
TYPE - PC Candidate Antibodies				
No.	Type	Weekday Group	10-min-interval Group	95.5 Percentile Threshold
Ab 1	PC	Mon	10:00	0.01
Ab 2	PC	Mon	10:10	0.00
Ab 3	PC	Mon	10:20	0.01
Ab 4	PC	Mon	10:30~11:20	0.01
Ab 5	PC	Mon	11:30~16:00	0.01
Ab 6	PC	Tue	10:00	0.03
Ab 7	PC	Tue	10:10	0.00
Ab 8	PC	Tue	10:20	0.02
Ab 9	PC	Tue	10:30~11:20	0.04
Ab 10	PC	Tue	11:30~16:00	0.02
Ab 11	PC	Wed	10:00	0.02
Ab 12	PC	Wed	10:10	0.00
Ab 13	PC	Wed	10:20	0.01
Ab 14	PC	Wed	10:30~11:20	0.01
Ab 15	PC	Wed	11:30~16:00	0.02
Ab 16	PC	Thu	10:00	0.03
Ab 17	PC	Thu	10:10	0.00
Ab 18	PC	Thu	10:20	0.01
Ab 19	PC	Thu	10:30~11:20	0.02
Ab 20	PC	Thu	11:30~16:00	0.00
Ab 21	PC	Fri	10:00	0.00
Ab 22	PC	Fri	10:10	0.00
Ab 23	PC	Fri	10:20	0.01
Ab 24	PC	Fri	10:30~11:20	0.01
Ab 25	PC	Fri	11:30~16:00	0.01

The data from 20/04/2001 to 24/04/2001 are used as input into the system day by day. At the same time the *Repertoire* is updated at the end of every day, so that the system adjusts the trading patterns by referring the every new day stock data. All the alerts launched by the AIAS are shown in Figure 9.

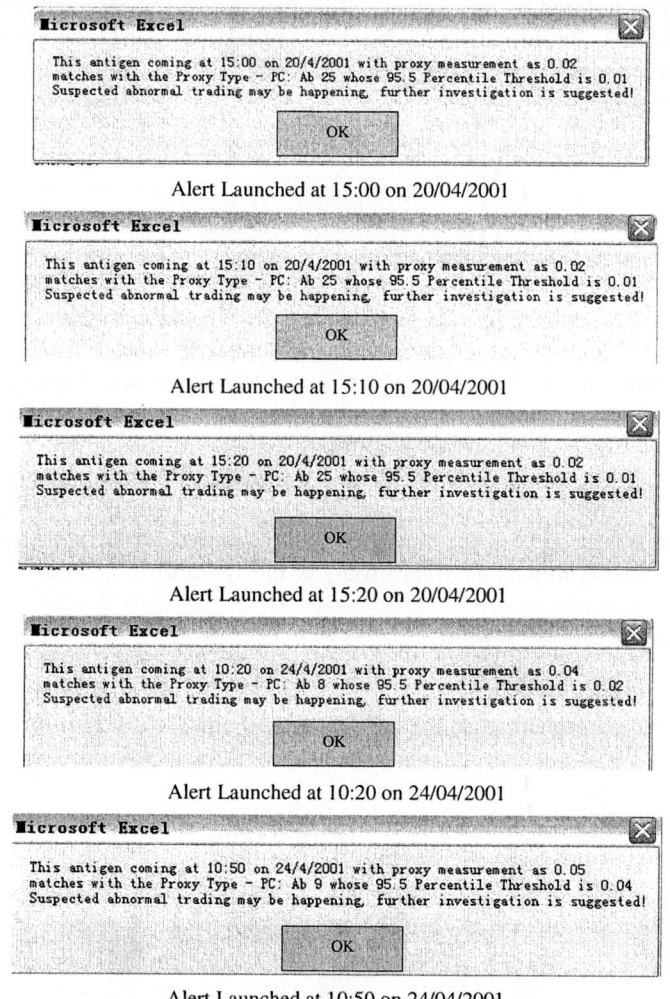

Fig. 9. The Various Alerts Launched by the AIAS from 20/04/2001 to 24/04/2001

As shown in Figure 9, there is no alert launched on 23/04/2001, but three on 20th and two on 24th respectively. Note: the *Rivkin*'s insider trading case happened on 24/04/2001 [12].

4 Conclusion

This paper explores the practicability of the AIS in the financial area. Through the literature reviews over the NIS and the financial area, the related NIS theory is applied to develop and construct the proposed *artificial immune abnormal-trading-detecting system* (AIAS). The core function of our proposed AIAS is to extract and update the trading patterns out of the proxy data from the stock market and generate related benchmarks (candidate antibodies) to detect abnormal trading on the market.

Compared with other software packages already used to surveil the stock market, the AIAS is able to identify/update trading patterns and generate benchmarks automatically without any personnel intervention. However, the algorithms applied in the system to these calculations are worth further researching in order to improve AIAS's real time performance. Meanwhile, the system can also compare the performance of different metrics and co-using of two or more metrics interactively so as to enhance the detection capability of the system. All those will be the objectives for the future study.

References

1. De Castro, L. N., Von Zuben, F. J.: Artificial Immune Systems: Part I – Basic Theory And Applications. Springer Press. (Dec.1999)
2. Anon., Asthma & Allergy. (2004) Glossary. CelebrateLove.com owned and operated by James Larry & CelebrateLove.com. Accessed October, Available from: http://www.celebratelove.com/asthmaglossary.htm
3. Dasgupta, D.: Artificial Immune Systems and Their Applications. Springer. (1998)
4. Timmis, J., Bentley, P., Hart, E.: Artificial Immune Systems, in Proceedings of Second International Conference, ICARIS 2003, Edinburgh, UK. (Sep. 2003)
5. Anon., Noise [Glossary online] (2004). IT Locus.com owned and operated by IT Locus. Accessed September, Available from: http://itlocus.com/glossary/noise.html
6. Black, F., "Noise", Journal of Finance, Vol. 41(3) (1985) 530 - 531
7. Gregoire, P., Huangi, H.: Insider Trading, Noise Trading and the Cost of Equity. (2001) 4
8. Aitken, M., Siow, A.: Improving the Effectiveness of the Surveillance Function in Securities Markets. Working paper. Department of Finance. University of Sydney. Sydney. (2003)
9. Anon., Issue 4 (2004) – The Importance of Benchmark Creation [Database online]. smarts.com owned and operated by SMARTS Limited. Accessed August 2004, Available from: http://www.smarts.com.au/discovery/09d_Discovery4.html
10. Anon., (2004), Access to software tools [Library online]. smarts.com owned and operated by SMARTS Limited. Accessed August, Available from:
http://www.smarts.com.au/new/library/issue.asp
11. Wilmott, Paul: Derivatives – The theory and practice of financial engineering. Chichester, England: John Wiley & Sons Ltd. (1998)
12. Anon., (2005) Archive of Corporate Law Bulletins of Melbourne University. Owned and operated by Centre for Corporate Law and Securities Regulation, Faculty of Law, The University of Melbourne. Accessed February, Available from: http://cclsr.law.unimelb.edu.au/bulletins/archive/Bulletin0052.htm

Adaptive Parameter Selection of Quantum-Behaved Particle Swarm Optimization on Global Level

Wenbo Xu and Jun Sun

School of Information Technology, Southern Yangtze University,
No.1800, Lihu Dadao, Wuxi Jiangsu 214122, China
xwb@sytu.edu.cn
sunjun_wx@hotmail.com

Abstract. In this paper, we formulate the philosophy of Quantum-behaved Particle Swarm Optimization (QPSO) Algorithm, and suggest a parameter control method based on the population level. After that, we introduce a diversity-guided model into the QPSO to make the PSO system an open evolutionary particle swarm and therefore propose the Adaptive Quantum-behaved Particle Swarm Optimization Algorithm (AQPSO). Finally, the performance of AQPSO algorithm is compared with those of Standard PSO (SPSO) and original QPSO by testing the algorithms on several benchmark functions. The experiments results show that AQPSO algorithm outperforms due to its strong global search ability, particularly in the optimization problems with high dimension.

1 Introduction

Particle Swarm Optimization (PSO), motivated by the collective behaviors of bird and other social organisms, is a novel evolutionary optimization strategy introduced by J. Kennedy and R. Eberhart in 1995 [1]. It has already been shown that PSO is comparable in performance with traditional optimization algorithms such as simulated annealing (SA) and the genetic algorithm (GA) [3], [4], [5], [7].

Since its origin in 1995, many revised versions of PSO have been proposed to improve the performance of the algorithm. In 1998, Shi and Eberhart introduced inertia weight w into evolution equation to accelerate the convergence speed [9] and therefore proposed the so-called Standard PSO (SPSO). In 1999, Clerc employed Constriction Factor K to guarantee convergence of the algorithm and release the limitation of velocity [8]. Ozcan in 1999 and Clerc in 2002 did trajectory analysis of PSO respectively [10], [13], and their works provide the golden rule of parameter selection. Other achievements in this field include Neighborhood Topology Structure [12], Selection Operator in PSO [3], Binary Version of PSO [14] and so forth.

In our previous work, we introduce quantum theory into PSO and propose a Quantum-behaved PSO based on Delta potential well (QPSO) algorithm [2], [15]. The experiment results testified that QPSO works better than standard PSO on several benchmark functions and is a promising algorithm due to its global convergence-guaranteed characteristic.

In this paper, we will propose an Adaptive Quantum-Behaved Particle Swarm Optimization algorithm based on Diversity-Guided Model, the parameter control of which is adaptive on global level and are able to overcome the problem of premature convergence efficiently. The paper is structured as follows. In Section 2, the philosophy of QPSO is formulated. Section 3 proposes a Diversity-Guided QPSO and Section 4 shows the numerical results of the algorithms. Some conclusion remarks are given in Section 5.

2   Quantum-Behaved Particle Swarm Optimization

In the Standard PSO model, each individual is treated as a volume-less particle in the D-dimensional space, with the position vector and velocity vector of the ith particle represented as $X_i(t)=(x_{i1}(t),x_{i2}(t),\cdots,x_{iD}(t))$ and $V_i(t)=(v_{i1}(t),v_{i2}(t),\cdots,v_{iD}(t))$. The particles move according to the following equation:

$$v_{id}(t+1) = w * v_{id}(t) + \varphi_1(P_{id} - x_{id}(t)) + \varphi_2(P_{gd} - x_{id}(t)) \ . \tag{1a}$$

$$x_{id}(t+1) = x_{id}(t) + v_{id}(t+1) \ . \tag{1b}$$

where φ_1 and φ_2 are random numbers whose upper limits are parameters of the algorithm that have to be selected carefully. Parameter w is the inertia weight introduced to accelerate the convergence speed of the PSO. Vector $P_i = (P_{i1}, P_{i2}, \cdots, P_{iD})$ is the best previous position (the position giving the best fitness value) of particle i called *pbest*, and vector $P_g = (P_{g1}, P_{g2}, \cdots, P_{gD})$ is the position of the best particle among all the particles in the population and called *gbest*.

In essence, the traditional model of PSO system is of linear system, if *pbest* and *gbest* are fixed as well as all random numbers are considered constant. Trajectory analysis [13] shows that, whatever model is employed in the PSO algorithm, each particle in the PSO system converges to its Local Point (LP) $p = (p_1, p_2, \cdots p_D)$, one and only local attractor of each particle, of which the coordinates are

$$p_d = (\varphi_1 P_{id} + \varphi_2 P_{gd})/(\varphi_1 + \varphi_2) \ . \tag{2}$$

so that the *pbests* of all particles will converges to an exclusive *gbest* with $t \to \infty$.

In the quantum model of a PSO, the state of a particle is depicted by wavefunction $\Psi(\bar{x},t)$, instead of position and velocity. The dynamic behavior of the particle is widely divergent from that of the particle in traditional PSO systems in that the exact values of X and V cannot be determined simultaneously. We can only learn the probability of the particle's appearing in position X from probability density function $|\psi(X,t)|^2$, the form of which depends on the potential field the particle lies in.

In our previous work [2], employed Delta potential well with the canter on point $p = (p_1, p_2, \cdots p_D)$ to constrain the quantum particles in PSO in order that the particle can converge to their local p without explosion.

For simplicity, we consider a particle in one-dimensional space firstly. With point p the center of potential, the potential energy of the particle in one-dimensional Delta potential well is represented as

$$V(x) = -\gamma\delta(x - p) . \tag{3}$$

Hence, we can solve the *Schrödinger equation* and obtain the normalized wavefunction as

$$\psi(x) = \frac{1}{\sqrt{L}} \exp(-\|p - x\|/L) . \tag{4}$$

The probability density function is

$$Q(x) = |\psi(x)|^2 = \frac{1}{L} \exp(-2\|p - x\|/L) . \tag{5}$$

and the distribution function is

$$D(y) = \int_{-\infty}^{y} Q(y) dy = \frac{1}{L} e^{-2|y|/L} . \tag{6}$$

The parameter L depending on energy intension of the potential well specifies the search scope of a particle. Using Monte Carlo Method, we can get the position of the particle

$$x(t+1) = p \pm \frac{L(t)}{2} \ln(1/u) \quad u = rand(0,1) . \tag{7}$$

where p is defined by equation (2) and u is random number distributed uniformly on (0,1). In [2], the parameter L is evaluated by

$$L(t) = 2 * \alpha * |p - x(t)| . \tag{8}$$

Thus the iterative equation of Quantum PSO is

$$x(t+1) = p \pm \alpha * |p - x(t)| * \ln(1/u) . \tag{9}$$

which replaces Equation (1) in PSO algorithm.

In [15], we employ a mainstream thought point to evaluate parameter L. The Mainstream Thought Point or Mean Best Position (*mbest*) is defined as the mean value of all particles' *pbest*s. That is

$$mbest = \frac{1}{M}\sum_{i=1}^{M} P_i = \left(\frac{1}{M}\sum_{i=1}^{M} P_{i1}, \frac{1}{M}\sum_{i=1}^{M} P_{i2}, \cdots, \frac{1}{M}\sum_{i=1}^{M} P_{id}\right) . \tag{10}$$

where M is the population size and p_i is the *pbest* position of particle i. The value of L is given by

$$L(t) = 2 * \beta * |mbest - x(t)| . \tag{11}$$

where β is called Contraction-Expansion Coefficient. Thus Equation (7) can be written as

$$x(t+1) = p \pm \beta * |mbest - x(t)| * \ln(1/u) . \tag{12}$$

This is the iterative equation of QPSO.

3 Diversity-Guided Model of QPSO

As we know, a major problem with PSO and other evolutionary algorithms in multi-modal optimization is premature convergence, which results in great performance loss and sub-optimal solutions. In a PSO system, with the fast information flow between particles due to its collectiveness, diversity of the particle swarms declines rapidly, leaving the PSO algorithm with great difficulties of escaping local optima. Therefore, the collectiveness of particles leads to low diversity with fitness stagnation as an overall result. In QPSO, although the search space of an individual particle at each iteration is the whole feasible solution space of the problem, diversity loss of the whole population is also inevitable due to the collectiveness.

Recently, R. Ursem has proposed a model called Diversity-Guided Evolutionary Algorithm (DGEA) [6], which applies diversity-decreasing operators (selection, recombination) and diversity-increasing operators (mutation) to alternate between two modes based on a distance-to-average-point measure. The performance of the DGEA clearly shows its potential in multi-modal optimization.

In 2002, Riget *et al* [7] adopted the idea from Usrem into the basic PSO model with the decreasing and increasing diversity operators used to control the population. This modified model of PSO uses a diversity measure to have the algorithm alternate between exploring and exploiting behavior. They introduced two phases: attraction and repulsion. The swarm alternate between these phases according to its diversity and the improved PSO algorithm is called Attraction and Repulsion PSO (ARPSO) algorithm.

Inspired by works undertaken by Ursem and Riget *et al*, we introduce the Diversity-Guided model in Quantum-behaved PSO. As Riget did, we also define two phases of particle swarm: attraction and repulsion. It can be demonstrated that when the Contraction-Expansion Coefficient satisfies $\beta \leq 1.7$, the particles will be bound to converge to its Local Point p, and the particles will diverge from p when $\beta > 1.8$. Consequently, the two phases is distinguished by the parameter β and defined as

- Attraction Phase: $\beta = \beta_a$, where $\beta_a \leq 1.7$;
- Repulsion Phase: $\beta = \beta_r$, where $\beta_r > 1.8$.

In attraction phase ($\beta = \beta_a$) the swarm is contracting, and consequently the diversity decreases. When the diversity drops below a lower bound, d_{low}, we switch to the repulsion phase ($\beta = \beta_r$), in which the swarm expands. Finally, when the diversity reaches a higher bound d_{high}, we switch back to the attraction phase. The result of this is a QPSO algorithm that alternates between phases of exploiting and exploring-attraction and repulsion-low diversity and high diversity, according to the diversity of the swarm measured by

$$diversity(S) = \frac{1}{|S| \cdot |A|} \cdot \sum_{i=1}^{|S|} \sqrt{\sum_{j=1}^{D} (x_{ij} - \overline{x_j})^2} \quad . \tag{13}$$

where S is the swarm, $|S| = M$ is the population size, $|A|$ is the length of longest the diagonal in the search space, D is the dimensionality of the problem, x_{ij} is the jth value of the ith particle and $\overline{x_j}$ is the jth value of the average point

The Quantum-behaved PSO algorithm with attraction and repulsion phases is called Adaptive Quantum-Behaved Particle Swarm Optimization (AQPSO) algorithm, which is described as following.

```
Initialize the population
Do
   find out the mbest of the swarm;
   measure the diversity of the swarm by equation (13);
   if (diversity<dlow)
      β=βa;
   if (diversity>dhigh) begin
      β=βr;
   for I=1 to population size M;
      if f(pi)<f(xi) then pi=xi
      pg=argmin(pi)
      for d=1 to dimension D
         fi=rand(0,1);
         p=fi*pid+(1-fi)*pgd;
         u=rand(0,1);
         if rand(0,1)>0.5;
            xid=p-β*abs(mbestd-xid)*log(1/u);
         else
            xid=p+β*abs(mbestd-xid)*log(1/u);
         end
   end
until the termination criterion is met.
```

In AQPSO, d_{low} and d_{high} is two parameters. The smaller the former, the higher the search precision of the algorithm is, and the larger the latter is, the strong the global search ability of AQPSO. Generally, d_{low} can be set to be less than 0.01, and d_{high} can be set to be between 0.1 and 1. As of Creativity Coefficient, $β_a$ must be less than 1.0, and $β_r$ should be 1.8 at least.

4 Experiment Results

To test the performance of AQPSO, seven benchmark functions are used here for comparison with SPSO in QPSO. The first function F_1 is Sphere function, the second function F_2 is called Rosenbrock function, the third function F_3 is the generalized Rastrigrin function, the fourth function F4 is generalized Griewank function, and the fifth function F5 is Shaffer's function. These functions are all minimization problems with minimum value zero. Their expressions are described in Table 1.

In all experiments, the initial range of the population listed in Table 1 is asymmetry and Table 1 also lists V_{max} and X_{ma} values for all the functions, respectively.

The fitness value is set as function value and the neighborhood of a particle is the whole population. We had 50 trial runs for every instance and recorded mean best fitness. In order to investigate the scalability of the algorithm, different population sizes M are used for each function with different dimensions. The population sizes are 20, 40 and 80. Generation is set as 1000, 1500 and 2000 generations corresponding to

Table 1. Expressions of benchmark functions and configuration of some parameters

	Functions	Initial Range	X_{max}	V_{max}
F_1	$\sum_{i=1}^{n} x_i^2$	(50, 100)	100	100
F_2	$\sum_{i=1}^{n}(100(x_{i+1} - x_i^2)^2 + (x_i - 1)^2)$	(15, 30)	100	100
F_3	$f(x)_3 = \sum_{i=1}^{n}(x_i^2 - 10\cos(2\pi x_i) + 10)$	(2.56, 5.12)	10	10
F_4	$f(x)_4 = \frac{1}{4000}\sum_{i=1}^{n}(x_i - 100)^2 - \prod_{i=1}^{n}\cos(\frac{x_i - 100}{\sqrt{i}}) + 1$	(300, 600)	600	600
F_5	$f(x)_5 = 0.5 + \frac{(\sin\sqrt{x^2 + y^2})^2 - 0.5}{(1.0 + 0.001(x^2 + y^2))^2}$	(30, 100)	100	100

the dimensions 10, 20 and 30 for first five functions, respectively, and the dimension of the last two functions is 2. We also test performance of the QPSO, in which the Contraction-Expansion Coefficient β decreases from 1.0 to 0.5 linearly when the algorithm is running. In the experiments to test AQPSO, we set β_a to be 0.73, β_r to be 2.0, d_{low} to be 0.005, and d_{high} to be 0.25.

The mean values and standard deviations of best fitness values for 50 runs of each function are recorded in Table 2 to Table 6. The average best fitness of the first four functions that indicate the convergence performance of the algorithms is shown in Figure 1. The numerical results show that the AQPSO works better on Rosenbrock benchmark functions than and has comparable performance on Rstringrin and Griewank functions with QPSO and SPSO. On Shaffer's function, the performance of AQPSO is better than QPSO. However, on Sphere function, AQPSO is less effective. Moreover, Figure 1 shows that AQPSO converge more rapidly that the other two algorithms in the early stage but slows down in later stage.

5 Conclusion

In this paper, based on the Quantum-behaved PSO, we formulate the philosophy of quantum-behaved PSO, which is not discussed in detail in [2]. And then, we set forth a diversity-guided parameter control and propose AQPSO algorithm. The AQPSO generally outperforms QPSO and SPSO in global search capability. In the AQPSO, the evaluation of parameter L depends on a global position, Mean Best Position (*mbest*), which is relatively stable as the population is evolving, and parameter β alternate between two phases (attraction and repulsion). In this model, the PSO system is an open system, and therefore the global search capability of the algorithm is enhanced.

Table 2. Sphere Function

M	Dim.	Gmax	SPSO		QPSO		AQPSO	
			Mean Best	St. Dev.	Mean Best	St. Dev.	Mean Best	St. Dev.
20	10	1000	3.16E-20	6.23E-20	2.29E-41	1.49E-40	8.53E-08	2.28E-07
	20	1500	5.29E-11	1.56E-10	1.68E-20	7.99E-20	3.77E-07	5.81E-07
	30	2000	2.45E-06	7.72E-06	1.34E-13	3.32E-13	5.58E-05	9.26E-05
40	10	1000	3.12E-23	8.01E-23	8.26E-72	5.83E-71	2.43E-09	5.22E-09
	20	1500	4.16E-14	9.73E-14	1.53E-41	7.48E-41	5.64E-08	7.05E-08
	30	2000	2.26E-10	5.10E-10	1.87E-28	6.73E-28	5.04E-06	1.23E-05
80	10	1000	6.15E-28	2.63E-27	3.1E-100	2.10E-99	2.14 E-10	4.50E-09
	20	1500	2.68E-17	5.24E-17	1.56E-67	9.24E-67	2.36 E-09	7.72E-09
	30	2000	2.47E-12	7.16E-12	1.10E-48	2.67E-48	4.14 E-07	6.33E-07

Table 3. Rosenbrock Function

M	Dim.	Gmax	SPSO		QPSO		AQPSO	
			Mean Best	St. Dev.	Mean Best	St. Dev.	Mean Best	St. Dev.
20	10	1000	94.1276	194.3648	59.4764	153.0842	35.9477	37.1740
	20	1500	204.337	293.4544	110.664	149.5483	91.8425	73.9530
	30	2000	313.734	547.2635	147.609	210.3262	78.0670	53.9014
40	10	1000	71.0239	174.1108	10.4238	14.4799	12.4419	8.3466
	20	1500	179.291	377.4305	46.5957	39.5360	45.0769	22.1808
	30	2000	289.593	478.6273	59.0291	63.4940	61.6228	69.3932
80	10	1000	37.3747	57.4734	8.63638	16.6746	13.7910	17.5702
	20	1500	83.6931	137.2637	35.8947	36.4702	33.1462	35.4184
	30	2000	202.672	289.9728	51.5479	40.8490	58.4388	65.3852

Table 4. Rastrigrin Function

M	Dim.	Gmax	SPSO		QPSO		AQPSO	
			Mean Best	St. Dev.	Mean Best	St. Dev.	Mean Best	St. Dev.
20	10	1000	5.5382	3.0477	5.2543	2.8952	4.9333	3.7982
	20	1500	23.1544	10.4739	16.2673	5.9771	26.592	21.8851
	30	2000	47.4168	17.1595	31.4576	7.6882	42.2754	19.2729
40	10	1000	3.5778	2.1384	3.5685	2.0678	4.6597	4.3427
	20	1500	16.4337	5.4811	11.1351	3.6046	20.7342	16.4114
	30	2000	37.2796	14.2838	22.9594	7.2455	35.2366	13.3873
80	10	1000	2.5646	1.5728	2.1245	1.1772	2.36765	1.4331
	20	1500	13.3826	8.5137	10.2759	6.6244	12.4195	10.2988
	30	2000	28.6293	10.3431	16.7768	4.4858	26.3222	6.3102

Table 5. Griewank Function

M	Dim.	Gmax	SPSO		QPSO		AQPSO	
			Mean Best	St. Dev.	Mean Best	St. Dev.	Mean Best	St. Dev.
20	10	1000	0.09217	0.08330	0.08331	0.06805	0.06182	0.04715
	20	1500	0.03002	0.03255	0.02033	0.02257	0.02660	0.02113
	30	2000	0.01811	0.02477	0.01119	0.01462	0.01523	0.01837
40	10	1000	0.08496	0.07260	0.06912	0.05093	0.05536	0.03694
	20	1500	0.02719	0.02517	0.01666	0.01755	0.02829	0.02484
	30	2000	0.01267	0.01479	0.01161	0.01246	0.01423	0.01715
80	10	1000	0.07484	0.07107	0.03508	0.02086	0.05751	0.03058
	20	1500	0.02854	0.02680	0.01460	0.01279	0.03050	0.02369
	30	2000	0.01258	0.01396	0.01136	0.01139	0.01243	0.01463

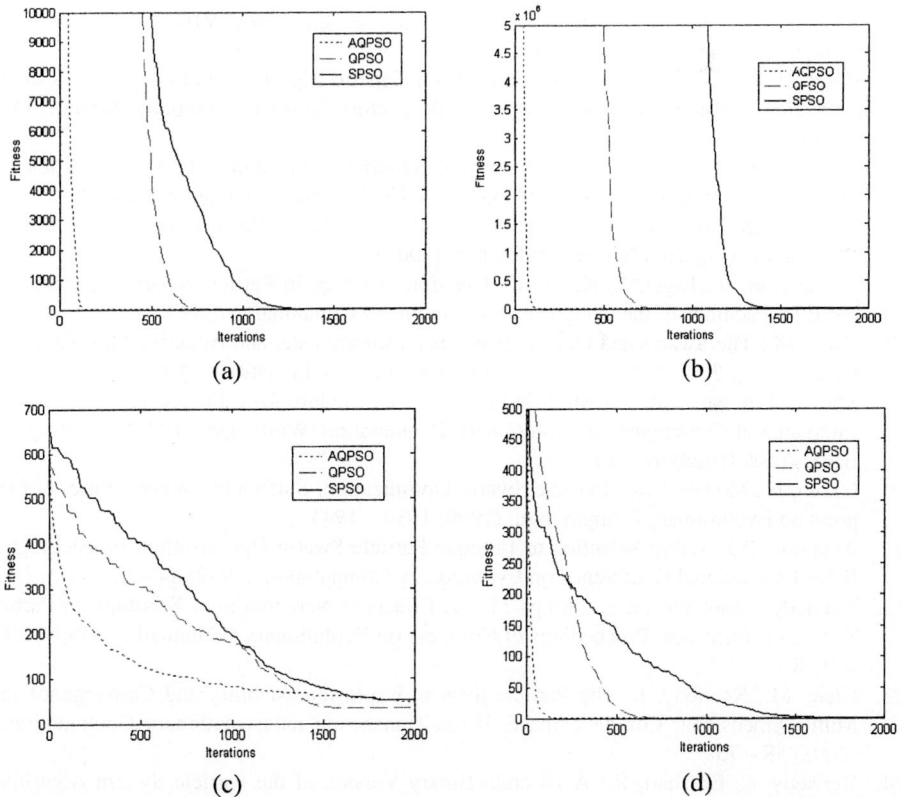

Fig. 1. The figure shows average best fitness of the benchmark functions with 30 dimensions. The population size is 20. (a) Sphere function; (b) Rosenbrock function; (c) Rastrigrin function; (d) Griewank function.

Table 6. Shaffer's Function

M	Dim.	Gmax	SPSO		QPSO		AQPSO	
			Mean Best	St. Dev.	Mean Best	St. Dev.	Mean Best	St. Dev.
20	2	2000	2.78E-04	0.000984	0.001361	0.003405	0.000758	0.000834
40	2	2000	4.74E-05	3.59E-05	3.89E-04	0.001923	1.53E-05	4.97E-05
80	2	2000	3.67E-10	3.13E-10	1.72E-09	3.30E-09	1.46E-10	5.67E-10

References

1. Kennedy, J., Eberhart, R.: Particle Swarm Optimization. Proceedings of IEEE Int. Conf. On Neural Network. (1995) 1942 - 1948
2. Sun, J., Feng, B., Xu, W.: Particle Swarm Optimization with Particles Having Quantum Behavior. IEEE Proc. of Congress on Evolutionary Computation. (2004)
3. Angeline, P.J.: Evolutionary Optimizaiton Versus Particle Swarm Opimization: Philosophyand Performance Differences. Evolutionary Programming VIII. Lecture Notes in Computer Science 1477. Springer. (1998) 601 – 610
4. Eberhart, R.C., Shi, Y.: Comparison between Genetic Algorithm and Particle Swarm Optimization. Evolutionary Programming VII. Lecture Notes in Computer Science 1447. Springer (1998) 611 – 616
5. Krink, T., Vesterstrom, J., Riget, J.: Particle Swarm Optimization with Spatial Particle Extension. IEEE Proceedings of the Congress on Evolutionary Computation. (2002)
6. Ursem, R.K.: Diversity-Guided Evolutionary Algorithms. Proceedings of The Parallel Problem Solving from Nature Conference. (2001)
7. Vesterstrom, J., Riget, J., Krink, T.: Division of Labor in Particle Swarm Optimization. IEEE Proceedings of the Congress on Evolutionary Computation. (2002)
8. Clerc, M.: The Swarm and Queen: Towards a Deterministic and Adaptive Particle Swarm Optimization. Proc. IEEE Congress on Evolutionary Computation. (1999) 1591 – 1597
9. Shi, Y., Eberhart, R.C.: A Modified Particle Swarm Optimizer. Proceedings of the IEEE International Conference on Evolutionary Computation. Washington. DC. Piscataway. NJ: IEEE Press. (1998) 69 – 73
10. Ozcan, E., Mohan, C.K.: Particle Swarm Optimization: Surfing the Waves. Proc. of Congress on Evolutionary Computation. (1999) 1939 – 1944
11. Angeline, P.J.: Using Selection to Improve Particle Swarm Optimization. Proceedings of IEEE International Conference on Evolutionary Computation. (1998) 84 – 89
12. Kennedy: Small Worlds and Mega-Minds: Effects of Neighborhood Topology on Particle Swarm Performance. Proceedings of Congress on Evolutionary Computation. (1999) 1931 – 1938
13. Clerc, M., Kennedy, J.: The Particle Swarm: Explosion, Stability and Convergence in a Multi-Dimensional Complex Space. IEEE Transaction on Evolutionary Computation. 6 (2002) 58 – 73
14. Kennedy, J., Eberhart, R.: A Discrete Binary Version of the Particle Swarm Algorithm. Proceedings of IEEE conference on Systems, Man and Cybernetics. (1997) 4104 – 4109
15. Sun, J., et al: A Global Search Strategy of Quantum-behaved Particle Swarm Optimization. Proceedings of IEEE conference on Cybernetics and Intelligent Systems. (2004) 111 - 116

New Method for Intrusion Features Mining in IDS[1]

Wu Liu, Jian-Ping Wu, Hai-Xin Duan, and Xing Li

Network Research Center of Tsinghua University, 100084 Beijing, P. R. China
liuwu@ccert.edu.cn

Abstract. In this paper, we aim to develop a systematic framework to semi-automate the process of system logs and databases of intrusion detection systems (IDS). We use both Ef-attribute based mining and Es-attribute based mining to mine effective and essential attributes (hence interesting patterns) from the vast and miscellaneous system logs and IDS databases.

1 Introduction

Intrusion prevention techniques, such as user authentication, authorization, and access control etc. are not sufficient [5]. Intrusion Detection System (IDS) is therefore needed to protect computer systems. Currently many intrusion detection systems are constructed by manual and ad hoc means. In [2] rule templates specifying the allowable attribute values are used to post-process the discovered rules. In [4] Boolean expressions over the attribute values are used as item constraints during rule discovery. In [3], a "belief-driven" framework is used to discover the unexpected (hence interesting) patterns. A drawback of all these approaches is that one has to know what rules/patterns are interesting or are already in the belief system. We cannot assume such strong prior knowledge on all audit data.

We aim to develop a systematic framework to semi-automate the process of building intrusion detection systems. We take a data-centric point of view and consider intrusion detection as a data analysis task. Anomaly detection is about establishing the normal usage patterns from the audit data, whereas misuse detection is about encoding and matching intrusion patterns using the audit data.

2 Basic Ideas of Our Methods for Intrusion Features Mining

We attempt to develop general rather than intrusion-specific tools in response to the challenges of IDS. The idea is to first compute the association rules and frequent episodes from audit data, which capture the intra-and inter-audit record patterns. These frequent patterns can be regarded as the statistical summaries of system activities captured in the audit data, because they measure the correlations among system features and temporal co-occurrences of events. Therefore, with user participation,

---

[1] This work is supported by grants from 973, 863 and the National Natural Science Foundation of China (Grant No. #90104002 & #2003CB314800 & #2003AA142080 & #60203044) and China Postdoctoral Science Foundation.

these patterns can be utilized, to guide the audit data gathering and feature selection processes.

We would like to stress that the class/style files and the template should not be manipulated and that the guidelines regarding font sizes and format should be adhered to. This is to ensure that the end product is as homogeneous as possible.

From [5], let A be a set of attributes, and I be a set of values on A, called items. Any subset of I is called an item-set. The number of items in an item-set is called its length. Let D be a database with n attributes (columns). We define support(X) as the percentage of transactions (records) in D that contain item-set X. An association rule is the expression

$$X \to Y, c, s \tag{1}$$

Here X and Y are item-sets, and $X \cap Y = \varnothing$. $s = support(X \cup Y)/support(X \cup Y)$ is the support of the rule, and c = support(X) is the confidence. For example, an association rule from the shell command history file of a user is

$$Trn \to rec.humor, 0.3, 0.1, \tag{2}$$

which indicates that 30% of the time when the user invokes Trn, he or she is reading the news in rec.humor, and reading this news group accounts for 10% of the activities recorded in his or her command history file. We implemented the association rules algorithm following the main ideas of Apriori [1]. Briefly, we call an item-set X a frequent item-set if $support(X) \geq minimum\ support$, observed that any subset of a frequent item-set must also be a frequent item-set. The algorithm starts with finding the frequent item-sets of length 1, then iteratively computes frequent item-sets of length k + 1 from those of length k. This process terminates when there are no new frequent item-sets generated. It then proceeds to compute rules that satisfy the minimum confidence requirement.

3   Our Methods for Intrusion Features Mining in IDS

3.1   Ef-Attribute Based Mining

Although we cannot know in advance what patterns, which involve actual attribute values, are interesting, we often know what attributes are more important or useful given a data analysis task, which we the effective attribute or *Ef-attribute*. Assume I is the interestingness measure of a pattern p, then:

$$I(p) = f(support(p); confidence(p)) \tag{3}$$

where f is some ranking function. We propose here to incorporate schema-level information into the interestingness measures. Assume IA is a measure on whether a pattern p contains the specified important (i.e. interesting) attributes, our extended interestingness measure is:

$$Ie(p) = fe(IA(p); f(support(p); confidence(p))) = fe(IA(p); I(p)) \tag{4}$$

where fe is a ranking function that first considers the attributes in the pattern, then the support and confidence values.

In the next section, we describe a schema-level characteristic of audit data, in the form of "what attributes must be considered", that can be used to guide the mining of relevant features.

3.2 Es-Attribute Based Mining

There is a partial "importance-order" among the attributes of an audit record. Some attributes are essential in describing the data, while others only provide auxiliary information. A network connection can be uniquely identified by

$$< timestamp; src\ host; src\ port; dst\ host; service > \tag{5}$$

that is, the combination of its start time, source host, source port, destination host, and service (destination port). These are the essential attributes when describing network data. We argue that the "relevant" association rules should describe patterns related to the essential attributes. Patterns that include only the unessential attributes are normally "irrelevant". For example, the basic association rules algorithm may generate rules such as:

$$src\_bytes = 200 \rightarrow flag = SF \tag{6}$$

These rules are not useful and to some degree are misleading. There is no intuition for the association between the number of bytes from the source, src bytes, and the normal status (flag = SF) of the connection, but rather it may just be a statistical correlation evident from the dataset.

Here, we call the essential attribute the *Es-attribute* when they are used as a form of item constraints in the association rules algorithm. During candidate generation, an item set must contain value(s) of the Es-attribute. We consider the correlations among non- essential attribute as not interesting. In other words,

$$I_A(p) = \begin{cases} 1 & if\ p\ \text{contains Es-attribute} \\ 0 & \text{otherwise} \end{cases} \tag{7}$$

In practice, we need not designate all essential attributes as the Es-attributes. For example, some network analysis tasks require statistics about various network services while others may require the patterns related to the hosts. We can use service as the Es-attribute to compute the association rules that describe the patterns related to the services of the connections.

It is even more important to use the Es-attribute to constrain the item generation for frequent episodes. The basic algorithm can generate serial episode rules that contain only the unimportant attribute values. For example

$$src\ bytes = 1000;\ src\ bytes = 1000 \rightarrow dst\ bytes = 1500;\ src\ bytes = 1000 \tag{8}$$

Note that here each attribute value is from a different connection record. To make matter worse, if the support of an association rule on non-Es-attributes, $X \rightarrow Y$, is high then there will be a large number of "useless" serial episode rules of the form $(X|Y)(,X|Y)* \rightarrow (X|Y)(,X|Y)*$, due to the following theorem:

Theorem 1. Let s be the support of the association $X \rightarrow Y$, and let N be the total number of episode rules on (X|Y), i.e., rules of the form:$(X|Y)(,X|Y)^* \rightarrow (X|Y)(,X|Y)^*$, then N is at least an exponential factor of s.

To avoid having a huge amount of "useless" episode rules, we extended the basic frequent episodes algorithm [5] to compute frequent sequential patterns in two phases: first, it finds the frequent associations using the Es-attribute; second, it generates the frequent serial patterns from these associations. That is, for the second phase, the items (from which episode item-sets are constructed) are the associations about the Es-attribute, and the Es-attribute values. An example of a rule is

$$(service = ftp; src\ bytes = 1000); \tag{9}$$

$$(service = http;\ src\ bytes = 1000) \rightarrow (service = ftp;\ src\ bytes = 1500) \tag{10}$$

Note that each item-set of the episode rule is an association. We in effect have combined the associations among attributes and the sequential patterns among the records into a single rule. This rule formalism not only eliminates irrelevant patterns, it also provides rich and useful information about the audit data.

4 Using the Mined Features

In this section we present our experience in mining the audit data and using the discovered patterns both as the indicator for gathering data and as the basis for selecting appropriate temporal statistical features.

4.1 Data Collecting

We posit that the patterns discovered from the audit data on a protected target (e.g., a network, system program, or user, etc.) corresponds to the target's behavior. When we gather audit data about the target, we compute the patterns from each new audit data set, and merge the new rules into the existing aggregate rule set. The added new rules represent variations of the normal behavior. When the aggregate rule set stabilizes, i.e., no new rules from the new audit data can be added, we can stop the data gathering since the aggregate audit data set has covered sufficient variations of the mined patterns. Our approach of merging rules is based on the fact that even the same type of behavior will have slight differences across audit data sets. Therefore we should not expect perfect (exact) match of the mined patterns. Instead we need to combine similar patterns into more generalized ones.

We merge two rules, r_1 and r_2, into one rule r if

- their right and left hand sides are exactly the same, or their right hand sides can be combined and left hand sides can also be combined; and
- the support values and the confidence values are close, i.e., within a user-defined threshold.

The concept of combining here is similar to clustering in [6] in that we also combine rules that are "similar" syntactically with regard to their attributes, and are "adjacent" in terms of their attribute values. That is, two left hand sides (or right hand sides) can be combined, if

- they have the same number of item-sets; and
- each pair of corresponding item-sets (according to their positions in the patterns) have the same Es-attribute value(s), and the same or adjacent non-essential attribute value(s).

As an example, consider combining the left hand sides and assume that the left hand side of r_1 has just one item-set,

$$(ax_1 = vx_1; a_1 = v_1) \tag{11}$$

Here ax_1 is an Es-attribute. The left hand side of r_2 must also have only one item-set,

$$(ax_2 = vx_2; a_2 = v_2) \tag{12}$$

Further, $ax_1 = ax_2$, $vx_1 = vx_2$, and $a_1 = a_2$ must hold. For the left hand sides to be combined, v_1 and v_2 must be the same value or adjacent bins of values. The left hand side of the merged rule r is

$$(ax_1 = vx_1; v_1 \leq a_1 \leq v_2) \tag{13}$$

assuming that v_2 is the larger value. For example,

$$(service = ftp; src\ bytes = 1000) \tag{14}$$

And

$$(service = ftp; src\ bytes = 1500) \tag{15}$$

can be combined into

$$(service = ftp; 1000 \leq src\ bytes \leq 1500) \tag{16}$$

To compute the statistically relevant support and confidence values of the merged rule r, we record support lhs and db size of r_1 and r_2 when mining the rules from the audit data. Here support lhs is the support of a LHS and db size is the number of records in the audit data. The support value of the merged rule r is

$$\sup port(r) = \frac{support(r_1) \times db\_size(r_1) + support(r_2) \times db\_size(r_2)}{db\_size(r_1) + db\_size(r_2)} \tag{17}$$

And the support value of LHS of r is

$$\sup port\_lhs(r) = \frac{support\_lhs(r_1) \times db\_size(r_1) + support\_lhs(r_2) \times db\_size(r_2)}{db\_size(r_1) + db\_size(r_2)} \tag{18}$$

And therefore the confidence value of r is

$$confidence(r) = \frac{\sup port(r)}{\sup port\_lhs(r)} \tag{19}$$

4.2 Feature Selection

An important use of the mined patterns is as the basis for feature selection. When the Es-attribute and Ef-attribute used as the class label attribute, features (the attributes) in the association rules should be included in the classification models.

In addition, the time windowing information and the features in the frequent episodes suggest that their statistical measures, e.g., the average, the count, etc., should also be considered as additional features.

5 Experimental Results

Here we test our hypothesis that the merged rule set can indicate whether the audit data has covered sufficient variations of behavior.

We obtained the TCP/IP network traffic data from CCERT (China Education and Research Network Computer Emergency Response Team) IDS, we hereafter refer it as the CCERTIDS dataset). We segmented the data by day. And for data of each day, we again segmented the data into four partitions: morning, afternoon, evening and night. This partitioning scheme allows us to cross evaluate anomaly detection models of different time segments that have different traffic patterns. It is often the case that very little (sometimes no) intrusion data is available when building an anomaly detector. A common practice is to use audit data (of legitimate activities) that is known to have different behavior patterns for testing and evaluation.

Here we describe the experiments and results on building anomaly detection models for the "weekday morning" traffic data on connections originated from CCERTIDS to the outside world. We compute the frequent episodes using the network service as the Es-attribute. Recall from our earlier discussion that this formalism captures both association and sequential patterns. For the first three weeks, we mined the patterns from the audit data of each weekday morning, and merged them into the aggregate rule set. For each rule we recorded merge count, the number of merges on this rule. Note that if two rules r_1 and r_2 are merged into one rule r, its merge count is the sum from the two rules. Merging count indicates how frequent the behavior represented by the merged rule is encountered across a period of time.

We call the rules with (20) the *frequent rules*.

$$\text{merge count} \geq \text{min frequency} \qquad (20)$$

Figure 1 shows how the rule set changes as we merge patterns from each new audit data set. We see that the total number of rules keeps increasing. We visually inspected the new rules from each new data set. In the first two weeks, the majority are related to new network services that have no prior patterns in the aggregate rule set. And for the last week, the majority is just new rules of the existing services. Figure 1 shows that the rate of change slows down during the last week. Further, when we examine the frequent rules (here we used min frequency = 2 to filter out the "one-time" patterns), we can see in the figure that the rule sets of all services as well as the individual services grow at a much slower rate and tend to stabilize.

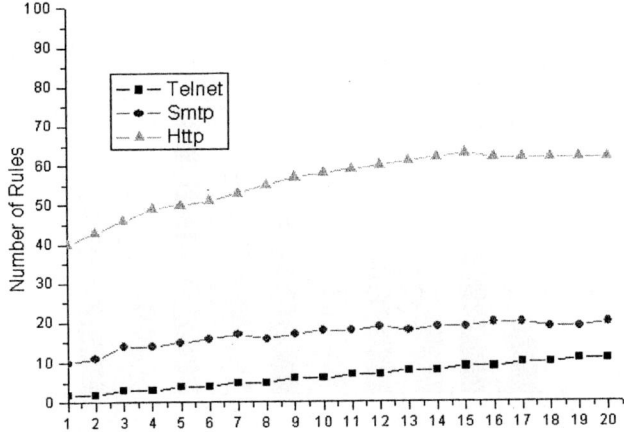

Fig. 1. The Number of Rules vs. The Number of Audit Data Sets

We used the set of frequent rules of all services as the indicator on whether the audit data is sufficient. We tested the quality of this indicator by constructing four classifiers, using audit data from the first 8, 10, 16, and 18 weekday mornings, respectively, for training. We used the services of the connections as the class labels, and included a number of temporal statistical features (the details of feature selection is discussed in the next session). The classifiers were tested using the audit data (not used in training) from the mornings and nights of the last 5 weekdays of the month, as well as the last 5 weekend mornings.

Figures 2 to 5 show the performance of these four classifiers in detecting anomalies (different behavior) respectively. In each figure, we show the misclassification rate (percentage of misclassifications) on the test data.

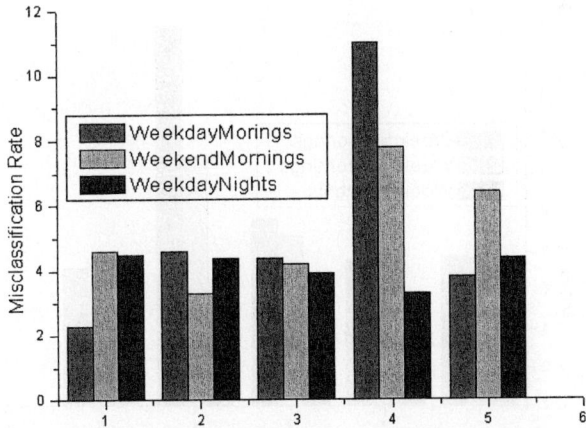

Fig. 2. Misclassification Rates of Classifier Trained on First 8 Weekdays

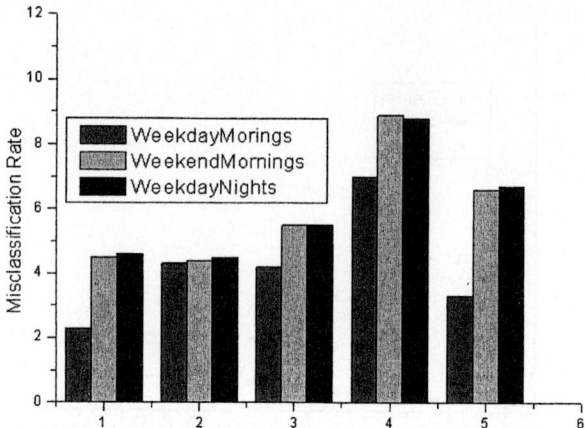

Fig. 3. Misclassification Rates of Classifier Trained on First 10 Weekdays

Since the classifiers model the weekday morning traffic, we wish to see this rate to be low on the weekday morning test data, but high on the weekend morning data as well as the weekday night data. The figures show that the classifiers with more training (audit) data perform better. Further, the last two classifiers are effective in detecting anomalies, and their performance are very close (see figures 4 and 5). This is not surprising at all because from the plots in figure 1, the set of frequent rules (our indicator on audit data) is growing in weekdays 8 and 10, but stabilizes from day 16 to 18. Thus this indicator on audit data gathering is quite reliable.

An important use of the mined patterns is as the basis for feature selection. When the Ef-attribute and Es-attribute are used as the class label attribute, features (the attributes) in the association rules should be included in the classification models.

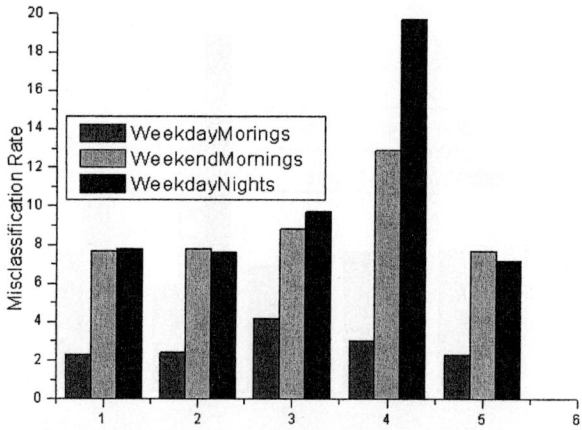

Fig. 4. Misclassification Rates of Classifier Trained on First 16 Weekdays

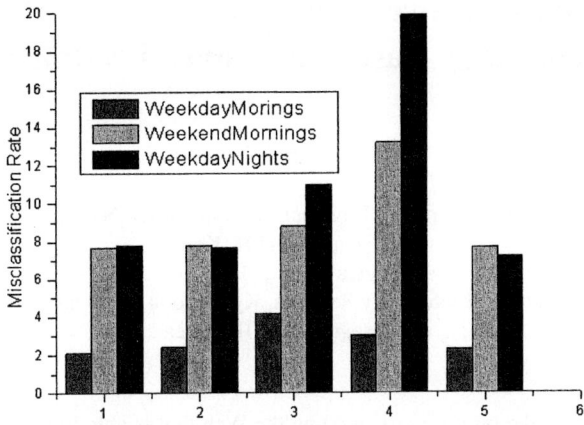

Fig. 5. Misclassification Rates of Classifier Trained on First 18 Weekdays

In addition, the time windowing information and the features in the frequent episodes suggest that their statistical measures, e.g., the average, the count, etc., should also be considered as additional features.

6 Conclusion

In this paper, we discussed data mining techniques of Ef-attribute based mining and Es-attribute based mining for mining effective and essential attributes (hence interesting patterns) from the vast and miscellaneous system logs and IDS databases.

References

1. Agrawal, R. and Srikant, R.: Fast algorithms for mining association rules. In Proceedings of the 25th VLDB Conference, Santiago, Chile, 2004
2. Klemettinen, M., Mannila, H., Ronkainen, P., Toivonen, H., and Verkamo, A. I.: Finding interesting rules from large sets of discovered association rules. In Proceedings of the 6rd International Conference on Information and Knowledge Management, Gainthersburg, MD, 2002
3. Padmanabhan, B. and Tuzhilin, A.: A belief-driven method for discovering unexpected patterns. In Proceedings of the 4th International Conference on Knowledge Discovery and Data Mining, New York, NY, August 1998
4. Srikant, R., Vu, Q., and Agrawal, R.: Mining association rules with item constraints. In Proceedings of the 8rd International Conference on Knowledge Discovery and Data Mining, Newport Beach, California, August 2001
5. Wu LIU, Study on Intrusion Detection Technology with Traceback and Isolation of Attacking Sources, PhD Thesis 2004
6. Lent, B., Swami, A., and Widom, J.: Clustering association rules. In Proceedings of the 13th International Conference on Data Engineering, Birmingham, UK, 1997

The Precision Improvement in Document Retrieval Using Ontology Based Relevance Feedback

Soo-Yeon Lim[1] and Won-Joo Lee[2]

[1] Department of Computer Engineering, Kyungpook National University,
Daegu, 702-701, Korea
nadalsy@sejong.knu.ac.kr
[2] Information Technology Services, Kyungpook National University,
Daegu, 702-701, Korea

Abstract. For the purpose of extending the Web that is able to understand and process information by machine, Semantic Web shared knowledge in the ontology form. For exquisite query processing, this paper proposes a method to use semantic relations in the ontology as relevance feedback information to query expansion. We made experiment on pharmacy domain. And in order to verify the effectiveness of the semantic relation in the ontology, we compared a keyword based document retrieval system that gives weights by using the frequency information compared with an ontology based document retrieval system that uses relevant information existed in the ontology to a relevant feedback. From the evaluation of the retrieval performance, we knew that search engine used the concepts and relations in ontology for improving precision effectively. Also it used them for the basis of the inference for improvement the retrieval performance.

1 Introduction

There are many search engines of general purpose for searching for web pages suitable for queries requested on the web. In order to obtain information suitable in a professional field, however, professional knowledge is required conforming thereto, which makes a user feel difficult. Most users have difficulty in making queries in a fixed form in a context without detailed medical knowledge or a specific environment for search in the field, prepared in advance. For efficient search, most users search for necessary information, while partially checking the result and correcting queries, repetitively. The purpose of this paper is to reduce such a user's difficulty, to check properly user's intention and to provide related information in order to enhance the quality of information retrieval. In particular, in order to reduce difficulties resulting from special terminologies or concepts such as medical information which general users encounter, this paper puts emphasis on semantic information retrieval by means of positive user interface agents on a terminology centered ontology basis.

This paper builds a domain ontology with the semantic relation that exists in an ontology by extracting a semantic group and hierarchical structure after classifying and analyzing the patterns of terminology that appeared in a Korean document as a type of compound noun[10]. A built ontology can be used in various fields. This paper proposes an ontology that classifies words related to a specific subject by using a hierarchical structure for a method to improve the performance of retrieving a document. A retrieval engine can be used as a base of inference to use the retrieval function using the concept and rule defined in the ontology. In order to experiment this retrieval engine, the texts that exists in a set of documents related to the pharmacy field are used as an object of the experiment.

The existing methods for extracting terminology can be largely classified by a rule based method and statistics based method. A rule-based method[3,4,6,8,9] builds a configuration pattern of terminology by hand or learning corpus and recognizes terminology by building a recognition pattern automatically by using it. Here, it presents a relatively exact result because people directly describe the rules using a noun or suffice dictionary. A statistics based method[5,11] uses some kind of knowledge, such as the hidden Markov model, maximum entropy model, word type, and vocabulary information in order to learn the knowledge of the recognition from a learning corpus. This paper extracts terminology by using a rule based method and configures a rule to extract terminology by analyzing the appearance patterns of the terminology.

In this paper, it is an object to propose a scheme for building a domain ontology using the analysis results of texts in a document set and to apply the built ontology in professional knowledge approach such as information retrieval. A test domain is defined as a pharmacy field, and the document set for building is for medicine manuals consisting of 21,113 documents

2 Scheme for Building Ontology

2.1 Steps of Building Ontology

The building process of the proposed ontology consists of four steps as follows. First, the web documents that exist in the related web will be collected to create a corpus and structurized through a document transformation process. Second, the verbs that will express the relation between the nouns that become concept and the extracted concepts will be extracted after passing a simple natural language process. Third, the terminology will be extracted from the extracted concepts and produce a hierarchical structure from the results of the analysis of the structure. Finally, the extracted relations will be added to the existing ontology with the concepts. Fig. 1 generally shows the ontology building process in four steps proposed in this paper.

For building an ontology, a designer decides a minor node, which is located in the upper level, and builds an ontology based on it and extends it. It is called by a base ontology. This paper configures a base ontology using 48 words. In order to configure this scheme, the concepts of the name of a disease, symptoms, drugs are defined as the highest node and its 45 hyponym nodes. The hyponym nodes consist of each node group; 20 nodes, which are defined by following the classification of a specific noun or suffices that form the name of a disease or symptom that exist in the pharmacy domain, and 15 nodes for the configured structures, and 10 nodes, which express a general noun that has a high frequency of appearance.

The stop words included in a document will be removed after the processes of morpheme analysis and tagging. Then, all nouns and verbs are extracted from the sentences of a document. In order to perform this process, the stop words list was

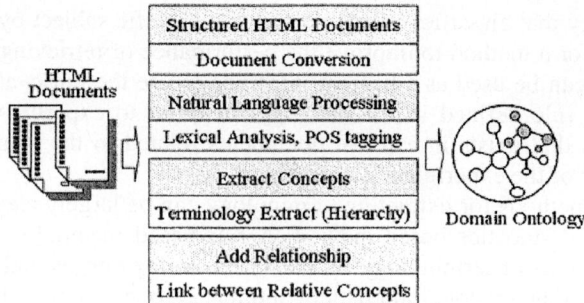

Fig. 1. A process of building ontology

made by using 181 words by considering the morphological characteristics of Korean language in which a part of suffices was excluded in the process of stemming because these suffices will be nicely used to extract terminology. The extracted nouns from an ontology that is a kind of network with a lot of words mean the concept of ontology. The tags and verbs present the relation between the concepts and act as a network, which plays a role in the connection of the concepts to each other.

There are some proper nouns and compound nouns in the documents what we have an interest in, such as the name of a disease, symptoms, ingredients, and other things. The proper nouns that present a major concept are processed the same way as a general noun. In addition, the terminology that appeared as a compound noun in the domain is extracted and hierarchicalized, and then is added into the ontology.

In order to give meaning to the extracted concepts in this paper, two methods are used. One is to use tag values attached to the head of a text and the other is to extract and use verbs existing in the text. Tag values in the structured document represent semantic relations to connect tagged word sets with higher concepts. Depending on the attached tag values in this paper, it was classified as 15 semantic relations. To define the relation among concepts, all verbs in the document were extracted and the verbs were classified as semantic patterns to set 18 semantic relations. The relationship between the extracted verbs and nouns can be verified by the co-occurrence information. If there is a relationship that exists between the nouns and verbs, otherwise it will compare a relationship between other nouns and verbs.

2.2 Extracting Terminology

Terminology is a set of words that has a specific meaning in a given domain and means a lexical unit that characterizes a subject by expressing the concept used in a domain. A language resource for terminology is important to perform effectively and precisely, such as a machine translation or retrieving information for a specific domain because this terminology is a necessary element to understanding a domain.

This paper analyzes an appearance type of terminology in order to extract the information automatically. The shape combining of terminology shows very varied ways. Almost all of the terminology appeared in an appropriate domain presented as a type of compound noun and can be classified as two types as follows. The one is a singleton term, that is, it has a simple shape of combining with one word that has no spacing words. The other is a multi-word term that has spacing words and is a kind of compound noun with two more words in which it has a semantic relation with the front element of a word.

The nouns and suffices that are configured by a singleton term terminology are classified by 20 kinds, such as yeom[1](염), jeung(증), tong(통), gyun(균), seong(성), jilwhan(질환), sok(속), yeomjeung(염증), jin(진), gam(감), jong(종), byeong(병), yeol(열), gweyang(궤양), seon(선), baekseon(백선), jeunghugun(증후군), hyeong(형), hwan(환), gun(균) in which it is appended by "hyponymOf" because it is almost a lower level word of a specific noun. A multi-word term terminology has almost a relation of modifier and keyword like "acute bronchitis" in which there are many cases that a keyword consists of terminology, which has a singleton term. This paper configures 5 semantic patterns and defines the semantic relation of an ontology according to these patters [10].

2.3 Extending Ontology

A built ontology can be extended by other resources, such as other ontologies, thesaurus, and dictionaries. It is possible to reduce the time and cost to extend an ontology by applying the predefined concepts and rules by using the existing resources.

The process in Fig.2 consists of 3 steps, such as the import, extract, and append. The step of import means the bringing and using of external resources. This paper uses two external resources(http://www.nurscape.net/nurscape/dic/frames.html, http://www. encyber.com/) In this step, the extraction was applied to terminology and its hyponym concepts for the results of the appropriate concept retrieving because the range of text was so wide. Then, the extracted concepts will be appended at a proper location by considering the upper and hyponym relations in the ontology.

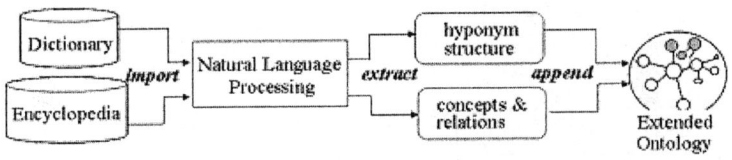

Fig. 2. A process of extending ontology

3 The Application of Document Retrieval

A built ontology can be used in various fields. This paper proposes a method that will improve the effect of retrieving a document using this ontology. A process of ontology selects a major set of documents in a specific field and extracts the concepts by analyzing these documents, and then appends these concepts by using a link. The objective of concept extraction is extracting the nouns that will represent the documents as well as possible. Especially, in the case of a retrieval or question-answering system using an ontology given by weights, it is helpful to the judgment of a user by presenting a selected minor information according to the weights.

---

[1] 'yeom' is Korean phonetic spelling of '염'.

This paper extracts the concepts using the extraction method proposed in this study and configures it as a node in the ontology. At this moment, the objective of this process is retrieving not only the concepts that are an inputted query in the ontology but also its hyponym concepts. A relevance feedback is well known as an effective way to process a reformation of a query that is an important part to access information. In the case of applying the relevance feedback to improve the traditional method of $tf \cdot idf$, it is well recognized that it can improve a precision rate if it is applied to a set of sample document of small scale.

This paper uses a hierarchical relation for the process of user relevance feedback in the ontology. The query will be extended by using the terminologies, which appeared as a hyponym information in the ontology that related to the inputted queries, and calculates the weights for the rewritten query. In this process, the hyponym retrieval level to retrieve the nodes in the ontology was set by 2. For instance, let us assume that the hyponym nodes of 'exudative otitis media' and 'acute exudative otitis media' for the node of 'otitis media' exist in the ontology. If a query [otitis media] is inputted as an input, the set of queries will be extended as [otitis media, secretory otitis media, acute secretory otitis media] after retrieving the ontology. Then, it will recalculate the similarity based on its weights. The most widely known method of (term-weights allocation strategy) will be used to grant the weights. The calculated weights will increase the retrieving speed and precision by storing it in order of the arrangement with a document number to the appeared document. Fig. 3 presents the configuration of a document retrieval system mentioned above in which it largely consists of a pre-processing module and retrieval module.

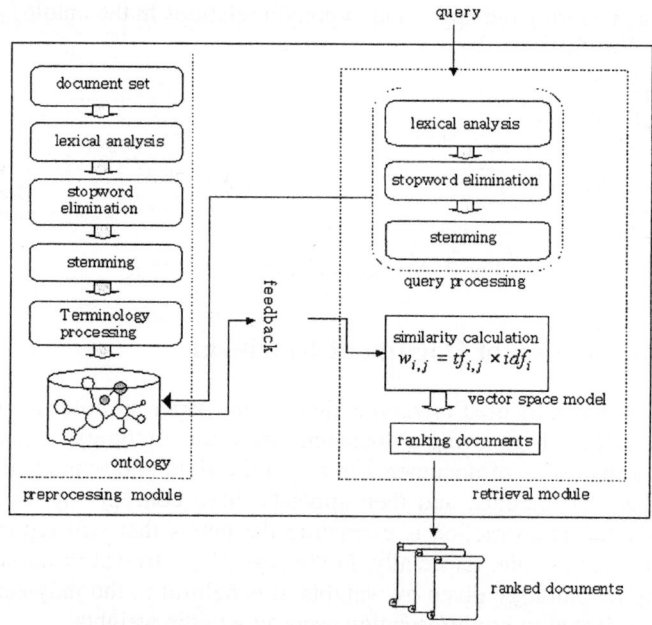

Fig. 3. configuration of the document retrieval system

First of all, a morpheme analysis process will be done in order to configure a set of index terms for the object documents in the preprocessing. From the results, the nouns are only extracted as a set of index terms in which the nouns will be mainly applied to get statistical information that represents a document for retrieving information and its classification. In this system, the ontology will be used as a set of index terms. In order to compare the retrieval performance, a set of the upper 30 correct answered documents for the 10 queries by introducing the 5 specialists' advices about 430 documents was configured. Then, the rates of recall and precision for each question were produced based on this configuration.

4 Experiments

From the results of the analysis of the morpheme of text in the formed corpus, the terminologies that were applied by specific expressions or patterns will be extracted from the extracted nouns. In the case of the failure of analyzing a morpheme due to the errors of word spacing or typing, the errors of the analyzing was modified.

The experiment was applied by the extraction method proposed in this paper for the text in the pharmacy domain. The experiment documents used in this experiment are 21,113. From the results of the expression analysis, the total numbers of the extracted nouns are 78,902. The numbers of terminology for the entire extracted nouns of 78,902 are 55,870. It covers about 70.8% for the total nouns. This means that the rate of terminology is very high in a specific domain.

Fig.4 presents the distribution according to the appearance types of terminology. There is good recognition for the terminologies that appeared and addition of 2,864 hyponym concepts after applying the extraction algorithm proposed in this paper in which the average level of the nodes, which existed in the ontology, is 1.8. The extracted terminologies can be investigated by three specialists by hand and evaluated by the precision. The precision of the extraction shows the ratio of the terminologies that are related by the correct relation for the extracted terminologies.

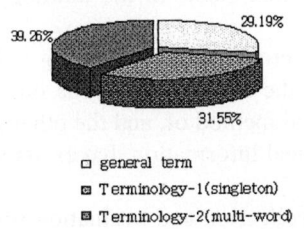

Fig. 4. distribution of terminologies

Fig. 5 shows the precision of the extraction of the terminologies that have a singleton term. From these results, the precision of the singleton term is 92.57% that shows a relatively good performance compared with the proposed algorithm. Fig. 6 presents the extraction precision of the multi-word term terminologies. From the results, the average precision of the multi-word term terminologies was 79.96%. In addition, 574 concepts were added.

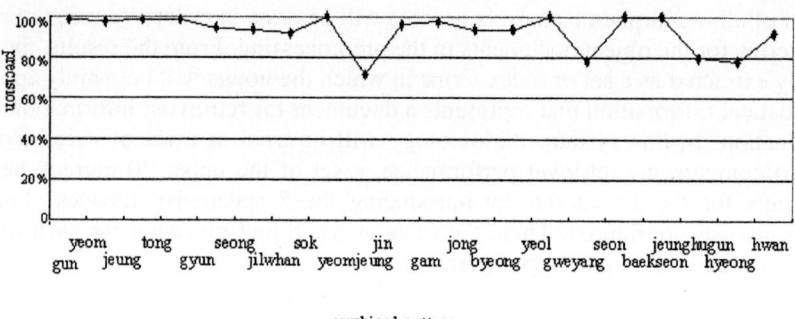

Fig. 5. precision of the extracted singleton term terminologies

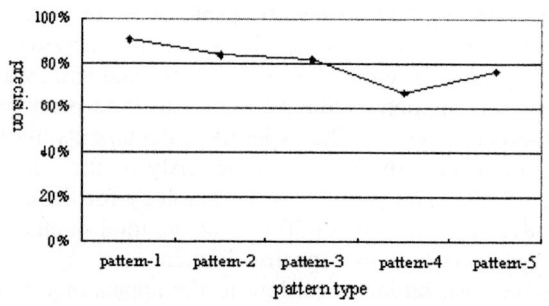

Fig. 6. precision of the extracted multi-word term terminologies

In order to verify the effectiveness of the built ontology, a keyword based document retrieval system that gives weights by using the traditional method of *tf·idf* will be compared and analyzed with an ontology based document retrieval system that uses a hyponym information that exists in the ontology to a relevant feedback and recalculates the weights.

In order to present the effectiveness for retrieving a document using the proposed method, this study compares the two methods. The one is a keyword based retrieval method by using the traditional method of, and the other is an ontology based retrieval method by using the hierarchical information that exists in the ontology to a relevance feedback.

An experiment reference collection and evaluation scale is used to evaluate an information retrieval system. An experiment reference collection consists of a set of literatures, information query examples, and set of relevant literatures for each information query. This paper collects 430 health/disease information documents from the home page of Korean Medical Association (http://www.kma.org) in order to configure a reference collection and configures an information query using 10 queries as follows. The experiment was carried for the 430 documents extraction. The objective of the experiment was to produce the recall and precision for the 10 queries in which a set of correct answers for each inputted question was defined in order of the documents set by the specialists.

An index for the texts was produced to increase the retrieval speed. It consists of two elements, such as the vocabulary and frequency. The vocabulary is a set of all words that exist in the text in which it has an appeared document vector for each word. An appeared document vector stores the location of the appeared document including the weights considered by the frequency. Fig. 7 and 8 present the comparison of the distributions of the precision and recall for the inputted queries using the two methods mentioned above. From the results, the ontology based document retrieval system that uses a hyponym information existed in the ontology to extend a question and grants the weights presented a high recall and precision by 0.78% and 4.79% respectively compared with the traditional method of $tf*idf$. This means that a hierarchical relation in the ontology that is used as a relevant feedback in a retrieval system will not largely affect to the recall but will affect the increase of the precision.

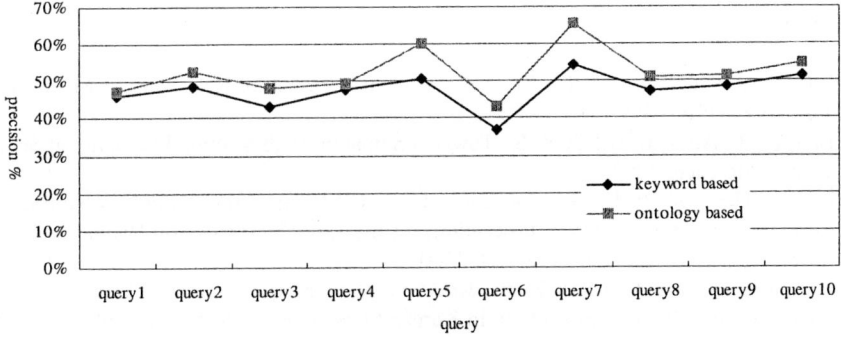

Fig. 7. comparison of the precision

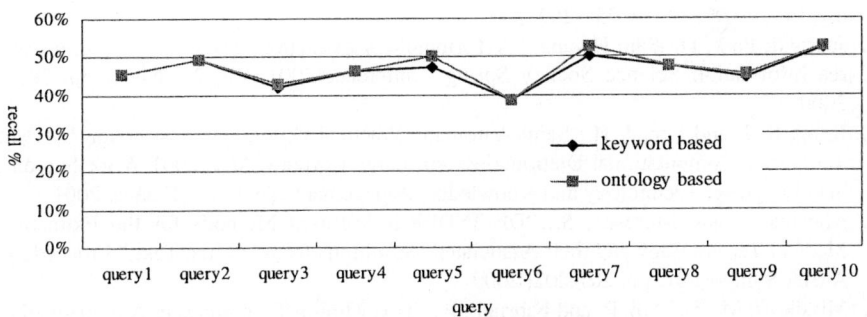

Fig. 8. comparison of the recall

5 Conclusions

The knowledge system, the web widely and easily shared by a plurality of users in any place and at any time can be considered a basic target pursued by human beings.

The semantic web ontology pursues wide development of a small-scale ontology. In this paper, we used the semantic relation information existing in the ontology for precisely handling queries when expanding queries for relevance feedback. As a result of comparing the performance of the keyword based document retrieval and the ontology based document retrieval, we could see that there was almost no change in the recall ratio and the precision ratio was improved by 4.79%. The medicine ontology to be tested is a domain ontology built using text mining technology with the result of analyzing terminology forms appearing in a related document in conformity to a specific domain.

Building of an ontology rich in semantics is a key purpose for enhancing retrieval efficiency. Accordingly, in addition to the 33 semantic relations established in the current medicine ontology, it is required to add necessary semantic relations and to expand concepts. It is considered that it is required to study ontology for various domains, not just for a specific domain.

References

1. Baeza-Yates, R. and Robeiro-Neto, B.: Modern Information Retrieval. ACM Press, New York, NY, USA, 1999
2. Bettina, B., Andreas, H., Gerd, S.: Towards Semantic Web Mining. International Semantic Web Conference, 2002
3. Gyeong-Hee Lee, Ju-HO Lee, Myeong- Choi, Gil-Chang Lim, "Study on Named Entity Recognition in Korean Text," Proceedings of the 12th Conference on Hangul and Korean Information Processing, pp. 292-299, 2000
4. Hyo-Shik Shin, Young-Soo Kang, Key-Sun Choi, Man-Suk Song, Computational Approach to Zero Pronoun Resolution in Korean Encyclopedia, Proceedings of the 13th Conference on Hangul and Korean Information Processing, pp. 239-243, 2001
5. JongHoon Oh, KyungSoon Lee, KeySun Choi, "Automatic Term Recognition using Domain Similarity and Statistical Methods," Journal of the Korea Information Science Society, Vol. 29, No. 4, pp. 258-269, 2002
6. Jung-Oh Park, Do-Sam Hwang, "A Terminology extraction system," Proceedings of Korea Information Science Society Spring Conference(2001),Vol 27, No 1, pp. 381-383, 2000
7. Kang, S. J. and Lee, J. H.: Semi-Automatic Practical Ontology Construction by Using a Thesaurus, Computational Dictionaries, and Large Corpora. ACL 2001 Workshop on Human Language Technology and Knowledge Management, Toulouse, France, 2001
8. Klavans, J. and Muresan, S., "DEFINDER:Rule-Based Methods for the Extraction of Medical Terminology and their Associated Definitions from On-line Text," Proceedings of AMIA Symposium, pp. 201-202, 2000
9. Missikoff, M., Velardi, P. and Fabriani, P., "Text Mining Techniques to Automatically Enrich a Domain Ontology," Applied Intelligence, Vol. 18, pp. 322-340, 2003
10. Soo-Yeon Lim, Mu-Hee Song, Sang-Jo Lee, "Domain-specific Ontology Construction by Terminology Processing," Journal of the Korea Information Science Society(B), Journal of the Korea Information Science Society Vol. 31, No. 3, pp. 353-360, 2004
11. Yi-Gyu Hwang, Bo-Hyun Yun, "HMM-based Korean Named Entity Recognition," Journal of the Korea Information Procissing Society(B), vol.10, No. 2, pp. 229-236, 2003

Adaptive Filtering Based on the Wavelet Transform for FOG on the Moving Base*

Xiyuan Chen

Department of Instrument Science and Engineering,
Southeast University, Nanjing City, 210096, P.R. China
chxiyuan@seu.edu.cn, chxiyuan@263.net

Abstract. An novel adaptive filtering method based on the wavelet transform is presented for a fiber optical gyroscope (FOG) on the moving base. Considering the performance difference of a FOG in different angular velocity, threshold values of different scales of wavelet coefficients are adjusted according to magnitude of FOG output signal, soft thresholding method is used to evaluate the wavelet coefficients, so effects of random signal noise and non-line of calibration factors of a FOG are removed at the maximum extent, and sensitivity of a FOG can be ensured. Filtering results of actual FOG show the proposed method has fine dynamic filtering effect.

1 Introduction

The concept of fiber optical gyros (FOG) was first proposed by Pircher and Hepner in 1967 and was given the first experimental demonstration at Utah University, Utah, USA, By Vali and Shorthill in 1976. So far, FOGs have been widely used because of its unique advantages such as high rotation-rate resolution, and a high zero-point stability, maintainance free usage, reliability, short warming-up time, etc.[1]-[3]. After extensive research on inertial navigation systems operating over long time intervals, it was found that noises and biasing drifts of a FOG represented the major error source. Especially the inertial measurement unit (IMU) integrated by inertial sensors such as three FOGs and three accelerometers is directly located under the gun base or beside the radar base for the strapdown inertial attitude system without any slider. Removing efficiently measurement error of a FOG is the one of key problems to ensure the attitude algorithm precision for the developing strapdown inertial attitude system based on FOGs. Up to the present, the causes of many noises and biasing drifts have been discovered and resolved [4]. However the problems of FOG's sensitivity to the environment still need to be studied further. Among the various factors, especially for FOGs in the dynamic environment, de-noising is necessary to remove the white noise, to overcdome the disturbance of periodic and non-line calibration factors, and to ensure the sensitivity of FOGs to angular velocity.

Filtering methods based on the wavelet transforms for the FOG signal on the static base are studied in [5][6][7]. Ref.[5] applied *db5* wavelet base and decomposition

---

* The work was supported by the southeast university excellent young teacher foundation (4022001002) and the national defense advanced research foundation (6922001019).

scale 5 to filter the FOG output signal. Although the biasing drifts are overcome, this method reduced the sensitivity of FOGs to angular velocity. However the sensitivity of FOGs to angular velocity is more important in the moving environment than in the static environment. Ref.[6] compared the filtering effect of a FOG in the static base when wavelet base *db1* and *db6* are at scale 2,3,4,5 respectively, but wavelet coefficients thresholding set at a different scale is not demonstrated. Ref.[7] proposed a norm maximum value algorithm controlled by noise to obtain fine filtering effect for FOGs in the static base. However, because of the complexity of FOG's dynamic performance, FOG output signal filtering on the moving base is still to be studied further and the signal processing strategy still needs to be verified.

This paper presents an novel adaptive filtering method based on the wavelet transform for a fiber optical gyroscope (FOG) on the moving base. Considering the performance difference of a FOG in different angular velocity, threshold values of different scales of wavelet coefficients are adjusted according to the magnitude of the FOG output signal, a soft thresholding method is used to evaluate the wavelet coefficients.

The organization of this paper is as follows: In Section 1, the researching backgrounds are provided. Section 2 introduces the wavelet filtering theory and de-noising method based on thresholding proposed by Donoho. An novel adaptive filtering method based on the wavelet transform for a fiber optical gyroscope (FOG) on the moving base and actual signal filtering of FOG experiments are presented in section 3, and finally, conclusions are given in section 4.

2 Wavelet Filtering Theory and De-noising Method

2.1 Wavelet Filtering Theory

The continuous wavelet transform was developed as a tool to obtain simultaneous, high-resolution time and frequency information about a signal using a variable sized window. The process of computing the Continuous Wavelet Transform of a signal is very similar to that of the Short-Time Fourier Transform. The wavelets are constructed by translating in time and dilation with the scale from a mother wavelet function. A wavelet is compared to a section at the beginning of a signal to calculate the time-scale wavelet coefficient that shows the degree of correlation between the wavelet and signal section. The wavelet is translated in time and the process is repeated until the whole signal is covered. The wavelet is scaled again and the previous process is repeated for all scales. Low scales correspond to compressed wavelets and high scales correspond to stretched wavelet. Since the scale is inversely related to the frequency, the wavelet transform allows the use of narrower wavelets where we require more precise high frequency information and stretched wavelets where we require low frequency information. We may then obtain the corresponding time-frequency wavelet coefficients of a signal.

In contrast to the continuous wavelet transform, the discrete wavelet transform calculates the wavelet coefficients at discrete intervals of time and scale, instead of at all scales. Like the Fast Fourier Transform, a fast algorithm of discrete wavelet transform (Dyadic Wavelet Transform) is possible if the scale parameter varies only along the dyadic sequence (2^j) .If we also want to impose the idea that the translation parameter varies along dyadic sequence (dyadic scales and positions), then more constraints must be imposed in order to construct orthogonal wavelets(Daubechies,1988) [8].The construction of these orthogonal bases can be related to multi-resolution signal approximations since orthogonal wavelets dilated by 2^j carry signal variations at the resolution 2^{-j}. Following this link leads us to an unexpected equivalence between wavelet bases and conjugate mirror filters used in discrete multirate filter banks. These filter banks implement a fast orthogonal wavelet transform and its inverse transform. An efficient way to implement this fast pyramid algorithm using filters was developed by Mallat (1989) [9]. The signal is passed through low-pass and high-pass filters and down-sampled (i.e. throwing away every second data point) to keep the original numbers of data points. Each level of the decomposition algorithm then yields low-frequency components of the signal (approximations) and high-frequency components (details). The reconstruction algorithm then involves up-sampling (i.e. inserting zeros between data points) and filtering with dual filters. By carefully choosing filters for the decomposition and reconstruction phases that are closely related (conjugate mirror filters), one can achieve perfect reconstruction of the original signal in the inverse orthogonal wavelet transform (Smith and Barnwell,1986 [10]; Daubechies,1988[8]). For a general introduction to discrete wavelet transform and filter banks, the reader is referred to several recent monographs (Strang and Nguyen,1997 [11]; Mallat,1998 [12]).

The Mallat algorithm is a fast, linear operation that operates on a data vector whose length is integer power of two, transforming it into a numerically different vector of the same length. Many wavelet families are available. However only orthogonal wavelets (such as Haar, Daubechies, Coiflet, and Symmlet wavelets) allow for perfect reconstruction of a signal by inverse discrete wavelet transform, i.e. the inverse transform is simply the transpose of the transform. For more detail wavelet filtering process, The reader is referred to Ref.[13].

Wavelet transform is widely used in filtering and de-nosing with its unique advantages. Many researchers have done convincing work. Fang H.T., Huang D.S. et al. did a series of research work in de-noising and digital filtering for lidar signals [14,15,16]. Ref.[14] proposed a new de-noising method for lidar signals based on a regression model and a wavelet neural network (WNN) that permits the regression model not only to have a good wavelet approximation property but also to make a neural network that has a self-learning and adaptive capability for increasing the quality of lidar signals. Specifically, the performance of the WNN for anti-noise approximation of lidar signals by simultaneously addressing simulated and real lidar signals was investigated. To clarify the anti-noise approximation capability of the WNN for lidar signals, they calculated the atmosphere temperature profile with the real signal processed by the WNN. To show the contrast, they also demonstrated the results of the Monte Carlo moving average method and the finite impulse response filter. A new method of the lidar signal acquisition based on the wavelet trimmed thresholding technique to increase the effective range of lidar measurements is presented [15]. The performance of this method is investigated by detecting the real signals in noise. They did some experiments to verify the proposed method which is superior to the traditional Butterworth filter. Ref.[16] proposed a new method of the Lidar signal acquisition based on discrete wavelet transform (DWT).This method can significantly improve the SNR so that the effective measured range of Lidar is increased. The performance for this method is investigated by detecting the simulating and real Lidar signals in white noise. To contrast, the results of Butterworth filter, which is a kind of finite impulse response (FIR) filter, are also demonstrated. Finally, the experimental results show that the proposed approach outperforms the traditional methods[16].

Considering the actual demand of the strapdown inertial attitude algorithm, denoising based on wavelet thresholding is suitable for the FOG output signal because of its high real time demand and unique advantages.

2.2 Wavelet De-noising

Unlike Fourier basis functions that characterize the entire time interval, wavelets employ basis functions whose support is contained within any interval length, no matter its duration (Mallat 1998) [12]. Therefore, compactly supported wavelet basis functions can model local signal behavior efficiently because they are not constrained by properties of the signal far away from the location of interest. This property makes wavelet analysis suitable for signals that have abrupt transitions or localized phenomena (Resnikoff and Wells 1998) [17]. Wavelet de-noising (Stein 1981[22]; Vidakovic 1999 [18]; Jansen 2001 [19]) is a term used to characterize noise rejection by setting thresholds for wavelet coefficients, which form the contribution of each wavelet basis to the signal. Wavelet coefficients smaller than the threshold are set to zero and other coefficients are shrunk by the threshold or left un-touched, depending on the algorithm used.

The discrete wavelet transform is linear and orthogonal, thus transforming white noise in the time space to white noise in the space of the wavelet coefficients [13]. It also enables compact coding, since the wavelet coefficients of the details possess high absolute values only in the intervals of rapid time series change. These properties led Donoho and Johnstone to propose denoising with thresholding [20]. They present a wavelet shrinkage de-noising algorithm used to suppress Gaussian noise and propose methods that select wave-let coefficient thresholds depending on the variance of the noise components in the signal. The efficiency of this denoising algorithm relies on the choice of wavelet basis, estimation of noise level, threshold selection method and parameters specific to the application, which consists of the following steps:

1) Select a wavelet base function ψ, a time series is transformed to the wavelet coefficients d_j, $j = 1,\cdots,J$, of the details.

2) The wavelet coefficients d_j of the details on each scale, $j = 1,\cdots,J$, are separately thresholded as

$$\hat{d}_j = \begin{cases} d_j, & |d_j| \geq \lambda_j \\ 0, & |d_j| < \lambda_j \end{cases} \qquad (1)$$

where the soft thresholding function[20]

$$\hat{d}_j = \begin{cases} sign(d_j)(|d_j| - \lambda_j), & |d_j| \geq \lambda_j \\ 0, & |d_j| < \lambda_j \end{cases} \qquad (2)$$

removes wavelet coefficients d of the details, that are absolutely smaller than the threshold λ_j, and reduces absolute values of those wavelet coefficients of the details which exceed the threshold.

3) From the modified wavelet coefficients $\hat{d}_j$, $j = 1,\cdots,J$, of the details the denoised time series is reconstructed.

Setting the thresholds is the essential part of de-noising. Assuming that most of the wavelet coefficients of the details contribute to the Gaussian white noise, Donoho and Johnstone set the thresholds to

$$\lambda_j = \sigma_j \sqrt{2\log(N)} \qquad (3)$$

where σ_j is the standard deviation of the wavelet coefficients of the details on the scale j and N is the number of all wavelet coefficients. Equation (3) is the basic formula to set the thresholds. This method, however, tends to underfit the data[21] and was therefore further enhanced by a criterion based on the Stein's unbiased risk estimate [22]. Called the SureShrink, this enhanced method is most widely used in de-noising with wavelets. Most methods proposed by other authors [23,24] differ from the foregoing in the way thresholds are estimated.

3 Adaptive Filtering Based on the Wavelet Transform for FOG on the Moving Base and Experiment Verification

3.1 Adaptive Filtering Based on the Wavelet Transform for FOG

According to the above description, how to select wavelet coefficients threshold is the key question during the wavelet de-noising process based on the wavelet transform for a FOG on the moving base. A new adaptive filtering approach towards threshold estimation and selection for FOGs is proposed.

The steps to determine wavelet coefficient threshold on each scale, $j = 1,\cdots,J$, for FOGs in dynamic environment are as follows in detail:

1) Select a wavelet base function ψ, a time series is transformed to the wavelet coefficients d_j, $j = 1,\cdots,J$, of the details. Here select *db4* as wavelet base function, $J = 6$;

2) Corresponding output domain $S_1[s_{1\min}, s_{1\max}]$ and $S_2[s_{2\min}, s_{2\max}]$ of FOG are obtained according to $\Omega[\Omega_{\min}, \Omega_{\max}]$, $\Omega'[\Omega'_{\min}, \Omega'_{\max}]$, which are maximum and secondary maximum angular velocity respectively and obtained by non-line maximum point detection method proposed in Ref.[25]. Maximum norm $|W_j(m)|_{\max}$ and $|W_j'(m)|_{\max}$ of corresponding wavelet coefficients in the two domain are obtained simultaneously, namely

$$\lambda_j(1) = |W_j(m)|_{\max}, s(t) \in S_1 \qquad (4)$$

$$\lambda_j(2) = \left|W_j{}'(m)\right|_{max}, s(t) \in S_2 \qquad (5)$$

3) Selection output sample data $\{s_{i,0}(n)\}$ of FOG on static base, where i ($i = 1, \cdots, N$) is test group number, 0 represents static base, n is sample length;
4) Wavelet coefficients $W_{i,0,j}(n)$ on each scale are obtained after every time series $s_{i,0}(n)$ is transformed, so corresponding scale maximum norm of wavelet coefficient $\left|W_{I,0,J}(m)\right|_{max}$ is obtained;
5) Wavelet coefficients threshold $\lambda_j(0)$ on each scale j, $j = 1, \cdots, J$, ($J = 6$) can be calculated according to equation (6):

$$\lambda_j(0) = \frac{1}{N}\sum_{i=0}^{N-1}\left|W_{i,0,j}(m)\right|_{max}, s(t) \notin S_1 \cup S_2 \qquad (6)$$

So the detailed steps of adaptive filtering based on wavelet transform for the FOG signal on moving base are as follows:

1) Select a wavelet base function *db4*, a time series is transformed to the wavelet coefficients d_j, $j = 1, \cdots, 6$, of the details.
2) The wavelet coefficients d_j of the details on each scale, $j = 1, \cdots, J$, are separately calculated by soft thresholding as

$$\hat{d}_j = \begin{cases} d_j, & |d_j| \geq \lambda_j(0) \\ 0, & |d_j| < \lambda_j(0) \end{cases}, s(t) \notin S_1 \cup S_2 \qquad (7)$$

$$\hat{d}_j = \begin{cases} d_j, & |d_j| \geq \lambda_j(1) \\ 0, & |d_j| < \lambda_j(1) \end{cases}, s(t) \in S_1 \qquad (8)$$

$$\hat{d}_j = \begin{cases} d_j, & |d_j| \geq \lambda_j(2) \\ 0, & |d_j| < \lambda_j(2) \end{cases}, s(t) \in S_2 \qquad (9)$$

removes wavelet coefficients d of the details, that are absolutely smaller than the threshold λ_j, and reduces absolute values of those wavelet coefficients of the details which exceed the threshold.

3) From the modified wavelet coefficients $\hat{d}_j$, $j = 1, \cdots, J$, of the details the denoised time series $\hat{s}(t)$ is reconstructed.

3.2 Experiment Verification

According to 3.1 description, an experiment for B-215 type FOG (fig.1) is done in the 3-axises simulation table (Fig.2) in lab.

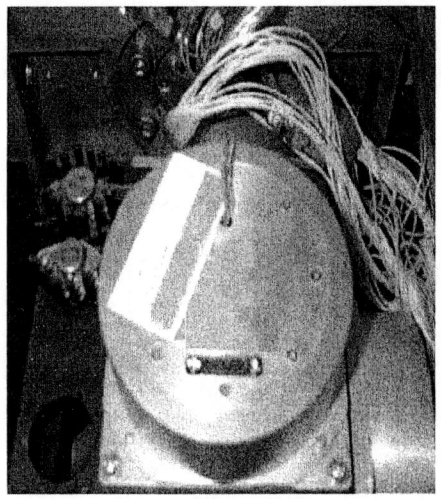 

Fig. 1. B-215 type FOG   **Fig. 2.** 3-axises simulation table

The threshold on each scale can be calculated as in table1.

Table 1. Threshold on each scale for B-215 type FOG

i	$\lambda_6(i)$	$\lambda_5(i)$	$\lambda_4(i)$	$\lambda_3(i)$	$\lambda_2(i)$	$\lambda_2(i)$
0	22.62	81.81	41.94	24.38	12.38	10.11
1	191.72	166.35	101.64	128.34	116.6	77.54
2	40.22	63.95	97.18	91.72	64.65	46.17

The actual signal of a FOG on the static base is processed by adaptive filtering according to above description, the result is shown in Fig.3, the yellow line represents the signal filtered.

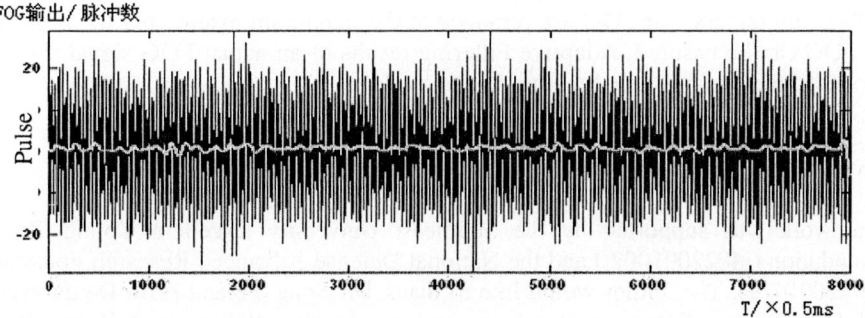

Fig. 3. the filtering result of actual signal of FOG(blue line represents original signal, yellow line represents the filtering signal)

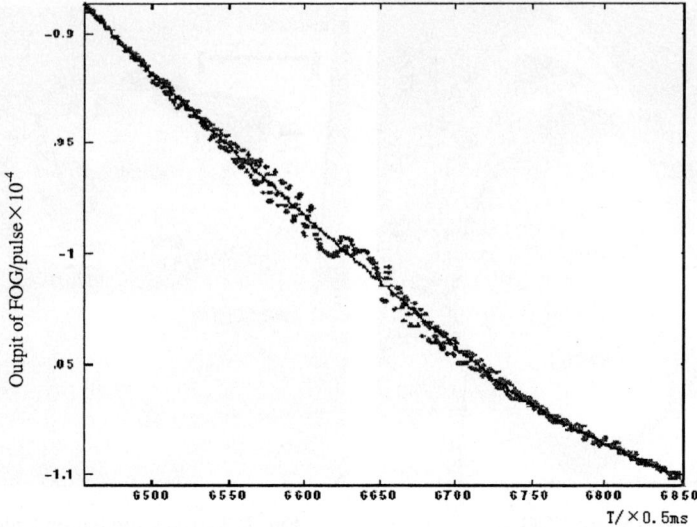

Fig. 4. the filtering result of the FOG signal when simulation table is in free pendulous rotation, the red scattered points represent the original signal, and the solid line represents the filtered signal

The actual signal of FOG when the simulation table is in free pendulum rotation is processed according to above description, the filtering signal is smooth and the adaptive filtering has a good effect. For the convenience of analysis, Fig.4 shows the filtering effect near the maximum odd angular velocity.

5 Conclusions

Considering the performance difference of a FOG at the different angular velocities, threshold values of different scales of wavelet coefficients are adjusted according to the magnitude of FOG output signal, the soft thresholding method is used to evaluate the wavelet coefficients, therefore the effects of random signal noise and non-line of calibration factors of FOG are removed at the maximum extent, and the sensitivity of FOG can be ensured. Adaptive Filtering results of an actual FOG signal show the proposed method has fine dynamic filtering effects.

Acknowledgement

The work was supported by the Southeast University Excellent Young Teacher Foundation (4022001002) and the National Defense Advanced Research Foundation (6922001019). The author would like to thank Dr. Song Ye and Prof. Dejun Wan of the Department of instrument science and engineering at Southeast University for helpful suggestions.

References

1. Hotate, K.: Fiber Optic Gyros. A Chapter in Optical Fiber Sensors Vol. IV. Norwood, MA. Artech House (1997)167-206
2. Lefevre, H.C.: The Fiber Optic Gyroscope. Norwood, MA. Artech House (1993)
3. Hotate, K.: Future Evolution of Fiber Optic Gyros. In Proc. SPIE Fiber Optic Gyros.20th Anniversary Conf., vol.2837, Denver, CO (1996)33-44
4. Chen,X.-Y.: Modeling Temperature Drift of FOG by Improved BP Algorithm and by Gauss-Newton Algorithm. Lecture Notes in Computer Science, Vol. 3174. Springer-Verlag, Berlin Heidelberg New York (2004) 805–812
5. Miao, L.-J.: Application of Wavelet Analysis in the Signal Processing of the Fiber Optic Gyro . Journal of Astronautics, Vol.21, No.1(2000)42-46
6. Qi, Y.-X., Gao X.P., Yuan R.M.: New Method for Eliminating Signal Zero Drift of Fiber Optic Gyro. Journal of Transducer Technology, Vol.22,No.10 (2003)57-59
7. Yuan, R.-M., Wei X.H., Li Z.Y.: De-noising Algorithm for Signal in FOG Based on Wavelet Filtering Using Threshold Value. Journal of Chinese Inertial Technology, Vol.11,No.5 (2003)43-47
8. Daubechies,I.: Ortho-normal Bases of Compactly Supported Wavelets. Communications on Pure and Applied Mathematics, No.41 (1988) 909-916
9. Mallat, S.:A Theory for Multi-resolution Signal Decomposition: The Wavelet Representation. IEEE Transaction on Pattern Analysis and Machine Intelligence,Vol.11,No.7 (1989) 674-693
10. Smith,M.J., Barnwell,T.P.: Exact Reconstruction for Tree-Structured Subband Coders. IEEE Transaction on Acoustics, Speech and Signal Processing, Vol.34, No.3 (1986) 431-441
11. Strang,G.,Nguyen,T.:Wavelets and Filter Banks. Wellesley-Cambridge Press, Wellesley, MA (1997)
12. Mallat,S.:A Wavelet Tour of Signal Processing. Academic Press, San Diego (1998)
13. Donoho,D.L.:Denoising by Soft-Thresholding. IEEE Trans. Inform. Theory,Vol.41, No.3 (1995) 613-627
14. Fang, H.-T., Huang, D.-S., Wu, Y.-H.: Antinoise Approximation of the Lidar Signal with Wavelet Neural Networks. Applied Optics, Vol.44, No.6 (2005) 1077-1083
15. Fang, H.T., Huang, D.-S.: Lidar Signal De-noising Based on Wavelet Trimmed Thresholding Technique. Chinese Optics Letters, Vol.2, No.1 (2004) 1-3
16. Fang, H.T., Huang, D.-S.: Noise Reduction in Lidar Signal Based on Discrete Wavelet Transform. Optics Communications. No.233 (2004) 67-76
17. Resnikoff, H.L., Wells, R.: Wavelet Analysis: The Scaleable Structure of Information. Springer-Verlag, New York (1998)
18. Vidakovic, B.: Statistical Modeling by Wavelets. Wiley, New York (1999)
19. Jansen, M.: Noise Reduction by Wavelet Thresholding. Springer-Verlag, New York (2001)
20. Donoho,D.L., Johnstone,I.M.: Ideal Spatial Adaptation by Wavelet Shrinkage. Biometrika,No.81 (1994) 425-455
21. Donoho,D.L., Johnstone,I.M.: Adapting to Unknown Smoothness via Wavelet Shrinkage.J.Am.Stat.Assoc.,Vol.90,No.432 (1995) 1200-1224
22. Stein,C.M.:Estimation of the Mean of a Multivariate Normal Distribution.Ann.Stat.,Vol.9,No.6 (1981) 1135-1151
23. Abramovich, F., Sapatinas, T., Silverman, B.W.: Wavelet Thresholding via Bayesian Approach. J.Roy.Stat.Soc.B, No.60 (1998) 723-749
24. Nason,G.P.:Wavelet Shrinkage Using Cross-Validation. J.Roy.Stat.Soc.B, No.58 (1996) 463-479
25. Ye,S.: Study on Data Processing and Fusion Technology in FOG Strapdown/GPS Integrated Attitude and Heading System [D]. Southeast University, Nanjing (2004)

Text Similarity Computing Based on Standard Deviation

Tao Liu and Jun Guo

School of Information Engineering, Beijing University of Posts and
Telecommunications, Beijing 100876, China,
sdlclt@sohu.com

Abstract. Automatic text categorization is defined as the task to assign free text documents to one or more predefined categories based on their content. Classical method for computing text similarity is to calculate the cosine value of angle between vectors. In order to improve the categorization performance, this paper puts forward a new algorithm to compute the text similarity based on standard deviation. Experiments on Chinese text documents show the validity and the feasibility of the standard deviation-based algorithm.

1 Introduction

Text categorization has recently become an active research topic in the area of information retrieval. The objective of text categorization is to assign free text documents to one or more predefined categories based on their content. Traditionally text categorization is performed manually by domain experts. Each incoming document is read and comprehended by the expert and then it is assigned a number of categories chosen from the set of prespecified categories. This process is very time-consuming and costly, thus limiting its applicability.

A promising way to deal with this problem is to learn a categorization scheme automatically from training collection. Once the categorization scheme is learned, it can be used for classifying future documents. It involves issues commonly found in machine learning problems. Since a document may be assigned to more than one category, the scheme also requires the assignment of multiple categories. There is a growing body of research addressing automatic text categorization. A number of statistical classification and machine learning techniques has been applied to text categorization, including regression models[1][2], nearest neighbor classifiers[3][4], Bayesian classifiers[5][6], decision trees[1][6][7], rule learning algorithms[8][9][10], neural networks[1], inductive learning techniques[11][12], Support Vector Machines[13], relevance feedback[14] and voted classification[15].

In order to improve the categorization performance, this paper puts forward a new algorithm to compute the text similarity based on standard deviation. Experiments show the validity and the feasibility of the standard deviation-based algorithm.

This paper contains 6 sections. In Section 2 we describe the vector space model; in Section 3 describe several typical methods that have been successfully applied to text feature selection and categorization; Section 4 introduces the proposed method applied to text similarity computing and categorization based on standard deviation and compares it with the classical method based on cosine similarity; experimental results and evaluation are given in Section 5; finally, we draw to a conclusion.

2 Vector Space Model

The most commonly used document representation is the so called vector space model (VSM)[16]. In the vector space model, each document can be represented by vector $\mathbf{v} = (\mathbf{w_1}, \mathbf{w_2}, \ldots, \mathbf{w_m})$, where w_i represents the corresponding weight of the ith feature t_i of the document and denotes the importance of t_i in describing the document's content. Therefore, the expression and matching issue of text information is converted to that of the vector in VSM [17]. Experiment shows that word is a better candidate for feature than character and phrase.

At present there are several ways of determining the weight w_i, Intuitively, w_i should express the two aspects as follows:

- The more often a word occurs in a document, the more effectively it is to reflect the content of the document.
- The more often the word occurs throughout all documents in the collection, the more poorly it discriminates between documents.

A well-known approach for computing word weights is the tf*idf weighting, which assigns the weight to word in document in proportion to the number of occurrences of the word in the document, and in inverse proportion to the number of documents in the collection for which the word occurs at least once.

Among several existing tf*idf formulas, we selected a commonly used one in our system:

$$W(t, \mathbf{d}) = \frac{tf(t, \mathbf{d}) \times \log\left(\frac{N}{n_t} + 0.01\right)}{\sqrt{\sum_{t \in \mathbf{d}} [tf(t, \mathbf{d}) \times \log\left(\frac{N}{n_t} + 0.01\right)]^2}} \quad (1)$$

where $W(t, \mathbf{d})$ is the weight of word t in document $\mathbf{d}$, $tf(t, \mathbf{d})$ is the frequency of word t in document $\mathbf{d}$, N is the number of documents in the training collection and n_t is the number of documents in the whole collection for which word t occurs at least once.

The main advantage of VSM is in the fact that it simplifies the documents' content to vectors comprising features and weights, which greatly decreases the complexity of the problem.

3 Typical Text Feature Selection and Categorization Methods

A major problem in text categorization is the high dimensionality of the feature space. Generally the feature space consists of hundreds of thousands words even for a moderated-size documents collection. Standard classification techniques can hardly deal with such a large feature set since processing is extremely costly in computational terms, and overfitting can not be avoided due to the lack of sufficient training data. Hence, there is a need for a reduction of the original feature set without decreasing the categorization accuracy, which is commonly known as dimensionality reduction in the pattern recognition literature.

Feature selection attempts to remove non-informative words from documents in order to improve categorization effectiveness and reduce computational complexity. Before feature selection, word segmentation which is necessary for Chinese text categorization has to be made because there is not apparent delimiter between the character in the text. In [18] a thorough evaluation of the five known feature selection methods: Document Frequency Thresholding, Information Gain, χ^2-statistic, Mutual Information and Term Strength is given.

In Document Frequency Thresholding, we computes the document frequency for each word in the training collection and removes those words whose document frequency is less than a predetermined threshold. The basic assumption is that rare words are either non informative for category prediction, or not influential in global performance. Information Gain is frequently employed as a termgoodness criterion in the field of machine learning [19][20]. It measures the number of bits of information obtained for category prediction by knowing the presence or absence of a word in a document. The information gain of a word t is defined to be:

$$IG(t) = -\sum_{j=1}^{n} P(C_j) \log P(C_j) + P(t) \sum_{j=1}^{n} P(C_j|t) \log P(C_j|t) \\ + P(\bar{t}) \sum_{j=1}^{n} P(C_j|\bar{t}) \log P(C_j|\bar{t}) \quad (2)$$

Wherein, n is the number of the category, $P(C_j)$ is the probability that class C_j occurs in the total collection and $P(t)$ is that of word t. $P(C_j|t)$ can be computed as the fraction of documents from class C_j that have at least one occurrence of word t and $P(C_j|\bar{t})$ as the fraction of documents from class C_j that does not contain word t. The information gain is computed for each word of the training collection, and the words whose information gain is less than some predetermined threshold are removed. The χ^2-statistic measures the lack of independence between word t and class C_j. It is given by:

$$\chi^2(t, C_j) = \frac{N \times (AD - CB)^2}{(A+C)(B+D)(A+B)(C+D)} \quad (3)$$

A set of tokens with the highest χ^2 measures are then selected as keyword features. A is the number of documents from class C_j that contains word t and B is the number of documents that contains t but does not belong to class C_j. C is the number of documents from class C_j that does not contain word t and D is the number of documents that belongs to class C_j nor contains word t. Mutual Information is a criterion commonly used in statistical modelling of word associations and related applications[21][22][23]. Term strength measures how informative a word is in identifying two related documents. The strength of a word t is defined as the probability of finding t in a document which is related to any document in which t occurs[24]. In our experiments, the method of χ^2-statistic is found to be the most effective.

There exists various text categorization algorithms, such as Rocchio's algorithm, Naive Bayes, K-nearest neighbor, neural network, support vector machine etc, wherein the first two are employed in most applications. The proposed method in this paper is in fact an improved version of Rocchio's algorithm. Here we only introduce the Naive Bayes while Rocchio's algorithm will be described in detail in the next section. The naive Bayes classifier estimate the probability of each class based on Bayes theory:

$$P(C_j|d) = \frac{P(C_j) P(d|C_j)}{P(d)} \quad (4)$$

$P(d)$ is same to all the caterogies and the assumption is made that the features are conditionally independent. This simplifies the computations yielding:

$$P(C_j|d) = P(C_j) \prod_{i=1}^{m} P(t_i|C_j) \quad (5)$$

$P(C_j)$ is the probability that class C_j occurs in the total collection. An estimate $\hat{P}(t_i|C_j)$ for $P(t_i|C_j)$ is given by:

$$\hat{P}(t_i|C_j) = \frac{1 + N_{ij}}{m + \sum_{l=1}^{m} N_{lj}} \quad (6)$$

N_{ij} denotes the number of times word t_i occurred within documents from class C_j in the training collection.

4 Proposed Method Based on Standard Deviation

Rocchio's algorithm is the classical method for document routing or filtering in information retrieval. In this method, a prototype vector $\mu_j = (\mu_{j1}, \mu_{j2}, ... \mu_{jm})$ is computed as the average vector over n_j training document vectors that belong to class C_j, where the feature $\mu_{ji} = \frac{1}{n_j} \sum_{k=1}^{n_j} w_{j,ki}$ is the mean of $w_{j,ki}$ and $w_{j,ki}$ is the weight of word i in document $\mathbf{d_k}$ of category C_j. A document $\mathbf{d_{test}}$ is classified by calculating the similarity between document vector $\mathbf{v_{test}} = (\mathbf{w_{test1}}, \mathbf{w_{test2}}, ..., \mathbf{w_{testm}})$ of $\mathbf{d_{test}}$ and each of the prototype vectors μ_j. The similarity can be computed as follows [25]:

$$Sim(\mathbf{d_{test}}, C_j) = \frac{\mathbf{v_{test}} \cdot \mu_j}{\|\mathbf{v_{test}}\| \cdot \|\mu_j\|} \quad (7)$$

Since (7) exactly denotes the cosine function of the angle between the two vectors, the similarity defined in (7) is usually called 'cosine similarity'. Consequently, the category into which the document $\mathbf{d_{test}}$ falls is determined by the equation:

$$c = \arg\max_j Sim\left(\mathbf{d_{test}}, C_j\right) \qquad (8)$$

As we can see from the analysis above, Rocchio's method has such advantage as simple categorization mechanism, rapid process rate while its main defect is due to the fact that it is difficult to roundly describe the characteristics of the category with the only information of samples' mean. A new method is proposed in this paper to overcome the main defect of classical method by describing the characteristics of category more precisely with not only the mean vector but also the standard deviation and classifying documents with new similarity rather than 'cosine similarity' employed. In our study, it is found that the standard deviation, which is a common used statistics reflecting the distribution of the samples in pattern recognition, of each feature in diverse category changes distinctly, while the fact is not concerned in classical Rocchio's method leading to degraded categorization result. The new method in this paper obtains better performance because of considering of the difference of the standard deviation.

Two vectors, mean vector $\mu_\mathbf{j}$ and standard deviation vector $\sigma_\mathbf{j} = (\sigma_{j1}, \sigma_{j2}, ...\sigma_{jm})$, are chosen as the prototype vectors of category C_j, wherein $\sigma_{ji} = \sqrt{\frac{1}{n_j - 1} \sum_{k=1}^{n_j} (w_{j,ki} - \mu_{ji})^2}$. A modified street distance is proposed to be a new similarity in text categorization to accommodate the two new prototype vectors,

$$Sim\left(\mathbf{d_{test}}, C_j\right) = -\sum_{i=1}^{m} \frac{\max\{|w_{testi} - \mu_{ji}| - \sigma_{ji}, 0\}}{\sigma_{ji}^2} \qquad (9)$$

As is shown, any document locates within the sphere determined by μ_{ji} and σ_{ji}^2 will be classified to category C_j. It is noted that because of the probably of being zero σ_{ji}^2 should be modified before used as denominator in (9).

In categorization experiments, mean vector $\mu_\mathbf{j}$ and standard deviation vector $\sigma_\mathbf{j}$, are obtained during the training process according to the training collection. Consequently, the similarity between document $\mathbf{d_{test}}$ to be categorized and each category is computed by (9). Finally, the categorization results is obtained by (8).

Compared with classical Rocchio's method based on 'cosine similarity', the advantage of the proposed method is illustrated in Fig.1. For the convenience of illustrating, we suppose each document has two features. Since in (7) $\mathbf{d_{test}}$ and C_j have been cosine normalized, the similarity described in (7) reflects the Euclidean distance between $\mathbf{d_{test}}$ and C_j. $\mathbf{v_{test}}$ is the document vector of the document $\mathbf{d_{test}}$ to be categorized, $\mu_\mathbf{A}$ and $\mu_\mathbf{B}$ are mean vectors of category A and B, respectively. σ_{A1} and σ_{A2} represent the two dimensional standard deviation of A, and σ_{B1} and σ_{B2} denote that of B. Since the distance D_A between $\mathbf{v_{test}}$ and $\mu_\mathbf{A}$ is greater than the distance D_B between $\mathbf{v_{test}}$ and $\mu_\mathbf{B}$, $D_A > D_B$, the categorization result is $\mathbf{d_{test}} \in B$ based on classical Rocchio's method. However, $\mathbf{d_{test}}$ locates in the field of category A not category B, so the probability of

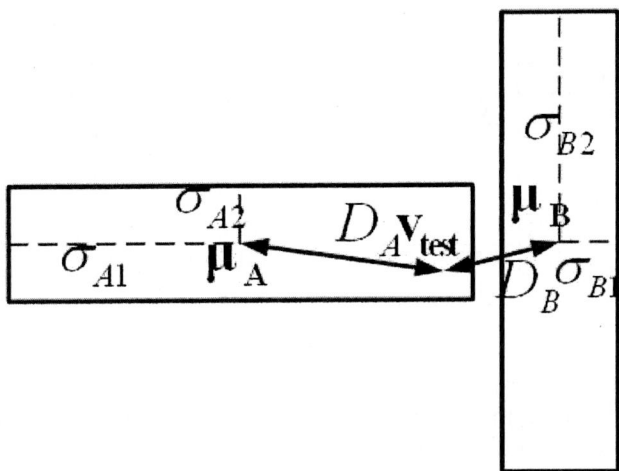

Fig. 1. Advantage of proposed method

$d_{test} \in A$ is larger than that of $d_{test} \in B$. It is more likely to put d_{test} in wrong category by classical Rocchio's method than the proposed method.

5 Experiment Results and Evaluation

Text categorization experiment results of 36 categories of Chinese text documents according to Chinese Library Classification (version 4) based on standard deviation and 'cosine similarity' are given in this section. We take Chinese Text Categorization (TC) Evaluation collection of Chinese Language Processing and Intelligent Human Machine Interface evaluation of the National High Technology Research and Development Program(HTRDP) in 2003 as the training collection and download 3134 text documents (labeled by the experts with reference to Chinese Library Classification) from the web. The HTRDP Evaluation of Chinese Language Processing and Intelligent Human Machine Interface, also called the '863' Evaluation, is a series of evaluation activities sponsored by China's National High Technology Research and Development Program (HTRDP, also called the '863' Program). The purpose of the HTRDP Evaluation is to provide infrastructural support for research and development on Chinese information processing and intelligent human-machine interface technology, to enhance interaction among industry, academia, and government, and to speed up the transfer of technology from research labs into commercial products.

Experiment results of 8 categories are shown in Table 1, where a represents the number of documents correctly assigned to this category; b represents the number of documents incorrectly assigned to this category; c represents the number of documents incorrectly rejected from this category. The category label is with reference to Chinese Library Classification.

Table 1. Experiment results of 8 categories

category	B	E	J	K	TU	G	TD	TP
a	94	69	29	55	57	27	70	77
b	7	5	28	25	38	80	18	5
c	26	50	21	20	34	3	14	50
Precision	93.07%	93.24%	50.88%	68.75%	60.00%	25.23%	79.55%	93.90%
Recall	78.33%	57.98%	58.00%	73.33%	62.64%	90.00%	83.33%	60.63%
F1	85.07%	71.50%	54.21%	70.97%	61.29%	39.42%	81.40%	73.68%
F1 Gain	2.12%	1.35%	8.83%	20.97%	22.97%	5.84%	2.91%	13.27%

Categorization effectiveness is measured in terms of the commonly used IR notions of precision and recall, adapted to the case of text categorization. Precision is defined as the probability that if a random document $\mathbf{d_{test}}$ is categorized under C_j, this decision is correct. Analogously, Recall is defined as the probability that, if a random document $\mathbf{d_{test}}$ should be categorized under C_j, this decision is taken. Another evaluation criterion that combines recall and precision is the F1 measure. The definitions of Precision, Recall and F1 are given below [26]:

$$Precision = a/(a+b) \qquad (10)$$

$$Recall = a/(a+c) \qquad (11)$$

$$F1 = (Precision \times Recall \times 2)/(Precision + Recall) \qquad (12)$$

Table 2. Measure comparison of two methods

	Macro-averaging Precision	Macro-averaging Recall	Macro-averaging F1
cosine	70.81%	70.71%	70.76%
deviation	74.21%	74.43%	74.32%

For evaluating performance average across categories, Macro-averaging performance scores are determined by first computing the performance measures per category and then averaging these to compute the global means and are calculated as follows:

$$Macro-averaging\ Precision = \frac{1}{n}\sum_{j=1}^{n} Precision_j \qquad (13)$$

$$Macro-averaging\ Recall = \frac{1}{n}\sum_{j=1}^{n} Recall_j \qquad (14)$$

$$Macro-averaging\ F1 = \frac{1}{n}\sum_{j=1}^{n} F1_j \qquad (15)$$

n is the number of the category. From Table 2, we can see that both the Macro-averaging Precision and Recall of proposed method is around 74%, which is greater than that of the traditional one. The validity and the feasibility of the standard deviation-based algorithm is validated by experiment results.

6 Conclusions

As Chinese text information available on the Internet continues to increase, there is a growing need for tools helping people better manage the information. Text categorization, the assignment of free text documents to one or more predefined categories based on their content, is an important component to achieve such task and attracts more and more attention.

A number of statistical classification and machine learning techniques has been applied to text categorization. In order to improve the categorization performance, this paper puts forward a new algorithm to compute the text similarity based on standard deviation. Experiments show the validity and the feasibility of the standard deviation-based algorithm.

References

[1] Fuhr, N., Hartmanna, S., Lustig, G., Schwantner, M.,Tzeras, K.: Air/x - a rule-based multistage indexing systems for large subject fields. Proceedings of RIAO'91. (1991) 606–623
[2] Yang, Y., Chute, C.G.: A Linear Least Squares Fit mapping method for information retrieval from natural language texts. Proceedings of 14th International Conference on Computational Linguistics (COLING'92).**II** (1992) 447–453
[3] Creecy, R.H., Masand, B.M., Smith, S.J., Waltz, D.L.: Trading MIPS and memory for knowledge engineering:classifying census returns on the connection machine. Comm. ACM.**35** (1992) 48–63
[4] Yang, Y., Chute, C.G.: An example-based mapping method for text classification and retrieval. ACM Transactions on Information Systems (TOIS) **12** (1994) 253–277
[5] Tzeras,K., Hartmann, S.: Automatic Indexing Based on Bayesian Inference Networks. In Proceedings of the 16th Annual International ACM SIGIR Conference on Research and Development in Information Retrieval(SIDIR'93). (1993) 22–34
[6] Lewis, D.,Ringuette, M.: A comparison of two learning algorithms for text clasification, In Third Annual Symposium on Document Analysis and Information Retrieval. (1994) 81–93
[7] Moulinier, I.: Is learning bias an issue on the text categorization problem? In Technical report, LAFORIA-LIP6, Universite Paris VI, (1997)
[8] Apte, C., Damerau, F., Weiss, S.: Towards language independent automated learning of text categorization models. In Proceedings of the Seventeenth Annual International ACM/SIGIR Conference. (1994)
[9] Wiener, E., Pedersen, J.O., Weigend, A.S.: A neural network approach to topic spotting. In Proceedings of the Fourth Annual Symposium on Document Analysis and Information Retrieval(SDAIR'95). (1995)

[10] Moulinier, I., Raskinis, G., Ganascia, J.: Text categorization: a symbolic approach. In Proceedings of the Fifth Annual Symposium on Document Analysis and Information Retrieval. (1996)
[11] William, W. C., Singer, Y.: Context-sensitive learning methods for text classification. In SIGIR'96: Proceedings of the 19th Annual International ACM SIGIR Conference on Research and Development in Information Retrieval. (1996) 307–315
[12] David, D. L., Robert E. S., Callan, J.P., Papka, R.: Training Algorithms for Linear Text Classifiers. In SIGIR '96:Proceedings of the 19th Annual International ACM SIGIR Conference on Research and Development in Information Retrieval. (1996) 298–306
[13] Joachims, T.: Text categorization with support vector machines: Learning with many relevant features. In Proc. 10th European Conference on Machine Learning (ECML) Springer Verlag. (1998)
[14] Rocchio, J.: Relevance feedback in information retrieval. In The SMART Retrieval System: Experiments in Automatic Document Processing. PrenticeHall Inc.. (1971) 313–323
[15] Weiss, S. M., Apte, C., Damerau, F. J.,Johnson, D.E., Oles, F.J., Goetz, T., Hampp, T.: Maximizing Text-Mining Performance. Intelligent Systems and Their Applications, IEEE [see also IEEE Intelligent Systems] 14 (1999) 63-69
[16] Salton, G., Lesk, M. E.: Computer evaluation of Indexing and text processing. Association for Computing Machinery **15** (1968) 8–36
[17] Salton, G., Wong, A., Yang, C. S.: A Vector Space Model for Automatic Indexing. Communications of ACM **18** (1975) 613–620
[18] Yiming, Y., Jan, P. P.: A comparative study on feature selection in text Categorization. In:Proceedings of ICML'97, 14th International Conference on Machine Learning. (1997) 412–420 Morgan Kaufmann
[19] Tom, M.: Machine Learning. McCraw Hill, 1996.
[20] Quinlan, J.: Induction of decision trees. Machine Learning, **1** (1986) 81–106
[21] Keeneth, W. C., Patric, H.: Word association norms, mutual information and lexicography. In Proceeding of ACL 27 (1989) 76–83 Vabcouver, Canada
[22] Fano, R.: Transmission of Information. MIT Press, Cambridge, MA, (1961)
[23] Wiener, E., Pedersen, J.O., Weigend, A.S.: A neural network apporach to topic spotting. In Proceedings of the Fourth Annual Symposium on Document Analysis and Information Retrieval(SDAIR'95) (1995)
[24] Yiming, Y.: Noise Reduction in a Statistical Approach to Text Categorization. ACM SIGIR Conference on Research and Development in Information Retrieval (SIGIR'95) (1995) 256–263
[25] Salton, G.: Automatic text processing: the transformation analysis and retrieval of information by Computer. Reading, Pennsylvania: Aoldison-wesley (1989)
[26] Bin, L., Tiejun, H., Jun, C., Wen, G.: A New Statistical-based Method in Automatic Text Classification. Journal of Chinese information processing. **16** (2002) 18–24

Evolving Insight into High-Dimensional Data

Yiqing Tu, Gang Li*, and Honghua Dai

School of Information Technology, Deakin University,
221 Burwood Highway, Vic 3125, Australia
Tel: +61(3)9251 7434
{ytu, gang.li, hdai}@deakin.edu.au

Abstract. ISOMap is a popular method for nonlinear dimensionality reduction in batch mode, but need to run its entirety inefficiently if the data comes sequentially. In this paper, we present an extension of ISOMap, namely I-ISOMap, augmenting the existing ISOMap framework to the situation where additional points become available after initial manifold is constructed. The MDS step, as a key component in ISOMap, is adapted by introducing Spring model and sampling strategy. As a result, it consumes only linear time to obtain a stable layout due to the Spring model's iterative nature. The proposed method outperforms earlier work by Law [1], where their MDS step runs within quadratic time. Experimental results show that I-ISOMap is a precise and efficient technique for capturing evolving manifold.

1 Introduction

Recently, one conceptually simple yet powerful method for *nonlinear* mapping has been introduced by Tenenbaum et al.: ISOMap [2]. This method basically operates in a batch mode. When new data points come, it is necessary to repeat running the batch version. With incremental methods, new data point can be inserted into the existing manifold directly so that new manifold can be quickly evolved by slightly adjusting the previous layout.

Other related methods to tackle the nonlinear dimensionality reduction problem include LLE [3] and Laplacian Eigenmaps [4]. However, similar to ISOMap, they both need to be repeatedly run whenever new data point is available. In this paper, we introduce an *Incremental ISOMap* (I-ISOMap) algorithm to compute evolving manifold with new coming data. The only similar work is done by Law [1], who proposed an incremental approach for manifold computation. However, the MDS step in Law's method is based on a numerical approach, which takes quadratic time. We introduce Spring model in MDS step for updating co-ordinates instead of the numerical approach adopted by Law. Thanks to iterative nature of Spring model, the adapted MDS step achieves linear time complexity, greatly contributing to the speeding up of I-ISOMap algorithm.

---
* Corresponding author.

2 ISOMap Algorithm

Firstly, we establish our notation for the rest of the paper. Vectors are columns, and denoted by a single underline. Matrices are denoted by a double underline. The size of a vector, or matrix is denoted by subscripts. An example is $\underline{\underline{A}}_{mn}$ - a matrix with m rows and n columns. Particular column vectors within a matrix are denoted by a superscript, and a superscript on a vector denotes a particular observation from a set of observations. For example, $\underline{\underline{A}}^i_{mn}$ is the ith column vector in matrix A. Also, as in convention, a^{ij} is a scalar, the element of matrix $\underline{\underline{A}}_{mn}$ in row i and column j. Finally, we denote matrices formed by concatenation using square brackets. Thus $[\underline{\underline{A}}_{mn} \underline{b}]$ is an $(m \times (n+1))$ matrix, with vector $\underline{b}$ appended to $\underline{\underline{A}}_{mn}$ as a last column.

ISOMap belongs to a category of techniques called *Local Embeddings* [5], and attempts to preserve topology of data points locally. For data lying on a nonlinear manifold, the distance between two data points is better described by geodesic distance on the manifold, i.e. the distance along the surface of the manifold, rather than the direct Euclidean distance. The main purpose of ISOMap is to find the intrinsic geometry of the data, as captured in pair-wise geodesic distances.

Formally speaking, consider $\underline{\underline{X}}_{MN}$ representing N observed data points. Especially, each observed data points $\underline{\underline{X}}^i_{MN}$ lies on an unknown manifold $\mathcal{M}$ smoothly embedded in an M-d observation space. Suppose this observation space is denoted by $\mathcal{X}$, and $\mathcal{Y}$ denotes an m-d Euclidean feature space $(M > m)$. The goal of manifold learning is to recover a mapping $f : \mathcal{X} \to \mathcal{Y}$, so that the intrinsic geometric structure of the observed data $\underline{\underline{X}}_{MN}$ are preserved as well as possible. The mapping f can be described implicitly in terms of the feature points $\underline{\underline{Y}}_{mN}$, where each column $\underline{\underline{Y}}^i_{mN}$ represents one data points in the feature space $\mathcal{Y}$.

The main steps of ISOMap are summarized in Alg. 1. The ISOMap algorithm first construct a weighted undirected neighborhood graph $\mathcal{G}$. In Graph $\mathcal{G}$, vertex i and j is connected if and only if point i is one of the neighbors of point j, or the other way round. The weight of edge between i and j is equal to distance d^{ij}. Next, the shortest path between point pairs are found. The length of the path is taken as an estimation of geodesic distance between data pairs. Finally, MultiDimensional Scaling(MDS) [6] is applied on the geodesic matrix $\underline{\underline{G}}_{NN}$ to configure points in low dimensional space.

3 Spring Model

Spring model proposed by Eades [7] is analogous to a system of steel rings connected by springs. The attractive and repulsive forces exerted by the springs iteratively improve the layout of objects, and finally the whole system is driven to a state with a minimal energy. Chalmer [8] presents an improved version of Spring model achieving higher computational efficiency. Using caching and stochastic sampling strategies, N iteration is only needed to obtain a stable configuration of objects in feature space, resulting quadratic time complexity.

Algorithm 1. ISOMap Algorithm

Input: $\underline{\underline{D}}_{NN}$ as the pairwise distances between N data points, K as the number of nearest neighbors, m as the desirable dimensionality of target feature space
Output: $\underline{\underline{Y}}_{NN}$ as coordinate of each data point in m-d feature space $\mathcal{Y}$
1: Compute the K-nearest neighbors for each data point, and construct neighborhood graph $\mathcal{G}$
2: Compute the shortest distance $\underline{\underline{G}}_{NN}$ between point pairs in graph $\mathcal{G}$ by Dijkstra's algorithm or Floyd's algorithm.
3: Calculate the data points' coordinates $\underline{\underline{Y}}_{mN}$ in feature space $\mathcal{Y}$ to describe intrinsic lower dimensional embedding.

This approach is extended in the MDS step of the proposed I-ISOMap, which will be described in details in Section 4.3.

4 The Incremental-ISOMap Framework

This section outlines the proposed I-ISOMap algorithm for dimensional reduction. The goal is to learn mapping $f:X \rightarrow Y$ incrementally. In other words, suppose mapping f is already recovered on N points, i.e. observed data $\underline{\underline{X}}_{MN}$ is represented in term of coordinates $\underline{\underline{Y}}_{mN}$ in m-d space. We want to find $\underline{\underline{Y}}'_{m,N+1}$ to accommodate one more observed point $\underline{x}^{N+1}$ so that the intrinsic structure for current observed data $\underline{\underline{X}}'_{M,N+1} = [\underline{\underline{X}}_{MN} \underline{x}^{N+1}]$ still remains best.

As in Alg. 2, the proposed I-ISOMap also consists of three steps. First, neighborhood graph $\mathcal{G}$ are updated. Second, based on updated graph $\mathcal{G}'$, geodesic distance $\underline{\underline{G}}'_{N+1,N+1}$ is estimated by calculating length of shortest path between each point pair. Finally, spring model is adapted for updating configuration in feature space. This MDS step rearranges the coordinates within constant iterations.

Algorithm 2. I-ISOMap Algorithm

Input: The previous coordinates $\underline{\underline{Y}}_{m,N}$, pair-wise distance matrix $\underline{\underline{D}}_{N,N}$, new data point $\underline{x}^{N+1}$
Output: The updated coordinates $\underline{\underline{Y}}'_{m,N+1}$
1: Update neighborhood Graph $\mathcal{G}$ due to introduction of new data point $\underline{x}^{N+1}$ as in Alg. 3.
2: Update geodesic distance $\underline{\underline{G}}_{N,N}$ into $\underline{\underline{G}}'_{N+1,N+1}$ using modified Dijkstra's algorithm as in Alg. 4
3: Update the coordinates in feature space from $\underline{\underline{Y}}_{mN}$ to $\underline{\underline{Y}}'_{m,N+1}$ as in Alg. 5

4.1 Updating Neighborhood Graph

The updating of neighborhood graph is based on previous neighborhood Graph $\mathcal{G}$ and distance vector $\underline{d}^i$, where $\underline{d}^i$ is the distance between new point and all remaining points. The other output is a set of deleted edges J. This set records all the edges that are deleted because of the adding of new points. As a result, only those geodesic distance affected by deleted edges will be recomputed in following steps. Neighbor lists $\underline{\underline{L}}_{KN}$ are employed to maintain a list of each point's K nearest neighbors. Points in $\underline{\underline{L}}_{KN}^i$ are ordered by their distance to point i. The neighbor lists are maintained when each new point is added.

Algorithm 3. Updating Neighborhood Graph

Input: the distance between new point and remaining points $\underline{d}^{N+1}$, previous geodesic distance $\underline{\underline{G}}_{NN}$
Output: new geometric distance $\underline{\underline{G}}'_{N+1,N+1}$, deleted-edge set J

1: **for** each neighbor list $\underline{\underline{L}}_{k,N}^i$ **do**
2:    **if** There are less than K elements in list $\underline{\underline{L}}_{KN}^i$ **then**
3:      Insert point $N+1$ into list $\underline{\underline{L}}_{KN}^i$
4:    **else**
5:      **if** point $N+1$ should be one of point i's K nearest neighbors **then**
6:        Insert new point $N+1$ into list $\underline{\underline{L}}_{KN}^i$
7:        Remove last element p from list $\underline{\underline{L}}_{KN}^i$
8:        Add edge related to point p into set J
9:      **end if**
10:   **end if**
11: **end for**
12: Initialize $\underline{\underline{G}}'$ so that $\underline{\underline{G}}'_{N+1,N+1} = \begin{vmatrix} \underline{\underline{G}}_{N,N} & \underline{d}^{N+1} \\ (\underline{d}^{N+1})' & 0 \end{vmatrix}$
13: **for** each edge $e_{ij} \in J$ **do**
14:   **if** point i is one of the neighbors of point j **then**
15:     Remove edge e_{ij} and e_{ji} from graph $\mathcal{G}$
16:   **else**
17:     Exclude edge e_{ij} and e_{ji} from set J
18:   **end if**
19: **end for**
20: $\underline{\underline{L}}_{K,N+1}^{N+1} \leftarrow$ Point $N+1$'s K nearest neighbors
21: **for** each $u \in \underline{\underline{L}}_{K,N+1}^{N+1}$ **do**
22:   **if** point $N+1$ is one of the neighbors of point u **then**
23:     Add edge $e_{N+1,u}$ and $e_{u,N+1}$ to $\mathcal{G}$
24:     Weight of edge $e_{N+1,u}$ and $e_{u,N+1} \leftarrow \underline{d}^{N+1,u}$
25:   **else**
26:     Add edge $e_{N+1,u}$ and $e_{u,N+1}$ to deleted-edge set J
27:   **end if**
28: **end for**

Alg. 3 involves two parts: updating neighborhood graph related to point $1, ..., N$ (Line 1-19), and the updating related to $(N+1)$th point (Line 20-28). First, when $(N+1)$th point is added, it is likely that it will take the place of some other points as one of K nearest neighbors of some point. Second part of the Alg. 3 performs updating regarding to $(N+1)$th point. Neighborhood graph are updated by adding edges between new point to its K nearest neighbors . Edges that are connecting new point and other far-away points are put into set J together with other deleted edges.

By using a proper data structure, such as linked list, to maintain $\underline{\underline{L}}_{K,N+1}^{i}$, the time complexity of Alg. 3 is

$$T_1 = O(N(c_1 + c_2) + Kc_3) \qquad (1)$$

where c_1 is the cost of determining whether $(N+1)$th point should be included in list $\underline{\underline{L}}_{K,N+1}^{i}$ (Line 5), c_2 is the cost of re-checking each edge in set J(Line 15) , and c_3 is the cost of finding ith nearest neighbor of point N. Note c_1, c_2 and c_3 all have upper bound K. As a result, the time complexity of Algorithm 3 is $T_1 = O(N)$.

4.2 Updating Geodesic Distance

The algorithm by [1] is employed in this part for updating geodesic distance based on neighborhood graph and previous geodesic distance (Alg. 4). At First, the point pairs whose geodesic distances need to be updated are determined. After this, geodesic distances between point i and j are computed. Using a modified Dijkstra's algorithm, high efficiency can be achieved compared to classic Dijkstra or Floyd algorithm employed by ISOMap [2]. At last, geodesic distances that are related to new data points are updated.

As for the time complexity of Alg. 4, the worst case bound for the time of running Alg. 4 is $O(N^2 \log N + N^2 K)$, where N is the number of data point and K is neighbourhood size [1].

Algorithm 4. Updating Geodesic distance

Input: deleted-edge set $V, sec: experiment$, neighborhood Graph $\mathcal{G}'$, previous geodesic distance $\underline{\underline{G}}_{NN}$
Output: new geodesic distance $\underline{\underline{G}}'_{N+1,N+1}$, set of moving data points J
1: Find those point pairs whose geodesic distance need to be updated.
2: Updating geodesic distance between point pairs obtained above.
3: Update geodesic distance related to the $(N+1)$th point.

4.3 Incremental MultiDimensional Scaling

Our main contribution is the introduction of incremental MDS step extended from Spring model. We adopt spring model instead of other statistical dimensional reduction techniques due to spring model's iterative nature. This nature enables an extra data point to be added onto a trained model by ad-

Algorithm 5. I-MDS for updating point coordinates in feature space

Input: the set of moving points V, neighbor list $\underline{\underline{L}}_{K,N+1}$, geodesic distance $\underline{\underline{G}}_{N+1,N+1}$, the previous coordinates of points in feature space $\underline{\underline{Y}}_{mN}$

Output: the new coordinates of points in feature space $\underline{\underline{Y}}'_{m,N+1}$

1: Initialize $\underline{\underline{Y}}^{i\,\prime}_{m,N+1}$ as $\underline{\underline{Y}}^i_{mN}$, i=1..., N
2: Initialize $\underline{\underline{Y}}^{N+1\,\prime}_{m,N+1}$ as Equation(2);
3: Initialize velocity of each points as 0;
4: **for** each $v \in V$ **do**
5:     **for** j= 1 to γ_1 **do**
6:         update position of point m according to force exerted
7:     **end for**
8: **end for**
9: **for** each point $i = 1$ to $N + 1$ **do**
10:     **for** j= 1 to γ_2 **do**
11:         update position of point i according to force exerted
12:     **end for**
13: **end for**

---

justing the model with only a few additional iterations. By contrast, other techniques, such as PCA, would require to be re-run its entirety for ongoing application.

The incremental MDS utilizes Chalmer's [8] sampling strategy. We shall show that the proposed method can compute a stable layout of data points in feature space with good preservation of current geodesic distance. More importantly, the algorithm can finish within linear time by constant number of iterations.

Estimating Initial Position of New Point. The position of new point $N + 1$ should be initialized at the beginning. The initial position of new point is calculated according to the distance between the point and its neighbors. Firstly, a point set of size d out of total N points is constructed. The set construction depends on neighborhood size K and low dimensionality m. If $K > m$, m nearest neighbors from the point's neighbor list is selected; otherwise, another $(m - K)$ points are selected randomly from non-neighbor points. Let m selected points be $h_1, ..., h_m$, and $\underline{y}^{h_i}$ be the coordinate of point h_i in feature space. $\underline{y}^0$ denotes desirable initial position of new point. $\underline{y}^0$ can be determined by solving the following system of equations:

$$|\underline{y}_0 - \underline{y}^{h_1}| = g^{N+1,h_1}$$
$$|\underline{y}_0 - \underline{y}^{h_2}| = g^{N+1,h_2}$$
$$\ldots$$
$$|\underline{y}_0 - \underline{y}^{h_d}| = g^{N+1,h_d} \qquad (2)$$

Algorithm 6. update position of each point i

Input: neighbor list $\underline{\underline{L}}^i_{K,N+1}{}'$, geodesic distance $\underline{\underline{G}}'_{N+1,N+1}$, the current point positions $\underline{\underline{Y}}_{mN}$ in feature space
Output: the new position $\underline{\underline{Y}}'_{m,N+1}$ of point i in feature space

1: Construct Neighborhood set Q from $\underline{\underline{L}}^i_{K,N+1}$
2: Construct Random set R using point randomly chosen from $\overline{Q}$ ($\overline{Q} = \{1, ..., N\} - Q$)
3: $P \leftarrow Q \cup R$
4: Initialize total force $\underline{F}_m$ as 0;
5: **for** each $p \in P$ **do**
6: $\quad \underline{F}_m \leftarrow \underline{F}_m +$ force exerted by point p on point i
7: **end for**
8: Update velocity of point i according to total Force $\underline{F}_m$
9: update position of point i according to its velocity

Updating Positions Iteratively. After estimating the initial position, the process of position updating involves two stages. At the first stage(Line 4-8), the positions of points only from set V are updated for γ_1 times. Note that set V records the points whose geodesic distance with others have changed because of introduction of new data points. By limiting scope of position updating, error rate is reduced more efficiently than by updating all data points at the same time. At the second stage(Line 9- 13), γ_2 iterations of further updating are on whole data set to refine the preliminary configuration. Note γ_1 should be to greater than γ_2 for the sake of efficiency.

At the both stages, Alg. 6(Line 6 and Line 11) are applied to update position of each point. Two sets, namely *Neighbour Set* and *Random Set*, are constructed. Neighbor Set Q includes K nearest neighbors of point i; whereas Random Set R includes fixed number of points randomly drawn from remaining data points. The two sets compose a pool P, where the force calculation is limited. The point's velocity is updated based on to the total force $\underline{F}_m$ on each point i, and positions of the points are in turn adjusted. It is interesting that, with the sampling strategy by Chalmer, fixed number of iterations (γ_1 and γ_2) for position updating is sufficient in Alg. 5.

The time complexity of Alg. 5 is

$$T_3 = O((m \times \gamma_1 + (N+1) \times \gamma_2) \times T'_3) \qquad (3)$$

where m is the size of moving-point set which have upper bound of N, and T'_3 is time for updating position of one data, which is $O(K + c_4)$. Here, K is the size of neighborhood, and c_4 is the size of random set. As a result, we obtain

$$T_3 = O((N \times \gamma_1 + (N+1) \times \gamma_2) \times (K + c_4)) \qquad (4)$$

Since γ_1, γ_2, K, and C_4 are all constant, the time complexity for I-MDS is $O(N)$.

5 Experiment

In this section, we present our experiment results on one synthetic data set and two image data sets. The neighborhood size is set to $K = 7$. The two iteration numbers in Alg. 6 are chosen as $\gamma_1 = 20$ and $\gamma_2 = 3$ respectively.

The proposed algorithm is tested on the Swiss roll data set [9]. As a starting point, a batch ISOMap was run, and an initial manifold with 100 data points was obtained. The geodesic distances and neighbor lists were all recorded as a base of incremental algorithm. The remaining 900 data points were added one by one using the proposed I-ISOMap. The final visualization result of the proposes method can be seen from Fig. 1(c). Comparing this result with that obtained using batch ISOMap (Fig. 1(b)), we can see these two layouts are similar.

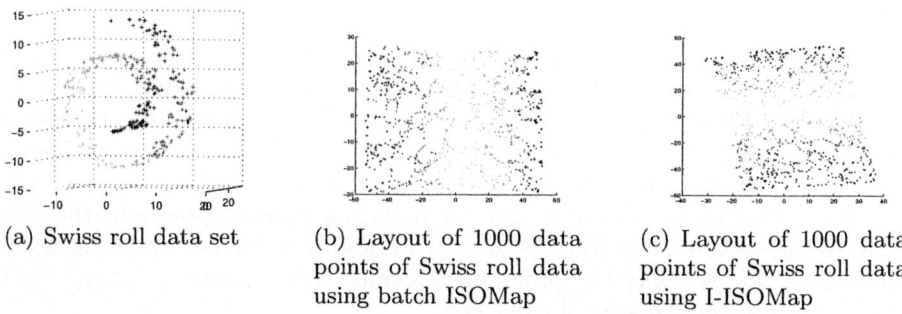

(a) Swiss roll data set (b) Layout of 1000 data points of Swiss roll data using batch ISOMap (c) Layout of 1000 data points of Swiss roll data using I-ISOMap

Fig. 1. Swiss roll data and Visualization Result

Accuracy - Residual Variance. We use residual variance to estimate accuracy of our algorithm [2]. Fig. 2(a) show the residual variance of I-ISOMap when data points are added in random order from $101st$ samples until we get 1000 samples. The X-axis of the figure represents the number of added point ranging from 101 to 1000; its Y-axis represents residual variance, reflecting the goodness of produced layout. The error rate by I-ISOMap is compared with error of batch ISOMap (Fig 2(b) and Fig 2(c)). Average residual variance by batch ISOMap and I-ISOMap are 0.2415 and 0.2175 respectively.

The algorithm was run with a Pentium 2.4 GHz PC with 512M memory. Most of the code was implemented with Matlab. Average time for accommodating one new data point is 1.55 minutes. It can be seen from the analysis of time complexity that proposed I-MDS is bound by linear time complexity, which contributes largely to the efficiency improvement of I-ISOMap.

Computational Time. For further comparison of time efficiency between the batch ISOMap and I-ISOMap, experiments were also carried out based on Swiss Roll data set in larger size. Starting from 1000 and 1500 data points respectively, 10 data points are appended one by one. Both time and residual error with I-ISOMap and baseline ISOMap are shown in Table 1. In the table, time for the

Table 1. Comparison of time and residual error on a large data set

	1001st ~ 1010th point		1501st ~ 1510th point	
	Time (min)	AvgError	Time (min)	AvgError
Batch	1.8	0.002	9.1	0.001
I-ISOMap	1.2	0.002	2.5	0.001

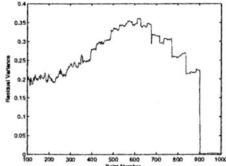 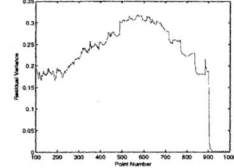 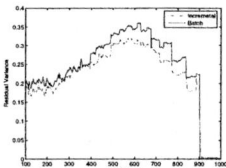

(a) Residual variance for Swiss roll data set by batch ISOMap

(b) Residual variance for Swiss roll data set by I-ISOMap

(c) Comparison of residual variance for Swiss roll data set by batch and I-ISOMap

Fig. 2. Error for Swiss roll data set using batch ISOMap and I-ISOMap

batch version is the time for computing configuration of 1000 and 1510 data points; whereas time for I-ISOMap is average time when 1001st to 1010th and 1501st to 1510th data points are appended. The result shows that I-ISOMap only spent roughly 1.3 more minutes for adding 1501th point than adding the 1001st point. Meanwhile, both version results in similar residual errors. This suggests the I-ISOMap is able to incorporate previous configuration efficiently when more data are available lately.

Image Data. We further test the proposed I-ISOMap using one digit image set [10], which are typically of high dimensionality. It includes 1000 binary images of digits from '0' to '9' at 16×16 resolution. The dimension of the digit images is reduced to $d = 7$ in following experiments.

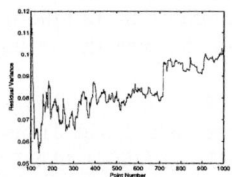 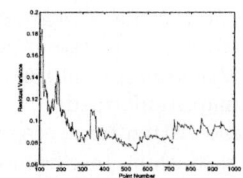 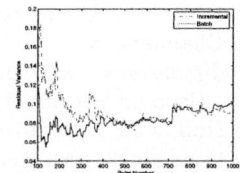

(a) Residual variance for digits images by batch ISOMap

(b) Residual variance for digit images by I-ISOMap

(c) Comparison of residual variance for digit images by batch and incremetal ISOMap

Fig. 3. Error for digit images using batch ISOMap and I-ISOMap

Again, we start with 101th data points based on result of batch version using 100 data points. The residual variance by batch ISOMap, I-ISOMap and comparison of error on digit images are shown in Fig. 3(a), Fig. 3(b) and Fig. 3(c). The figures reveal that the result with batch and incremental ISOMap are comparable on accuracy. Average residual variance of 900 points by batch and incremental algorithm on digit images are 0.0835 and 0.0929 respectively.

6 Conclusion

We have extended ISOMap into an incremental algorithm, I-ISOMap, for dimensional reduction and visualization. I-ISOMap can accommodate new coming data in a few constant number of iterations, and generates a stable layout embedding in a lower dimensional space. Compared with ISOMap, the proposed I-ISOMap is more efficient, and can produce comparable layout to that by batch ISOMap. The efficiency of our method is mainly gained from the iterative nature of Spring model. In comparison to earlier work in [1], our method further reduce time complexity to $O(N)$ instead of quadratic time complexity in MDS step.

References

[1] Law, M., Zhang, N., Jain, A.: Nonlinear manifold learning for data stream. In: Saim Data Mining(SDM). (2003)
[2] Tenenbaum, J., de Silva, V., Langford, J.: A global geometric framework for nonlinear dimensionality reduction. Science **290** (2000) 2319–2323
[3] Roweis, S., Saul, L.: Nonlinear dimensionality reduction by locally linear embedding. Science **290** (2000) 2323–2326
[4] Belkin, M., Niyogi, P.: Laplacian eigenmaps and spectral techniques for embedding and clustering. In: Advances in Neural Information Processing Systems (NIPS). (2002)
[5] Vlachos, M., Domeniconi, C., Gunopulos, D.: Non-linear dimensionality reduction techniques for classification and visualization. In: Proceedings of the 8th ACM SIGKDD International Conference on Knowledge Discovery and Data Mining. (2002)
[6] Cox, T., Cox, M.: Multidimensional Scaling. London:Chapman and Hall (1994)
[7] Eades, P.: Aheuristic for graph drawing. Congressus Numerantium **42** (1984)
[8] Chalmers, M.: A linear iteration time layout algorithm for visualising high-dimensional data. In: IEEE Visualization. (1996) 127–132
[9] : (Isomap website:http://isomap.stanford.edu/)
[10] Hull, J.J.: A database for handwritten text recognition research. IEEE Transactions on Pattern Analysis and Machine Intelligence (PAMI) **16** (1994) 49–67

SVM Based Automatic User Profile Construction for Personalized Search*

Rui Song, Enhong Chen, and Min Zhao

Department of Computer Science and Technology,
University of Science and Technology of China,
Hefei, Anhui, 230027, P.R. China
{ruisong, zhaomin}@mail.ustc.edu.cn
cheneh@ustc.edu.cn

Abstract. The number of accessible Web pages has been growing fast on the Internet. It has become increasingly difficult for users to find information on the Internet that satisfies their individual needs. This paper proposes a novel approach and presents a prototype system for personalized information retrieval based on user profile. In our system, we return different searching results to the same query according to each user's profile. Compared with other personalized search systems, we learn the user profile automatically without any effort from the user. We use the method of support vector machine to construct user profile. A profile ontology is introduced in order to standardize the user profile and the raw results returned by the search engine wrapper. Experiments show that the precision of the returned web pages is effectively improved.

1 Introduction

Web search engine plays an important role in information retrieval on the Internet. However most search engines return the same results to the same query not caring about whom the query is from. In other words, the existing search engines do not take the user's profile into account, which in some extent cause lower precision in information retrieval. In order to solve this problem, several approaches applying data mining techniques to extract usage patterns from Web logs are put forward. However, most of these approaches ask users to construct their individual profile either manually or semi automatically which makes the user profile construction an energy consuming task. How to construct the user profile automatically without any effort from the user is becoming a significant problem to be solved. Therefore, in this paper, we propose a method that can be used to obtain user profile automatically according to user's recently accessed documents. The constructed user profile can then be used to adapt the user's query results to his need. We use the method of Support Vector Machine (SVM) to construct the user profile based on the profile ontology. Search Engine Wrapper (SEW) collects the user's queries and passes them to several search engines. The Raw Results returned by those search engines are processed according to user profile and each result is given a score to form a ranked list.

---

* This work was supported by Nature Science Foundation of Anhui Province (No.050420305) and National "973" project (No. 2003CB317002).

The rest of this paper is organized as follows: In Section 2, we review related work focusing on personalized search system. In Section 3, we propose our approach of constructing user profile automatically from user's recently accessed documents based on profile ontology. We describe the whole query process of returning different results according to user's individual profile as well. In Section 4, we present the experimental results for evaluating our proposed approaches. Finally, we conclude the paper with a summary and directions for future work in Section 5.

2 Related Work

Standard search engines, Google, Yahoo, for example, do not take the user's individual interest into account. To solve this problem, several types of search systems that provide users with information more relevant to their individual needs have been developed. A novel approach to enhance the search engines by retrieving the semantic content of a page and by comparing it to the one specified by the user itself is proposed in [1]. In [3], the integration of agent technology and ontology is proposed as a significant impact on the effective use of the web services. But in most of these systems, users have to register personal information beforehand or to provide feedback on relevant or irrelevant judgments. The OBIWAN system [6], for example, asks the user to submit a hierarchical tree of concepts that represents their view of the world. These types of registration can become energy consuming for users. Therefore, this paper proposes an approach to construct user profile automatically using SVM based on user profile ontology. The constructed user profile can be used by SEW to analyze the web pages returned by several standard search engines.

3 System Architecture

Figure 1 presents a high level view of the proposed architecture for user profile constructing, updating and query processing.

The user is asked to submit keywords to the system by means of a graphical user interface the same as the ordinary search engines. His query is submitted to several standard search engines by SEW. The returned searching results are collected as Raw Results. Then each page is analyzed and scored by Document Processor based on the user profile. The user profile is constructed using SVM classifier based on the profile ontology. The construction process involves two phases: classifying user's recently accessed documents and constructing user profile vector. Both of these two phases are implemented without any effort from the user.

3.1 Constructing User Profile

In our system, user only has to provide a directory in which there are some documents he is interested in. These documents can be the resources downloaded recently from the internet or the materials written by him. We provide a profile ontology organized as a hierarchy tree with its leaves as the elementary categories. The documents given

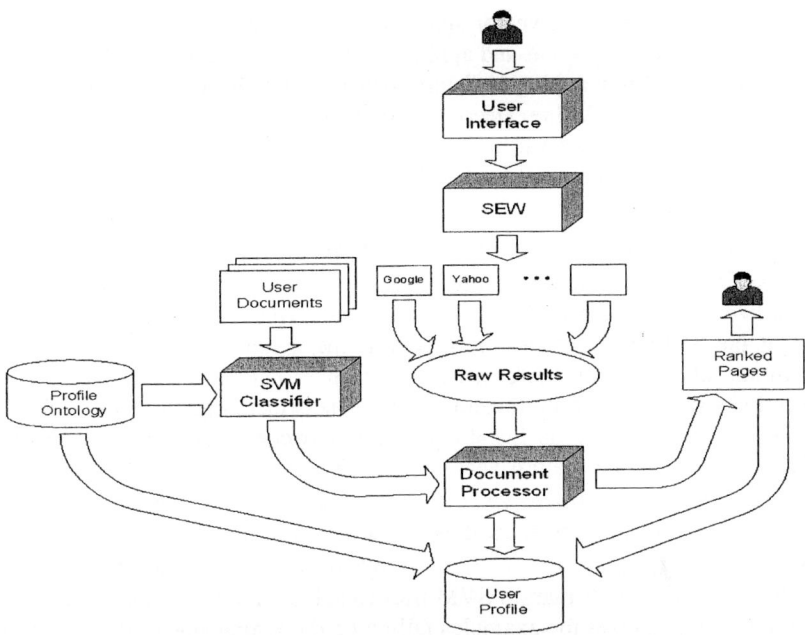

Fig. 1. Framework of Personalized Search System with User Profile

in the user directory are classified into those elementary categories. We do not use the method of clustering because using classification we can easily find the relationships between every two interest categories in the hierarchy tree during the retrieval process and we can obtain a relatively normalized user profile vector. In our system, we aims at maximally automating the process of user profile construction to significantly reduce the cost of users. So we can not ask users to give us a lot of documents. We choose SVM as the classification method to map the document feature space to the user profile space because SVM has overwhelming effect in learning from small training sets.

3.1.1 Using Latent Semantic Indexing for Dimension Reduction

The documents in the user document collection are represented by a term-document matrix, which contains the values of the index terms t occurring in each document d. We denote m as the number of index terms in a collection of documents and n as the total number of documents. Formally, we let M denote a term-document matrix with n rows and m columns and let $w_{i,j}$ be an element (i, j) of M. Each $w_{i,j}$ is assigned a weight associated with the term-document pair (d_i, t_j), where d_i ($1 \leq i \leq n$) represents the i-th document and t_j ($1 \leq j \leq m$) represents the j-th term. For example, using a tf-idf representation, we have $w_{i,j} = tf(t_j\text{-}d_i)idf(t_j)$. Thus, given the values of $w_{i,j}$, the term-document matrix represents the whole document collection.

Generally the matrix is very sparse because there are so many terms in the document collection. We use the method of Latent Semantic Indexing (LSI) [7, 8] to delete the terms with lower expressive force. LSI is an approach that maps documents as well as terms to a representation in the so-called latent semantic space. It usually takes the high dimensional vector space representation of documents based on term frequencies as a starting point and applies a dimension reducing linear projection. The specific form of this mapping is determined by a given document collection and is based on a Singular Value Decomposition of the corresponding term/document matrix. In our system, we use LSI for dimension reduction. We use the dimension reduced matrix M' as the input for SVM method to construct the user profile.

3.1.2 Using SVM to Construct User Profile Based on Profile Ontology

In our system, we provide a profile ontology referenced from Yahoo directories to standardize user's interests. It is represented as a hierarchy tree. We use M' as the input of the SVM method and the leaves as the given categories. SVM has shown outstanding classification performance in practice. It is based on a solid theoretical foundation-structural risk minimization [10]. The decision function of an SVM is $f(x) = <w \cdot x>$, where $<w \cdot x>$ is the dot product between w (the normal vector to the hyperplane) and x (the feature vector representing an example). The margin for an input vector x_i is $y_i f(x_i)$ where $y_i \in \{-1,1\}$ is the correct class label for x_i. Seeking the maximum margin can be expressed as a quadratic optimization problem: minimizing $<w \cdot w>$ subject to $y_i(<w \cdot x_i>+b) \geq 1$, $\forall$ i. When positive and negative examples are linearly inseparable, soft-margin SVM tries to solve a modified optimization problem that allows but penalizes the examples falling on the wrong side of the hyperplane.

We use SVM to classify the documents in the user given document collection into the elementary categories. The number of documents in each category is labeled as each leave's weight in the tree data structure. We obtain the other nodes' weight in the tree by making each node's weight equaling to the sum of its sub nodes' weight. Thus all the nodes in the tree have their weights for describing their relative importance in the user profile. We choose the n biggest nodes to be normalized to form the user profile vector.

The algorithm of user profile construction is shown in Fig. 2. In the algorithm, the user profile structure is represented as a hierarchy tree. Suppose it has r nodes totally, including k leaf nodes, which are sub_1, sub_2, ..., sub_k from left to right. Every node has the attributes of w and w', w is the weight, and w' is the temporary weight for updating. The user document collection D has been classified into the elementary categories using SVM. For leaf nodes sub_1, sub_2, ..., sub_k, there are d_1, d_2, ..., d_k documents respectively. The total number of documents in D is $d=d_1 + d_2 + ... + d_k$.

3.4 Processing the User's Queries

The user submits a query to our system through graphical user interface. SEW submits the user's query to several standard search engines. The returned pages are preprocessed before downloading to form the Raw Results. Then each page is classified into the elementary categories and scored according to its similarity to the user profile by Document Processor. We present the results to the user in the descending order of score. The algorithm of page scoring is as follows:

Algorithm 1 - User Profile Construction
Input: Profile Tree T
Output: User Profile Vector U

```
(1) Traverse the profile tree from leaf to root
(2) for each node_i, 1 ≤ i ≤ r do begin
        if (node_i is sub_j)
           node_i.w = d_j
        else
```
$$node_i.w = \sum_{j=1}^{ni} node_i.child_j.w, \text{ where } ni \text{ is the}$$
```
           number of child nodes of node_i
    node_i.w' = node_i.w
    end
(3) Sort node_1, node_2, ..., node_r in the descending order
    of weight
(4) Get (l_1,v_1), (l_2,v_2), ..., (l_n,v_n) of the first n
    nodes from node_1, node_2, ..., node_r, l_i is the
    category of the node, and v_i is the weight
(5) U={ (l_1,w_1), (l_2,w_2), ..., (l_n,w_n) },
```
$$(l_i, w_i) = (l_i, v_i / \sum_{k=1}^{n} v_k), 1 \leq i \leq n,$$
```
    U is user profile vector.
```

Fig. 2. Algorithm of User Profile Construction

Algorithm 2 - Scoring Page P Returned by SEW
Input: Page P
Output: Score for P

```
Use SVM to classify the returned page P. Suppose P
belongs to l_q with the probability prob_q, l_q is the
category of leaf sub_q, 1 ≤ q ≤ k.
(1) if (l_q=l_i, l_i is one category in U, 1 ≤ i ≤ n)
       Score (P) = w_i * prob_q
    else if (∃ node_a, l_a=l_j, l_j is one category in U,
    1 ≤ j ≤ n (node_a is the ancestor of sub_q))
       Score (P) = w_j * prob_q
       else
          Score (P) = 0
(2) From sub_q to the root, update w' of each node using
    the same method as calculating w.
```

Fig. 3. Algorithm of Scoring a Web Page P Returned by SEW

3.5 Updating User Profile According to User's Queries

During the page scoring process each node's w' is calculated. We compare the maximum of w' with the minimum weight of the user profile vector items. The bigger one will be included in the user profile vector. The user profile updating algorithm is as follows:

Algorithm 3 – Updating User Profile
Input: Initial User Profile Vector, m representing the number of pages we want to use
for updating
Output: Updated User Profile Vector
```
(1) Sort the returned pages in the descending order of
    Score.
(2) If there are only m' pages whose Score>0 and m'<m
    m = m'.
(3) Get the first m pages P₁...Pₘ.
(4) if m > 0
    {
      for each Pᵢ,1 ≤ i ≤ m
        Suppose Pᵢ's category is lⱼ in U, update (lⱼ,wⱼ)
          with (lⱼ,wⱼ')
    }
    else
    {
      Suppose w=min{wᵢ|1≤i≤n},and its corresponding
        category l∈{l₁,l₂,...,lₙ}, pₘₐₓ is the page
        whose weight is maximum, wₘₐₓ is the weight of
        pₘₐₓ, and lₘₐₓ is the corresponding category.
        if wₘₐₓ > w
          (lₘₐₓ,wₘₐₓ)→(l,w)
    }
(5) U={ (l₁,w₁),(l₂,w₂),...,(lₙ,wₙ) },
```
$$(l_i, w_i) = (l_i, w_i / \sum_{k=1}^{n} w_k), 1 \leq i \leq n$$

Fig. 4. Algorithm of User Profile Updating

4 Experiment and Results

We provide an ontology in a tree hierarchy of 5 layers and 16 leaf nodes as the basis for user profile construction. We use C-SVC in LIBSVM [11] as the SVM tool and choose RBF as the kernel function. Our SVM model is trained by the categories and the corresponding documents referenced from yahoo directory. As to the main parameters in SVM, we set $\gamma = 0.15$ and $C = 100$.

In this paper we give two users as an example to analyze the performance of our system. Their interests are different from each other. User A is a young girl. User B is a forty-years-old man. We do not know their interests explicitly. They both are asked to provide a directory. User A gave us a directory including 32 documents in her interesting field. User B provided us 48 documents. We parse the documents and represent User A's and B's documents as Matrix A and Matrix B. After dimension reduction we use the trained SVM classifier to classify those documents. User A's documents are classified into 6 categories. They are travel, food, TV, movie, digital, and animal. We calculate the weight of each node in the hierarchy tree and choose 4 nodes as her interesting field according to the nodes' weights. To avoid the profile vector being too general, we do not choose the nodes in the first two layers and we choose the child node instead of the parent node if their weights are equal. The 4 nodes from user A are TV, food, animal and travel. Her profile vector is constructed as {('TV', 0.375), ('food', 0.281), ('animal', 0.219) ('travel', 0.09)} using the algorithm above. User B's documents are classified into 8 categories as automotive, plant, travel, painting, tea, god, Internet and photography. His profile vector is constructed as {('plant', 0.313), ('automotive', 0.188), ('Internet', 0.146), ('astronomy', 0.104)}. Each user enters 5 queries. They are "spider", "bean", "mars", "jaguar" and "huang mountain". Take the query "spider" for an example: Spider is the name of a type of arthropods and also a type of search engine. From the two users' interests, we know User A may want to learn some knowledge of the arthropod spider, as she is interested in animal. User B may want to learn the knowledge of spider search engine. Obviously, they have distinct intent although they enter the same query. Using standard search engines, they will obtain the same returned pages. As to the last query "huang mountain", generally we mean the Mountain itself. But the ordinary search engines will return various pages such as huang mountain tea as the tea from Huang Mountain is as noted as the Mountain itself. It is the same with the other queries.

We give our results comparing with some standard search engines in Table 1 and Table 2 (the size of profile vector is 4). Results are reported in the form $x_1/10$ and $x_2/30$: x_1 represents the number of relevant links found in the first result page and x_2 represents the number of relevant links found in the first three result pages (10 links per page). Investigation demonstrates that users only care about the first 30 links.

The column R reports recall of each query and the column P reports the precision of each query in our system. R and P are defined as:

$$R = Num_c / Num_t . \quad (1)$$

$$P = Num_c / Num_r . \quad (2)$$

Where Num_c denotes the number of correct pages returned by our system, Num_t denotes the total number of correct pages in the first 50 links returned by Google and Yahoo each, and Num_r denotes the number of relevant pages returned by our system.

BK is a composite measure proposed by Borko. (BK = P+R-1) We use it to reflect the influence of the varying size of profile vector to our system. In our experiment, the size of profile vector changes from 3 to 5.

We can see * in the row of the query "huang mountain" of User B. It is because huang mountain does not belong to any category in the user profile vector. We can not decide whether the returned results are relevant to his interest or not.

Table 1. Results returned by our system comparing with Google and Yahoo for User A

query	Google		Yahoo		Our System		R	P	BK
	$X_1/30$	$X_2/10$	$X_1/30$	$X_2/10$	$X_1/30$	$X_2/10$			
spider	10/30	2/10	7/30	1/10	10/30	4/10	71.4%	76.9%	0.483
bean	4/30	3/10	4/30	5/10	6/30	6/10	75%	64.5%	0.395
mars	2/30	2/10	1/30	0/10	2/30	2/10	66.7%	50%	0.167
Jaguar	5/30	1/10	4/30	2/10	8/30	5/10	66.7%	61.5%	0.282
Huang mountain	22/30	7/10	19/30	6/10	24/30	8/10	80%	82.8%	0.628

Table 2. Results returned by our system comparing with Google and Yahoo for User B

query	Google		Yahoo		Our System		R	P	BK
	$X_1/30$	$X_2/10$	$X_1/30$	$X_2/10$	$X_1/30$	$X_2/10$			
spider	2/30	4/10	2/30	2/10	6/30	4/10	40%	66%	0.06
bean	6/30	3/10	7/30	1/10	8/30	4/10	80%	63.3%	0.433
mars	8/30	8/10	9/30	10/10	12/30	10/10	66.7%	70.1%	0.368
Jaguar	5/30	6/10	4/30	6/10	8/30	7/10	80%	70.1%	0.501
Huang mountain	*								

From Figure 5 and Figure 6 we can see that the results returned by our system have higher precision than Google and Yahoo based on a small training set.

Let us see the difference of the returned results when the size of profile vector n changes. If n equals 3, user A's profile vector is {('TV', 0.375), ('food', 0.281), ('animal', 0.219)}. The results are the same as in table 1 except for the query "huang mountain". The row of "huang mountain" must be remarked as "*" because "travel" is not included in the user's profile. And it is the same with user B's third query "mars" as User B's profile vector is {('plant', 0.313), ('automotive', 0.188), ('Internet', 0.146)}. If n equals 5, user A's profile vector is {('TV', 0.375), ('food', 0.281), ('animal', 0.219), ('travel', 0.093), ('digital', 0.031)}. The results do not change because all the returned pages have low probability to be categorized to the new category "digital". User B's profile vector is {('plant', 0.313), ('automotive', 0.188), ('Internet', 0.146), ('astronomy', 0.104), ('travel', 0.083}. Since the categorization "travel" is included in his profile, the fifth query "huang mountain" will have the same results as in Table 1.

Each query has a BK value and we calculate the average value in Table 3. We use this composite measure to consider the recall and precision change of our system while the size of profile vector n changes. From Table 3 we can see that our system has better performance with the augment of n. It is because our system has more information to be based on to choose the appropriate pages. However n can not be too big because it will make the profile too general and the computation too complex. We must choose the appropriate n according to the scale of the ontology.

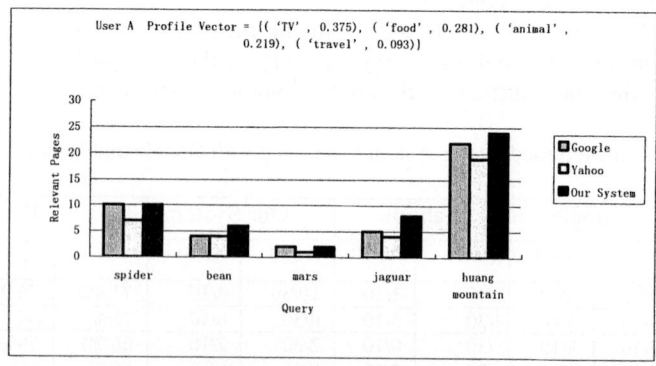

Fig. 5. Results returned by our system comparing with Google and Yahoo for User A

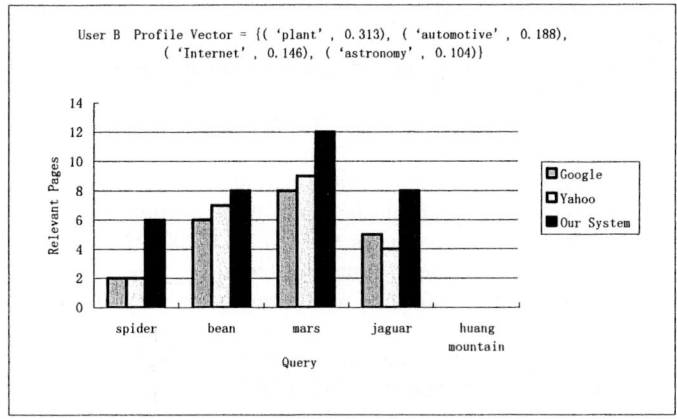

Fig. 6. Results returned by our system comparing with Google and Yahoo for User B

Table 3. BK value of our system with different size of profile vector n

n	BK of User A	BK of User B
3	0.348	0.287
4	0.391	0.341
5	0.391	0.398

5 Conclusion and Future Work

In this paper, we presented the idea of constructing user profile automatically and using it for personalized information retrieval on the Internet. This was accomplished based on a profile ontology. We use SVM as the method of categorization to process the user given document collection and also the web pages returned by the Search Engine Wrapper. Compared with other systems, we release the user from building personal profile by themselves or spending time interacting with the system. The results of our experiment show that the precision of the returned web pages is effectively improved. Since this paper focuses on the content-based personalized information retrieval, in the future, we can consider about integrating this with collaborative methods. We can also combine user's long-term interest with short-term interest to update user profile more effectively.

References

1. Carmine, C., Antonio, A., Antonio, P.: An Intelligent Search Agent System for Semantic Information Retrieval on the Internet. Workshop on Web Information and Data Management (2003) 111-117
2. Kazunari, S., Kenji, H., Masatoshi, Y.: Adaptive Web Search Based on User Profile Constructed without Any Effort from Users. Proc. International WWW Conference, New York, USA (2004)

3. Weihua, L.: Ontology Supported Intelligent Information Agent. Proc. on the First Int.IEEE Symp. on Intelligent Systems (2002) 383-387
4. Weifeng, Z., Baowen, X., Lei, X.: Personalize Searching Results using Agent. Mini-Micro Systems, Vol. 22 (2001) 724-727
5. Xiaolan, Z., Susan, G., Lutz, G., Nicholas, K., Alexander, P.: Ontology-Based Web Site Mapping for Information Exploration. Proc. of the 1999 ACM CIKM International Conference on Information and Knowledge Management, USA (1999) 188-194
6. Jason, C., Susan, G.: Personal Ontologies for Web Navigation. Proc. 9th Intl. Conf. on Information and Knowledge Management McLean VA (2000) 227-234
7. Thomas, H.: Probabilistic Latent Semantic Indexing. Proc. of the 22nd Annual International ACM SIGIR Conference on Research and Development in Information Retrieval, USA (1999) 50-57
8. Noriaki, K.: Latent Semantic Indexing Based on Factor Analysis. Soft Computing Systems - Design, Management and Applications, Chile (2002)
9. Dell, Z., Weesun, L.: Web Taxonomy Integration using Support Vector Machines. Proc. of the Sixteenth International Conference on Machine Learning (1999) 200-209
10. Vapnik, V.N.: The Nature of Statistical Learning Theory. Springer-Verlag, New York, NY, (2000)
11. Chih-Chung, C., Chih-Jen, L.: LIBSVM, http://www.csie.ntu.edu.tw/~cjlin/libsvm/

Taxonomy Building and Machine Learning Based Automatic Classification for Knowledge-Oriented Chinese Questions[1]

Yunhua Hu, Qinghua Zheng, Huixian Bai, Xia Sun, and Haifeng Dang

Computer Department of Xi'an Jiaotong University, Xi'an, Shaanxi, China, 710049
{yunhuahu, qhzheng, sx}@mail.xjtu.edu.cn
{baihuixian, xjtu_hfdang}@163.com

Abstract. In this paper, we propose a taxonomy for knowledge-oriented question, and study the machine learning based classification for knowledge-oriented Chinese questions. By knowledge-oriented questions, we mean questions carrying information or knowledge about something, which cannot be well described by previous taxonomies. We build the taxonomy after the study of previous work and analysis of 6776 Chinese knowledge-oriented questions collected from different realistic sources. Then we investigate the new task of knowledge-oriented Chinese questions classification based on this taxonomy. In our approach, the popular SVM learning method is employed as classification algorithm. We explore different features and their combinations and different kernel functions for the classification, and use different performance metrics for evaluation. The results demonstrate that the proposed approach is desirable and robust. Thorough error analysis is also conduced.

1 Introduction

Question classification is a process of identifying the semantic classes of questions and assigning labels to them based on expected answer types. It plays an important role in Question Answering (QA) system. It would be used to reduce the search space for answers, and also can provide the information for downstream processing of answer extraction [1, 2, 3].

Nowadays the question type taxonomies for fact-based and task-oriented [4] questions are well studied. However, we find that previous taxonomies can not meet the requirements of knowledge-oriented questions. By knowledge-oriented questions, we mean questions carrying information or knowledge about something, which are more realistic and more comprehensive. For example, "Why can Bill Gates be the richest person in the world?" and "What is the difference between face-based questions and task-oriented questions?" are knowledge-oriented questions. As we studied, further

---

[1] This work was supported by NSF of China (Grant No. 60473136), the National High Technology Research and Development Major Program of China (863 Program) (Grant No. 2004AA1Z2280), the Doctoral Program Foundation of the China Ministry of Education (Grant No. 20040698028) and the Project of Tackling Key Problems in Science and Technology of Shaanxi province in China (Grant No. 2003K05-G25).

researches are needed for question taxonomy for the following three reasons. Firstly, previous taxonomies can't cover nearly 30% realistic questions based on our survey. Secondly, the different criteria between taxonomies make question classification more difficult. Fact-based questions are distinguished by semantic type. While Task-oriented questions usually are classified by their degree of knowledge. Thirdly, these taxonomies overlapped partially. It is worthwhile to work on the construction of a uniform taxonomy for knowledge-oriented questions

Knowledge-Oriented question classification is a new task. Previous work has been conducted on the automatic question classification task for fact-based questions in English, Japanese, and other kind of languages. But how to conduct the automatic classification for knowledge-oriented questions is still a problem. Moreover, classification for Chinese questions is more challenge than it in other languages for the complexities of Chinese. It is worthwhile to investigate this problem.

In this paper, we propose a taxonomy consists of 25 coarse classes for 6776 Chinese knowledge-oriented questions collected from different realistic sources. Previous taxonomies are well organized into this uniform architecture. Then we study the automatic classification problem for knowledge-oriented Chinese questions. In our approach, we manually annotate training questions in the collected data set and take them as training data, train machine learning models, and perform question type classification using the trained models. We explore different features and their combinations and different kernel functions for the classification, and use different performance metrics for evaluation. The results demonstrate that the proposed approach is desirable and robust. Thorough error analysis is also conduced.

The rest of the paper is organized as follows. In Section 2, we introduce related work, and in Section 3, we show the building of knowledge-oriented question taxonomy. Section 4 describes our method of automatic classification for knowledge-oriented questions. Section 5 gives our experimental results and analysis of misclassification cases. Section 6 summarizes our work in this paper.

2 Related Work

Several question taxonomies for classification have been proposed till now. Wendy Lehnert [5] proposed a conceptual taxonomy for questions. ISI built a taxonomy with 94 classes of questions [6]. Li and Roth [7] simplified the taxonomy of ISI and construct a taxonomy for face-based questions in TREC QA system. This taxonomy is consists of 6 coarse grained classes and 50 fine grained classes. It's followed by many researchers as standard in the research work. Suzuki, et al [8], proposed a similar taxonomy for Japanese fact-based questions.

The above taxonomies mainly focus on the fact-based questions. For task-oriented questions, Bloom's taxonomy is used [9]. Bloom's taxonomy consists of six categories namely knowledge, comprehension, application, analysis, synthesis and evaluation. This taxonomy is to reflect the degree of knowledge that students mastered during their study. And questions in this taxonomy are partially overlapped with questions in fact-oriented taxonomies.

Approaches for QC fall into two categories, rule based approach and machine learning based approach.

Singhal et al. [10], used hand craft rules for questions classification. The rule based approach can achieve good performance if the rules were well defined. But unfortunately, it has many disadvantages such as it's difficult to be constructed and adapted to new taxonomy or new dataset.

Li and Roth introduced machine learning algorithm in their question classification task and took syntactic and semantic information as features. Zhang et al [11], used SVM as algorithm for classification in the same task and a special tree kernel was developed to improve the performance. Suzuki et al [8], addressed the Japanese question classification problem in the similar way. The main difference is that they designed a HDAG kernel to combine syntactic and semantic features. To verify the robust of machine learning approaches, Metzler et al [12], investigate the question classification problem in different data sets and different questions with different question classification standards. They all obtained desirable results of classification.

Previous work shows that machine learning based approaches works well in fact-based question classification. However, to the best of our knowledge, no related work on the classification of knowledge-oriented questions has conducted.

3 Taxonomy Building for Knowledge-Oriented Question

Based on previous work and analysis of many questions collected from different realistic sources, we construct a knowledge oriented question taxonomy.

In out dataset, 6776 questions were collected from three different realistic sources. The first source is the database of our QA system for e-learning [13]. More than 1500 questions came from this source. The second source is the open Web. More than 1000 questions were selected from FAQ, forum, mail list, and other Web pages in internet. These questions involve various topics, such as government documents, programming skills, introduction of products, etc. The third source we used is the large corpus named CWT100g (the Chinese Web Test collection with 100 GB web pages). This corpus is provided by network group of Peking University in China [14]. The rest questions in our data set are all extracted from this corpus by heuristic rules and filtered by hand. The questions which carry no information were ignored. For example, question "And then?" is the ignored case.

Table 1. Distribution of 6776 questions in taxonomy for knowledge-oriented questions

Name	Perc.	Name	Perc.	Name	Perc.
degree	0.010	purpose	0.009	choice	0.021
place	0.021	distinction	0.016	evolution	0.009
definition	0.063	person	0.049	application	0.007
method	0.209	time	0.018	theory	0.005
classification	0.011	entity	0.073	reason	0.162
relation	0.014	yesno	0.112	attribute	0.026
attitude	0.018	number	0.026	function	0.013
structure	0.008	condition	0.024		
description	0.065	position	0.011		

Table 2. Example of knowledge-oriented questions with question types

Type	Example
method	How to resolve this problem?
place	Where is the capital of China?
degree	How hot is the core of the earth?
relation	What is the relation between TCP and UDP?
purpose	What is her purpose to do this?

Although many question taxonomies are organized as hierarchy, we only consider the coarse grained classes of knowledge-oriented question taxonomy in this paper. The taxonomy we build is based on the semantic analysis of questions. The distribution of questions in the building taxonomy is shown in table 1 (Perc. means percentage).Table 2 shows some examples in this taxonomy.

From table 1 and table 2 we can find more than 20% questions are talking about the process of doing something. They called task-oriented questions. And about 50% questions are fact-based questions such as type 'place'. But their distribution is quite different from that in TREC. And we can also find that there are nearly 30% questions which can be included in neither fact-based nor task-oriented questions taxonomy such as 'relation'. It shows that our work is necessary in some sense.

The meanings and ranges of question types in our taxonomy are quite different from those in other taxonomies. We keep some fact-based questions in our taxonomy the same as them in taxonomy proposed by Li and Roth. But some types are redefined. For example, we split the 'number' classes in Li's taxonomy into three classes, namely, 'time', 'number' and 'degree'. And some classes in conceptual taxonomy and Bloom's taxonomy are also considered in our taxonomy such as class 'cause' and 'purpose'.

4 Question Classification for Knowledge-Oriented Chinese Questions

Previous work on question classification can fall into two categories, rule based approach and machine learning based approach. We used machine learning approach in knowledge-oriented Chinese questions classification in this paper. Machine learning approach outperforms rule based approach for several reasons. Firstly, the construction of rules is a tedious and time consuming work. Secondly, it's difficult to weight the importance of each rule by hand. Thirdly, the consistence between rules is difficult to keep especially while the question taxonomy is expanded. Fourthly but not lastly, the rules can not be adapted to new taxonomy or new domain conveniently. Machine learning based approach can overcome this difficulty well.

Question classification is the same as text classification in some sense. Although it's difficult to distinguish the semantic difference between questions, we can consider question classification as text classification problem and take advantage of experience of them after advance analysis of questions.

4.1 Outline

Question classification based on machine learning consists of training and prediction.

In learning, the input is sequence of questions with corresponding question type. We take labeled questions in the sequences as training data and construct models for identifying whether a question has a correct question type. One-against-rest strategy is used for model training.

In prediction, the input is questions with unknown question types. We employ models for every specific question types to identify whether a question has such semantic type. We can take the type with maximum output score as final result we are desired.

4.2 Features

The following are features used in our approach. The former three features are syntactic and the later one is semantic.

SingleWord: This feature consists of bag-of-words. Some words which are taken as stop words in some tasks are really helpful for question classification, such as 'what', 'consist', 'reason', and so on.

BigramWords: All two continuous word sequences in the question. For example, combination of '定义(definition)' and '是(is)' will be useful for identifying the definition questions.

WordAndPos: Combination of word and POS (part of speech tag) of two continuous words. For example, in sequence '定义/n 是/v (definition/n is/v)' the WordAndPos features are '定义v' and 'n是'. It will be related to '定义/n 为/v (definition/n is/v)' by feature '定义v'. This is a useful feature in our approach since many Chinese words have the similar meaning when they are similar in structure.

Semantic: WordNet like semantic information.

Semantic information we used here are derived from HowNet, a Chinese version semantic dictionary [15]. The HowNet consists of several categories of semantic information such as event, entity, attribute, quantity, and so on. In each category meta-semantic elements are organized as a tree. Different word perhaps contains the same types of meta-semantic elements. For example, the word 'definition' contains the semantic of 'information/信息' as well as the word 'concept' It means that these two words are similar in semantic although they are different.

4.3 Algorithm

SVM perform state-of-art in text classification and are confirmed helpful to fact-based question classification [11]. So we chose SVM as algorithm in our knowledge-oriented question classification scenario.

A classification task usually involves with training and testing data which consist of some data instances. Each instance contains one classes labels and some features. The goal of classification is to produce a model which predicts class labels of data instance in testing set which are given only the features. Support Vector Machines are such discriminative model. The SVM is the form of $f(x)=<w \cdot x>$, where $<w \cdot x>$ is the inner product between input feature vector and the weight vector of features. The class label will be positive if f(x) > 0 otherwise the label will be negative.

The main idea of SVM is to map input feature space into a higher dimensional space and select a hyperplane that separates the positive and negative examples while maximizing the minimum margin. Given a training set of instance-label pairs (x_i, y_i), $i=1,...,l$ in which $x_i \in R^n$ and $y_i \in \{1,-1\}^l$, the SVM conduct the optimization problem as the following [16]:

$$\min_{w,b,\xi} \tfrac{1}{2} w^T w + c \sum_{i=1}^{l} \xi_i$$
$$\text{subject to} \quad yi(w^T \phi(x_i) + b) \geq 1 - \xi_i \quad (1)$$
$$\xi_i \geq 0.$$

Here training vectors of x_i are mapped into a higher dimensional space by the function $\Phi(x_i)$. And the penalty of minimization of structure risk is controlled by parameter C > 0. $K(x_i, x_j) \equiv \phi(x_i)^T \phi(x_j)$ is so called the kernel function. We used the following four basic kernels in our experiments.

linear: $\quad K(x_i, x_j) = x_i^T x_j.$

polynomial: $\quad K(x_i, x_j) = (\gamma x_i^T x_j + r)^d, \gamma > 0.$

radial basis function (RBF): $\quad K(x_i, x_j) = \exp(-\gamma \|x_i - x_j\|^2), \gamma > 0.$

sigmoid: $\quad K(x_i, x_j) = \tanh(\gamma x_i^T x_j + r).$

Here, γ, r, and d are kernel parameters.

We used SVM light which is implemented by Joachims[17] in this paper.

5 Experimental Results and Error Analysis

5.1 Data Set and Evaluation Measures

As mentioned in section 3, we used the dataset consists of 6776 Chinese questions collected from three different sources. We call the data set as CKOQ (Chinese Knowledge-Oriented Questions) in the following. The characteristic of these questions

Table 3. Difference between CKOQ and TREC questions

Name	CKOQ	TREC
Language	Chinese	English
Question Style	Knowledge-oriented	Fact-based
Length of sentence (count by words)	12.62	9.98
Average vocabulary for each question	1.74	1.51

are different from TREC questions. For TREC questions, we mean the 5952 fact-based questions used by Li and Roth. Table 3 shows the difference of characteristic between questions in CKOQ and TREC. We see that the knowledge-oriented questions are quite different from those in TREC. Questions in CKOQ are more complicated than in TREC.

In our experiments, we conducted evaluations in terms of accuracy. Given a set of N questions, their corresponding semantic types, and a ranked list of classification results, the evaluation measures can be defined as follows:

$$Accuracy = \frac{1}{N}\sum_{i=1}^{N}\delta(rank_i,1) \qquad (2)$$

Where δ is the Kronecker delta function defined by:

$$\delta(a,b) = \begin{cases} 1 & if \quad a=b \\ 0 & otherwise \end{cases} \qquad (3)$$

And $rank_i$ means the rank of correct type in the results list returned by classifiers.

One purpose of question classification is to reduce the search space of answers. However, the metric defined above only consider accuracy of the first result. To investigate the credit as long as the correct semantic type given by classifier is found anywhere in the top n ranked semantic types, we used another less common but generalized version of accuracy in this paper. It is defined as:

$$P_{\leq m} = \frac{1}{N}\sum_{k=1}^{m}\sum_{i=1}^{N}\delta(rank_i,k) \qquad (4)$$

It is the relaxed version of accuracy in top m. A system taking use of the generalized accuracy of question classification will improve the performance of answer retrieval in some sense. For example, if the accuracy of question classification in one system is $P_{\leq 1} = 0.7$ and $P_{\leq 2} = 0.9$ respectively, it's desirable to consider the top two types of question classification. Although more noise answers will be retrieved, it will improve the overall performance of the system.

5.2 Comparison Between Different Features and Feature Combinations

We investigated how features perform respectively and how their combinations work. Table 4 to Table 5 shows the experimental results. The kernels mean kernel function of SVM described in section 4.3. In table 4, we can find the feature WordAndPos performs better than other features even than BigramWord. This indicates the usefulness of combination of continuous words and POS tags. The semantic information itself can achieve more than 60 percentage accuracy in question classification. And we observed that the accuracy of correct question type in top 2 returned results is nearly 10 percentages than in top 1.

Table 4. Performance of SVM with different features and kernels

Feature / Kernel	WordAndPos		BigramWords		SingleWord		Semantic	
	$P_{\leq 1}$	$P_{\leq 2}$	$P_{\leq 1}$	$P_{\leq 2}$	$P_{\leq 1}$	$P_{\leq 2}$	$P_{\leq 1}$	$P_{\leq 2}$
Linear	0.703	0.800	0.651	0.755	0.679	0.736	0.625	0.757
Polynomial	0.700	0.796	0.647	0.749	0.642	0.773	0.626	0.761
RBF	0.700	0.798	0.649	0.750	0.326	0.469	0.631	0.761
Sigmoid	0.532	0.634	0.532	0.634	0.448	0.526	0.520	0.652

Table 5. Performance of SVM with different features combinations and kernels

Features / Kernel	WordAndPos		+BigramWords		+ SingleWord		+Semantic	
	$P_{\leq 1}$	$P_{\leq 2}$	$P_{\leq 1}$	$P_{\leq 2}$	$P_{\leq 1}$	$P_{\leq 2}$	$P_{\leq 1}$	$P_{\leq 2}$
Linear	0.703	0.800	0.753	0.849	0.741	0.832	0.720	0.825
Polynomial	0.700	0.796	0.752	0.850	0.671	0.787	0.658	0.777
RBF	0.700	0.798	0.749	0.847	0.274	0.428	0.231	0.390
Sigmoid	0.532	0.634	0.466	0.544	0.469	0.548	0.457	0.511

Table 5 shows the performance of features combinations. Each column with '+' before the feature means the incremental features based on the former feature combination. For example, the last column with features '+Semantic' means the features combination of 'WordAndPos+BigramWord+Word+Semantic'. To our surprise, the adding of the last two features, Word and Semantic, will harm the performance. We see from table 4 that the top 2 features are WordAndPos and Word. But when we conduct the experiment of features combination of these two features, we find the experimental result is less than that in the combination of WordAndPos and BigramWord. And when we add the feature 'Word' based on the former combination, the result becomes worse. The reason mainly lies in the comprehension of knowledge-oriented questions and the dirty data. Too many unrelated words will harm the performance of question classification.

5.3 Comparison Between Different Kernels of SVM

We conduct the experiments to show how different kernels of SVM perform. The experimental result is shown in table 4 to table 5. All the parameters of SVM are set as default. Although the kernel of sigmoid works worst, there are no significant differences between the former three kernels. That is to say, the result is consistent and machine learning approach we proposed in this paper is robust.

5.4 Error Analysis and Discussion

We conduct an error analysis on the results of SVM with linear kernel and feature combination of WordAndPos and Bigram, which is the best result in our approach. We found that the errors mainly fell into three categories.

5.4.1 Ambiguous Labeled Data

Although we build a taxonomy for knowledge-oriented questions carefully, there are still some problems in the data set. Firstly, some questions are difficult to label without context information. For question "你知道北京在哪吗？ (Did you know where is Beijing?)", the type of it will be 'location' if it is asked by a traveler. But if this question appears in a questionnaire which only want know how people are familiar with Beijing, it become a 'YesNo' question. Secondly, there are maybe more than one question types for a question. Thirdly, questions in our dataset are more realistic than those in TREC. Some time it's even difficult for human to label them.

5.4.2 Typical Difficulty in Chinese Questions

In Chinese some words have many meaning. For example "谁(Who)" in Chinese can indicate human, organization and other entities. And the word "什么(What)" can be taken as anything depending on the context it appears. And the omit content in questions make the classification of semantic type difficult too. There are three examples as the following.

是什么导致了恐龙的消失?(What cause the disappearing of dinosaur?)	reason
猫和狗谁更聪明?(Which one is cleverer between cat and dog?)	choice
UDP是传输层协议, TCP呢? (UDP is a protocol of transfer layer, how about TCP?)	description

5.4.3 The Errors in Segmentation and POS

Unlike English sentence, Chinese sentence is a continuous character without space to separate each word. We should conduct word segmentation during preprocessing. The errors in segmentation of questions cause downstream errors. Further more, the semantic information extracted from HowNet depends on the word and its POS. Either the error of segmentation or the error of POS will cause the semantic information error. The simple way we derived the semantic information is another cause of why semantic information doesn't work in our approach. We ignored the structure of meta-semantic elements in semantic categories.

6 Conclusion

In this paper, we have investigated and built a taxonomy for knowledge-oriented questions. And then we have tried using a machine learning approach to address the automatic classification problem for Chinese knowledge-oriented questions. The main contributions of this paper are as follows.

(1) A taxonomy consists of 25 coarse classes is built based on 6776 Chinese knowledge-oriented questions collected from different realistic sources. Fact-based questions, task-oriented questions and other questions are well organized into a uniform architecture based on their semantic relations.

(2) We study the classification problems with different features and features combinations, different kernel functions, and different performance metrics in detail. The results demonstrated that our approach is desirable and robust. We found that the

syntactic structures of questions are really helpful for question classification. Semantic information, however, shows no improvement in our approach.

(3) We conduct the error analysis carefully and the analysis shows the insight into ways these problems may be overcome.

Question classification of knowledge-oriented question classification is a novel task and worth to conduct further research. We are planning to use advanced Natural Language Processing technologies to conduct deep semantic and syntactic analysis in the feature. The effective way to use of semantic information is need to be conducted. And the multi-class classification strategy for question classification is also an interesting approach to improve the classification performance.

References

1. Hermjakob, U.: Parsing and Question Classification for Question Answering. In Proceedings of the Association for Computational Linguistics 2001 Workshop on Open-Domain Question Answering (2001) 17-22
2. Hovy, E., Hermjakob, U., Ravichandran, D.: A Question/Answer Typology with Surface Text Patterns. In Proceeding of the Human Language Technology Conference (2002)
3. Roth, D., Cumby, C., Li, X., Morie, P., Nagarajan, R., Rizzolo, N., Small, K., Yih, W.: Question-Answering via Enhanced Understanding of Questions. In Proceeding of the 11th Text REtrieval Conference (2002)
4. Kelly, D., Murdock, V., Yuan, XJ., Croft, WB., Belkin, NJ.: Features of Documents Relevant to Task- and Fact- Oriented Questions. In Proceeding of the Eleventh International Conference on Information and Knowledge Management (2002) 645-647
5. Lehnert, W. G.: A Conceptual Theory of Question Answering. Natural Language Processing (1986) 651-658
6. Hovy, E., Gerber, L., Hermjakob, U., Junk, M., Lin C.: Question Answering in Webclopedia. In Proceedings of the Ninth Text REtrieval Conference (2002) 655-664
7. Li, X., Roth, D.: Learning Question Classifiers. In Proceedings of the 19th International Conference on Computational Linguistics (2002) 556-562
8. Suzuki, J., Taira, H., Sasaki, Y., Maeda, E.: Question Classification using HDAG Kernel. In Proceeding of the 41st Annual Meeting of the Association for Computational Linguistics (2003) 61-68
9. http://www.broward.k12.fl.us/learnresource/Info_literacy/Bloom's_Taxonomy.pdf
10. Singhal, A., Abney, S., Bacchiani, M., Collins, M., Hindle, D., and Pereira, F.: AT&T at TREC-8. In Proceedings of the 8th Text REtrieval Conference (2000) 317-330
11. Zhang, D., Lee, W.S.: Question Classification using Support Vector Machines. In Proceedings of the 26th Annual International ACM SIGIR Conference on Research and Development in Information Retrieval (2003) 26-32
12. Metzler, D. and Croft, W. B. Analysis of Statistical Question Classification for Fact-based Questions. In Journal of Information Retrieval (2005) 481-504
13. Zheng, Q., Hu, Y., Zhang, S.: The Research and Implementation of Nature Language Web Answer System. Mini-Micro Systems (2005) 554-560
14. http://www.cwirf.org/
15. http://www.keenage.com
16. Cortes, C., Vapnik, V.: Support-Vector Networks. Machine Learning, (1995) 273-297
17. Joachims, T.: Estimating the Generalization Performance of a SVM Efficiently. In Proceeding of the International Conference on Machine Learning (2000) 431-438

Learning TAN from Incomplete Data

Fengzhan Tian, Zhihai Wang, Jian Yu, and Houkuan Huang

School of Computer & Information Technology,
Beijing Jiaotong University, Beijing 100044 P. R. China
{fztian, zhwang, jianyu, hkhuang}@center.njtu.edu.cn

Abstract. Tree augmented Naive Bayes (TAN) classifier is a good tradeoff between the model complexity and learnability in practice. Since there are few complete datasets in real world, in this paper, we develop research on how to efficiently learn TAN from incomplete data. We first present an efficient method that could estimate conditional Mutual Information directly from incomplete data. And then we extend basic TAN learning algorithm to incomplete data using our conditional Mutual Information estimation method. Finally, we carry out experiments to evaluate the extended TAN and compare it with basic TAN. The experimental results show that the accuracy of the extended TAN is much higher than that of basic TAN on most of the incomplete datasets. Despite more time consumption of the extended TAN compared with basic TAN, it is still acceptable. Our conditional Mutual Information estimation method can be easily combined with other techniques to improve TAN further.

1 Introduction

Classification is a very common and important task in the real-world applications. Bayesian classifiers have become very important approaches used for data mining, machine learning as well as pattern recognition [6,7,12]. When using MAP (Maximum A Posteriori) Principle, Bayesian classifiers are optimal classifiers.

The early Naive Bayes are very simple, that are surprisingly effective [10]. Many research showed that Naive Bayes predicts just as well as C4.5 [10,11]. Domingos gives a good explanation why a naive Bayes works surprisingly well despite its strong independence assumption [3]. However, the assumption rarely ever holds in real applications. One solution to the problem is to add "augmented edges" among attribute variables. Followed this idea, Semi-Naive Bayes, Tree Augmented Naive Bayes (TAN) [4], Bayesian Network Augmented Naive Bayes (BAN), and General Bayesian Network classifiers (GBN) were proposed successively [1,4,5,9]. Unfortunately, learning an optimal Augmented Naive Bayes (ANB) is intractable [4]. TAN is a good trade-off between the model complexity and learnability in practice.

After TAN is proposed, a large number of TAN learning algorithms have been developed. However, the above methods mainly focus on learning Bayesian classifiers from complete datasets. Unfortunately, in practice datasets are rarely complete: unreported, lost and corrupted data are a distinguished feature of the real world datasets. Simple solutions to handle missing values are either to ignore the cases including unknown entries or to ascribe these entries to an ad hoc dummy state of the respective variables. Both these solutions are known to introduce potentially dangerous biases in the estimates [8]. Friedman et al. suggest the use of the EM algorithm, gradient descent or Gibbs sampling so as to complete the incomplete datasets, and learn Naive Bayes from complete data [4]. Ramoni and Sebastiani introduced a method that constructs Naive Bayes directly from the incomplete data [13].

Another way to learn a TAN is by the use of the algorithms for learning Bayesian networks from incomplete datasets. But it is usually does not work. This is because of the following two reasons. First, [4] shows theoretically that the general Bayesian network learning methods may result in poor classifiers due to the mismatch between the objective function used (likelihood or a function thereof) and the goal of classification (maximizing accuracy or conditional likelihood). Second, the existing learning algorithms handling missing values are all search & scoring based methods, which are often less efficient. Thereafter, how to learn efficiently TAN from incomplete data still needs further study.

In this paper, we will first present an efficient method that could estimate conditional Mutual Information directly from incomplete data. And then we will extend basic TAN learning algorithm to incomplete data using our conditional Mutual Information estimation method. Finally, we carry out experiments to evaluate the extended TAN and compare it with basic TAN.

2 Background

TAN classifier is an extended tree-like Naive Bayes, in which the class node directly points to all attribute nodes and an attribute node can have only one parent from another attribute node (in addition to the class node). Learning such structures from complete datasets can be easily achieved by using a variation of the Chow-Liu's tree construction algorithm [2]. Given an attribute set $X_1, X_2, \cdots, X_n$ and a complete dataset D, the learning algorithms for TAN can be described as follows:

1. Compute conditional mutual information $I_{P_D}(X_i, X_j | C)$ between each pair of attributes X_i and X_j, where $i \neq j$.
2. Build a complete undirected graph in which nodes are attributes $X_1, X_2, \cdots, X_n$. Annotate the weight of an edge connecting X_i to X_j by $I_{P_D}(X_i, X_j | C)$.
3. Build a maximum weighted spanning tree.
4. Transform the resulting undirected tree to a directed one by choosing a root attribute and setting the direction of all edges to be outward from it.
5. Construct a TAN model by adding a node labeled by C and adding an arc from C to each X_i.

In above procedure, conditional mutual information can be defined as follows.

Definition 1. *Let X, Y, Z are three variables, then the conditional mutual information between X and Y given Z is defined by the following equation* [2].

$$I_P(X, Y | Z) = \sum_{x,y,z} P(x, y, z) \log \frac{P(x, y, z) P(z)}{P(x, z) P(y, z)}. \tag{1}$$

Roughly speaking, conditional mutual information measures how much information that Y provides about X when the value of Z is known.

When datasets contains missing values, we have to compute approximately conditional mutual information with incomplete data. Based on Equation (1), we define conditional mutual information estimate on incomplete datasets as follows.

Definition 2. *Let X, Y, Z are three variables, then the conditional mutual information estimate between X and Y given Z is defined by the following equation.*

$$I'_P(X,Y|Z) = \sum_{x,y,z} P'(x,y,z) \log \frac{P'(x,y,z)P'(z)}{P'(x,z)P'(y,z)} . \tag{2}$$

According to Definition 2, to estimate conditional mutual information, we need to estimate the joint probabilities $P'(x, y, z)$. This work will be depicted in next section.

3 Learning TAN from Incomplete Data

In order to efficiently learn TAN from incomplete data, we must estimate the conditional mutual information used in the learning process from incomplete data. For this purpose, we present a method for efficiently estimating conditional mutual information directly from data with missing values. This method is mainly composed of three steps: *Interval Estimation*, *Point estimation* and *Conditional Mutual Information Estimation*. We will first discuss how to get a joint probability's interval estimate, whose extreme points are the maximum and minimum Bayes estimates that would be inferred from all possible completions of the dataset. Then, we will describe the approach to obtain a point estimate using these interval estimates, given the data missing mechanism. Next, we can estimate Conditional mutual information using these point estimates by Equation (2), and extend basic TAN learning algorithm to incomplete data field.

3.1 Interval Estimation

Interval estimation computes, for each parameter, a probability interval whose extreme points are the minimal and the maximal Bayes estimate that would have been inferred from all possible completions of the incomplete dataset.

Let **X** be a variable set, having r possible states $x_1, x_2, \cdots, x_r$, and the parameter vector $\boldsymbol{\theta} = \{\theta_1, \theta_2, \cdots, \theta_r\}$, which satisfies the assumption of having a prior Dirichlet distribution, be associated to the probability $\theta_i = P(\mathbf{X} = x_i), i = 1, 2, \cdots, r$, so that $\sum_i \theta_i = 1$. Hence the prior distribution of $\boldsymbol{\theta}$ is as follows:

$$P(\boldsymbol{\theta}) = Dir(\boldsymbol{\theta} | \alpha_1, \alpha_2, \cdots, \alpha_r) = \frac{\Gamma(\alpha)}{\prod_{k=1}^{r} \Gamma(\alpha_k)} \prod_{k=1}^{r} \theta_k^{\alpha_k - 1} ,$$

where $\alpha_k = 1, k = 1, 2, \cdots, r$ are super parameters and $\alpha = \sum_{k=1}^{r} \alpha_k$.

Let $N(\mathbf{x}_i)$ be the frequency of complete cases with $\mathbf{X} = \mathbf{x}_i$ in the dataset and $N^*(\mathbf{x}_i)$ be the artificial frequency of incomplete cases with $\mathbf{X} = \mathbf{x}_i$, which can be obtained by completing incomplete cases of the variables set $\mathbf{X}$. An illustration of how to compute the frequencies of complete cases and artificial frequencies of incomplete cases is given in Table 1. In Table 1, as for variable X_1, there are two missing values in the four cases. Both of them can be completed as true or false. So the possible maximal numbers of the artificially completed cases corresponding to the two values are all 2. Therefore the artificial frequencies of variable X_1 are also all 2, i.e. $N^*(X_1 = true) = 2$ and $N^*(X_1 = false) = 2$. So are the artificial frequencies of variable X_2. As for variable set $\mathbf{X} = \{X_1, X_2\}$, there are 3 incomplete cases all together. The possible maximal numbers of the artificially completed cases corresponding to the four values, (*true, true*), (*true, false*), (*false*, true) and (*false, false*) are 2, 1, 3 and 2 respectively. Hence, we can get $N^*(X_1 = true, X_2 = true) = 2$, $N^*(X_1 = true, X_2 = false) = 1$, $N^*(X_1 = false, X_2 = true) = 3$ and $N^*(X_1 = false, X_2 = false) = 2$, as shown in Table 1. Following the same way, we can calculate the frequencies $N(\mathbf{x}_i)$ of complete cases corresponding to various values $\mathbf{x}_i$.

Table 1. Compute the frequencies and artificial frequencies from an simplified incomplete dataset

	X1	X2
Case1	true	true
Case2	?	true
Case3	false	?
Case4	?	?

N(X1=true)=1
N(X1=false)=1
N(X2=true)=2
N(X2=false)=0
N(X1=true, X2=true)=1
N(X1=true, X2=false)=0
N(X1=false, X2=true)=0
N(X1=false, X2=false)=0

N*(X1=true)=2
N*(X1=false)=2
N*(X2=true)=2
N*(X2=false)=2
N*(X1=true, X2=true)=2
N*(X1=true, X2=false)=1
N*(X1=false, X2=true)=3
N*(X1=false, X2=false)=2

Assume that we have completed the dataset by filling all possible incomplete cases of $\mathbf{X}$ with value $\mathbf{x}_i$, and denote the completed dataset as $D_\mathbf{X}^i$. Then given $D_\mathbf{X}^i$, the prior distribution of θ is updated into the posterior distribution using Bayes' theorem:

$$P_i(\theta \mid D_\mathbf{X}^i) = Dir(\theta \mid \alpha_1', \alpha_2', \cdots, \alpha_r'), i = 1, 2, \cdots, r, \quad (3)$$

where $\alpha_i' = \alpha_i + N(\mathbf{X}_i) + N^*(\mathbf{X}_i)$ and $\alpha_k' = \alpha_k + N(\mathbf{X}_k)$ for all $k \neq i$. From Equation (3), we obtain the maximal Bayes estimate of $P(\mathbf{X} = \mathbf{x}_i)$:

$$P^*(\mathbf{X} = \mathbf{x}_i) = E(\theta_i) = \frac{\alpha_i + N(\mathbf{x}_i) + N^*(\mathbf{x}_i)}{\alpha + N^*(\mathbf{x}_i) + \sum_{k=1}^{r} N(\mathbf{x}_k)}. \quad (4)$$

In fact, Equation (3) also identifies a unique probability for other states of $\mathbf{X}$:

$$P_{i*}(\mathbf{X} = \mathbf{x}_l) = E(\theta_l) = \frac{\alpha_l + N(\mathbf{x}_l)}{\alpha + N^*(\mathbf{x}_i) + \sum_{k=1}^{r} N(\mathbf{x}_k)}, \quad (5)$$

where $l = 1, 2, \cdots, r$ and $l \neq i$. Now we can define the minimal Bayes estimate of $P(\mathbf{X} = \mathbf{x}_i)$ as follows:

$$P_*(\mathbf{X} = \mathbf{x}_i) = \min_k (P_{k*}(\mathbf{X} = \mathbf{x}_i)) = \frac{\alpha_i + N(\mathbf{x}_i)}{\alpha + \max_{l \neq i} N^*(\mathbf{x}_l) + \sum_{k=1}^{r} N(\mathbf{x}_k)}, \quad (6)$$

$i = 1, 2, \cdots, r$. Note that when the dataset is complete, we will have

$$P^*(\mathbf{X} = \mathbf{x}_i) = P_*(\mathbf{X} = \mathbf{x}_i) = \frac{\alpha_i + N(\mathbf{x}_i)}{\alpha + \sum_{k=1}^{r} N(\mathbf{x}_k)}, \quad i = 1, 2, \cdots, r,$$

which is the Bayes estimate of $P(\mathbf{X} = \mathbf{x}_i)$ that can be obtained from the complete dataset.

The minimal and maximal Bayes estimates of $P(\mathbf{X} = \mathbf{x}_i)$ comprise the lower and upper bounds of its interval estimates. The above calculation does not rely on any assumption on the distribution of the missing data because it does not try to infer them: the available information can only give rise to constraints on the possible estimates that could be learned from the dataset. Furthermore, it provides a new measure of information: the width of the interval represents a measure of the quality of the probabilistic information conveyed by the dataset about a parameter. In this way, intervals provide an explicit representation of the reliability of the estimates.

3.2 Point Estimation

Next we will discuss how to compute the point estimates from these interval estimates using a convex combination of the lower and upper bounds of this interval.

Suppose that some information on the pattern of missing data is available, for each incomplete case, the probability of a completion is as follows:

$$P(\mathbf{X} = \mathbf{x}_i \mid \mathbf{X} = ?, D_{inc}) = \lambda_i, \quad i = 1, 2, \cdots, r, \quad (7)$$

where D_{inc} denotes the original incomplete dataset, ? denotes the incomplete values in D_{inc}, and $\sum_{i=1}^{r} \lambda_i = 1$. Such information can be used to summarize the interval estimate into a point estimate via a convex combination of the extreme probabilities:

$$P'(\mathbf{X} = \mathbf{x}_i) = \lambda_i P^*(\mathbf{X} = \mathbf{x}_i) + \sum_{k \neq i} \lambda_k P_{k*}(\mathbf{X} = \mathbf{x}_i), \quad (8)$$

where $i = 1, 2, \cdots, r$. This point estimate is the expected Bayes estimate that would be obtained from the complete dataset when the mechanism generating the missing data in the dataset is described by Equation (7). When no information on the mechanism generating missing data is available and therefore any pattern of missing data is equally likely, then

$$\lambda_k = \frac{1}{r}, \quad k = 1, 2, \cdots, r. \quad (9)$$

This case corresponds to the assumption that data is *Missing Completely at Random (MCAR)* [14]. When data is *Missing at Random (MAR)* [14], we will have:

$$\lambda_k = \frac{\alpha_i + N(\mathbf{x}_k)}{\alpha + \sum_{l=1}^{r} N(\mathbf{x}_l)}, \; k=1,2,\cdots,r. \tag{10}$$

3.3 Extended TAN Learning Algorithm

Suppose now that we have obtained the Bayes point estimates of the joint probabilities interested from Equation (8), we can estimate Conditional mutual information using these point estimates by Equation (2). The computation depends only on the frequencies of complete entries and artificial frequencies of incomplete entries in the dataset, both of which can be obtained after only one dataset scan. Assume that we have obtained these frequencies, from Equation (2) we can see that the computation of the estimate of conditional mutual information requires at most $O(r^2)$ basic operations (such as logarithm, multiplication, division), where r equals the number of the possible states of the variable set of $\{X, Y, Z\}$.

Once we have got all the needed conditional mutual information estimates, we can extend basic TAN learning algorithm. The extension can be simply achieved just by replacing the exact computation of conditional mutual information with our approximate estimation of conditional mutual information.

4 Experiments

The main aim of the experiments in this section is to evaluate the extended TAN. For this aim, we compared the extended TAN with basic TAN through experiments in terms of classification accuracy and efficiency.

4.1 Materials and Methods

We run our experiments on 27 datasets from the UCI repository, listed in Table 2. Where 12 datasets contain missing values, and the other 15 ones do not. As for the missing values in the twelve datasets, basic TAN treated them as single particular values as if they are normal. If there existed continuous attributes in the datasets, we applied a pre-discretization step before learning TAN that was carried out by the MLC++ system[1].

As for larger datasets, the holdout method was used to measure the accuracy, while for the smaller ones, five-fold cross validation was applied. And the accuracy of each classifier is based on the percentage of successful predictions on the test sets of each dataset. We used the MLC++ system to estimate the prediction accuracy for each classifier, as well as the variance of this accuracy. For the convenience of comparison, basic TAN and extended TAN have been implemented in a Java based system. All the experiments were conducted on a Pentium 2.4 GHz PC with 512MB of RAM running under Windows 2000. The experimental outcome was the average of the results running 10 times for each level of missing values of each dataset.

---

[1] http://www.sgi.com/tech/mlc/

Table 2. Datasets used in experiments

No.	Datasets	Attributes	Classes	Instances train	Instances test	Missing values
1	Car	6	4	1728	CV5	No
2	Chess	36	2	2130	1066	No
3	Contact-Lenses	5	3	24	CV5	No
4	Diabetes	8	2	768	CV5	No
5	Flare-C	10	8	1389	CV5	No
6	German	20	2	1000	CV5	No
7	Iris	4	3	150	CV5	No
8	LED	10	7	1000	CV5	No
9	Letter	16	26	15000	5000	No
10	Nursery	8	5	8640	4320	No
11	Promoters	57	2	106	CV5	No
12	Satimage	36	6	4435	2000	No
13	Segment	19	7	1540	770	No
14	Tic-Tac-Toe	9	2	958	CV5	No
15	Zoology	16	7	101	CV5	No
16	Annealing	38	6	798	CV5	Yes
17	Australian	14	2	690	CV5	Yes
18	Breast	10	2	683	CV5	Yes
19	Crx	15	2	653	CV5	Yes
20	Hepatitis	19	2	80	CV5	Yes
21	Horse Colic	21	2	300	68	Yes
22	House Vote	16	2	435	CV5	Yes
23	Lung Cancer	56	3	32	CV5	Yes
24	Marine Sponges	45	12	76	CV5	Yes
25	Mushroom	22	2	5416	2708	Yes
26	Primary Tumor	17	22	339	CV5	Yes
27	Soybean Large	35	19	683	CV5	Yes

4.2 Experimental Results

To evaluate the mutual information estimation method and the extended TAN based on that, we run experiments of the extended TAN in terms of complete datasets and incomplete datasets created from the complete ones. We throw away respectively 5%, 10% and 20% values from each complete dataset at random, thus we get 15 incomplete datasets, each contains three percentages of missing values. These incomplete datasets follow the assumption that data is *Missing Completely at Random*. This means that we can compute the weights λ_i by Equation (9). Because when running on complete datasets, the extended TAN degenerates to basic TAN and the results are exactly as same as that of basic TAN, we will not run basic TAN on complete datasets.

Table 3 shows the experimental results of the extended TAN in terms of complete datasets and incomplete datasets with various percentages of missing values. From Table 3, it is unsurprising that all the results on complete datasets are better than that on incomplete datasets with various percentages of missing values, and as the percentage of missing values increases, the accuracy of extended TAN decline gradually. Nevertheless, we find that the classification accuracy on datasets with 5% missing values is almost the same as that on the complete ones, the average accuracy on datasets with 10% missing values is about 1% lower than that on complete ones, and even as for datasets with 20% missing values, compared with that on complete ones, the average accuracy declines about 2%. This indicates that our mutual information estimation method is quite accurate.

Table 3. Experimental results of the extended TAN on complete datasets and incomplete datasets

Datasets	Extended TAN			
	Complete	5% missing	10% missing	20% missing
Car	90.89±0.37	91.41±0.43	91.07±0.73	89.98±0.95
Chess	93.14±0.58	93.13±0.36	92.72±0.79	91.92±0.88
Contact-Lenses	66.26±3.14	65.81±3.89	64.65±4.68	63.29±5.72
Diabetes	75.52±1.11	74.91±1.42	74.39±2.01	72.56±2.63
Flare-C	82.97±0.73	82.97±0.61	82.45±0.78	81.49±0.92
German	72.89±1.42	72.67±1.76	71.84±1.93	70.38±2.32
Iris	91.92±1.83	91.69±1.53	90.58±1.94	89.27±2.07
LED	73.88±0.56	73.84±0.52	72.83±0.86	71.85±0.81
Letter	85.67±0.53	85.57±0.63	84.79±0.69	83.61±1.34
Nursery	92.84±0.41	92.39±0.28	91.62±0.49	90.68±0.94
Promoter	83.06±3.74	82.88±4.16	82.03±3.98	80.92±3.89
Satimage	87.13±0.83	87.04±0.79	86.47±0.92	85.01±1.24
Segment	95.42±0.85	94.88±0.67	94.53±0.96	93.19±1.58
Tic-Tac-Toe	74.29±1.31	73.93±1.29	73.06±1.42	71.75±1.95
Zoology	95.38±0.72	95.26±0.59	94.28±0.95	92.82±1.52

Table 4 shows the experimental results of basic TAN and the extended TAN on the twelve incomplete datasets. We run the extended TAN under the assumption that the incomplete data follows *Missing at Random*. This means that we can compute the weights λ_i by Equation (10). In Table 4, the boldface items indicate the higher accuracy of the two classifiers, from that we could find that the extended TAN outperforms basic TAN at an average significance level more than 0.05. The accuracy of the extended TAN is much higher than that of basic TAN on ten out of twelve datasets, and is less than or almost equal to that of basic TAN only on the other two datasets. This result proves that the extended TAN could make exhaustive use of the information contained in incomplete data to heighten the classification accuracy. The lower accuracy of the extended TAN on the two datasets is because the missing values in them are relatively few, and the information supplement of the incomplete data probably could not compensate the information loss resulted from the mutual information estimate.

Table 4. Experimental comparison of basic TAN and the extended TAN on incomplete datasets

Datasets	Basic TAN		Extended TAN	
	Accuracy	Run Time(s)	Accuracy	Run Time(s)
Annealing	92.75±2.62	0.25	**96.08±0.54**	15.24
Australian	84.31±1.37	0.23	**89.57±1.14**	1.19
Breast	**96.87±1.45**	0.24	96.75±1.32	0.37
Crx	84.75±1.51	0.21	**89.14±1.26**	0.85
Hepatitis	90.28±2.26	0.16	**94.21±1.87**	0.72
Horse Colic	79.96±3.72	0.19	**87.28±2.49**	8.96
House Vote	91.87±0.64	0.13	**94.75±0.73**	0.83
Lung Cancer	**46.69±3.21**	0.23	45.44±3.26	0.25
Marine Sponges	67.29±2.94	0.16	**76.37±2.39**	0.56
Mushroom	97.41±0.28	0.20	**98. 69±0.52**	6.58
Primary Tumor	76.39±1.92	0.18	**84.16±1.58**	2.48
Soybean Large	91.36±0.36	0.61	**94.27±0.75**	4.39

Table 4 also gives out the learning time of the two classifiers on the twelve incomplete datasets. From that we could find that the extended TAN learning algorithm have to spend much more time than basic TAN learning algorithm. The ratio of the learning time of the two classifiers could reach several decades, such as on dataset *Annealing, Horse Colic* and *Mushroom*. The very low efficiency of the extended TAN on these three datasets is because they contain either very large number of missing data (*Annealing* and *Horse Colic*), or very large number of instances (*mushroom*). Even so, the biggest time consumption of the extended TAN on all datasets is only 15.24 seconds, which is far less than what could not be accepted.

5 Conclusion

In this paper, we introduce a method that estimates the Conditional Mutual Information directly from incomplete datasets. And we extend basic TAN learning algorithm to incomplete data field by this method. The experimental results show that not only our conditional mutual information estimation method is quite accurate, but also the accuracy of the extended TAN is much higher than that of basic TAN on most of the incomplete datasets.

Additionally, it is worth noting that the method in the paper is a general method to estimate conditional Mutual Information. On one hand, other techniques for learning TAN could be combined with the method easily so as to improve its performance further. On the other hand, the method in the paper can be applied to any other tasks related to mutual information calculation.

Acknowledgments

This work is supported by NSF of China under Grant No. 60303014 and Science Foundation of Beijing Jiaotong University under Grant No. 2004RC063.

References

1. Cheng, J., Greiner, R., Liu, W.: Comparing Bayesian network classifiers. Fifth Conf. on Uncertainty in Artificial Intelligence (1999) 101–107
2. Chow, C.K., Liu, C.N.: Approximating discrete probability distributions with dependence trees. IEEE Transactions on Information Theory 14 (1968) 462–467
3. Domingos, P., Pazzani, M.: On the optimality of the simple Bayesian classifier under zero-one loss. Machine Learning 29 (1997) 103–130
4. Friedman, N., Geiger, D., Goldszmidt, M.: Bayesian network classifiers. Machine Learning, 29 (1997) 131–161
5. Friedman N., Goldszmidt, M.: Building classifiers using Bayesian networks. AAAI/IAAI, Vol. 2 (1996) 1277–1284
6. Gyllenberg, M., Carlsson, J., Koski, T.: Bayesian network classification of binarized DNA fingerprinting patterns. In: Capasso, V. (eds.): Mathematical Modeling and Computing in Biology and Medicine, Progetto Leonardo, Bologna (2003) 60-66
7. Karieauskas, G.: Text categorization using hierarchical Bayesian network classifiers. http://citeseer.ist.psu.edu/karieauskas02text.html (2002)
8. Kohavi, R., Becker, B., Sommerfield, D.: Improving simple Bayes. Invan Someren, M., Widmer, G. (eds.) Proceedings of ECML-97 (1997) 78–87
9. Kononenko, I.: Semi-naive Bayesian classifier. Proceedings of the Sixth European Working Session on Learning, Springer-Verlag (1991) 206–219
10. Langley, P., Iba, W., Thompson, K.: An analysis of Bayesian classifiers. Proceedings of AAAI-92 (1992) 223–228
11. Pazzani, M.J.: Searching for dependencies in Bayesian classifiers. Learning from Data: Artificial intelligence And Statistics V, New York: Springer-Verlag (1996) 239–248
12. Pham, H.V., Arnold, M.W., Smeulders, W.M.: Face detection by aggregated Bayesian network classifiers. Pattern Recognition Letters 23 (2002) 451–461
13. Ramoni, M., Sebastiani, P.: Robust Bayes classifiers. Artificial Intelligence 125 (2001) 209–226
14. Singh, M.: Learning Bayesian networks from incomplete data. The 14th National Conf. on Artificial Intelligence (1997) 27-31

The General Method of Improving Smooth Degree of Data Series

Qiumei Chen and Wenzhan Dai

College of Information and Electronic Engineering,
Zhejiang Gongshang University, 310033, HangZhou, China
xiaoqiu@pop.hzic.edu.cn
dwzhan@zist.edu.cn

Abstract. Increasing the smooth degree of data series is key factor of grey model's precision. In this paper, the more general method is put forward on the basis of summarizing several kinds of ways to improve smooth degree of data series, and a new transformation is represented. The practical application shows the effectiveness and superiority of this method.

1 Introduction

The grey system theory has caught great attention of researchers since 1982 and has already been widely used in many fields. Theoretically and practically it is proved that model's precision can be increased greatly if the smooth degree of initial data series is better. Therefore, improving the smooth degree of initial data series has great signification for increasing grey model's precision. In order to enhance the smooth degree of data series, logarithm function transformation, exponent function transformation and power function transformation are put forward in papers [2-4] respectively, and a great achievement has been made in practical application.

In this paper, the more general method for improving smooth degree of data series is put forward. It is proved that logarithm function transformation, exponent function transformation and power function transformation all accord with the more general theorem proposed in this paper to improve smooth degree of data series. According to the theorem, a new transformation is put forward and practical application shows the effectiveness of this method.

2 The General Method of Improving Smooth Degree of Data Series

Definition[2] let $\{x^{(0)}(k), k=1,2,\cdots n\}$ be non-negative data series, for $\forall \varepsilon > 0$, if there exists a k_0, when $k > k_0$, the follow equation$^{(1)}$ is held,

$$\frac{x^{(0)}(k)}{\sum_{i=1}^{k-1} x^{(0)}(i)} = \frac{x^{(0)}(k)}{x^{(1)}(k-1)} < \varepsilon \qquad (1)$$

Then series $\{x^{(0)}(k), k=1,2,\cdots n\}$ is defined as smooth data series.

Lemma[2] The necessary and sufficient condition for $\{x^{(0)}(k), k=1,2,\cdots n\}$ to be the smooth data series is that the function $\dfrac{x^{(0)}(k)}{\sum_{i=1}^{k-1} x^{(0)}(i)}$ is decrement with k.

Theorem. Let the non-negative original data series be denoted by $x^{(0)}(k)$, if there exists non-negative and decrement function $f(x^{(0)}(k), k)$ with k and the transfer function F that can be written in the following form

$$F(x^{(0)}(k)) = x^{(0)}(k) \cdot f(x^{(0)}(k), k) \qquad (2)$$

Then the smooth degree of data series $F(x^{(0)}(k))$ is better than that of data series $x^{(0)}(k)$.

Proof: Because $f(x^{(0)}(k), k)$ is non-negative and decrement strictly with k, so

$$0 < f(x^{(0)}(k), k) < f(x^{(0)}(i), i), \ i=1,2,\cdots,k-1$$

According to the non-negative transfer $F(x^{(0)}(k)) = x^{(0)}(k) \cdot f(x^{(0)}(k), k)$, we can obtain

$$F(x^{(0)}(1)) = x^{(0)}(1) \cdot f(x^{(0)}(1),1) > x^{(0)}(1) \cdot f(x^{(0)}(k), k)$$
$$F(x^{(0)}(2)) = x^{(0)}(2) \cdot f(x^{(0)}(2),2) > x^{(0)}(2) \cdot f(x^{(0)}(k), k)$$
$$\cdots$$
$$F(x^{(0)}(k-1)) = x^{(0)}(k-1) \cdot f(x^{(0)}(k-1),(k-1)) > x^{(0)}(k-1) \cdot f(x^{(0)}(k), k)$$

Therefore

$$\sum_{i=1}^{k-1} F(x^{(0)}(k)) = \sum_{i=1}^{k-1}(x^{(0)}(i) \cdot f(x^{(0)}(i), i)) > f(x^{(0)}(k), k) \cdot \sum_{i=1}^{k-1} x^{(0)}(i)$$

While $F(x^{(0)}(k)) = x^{(0)}(k) \cdot f(x^{(0)}(k), k)$

Then $\dfrac{F(x^{(0)}(k))}{\sum_{i=1}^{k-1} F(x^{(0)}(i))} < \dfrac{x^{(0)}(k) \cdot f(x^{(0)}(k), k)}{f(x^{(0)}(k), k) \cdot \sum_{i=1}^{k-1} x^{(0)}(i)} = \dfrac{x^{(0)}(k)}{\sum_{i=1}^{k-1} x^{(0)}(i)}$

By deducting the two above equations, the theorem is easily proved completely.

The general method to improve smooth degree of data series has been put forward according to the theorem. In fact, the methods proposed in papers[2~4] can be considered as examples of the theorem. For example:

(1) Logarithm function transfer [2].

Let $x^{(0)}(k)$ be an increment data series with k and $x^{(0)}(1) > e$

Make transfer $F(x^{(0)}(k)) = \ln[x^{(0)}(k)]$, it is easy to prove that $f(x^{(0)}(k), k)$ is a non-negative and decrement function strictly with

$$f(x^{(0)}(k), k) = \frac{\ln[x^{(0)}(k)]}{x^{(0)}(k)}.$$

Proof: Because $x^{(0)}(k)$ is an increment series with $x^{(0)}(1) > e$, we can obtain

$$f(x^{(0)}(k), k) = \frac{\ln[x^{(0)}(k)]}{x^{(0)}(k)} > 0. \text{ It is to say, } f(x^{(0)}(k), k) \text{ is non-negative.}$$

Because $(f(x^{(0)}(k), k))' = (\frac{\ln[x^{(0)}(k)]}{x^{(0)}(k)})' = \frac{1 - \ln[x^{(0)}(k)]}{(x^{(0)}(k))^2} \cdot (x^{(0)}(k))' < 0$

then $f(x^{(0)}(k), k)$ is decreased strictly.

(2) Exponent function transfer [3].

Let $x^{(0)}(k)$ be an increment series and $x^{(0)}(1) > 0$

Make transfer $F(x^{(0)}(k)) = x^{(0)}(k)^a$ ($0 < a < 1$), it is easy to prove that $f(x^{(0)}(k), k)$ is a non-negative and decrement function with

$$f(x^{(0)}(k), k) = \frac{x^{(0)}(k)^a}{x^{(0)}(k)} = x^{(0)}(k)^{a-1}.$$

Proof: Because $x^{(0)}(k)$ is an increment series with $x^{(0)}(1) > 0$ and $0 < a < 1$, we can obtain $f(x^{(0)}(k), k) = \frac{x^{(0)}(k)^a}{x^{(0)}(k)} = x^{(0)}(k)^{a-1} > 0$. It is to say, $f(x^{(0)}(k), k)$ is non-negative.

Because $(f(x^{(0)}(k), k))' = (x^{(0)}(k)^{a-1})' = (a-1)x^{(0)}(k)^{a-2}(x^{(0)}(k))' < 0$

then $f(x^{(0)}(k), k)$ is decreased strictly.

(3) Power function transfer [4].

Let $x^{(0)}(k)$ be an increment series and $x^{(0)}(1) > 0$

Make transfer $F(x^{(0)}(k)) = a^{-x^{(0)}(k)}$ ($a > 1$), it is easy to prove that $f(x^{(0)}(k), k)$ is a non-negative and decrement function with $f(x^{(0)}(k), k) = \frac{a^{-x^{(0)}(k)}}{x^{(0)}(k)}.$

Proof: Because $x^{(0)}(k)$ is an increment series with $x^{(0)}(1) > 0$ and $a > 1$, we can obtain $f(x^{(0)}(k),k) = \dfrac{a^{-x^{(0)}(k)}}{x^{(0)}(k)} > 0$. It is to say, $f(x^{(0)}(k),k)$ is non-negative.

Because

$$(f(x^{(0)}(k),k))' = (\dfrac{a^{-x^{(0)}(k)}}{x^{(0)}(k)})' = \dfrac{-\ln(a) \cdot x^{(0)}(k) \cdot a^{-x^{(0)}(k)} - 1}{(x^{(0)}(k))^2} \cdot ((x^{(0)}(k))' < 0$$

then $f(x^{(0)}(k),k)$ is decreased strictly.

Only three kinds of transformation functions for improving smooth degree of data series have been put out above. It is also has been proved that all these transfer functions accord with theorem proposed in this paper. Here, the author puts forward another kind of new transfer function, that is $F(x^{(0)}(k)) = x^{(0)}(k) \cdot k^{-d} (d > 0)$. It is easy to prove that $f(x^{(0)}(k),k)$ is a non-negative and decrement function with $f(x^{(0)}(k),k) = k^{-d}$.

Proof: Because of $x^{(0)}(k) > 0$ and $d > 0$, we can obtain $k^{-d} > 0$ and $f(x^{(0)}(k),k) = k^{-d} > 0$.

$(f(x^{(0)}(k),k))' = (k^{-d})' = -d \cdot k^{-d-1} < 0$.

It is to say, $f(x^{(0)}(k),k)$ is both non-negative and decrement.

3 The Modeling Mechanism of Grey Model with a Transformation Function $y \cdot k^{-d} (d > 0)$

GM (1,1) is the most frequently used grey model. It is formed by a first order differential equation with a single variable. Its modeling course is as follows:

(1) Let non-negative and increment original series be denoted by

$$Y^{(0)} = \{y^{(0)}(1), y^{(0)}(2),..., y^{(0)}(n)\}, \ y^{(0)}(i) > 1, i = 1,2,...,n \quad (3)$$

(2) The original data series are transformed by $y \cdot k^{-d} (d > 0)$:

$$X^{(0)} = \{x^{(0)}(1), x^{(0)}(2),..., x^{(0)}(n)\} \quad (4)$$
$$x^{(0)}(k) = [y^{(0)}(k)] \cdot k^{-d}, i = 1, 2,..., n$$

(3) The AGO (accumulated generation operation) of original data series is defined as:

$$X^{(1)} = \{x^{(1)}(1), x^{(1)}(2),..., x^{(1)}(n)\}, \quad (5)$$

$$x^{(1)}(k) = \sum_{i=1}^{k} x^{(0)}(i), k = 1,2,...,n$$

(4) The grey model can be constructed by establishing a first order differential equation as following:

$$\frac{dx^{(1)}(t)}{dt} + ax^{(1)}(t) = u \tag{6}$$

The difference equation of GM (1,1) model:

$$x^{(0)}(k) + az^{(1)}(k) = u, k = 2,3,...,n \tag{7}$$

Unfolding equation (7), we can obtain:

$$\begin{bmatrix} x^{(0)}(2) \\ x^{(0)}(3) \\ \vdots \\ x^{(0)}(n) \end{bmatrix} = \begin{bmatrix} -z^{(1)}(2) & 1 \\ -z^{(1)}(3) & 1 \\ \vdots & \vdots \\ -z^{(1)}(n) & 1 \end{bmatrix} \times \begin{bmatrix} a \\ u \end{bmatrix} \tag{8}$$

Let $Y = [x^{(0)}(2), x^{(0)}(3),..., x^{(0)}(n)]^T$, $\Phi = [a \quad u]^T$,

$$B = \begin{bmatrix} -z^{(1)}(2) & 1 \\ -z^{(1)}(3) & 1 \\ \vdots & \vdots \\ -z^{(1)}(n) & 1 \end{bmatrix} \tag{9}$$

The background $z^{(1)}(k)$ in equation (7) is defined as:

$$z^{(1)}(k+1) = \frac{1}{2}[x^{(1)}(k+1) + x^{(1)}(k)], \ k = 1,2,...,n-1$$

(5) The estimation value of parameter Φ by using least squares is

$$\hat{\Phi} = (B^T B)^{-1} B^T Y$$

(6) The discrete solution of equation (6):

$$\hat{x}^{(1)}(k+1) = [x^{(1)}(1) - \frac{u}{a}]e^{-ak} + \frac{u}{a} \tag{10}$$

(7) Revert $\hat{x}^{(1)}(k+1)$ into

$$\hat{x}^{(0)}(k+1) = \hat{x}^{(1)}(k+1) - \hat{x}^{(1)}(k)$$
$$= (1-e^a)[x^{(1)}(1) - \frac{u}{a}]e^{-ak};$$
$$\hat{y}^{(0)}(k) = \hat{x}^{(0)}(k)/k^{-d}$$

4 Example

4.1 Example 1

The average number of city hospital berth is an important index of the development of national health care industry. To build the model of average number of city hospital berth and forecast its trend will have great significance. Now let build the model based on 《Statistic almanac of China-2002》 from 1990 to 1999 and predict berth number of 2000 and 2001.

The traditional GM (1,1) model of these data is as following:

$$\hat{x}^0(k+1) = 145.1334e^{0.0323k} \quad k>1$$
$$x^0(1) = 138.7$$

The model by using the method proposed in this paper is as following:

$$\hat{y}^0(k+1) = 127.842e^{0.0017k}/k^{-d}$$
$$d = 1/6, k>1, \hat{y}^0(1) = 138.7$$

Table 1 gives comparison of two modeling methods. Figure 1 is the fitted curve.

Table 1. Comparison of two modeling methods (Unit: Ten thousand)

Year	No.	Number of city hospital berth	Traditional GM (1,1)		Method proposed in this paper	
			Model values	Relative error(%)	Model values	Relative error(%)
1990	1	138.7	139.57	0	138.7	0
1991	2	144.8	149.90	-3.52	143.74	0.73
1992	3	152.4	154.83	-1.59	154.04	1.08
1993	4	159.6	159.92	-0.20	161.88	1.43
1994	5	170.7	165.17	3.23	168.30	1.40
1995	6	174.0	170.60	1.95	173.78	0.12
1996	7	179.1	176.21	1.61	178.60	0.27
1997	8	184.2	182.00	1.19	182.93	0.68
1998	9	187.2	187.98	-0.42	186.87	0.17
1999	10	188.7	194.16	-2.89	190.50	-0.95
2000*	11	191.4	207.04	-8.18	194.23	-1.48
2001*	12	195.9	213.84	-9.16	197.42	-0.77

(* Forecasting value)

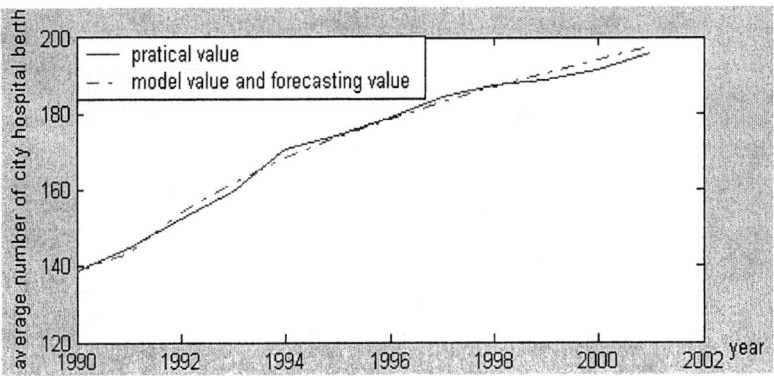

Fig. 1. The fitted curve

From table 1, it is easy to see that the absolute value of the relative error of the model proposed by this paper is almost less than 1.5%. The error inspection of post-sample method can be used to inspect quantified approach. The post-sample error $c = S_1 / S_0$ of the model proposed by this paper is 0.0827 (where S_1 is variation value of the error and S_0 is variation value of the original series), while the post-sample error of traditional GM (1,1) model is 0.2002. The probability of the small error $p = \{|e^{(0)}(i) - e^{-(0)}|\} < 0.6745S_0 = 1$. The post-sample error of this model is far less than 0.3640. In a word, it is obvious that the method proposed by this paper has improved the fitted precision and prediction precision.

4.2 Example 2

The steel and iron industry is one of the basic industries in the country. The per capita output of steel is an important index of the development of the country. To build the mode of the per capita output of steel can forecast its trend. Now let build the model based on 《Statistic almanac of China-2002》 from 1981 to 1999 and predict per capita output of 2000 and 2001.

The traditional GM (1,1) model of these data is as following:

$$\hat{x}^0(k+1) = 36.0458e^{0.0566k}$$
$$x^0(1) = 35.82 \quad k > 1$$

The model by using the method proposed in this paper is as following:

$$\hat{y}^0(k+1) = 33.1299e^{0.0499k} / k^{-d}$$

$$d = 1/15, k > 1, \hat{y}^0(1) = 35.82$$

Table 2 is Comparison of two modeling methods. Figure.2 is the fitted curves.

Table 2. Comparison of two modeling methods (Unit: Kilogram)

Year	No.	Per Capita output of steel	Traditional GM (1,1)		Method proposed in this paper	
			Model values	Relative error(%)	Model values	Relative error(%)
1981	1	35.82	35.8200	0	35.8200	0
1982	2	36.84	38.1463	-3.5458	36.4728	0.9967
1983	3	39.11	40.3691	-3.2195	39.3902	-0.7165
1984	4	41.93	42.7216	-1.8878	42.2083	-0.6637
1985	5	44.52	45.2110	-1.5522	45.0338	-1.1540
1986	6	48.93	47.8456	2.2162	47.9178	2.0687
1987	7	51.92	50.6337	2.4775	50.8909	1.9821
1988	8	53.95	53.5842	0.6780	53.9742	-0.0448
1989	9	55.50	56.7067	-3.0094	56.6337	-2.8768
1990	10	58.54	60.0111	-2.6709	59.9505	-2.5671
1991	11	61.70	63.5081	-2.9305	62.8052	-1.7912
1992	12	69.74	67.2089	3.2548	67.7088	2.5352
1993	13	76.00	71.1253	6.4141	73.8354	2.8482
1994	14	77.70	75.2700	3.1275	75.5906	2.7148
1995	15	79.15	79.6561	-0.6394	79.8262	-0.8543
1996	16	83.15	84.2979	-1.3805	84.2740	-1.3518
1997	17	88.57	89.2101	-0.7227	88.9465	-0.4251
1998	18	93.05	94.4086	-1.4601	93.8563	-0.8666
1999	19	99.12	99.9100	-0.7970	99.0168	0.1041
2000*	20	101.77	105.6562	-3.8186	104.4023	-2.5865
2001*	21	119.22	111.8088	6.2164	116.0946	-2.6215

(* Forecasting value)

From table 2, it is obvious that the absolute value of the relative error of the model proposed by this paper is less than 3 %. The error inspection of post-sample method can be used to inspect quantified approach. The post-sample error $c = S_1 / S_0$ of the model proposed by this paper is 0.0705 (where S_1 is variation value of the error and S_0 is variation value of the original series), while the post-sample error of traditional GM (1,1) model is 0.0802. The probability of the small error $p = \{|e^{(0)}(i) - e^{-(0)}|\} < 0.6745 S_0 = 1$. Then we can come to conclusion that the method proposed by this paper has improved the fitted precision and prediction precision.

The two examples above show that the method proposed by this paper has increased the smooth degree of data series. Therefore, the method improves the fitted precision and prediction precision of models greatly.

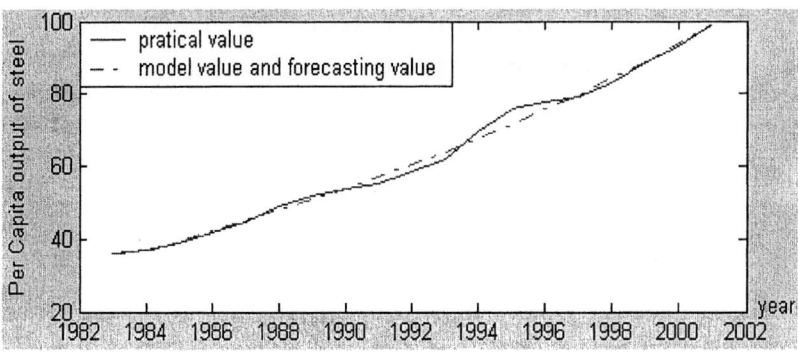

Fig. 2. The fitted curve

5 Conclusions

Increasing the smooth degree of data series is key factor of enhancing grey model's precision. In this paper, the more general method is put forward on the basis of summarizing several kinds of ways to improve smooth degree of data series, and a new transformation is represented. The practical application shows the effectiveness and superiority of this method.

Acknowlegement

This paper is supported by the Natural Science Fundation of Zhejiang Province, P.R.China(NO : 602016)

References

1. Liu Sifeng, Guo Tianbang, Dang Yaoguo,: Theory and Application of Grey System, BeiJing, Science Publishing Company, 1999. 10
2. Chen Taojie: An Expansion of Grey Prediction Model, System Engineering, 1990.8(7) : 50~52
3. Li Qun: The Further Expansion of Grey Prediction Model, Systems Engineering—Theory & Practice, 1993.13(1) : 64~66
4. He Bin, Meng Qing: Research on Methods of Expending Grey Prediction Model, Systems Engineering—Theory & Practic, 2002,22(9)□138~141
5. Xiang Yuelin: Research on GIM (1) Model for Investment on Environmental Protection, Environmental Protection Science 1995, 2l(2) : 72~76
6. Xiang Yuelin: GIM(1) Model of Predicting Regional Ambient Noise, SiChuan Environment, 1996.15 (1) : 68—71
7. Chen Yaokai and Tan Xuerui: Grey Relation Analysis on Serum Makers of Liver Fribrosis, The Journal of Grey System, 1995.7(1) : 63~68
8. Liu Sifeng, Deng Julong: The Range Suitable for GM(1,1), The Journal of Grey System, 1999.11(1) : 131~138
9. Kendrick, J., David, A.: Stochastic Control for Economic Models, New York, McGraw-Hill, 1981.
10. Deng Julong, Zhou Chaoshun: Sufficient Conditiond for the Stability of a Class of Interconnected Dynamic System, System and Control Letters, 1986.5(2) : 105~108

A Fault Tolerant Distributed Routing Algorithm Based on Combinatorial Ant Systems

Jose Aguilar[1] and Miguel Labrador[2]

[1] CEMISID. Dpto. de Computación, Facultad de Ingeniería. Universidad de los Andes
Av. Tulio Febres. Mérida, 5010, Venezuela
aguilar@ula.edu

[2] Department of Computer Science and Engineering University of South Florida. Tampa, FL 33620 USA
labrador@cse.usf.edu

Abstract. In this paper, a general Combinatorial Ant System-based fault tolerant distributed routing algorithm modeled like a dynamic combinatorial optimization problem is presented. In the proposed algorithm, the solution space of the dynamic combinatorial optimization problem is mapped into the space where the ants will walk, and the transition probability and the pheromone update formula of the Ant System is defined according to the objective function of the communication problem.

1 Introduction

The problem to be solved by any routing system is to direct traffic from sources to destinations while maximizing some network performance metric of interest. Depending on the type of network, common performance metrics are call rejection rate, throughput, delay, distance, and energy, among the most important ones. Routing in communication networks is necessary because in real systems not all nodes are directly connected. Currently, routing algorithms face important challenges due to the increased complexity found in modern networks. The routing function is particularly challenging in modern networks because traffic conditions, the structure of the network, and the network resources are limited and constantly changing. The lack of adaptability of routing algorithms to frequent topological changes, node capacities, traffic patterns, load changes, energy availability, and others, reduces the throughput of the network. This problem can be defined as a distributed time-variant dynamic combinatorial optimization problem [2, 11].

Artificial Ant Systems provide a promising alternative to develop routing algorithms for modern communication networks. Inherent properties of ant systems include massive system scalability, emergent behavior and intelligence from low complexity local interactions, autonomy, and stigmergy or communication through the environment, which are very desirable features for many types of networks. In general, real ants are capable of finding the shortest path from a food source to their nest by exploiting pheromone information [1, 3, 4, 5, 6]. While walking, ants deposit pheromone trails on the ground and follow pheromone previously deposited by other ants. The above behavior of real ants has inspired the Ants System (AS), an algorithm in which a set of artificial ants cooperate to the solution of a problem by exchanging information via pheromone deposited on a graph.

Ants systems have been used in the past to solve other combinatorial optimization problems such as the traveling salesman problem and the quadratic assignment problem, among others [3, 4, 5, 6, 7]. We have proposed a distributed algorithm based on AS concepts, called the Combinatorial Ant System (CAS), to solve static discrete-state and dynamic combinatorial optimization problems [1,2]. The main novel idea introduced by our model is the definition of a general procedure to solve combinatorial optimization problems using AS. In our approach, the graph that describes the solution space of the combinatorial optimization problem is mapped on the AS graph, and the transition function and the pheromone update formula of the AS are built according to the objective function of the combinatorial optimization problem. In this paper, we present a routing algorithm based on CAS. Our scheme provides a model for distributed network data flow organization, which can be used to solve difficult problems in today's communication networks. The remaining of the paper is organized as follows. Section 2 presents the Combinatorial Ant System and the Routing Problem. Section 3 presents the general distributed routing algorithm based on the CAS. Then, Section 4 presents and evaluates the utilization of this algorithm on communication networks. Finally, conclusions are presented.

2 Theoretical Aspects

2.1 The Combinatorial Ant System (CAS)

Swarm intelligence appears in biological swarms of certain insect species. It gives rise to complex and often intelligent behavior through complex interaction of thousands of autonomous swarm members. Interaction is based on primitive instincts with no supervision. The end result is the accomplishment of very complex forms of social behavior or optimization tasks [1,3,4,5,7]. The main principle behind these interactions is the autocatalytic reaction like in the case of Ant Systems where the ants attracted by the pheromone will lay more of the same on the same trail, causing even more ants to be attracted.

The Ant System (AS) is the progenitor of all research efforts with ant algorithms, and it was first applied to the Traveling Salesman Problem (TSP) [5, 6]. Algorithms inspired by AS have manifested as heuristic methods to solve combinatorial optimization problems. These algorithms mainly rely on their versatility, robustness and operations based on populations. The procedure is based on the search of agents called "ants", i.e. agents with very simple capabilities that try to simulate the behavior of the ants.

AS utilizes a graph representation (*AS graph*) where each edge *(r,s)* has a desirability measure γ_{rs}, called *pheromone*, which is updated at run time by artificial ants. Informally, the following procedure illustrates how the AS works. Each ant generates a complete tour by choosing the nodes according to a probabilistic state transition rule; ants prefer to move to nodes that are connected by short edges, which have a high pheromone presence. Once all ants have completed their tours, a global pheromone updating rule is applied. First, a fraction of the pheromone evaporates on all edges, and then each ant deposits an amount of pheromone on the edges that belong to its tour in proportion to how short this tour was. At his point, we continue with a new iteration of the process.

There are two reasons for using AS on the TSP. First, the TSP graph can be directly mapped on the AS graph. Second, the transition function has similar goals to the TSP. This is not the case for other combinatorial optimization problems. In [1, 2], we proposed a distributed algorithm based on AS concepts, called the CAS, to solve any type of combinatorial optimization problems. In this approach, each ant builds a solution walking through the AS graph using a transition rule and a pheromone update formula defined according to the objective function of the combinatorial optimization problem. This approach involves the following steps:

1. *Definition of the graph that describes the solution space of the combinatorial optimization problem (COP graph).* The solution space is defined by a graph where the nodes represent partial possible solutions to the problem, and the edges the relationship between the partial solutions.
2. *Building the AS graph.* The COP graph is used to define the *AS graph,* the graph where the ants will finally walk through.
3. *Definition of the transition function and the pheromone update formula of the CAS.* These are built according to the objective function of the combinatorial optimization problem.
4. Executing the AS procedure described before.

2.1.1 Building the AS Graph

The first step is to build the COP graph, then we define the AS graph with the same structure of the COP graph. The AS graph has two weight matrices. The first matrix is defined according to the COP graph and registers the relationship between the elements of the solution space (COP matrix). The second one registers the pheromone trail accumulated on each edge (pheromone matrix). This weight matrix is calculated/updated according to the pheromone update formula. When the incoming edge weights of the pheromone matrix for a given node become high, this node has a high probability to be visited. On the other hand, if an edge between two nodes of the COP matrix is low, then it means that, ideally, if one of these nodes belongs to the final solution then the other one must belong too. If the edge is equal to infinite then it means that the nodes are incompatible, and therefore, they don't belong to at the same final solution.

We define a data structure to store the solution that every ant k is building. This data structure is a vector (A^k) with a length equal to the length of the solution, which is given by n, the number of nodes that an ant must visit. For a given ant, the vector keeps each node of the AS graph that it visits.

2.1.2 Defining the Transition Function and the Pheromone Update Formula

The state transition rule and the pheromone update formula are built using the objective function of the combinatorial optimization problem. The transition function between nodes is given by:

$$Tf(\gamma_{rs}(t), Cf^k_{r->s}(z)) = \frac{\gamma_{rs}(t)^\alpha}{Cf^k_{r->s}(z)^\beta}$$

where $\gamma_{rs}(t)$ is the pheromone at iteration t, $Cf^k_{r->s}(z)$ is the cost of the partial solution that is being built by ant k when it crosses the edge (r, s) if it is in the position r, z-1 is the current length of the partial solution (current length of A^k), and, α and β are two adjustable parameters that control the relative weight of trail intensity $(\gamma_{rs}(t))$ and the cost function.

In the CAS, the transition probability is as follows: an ant positioned at node r chooses node s to move to according to a probability $P^k_{rs}(t)$, which is calculated according to Equation 1:

$$P^k_{rs}(t) = \begin{cases} \dfrac{Tf(\gamma_{rs}(t), Cf^k_{r->s}(z))}{\sum_{u \in J^k_r} Tf(\gamma_{ru}(t), Cf^k_{r->u}(z))} & \text{If } s \in J^k_r \\ 0 & \text{Otherwise} \end{cases} \quad (1)$$

where J^k_r is the set of nodes connected to r that remain to be visited by ant k positioned at node r. When $\beta=0$ we exploit previous solutions (only trail intensity is used), and when $\alpha=0$ we explore the solution space (a stochastic greedy algorithm is obtained). A tradeoff between quality of partial solutions and trail intensity is necessary. Once all ants have built their tours, pheromone, i.e. the trail intensity in the pheromone matrix, is updated on all edges according to Equation 2 [1, 2, 3, 4, 5, 6]:

$$\gamma_{rs}(t) = (1-\rho)\gamma_{rs}(t-1) + \sum_{k=1}^{m} \Delta\gamma^k_{rs}(t) \quad (2)$$

where ρ is a coefficient such that $(1 - \rho)$ represents the trail evaporation in one iteration (tour), m is the number of ants, and $\Delta\gamma_{rs}^k(t)$ is the quantity per unit of length of trail substance laid on edge (r, s) by the kth ant in that iteration

$$\Delta\gamma^k_{rs}(t) = \begin{cases} \dfrac{1}{C^k_f(t)} & \text{If edge } (r, s) \text{ has been crossed by ant } k \\ 0 & \text{Otherwise} \end{cases} \quad (3)$$

where $C^k_f(t)$ is the value of the cost function (objective function) of the solution proposed by ant k at iteration t. The general procedure of our approach is summarized as follows:

1. Generation of the AS graph.
2. Definition of the state transition rule and the pheromone update formula, according to the combinatorial optimization problem.
3. Repeat until system reaches a stable solution
    3.1. Place m ants on different nodes of the AS graph.
    3.2. For i=1, n

3.2.1. For j=1, m
  3.2.1.1. Choose node *s* to move to, according to the transition probability (Equation 1).
  3.2.1.2. Move ant *m* to the node *s*.
 3.3. Update the pheromone using the pheromone update formula (Equations 2 and 3).

2.2 The Routing Problem

Routing is the function that allows information to be transmitted over a network from a source to a destination through a sequence of intermediate switching/buffering stations or nodes. Routing is necessary because in real systems not all nodes are directly connected. Routing algorithms can be classified as static or dynamic, and centralized or distributed [10]. Centralized algorithms usually have scalability problems, and single point of failure problems, or the inability of the network to recover in case of a failure in the central controlling station. Static routing assumes that network conditions are time-invariant, which is an unrealistic assumption in most of the cases. Adaptive routing schemes also have problems, including inconsistencies arising from node failures and potential oscillations that lead to circular paths and instability. Routing algorithms can also be classified as minimal or non-minimal [10]. Minimal routing allows packets to follow only minimal cost paths, while non-minimal routing allows more flexibility in choosing the path by utilizing other heuristics. Another class of routing algorithms is one where the routing scheme guarantees specified QoS requirements pertaining to delay and bandwidth [10].

Commonly, modern networks utilize dynamic routing schemes in order to cope with constant changes in the traffic conditions and the structure or topology of the network. This is particularly the case of wireless ad hoc networks where node mobility and failures produce frequent unpredictable node/link failures that result in topology changes. A vast literature of special routing algorithms for these types of networks exist [13, 14], all of them with the main goal of making the network layer more reliable and the network fault tolerant and efficient. However, maximizing throughput for time-variant load conditions and network topology is a NP-complete problem. A routing algorithm for communication networks with these characteristics can be defined as a dynamic combinatorial optimization problem, that is, like a distributed time-variant problem. In this paper, we are going to use our model to propose a routing algorithm for these networks, which support multiple node and link failures.

3 The General CAS-Based Distributed Routing Algorithm

There are a number of proposed ant-based routing algorithms [3, 9, 12, 13, 14]. The most celebrated one is AntNet, an adaptive agent-based routing algorithm that has outperformed the best-known routing algorithms on several packet-switched communication networks. Ant systems have also been applied to telephone networks. The Ant-Based Control (ABC) scheme is an example of a successful application. We are going to propose a new routing algorithm based on our CAS that can be used in different networking scenarios, such as networks with static topologies, networks with constant topology changes, and network with energy constraints.

We can use our approach for point to point or point to multipoint requests. In the case of point to point, one ant is launched to look for the best path to the destination. For a multipoint request with m destinations, m ants are launched. The route where intermediate nodes have large pheromone quantities is selected. For this, we use the local routing tables of each node like a transition function to its neighbours. Thus, according to the destination of the message, the node with highest probability to be visited corresponds to the entry in the table with the larger amount of pheromone. Then, the local routing table is updated according the route selected. Our algorithm can work in combinatorial stable networks (networks where the changes are sufficiently slow for the routing updates to be propagate to all the nodes) or not, because our approach works with local routing tables and the changes only must be propagated to the neighbours.

3.1 Building the AS Graph

We use the pheromone matrix of our AS graph like the routing table of each node of the network. Remember that this matrix is where the pheromone trail is deposited. Particularly, each node i has k_i neighbors, is characterized by a capacity C_i, a spare capacity S_i, and by a routing table $R_i=[r^i_{n,d}(t)]_{ki,N-1}$. Each row of the routing table corresponds to a neighbor node and each column to a destination node. The information at each row of node i is stored in the respective place of the pheromone matrix (e.g., in position i, j if k_i neighbor = j). The value $r^i_{n,d}(t)$ is used as a probability. That is, the probability that a given ant, where the destination is node d, be routed from node i to neighbor node n. We use the COP matrix of our AS graph to describe the network structure. If there are link or node failures, then the COP graph is modified to show that. In addition, in each arc of the COP graph, the estimation of the trip time from the current node i to its neighbor node j, denoted $\Gamma_i=\{\mu_{i->j}, \sigma^2_{i->j}\}$ is stored, where $\mu_{i->j}$ is the average estimated trip time from node i to node j, and $\sigma^2_{i->j}$ is its associated variance. Γ_i provides a local idea of the global network's status at node i. Finally, we define a cost function for every node, called $C_{ij}(t)$, that is the cost associated with this link. It is a dynamic variable that depends on the link's load, and is calculated at time t using Γ_i.

3.2 Defining the Transition Function and the Pheromone Update Formula

We have explained that in our decentralized model each node maintains a routing table indicating where the message must go in order to reach the final destination. Artificial ants adjust the table entries continually affecting the current network state. Thus, routing tables are represented like a pheromone table having the likelihood of each path to be followed by artificial ants. Pheromone tables contain the address of the destination based on the probabilities for each destination from a source. In our network, each ant launched influences the pheromone table by increasing or reducing the entry for the proper destination.

In our model, each node of the network is represented as a class structure containing various parameters (identification of the node, adjacent nodes, spare capacity, number of links), and Equation 3 has the following meaning: $C_f^k(t)$ is the cost of kth ant's route, $\Delta\gamma_{is}^k(t)$ is the amount of pheromone deposited by ant k if edge (i, s) belongs to the kth ant's route (it is used to update the routing table R_i in each node), and $P_{ij}^k(t)$ is the probability that ant k chooses to hop from node i to node j (it is calculated from the routing table R_i). In this way, ants walk according to the probabilities given in the pheromone tables and they visit one node every time. Ant k updates its route cost each time it traverses a link $C_f^k(t) = C_f^k(t) + C_{ij}(t)$ if $i,j \in$ path followed by ant k. In this way, an ant collects the experienced queues and traffic load that allows it to define information about the state of the network. Once it has reached the destination node d, ant k goes all the way back to its source node through all the nodes visited during the forward path, and updates the routing tables (pheromone concentration) and the set of estimations of trip times of the nodes that belong to its path (COP graph), as follows:

- The times elapsed of the path i->d ($T_{i->d}$) in the current k^{th} ant's route is used to update the mean and variance values of Γ_i of the nodes that belong to the route. $T_{i->d}$ gives an idea about the goodness of the followed route because it is proportional to its length from a traffic or congestion point of view.
- The routing table R_i is changed by incrementing the probability $r_{j,d}^i(t)$ associated with the neighbor node j that belongs to the kth ant's route and the destination node d, and decreasing the probabilities $r_{n,d}^i(t)$ associated with other neighbor nodes n, where $n \neq j$ for the same destination (like a pheromone trail).

The values stored in Γ_i are used to score the trip times so that they can be transformed in a reinforcement signal $r = 1/\mu_{i>j}$, $r \in [0,1]$. r is used by the current node i as a positive reinforcement for the node j:

$$r_{i-1,d}^i(t+1) = r_{i-1,d}^i(t)(1-r) + r$$

and the probabilities $r_{n,d}^i(t)$ for destination d of other neighboring nodes n receive a negative reinforcement

$$r_{n,d}^i(t+1) = r_{n,d}^i(t)(1-r) \qquad \text{for } n \neq j$$

In this way, artificial ants are able to follow paths and avoid congestion while balancing the network load. Finally, $C_{ij}(t)$ is updated using Γ_i and considering the congestion problem (we must avoid congested nodes):

$$C_{ij}(t+1) = Ce^{-ds_j(t)} \frac{\mu_{i \to j}}{\sigma_{i \to j}^2} \qquad (4)$$

where C and d are constants, and $s_j(t)$ is the spare capacity of the node j at time t. The incorporation of delay ($Ce^{-ds_j(t)}$) reduces the ant flow rate to congested nodes, permitting other pheromone table entries to be updated and increased rapidly (negative backpropagation). In the case of link failures, the algorithm avoid those nodes according to the following formula:

$$C_{ij}(t+1) = \infty \qquad \text{(node } j \text{ with failures)} \qquad (5)$$

4 Performance Evaluation of the CAS Algorithm

In this section we test our approach considering three cases, networks with static topologies (no failures), networks with constant topology changes due to node and link failures, and networks with energy constraints with and without failures.

In this experiment, we evaluate our algorithm in networks with constant topology changes introducing link and node failures. Here, if a link failure occurs and the node has more than one linkage, then the node can be reached via other path. If the node has no other link to any node in the network then a node failure occurs. We assume that link failures follow a uniform distribution and do not exceed 10% of the total number of links in the network. In the presence of a link failure, the cost of a call from source node i to destination node j will be defined as infinite (see Eqn. 5), and the probability in the proper column and row in the pheromone table is set to zero.

As in [8], we also consider the incorporation of additive noise in order to handle the so-called *shortcut* and *blocking* problems. The shortcut problem occurs when a shorter route becomes suddenly available while the blocking problem occurs when an older route becomes unavailable. In both situations, artificial ants have difficulties finding new routes, as they work guided by the pheromone tables and don't have an adequate dynamic reaction. With the inclusion of the noise factor f, ants select a purely random path with probability f and a path guided by the pheromone table with probability $(1-f)$. As shown in [8, 9, 12], the noise factor must not exceed 5%, because a noise factor greater that 5% makes the system unstable, reducing the network throughput and the performance of the routing method.

We performed simulations and compared our algorithm with the approach presented in [8] using the same partially meshed Synchronous Digital Hierarchy (SDH) network. The network has 25 nodes partially connected and all links have a capacity of 40 calls. We make random selection of call probabilities, link failure random generations, and collect data to evaluate the performance of the schemes in terms of throughput and mean delay per node. Figures 1 and 2 show these results.

In Figure 1, we show that our approach provides better performance than [8] in the presence of link failures. The mean delay per node is considerably better because we consider the congested node problem. Similarly, Figure 2 shows that the throughput response of the proposed system is better, as it handles the incoming call variations and simultaneous link failures better than [8]. Link failures essentially form a constantly changing network topology to which our agent-based algorithm seems to adapt particularly well. This actually means that the proposed routing algorithm is a good candidate for networks with constant topology changes such as mobile wireless ad hoc networks, where node mobility causes constant link failures.

We also compared our model with the traditional *Link State* routing scheme described in [10] and the Ant-Based approach proposed in [9] using the same network. In Figure 3, it is shown that our CAS scheme provides substantially better throughput performance in the presence of multiple link/node failures.

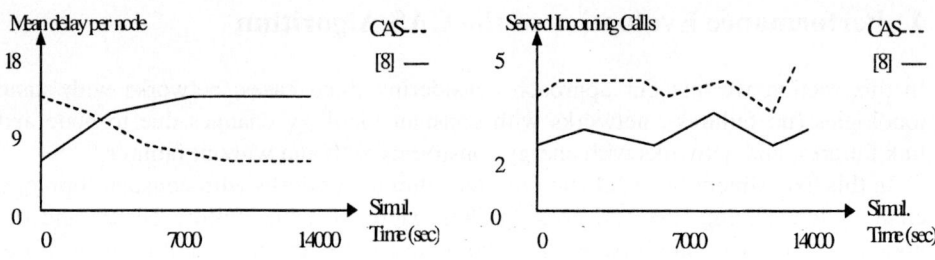

Fig. 1. Mean delay per node **Fig. 2.** Throughput response

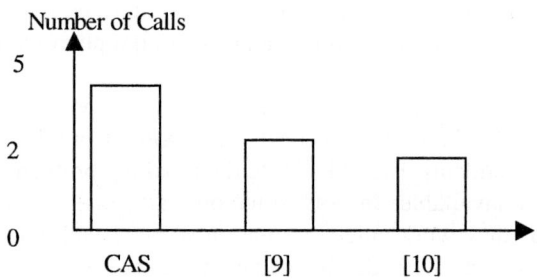

Fig. 3. Throughput response

5 Conclusions

In this work we propose a General Combinatorial Ant System-based Distributed Routing Algorithm for wired, wireless ad hoc and wireless sensor networks based on the Combinatorial Ant System. This work shows the versatility of our routing algorithm exemplified by the possibility of using the same model to solve different telecommunication problems, like dynamic combinatorial optimization problems of various sizes. Our approach can be applied to any routing problem by defining an appropriate graph representation of the solution space of the problem considered, the dynamic procedure to update that representation, and an objective function that guides our heuristic to build feasible solutions. In our approach, the dynamic environment of the combinatorial optimization problem is defined through the Combinatorial Optimization Problem matrix that forms part of the space through which the ants will walk (AS graph). Ants walk through this space according to a set of probabilities updated by a state transition and a pheromone update rule defined according to the objective function of the combinatorial optimization problem considered. Messages between nodes are replaced by ants simultaneously biasing the network parameters by laying pheromone on route from source to destination. The results show that our approach obtains good performance in the presence of multiple failures (links, nodes), contributes to congestion avoidance (load balancing), and keeps the network throughput stable.

Reference

1. Aguilar, J., Velásquez, L.: Pool M. The Combinatorial Ant System. Applied Artificial Intelligence, Vol. 18, (2004) 427-446.
2. Aguilar, J.: The Combinatorial Ant System for Dynamic Combinatorial Optimization Problems. Revista de Matematica: Teoria y Aplicaciones (to appear), Universidad de Costa Rica, San José, Costa Rica, (2005).
3. Bonabeau, E., Dorigo, M., Theraulaz, G.: Swarm Intelligence: from Natural to Artificial Swarm Systems. Oxford University Press, 1999.
4. Corne, D., Dorigo, M., Glover, F.: New Ideas in Optimization. McGraw Hill, 1999.
5. Dorigo, M., Maniezzo, V., Coloni, A.: The Ant System: Optimization by a Colony of Cooperating Agents. IEEE Trans. of Systems, Man, Cybernetics, Vol. 26, (1996) 29-41.
6. Dorigo, M., Gambardella, L.: Ant Colony System: A Cooperative Learning Approach to the Traveling Salesman Problem. IEEE Trans. on Evol. Computation, Vol. 1, (1997) 53-66.
7. Hidrobo, F., Aguilar, J.: Toward a Parallel Genetic Algorithm Approach Based on Collective Intelligence for Combinatorial Optimization Problems. Proc. of IEEE International Conference on Evolutionary Computation (1998) 715-720.
8. Mavromoustakis, C., Karatza, H.: Agent-Based Throughput Response in Presence of Node and/or Link Failure (on demand) for Circuit Switched Telecommunication Networks. Computer Communications, Vol. 27 (2004) 230-238.
9. Schoonderwoerd, R., Holland, O., Bruten, J., Rothkrantz, L.: Ant-based Load Balancing in Telecommunications Networks, Adaptive Behavior, Vol. 5 (1997) 169-207.
10. Bertsekas, D., Gallager, R.: Data Networks. Prentice Hall, 1987.
11. Aguilar, J., Ramirez, W.: Some Results of CAS for Ad Hoc Networks", *Technical Report*, CEMISID, Facultad de Ingenieria, Universidad de los Andes, 2004.
12. Appleby, S., Stewart, S.: Mobile Software Agents for Control in Telecommunication Networks. BT Technology Journals, Vol. 12, (1994) 104-113.
13. Caro, G.D., Dorigo, M.: AntNet: Distributed Stigmergetic Control for Communication Networks. Journal of Artificial Intelligence Research, Vol. 9 (1998) 317-365.
14. Gabber, E., Smith, M.: Trail Blazer: A Routing Algorithm Inspired by Ants. Proc. of the 12th IEEE International Conference on Network Protocols (2004).

Improvement of HITS for Topic-Specific Web Crawler

Xiaojun Zong, Yi Shen, and Xiaoxin Liao

Department of Control Science and Engineering,
Huazhong University of Science and Technology, Wuhan, Hubei, 430074, China
Yishen64@163.com

Abstract. The rapid growth of the World-Wide Web poses unprecedented scaling challenges for general-purpose crawlers. Topic-specific web crawler is developed to collect relevant web pages of interested topics form the Internet. Based on the analyses of HITS algorithm, a new P-HITS algorithm is proposed for topic-specific web crawler in this paper. Probability is introduced to select the URLs to get more global optimality, and the metadata of hyperlinks is appended in this algorithm to predict the relevance of web pages better. Experimental results indicate that our algorithm has better performance.

1 Introduction

The Internet is undoubtedly a huge source of information. We can get multifarious information from the World Wide Web on any topic via web pages. Therefore, more and more people use the Web for information searching. Generally the popular portals or search engines like Yahoo and Google are used for gathering information on the World Wide Web. The search engine serves as a tool providing desired web pages. It will gather web pages from the Internet by using a web crawler. After collecting web pages, search engine will allow users to find what they want from those colleted web pages by simply entering the key words and browsing.

However, with the explosive growth of the Web, searching information on the Web is becoming an increasing difficult task. On the one hand the total web in the Internet is composed of several billion pages [1]. This leads to a large number of web pages being retrieved for most queries, and the returned results are presented as pages of scrolled lists. Going through these pages to find the relevant information is tedious. On the other hand, the web continues to grow rapidly at a million pages per day [2]. The largest crawls cover only 30-40% of the web, and the refreshes take weeks to months [2], [3]. During the refreshing time, some new web pages are progressively produced, and some are disappeared. It is impossible to collect all of them on time.

Because of those problems, topic-specific web crawler has been developed to collect web pages from the internet by choosing to gather only particular pages related to a specific topic. Being differed from the generic crawlers, this kind of web crawler does not need to gather every web page from the Internet. Apparently, the core of topic-specific web crawler is the gathering of related information. In this paper, we analyze some present algorithms, and propose an improved one. Experimental results indicate that the new algorithm has better performance. Here, we categorize some knowledge bases of the topic-specific web crawler as follows, and these knowledge bases are considered as the experience of the crawler:

- *Topic keywords*. As the crawler must compare a topic of interest with the content of collected web pages, it should have proper keywords to describe the topic of interest for comparison. Sometimes topic keywords are sent to a search engine in order to build an initial set of starting URLs.
- *Starting URLs(Seed URLs)*. As the crawler must collect as many as relevant web pages as possible, it needs a set of good URLs, which point to a large number of relevant web pages as the starting point.
- *URL weight*. As the crawler must pre-order the URLs for un-downloaded web pages, it should predict the relevancy of those pages before downloading. The crawler will calculate the URL weight by using the seen content of web pages obtained from previous crawling attempts.

We organize this paper in the following way. Section 2 reviews related works of topic-specific web crawlers. Section 3 gives some necessary techniques that a topic-specific crawler must have and analyses HITS algorithm for the topic-specific crawler. Then we further to the improvement of HITS and propose an improved one. Section 4 presents and analyzes our experiences with topic-specific web crawler. Finally, section 5 concludes the paper.

2 Related Works

A working process of a topic-specific web crawler is composed of two main steps. The first step is to determine the starting URLs or the starting point of a crawling process. The crawler is unable to traverse the Internet without starting URLs. Moreover, the crawler cannot discover more relevant web pages if starting URLs are not good enough to lead to target web pages.

The second step in a topic-specific web crawling process is the crawling method. In theoretical point of view, a topic-specific web crawler smartly selects a direction to traverse the Internet. However, the crawler collects web pages from the Internet, extracts URLs from those web pages, and puts the result into a queue. Therefore, a clever route selection method of the crawler is to arrange URLs so that the most relevant ones can be located in the first part of the queue. The queue will then be sorted by relevancy in descending order.

In recent time there has been much interest in topic-specific web crawler. The emphasis is the algorithm for collecting web pages related to a particular topic.

The fish-search algorithm for collecting topic-specific pages is initially proposed by P.DeBra et al.[4]. The crawler dynamically maintains a list of uncollected URLs that ordered by priority of being gathered and downloads the web pages according to the queue. In the processing, these URLs in the relevant web pages are given higher priorities than the others in the irrelevant web pages.

Based on the improvement of fish-search algorithm, M.Hersovici et al. proposed the shark-search algorithm [5]. The effects of those characters in the hyperlinks are considered when the priority of URLs being calculated in this algorithm. At the same time, the vector space model is introduced to calculate the similarity factor of web pages.

The neural network is adopted in the Info Spider designed by F.Menczer et al.[6]. The information in the hyperlinks of web pages is extracted as the input of the neural network, and the output is considered as the guidance to farther collection.

The VTMS (Web Topic Management System) crawler, designed by S.Mukherjes[7], downloads the seed URLs that relevant to the topic and creates a representative document vector (RDV) based on the frequently occurring keywords in these URLs. The crawler then downloads the web pages that are referenced from the seed URLs and calculates their similarity to the RDV using the vector space model [8]. If the similarity is above a threshold, the links form the pages are added to a queue. The crawler continues to follow the out-links until the queue is empty or a user-specified limit is reached. Meanwhile, the crawler uses heuristics to improve performance.

All these works present algorithms that enable their crawlers to select web pages related to a particular topic. The main purpose of those algorithms is to gather as many relevant web pages as possible. Therefore, the efficiency of topic-specific web crawling is measured in terms of a proportion of the number of relevant web pages and the total number of downloaded web pages. If this proportion is high, it means that the crawler can collect more relevant web pages than irrelevant ones. This is a result we expect for the web crawler.

3 Algorithms and Strategy

The mission for the topic-specific crawler is gathering more related information as soon as possible. We must provide the means of justifying the relevancy of a web page and the means of finding the best route to go to other relevant items. In this section, we first give the necessary techniques that a topic-specific crawler must have and analyses HITS algorithm for the topic-specific crawler. Then we further to the improvement of HITS and propose an improved one.

3.1 Authority and Hub

There are two kinds of important web pages in the Internet. One is authority, and another is called hub. An authority is linked by a set of relevant web pages. Generally, the authority is an authoritative web page, and its similarity to the topic is high. A hub is one or a set of web pages that provides collections of links to authorities. Hub pages may not be prominent themselves, or there may exist few links pointing to them; however, they provide links to a collection of prominent sites on a common topic. In generally, a good hub is a page that points to many good authorities; a good authority is a page pointed to by many good hubs. Such a mutual reinforcement relationship between hubs and authorities helps the gathering of topic-specific web pages.

The links of web pages is similar to using citations among journal articles. Based on this, J.Kleinberg proposed the concepts of authority and hub for web pages [9]. The authority and hub weights are updated based on the following equations:

$$\text{Authority}(p) = \sum_{(q \to p)} \text{Hub}(q) \ . \tag{1}$$

$$\text{Hub}(p) = \sum_{(q \leftarrow p)} \text{Authority}(q) \ . \tag{2}$$

Equation (1) implies that if a page is pointed by many good hubs, its authority weight should increase (i.e., it is the sum of current hub weights of all of the pages pointing to it). Equation (2) implies that if a page is pointing to many good authorities, its hub weight should increase (i.e., it is the sum of the current authority weights of all of the pages it pints to).

3.2 Topic Similarity

The analysis of topic similarity is the most important stage for a topic-specific web crawling. Since the crawler cannot make a decision that which web page is relevant to the topic of interest, a similarity score should be used to determine the relevancy of a web page. At the beginning of the first crawling, the crawler dose not know the content of web pages represented by the starting URLs, it must collect those web pages to investigate their contents and extract new URLs inside them. To choose new URLs to visit, the crawler should calculate the relevancy of web pages represented by those URLs. Normally, authors of most web pages often include some links, as well as anchor texts, which relate to page contents. Thus, the crawler can use both combination of a parent page's similarity score and an anchor text's similarity score to calculate the weight of those new URLs, then estimates whether URLs should be crawled.

There are many ways to compute the topic similarity, but here we only focus on a vector space approach. The vector space is widely used in information retrieval research. The main idea of the vector space is to represent a web page as a vector. Each distinct word in that page is considered to be an axis of a vector in a multidimensional space. The direction of a vector corresponds to the content of the web page. Two web pages are assumed to be relevant, i.e. relating to the same topic, if their vectors point the same direction. Mathematically, if two vectors point the same direction, this means that they have an angle of zero degree and their cosine value becomes 1. Therefore, we define $sim(k,d)$ to be the page similarity function with the formula written in Equation (3)[10]

$$sim(k,d) = \frac{\sum_{t \in (k \cap d)} f_{tk} f_{td}}{\sqrt{\sum_{t \in k} f_{tk}^2 \sum_{t \in d} f_{td}^2}} \ . \tag{3}$$

Here, k is a set of keywords describing an interest topic, d is the web page which we want to compare, f_{tk} and f_{td} are the number of terms t in the set of keywords k and a web page d, respectively. The range of $sim(k,d)$ value lies between 0 and 1, and the similarity increases as this value increases.

3.3 HITS (Hyperlink-Induced Topic Search)

An algorithm using hubs, called HITS (Hyperlink-Induced Topic Search), was discussed as follows.

First, HITS uses the query terms to collect a starting set of pages from an index-based search crawler. These pages form the root set. Since many of these pages are presumably relevant to the search topic, some of them should contain links to most of the prominent authorities. Therefore, the root set can be expanded into a base set by including all of the pages that the root-set pages link to.

Second, a weight-propagation phase is initiated. This is an iterative process that determines numerical estimates of authority and hub weights. A nonnegative authority weight a_p and a nonnegative hub weight h_p are associated with each page p in the base set, and all a and h values are initialized to a uniform constant. The weights are normalized and an invariant is maintained that the squares of all weights sum to 1. Equation (1) and equation (2) are used for calculating the weights of authority and hub.

Finally, the HITS algorithm outputs a short list of the pages with large hub weights, and the pages with large authority weights for the given search topic.

Systems based on the HITS algorithm include Clever and another system, Google, based on a similar principle. By analyzing Web links and textual context information, it has been reported than such systems can achieve better quality search results than those generated by term-index engines such as AltaVista and those created by human ontologists such as Yahoo.

Although many experiments have shown that HITS provides good search results for a wide range of queries, the method may encounter some difficulties by ignoring textual contexts. Sometimes HITS drifts when hubs contain multiple topics. It may also cause "topic hijacking" when many pages from a single web site point to the same single popular site. Such problems can be overcome by using metadata of hyperlink to adjust the weight of the links at a certain extent.

3.4 Metadata of Hyperlink

Hyperlink is the important portion of network information. The designer could organize the whole frame of web pages by using the hyperlinks. Convenient for the continued browsing, the text content indicated in hyperlink is concise, which we called metadata of hyperlink. Metadata is not only a good guide to browse web pages, but also a better clue to topic-specific crawling. So the information in metadata can do help to select the searching path in the process of topic-specific crawling.

Metadata is composed of anchor text and HREF information. The information of HREF should be condensed for disposing of the useless segments. For example,

Table 1. Metadata of Hyperlink

Hyperlink		http://service.symantec.com/subscribe
Described		<a href="http://service.symantec.com/subscribe">"How do I Purchase and Activate my Definition Subscription?"</a>
Meta -data	Anchor text	How do I Purchase and Activate my Definition Subscription
	HREF	Service Symantec Subscribe

3.5 P-HITS

The HITS algorithm is a heuristic strategy, and those URLs to be selected are ranked on the principle of priority. The priorities are judged by analyzing those web pages that have been collected. So those priorities are local and this algorithm can easily fall in the trap of local optimality [11].

Considering more global optimality, probability is introduced to select the URLs based on the HITS. So we call it P-HITS algorithm. In P-HITS, a buffer is added in when URLs are selected from the queue for next crawling. If the size of buffer is m, it means that the number of anterior URLs with high weight we select form the queue is m. The strategy of selecting the next URL to crawl may accord as equation (4) below.

$$P(URL_i) = 1 - \lambda \frac{\sum_{j \neq i} W(URL_j)}{\sum W(URL_j)} \quad i, j = 1, 2, \ldots, m \, ; \, 0 \leq \lambda \leq 1. \quad (4)$$

Here, $W(URL_i)$ is the weight of URL in the buffer. It can be easily comprehended that those URLs with low weight may get more chances to be selected for the next crawling. The variable m is degressive along with the process of crawling. When m decreases to 1 or λ is null, the corresponding P is equal to 1. It means an ordinal selection. The P-HITS algorithm is described as below.

```
Seed_URLs: = Search Engine (topic_keyword);
For all URL in Seed_URLs do
    url_weight := sim(url.page.description, topic_keyword);
    Enqueue (url, url_weight, url_queue);
End for
While url_queue not empty do
    url_buffer_size := m;
    enbuffer (url_with_high_weight, m);
    While m > 1 do
        url := Select_url(P, enbuffer);
        page := Download_document (url);
        page_weight := HITS (page.urls);
        For all un-crawled link in page.links do
            link_metadata _weight := sim (link.metadata,
                topic_keyword);
            link_weight := merge (page_weight,
                link_metadata _weight);
            Enqueue (link, link_weight, url_queue);
        End for
        m--;
    End while
End while
```

First some topic keywords are sent to a search engine in order to build an initial set of starting URLs. The crawler computes the weights to pre-order the URLs in that set. Then, it will roughly select URL as equation (4) from a url-buffer, download the page and compute the page weight, extract un-crawled links and compute the weights. New extracted URLs will be ordered by merging page weights and metadata weights.

4 Evaluation

In this section we will present our experiences with topic-specific web crawler. In contrast, both the HITS and P-HITS are tested.

4.1 Experimental Setup

Our web crawler is implemented by JAVA language. We run the crawler on a windows OS machine equipped with Pentium-IV 2000 MHz and 512 MB RAM. This machine is connected with the public network through an Ethernet network card. We illustrate a crawled result on a two-dimension graph where x-axis is the number of crawled web pages and y-axis is a relevance ratio calculated from equation (5) below.

$$\text{Relevance ration} = \frac{\text{number of relevant pages}}{\text{number of pages crawled}} \times 100 \ . \tag{5}$$

In our experiments, we choose 'network security' as the topic to the performance of our algorithm. We first assign the initial set of keywords as 'network', 'security', 'antivirus' and 'firewall' to describe the topic 'network security'. We then perform a search on the Google and select the top 100 resulting URLs to be the initial set of starting URLs. For easily judging the relevance of the crawl by human inspection, we will collect and study only the first 1000 web pages for each crawling attempt.

4.2 Experimental Results

In our experiments, the value of m in equation (4) is initially set as 20, and decreases by unity after crawling every 100 Web pages till it becomes 1. Some available results can be drawn from Fig. 1 and Fig. 2. B-spline curve is adopted to describe the trendline in both of these figures.

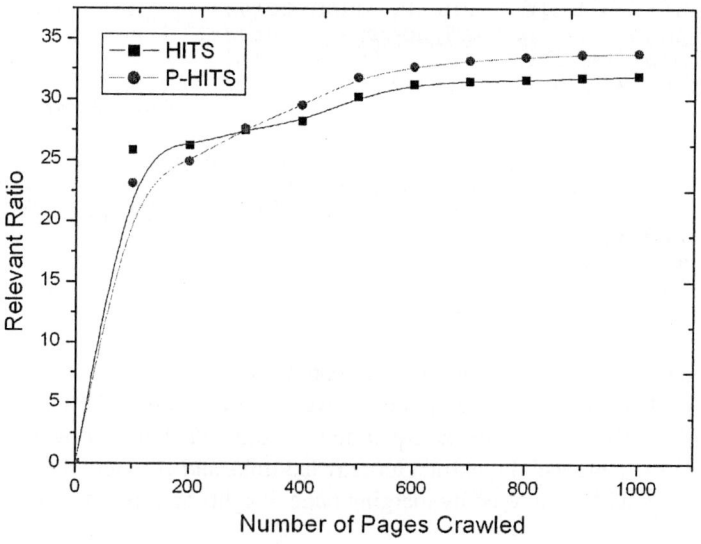

Fig. 1. The algorithm's characteristics of HITS and P-HITS

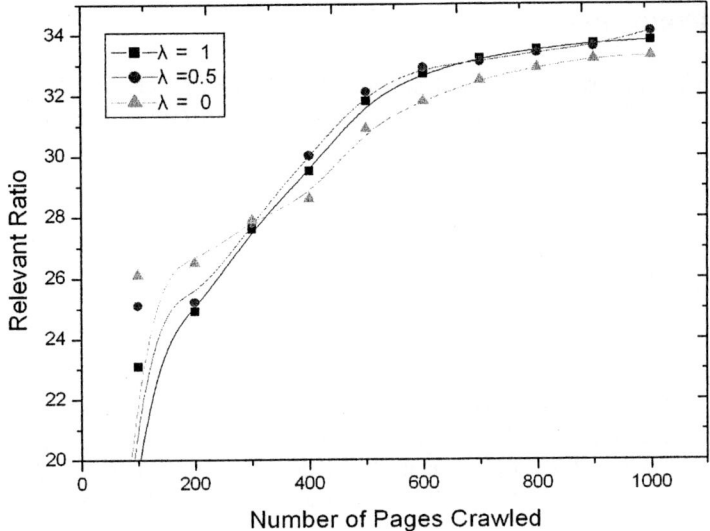

Fig. 2. The algorithm's characteristics of P-HITS with different variable λ

Fig. 1 shows the algorithm's characteristics of HITS and P-HITS. The value of λ is initially set as 1 in this comparative trial. In early crawling time, the relevance ration of HITS is a little higher than that of P-HITS. With the running of crawler, the relevance ration of P-HITS preponderates over that of HITS and maintains a stabile value. It means that P-HITS is superior to HITS in topic-specific crawling.

Fig. 2 shows the algorithm's characteristics of P-HITS with different variable λ. For distinctly describing the change of relevant ratio, the scale of y-axis begins from 20. We can see that in early crawling time, the relevance ration is better with smaller λ because smaller λ means the URL selection is closer to the optimality. With the running of crawler, the relevance ration with higher λ increases rapidly and maintains a stabile value. It means that those Web pages with low weights get more chance to be gathered in the P-HITS algorithm for topic-specific crawling; so higher relevance ration is gained. The value of m may have an effect on the relevance ration, but more creditable conclusions could be shown only quite a few Web pages have been crawled.

5 Conclusion

In this paper, we present a new algorithm to build an effectual topic-specific web crawler, which we called P-HITS. Probability is introduced to select the URLs to get more global optimality when crawling the Web pages, and the metadata of hyperlinks is appended in this algorithm to predict the relevance of web pages better. Our experiments have shown that this algorithm provides good results. However, there are some limitations in our study, e.g. the volume of examined web pages is too small. We expect to do more extensive test with large volume of web pages, later. Some specifically automatic classifier would be used to judge the relevance of crawled web pages.

Acknowledgments

The work was supported by Natural Science Foundation of Hubei (2004ABA055) and National Natural Science Foundation of China (60274007,60074008).

References

1. Murry BH., Moore A.: Sizing the Internet. A White Paper. Cyveillance, Inc (2000)
2. Bharet K., Broder A.: A Technique for Measuring the Relative Size and Overlap of Public Web Search Engines. Computer Networks and ISDN Systems, Special Issue on the 7th Int. World Wide Web Conf., Brisbane, 30 (1998) 1–7
3. Lawrence S., Giles C.L.: Searching the World Wide Web. Science, 280 (1998) 98–100
4. Debra P., Post R.: Information Retrieval in the World Wide Web: Making Client – Based Searching Feasible. Proc. 1st International World Wide Web Conference (1994)
5. Michael Hersovici, Michal Jacov et al.: The Shark-search Algorithm-An Application: Tailored Web Site Mapping. Proc. 7th International World Wide Web Conference, Brisbane, Australia (1998) 317–326
6. Menczer F., Monge AE.: Scalable Web Search by Adaptive Online Agents: An Infospider Case Study. Intelligent Information Agents: Agent-based Information Discovery and Management on the Internet, Berlin (1999) 323–347
7. Sougata Mukherjea.: WTMS: A System for Collecting and Analyzing Topic-specific Web Information. Computer Networks, 33 (2000) 457–471
8. Salton G., McGill M.: Introduction to Modern Information Retrieval. McGraw-Hill, New York (1983)
9. Kleinberg J.: Authoritative Sources In a Hyperlinked Environment. Proc. 9th Ann. ACM-SIAM Symp. Discrete Algorithms, ACM Press, New York (1998) 668–677
10. 10.Rungsawang A., Angkawattanawit N.: Learnable Topic-specific Web Crawler. Network and Computer Applications, 28 (2005) 97-114
11. 11.Menczer F.: Complementing Search Engines With Online Web Mining Agents. Decision Support Systems, 35 (2003) 195–212

Geometrical Profile Optimization of Elliptical Flexure Hinge Using a Modified Particle Swarm Algorithm

Guimin Chen, Jianyuan Jia, and Qi Han

School of Electronical and Mechanical Engineering,
Xidian University, Xi'an, China 710071
efoxxx@126.com, jyjia@xidian.edu.cn, hanqi_aa@sina.com

Abstract. Elliptical flexure hinges are one of the most widely used flexure hinges for its high flexibility. To design elliptical flexure hinges of best performance, the author proposed a modified particle swarm optimization (MPSO) search method, where an exponentially decreasing inertia weight is deployed instead of a linearly decreasing inertia weight. Simulations indicate that the MPSO method is very effective. The optimal design parameters including the cutout configuration and the minimum thickness are obtained.

1 Introduction

The flexure hinge is a kind of micro transmission mechanism. For its several outstanding advantages, such as small in dimensions, no mechanical friction, no hysteresis, no gap, and high motion sensitivity, the flexure hinge is widely used in many fields to achieve higher precision and displacement resolution of the whole system. In recent years, it has more applications of using piezoelectric element to drive and flexure hinge to realize fine positioning, such as stability control of light-beam tracking-pointing [1], micro-robots [2][3], control of the beam line of synchrotron radiation [4], laser welding, etc. Nanopositioning technology is the base of the realization of nanomachining and nano measurement, the flexure hinge also has momentous applications in this area.

Flexure hinges of single-axis can be divided into two main categories: leaf and notch type hinges. Because of relative low rotation precision and stress concentration, leaf type hinge is seldom adopted. In 1965, Paros and Weisbord [5] introduced the first notch hinge — right-circular flexure hinge, which incorporates a circular cutout on either side of the blank to form a necked-down section. The common feature of these two types is easy to manufacture. With the advent of CNC milling machines and particularly CNC wire electrodischarge machining (WEDM), hinges of arbitrary shape can now be readily produced. Therefore, researchers turned their attention to hinges of other cutout profiles, which could provide precision rotation in an even larger angular range. Smith *et al.* [6] introduced the elliptical flexure hinge, which becomes one of the most widely used flexure hinges for its high flexibility. In fact, it becomes a right-circular flexure hinge when minor axis equals to major axis.

It is approved that the elliptical cutout dimensions and the minimum thickness affect the performance of flexure hinges greatly. In this paper, to achieve optimal profile of elliptical flexure hinge, the author proposed a modified particle swarm optimization (MPSO) search method.

The particle swarm optimization (PSO) was introduced by Kennedy and Eberhart in 1995 [11] and has been compared favorably to genetic algorithms. Particle Swarm Optimization (PSO) is one of the evolutionary computation techniques. It was developed through simulation of a simplified social system, and has been found to be robust in solving continuous nonlinear optimization problems. The PSO technique can generate high-quality solutions within shorter calculation time and stable convergence characteristic than other stochastic methods.

2 Formal Problem Definition

The flexibility and maximum stress are two of the most important parameters to assess the performance of flexure hinges. The optimization objective is to find the best profile that could make an elliptical hinge more flexible while less stress concentration.

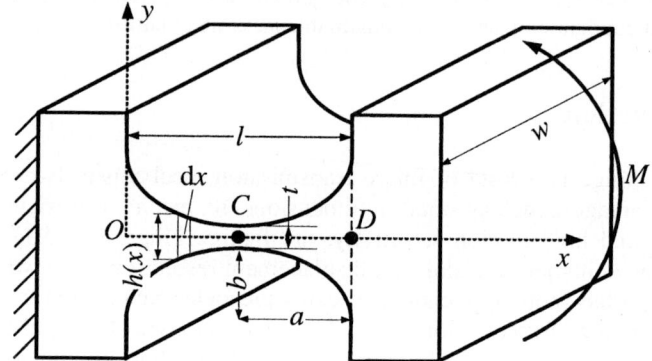

Fig. 1. Elliptical flexure hinge and its coordinate system

2.1 Compliance Model

Compliance is the most important parameter for flexure hinge design. To calculate the compliance of a hinge, the bending theory of Euler-Bernoulli beam is adopted generally [7] [8].

From simple bending theory of mechanics of materials, the curvature radius of the neutral plane of a flexure hinge can be expressed as:

$$\frac{1}{\rho} = \frac{M(x)}{EI(x)} \qquad (1)$$

where, E is the elastic modulus, $M(x)$ is the moment applied on dx, and $I(x)$ is the second moment of the cross-section on dx.

For the curvature radius of curve $y = f(x)$ is given by

$$\frac{1}{\rho} = \frac{\dfrac{d^2y}{dx^2}}{\left[1 + (\dfrac{dy}{dx})^2\right]^{3/2}} \quad (2)$$

the bending equation of the flexure hinge can be expressed as:

$$\frac{\dfrac{d^2y}{dx^2}}{\left[1 + (\dfrac{dy}{dx})^2\right]^{3/2}} = \frac{M(x)}{EI(x)} \quad (3)$$

Generally, the length of the cutout is far less than other dimensions of the flexure hinge. Therefore, the variation of the bending moment along the hinge can be ignored, i.e., the $M(x)$ can be considered as a constant. Moreover, $dy/dx \ll 1$ for the deflection of the flexure hinge is very little, so the equation above can be simplified as:

$$\frac{d^2y}{dx^2} = \frac{M}{EI(x)} \quad (4)$$

Deflections out of the cutout section can be ignored because most deflections of flexure hinge come from the cutout section. Substituting $I(x) = \dfrac{b \times [h(x)]^3}{12}$ into Equation (5) and integrating it from point O to point D, an approximate expression for the angular deflection about the neutral axis can be obtained:

$$\theta = \int_0^{2a} \frac{d^2y}{dx^2} dx = \int_0^{2a} \frac{12M}{Eb[h(x)]^3} dx \quad (5)$$

where, the height $h(x)$ at a position x on the hinge is given by:

$$h(x) = 2b + t - 2b\sqrt{1 - \frac{(a-x)^2}{a^2}} \quad (6)$$

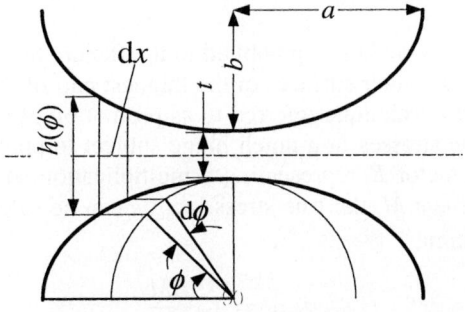

Fig. 2. Diagram of infinitesimal

Considering the elliptical cutout profile in **Fig. 2**. By introducing the centrifugal angle ϕ (which is from 0 to π) as the integral variable, to give

$$h(\phi) = 2b + t - 2b \sin \phi \tag{7}$$

For $x = a - a \cos \phi$, to give

$$dx = a \sin \phi d\phi \tag{8}$$

Substituting Equation (4), (5), (6) and (7) into Equation (3), to give

$$\theta = \frac{12Ma}{Ew} \int_0^\pi \frac{\sin \phi}{(2b + t - 2b \sin \phi)^3} d\phi \tag{9}$$

Let

$$s = b/t \tag{10}$$

substitute it into Equation (9) and solve the definite integrals to give

$$\theta = 24Ma\gamma / Ewt^3 \tag{11}$$

where,

$$\gamma = \frac{6s(8s^3 + 12s^2 + 6s + 1)}{(2s+1)^2(4s+1)^{5/2}} \text{arctg} \frac{2s}{\sqrt{4s+1}} + \\ \frac{6s(2s+1)}{(4s+1)^{5/2}} \text{arctg} \frac{1}{\sqrt{4s+1}} + \frac{12s^3 + 14s^2 + 6s + 1}{(2s+1)^2(4s+1)^2} \tag{12}$$

Based on Equation (9), the compliance equation of elliptical flexure hinge can be expressed as

$$1/k = \frac{\theta}{M} = 24a\gamma / Ewt^3 \tag{13}$$

2.2 Stress Model

When a pure bending moment being applied to the flexure hinge, the maximum stress will occur at each of the outer surfaces of the thinnest part of the notch. Using Fourier integral methods and a technique referred to as promotion of rank, Ling has obtained a full solution for the stresses in a notch hinge subject to pure bending in terms of a stress concentration factor K_t representing a multiplication so that [6][9][10], for an applied bending moment M, the true stress, σ_{max}, can be calculated from a nominal stress using the equation

$$\sigma_{max} = K_t \frac{6M}{t^2 w} \tag{14}$$

with the stress concentration factor given by

$$K_t = (1 + \frac{t}{2a})^{9/20} = (1 + \frac{x_1}{2})^{9/20} \tag{15}$$

where,

$$x_1 = t/a, \; (0 < x_1 < 4.6) . \tag{16}$$

2.3 Optimization Model

Optimization of the profile by any of the optimization techniques corresponds to searching for an elliptical cutout and a maximum thickness that could make a hinge more flexible while less stress concentration.

Let $x_2 = b/a$, to give $s = x_2/x_1$. Then the mathematical model of the problem can be described as the following:

$$\text{Min} \quad J(x_1, x_2) = \frac{\sigma_{\max}}{E \cdot \theta} = \frac{x_1 K_t}{4\gamma} \tag{17}$$
$$\text{Subject to} : 0 < x_1 \leq 4.6 \text{ and } 0 < x_2 \leq 1$$

where γ and K_t is expressed by **(12)** and **(15)** respectively.

3 Particle Swarm Optimization

Similar to other population-based optimization methods such as genetic algorithms, the particle swarm algorithm starts with the random initialization of a population of individuals (particles) in the search space. The PSO algorithm works on the social behavior of particles in the swarm. Therefore, it finds the global best solution by simply adjusting the trajectory of each individual toward its own best location and toward the best particle of the entire swarm at each time step (generation) [12].

3.1 PSO Algorithm

In the particle swarm algorithm, the trajectory of each individual in the search space is adjusted by dynamically altering the velocity of each particle, according to its own flying experience and the flying experience of the other particles in the search space.

The basic elements of PSO technique are briefly stated and defined as follows [13] [14] [15] [16]:

1. *Particle, X(t)*: It is a candidate solution represented by an m-dimensional vector, where m is the number of optimized parameters. At time t, the jth particle $X_j(t)$ can be described as $X_j(t)=[x_{j,1}(t), ...,x_{j,m}(t)]$, where xs are the optimized parameters and $x_{j,k}(t)$ is the position of the jth particle with respect to the kth dimension, i.e. the value of the kth optimized parameter in the jth candidate solution.
2. *Population, pop(t)*: It is a set of n particles at time t, i.e. $pop(t) = [X_1(t), ..., X_n(t)]^T$.

3. *Swarm*: It is an apparently disorganized population of moving particles that tend to cluster together while each particle seems to be moving in a random direction.
4. *Particle velocity*, $V(t)$: It is the velocity of the moving particles represented by an m-dimensional vector. At time t, the jth particle velocity $V_j(t)$ can be described as $V_j(t) = [v_{j,1}(t), ..., v_{j,m}(t)]$, where $v_{j,k}(t)$ is the velocity component of jth particle with respect to the kth dimension.
5. *Inertia weight*, $w(t)$: It is a control parameter that is used to control the impact of the previous velocities on the current velocity. Hence, it influences the trade-off between the global and local exploration abilities of the particles. For initial stages of the search process, large inertia weight to enhance the global exploration is recommended while, for last stages, the inertia weight is reduced for better local exploration.
6. *Individual best*, $X^*(t)$: As a particle moves through the search space, it compares its fitness value at the current position to the best fitness value it has ever attained at any time up to the current time. The best position that is associated with the best fitness encountered so far is called the individual best, $X^*(t)$. For each particle in the swarm, $X^*(t)$ can be determined and updated during the search. In a minimization problem with objective function J, the individual best of the jth particle $X_j^*(t)$ is determined such that $J(X_j^*(t)) \leq J(X_j(\tau))$, $\tau \leq t$. For simplicity, assume that $J_j^* = J(X_j^*(t))$. For the jth particle, individual best can be expressed as $X_j^*(t) = [x_{j,1}^*(t), ..., x_{j,m}^*(t)]$.
7. *Global best*, $X^{**}(t)$: It is the best position among all individual best positions achieved so far. Hence, the global best can be determined such that $J(X^{**}(t)) \leq J(X_j^*(t)), j = 1, ..., n$. For simplicity, assume that $J^{**} = J(X^{**}(t))$.
8. *Stopping criteria*: These are the conditions under which the search process will terminate. In this study, the search will terminate if one of the following criteria is satisfied: (i) the number of iterations since the last change of the best solution is greater than a prespecified number of (ii) the number of iterations reaches the maximum allowable number.

3.2 Modified PSO Algorithm

In 1999, Shi and Eberhart [17] have found a significant improvement in the performance of the PSO method with a linearly varying inertia weight over the generations (LPSO). The mathematical representation of this concept is given by (**18**), (**19**) and (**20**)

$$v_{j,k}(t) = w(t) \times v_{j,k}(t-1) + c_1 \times \text{Rand}_1 \times (x_{j,k}^*(t-1) - x_{j,k}(t-1)) \\ + c_2 \times \text{Rand}_2 \times (x_{j,k}^{**}(t-1) - x_{j,k}(t-1)) \qquad (18)$$

$$x_{j,k}(t) = x_{j,k}(t-1) + v_{j,k}(t) \qquad (19)$$

where $w(t)$ is given by

$$w(t) = (w_{\max} - w_{\min}) \times \frac{\text{MAXITER} - t}{\text{MAXITER}} + w_{\min} \qquad (20)$$

c_1 and c_2 are constants known as acceleration coefficients, c_3 is constants known as velocity coefficient, t is the current iteration number, MAXITER is the maximum number of allowable iterations, Rand_1 and Rand_2 are two separately generated uniformly distributed random numbers in the range [0, 1].

The first part of (18) represents the inertia, which provides the necessary momentum for particles to roam across the search space. The second part, known as the "cognitive" component, represents the personal thinking of each particle. The cognitive component encourages the particles to move toward their own best positions found so far. The third part is known as the "social" component, which represents the collaborative effect of the particles, in finding the global optimal solution. The social component always pulls the particles toward the global best particle found so far. By decreasing the inertia weight gradually from a relative large value (empirically $w_{\max} = 0.9$) to a small value (empirically $w_{\min} = 0.4$) through the course of LPSO run, LPSO tends to have more global search ability at the beginning of the run while having more local search ability near the end of the run.

To improve the performance of LPSO, the authors proposed a modification to the strategy of decreasing inertia weight. In this modified PSO algorithm (MPSO), instead of a linearly decreasing inertia weight, an exponentially decreasing inertia weight is deployed. The mathematical representation of $w(t)$ is given by

$$w(t) = w_{\min}\left(\frac{w_{\max}}{w_{\min}}\right)^{1/(1+t/c_3)}. \qquad (21)$$

Equation (18), (19) and (21) form the MPSO algorithm together.

In this MPSO algorithm, the population has n particles and each particle is an m-dimensional vector, where m is the number of optimized parameters. Incorporating the above modifications, the computational flow of PSO technique can be described in the following steps.

- *Step 1 (Initialization)*: Set the time counter $t = 0$ and generate randomly n particles, $\{X_j(0), j = 1, \ldots, n\}$, where $X_j(0)=[x_{j,1}(0), \ldots, x_{j,m}(0)]$. $x_{j,k}(0)$ is generated by randomly selecting a value with uniform probability over the kth optimized parameter search space $[x_k^{\min}, x_k^{\max}]$. Similarly, generate randomly initial velocities of all particles, $\{V_j(0), j = 1, \ldots, n\}$, where $V_j(0)=[v_{j,1}(0), \ldots, v_{j,m}(0)]$. $v_{j,k}(0)$ is generated by randomly selecting a value with uniform probability over the kth dimension $[-v_k^{\max}, v_k^{\max}]$. Each particle in the initial population is evaluated using the objective function, J. For each particle, set $X_j^*(0) = X_j(0)$ and $J_j^* = J_j$, $j = 1, \ldots, n$. Search for the best value of the objective function J_{best}. Set the particle associated with J_{best} as the global best, $X^{**}(0)$, with an objective function of J^{**}. Set the initial value of the inertia weight $w(0)$.
- *Step 2 (Time updating)*: Update the time counter $t = t+1$.
- *Step 3 (Weight updating)*: Update the inertia weight $w(t)$ according to (21).
- *Step 4 (velocity updating)*: Using the global best and individual best of each particle, the jth particle velocity in the kth dimension is updated according to (18).
- *Step 5 (Position updating)*: Based on the updated velocities, each particle changes its position according to (19). If a particle violates its position limits in any dimension, set its position at the proper limit.

Step 6 (Individual best updating): Each particle is evaluated according to its updated position. If $J_j < J_j^*$, $j = 1, ..., n$, then update individual best as $X_j^*(t) = X_j(t)$ and $J_j^* = J_j$ and go to Step 7; else go to Step 7.

Step 7 (Global best updating): Search for the minimum value $J_{\min}$ among J_j^*, where min is the index of the particle with minimum objective function, i.e. min $\in \{j; j = 1, ..., n\}$. If $J_{\min} < J^{**}$, then update global best as $X^{**}(t) = X_{\min}(t)$ and $J^{**} = J_{\min}$ and go to Step 8; else go to Step 8.

Step 8 (Stopping criteria): If one of the stopping criteria is satisfied then stop; else go to Step 2.

3.3 MPSO Implementation

The performance of the newly developed MPSO method applied to profile optimization of elliptical flexure hinges was observed in comparison with the LPSO method. In our implementation, the maximum inertia weight $w_{\max} = 0.9$, and the minimum $w_{\min} = 0.4$. The maximum allowable velocity $v_1^{\max} = 4.6$, and $v_2^{\max} = 1$. Other parameters are selected as: number of particles $n = 20$, accelerate coefficients $c_1 = c_2 = 2.0$, maximum number of allowable iterations MAXITER = 500, and velocity coefficients $c_3 = 30$.

Fig. 3 shows variation of the average particle fitness of 50 runs with iterations. From the results, it is clear that MPSO method can find the optimal solution faster than LPSO. The modification of the strategy of decreasing inertia weight significantly boost the performance without much added complexity. **Fig. 4** shows the optimal profile found by MPSO method.

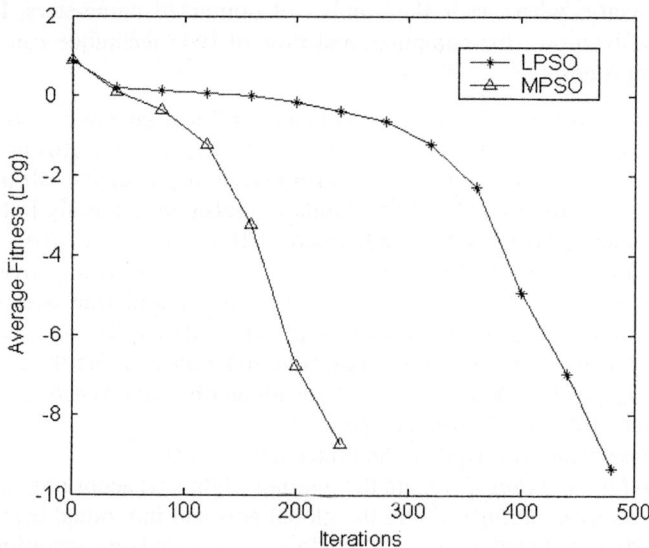

Fig. 3. The average particle fitness versus the number of iterations. We observe that the average solution quality in terms of the fitness value is significantly improved with the variation of iterations, meaning that the particles are resorting to high quality solutions as the swarm converges.

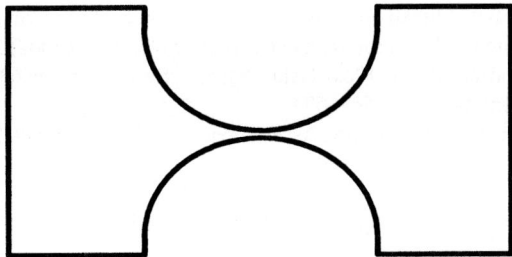

Fig. 4. The optimal profile ($x_1 = 0.0771$ and $x_2 = 0.8626$)

4 Conclusion

In this paper, we proposed a modified particle swarm optimization (MPSO) search method, where an exponentially decreasing inertia weight is deployed instead of a linearly decreasing inertia weight. By using this method, optimal design parameters of an elliptical flexure hinge including the cutout configuration and the minimum thickness are obtained. The MPSO algorithm is very efficient to solve global optimization problems with continuous variables.

References

1. Sweeney M., Rynkowski G., Ketabchi M., Crowley R.: Design Considerations for Fast Steering Mirror (FSMs). Proceedings of SPIE Vol. 4773 (2002) 1–11
2. Yi B. J., Chung G. B., Heung N. Y.: Design and Experiment of a 3-DOF Parallel Micromechanism Utilizing Flexure Hinges. IEEE Transactions on Robotics and Automation 19 (2003) 604–612
3. John E. McInroy, and Jerry C. Hamann: Design and Control of Flexure Jointed Hexapods. IEEE Transactions on Robotics and Automation 16 (2000) 372–381
4. Fu X., Zhou R. K., Zhou S. Z.: Flexure Hinge in Beam Line of Synchrotron Radiation. Optics and Precision Engineering 9 (2001) 67–70
5. Paros J. M, Weisboro L.: How to Design Flexure Hinges. Machine Design 37 (1965) 151–157
6. Smith S. T., Badam I. V. G., Dale J. S.: Elliptical Flexure Hinges. Review of Scientific Instruments 68 (1997) 1474–1483
7. Gou Y. J., Jia J. Y., Chen G. M.: Study on the Accuracy of Flexure Hinge. Proceedings of the First International Symposium on Mechatronics (2004) 108–112
8. Jia J. Y., Chen G. M., Liu X. Y.: Design Calculation and Analysis of Elliptical Flexure Hinges. Engineering Mechanics 22 (2005) 136–140
9. Timoshenko S.: Strength of Materials (Advanced theory and problems). D. Van Nostrand Company (1957) 279–280
10. Tian Z. S., Liu J. S., Ye L.: Studies of Stress Concentration by Using Special Hybrid Stress Elements. International Journal for Numerical Methods in Engineering 40 (1997) 1399–1411
11. Kennedy J., Eberhart R.: Particle Swarm Optimization. Proceeding of IEEE International Conference on Neural Networks (1995) 1942–1948

12. Ayed Salman, Imtiaz Ahmad and Sabah Al-Madani. Particle Swarm optimization for Task Assignment Problem. Microprocessors and Microsystems 26 (2002) 363–371
13. Abido M. A.: Optimal Power Flow Using Particle Swarm Optimization. Electrical Power and Energy System 24 (2002) 563–571
14. Elegbede C.: Structural Reliability Assessment Based on Particles Swarm Optimization. Structural Safety 27 (2005) 171–186
15. Clerc M., Kennedy J.: The Particle Swarm: Explosion, Stability, and Convergence in a Multi-dimensional Complex Space, IEEE Transactions on Evolutionary Computation 8 (2002) 58–73
16. Mendes R., Kennedy J., Neves J.: The Fully Informed Particle Swarm: Simpler, Maybe Better. IEEE Transactions on Evolutionary Computation 8 (2002) 58–73
17. Shi Y., Eberhart R. C.: Empirical Study of Particle Swarm Optimization. Proceeding of IEEE International Conference on Evolutionary Computation (1999) 101–106

Classification of Chromosome Sequences with Entropy Kernel and LKPLS Algorithm

Zhenqiu Liu[1] and Dechang Chen[2]

[1] Department of Statistics, The Ohio State University,
1958 Neil Avenue, Columbus, OH 43210, USA
[2] Preventive Medicine and Biometrics,
Uniformed Services University of the Health Sciences,
4301 Jones Bridge Road, Bethesda, MD 20814, USA
liu@stat.ohio-state.edu
dchen@usuhs.mil

Abstract. Kernel methods such as support vector machines have been used extensively for various classification tasks. In this paper, we describe an entropy based string kernel and a novel logistic kernel partial least square algorithm for classification of sequential data. Our experiments with a human chromosome dataset show that the new kernel can be computed efficiently and the algorithm leads to a high accuracy especially for the unbalanced training data.

1 Introduction

A fundamental problem in computational biology is the classification of proteins into different classes based on evolutionary similarity of protein sequences. Popular algorithms for aligning sequences include the Needleman Wunsch algorithm for optimal global alignment and the Smith-Waterman algorithm for optimal local alignment[1]. Both algorithms give natural similarity scores related to the distance. However, standard pairwise alignment scores are not valid kernels [2]. Many string kernels such as the counting kernel [3,4] came from the text mining field. These kernels work well for the text mining but may not be appropriate for biological sequences because of the frequent context changes. [4] has proposed a marginal kernel through adding hidden contents to the counting kernel and shown that the Fisher kernel [5] is a special case of the marginalized kernels. Other kernels such as spectrum kernels, mismatch string kernels, and profile kernels have been proposed recently [6] and claimed to be successful in protein sequence classification. For most experiments performed, binary classification is considered and support vector machines (SVMs) are used for the classification task. However, SVM, a maximal margin classifier, is not very accurate when data are completely unbalanced. Unbalanced data often occur for multi-class problems when using the "one-against-others" scheme. Logistic regression, on the other hand, performs well for unbalanced data. In this paper, we propose an entropy kernel and the logistic kernel partial least squares algorithm (LKPLS) for classification of the multi-class sequences. LKPLS combines the kernel partial least squares with logistic regression in a natural way. LKPLS involves three steps: a kernel computation step, a dimension reduction step, and a classification step. The proposed entropy kernel and LKPLS have been used to classifying human chromosomes, and their effectiveness has been demonstrated by the experimental results.

This paper in organized as follows. In Section 2, we discuss the string kernels. Partial least squares (PLS) method and LKPLS algorithm are introduced in Section 3. Computational results are given in Section 4. Conclusions and remarks are provided in Section 4.

2 Entropy Kernel

Kernel methods including support vector machines have been a hot topic in machine learning and statistical research. To define a kernel function, we need to first transform the input into a feature space with a nonlinear transform function $\mathbf{x} \to \Phi(\mathbf{x})$. A kernel function is then defined as the inner product into a feature space: $K(\mathbf{x}_i, \mathbf{x}_j) = \Phi(\mathbf{x}_i)'\Phi(\mathbf{x}_j)$. The following is a simple example to illustrate the concept. Suppose we have a two dimensional input $\mathbf{x} = (x_1, x_2)'$. Let the nonlinear transform be

$$\mathbf{x} \to \Phi(\mathbf{x}) = (x_1^2, x_2^2, \sqrt{2}x_1 x_2, \sqrt{2}x_1, \sqrt{2}x_2, 1)'. \tag{1}$$

Then, given two points $\mathbf{x}_i = (x_{i1}, x_{i2})'$ and $\mathbf{x}_j = (x_{j1}, x_{j2})'$, the inner product (kernel) is

$$\begin{aligned} &K(\mathbf{x}_i, \mathbf{x}_j) \\ &= \Phi(\mathbf{x}_i)'\Phi(\mathbf{x}_j) \\ &= x_{i1}^2 x_{j1}^2 + x_{i2}^2 x_{j2}^2 + 2x_{i1}x_{i2}x_{j1}x_{j2} + 2x_{i1}x_{j1} + 2x_{i2}x_{j2} + 1 \\ &= (1 + x_{i1}x_{j1} + x_{i2}x_{j2})^2 = (1 + \mathbf{x}_i'\mathbf{x}_j)^2. \end{aligned} \tag{2}$$

This is a second order polynomial kernel. The above equation (2) shows that the kernel function is an inner product in the feature space that can be evaluated even without explicitly knowing the feature vector $\Phi(\mathbf{x})$.

In general, a kernel can be defined as a measure of *similarity*. The value of the kernel is greater if two samples are more similar and becomes zero if two samples are independent. Mathematically, given two samples $\mathbf{x}$ and $\mathbf{y}$, a kernel function $K(\mathbf{x}, \mathbf{y})$ has to meet the following properties:

1. $K(\mathbf{x}, \mathbf{y}) \geq 0$
2. $K(\mathbf{x}, \mathbf{y}) = K(\mathbf{y}, \mathbf{x})$
3. $K(\mathbf{x}, \mathbf{x}) \geq K(\mathbf{x}, \mathbf{y})$
4. the distance between $\mathbf{x}$ and $\mathbf{y}$ can be defined as

$$d(\mathbf{x}, \mathbf{y}) = \sqrt{K(\mathbf{x}, \mathbf{x}) + K(\mathbf{y}, \mathbf{y}) - 2K(\mathbf{x}, \mathbf{y})}. \tag{3}$$

Similarity can be evaluated by information measures. An information measure is a measure of evidence that a data set provides about a parameter in a parametric family. Fisher information and Kullback-Leibler (KL) information [7,8] are two such popular measures. In this paper, we concentrate on the KL mutual information. Mutual information is defined for information between distributions. It can lead to a kernel function after some revision. In the following, we define an entropy kernel based on the mutual information between two sequences.

Given two sequences **x** and **y**, the normalized mutual information is defined as

$$I(\mathbf{x},\mathbf{y}) = \frac{\sum_{i,j} p(x_i, y_j) \log \frac{p(x_i, y_j)}{p(x_i)p(y_j)}}{\min\{h(\mathbf{x}), h(\mathbf{y})\}}. \tag{4}$$

where $h(\mathbf{x}) = -\sum_i p(x_i) \log \frac{1}{p(x_i)}$ is the entropy for sequence **x**, $h(\mathbf{y})$ is the entropy for sequence **y**, and $0 \leq I(\mathbf{x},\mathbf{y}) \leq 1$. Using the normalized mutual information, we define the kernel matrix K by

$$K(\mathbf{x},\mathbf{y}) = e^{-\beta d^2}. \tag{5}$$

where β is a free positive parameter and the distance d is such that

$$d^2 = 2 - 2I(\mathbf{x},\mathbf{y}). \tag{6}$$

We see that the entropy kernel K meets all the properties of a kernel. For the sequences of different lengths, we can use a moving forward window and choose the one with maximal mutual information I.

For a comparison purpose, here we briefly introduce the mismatch kernels [6]. A mismatch kernel is a string kernel defined in terms of feature maps. For a fixed k-mer $\gamma = \gamma_1\gamma_2\ldots\gamma_k$, where γ_i is a character in an alphabet Σ, the (k,m) neighborhood, denoted by $N_{(k,m)}(\gamma)$, is the set of all k-length sequences α from Σ that differ from γ by at most m mismatches. For a k-mer γ, the feature map is defined as

$$\Phi_{(k,m)}(\gamma) = (\phi_\alpha(\gamma))_{\alpha \in \Sigma^k}. \tag{7}$$

where $\phi_\alpha(\gamma) = 1$ if $\alpha \in N_{(k,m)}(\gamma)$ and 0 otherwise. For a sequence **x** of any length, the feature map is defined as the sum of $\Phi_{(k,m)}(\gamma)$ over the k-mers contained in **x**:

$$\Phi_{(k,m)}(\mathbf{x}) = \sum_{k-\text{mers } \gamma \in \mathbf{x}} \Phi_{(k,m)}(\gamma). \tag{8}$$

The (k,m) mismatch kernel is the inner product:

$$K_{(k,m)}(\mathbf{x},\mathbf{y}) = \langle \Phi_{(k,m)}(\mathbf{x}), \Phi_{(k,m)}(\mathbf{y}) \rangle. \tag{9}$$

he normalized kernel is then given by

$$K(\mathbf{x},\mathbf{y}) \leftarrow \frac{K(\mathbf{x},\mathbf{y})}{\sqrt{K(\mathbf{x},\mathbf{x})}\sqrt{K(\mathbf{y},\mathbf{y})}}. \tag{10}$$

3 PLS and LKPLS Algorithms

PLS method is based on linear transition from a number of original descriptors (independent variables) to a new variable space based on a small number of orthogonal factors (latent variables). This technique is especially useful in cases where the number of descriptors is comparable to or greater than the number of compounds (data points) and/or there exist other factors leading to correlations between variables. In these cases, the solution of classical least squares method does not exist or is unstable and unreliable. On the other hand, the PLS approach leads to stable, correct and highly predictive models even for correlated descriptors [9]. The latent factors are mutually independent (orthogonal) linear combinations of original descriptors. Unlike principal component analysis, latent variables in PLS are chosen in such a way that the maximum correlation with the dependent variable is obtained. Thus, a PLS model contains the smallest number of necessary factors. When increasing the number of factors, a PLS model converges to an ordinary multiple linear regression model (if one exists). In addition, the PLS approach allows one to detect the relationship between activity and descriptors even if key descriptors have little contribution to the first few principal components. Figure 1 provides an example to illustrate the concept. Given two independent variables x_1 and x_2 and one dependent variable y, the first latent component given by PCA and PLS is the line in the left and right panel of the figure, respectively.

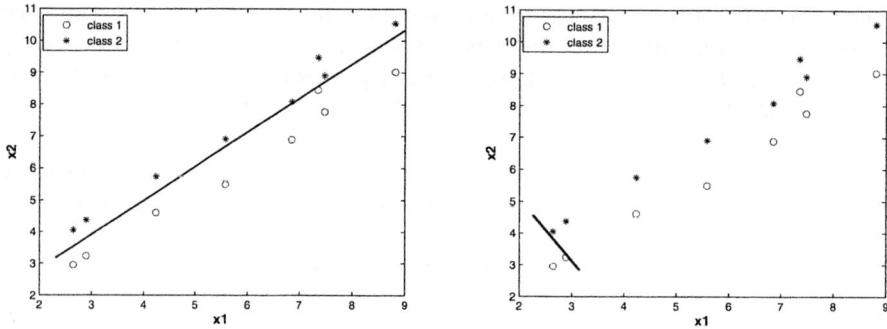

Fig. 1. The first latent component given by PCA (left) and PLS (right)

Figure1 shows that for this simple problem, PCA selects the poor latent variable which can not be used to separate the two classes, while the PLS component can be used to separate the two class efficiently.

Kernel partial least squares (KPLS) method is a generalization and nonlinear version of the partial least squares method. Only recently PLS has been extended to nonlinear regression through the use of kernels [10]. The general approach for KPLS is first to map each point nonlinearly to a higher dimensional feature space with a transform $x \rightarrow \Phi(x)$, and then to perform a linear PLS in the mapped space, which corresponds to a nonlinear PLS in the original input space. Kernel partial least squares method was mainly designed for the regression task originally. Recently, Liu and Chen have combined the KPLS and logistic regression to obtain LKPLS for classification of microarray data [11]. The algorithm is given as follows.

LKPLS Algorithm:

Given the training sequence data $\{\mathbf{x}_i\}_{i=1}^n$ with class labels $\{y_i\}_{i=1}^n$ and the test sequence data $\{\mathbf{x}_t\}_{t=1}^{n_t}$ with labels $\{y_t\}_{t=1}^{n_t}$

1. Compute the kernel matrix, for the training data, $K = [K_{ij}]_{n \times n}$, where $K_{ij} = K(\mathbf{x}_i, \mathbf{x}_j)$. Compute the kernel matrix, for the test data, $K_{te} = [K_{ti}]_{n_t \times n}$, where $K_{ti} = K(\mathbf{x}_t, \mathbf{x}_i)$.
2. Centralize K and K_{te} using

$$K = \left(\mathbf{I}_n - \frac{1}{n}\mathbf{1}_n\mathbf{1}_n'\right) K \left(\mathbf{I}_n - \frac{1}{n}\mathbf{1}_n\mathbf{1}_n'\right). \tag{11}$$

and

$$K_{te} = \left(K_{te} - \frac{1}{n}\mathbf{1}_{n_t}\mathbf{1}_n' K\right)\left(\mathbf{I}_n - \frac{1}{n}\mathbf{1}_n\mathbf{1}_n'\right). \tag{12}$$

where $\mathbf{1}_n$ is the vector of n 1's.
3. Call KPLS algorithm to find k component directions [10]:
   (a) for $i = 1, \ldots, k$
   (b) initialize $\mathbf{u}^i$
   (c) $\mathbf{t}^i = \Phi' \Phi \mathbf{u}^i = K \mathbf{u}^i$, $\mathbf{t}^i \leftarrow \mathbf{t}^i / \|\mathbf{t}^i\|$
   (d) $\mathbf{c}^i = \mathbf{y}^i \mathbf{t}^i$
   (e) $\mathbf{u}^i = \mathbf{y} \mathbf{c}^i$, $\mathbf{u}^i \leftarrow \mathbf{u}^i / \|\mathbf{u}^i\|$
   (f) repeat steps (b) -(e) until convergence
   (g) deflate K, $\mathbf{y}$ by $K \leftarrow (I - \mathbf{t}^i \mathbf{t}^{i\prime}) K (I - \mathbf{t}^i \mathbf{t}^{i\prime})$ and $\mathbf{y} \leftarrow \mathbf{y} - \mathbf{t}^i \mathbf{t}^{i\prime} \mathbf{y}$
   (h) obtain the component matrix $U = [\mathbf{u}^1, \ldots, \mathbf{u}^k]$
4. Find the projections $\mathbf{V} = KU$ and $\mathbf{V}_{te} = K_{te}U$ for the training and test data, respectively.
5. Build a logistic regression model using $\mathbf{V}$ and $\{y_i\}_{i=1}^n$ and test the model performance using $\mathbf{V}_{te}$ and $\{y_t\}_{t=1}^{n_t}$.

We can show that the above LKPLS classification algorithm is a nonlinear version of the logistic regression. In fact, from our LKPLS classification algorithm, we see that the probability of the label y given the projection $\mathbf{v}$ may be expressed as

$$P(y|\mathbf{w}, \mathbf{v}) = g\left(b + \sum_{i=1}^{k} w_i v_i\right). \tag{13}$$

where the coefficients $\mathbf{w}$ are adjustable parameters and g is the logistic function

$$g(u) = (1 + \exp(-u))^{-1}. \tag{14}$$

Given a data point $\Phi(\mathbf{x})$ in the transformed feature space, its projection v_i ($i = 1, ..., k$) can be written as

$$v_i = \Phi(\mathbf{x})\Phi'\mathbf{u^i} = \sum_{j=1}^{n} u_j^i K(\mathbf{x}_j, \mathbf{x}). \tag{15}$$

Therefore,

$$P(y|\mathbf{w}, \mathbf{v}) = g\left(b + \sum_{j=1}^{n} c_j K(\mathbf{x}_j, \mathbf{x})\right). \tag{16}$$

where

$$c_j = \sum_{i=1}^{k} w_i u_j^i, \quad j = 1, \cdots, n.$$

When $K(\mathbf{x}_i, \mathbf{x}_j) = \mathbf{x}_i'\mathbf{x}_j$, equation (16) becomes a logistic regression. Therefore, LKPLS classification algorithm is a generalization of logistic regression.

Similar to support vector machines, LKPLS algorithm was originally designed for two class classification. However, we can deal with the multi-class classification problem through the popular "one against all others" scheme. What we do is first to classify each class against all others, and then send each sequence to the class with the highest probability.

Computational Issues: LKPLS algorithm requires that the user specify the number of latent variables. The number of latent variables can be determined by either tuning with the validation sets or using Akaike's information criteria (AIC):

$$AIC = -2\log(\hat{L}) + 2(k+1), \tag{17}$$

where k is the dimension of the projection and $\hat{L}$ is the maximum likelihood calculated according to

$$\hat{L} = \prod_{i=1}^{n} \left(p(y|\mathbf{w}, \mathbf{v})\right)^y \left(1 - p(y|\mathbf{w}, \mathbf{v})\right)^{1-y} \tag{18}$$

The k that minimizes AIC should be selected. Our experience shows that for certain types of chromosome data, the prediction performance is not extremely sensitive to the choice of the number of latent variables.

4 Computational Results

Chromosome analysis is important in many applications such as detecting malignant diseases. In this section, we present the experimental results on classification of humane chromosomes when using LKPLS and the entropy kernel described previously. The data used in the experiments were extracted from a database of approximately 7,000 chromosomes classified by experts [12]. Each digitized chromosome image was automatically transformed into a string. The initial string composed of symbols from the alphabet $\{1,2,3,4,5,6\}$. The string is the difference coded to represent signed differences of successive symbols, with the alphabet $\Sigma = \{e,d,c,b,a,=,A,B,C,D,E\}$, where "=" is used for a difference of 0, "A" for +1, "a" for -1, and so on. Totally, there are 22 homologous pairs or 'autosomes' and one pair of sex chromosomes. Only the 22 autosomes were considered. The dataset contains 200 string samples from each of the 22 non-sex chromosome types, providing a total of 4400 samples. Samples of raw chromosome data pieces are given in Table 1.

Table 1. Samples of chromosome pieces

```
A=A=a===B==a====D==d=====D==e======B==b====B==b
A=B====a==A==a==D==d=====D==d======C==b===A===c
A===B==a==C==a==A==c===D===d=======C===b==B===d
A==B==a=A===a===B===b==D====e====A===a==A==a=Aa=A
A=B=aB==a=A==a==C===c==C=====d=====C==b===A===c===A
A==B==a==A==a====B==b==D====c=====B==b==A==c====A
A=A=a==A==a====A==a====E===d====B====b==A===b==A
A===C==a==A==a==A==a===C===c====A===a===A===c===A
```

Table 2. Error rates and their corresponding 95% intervals

Methods	Error (%)	95% Interval
LKPLS(entropy)	3.1	(2.4-3.9)
LKPLS(mismatch)	2.9	(2.5-3.7)
SVM(entropy)	4.3	(3.8-4.9)
SVM(mismatch)	3.8	(3.4-4.3)
k-NN	3.9	(3.3-4.5)
Constrained Markov Nets	7.3	(6.5-8.1)
ECGI Algorithm	7.5	(6.7-8.3)
Multilayer Perceptron	9.1	(8.3-10)
Hidden Markov Model	9.3	(8.5-10.2)

An additional piece of information contained in the data is the location of the centromere in each chromosome string. However, we decided not to use this information in this work. The task of chromosome classification is to detect deviations from the standard Karyotype, through labelling each chromosome with its type in order to identify missing, extra, or distorted chromosomes. Much work has been done with this dataset [13,14].

Our experiments were focused on demonstration of the performance of the LKPLS algorithm and the proposed entropy kernel. We randomly partitioned the data into two equally sized groups. Each group was composed of 2200 samples with 100 samples from each of the chromosome classes. With the cross-validation framework, we needed two runs: training the model with one group and testing using the other. The average error rates were computed. Our results and those reported in the literature are presented in Table 2.

In Table 2, $\beta = 4$ and 25 components were used for the combination of LKPLS and the entropy kernel, and $(k, m) = (5, 1)$ and 30 components were used for the combination of LKPLS and the mismatch kernel. The results on k-NN, constrained Markov nett, multilayer perceptron, and HMM are from the work in [13], [14], and [15]. It is seen that the performance of the entropy kernel and LKPLS is comparable to the best performance from the mismatch kernel and LKPLS.

Figure 2 is the plot for the error rate (%) and number of components used in the LKPLS algorithm. The plot shows that we need at least 20 components to achieve a better performance.

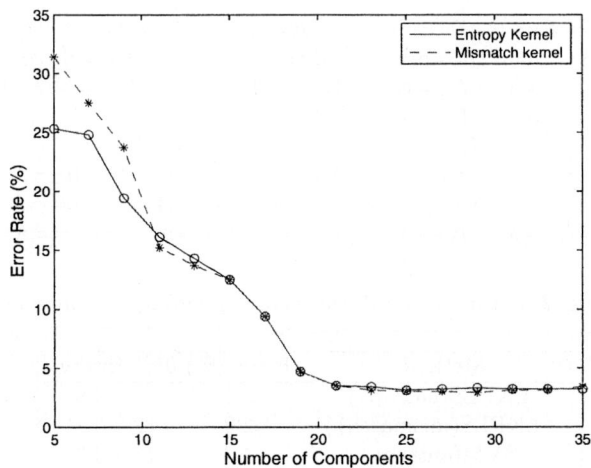

Fig. 2. Number of components and error rates of LKPLS algorithm

5 Conclusions and Discussion

The results of the experiments carried out in this paper demonstrate that our algorithm LKPLS and the entropy kernel performed very well for the given dataset. It is expected that LKPLS should work well when dealing with sequential and time series classification problems, as it is based solely on a kernel matrix. Usually, the design of kernels influences the performance of kernel related algorithms. How to design new kernel functions is a hot topic in the machine learning community. The entropy based kernel function proposed in this paper measures the similarity between sequences without requiring any alignment. Our experiments provided in this paper indicate that calculation of the entropy based kernel is faster than other kernels and that the performance of the entropy based kernel is comparable to those of some popular string kernels. To generalize these, one may need to conduct more experiments on different datasets and make thorough comparisons. It is also expected that the use of LKPLS in conjunction with the entropy kernel would be promising in classification tasks such as protein and DNA sequence classification. These are our future works.

Acknowledgement

D. Chen was supported by the National Science Foundation grant CCR-0311252. The authors wish to thank Dr. Jens Gregor for providing the preprocessed chromosome data.

Note: The opinions expressed herein are those of the authors and do not necessarily represent those of the Uniformed Services University of the Health Sciences and the Department of Defense.

References

1. Waterman, M. S., Joyce, J., and Eggert, M.: Computer Alignment of Sequences. In *Phylogenetic Analysis of DNA Sequences*, Oxford University Press(1991) 59-72
2. Vert, J.-P., Saigo, H., and Akutsu, T.: Local Alignment Kernels for Biological Sequences. In *Kernel Methods in Computational Biology*, MIT Press (2004)
3. Frakes, W. B. and Baeza-Yates, R.(Eds.): *Information Retrieval: Data Structures and Algorithms*, Prentice Hall (1992)
4. Tsuda, K., Kin, T., and Asai, K.: Marginalized Kernels for Biological Sequences. *Bioinformatics*, 18(2002) S268-S275
5. Jaakkola, T. and Haussler, D.: Exploiting Generative Models in Discriminative Classifie. In *Advances in Neural Information Processing Systems*, 11(1999) 487-493
6. Leslie, C., Eskin, E., Cohen, A., Weston, J., and Noble, W. S.: Mismatch String Kernels for Discriminative Protein Classification. *Bioinformatics*, 20(4) (2004)467-76
7. Kullback, S. and Leibler, R. A.: On Information and Sufficiency. *The Annals of Mathematical Statistics*, 22 (1951) 76-86
8. Topsoe, F. : Some Inequalities for Information Divergence and Related Measures of Discrimination.*IEEE Transaction on Information Theory*, 46(2000)1602-1609
9. Martens, H. and Naes, T.: *Multivariate Calibration*. Wiley, Chichester (1989)
10. Rosipal, R. and Trejo, L. J.: Kernel Partial Least Squares Regression in RKHS, Thoery and Empirical Compariso. Technical report, University of Paisley, UK (2001)
11. Liu, Z. and Chen, D.: Gene Expression Data Classification with Revised Kernel Partial Least Squares Algorithm. *Proceedings of the 17th International FLAIRS conference* (2004)104-108
12. Lundsteen, C., Philip, J., and Granum, E.: Quantitative Analysis of 6895 Digitized Trypsin G-banded Human Metaphase Chromosomes. *Clinical Genetics*, 18(1980) 355-373
13. Gregor, J. and Thomason, M. G.: A Disagreement Count Scheme for Inference of Constrained Markov Networks. In *Grammatical Inference: Learninf Syntax from Sentences* (1996)168-178
14. Martinez-Hinarejos, C. D., Juan, A., and Casacuberta, F.: Prototype Extraction for k-NN Klassifier Using Median Strings, In *Pattern Recognition and String Matching* (D. Chen and X. Cheng, eds.), Kluwer Academic Publishers (2002)465-476
15. Vidal, E. and Castro, M. J.: Classification of Banded Chromosomes Using Error-correcting Grammatical Inference (ECGI) and Multilayer Perceptron (MLP). *Proceeding of the VII Simposium Nacional de Reconocimiento de Forms y Análisis de Imágenes*, 1(1997) 31-36

Adaptive Data Association for Multi-target Tracking Using Relaxation

Yang-Weon Lee

Department of Information and Communication Engineering, Honam University,
Seobongdong, Gwangsangu, Gwangju, 506-714, South Korea
ywlee@honam.ac.kr

Abstract. This paper introduces an adaptive algorithm determining the measurement-track association problem in multi-target tracking. We model the target and measurement relationships and then define a MAP estimate for the optimal association. Based on this model, we introduce an energy function defined over the measurement space, that incorporates the natural constraints for target tracking. To find the minimizer of the energy function, we derived a new adaptive algorithm by introducing the Lagrange multipliers and local dual theory. Through the experiments, we show that this algorithm is stable and works well in general environments.

1 Introduction

The primary purpose of a multi-target tracking(MTT) system is to provide an accurate estimate of the target position and velocity from the measurement data in a field of view. Naturally, the performance of this system is inherently limited by the measurement inaccuracy and source uncertainty which arises from the presence of missed detection, false alarms, emergence of new targets into the surveillance region and disappearance of old targets from the surveillance region. Therefore, it is difficult to determine precisely which target corresponds to each of the closely spaced measurements. Although trajectory estimation problems have been well studied in the past, much of this previous work assumes that the particular target corresponding to each observation is known. Recently, with the proliferation of surveillance systems and their increased sophistication, the tools for designing algorithms for data association have been announced.

Generally, there are three approaches in data association for MTT : non-Bayesian approach based on likelihood function [1], Bayesian approach [2,3], and neural network approach [4,5]. The major difference of the first two approaches is how to treat the false alarms. The non-Bayesian approach calculates all the likelihood functions of all the possible tracks with given measurements and selects the track which gives the maximum value of the likelihood function. Meanwhile, the tracking filter using Bayesian approach predicts the location of interest using *a posteriori* probability. The two approaches are inadequate for real time applications because the computational complexity is overwhelming even for relatively large targets and measurements and a computationally efficient substitute based on a careful understanding of its properties is lacking.

As an alternative approach, Sengupta and Iltis [4] suggested a Hopfield neural probabilistic data association (HNPDA) network to approximately compute *a posteriori* probability in the joint probabilistic data association filter(JPDAF)[6] as a constrained minimization problem. It might be a good alternative to reduce the computation comparing with *ad hoc* association rules [7]. However, the neural network developed in [4] has been shown to have improper energy functions. Since the value of *a posteriori* probability in the original JPDAF are not consistent with association matrix of [4], these dual assumptions of no two returns from the same target and no single return from two targets should be used only in the generation of the feasible data association hypotheses, as pointed out in [8]. This resulted from misinterpretations of the properties of the JPDAF which the network was supposed to emulate. Furthermore, heuristic choices of the constant coefficients in the energy function in [4] didn't guarantee the optimal data association.

In contrast to the HNPDA, the MTT investigated here is a nonlinear combinational optimization problem irrespective of JPDAF. Its objective function consists of the distance measure based on the nearest neighbor weighted by target trajectory and matching factor which is getting from the validation and feasible matrix. We derived a method to solve the nonlinear MTT by converting the derived objective function with constraints into the minimization problem of energy function by MAP estimator [9]. The constrained energy function is also converted by an unconstrained energy equation using the Lagrange multipliers [10] and local dual theory [11]. To compute the feasible matrix from the energy function, we introduced a parallel relaxation scheme which can be realized on an array processor.

2   Problem Formulation and Energy Function

2.1   Representing Measurement-Target Relationship

Fig. 1 shows the overall scheme. This system consists of three blocks: acquisition, association, and prediction. The purpose of the acquisition is to determine the initial starting position of the tracking. After this stage, the association and prediction interactively determine the tracks. Our primary concern is the association part that must determine the actual measurement and target pairs, given the measurements and the predicted gate centers.

Let m and n be the number of measurements and targets, respectively, in a surveillance region. Then, the relationships between the targets and measurements are efficiently represented by the *validation matrix* ω [6]:

$$\omega = \{\omega_{jt} | j \in [1, m], t \in [0, n]\}, \tag{1}$$

where the first column denotes clutter and always $\omega_{j0} = 1$. For the other columns, $\omega_{jt} = 1$ ($j \in [1, m], t \in [1, n]$), if the validation gate of target t contains the measurement j and $\omega_{jt} = 0$, otherwise.

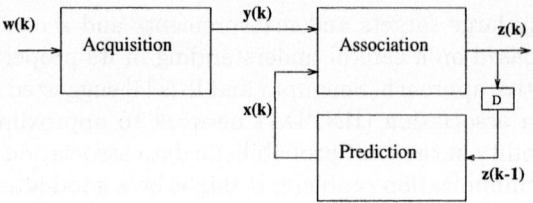

Fig. 1. An overall scheme for target tracking

Based on the validation matrix, we must find *hypothesis matrix* [6] $\hat{\omega}(=\{\hat{\omega}_{jt}|j \in [1,m], t \in [0,n]\})$ that must obey the data association hypothesis(or feasible events [6]):

$$\begin{cases} \sum_{t=0}^{n} \hat{\omega}_{jt} = 1 & \text{for } (j \in [1,m]), \\ \sum_{j=1}^{m} \hat{\omega}_{jt} \leq 1 & \text{for } (t \in [1,n]). \end{cases} \quad (2)$$

Here, $\hat{\omega}_{jt} = 1$ only if the measurement j is associated with clutter ($t = 0$) or target ($t \neq 0$). Generating all the hypothesis matrices leads to a combinatorial problem, where the number of data association hypothesis increases exponentially with the number of targets and measurements.

2.2 Constraining Target Trajectories

To reduce the search space further, one must take advantage of additional constraints. Let's consider a particular situation of radar scan like Fig. 2.

In this figure, the position of the gate center of target t at time k is represented by $\mathbf{x_t}(k)$. Also, $\mathbf{y_j}(k)$ means the coordinate of the measurement j at time k. Among the measurements included in this gate, at most one must be chosen as an actual target return. Note that the gate center is simply an estimate of this actual target position obtained by a Kalman filter [15].

Since the target must change its direction smoothly, a possible candidate must be positioned on the gate which is as close to the trajectory as possible.

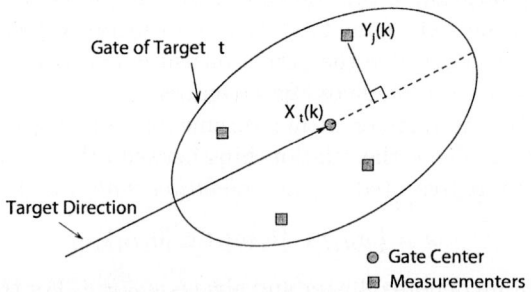

Fig. 2. Target trajectory and the measurements

As a measure of this distance, one can define the distance measure r_{jt} from the target trajectory line to the measurement point, $\mathbf{y_j} = (x_j, y_j)$ as

$$r_{jt}^2 \triangleq \frac{(x_j d_{y_t} - y_j d_{x_t})^2}{(d_{x_t}^2 + d_{y_t}^2)}. \tag{3}$$

where d_{x_t} and d_{y_t} are target's x, y axis directional distance over times k and $k+1$ which is calculated by using the estimation and prediction filter and x_j and y_j are relative positions from the gate center to the measurement position. Notice that this is the normal from the observation to the target trajectory.

As shown in Fig. 2, the distance from the target to the measurement is weighted by (3). In this case, the system tries to associate the target with the measurement which is located near the estimated point and target trajectory line within the gate. This will make one-to-one association decisions for track updating. But the JPDA [8] approach does not perform one-to-one associations of returns to tracks since the essence of the JPDA method is the computation of association probabilities for every track with every measurement in the present scan, and the subsequent use of those probabilities as weighting coefficients in the formation of a weighted average measurement for updating each track.

2.3 MAP Estimates for Data Association

The ultimate goal of this problem is to find the hypothesis matrix $\hat{\omega} = \{\hat{\omega}_{jt} | j \in [1,m], t \in [0,n]\}$, given the observation $\mathbf{y}$ and $\mathbf{x}$, which must satisfy (2). From now on, let's associate the realizations- the gate center $\mathbf{x}$, the measurement $\mathbf{y}$, the validation matrix ω, and $\hat{\omega}$- to the random processes- $X, Y, \Omega,$ and $\hat{\Omega}$.

Next, consider that $\hat{\Omega}$ is a parameter space and (Ω, Y, X) is an observation space. Then, *a posteriori* can be derived by the Bayes rule:

$$P(\hat{\omega}|\omega, \mathbf{y}, \mathbf{x}) = \frac{P(\omega|\hat{\omega})P(\mathbf{y}, \mathbf{x}|\hat{\omega})P(\hat{\omega})}{P(\omega, \mathbf{y}, \mathbf{x})}. \tag{4}$$

Here, we assumed that $P(\omega, \mathbf{y}, \mathbf{x}|\hat{\omega}) = P(\omega|\hat{\omega})P(\mathbf{y}, \mathbf{x}|\hat{\omega})$, since the two variables ω and $(\mathbf{x}, \mathbf{y})$ are separately observed. This assumption makes the problem more tractable as we shall see later. This relationship is illustrated in Fig. 3.

Given the parameter $\hat{\Omega}$, Ω and (X, Y) are observed. If the conditional probabilities describing the relationships between the parameter space and the observation spaces are available, one can obtain the MAP estimator:

$$\omega^* = \arg\max_{\hat{\omega}} P(\hat{\omega}|\omega, \mathbf{y}, \mathbf{x}). \tag{5}$$

2.4 Representing Constraints by Energy Function

As a system model, we assume that the conditional probabilities are all Gibbs distributions:

$$\begin{cases} P(\mathbf{y}, \mathbf{x}|\hat{\omega}) \triangleq \frac{1}{Z_1} \exp\{-E(\mathbf{y}, \mathbf{x}|\hat{\omega})\}, \\ P(\omega|\hat{\omega}) \triangleq \frac{1}{Z_2} \exp\{-E(\omega|\hat{\omega})\}, \\ P(\hat{\omega}) \triangleq \frac{1}{Z_3} \exp\{-E(\hat{\omega})\}, \end{cases} \tag{6}$$

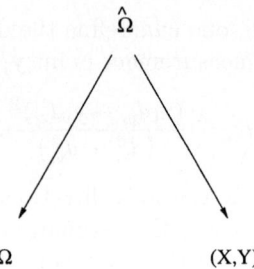

Fig. 3. The parameter space and the observation space

where Z_s ($s \in [1, 2, 3]$) is called partition function:

$$Z_s = \int_{\hat{\omega} \in \mathcal{E}} \exp\{-E(\hat{\omega})\} d\hat{\omega}. \quad (7)$$

Here, E denotes the energy function. Substituting (6) into (4), (5) becomes

$$\hat{\omega}^* = \arg\min_{\hat{\omega}}[E(\mathbf{y}, \mathbf{x}|\hat{\omega}) + E(\omega|\hat{\omega}) + E(\hat{\omega})]. \quad (8)$$

Since the optimization is executed with respect to $\hat{\omega}$, the denominator in (4) is independent of $\hat{\omega}$ and therefore irrelevant for its minimization.

The energy functions are realizations of the constraints both for the target trajectories and the measurement-target relationships. For instance, the first term in (8) represents the distance between measurement and target and could be minimized approximately using the constraints in (3). The second term intends to suppress the measurements which are uncorrelated with the valid measurements. The third term denotes constraints of the validation matrix and it can be designed to represent the two restrictions as shown in (2). The energy equations of each term are defined respectively:

$$\begin{cases} E(\mathbf{y}, \mathbf{x}|\hat{\omega}) \triangleq \sum_{t=1}^{n} \sum_{j=1}^{m} r_{jt} \hat{w}_{jt}, \\ E(\omega|\hat{\omega}) \triangleq \sum_{t=1}^{n} \sum_{j=1}^{m} (\hat{w}_{jt} - w_{jt})^2, \\ E(\hat{\omega}) \triangleq \sum_{t=1}^{n} (\sum_{j=1}^{m} \hat{w}_{jt} - 1) + \sum_{j=1}^{m} (\sum_{t=0}^{n} \hat{w}_{jt} - 1). \end{cases} \quad (9)$$

Putting (9) into (8), one gets

$$\hat{\omega}^* = \arg\min_{\hat{\omega}} \Big[\alpha \sum_{t=1}^{n} \sum_{j=1}^{m} r_{jt}^2 \hat{w}_{jt} + \frac{\beta}{2} \sum_{t=1}^{n} \sum_{j=1}^{m} (\hat{w}_{jt} - w_{jt})^2 \\ + \sum_{t=1}^{n} (\sum_{j=1}^{m} \hat{w}_{jt} - 1) + \sum_{j=1}^{m} (\sum_{t=0}^{n} \hat{w}_{jt} - 1) \Big], \quad (10)$$

where α and β are a coefficient of the weighted distance measure and the matching term respectively.

Using this scheme, the optimal solution is obtained by assigning observations to tracks in order to minimize the weighted total summed distance from all observations to the tracks to which they are assigned. This is thought of as a version of the well-known assignment problem for which optimal solutions have been developed with constraints [12][16].

3 Relaxation Scheme

The optimal solution for (10) is hard to find by any deterministic method. So, we convert the present constrained optimization problem to an unconstrained problem by introducing the Lagrange multipliers and local dual theory [11,10]. In this case, the problem is to find $\hat{\omega}^*$ such that $\hat{\omega}^* = \arg\min_{\hat{\omega}} L(\hat{\omega}, \lambda, \epsilon)$ where

$$L(\hat{\omega}, \lambda, \epsilon) = \alpha \sum_{t=1}^{n} \sum_{j=1}^{m} r_{jt}^2 \hat{\omega}_{jt} + \frac{\beta}{2} \sum_{t=1}^{n} \sum_{j=1}^{m} (\hat{\omega}_{jt} - \omega_{jt})^2$$
$$+ \sum_{t=1}^{n} \lambda_t (\sum_{j=1}^{m} \hat{\omega}_{jt} - 1) + \sum_{j=1}^{m} \epsilon_j (\sum_{t=0}^{n} \hat{\omega}_{jt} - 1), \qquad (11)$$

Here, λ_t and ϵ_j are the Lagrange multipliers. Note that (11) includes the effect of the first column of the association matrix, which represents the clutter as well as newly appearing targets. In general setting, we assume $m > n$, since most of the multitarget problem is characterized by many confusing measurements that exceed far over the number of original targets.

Let's modify (11) so that each term has equal elements:

$$L(\hat{\omega}, \lambda, \epsilon) = \alpha \sum_{t=0}^{n} \sum_{j=1}^{m} r_{jt}^2 \hat{\omega}_{jt}(1 - \delta_t) + \frac{\beta}{2} \sum_{t=0}^{n} \sum_{j=1}^{m} (\hat{\omega}_{jt} - \omega_{jt})^2$$
$$+ \sum_{t=0}^{n} \lambda_t \left\{ \sum_{j=1}^{m} \hat{\omega}_{jt} - 1 - d_{mn}\delta_t \right\} + \sum_{j=1}^{m} \epsilon_j (\sum_{t=0}^{n} \hat{\omega}_{jt} - 1), \qquad (12)$$

where $d_{mn} \triangleq m - n - 1$.

We now look for a dynamical system of ordinary differential equations. The state of this system is defined by $\hat{\omega} = \{\hat{\omega}_{jt}\}$ and the energy equation L is continuously differentiable with respect to $\hat{\omega}_{jt}, (j = 1, \cdots, m$ and $t = 0, \cdots, n)$. Since we are dealing with a continuous state problem, it is logical to use the Lagrange multipliers in the differential approach:

$$\begin{cases} \frac{d\hat{\omega}}{dt} = -\eta(\hat{\omega}) \frac{\partial L(\hat{\omega}, \lambda, \epsilon)}{\partial \hat{\omega}} \\ \frac{d\lambda_t}{dt} = \frac{\partial L(\hat{\omega}, \lambda, \epsilon)}{\partial \lambda_t} \\ \frac{d\epsilon_j}{dt} = \frac{\partial L(\hat{\omega}, \lambda, \epsilon)}{\partial \epsilon_t}, \end{cases} \qquad (13)$$

where $\eta(\hat{\omega})$ is a modulation function that ensures that the trajectories of (13) are in a state space contained in Euclidean nm-space. Performing gradient ascent on

λ and ϵ have been shown [16] to be very effective in the resolution of constrained optimization problems.

To find a minimum of this equation by iterative calculations, we can use the gradient descent method :

$$\begin{cases} \hat{\omega}^{n+1} = \hat{\omega}^n - \eta(\hat{\omega})\nabla_{\hat{\omega}}L(\hat{\omega},\lambda,\epsilon)\Delta t, \\ \lambda^{n+1} = \lambda^n + \nabla_{\lambda}L(\hat{\omega},\lambda,\epsilon)\Delta t, \\ \epsilon^{n+1} = \epsilon^n + \nabla_{\epsilon}L(\hat{\omega},\lambda,\epsilon)\Delta t, \end{cases} \quad (14)$$

where $\hat{\omega}^0$, λ^0, and ϵ^0 are initial states, Δt is the unit step size for each iteration, and $\nabla_{\hat{\omega}}$, ∇_{λ}, ∇_{ϵ} are gradients. The trajectory of this dynamical equation is chosen in such a way that the energy $L(\hat{\omega},\lambda,\epsilon)$ decreases steadily along the path. Hence, $L(\hat{\omega},\lambda,\epsilon)$ is a *Lyapunov function* for this dynamical equation. Note that this algorithm converges to a minimum point nearest to the initial state. In general, the gradient search method has the property of converging to one of the local minima depending on the initial states.

We assume that the energy is analytic and also that the energy is bounded below, *i.e.*, $L \geq 0$. A complete form of the relaxation equations are given by

$$\begin{cases} \hat{\omega}_{jt}^{n+1} = \hat{\omega}_{jt}^n - \Delta t\left[\alpha r_{jt}^2(1-\delta_t) + \beta(\hat{\omega}_{jt}^n - \omega_{jt}) + \lambda_t^n + \epsilon_j^n\right], \\ \lambda_t^{n+1} = \lambda_t^n + \Delta t\left[\sum_{j=1}^m \hat{\omega}_{jt}^n - 1 - d_{mn}\delta_t\right], \\ \epsilon_j^{n+1} = \epsilon_j^n + \Delta t\left[\sum_{t=0}^n \hat{\omega}_{jt}^n - 1\right]. \end{cases} \quad (15)$$

This equation can be computed by an array processor. A processing element in this array stores and updates the states by using information coming from nearby processors, together with their previous states. To terminate the iteration, we can define in advance either the maximum number of iterations or a lower bound of the change of $\hat{\omega}$, ϵ and λ in successive steps.

The MAP estimate adaptive data association scheme's computational complexity of basic routine per iteration require $\mathcal{O}(nm)$ computations. When we assume the average iteration number as $\bar{k}$, the total data association calculations require $\mathcal{O}(\bar{k}nm)$ computations. Therefore, even if the tracks and measurements are increased, the required computations are not increasing exponentially. However JPDAF as estimated in [4] requires the computational complexity $\mathcal{O}(2^{nm})$, so its computational complexity increases exponentially depending on the number of tracks and measurements.

4 Experimental Results

In this section, we present some results of the experiments comparing the performance of the proposed MAP estimate adaptive data association(MAPADA) with that of the JPDA [6]. We just used the standard Kalman filter [15] for the estimation part once feasible matrix $\hat{\omega}$ is computed. The performance of the MAPADA is tested in two separate cases in the simulation. In the first case, we consider two crossing and parallel targets for testing the track maintenance

Table 1. Initial Positions and Velocities of 10 targets

Target	Position (km)		Velocity (km/s)	
i	x	y	$\dot{x}$	$\dot{y}$
1	-4.0	1.0	0.2	-0.05
2	-4.0	1.0	0.2	0.05
3	-6.0	-5.0	0.0	0.3
4	-5.5	-5.0	0.0	0.3
5	8.0	-7.0	-0.4	0.0
6	-8.0	-8.0	0.4	0.0
7	-5.0	9.0	0.25	0.0
8	-5.0	8.9	0.25	0.0
9	0.5	-3.0	0.1	0.2
10	9.0	-9.0	0.01	0.2

Table 2. The track performance based on the crossing and parallel targets

Clutter density ($/km^2$)	Position error (km)		Velocity error (km/s)		Track maintenance (%)	
	JPDA	MAPADA	JPDA	MAPADA	JPDA	MAPADA
0.2	0.040	0.043	0.018	0.019	100	100
0.4	0.056	0.059	0.033	0.034	80	80
0.6	0.079	0.061	0.047	0.045	48	63
0.8	0.117	0.063	0.072	0.059	35	53
Average	0.073	0.057	0.043	0.039	66	74

and accuracy in view of clutter density. In the second case, all the targets as listed in Table 1 are used for testing the multi-target tracking performance. The dynamic models for the targets have been used by the Singer model developed in [14]. Target 8 and 9 in Table 1 were given acceleration $20m/sec^2$ and $10m/sec^2$ between 15 and 35 turn period, respectively.

The crossing and parallel targets whose initial parameters are taken from target 1,2,3,and 4, respectively in Table 1 are tested. The rms estimation errors and track maintenance capability from the filtering based on the crossing and parallel targets are listed in Table 2. We note that an obvious trend in the results is making it harder to maintain tracks by increasing the clutter density. We also note that, although we have simulated just two scenarios: parallel and crossing targets, the performance of the MAPADA is almost same as that of the HNPDA in view of both tracking accuracy and maintenance.

Table 3 summarizes the rms position and velocity errors for each target in the second test. From Table 3, we note that MAPADA's track maintenance capability is higher than JPDA, even if the rms error of each track is a little larger than the JPDA. This result comes from the choosing the course weighted distance measure. We also note that the general performance of the MAPADA is almost equivalent to that of JPDA. The MAPADA appears to be a alternative to

Table 3. RMS Errors in the case of ten targets

target i	Position error (km)		Velocity error (km/s)		Track maintenance (%)	
	JPDA	MAPADA	JPDA	MAPADA	JPDA	MAPADA
1	0.039	0.042	0.017	0.018	100	100
2	0.038	0.043	0.021	0.019	100	100
3	0.039	0.042	0.016	0.018	100	100
4	0.044	0.042	0.021	0.018	93	100
5	0.040	0.044	0.016	0.019	100	100
6	0.040	0.042	0.011	0.045	100	100
7	0.040	0.042	0.011	0.045	100	100
8	0.251	0.295	0.072	0.118	65	53
9	0.056	0.052	0.024	0.020	85	93
10	0.040	0.044	0.017	0.018	100	98

the JPDA instead of HNPDA. Also, It could replace the sequential computations required for the JPDA with a parallel scheme. But the difficulty in adjusting the parameters still exist.

5 Conclusion

The purpose of this paper was to explore an adaptive data association method as a tool for applying multi-target tracking. It was shown that it always yields consistent data association, in contrast to the HNPDA, and that these associated data measurements are very effective for multi-target filters. Although the MAPADA finds the convergence recursively, the MAPADA is a general method for solving the data association problems in multi-target tracking. A feature of our algorithm is that it requires only $\mathcal{O}(nm)$ storage, where m is the number of candidate measurement associations and n is the number of trajectories, compared to some branch and bound techniques, where the memory requirements grow exponentially with the number of targets. The experimental results show that the MAPADA outperforms the HNPDA and is almost equivalent to the JPDA in terms of both rms errors and track maintenance rate. This algorithm has several applications and can be effectively used in long range multitarget tracking systems using surveillance radar.

References

1. Alspach D. L.: A Gaussian sum approach to multi-target identification tracking problem, Automatica, Vol. 11 (1975) 285-296
2. Reid D. B.: An algorithm for tracking multiple targets, IEEE Trans. on Automat. Contr., Vol. 24 (1979) 843-854 (J. Basic Eng.,) Vol.82 (1960) 34-45

3. Bar-Shalom Y.: Extension of probabilistic data associatiation filter in multitarget tracking, in Proc. 5th Symp. Nonlinear Estimation Theory and its Application (1974) 16-21
4. Sengupta, D., and Iltis, R. A.: Neural solution to the multitarget tracking data association problem, IEEE Trans. on AES, AES-25 (1989) 96-108
5. Kuczewski R.: Neural network approaches to multitarget tracking, In proceedings of the IEEE ICNN conference (1987)
6. Fortmann T. E., Bar-Shalom Y., and Scheffe M.: Sonar Tracking of Multiple Targets Using Joint Probabilistic Data Association, IEEE J. Oceanic Engineering, Vol. OE-8 (1983) 173-184
7. Fitzgerald R.J.: Development of practical PDA logic for multitarget tracking by microprocessor, In Proceedings of the American Controls Conference, Seattle, Wash. (1986) 889-898
8. Fortmann T. E., Bar-Shalom Y.: Tracking and Data Association, Orland Acdemic Press (1988)
9. Lee Y. W., and Jeong H.: A Neural Network Approach to the Optimal Data Association in Multi-Target Tracking, Proc. of WCNN'95, INNS Press
10. Luenberger D. G.: Linear and Nonlinear Programming, Addition-wesley Publishing Co. (1984)
11. Hiriart-Urruty J. B. and Lemarrecchal C.: Convex Analysis and Minimization Algorithms I, Springer-Verlag (1993)
12. Cichocki A., and Unbenhauen R.: Neural networks for optimization and signal processing, Wiley, New York (1993)
13. Emile Aarts and Jan Korst: Simulated annealing and Boltzmann Machines, Wily, New York (1989)
14. Singer, R.A.: Estimating optimal tracking filter performance for manned maneuvering targets, IEEE Transactions on Aerospace and Electronic Systems, Vol. 6 (1970) 473-483
15. Kalman R.E.: A new approach to linear filtering and prediction problems, Trans. ASME, (J. Basic Eng.,) Vol.82 (1960)
16. Platt J.C., and Barr A.H.: Constrained Differential Optimization, Proceedings of the 1987 IEEE NIPS Conf., Denver (1987)

Demonstration of DNA-Based Semantic Model by Using Parallel Overlap Assembly

Yusei Tsuboi[1], Zuwairie Ibrahim[1,2], and Osamu Ono[1]

[1] Institute of Applied DNA Computing,
Graduate School of Science & Technology, Meiji University,
1-1-1 Higashi-mita, Tama-ku, Kawasaki-shi,
Kanagawa-ken, 214-8571 Japan
Phone: +81-44-934-7289, Fax: +81-44-934-7909
{tsuboi, zuwairie, ono}@isc.meiji.ac.jp
http://www.isc.meiji.ac.jp/~i3erabc/IADC.htm
[2] Faculty of Electrical Engineering, Universiti Teknologi Malaysia,
81310 UTM Skudai, Johor Darul Takzim, Malaysia
zuwairie@fke.utm.my

Abstract. In this paper, we propose a novel approach to DNA computing-inspired semantic model. The model is theoretically proposed and constructed with DNA molecules. The preliminary experiment on construction of the small test model was successfully done by using very simple techniques: parallel overlap assembly (POA) method, polymerase chain reaction (PCR), and gel electrophoresis. This model, referred to as *'semantic model based on molecular computing'* (SMC) has the structure of a graph formed by the set of all (attribute, attribute values) pairs contained in the set of represented objects, plus a tag node for each object. Each path in the network, from an initial object-representing tag node to a terminal node represents the object named on the tag. Input of a set of input strands will result in the formation of object-representing dsDNAs via parallel self-assembly, from encoded ssDNAs representing both attributes and attribute values (nodes), as directed by ssDNA splinting strands representing relations (edges) in the network. The proposed model is very suitable for knowledge representation in order to store vast amount of information with high density.

1 Introduction

In 1994, L. Adleman's [7] ground-breaking work demonstrated the way to use DNA molecules for computational purposes. This experience also contributed into a better understanding where to go with DNA machines, namely, to try to develop memory machines that are machines with very large memory that implements rather simple search operations. Such computation with DNA is, what is called, DNA computing or biomolecular computing. DNA computing has showed its potential by solving several mathematical problems, such as a graph and satisfiability problems [7, 11]. It appears, however, that current technology is not capable of the level of control of biomolecules that is required for large, complex computations [8]. One of the DNA's most attractive applications is a memory unit, using its advantages such as the vast parallelism, exceptional energy efficiency and extraordinary information density. Baum [1] proposed an idea of constructing DNA-based memory, and then some experimental work on this has been reported. In their reports, instead of encoding data into {0, 1} bit in case of conventional memory, the data is encoded as {A, T, C, G} base sequences then it is stored in DNA strands. Reif [5] reported these characteristics of DNA-based computing as follows. The extreme compactness of DNA as data storage is nothing short of incredible. Since a mole contains 6.02×10^{23} DNA base monomers, and the mean molecular weight of a monomer is approximately 350 grams/mole, then 1 gram of DNA contains 2.1×10^{21} DNA bases. Since there are 4 DNA bases can encode 2 bits, and it follows that 1 gram of DNA can store approximately 4.2×10^{21} bits. In contrast, conventional storage technologies can store at most roughly 10^9 bits per gram, so DNA has the potential of storing data on the order of 10^{12} more compactly than conventional storage technologies. However, they never sufficiently compensate for the way to arrange knowledge information for representing an object in DNA strands, which makes it difficult for human to understand any information instinctively.

Semantic networks are graphic notations for representing knowledge in patterns of interconnected nodes and edges. Computer implementations of semantic networks were first developed for artificial intelligence (AI) and machine translation, but earlier versions have long been used in philosophy, psychology, and linguistics. Brain information processing often involves comparing concepts. There are various ways of assessing concept similarity, which vary depending on adopted models of knowledge representation. In featural representations, concepts are represented by sets of features. In Quillian's model of semantic memory [10], concepts are represented by the relationship name via links. Links are labeled by the name of the relationship and are assigned criteriality tags that attest to the importance of link. In artificial computer implementations, criteriality tags are numerical values the represent the degree of association between concept pairs (i.e., how often the link is traversed), and the nature of the association. There are many variations of semantic networks, but all can be categorize into either one of the six categories: definitional, assertional, implicational, executable, learning, and hybrid [3]. The detail of such categories is not covered in this paper. The point of our study is in the structure of semantic networks.

In semantic networks, a basic structure of relations between two objects is described with nodes, directed edges and labels as shown in Figure 1. The nodes denote "object" and the directed edge denotes the relations. Semantic network is a set of objects described by such structures. The nodes and edges are changeable under the situations of various representations. It is easy for human to intuitively understand meanings of the object with network structures. The graph of the semantic network potentially realizes reasonable knowledge representation.

It is considered that one method of approaching a memory with power near to that of human is to construct a semantic network model based on molecular computing. However it is difficult to apply existing semantic networks when we build DNA based semantic memory because duplex DNA structures forced them to limit knowledge representation abilities. It seems a new semantic model like a network will be required for DNA computing approach.

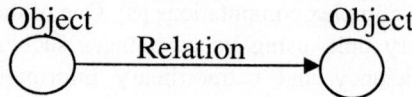

Fig. 1. A standard structure of a semantic network: the nodes denote object and the directed edge denote relation between the two objects

In this paper, we propose a novel semantic model derived from existing semantic networks for implementation with DNA. A set of semantic information is stored in single-stranded (ss) or double-stranded (ds) DNAs. These strands will form linear dsDNA by using parallel overlap assembly (POA) method. POA has some advantages over the hybridization/ligation method by Adleman [7] in terms of generation speed and material consumption [6]. At the POA process, DNA strands were overlapped during an annealing step in the assembly process while the remaining parts of the DNA strands were extended by dNTP incorporation by polymerase chain reaction (PCR) in that it repeats the denaturation, annealing, and extension.

2 Semantic Model Based on Molecular Computing

2.1 Model

A number of researches have engaged in solving simple problems on semantic networks using AI technique. Therefore it was conceivable that early semantic network was done using papers and pencils. When researchers realize the need to represent bigger problem domain, computers software were designed to overcome the limitation inherent in the conventional methods. These graphical notions were ultimately transferred on to computers using software specific data structures to represent nodes and edges. We produce another way by using DNA-based computers to implement such semantic networks.

In our consideration, however, it is almost impossible to implement existing semantic networks by using DNA computing techniques, because DNA's duplex structure limits representations of them heavily. Thus, instead of using the standard existing semantic network described in the previous section, we create a new semantic model such a network in order to enable to use the DNA computing techniques. Such a model has to be created to maintain the concept of an existing semantic network as much as possible. The way to make the semantic model is described as follows.

First, a tag as a name of an object is set to an initial node in the graph. After we determine the number and the kinds of the attribute of the object, both the attribute and attribute value are sequentially set to another node following by the tag node. Second, a directed edge is connected between (attribute, attribute value) pair nodes. Figure 2 shows a basic structure of this. It is imperative to transform complicated graphs into simpler ones. An AND/OR graph enables the reduction of graph size, and facilitates easy understanding. The relation between the nodes and edges is represented using a new defined AND/OR graph. A directed edge in the terminal direction is sequentially connected between the nodes except for the following case (AND). If there are two nodes which have same attributes but different attribute values, each of directive edges

Attribute, Attribute Value   Attribute, Attribute Value

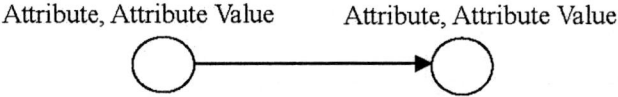

Fig. 2. A basic structure of the semantic model proposed: the nodes except for the tag node denote both attributes and attribute values. The directed edge merely is connected between the two nodes.

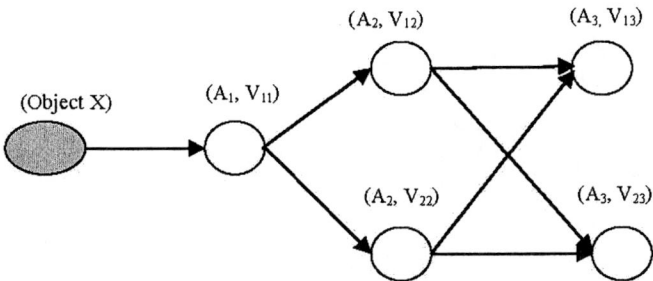

Fig. 3. A simple object model of X; the three determind attributes are A_1, A_2, and A_3

is connected in parallel (OR). Each edge denotes only connection between the nodes in the directive graph. Finally, labels are attached to the nodes, such as '(Tag: O)' and '(Attribute: A, Attribute Value: V)'.

The nodes denote either a name of the object or both the attribute and the attribute values. In short, one path from an initial node to a terminal node means one object named on the tag. We define this graph as a knowledge representation model. The model represents an object, as reasoned out by the combinations between the nodes connected by the edges. For example, Figure 3 illustrates this object representation in the context of object X (named via the tag). An overall graph is then formed by the union of a set of such basic objects, each of which is described in similar, simple fashion. Figure 4 shows an example of such a network. We name such a graph a *semantic model based on molecular computing* (SMC). An SMC contains all attributes common to every object as well as each attribute value. Attribute layers consist of attribute values, lined up. If an object has no value of a certain attribute, the attribute value is assigned '*no value*', such a question of 'What is water's shape?' An object is expressed by the list representation style as follows,

$$\{<O, A_i, V_{ji}> | i=1, 2,..., m; j=1,2,..., n\}$$

Although an attribute generally corresponds to an attribute value with one to one, in the SMC the attribute is allowed to have one or more attribute values.

For example, object X such as Figure 3 is,

<Object X, A_1, V_{11}>
< Object X, A_2, V_{12}>
< Object X, A_2, V_{22}>
< Object X, A_3, V_{13}>
<Object X, A_3, V_{23}>

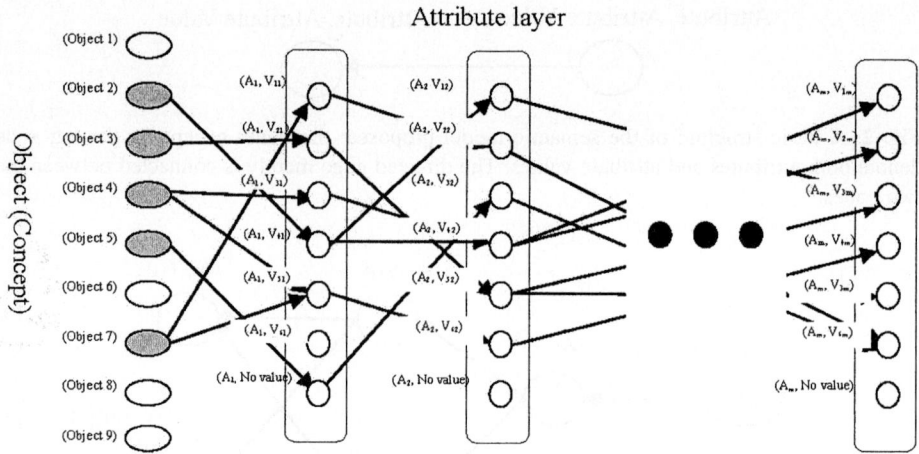

Fig. 4. A semantic model based on molecular computing (SMC), which collectively models a set of objects, given a total number of attribute layers, *m*.

A_2 has two attribute values, V_{12} and V_{22}. And also, A_3 has two attribute values, V_{13} and V_{23}.

2.2 DNA Representation

Each of the nodes and edges within an SMC may be represented by a DNA strand, as follows. First, each node is mapped onto a unique, ssDNA oligonucleotide, in a DNA library of strands. In the DNA library, a row shows attributes, a column shows attribute values and each DNA sequence is designed. The DNA sequences are assigned by 20 oligonucleotides. Here, an important thing is that every sequence is designed according to these relations to prevent mishybiridization via other unmatching sequences. The sequences used are designed by DNA Sequence Generator [2] that is a program for the design of DNA sequences useful for DNA computing, nanotechnology and the design of DNA Arrays. Second, each edge is mapped by following the Adleman [7] scheme. Finally, the overall strands are respectively represented by the size which suits the end of the DNA pieces of the initial or the terminal exactly. In this way, the semantic model of each object is represented by DNA strands. Figure 5 shows one of paths shown for object X model in Figure 3, as represented by a set of ssDNA ((Object X)→(A_1, V_{11})→(A_2, V_{12})→(A_3, V_{13})).

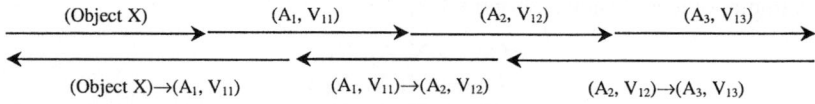

Fig. 5. One of the DNA paths of the graph (Object X) in Figure 3. The arrowhead indicates the 3' hydroxyl end.

3 Experimental Construction with POA

The SMC is theoretically described in the section 2. It is essential to demonstrate SMC with DNA molecules to experimentally verify our theory. The target model is Figure 3, a small SMC graph. Every path from the initial node to the terminal node in this model

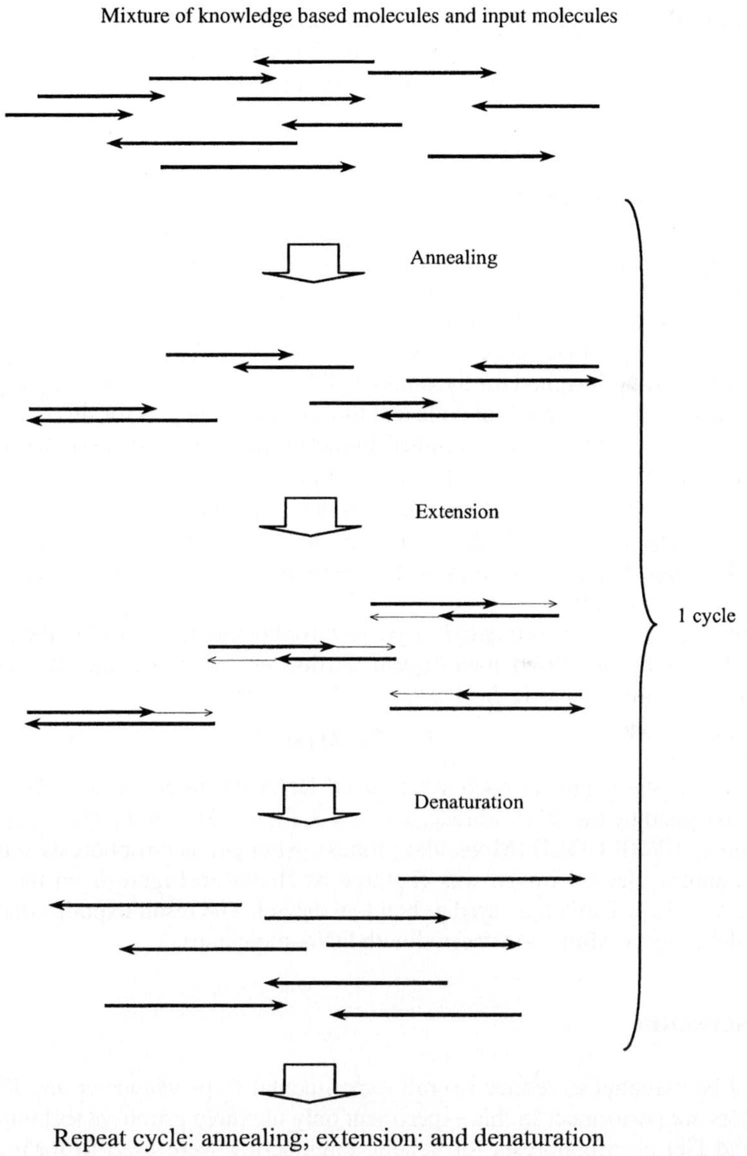

Fig. 6. Schematic diagram of POA steps. The thick arrows denote the ssDNA which participate in each step of the reaction. The thin arrows denote elongated part by dNTPs incorporation.

is represented by a set of ssDNA. The ssDNAs of tag node and the edges within the set are synthesized as *knowledge based molecules*. The ssDNAs representing label (A_1, V_{11}), (A_2, V_{12}), and (A_3, V_{13}) is synthesized as an *input molecule* to be combined with the DNA library.

Knowledge based molecules are first inserted into a test tube, followed by addition of the input molecules, were mixed. The schematic diagram of POA steps is shown in Figure 6. The mixture contained each 1μl DNA oligonucletoides, 67.5 μl distilled water, 10μl dNTP (TOYOBO, Japan), 10μl 10×KOD dash buffer (TOYOBO, Japan), 0.5μl KOD dash (TOYOBO, Japan). The total reaction volumes was 99μl .The POA was processed for 25 cycle at 94 for 30 seconds, at 55 for 30 seconds, and at 74 seconds for 10 seconds.

4 Experimental Result

POA product in the test tube was checked to confirm whether our necessary dsDNAs exist, with two methods, PCR and Gel electrophoresis. The only DNAs representing the test model are efficiently amplified with PCR to surely obtain correct dsDNA length for a next operation, gel electrophoresis. Although the PCR needs two primers of sequences properly designed for the template DNAs, it is possible to design them by the sequences assigned by the initial node and the terminal node. As for this experiment the sequences used as primers are designed based on the sequences assigned by the two nodes, (object X) and (A_3, V_{13}). The PCR is performed in a 25μl solution including 13.875μl distilled water, 2.5μl for each primers, 1μl template, 2.5μl 10×KOD dash buffer (TOYOBO, Japan), 0.125μl KOD dash (TOYOBO, Japan), and 2.5μl dNTP (TOYOBO, Japan) for 25 cycle at 94 for 30 seconds, at 55 for 30 seconds, and at 74 seconds for 10 seconds.

The PCR products are subjected to gel electrophoresis to determine the presence of the correct dsDNAs, which then appear as discrete bands on the gel. The correct dsDNA length, denoted as L_S, is given by the simple relation:

$$L\_S = L\_D\ (m+1)$$

where L_D is the length of each synthesized DNA fragment. Now L_D is length of 20-mer assigned by the DNA library, and $m = 3$, then L_D is 80 bp (base pair). The gel is stained by SYBR GOLD (Molecular probes). After gel electrophoresis was executed for 35 minutes, the gel image was captured as shown in Figure 6. In the lane 1 the dsDNAs of 80 bp clearly appeared as band on the gel. The result explains that the small test model is successfully constructed with DNA molecules.

5 Discussion

It would be essential to reduce overall experimental steps whenever any DNA-based computers are performed. In this experiment only the three primitive techniques: POA, PCR, and Gel electrophoresis for genetic engineering were used to form and check necessary dsDNAs. It is very simple experiment for implementation of DNA-based model. The total of the experiment time is approximately 3 hours.

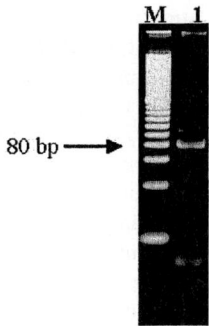

Fig. 7. The result of the gel electrophoresis on 10% polyacrylamide gel. The POA product amplified by PCR is in lane 1. Lane M denotes 20 bp ladder.

In AI research fields, some work has been focusing on how to arrange and store a set of knowledge in memories. By semantic models composed of nodes and edges, not logical rules, a certain object is administered by a tag node linking (attribute, attribute value) pair nodes. It might have difficulties treating complicated representation only by (attribute, attribute value) pairs. However, the proposed model has much advantage over the logical rule approaches. By picking the nodes up, it is possible to reach for the object at a stretch. Human can understand and recognize the context of the object smoothly compared with other models [1, 4, 5, 9] when he sees the model. If individual objects have common (attribute, attribute value) pair nodes, they can share them, which enable to avoid the synthesis of such overlap sequences. Moreover, knowledge representation abilities of this model are very higher, because the SMC is represented by both attributes and attribute values with a concept of AND/OR graph. Therefore, we believe that the proposed one is very suitable for semantic memories with DNA molecules. In the future work, the optimization of some parameters, such as reaction-temperature, oligonucleotide concentrations, reaction time, etc. will be experimentally tested to achieve reliable performance, when a large complex model is tread. The proposed model will appears as an interaction between AI and biomolecular computing research fields, and will be further extend for several AI applications.

Acknowledgement

The research was financially supported by the Sasakawa Scientific Research Grant from The Japan Science Society. The second author is very thankful to Universiti Teknologi Malaysia for a study leave.

References

1. Baum, E. B.: How to Build an Associative Memory Vastly Larger than the Brain, Science 268 (1995), pp.583-585
2. Udo, F., Sam, S., Wolfgang, B., and Hilmar, R.: DNA Sequence Generator, A program for the Construction of DNA Sequences. In N. Jonoska and N. C. Seeman (editors), Proc. of the Seventh International Workshop on DNA Based Computers, pp.21-32

3. Sowa, J. F.: Semantic Networks, http://www.jfsowa.com/pubs/semnet.htm (2005)
4. Rief, J. H.: Parallel Molecular Computation, Models and Simulations, Proc. of the seventh Annual Symposium on Parallel Algorithms and Architectures (1995), pp.213-223
5. Reif, J. H., LaBean, H. T., Pirrung, M., Rana, V. S., Guo, B., Kingsford, C., Wickham, G. S.: Experimental Construction of Very Large Scale DNA Databases with Sssociative Search Capability, The tenth International Workshop on DNA Based Computers, Revised Papers, Lecture Notes in Computer Science 2943 (2002), pp.231-247
6. Lee, J.Y., Lim, H-W., Yoo, S-I., Zhang, B-T., and Park, T-H.: Efficient Initial Pool Generation for Weighted Graph Problems Using Parallel Overlap Assembly, Preliminary Proc. of the Tenth International Meeting on DNA Based Computers (2004), pp.357-364
7. Adleman, L.M.: Molecular Computation of Solutions to Combinatorial Problems, Science 266 (1994), pp.583-585
8. Adleman, L.M.: DNA computing FAQ., http://www.usc/dept/melecular-science/ (2004)
9. Arita, M., Hagiya, M. and Suyama, A.: Joining and Rotating Data with Molecules, Proc. of IEEE International Conference on Evolutionary Computation (1997), pp.243-248
10. Quillian, M.R. and Minsky, M.: Semantic Memory, Semantic Information Processing, MIT Press, Cambridge, MA. (1968), pp.216-270
11. Lipton, R.J.: DNA Solution of Hard Computation Problems, Science 268 (1995), pp.542-545

Multi-objective Particle Swarm Optimization Based on Minimal Particle Angle

Dun-Wei Gong, Yong Zhang, and Jian-Hua Zhang

School of Information and Electrical Engineering,
China University of Mining and Technology,
Xuzhou, Jiangsu, 221008, P. R. China
dwgong@vip.163.com

Abstract. Particle swarm optimization is a computational intelligence method of solving the multiobjective optimization problems. But for a given particle, there is no effective way to select its globally optimal particle and locally optimal particle. The particle angle is defined by the particle's objective vector. The globally optimal particle is selected according to the minimal particle angle. Updating the locally optimal particle and particle swarm is based on the Pareto dominance relationship between the locally optimal particle and the offspring particles and the particle's density. A multiobjective particle swarm optimization based on the minimal particle angle is proposed. The algorithm proposed is compared with sigma method ,NSPSO method and NSGA- II method on four complicated benchmark multiobjective function optimization problems. It is shown from the results that the Pareto front obtained with the algorithm proposed in this paper has good distribution, approach and extension properties.

1 Introduction

Particle swarm optimization (PSO) is a population-based stochastic optimization technique proposed by Kennedy et al which imitates the social behavior of a bird flock flying around and sitting down on a pylon [1]. PSO was initially proposed to deal with single objective optimization problems. The traditional PSO is not suitable for multi-objective optimization problems whose objective functions may conflict with each other and so it needs an improvement in the following two aspects: selecting the globally optimal particle and selecting the locally optimal particle.

On selecting the globally optimal particle, Hu presented a multi-objective particle swarm optimization in which dynamic neighborhood strategy is adopted to select the globally optimal particle [2]. Selecting globally optimal particles depends on just one of the objectives in nature. Coello presented a method based on dividing the objective space [3], which is a variant of the roulette wheel selection in nature. Fieldsend et al applied the archive to selecting the globally optimal particle, which selects the globally optimal particle from the archive by comparing with the composite points in a structure tree [4]. Although this method can find the globally optimal particle favorable for particle evolution, it may make most particles only point to some special ones of the archive, hence producing local convergence. Mostaghim et al put forward a sigma method to select the globally optimal particle[5]. Although the method can avoid the deficiencies existed in the method proposed by Fieldsend et al, there still exist the following deficiencies: (1) It requires all the values of the objective functions to be positive. (2) The convergence property of the method is not good when it is applied in optimizing functions with 3 objectives or more.

Multi-objective particle swarm optimization selects the locally optimal particle via simple comparison [1]. However in the method, when the new generated particle may be better than other particles or other locally optimal particles, it is easy to lose the obtained optimal information, thereby resulting in low evolutionary efficiency. In NSPSO[6],Li presented a new method of selecting the locally optimal particle which selects among the locally optimal particles and the offspring particles of the N particles. The method avoids the deficiency of the traditional selection operators effectively, however there also exists a deficiency: the computation is very complicated.

In this paper the particle angle is defined and the globally optimal particle is chosen according to the minimal particle angle. Updating the locally optimal particle and particle swarm is determined according to the Pareto dominance relationship between the locally optimal particle and the offspring particles and the particle's density. A multi-objective particle swarm optimization based on the minimal particle angle is proposed. Finally the algorithm proposed in this paper is compared with sigma method ,NSPSO method and NSGA-II method on benchmark functions

In the remainder of the paper, particle swarm optimization is simply introduced in section 2. In section 3, the particle angle is defined and a novel method of selecting the globally optimal particle based on the above definition is put forward. In section 4, the particle's density are defined and subsequently a novel strategy for updating the locally optimal particle based on the above definitions is proposed. In section 5, a multi-objective particle swarm optimization based on the minimal particle angle is presented. In section 6, the algorithm proposed in this paper is compared with other three methods. Finally, the conclusions for this paper are drawn.

2 Particle Swarm Optimization

Particle swarm optimization searches for the solution space by using a particle swarm, each particle represents a solution of the optimized problem, and each particle has a velocity corresponding to its flying direction and distance. Each particle x_i updates itself by tracking two optima, the one is its locally optimal particle $P_i = (p_{i1}, p_{i2}, \ldots, p_{id}) \in R^d$, which is the best position that x_i has visited so far, the other is the globally optimal particle $P_g = (p_{g1}, p_{g2}, \ldots, p_{gd}) \in R^d$, which is the best position that the whole swarm has visited so far. Each particle updates itself according to the following formula:

$$v_{ij}(t+1) = wv_{ij}(t) + c_1 r_1 (p_{ij}(t) - x_{ij}(t)) + c_2 r_2 (p_{gj}(t) - x_{ij}(t)) \tag{1}$$
$$x_{ij}(t+1) = x_{ij}(t) + v_{ij}(t+1)$$

where $j=1,2\ldots\ldots d$, $i=1,2,\ldots,N$, N is the swarm size, c_1,c_2 are two positive constants, r_1,r_2 are two random numbers within $[0,1]$, w is an inertia weight to control exploration in the search space, $v_{ij} \in [-v_{max}\ v_{max}]$, where v_{max} is set by the user.

3 Selecting Globally Optimal Particle

In order to give how to select the globally optimal particle based on the minimal particle angle, the concept of the particle angle is firstly defined.

Considering two particles x_i, x_j in a swarm, define

$$\delta_{ij} = \delta(x_i,x_j) = \arccos \frac{f(x_i) \cdot f(x_j)}{|f(x_i)| \cdot |f(x_j)|} \quad (2)$$

as the particle angle between x_i and x_j in objective space, namely, the particle angle for short, where $f(x_i)$ and $f(x_j)$ are vector values of the objective function of x_i and x_j.

The following is the method of selecting the globally optimal particle of x_i from archive A. Firstly compute the particle angle $\delta(x_i,a_j)$ between x_i and particle a_j in A, where $j=1,2,\ldots,|A|$ and $|A|$ is the number of the elements in A. Secondly look for a particle a_k in A satisfying $\delta(x_i,a_k) = \min_{j \in \{1,2,\ldots,|A|\}} \delta(x_i,a_j)$. Then a_k is the globally optimal particle of x_i. The geometric explanation is shown as Fig. 1. It can be seen from Fig. 1 that selecting the globally optimal particle of x_i based on the minimal particle angle actually selects the particle such that the particle angle between it and x_i is minimal.

It can be seen from formula (2) that the method doesn't require the objective vectors to be positive and the computation is more direct.

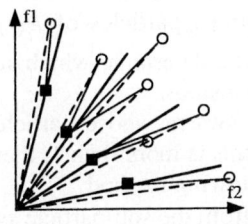

Fig. 1. Geometric explanation of selecting globally optimal particle, where ■ stands for one particle in A; ○ stands for one particle in particle swarm

4 Updating Particle and Its Locally Optimal Particle

4.1 Sub-particle Swarm and Particle's Density

According to the above method of selecting the globally optimal particle based on the minimal particle angle, each particle in the swarm can find its globally optimal particle in A. In other words, each particle in A can find a certain number of corresponding particles from the particle swarm. If it is not, let the number of corresponding particles be zero.

For an arbitrary $a_j \in A$, $j = 1, 2, \ldots, |A|$, the set composed of the particles whose globally optimal particle is a_j is called the sub-particle swarm of a_j

The number of particles in a sub-particle swarm is called the particle's density of the sub-particle swarm or called the particle's density for short.

According to the above definition, the number of the sub-particle swarm is $|A|$. It is known from the definition of the particle's density that the more particles a sub-particle swarm has, the bigger the particle's density is and vice versa. The maximum and the minimum of the particle's density are N and 0 respectively.

4.2 Updating the Particle and Its Locally Optimal Particle

The methodology of updating the particle and its locally optimal particle is as follows. If there exists the Pareto dominance relationship between a particle's locally optimal particle and its offspring particle, then the particle's locally optimal particle is determined according to the Pareto dominance relationship, otherwise update the particle and its locally optimal particle according to the Pareto dominance relationship and the particle's density.

Now let's discuss the following two cases according to the above methodology.

The first case is that the locally optimal particle $P_i(t)$ of particle $x_i(t)$ is non-dominated to its offspring particle $x_i(t+1)$ and vice versa.

Step 1 Save $P_i(t)$ to the particle swarm and initialize the locally optimal particles of $P_i(t)$ and $x_i(t+1)$ as themselves.

Step 2 Find the globally optimal particles of $P_i(t)$ and $x_i(t+1)$ according to the method mentioned in section 3 and determine which sub-particle swarm they belong to respectively and update their densities.

Step 3 Find the sub-particle swarm whose particle's density is maximal. If the number of such sub-particle swarms is more than 1, then randomly select one of them and update it according to the following method.

Step3.1 Select a particle $x_r(t)$ from the sub-particle swarm randomly.

Step3.2 Compare $x_r(t)$'s locally optimal particle $P_r(t)$ with other particles' locally optimal particles, if $P_r(t)$ dominates to the latter, then delete the latter and its corresponding particles, otherwise delete $x_r(t), P_r(t)$.

The second case is that there exists the Pareto dominance relationship between $P_i(t)$ and $x_i(t+1)$.

If $P_i(t)$ dominates to $x_i(t+1)$, then still select $P_i(t)$ as $x_i(t)$'s locally optimal particle, otherwise select $x_i(t+1)$ as $x_i(t)$'s locally optimal particle.

4.3 Further Explanation

Further explanation for the above strategy is given in this subsection. The process of updating the particle and its locally optimal particle according to the above strategy is depicted as Fig. 2.

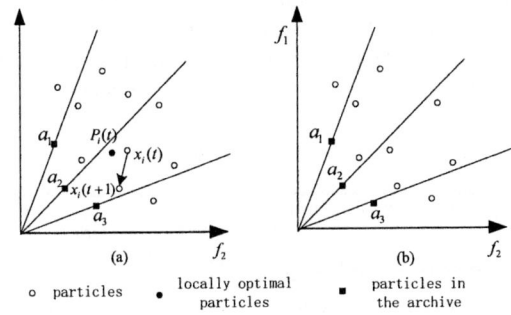

Fig. 2. Updating particle and its locally optimal particle when $P_i(t)$ and $x_i(t+1)$ are not dominated to each other

When $P_i(t)$ and $x_i(t+1)$ are not dominated to each other, it can be seen from figure 2(a) that the number of particles whose globally optimal particles are a_1, a_2 and a_3 is 3,4 and 2, respectively, hence the densities of the sub-particle swarms determined by a_1, a_2 and a_3 are 3,4 and 2, respectively. According to the above strategy for updating the particle and its locally optimal particle, $P_i(t)$ and $x_i(t+1)$ are saved to the particle swarm and the density of each sub-particle swarm is updated. Consequently the densities of the sub-particle swarms determined by a_1, a_2 and a_3 are changed into 3,4 and 3, respectively. The sub-particle swarm whose particle's density is maximal is chosen, which is the sub-particle swarm determined by a_2, from which the worst particle as well as its locally optimal particle are deleted. Finally the densities of the sub-particle swarms determined by a_1, a_2 and a_3 become 3,3 and 3, respectively, which is shown as the figure 2(b).

From the above process of updating the particle and its locally optimal particle, it can be seen that the strategy mentioned in subsection 4.2 makes the particles have good distribution in objective space, sequentially obtaining the Pareto front with good distribution. Furthermore, the better offspring particles have more chances to be conserved by deleting bad particles and their locally optimal particles, which makes the particle swarm approach to the Pareto front further, hence improving the quality of the Pareto optimal solutions.

For a particle swarm whose swarm size is N, the archive size is |A|, O(|A|+mN) computations are required for updating each particle when its locally optimal particle and its offspring particle are not dominated to each other. In the worst case when each particle's locally optimal particle and its offspring particle are not dominated to each other, the computational complexity of updating the particle swarm is O(|A|N+mN^2).

5 Multi-objective Particle Swarm Based on Minimal Particle Angle

A novel multi-objective particle swarm optimization, namely, multi-objective particle swarm optimization based on the minimal particle angle is proposed when the method of selecting the globally optimal particle given in section3 and the strategy for updating the particle and its locally optimal particle given in section 4 are applied to multi-objective particle swarm optimization, whose steps are described as follows.

Table 1. Test functions

Test functions	Description of objective functions	Number of variables	Variables' range
function 1 (ZDT1)	$g(x_2,x_3,\ldots,x_n)=1+9(\sum_{i=2}^{n}x_i)/5$ $h(f_1,g)=1-\sqrt{f_1/g(x)}$ $f_1(x_1)=x_1$ $f_2(x)=g(x_2,x_3,\ldots,x_n)\cdot h(f_1,g)$	30	[0,1]
function 2 (ZDT2)	$g(x_2,x_3,\ldots,x_n)=1+9(\sum_{i=2}^{n}x_i)/5$ $h(f_1,g)=1-(f_1/g(x))^2$ $f_1(x_1)=x_1$ $f_2(x)=g(x_2,x_3,\ldots,x_n)\cdot h(f_1,g)$	30	[0,1]
function 3 (ZDT3)	$g(x)=1+9(\sum_{i=2}^{n}x_i)/(n-1)$ $f_1(x)=x_1$ $f_2(x)=g(x)[1-\sqrt{\frac{x_1}{g(x)}}-\frac{x_1}{g(x)}\sin(10\pi x_1)]$	30	[0,1]
function 4 (DTLZ2)	$f_1(x)=(1+g(x))\cos(\pi x_1/2)\cos(\pi x_2/2)$ $f_2(x)=(1+g(x))\cos(\pi x_1/2)\sin(\pi x_2/2)$ $f_3(x)=(1+g(x))\sin(\pi x_1/2)$ $g(x)=\sum_{i=3}^{12}(x_i-0.5)^2$	12	[0,1]

Step1 Initialize the position, velocity and locally optimal particle of each particle in the particle swarm and the archive.

Step2 Update the archive by comparing the dominance relationship between each pair of the locally optimal particles.

Step3 Select the globally optimal particle of each particle according to the method given in section 3.

Step4 Generate offspring particles via formula (1).

Step5 Update the locally optimal particles and the particle swarm according to the method given in section 4.

Step6 Judge whether the termination condition is satisfied or not, if yes, stop the algorithm, otherwise go to step 2.

Notation: In the above method, the crowded degree method proposed by Deb et al is applied to update the archive in order to guarantee the particles in the archive having good distribution and extension properties [7].

Table 2. Parameters Setting

Parameters	Values of parameters
N	100 in function1, 2 and 3, 200 in function4
T_{max}	100 in function1, 2,3 and 4
w	0.4

6 Experiments and Analysis

6.1 Test Functions and Parameter Settings

The test functions from literature[8],[9] listed in table 1 are used to test the performance of the algorithm, the values of the pertinent parameters are set as table 2.

6.2 Experimental Results and Comparison

The algorithm proposed in this paper, for short MAPSO method, is compared with sigma method, NSPSO method and NSGA-II method on optimizing the functions listed in Table. 1. For equitable comparison, NSPSO method and sigma method are set the same parameter as MAPSO method. However, for NSGA-II method, population size and running generations are set the same as MAPSO method, the crossover probability is set to 0.8 and the mutation probability is set to 0.05.

Fig 3 and 4 show the results of four methods optimizing the test functions ZDT1 and ZDT2. It is shown from the figures that MAPSO method, sigma method and NSPSO method have better convergence than NSGA-II method. For sigma method, the strategy of updating the particle and its locally optimal particle reduces the computational complexity, but its results' diversity is not better than MAPSO method and NSPSO method. For NSPSO method, MAPSO method isn't able to get better convergence than NSPSO method, but the diversity of its results is slightly better than that of NSPSO method and its computation ($O(|A|N+mN^2)$) is less complicated than NSPSO method($O(mN^3)$). Although the C metric[10] is not a proper quantitative metric for comparing [11],we try to use this metric to make Figure 3 and 4 more clear . The average C-metric of 30 times runs is listed in table 3,where M,S,NP and NG stand for MAPSO method, sigma method, NSPSO method and NSGA-II method. It can be seen from table 3 that the solutions obtained with MAPSO method is good.

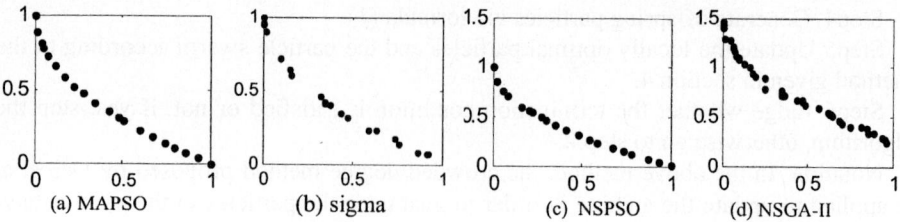

Fig. 3. Results of different method optimizing ZDT1

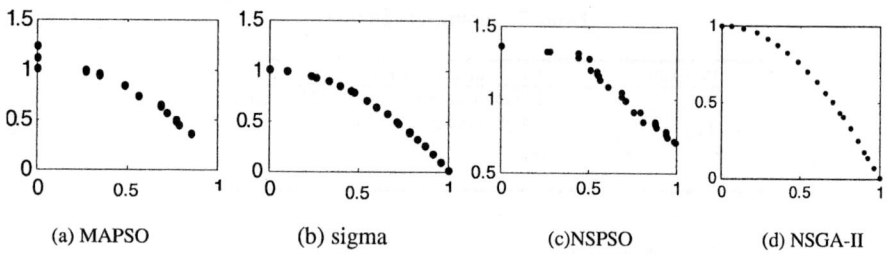

Fig. 4. Results of different method optimizing ZDT2

Table 3. Average C-metric of four methods on ZDT1 and ZDT2

Functions	C(M,S) (C(S,M))	C(M,NP)(C(NP,M))	C(M,NG) (C(NG,M))	C(NS,S) (C(S,NS))
ZDT1	0.2500 (0.0000)	0.2000 (0.0000)	1.0000 (0.0000)	0.1500 (0.0750)
ZDT2	0.9000 (0.00000)	0.0500 (0.0000)	1.0000 (0.0000)	0.8750 (0.0000)

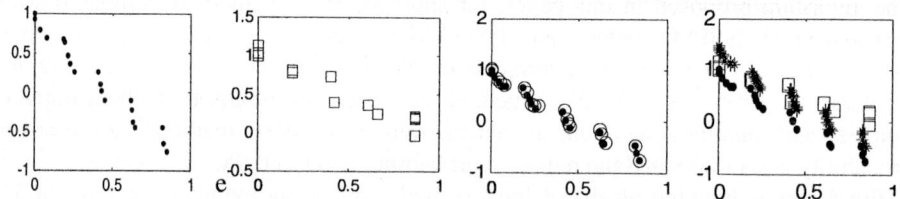

Fig. 5. Results of different method optimizing ZDT3, where·, □, ○, * stand for MAPSO method, sigma method, NSPSO method and NSGA-II.

Fig 5 shows the results of function ZDT3. It can be seen from Fig 5 that MAPSO method and NSPSO method have better convergence than the other two methods. However the faults of sigma method is that it can't deal with objective functions of positive value, make it only obtain several solutions which have negative y-axis values, and effect its convergence. NSGA-II method is able to get solutions with better diversity, but its convergence is worse than MAPSO method and NSPSO method.

Fig 6 shows the results of optimizing DTLZ2. It is shown from the figure, MAPSO method and NSPSO method have better diversity than the other two methods. For DTLZ2, $f_1^2(x_i) + f_2^2(x_i) + f_3^2(x_i) = 1$ means that particles x_i is on the Pareto front. So the performance in convergence of the methods is evaluated as follows:

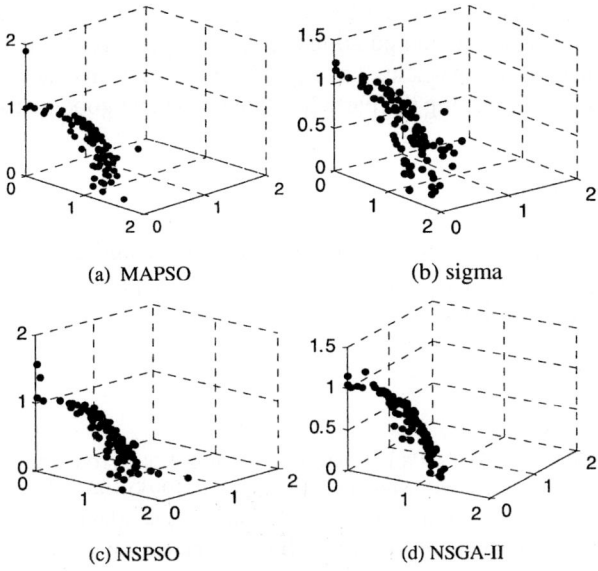

Fig. 6. Results of different methods optimizing DTLZ2

Table 4. Number of particles in different *err* values of four methods

Err	0	0.1	0.2	0.3	0.4	0.5	1.0
MAPSO	0	43	79	85	89	93	97
Sigma	0	12	41	71	76	80	91
NSPSO	0	39	73	87	89	93	95
NSGA2	0	35	70	81	91	92	99

$$err = 1 - f_1^2(x_i) + f_2^2(x_i) + f_3^2(x_i), \forall x_i \in A \qquad (3)$$

The particles which have an $err = 0$ are on the Pareto optimal front. Table 4 lists the number of particles in different *err* values of four methods. When $err = 0$, four methods can not find the corresponding particles, namely, all the four methods can't find Pareto optimal solutions. However when $err < 0.1$, MAPSO method, NSPSO method and NSGA-II method can find more particles than sigma method. In other words, for sigma method, although it take less time than the other three methods in the same individual evaluation, its convergence is worse than the other three methods.

7 Conclusion

In this paper ,a multi-objective particle swarm optimization based on the minimal particle angle is proposed. The performance of the algorithm proposed in this paper is compared with the other three methods. It can be shown from the results that: (1)Selecting the globally optimal particle based on the minimal particle angle not only avoids the deficiency of concentratively selecting the globally optimal particle, but also the value of the particle's objective function is not required to be positive.

(2) Updating the particle and its locally optimal particle proposed in this paper not only reduces the algorithms computational complexity, increases the convergence of solutions, but also makes particles have good distribution and gets the Pareto front with good distribution, making the particles of swarm have good distribution.

Acknowledgments

Project Research on Key Issues of Co-interactive Evolutionary Computation and Their Implementation based on Internet (60304016) supported by National Natural Science Foundation of China.

References

1. Kennedy J., Eberhart R.C.: Particle swarm optimization. Proceedings of IEEE International Conference on Neural Networks, Piscataway (1995) 1942-1948
2. Hu X.: Multi-objective optimization using dynamic neighborhood particle swarm optimization. Proceedings of the 2002 IEEE Congress on Evolutionary Computation, Honolulu (2002)
3. Coello C. C. A., Lechuga M.S.: MOPSO:A proposal for multiple objective particle swarm optimization. Proceedings of the 2002 IEEE Congress on Evolutionary Computation, Honolulu (2002) 1051-1056
4. Fieldsend J., Singh S.: A multi-objective algorithm based upon particle swarm optimization. Proceedings of The U.K. Workshop on Computational Intelligence (2002) 34-44
5. Mostaghim S., Teich J.: Strategies for finding good local guides in multi-objective particle swarm optimization (MOPSO). Proceedings of the 2003 IEEE Swarm Intelligence Symposium, Indianapolis (2003) 26-33
6. LI X. D.: A Non-dominated Sorting Particle Swarm Optimizer for Multi-objective Optimization. Proceedings of Genetic and Evolutionary Computation. Springer (2003) 37-48
7. Deb K., Pratap A.: A fast elitist multi-objective genetic algorithm: NSGA-II.IEEE Transactions on Evolutionary Computation, 6(2) (Apr,2002) 182-197
8. Zitller E., Deb K., Thiele L.: Comparison of multi-objective evolutionary algorithms: empirical results. Evolutionary Computation, 8(2) (Apr,2000) 173-195
9. Deb K., Thiele L., Laumanns M.: Scalable test problems for evolutionary multi-objective optimization. Technical Report 112, Computer Engineering and Networks Laboratory (TIK), Swiss Federal Institute of Technology (ETH), Zurich, Switzerland (2001)
10. Schon J. R.: Fault tolerant design using single and multi-criteria genetic algorithm optimization. Master's thesis. Massachusetts Institute of Technology, Cambridge (1995)
11. Knowles J., Corne D.: On metrics for comparing nondominated sets. IEEE Proceedings, World Congress on Computational Intelligence (CEC2002) 711-716

A Quantum Neural Networks Data Fusion Algorithm and Its Application for Fault Diagnosis*

Daqi Zhu[1], ErKui Chen[1], and Yongqing Yang[2]

[1] Research Centre of Control Science and Engineering, Southern Yangtze University,
214122 Wu Xi, Jiangshu Province, China
zdq367@yahoo.com.cn
http://www.cc.sytu.edu.cn
[2] Department of mathematics, Southern Yangtze University,
214122 Wu Xi, Jiangsu Province, China
yqyang@sytu.edu.cn

Abstract. An information fusion algorithm based on the quantum neural networks is presented for fault diagnosis in an integrated circuit. By measuring the temperature and voltages of circuit components of mate changing circuit board of photovoltaic radar, the fault membership functional assignment of two sensors to circuit components is calculated, and the fusion fault membership functional assignment is obtained by using the 5-level transfer function quantum neural network (QNN). Then the fault component is precisely found according to the fusion data. Comparing the diagnosis results based on separate original data、DS fusion data、BP fusion data with the ones based on QNN fused data, it is shown that the quantum fusion fault diagnosis method is more accurate.

1 Introduction

The analog circuit fault diagnosis can be divided into two types. One is called pre-test simulation [1] method and the other is post-test simulation [2] method. In these methods, no matter pre-test simulation diagnosis or post-test simulation method, generally, we must analyze the working principle and detailed circuit structure, and know some information about the circuit, then undergo the diagnosis. However, in many cases, it's very difficult for us to get such information, and the effectiveness of these diagnosis methods is limited; On the other hand, when a fault in an analog circuit occurs, both the fault component's output signal and the neighboring components corresponding signals will be distorted, due to the mutual interaction between the neighboring components. So it's still difficult to find precisely fault components by using normal fault diagnosis methods.

In this paper the multi-sensor information fusion technique [3-5] based on QNN(Quantum neural network) is introduced for integrated circuit fault diagnosis. By

---

* This project is supported by JiangSu Province Nature Science Foundation (NO. BK 2004021) and the Key Project of Chinese Ministry of Education.(NO.105088).

using multi-dimension signal processing method of information fusion and the inherently fuzzy architecture of quantum neural network [6], it can reduce the uncertainty of analog electronic components fault diagnosis, and exactly recognize the fault components. Moreover, it's unnecessary to know the principle and structure of the circuit. This is a blind diagnosis method.

This paper is organized as follows. The algorithm of multi-sensor information fusion based on the quantum neural networks is introduced in section 2, followed by its application to certain type plane photovoltaic radar electronic components fault diagnosis system in section 3. In order to demonstrate and illustrate the capability of our proposed fault detection approach, the comparison of diagnosis results by single sensor、DS fusion 、BP and QNN fusion is performed. Finally, conclusions are provided in section 4.

2 Quantum Neural Networks Information Fusion Fault Diagnosis

Figure 1 shows the block diagram of a multi-sensor information fusion system. Two sensors measure voltage and temperature respectively. The voltage of each circuit component is obtained by using probes, the temperature of each component is measured by using thermal image instrument. According to fuzzy logic theory, for each sensor, the possibility of fault on tested component can be described by a set of fault membership function values. Then two fuzzy sets of fault membership function values can be gathered. Two cases may occur. The first one is for the mutual interference among components in an electric circuit, which may cause misjudgment when using a signal sensor fault membership function values to distinguish a fault component. The second one is when two sensors give different fault membership function values for the same component. This case makes it very difficult to find the fault component. To solve the problems in the two cases described above, this paper presents an approach that uses a fuzzy neural network information fusion fault classification model. This model combines fuzzy logic with quantum neural network, and the input and output data of the neural network are the meaningful fault membership function values. During fault diagnosis, the fault membership function values of voltage and temperature are taken as the input data of quantum neural network, and output data of the network are the fusion fault membership function values. By using the fusion fault membership function values, the fault component can be determined on certain fault determination criteria. In Figure 1, $\mu_{11}, \mu_{12}, ..., \mu_{1n}$

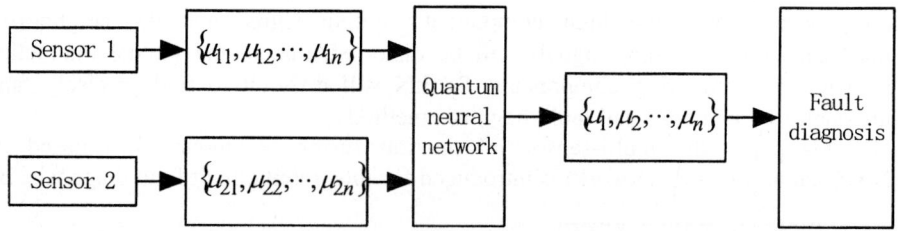

Fig. 1. Fault diagnosis based on quantum neural network information fusion

are fault membership function values of tested components $1,2,...,n$ respectively by sensor 1, $\mu_{21},\mu_{22},...,\mu_{2n}$ are fault membership function values of tested components $1,2,...,n$ respectively by sensor 2, $\mu_1,\mu_2,...,\mu_n$ are fault membership function values of tested components $1,2,...,n$ respectively by the quantum neural network information fusion.

2.1 Forms of Fault Membership Function

Fault membership function is designed by the working characteristics of the sensors and the measured parameters. For a certain component in the electronic system, when the system is working properly, the voltages of components should be stable and the temperature should be fixed value relatively. If there are some fault components in the system, generally the voltage and temperature values will deviate from the normal range. The more deviation is, the higher possibility of fault is. For convenience, the distribution of fault membership function μ_{ij} is defined as shown in Figure 2, where x_{0ij} is the standard parameter value of the tested component when the electronic circuit is working properly; e_{ij} is the normal changing range of the tested component parameters; t_{ij} is the maximum deviation of the parameter of the component to be tested; μ_{ij} is the fault membership function of the component j tested by the sensor i; x_i is the real measured value of the sensor i; α is correction coefficient. The formula (1) is the distribution of fault membership function μ_{ij}.

$$\mu_{ij} = \begin{cases} 1 & x_i \leq 0 \\ -\alpha(x_i - x_{0ij} + e_{ij})/(t_{ij} - e_{ij}) & 0 < x_i \leq x_{0ij} \\ 0 & x_i = x_{0ij} \\ \alpha(x_i - x_{0ij} - e_{ij})/(t_{ij} - e_{ij}) & x_{0ij} < x_i \leq 2x_{0ij} \\ 1 & x_i > 2x_{0ij} \end{cases} \quad (1)$$

2.2 Quantum Neural Network Model

The multi-transfer function quantum neural network [7-9] was presented by Karayiannis N.B. in 1997, and it's widely used in data classification, pattern recognition and so on [11-13]. In this paper, the quantum neural network was used in information fusion to recognize the fault components. The quantum neural network has an inherently fuzzy architecture which can encode the sample information into discrete levels of certainty/uncertainty [10]. The goal is accomplished by using quantum neurons in the hidden layer of the neural network. The transfer function

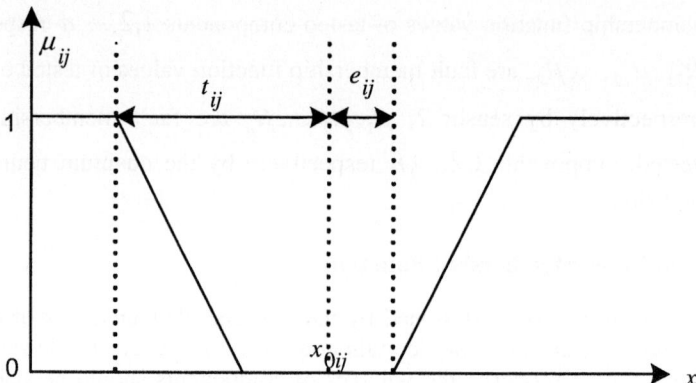

Fig. 2. Fault membership function distribution

(activation function) of quantum neuron has the ability to form graded partitions instead of crisp linear partitions of the feature space. One possibility of obtaining this kind of transfer function is to take the superposition of ns sigmoidal functions, each shifted by quantum interval θ_s ($s = 1,2,...,ns$), where ns is called the total number of quantum levels. The output of quantum neuron can be written as:

$$\frac{1}{ns}\sum_{s=1}^{ns} sig(\beta *(W^T X - \theta_s)) \qquad (2)$$

where $sig(x) = 1/(1+\exp(-x))$, W is weight vector, X is the input vector. β is a slope factor. The quantum intervals of QNN will be determined by training. No a priori fuzzy measure needs to be pre-defined. Given a suitable training algorithm, the uncertainty in the sample data will be adaptively learned and quantified. If the feature vector lies at the boundary between overlapping classes, the quantum neural network will assign it partially to all related classes. If no uncertainty exists, QNN will assign it to the corresponding class.

QNN has three layer neurons. When applying the quantum neural network information fusion to recognize the fault pattern, the input and output of QNN are values that have certain physical meanings. In the paper, the input layer corresponds to the fault feature (the fault membership function value of voltages and temperature), and the output layer corresponds to fault reasons (i.e., fault component). In practical information fusion fault diagnosis of electric system, the applied network model has $2*n$ input nodes to represent fault membership function values of the components tested by the two sensors. There are $n+1$ nodes in the output layer to represent the membership function values of the fusion data. There is one hidden layer.

2.3 Training Algorithms

The gradient-descent-based algorithm is used to train the quantum neural network. In each training epoch. The training algorithm updates both connectivity weights among different layers and quantum intervals of the hidden layer. Weight updating is carried out by the standard back-propagation algorithm. Once the synaptic weights have been obtained, quantum intervals can be learned by minimizing the class-conditional variances [14] at the outputs of the hidden units.

The variance of the output of the i-th hidden unit for class C_m is

$$\sigma_{i,m}^2 = \sum_{x_k \in C_m}(<O_{i,m}> - O_{i,k})^2 \tag{3}$$

where $O_{i,k}$ is the output of the i-th hidden unit with input vector x_k,
$<O_{i,m}> = \frac{1}{|C_m|}\sum_{x_k \in C_m} O_{i,k}$, $|C_m|$ denotes the cardinality of C_m.

By minimizing $\sigma_{i,m}^2$, we can get the update equation for $\theta_{i,s}$ as follows[8], for each hidden unit i and its s-th quantum level ($s = 1, 2, \ldots, ns$):

$$\Delta\theta_{i,s} = \eta\frac{\beta}{ns}\sum_{m=1}^{n}\sum_{x_k \in C_m}(<O_{i,m}> - O_{i,k}) * (<v_{i,m,s}> - v_{i,k,s}) \tag{4}$$

Where η is the learning rate, n is the number of output notes, i.e. the number of classes.

$$<v_{i,m,s}> = \frac{1}{|C_m|}\sum_{x_k \in C_m} v_{i,k,s} \ ; v_{i,k,s} = O_{i,k,s}(1 - O_{i,k,s}) \tag{5}$$

$O_{i,k,s}$ is the output of the s-th quantum level of the i-th hidden unit with input vector x_k.

2.4 Fault Component Judge Criterion

To each fused component fault membership function value, the following rules can be used to identify the fault pattern:

① Object patterns to be judged should have maximal fault membership function value; moreover, it should be larger than a certain value. Generally, this value should be at least more than $\frac{1}{n}$, where n stands for the number of components to be tested. The larger the threshold value is, the more precise the judgment should be. However, if the threshold value is too large, fault membership function value from testing will not satisfy the requirement. Therefore, we should, according to the real situation, choose a moderate one, such as 0.80 in the diagnosis example given in this paper.

② The difference of the fault membership function value between object patterns with other patterns should be larger than a certain threshold value, such as 0.60 in the example of this paper.

3 The Fault Diagnosis for Integrated Circuit

In a certain type plane photovoltaic radar electronic components fault diagnosis system, the "fault tree analysis" method is adopted to search fault components [15]. By measuring the voltage of components on the specific circuit board and comparing it with the normal signals, we can make a proper judgment. Although this method is simple and convenient, the diagnosis accuracy is poor. The reason lies in two aspects: when a fault in an analog circuit occurs, both the fault component's output signal and the neighboring components corresponding signals will be distorted, due to the mutual interaction between the neighboring components, in other words, the fault feature vector lies at the boundary between overlapping fault classes; On the other hand, for some components, we don't know their circuit structure, the "fault tree" may not be well rationalized.

So we introduce the quantum neural network to fault components hunting. By using multi-dimension signal processing method of information fusion and the inherently fuzzy architecture of quantum neural network, it can reduce the uncertainty of analog electronic components fault diagnosis, and exactly identify the fault components. The fault diagnosis example is a mate changing circuit board of photovoltaic radar electronic components in the paper, Figure 3 is its theory circuit chart.

Its major function is to change analog voltage into a digital signal. There are four electric components (A_1, A_2, A_3, A_4) for diagnosis. V_1, V_2, V_3 and V_4 are the voltages of the component 1,2,3 and 4 respectively. Firstly, the temperature of each component is measured by a thermal image instrument (inframetrics 600) when the

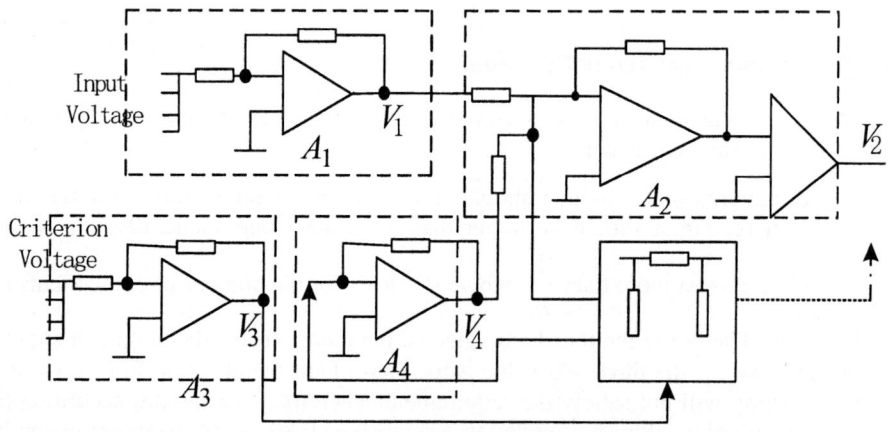

Fig. 3. The mate changing circuit of photovoltaic radar diagnosed

circuit is working properly. Secondly, if there is a fault in the circuit, the fault component's temperature will change (increase or decrease), so the new temperature of each component is measured, and the temperature fault membership function value to all suspected fault components can be calculated using the formula given before. In addition, the voltage of each component can be measured by using probes, and the voltage fault membership function values can be obtained with same calculating method.

3.1 Structure of Quantum Neural Network and Its Training Sample

In this fusion experiment, there are four object components and two sensors, so the network has 8 input neurons, 5 output neurons(four outputs stand for four fault components, the fifth output is fault uncertainty), and 12 neurons in the middle hidden layer. The total number of quantum levels is $ns = 4$; the initialized weights and quantum interval (θ_s) are normalized random values. Fault membership function is defined mainly by the characteristics of sensors and object parameters. The distribution of fault membership function μ_{ij} can be illustrated by formula (1). To simplify the problem, while keeping the fault characteristics unchanged, we choose $e_{ij} = 0, t_{ij} = x_{0ij}$, voltage sensor $\alpha = 1/3$, temperature sensor $\alpha = 2.8$, and set fault location on different components, under different input voltage of A/D conditions, calculate the fault membership function value of each component tested by the two sensors. Each group of sensor membership function value is normalized and then used as the input vector of network training sample; the output vector is defined by the fault components. That is, for the real fault component, its corresponding neuron output is 1, and the outputs for other components and the uncertainty are 0s. In the example, 15 samples are used. These 15 samples are typical cases for all kinds of fault instances, in fact, under different input voltage of A/D conditions, a lot of samples can be gotten by the same computation method.

3.2 Discussion of Fault Diagnosis Results

According to the above fusion algorithm and fault judgment rules, by using BP network and the multi-transfer function quantum neural network information fusion, the fault diagnosis result of the circuit is shown in Table 1. The first and the second groups of the table are respectively the fault membership function values of components tested by temperature sensor and voltage sensors. The third, the fourth and fifth group show the fault membership function values of each components after DS evidence theory [16], BP and QNN fusion. Fusion fault membership function values are respectively the outputs of BP and QNN trained by samples, when the inputs of BP and QNN are respectively fault membership function values of components tested by temperature sensor and voltage sensors (table 1).

Table 1 shows distinctly that fault membership function values obtained by the two sensors respectively are sometimes very close for the four components diagnosed in photovoltaic radar electric circuit system, the single sensor can not identify the fault

Table 1. comparison of diagnosis results by single sensor and DS、BP、QNN fusion

real fault comp.	sensor	Fault membership function value					Diagnosis results
		A_1	A_2	A_3	A_4	uncertainty	
A_3	voltage	.0327	.2172	.3022	.2943	.1536	uncertain
	temperature	.1647	.1957	.2771	.1565	.2060	uncertain
	DS fusion	.0740	.2320	.3729	.2585	.0626	uncertain
	BP fusion	.0634	.0111	.7280	.1531	.0446	uncertain
	QNN fusion	.0524	.0221	.8481	.0732	.0044	A_3 failure

component correctly using the judgment criterion given before, but quantum neural network fusion membership function values can identify the fault component exactly. In the example, component 3(A_3) is the real fault component, the fault membership function values of component 2,4 and ones of component 3 from voltage sensor are very close, so we can not decide the fault component. If simply taking the single maximum for a fault indicator, the component 3 would be regarded as the fault component, it is a wrong result. But after QNN fusion, the fault membership function value of component 3 increases significantly and is much larger than the ones of the other components, and therefore we can point out the fault component correctly. In other words, the QNN information fusion algorithm increases the objective membership function value assignment and decreases the membership function values of the other components. This makes the uncertainty of the diagnosis system decrease effectively, and error diagnosis has been removed from the inadequacy of single sensor information. In the example, the correction rate of fault diagnosis of quantum neural network information fusion can reach 100%. So multi-sensor information fusion based on quantum neural network and fuzzy logic theory improves the analyzability of the system and the identification ability of fault models, and greatly increases the correction rate of fault component decision.

Comparing DS evidence theory、BP network with the multi-transfer function quantum neural network, it can be found that quantum neural network fusion is more effective than BP and DS evidence theory fusion. For example, when maximal fault membership function threshold value is 0.35, both information fusion results of BP network、DS evidence theory and the multi-transfer function quantum neural network can all find the fault component correctly. But when the maximal fault membership function threshold value is 0.7, the DS fusion can not always identify the fault component correctly. BP network can not always identify the fault component correctly, when the maximal fault membership function threshold value is 0.8.

The larger the threshold value is, the more precise the judgment should be. However, if the threshold value is too large, fault membership function value from testing will not satisfy the requirement. Therefore, according to the real situation, we choose a moderate one, such as 0.80 in the diagnosis example given in this paper. In this fusion experiment, the correction rate of fault diagnosis can reach 100% for QNN fusion

It can be seen that the accuracy of multi-sensor information fusion is the highest grades. The quantum neural network information fusion is more effective than BP network fusion and DS fusion.

4 Conclusion

We can see from the experiment results that, as long as the components under diagnosis are properly chosen, and the measured signals are precise, then the fault components can be precisely recognized by using the multi-sensor quantum neural network information fusion method. In addition, the quantum neural network information fusion is more effective than BP network fusion and DS fusion.

References

1. Chakrabarti, S., Chatterjee, A.: Compact fault dictionary construction for efficient isolation, Proc. 20th Anniversary Conf. Advanced Research in VLSI, (1999) 327-341.
2. Chen, Y. Q.: Experiment on fault location in large-scale analogue circuits, IEEE Trans. On IM. 1(1993)30-34.
3. Luo, R.C., Scherp, R.S.: Dynamic multi-sensor data fusion system for intelligent robots. IEEE J. Robotics and Automation, 4(1998) 386-396.
4. Gao, P., Tantum, S., Collins, L.: Single sensor processing and sensor fusion of GPR and EMI data for landmine detection, Proc. SPIE Conf. Detection Technologies Mines Mine-like Targets, Orlando, (1999)1139-1148.
5. Milisavljevic, N., Bloch, I.: Sensor fusion in anti-personal mine detection using a two-level belief function model, IEEE Trans. On Systems, Man, and Cybernetics, 2(2003)269-283.
6. Huang, D.S., Ma, S.D.: Linear and nonlinear feedforward neural network classifiers: A comprehensive understanding, Journal of Intelligent Systems, 1(1999)1-38.
7. Karayiannis, N.B., Purushothaman, G.: Fuzzy pattern classification using feed forward neural networks with multilevel hidden neurons. IEEE International Conference on neural networks,1(1994)127-132.
8. Gopathy, P., Nicolaos, B., Karayiannis N.B.: Quantum Neural networks: Inherently fuzzy feedforward neural networks. IEEE Transactions on neural networks, (1997)679-693.
9. Behman, E.C., Chandrashkar, V.G., Wang, C.K.: A quantum neural network computes entanglement. Physical Review Letters, 1(2002)152-159.
10. Narayanan, A., Menneer, T.: Quantum artificial neural network architectures and components, Infirmation Sciences, 3(2000)231-255.
11. Zhou, J., Qing, G., Adam, Krzyzak.: Recognition of handwritten numerals by quantum neural network with fuzzy features, IJDAR, 1(1999)30-36.
12. Li, F., Zhao, S.G., Zheng, B.Y.: Quantum neural network in speech recognition, 6th International Conference on Signal Processing, (2002)1234-1240.
13. Shiyan H.: Quantum neural network for image watermarking, Lecture Notes in Computer Science, Springer-Verlag, (2004)669-674.

14. Duda, R.O., Hart P.E.: Pattern classification and scene analysis, John Wiley & Sons, New York, 1973.
15. Zhu, daqi: new technology study of fault diagnosis for aeronautics electronic equipment, doctor thesis, Nanjing university of aeronautics and astronautics, 2002. 6.
16. Zhu, daqi, Yang, Yongqing, Yu, shenglin: Dempster-Shafer information fusion algorithm of electronic equipment fault diagnosis, Control Theory and Applications, 4(2004), 559-663.

Statistical Feature Selection for Mandarin Speech Emotion Recognition

Bo Xie, Ling Chen, Gen-Cai Chen, and Chun Chen

College of Computer Science,
Zhejiang University, Hangzhou 310027, P.R. China
lingchen@cs.zju.edu.cn

Abstract. Performance of speech emotion recognition largely depends on the acoustic features used in a classifier. This paper studies the statistical feature selection problem in Mandarin speech emotion recognition. This study was based on a speaker dependent emotional mandarin database. Pitch, energy, duration, formant related features and some velocity information were selected as base features. Some statistics of them consisted of original feature set and full stepwise discriminant analysis (SDA) was employed to select extracted features. The results of feature selection were evaluated through a LDA based classifier. Experiment results indicate that pitch, log energy, speed and 1st formant are the most important factors and the accuracy rate increases from 63.1 % to 76.5 % after feature selection. Meanwhile, the features selected by SDA are better than the results of other feature selection methods in a LDA based classifier and SVM. The best performance is achieved when the feature number is in the range of 9 to 12.

1 Introduction

If a computer can recognize emotion, it could respond to user differently according to his / her emotional state. Emotion recognition in speech is one research field for emotional human-computer interaction or affective computing [1]. Performance of emotion recognition largely depends on acoustic features used in a classifier. Murray [2] summarized the relationship between emotion and acoustic features including pitch, intensity, rate and voice quality etc. Later researchers added formants, LPC and MFCC etc. to combine phonetic and prosodic features together in emotion recognition. For a fixed-length feature vector, researchers computed derived features and statistics including range, mean, standard deviation, and so on. But such large number of features is not suitable for classification. Because the accuracy rate will not increase along with the feature number and the generalization of the classifier will decrease while in high dimension space, feature selection is necessary to achieve high recognition performance. In addition, emotion expressed in speech will change with language, culture and territory, so the feature selection results of one language will not suit others.

The commonly used feature selection methods for speech emotion recognition are: promising first selection (PFS); sequential forward selection (SFS); sequential backward selection (SBS). Frank and Polzin [3] firstly used PFS and SFS for emotion recognition. They selected 5 features from 17, and error rate decreased by 3.5 %. It was reported that the most salient features which represent the acoustical correlates of emotion are maximum, minimum and median of fundamental frequency and the mean positive derivative of the regions where the F0 curve is increasing. Lee [4] used those two methods, and error rate decreased by 10 % on average. Kwon [5] selected features according to a two dimensional ranking figure result from SFS and SBS methods. Because mandarin speech is a tonal language and its characteristics are different from English, feature selection should be carried out for it separately. Wang [6] used fuzzy entropy theory to measure the effectiveness of emotional features of mandarin. To our knowledge, there are no other results about feature selection for mandarin speech emotion recognition.

This paper studied the statistical feature selection problem in mandarin speech emotion recognition. This study was based on a speaker dependent emotional mandarin database. Pitch, energy, duration, formant related features and some velocity information were selected as base features. Some statistics of them consisted of original feature set. Because full stepwise discriminant analysis (SDA) can track the previous selected feature using F test for the distinguish ability of it might reduce while new feature added, it was used to select features. And the results of feature selection were evaluated through a LDA based classifier. In addition, the SDA method was compared with PFS, SFS and SBS methods with varying feature number using a LDA based classifier and support vector machine (SVM).

The contribution of our work is to extend feature selection study for speech emotion recognition into three aspects: used statistical feature selection to search for efficient features for mandarin speech emotion recognition; studied the effect of feature number variety in feature selection; compared SDA with other popular feature selection methods in LDA based classifier and SVM.

2 Feature Extraction and Selection

Emotional feature selection and analysis described in this paper has four stages: building an emotional speech database; feature extraction; feature selection and analysis; performance evaluation.

2.1 Emotional Mandarin Speech Database

Emotional mandarin speech was collected through studio recording. This speech was speaker dependent corpus recorded from 10 Chinese professional actors of different gender (five male and five female). There were five text scripts used in the recording and actors were asked to speak every script with all five emotions: angry; fear; happy; sad and neutral. The emotional mandarin speech database used in our research consisted of 696 utterances and the recording format is mono, 16-bit, 16 KHz.

2.2 Feature Extraction

We selected the pitch, log energy, energy and formant as the base features based on the previous study results [8, 9] and our preliminary result. The frame shift in feature extraction was 5ms. We first segmented only speech parts from an input utterance by using an endpoint detector based on zero crossing rate (ZCR) and frame energy. For each frame of speech signal, we estimated F0, energy, three formant frequencies and bandwidth. PRAAT tools [10] were used to estimate F0 and formant. We also added velocity information for pitch and energy, respectively, to take the rate of speaking into account and model the dynamics of the corresponding temporal change of pitch and energy. Hence we have 12 feature streams (including velocity components). These streams were the base of feature selection and analysis.

Because emotion was expressed mainly at the utterance-level, it is crucial to normalize the feature. Although a back-end classifier working with a fixed-length feature vector is used for classification, it is also important to convert feature streams into a representative fixed-length feature vector. Therefore, some statistics (including mean, max, min, range and standard deviation) of above feature streams were calculated to get a fixed-length feature vector. We also add speed information for duration that concerned about how many words were pronounced in one second. The dimension of the feature vector, the input vector of the feature selection, was 31 for each utterance. Tab. 1 shows the base features of emotion speech. Each feature has its own unit and samples must be normalized (mean is 0 and standard deviations is 1).

Table 1. Base features of emotion speech

Category	Streams (abbreviation)	Statistics (abbreviation)
Pitch	Pitch (F0)	Mean (Mean) Maximum (Max)
	Pitch velocity (F0Vel)	Minimum (Min)
		Range (Range) (except velocity)
		Standard deviation (Sd) (except velocity)
Energy	Log energy (LogEng)	Mean (Mean) Maximum (Max)
	Log energy velocity (LogEngVel)	Minimum (Min)
	Energy (Eng)	Range (Range) (except velocity)
	Energy velocity (EngVel)	Standard deviation (Sd) (except velocity)
Voice quality	The first formant (F1)	Mean (Mean)
	The second formant (F2)	Bandwidth mean (bwMean)
	The third formant (F3)	
Duration		Speed (Speed)

2.3 Feature Selection

The purpose of feature selection is to identify features that contribute the most in classification. The results would tell us what features and properties of the speech are important in distinguishing emotion. We could then add more relevant features accordingly to improve classification accuracy. However, it is prohibitively time consuming to perform exhaustive search for the subset of features that give the best classification. Instead, we use SDA to rank the features and identify the subset that contributes the most in classification.

SDA used Wilks' lambda and F ratio as a measure of effectiveness. Wilks' lambda is a basic quantity in testing the significance of the overall difference among several class centroids. It is a ratio of the determinants of the within-classes and the total sums of squares and cross product (SSCP) matrices. Wilks' lambda Λ is given by:

$$\Lambda = \frac{|W|}{|T|} = \frac{|W|}{|W+B|} \tag{1}$$

where W, B and T are within, between and total SSCP matrix respectively. In emotion speech recognition, Wilks' lambda Λ can be used to identify the emotion distinguishing ability of a feature set. The more Wilks' lambda decreases, the more the distinguishing ability of feature set increases.

The following is the procedure of SDA feature selection:

(1) Select one feature from the original feature set that minimizes the Wilks' lambda value.
(2) Select the next feature, suppose the rth feature, from unselected original feature set, which combines the selected features, and minimizes the Wilks' lambda value. Perform a F test to show significance of the new selected feature. If significant then go to (3), else go to (4).
(3) Perform a F test to show significant of the previous r selected features one by one after the $(r+1)^{th}$ feature was selected. If the previous r selected features are not significant, select a feature whose F ratio is minimum, and remove it from the selected feature set, until all the selected features are significant.
(4) If no one feature can be selected from original feature set, and no one feature can be removed from selected feature set, the feature selection procedure is finishes.

The procedure that minimizes Wilks' lambda will maximize the total spread of the data while minimizing the spread of individual classes. Thus, the procedure increases the class separability.

3 Experiment and Results

We used LDA classifier to evaluate the performance of SDA within the emotional mandarin speech database mentioned in section 2.1. SDA and LDA were implemented with SPSS software. After setting the minimum / maximum partial F to enter / remove, we perform SDA for every speaker's original feature set. Tab. 2 shows the feature selection results for a female speaker. Wilks' lambda value indicates that the distinguishing ability increased step by step.

After feature selection, the selected feature set of each speaker had about 7 - 15 features. LDA was used as classifier for emotion recognition. For each speaker we divided the corpus into training and testing sample data with a 7 : 3 ratio. Fig. 1 gives a scatter plot where the x - y coordinate is two discriminant functions (LDA uses four discriminant functions in this case) for a female speaker.

Tab. 3 shows the confusion matrix of LDA. It achieved an overall accuracy of 76.5 % after feature selection for 10-speaker average. While using the original feature set, LDA gets an overall accuracy of 63.1%. Therefore, LDA could efficiently classify emotional speech.

Fig. 2 shows mean accuracy rate over 5 – class emotion. The mean accuracy rate of the selected feature set and the original feature set was compared and SDA based feature selection increased mean accuracy rate by 13 points.

Table 2. Stepwise discriminant procedure, F to enter is 3.84, F to remove is 2.71

Step	Entered	Wilks' Lambda
1	F0Mean	.172
2	EngMean	.036
3	LogEngMean	.022
4	LogEngSd	.016
5	LogEngRange	.010
6	F3Mean	.007
7	F2Mean	.005
8	EngSd	.004
9	F0VelMean	.003
10	F2bwMean	.003
11	F3bwMean	.002
12	F0VelMax	.002
13	LogEngVelMean	.002
14	LogEngVelMax	.001

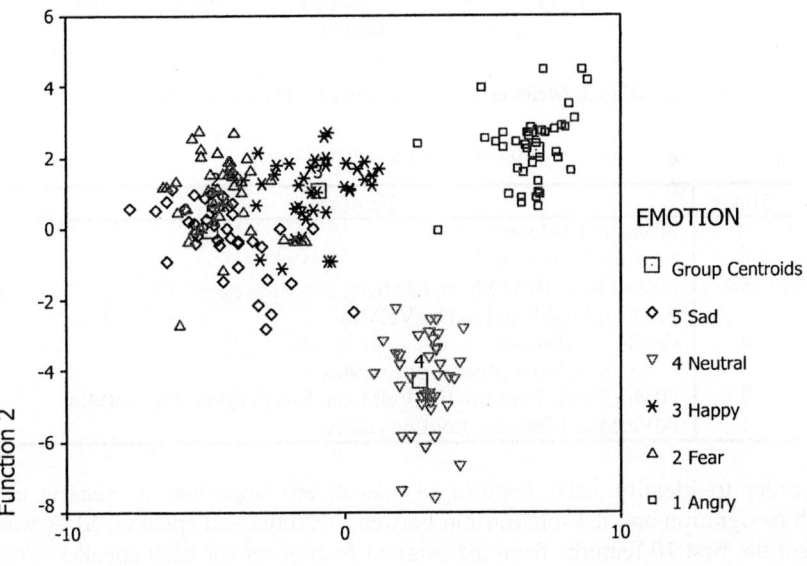

Fig. 1. Scatter plot of discriminant function 1&2

Table 3. Confusion matrix of 5-class emotion recognition using LDA after SDA feature selection

	Angry	Fear	Happy	Neutral	Sad
Angry	70.0	2.8	24.4	0.0	2.8
Fear	0.0	78.9	11.9	0.0	9.3
Happy	17.4	5.6	73.5	2.8	0.8
Neutral	2.8	5.0	0.0	89.4	2.8
Sad	3.7	19.0	5.3	0.8	71.2

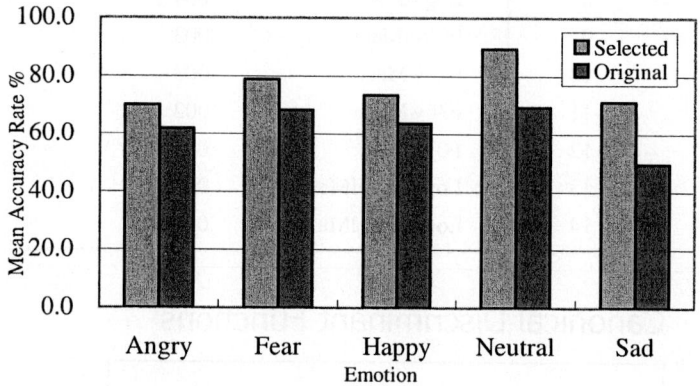

Fig. 2. Mean accuracy rate over 5 - class emotion

Table 4. Feature selected times

Times	Features
8	F0Mean, F1Mean
6	LogEngMean, Speed
5	F0VelMean, F1bwMean, F2Mean, F3Mean, EngMean, LogEngVelMax
4	EngSd, LogEngSd
3	F2bwMean, F3bwMean, LogEngMax
2	F0Min, F0Sd, EngMin, EngVelMean, EngVelMax, EngVelMin
1	F0VelMin, F0Range, LogEngVelMin

In order to identify what features of speech are important in generic emotion speech recognition and the relationship between features and speaker, SDA was used to select the first 10 features from the original feature set for each speaker. There are 24 selected features and Tab. 4 gives the selected times of them. F0 mean and F1 mean are selected 8 times; log energy mean and speed are selected 6 times.

The results showed that F0 mean, F1 mean, log energy mean and speed got high selection times than others, and were the effective features in generic emotion speech recognition. Meanwhile, the selected features covered 77.4% of the original feature set, and they were related to different speakers. So the different speakers express emotion using both common acoustic features and their own styles.

4 Discussion

As a reference for comparing the results of our feature selection method using SDA, we studied PFS, SFS and SBS methods. There are a large variety of measures that can be used to evaluate the goodness of a feature subset. KNN classification and leave-one-out cross validation error rate were used to select features by these methods. In PFS, original features were ordered by increasing error rate, thus adding a new feature from the original feature set to the selected feature set successively each time. SFS adds one feature at a time by choosing the next one that least decreases error rate, while SBS starts with all original features and sequentially deletes the next feature that least increases error rate.

SDA and these three methods were used to perform feature selection for each speaker with feature number ranging from 1 to 28. Fig. 3 shows the mean accuracy rate of the four feature selection methods using LDA based classifier. As shown in Fig. 3, SDA achieved best performance while feature number is in the range of 9 to 12 and it exceeded other methods in most cases. When the feature number exceeds a given value (e.g. 12 for SDA), the generalization accuracy is hurt.

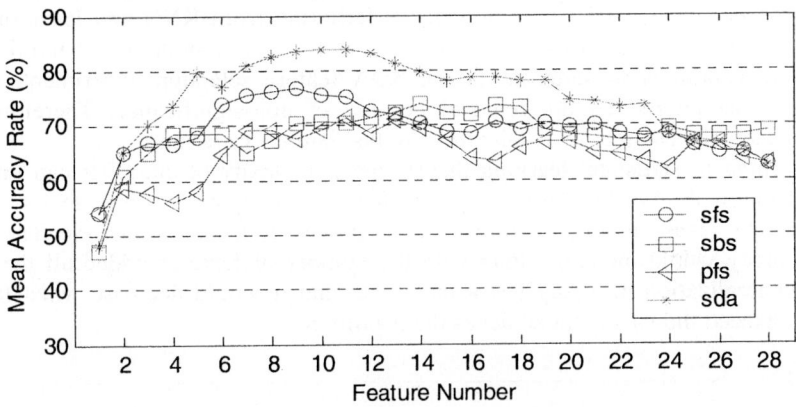

Fig. 3. Mean accuracy rate change with feature number using LDA for speaker dependent

For the text and speaker-independent task, the performance of those feature selection methods is shown in Fig. 4. The SDA still achieved top performance in those four feature selection methods with the feature number range from 3 to 25, but the generalization accuracy always increases with the feature number. Compared with Fig. 3 the original feature set is not good enough to get optimum generalization accuracy, while the original feature set is redundancy for the speaker-dependent task, there are exist optimum feature number less than 31.

To verify that the feature selection methods are independent with the classifier, we also employed support vector machine (SVM) as a classifier to evaluate the performance of those 4 feature selection methods. As the Fig. 5 and 6 shown, the SDA still

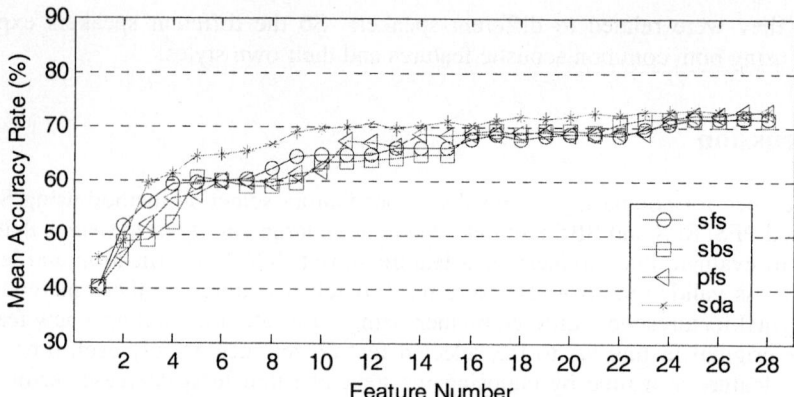

Fig. 4. Mean accuracy rate change with feature number using LDA for text and speaker-independent

has better performance than the other 3 methods and the curve of the accuracy rate is similar to the LDA one.

The reason that SDA gets the best performance in classification maybe result from two aspects: (1) SDA can track the selected features using F test. It is crucial to track the previous selected feature, because the distinguishing ability of selected feature might reduce while new feature added; (2) Differing from KNN and leave-one-out cross validation error rate, Wilks' lambda is based on an assumption of multivariate normal distributions in samples. In the SDA feature selection experiment, a few speakers encountered feature removal because of non-significance. Therefore, the performance of SDA mainly results from Wilks' lambda criterion.

According to statistical learning theory, the complexity of classifier can hurt the generalization ability of pattern recognition for small samples. It can be seen from the experimental results that the feature selection procedure has the same rule. The training accuracy would increase along with the number of features added all the time, while generalization accuracy has a maximum and it would decrease when feature number exceed the value that achieves the maximum.

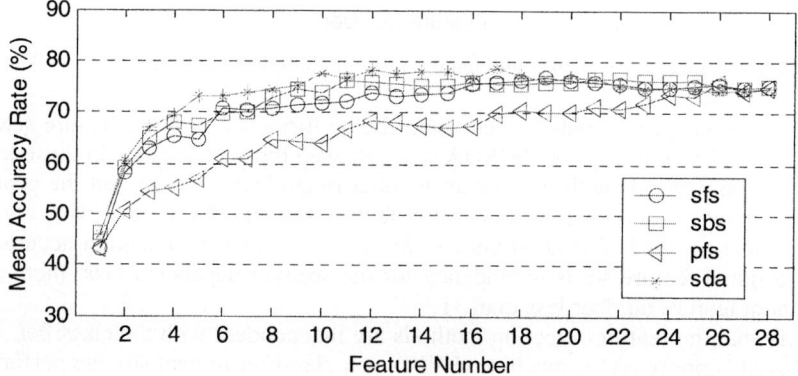

Fig. 5. Mean accuracy rate change with feature number using SVM for speaker-dependent

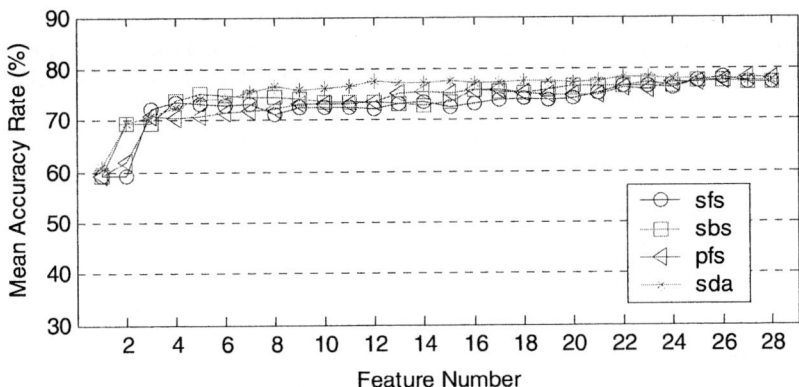

Fig. 6. Mean accuracy rate change with feature number using SVM for text and speaker-independent

5 Conclusions

This paper studied the effects of SDA based feature selection on mandarin speech emotion recognition. Experiment results indicate that F0, F1 mean, energy and speed play a major role in recognizing emotion and the importance of other features would change with different speakers. A LDA based classifier was used to evaluate the selection results of SDA with speaker dependent emotional mandarin speech. Recognition results show that the classification accuracy of LDA significantly increased in all 5 - class emotion while SDA feature selection was used. In addition, the comparison of feature selection methods showed SDA to perform better than PFS, SFS and SBS, and the best performance is achieved while feature number is in the range of 9 to 12.

We propose some future directions in this area: (1) explore new features better representing emotion; (2) develop a set of new statistics of feature to reduce the variability in F0, energy, formant and duration due to speaker, gender and phonetic dependency.

References

1. Picard, R.W.: Affective Computing. The MIT Press (1997)
2. Murray, I.R., Arnott, J.L.: Toward the Simulation of Emotion in Synthetic Speech: A Review of the Literature on Human Vocal Emotion. Journal of the Acousttcal Society of America, Vol. 93, No. 2, (1933) 1097-1108
3. Dellaert, F., Polzin, T., Waibel, A.: Recognizing Emotion in Speech. Proceedings of International Conference on Spoken Language Processing (1996) 1970-1973
4. Lee, C.M., Narayanan, S., Pieraccini, R.: Recognition of Negative Emotions from the Speech Signal. Proceedings of IEEE Workshop on Automatic Speech Recognition and Understanding (2001) 240-243
5. Kwon, O.W., Chan, K., Hao, J., Lee, T.W.: Emotion Recognition by Speech Signals. Proceedings of EUROSPEECH (2003) 125-128
6. Wang, Z.P., Zhao, L., Zou, C.R.: Emotion Recognition of Speech using Fuzzy Entropy Effectiveness Analysis. Journal of circuits and systems, Vol. 8, No. 3 (2003) 109-112

7. James, M.: Classification Algorithms. John Wiley & Sons, London (1985)
8. Cowie, R., Douglas-Cowie, E., Tsapatsoulis, N., et al.: Emotion Recognition in Human-computer Interaction. IEEE Signal Processing Magazine, Vol. 18, No. 1 (2001) 32-80
9. Cai, L.L., Jiang, C.H., Wang, Z.P.: A Method Combining the Global and Time Series Structure Features for Emotion Recognition in Speech. Proceedings of International Conference on Neural Networks and Signal Processing (2003) 904–907
10. Boersma, P., Weenink, D.: Praat Speech Processing Software. Institute of Phonetics Sciences of the University of Amsterdam. http://www.praat.org

Reconstruction of 3D Human Body Pose Based on Top-Down Learning

Hee-Deok Yang[1], Sung-Kee Park[2], and Seong-Whan Lee[1]

[1] Department of Computer Science and Engineering, Korea University,
Anam-dong, Seongbuk-gu, Seoul 136-713, Korea
{hdyang, swlee}@image.korea.ac.kr
[2] Intelligent Robotics Research Center, Korea Institute of Science and Technology,
P.O. Box 131, Cheongryang, Seoul 130-650, Korea
skee@kist.re.kr

Abstract. This paper presents a novel method for reconstructing 3D human body pose from monocular image sequences based on top-down learning. Human body pose is represented by a linear combination of prototypes of 2D silhouette images and their corresponding 3D body models in terms of the position of a predetermined set of joints. With a 2D silhouette image, we can estimate optimal coefficients for a linear combination of prototypes of the 2D silhouette images by solving least square minimization. The 3D body model of the input silhouette image is obtained by applying the estimated coefficients to the corresponding 3D body model of prototypes. In the learning stage, the proposed method is hierarchically constructed by classifying the training data into several clusters recursively. Also, in the reconstructing stage, the proposed method hierarchically reconstructs 3D human body pose with a silhouette image or a silhouette history image. We use a silhouette history image and a blurring silhouette image as the spatio-temporal features for reducing noise due to extraction of silhouette image and for extending the search area of current body pose to related body pose. The experimental results show that our method can be efficient and effective for reconstructing 3D human body pose.

1 Introduction

There has been a growing interest in improving the interaction between humans and computers. It is argued that to achieve comfortable human-computer interaction(HCI), there is a need for the computer to be able to interact naturally with the human, similar to the way where human-human interaction takes place. Humans interact with each other through gesture, speech, etc.

Recognizing body gesture by estimating human body pose is one of the most difficult and commonly occurring problems in computer vision system. A number of researches have been developed for estimating and reconstructing 2D or 3D body pose [3, 6, 7, 8, 9, 10]. Bowden et al. [4] used a non-linear statistical model consisting of the positions of 2D shape contour and its corresponding 3D skeleton vertices. The 2D features are fused with the 3D model and learnt using hierarchical PCA(Principal Component Analysis). They used position of head and hands and body contour to reconstruct upper body pose. Song et al. [10] modeled the manifold of human body pose by a hidden Markov model and learnt by entropy minimization. They used dynamic programming to calculate the global labeling of the joint probability density function of the position and velocity of body features. It was assumed that it is possible to track these features for continuous frames. Rosales et al. [9] used the Specialized Mapping Architecture(SMA). The SMA's fundamental components are a set of specialized mapping functions and a single feedback matching function. All of these functions are estimated directly from training data. Howe et al. [5] used a learning based approach for estimating human body pose, where a statistical approach was used to estimate the three-dimensional motions of human body pose.

In this paper, we solve the problem of reconstructing 3D human body pose using a hierarchical linear learning method [4]. The proposed method is related to machine learning models [3, 8] that use a hierarchical method to reduce the complexity of the learning problem by splitting it into several simpler ones. The differences of our approach are: the 2D silhouette images and their corresponding 3D positions of body components are used to learn; the spatio-temporal features are used to reduce noise which occurs in the extraction of silhouette images; the training and testing data are normalized by an appropriate method.

2 3D Human Model

The 3D human modeling includes human body representation and kinematics. For human body representing, we build a 3D human model consists of body segments, joints, and a perfect coordination among them. It has 17 body parts with 17 joints and 37 DOF(Degree of Freedom) and has 5 additional joints which are end-effectors used to calculate angle of each body segment in inverse kinematics [1]. Our 3D human model and the corresponding tree structure are shown in Fig. 1. Each component of 3D human model built from super-quadric ellipsoids with 3 parameters. In Eq. (1). An is the angular joint parameter consisted of three values, abduction, flexion and rotation of a body segment, Sh is the deformable shape parameter consisted of two values, volume and height of a body segment and Po is the position of a body segment used to calculate angle of each body segment in inverse kinematics. 3D human model is represented such as:

$$M = ((An_1, Sh_1, Po_1),..., (An_n, Sh_n, Po_n)) \qquad (1)$$

where n is the number of human body segments.

Although this 3D human model is far from real human body shape, it suffices to represent human body pose and to detect the collision of body segments.

For human body kinematics, we build simple forward and inverse kinematics. Transformation created in one joint spreads to all child segments and joints. Forward kinematics gets the position and orientation of last segment in a kinematics chain by calculating angles for every joint. The forward transformation of the segment is calculated by Eq. (2).

$$Rf_i = (R_1 \times ,...,\times R_i) \qquad (2)$$

where R_1 is the local transformation of the upper torso of the human model and R_i is the local transformation of ith joint of the 3D human model in a kinematics chain.

On the other hand, inverse kinematics computes the joint angles of a kinematics chain based on the position and orientation of last kinematics segment.

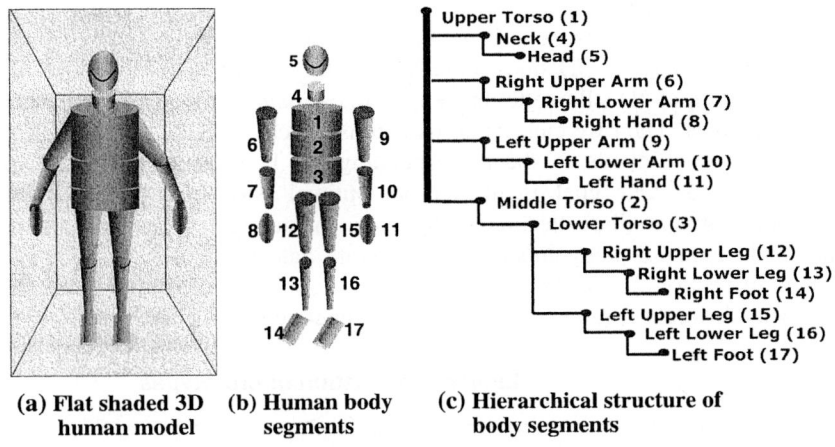

(a) Flat shaded 3D human model (b) Human body segments (c) Hierarchical structure of body segments

Fig. 1. 3D human model

3 Top-Down Learning

3D human body pose can be defined as a collection of a 2D silhouette image with a 3D body model. A silhouette image is represented by a linear combination of prototypes of 2D silhouette images. We estimate optimal coefficients for a linear combination of prototypes of 2D silhouette images and their corresponding 3D body models by solving least square minimization. The proposed method is constructed with top-down learning by classifying the training data into several clusters recursively. The silhouette images used in this paper are normalized by the method described in Section 3.1.

3.1 Data Normalization

We generate silhouette images and their corresponding 3D body models using 3D human model described in Section 2. The generated silhouette images have various human body poses. For each silhouette image, we apply PCA for the foreground pixels to obtain two eigenvectors. To calculate the mean of foreground region, a set of foreground pixels is created. Let X is the foreground pixels:

$$X = ((x_1, y_1), ..., (x_n, y_n))^T \tag{3}$$

where x, y are a position of a pixel in the 2D world and n is the number of foreground pixels in the silhouette image.

To normalize data, a distance between the pixels in a set X and the longest eigenvector is calculated. The pixels in range of threshold are selected. And the silhouette image is rotated by the longest eigenvector to have same angle in all data.

3.2 3D Gesture Representation

In order to reconstruct 3D human body pose from continuous silhouette images, we used a learning based approach. If we have sufficiently large amount of pairs of a silhouette image and its 3D body model as prototypes of human gesture, we can reconstruct an input 2D silhouette image by a linear combination of prototypes of 2D silhouette images. Then we can obtain its reconstructed 3D body model by applying the estimated coefficients to the corresponding 3D body model of prototypes as shown in Fig. 2. Our goal is to find an optimal parameter set α which best reconstructs a given silhouette image. To make various prototypes of 2D silhouette images and theirs 3D body models, we generate data using the 3D human model described in Section 2. The parameters of An in Eq. (1) are randomly changed. As a result, many different D human body poses are generated.

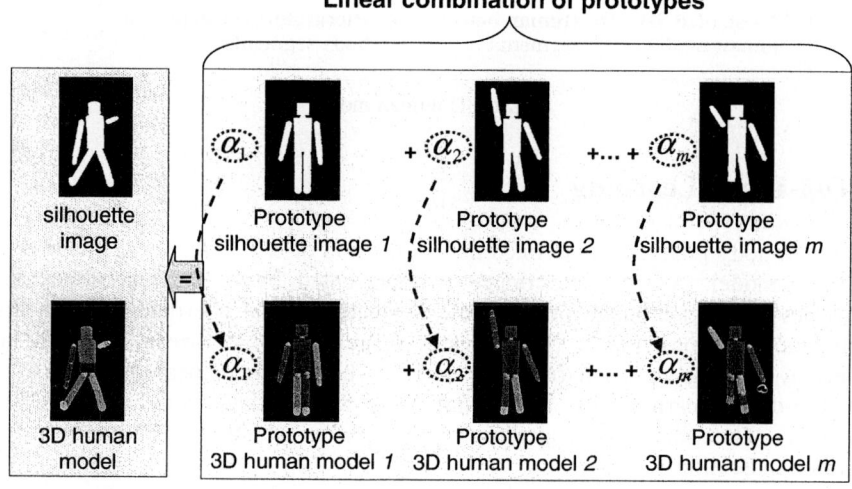

Fig. 2. Basic idea of the proposed method

The proposed method is based on the statistical analysis of a number of prototypes of the 2D images are projected from 3D human model. The silhouette image is represented by a vector $s = (s'_1, ..., s'_n)^T$, where n is the number of pixels in the image and s' is the intensity value of a pixel in the silhouette image. The 3D body model is represented by a vector $p = ((x_1, y_1, z_1), ..., (x_q, y_q, z_q))^T$, where x, y and z are the position of body joint in the 3D world and q is the number of joints in 3D human model. Eq. (4) explains training data.

$$S = (s_1, ..., s_m), \quad P = (p_1, ..., p_m) \qquad (4)$$

where m is the number of prototypes.

A 2D silhouette image is represented by a linear combination of a number of prototypes of 2D silhouette images and its 3D body model represented by estimated coefficients to the corresponding 3D body model of prototypes by such as:

$$\tilde{S} = \sum_{i=1}^{m} \alpha_i s_i \, , \quad \tilde{P} = \sum_{i=1}^{m} \alpha_i p_i \qquad (5)$$

3.3 Hierarchical Statistical Model

In order to reduce search area, we construct our algorithm hierarchically. Given a set of silhouette images and their 3D body models for training, we classify them into several clusters. A set of cluster is built in which each has similar shape in 2D silhouette image space. Then, for each of the cluster, we divide it into several sub-clusters recursively. To divide training data into sub-clusters, we apply K-means algorithm. Our hierarchical model has three-levels as shown in Fig. 3.

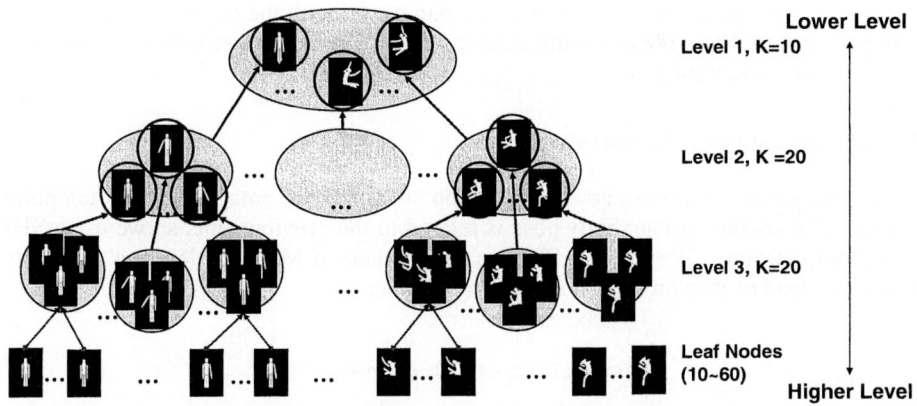

Fig. 3. Building a hierarchical statistical model

The lower level is the mean value of the data in higher-cluster. In our model, the value of the first level is the mean value of each cluster in the second level, the values of the second and third level are the mean value of each cluster in the higher level respectively, and all leaf nodes are about 100,000 and each cluster in the third level has about 10~60 leaf nodes. Each cluster in the hierarchical model is presented by one gesture representation described by Eq. (5). In other words, the element of S is the mean values of its each sub-cluster, so the number of elements of S is the number of its sub-clusters. Also, a matrix P of each cluster is the 3D body models in terms of the position of body joints which are the mean values of its each sub-cluster, so the number of elements of P is the number of its sub-clusters. The clusters in level three are presented by a linear combination of the leaf nodes instead of the mean values of their sub-clusters.

4 Reconstruction of 3D Human Body Pose

To reconstruct 3D human body pose, we use three-level hierarchical model. In the first level, we estimate 3D human body with a silhouette history image(SHI) applied spatio-temporal features which include continuous silhouette information. We compare the silhouette image reconstructed by a linear combination of prototypes of 2D silhouette images with the prototypes of 2D silhouette images, and select the prototype has the min distance. We use template matching to compare two silhouette images. After the first level, we reconstruct 3D human body with a silhouette image in the sub-cluster of current level. Our reconstruction process consists of five-steps.

Step 1. Make a SHI applied spatio-temporal features from continuous silhouette images and normalize input data.
Step 2. Estimate a parameter set α to reconstruct silhouette image from the given silhouette image.
Step 3. Reconstruct a 3D human model with the parameter set α estimated at Step 2.
Step 4. Compare the reconstructed 3D human model with the training data.
Step 5. Repeat Step 2, 3 and 4 for all levels of the hierarchical statistical model from top to bottom level.

4.1 Spatio-temporal Features

To reduce noise which occurs in extraction of silhouette image, we use temporal feature. The current human body pose is related to the previous one, so we use a SHI as temporal feature. We make a SHI use the method of MHI [2]. We use silhouette images instead of motion images making a SHI such as:

$$H_t(x, y, t) = \begin{cases} \tau & \text{if } D(x, y, t) = 1 \\ \max(0, H_t(x, y, t-1)) & \text{otherwise} \end{cases} \quad (6)$$

where τ is the duration of temporal extension to previous silhouette image and $D(x,y,t)$ is a silhouette image indicating the region of human at time t.

By accumulating silhouette image, the weight of noise data is reduced and the search area of current human body pose is expanded to related human body pose.

Another feature used to reduce noise is spatial information. The contour of silhouette image has noise such as a steep line, an unexpected concave or curvature. By blurring silhouette image, the weight of noise is reduced.

4.2 Estimating Reconstruction Parameter Set

To reconstruct 3D human body pose, we calculate the inverse matrix of S in Eq. (5). The inverse S^{-1} of a matrix S exists only if S is square. However, a matrix S is not square. In this case, we apply least square minimization. The pseudo inverse S^+ is a generation of the inverse for a matrix S. The pseudo inverse is defined such as:

$$S^+ = (S^T S)^{-1} S^T \quad (7)$$

eAnd, the solution $\alpha S = \tilde{S}$ can be rewritten such as:

$$\alpha = S^+ \tilde{S} \qquad (8)$$

After calculating $\alpha S = \tilde{S}$, we solve a set of coefficients of prototypes by Eq. (8). The silhouette image is calculated such as:

$$\tilde{S}_i = \sum_{k=1}^{m} \alpha_k S_k \qquad (9)$$

Then, the position of each segment of 3D human model is calculated by Eq. (10).

$$\tilde{P}_i = \sum_{k=1}^{m} \alpha_k P_k \qquad (10)$$

5 Experimental Results and Analysis

5.1 Experimental Environment

For training the proposed method, we generated approximately 100,000 pairs of silhouette images and their 3D human models. The silhouette images are 170 x 190 pixels and their 3D human models are 17 joints in the 3D world. Fig. 4 shows examples generated by 3D human model. We use a perspective projection transform to achieve silhouette images. Fig. 4(a) is the silhouette image projected from 3D human model at front view, and Fig. 4(b) is front, left side and right side view of 3D human model.

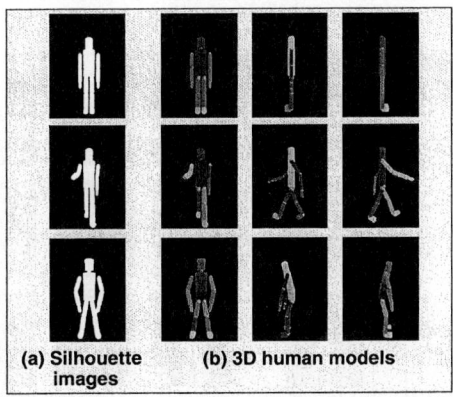

Fig. 4. Examples of 3D human body pose generated by 3D human model

For testing the performance of our method, we use two data sets. One is synthetic human bodies generated by 3D human model. The other is real data such as walking, sitting on chair, hand raising and bending captured with digital camera, Sony PC-100 with the resolution of 320 x 240. The real data were captured for 2 humans at front and right side view at the same time. The sequence lengths of data are about 120~180 frames for each sequence.

5.2 Experimental Results

Fig. 5 shows the reconstructed results obtained in several images come from video sequences at front view. Even though the characteristics of the human body different from the ones used for testing, good result is achieved. In Fig. 5, the input image is silhouette image at front view and Fig. 5(a) is side view of Fig. 5(b). Fig. 5(c) represents the reconstructed 3D human model of Fig. 5(b). The images in Fig. 5(c) represents silhouette image, front view, left side view and right side view of reconstructed 3D human body respectively. Fig. 6 shows the reconstructed 3D human body pose with a sitting on chair sequence.

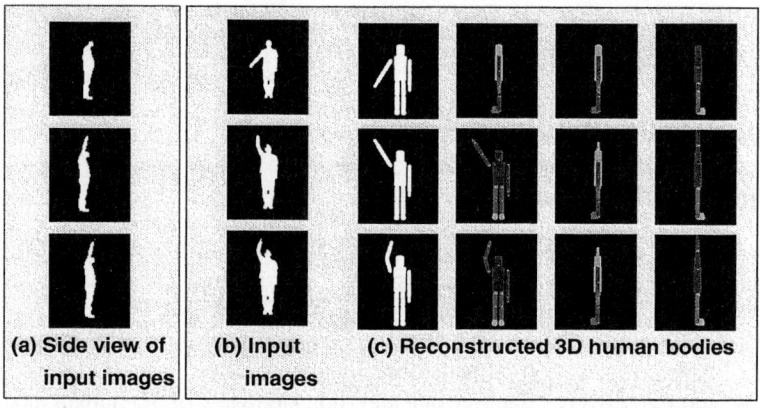

Fig. 5. Examples of the reconstructed 3D human body pose with a hand raising sequence

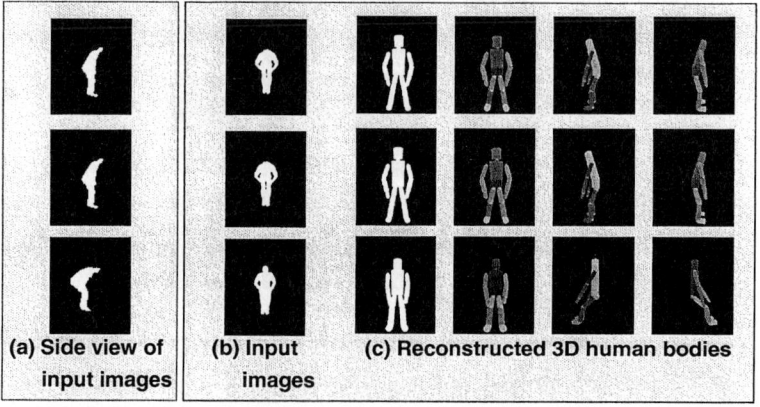

Fig. 6. Examples of the reconstructed 3D human body pose with a sitting on chair sequence

Fig. 7 shows examples containing errors. The second row shows one example of ambiguity of hand position. The hand of the input image is located at the front part of the body, but the reconstructed 3D human body shows that one is at front part of the body, the other is at back of the body. It stems from that a silhouette image is related to different human body poses. In other words, a silhouette image is generated by different human's pose each other.

Fig. 8 shows the reconstructed results obtained in test set generated by 3D human model described in Section 2. We generated 1,000 pairs of silhouette images and theirs 3D human models for testing.

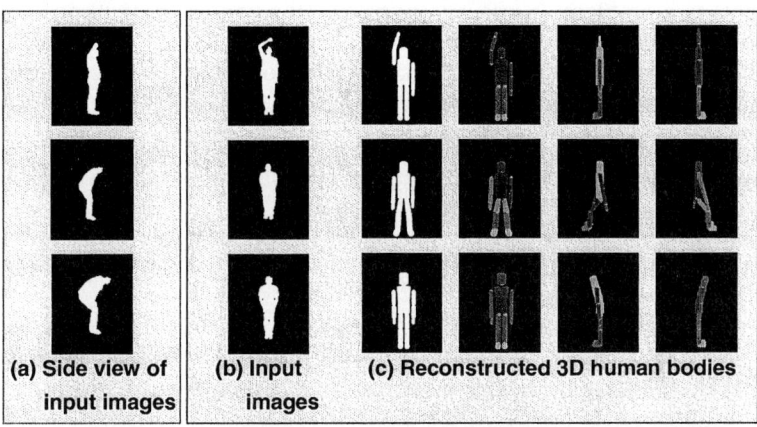

Fig. 7. Examples of the reconstructed 3D human body pose containing errors

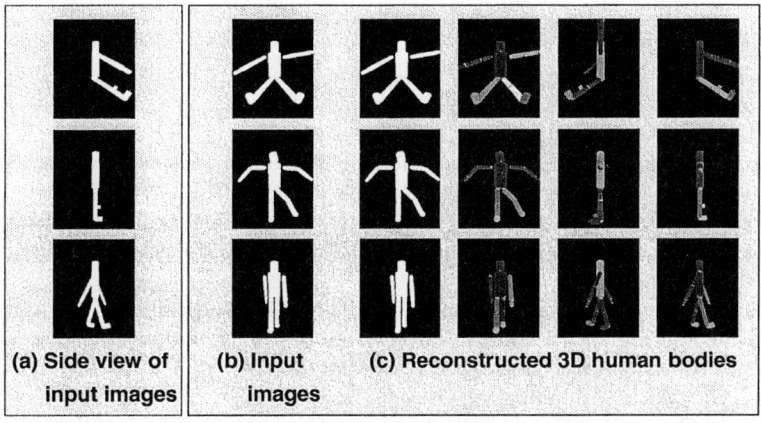

Fig. 8. Examples of the reconstructed 3D human body pose using silhouette images generated by 3D human model

6 Conclusion and Further Research

In this paper, we proposed an efficient method for reconstructing 3D human body pose from monocular image sequence using top-down learning. Human body pose is represented by a linear combination of prototypes of 2D silhouette images and their corresponding 3D body models in terms of the position of a predetermined set of joints. With the 2D silhouette images and their corresponding 3D body models, we can estimate optimal coefficients for a linear combination of prototypes of the 2D silhouette images and their corresponding 3D body models by solving least square minimization.

The performance of the presented method shows that reconstructing 3D human body pose from visual features obtained from a single image is possible. But, a silhouette image is related to different human body poses. Because a silhouette image is generated by different human's pose each other. The depth of human body components is a problem. In order to overcome the confusion of ill-posed problem, we use low-level information extracted from continuous stereo images. Using additive low-level information such as depth, color information, we can analyze the relationship of human body components.

Acknowledgements

This research was supported by the Intelligent Robotics Development Program, one of the 21st Century Frontier R&D Programs funded by the Ministry of Commerce, Industry and Energy of Korea.

References

1. Ahmad, M., Lee, S.-W.: 3D Human Body Modeling for Gesture Recognition. Technical report CAVR-TR-2004-12, Center for Artificial Vision Research, Korea University (2004)
2. Bobick, A. F., Davis, J.: The Representation and Recognition of Action Using Temporal Templates. IEEE Trans. on Pattern Analysis and Machine Intelligence, Vol. 23, No. 3, (2001) 257-267
3. Bowden, R., Mitchell, T. A., Sarhadi, M: Non-linear Statistical Models for 3D Reconstruction of Human Pose and Motion from Monocular Image Sequences. Image and Vision Computing, Vol. 18, No. 9, (2000) 729-737
4. Heap, T., Hogg, D.: Improving Specificity in PDMs Using a Hierarchical Approach. Proc. of 8th British Machine Vision Conference, Colchester, UK (Sep. 1997) 590-599
5. Howe, N., Leventon, M., Freeman, M.: Bayesian Reconstruction 3D Human Motion from Single-Camera Video. Advances in Neural Information Processing System, Vol. 12, (2000) 820-826
6. Mori, G. et al.: Recovering Human Body Configurations: Combining Segmentation and Recognition. Proc. of the IEEE Computer Society Conference on Computer Vision and Pattern Recognition, Washington D.C., USA (July 2004)
7. Mori, G., Malik, J.: Estimating Human Body Configurations using Shape Context Matching. Proc. of 7th European Conference on Computer Vision, Copenhargen, Denmark (May 2002) 666-680
8. Ong, E. J., Gong, S.: A Dynamic Human Model Using Hybrid 2D-3D Representations in Hierarchical PCA Space. Proc. of 10th British Machine Vision Conference, Nottingham, UK (Sep. 1999) 33-42
9. Rosales, R., Sclaroff, S.: Specialized Mapping and the Estimation of Human Body Pose from a Single Image. Proc. of IEEE Workshop on Human Motion, Texas, USA (Dec. 2000) 19-24
10. Song, Y., Feng, X., Perona. P.: Towards Detection of Human Motion. Proc. of the IEEE Computer Society Conference on Computer Vision and Pattern Recognition, South Carolina, USA, (June 2000) 810-817

2D and 3D Full-Body Gesture Database for Analyzing Daily Human Gestures

Bon-Woo Hwang, Sungmin Kim, and Seong-Whan Lee[*]

Department of Computer Science and Engineering, Korea University,
Anam-dong, Seongbuk-ku, Seoul 136-713, Korea
{bwhwang, smkim, swlee}@image.korea.ac.kr

Abstract. This paper presents a database of 14 representative gestures in daily life of 20 subjects. We call this database the 2D and 3D Full-Body Gesture (FBG) database. Using 12 sets of 3D motion cameras and 3 sets of stereo cameras, we captured 3D motion data and 3 pairs of stereo-video data at 3 different directions for each gesture. In addition to these, the 2D silhouette data is synthesized by separating a subject and background in 2D stereo-video data and saved as binary mask images. In this paper, we describe the gesture capture system, the organization of database, the potential usages of the database and the way of obtaining the FBG database. We expect that this database would be very useful for the study of 2D/3D human gestures.

1 Introduction

Human gesture recognition has been an interesting topic in the research of human behavior understanding, human-machine interaction, machine control, surveillance, etc. For the efficient and reliable study in these fields, gesture database is desperately required for observing or analyzing the characteristics of human gesture and verifying or evaluating the developed algorithms. Several gesture databases have been constructed for these purposes[1][2][3][7][9]. They can be classified according to the target components of human body and the dimension of obtained data. In the classification of target component, hand gesture database is commonly collected and used for studying algorithm and developing system for hand gesture and sign language recognition[10][13]. The gesture data of upper-body including hands, arms, shoulders and head is exploited for the research of tracking the human body parts and reconstructing human pose[6][11]. However, there are few databases, in which upper body gestures are systematically collected. Capturing lower-body data mainly focuses on human gait for individual identification[5][8]. Gait data are carefully captured at different directions under indoor or outdoor environments[2][7][9][12]. In this case, the motion and appearance of not only lower body, but also full-body are often captured as simple 'waking' gesture.

In spite of many restrictions such as view dependency and self occlusion among body components, many studies of gesture have been performed on 2 dimensional data due to the low complexity and the easiness of data acquisition[5][8][13]. The 2D

---

[*] To whom all correspondence should be addressed.

gesture data is commonly captured with video cameras on blue screen or plain background for extracting body parts of subjects from background. Although some 3D full-body gesture databases are constructed for studying 3D pose estimation and gesture recognition[1][3], they rarely contain gestures which are performed by many subjects. In the 'Motion Capture Database' at CMU Graphics Lab, although a subject performs same gesture in many times (2~15 times), gestures that more than 5 subjects perform seldom exist except 'walk' gesture[1]. In the 'ICS Action Database' at University of Tokyo, 3D motion data of 25 gestures for 5 subjects are opened in public[3]. Anyway, there has been no database that tried to contain various full-body gestures performed by many subjects and obtained by 2D and 3D capture system at the same time.

This paper presents the 2D and 3D Full-Body Gesture (FBG) database for analyzing 2D and 3D gesture and its related studies. In the FBG database, each of 14 gestures is performed by 20 subjects and captured by 3D motion capture system and 3 sets of stereo camera systems, simultaneously. In this paper, we describe the gesture capture system, the organization of database, the potential usages of the database and the way of obtaining the FBG database.

Fig. 1. Studio for capturing 2D and 3D gesture data

2 Database Overview

The FBG database contains 14 representative full-body gestures in daily life for 10 male and 10 female subjects of 60~80 ages. Table 1 shows the distribution of subjects by sex and age. This database consists of major three parts: (1) 3D motion data, (2) 2D stereo-video data and (3) 2D silhouette data. Using 12 sets of 3D motion cameras and 3 sets of stereo cameras, the 3D motion data and three pairs of 2D stereo-video data are obtained for each gesture in a capture studio(Figure 1). After capturing, we manually construct the correspondence between 3D motion data and 2D video data by using a synchronization program. An program operator finds the matched one among rendered skeleton images from 3D motion data to each 2D gesture image in 2D video data. The frame number of the matched image is recorded as synchronization data for 2D video data. The 2D silhouette data is synthesized by separating a subject and background from the 2D stereo-video data. All data are saved according to the file naming rule(Table 2) in order that user can easily recognize information of subject, capture condition and file.

Table 1. Distribution of subjects by age and sex

Sex \ Age	60~69(50%)	70~79(35%)	More than 80(15%)
Male (10 subjects)	5	4	1
Female (10 subjects)	5	3	2

3 Capture System

3.1 3D Gesture Capture System

In order to obtain 3D motion data of various gestures, we exploit the *Eagle Digital System* of *Motion Analysis Co*. The *Eagle Digital System* consists of *Eagle Digital Cameras*, the *EagleHub*, and *EVaRT* software, which can capture subject's motion with high accuracy(Figure 2). The motion capture camera, *Eagle Digital Camera* supports a resolution of 1280 x 1024 pixels at up to 500 frames per second. The 3 sets of Digital Cameras are located in each side of the studio, the volume of which is 10.1m (width) x 11.1m (length) x 5.1m (height). We totally positioned the 12 cameras at 4.6m heights(Figure 4). The hub system, *EagleHub* consists of multi-port Ethernet switch (100 Mbps) and provides power for the 12 digital cameras by connecting with them. Using real-time motion capture software, *EVaRT*, we can install system, calibrate cameras, capture motion in real-time, edit and save data in the various formats. All subjects wear a black and blue color suit, on which 33 markers reflecting

Fig. 2. 3D motion capture system of *Motion Analysis Co.*: Digital camera (left), Hub system (center) and real-time motion capture software (right)

Fig. 3. Body suit for motion capture

light from LED of 3D cameras are attached(Figure 4). All 3D cameras are synchronized and the 3D position of makers is obtained at 60 frames per second. Figure 3 shows a body suit with makers.

3.2 2D Gesture Capture System

In contrary to the previous gesture databases, we captured 2D and 3D gesture data, simultaneously. Especially, 2D video data is captured with stereo camera system (STH-MDCS2) made by *Videre Design*. 2D stereo camera systems are 4m away from a subject and placed at +45, -45, 0 degrees for obtaining gestures at 3 different directions. This system includes 6.0mm, F 1.4, C mount lenses and the stereo baseline of camera is 9cm. It captures uncompressed video at 320 x 240 resolution, color and 30 frames per second. The 3 pairs of stereo-video data are captured by progressive scan mode and saved in uncompressed 'AVI' file format. Before capturing gestures images, each camera is calibrated with black and white pattern, which is usually exploited for calibration process. In order to easily separate subject from background, several pieces of white fabric are used. Those cover right, left and rear sides of studio, which appear in views of 3 stereo cameras(Figure 4).

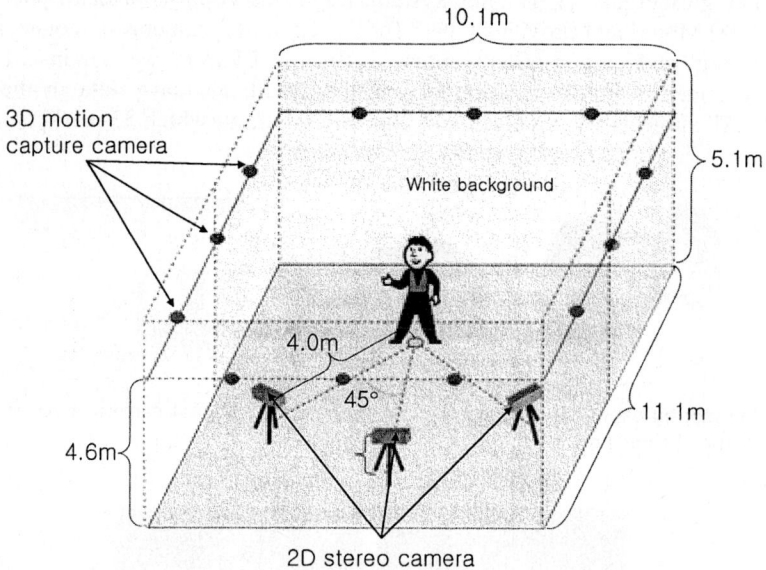

Fig. 4. Studio overview

4 Database Organization

4.1 Captured Gestures

Although the human gestures in everyday life are a wide variety, we define the most common 14 gestures: (1) sitting on a chair, (2) standing up from a chair, (3) walking at a place, (4) touching a knee and a waist, (5) raising a right hand, (6) sticking out a hand, (7) bending a waist, (8) sitting on the floor, (9) getting down on the floor, (10) lying down on the floor, (11) waving a hand, (12) running at a place, (13) walking forward, and (14) walking circularly. We ask a subject to behavior with his/her own style and captured 14 gestures with 3D motion capture cameras and 3 sets of stereo cameras at 3 different directions. Figure 5 shows the examples of each gesture.

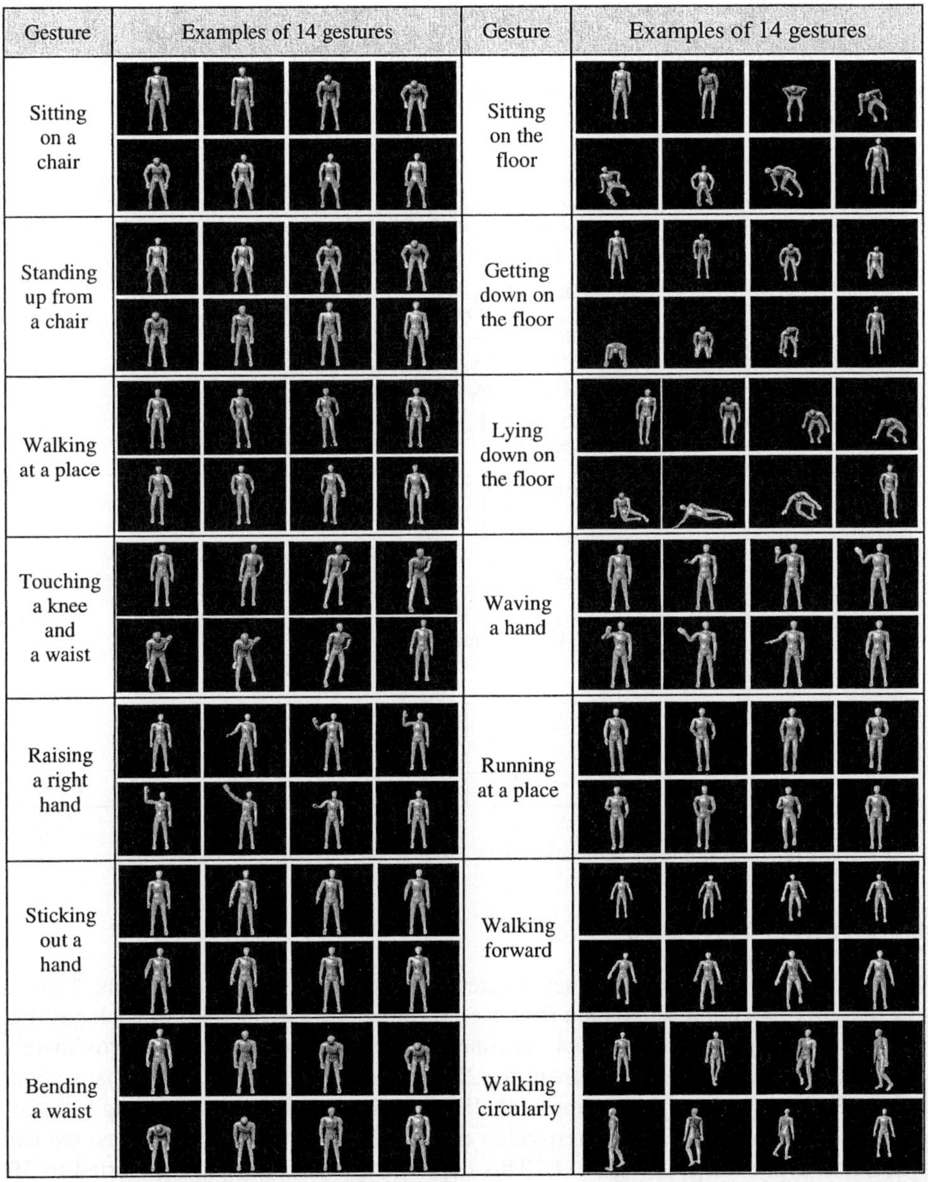

Gesture	Examples of 14 gestures	Gesture	Examples of 14 gestures
Sitting on a chair		Sitting on the floor	
Standing up from a chair		Getting down on the floor	
Walking at a place		Lying down on the floor	
Touching a knee and a waist		Waving a hand	
Raising a right hand		Running at a place	
Sticking out a hand		Walking forward	
Bending a waist		Walking circularly	

Fig. 5. Examples of 14 gestures

4.2 3D Motion Data

For representing the captured 3D motion data, we need a 3D human body model. Our 3D human body model consists of 21 segments, which are shown in Figure 6[4]. 3D motion data is saved in *Motion Analysis* 'HTR (Hierarchical Translation Rotation)' format, which contains the hierarchy of 3D human body model, the base position of each segment and 3D translation and rotation difference between adjacent frames at each segment. The storage required per subject is approximately 20MB. Thus, the total storage requirement for 20 subjects is about 400MB. Figure 7 represents front and side views of 3D motion data of 'lying down on the floor' gesture.

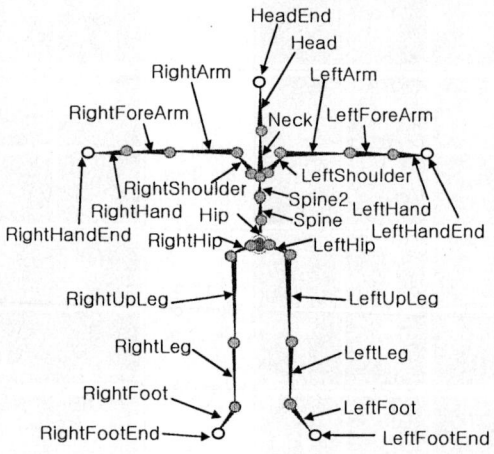

Fig. 6. 3D human body model

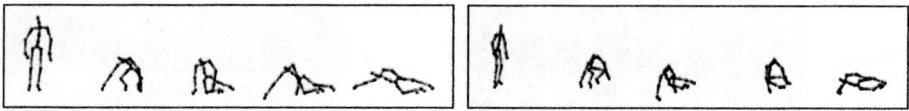

Fig. 7. Example of 3D motion data: front view(left) and side view(right)

4.3 2D Stereo-Video Data

We captured 14 gestures from 3 sets of stereo cameras at different directions. Thus, 6 sets of 320 x 240 color video data from stereo cameras are generated for each gesture. The storage required per subject for uncompressed 'AVI' files is approximately 4.5GB. The total storage requirement for 2D stereo-video data of 20 subjects is about 90GB. Figure 8 shows the example of 2D stereo-video data for 'Bending a waist' gesture. From a pair of 2D stereo-video data captured with stereo cameras, we can synthesize depth images (Figure 9). This depth image can be easily converted to 3D data in x, y, z space.

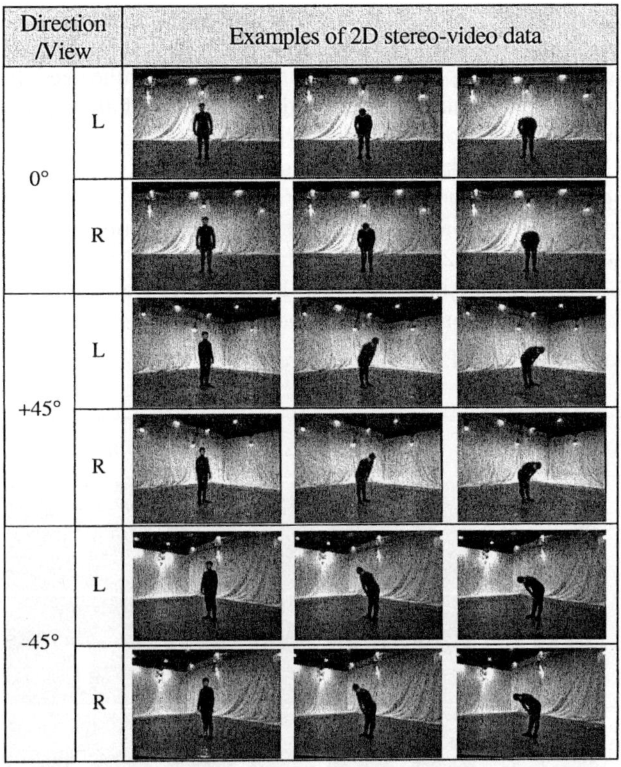

2D and 3D Full-Body Gesture Database for Analyzing Daily Human Gestures   617

Fig. 8. Examples of 2D stereo-video data captured with 3 sets of stereo cameras

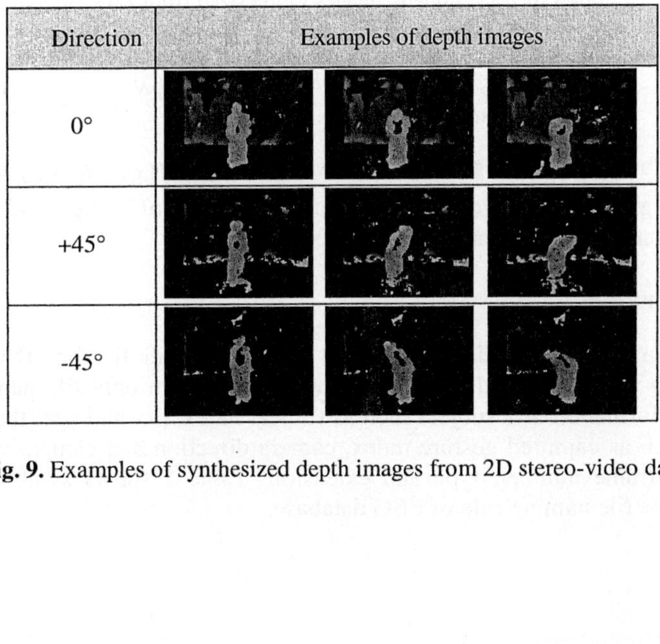

Fig. 9. Examples of synthesized depth images from 2D stereo-video data

4.4 2D Silhouette Data

2D silhouette data is not captured from cameras, but synthesized from 2D stereo-video data. Therefore, the number, size and frame rate of the synthesized data are same to those of the 2D stereo-video data. When the 2D stereo-video data is captured, more than 500 frames of background video data without a subject are saved for each subject. For separating a subject who performs 'sitting on a chair' and 'standing up from a chair' gestures, the background video data with only a chair and without a subject is also captured. The 2D stereo-video data and the background video data are converted to 8 bit gray-level images. From Equation 1, we synthesize 2D silhouette data, $S(x_i)$, in which the region of background is '0' and the region of a subject is '1', and save them as binary 'BMP' format(Figure 10).

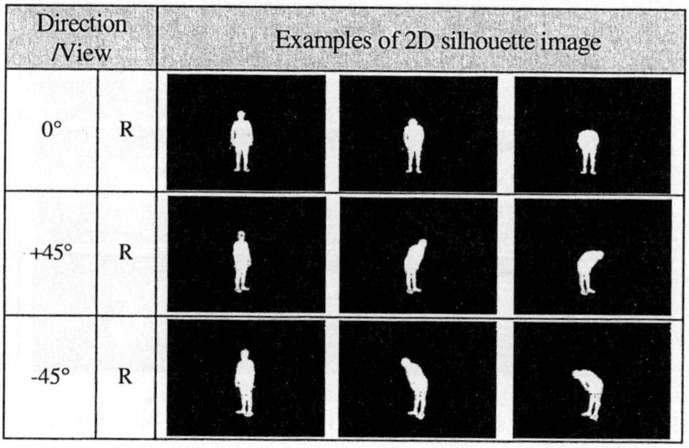

Fig. 10. Examples of 2D silhouette images synthesized from 2D stereo-video data

$$S(x_i) = \begin{cases} 1 & if \ |I(x_i) - B^M(x_i)| \geq k \cdot B^{Std}(x_i) \\ 0 & otherwise \end{cases}, \ i=1,...,N \quad (1)$$

where N is the number of pixels in 2D video image. $I(x_i)$, $B^M(x_i)$ and $B^{Std}(x_i)$ represent the gray-level image of 2D video data, the mean of background data and the standard deviation of background data, respectively.

4.5 File Naming Rule

For user's convenience, we define a simple file naming rule for the 3D motion data, the 2D stereo-video data and the 2D silhouette data. From only file name, user can realize the information of a subject such as subject index, sex and age; that of capture condition such as captured gesture index, camera direction and camera view; that of file such as frame number, type and extension. Table 2 shows symbols and their meaning in the file naming rule of FBG database.

Table 2. File naming rule for gesture data

\multicolumn{2}{c}{Inx-E-SAg_Gi_CdV_TFram.Ext}
Symbol	Meaning
Inx	Index {000, ..., 999}
E	sEssion {1, 2, ..., 9}
S	Sex {M, F}
Ag	Age {00,..., 99}
Gi	Gesture index {00,..., 99}
Cd	**Camera direction**: {00: 3D motion data, 01(0°), 02(45°), 03(-45°): 2D stereo-video data, 04: synchronization data }
V	**camera View** {D: 3D motion data, L: left view, R: right view, C: synchronization data}
T	**file Type** {D: 3D motion data, V: 2D stereo-video data, S: 2D silhouette data, C: synchronization data}
Fram	**Frame number** {dddd: 3D motion data, vvvv: video data, 0000, ..., 9999: image data, cccc: synchronization data }
Ext	file Extension {HTR, AVI, BMP}

5 Potential Usages of the Database

At present, this database is used in human gesture studies such as 3D gesture recognition and 3D body components estimation at the Center for Artificial Vision Research, Korea University. Including these, we present some of the potential usages of the 2D and 3D FBG database:

- Development of 3D gesture recognition algorithms that use depth information from stereo cameras.
- Evaluation of 2D/3D body components estimation algorithms from color, gray or silhouette images of various gestures.
- Evaluation of 2D gesture recognition algorithms under pose variations: i.e. algorithms for which the gallery and probe images have different poses.
- Evaluation of 2D gesture recognition algorithms that use multiple images across pose.
- Evaluation of 3D gesture recognition algorithms from position or angle of body components.

6 Obtaining the FBG Database

Anyone who is interested in downloading the database must contact the third author by e-mail at swlee@image.korea.ac.kr or visit the FBG database web site at http://image.korea.ac.kr/projects/gesture.html.

Acknowledgements

This research was supported by the Intelligent Robotics Development Program, one of the 21st Century Frontier R&D Programs funded by the Ministry of Commerce, Industry and Energy of Korea.

References

1. The Motion Capture Database at CMU Graphics Lab, http://mocap.cs.cmu.edu/
2. The Gait Recognition Database at Georgia Tech, http://www.cc.gatech.edu/cpl/projects/hid/Description.html.
3. The ICS Action Database at University of Tokyo, http://www.ics.t.u-tokyo.ac.jp/action/ (in Japanese)
4. Ahmad, M., Lee, S.-W.: 3D Human Body Modeling for Gesture Recognition. Technical report CAVR-TR-2004-12, Center for Artificial Vision Research, Korea University, (2004)
5. Bobick, A.F., Johnson, A.Y.: Gait Recognition Using Static, Activity-Specific Parameters. Proc. of the IEEE Computer Society Conference on Computer Vision and Pattern Recognition, Vancouver, Canada, (July 2001) 423-430
6. Bowden, R., Mitchell, T., Sarhadi, M.: Non-Linear Statistical Models for the 3D Reconstruction of Human Pose and Motion from Monocular Image Sequences. Image and Vision Computing, Vol. 18. No. 9. (2000) 729-737
7. Lee, L., Grimson, W.: Gait Analysis for Recognition and Classification. Proc. of the IEEE Conference on Face and Gesture Recognition, Washington, DC, USA. (May 2002) 155-161
8. Foster, J., Nixon, M., Prugel-Bennett, A.: Automatic Gait Recognition Using Area-Based Metrics. Pattern Recognition Letters, Vol. 24. No. 14. (2003) 2489-2497
9. Gross, R., Shi, J.: The CMU Motion of Body (MoBo) Database. Technical Report CMU-RI-TR-01-18, Robotics Institute, Carnegie Mellon University, (2001)
10. Martinez, A., Wilbur, R., Shay, R., Kak, A.: Purdue RVL-SLLL ASL Database for Automatic Recognition of American Sign Language. Proc. of IEEE International Conference on Multimodal Interfaces, Pittsburgh, USA, (October 2002) 167-172
11. Ong, E.-J., Gong, S.: The Dynamics of Linear Combinations: Tracking 3D Skeletons of Human Subjects. Image and Vision Computing, Vol. 20. No. 5-6. (2002) 397-414
12. Shutler, J., Grant, M., Nixon, M., Carter, J.N.: On a Large Sequence-Based Human Gait Database. Proc. 4th International Conference on Recent Advances in Soft Computing, Nottingham, UK, (December 2002) 66-71
13. Triesch, J., von der Malsburg, C.: A System for Person-Independent Hand Posture Recognition against Complex Backgrounds. IEEE Trans. on Pattern Analysis and Machine Intelligence, Vol. 23. No. 12. (2001) 1449-1453

Sound Classification and Function Approximation Using Spiking Neural Networks

Hesham H. Amin and Robert H. Fujii

The University of Aizu, Aizu-Wakamatsu, Fukushima, Japan
{d8042201, fujii}@u-aizu.ac.jp

Abstract. The capabilities and robustness of a new spiking neural network (SNN) learning algorithm are demonstrated with sound classification and function approximation applications. The proposed SNN learning algorithm and the radial basis function (RBF) learning method for function approximation are compared. The complexity of the learning algorithm is analyzed.

1 Introduction

Spiking neural networks (SNNs) model biologically inspired neural networks [1]. An SNN is composed of spiking neurons as processing elements which utilize interconnection synapses to exchange information with each other [10]. A spiking neuron receives spikes at its inputs and fires an output spike at a time dependent on the inter-spike times of the input spikes. Thus, SNNs use temporally coded input for information processing. A sequence of spikes with various inter-spike interval (ISI) times forms a spike train.

In this paper, the usefulness and practicality of a new proposed learning algorithm for SNN which process spike trains are shown with two applications. The spike train mapping scheme described in [3] is used in conjunction with the proposed SNN learning algorithm.

2 Spiking Neuron Model

The spiking neuron model employed in this paper is based on the Spike Response Model (SRM) [6] with some modifications. Input spikes come at times $\{t_1...t_n\}$ into the input synapse(s) of a neuron. The neuron outputs a spike when the internal neuron membrane potential $x_j(t)$ crosses the threshold potential ϑ from below at firing time $t_j = min\{t : x_j(t) \geq \vartheta\}$. The threshold potential ϑ is assumed to be constant for the neuron.

The relationship between input spike times and the internal potential of neuron j (or Post Synaptic Potential (PSP)) $x_j(t)$ can be described as follows:

$$x_j(t) = \sum_{i=1}^{n} W_i . \alpha(t - t_i), \quad \alpha(t) = \frac{t}{\tau} e^{1-\frac{t}{\tau}}. \tag{1}$$

W_i is the ith synaptic weight variable which can change the amplitude of the neuron potential $x_j(t)$, t_i is the ith input spike arrival-time, $\alpha(t)$ is the spike response function, and τ represents the membrane potential decay time constant.

In this paper, the $\alpha(t)$ function is approximated as a linear function for $t \ll \tau$. It then follows that the internal neuron potential Equation 1, can be re-written as:

$$x_j(t) = \frac{t}{\tau_1} \sum_{i=1}^{n} W_i.u(t-t_i); \quad t \ll \tau_1. \tag{2}$$

$u(t)$ is the Heaviside function and $\tau_1 = \frac{e}{\tau}$.

3 Mapping-Learning Scheme for Spiking Neural Networks

A one-to-one correspondence between input spike trains and output spike firing times is necessary for the learning algorithm proposed in this paper. By selecting an appropriate set of synaptic weights for a neuron, a particular spike train or a set of spike trains which belong to the same class can be distinguished by the output firing time of the neuron because of the one-to-one correspondence between the input and output. The combined mapping-learning organization is shown in Figure 1.

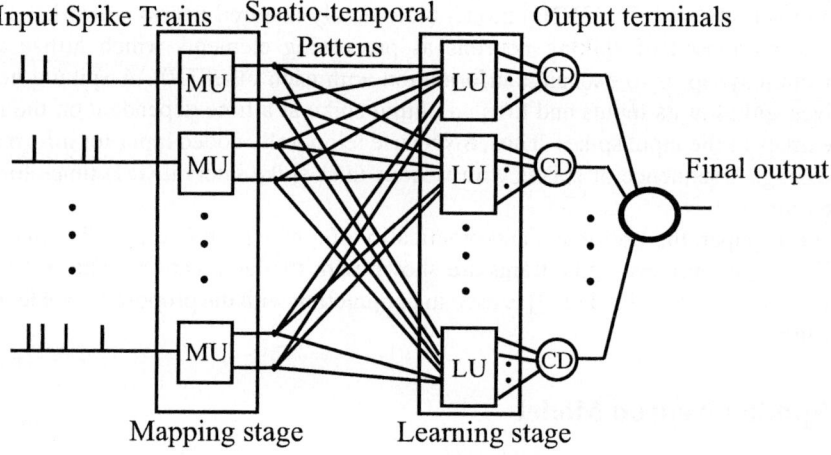

Fig. 1. Combined mapping-learning organization

Learning is performed in two stages: (1) The *mapping stage* is composed of neural mapping units (MUs) as shown in Figure 1. This stage was described in [2],[3] and it is used for mapping the input spike train(s) into unique spatio-temporal output patterns. The one-to-one relationship between the inputs and outputs of the mapping stage was proved in Appendix A of [3]. (2) The *learning stage* consists of several learning units (LUs) as shown in Figure 1. The learning stage receives the spatio-temporal output pattern produced by the mapping stage. Each learning unit is composed of sub-learning units as shown in Figure 2(A). Each sub-learning unit (e.g LUA1) takes inputs from one mapping unit (MU) as shown in Figure 2(B). As shown in Figure 2(B), the outputs t_1 and t_2 from the mapping unit are input into the sub-learning unit ISI blocks. The ISI block performs the same function as the ISI block used in the mapping units used in [2],[3]; the learning unit ISI block input synaptic weights are assigned using $W_i = \beta \cdot t_i$ and $W_i = \frac{\beta}{t_i}$ for the ISI1 and ISI2 blocks respectively. It should be noted that in a learning unit there are $2n$ ISI blocks where n is the number of input spike trains. The t_r reference time input shown in Figure 2(A) is used as a local reference signal for the combined mapping-learning organization shown in Figure 1. The coincidence generation (CG) neurons in a sub-learning unit perform the function of aligning their output spike times. When all CG neurons in an LU fire simultaneously, the coincidence detection (CD) neuron fires.

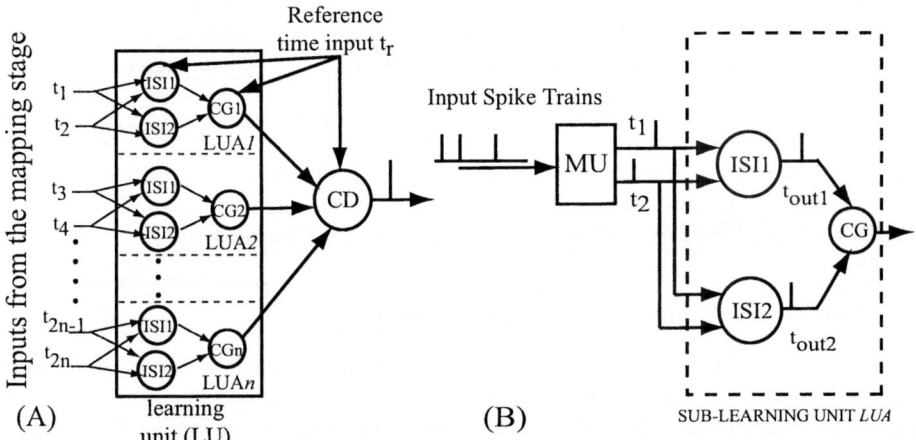

Fig. 2. (A) Details of Learning Unit (LU) (B) MU and sub-learning unit

Past learning algorithms for spiking neural networks such as back-propagation (SpikeProp) [5], self-organizing map (SOM) [12], and radial basis function (RBF) [11] used synaptic weights and delays as well as multiple sub-synapses as the learning parameters. The learning algorithm proposed in this paper can perform learning in one step and utilizes only synaptic weights for learning. Hence, the proposed algorithm is simpler than past approaches and more practical to implement in hardware.

3.1 The Learning Algorithm

The spatio-temporal patterns generated by the ISI1 and ISI2 blocks in the mapping stage, described in [2],[3], are used as inputs for the learning stage where a supervised learning method is used to classify input patterns. Clustering of input patterns which belong to the same class is achieved by setting the synaptic weights for a learning unit (LU) so that its output fires at approximately the same time for as many input spike trains as possible that belong to the same class. In this learning scheme, all input spike train samples used for learning must be known a priori.

The supervised learning algorithm works as follows:

1. Choose an input pattern vector (say P_A) at random from the set of $P_l = (P_A, P_B, ...)$ pattern vectors to be used for the learning phase. The randomly chosen pattern P_A is used to assign weights to all the ISI blocks in a learning unit. This learning unit will represent the class to which pattern P_A belongs. Once the weights have been assigned, they are temporarily fixed. The weights selected for the initial input pattern works as a center vector which can later be modified slightly to accommodate more than one input pattern.
2. Another input pattern (P_B) belonging to the same class as pattern P_A is selected. This new pattern is applied to the learning unit for P_A and the output of the ISI blocks times for $P_B\{t_{out1}, t_{out2}, ..., t_{out2n}\}$ are compared against the output times for P_A $\{t^*_{out1}, t^*_{out2}, ...t^*_{out2n}\}$. This new pattern ($P_B$) is assigned to the learning unit (e.g. P_A) with which each of the output times differ by less than ϵ.

$$|t^*_{out1} - t_{out1}| \leq \epsilon \quad , |t^*_{out2} - t_{out2}| \leq \epsilon \quad , ... \quad \text{and} \quad |t^*_{out2n} - t_{out2n}| \leq \epsilon. \quad (3)$$

ϵ is a small error value determined empirically. If the error is larger than ϵ for any one of the error conditions in Equation 3, a new learning unit is added.
3. Steps 1 and 2 are repeated for all input patterns in the learning set P_l.

3.2 Learning Unit Input-Output Time Relationship

A one-to-one relationship between inputs and outputs for each of the learning units must be achieved in order to guarantee that each learning unit outputs a spike at a time which is different from the output times corresponding to other inputs. This one-to-one relationship will be shown using one MU and one sub-learning unit. When a new pattern (e.g. pattern P_B with MU output times t_1^B and t_2^B) is input into a sub-learning unit within an LU which had its synaptic weights fixed during the learning of pattern P_A, the following will result: $(\{t^A_{out1}, t^A_{out2}\} \neq \{t^B_{out1}, t^B_{out2}\}$, where t_{out} is the output firing time of an ISI block. This can be proved by contradiction:

Assume that P_B produces the same t_{out1} or t_{out2} as P_A. For the moment, t_{out1} and t_{out2} will not be distinguished and they will simply be referred to as t_{out}. Then the internal neuron potentials $x_j(t)$ (Equation 2) for P_A and P_B at time t_{out} can be written as follows:

$$\sum_{i=1}^{2} W_i^A . u(t - t_i^A) = \sum_{i=1}^{2} W_i^A . u(t - t_i^B). \quad (4)$$

W_i^A's are the synaptic weights which have been fixed for the learning unit P_A. It can be shown from Equation 4, if the ISI1 block outputs a spike at the same t_{out} time for both P_A and P_B, the ISI2 output times will be not equal and vice versa.

3.3 Firing of Only One Learning Unit

Assume that patten P_A was learned by the learning unit A(LUA) and that patten P_B was learned by the learning unit B (LUB). Assume that the sub-learning units LUA1 and LUB1 get inputs from the same mapping unit (MU). If pattern P_A is input into both LUA1 and LUB1, the neuron internal potentials for LUA's ISI1 or ISI2 and LUB's ISI1 or ISI2 will increase according to equation 2. If $t^A_{out1} = t^B_{out1} = t_{out1}$ and $t^A_{out2} = t^B_{out2} = t_{out2}$ are assumed, the only way for LUA1 and LUB1 to produce an output spike at the same $t_{out1}(t_{out2})$ time is to have the following condition satisfied:

$$\sum_{i=1}^{2} W_i^A = \sum_{i=1}^{2} W_i^B. \tag{5}$$

Thus, if the condition specified by Equation 5 is not satisfied by any one of the sub-learning units, only one of the learning units will respond to an input pattern.

In order to have only one learning unit fire for a given input pattern, output times of the CG neurons in the sub-learning units (Figure 2(A)) have to be made coincident by changing the input synaptic weight values of the coincidence generation (CG) neurons. The coincidence detection neuron (CD), shown in Figure 2(A), uses the exponential response function (Equation 1) of a spiking neuron.

The outputs of the ISI1 and ISI2 blocks of each sub-learning unit (Figure 2(A)) fire at certain times according to the assigned synaptic weight centers. The other patterns which have been joined to the same learning unit cause the outputs to fire at times which are close to the ones corresponding to the center pattern. The coincidence detection neuron threshold value ϑ is adjusted so as to allow some fuzziness in the input spike times.

3.4 Local Reference Time

In section 3.2 it was proved that the output combination $\{t_{out1}, t_{out2}\}$ for the ISI1 and ISI2 blocks will be unique for each sub-learning unit; however, the relative time $|t_{out1} - t_{out2}|$ should also be considered for all the sub-learning units of different learning units (LUs). In other words, two different sub-learning units in two different learning units can fire at different output times, t_{out1} and t_{out2}, but the relative time $|t_{out1} - t_{out2}|$ may be the same; this would lead to two (or more) learning units firing outputs for the same input pattern. Thus, a reference time (bias) t_r input is necessary to differentiate these outputs as shown in Figure 2(A). This reference time t_r is the time when the first input spike arrives at one of the mapping stage inputs.

4 Complexity of the Learning Algorithm

The complexity of the proposed learning algorithm is calculated for a sequential machine. The learning algorithm complexity is $O(4nkp + kp^2)$ where k is the number of input spike trains, n is the number of spikes per spike train, and p is the number of learning samples. The complexity order was calculated for the worst case number of clusters needed for classification which occurs when the number of the clusters is equal to the number of learning samples p. If the number of learning patterns p is assumed to be the dominant factor in this learning algorithm, then the learning algorithm complexity can be approximated to $O(kp^2)$.

5 SNN Applications

Two applications were used to evaluate the classification capability and robustness of the proposed learning algorithm. First, a spatio-temporal input application example for a non-linear function approximation was carried out; in this case, the input mapping stage (described in [2],[3]) was not used. Instead a different pre-processing method which can encode a continuous input value into a spatio-temporal outputs described in section 5.1 below, was used. Second, the classification of sounds produced when a glass ball struck different materials of various shapes and sizes was carried out. In this case, spike trains were generated by a sound signal pre-processing unit described in [9] and then the spike trains were mapped by the mapping stage described in [3] into a spatio-temporal pattern; the learning stage was used to classify the different materials. All application simulations were carried out using Matlab® version 7.0.

5.1 Non-linear Function Approximation

An encoding scheme [5] based on an array of overlapped basis functions was employed for transforming a continuous input variable x into spike times $\{t_1, t_2, ..., t_n\}$ as shown in Figure 3. In this encoding scheme a number of gaussian functions are used to represent an input value. For the simulations, the input variable x was encoded using 8 gaussian functions. Improved representation accuracy for a particular variable can be achieved by making the width of each basis function narrower and increasing the number of basis functions. This encoding scheme has been applied successfully in unsupervised and supervised learning [4],[5].

The proposed SNN learning algorithm was used to approximate the following non-linear function $f(x) = e^{-x}\sin(3x)$ in the interval $[0, 4]$ as shown on Figure 4. The interval $[0, 4]$ was sampled at 41 points with an interval spacing of 0.1. The learning algorithm, described in section 3.1, was used to train the neural network to assign cluster

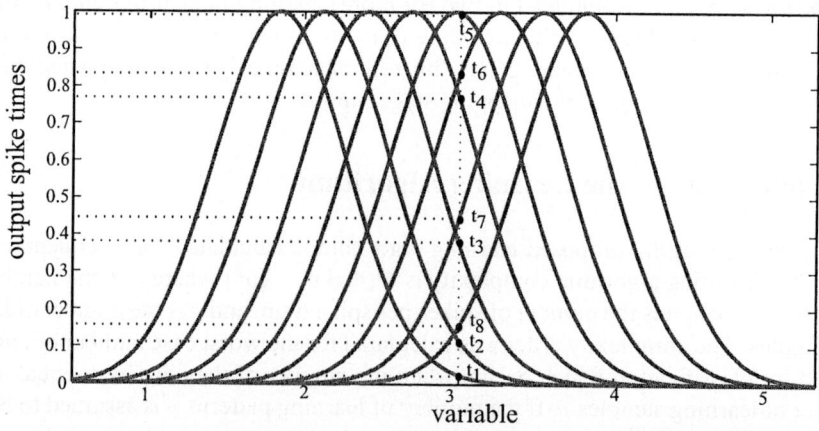

Fig. 3. Input variable x encoded into 8 spike times using 8 gaussian basis functions

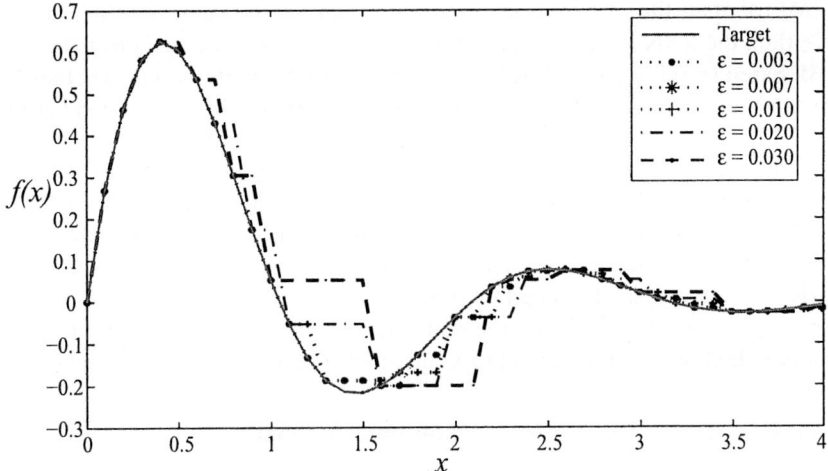

Fig. 4. Function approximation using various learning tolerances ε

centers using the PSP $x_j(t)$ function variables set to $\vartheta = 0.3$, $\beta = 0.5$, and $\tau = 5.0$ in the simulations.

To test the generalization capability of the network after training, the same interval was sampled at 401 points, at intervals of 0.01, in order to generate the test data for the neural network.

Table 1 shows the proposed SNN learning results together with the radial basis function (RBF) [7] based learning results. It can be observed from these results that as the number of clusters increases (achieved by decreasing the learning tolerance ε value in the proposed learning algorithm described in section 3.1) the learning accuracy is improved. It can also be observed that RBF learning produces a smaller maximum fit

Table 1. Comparison of the proposed SNN and RBF learning algorithms for function approximation

ε	No. of learning clusters	The Proposed SNN Learning Algorithm		RBF Algorithm (the same number of clusters)	
		Mean squared error	Max fit error	Mean squared error	Max fit error
0	41	0	0	0	0
0.003	36	0.0012	0.0320	0.0002	0.0227
0.007	28	0.0059	0.0453	0.0022	0.0415
0.010	23	0.0118	0.0867	0.0103	0.0490
0.020	17	0.0332	0.1656	0.0368	0.0789
0.030	14	0.0530	0.2700	0.0752	0.0281

error (MFE)[1] than the proposed learning algorithm for the same ε values. However, for $\varepsilon \geq 0.02$ the SNN learning algorithm had a lower mean squared error (MSE) than the RBF based learning method. RBF learning needed many iterations to achieve equal MSE values while the proposed learning algorithm only requires a one step learning. The RBF Learning method is used as a comparison because it employs supervised learning as in the proposed spiking neural learning algorithm and the number of RBF hidden layer neurons (basis function neurons) can be made equal to the number of learning units (clusters) used in the proposed SNN learning algorithm. The RBF algorithm adjusts the hidden layer neuron weights (basis function neurons) so that the summation of the basis function outputs for a certain input value produces the desired output. For the proposed SNN learning algorithm the coincidence detection neuron (CD) makes only one learning unit fire for a set of appropriate inputs belonging to the same class.

5.2 Classification of Materials Based on Impact Sounds

A sound classification experiment using actual sounds produced when a small glass ball struck different materials was performed. Sounds were recorded and then classified using the proposed SNN learning algorithm. The impacted materials were of different sizes and shapes. The materials consisted of sheets of steel (S), sheets of copper (C), and pieces of wood (D). For example C_1, C_2 in Table 2 represent two sheets of copper of different thicknesses and sizes.

Table 2. Impact sound based material classification accuracy

Material type	No. of learning patterns	No. of clusters	No. of testing patterns	Classification accuracy
S	30	18	10	70%
C_1	15	11	5	80%
C_2	15	10	5	80%
D_1	45	8	15	87%
D_2	30	13	10	70%

To encode each sound into spike trains, the method used in [8],[9] was employed. The method used in [8],[9] utilized a frequency tuned bank of filters and signal processing which produced three features for each of the filters: onset, offset, and peak times. In this experiment, all onset, offset, and peak output times of each filter belonging to the filter bank consisting of 20 band-pass filters were used as shown in Figure 5. The 20 spike trains generated by the outputs of the filter bank were mapped into a spatio-temporal pattern containing 40 spikes (two output spikes for each mapping unit) using the mapping stage described in [3]. The filter bank center frequencies ranged from 100Hz to 4000 Hz, with each filter having a bandwidth of 200 Hz. The spatio-temporal patterns for the various impact impact sounds were then used as input patterns for the learning stage as shown in Figure 1.

---

[1] MFE measures the maximum absolute error between the desired and actual outputs of the trained function.

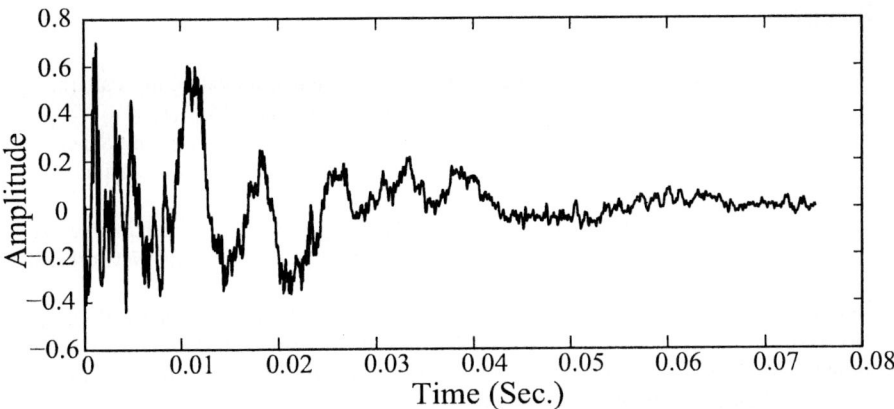

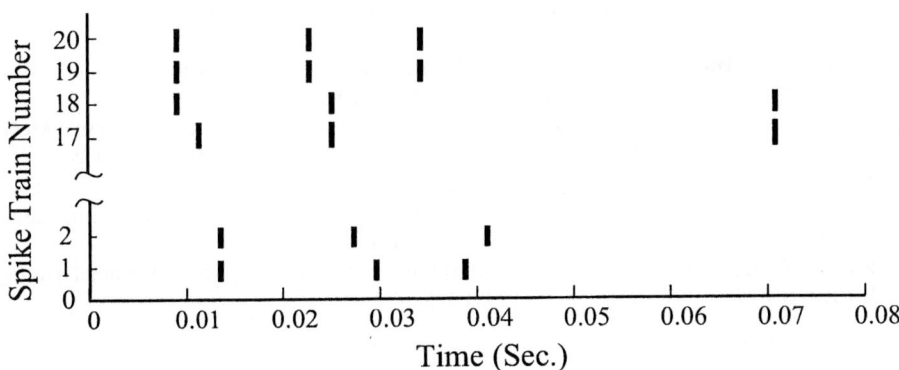

Fig. 5. Steel plate impact sound waveform and its corresponding 20 spike trains

As can be seen in Table 2, each material could be correctly classified with relatively good accuracy in the testing phase. It should be noted that the learning phase can achieve 100% learning of the learning set because learning units can be added incrementally as needed. Better classification accuracy during the testing phase may require a better way to pre-process sound signals as well as a larger learning set.

6   Conclusions

The effectiveness of a new spiking neural network learning algorithm was shown with two applications: function approximation and sound classification.

For both applications, reduction of the number of input pattern clusters needs to be addressed in order to optimize neural network size.

References

1. Abeles, M., Bergman, H., Margalit, E., Vaadia, E. : Spatiotemporal Firing Patterns in the Frontal Cortex of Behaving Monkeys. J. Neurophysiol. **70** (1993) 1629–1658
2. Hesham H. Amin, Robert H. Fujii: Input Arrival-Time-Dependent Decoding Scheme for a Spiking Neural Network. Proceeding of the 12th European Symosium of Artificial Neural Networks (ESANN'2004). (2004) 355–360
3. Hesham H. Amin, Robert H. Fujii: Spike Train Decoding Scheme for a Spiking Neural Network. Proceedings of the 2004 International Joint Conference on Neural Networks. IEEE. (2004) 477–482
4. Bohte, S.M., La Poutré, H., Kok, J.N.: Unsupervised Classification in a Network of Spiking Neurons. IEEE Transactions on Neural Networks. **13(2)** (2002) 426–435
5. Bohte, S.M., Kok, J.N., La Poutré, H.: Spike-Prop: Error-Backprogation in Multi-Layer Networks of Spiking Neurons. Proceedings of the European Symposium on Artificial Neural Networks (ESANN'2000). (2000) 419–425
6. Gerstner, W., Kistler, W.: Spiking Neuron Models. Single Neurons. Populations. Plasticity. Cambridge University Press. (2002)
7. Simon Haykin: Neural Networks, A Comprehensive Foundation. Prentice Hall International Inc. (1999)
8. Hopfield, J.J., Brody, C.D.: What Is a Moment? Cortical Sensory Integration Over a Brief Interval. Proc. Natl. Acad. Sci. **97(25)** (2000) 13919–13924
9. Hopfield, J.J., Brody, C.D.: What Is a Moment? Transient Synchrony as a Collective Mechanism for Spatiotemporal Integration. Proc. Natl. Acad. Sci. **98(3)** (2001) 1282–1287
10. Maass, W., Bishop, C., editors: Pulsed Neural Networks. MIT press. Cambridge (1999)
11. Berthold Ruf: Computing and Learning with Spiking Neurons - Theory and Simulations. Chapter (8). *Doctoral Thesis*. Technische Universitaet Graz. Austria (1997)
12. Ruf, B., Schmitt, M.: Self-Organization of Spiking Neurons Using Action Potential Timing. IEEE Trans. Neural Networks. **9(3)** (1998) 575–578

Improved DTW Algorithm for Online Signature Verification Based on Writing Forces

Ping Fang[1,2], ZhongCheng Wu[1], Fei Shen[1], YunJian Ge[1], and Bing Fang[3]

[1] State Key Laboratories of Transducers Technology,
Institute of Intelligent Machines, CAS, Hefei, 230031 Anhui, China
[2] Department of AutomationUniversity of Science Technology of China
Hefei, 230026 Anhui, China
[3] Department of Computer Science, Hong Kong Baptist University,
Hong Kong, P.R. China
pingfang@ustc.edu,
{zcwu, shenfei, yjge}@iim.ac.cn, fangb@comp.hkbu.edu.hk

Abstract. Writing forces are important dynamics of online signatures and they are harder to be imitated by forgers than signature shape. An improved DTW (Dynamic Time Warping) algorithm is put forward to verify online signatures based on writing forces. Compared to the general DTW algorithm, this one deals with the varying consistency of signature point, signing duration and the different weights of writing forces in different direction. The iterative dexperiment is introduced to decide weights for writing forces in different direction and the classification threshold. A signature database is constructed with F_Tablet and the equal error rate of 1.4% is realized with the improved algorithm.

1 Introduction

When a person signs his name, he writes not only the characters, but also his identity, which is implied in the dynamic writing process and the static signature image. Computer based online and offline signature verification approaches have been developed to extract the identity. Compared to the static handwriting image of offline approach, online one uses those dynamics during signing and has relatively higher classification rate [1], [2], [3].

The signature is the trajectory of the writing pen's contact movement on the writing surface driven by writing forces. So writing forces are one of the most important information of writing dynamics and many researches have been done on them. Crane and Ostrem developed a three-dimension force-sensitive pen to get the writing forces [4]. With input device of the SmartPen, Martens and Claesen devised an online signature verification system based on three-dimension forces [5]. Tanabe studied signature verification based on the pressure with digital pen device [6]. Sakamoto did research on signature verification incorporating pen position, pen pressure and pen inclination with WACOM Tablet [7]. Shimizu developed an electrical pen using two-dimensional optical angle sensor to get writing forces [8].

Although all kinds of writing pen devices are used to get the writing forces, they can't get the forces accurately, because a writing pen may be rotated during writing which would change the measure coordination. And WACOM Tablets can only get the writing pressure. A new writing tablet, named F_Tablet, is used here to capture the three-dimension writing forces. And an improved DTW algorithm is used to verify those signatures. Compared to the general DTW algorithm, this one deals with the varying consistency of signature point, signing duration and weights of writing forces in different direction. The F_Tablet signature capturing device and the signature database are introduced in the next chapter. Chapter 3 discusses the improved DTW algorithm and iterative experiment in detail. The experimental dresults are given in chapter 4 and conclusions are made in the last chapter.

2 Signature Acquisition

2.1 The F_Tablet

This F-Tablet is capable of capturing three perpendicular forces of the pen-tip to the contacting plane and two dimension torques directly because of the core part of a multi-dimension force/torque sensor. With the specially designed structure, the static trajectory of the pen-tip and other dynamic signals such as velocities, accelerations and writing angles can also be calculated indirectly [9]. The input tablet of $70 \times 70 mm^2$ is on the up-left side of the F_Tablet, as shown in Fig.1(a). The device is connected to computer via USB interface with a maximum sample rate of 120Hz. And there is no special requirement on the writing pen. Fig.1(b) explains the coordinate of the F_Tablet. The coordinate is fixed during design and it won't change no matter how the writing pen is rotated. Here the device is used to get the three-dimension writing forces F_x, F_y and F_z.

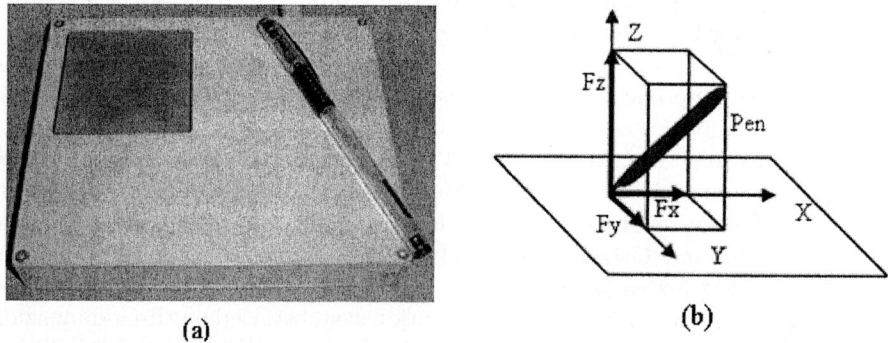

Fig. 1. (a)Photo of the F_Tablet;(b)Coordinate of the F_Tablet

2.2 Signature Database Construction

The database is constructed with 30 persons involved over a one-month period. Each subject donated 40 signatures with 10 ones every week. At the same time, each subject is told to practice and imitate other subject's signature after the static signature images as simple forgeries. And 10 persons are recruited to make skilled forgeries. Before the skilled forgeries were collected, each subject can view the signing process of the signature to be imitated with a special program and practice for a period of time what ever long he wants. And a signature database with 1200 genuine signatures, 600 simple forgeries and 300 skilled forgeries is constructed.

2.3 Signature Preprocessing

To improve the classification result, signatures are preprocessed before calculation. The preprocessing methods taken here are filtering, direction adjustment and dehooking.

1) *Filtering* To remove the noise in the signature data, the Gaussian filter is applied to filter the three dimension forces respectively.

2) *Direction Adjustment* The posing of the writer or the position of the F_Tablet may change when the signatures are collected in several batches, which results in the inconsistency of the signature direction. So force direction adjustment is introduced to adjust the force direction in X and Y direction.

3) *Dehooking* As the writing surface of the F_Tablet is a little smoother than the general paper, so a jerk may occur when a pen collide with the tablet, which will cause the wrong judgement of the pen-down or pen-up status. So dehooking is taken to remove the jerk.

3 Improved DTW Algorithm

Compared to the general DTW algorithm, this improved one takes the varying consistency of different signature point, the signing duration and weights of writing forces in different direction into account. First, multiple signature templates are generated with general DTW algorithm. Then weights of different template points are calculated. The weights of writing forces in different and classification threshold are decided with iterative experiment.

3.1 Multiple Templates Generation

Concerned to the fact that even genuine signatures from the same subject may have different stroke numbers, multiple templates are generated for signature verification. The templates can be generated with the following 6 steps:

Step 1 Segmented the registered signatures into strokes according to F_z, and classified them into different groups according to stroke number;

Step 2 Take one group and calculate each corresponding stroke's average length as the its template's stroke length, including pen-up strokes and pen-down strokes;

Step 3 Take one signature from the group and resample each pen-down stroke according to the template's corresponding stroke's length, and the result is taken as template T;

Step 4 Take another signature from the group and resample it as step 3, then aligned every pen-down stroke to the corresponding stroke of T with general DTW algorithm , and update the aligned signature point's writing force value of T. The DTW alignment is calculated with (1).

$$D_k(T_i, S_j) = \min \begin{cases} D_k(T_{i-2}, S_{j-1}) + \frac{1}{2}[d(T_{i-1}, S_j) + d(T_i, S_j)] \\ D_k(T_{i-1}, S_{j-1}) + d(T_i, S_j) \\ D_k(T_{i-1}, S_{j-2}) + \frac{1}{2}[d(T_i, S_{j-1}) + d(T_i, S_j)] \end{cases} \quad (1)$$

$$d(T_i, S_j) = \left[w_x(F_x^{(i)} - F_x^{(j)})^2 + w_y(F_y^{(i)} - F_y^{(j)})^2 + w_z(F_z^{(i)} - F_z^{(j)})^2 \right]^{1/2}. \quad (2)$$

where w_x, w_y and w_z are weights of F_x, F_y and F_z respectively, and $d(T_i, S_j)$ is the distance between signature point of template and that of sample. When the distance $D_k(T_i, S_j)$ between two corresponding strokes are calculated, then the alignment path is decided at the same time. The value of the signature point is updated as (3).

$$T_i^{(j)} = \frac{r_i^{(j-1)} * T_i^{(j-1)} + S_{i,1} + S_{i,2} + \cdots + S_{i,k}}{r_i^{(j-1)} + k}. \quad (3)$$

$$r_i^{(j)} = r_i^{(j-1)} + k. \quad (4)$$

where $T_i^{(j)}$ is the ith template point value which has been updated with j samples; $r_i^{(j)}$ is the number of times of the ith template point being aligned with j samples; $S_{i,k}$ is the kth sample point that has been aligned to the ith template point. The three dimension forces are updated respectively with (3) and (4).

Step 5 Repeat step 4 until all signatures in the group have been used and the result is the template of the group;

Step 6 Repeat step 2 ~ 5 calculation on all groups.

The calculated results are the multiple signature templates [10].

3.2 Template Signature Point Weight Calculation

Different subject has different signature consistency. And the signature points of the same subject also have different consistency. Put more weight on the points with better consistency can be sure to improve the classification result, so different weight is set for each signature point of the templates according to its consistency.

The consistency of each signature point can be described with the distribution of the corresponding aligned sample points to the template signature point. As the signature template is generated from the samples with DTW algorithm, it can be viewed as made up of the average value of the sample signature points. So for a specific signature point, if the corresponding aligned sample signature points fall far from the template point, then its signature point consistency is bad; otherwise, the specific signature point has a good consistency. That is, the consistency of a specific signature point is inverse proportion to the distances between the corresponding aligned sample points and the template signature point. The average value of the distances will be calculated to decide the template signature point weight. After the multiple templates are generated, each template and the samples of the corresponding group are used to decide each template signature point weight. First, each sample signature is aligned to the template with DTW algorithm with (1) and (2). From (1), we can see that every template signature point may have one or two aligned point pairs for each sample in the group.

After every sample of the group is calculated, we can get the following point pairs for every template point as (5).

$$(T_i, S_{1,1}), [(T_i, S_{1,2})], (T_i, S_{2,1}), [(T_i, S_{2,2})], \cdots (T_i, S_{n,1}), [(T_i, S_{n,2})] . \quad (5)$$

where T_i denotes the ith template point; $S_{n,1}$ and $S_{n,2}$ denote the first and second aligned point of the nth sample; and the point pairs in the brackets may not exist. The distance between every point pair can be calculated with (2). And the average value of the distances can be expressed as (6).

$$avgd_i = \sum_{j=1}^{n} (d_{(T_i, S_{j,1})} + [d_{(T_i, S_{j,2})}]) \Big/ M . \quad (6)$$

where M denotes the number of the aligned point pairs; and still the content in the bracket may not exist. Then the point weight can be set as inverse proportion to the average distance.

3.3 Distance Calculation

The difference between the general DTW algorithm and this improved one lies on the distance calculation. This one takes weight of writing forces in different direction, weight of different signature point and signing duration into account.

A sample signature is first segmented into strokes. If the number of strokes is different from any of the multiple templates', the distance between the sample and templates is considered to be infinite, and the sample is rejected as a forgery directly. Otherwise, a template with the same number of strokes is selected to calculate the distance between the sample and the multiple templates. First, each of the sample's pen-down stroke is resampled according to the length of the template's corresponding stroke. Then, the distance between the corresponding stroke of template and that of the sample is calculated with (7). The difference between (1) and (7) is that template point weight is used in (7). And Euclidean distance is used to measure the difference between the point pair as (2).

$$D_k(T_i, S_j) = \min \begin{cases} D_k(T_{i-2}, S_{j-1}) + \frac{1}{2}[w_{i-1}d(T_{i-1}, S_j) + w_i d(T_i, S_j)] \\ D_k(T_{i-1}, S_{j-1}) + w_i d(T_i, S_j) \\ D_k(T_{i-1}, S_{j-2}) + \frac{1}{2}[w_i d(T_i, S_{j-1}) + w_i d(T_i, S_j)] \end{cases} \quad (7)$$

Compared to the image of a signature, the writing velocity is sure to be more difficult to imitate. So the signing duration is taken into distance calculation. Signing duration can be divided into two parts, the pen-down stroke time and the pen-up stroke time. The pen-up stroke time is used to think about how to write the next stroke and adjust the pen tip position, so the pen-up stroke time and the following pen-down stroke time can be viewed as a whole stroke time. For the first pen-down stroke of a signature, the foregoing pen-up stroke time is zero. Two kinds of signing durations can be used to update the distance between the corresponding strokes. One is that only the pen-down stroke time is used, expressed as (8); the other is that both the pen-up and the following pen-down stroke time are considered as the stroke time, expressed as (9).

$$D'_k(T, S) = D_k(T, S)\left(1 + \frac{|t_{s,down} - t_{t,down}|}{t_{t,down}}\right). \quad (8)$$

$$D'_k(T, S) = D_k(T, S)\left(1 + \frac{|t_{s,down} + t_{s,up} - t_{t,down} - t_{t,up}|}{t_{t,down} + t_{t,up}}\right). \quad (9)$$

where $D_k(T, S)$ is the kth stroke distance between the sample and the weighted template and $D'_k(T, S)$ is the distance which take signing duration into account; $t_{s,down}$ and $t_{s,up}$ are the pen-down stroke point number and pen-up stroke point number of the sample before being resampled; $t_{t,down}$ and $t_{t,up}$ are those of the template.

So the final distance between the sample signature and templates is the sum of the distances between the corresponding strokes, expressed as (10).

$$D(T, S) = \sum_k D'_k(T, S). \quad (10)$$

3.4 Threshold Setting

Because the consistencies of signing of different subjects are sure to be different, so the threshold is set to reflect both personal and global signing characteristics. The threshold D_i^{th} of subject i is expressed as (11).

$$D_i^{th} = \mu_i + f\sigma_i. \quad (11)$$

where μ_i and σ_i are the average value and standard deviation of the distances between the registered signatures and templates respectively; while f is the global coefficient to make sure to get an optimistic classification performance for all subjects.

3.5 Iterative Experiment

Iterative experiment is introduced to decide the global threshold coefficient f and the ratio between force weight in different direction, that is the ration between w_x, w_y and w_z in (2). Fig.2 gives the charts of writing force F_x, F_y and F_z of two different Chinese characters. As F_x and F_y are forces parallel to the writing tablet, and we can see that their amplitudes are comparable, while that of F_z is much more bigger. So F_x and F_y are set to have same weight in distance calculation. In the experiment, the ratio between w_x, w_y and w_z is set to be $w : w : (3 - 2 \times w)$. Then the goal of the experiment is to decide the values of f and w to get an optimum signature classification result.

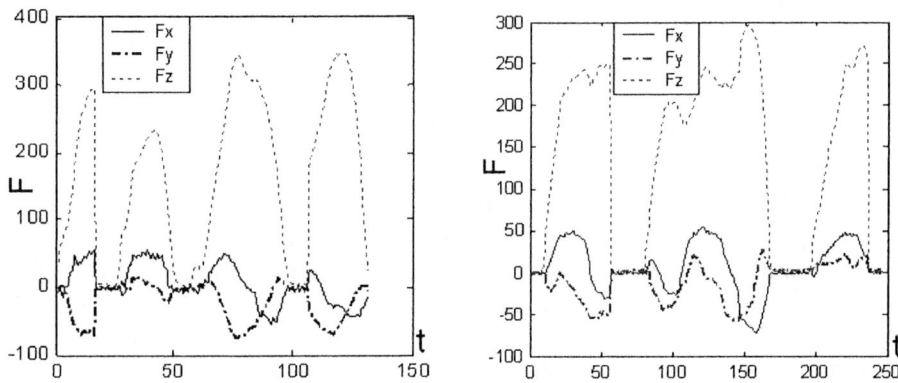

Fig. 2. The three-dimension writing forces of two different characters

First initial values are set for f and w, the three dimension writing forces are set to have the same weight. Then, let f changes from 1.0 to 3.0 and record the classification result. It is sure there is a value of f where FAR equals FRR. Then fix the value of f, change w from 0.1 to 1.5, and find the value of w where FAR equals FRR. Parameters f and w are alternatively changed to do signature classification experiments on the signature database to find the other one where FAR equals FRR, until the difference between the values of parameter f of two successive experiments is small enough. The flow chart of the iterative experiment is explained with Fig.3.

4 Experimental Results

Experiments are carried out on the constructed signature database. Four kinds of algorithms are used to compare their effects. They are the general DTW algorithm, weighted DTW algorithm, and the two improved DTW algorithm, which takes the pen-down stroke time and pen-down/pen-up stroke time together into account respectively. First the iterative experiments are carried out to get

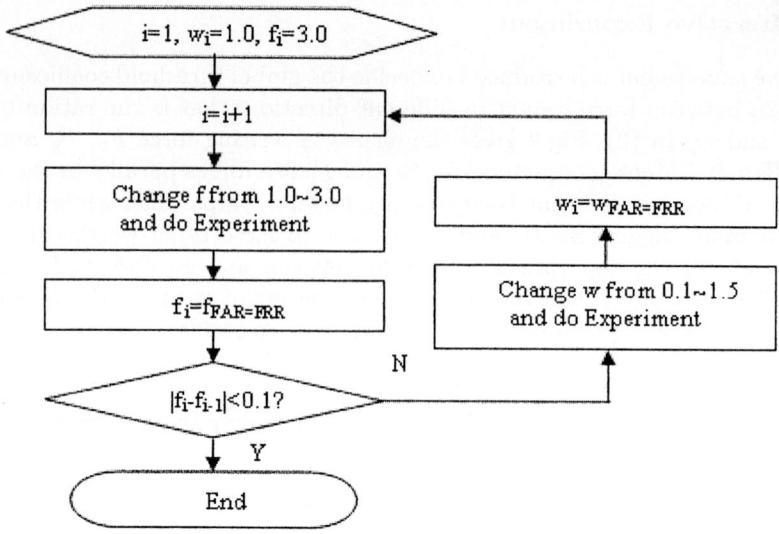

Fig. 3. Flow chart of the iterative experiment

the optimum ratio between w_x, w_y and w_z, and the optimum ratio is found to be 7 : 7 : 1. With the same force direction weight ratio, the global threshold coefficients are found to be different from each other, the main reason is that with different kinds of algorithms, different distribution of the distance are calculated. And the experimental results of the four kinds of algorithms are depicted in Fig.4. The equal error rate of the general DTW algorithm is 2.6%, not good as that of the weighted one, 2.1%. This reveals that put more weight on those signature points with better consistency can improve the classification results. But they are obviously inferior to those of the two improved DTW algorithms, 1.6% and 1.4% respectively. And the one which takes both the pen-up stroke time and pen-up stroke time gives better result. This is because it is very difficult to imitate a signature's shape, writing forces and writing velocity at the same time. And the pen-up stroke time, which is used to adjust the nib position and think about how to write the following stroke, is even harder to imitate.

Therefore, with the improved DTW algorithm, we get the optimum equal error classification rate of 1.4% where the global threshold coefficient f is 2.30 and the ratio between w_x, w_y and w_z is 7 : 7 : 1.

5 Conclusion

An improved DTW algorithm is put forward to verify online signature based on writing forces. The F_Tablet is used to capture the three perpendicular writing forces. Compared to the general DTW algorithm, this one deals with the varying consistency of signature point, signing duration and the different weights of writing forces in different direction. And iterative experiment is introduced

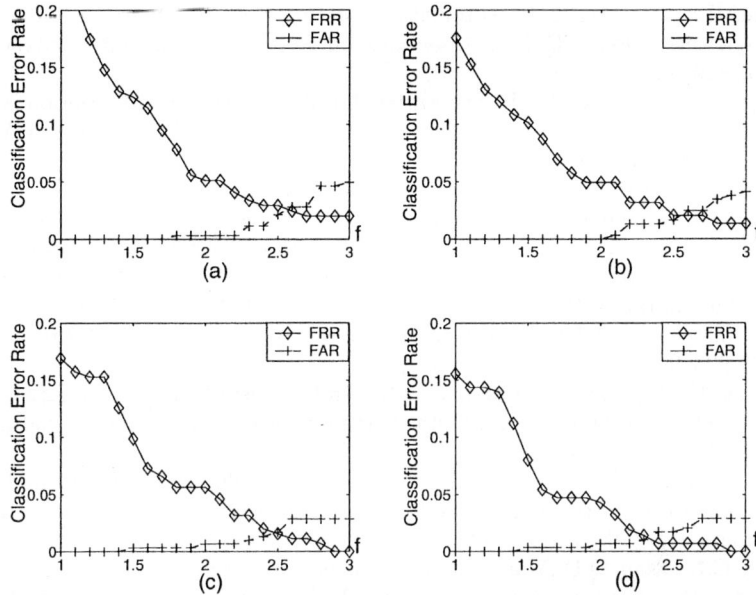

Fig. 4. Classification error rate to the varying threshold coefficient f. (a)the general DTW algorithm; (b)the weighted DTW algorithm; (c)the improved DTW algorithm (pen-down time); (d)the improved DTW algorithm (pen-down/pen-up time).

to decide weights for writing forces in different direction and the classification threshold. With this algorithm, the optimum equal error classification rate of 1.4% is realized based on the constructed signature database, where the ratio between w_x, w_y and w_z is $7:7:1$.

Although this improved DTW algorithm has better performance, it does have its deficiency. This algorithm consumes more computation time and memory space and these problems are left for the future work.

Acknowledgment

The work is supported by National Natural Science Foundation of China (NSFC No.60375027 and No.60475005).

References

1. Plamondon,R., Lorette,G.: Automatic Signature Verification and Writer Identification: The state of the art. Pattern Recognition.**22,2**) (1989) 107–131
2. Leclerc, F., Plamondon, R.: Automatic Signature Verification: The state of the art 1989-1993. IJPRAI. **8,3** (1994) 643–660
3. Nalwa, V.S.: Automatic On-Line Signature Verification. Proceedings of the IEEE. **85** (1997) 215–239

4. Crane, H.D., Ostrem, J.S.: Automatic Signature Verification Using a Three-Axis Force-Sensitive Pen. IEEE Transactions on Systems, Man, and Cybernetics. **3** (1983) 329–33
5. Martens, R., Claesen, L.: Incorporating Local Consistency Information into the Online Signature Verification Process. International Journal on Document Analysis and Recognition. **1** (1998) 110–115
6. Tanabe, K., Yoshihara, M., Kameya S. et al: Automatic Signature Verification Based on the Dynamic Feature of Pressure. Proceedings of Sixth International Conference on Document Analysis and Recognition. **1** (2001) 1045–1049
7. Sakamoto, D., Morita, H., Ohishi, T. et al: On-Line Signature Verification Algorithm Incorporating Pen Position, Pen Pressure and Pen Inclination Trajectories. IEEE International Conference on Acoustics, Speech, and Signal Processing. **2** (2001) 993–996
8. Shimizu, H., Kiyono, S., Motoki, T. et al: An Electrical Pen for Signature Verification Using a Two-Dimensional Optical Angle Sensor. Sensors and Actuators. **111** (2004) 216–221
9. Ping, F., Zhong, C. W., Ming, M. et al: A Novel Tablet for On-Line Handwriting Signal Capture. Proceedings of the 5th World Congress on Intelligent Control and Automation. **6** (2004) 3714–3717
10. Sheng, C.L., Xiao, Q.D.,Yan, C.: On Line Signature Verification Based on Weighted Dynamic Programming Matching. Journal of Tsinghua University. **39**,9 (1999) 61–64

3D Reconstruction Based on Invariant Properties of 2D Lines in Projective Space

Bo-Ra Seok, Yong-Ho Hwang, and Hyun-Ki Hong

Dept. of Image Eng., Graduate School of Advanced Imaging Science, Multimedia and Film,
Chung-Ang Univ., 221 Huksuk-dong, Dongjak-ku, Seoul, 156-756, Korea
{seokbr, hwangyh}@wm.cau.ac.kr, honghk@cau.ac.kr

Abstract. Projective reconstruction is known to be an important step for 3D reconstruction in Euclidean space. In this paper, we present a new projective reconstruction algorithm based on invariant properties of the line segments in projective space: collinearity, order of contact, intersection. Points on each line segment in the image are reconstructed in projective space, and we determine the best-fit 3D line from them by Least-Median-Squares (LMedS). Our method regards the points unsatisfying collinearity as outliers, which are caused by false feature detection and tracking. In addition, both order of contact and intersection in projective space are considered. By using the points that are the orthogonal projection of outliers onto the 3D line, we iteratively obtain more precise projective matrix than the previous method.

1 Introduction

The seamless integration of virtual objects with an image of a real scene has long been one of the central topics in computer vision and computer graphics. To generate a high quality synthesized image, consistency of geometry, illumination, and time has to be taken into account [1]. Reliable 3D reconstruction and camera recovery from images can guarantee geometric consistency to add synthetic objects into a real-world scene. In addition, building 3D models of outdoor scenes is widely used for virtual environment and augmented reality, but has always been a difficult problem.

Since 3D modeling from un-calibrated images has little calibration constraints and may be applied to various images, many researches have been presented up to now [2~7]. These are mainly focused on improving auto-calibration algorithm, applying to long sequence with key frame selection, and using user's knowledge for complete 3D model. However, there were few studies on more precise projective reconstruction, which is needed as a preceding step for 3D reconstruction in metric space.

This paper presents a new projective reconstruction algorithm based on invariant properties of the linear components in projective space: collinearity, order of contact, intersection. Collinearity means any points located on the 2D imaged line should lie on the reconstructed projective line. Therefore, we regard the points unsatisfying collinearity as outliers caused by false feature detection and tracking. In addition, the order of contact and the intersection points of the line segments in projective space are considered.

First, we establish correspondence over images and estimate a fundamental matrix (F-matrix) to determine the set of inlying feature tracks. After the points on each line segment in the image are transformed into projective space, we examine whether they satisfy linear properties in projective. More specifically, after a new 3D line is determined from the reconstructed points by Least-Median-Squares (LMedS), we iteratively obtain more precise projective matrix by using the points that are the orthogonal projection of outliers onto the line. Our method can alleviate the effect of false correspondences. This work is targeted on architectural scenes with a polyhedral shape. Because many man-made objects are constructed by using predefined rules, they often have many line segments.

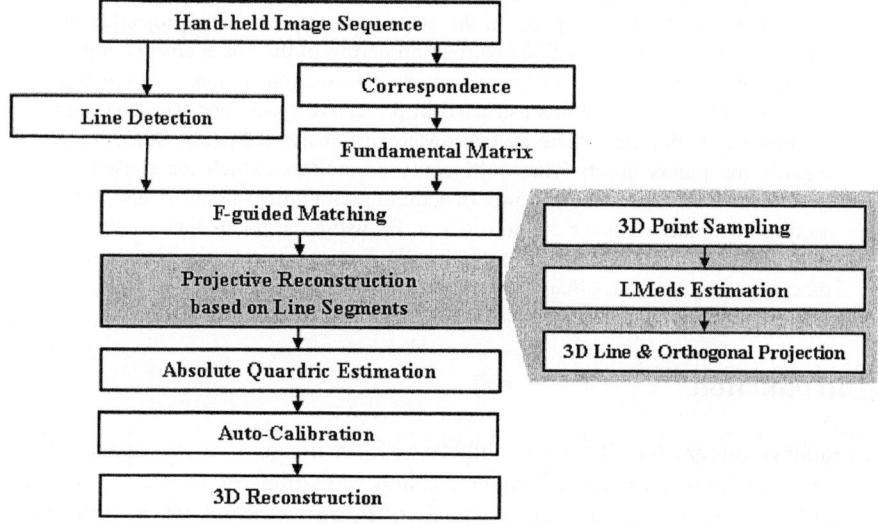

Fig. 1. Proposed 3D reconstruction algorithm

The remainder of this paper is structured as follows: Sec. 2 presents how to detect and match the line segments over views. Sec. 3 discusses projective reconstruction, and the proposed algorithm is detailed in Sec. 4. After comparisons of the experimental results are given in Sec. 5, the conclusion is described in Sec. 6.

2 Detecting and Matching Line Segments

When the points consisting of the line segment are transformed into projective space, those linear invariant properties are preserved. This paper uses the line segments for more precise calibration, so we have to detect and match them over two views.

The lines are obtained by Hough transform, which is a very popular algorithm to detect lines [8]. Since Hough transform finds the lines on a parametric space, they always do not coincide with the edges in the image. Therefore, our method applies Canny edge operator to extract the line candidates from the images. By comparing a distance between the lines and the edges, we can select the line segments located on the edges in the image.

Given a line l in one image and a corresponding line l' in the second image, we can find a correspondence on the epipolar line (l^e). Epipolar geometry is a fundamental constraint used whenever two images of a static scene are to be registered. In Fig. 2, the epipolar line, F-matrix (F) and two points (x, x') are satisfying [9, 10]:

$$x' = l' \times l'^e = l' \times (Fx) . \quad (1)$$

The corresponding points satisfying Eq. (1) on two views are located on two lines (l and l'), and we can obtain the corresponding lines. Fig. 3 shows the corresponding line segments over two views of the cube.

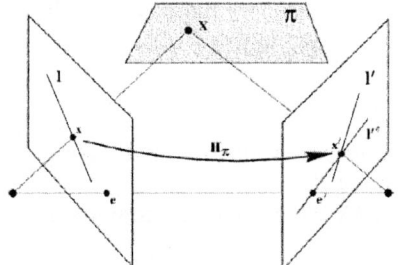

Fig. 2. Relation of lines and points on two views

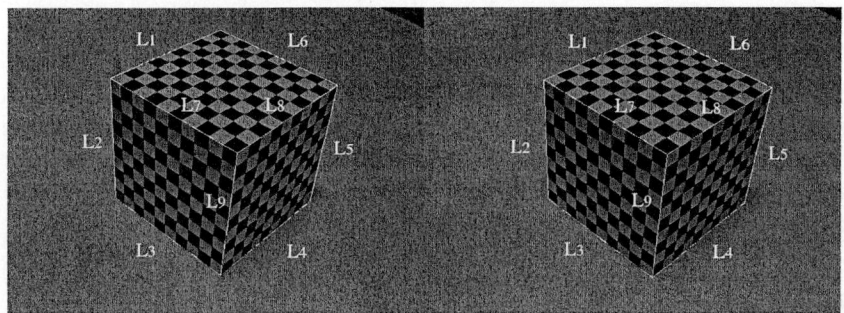

Fig. 3. Corresponding line segments over two views

3 Projective Camera and Reconstruction

Projective reconstruction is necessary for auto-calibration and 3D reconstruction from un-calibrated images [2]. Without some knowledge of a scene's placement with respect to 3D coordinate frame, it is impossible to reconstruct the absolute position or orientation of a scene from a pair of views. Therefore, the first camera is assumed to be located at the origin of a Euclidean coordinate system, and the projective matrix of the second in the camera coordinate is derived from F-matrix as follows:

$$P_1 = [I \mid 0], \quad P_2 = \left[[e']_\times F \mid e' \right], \tag{2}$$

where e' and P_n are the epipole of the second image and the projection matrix of the nth camera, respectively.

The image points are inversely projected from each camera center, and then the point in Euclidean 3D space is reconstructed using the intersection point on the epipolar plane. We can derive the linear equations for the camera projective matrices, the image points (x), and the points in 3D space (X). In each image we have a measurement $x = PX$, $x' = P'X$ in homogeneous, and these equations can be combined into a form $AX = 0$, which is an equation linear in X.

The homogeneous scale factor is eliminated by a cross product to give three equations for each image point, of which two are linearly independent. For the first image, $x \times (PX) = 0$ and writing this out gives:

$$x(p^{3T}X) - (p^{1T}X) = y(p^{3T}X) - (p^{2T}X) = x(p^{2T}X) - y(p^{1T}X) = 0, \tag{3}$$

where p^{nT} represent the transposed nth row of P. These equations are linear in the components of the world point X.

An equation of the form $AX = 0$ can then be composed as follows:

$$A = \begin{bmatrix} xp^{3T} - p^{1T} \\ yp^{3T} - p^{2T} \\ x'p'^{3T} - p'^{1T} \\ y'p'^{3T} - p'^{2T} \end{bmatrix}, \tag{4}$$

where two equations have been included from each image, giving a total of four equations in four homogeneous unknowns. This is a redundant set of equations, since the solution is determined only up to scale. After setting up the linear equation for the camera matrices and the corresponding points in two views, we can determine 3D points by linear method such as Singular Value Decomposition (SVD).

4 Proposed Algorithm

4.1 Determining 3D Lines in Projective Space

Fig. 4 presents the linear segments on a plane of the cube and the transformed lines in projective space. Though the shape of the plane and the slopes of the line segments are distorted, their linear properties are preserved as shown in Fig. 4. For example, any points located on the 2D imaged line must be on the reconstructed projective line because of collinearity. In general, since it is difficult to establish correspondences between two views, some segments may get disappeared as (b).

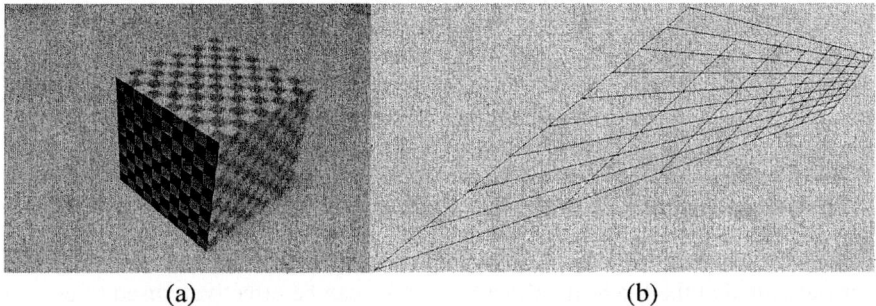

Fig. 4. Line segments (a) on a plane and (b) in projective space

Reconstruction of the points in the projective space is followed by refinement of the projective matrix. In order to refine the projective matrix, we determine 3D lines accurately in projective by LMedS based on random sampling. After the image points on each linear segment are transformed into projective space, we select any two points, $p(x_1, y_1, z_1)$ and $q(x_2, y_2, z_2)$, among them. Our method examines the distances between 3D line and other projective points by the vector operations as follows:

$$error = \sum_j^n sqrt\left(\frac{v_1^2 + v_2^2 + v_3^2}{(x_2 - x_1)^2 + (y_2 - y_1)^2 + (z_2 - z_1)^2}\right),$$
$$v_1 = (y - y_1) \times (z_2 - z_1) - (z - z_1) \times (y_2 - y_1)$$
$$v_2 = (z - z_1) \times (x_2 - x_1) - (x - x_1) \times (z_2 - z_1) \quad (5)$$
$$v_3 = (x - x_1) \times (y_2 - y_1) - (y - y_1) \times (x_2 - x_1)$$

The threshold value to discriminate the points with high errors is computed as follows:

$$r = 2.5 \times 1.4826 \times \left(\frac{1 + 5.0}{n - 2}\right)\sqrt{median}, \quad (6)$$

where *median* and n are a minimal median value, and the number of points consisting in the 2D line segment, respectively. As removing the outliers by false feature detection and tracking, LMeds based method determines iteratively an optimal 3D line accurately. Then, the outliers within some distances are orthogonally projected onto the 3D line. That means the points on the lines in projective space are moved so that they satisfy linear invariance.

As described in the previous, both 3D points on the lines in projective space and 2D image points are used to estimate the camera projective matrix. The 3D points are back-projected to 2D images, and compute each residual as follows:

$$residual = (q_x - (P_sQ)_x)^2 + (q_y - (P_sQ)_y)^2, \quad (7)$$

where P_s, q, and Q are the camera projective matrix by 6 points pair, the image point, and 3D points, respectively. LMedS based method obtains an optimal projective matrix that minimizes residual (Eq. 7). The threshold for rejecting the camera that causes projective matrix estimation to fail can be computed as follows:

$$r = 2.0 \times 1.4826 \times \left(\frac{1+5.0}{n-6}\right)\sqrt{median},\qquad(8)$$

where n is the number of correspondence pairs between the image points and 3D points. In final, we can determine a precise camera projective matrix iteratively.

4.2 Reconstruction of Planes Based on Linear Invariance

It is difficult to establish correspondences of every points and lines over views. This paper presents that the linear invariant properties can be effectively used to cope with missing correspondences.

The surface equation of a 3D plane is obtained from the cross product of the direction vectors of 3D lines, and we can classify 3D lines on the same plane. By examining iteratively whether two lines are located on a plane, 3D planes in projective space are reconstructed. Our method makes 3D lines longer on a plane, and hypothesizes the intersection points based on linear invariance to ascertain if they are the missing correspondences through views. This verification process can be used for more precise feature detection and tracking in the image sequence.

5 Experimental Results

We have experimented on 2 synthetic images (640×480) of the barn with a checkered pattern. The internal parameters of the projective cameras by the proposed algorithm and the previous are compared in table 1. The previous method uses only the corresponding points to obtain the projective matrix. On the contrary, our algorithm re-estimates the projective matrix based on linear invariance.

In order to evaluate an accuracy of the camera matrix, 3D points are back-projected into 2D images, and we compute the squared average errors that are the distances between the projected points and the real image points. The distance errors by the proposed algorithm and the previous are 0.65001 and 0.65432, respectively.

In addition, we have experimented on three real images (640×480) of the cube with a checkered pattern. The internal parameters of the projective cameras by the proposed algorithm and the previous are compared in table 2.

The distance errors by the proposed algorithm and the previous are 2.4978 and 2.5776, respectively. Fig. 7 shows the reconstructed cube with 3D lines, planes, and textured surfaces. The results showed that the proposed method can estimate precisely the camera parameters and reconstruct 3D model from un-calibrated images.

Fig. 8 and 9 show the input sequence and an accumulation error of the internal camera parameters, respectively. Merging-based projective method estimates the projective matrix of the second camera from that of the first by using F-matrix. Merging methods successively obtain the projective matrices and combine them over image sequences [11]. Comparing the squared average errors of the internal parameters - focal length ratio, the principal point, and the skew - by the previous merging-based method, we ascertain that a reduction of 84, 77, and 85% percent of their averaged errors is achieved, respectively.

Table 1. Internal parameters of the projective camera

	1st camera (Ideal)	Previous method (error)	Proposed method (error)
Focal length ratio	1.00	0.858 (-0.142)	0.872 (-0.128)
Principal point	0.00	0.413 (+0.413)	0.395 (+0.395)
Skew	0.00	0.344 (+0.344)	0.322 (+0.322)

Table 2. Internal parameters of the projective camera

	1st camera	Previous method		Proposed method	
		2nd camera	3rd camera	2nd camera	3rd camera
Focal length ratio (error)	1.00	1.0056 (+0.0056)	0.7320 (-0.2680)	1.0048 (+0.0048)	0.943 (-0.057)
Principal point (error)	0.00	0.0053 (+0.0053)	0.6150 (+0.6150)	0.0050 (+0.0050)	0.411 (+0.411)
Skew (error)	0.00	-0.040 (-0.040)	-0.040 (-0.040)	-0.02219 (-0.02219)	0.040 (+0.040)

(a) Previous method (340 corresponding points)

(b) Proposed method (36 corresponding lines)

Fig. 5. Input synthetic images (modeling by 3D Max) at 2-view

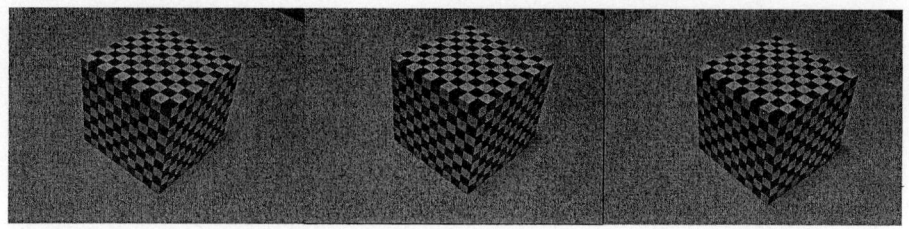

Fig. 6. Input images

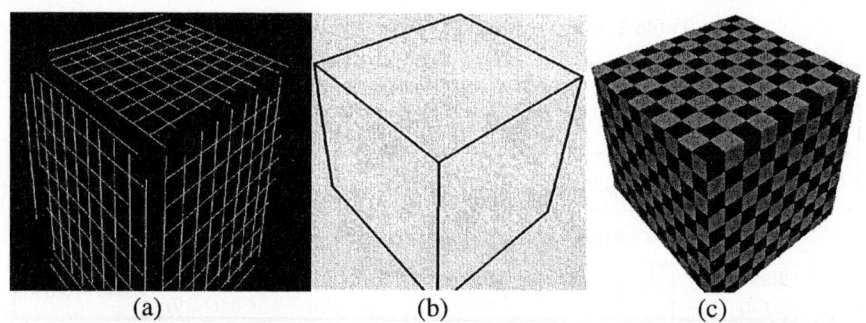

Fig. 7. Reconstructed (a) lines, (b) planes, and (c) textured surfaces

Fig. 8. Input sequence: 1, 3, 6 frames

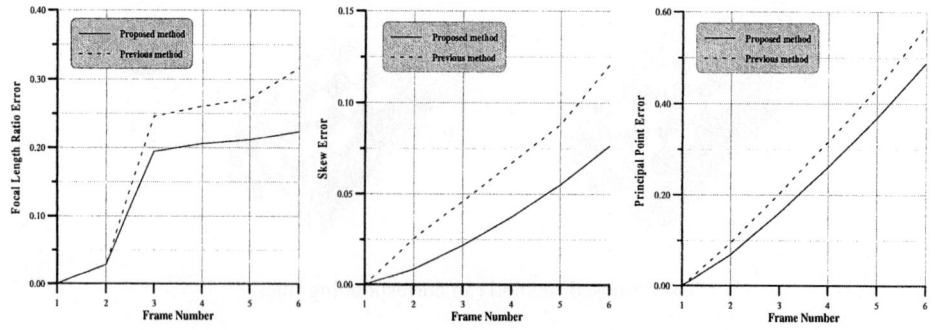

Fig. 9. Accumulation error graph of the focal length ratio

6 Conclusion

This paper presents a new projective reconstruction algorithm based on linear invariance. In order to evaluate the performance of the method, we estimate the internal parameters of the projective camera. In addition, after 3D points in projective space are back-projected to the image, their squared average errors are computed. By comparing the proposed method with the previous, we ascertained that our method can cope with the effects of outliers and recover the camera parameters precisely. Our method is a suitable for architectural scenes with many line segments and planes.

Further study will include more performance evaluations on various images, and build lighting representation from the reconstructed scene structure for generating photo-realistic rendering images.

Acknowledgement

This work was supported by Korea Research Foundation Grant(KRF-2004-041-D00620)

References

1. I. Sato, Y. Sato, and K. Ikeuchi, "Acquiring a Radiance Distribution to Superimpose Virtual Objects onto a Real Scene," *IEEE Trans. on Visualization and Computer Graphics*, vol. 5, no. 1 (1999), pp. 1-12
2. R. Hartley and A. Zisserman, *Multiple View Geometry in Computer Vision*, Cambridge University Press. (2000)
3. P. Debevec, C. Taylor, and J. Malik, "Modeling and rendering architecture from photos: a hybrid geometry and image-base approach," *SIGGRAPH* (1996), pp.11-20
4. C. Baillard, and A. Zisserman., "Automatic reconstruction of piecewise planar models from multiple views," *In proc. of the IEEE Conference on Computer Vision and Patter Recognition* (1999), pp. 559-565
5. R. Hartley, "A linear method for reconstruction from lines and points," *In proc. of IEEE International Conference on Computer Vision* (1995), pp. 882-887
6. S. Gibson, J. Cook, T. Howard, R. Hubbold, and D. Oram, "Accurate camera calibration for off-line, video-based augmented reality," *In proc. of IEEE and ACM ISMAR* (2002), pp. 37-46
7. S. Gibson, R. Hubbold, J. Cook, and T. Howard, "Interactive reconstruction of virtual environments from video sequences," *Computer Graphics*, vol. 27 (2003), pp. 293-301
8. E. Trucco and A. Verri, *Introductory Techniques for 3-D Computer Vision*, Prentice Hall (1998)
9. C. Schmid and A. Zisserman, "Automatic Line Matching across Views," *In proc. of IEEE Computer Vision and Pattern Recognition* (1997), pp. 666-671
10. Z. Zhang, R. Deriche, O. Faugeras, and Q. T. Luong, "A Robust Technique for Matching Two Uncalibrated Images through the Recovery of the Unknown Epipolar Geometry," *Artificial Intelligence Journal*, vol. 78, no. 1-2 (1995), pp. 87-119
11. A. Fitzgibbon and A. Zisserman, "Automatic Camera Recovery for Closed or Open Image Sequences," *In proc. of European Conference on Computer Vision* (1998) pp. 311-326

ANN Hybrid Ensemble Learning Strategy in 3D Object Recognition and Pose Estimation Based on Similarity

Rui Nian[1], Guangrong Ji[1], Wencang Zhao[1,2], and Chen Feng[1]

[1] College of Information Science and Engineering, Ocean University of China,
Qingdao, 266003, China
[2] College of Automation and Electronic Engineering,
Qingdao University of Science & Technology, Qingdao, 266042, China
nianrui_80@163.com, grji@mail.ouc.edu.cn,
wencangzhao@mail.edu.cn, fccjg@sdu.edu.cn

Abstract. In this paper, we present an ANN hybrid ensemble scheme for simultaneous object recognition and pose estimation from 2D multiple-view image sequence, and realized human vision simulation within an intelligent machine. Based on the notion of similarity measure at various metrics, the paradox between information simplicity and accuracy is balanced by a model view generation procedure. An ANN hierarchical hybrid ensemble framework, much like a decision tree, is then set up, with multiple weights and radial basis function neural networks respectively employed for different tasks. The strategy adopted not only determines object identity by spatial geometrical cognition and omnidirectional accumulation through connectivity, but also assigns an initial pose estimation on a viewing sphere in a coarse to fine process. Simulation experiment has achieved encouraging results, proved the approach effective, superior and feasible in large-scale database and parallel computation.

1 Introduction

Object recognition, the course of classification, description, judgment, estimation, recognition and comprehension, plays a main role in computer vision system, which is the most enhancements to an intelligent machine [1]. Most approaches developed so far can be categorized as geometry-based or appearance-based [1]. The former explicitly stores volume or surface representation relying on 3D geometry, either by complete specification or by description in form of certain features, invariants, parts, ridges and surface patches [1]. The latter directly compares and matches 2D images rather than 3D objects by similarity measure based on intensity, geometry, topology or their combination, results in significant reduction in dimensionality [1-4]. Recently, Artificial Neural Networks (ANN), with merits of robustness, fault tolerance, and powerful capacities in function approximation and data fitting [5, 6], is increasingly widely used in object recognition and pose estimation [3, 6-9].

Human vision system is inherently 2D, but is exceptionally adept at 3D object recognition and pose estimation [10]. The usual way that humans recognize objects is to observe from some distinct angles, which can be seen as selecting strategic locations on the spherical viewing surface. When taking cues from relatively intricate knowledge in 3D objects, only a few distinct views could usually suffice for humans to get the whole picture. If any kind of 3D structural information could not be utilized, all possible images needs to learn for reliable object recognition and pose estimation. With proper incorporated 3D hints, it is reasonable to implement human vision simulation on 2D multiple-view sequence in an intelligent machine [7, 8].

In this paper, a general framework for object recognition and pose estimation in parallel computation and large-scale database is constructed. Based on the notion of similarity measure at various metrics [11-13], a model view generation procedure is involved to balance the paradox between information simplicity and accuracy so that model views as few as possible are utilized for learning. The hierarchical hybrid ANN ensemble scheme, not only determines object identity in optimal spatial coverage cognition or omnidirectional accumulation through connectivity [6, 9, 14], but also assigns an initial pose estimation on a viewing sphere in a coarse to fine process, with three layers respectively employing multiple weights [15] and radial basis function neural networks. The approach adopted imposes a local constraint on images while maintaining a global structure. Spatial relationships or correlation feature among view points are iteratively fitted to images, obtaining a mapping from images to poses.

2 Similarity Measure for Simplicity and Accuracy Tradeoff

2D multiple-angle views give us overall and thorough information for appearance-based cognition on 3D object. However, information simplicity and accuracy is a pair of paradox. On the one hand, we try to acquire adequate information, include nearly all conditions, to construct spatial coverage. On the other hand, we pursue concise and efficient learning without redundancy, in a size as small as possible. So high quality recognition, together with a low quantity of samples, is all we want to accomplish.

Let there be N 3D objects $\{O_1, O_2, \cdots, O_n, \cdots, O_{N-1}, O_N\}$, each composed of M 2D images sampling the viewing sphere, $\{I_1^1, \cdots, I_m^n, \cdots, I_M^N\}$, with I_m^n denoting the m th image of object O_n, so the whole image database consists of $N \bullet M$ images.

2.1 Model View Generation

In order to balance the competing aims, we prefer to employ relatively few views, each representing a moderate range of possible appearances. Instead of learning the full set of images, a model view generation procedure is introduced to improve economy and accuracy tradeoff, minimize view set required for objects, and reduce complexity in matching process, shown in Fig.1. Preprocessed images are clustered into groups. Members in each group are then generalized to form a model view, characteristic of neighboring images. In the meantime, a hierarchical image representation system, vital for object recognition and pose estimation, is also established. Model view learning procedure keeps spatial relationships intact, and training sets were formed with different similarity distances between every two adjoining model views. View acquisition frequency is an important parameter which in conjunction to shape complexity determines the overall recognition accuracy.

Let C_m^n be a cluster of object O_n, with a collection of images ranging from $I_{m-k^-}^n$ to $I_{m+k^+}^n$ represented by characteristic view I_m^n, where (k^-, k^+) are left and right radius, $C_m^n = \{I_{m-k^-}^n, \cdots, I_{m-1}^n, I_m^n, I_{m+1}^n, \cdots, I_{m+k^+}^n\}$. The dissimilarity is denoted as the distance between two images I_m^n and I_i^j, i.e., $d(I_m^n, I_i^j)$.

While forming model views, several criteria are imposed to maintain successful object recognition and pose estimation [4, 6, 9]. A pair of cluster boundaries could be derived from those criteria. An iterative scheme, much like 'seed region growing' method, determines cluster segmentation and model view shape. Each image is initially considered a distinct model view, and then two images with the lowest global distance are chosen as the candidates to be merged as long as the boundaries are not violated. Model view is what minimizes the distance to all others in a cluster. Cluster number depends on object complexity as well as dissimilarity metric sensitivity.

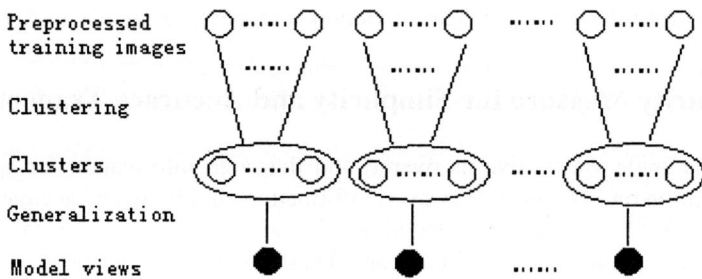

Fig. 1. Model view learning procedure in one object

Homologous Continuity Law. Suppose that the difference between two images from identical object is changing gradually or not determined by quantum, then there must be at least one course, where all the images in transition belong to the same object. Topology nature is used as pre-acquired knowledge [6, 9]. Let I be a set including all images of an object. Given $x, y \in I$ and $\varepsilon > 0$, there must be a set J,

$$J = \{x_1 = x, x_2 \cdots x_{n-1}, x_n = y | \rho(x_m, x_{m+1}) < \varepsilon, \forall m \in [1, n-1], m \in N\} \subset I \quad (1)$$

Local Monotonicity. Images monotonically change within each cluster starting from model view [4]. For each characteristic view I_m^n, there exists an integer $a > 0$ such that

$$d(I_m^n, I_{m\pm i}^n) \leq d(I_m^n, I_{m\pm j}^n) \quad \forall i < j \leq a \quad (2)$$

As relative visual angle increases, the dissimilarity increases for some range of angles. It relies on the sampling sufficiency assumption that a sampling rate must be high enough so that the monotonicity condition still holds, or satisfies for the unsampled images between samples.

Cluster Distinctiveness. The distance from the model view to images in the same cluster must be smaller than any distance to the images outside the cluster [4]. For each cluster C_m^n with model view I_m^n,

$$\max_{I_i^n \in C_m^n} d(I_m^n, I_i^n) \leq \min_{I_j^k \notin C_m^n} d(I_m^n, I_j^k) \tag{3}$$

Cluster Separability. For any pair of model views I_m^n and I_q^p from distinct clusters,

$$d(I_m^n, I_q^p) > 2 R(I_m^n) \tag{4}$$

where cluster radius is defined as

$$R(I_m^n) = \arg \max_{I_i^n \in C_m^n} d(I_i^n, I_m^n) = \arg \max_{\pm} d(I_{k\pm}^n, I_m^n) \tag{5}$$

Given any image I_i^j conformable to cluster separability, object identity j and cluster i could be correctly identified by the model view with minimum distance to I_i^j [4].

$$j = \arg \min_{I_m^n} d(I_i^j, I_m^n) \tag{6}$$

2.2 Similarity Metric

The feature of similarity, or correlation among view points is the key in separability. Different similarity metrics are employed to measure distance [11-13], such as Euclidean metric, Chamfer metric, Hausdorff metric, generalized Mahalanobis distance, Cityblock and Chessboard distance, Curve and Shock metric [12]. Due to each metric nature and relative shape weighting, model view generation course results in different selection of prototypes or characteristic views. Similarity metric is also essential to the knowledge expansion course on how to effectively combine newly increased untrained samples with already cognized data. The selection or combination of proper similarity metrics still need to pay more attention to in the future. We simply made some primary attempts at distance metrics for verification.

3 ANN Hybrid Ensemble Learning Strategy

An ANN hybrid ensemble, with three layers or more, is set up for object recognition and pose estimation. The first layer is assigned to recognize objects, the second to determine the cluster inside each object by model views, and the third or more to decide detailed and subtle inner relationship by images belonging to the same cluster. A hierarchical hybrid ensemble framework, much like a decision tree [16] is then shaped, with layers respectively employing various neural networks such as multiple weights (MWNN) [15] and radial basis function (RBF) neural networks. An input image flows from image database to object nodes in MWNN ensemble, then to model view node and at last to image node in RBF ensemble. Fig.2 is the spatial geometrical coverage in one object. Let there be $P(P \geq 3)$ layers $\{L_1, L_2, \cdots, L_P\}$ in the hybrid ensemble, each layer composed of a number of neural networks, NN_q^p and $E_q^p(I)$ respectively denoting the qth neural networks and the output from input I in layer L_p.

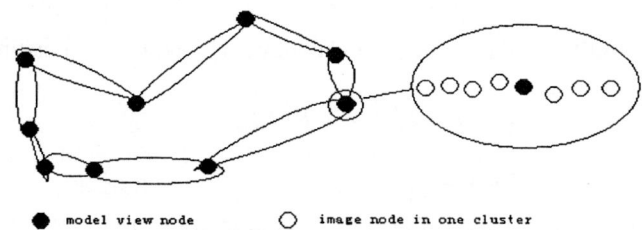

● model view node   ○ image node in one cluster

Fig. 2. Spatial geometrical coverage for one object in MWNN and RBF ensemble

3.1 Object Recognition

Object recognition emphasizes most on knowledge cognition one by one, rather than optimal interface classification simultaneously [9]. With high dimensional manifold as topology nature and homologous connectivity law as pre-acquired knowledge, an optimal spatial geometrical coverage is intended to establish for each object [6, 9, 14]. So an MWNN [15] ensemble is selected for object geometrical shape formation.

General neuron models [6, 17] with m weights in MWNN [15] can be denoted as

$$Y = f[\sum_{i=1}^{n} \Phi(W_{i1}, W_{i2}, \cdots, W_{im}, X_i) - \theta] \qquad (7)$$

From the perspective of geometry analysis in high dimensional space, neurons can be considered to be an (n-1)-dimensional hyper plane or curved surface in an n-dimensional space [6]. One kind of neurons like hyper sausages is chosen here. Let

$$d(x, \overline{x_1 x_2}) = \min_{\alpha \in [0,1]} d(x, \alpha x_1 + (1-\alpha) x_2) \qquad (8)$$

be the distance of x and line segment $\overline{x_1 x_2}$, and hyper sausage set is

$$S(x_1, x_2; r) = \{x | d^2(x, \overline{x_1 x_2}) < r^2\} \qquad (9)$$

The input-output transfer function is

$$f(x; x_1, x_2) = \phi(d(x, \overline{x_1 x_2})) \qquad (10)$$

where $\phi(\cdot)$ is the nonlinear threshold function, a variant of MWNN, $x \in R^n$ is input vector and $x_1, x_2 \in R^n$ are two centers. The neural networks consist of three layers: an input layer, a single hidden layer and an output layer of linear weights.

An MWNN ensemble is set up, every object an individual neural network. Training set with transition in explicitly temporal order, is reasonably selected to learn object geometrical shape. During test, Input images will be judged as the class with the maximal MWNN output, denoting corresponding 3D object. If $n = \arg \max\limits_{q=1}^{N} E_q^1(I)$, then $I \in O_n$. A series of comparison are made to validate recognition correctness, and decision on whether spatial coverage needs to be improved and adjusted also could be made. If untrained images can be justified as their counterparts by existing system, there is no need to modify anything. Otherwise, complicated training set will be adopted. Recognition results of multiple images from the same object were input into a working memory for evidence accumulation over time so that ultimate prediction could be achieved. In this way, an entire information representation, or enclosed coverage set will eventually be constructed in space. The course for knowledge expansion goes without affecting pre-acquired knowledge coming from other classes.

In fact, a $\omega_i / \overline{\omega}_i$ problem is solved here. With $d(x)$ the decision function and N the class number, when $d_i(x) > 0$, $d_j(x) \leq 0$, $j = 1, 2, \cdots, N, j \neq i$, then $x \in \omega_i$. We construct a spatial geometrical coverage inclusive of embedding set I in feature space, nearly close to the combination of all with arbitrary point in multiple dimension manifold in set I as the center and the constant k as the radius, i.e., topological product between set I and n dimension hyper sphere[14]. In practice, topological covering set V formed by set I is defined mathematically below, I_i is one of the samples in set I.

$$V = \bigcup_i V_i, V_i = \{x | \rho(x, y) \leq k, y \in J_i, x \in R^n\}, J_i = \{x | x = \alpha I_i + (1-\alpha)I_{i+1}, \alpha = [0,1]\} \quad (11)$$

When indefinite superposition region encounters, more detailed training set is needed in order to subdivide.

3.2 Pose Estimation

3D Objects are two degrees of freedom on the viewing sphere, corresponding to elevation and azimuth angles respectively. Each 2D image can be thought of being generated by a visual sensor, aimed towards the sphere center and traveling along the longitudes and latitudes. Pose estimation means to determine viewing parameters or relative position and orientation of 3D objects from unknown 2D images with respect to a reference camera system [7, 8].

The iterative hierarchical hybrid ensemble framework present here not only determines object identity, but also gives an initial pose estimation by identifying cluster as the focus region that images belong to in each object. The viability of pose estimation is created in a coarse to fine process, much like a decision tree. Two or more layers, employing RBF neural networks, are shaped in the ANN ensemble, which are assigned to determine concrete pose information from clusters as well as subtle inner relationship inside clusters. The viewing sphere is initially coarsely located in the second layer, at big increments in virtue of model views. Further pose estimation is gradually acquired in the following layers, where a higher level match is conducted at each stage. Further subdivision could be carried out by repeating this process, until accuracy determined by metric is reached at last. The advantage of this method is that it avoids initialization difficulty in 2D-3D registration methods by an efficient global search based on similarity. However, similarity metrics are not uniformly distributed on the viewing sphere, which implies the hierarchical approach should be further improved. Fig.3 shows pose estimation in a coarse to fine process with two degrees of freedom.

$$\begin{aligned} &\textbf{If } m = \arg\max_q E_q^2(I), \textbf{ then } I \in C_m^n. \\ &\textbf{If } k = \arg\max_q E_q^p(I) \textbf{ and } 2 \le p \le P, \textbf{ then } I \in I_k^n \in C_m^n. \end{aligned} \quad (12)$$

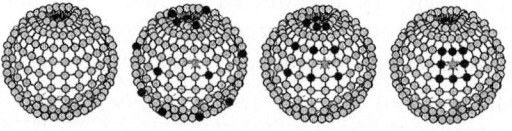

Fig. 3. Pose estimation in a coarse to fine process with two degrees of freedom

4 Simulation Experiment

Image database consists of multiple video images of sorts of objects. With camera fixed, each object model was mounted onto a precise revolver against a common light background for the benefit of figure-ground segmentation, rotated in increments of 5° through 360° in the horizontal plane. For each visual angle, camera was spun and frames resulted in many images. The images were 128×128 pixel color photos originally, and then transferred to gray ones, thresholded and binarized. Both noise-free and noisy images introduced by an additive noise process, were collected, covering a sphere surrounding object. Based on global similarity with overall quality to guide decisions and on the assumption of sampling sufficiency, several criteria were imposed to a clustering and generalization procedure to form model views and maintain successful cognition. With different distances between adjoining model views, training sets were formed. Images from both trained and untrained objects were taken for error test, with trained ones also for correct rate calculation. Fig.4 shows some example images in database.

Some preprocessing, such as illuminant discounts, noise suppressing, boundary segmentation, invariance, coarse coding and so on, was done in advance before formal operation for feature extraction. Smoothed edges, delineating natural outline of interest, were completed with the help of dilation and erosion in morphography. View information was transformed into an invariant presentation by log-polar. With shift parameters exactly recorded down, an optional inverse transform could be taken to revert into initial state. Coarse coding or dimension reduction, necessary to speed up calculation and compensate for modest 3D foreshortening, was also carried out. Fig.5 shows an invariance example of four edge views identical except for translation, rotation and scale, images become much similar in centered log-polar domain.

Object recognition and pose estimation was performed in aid of ANN hierarchical hybrid ensemble architecture with MWNN and RBF neural networks respectively employed. Several parallel neural networks denoting different categories, were set up to organize images into object nodes, view clusters and subtle inner relationship. Their outputs combine into or converge at a clear response to the geometrical shape that certain kind of category occupies or covers in space. For object recognition, multiple images from the same object were input into a working memory for evidence accumulation over time to improve effect, where each occurrence (nonoccurrence) of an image increases (decreases) the corresponding object node's activity and the maximally active object node is used for prediction or judgement. For pose estimation, every image needed to go deep through the ensemble to the second, the third or more layers to acquire exact pose information, much like decision tree theory.

Fig. 4. Example images

Fig. 5. Invariance example. At the top, preprocessed results of one kangaroo image; in the middle and at the bottom, Column 1,2,3,4 respectively edge, centered, log-polar transformed, centered log-polar images from left to right.

5 Result Analysis

At object recognition stage, input images were recognized as different object classes via MWNN ensemble. Training recognition rates were all 100%. With properly varied parameters, error rate (mistaken recognition) for images from unlearned objects could be 0%, i.e., unknown objects were all rejected without incorrect recognition. Besides MWNN ensemble, BP, RBF ensemble were also adopted as options. For images from learned objects, the performance contributed to average correct recognition rate is compared in Fig. 6, and various fractions between images at trained visual angles and all are listed. Training time in MWNN ensemble is much faster than the others.

At pose estimation stage, input images entered the second and the third layers in ANN ensemble with RBF neural networks, which were first classified into a cluster by model views, and then further located into concrete position inside that cluster. A series of object recognition and pose estimation results are shown in Fig. 7.

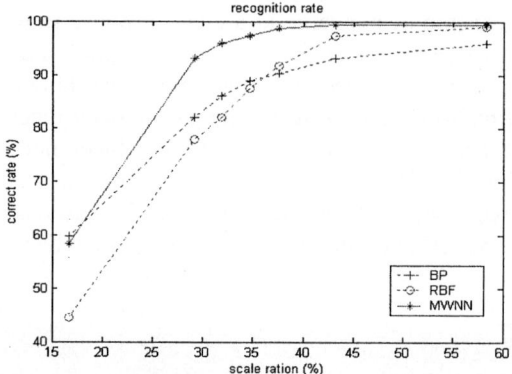

Fig. 6. Performance comparison to recognition rate in BP, RBF and MWNN ensemble

Fig. 7. A series of object recognition and pose estimation results to one kangaroo input. From input image, object node, cluster node and model view, to concrete position inside cluster.

6 Conclusions

In this paper, in aid of ANN hybrid ensemble, we present a novel scheme for simultaneous object recognition and pose estimation from 2D multiple-angle image sequence with proper embedded 3D hints, as much as the way that human being does. Correlation feature among view points is the key in human vision simulation. Based on the notion of similarity measure at various metrics, the paradox between information simplicity and accuracy is balanced by a model view generation procedure in an iterative method like 'seed region growing'. A hierarchical hybrid ensemble architecture, similar to a decision tree, imposes a local constraint on images close to each other, while maintaining a global structure, with three layers employing various neurons like multiple weights and radial basis function for different tasks. The strategy adopted not only determines object identity by optimal spatial coverage cognition and omnidirectional accumulation through connectivity, but also assigns an initial pose estimation on a viewing sphere in a coarse to fine process, with as few model views as possible learned. A clear response to the optimal geometrical coverage and overall information representation is gradually constructed and cognized one by one. Ensemble framework makes time cost lessened and accuracy improved in parallel computation, not affecting what has been acquired before in other objects in knowledge expansion course. Encouraging results have been achieved in simulation experiment and proved effective, superior and feasible in the approach proposed.

Acknowledgements

The National 863 Natural Science Foundation of P. R. China (2001AA635010) fully supported this research.

References

1. Besl, P., Jain, R.: Three-dimensional object recognition .Surveys, 17(1), (1985) 75–145
2. Pope, A., Lowe, D.: Learning object recognition models from images. In ICCV93, (1993) 296–301
3. Bradski, G., Grossberg, S.: Fast-Learning VIEWNET Architectures for Recognizing Three-dimensional Objects from Multiple Two-dimensional Views. Neural Networks, Vol.8, (1995) 1053-1080
4. Cyr, C. M., Kimia, B. B.: A Similarity-Based Aspect-Graph Approach to 3D Object Recognition. International Journal of Computer Vision, Volume 57 Issue 1 (2004)
5. Fauselt, L.: Fundamentals of Neural Networks. Prentice-Hall, International Inc., Englewood CliGs, NJ, (1994)
6. Huang, D.S.: Systematic Theory of Neural Networks for Pattern Recognition. Publishing House of Electronic Industry of China, Beijing (1996) 70-78
7. Wunsch, P., Winkler, S., Hirzinger, G.: Real-time pose estimation of 3-D objects from camera images using neural networks. In Proc. IEEE International Conference on Robotics and Automation, vol. 4, (1997) 3232-3237
8. Yuan, C., Niemann, H.: Neural networks for the recognition and pose estimation of 3D objects from a single 2D perspective view. Image and Vision Computing, Vol.19, Issue 9-10, (2001) 585-592
9. Wang, S. J.: Biomimetics pattern recognition. INNS, ENNS, JNNS Newletters Elseviers, (2003)
10. Liu, Z., Kersten, D.: 2D observers for human 3D object recognition? In: Advances in Neural Information Processing Systems. Jordan, M. I., Kearns, M. J., Solla, S. A., eds., vol. 10, (1997)
11. Basri, R., Weinshall, D.: Distance Metric between 3D Models and 2D Images for Recognition and Classification. Mathematics & Computer Science, (1994)
12. Klein, P., Sebastian, T., and Kimia, B.: Shape matching using edit-distance: an implementation. In: Twelfth Annual ACM-SIAM Symposium on Discrete Algorithms (SODA), (2001) 781-790

13. Xing, E.P., Ng, A.Y., Jordan, M.I., and Russell, S.: Distance metric learning, with application to clustering with side information. Advances in Neural Information Processing Systems, (2002)
14. Wang, S. J., Wang, B.N.: Analysis and theory of high-dimension spatial geometry for Artificial Neural Networks. Acta Electronica Sinica, Vol. 30 No.1, (2002) 1-4
15. Wang, S. J., Xu, J., Wang, X.B., Qin, H., Multi-camera human-face personal identification system based on the biomimetics pattern recognition. Acta Electronica Sinica, Vol. 31 No.1, (2003) 1-4
16. Katsuhiko,T., Shogo,N.: Implementation and refinement of decision trees using neural networks for hybrid knowledge acquisition. Artificial Intelligence in Engineering Vol. 9, Issue 4, (1995) 265-276
17. Wang, S. J., Li, Z. Z., Chen, X. D., Wang, B.N.: Discussion on the basic mathematical models of Neurons in General purpose Neurocomputer. Acta Electronica Sinica, Vol. 29 No.5, (2001)

Super-Resolution Reconstruction from Fluorescein Angiogram Sequences*

Xiaoxin Guo, Zhiwen Xu, Yinan Lu, Zhanhui Liu, and Yunjie Pang

Key Laboratory of Symbol Computation and Knowledge Engineering of the Ministry of Education, College of Computer Science and Technology, Jilin University, Qianjin Road 10#, Changchun, 130012, P.R.China
xiaoxin@mail.jl.cn

Abstract. Intensity degradations are a familiar problem for fluorescein angiogram sequences. In this paper, we attempt to super-resolve a fluorescein angiogram, and to keep the high intensity pixels from degrading. To this end, we incorporate a new constraint, called intensity constraint, to Miller's regularization formulation with a smoothness constraint. Considering the specified requirement for fluorescein angiograms, we also modify the Q-th order converging algorithm for implementation purpose. In our scheme, including its formulation and implementation, super-resolution reconstruction can not only handle the traditional problems, such as blur, decimation, and noise, but also achieve an important feature, intensity preservation. The experiments show that our approach has satisfactory results in the two aspects.

1 Introduction

Super-resolution (SR) reconstruction is the process of combining a sequence of undersampled and degraded low resolution (LR) images in order to produce a still image at a higher resolution. The technique aims to overcome the inherent hardware limitations of a camera system, i.e., to increase image resolution by capturing the additional new detail revealed in each frame of the sequence of images portraying the same scene.

Tsai and Huang [1] were the first to demonstrate that unique information in a sequence of translated and aliased images can be exploited to produce an enhanced resolution image. In their early work, the LR images were neither blurred nor noisy, and they assumed that the shifts between the LR images were known. Since then, several other approaches have been proposed in the literature, and the formulation has been extended to include noise, blur, and more complex motion models. These approaches include weighted least-squares [2][3], an iterative technique similar to back-projection [4], projection onto convex sets (POCS) [5]-[8], and MAP and least squares regularization formulations solved iteratively using gradient-decent optimization [9][10].

---

* This work was supported by the foundation of science and technology development of Jilin Province, China under Grant 20040531.

Fluorescein angiography is used as a physiological assay for retinal function since the dye-dilution technique with fluorescein angiography allows the characterization of retinal blood flow from all the major retinal vessels in the field of view simultaneously. Fluorescein angiograms are taken using a scanning laser ophthalmoscope (SLO) [11][12] after the injection of fluorescein. The angiograms can record the process of the flowing of fluorescent dye and the changing of the fluorescence intensity. There are two fluorescence intensity peaks appearing in artery and vein of the fundus of eyes, respectively. The interval between two peaks is relatively short, and most of time is in the phases when dye is arriving or exhausting. In this case, the locations where dye has not arrived or has exhausted appear dark and beyond recognition. In medical practice, the locations of vasculature bed full of fluorescent dye in the angiograms are the ones that clinicians are interested in. Clinicians expect to obtain a single higher resolution image, showing finer vasculature details taken from the angiogram sequence at different time. The resulting image may be a warrant for the correct diagnosis.

In this paper, considering the specified requirement for fluorescein angiograms, we propose an improved SR reconstruction scheme. In order to preserve high intensity in the resulting reconstruction for medical analysis and diagnosis, our scheme imposes a novel constraint, called intensity constraint, on the traditional minimization problem for SR reconstruction. With the definition of the constant vector, called intensity template, we can guide the minimization process to our desirable goal. This goal is just intensity preservation. For the same purpose, we employ a bilateral filter as a pre-filtering process. In addition to noise suppression, the filter can provide us with the intensity template mentioned above for the reconstruction process. Correspondingly, we make the modification to the Q-th order converging algorithm for our specified requirement. The algorithm with a high convergence rate incorporates the projection operator, resulting in considerable generality. Prior to experiments, a novel criterion of the performance estimation for intensity contrast, called mean gray level (MGL) ratio is introduced, and will be used in the subsequent experiments. The experimental results present a comparison among the intensity constraints with different regularization parameters, and shows better applicability and robustness in intensity preservation.

In Section 2, we first model the SR reconstruction problem as a minimization problem, and present a new constraint for the minimization problem. In Section 3, we describe the method used to implement the solution of the minimization problem, including the preprocessing, the modified algorithm and the assessment criterions of intensity. Experiments and results are presented in Section 4 for test purposes. Conclusions are drawn in Section 5.

2 Modeling

In many practical situations image degradations may be modeled by a distortion and an additive noise term which is uncorrelated with the signal. The distorted LR image can then be described by the following algebraic model:

$$\mathbf{g} = \mathbf{A}\mathbf{f} + \mathbf{n} \ . \qquad (1)$$

where the distortion operator $\mathbf{A}$ is known or can be satisfactory identified, combining the effect of the warp, the decimation and the blur. The original and distorted images are denoted by the lexicographically ordered vectors $\mathbf{f}$ and $\mathbf{g}$, respectively. The characteristics of the noise term $\mathbf{n}$ are only partially known in practice; hence the exact original image cannot be computed from the distorted version. SR reconstruction concentrates on removing the degradations caused by the blur, decimation and the noise to obtain an improved image $\hat{\mathbf{f}}$ which is an acceptable approximation to the original image. Since the inverse problem formulated in Eq. (1) is ill-posed, the solution method has to be regularized in order to obtain physically meaningful solutions. To this end we require the reconstructed image $\hat{\mathbf{f}}$ to satisfy the following conditions in which adaptivity and a deterministic constraint are introduced to achieve both noise suppression and ringing reduction [13][14]:

$$\left\| \mathbf{g} - \mathbf{A}\hat{\mathbf{f}} \right\|_{\mathbf{S}}^2 = (\mathbf{g} - \mathbf{A}\hat{\mathbf{f}})^T \mathbf{S}(\mathbf{g} - \mathbf{A}\hat{\mathbf{f}}) \leq \varepsilon^2 \ . \qquad (2)$$

where $\mathbf{S}$ is a diagonal weight matrix which locally regulates the restoration process.

$$\left\| \mathbf{L}\hat{\mathbf{f}} \right\|_{\mathbf{V}}^2 = (\mathbf{L}\hat{\mathbf{f}})^T \mathbf{V}(\mathbf{L}\hat{\mathbf{f}}) \leq E^2 \ . \qquad (3)$$

where $\mathbf{L}$ represents a high-pass filter. Eq. (3) imposes a smoothness condition on the reconstructed image $\hat{\mathbf{f}}$ which is locally regulated by the weight matrix $\mathbf{V}$.

Besides, we impose a new constraint specified for fluorescein angiograms. Because of low gray level and liability to noise interference, the dark frames cannot be analyzed as reliable data, and their information is relatively secondary in importance. For the bright frames, it is just the reverse. Therefore, it is desirable that the resulting image after reconstructing can preserve high intensity pixels, which are regions of interest, and reduce or remove the influence of dark pixels on bright ones at the identical spatial location. Therefore, considering the above reasons, the following constraint related to fluorescence intensity is a more intuitive and reasonable choice.

$$\left\| \hat{\mathbf{f}} - \hat{\mathbf{f}}_t \right\|_{\mathbf{U}}^2 = (\hat{\mathbf{f}} - \hat{\mathbf{f}}_t)^T \mathbf{U}(\hat{\mathbf{f}} - \hat{\mathbf{f}}_t) \leq \delta^2 \ . \qquad (4)$$

Eq. (4) imposes an intensity constraint (or intensity penalty in regularization theory) on the reconstructed image $\hat{\mathbf{f}}$ which is locally regulated by the weight matrix $\mathbf{U}$. $\hat{\mathbf{f}}_t$ is a constant vector, called intensity template, representing an estimated image with desirable intensities. The intuition behind this constraint is to determine the upper and the lower bound of the resulting image intensities. Clearly, when the constant vector is a zero vector, the above intensity constraint reduces to an amplitude constraint.

In addition to Eq. (2), (3) and (4), the reconstructed image $\hat{\mathbf{f}}$ satisfies the constraints C, representing certain deterministic *a priori* information about the original image **f**. The nonexpansive projection onto the closed convex sets described by C is denoted by **P**.

To compute a solution $\hat{\mathbf{f}}$ satisfying the above conditions, there exist two alternative approaches. The first approach is to add the constraint in Eq. (4) to convex sets C, and then combine Eq. (2) and (3) into a single quadrature formula to form Miller's regularization formulation [21]

$$\Phi(\hat{\mathbf{f}}) = \left\|\mathbf{g} - \mathbf{A}\hat{\mathbf{f}}\right\|_{\mathbf{S}}^2 + \alpha \left\|\mathbf{L}\hat{\mathbf{f}}\right\|_{\mathbf{V}}^2 . \tag{5}$$

where the regularization parameter has the fixed value $\alpha = (\epsilon/E)^2$. The solution to the SR problem is given by the vector $\hat{\mathbf{f}}$ which minimizes the functional $\Phi(\hat{\mathbf{f}})$ subject to the deterministic constraints C.

The second option is to compute the following minimization problem subject to the deterministic constraints C, which excludes the constraint in Eq. (4):

$$\Phi(\hat{\mathbf{f}}) = \left\|\mathbf{g} - \mathbf{A}\hat{\mathbf{f}}\right\|_{\mathbf{S}}^2 + \alpha \left\|\mathbf{L}\hat{\mathbf{f}}\right\|_{\mathbf{V}}^2 + \gamma \left\|\hat{\mathbf{f}} - \hat{\mathbf{f}}_t\right\|_{\mathbf{U}}^2 . \tag{6}$$

where the regularization parameter has the fixed value $\gamma = (\delta/E)^2$. The parameter γ and α play an important part in the relationship of the smoothness, the faithfulness to the data and the intensity preservation. γ has the ability to control and adjust the strength of the three aspects. Especially, when $\gamma = 0$, the resulting image reduces to a direct reconstruction without intensity constraint as Miller's regularization formulation in Eq. (5).

3 Implementation

3.1 Preprocessing

In the stage of preprocessing, the most important step is to determine the intensity template $\hat{\mathbf{f}}_t$ in Eq. (4) according to the specified requirement of fluorescein angiograms. To this end, we employ the bilateral filter. The idea of the bilateral filter was first proposed in [15] as a very effective one pass filter for denoising purposes while keeping sharp edges. The bilateral filter defines the closeness of two pixels not only based on geometric distance but also based on photometric distance. Considering 2-D case, the result of applying bilateral filter in estimating pixel $\hat{\mathbf{g}}(i,j)$ is:

$$\hat{\mathbf{g}}(i,j) = \left(\sum_{(m,n,k)\in\Omega} \mathbf{w}_{(i,j)}(m,n,k)\mathbf{g}_k(m,n)\right) \bigg/ \left(\sum_{(m,n,k)\in\Omega} \mathbf{w}_{(i,j)}(m,n,k)\right) . \tag{7}$$

where $g_k(m,n)$ is the gray level value of degraded pixel (i,j) in the kth LR frame, Ω is a 3-D support centered at pixel (i,j). $\hat{\mathbf{g}}(i,j)$ is an output after weighting the pixels within the support Ω. The weights of the bilateral filter consider both photometric and spatial difference of pixel (i,j) in LR frame g_k from its neighbors to define the value of the estimated vector $\hat{\mathbf{g}}$, given by

$$\mathbf{w}_{(i,j)}(m,n,k) = \mathbf{w}_{S(i,j)}(m,n,k) \cdot \mathbf{w}_{P(i,j)}(m,n,k) . \tag{8}$$

where the spatial and photometric difference weights were defined for $m = -q,\ldots,q; n = -q,\ldots,q; k = 1,\ldots,N$ as

$$\mathbf{w}_{S(i,j)}(m,n,k) = \exp\left(-\frac{1}{2}\frac{(m-u)^2+(n-v)^2}{c_2\sigma_d^2}\right) . \tag{9}$$

$$\mathbf{w}_{P(i,j)}(m,n,k) = \exp\left(-\frac{1}{2}\frac{(\mathbf{g}_k(i+m,j+n)-\mathbf{g}_t(i+u,j+v))^2}{c_1\sigma_r^2}\right) . \tag{10}$$

The quantity $c_1(c_1 > 0)$, $c_2(c_2 > 0)$ and $q(q \geq 0)$ is the parameters of the filter. g_t denotes the gray level value of the pixel of interest (POI) within the support Ω. Which pixel is defined as the POI within the support can reflect what kind of values of signal (or pixels) is desirable to be preserved after filtering. q controls the bilateral kernel size. Parameters σ_r^2 and σ_d^2 control the strength of spatial and photometric property of the filter.

The bilateral filter enhances the ability to control the weights of pixels within the support Ω. The filter allows us to adjust the parameters c_1, c_2 and q to control the weight distribution as we require. If we limit window radius $q = 0$, then the filter reduces to a temporal filter. If c_1 is large enough, then the weights of w_p obtain almost the same value, and the filter becomes an averaging filter. In contrast, when c_1 approaches to zero, the filter turns out to be an impulse filter, whose output is the predefined POI g_t. Clearly, c_2 is the same situation for w_s.

Besides the similarity to parameter c_1, the use of parameter c_2 lends itself to supporting motion compensation. It is well known that there is a trade-off between the amount of noise removal and blurring in noise filtering. At relatively high SNR levels, motion-compensated adaptive temporal filters can provide effective noise reduction without introducing spatial blurring if the estimated motion trajectories are sufficiently accurate. At low SNR levels, however, the number of image points within the temporal support may not be large enough to achieve sufficient noise reduction. By adjusting parameter c_2, we can control the number of image points within the filter support and the strength of motion compensation.

On the other hand, by allowing defining POI g_t, the bilateral filter extends the ability of controlling the output. As mentioned above, when the filter corresponding to w_s is an impulse one, the result is just g_t, a predefined POI. The procedure of selecting the POI can be considered as a pre-filtering stage. This property can be utilized when we design desirable g_t according to the specified requirements. If we change the POI, actually we also alter the center of the weight distribution. Therefore, the use of the POI makes it more flexible for the bilateral filter to achieve a controlled output. In the case of fluorescein angiogram sequences, in order to achieve the two goals, suppressing noise as well as preventing the high intensities from being degraded, we define the POI using aggregate functions, given by

$$\mathbf{g}_t(i+u, j+v) = \max\{\mathbf{g}_k(i+m, j+n)\} \ . \tag{11}$$

for $m = -q, \ldots, q; n = -q, \ldots, q; k = 1, \ldots, N$, where $\max\{\cdot\}$ is a maximum operator, (u, v) is the offset of the pixel with the highest intensity within the support relative to the current pixel (i, j). Through the definition, we see more high-intensity values can be preserved after filtering. After appropriate interpolations, we obtain $\hat{\mathbf{f}}_t$ from the estimated vector $\hat{\mathbf{g}}$, and the resulting intensity template is used in Eq. (6).

Summarily, the bilateral filter supplies some possibilities to adjust itself. The following factors strengthened the flexibility and the generality of the filter: (1) the parameters c_1 and c_2 to control the distribution of the weights and the ability of motion compensation; (2) the POI to position the center of weight distribution and define a desirable output value; (3) the parameters q to limit the size of the support.

3.2 A Q-th Order Converging Algorithm

Because the minimization problem in Eq. (6) is nonlinear and space-variant, an iterative solution method called Q-th order converging algorithm is used. In [16] Singh et al. proposed this iterative technique with a quadratic convergence rate, which Morris et al. [17] generalize and extend to a regularized iterative SR reconstruction algorithm with a Q-th order convergence rate ($Q = 2, 3, \ldots$). Consider the minimization of the functional $\Phi(\hat{\mathbf{f}})$ in Eq. (6). The solution $\hat{\mathbf{f}}$ of this problem can formulated as

$$\left(\mathbf{A}^T \mathbf{S} \mathbf{A} + \alpha \mathbf{L}^T \mathbf{V} \mathbf{L} + \gamma \mathbf{U}\right) \hat{\mathbf{f}} = \mathbf{A}^T \mathbf{S} \mathbf{g} + \gamma \mathbf{U} \hat{\mathbf{f}}_t \ . \tag{12}$$

In this case, the Q-th order converging algorithm is written as follows

$$\mathbf{f}_0 = \beta(\mathbf{A}^T \mathbf{S} \mathbf{g} + \gamma \mathbf{U} \hat{\mathbf{f}}_t), \mathbf{B}_0 = \mathbf{I} - \beta(\mathbf{A}^T \mathbf{S} \mathbf{A} + \alpha \mathbf{L}^T \mathbf{V} \mathbf{L} + \gamma \mathbf{U}) \ . \tag{13}$$

$$\mathbf{f}_{k+1} = \sum_{j=0}^{Q_k - 1} \mathbf{B}_k^j \mathbf{f}_k \ . \tag{14}$$

$$\mathbf{B}_{k+1} = \mathbf{B}_k^{Q_k} \quad (Q_k \geq 2) \ . \tag{15}$$

where $\mathbf{f}_0$ is an initial guess, Q_k determines the degree of convergence at iteration step k. The efficiency of the algorithm depends on the choice of the convergence rate parameter Q. To ensure the convergence of the iterations to the solution of the minimization of Eq. (6), $\mathbf{B}_k$ has to be a contraction mapping, yielding the below conditions for β.

$$0 < \beta < 2 \left\| \mathbf{A}^T \mathbf{S} \mathbf{A} + \alpha \mathbf{L}^T \mathbf{V} \mathbf{L} + \gamma \mathbf{U} \right\|^{-1} \ . \tag{16}$$

In order to incorporate the projection operator $\mathbf{P}$ in the Q-th order converging algorithm, the simplest and most direct method is to project the iterates as was proposed in [16]

$$\mathbf{f}_{k+1} = \mathbf{P}\left[\sum_{j=0}^{Q_k-1} \mathbf{B}_k^j \mathbf{f}_k\right] . \qquad (17)$$

This extension will, however, inevitably lead to diverging iterations and erroneous results since the unaltered iterations on the matrix $\mathbf{B}_k$ in Eq. (15) would progress independently of the projection operator. Consequently, the incorporation of the projection operator in Eq. (17) must be followed by a modification of Eq. (15) as well, as follows.

$$\mathbf{P}(\mathbf{I} - \mathbf{B}_k^{Q_k})\mathbf{f} = (\mathbf{I} - \mathbf{B}_{k+1})\mathbf{f} . \qquad (18)$$

For a linear projection operator Eq. (18) reduces to

$$\mathbf{B}_{k+1} = \mathbf{PB}_k^{Q_k} + (\mathbf{I} - \mathbf{P}) . \qquad (19)$$

which includes Eq. (15) as a special case. Eq.(18) cannot be solved for a nonlinear projection operator, therefore the extension in Eq. (17) holds merely for linear projections. In practice, most projections can satisfy this condition.

3.3 Assessment Criterions of Intensity

In order to obtain a quantified index of the intensities of images, the assessment criterions of intensity are introduced here. MGL [18] is one of the criterions to assess the intensities of images, given by

$$MGL(G) = \frac{IOD(G)}{A(G)} = \left(\int_G^\infty DH(D)dD\right) \Big/ \left(\int_G^\infty H(D)dD\right) . \qquad (20)$$

where G is a threshold of gray level, $A(G)$ is an area where the intensities are not less than the threshold G; $IOD(G)$ is the integral of the product of the gray level and its probability density with respect to the gray level from the threshold G to infinite, called integrated optical density; $H(D)$ is the histogram value of the gray level D. $IOD(0)$ and $A(0)$ can be simply denoted as IOD and A.

We introduce a novel assessment criterion used in the experiments later, called MGL Ratio, to assess the intensity contrast, given by

$$MGLR(G) = \frac{MGL(G) \cdot A - IOD(G)}{IOD - IOD(G)} . \qquad (21)$$

where $MGLR(G)$ is a ratio of the average gray level of the pixels, whose intensities are not less than the threshold G, to that less than the threshold G. When the threshold G is an optimal threshold, which results in that the smallest number of pixels in the object and the background is missegmented [19], MGL Ratio can be considered as the intensity contrast of the object and the background if the object is brighter than the background as is shown in the fluorescein angiograms; and vice versa. Obviously, MGL Ratio is always greater than 1.

4 Experiments

In this section, we present the results of experiments using real images. For comparison, the resulting images are reconstructed based on the minimization of Eq. (6) with different regularization parameters γ. A 2-D Laplacian filter [20] is then used as a high-pass filter $\mathbf{L}$, and the regularization parameter α of the smoothness constraint is set to 0.01. The other parameters include $\beta = 1.9$, $Q = 4$, $\mathbf{S} = \mathbf{V} = \mathbf{U} = Identity$.

We use fluorescein angiographic video sequences as our data source. The SLO took 1208 frames in 40 seconds. Fig. 1 (a) and (d) show 2 frames out of the sequences, to be exact, 2 original image fragments with 128×128 pixel resolution. The resulting reconstructions are presented in Fig. 1. Fig. 1 (b) and (e) show the reconstructions from 65 frames and 216 frames, respectively, when $\gamma = 0$; Fig. 1 (c) and (f) are the reconstructions from 65 frames and 216 frames, respectively, when $\gamma = 1$.

In Fig. 1 (b) and (d), we note that when the number of the LR frames required for reconstruction increases, the intensities of the resulting images using Miller's regularization approach (i.e. $\gamma = 0$) degrade to a great degree. The reason causing the phenomenon is primarily that the intensities of images are not constant along temporal axis, i.e., not temporally homogeneous. In the most of time, the fluorescence intensity is dark. In the process of reconstruction with the increase of frames, the intensities are inevitably influenced by more "dark" frames, even degraded obviously if excessive frames are used. In contrast, visual inspections

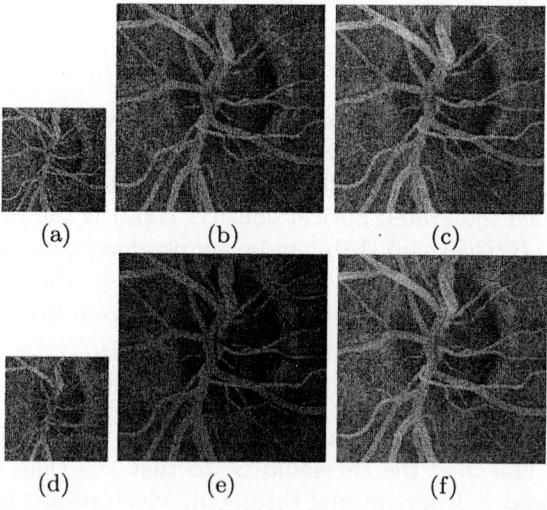

Fig. 1. Experimental results. (a) and (d) are two original image fragments (128×128 pixel resolution); (b) and (c) are two reconstructed images from 65 frames with the parameters $\gamma = 0$ and $\gamma = 1$, respectively (256×256 pixel resolution); (e) and (f) are two reconstructed images from 216 frames with the parameters $\gamma = 0$ and $\gamma = 1$, respectively (256×256 pixel resolution).

Table 1. Experimental results of PSNR and MGL Ratio of the reconstructed images with different regularization parameters γ

γ	0	0.0625	0.125	0.25	0.5	1
PSNR	30.792	30.442	29.039	28.971	28.906	26.918
MGL Ratio	1.4182	1.5597	1.5639	1.5644	1.5748	1.5859

show that the reconstruction with intensity constraint is obviously superior to the direct reconstruction from the aspect of high-intensity preservation. In addition, the proposed scheme is comparable to the direct one in noise suppression, and produces a satisfactory result.

We also quantitatively measure PSNR and MGL Ratio among the different resulting images, with results summarized in Table 1. Note that when γ increases, MGL Ratio increases, whereas PSNR decreases. It means that the ability of keeping the intensities from degrading increases, at the expense of reducing the ability of noise suppression. The closer to the desirable intensity template is the enhanced image, the weaker is the ability to control data faithfulness and solution smoothness. Since we set regularization parameters α to a fixed value ($\alpha = 0.01$), the control capability for α isn't considered and discussed in this paper. When γ approaches zero, the situation is just the reverse. Generally, Table 1 supports the following conclusion that the proposed scheme can rival the direct scheme in respect of SR reconstruction, and shows better applicability and robustness in intensity preservation.

5 Conclusions

In this paper, we propose a SR reconstruction scheme for fluorescein angiogram sequence, which combine the new intensity constraint to Miller's regularization formulation to achieve the ability of preserving high-intensity. In implementation, we use a modified Q-th order converging algorithm to solve the minimization problem subject to both the smoothness and the intensity constraint. An intensity template derived from a pre-filtering process is employed in the intensity constraint, representing an image with the desirable intensities. The estimated SR reconstruction may be influenced by the regularization parameter of the intensity constraint as does the regularization parameter of the smoothness constraint. We also introduce a novel assessment criterion used in experiments, called MGL Ratio, to assess the performance of high-intensity preservation. In the experiments, we show that the proposed scheme obtain the good results in high-intensity preservation and SR reconstruction.

References

1. Huang, T.S., Tsai, R.Y.: Multiple Frame Image Reconstruction and Registration, in Advances in Computer Vision and Image Processing, T.S. Huang, Ed. Greenwich, CT: JAI, ch. 7, 1984

2. Kim, S.P., Bose, N.K., Valenzuela, H.M.: Recursive Reconstruction of High Resolution Image from Noisy Undersampled Multiframes, IEEE Trans. Acoust., Speech, Signal Proc., vol. 38, no.6, pp. 1013–1027, June, 1990
3. Kim, S., Su, W.Y.: Recursive High-Resolution Reconstruction of Blurred Multiframe Images, IEEE Trans. Image Processing, vol. 2, pp. 534–539, Oct. 1993
4. Irani, M., Peleg, S.: Improving Resolution by Image Registration, CVGIP: Graph., Models, Image Processing, vol. 53, pp. 231–239, May 1991
5. Stark, H., Oskoui, P.: High-resolution Image Recovery from Image-Plane Arrays Using Convex Projections, J. Opt. Soc. Amer. A, vol. 6, pp. 1715–1726, 1989
6. Tekalp, A.M., Ozkan, M.K., Sezan, M.I.: High-Resolution Image Reconstruction from Lower-Resolution Image Sequences and Space-Varying Image Restoration, in Proc. IEEE Int. Conf. Accoust., Speech, Signal Processing, San Francisco, CA, pp. III-169 to III-172, 1992
7. Patti, A.J., Sezan, M.I., Tekalp, A.M.: Superresolution Video Reconstruction with Arbitrary Sampling Lattices and Non-Zero Aperture Time, IEEE Trans. Image Processing, vol. 6, pp. 1064–1076, Aug. 1997
8. Elad, M., Feuer, A.: Restoration of a Single Superresolution Image from Several Blurred, Noisy, and Undersampled Measured Images, IEEE Trans. Image Processing, vol. 6, pp. 1646–1658, Dec. 1997
9. Schultz, R.R., Stevenson, R.L.: Extraction of High-Resolution Frames from Video Sequences, IEEE Trans. Image Processing, vol 3., pp. 233–242, May 1994
10. Borman, S., Stevenson, R.L.: Simultaneous Multiframe MAP Super-Resolution Video Enhancement Using Spatio-temporal Priors, Proc. IEEE Int. Conf. in Image Processing, vol. 3, pp. 469–473, 1999
11. Manivannan, A., Sharp, P.F., Phillips, R.P. et al.: Digital Fundus Imaging Using a Scanning Laser Ophthalmoscope, Physiological Measurement 14, pp. 43–56, 1993
12. Webb, R.H., Hughes, G.W.: Scanning Laser Ophthalmoscope, IEEE Trans. Biomedical Engineering vol. 28, pp.488–492, 1981
13. Biemond, J., Lagendijk, R.L.: Regularized Iterative Image Restoration in a Weighted Hilbert Space, ICASSP'86
14. Lagendijk, R.L., Biemond, J., Boekee, D.E.: Regularized Iterative Image Restoration with Ringing Reduction, IEEE trans. on Acoust., Speech and Signal Processing, 1987
15. Tomasi, C., Manduchi, R.: Bilateral Filtering for Gray and Color Images, in Proceedings of IEEE Int. Conf. on Computer Vision, Jan., pp. 836–846, 1998
16. Singh, S., Tandon, S.N. Gupta, H.M.: An Iterative Restoration Technique, Signal Processing, vol. 11, 1986
17. Morris, C.E., Richards, M.A., Hayes, M.H.: An Iterative Deconvolution Algorithm with P-th Order Convergence, Proc. of the 1986 DSP Workshop, Chatham, Mass., 1986
18. Castleman, K.R.: Digital Image Processing. Upper Saddle River, New Jersey: Prentice Hall International, Inc., pp. 64–65, 1998
19. Gonzalez, R.C., Woods, R.E.: Digital Image Processing. 2rd. ed., Addison-Wesley, pp. 602–607, 1987
20. Andrews, H.C., Hunt, B.R.: Digital Image Restoration. Englewood Cliffs, NJ: Prentice-Hall, 1977
21. Miller, K.: Least-Squares Method for Ill-Posed Problems with a Prescribed Bound, SIAM J. Math. Anal., vol. 1, pp. 52–74, Feb. 1970

Signature Verification Using Wavelet Transform and Support Vector Machine*

Hong-Wei Ji** and Zhong-Hua Quan

Institute of Intelligent Machines, Chinese Academy of Sciences, P.O. Box 1130,
Hefei Anhui 230031, China
{njnujhw4, quanzhonghua}@iim.ac.cn

Abstract. In this paper, we propose a novel on-line handwritten signature verification method. Firstly, the pen-position parameters of the on-line signature are decomposed into multiscale signals by wavelet transform technique. For each signal at different scales, we can get a corresponding zero-crossing representation. Then the distances between the input signature and the reference signature of the corresponding zero-crossing representations are computed as the features. Finally, we build a binary Support Vector Machine (SVM) classifier to demonstrate the advantages of the multiscale zero-crossing representation approach over the previous methods. Based on a common benchmark database, the experimental results show that the average False Rejection Rate (FRR) and False Acceptance Rate (FAR) are 5.25% and 5%, respectively, which illustrates such new approach to be quite effective and reliable.

1 Introduction

Handwritten signature verification, as one type of biometrics technology, has been extensively studied and now become more and more important in authorizing electronic transaction and authenticating a person's identity due to its availability and acceptability [1-3].

In general, handwritten signature verification can be divided into two main areas depending on the data acquisition method: off-line and on-line signature verification. Off-line signature verification deals with the digital images which are acquired by scanning a document into the computer, while on-line signature verification uses a digitizing tablet and a pressure-sensitive pen to capture the dynamic signature information, such as pen-position, pen-pressure and pen-orientation. A variety of advanced methods has been developed in recent years [4-9]. Many techniques have been used for signature verification, such as Regional Correlation, Dynamic Time Warping and Skeletal Tree Matching, etc.. In 1990, Parizeau and Plamondon [10] reported a comparative study of the above three different matching algorithms in the

---

* This work was supported by the National Natural Science Foundation of China (Nos.60472111 and 60405002).
** Corresponding author.

context of signature verification. Verification errors showed that no algorithm consistently outperformed the others in all circumstances. Hidden Markov Models (HMMs), well known for their success in speech recognition, have also been applied to signature verification [11].

Recently, Wavelet Transform (WT), due to its capability of multiresolution analysis, has been successfully applied to signature verification [12-15]. This paper proposes a novel on-line signature verification method based on the multiscale zero-crossing representation of wavelet transforms. This approach can take full advantage of multiscale zero-crossings of wavelet transforms. Firstly, the pen-position parameters (X-coordinate function and Y-coordinate function) of the on-line signature are decomposed into multiscale signals by wavelet transform technique. For each signal at significant scales, we can obtain a corresponding zero-crossing representation described by a sequence of integral values. Then the distances between the input signature and the reference signature of the corresponding zero-crossing representations are computed using Dynamic Time Warping [9]. Taking these distances as the features, we build a binary Support Vector Machine (SVM) classifier to demonstrate the advantages of the multiscale zero-crossing representation approach over the previous methods. Based on a common benchmark database, the experimental results show that the average False Rejection Rate (FRR) and False Acceptance Rate (FAR) are 5.25% and 5%, respectively, which illustrates such new approach to be quite effective and reliable.

The remainder of this paper is organized as follows. Section 2 describes and discusses in detail the feature extraction method based on the zero-crossing representation of wavelet transforms. In Section 3, a binary SVM classifier is employed to verify the signature. Experimental results are reported in Section 4 and Section 5 concludes this paper.

2 Feature Extraction of Signatures Based on the Multiscale Zero-Crossing Representation of Wavelet Transforms

2.1 Preprocessing

Considering that only the pen-position information can be acquired by small pen-based input devices such as Personal Digital Assistants (PDA), we just use the coordinate information contained in the signature. In order to decrease the personal fluctuation, we must normalize the pen-position parameters. The normalized pen-position parameters can be defined as

$$x'(t) = \frac{x(t) - x_{min}}{x_{max} - x_{min}} \quad (x_{min} \le x(t) \le x_{max}) . \tag{1}$$

$$y'(t) = \frac{y(t) - y_{min}}{y_{max} - y_{min}} \quad (y_{min} \le y(t) \le y_{max}) . \tag{2}$$

where $x(t)$ and $y(t)$ are the original X-coordinate function and Y-coordinate function, respectively. An example of the handwritten signature and its corresponding normalized pen-position parameters are shown in Fig. 1.

(a) An example of the handwritten signature

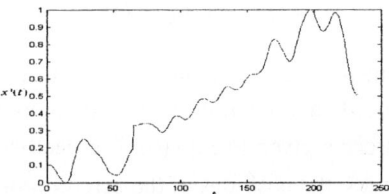

(b) The normalized X-coordinate function $x'(t)$

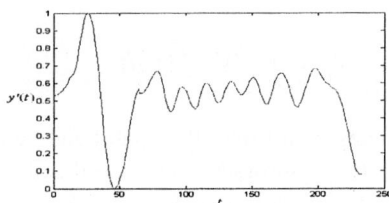

(c) The normalized Y-coordinate function $y'(t)$

Fig. 1. An example of the handwritten signature and its corresponding normalized pen-position parameters

2.2 Feature Extraction

Since WT has the capability of multiresolution analysis [16], we here apply WT to build the signature verification system. The mother wavelet employed in this paper is the second derivative of the Gaussian function. For one dimensional signal, the mother wavelet can be written as

$$\psi(t) = \frac{d^2 g(t)}{dt^2} = (1-t^2)e^{-\frac{t^2}{2}}. \tag{3}$$

where $g(t)$ is a Gaussian function. We denote the dilation of $\psi(t)$ by $\psi_s(t)$:

$$\psi_s(t) = \frac{1}{s}\psi(\frac{t}{s}) . \qquad (4)$$

where s is the scale factor. The wavelet transform of a function $f(t)$ at the scale s is given by the convolution product

$$W_s f(t) = f * \psi_s(t) . \qquad (5)$$

For practical applications the scale factor must be discretized. Here we define $s = 2^j$, therefore the wavelet transform at the scale 2^j can be represented as

$$W_{2^j} f(t) = f * \psi_{2^j}(t) . \qquad (6)$$

After the preprocessing step, the normalized pen-position parameters (X-coordinate function $x'(t)$ and Y-coordinate function $y'(t)$) are then decomposed into multiscale signals by wavelet transforms given in Eqn.(6). For each signal at different scales, we extract two sets of attributes: the positions of the zero-crossings $(z_0, z_1, ..., z_n)$ and the integral values between two consecutive zero-crossings $(e_0, e_1, ..., e_n)$, where e_n is defined by

$$e_n = \int_{z_{n-1}}^{z_n} W_{2^j} f(t) dt . \qquad (7)$$

The two sets of attributes constitute the stabilized zero-crossing representation explained by Mallat [17]. Fig. 2 shows Mallat's stabilized zero-crossing representation of a wavelet transform.

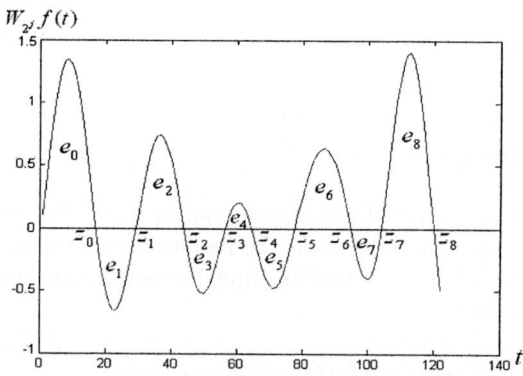

Fig. 2. Mallat's stabilized zero-crossing representation of a wavelet transform

In fact, Mallat has proven that the original signal can be reconstructed from its stabilized zero-crossing representation of wavelet transforms [17]. Furthermore, we normalize the integral values with the following equation:

$$e'_n = \frac{e_n}{z_n - z_{n-1}}. \tag{8}$$

So, for each signal at different scales, we can obtain a corresponding zero-crossing representation described by a sequence of normalized integral values $(e'_0, e'_1, ..., e'_n)$. Here we only utilize the normalized integral sequences at scales of $2^1, 2^2, 2^3, 2^4$. The reason for doing so is that when the scale is larger than 2^4, the number of the zero-crossings is close to zero, thus it is meaningless for the zero-crossing representation. Since two functions (X-coordinate function $x'(t)$ and Y-coordinate function $y'(t)$) are decomposed by wavelet transforms, total eight sequences of normalized integral values are acquired. Then Dynamic Time Warping (DTW) [9] is employed to compute the distances between the input signature and the reference signature of the corresponding normalized integral sequences. DTW is a pattern matching technique, which can compare two sequences of different lengths. For more details about DTW, readers can refer to the work of Sakoe and Chiba [9]. We take these distances as the features. Eventually, total eight features are extracted for signature verification.

3 Signature Verification Using SVM

In this section, a binary SVM classifier is employed to verify the signature. SVM is a new and promising classification technique developed by Vapnik [18]. It has a good generalization performance even under the conditions of small training set. For signature verification problem, both genuine signatures and forgeries used for training are very limited, so we choose SVM to distinguish the genuine signature from its forgery.

Usually, SVM is to solve the following optimization problem:

$$\min_{w,b,\xi} \frac{1}{2} w^T w + C \sum_{i=1}^{l} \xi_i \tag{9}$$

subject to $y_i(w^T \phi(x_i) + b) \geq 1 - \xi_i$, $C > 0$, $\xi_i \geq 0$.

where the original input space is mapped into a higher dimensional feature space by the function ϕ. Then SVM can find an optimal linear separating hyperplane with the maximal margin in this higher dimensional feature space. C is the penalty parameter of the error term. For a two-class problem, the nonlinear decision function derived from the SVM classifier can be formulated as

$$f(x) = sign(\sum_{i=1}^{l} \lambda_i y_i K(x, x_i) + b). \tag{10}$$

where $K(x, x_i) = \phi(x)^T \phi(x_i)$ is called the kernel function.

Thus, the whole training and verification procedure using SVM for signature verification can be described by the following steps:

Step 1: For the signatures used for training, the features are extracted using the method described in Section 2.
Step 2: Consider the Gaussian kernel $K(x, x_i) = \exp(-\gamma \| x - x_i \|^2)$, $\gamma > 0$.
Step 3: Cross-validation [19] based on subsets of the training feature set is employed to find the best SVM parameters C and γ.
Step 4: The SVM classifier is trained using the whole training feature set.
Step 5: During the verification phase, the feature vector of an input signature is submitted to the well-trained SVM classifier to determine whether the signature is genuine or forged.

4 Experimental Results

The common benchmark signature database used in this paper is obtained from the First International Signature Verification Competition (SVC2004) held in Hong Kong [20]. This database has 40 sets of signature data which belong to 40 users, respectively. Each set contains 20 genuine signatures from one signature contributor and 20 skilled forgeries from five other contributors. For each set of signature data, we randomly selected 10 genuine signatures and 10 skilled forgeries for training, and the others for testing.

Firstly, the pen-position parameters of all the signatures were decomposed into multiscale signals by wavelet transforms. For each signature, eight sequences of normalized integral values were obtained from its multiscale zero-crossing representation of wavelet transforms. Next, the distances between the input signature and the reference signature of the corresponding integral sequences are computed as the features (Among the training signatures, the genuine signature which has a moderate number of zero-crossings at different scales, is regarded as the reference signature). For each user, a binary SVM classifier was then trained on the training feature set. Finally we applied the SVM classifier on the testing feature set.

In order to demonstrate the advantages of our multiscale method, we compared it with the single scale method (only using the features obtained from a certain scale). Table 1 shows the detailed experimental results.

From Table 1, it can be found that our multiscale method, which achieves 5.25% FRR and 5% FAR, obviously outperforms the single scale method. Therefore, the average error rate we achieved is 5.125%. In SVC2004, total 12 teams submitted their programs for Task 1. The average error rates obtained by the top five teams are 5.50%,

Table 1. Experimental results based on the common benchmark signature database from SVC2004

Scale	2^1	2^2	2^3	2^4	Multiscale
FRR	25%	8.25%	18.25%	5.75%	**5.25%**
FAR	18.25%	21.75%	15%	31.75%	**5%**

6.45%, 7.33%, 9.80% and 11.10%, respectively. Compared with these results, we are ranked 1st, which illustrates that our approach is quite satisfying.

5 Conclusions

In this paper, a novel on-line signature verification method based on the multiscale zero-crossing representation of wavelet transforms was proposed. Experimental results based on a common benchmark signature database showed that our approach is quite effective and reliable. In the future works, we shall combine the features extracted by the proposed wavelet-based method with the global statistical or geometrical features, to improve the overall verification performance and build a more powerful on-line handwritten signature verification system for practical applications.

References

1. Plamondon, R. and Lorette, G.: Automatic signature verification and writer identification-the state of the art. Pattern Recognition, 22(2):107-131, 1989
2. Leclerc, F. and Plamondon, R.: Automatic signature verification: the state of the art 1989-1993. International Journal of Pattern Recognition and Artificial Intelligence, 8(3):643-660, 1994
3. Plamondon, R. and Srihari, S.N.: On-line and off-line handwriting recognition: a comprehensive survey. IEEE Trans. on Pattern Anal. Mach. Intel. 22 (1):63-84, 2000
4. Huang, D.S.: Systematic Theory of Neural Networks for Pattern Recognition. Publishing House of Electronic Industry of China, Beijing, 1996
5. Huang, D.S.: The local minima free condition of feedforward neural networks for outer-supervised learning. IEEE Trans on Systems, Man and Cybernetics, Vol.28B, No.3, 1998,477-480
6. Huang, D.S. and Ma, S.D.: Linear and nonlinear feedforward neural network classifiers: A comprehensive understanding. Journal of Intelligent Systems, Vol.9, No.1, 1-38,1999
7. Huang, D.S.: Application of generalized radial basis function networks to recognition of radar targets. International Journal of Pattern Recognition and Artificial Intelligence, Vol.13, No.6, 945-962,1999
8. Huang, D.S.: Radial basis probabilistic neural networks: Model and application. International Journal of Pattern Recognition and Artificial Intelligence, 13(7), 1083-1101,1999
9. Sakoe, H. and Chiba, S.: Dynamic programming algorithm optimization for spoken word recognition. IEEE Trans. Acoust., Speech, Signal Processing, vol. ASSP-26, no. 1, pp. 43-49, Feb. 1978
10. Parizeau, M. and Plamondon, R.: A comparative analysis of regional correlation, dynamic time warping, and skeletal tree matching for Signature Verification. IEEE Transactions on Pattern Analysis and Machine Intelligence, Vol. 12, Iss. 7, July 1990
11. Yang, L., Widjaja, B.K. and Prasad, R.: Application of hidden markov models for signature verification. Pattern Recognition, Vol. 28, No. 2, pp. 161-170, 1995
12. Qi, Y. and Hunt, B.R.: A multiresolution approach to computer verification of handwritten signature. IEEE Trans. Image Processing, vol. 4, pp. 870-874, June 1995

13. Wen, C.J., Jeng, B.S. and Yau, H.F.: Tremor detection of handwritten Chinese signatures based on multiresolution decomposition using wavelet transforms. Electronic Letters, vol. 32, no. 3, pp. 204-206, July 1996
14. Deng, P.S., Liao, H.M., Ho, C.W. and Tyan, H.: Wavelet-based off-line signature verification. Proc. IEEE, 1997
15. Nakanishi, I., Nishiguchi, N., Itoh, Y. and Fukui, Y.: On-line signature verification method utilizing feature extraction based on DTW. Proceedings of the 2003 International Symposium on Circuits and Systems, 2003
16. Mallat, S.G.: A theory for multiresolution signal decomposition: The wavelet representation. IEEE Trans. Pattern Anal. Machine Intell., vol. 11, pp. 674–693, July 1989
17. Mallat, S.G.: Zero-crossings of a wavelet transform. IEEE Trans. Information Theory, vol. 37, no. 4, pp. 1019-1033, July 1991
18. Vapnik, V.: The nature of statistical learning theory. Springer Verlag, 1995
19. Hsu, C.W., Chang, C.C. and Lin., C.J.: A practical guide to support vector classification, 2001
20. The First International Signature Verification Competition (http://www.cs.ust.hk/svc2004), 2004

Visual Hand Tracking Using Nonparametric Sequential Belief Propagation

Wei Liang, Yunde Jia, and Cheng Ge

Department of Computer Science and Engineering,
Beijing Institute of Technology, Beijing 100081, P.R. China
{liangwei, yjiar, gecheng}@bit.edu.cn

Abstract. Hand tracking is a challenging problem due to the complexity of searching in a 20+ degrees of freedom space for an optimal estimate. This paper develops a statistical method for robust visual hand tracking, in which graphical model decoupling different hand joints is performed to represent the hand constraints. Each node of the graphical model represents the position and the orientation of each hand joint in world coordinate. Then, the problem of hand tracking is transformed into an inference of graphical model. We extend Nonparametric Belief Propagation to a sequential process to track hand motion. The Experiment results show that this approach is robust for 3D hand motion tracking.

1 Introduction

Gesture is one of the most common and natural communication means among human beings. Rather than the mouse and the keyboard, hand gesture should be used as a more natural and convenient way for human-computer interaction. Cyber-glove was firstly used to capture hand motion in real time, but it is not natural and friendly. Recent years, many researchers in computer vision try to seek more effective methods to solve these problems. However, highly articulated human hand motion always presents complex rotation, translation and self-occlusion, which make hand hard to be tracked. Robust tracking of human hands is difficult in general for three reasons. Firstly, hand is deformable, self-occlusion and highly articulated with 20+ degrees of freedom. Secondly, many applications require hands to be tracked in cluttered scenes and under poorly controlled varying lighting conditions. Thirdly, tracking hands requires a solution to the temporal problem.

A strategy way to address the high dimensionality of articulated tracking problem is bottom-up approaches, and graphical model is a good choice. Because the articulated objects consist of a number of articulated parts, the increase of dimensionality will incur exponential increase of computation. The advantage of graphical model is that it decouples different hand joints, and simplifies some tracking problem. Thus the problem of hand tracking is transformed into an inference of graph model. And Belief Propagation (BP) has been used to perform inference in graphical model. Recently, two approaches have been used widely in computer vision to perform inference of graphical model: loopy belief propagation [6] and particle filter [2] [4]. When there is no loop in the graph model, BP could obtain the exact inference more efficiently through a local message passing process. The Nonparametric BP [2] and the PAMPAS algorithm [4] combine the BP algorithm with MCMC technique to implement the inference. Also, a mean field Monte Carlo algorithm (MFMC) has been proposed in [3] for tracking articulated body by integrating sequential Monte Carlo technique with the mean field variational method.

In this paper, we represent the hand by graphical model including kinematics constraints and structure constraints. Then the problem of hand tracking is performed by graphical model inference. We extend Nonparametric Belief Propagation to a sequential process to track hand motion. Like Monte Carlo, NSBP (nonparametric sequential belief propagation) approximates the posterior distribution of hand configurations as a collection of samples. In Section 2 we describe the hand model represented by graphical model. Observation model is presented in Section 3. Section 4 gives an outline of the nonparametric sequential belief propagation with particle filter to perform the inference. Finally, experiment results are shown in Section 5 and conclusion are drawn in Section 6.

2 Hand Modeling

In this paper, hand is represented by graphical model in which each node associated with a configuration vector defining the joint's position and orientation in word coordinates. Placing each joint in a global coordinate enables the joints observation to operate independently while the whole hand is assembled by the process of graphical model inference. Edges in the graphical model correspond to constraints between adjacent joints, as illustrated in Fig.1. Just as standard graphical models, we assume the variables in a node are independent of those in non-neighboring nodes.

In this paper, each hand joint is modeled by a cylinder with 2 fixed (person specific) and 6 estimated parameters. The fixed parameters are denoted as $f_i = (L_i, W_i)$, where L_i and W_i are the joint's length and width, respectively, and the estimated parameters are written as $e_i = (X_i, \theta_i)$ representing the position and orientation of joint i in word coordinates, where $X_i \in R^3$ and $\theta_i \in SO(3)$. Then the configuration of the whole hand is written as $E = \{e_1, \cdots, e_{16}\}$. It is obviously that the representation is redundant and there are dependencies among the elements of e_i. Each edge between parts i and j has an associated potential function $\psi_{i,j}(e_i, e_j)$ that encodes the compatibility between pairs of part configurations and intuitively can be thought of as the probability of configuration e_j of part j conditioned on the part i. Fig.1 shows the structure constraints and the kinematics constraints. The structure constraints $\psi^S_{i,j}(e_i, e_j)$ are defined by Eq.(1) as described in [8].

$$\psi^S_{i,j}(e_i, e_j) = \begin{cases} 1 & \| e_i(X) - e_j(X) \| > \delta_{i,j} \\ 0 & otherwise \end{cases} \quad (1)$$

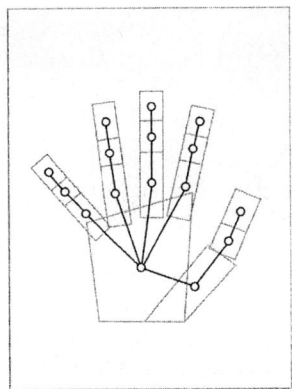 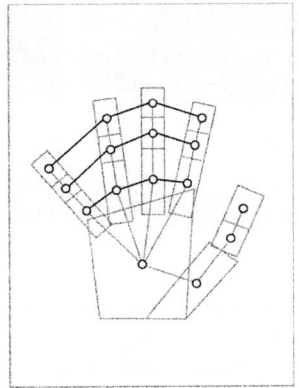

Fig. 1. Hand model represented by graphical model (a) kinematics constraints (b)structure constraints

where $\delta_{i,j}$ is a threshold determined by fixed parameters f_i and f_j. And the kinematics constraints $\psi_{i,j}^K(e_i, e_j) = 1$ if the pair (e_i, e_j) is valid rigid body configurations associated with some setting of the angles of joint (i, j) and zero otherwise. For notational convenience, we define an order: from the fingertip to the palm is 'forward' and from the palm to the fingertip is 'backward'.

Most of the typical tracking methods are based on Bayesian system that the state of the object at time t is denoted by x_t, and its history is $X_t = \{x_1, \cdots, x_t\}$. Similarly the set of image features at time t is z_t with history $Z_t = \{z_1, \cdots, z_t\}$. All these methods estimate x_{t+1} at time $t+1$ through Bayesian formulation shown by Eq. (2).

$$p(x_{t+1} | Z_{t+1}) \propto p(z_{t+1} | x_{t+1}) p(x_{t+1} | Z_t). \tag{2}$$

Where $p(x_{t+1} | Z_t) = \int_{x_t} p(x_{t+1} | x_t) p(x_t | Z_t) dx_t$. A general assumption is made for the probabilistic framework that the object dynamics form a temporal Markov chain so the new state is conditioned directly only on the immediately preceding state, independent of the earlier history. And we assume that our dynamical model obey the Gaussian distribution for each component at time t.

3 Observe Model

The observation process is defined by $p(z_t | x_t)$ and z_t represents the set of entire image features at time t. In this paper, we employ edges and foreground silhouette to construct the observation model. The cylinder model of each joint with the hypothetical configuration is projected in the real image plan, then a gradient based edge detection mask is used to detect edges of the real image and the result is thresholded to eliminate spurious edges, smoothed with Gaussian mask and remapped between 0 and 1 as shown in Fig.2(b). This method produces a pixel map which each pixel is as

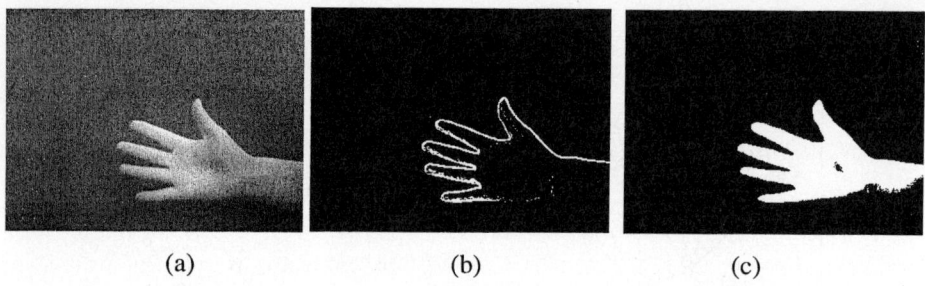

(a)   (b)   (c)

Fig. 2. Feature extraction. A gradient based edge detection mask is used to find edges. (a) is the real image. The result is thresholded to eliminate spurious edges and smoothed using a Gaussian mask to produce map (b) in which the value of each pixel is related to it's proximity to an edge. The foreground is segmented using thresholded background subtraction to produce the pixel map (c) used in the weighting function.

signed a value related to proximity to an edge. At last, a SSD (Sum-Squared Difference) function is computed using Eq.(3).

$$\sum\nolimits^e (X,Z) = \frac{1}{N} \sum_{i=1}^{N} (1 - p_i^e(X,Z))^2 . \qquad (3)$$

Where X is the model's configuration vector and Z is the image from which the pixel map is derived. $p_i(X,Z)$ is the values of the edge pixel map at the N sampling points taken along the model's edge. So the edges observe model is written as Eq.(4).

$$p(z_t | x_t) \propto \exp{-\sum\nolimits^e (X,Z)} . \qquad (4)$$

Thresholds background subtraction is used here to separate the hand from the background and the result is shown in Fig.2(c). And a SSD is computed as Eq.(5).

$$\sum\nolimits^s (X,Z) = \frac{1}{N} \sum_{i=1}^{N} (1 - p_i^s(X,Z))^2 . \qquad (5)$$

where $p_i(X,Z)$ are the values of the foreground pixel map at the N sampling points taken from the interior of the cylinder sections. So the foreground silhouette observe model is written as Eq.(6).

$$p_S(z_t | x_t) \propto \exp{-\sum\nolimits^s (X,Z)} . \qquad (6)$$

For the means of graphical model representation, each hand component is projected to a disjoint part, so the likelihoods are estimated independently for each joint by projecting the 3D model into the corresponding image projection plane. And the observe model will be decomposed as Eq.(7) and (8) respectively.

$$p_E(z_t | x_t) \propto \prod_{i=1}^{16} p_E(z_t | x_{t,i}) . \qquad (7)$$

$$p_S(z_t \mid x_t) \propto \prod_{i=1}^{16} p_S(z_t \mid x_{t,i}). \tag{8}$$

4 Sequential Belief Propagation with Particle Filter

Hand tracking could be considered as a graphical model inference problem. In this paper, we perform sequential belief propagation (SBP) via message passing between the nodes of the graph to handle the inference problem which is extended from standard BP [1].

4.1 Sequential Belief Propagation

An undirected graph ξ is defined by a set of nodes v and a corresponding set of edges ε. The neighborhood of a node $j \in v$ is defined as $\Gamma(j) = \{i \mid (j,i) \in \varepsilon\}$ with each node $j \in v$ associatting with an observation z_j. For continuous valued marginal distributions, the message $m_{ij}(x_j)$ which is sent to each neighboring nodes $j \in \Gamma(i)$ from each $i \in v$ is written as:

$$m_{ij}^n(x_j) \leftarrow \int \varphi_{ji}(x_j, x_i) \phi_i(x_i) \prod_{u \in \Gamma(i) \setminus j} m_{ui}^{n-1}(x_i) dx_i. \tag{9}$$

where $\varphi_{ji}(x_j, x_i)$ is the correlation function between i and j, and $\phi_i(x_i)$ is the observation at node i. The message, which is a vector with the same dimensionality as x_j, is proportional to how likely node i considers it is that node j will be in the corresponding state. The marginal distribution (belief) at node j is approximated by combining all the messages coming into node j with the local observation which is shown in Eq.(10).

$$b(x_j) \propto \phi_j(x_j) \prod_{i \in \Gamma(j)} m_{ij}^n(x_j). \tag{10}$$

For the sequential hand tracking, the BP method is extended by rewritten Eq.(9) of the message update process at time t as Eq.(11).

$$m_{t,ij}^n(x_{t,j}) \leftarrow \int \varphi_{j,i}(x_{t,j}, x_{t,i}) \phi_i(x_{t,i})$$

$$[\int_{x_{t-1,i}} p(x_{t,i} \mid x_{t-1,i}) p(x_{t-1,i} \mid Z_{t-1}) dx_{t-1,i}] \prod_{u \in \Gamma(i) \setminus j} m_{ui}^{n-1}(x_{t,i}) dx_{t,i} \tag{11}$$

And the marginal distribution at time t is given by:

$$b(x_{t,j}) \propto \phi_j(x_{t,j}) \prod_{i \in \Gamma(j)} m_{ij}^n(x_{t,j}) \int_{x_{t-1,j}} p(x_{t,j} \mid x_{t-1,j}) p(x_{t-1,i} \mid Z_{t-1}) dx_{t-1,j}. \tag{12}$$

For hand tracking, the variables x_j take continuous values. Just as most computer vision problems, accurate discretization of the multiple degrees of freedom at each

> For each node $j=1\cdots 16$, generate $\{s_{t,j}^{(n)}, \omega_{t,j}^{(i,n)}\}_{n=1}^N$ and $\{s_{t,j}^{(n)}, \pi_{t,j}^{(n)}\}_{n=1}^N$ at time t.
> 1. Initialization
>   For $j=1\cdots 16$
>   1.1 Resampling: Resample from $\{s_{t-1,j}^{(n)}, \pi_{t-1,j}^{(n)}\}_{n=1}^N$ according to the weights $\pi_{t-1,j}^{(n)}$ and set $\pi_{t-1,j}^{(n)} = \frac{1}{N}$, then get $\left\{\tilde{s}_{t-1,j}^{(n)}, \frac{1}{N}\right\}_{n=1}^N$
>   1.2 Prediction: For each sample in $\left\{s_{t-1,j}^{(n)}, \frac{1}{N}\right\}_{n=1}^N$, sampling to get new samples $\left\{s_{t,j,k}^{(n)}\right\}_{n=1}^N$ based on $p(x_{t,j} | x_{t-1,j})$
>   1.3 Initialization: Assign weight $\omega_{t,j,k}^{(i,n)} = \frac{1}{N}$, $\pi_{t,j,k}^{(n)} = p_j(z_{t,j,k}^{(n)} | s_{t,j,k}^{(n)})$ and normalize them.
> 2. Message update
> $k \leftarrow k+1$
>   2.1 Importance Sampling: Sample $\{s_{t,j,k}^{(n)}\}_{n=1}^N$ from $p(x_{t,j} | x_{t-1,j})$
>   2.2 Re-weight: For each sample $\{s_{t,j,k}^{(n)}\}_{n=1}^N$ and each $i \in \Gamma(j)$ set the weight
> $$\omega_{t,j,k}^{(i,n)} = G_{t,j}^{(i)}(s_{t,j,k}^{(n)}) / I_{t,j}(s_{t,j,k}^{(n)})$$
>   where $G_{t,j}^{(i)}(s_{t,j,k}^{(n)}) = \sum_{m=1}^N \{\pi_{t,i,k-1}^{(m)} p_i(z_{t,i,k-1}^{(m)} | s_{t,i,k-1}^{(m)}) \prod_{l \in \Gamma(i) \setminus j} \omega_{t,i,k-1}^{(l,m)}$
> $\times [\frac{1}{N} \sum_{r=1}^N p(s_{t,j,k-1}^{(n)} | s_{t-1,i}^{(r)})] \cdot \psi_{i,j}(f_i(s_{t,i,k-1}^{(m)}), f_j(s_{t,j,k}^{(m)}))\}$
> $I_{t,j}(s_{t,j,k}^{(n)}) = (\frac{1}{N} \sum_{r=1}^N p(s_{t,j,k}^{(n)} | s_{t-1,j}^{(r)}))$
>   2.3 Normalization: Normalize $\omega_{t,j,k}^{(i,n)}$, and set
> $\pi_{t,j,k}^{(n)} = p_j(z_{t,j}^{(n)} | s_{t,j}^{(n)}) \prod_{l \in \Gamma(j)} \omega_{t,j}^{(l,n)} \times \sum_r (s_{t,j,k}^{(n)} | s_{t-1,j}^{(r)})$ and normalize it.
>   Then, $\{s_{t,j,k}^{(n)}, \omega_{t,j,k}^{(i,n)}\}_{n=1}^N$ and $\{s_{t,j,k}^{(n)}, \pi_{t,j,k}^{(i,n)}\}_{n=1}^N$ are obtained.
>   2.4 Iterate 2.1-2.3 until convergence.

Fig. 3. Nonparametric sequential Belief Propagation message update with Monte Carlo in hand tracking

joint is intractable. So the generalization of particle filtering method is adopted. This generalization is achieved by treating the weighted particles set which is propagated in a standard particle filter as an approximation to the "message" used in the belief propagation algorithm, and replacing the conditional distribution from the previous time step by a product of incoming message sets. In Section 4.2, belief propagation with Monte Carlo method is extended to a sequential process to perform the inference of graphical model.

4.2 Message Propagation

Using particle sets to perform belief propagation concentrates on how to approximate the integral with Monte Carlo. Different from NBP [8] and PAMPAS [9] modeling the message in BP as Gaussian mixtures and using complex MCMC to sample the new Gaussian mixture kernels of the updated messages, we take SBPMC (Sequential Belief Propagation with Monte Carlo) avoiding complex MCMC samplers to update the messages. In fact, NBP and PAMPAS are semi-parametric, for assuming that the message is a mixture of Gaussian. For hand tracking, there is no loop in the model, so BP can get exact inference result better than MFMC only getting an approximate result. On the other hand, the method of SBPMC takes sequential into account. In this paper, each message and belief are represented by a set of weighted samples respectively:

$$m_{ij}(x_j) \sim \{(s_{t,j}^{(n)}, \omega_{t,j}^{(i,n)})\}_{n=1}^N. \tag{13}$$

where $i \in \Gamma(j)$ and $j \in v$. And

$$b(x_{t,j}) \sim \{(s_{t,j}^{(n)}, \pi_{t,j}^{(n)})\}_{n=1}^N. \tag{14}$$

$s_j^{(n)}$ and $\omega_j^{(i,n)}$ in Eq.(13) denote the samples drawn from m_{ij} and their weights respectively. $s_j^{(n)}$ in Eq.(14) is the same as Eq.(13) and $\pi_j^{(n)}$ is the belief. The message update process is shown in Fig. 3.

Both step 2.2 and 2.3 in Fig.3 sample from the message product. When the numbers of the message are more than 2, Gibbs sampler described in [7] is used to form the message products.

5 Experiment Results

Although for hand tracking with bottom-up algorithm, the state dimension has increased from 26 DOF to 96 DOF at first glance, local encoding of the model state greatly simplifies many other aspects of the tracking problem. Each hand joint placed in global coordinate frame enables the joints observation to operate independently while the whole hand is assembled by the process of graphical model inference. For nonparametric sequential belief propagation, the message update order influences the tracking results. We have defined backward and forward order in Section 2. In this paper, we perform message update forward and then backward. The temporal constraints of the hand tracking in sequential images are modeled by Gaussian distribution.

We take a group of image sequence by calibrated camera and all experiments are performed in a cluttered background. Messages are represented by 200 samples and Fig. 4 shows the iteration of the message update. Instead of showing all the samples in the image, we only give the estimation result of each iteration for clarity. Fig. 5 shows

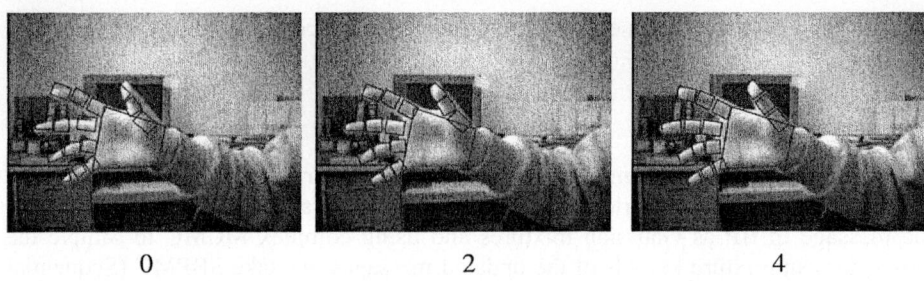

Fig. 4. An example nonparametric sequential belief propagation message update. We show the results of 0,2 and 4 iterations.

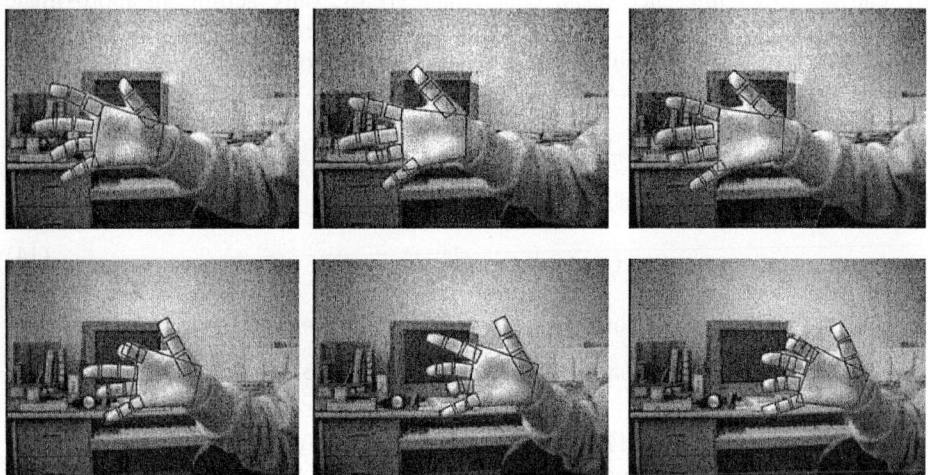

Fig. 5. Hand motion configurations from sequential belief propagation

the result of the sequential belief propagation. Sudderth *et al* [8] also performed Nonparametric Belief Propagation (NBP) to track hand. But in fact, it is a semi-parametric because it models the message as Gaussian mixtures and using complex MCMC samplers to update the messages. The advantage of our method is obviously that it avoids complex MCMC samplers. On the other hand, the method is sequential.

6 Conclusions

Human hand consists of a number of articulated parts and the high dimensionality incurs exponential increase of computation. In this paper, graphical model is used to represent the hand, which decoupling different parts and simplifying some tracking problem. Each node of the graphical model represents the position and the orientation of each hand joint in world coordinate. The problem of hand tracking is then transformed into an inference of graphical model. We extend Belief Propagation to a sequential process to track hand motion. In our tracking method, the message and belief are both represented by a set of weighted particles. And the advantage is obvious that it is nonparametric and sequential. Furthermore, our method allows edges and foreground silhouette as likelihood measures.

Acknowledgements

This work was partially supported by Grant no. (60473049) from the Chinese National Science Foundation

References

1. Yedidia, J. S., Freeman, W. T., Weiss,Y.: Constructing Free Energy Approximations and Generalized Belief Propagation Algorithms. MERL TR 2002- 35, (2002)
2. Sudderth, E. B., Ihler, A. T., Freeman, W. T., Willsky, A. S.: Nonparametric belief propagation. In Proc. IEEE Conf. on Computer Vision and Pattern Recognition, Madison, Wisconsin, June (2003) 605–612
3. Wu, Y., Hua, G., Yu, T.: Tracking Articulated Body by Dynamic Markov Network. In Proc. IEEE International Conference on Computer Vision, Nice,Cˆote d'Azur,France, Octobor (2003) 1094–1101
4. Isard, M.: PAMPAS: Real-valued Graphical Models for Computer Vision. In Proc. IEEE Conf. on Computer Vision and Pattern Recognition, Madison, Wisconsin, June (2003) 613–620
5. Gang Hua, Ying Wu: Multi-scale Visual Tracking by Sequential Belief Propagation. In Pro. IEEE Conf. on Computer Vision and Pattern Recognition, (2004)
6. Yedidia, J.S., Freeman, W.T., Weiss, Y.: Understanding Belief Propagation and its Generalizations. Technical Report TR2001-22, MERL, (2001)
7. Geman, S. and Geman, D.: Stochastic Relaxation, Gibbs Distributions, and the Bayesian Restoration of Images. IEEE Trans. PAMI, 6(6): Nov. (1984) 721–741
8. Sudderth, E. B., Mandel, M. I., Freeman, W. T.: Vision Hand Tracking Using Nonparametric Belief Propagation. IEEE CVPR Workshop on Generative Model Based Vision, (2004)
9. Silverman, B. W.: Density Estimation for Statistics and Data Analysis. London: Chapman & Hall, (1986)

Locating Vessel Centerlines in Retinal Images Using Wavelet Transform: A Multilevel Approach

Xinge You[1,2], Bin Fang[1,3], Yuan Yan Tang[1,2], Zhenyu He[2], and Jian Huang[2]

[1] Faculty of Mathematics and Computer Science, Hubei University, P.R. China
[2] Department of Computer Science, Hong Kong Baptist University
{xyou, bfang, yytang, zyhe, jhuang}@comp.hkbu.edu.hk
[3] Chongqing University, Chongqing, P.R. China

Abstract. Identifying centerlines of vessels in the retinal image is helpful to provide useful information in diagnosis of eye diseases and early signs of systemic disease. This paper presents a novel thinning method based on the wavelet transform with a multilevel scheme. The development of the method is inspired by the favorable characteristics of wavelet transform moduli. Mathematical analysis is given to show that the vessel edge and centerline can be detected efficiently by computing the maxima and minima of wavelet transform moduli. The implementation is performed by applying various scale sizes of the wavelet transform to thin the multiple-pixel-wide ribbon-like vessel gradually to be one-pixel-wide centerline. Experiment results show that the identified centerline of vascular trees are accurate by visual inspection and are useful for further applications.

1 Introduction

Ocular fundus image can provide useful information on pathological changes caused by eye diseases and on early signs of systemic diseases, such as diabetes and hypertension [3,7,14]. Identifying vascular tree features of blood vessels has been an important processing step in diagnosis of eye diseases because these features can be used to reveal status of relative diseases in terms of abnormalities in diameter and formulate the basis for registration. Much effort has been devoted to automate this process and different types of detectors have been explored for vessel identification [3,7,13,10]. However, existing methods suffer from computation complexity due to multiple operations of the detector at various directions to suit with the topology of vascular tree. In addition, the centerline of the vascular tree instead of the core body of vessels is more favorable in some special applications such as registration.

Different approaches are used to extract the centerline structure. They can be classified in the following categories: (i) thresholding and then object connectivity, (ii) thresholding followed by a thinning procedure, and (iii) extraction based on graph description. An interesting overview and several references on different definitions of centerline using paradigms from geometry and mechanics can be found in [1,9].

An interesting overview and several references on different definitions of centerline (or skeleton) using paradigms from geometry and mechanics can be found in [9]. Relation between evolutes and centerlines are presented in [1]. A theoretical mathematical discussion of the medial axis transform in the context of differential geogmetry can be found in [8]. There exists a wide variety of sequential and parallel thinning algorithms for continuous and discrete 2D and 3D data applying one or more of the definitions in [2] on given data. Algorithms are classified by method (topological thinning, distance map based, or based on Voronoi doagrams) or by input data (continuous, polygonal, discrete).

As we focus on techniques for digital retinal fundus images, only thinning algorithms for discrete data will be discussed here. Nytrom and Chen et al. [4] list some criteria for a good thinning approximation in discrete data: preserve the topology of the original shape, approximate the central axis, centerline is thin, smooth, and continuous. In general, the direct application of the definitions mentioned above on discrete data leads to problems: noisy data produces centerline with too many branches, and quality and shape of the approximated centerline depends highly on the chosen discrete metric. In addition, the centerline is not uniquely defined due to the finite resolution of the underlying grid and special care has to be taken to preserve connectivity.

The recently proposed technique based on maximum modulus symmetry of wavelet transform (MMSWT) [17,16] Improved greatly in these respects. Anyway, it has a poor performance with wide-structure objects, such as, centerline extraction of blood vessels in retinal image. As mentioned in [16], the MMSWT-based algorithm is not applicable for computing centerline of object with fairly contrast of wide structure, such as blood vessels in the retinal image, which some small branches of the vessel tree are much thinner than the main vessels. Thus, structure widths of objects in the same image have the strong difference the ideal scale of wavelet transform which is suitable for thinning objects with various different width is not decided easily even does not exists at all. As a result, some natural centerline points of the shape can not be extracted completely. In addition, as we mentioned in our previous discussion [16], it is still puzzled how to choose proper scale for wavelet transform according to the width of shape in practice. To overcome the above problems, a novel wavelet-based method is presented in this paper. In this way, a new wavelet function which has been constructed in our previous works [16] is applied. Some significant characteristics of the local minima of wavelet transform modulus are mathematically investigated. A novel algorithm which is capable of extracting centerlines of blood vessels in the retina images benefited from some significant characteristics of the local minima of wavelet transform is developed.

The rest of the paper is organized as follows: In Section 2, thinning analysis based on wavelet transform moduli of images is presented. In Section 3, a multilevel-based thinning algorithm is developed. Several experiments are illustrated in Section 4 and demonstrate that the centerline produced by the algorithm can be used for representing blood vessels in the retina image efficiently. Finally, the conclusions and remarks are made in Section 5.

2 Thinning Analysis Based on Wavelet Transform Modulus

Let $L^2(R^2)$ be the Hilbert space of all square-integrable 2-D functions on plane R^2, $\psi \in L^2(R^2)$ is called a wavelet function, if

$$\int_R \int_R \psi(x,y) dx dy = 0, \quad (1)$$

For $f \in L^2(R^2)$ and scale $s > 0$, the scale wavelet transform of $f(x,y)$ is defined by

$$W_s f(x,y) := (f * \psi_s)(x,y) = \int_R \int_R f(u,v) \frac{1}{s^2} \psi(\frac{x-u}{s}, \frac{y-v}{s}) du dv, \quad (2)$$

where * denotes the convolution operator, and $\psi_s(u,v) := \frac{1}{s^2} \psi(\frac{u}{s}, \frac{v}{s})$.

The general theory of the scale wavelet transform can be found in [5,6].

The centerline of blood vessels in the retinal image can be viewed as transient components with low frequencies. They are highly localized in the space, which makes their analysis and processing with the classical Fourier transforms ineffective. Wavelet analysis is used to alleviate this deficiency.

As well known, edge points of shape in the image are often located in such a region, where the image intensity has sharp transition. The local maxima of the absolute value of the first derivative of $f * \theta_s(x)$ are the sharp variation points of $f(x)$). Where, $\theta(x)$ is a real smoothing function, and it satisfiies $\theta(x) = O(\frac{1}{(1+x^2)})$ and whose integral is nonzero. It can be viewed as the impulse response of a low-pass filter. Let $\theta_s(x) = (\frac{1}{s})\theta(\frac{x}{s})$ and $f(x) \in L^2(R)$. Singular points (such as the edges in images) at the scale s are defined as the local sharp variation points of $f(x)$ smoothed by $\theta_s(x)$. Whereas the skeleton point of underlying shape should be midpoint between the two edge points along the gradient and where the image intensity of shape has the slowest transition.

Hence the skeleton points of the underlying shape correspond to the local minima of the wavelet transform modulus $|\nabla W f(s,x)|$ which is called "*wavelet maxima*". From a viewpoint of the mathematical analysis, there should be a local minimum locating between the two consecutive local maxima of $|\nabla W f(s,x)|$. Namely, the wavelet maxima may provide a localized information on $f(x)$ [11,12,15,16]. Therefore, the description of underlying shape by using the wavelet transform modulus minima also preserve most of the structural information in an image. Thus, for the fixed scale s and some "fine" wavelet, , the wavelet transform modulus minima point may locate at the center between two consecutive modulus maxima points and is independent of the scale. Further, all these minima points form the centerline of the underlying shape. Hence, the centerline of the underlying shape in an gray image can be measured by computing wavelet transform modulus minima.

On the other hand, there exist many different wavelet functions. Hence, it is very important to select a suitable one for this particular application. It is easy to see that the partial derivatives of a low-pass function can be the candidates for wavelet functions. In this paper, we use a novel wavelet constructed in our previous work [17,16] below:

Let the function $\phi(x) := \int_0^x \psi(u)du$ is an even function, and compactly supported on $[-1, 1]$. The wavelet function $\psi(x)$ is defined as follows:
Let

$$\psi^+(x) := \begin{cases} \frac{2}{\pi}(4x \ln \frac{(1-8x^2+2\sqrt{1-16x^2})(1+\sqrt{1-x^2})}{9x-8x^2+3\sqrt{9-16x^2}} \\ \quad -\frac{1}{2x}(\sqrt{1-16x^2}-3\sqrt{9-16x^2}+8\sqrt{1-x^2})) & x \in (0, \frac{1}{4}) \\ \frac{2}{\pi}(4x \ln \frac{8x(1+\sqrt{1-x^2})}{9+3\sqrt{9-16x^2}-8x^2} + \frac{1}{2x}(3\sqrt{9-16x^2}-8\sqrt{1-x^2})) & x \in [\frac{1}{4}, \frac{3}{4}) \\ \frac{2}{\pi}(4x \ln \frac{1+\sqrt{1-x^2}}{x} - \frac{4}{x}\sqrt{1-x^2}) & x \in [\frac{3}{4}, 1), \\ 0 & x \in [1, \infty), \end{cases}$$

Set $\psi^-(x) := -\psi^+(-x)$, then we have

$$\psi(x) := \psi^+(x) + \psi^-(x).$$

The 2D wavelets are defined by

$$\begin{cases} \psi^1(x, y) := \phi'(\sqrt{x^2+y^2})\frac{x}{\sqrt{x^2+y^2}}, \\ \psi^2(x, y) := \phi'(\sqrt{x^2+y^2})\frac{y}{\sqrt{x^2+y^2}}. \end{cases} \quad (3)$$

With the wavelets $\psi^1(x, y)$ and $\psi^2(x, y)$ defined as above, the scale wavelet transforms become

$$\begin{cases} W_s^1 f(x, y) = (f * \psi_s^1)(x, y), \\ W_s^2 f(x, y) = (f * \psi_s^2)(x, y). \end{cases} \quad (4)$$

The gradient direction and the amplitude of the 2D wavelet transform are defined as in our previous work [17,16].

We prove mathematically the fact that it particularly suitable to characterize Dirac-structure edges in an image and its corresponding wavelet transform modulus provides with the four significant characteristics below:

- Symmetry: For different scales of the wavelet transform, if the scale of the wavelet transform is much larger than the width of the tubular object, the local minimum points exist uniquely and can form the skeleton lines of the tubular object. Further, the two contours obtained from the wavelet minima are symmetric with respect to this local centerline along the gradient direction.
- Slope invariant: the local maximum modulus of the wavelet transform of a Dirac-structure edge is independent on the slope of the edge.
- Grey-level invariant: the local maximum modulus of the wavelet transforms with respect to a Dirac-structure edge takes place at the same points when the images with different grey-levels are to be processed.

The above properties imply that the locations of wavelet minima cover exactly the inherent central line of the shape. Meanwhile, the location of maxima of the wavelet transform moduli, which are located nearly at the points of the original boundary, form the two new lines and they are symmetrical with respect to the inherent central line of the blood vessel, which can be identified by the local minima points. Therefore, the connective curve of all minimum points of the wavelet transform modulus is defined as the primary centerline of the blood vessel. Thus, a simple and direct strategy for extracting the centerline of the blood vessel will be developed.

3   Multilevel-Based Thinning Algorithm to Vessel Extraction

In practice, the detection of the wavelet minima in the discrete domain can be implemented analogously as the local maxima of the wavelet transform moduli [17,16]. Ideally, the centerlines of the blood vessels in the retina images are represented by a set of thin curves which consist of single pixels rather than a tubular object. As we previously mentioned [16], if and only if the scale of the wavelet transform matches well with the width of the blood vessel, namely, its value is much bigger than the width of the blood vessel, the modulus minimum points between two homologous modulus maximum points exist uniquely. Otherwise, there maybe exist numerous modulus minimum points and resulting centerline is more than one pixel wide.

In practice, it is difficult to choose the suitable scale of the wavelet transform according to the width of the blood vessel structure so that the centerline obtained from the wavelet minima contains single pixels. For most cases, the primary centerline obtained from the modulus-minimum-based algorithm is generally the bandwidth ribbon consisting of multiple pixels than the perfect centerline containing a single pixel. Although a relatively large scale is favorable to process on wide blood vessel and a single pixel centerline may be produced, but it usually suffers from much heavy computational cost.

To solve this problem, the following mulitilevel-based approach is proposed. Our basic idea is as follows: For each input image, we choose a reasonable scale of the wavelet transform and extract the corresponding center ribbon of the underlying blood vessels in the retina image by computing all wavelet minima. As a result, all these local minimum points produce a primary center ribbon of the underlying blood vessel, which may be multiple pixels wide. Obviously, these primary center ribbons are apparently thinner than the original blood vessel and preserve exactly the topological properties of the original blood vessel. Then, we choose a smaller scale than the prior one to perform the second wavelet transform on the image, which contains generated center ribbons, and compute the second level center ribbon. The above procedure is iterated until the central curves, which consist of single pixels, is eventually extracted from the underlying blood vessels in the retina image. This procedure is called Multilevel-based Algorithm (MA).

It has been observed that distribution of the gray level throughout the whole image is not balanced. Intensity contrast at one location may be rather weak than others. In order to deal with the problem, we come up with the idea to segment the image into a number of blocks for wavelet transform. The detail is as follows. We divide the image into blocks of size $W \times W$. By computing the modulus maxima of wavelet transform, we detect all edge points of blood vessels and record their intensity value. Next, we compute the mean of the intensity value of all edge point for every block as threshold value. By comparing the intensity of every pixel in the block with this threshold, we may segment the foreground from the background in the block. In summary, the algorithm is designed as follows:

Algorithm 1 *(Multilevel Thinning Algorithm)*

Step 1. *Input a retina image, and select randomly a scale for wavelet tranform*

Step 2. *Perform WT with the wavelets defined by Eq. (3) on the input image.*

Step 3. *Calculate the moduli of wavelet transforms $|\nabla W_s f(x,y)|$.*

Step 4. *Detect edge points by computing the modulus maxima of the wavelet transform.*

Step 5. *Compute the mean of edge points in the block as threshold T_{mean} and segment the foreground and background points based on the threshold.*

Step 6. *Extract the candidates of primary center ribbon points by computing the modulus minima of wavelet transform.*

Step 7. *Select all points which are those foreground points with modulus minima in the modulus image as the initial centerline points.*

REPEAT { the obtained the image which contained center ribbon }
If {the obtained centerline consist of multiple pixels }
Then { Perform the second wavelet with new scale for the previous center ribbon image, and compute modulus minima as new centerline; }
Otherwise { the obtained centerline consist of single pixels }
Exit loop
UNTIL { The obtained centerline consist of single pixels. }
END

4 Experiments

We use a number of gray level fundus images of both left and right eyes of patients to evaluate the performance of the proposed method. The image size is $1096 \times 960 \times 8$ bits. In the experiment, one result is illustrated in Fig. 1 and 2 to show the full procedure of the proposed thinning algorithm step by step. An original grey scale retina image consisting of the distracting background is

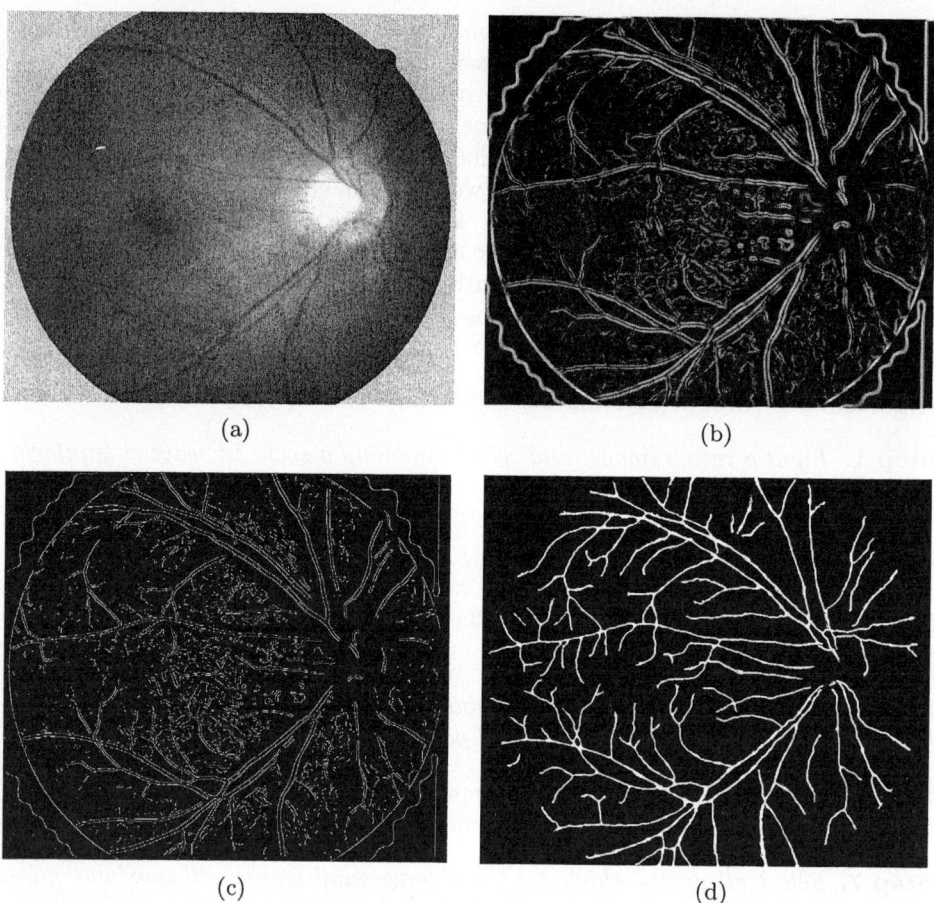

Fig. 1. (a) The original retina image; (b) the raw output of modulus obtained from WT with scale $s = 6$; (c) the modulus maxima image; (d) the primary center ribbon from the first WT

shown in Fig. 1(a). As mentioned previously, most of existing algorithms often fail to directly process these gray images. The modulus image obtained from the WT with scale s = 6 is presented in Figs. 1(b) to segment the foreground from the back-ground, detect the edge by computing the modulus maxima. The corresponding modulus image is illustrated in Fig. 1(c). The primary center ribbon from the modulus minima of the first WT is shown in Fig. 1(d). Obviously, these center ribbon consists multiple pixels. Next, for the primary center ribbon from Fig. 1(d), we perform the second WT with scale s = 6. By computing the modulus minima of the wavelet transform in Fig. 2(a), the final centerlines of blood vessels are obtained, as shown in Fig. 2(b) and it meets the requirements for centerline detection of blood vessels in the retina image. It is clear by visual in-spection that the extracted vascular trees are accurate, practical and ready for further applications.

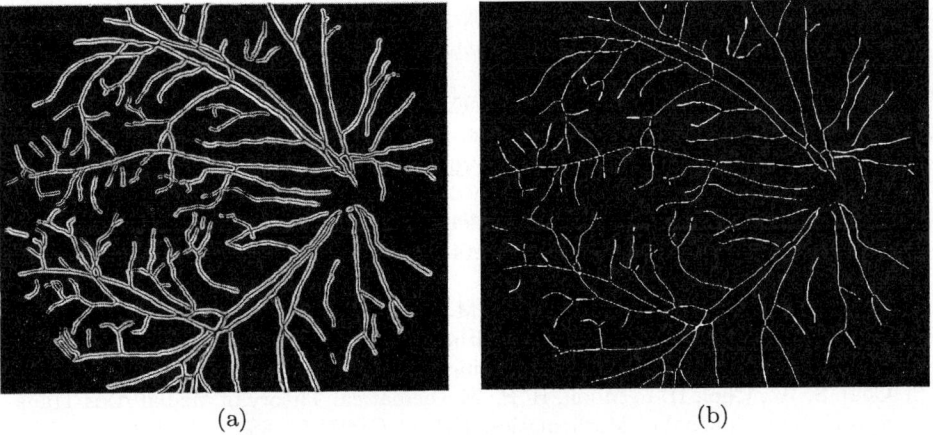

Fig. 2. (a) The raw output of modulus obtained from the second WT with scale $s = 6$; (b) the final skeleton from the second WT with scale $s = 6$

5 Conclusions

In this paper, a new wavelet-based method for extracting centerline of blood vessels in the retina fundus images is proposed. It is based on the favorable properties of a new constructed wavelet function which allow us to solve the problem of computing centerlines of tubular objects from a unique point of view. By applying the wavelet transform and finding corresponding minima, we are able to directly extract centerlines of blood vessels with a multilevel scheme. The skeletonization method can be implemented very efficiently with relatively lower computational time compared with existing methods. The identified centerline of vascular trees are accurate and helpful to form the basis for further applications such as feature based registration.

Acknowledgments

This research was partially supported by a grant (60403011) from National Natural Science Foundation of China, and grants (2003ABA012) and (20045006071-17) from Science &Technology Department, Hubei Province and the People's Municipal Government of Wuhan respectively, China. This research was also supported by the grants (RGC and FRG) from Hong Kong Baptist University and Hubei University.

References

1. Yoshizawa Alexander, S., Belyaev, G.: On Evolute Cusps and Skeleton Bifurcations. In Int. Conf. on Shape Modeling ad Its Applications. (2001) 134–141
2. Katja Bühler, Petr Felkel., Alexandra, L. C.: Geometric Methods for Vessel Visualization and Quantification- A Survey. In H. Müller G. Brunnett, B. Hamann, editor, Geometric Modelling for Scientific Visualization. Springer (2003) 399–421

3. Chaudhuri, S., Chatterjee, S., Katz, N. Goldbaum, M.: Detection of Blood Vessels in Retinal Images Using Two-dimensional Matched Filters. IEEE Trans. Med. Imag. **3** (1989) 263–269
4. Liang, Z. R., Wan M., Kaufman, A. E., Chen, W. M., Li, D. Q.: Tree-branch-searching Multiresolution Approach to Skeletonization for Virtual Endoscopy. Proc. Medical Imaging 2000: Image Processing, SPIE2000, Kenneth M. Hanson; Eds. **3979** (2000) 726–734
5. Chui, C. K.: An Introduction to Wavelets. Academic Press, Boston, (1992)
6. Daubechies, I.: Ten Lectures on Wavelets. Society for Industrial and Applied Mathemathics, Philadelphia. (1992)
7. Luo, G, Chutatape, O., Krishnan, S. M.: Detection and Measurement of Retinal Vessels in Fundus Images Using Amplitude Modified Seconder-order Gaussian Filter. IEEE Trans. Biomedical Engineering. **49(2)** (2002)168–172
8. Choi, S. W., Choi, H. I., Moon, H. P.: Mathematical Theory of Medial Axis Transform. Pacific Journal of Mathematics. **181(1)** (1997) 57–88
9. Hoffmann, C. M. : Computer Vision, Descriptive Geometry, and Classical Mechanics. Technical Report CSD-TR-91-073, Purdue University, CS Dept., West Lafayette, IN 47907, USA. (1991)
10. Hoover, A., Kouznetsova, V., Goldbaum, M.: Locating Blood Vessels in Retinal Images by Piecewise Threshold Probing of a Matched Filter Reponse. IEEE Trans. Med. Imag. **19(3)** (2000) 203–210
11. Mallat, S.: Wavelet Tour of Signal Processing. Academic Press, San Diego, USA, (1998)
12. Mallat, S. G., Hwang, W. L.: Singularity Detection and Processing with Wavelets. IEEE Transactions on Information Theory. **38(2)** (1992) 617–643
13. Yehia, A. B., Solouma, H. N., Youssef, Abou-Bakr M., Kadah, Y. M.: A New Real-time Retinal Tracking System for Image-guided Laser Treatment. IEEE Trans. Biomedical Engineering. **49(9)** 1059–1067
14. American Academy of Ophthalmology. Retina and Vitreous, Basic and Clinical Science Course, **Section 11** (1991) 13-27, 31-39
15. Tang, Y. Y., Yang, L. H., Liu, J. M.: Characterization of Dirac-Structure Edges with Wavelet Transform. IEEE Trans. Systems, Man, Cybernetics (B). **30(1)** (2000) 93–109
16. Tang, Y. Y., You, X. G.: Skeletonization of Ribbon-like Shapes Based on A New Wavelet Function. IEEE Transactions on Pattern Analysis and Machine Intelligence. **25(9)** (2003) 1118–1133
17. Yang, L. H., You, X.G., Haralick, R. M., Phillips, I. T., Tang, Y.Y.: Characterization of Dirac Edge with New Wavelet Transform. Proc. 2th Int. Conf. Wavelets and its Application. Lecture Notes in Computer Science, Springer. **2251** (2001) 872–878

Stability Analysis on a Neutral Neural Network Model

Yumin Zhang, Lei Guo, and Chunbo Feng

Research Institute of Automation, Southeast University Nanjing 210096, China
zhyminus@sohu.com, l.guo@seu.edu.cn

Abstract. To describe the complicated neural dynamics of cerebra with time delays, a new type of model called generalized cellular neutral neural networks (GCNNNs) is studied in this paper. It is noted that the GCNNNs reduce to generalized cellular neural networks (GCNNs) in the absence of the neutral term in systems. Some criteria for mean square exponential stability and asymptotic stability of GCNNNs are established and the relationship between the neutral item and the whole system is analyzed. Simulation results are given to show the effectiveness of the proposed analysis algorithms.

1 Introduction

During the past two decades, artificial neural networks theory has been received considerable attention since the Hopfield neural network (NN) was introduced in [1] (see [2], [3] for surveys). Along with the rapid development of the neutron sciences, some new type neural network models have been established to study different modes of the neural dynamic such as the Bolzmann machines model bases on simulated annealing (SA) algorithm [5], the adaptive resonance theory (ART) model bases on adaptive resonance theory [6], and the cellular neural network in case that all of the neural cells are not connected entirely [7], *et al.* The above neural network models have been successfully used in various practical processes.

On the other hand, based on the neurological theory and experiments, it can be seen that most single neural network should be a dynamic system with time delay, which has been focused on by most of researchers (see [3], [4], [8]). It is well known that the time delay may result in complicated system dynamic. Since the system dynamic is related with states in the past, it should be certainly related with the change rate of states in the past, which can result a more complicated neutral system dynamic. Basing on such ideal, the complicated dynamic of the neural cells can be farther researched, which leads to the motivation of this paper, where a new type of NN models named neutral neural network (NNN) is set up and analyzed.

Three reasons for us to introduce the neutral-type operation into the conventional NN models. First, based on biochemistry experiments, neural information may transfer across chemical reactivity, which results in a neutral-type process [9]. Then, in view of electronics, it has been shown that neutral phenomena exist in large-scale integrated (LSI) circuits [10], [11], [12]. Last, the key point is that cerebra can be considered as a super LSI circuit with chemical reactivity, which reasonably implies that the neutral dynamic behaviors should be included in neural dynamic systems.

However, to our best knowledge, there is only a few research up to date concentrating on neural network models using neutral dynamics, [14] begins to initiate the study on NNN models and gives one stable criteria of stochastic Hopfield NNN. Due to that the (generalized) cellular neural networks can be realized by electronic circuits (see [7], [13]), the generalized cellular NNN (GCNNN) will be analyzed in this paper. The remainder of this paper is organized as follows: In Section 2, the GCNNN model is given to characterize the neutral dynamic behavior of cerebra and the stability criteria are proposed. Simulation example is given in Section 3 to illustrate our results and conclusion is given in section 4.

2 Stability Analysis for Stochastic GCNNN

Consider a special case of system (2) as follows,

$$\begin{cases} d(x-G(x_t)) = [-Ax + Wf(x) + W_\tau f(x_t)]dt + \sigma(t,x,x_t)dw(t) \\ x(s) = \xi(s), \quad -\tau \leq s \leq 0 \end{cases} \quad (1)$$

where $A = \text{diag}(a_i)_{n \times n} \in R^{n \times n}$, $W = (w_{ij})_{n \times n} \in R^{n \times n}$, $W_\tau = (w_{ij}^\tau)_{n \times n} \in R^{n \times n}$ are weighted matrices. Let $(\Omega, F, \{F_t\}_{t \geq 0}, P)$ be a complete probability space with a filtration $\{F_t\}_{t \geq 0}$ (i.e. $F_t = \sigma\{w(s) : 0 \leq s \leq t\}$). It is suppose that $\{F_t\}_{t \geq 0}$ is right continuous and F_0 contains all P-null sets (see [4], [12]]). And $w(t) = (w_1(t), w_2(t), \cdots, w_m(t))^T$ is a m-dimension Brownian motion. Assume that $\xi \in C([-\tau,0]; R^n)$ is F_0- measurable and satisfies $\int_{-\tau}^0 E|\xi(s)|^2 ds < \infty$ (or $\xi \in L_{F_0}^2([-\tau,0]; R^n)$). Denote $x(t) = (x_1(t), x_2(t), \cdots, x_n(t))^T$ and $x_t = (x_1(t-\tau_1), x_2(t-\tau_2), \cdots, x_n(t-\tau_n))^T$ $(-\tau \leq -\tau_i \leq 0,)$,

where $\tau_i, i = 1, 2, \cdots, n$ are time delays. The related nonlinear mappings are denoted as follows: $\sigma : t \times R^n \times R^n \to R^{n \times m}$, $\sigma(t, x, x_t) = (\sigma_{ij}(t, x_i, x_i(t-\tau_i)))_{n \times m}$

$$f(x(t)) = (f_1(x_1(t)), f_2(x_2(t)), \ldots, f_n(x_n(t)))^T,$$

$$f(x_t) = (f_1(x_1(t-\tau_1)), f_2(x_2(t-\tau_2)), \ldots, f_n(x_n(t-\tau_n)))^T$$

In this paper, $f(x_i(t)) = 0.5M(|x_i(t)+1| - |x_i(t)-1|)$ $(i=1,2,\ldots,n)$ are not differentiable, where $M \in N$ is a finite constant. Then, system (1) can be rewritten into the following stochastic *generalized cellular neutral neural network* (GCNNN) model

$$\begin{cases} d(x-G(x_t)) = [-Ax + WH(x)x + W_\tau H(x_t)x_t]dt + \sigma(t,x,x_t)dw(t) \\ x(s) = \xi(s), \quad -\tau \le s \le 0 \end{cases} \quad (2)$$

where $H(x) = \text{diag}(H_i(x_i))_{n \times n}$, $H_i(x_i) = x_i^{-1} f_i(x_i)$, $H(u_\tau) = \text{diag}(H_i(x_t^i))_{n \times n}$, $H_i(x_t^i) = (x_t^i)^{-1} f_i(x_i(t-\tau_i))$, and $\|H\| \le M$.

Throughout this paper, we assume that $\sigma(t,\cdot,\cdot)$ is Lipschitz continuous and satisfy the nonlinear incremental conditions:

Assumption 1: *There exist constant* b_{ik} *and* c_{ik}, $i=1,2,\cdots,n$, $k=1,2,\cdots,m$ *such that* $\sigma(t,\cdot,\cdot)$ *satisfies* $\sigma_{ik}^2(t,x_i(t),x_i(t-\tau_i)) \le b_{ik}x_i^2(t) + c_{ik}x_i^2(t-\tau_i)$.

With *Assumption 1*, it is known (see [15]) that system (5) has a unique global solution on $t \ge 0$, denoted by $x(t;\xi)$. Moreover, it is assumed that $\sigma(t,0,0) \equiv 0$ so that system (5) admits an equation solution $x(t;0) \equiv 0$. For convenience, denote $|x|$ as the Euclidean norm of x, $\|W\| = \{\sup Wu : |u|=1\} = \sqrt{\lambda_{\max}(W^T W)}$ as the matrix norm of W, and $b = \max\left\{\sum_{k=1}^{m} b_{ik} : i=1,2,\cdots,n\right\}$, $c = \max\left\{\sum_{k=1}^{m} c_{ik} : i=1,2,\cdots,n\right\}$.

The neutral-iterm function $G(\cdot)$ is required to satisfy the following assumption.

Assumption 2: *There is* $k \in (0,1)$ *such that*

$$E|G(\varphi)|^2 \le k \sup_{-\tau \le \theta \le 0} E|\varphi(\theta)|^2, \quad \varphi \in L^2_{F_\infty}([-\tau,0]; R^n) \quad (3)$$

To study the stability, the following notations are useful for derivative procedures. Let $C^{1,2}([-\tau,\infty) \times R^n, R_+)$ denote the family of all functions $V(t,x):[-\tau,\infty) \times R^n \to R_+$, which are continuously twice differentiable in x and once in t. For any $V(t,x) \in C^{1,2}([-\tau,\infty) \times R^n, R_+)$, denote

$$V_x(t,x) = (V_{x_1}(t,x), V_{x_2}(t,x), \cdots, V_{x_n}(t,x)), \quad V_{xx}(t,x) = (V_{x_i x_j}(t,x))_{n \times n}$$

and denote $LV : R_+ \times C([-\tau,0]; R^n) \to R$ as

$$LV(t,\varphi) = V_t(t,\varphi(0)-G(\varphi)) + V_x(t,\varphi(0)-G(\varphi))f(t,\varphi(0),\varphi)$$
$$+\frac{1}{2}\text{trace}\left[g^T(t,\varphi(0),\varphi)V_{xx}(t,\varphi(0)-G(\varphi))g(t,\varphi(0),\varphi)\right] \quad (4)$$

In this paper, it is supposed that Assumptions 1 and 2 hold.

Theorem 1: Let $q > \left(1-k^{1/2}\right)^{-2}$. For a definite positive matrix Q and a Lyapunov candidate $V(t,\varphi) = \varphi^T Q \varphi$. Suppose that for all $t \geq 0$ and $\varphi \in L^2_{F_t}([-\tau,0];R^n)$ satisfying $|\varphi(0)| \leq |\varphi(\theta)|$, the following inequality

$$E|\varphi(\theta)|^2 \leq \frac{q}{\lambda_{\min}(Q)} EV(t,\varphi(0)-G(\varphi)), \quad -\tau \leq \theta \leq 0 \quad (5)$$

holds, then for all $\xi \in C^b_{F_0}([-\tau,0];R^n)$, we have

$$\limsup_{t\to\infty}\frac{1}{t}\log\left(E|x(t,\xi)|^2\right) \leq -\frac{1}{\tau}\log\frac{q}{\left[1+(kq)^{1/2}\right]^2} := -\gamma \quad (6)$$

That is, the zero-solution of GCNNN system (2) is mean square exponentially stable. In (5) and (6), q is the unique positive root of the following algebraic equation

$$-\lambda_{\min}\left(Q^{\frac{1}{2}}AQ^{-\frac{1}{2}} + Q^{-\frac{1}{2}}A^T Q^{\frac{1}{2}}\right) + 2\sqrt{\frac{kq\|Q\|}{\lambda_{\min}(Q)}}\|A\| + 2M\sqrt{\frac{q\|Q\|}{\lambda_{\min}(Q)}}\|W\|$$

$$+2M\sqrt{\frac{q\|Q\|}{\lambda_{\min}(Q)}}\|W_\tau\| + \frac{q\|Q\|(b+c)}{\lambda_{\min}(Q)} = -\gamma \quad (7)$$

Proof: By using of operation (4),

$$LV(t,\varphi(0),\varphi) = V_t(t,\varphi(0)-G(\varphi)) + V_x(t,\varphi(0)-G(\varphi))\left[-A\varphi(0)\right.$$
$$\left. + WH(t,\varphi(0))\varphi(0) + W_\tau H(t,\varphi)\varphi\right]$$
$$+\frac{1}{2}\text{trace}\left[\sigma^T(t,\varphi(0),\varphi)V_{xx}(t,\varphi(0)-G(\varphi))\sigma(t,\varphi(0),\varphi)\right]$$
$$= -2(\varphi(0)-G(\varphi))^T Q A(\varphi(0)-G(\varphi))$$

$$+2(\varphi(0)-G(\varphi))^T Q [AG(\varphi) + WH(t,\varphi(0))\varphi(0)$$
$$+W_\tau H(t,\varphi)\varphi] + \text{trace}[\sigma^T(t,\varphi(0),\varphi)Q\sigma(t,\varphi(0),\varphi)] \quad (8)$$

Furthermore, it can be estimated that

$$-2(\varphi(0)-G(\varphi))^T Q A(\varphi(0)-G(\varphi))$$
$$\leq -\lambda_{\min}\left(Q^{\frac{1}{2}}AQ^{-\frac{1}{2}} + Q^{-\frac{1}{2}}A^T Q^{\frac{1}{2}}\right)V(t,\varphi(0)-G(\varphi)) \quad (9)$$

$$2(\varphi(0)-G(\varphi))^T Q AG(\varphi)$$
$$\leq \varepsilon_1 V(t,\varphi(0)-G(\varphi)) + \frac{1}{\varepsilon_1}\|Q\|\|A\|^2 |G(\varphi)|^2 \quad (10)$$

$$2(\varphi(0)-G(\varphi))^T Q WH(t,\varphi(0))\varphi(0)$$
$$\leq \varepsilon_2 V(t,\varphi(0)-G(\varphi)) + \frac{1}{\varepsilon_2}\|Q\|\|W\|^2 M^2 |\varphi(0)|^2 \quad (11)$$

$$2(\varphi(0)-G(\varphi))^T Q W_\tau H(t,\varphi)\varphi$$
$$\leq \varepsilon_3 V(t,\varphi(0)-G(\varphi)) + \frac{1}{\varepsilon_3}\|Q\|\|W_\tau\|^2 M^2 |\varphi|^2 \quad (12)$$

$$\text{trace}\left[\sigma^T(t,\varphi(0),\varphi)Q\sigma(t,\varphi(0),\varphi)\right]$$
$$\leq \|Q\|\text{trace}\left[\sigma^T(t,\varphi(0),\varphi)\sigma(t,\varphi(0),\varphi)\right]$$
$$\leq \|Q\|\left(b|\varphi(0)|^2 + c|\varphi|^2\right) \quad (13)$$

Substituting (9)-(13) into (8) yields

$$LV(t,\varphi(0),\varphi) \leq -\lambda_{\min}\left(Q^{\frac{1}{2}}AQ^{-\frac{1}{2}} + Q^{-\frac{1}{2}}A^T Q^{\frac{1}{2}}\right)V(t,\varphi(0)-G(\varphi))$$
$$+(\varepsilon_1+\varepsilon_2+\varepsilon_3)V(t,\varphi(0)-G(\varphi)) + \frac{1}{\varepsilon_1}\|Q\|\|A\|^2 |G(\varphi)|^2$$

$$+\frac{1}{\varepsilon_2}\|Q\|\|W\|^2 M^2 |\varphi(0)|^2 + \frac{1}{\varepsilon_3}\|Q\|\|W_\tau\|^2 M^2 |\varphi|^2$$

$$+\|Q\|\left(b|\varphi(0)|^2 + c|\varphi|^2\right) \tag{14}$$

Choose $\varepsilon_1 = \sqrt{\dfrac{kq\|Q\|}{\lambda_{\min}(Q)}}\|A\|$, $\varepsilon_2 = \sqrt{\dfrac{q\|Q\|}{\lambda_{\min}(Q)}}\|W\|M$,

$\varepsilon_3 = \sqrt{\dfrac{q\|Q\|}{\lambda_{\min}(Q)}}\|W_\tau\|M$. Based on conditions (3), (5) and (7), (14) implies that

$$ELV(t,\varphi(0),\varphi) \le -\gamma\, EV(t,\varphi(0) - G(\varphi)) \tag{15}$$

Based on [12], it is seen that system (2) is mean square exponential stable. □

Theorem 1 provides a mean square exponential stability condition. Similarly it can be verified that GCNNN model (2) is also asymptotically stable with condition (7) changed as in the following Corollary.

Corollary 1: Suppose that condition (7) in Theorem 1 holds, where q satisfies the following inequality

$$-\lambda_{\min}\left(Q^{\frac{1}{2}}AQ^{-\frac{1}{2}} + Q^{-\frac{1}{2}}A^T Q^{\frac{1}{2}}\right) + 2\sqrt{\frac{kq\|Q\|}{\lambda_{\min}(Q)}}\|A\| + 2\sqrt{\frac{q\|Q\|}{\lambda_{\min}(Q)}}\|W\|M$$

$$+2\sqrt{\frac{q\|Q\|}{\lambda_{\min}(Q)}}\|W_\tau\|M + \frac{q\|Q\|(b+c)}{\lambda_{\min}(Q)} < 0 \tag{16}$$

then GCNNN system (2) is asymptotically stable.

To simplify the computation, Q can be chosen as I (identity matrix) in (7), which leads to the following result.

Corollary 2: By selecting $Q = I$ in Theorem 1, if the following equation has the unique solution,

$$-\lambda_{\min}(A+A^T) + 2\sqrt{q}\left(\sqrt{k}\|A\| + \|W\|M + \|W_\tau\|M\right) + q(b+c) = -\gamma \tag{17}$$

inequality (8) also holds for the GCNNN model.

Remark 2: If $\sigma(\cdot,\cdot,\cdot) \equiv 0$, system (2) turns to be a deterministic model of GCNNN and corresponding results to Theorem 1, Corollaries 1 and 2 can be obtained. If

$G(x_t) \equiv 0$, the neutral term is eliminated and model (5) is reduced to the conventional GCNN model (see [7], [13]) for which some existing results can be also generalized by the approaches in Theorem 1 and Corollaries 1 and 2. When $M = 1$ and $G(\cdot) \equiv 0$, Theorem 1 is also effective for the conventional cellular neural networks introduced by L. O. Chua (see [7]).

3 Sensitivity of Neutral Item in NNN: A Simulation Result

It is noted that inequality $q > \left(1 - k^{1/2}\right)^{-2}$ connects k with q in (11). When $k \in [0,1)$ is increased, solution q in (7) is changed in a certain field. On the other hand, the exponential rate $\gamma = \tau^{-1} \log\left\{q\left[1 + (kq)^{1/2}\right]^{-2}\right\}$ is decreased rapidly if k is increased. Generally, existence of condition $q > \left(1 - k^{1/2}\right)^{-2}$ may lead to the confinement of q within a small field, which implies that the neutral term is sensitive to the stability of the whole system. In the following, we consider a simulation result to illustrate the sensitivity.

Consider the following 2-dimension stochastic (G)CNNN model

$$d\left(x_1(t) - lx_1(t-0.1)\right)$$
$$= \left[-4x_1(t) + f_1(x_1(t)) + f_2(x_2(t))\right] dt$$
$$+ 0.4 f_1(x_1(t-0.1)) dw_1(t) + 0.2 f_2(x_2(t-0.1)) dw_2(t)$$
$$d\left(x_2(t) - lx_2(t-0.1)\right)$$
$$= \left[-6x_1(t) + f_1(x_1(t)) + f_2(x_2(t))\right] dt$$
$$- 0.2 f_1(x_1(t-0.1)) dw_1(t) + 0.3 f_2(x_2(t-0.1)) dw_2(t)$$

where $w_1(t)$ and $w_2(t)$ are independent identity distribution Weiner processes, $t \geq 0$, $l = k^{\frac{1}{2}}$, $A = \begin{bmatrix} 4 & 0 \\ 0 & 6 \end{bmatrix}$, $W = \begin{bmatrix} 1 & 1 \\ 1 & 1 \end{bmatrix}$, $W_\tau = 0$, and $f(x_i(t)) = 0.5(|x_i(t)+1| - |x_i(t)-1|)$ $(i=1,2)$ are sigmoid functions. It can be verified that $M \leq 1 (i=1,2)$, $\lambda_{\min}(A) = 4$, $\|A\| = 6$, $\|W\| = 2$, and

$$\text{trace}(\sigma^T \sigma) \leq 0.2 f_1^2(x_1(t-0.1)) + 0.13 f_2^2(x_2(t-0.1)).$$

Based on (17), it can be seen that

$$-4+(6l+2)q^{\frac{1}{2}}+0.1q=-5\left[\log q-2\log\left(1+lq^{\frac{1}{2}}\right)\right]$$

When $l \in [0, 0.3000]$, the above equation has unique positive solution. The following table is given to show the relationships of l, q and γ.

Table 1. Relationships of l, q and γ

l	q	γ
0.00000	1.35874796	1.53281830
0.00001	1.35875801	1.53273873
...	...	...
0.10000	1.45079121	0.72325221
0.30000	2.06102318	0.03445881
0.31000	2.00733973	-0.15689127

The first line is the case of conventional cellular neural networks. It is clear that the *Corollary 2* is invalid when $l \geq 0.31$ because $\gamma \leq -0.15689127$. In *Figures 1-3*, both the trajectories and the phases of the system are demonstrated, respectively. In *Figure 1*, the trajectories of (18) when $l = 0.30$ show the exponentially stability. The state responses of (18) are shown in *Figure 2* when $l \leq 0.70$ to demonstrate the asymptotical stability. In *Figure 3*, it is shown that the system is unstable where l is selected as 0.94. The simulation results show the efficiency of the proposed criteria. However, it is noted that the proposed results are conservative and improved approaches have to be discussed in the future research.

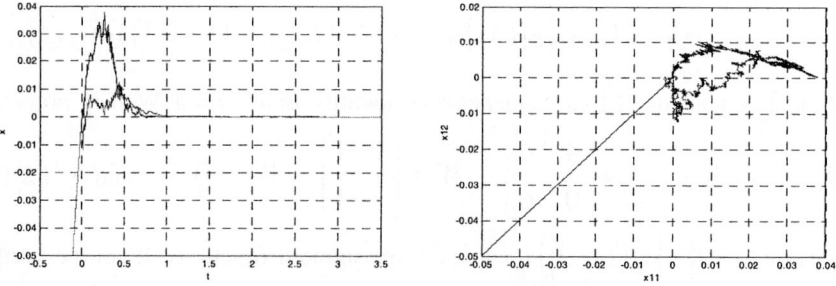

Fig. 1. Trajectories and phases when *l*=0.30

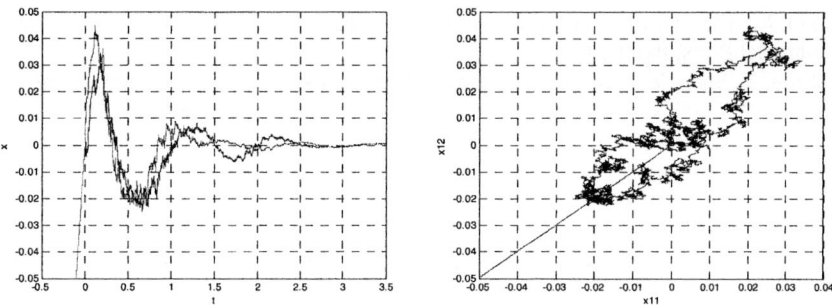

Fig. 2. Trajectories and phases when $l=0.70$

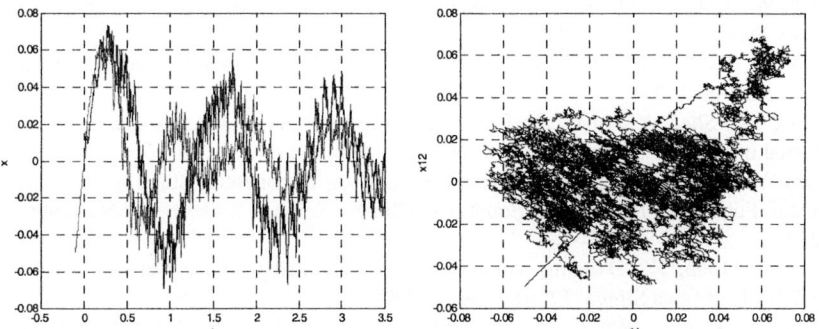

Fig. 3. Trajectories and phases when $l=0.94$

4 Conclusions

A new type model called generalized cellular neutral neural networks (GCNNNs) is studied in this paper. Some criteria for mean square exponential stability and asymptotic stability of GCNNNs are provided and the relationship between the neutral item and the whole system is also analyzed. Simulation results show the effectiveness of the proposed analysis algorithms. Further researches include more elegant stability analysis approaches with less conservativeness and some control strategies for the neutral differential systems.

Acknowledgement. This work is partially supported by the NSF of China (No. 60474050), the China Post-Doctoral Foundation and the Jiangsu Post-Doctoral Foundation of China.

References

1. Hopfield, J. J.: Neural Networks and Physical Systems with Emergent Collect Computational Abilities. Proc. Natl. Acad. Sci. USA 79(1982) 2554-2558
2. De Wilde, P.: Neural Networks Models, An Analysis. Springer-Verlag (1996)

3. Liao, X.: Theory and Applications of Stability for Dynamical Systems. National Defensive Industrial Press, Beijing (2000)
4. Zhang, Y, Shen, Y, Liao, X, et al.: Novel Criteria Stability for Stochastic Interval Delayed Hopfield Neural Networks. Advances in Syst. Sci. and Appl. 2(1) (2002) 37-41
5. Ackley, D. H., Hinton, G. E. and Sejinowski, T. J.: A lining Algorithm for Bolzmann Machines. Cognitive Science. 9(1985) 147-169
6. Grossberg, S.: Competitive Learning: From Interactive Activation To Adaptive Resonance. Cognitive Science. 11 (1987) 23-63
7. Chua, L. O., Yang, L.: Cellular Neural Network: Theory. IEEE Trans. Circuits Syst. 35(1988) 1257-1272
8. Zeng, Z., Wang, J., Liao, X.: Stability Analysis of Delayed Cellular Neural Networks Described Using Cloning Templates. IEEE Transactions on Circuits and Systems I: Fundamental Theory and Applications. 51(2004) 2313-2324
9. Curt, W.: Reactive Molecules: the Neutral Reactive Intermediates in Organic Chemistry. Wiley Press, New York (1984)
10. Salamon, D.: Control and Observation of Neutral Systems. Pitman Advanced Pub. Program, Boston (1984)
11. Zhang, Y., Shen, Y., Liao, X.: Robust Stability of Neutral Stochastic Interval Systems. Journal of Control Theory and Applications. 2(2004) 82-84
12. Shen, Y., Liao, X.: Razumikhin-type Theorems on Exponential Stability of Neutral Stochastic Functional Differential Equations. Chinese Science Bulletin. 44(24) (1999) 2225-2228
13. Shen, Y., Liao, X.: Dynamic Analysis for Generalized Cellular Networks with Delay. ACTA Electronica Sinica. 27(10) (1999) 62-64
14. Zhang, Y., Guo, L., Wu, L., et al.: On Stochastic Neutral Neural Networks. Lecture Notes in Computer Science. Vol. 3496 (2005) 69-74
15. Mao, X.: Exponential Stability of Stochastic Differential Equations. Marcel Dekker, New York (1994)

Radar Emitter Signal Recognition Based on Feature Selection and Support Vector Machines*

Gexiang Zhang[1], Zhexin Cao[2], Yajun Gu[3], Weidong Jin[4], and Laizhao Hu[5]

[1] School of Electronic Engineering, University of Electronic Science and Technology of China,
Chengdu 610054 Sichuan, China
gxzhang@ieee.org
[2] College of Profession and Technology, Jinhua 321000 Zhejiang, China
[3] School of Computer Science, Southwest University of Science and Technology,
Mianyang 621002 Sichuan, China
[4] School of Electrical Engineering, Southwest Jiaotong University,
Chengdu 610031 Sichuan, China
[5] National EW Laboratory, Chengdu 610036 Sichuan, China

Abstract. One of the intelligent aspects of human beings in pattern recognition is that man identifies an object in real world using Marked Characteristic Principle (MCP). This paper proposes a humanoid recognition method for radar emitter signals. The main points of the method include feature ordering and an improved one-versus-rest multiclass classification support vector machines. According to MCP, an approach for computing marked characteristic coefficients is presented to obtain the most marked feature of every radar emitter signal. Subsequently, a support vector network is designed using the improved one-versus-rest combination approach of several binary support vector machines. Experimental results show that the introduced method has faster recognition speed and better classification capability than conventional recognition approaches.

1 Introduction

Radar emitter signal recognition is one of the key procedures of signal processing in electronic intelligence, electronic support measures and radar warning receiver systems in electronic warfare [1-4]. It is also the precondition and foundation of electronic interfering. The state of the art of radar emitter signal recognition corresponds to the technical merit of electronic reconnaissance equipment. As countermeasure activities in modern electronic warfare become more and more drastic, advanced radars increase rapidly and become the main component of radars gradually [1-4]. Complex and changeful signal waveform destroys the laws for identifying radar emitters in traditional counter-radar equipment. It is difficult to know the potential menace correctly only using conventional parameters. So it is very

---

* This work was supported by the National EW Laboratory Foundation (No.NEWL51435 QT220401).

urgent to explore new recognition methods to improve signal-processing techniques in existing radar reconnaissance systems.

In recent years, although radar emitter signal recognition is paid much attention and some recognition methods were presented [5,6], using conventional parameters, traditional recognition methods and their improved methods encounter serious difficulties in identifying advanced radar emitter signals [2,3,7,8]. Unfortunately, the existing intrapulse characteristic extraction approaches only analyze qualitatively two or three radar emitter signals without considering the effects of noise nearly [7,8]. So the approaches cannot meet the intelligentized requirements of modern information warfare for electronic warfare reconnaissance systems. For the difficult problem of recognizing complicatedly and changefully advanced radar emitter signals, multiple methods need be used to extract the valid features from radar emitter signals from multiple different views firstly. The features construct a high-dimensional feature vector. Then, some feature selection methods are presented to select the most discriminatory features and to eliminate redundant features so as to lower the dimensionality of the feature vector. Finally, efficient classifiers are designed to fulfill automatic recognition of radar emitter signals. This paper discusses mainly the latter two problems.

In this paper, a humanoid recognition method is proposed to identify radar emitter signals. Firstly, a new feature selection based on Marked Characteristic Principle (MCP) is presented to compute marked characteristic coefficients and to obtain the most marked feature of every radar emitter signal. Then, according to the characteristics of the feature selection, an improved combination approach of binary support vector machines (SVMs) based on one-versus-rest (OVR) is used to solve the multiclass classification problem in radar emitter signals. Finally, simulation experiments are made to test the validity of the introduced method and conclusions are also drawn.

2 Feature Ordering

Supported by fast-speed and strong-function digital computers, various pattern recognition methods and equipments are used to recognize signals, images, speeches, and so on, but the methods and equipments still looks very clumsy and very slow, comparing with human beings [9,10,11]. For example, a one-year to two-year child can recognize exactly his parent from a large number of persons in an instant. If this task is finished by a computer, a series of time-cost and troublesome steps, including image input, noise filtering, template matching and various feature distance computing, need be performed and finally someone can be recognized by the computer only in a percent probability [9,10]. To quicken recognition speed, Jin pointed out that the intelligent aspect of human brain in pattern recognition is that man identifies an object using MCP, when he studied how man recognizes the complex things in real world [9,10]. According to MCP, the characteristics of the samples that man can recognizes are deposited beforehand in his memory and when man faces a sample to recognize, it is not all characteristics of the sample that are used to compare with those in his memory, but only the most marked characteristics of the sample are employed to match the corresponding ones in his memory. In this way, search space can decrease greatly and recognition time can save.

In this paper, MCP is introduced into radar emitter signal recognition to select the most marked features of every radar emitter signal. That is, a low-dimensional feature vector with large marked characteristic coefficient is firstly used to identify a certain radar emitter signal. If the radar emitter signal cannot be identified by using the low-dimensional feature vector, the high-dimensional feature vector is used. So the feature selection method is called feature ordering. Using the features obtained by the feature selection method based on MCP, radar emitter signals can be recognized at a fast speed.

The key of feature ordering is how to compute marked characteristic coefficients of signal features. The feature values of multiple radar emitter signals do not usually order a certain probability distribution, though the overlapping of the features in feature space has a direct influence on recognition results. According to the overlapping of signal features in feature space, a novel method for computing marked characteristic coefficient is given in the following description.

For a certain feature, the marked characteristic coefficient of the ith signal is represented with Df_i.

$$Df_i = \min(Df_i^l, Df_i^r), \quad (1)$$

Where Df_i^l and Df_i^r are respectively

$$Df_i^l = \frac{|E(X_{i-1}) - E(X_i)|}{\max_{k=1,2,\cdots,M}\{|x_{(i-1)k} - E(X_{i-1})|\} + \max_{k=1,2,\cdots,M}\{|x_{ik} - E(X_i)|\}}, \quad (2)$$

$$Df_i^r = \frac{|E(X_i) - E(X_{i+1})|}{\max_{k=1,2,\cdots,M}\{|x_{ik} - E(X_i)|\} + \max_{k=1,2,\cdots,M}\{|x_{(i+1)k} - E(X_{i+1})|\}}. \quad (3)$$

Where M is the number of feature values; $E(.)$ is a function for computing mathematic expectation; X_{i-1}, X_i and X_{i+1} are the sample vectors of the $(i-1)$th, ith and $(i+1)$th signals, respectively; x_{i-1}, x_i and x_{i+1} are the kth sample values of the $(i-1)$th, ith and $(i+1)$th signals, respectively.

In equation (1), the case of $Df_i \leq 1$ denotes there are some overlaps between the samples of the ith signal and the samples of neighboring signal of the ith signal. The smaller Df_i is, the more serious the overlapping is. The case of $Df_i > 1$ denotes there are some intervals between the samples of the ith signal and the samples of neighboring signal of the ith signal. The larger Df_i is, the bigger the interval is. Thus, the feature with the largest marked characteristic coefficient is the most marked feature of a certain signal. If the marked characteristic coefficient of a certain signal in a lower dimensional feature space is less than or equal to 1, the signal cannot be recognized well only using the lower dimensional feature vector. Obviously, the dimensionality of feature vector must increase and the marked characteristic coefficient of the signal is computed again in a higher dimensional feature space.

According to equation (1), the algorithm of feature ordering is given as follows.

Step 1. Determining the number N of signals, the number M of feature samples and the number H of features.

Step 2. For the jth feature, expectation values of feature samples of N signals are ordered from small to large.

Step 3. Computing marked characteristic coefficient of the jth feature of the ith signal.

Step 4. Repeating step 2 and step 3 till $j>H$.

Step 5. Determining the most marked feature of the ith signal. The most marked feature is the feature whose marked characteristic coefficient is the largest among H features of the ith signal and the marked characteristic coefficient must be more than 1. If the marked characteristic coefficient exists, the algorithm goes to the next step, otherwise, the dimensionality of the feature vector adds 1 and the algorithm goes to step 2. The method for increasing the dimensionality of the feature vector is that the feature with the largest marked characteristic coefficient among the rest (H-1) marked characteristic coefficients of the ith signal is firstly added into the prior feature vector.

Step 6. Repeating step 2 to step 5 till $i>N$.

Step 7. The most marked features of all signals are obtained.

3 Classifier Designing

According to the feature ordering method, this section gives a classifier design method corresponding to the feature selection using support vector machines. Many conventional signal recognition techniques including k-nearest neighboring classification and template matching rely on algorithms which are computationally intensive and require a key man to validate and verify the analysis [12]. Also, these techniques are inefficient and time-consuming for solving radar emitter signal recognition problems, and they often fail to identify signals under high signal density environment, especially, in near real time [12]. In recent years, neural networks are used to recognize different radar emitter signals so as to solve the above problems [12, 13]. However, neural networks have some internal shortcomings, such as difficulty for determining the structure of neural network, over-learning and getting into local extremes easily [14]. SVMs, developed principally by Vapnik [15], are considered as a good learning method that can overcomes the internal drawbacks of neural network [14]. An interesting property of SVMs is that it is an approximate implementation of the structure risk minimization induction principle that aims at minimizing a bound on the generation error of a model, rather than minimizing the mean square error over the data set [16, 17]. The subject of SVM covers emerging techniques which have proven successful in many traditionally neural network-dominated applications [17]. In recent years, SVMs were used successfully in many practical applications such as pattern matching and classification [18, 19], function approximation [20], data clustering [21] and forecasting [22].

SVMs were originally designed for binary classification. How to effectively extend it for multiclass classification is still an ongoing research issue [23]. Currently, OVR is one of main approaches for combining multiple binary-SVMs to solve the multiclass classification problems. Based on the main idea of OVR [15, 23, 24], this paper presented an improved OVR (IOVR) method to design classifiers for multiclass classification problems.

The structure of SVM classifier is given in Fig.1. Suppose that there are N signals to be classified, IOVR need constructs N SVM models, i.e. N binary SVMs. The ith SVM is trained with the examples in the ith class with positive labels, and the other examples with negative labels. From the feature ordering described in prior Section, every signal has its own the most marked features. So different binary-SVM has different training samples and different dimensionality of training feature vector. For example, if the 4th feature is the most marked feature of the 1st signal, in the 1st binary-SVM, the training samples are only from the 4th feature, that is, the dimensionality of the training feature vector is equal to 1. So in Fig.1, the training samples of SVM$_1$, SVM$_2$, ... , SVM$_N$ are different. This is different aspect between IOVR and OVR. In OVR, all binary SVM models have the same dimensionality of the training feature vector and have the training samples from the same features [15, 23, 24].

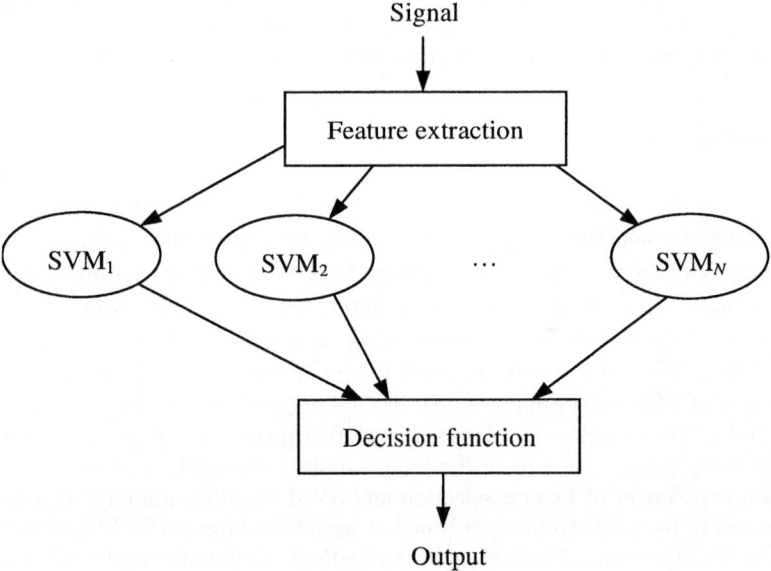

Fig. 1. The structure of SVM classifier

In testing phase, suppose that $g_1(x_1), g_2(x_2), \cdots, g_N(x_N)$ represent the decision functions of SVM$_1$, SVM$_2$, ... , SVM$_N$, respectively, where $x_1, x_2, \cdots, x_N$ are the testing sample vector of $g_1(x_1), g_2(x_2), \cdots, g_N(x_N)$, respectively. In OVR, the final decision function f_1 is

$$f_1 = \arg\{\max(g_1(x_1), g_2(x_2), \cdots, g_N(x_N))\} . \qquad (4)$$

In equation (4), $x_1, x_2, \cdots, x_N$ are identical. While in IOVR, the dimensionality of x_i $(i = 1, 2, \cdots, N)$ is the same as that of the training feature vector of binary SVM_i $(i = 1, 2, \cdots, N)$ model [15, 23, 24]. In IOVR, the final decision function f_2 is

$$f_2 = \{g_1(x_1) = 0, g_2(x_2) = 0, \cdots, g_i(x_i) = 1, \cdots, g_N(x_N) = 0\} \\ i = 1, 2, \cdots, N \tag{5}$$

The structure of SVM classifier shown in Fig.1 has the following advantages:

1) The dimensionality of the marked feature vector is usually very low, that is, the dimensionality of training sample vector is very low, so the training time of IOVR is usually much shorter than that of OVR;
2) In IOVR, the dimensionality of the testing feature vector is the same as the dimensionality of training sample vector. IOVR usually need smaller testing time than OVR;
3) What is more, the features obtained by using feature ordering method can separate easily one signal from the rest. So IOVR is very suitable for classifying multiple signals using the features obtained by using feature ordering method. IOVR has much better classification capability than OVR.

4 Simulations

To test the validity of the introduced method, 10 radar emitter signals are chosen to make simulation experiments. The 10 signals are represented with $x_1, x_2, \cdots, x_{10}$, respectively. In our prior work, 16 features have been extracted from the 10 radar emitter signals [2, 3, 25-27]. The 16 features are represented with $a_1, a_2, \cdots, a_{16}$, respectively. In the experiment, for every radar emitter signal, 150 feature samples are generated in each signal-to-noise rate (SNR) point of 5dB, 10dB, 15dB and 20dB. Thus, 600 samples of each radar emitter signal in total are generated when SNR varies from 5dB to 20dB. The samples are classified into two groups: training group and testing group. Training group, one half of the total samples generated, is applied to make the simulation experiment of feature selection and SVM classifier training. Testing group, the other half of the total samples generated, is used to test trained SVM classifiers.

We use the algorithm of feature ordering method to compute marked characteristic coefficients of the 10 radar emitter signals. Experimental results are given in Table 1. All values in Table 1 are obtained using equation (1). According to the definition of marked features, Table 1 shows that 9 radar emitter signals $x_1, x_3, \cdots, x_{10}$ have one-dimensional marked features and their marked features are respectively a_7, a_{14}, a_1, a_7, a_{12}, a_1, a_2, a_7 and a_7. The 2nd radar emitter signal has no one-dimensional marked feature, however, because the rest 9 radar emitter signals $x_1, x_3, \cdots, x_{10}$ have one-dimensional marked features, when IOVR classifier is used to classify the 10 radar emitter signals, the 2nd radar emitter signal can also be recognized. It is not necessary to compute the two-dimensional marked feature vector of the 2nd radar emitter signal. In this paper, we use a_1, a_2, a_7, a_{12} and a_{14} to construct a feature vector to recognize the 2nd radar emitter signal.

Table 1. Marked characteristic coefficients of radar emitter signals (FT stands for features)

FT	x_1	x_2	x_3	x_4	x_5	x_6	x_7	x_8	x_9	x_{10}
a_1	0.33	0.21	0.21	35.53	0.33	3.99	55.24	3.11	0.33	2.65
a_2	0.37	0.14	0.14	3.65	2.45	6.17	13.57	12.90	0.45	0.45
a_3	0.14	0.39	0.24	0.23	0.19	0.02	0.19	0.01	0.23	0.01
a_4	0.02	0.02	0.80	0.09	0.09	3.53	3.53	0.95	0.80	0.56
a_5	0.01	0.12	0.35	0.03	0.14	0.03	0.10	0.01	0.07	0.07
a_6	0.69	0.09	0.12	1.27	0.20	0.79	0.79	0.09	0.12	0.20
a_7	4.28	0.57	0.57	3.85	3.65	6.27	0.58	3.65	14.74	3.65
a_8	0.19	0.19	0.82	0.31	1.65	0.31	0.86	1.86	0.86	0.73
a_9	0.78	0.79	1.38	0.28	0.28	18.00	18.00	0.78	1.38	1.51
a_{10}	1.60	0.73	3.23	1.00	1.00	20.52	20.52	0.73	3.23	2.33
a_{11}	0.86	0.82	2.67	0.02	0.02	8.27	8.27	0.82	2.47	0.06
a_{12}	0.62	0.62	1.33	0.72	0.72	20.01	20.01	4.85	1.33	2.48
a_{13}	0.56	0.56	1.74	0.05	0.05	14.57	14.57	0.92	5.02	0.24
a_{14}	0.37	0.37	4.13	0.01	0.01	3.83	3.83	2.29	3.23	0.04
a_{15}	0.75	0.15	0.09	0.02	0.02	15.46	15.46	0.15	0.09	0.08
a_{16}	0.29	0.29	3.69	0.50	0.50	4.92	4.92	10.51	3.69	2.23

To bring into comparison, the features selected by the satisfactory feature selection method [28] and the resemblance coefficient feature selection method [29] are construct a feature vector to be input of OVR SVM classifiers to recognize the 10 radar emitter signals.

In the experiments, we use recognition error rate and recognition efficiency to evaluate the performances of IOVR and OVR. Recognition efficiency includes training time and testing time. The statistical results of many experiments are given in Table 2.

Table 2. Comparison of OVR and IOVR

Signals	OVR	IOVR
a_1	100.000 %	0.000 %
a_2	17.410 %	0.000 %
a_3	0.000 %	0.000 %
a_4	0.000 %	0.410 %
a_5	0.000 %	0.670 %
a_6	0.000 %	0.000 %
a_7	0.000 %	0.000 %
a_8	100.00 %	0.000 %
a_9	0.000 %	0.190 %
a_{10}	0.000 %	0.040 %
Average error rate	21.741 %	0.131 %
Training time (Sec.)	1386.666	1120.691
Testing time (Sec.)	294.950	257.914

In [28, 29], the selected feature subset is composed of a_5, a_{10} and the classifier uses neural networks. The average recognition error rate in [28, 29] is 1.340%, which is higher 1.209% than 0.131% that is obtained by the proposed method in this paper. From Table 2, IVOR achieves only 0.131% average recognition error rate, which is much lower than OVR with 21.741% average recognition error rate. Moreover, IVOR costs much shorter training time and testing time than OVR. So these experimental results show that IVOR is superior to OVR greatly in classification capability and recognition efficiency. This conclusion is consistent to the prior analysis in Section 3. Simultaneously, experimental results also indicate that the neural network classifier in [28, 29] has better classification performance than OVR.

IOVR is superior to OVR and neural network in classification capability. OVR is inferior to neural network in classification capability. Because IOVR is a classifier corresponding to feature ordering method, these experimental results also indicate that the introduction of feature ordering method proposed in this paper can enhance classification capability of OVR greatly.

5 Conclusions

This paper proposes a new radar emitter signal recognition approach called humanoid recognition method. The main points of the introduced method include a new feature selection called feature ordering and an improved one-versus-rest combination of multiple binary support vector machines. In feature ordering, an algorithm for computing marked characteristic coefficient is presented to obtain the marked features of every radar emitter signal. The most advantage of the feature selection is that low dimensional feature vector is usually obtained, which can quicken the training and recognition speed greatly. Corresponding to the feature selection, an improved one-versus-rest combination for support vector machines is used to design a support vector network to implement automatic recognition of radar emitter signals. Experimental results show that the proposed method can achieve good classification performance and good recognition efficiency. It is proven to be a practical and valid method.

This paper only uses an application example of radar emitter signal recognition to test the performances of the introduced method. In further study, we will use some standard datasets from UCI Machine Learning Repository and some engineering applications such as image recognition, speech recognition and human face recognition to verify the validity of the proposed method. Besides, what needs to point out is that the validity of the introduced algorithm for computing marked characteristic coefficients is only proven using the recognition issue in which the features are real numbers instead of discrete values.

References

1. Schroer, R.: Electronic Warfare. IEEE Aerospace Electronic Systems Magazine, Vol.18, No.7 (2003) 49-54
2. Zhang, G.X., Hu, L.Z., and Jin, W.D.: Resemblance Coefficient Based Intrapulse Feature Extraction Approach for Radar Emitter Signals. Chinese Journal of Electronics, Vol.14, No.2 (2005) 337-341

3. Zhang, G.X., Hu, L.Z., and Jin, W.D.: Intra-pulse Feature Analysis of Radar Emitter Signals. Journal of Infrared and Millimeter Waves, Vol.23, No.6 (2004) 477-480
4. Granger, E., Rubin, M.A., Grossberg, S., and Lavoie, P.: A What-and-Where Fusion Neural Network for Recognition and Tracking of Multiple Radar Emitters. Neural Networks, Vol.14, No.3 (2001) 325-344
5. Langley, L.E.: Specific Emitter Identification (SEI) and Classical Parameter Fusion Technology. Proceedings of the WESCON, (1993) 377-381
6. Shieh, C.S., Lin, C.T.: A Vector Network for Emitter Identification. IEEE Transaction on Antennas and Propagation, Vol.50, No.8 (2002) 1120-1127
7. Liu, W.K., Zhu, D.J., and Zhang, C.H.: The Extraction of Modulation Characteristics of Radar Signal Using Wavelet Transform. Proceedings of the 4th International Conference on Signal Processing, (1998) 288-291
8. Dudczyk, J., Matuszewski, J., and Wnuk, M.: Applying the Radiated Emission to the Specific Emitter Identification. Proceedings of 15th International Conference on Microwaves, Radar and Wireless Communications, Vol.2, (2004) 431-434
9. Jin, F., Fan, J.B.: A Humanoid Intelligent System for High-speed Recognitions. Proceedings of International Conference on Intelligent Information Processing and System, (1992) 412-414
10. Jin, F.: Intelligent Foundation of Neural Computing: Principles & Methods. Southwest Jiaotong University Press, Chengdu (2000)
11. Ren, L.Y., Lu, X.L.: Study and Implementation of BP Network Based on Serial-parallel Computing Topology. Journal of UEST of China, Vol.29, No.2 (2000) 197-200
12. Shieh, C.S., Lin, C.T.: A Vector Network for Emitter Identification. IEEE Transaction on Antennas and Propagation, Vol.50, No.8 (2002) 1120-1127
13. Zhang, G.X., Hu, L.Z., and Jin, W.D.: Radar Emitter Signal Recognition Based on Feature Selection Algorithm. Lecture Notes in Artificial Intelligence, Vol.3339 (2004) 1108-1114
14. Bian, Z.Q., Zhang, X.G.: Pattern Recognition (2nd Edition). Tsinghua University Press, Beijing (2000)
15. Vapnik, V.: The Nature of Statistical Learning Theory. Springer-Verlag, New York. (1995)
16. Cristianini, N., Shawe-Taylor, J.: An Introduction to Support Vector Machines and Other Kernel-Based Learning Methods. Translated by Li, G.Z., Wang, M., and Zeng, H.J., Publishing House of Electronics Industry, Beijing (2004)
17. Dibike, Y.B., Velickov, S., and Solomatine, D.: Support Vector Machines: Review and Applications in Civil Engineering. Proceedings of the 2nd Joint Workshop on Application of AI in Civil Engineering, (2000) 215-218
18. Osareh, A., Mirmehdil, M., Thomas, B., and Markham, R.: Comparative Exudate Classification Using Support Vector Machines and Neural Networks. Lecture Notes in Computer Science, Vol.2489. (2002) 413-420
19. Foody, G.M., Mathur, A.: A Relative Evaluation of Multiclass Image Classification by Support Vector Machines. IEEE Transactions on Geoscience and Remote Sensing, Vol.42, No. 6. (2004) 1335-1343
20. Ma, J.S., Theiler, J., and Perkins, S.: Accurate On-line Support Vector Regression. Neural Computation, Vol.15, No.11. (2003) 2683-2703
21. Ben-Hur, A., Horn, D., Siegelmann, H.T., and Vapnik, V.: Support Vector Clustering. Journal of Machine Learning Research, Vol.2, No.2. (2001) 125-137
22. Kim, K.J.: Financial Time Series Forecasting Using Support Vector Machines. Neurocomputing, Vol.55, No.1. (2003) 307-319

23. Hsu, C.W., Lin, C.J.: A Comparison of Methods for Multiclass Support Vector Machines. IEEE Transaction On Neural Networks, Vol.13, No.2, (2002) 415-425
24. Rifkin, R., Klautau, A.: In Defence of One-Vs-All Classification. Journal of Machine Learning Research, Vol.5, No.1. (2004) 101-141
25. Zhang, G.X., Hu, L.Z., and Jin, W.D.: Quantum Computing Based Machine Learning Method and Its Application in Radar Emitter Signal Recognition. Lecture Notes in Artificial Intelligence, Vol.3131. (2004) 92-103
26. Zhang, G.X., Rong, H.N., Hu, L.Z., and Jin, W.D.: Entropy Feature Extraction Approach of Radar Emitter Signals. Proceedings of International Conference on Intelligent Mechatronics and Automation, (2004) 621-625
27. Zhang, G.X., Jin, W.D., and Hu, L.Z.: Application of Wavelet Packet Transform to Signal Recognition. Proceedings of International Conference on Intelligent Mechatronics and Automation, (2004) 542-547
28. Zhang, G.X., Jin, W.D., and Hu, L.Z.: A Novel Feature Selection Approach and Its Application. Lecture Notes in Computer Science, Vol.3314. (2004) 665-671
29. Zhang, G.X., Jin, W.D., and Hu, L.Z.: Resemblance Coefficient and a Quantum Genetic Algorithm for Feature Selection. Lecture Notes in Artificial Intelligence. Vol.3245. (2004) 155-168

Methods of Decreasing the Number of Support Vectors via k-Mean Clustering

Xiao-Lei Xia[1], Michael R. Lyu[2], Tat-Ming Lok[3], and Guang-Bin Huang[4]

[1] Institute of Intelligent Machines, Chinese Academy of Sciences,
P.O.Box 1130, Hefei, Anhui, China
xlxia@iim.ac.cn
[2] Computer Science & Engineering Dept., The Chinese University of Hong Kong,
Shatin, Hong Kong
[3] Information Engineering Dept., The Chinese University of Hong Kong,
Shatin, Hong Kong
[4] School of Electrical and Electronic Engineering, Nanyang Technological University

Abstract. This paper proposes two methods which take advantage of k-mean clustering algorithm to decrease the number of support vectors (SVs) for the training of support vector machine (SVM). The first method uses k-mean clustering to construct a dataset of much smaller size than the original one as the actual input dataset to train SVM. The second method aims at reducing the number of SVs by which the decision function of the SVM classifier is spanned through k-mean clustering. Finally, Experimental results show that this improved algorithm has better performance than the standard Sequential Minimal Optimization (SMO) algorithm.

1 Introduction

Support Vector Machine (SVM) [1] is a new class of approaches for classification and regression problems. Currently, SVMs are gaining popularity due to attractive features and have been successfully applied to various fields. Unlike previous machine learning algorithms such as traditional neural network models [2, 3, 4, 5], the SVM developed by Vapnik is derived from statistical learning theory and employs the structural risk minimization (SRM) principle [1], which can significantly enhance SVM's generalization capability. With a clear geometrical interpretation, the training of the SVM is guaranteed to find the global minimum of the cost function.

In general, training an SVM requires the solution of a very large quadratic programming (QP) optimization problem. The large size of the training sets typically used in applications is a formidable obstacle to a direct use of standard quadratic programming techniques [6]. Recently, many algorithms have been developed to solve the problem [7, 8, 9]. The most typical one is John Platt's Sequential Minimal Optimization (SMO) [10], which breaks a large QP problem into a series of smallest possible QP problems. SMO is generally fast and efficient for linear SVMs and sparse data sets. However, the number of support vectors (SVs) that SMO produces is too large in proportion to the size of the input dataset for training SVM. It is shown that if the training vectors are separated without errors the expectation value of the probability of committing an error on a test example is bounded by the ratio between the expectation value of the number of support vectors and the number of training vectors:

$$E[\Pr(error)] \leq \frac{E[number\ of\ support\ vectors]}{number\ of\ training\ vectors}. \tag{1}$$

From inequality (1), it can be drawn that a small number of support vectors can lead to a small testing error and also a SVM with a better generalization capability.

In this paper k-mean clustering [11, 13] provides two methods to suppress the number of support vectors based on SMO algorithm in the training of SVM. For the first method, k-mean clustering method helps pick a set smaller than the original dataset to train SVM, which dramatically reduce the number of SVs without reducing the training correctness. It also can be concluded that with the decrease in the number of training examples the computational time that SMO requires greatly falls. The other application of k-mean clustering aims at finding less support vectors to describe the normal vector of the optimal hyperplane of SVM. The normal vector is spanned by a number of the mapping of SVs from input space into feature space where the kernel trick plays an essential role [6, 12]. k- mean clustering can help find a certain number of support vectors whose feature space image would well approximate the expansion. The two methods can suppress the number of SVs and result in a SVM with significant efficiency and outstanding generalization ability.

The paper is organized as follows. Section 2 gives a brief introduction to the theoretical background with reference to classification principles of SVM. Section 3 describes the two methods for the decrease in the number of SVs. Experimental results is demonstrated in Section 4 to illustrate the efficiency and effectiveness of our algorithm. Conclusions are included in Section 5.

2 Support Vector Machine

Consider the problem of separating the set of N training vectors belonging to two classes, where $x_i \in R^n$ is the i th input data and $y_i \in R$ is the i th output data

$$(x_1, y_1),...,(x_N, y_N) \in R^n \times Y, \quad Y = \{-1,+1\} \tag{2}$$

with a hyperplane

$$H: \quad \langle w, x \rangle + b = 0 \tag{3}$$

where w is normal to the hyperplane and $b/\|w\|$ is the perpendicular distance from the hyperplane to the origin. The hyperplane is regarded as optimal if all the training vectors are separated without error and the margin (i.e. the distance from the closest vector to the hyperplane) is maximal. Without loss of generality, it is appropriate to consider a canonical hyperplane, acquired by rescaling w and b so that the vectors x_i ($i = 1,..., N$) closest to the hyperplane satisfy:

$$|\langle w, x_i\rangle + b| = 1. \qquad (4)$$

Hence, the margin is $2/\|w\|$. Thus the hyperplane $<w,b>$ is given by the solution to the following optimization problem:

$$\min_{w,b} \frac{1}{2}w^T w \qquad (5)$$
$$\text{subject to} \quad y_i\left(w^T x_i + b\right) \geq 1$$
$$\forall i = 1, 2, ..., N$$

The training vectors for which the equation (4) holds are termed as support vectors (SV). The equivalent dual problem to equation (5) can be written as:

$$\max Q(\alpha) = \sum_{i=1}^{N} \alpha_i - \frac{1}{2}\sum_{i=1}^{N}\sum_{j=1}^{N} \alpha_i \alpha_j y_i y_j \langle x_i, x_j\rangle \qquad (6)$$
$$\text{subject to} \quad \sum_{i=1}^{N} \alpha_i y_i = 0$$
$$\forall i = 1, 2, ..., N$$

where the α_i are the Lagrange multipliers and constrained to be non-negative. The linear SVM classifier can be denoted as:

$$f(x) = \text{sgn}(\langle w, x\rangle + b). \qquad (7)$$

With respect to the case of nonlinearly separable datasets, SVM employs a kernel function K to implement the dot product between the functions $\phi(x_i)$ which can map the data from input space to a high dimensional feature space H, i.e.:

$$K(x_i, x_j) = \langle \phi(x_i), \phi(x_j)\rangle. \qquad (8)$$

The theory of functional analysis suggests that an inner product in feature space correspond to an equivalent kernel operator in input space provided that K satisfies Mercer's condition.

Furthermore, variables ξ_i $(i = 1, 2, ..., N)$ are introduced to allow the margin constraints to be violated while C determines the tradeoff between error and margin. Then a quadratic optimization problem is introduced as follows:

$$\min \frac{1}{2}w^T w + C\left(\sum_{i=1}^{l} \xi_i^2\right) \qquad (9)$$
$$\text{subject to} \quad y_i\left(w^T \phi(x_i) + b\right) \geq 1 - \xi_i$$

The decision function of the nonlinear classifier is:

$$f(x) = \text{sgn}\left(\sum_{i=1}^{n} y_i \alpha_i K(x_i, x) + b\right).\tag{10}$$

3 Reducing the Number of Support Vectors via k-Mean Clustering

3.1 The First Method Using k-Mean to Reduce the Number of SVs

The first method used to modify the conventional SMO is to employ the k-mean clustering to choose a set which reflects the general features of the full input dataset but has much fewer data points. The approach is based on the observation that in many cases a large proportion of the original input dataset is redundant for training SVM. A good SVM classifier could be generated from a small portion of the input dataset provided they outline the whole dataset approximately. The advantages of the approach lies in the fact that the smaller the input dataset is, the fewer SVs would be yielded and that it would require less CPU time and memory .Hence, k-mean clustering is introduced to choose the actual training set. k-mean clustering is applied respectively to the two groups into which the input datasets are divided according to the values of the output data y_i in order to generate two sets of centers. Centers are chosen such that the points are mutually farthest apart, which would well reflect the relative position of point-clusters of input dataset and thus characterize the outline of the full dataset. To achieve an optimal k, i.e. the number of centers, which would describe the full input dataset well, a portion of data will be removed as tuning set to adjust the number of centers to reach the best training precision.

As a result, the procedures for determining k and the set of smaller size can be summarized as follows:

Step 1. Remove a certain portion of an input set as the tuning set and divide the input dataset into two groups according to their labels y_i.

Step 2. Start with a small k, which is around 5% of the whole input set.

Step 3. Apply k-mean algorithm to the two groups respectively to produce a center set as the real dataset to train SVM.

Step 4. Apply the standard SMO algorithm to the training set in order to produce a classifier.

Step 5. Compute the correctness of the classifier on the tuning set

Step 6. If the correct rate of the tuning set is small enough, terminate the loop. Otherwise, increase k and continue from *Step 3*.

3.2 The Second Method Using k-Mean to Reduce the Number of SVs in the Decision Function of SVM Classifier

The second modification to the standard SMO aims at simplifying the decision function of the SVM classifier to strength its generalization capability. It has been noted that the number of SVs that nonlinear separable datasets generate makes up a large proportion of the input set, which would result in a high risk of poor performance on testing examples thus weak generalization capability according to inequality (1). To avoid problems mentioned above, k-mean clustering is employed after the training phase of SVM to suppress the number of SVs.

In the decision function of the SVM classifier, the normal vector of the optimal hyperplane is described by the kernel expansion.

$$w = \sum_{i=1}^{N} \alpha_i y_i \phi(x_i) = \sum_{i=1}^{N} \lambda_i \phi(x_i) \cdot \quad (11)$$

Now we wish to find a new solution:

$$w^* = \sum_{i=1}^{m} \beta_i \phi(s_i), \quad (12)$$

so that the kernel expansion would be shorter, i.e. $1 \leq m \ll N$ and well approximate the original expansion.

To simplify the problem, set both m and β as 1 and the problem of finding the new expansion of the normal vector can be formulated as the following optimization task:

$$\begin{aligned} s &= \arg\min_{s'} \| w - w^* \|^2 \\ &= \arg\min_{s'} \| \phi(s) - \sum_{i=1}^{N} \lambda_i \phi(x_i) \|^2 \quad (13) \\ &= \arg\min_{s'} k(s,s) - 2\sum_{i=1}^{N} \lambda_i k(s,x_i) + \sum_{i=1}^{N}\sum_{j=1}^{N} \lambda_i \lambda_j k(x_i,x_j) \end{aligned}$$

For Gaussian radial basis function (RBF) as shown in the equation below:

$$K(x,y) = \exp\left(-\|x-y\|^2 / (2\sigma^2)\right), \quad (14)$$

optimization task (13) leads to:

$$s = \arg\max_{s'} \sum_{i=1}^{N} \lambda_i \exp(-\|x_i - s\|^2 / \sigma^2). \quad (15)$$

For $m \geq 1$, the problem described by Equation (12) with RBF kernel can be converted into m optimization tasks of (15) which aims at finding an input vector s of the input dataset.

To solve the optimization task (15), k-mean is again used to cluster the points of the input dataset. And the algorithm of finding a shorter kernel expansion can be summarized as follows:

Step 1. Start with a small k, around 5% of the size of the input dataset.
Step 2. Apply the k-mean algorithm to yield k centers.
Step 3. Employ the set of these k centers as the actual input dataset and pick up the center which solve the optimization task (15)
Step 4. Compute the deviation between the two expansions of normal vector of the hyperplane
Step 5. If the standard deviation is small enough, terminate the loop. Otherwise increase k and start from *Step 2*.

4 Experimental Results

To verify the effectiveness and efficiency of the novel SMO combined with k-mean clustering, we use Riply's training dataset [14], which contains 250 points, and checkerboard's dataset [15] of 1000 points to test the proposed algorithms. All experiments are conducted on a platform of a machine with a Pentium 4 2.6GHz processor and 265 megabytes of memory.

4.1 Experiments of Combing the First Method with Standard SMO Algorithm

The first experiment on Riply's dataset uses Gaussian radial basis function (RBF) as the kernel function. Figure 1 depicts the decision boundaries of standard SMO with the red solid line and the novel SMO using k-mean with the blue line. The comparison of the novel SMO with the original one is demonstrated by Table 1. Parameter setting for standard SMO is C=30 in Equation (9) and σ=1 in Equation (14) after model selection. For the SMO using k-mean, C=5 and σ=1 in Figure 1.

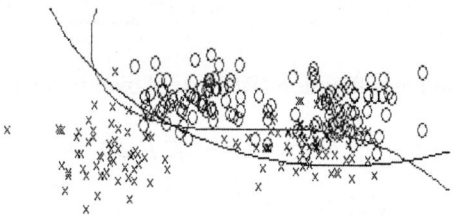

Fig. 1. Decision boundaries built from two SMO algorithms with 8 points (k=4)

Figure 1 illustrates two decision boundaries which bear several similarities to each other. However, the SMO using k-mean clustering shown in Figure 1 only employs 5 SVs while the standard SMO 94 SVs according to Table 1. It shows that the training of SVM has been sped up with the combination of k-mean clustering.

The second experiment, using Gaussian RBF kernel, is to classify the checkerboard dataset. Figure 2 illustrate the training results of the SMO using k-mean with the training set of only 16 points.

Table 1. Performance comparison between the two SMO algorithms 1=standard SMO ; 2 =SMO using k-mean

	k	Training Error	Testing Error	Number of SVs	Time (CPU sec)
1		0.128	0.096	91	**1.178**
	4	0.148	0.094	5	**0.016**
	8	0.192	0.166	10	**0.016**
2	16	0.140	0.099	22	**0.031**
	32	0.148	0.099	31	**0.031**
	64	0.128	0.095	57	**0.063**

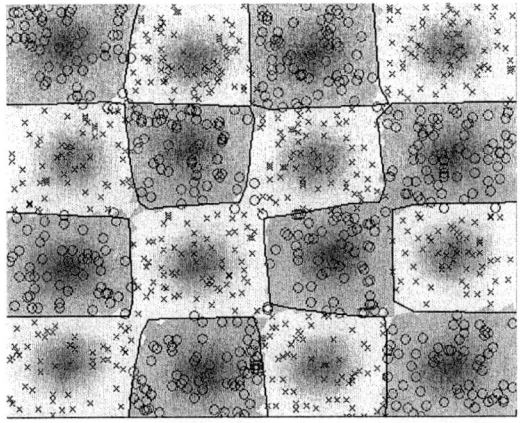

Fig. 2. Performance of SMO using k-mean on checkerboard with 16 training points (k=8) with the parameter setting: C=20 and σ=8

Table 2. Performance comparison between the two SMO algorithms 1=standard SMO; 2 =SMO using k-mean

	k	Training Error	Number of SVs	Time (CPU sec)
1		0.000	285	**226.2**
	8	0.034	16	**0. 11**
	16	0.105	32	**0. 15**
2	32	0.112	61	**0. 67**
	64	0.044	118	**0.75**
	128	0.030	192	**0.92**

From Table 2, it can be drawn that the classifier in Figures 2 which gives pretty good a representation of the checkerboard dataset are built on only 1.6% input data and 96.6% input data are classified correctly. A deeper comprehension of the advantages of the SMO using k-mean over the standard SMO can be also seen from the computational time.

4.2 Experiments of Combing the Second Method with Standard SMO Algorithm

The first experiment to verify the effectiveness of the second approach is implemented on Riply's dataset uses Gaussian radial basis function (RBF) as the kernel function with parameters: $C=20$ and $\sigma=1$.

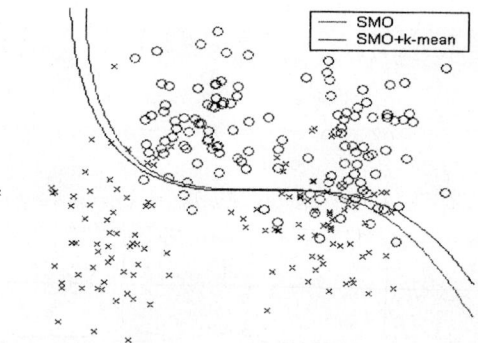

Fig. 3. Decision boundaries built from two SMO algorithms (k=8)

With Figure 3, it is shown that the decision boundaries built from two SMO algorithms are very similar to each other. However, the SVM classifier built with the second method to suppress the number of SVs only employs 8 SVs while the classifier built with standard SMO has 88 SVs.

The second experiment is to classify the checkerboard dataset using Gaussian RBF kernel. After the second method is applied, the normal vector of the hyperlane is spanned with 91 SVs while the original expansion of the normal vector has 273 SVs. The employment of the second method to reduce the number of SVs decreases the expectation value of the probability of committing an error on a test example and enhances SVM's generalization capability. Figure 4 illustrates the training results of the SVM using the second method with $k=91$.

5 Conclusions

This paper proposes and implements two methods which are intensely involved with k-mean clustering algorithm to suppress the number of SVs. It is shown with experiments that the first method of integrating k-mean clustering into the standard SMO algorithm significantly speeds up the training process and greatly decrease the

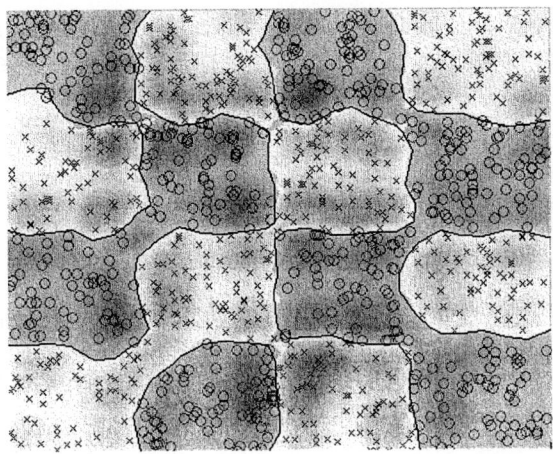

Fig. 4. Performance of the SMO using the second method (k =91)

number of SVs and the second method of combining k-mean clustering with standard SMO makes the number of SVs which span the decision function of the SVM classifier smaller and improves SVM's generalization capability. Future works involves applying the two methods to more real-world problems and modifying k-mean clustering algorithm so that the optimum value for the number of centers can be found.

Acknowledgements

The research is supported by the National Science Foundation of China (Nos.60472111 and 60405002), RGC Project No.CUHK 4170/04E, RGC Project No. CUHK4205/04E and UGC Project No.AoE/E-01/99.

References

1. V.,Vapnik: The nature of statistical learning theory. Springer Verlag, (1995)
2. D.S. Huang, Horace, H.S.Ip, Law Ken, C.K., Zheru Chi: Zeroing polynomials using modified constrained neural network approach. IEEE Trans. On Neural Networks, vol.16, no.3 (2005) 721-732
3. D.S. Huang, Horace, H.S.Ip, Zheru, Chi: A neural root finder of polynomials based on root moments. Neural Computation, Vol.16, No.8 (2004) 1721-1762
4. D.S. Huang: A constructive approach for finding arbitrary roots of polynomials by neural networks. IEEE Transactions on Neural Networks, Vol.15, No.2 (2004) 477-491
5. D.S. Huang: Systematic Theory of Neural Networks for Pattern Recognition. Publishing House of Electronic Industry of China, Beijing (1996)
6. N. Cristianini, J. Shawe-Taylor: An Introduction to Support Vector Machines and Other Kernel-based Learning Methods. Cambridge University Press(2000)

7. Bing-Yu Sun, D.S. Huang, Hai-Tao Fang: Lidar signal de-noising using least squares support vector machine. IEEE Signal Processing Letters, vol.12, no.2 (2005) 101-104
8. Bing-Yu Sun, D.S. Huang: Least squares support vector machine ensemble. The 2004 International Joint Conference on Neural Networks (IJCNN2004), Budapest Hungary (2004) 2013-2016.
9. T.,Joachims.: Making large-scale support vector machine learning practical. Advances in kernel methods: support vector learning. MIT Press (1999) 169-184
10. J. Platt.: Sequential Minimal Optimization: A Fast Algorithm for Training Support Vector Machines. Advances in kernel methods: support vector learning, MIT Press, (1999) 185-208
11. Anil K.Jain, Richard C. Dubes: Algorithms for Clustering Data. Prentice Hall (1988)
12. B. Scholkopf, A.J.,Smola: Learning with Kernels.The MIT Press, MA (2001)
13. J. H. Friedman, F. Baskett, and L. J. Shustek: An algorithm for finding nearest neighbours. IEEE transactions on Computers C-24 (1975)1000-1006
14. B. D. Riply: Neural networks and related methods for classifications. J.Royal Statistical Soc. Series B,56 (1994)409-456
15. ftp://ftp.cs.wisc.edu/math-prog/cpo-dataset/machine-learn/checker

Dynamic Principal Component Analysis Using Subspace Model Identification

Pingkang Li[1], Richard J. Treasure[2], and Uwe Kruger[3]

[1] School of Mechanical, Electrical and Control Engineering,
Beijing Jiaotong University, Beijing 100044, P.R. China
pkli@center.njtu.edu.cn
[2] Control Systems Research Group, University of Western Australia, Crawley 6009
rtreasure@ee.uwa.edu.au
[3] Intelligent Systems and Control Group, Queen's University Belfast, BT9 5AH, UK
uwe.kruger@ee.qub.ac.uk

Abstract. This work analyses a recently proposed statistically based technique for monitoring complex dynamic process systems [17]. The technique utilises a state space model that is cast into the multivariate statistical process control framework (i) to define a set of state variables that can describe dynamic process behaviour, (ii) to generate univariate statistics that can monitor dynamic process behaviour and (iii) to construct contribution plots from these statistics that can diagnose anomalous process behaviour. The presented analysis reveals that the size of the state space monitoring model can be reduced. The utility of the improved dynamic monitoring technique is demonstrated using an industrial application study to a glass-melter process.

1 Introduction

Over the past two decades, a number of statistically based techniques have been applied to enhance the monitoring of complex multivariate processes systems. Such techniques, collectively referred to as multivariate statistical process control, aim at exploiting the often observed high degree of correlation in a typically large set of the recorded process variable [11,12,23,13,16,8]. Principal component analysis (PCA) is one of the most popular techniques, as it can produce a significantly reduced set of statistically independent score variables from the recorded variable set [7,20,3,1,18]. This reduced set is accordingly simpler to monitor than the original variable set.

In recent years, dynamic extensions of PCA (DPCA) have gained attention [9,15,19,4]. Although a number of application studies have demonstrated the usefulness of DPCA in monitoring dynamic process behavior [9,2], [17] showed that DPCA may produce a significant number of time-lagged variables that need to be analysed. To reduce the number of original variables to be included in the dynamic condition monitor, Treasure et al. [17] incorporated a state space model structure into the conventional PCA technique. This resulted in a twofold reduction that (i) produces a reduced set of state variables that represented dynamic process behaviour, (ii) generates PCA based statistics to detect anomalous process behaviour and (iii) allows the use of contribution charts to diagnose such behaviour.

This paper analyses the technique by Treasure et al. [17] and shows that the dynamic model that is required to monitor the process can be further reduced in size. Since the work by Treasure et al. [17] only included a simulation study, a further contribution of the presented work is the application of the enhanced dynamic monitoring technique to an industrial application study.

2 Dynamic Process Monitoring

This section briefly summarises the dynamic monitoring technique by Treasure et al. [17] and is divided as follows. The estimation of the state sequences is discussed next. This is followed by showing how PCA is used to estimate the statical monitoring model for on-line monitoring.

2.1 Estimation of the State Sequences

Throughout this work, the N4SID algorithm [21] is utilised for subspace model identification. Assuming that the process has a total of n_N input and n_M output variables, the subspace monitoring technique determines n_S^{th} order state sequences $\mathbf{X} = [\mathbf{x}(1)\ \mathbf{x}(2)\ \cdots\ \mathbf{x}(K)]$ and $\mathbf{X}^+ = [\mathbf{x}^+(1)\ \mathbf{x}^+(2)\ \cdots\ \mathbf{x}^+(K)] \in \mathbb{R}^{n_S \times K}$ of length K and a state space model:

$$\begin{bmatrix}\widehat{\mathbf{X}}^+ \\ \widehat{\mathbf{Y}}\end{bmatrix} = \begin{bmatrix}\widehat{\mathbf{A}} & \widehat{\mathbf{B}} \\ \widehat{\mathbf{C}} & \widehat{\mathbf{D}}\end{bmatrix} \begin{bmatrix}\widehat{\mathbf{X}} \\ \widehat{\mathbf{U}}\end{bmatrix}, \qquad (1)$$

where $\widehat{\mathbf{U}} \in \mathbb{R}^{n_N \times K}$, $\widehat{\mathbf{Y}} \in \mathbb{R}^{n_M \times K}$, $\widehat{\mathbf{X}}^+$ and $\widehat{\mathbf{X}}$ are matrices storing the estimates of the input and output readings of the K observation and the state sequences, respectively. The matrices $\widehat{\mathbf{A}} \in \mathbb{R}^{n_S \times n_S}$, $\widehat{\mathbf{B}} \in \mathbb{R}^{n_S \times n_N}$, $\widehat{\mathbf{C}} \in \mathbb{R}^{n_M \times n_S}$ and $\widehat{\mathbf{D}} \in \mathbb{R}^{n_M \times n_N}$ are estimates of the state space matrices. A complete description of the above identification algorithm is available in references [22,17].

The estimation of the state sequences relies on block Hankel matrices $\mathbf{U}_f$, $\mathbf{U}_p \in \mathbb{R}^{N \cdot n_u \times K}$, $\mathbf{Y}_f$ and $\mathbf{Y}_p \in \mathbb{R}^{N \cdot n_y \times K}$, containing N time lagged block rows of the measured input and output sequences of length K, where the indices 'p' and 'f' refer to past and future. Next, the matrix $\mathbf{Y}_f$ is predicted using a linear regression involving $\mathbf{Y}_p$, $\mathbf{U}_p$ and $\mathbf{U}_f$:

$$\mathbf{Y}_p = \begin{bmatrix}\mathbf{R}_1 & \mathbf{R}_2 & \mathbf{R}_3\end{bmatrix}\begin{bmatrix}\mathbf{Y}_p \\ \mathbf{U}_p \\ \mathbf{U}_f\end{bmatrix} + \mathbf{G}, \qquad (2)$$

where $\mathbf{R}_1$, $\mathbf{R}_2$ and $\mathbf{R}_3$ are regression matrices and $\mathbf{G}$ is a residual matrix. On the basis of the above regression equation, the state sequences are estimated as follows:

$$\widehat{\mathbf{Y}}_f = \mathbf{R}_1 \mathbf{Y}_p + \mathbf{R}_2 \mathbf{U}_p = \mathbf{\Gamma} \mathbf{X}, \qquad (3)$$

where Γ is the extended observability matrix that are estimated up to a similarity transformation using a singular value decomposition of $\widehat{\mathbf{Y}}_f$. The second state sequence $\mathbf{X}^+$ is obtained by rearranging the rows in $\mathbf{U}_f$, $\mathbf{U}_p$, $\mathbf{Y}_f$ and $\mathbf{Y}_p$ and reapplying Equations (2) and (3) [21].

2.2 Determining the Dynamic Monitoring Model

Multivariate statistics are developed on the basis of a PCA decomposition of the following matrix:

$$\mathbf{Z} = \begin{bmatrix} \mathbf{X}^+ \\ \mathbf{Y} \\ \mathbf{X} \\ \mathbf{U} \end{bmatrix} = \mathbf{PT} + \mathbf{E}, \qquad (4)$$

where $\mathbf{P} \in \mathbb{R}^{2 \cdot n_S + n_N + n_M \times n_t}$ is a loading matrix, $\mathbf{T} \in \mathbb{R}^{n_t \times K}$ is a score matrix, $\mathbf{E} \in \mathbb{R}^{2 \cdot n_S + n_N + n_M \times K}$ is a residual matrix, and n_t represents the number of retained principal components (PCs).

On the basis of the above decomposition, the univariate (Hotelling's) T^2 and Q statistics can be defined as follows:

$$T^2(k) = \mathbf{t}^T(k)\boldsymbol{\Lambda}^{-1}\mathbf{t}(k) \qquad Q(k) = \mathbf{e}^T(k)\mathbf{e}(k), \qquad (5)$$

where k represents the sampling index, $T^2(k)$ and $Q(k)$ are the values of the T^2 and Q statistics, respectively, $\mathbf{t}(k) = \mathbf{P}^T\mathbf{z}(k)$ and $\mathbf{e}(k) = \mathbf{z}(k) - \mathbf{P}\mathbf{t}(k)$ are column vectors storing the values of the score variables and the residuals of the PCA decomposition, respectively, $\boldsymbol{\Lambda}$ is a matrix storing the variances of the retained PCs and $\mathbf{z}^T(k) = \left(\mathbf{x}^{+T}(k)\ \mathbf{y}^T(k)\ \mathbf{x}^T(k)\ \mathbf{u}^T(k) \right)$. The contribution of the individual variables to both univariate statistics can be obtained as discussed in Miller et al. [14] and Russel et al. [16] for example.

3 Analysis of the Dynamic Monitoring Technique

This section analyses the dynamic monitoring technique discussed in the previous section. Treasure et al. [17] showed that this technique requires considerably fewer variables to be analysed compared to DPCA. This is a direct result of arranging highly correlated process variables to represent an autoregressive model structure that is decomposed using PCA. In contrast, the state space model determines a minimum set of state variables to capture the dynamic process behaviour and hence, represents this behaviour using a reduced variable set.

Ljung [10] argued that it *"it is preferable to work with state space models in the multivariate case, since the model structure complexity is easier to deal with"*. In addition, George et al. [5] highlighted that general multivariate AR structures may be difficult to handle, as the complexity of such structures grows rapidly with an increasing model order, i.e. an increasing number of lagged terms.

Despite the benefits of the dynamic state space monitoring technique in a condition monitoring context, the following question can be raised: *how does* the information encapsulated in $\mathbf{x}(k+1)$ contribute to the monitoring model? According to Equation (1), the estimates of the state and output sequences, $\widehat{\mathbf{X}}^+$ and $\widehat{\mathbf{Y}}$, are linearly dependent upon the state and input sequences, $\widehat{\mathbf{X}}$ and $\widehat{\mathbf{U}}$. In addition, the PCA decomposition of $\mathbf{Z}$ in Equation (4) produces a matrix $\mathbf{PT}$, which relates to the T^2 statistic and a matrix $\mathbf{E}$ that relates to the Q statistic. By partitioning $\mathbf{P}^T = \begin{bmatrix} \mathbf{P}_{X+}^T & \mathbf{P}_Y^T & \mathbf{X}_X^T & \mathbf{P}_U^T \end{bmatrix}$, the estimation of the input, output and state sequences are given by:

$$\widehat{\mathbf{X}}^+ = \mathbf{P}_{X+}\mathbf{T} \quad \widehat{\mathbf{Y}} = \mathbf{P}_Y\mathbf{T} \quad \widehat{\mathbf{X}} = \mathbf{P}_X\mathbf{T} \quad \widehat{\mathbf{U}} = \mathbf{P}_U\mathbf{T} \tag{6}$$

Comparing Equations (1) and (6) reveals that the row spaces of $\mathbf{T}$ and $\begin{bmatrix} \widehat{\mathbf{X}} \\ \widehat{\mathbf{U}} \end{bmatrix}$ are identical. This implies, however, that the row spaces of $\widehat{\mathbf{X}}^+$ and $\widehat{\mathbf{Y}}$ are both in the row space of $\mathbf{T}$. Hence, the state sequences $\widehat{\mathbf{X}}^+$ represent redundant information in a process monitoring context. Furthermore, the state sequences $\mathbf{X}$ and $\mathbf{X}^+$ are both dependent upon the block Hankel matrices $\mathbf{Y}_f$, $\mathbf{Y}_p$, $\mathbf{U}_f$ and $\mathbf{U}_p$ and are both computed from variations in the process input and output variables. More precisely, a simplified monitoring model can be established as follows:

$$\widehat{\mathbf{Y}} = \begin{bmatrix} \widehat{\mathbf{C}} & \widehat{\mathbf{D}} \end{bmatrix} \begin{bmatrix} \widehat{\mathbf{X}} \\ \widehat{\mathbf{U}} \end{bmatrix}. \tag{7}$$

Estimates of the state space matrices $\widehat{\mathbf{C}}$ and $\widehat{\mathbf{D}}$ are given by:

$$\mathbf{P}_Y\mathbf{T} = \begin{bmatrix} \widehat{\mathbf{C}}\mathbf{P}_X & \widehat{\mathbf{D}}\mathbf{P}_U \end{bmatrix}\mathbf{T} \Rightarrow \begin{bmatrix} \widehat{\mathbf{C}} & \widehat{\mathbf{D}} \end{bmatrix} = \mathbf{P}_Y \begin{bmatrix} \mathbf{P}_X \\ \mathbf{P}_U \end{bmatrix}^\dagger, \tag{8}$$

where $[\cdot]^\dagger$ represents a generalised inverse. The score sequences $\mathbf{T}$ are given by $\mathbf{T} = \begin{bmatrix} \mathbf{P}_Y^T & \mathbf{P}_X^T & \mathbf{P}_U^T \end{bmatrix} \begin{bmatrix} \mathbf{Y} \\ \mathbf{X} \\ \mathbf{U} \end{bmatrix}$. Using the above PCA decomposition, the T^2 and Q statistic can be computed as shown in Equation (5).

4 Industrial Application Study

This section presents an application studies of the improved dynamic monitoring technique to an industrial melter process. The process is introduced first. This is followed by detailing the identification of the dynamic monitoring model prior to the analysis of a fault condition.

4.1 Process Description

The melter process is part of a disposal procedure. Waste material is preprocessed by an evaporation treatment to produce a powder that is then clad by glass as part of the melter process. The melter vessel is continuously filled with powder and raw glass is discretely introduced in the form of glass frit. This binary composition is heated by four induction coils, which are positioned around the vessel. Because of the heating procedure, the glass becomes molten homogeneously. The process of filling and heating continues until the desired height of the liquid column is reached. Then, the molten mixture is poured out through an exit funnel. After the content of the vessel is emptied to the height of the nozzle, the next cycle of filling and heating is carried out.

Measurements of 8 temperatures, the power in the 4 induction coils and voltage were recorded every five minutes. The filling and emptying cycles represented a dynamic relationship between the temperatures (process output variables), power in the induction coils and voltage (process input variables). The variable set therefore comprised $n_M = 8$ output and $n_N = 5$ input variables, which are listed in Table 1. Two data sets formed the basis of the analysis. A reference data set containing $K = 1000$ samples and a second data set of 50 samples describing a crack in the melter vessel. Prior to the identification of the dynamic monitoring model, the variables of the reference data set were normalized.

Table 1. Process variables of the industrial melter process

VARIABLE	DESCRIPTION	FUNCTION
T1	Temperature measurement	output variable
T2	Temperature measurement	output variable
T3	Temperature measurement	output variable
T4	Temperature measurement	output variable
T5	Temperature measurement	output variable
T6	Temperature measurement	output variable
T7	Temperature measurement	output variable
T8	Temperature measurement	output variable
P1	Power measurement	input variable
P2	Power measurement	input variable
P3	Power measurement	input variable
P4	Power measurement	input variable
V	Voltage measurement	input variable

4.2 Establishing a Dynamic Monitoring Model

After constructing the block Hankel matrices $\mathbf{U}_f$, $\mathbf{U}_p$, $\mathbf{Y}_f$ and $\mathbf{Y}_p$ using the reference data, the regression matrices $\mathbf{R}_1$, $\mathbf{R}_2$ and $\mathbf{R}_3$ were identified (Equation (2)), followed by a singular value decomposition of $\widehat{\mathbf{Y}}_f$ (Equation (3)). This yielded 3 dominant singular values, as shown in Figure 1. The state variables were then scaled to unit variance and included in the dynamic data matrix $\mathbf{Z}^T = \begin{bmatrix} \mathbf{Y}^T & \mathbf{X}^T & \mathbf{U}^T \end{bmatrix}$. Figure 2 shows that a PCA decomposition of $\mathbf{Z}$ suggested the retention of 4 PCs (eigenvalues of the correlation matrix $\frac{1}{K-1}\mathbf{Z}\mathbf{Z}^T$. The application of Equation (8) involving the partitioned loading matrix produced the following state space matrices $\mathbf{C}$ and $\mathbf{D}$ detailed in Tables 2 and 3, respectively. The notation x_1, x_2 and x_3 refers to the three state variables.

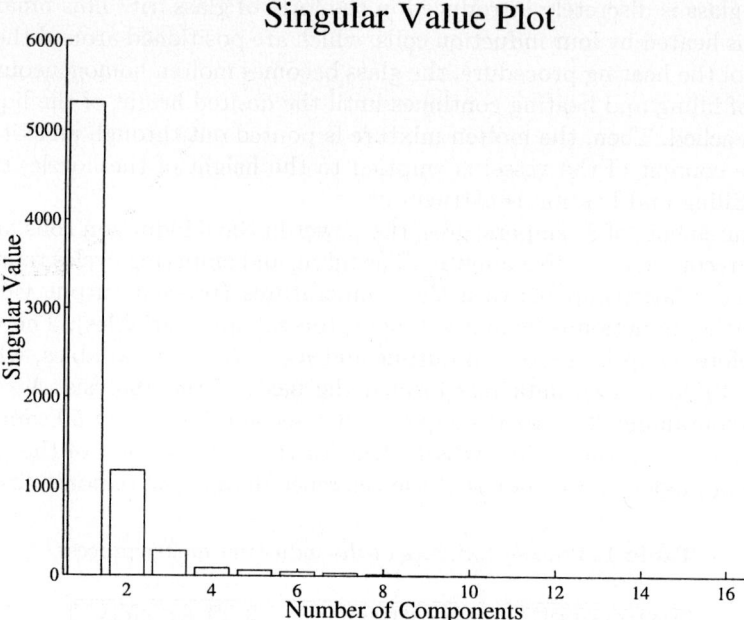

Fig. 1. Estimation of the number of state variables

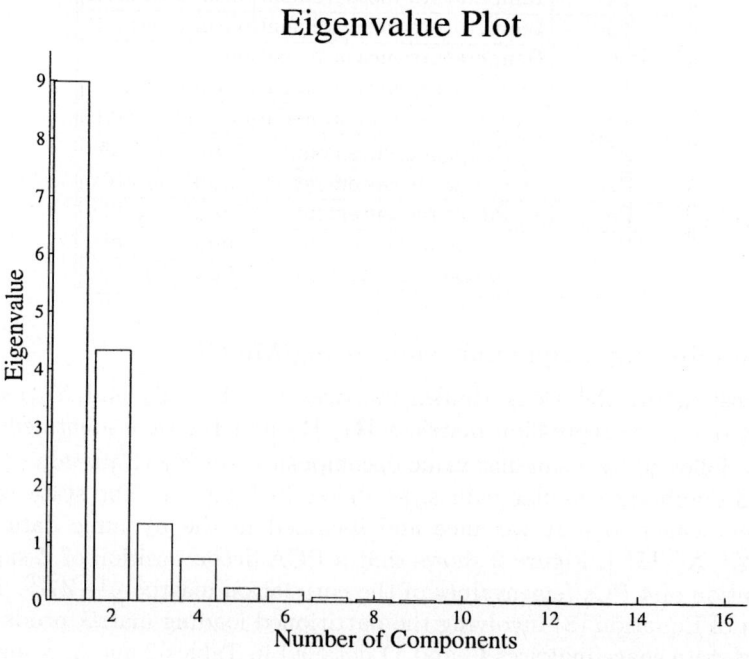

Fig. 2. Estimation of the number of retained PCs

Table 2. Elements of the state space matrix **C**

	x_1	x_2	x_3
T1	-0.6411	-0.1660	0.1187
T2	-0.5978	-0.1540	0.0692
T3	-0.1491	-0.1662	-0.0659
T4	-0.2671	-0.1887	-0.3293
T5	0.2023	-0.2890	0.0279
T6	0.0812	-0.1235	0.0424
T7	-0.6828	-0.0749	0.1783
T8	-0.0539	-0.0658	-0.0263

Table 3. Elements of the state space matrix **D**

	P_1	P_2	P_3	P_4	V
T1	-0.2507	0.0188	0.2448	0.2648	-0.0827
T2	-0.1759	0.0497	0.2591	0.2926	-0.0838
T3	0.1686	0.2009	0.2497	0.2448	0.0805
T4	-0.0671	0.1909	0.2632	0.2444	0.0115
T5	0.1099	0.2175	0.1439	-0.0260	0.3376
T6	0.3519	0.2069	0.1837	0.1733	0.1497
T7	-0.3103	-0.0769	0.1678	0.2264	-0.1701
T8	0.3554	0.1937	0.2157	0.2720	0.0363

4.3 Analysis of a Fault Condition

The melter vessel is made of graphite, which is a brittle material. As a consequence of the strong variations in temperature that the vessel is frequently exposed to, cracks along regions of high stress can occur. These cracks not only damage the shell of the melter, they also allow molten material to escape. It is therefore necessary to detect such cracks at the earliest possible time.

The identified dynamic monitoring model was applied to analyse a second data set that described the development of such a crack. The upper plot in Figure 3 shows the univariate monitoring statistics and their 99% confidence limits, computed as proposed in [6]. The middle and lower plot in Figure 3 details the model residuals of the state space model to the input and output variables, respectively, and highlights that variable T6, i.e. the 6th temperature sensor and the input variables P1 and V most significantly responded to this event. This made sense physically, as the crack developed in the vicinity of temperature sensor #6. The control system, in turn, responded to the increase in temperature by adjusting the power in one of the induction coils, i.e. P1 which is most dominantly influenced by the temperature sensors in the vicinity of the crack. As a consequence of the reduction in power, the voltage also dropped. At the end of the recorded data set, the temperature sensor T6 malfunctioned, which can be noticed by the sharp drop in the sensor reading and the sharp increase by P1 and V.

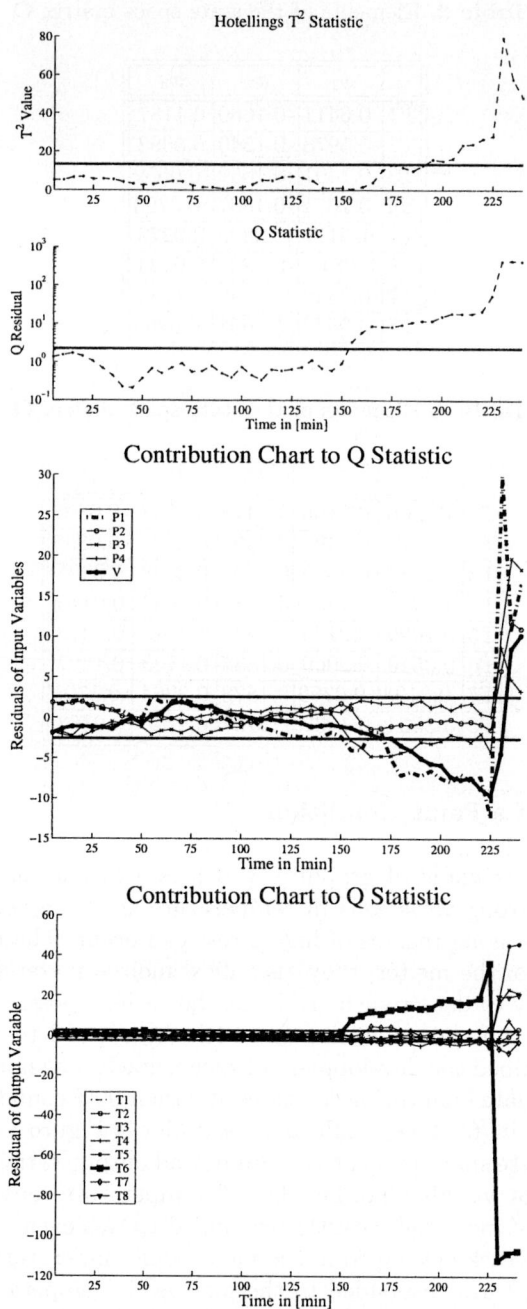

Fig. 3. Detection and diagnosis of the crack in the melter vessel. Upper plot → T^2 and Q monitoring statistics, middle and lower plot → contributions of the input and output variables to the Q statistic).

This application study therefore demonstrated that the improved dynamic monitoring technique is able to detect and correctly diagnose abnormal process behaviour, which went unnoticed by plant operators until the impact of the fault condition produced a very substantial increase in power consumed by induction coil 1 and voltage applied.

5 Conclusions

This article studied the recently proposed dynamic monitoring technique by Treasure et al. [17]. It was found that, although this technique can considerably reduce the number of variables to describe dynamic process behaviour, it can be reduced further. The state space model, obtained using subspace identification, does only require the input/output relationship in a condition monitoring context. The utility of the improved monitoring technique was demonstrated by analysing recorded data from an industrial melter process, where a crack in the melter vessel was detected and correctly diagnosed.

References

1. Chen,Q., Wynne,R., Goulding,P.R., Sandoz,D.J.: The Application of Principal Component Analysis and Kernel Density Estimation to Enhance Process Monitoring, Control Engineering Practice 8 (5) (2000) 531 - 543
2. Chen,J., Liu,K.: On-line Batch Process Monitoring using Dynamic PCA and Dynamic PLS Models, Chemical Engineering Science 57 (2002), 63 - 75
3. Cinar,A.,Undey, C.: Statistical process and controller performance monitoring: A tutorial on current methods and future direction, Proceedings of American Control Conference 4 (1999) 2625 - 2630
4. Callao,M.P., Rius,A.: Time Series: A Complementary Technique to Control Charts for Monitoring Analytical Systems, Chemometrics and Intelligent Laboratory Systems 66 (1) (2003) 79 - 87
5. George, B., Gwilym,M.J., Gregory,R.: Time series analysis: Forecasting & Control (3rd edition), Prentice Hall: Englewood Clifs, (1994)
6. Jackson,J.E.: Principal Components and Factor Analysis: Part I- Principal Components, Journal of Quality Technology 12 (4) (1980) 201 - 213
7. Kourti,T., MacGregor J.F.: Multivariate SPC Methods for Process and Product Management, Journal of Quality Technology 28 (1996) 409 - 428
8. Kruger,U., Chen, Q., Sandoz,D.J.,McFarlane,R.C.: Extended PLS Approach for Enhanced Condition Monitoring of Industrial Processes, AIChE journal 47 (9) (2001) 2076 - 2091
9. Ku,W., Storer,R.H., Georgakis,C.: Disturbance Detection and Isolation by Dynamic Principal Component Analysis, Chemometrics and Intelligent Laboratory Systems 30 (1995) 179 - 196
10. Ljung, L.: System Identification Toolbox - for Use with MATLAB, Mathwork: Natick, 2002
11. MacGregor,J.F., Marlin,T.E.,Kresta,J.,Skagerberg,B.: Multivariate Statistical Methods in Process Analysis and Control, AIChE Symposium of the 4th International Conference on Chemical Process Control, AIChE Publ. P-67, New York, (1991) 79 - 99

12. MacGregor,J.F., Kourti,T.: Statistical Process Control of Multivariate Processes, Control Engineering Practice 3 (3) (1995) 403 - 414
13. Martin,E.B., Morris,A.J.: An Overview of Multivariate Statistical Control in Continuous and Batch Process Performance Monitoring, Transactions of the Institite of Measurement and Control, 18 (1) (1996) 51 - 60
14. Miller,P.,Swanson,R.E.,Heckler,C.F.: Contribution Plots: A Missing Link in Multivariate Quality Control, Applied Mathematics and Computer Science 8 (4) (1998) 775 - 792
15. Negiz,A., Cinar,A.: Statistical Monitoring of Multivariate Dynamic Processes with State-Space Models, AIChE Journal 43 (1997) 2002 - 2012
16. Russel,E.L. ,Chiang,L.H., Braatz,R.D.: Data-Driven Techniques for Fault Detection and Diagnosis in Chemical Processes, Springer, London, (2000)
17. Treasure,R. , Kruger,U. , Cooper,J.E.: Dynamic Multivariate Statistical Process Control using Subspace Identification, Journal of Process Control, 14 (3) (2004) 279 - 292
18. Simoglou, A., Martin, E.B., Morris, A.J.: Multivariate Statistical Process Control of an Industrial Fluidised-Bed Reactor, Control Engineering Practice 8 (8) (2000) 893 - 909
19. Simoglou, A., Martin, E.B., Morris, A.J.: Statistical Performance Monitoring of Dynamic Multivariate Processes Using State Space Modeling, Computers and Chemical Engineering 26 (6) (2002) 909 - 920
20. Vedam,H., Venkatasubramanien,V.: PCA-SDG based Process Monitoring and Fault Detection, Control Engineering Practice 7 (7) (1999) 903 - 917
21. van Overschee,P., de Moor,B.: System Identification for the Identification of Combined Deterministic-Stochastic Systems, Automatica 30 (1) (1994) 75 - 93
22. van Overschee,P., de Moor,B.: Subspace Identification for Linear Systems, Kluwer Academic Publishers, Boston, (1996)
23. Wise,B.M., Gallagher,N.B.: The Process Chemometrics Approach to Process Monitoring and Fault Detection, Journal of Process Control 6 (6) (1996) 329 - 348

Associating Neural Networks with Partially Known Relationships for Nonlinear Regressions*

Bao-Gang Hu[1,2], Han-Bing Qu[1,2], and Yong Wang[1,2]

[1] NLPR, Institute of Automation
[2] Beijing Graduate School, Chinese Academy of Sciences,
P.O. Box 2728, Beijing, 100080 China
{hubg, hbqu, yongwang}@nlpr.ia.ac.cn

Abstract. In many regression applications, there exist common cases for users to know qualitatively, yet partially, about nonlinear relationships of physical systems. This paper presents a novel direction for constructing feedforward neural networks (**FNNs**) which are subject to the given nonlinear relationships. The *"Integrated models"*, associating FNNs with the given nonlinear functions, are proposed. Significant benefits will be obtained over the conventional FNNs by using these models. First, they add a certain degree of comprehensive power for nonlinear approximators. Second, they may provide better generalization capabilities. Two issues are discussed about the improved approximation and the estimation of the real parameters to the partially known function in the proposed models. Numerical studies are given in comparing with the conventional FNNs.

1 Introduction

The technique of neural networks has received much attention from every engineering field, since it has evolved into a general tool for various applications, such as modeling, control, pattern recognition, nonlinear regression, *etc* [1][2]. However, the models constructed from this technique suffer a difficulty in preserving the physical explanations to the real world problems. Usually, we consider neural networks to be a *"black-box"* approach. Due to this feature, neural networks have been greatly constrained for further developments.

We believe that one of the important directions for advancing neural networks lies on how to overcome the difficulty of *"non-transparency"* (also *"non-comprehensibility"* or *"non-interpretability"*) carried by the conventional neural networks. Up to now, several different paradigms of investigations towards this direction have been reported, namely:

I. Building prior information into neural network design [1][3].
II. Extracting rules embedded within neural networks [4][5].
III. Integrating neural network with fuzzy inferences [6] or knowledge bases [7].

---

* This work is supported in part by National Science Foundation of China (#60275025, #60121302) and Chinese 863 Program (#2002AA241221).

In this work, we propose a new scheme, falling into the first paradigm, but called as "*Integrating neural networks with partially known nonlinear functions*". This scheme is significantly different from the other reported schemes, say, using prior knowledge to set the connectivity of neural networks [1], or to initiate the hypothesis [8]. The importance behind this new scheme is initially justified by real world applications. Take plant growth modeling for an example. The yield of a plant is a nonlinear function in respect to many environmental variables, such as temperature, light intensity and distribution, water and fertilizer supplies, *etc*. Generally, partially known relationships to this nonlinear function can be obtained by the domain knowledge, say, the yield of a plant is proportional to $\sum (T - T_0)$, where T and T_0 are temperature and critical temperature, respectively. It is understandable that any model is better to incorporate and reflect domain knowledge as much as possible. However, some plant growth models based on neural networks fail to provide eco-physiological knowledge explicitly for mechanism explanations, although they may present reasonable predictions for the development and growth processes of plants.

Most existing neural networks act as completely-black boxes. Few investigations are reported for associating the partially known relationships in their designs. To simplify the discussion in this work, we focus our study on feed-forward neural networks (**FNNs**) for applications of only nonlinear regressions. This paper is organized as follows. Section 2 proposes a new way to formalize regression problem. A new scheme, given by so-called "*Integrated models*", is proposed in the design of FNN integrating with partially known nonlinear relationships in Section 3. Numerical examples are presented in Section 4. Finally, some remarks are given in Section 5.

2 A New Way to Formalize Regression Problem

Without loss of generality, we restrict the modeling case for a multiple-input-single-output (**MISI**) regression problem. In this case, it is supposed that

$$y = f(\mathbf{x}) . \qquad (1)$$

where y is the output of a nonlinear function with n-dimensional independent variables $\mathbf{x} = \{x_1, x_2, ..., x_n\}$; f is a function to be estimated. In the proposed scheme of handling regression problems, we assume that some relationships are partially known to the nonlinear function. Table 1 lists some examples for the partially known relationships in the applications of regression problems. With the partially known relationships, we formalize a regression problem by the following expression:

$$\begin{aligned}&min\|y - g(\mathbf{x}, \theta)\|\\ &\text{subject to} \quad g(\mathbf{x}, \theta) \propto \mathbf{R}_i(f), \quad i = 1, 2, \ldots\end{aligned} \qquad (2)$$

where $z = g(\mathbf{x}, \theta)$ is the approximate function with a parameter set θ; and $\mathbf{R}_i(f)$ is the ith relationship which is known or partially known of the given nonlinear function. This function is usually specified by users from their prior knowledge.

Table 1. Examples of partially known relationships about nonlinear functions in regression problems

Known Items/Aspects	Explanations	Examples
Constants or coefficients	A constant or coefficient is engaged but its nonlinear value is unknown in the function.	$(x - c)$, cx_1x_2
Components	A more complex form than a single variable or constant is known in a nonlinear function.	$x_1x_2x_4^3$, $exp[(ax_1 - c)/b]$
Superposition	The components are known for the superposition in the nonlinear function.	$x_1x_2 + sin(x_2 - c)$
Multiplication	A function is proportional or inversely proportional to some variables or components.	$log(x_1) * sin(x_2 - c)$, $x_2/log(x_1 + c)$
Boundary Conditions	A function is restricted within boundaries on its variables	$(x_i)_{min} < x_i < (x_i)_{max}$
Equality Conditions	A function is known on some key points.	$f(\mathbf{x}_0) = y_0$
Inequality Conditions	A function is constrained with inequality conditions.	$y_{min} < f(\mathbf{x}) < y_{max}$
Derivatives and Integrals	A function is restricted with its specified derivatives or integrals.	$f'(x) > 0$, $\int_a^b f(x)dx = 1$

The symbol "∝" represents as "*compatible to*". Different with the conventional neural networks or other nonlinear regression tools, eq. (2) describes that neural networks should not only approximate the desired signal y in a minimum sense, but also satisfy the known relationships of nonlinear functions in a certain degree of accuracy. Up to now, it seems few studies have been reported on the problems defined by eq. (2), particularly for neural networks.

3 Design of Neural Networks for Partially Known Relationships

In fact, there exist an infinite number of cases in representing the partially known relationships, or $R_i(f)$, from real world problems. Theoretically, the design of neural networks for partially known relationships is more problem dependent. Fig. 1 shows a schematic diagram of integrating feedforward neural networks with partially known nonlinear relationships. This model is called as "*Integrated model*". While the upper part in the integrated model in Fig. 1 represents $\mathbf{R}(f)(= \{R_1(f), R_2(f), \dots \})$ for the known relationships, the lower part will be the typical neural networks. A coupling exists between two parts for the reason of performing eq. (2). However, it seems that a generic or uniform approach for the integration, or the coupling, does not exist.

It is still an open problem in exploring heuristic guidelines for associating neural networks with partially known nonlinear relationships. In this work, we only discuss two simple cases (or *coupling operators*), i.e., "superposition" and "multiplication". The mathematical expressions for these two cases are:

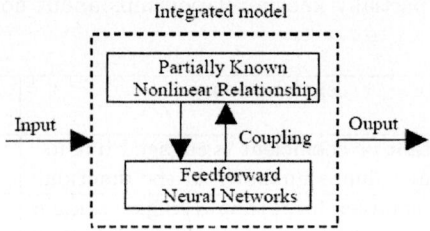

Fig. 1. Schematic diagram of integrating feedforward neural networks with partially known nonlinear relationships

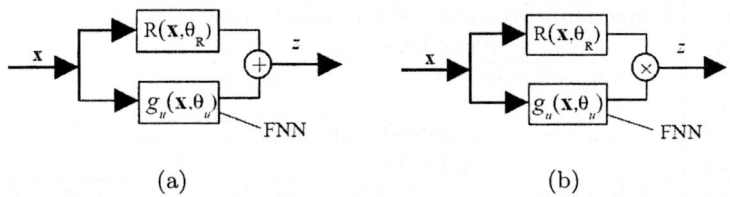

Fig. 2. Integrated models of feedforward neural networks with two cases of coupling: (a) for superposition, (b) for multiplication

$$\text{For superpositon:} \quad f(\mathbf{x}) = f_u(\mathbf{x}) + \mathbf{R}(\mathbf{x}) \ . \tag{3}$$

$$\text{For multiplication:} \quad f(\mathbf{x}) = f_u(\mathbf{x}) * \mathbf{R}(\mathbf{x}) \ . \tag{4}$$

where $f_u(\mathbf{x})$ is an unknown function and $\mathbf{R}(\mathbf{x})$ is a function which is known *fully* or *partially*. The partially known $\mathbf{R}(\mathbf{x})$ means that some parameter(s) for the function is not given. The integrated models of the FNNs for "superposition" and "multiplication" are proposed in Fig. 2, respectively, based on institutions. While $g_u(\mathbf{x}, \theta_u)$ models $f_u(\mathbf{x})$ by a conventional FNN, $\mathbf{R}(\mathbf{x}, \theta_R)$ is a mathematical expression. The parameter set will include two subsets:

$$\theta = \{\theta_u, \theta_R\} \ . \tag{5}$$

where $f_u(\mathbf{x})$ and $\mathbf{R}(\mathbf{x})$ are parameter sets for $f_u(\mathbf{x})$ and $\mathbf{R}(\mathbf{x})$, respectively. If $\mathbf{R}$ is fully, or exactly, known, θ_R will be an empty set. If $\mathbf{R}$ is partially known, θ_R can be obtained in a similar approach for the conventional FNN. However, two questions will arise about the integrated models:

I. *"Can the integrated models guarantee to provide a better regression performance than the conventional FNNs"*?
II. *"Can the integrated models estimate θ_R like a conventional approach for parameter identification"*?

The first question is important since one will always expect a better approximation from using prior knowledge. However, from a theoretical analysis, one

Table 2. Conditions for guaranteed better performance of the integrated models over the conventional FNNs in two cases

Case	Conditions
Superposition	$\|e_U\| < \|e_C\|$ and $\mathbf{R}(\mathbf{x})$ is exactly known
Multiplication	$\|e_U\| < \|e_C\|$, $\mathbf{R}(\mathbf{x})$ is exactly known, and $\|\mathbf{R}(\mathbf{x})\| \leq 1$

will fail to arrive at a generally "yes" answer to the first question. Therefore, it seems necessary to know under which conditions one can receive a better performance from using the integrated models. Assuming $\mathbf{R}(\mathbf{x})$ is fully known, the approximation errors of the integrated models are given by:

For superposition:
$$\| e_S \| = \| \mathbf{R}(\mathbf{x}) \| + \| f_u(\mathbf{x}) - g_u(\mathbf{x}, \theta_u) \| \ . \tag{6}$$

For multiplication:
$$\| e_M \| = \| \mathbf{R}(\mathbf{x}) \| * \| f_u(\mathbf{x}) - g_u(\mathbf{x}, \theta_u) \| \ . \tag{7}$$

and the error for a conventional model of using FNNs is

$$\| e_C \| = \| f(\mathbf{x}) - g(\mathbf{x}, \theta) \| \ . \tag{8}$$

Suppose that
$$\| e_U \| = \| f_u(\mathbf{x}) - g_u(\mathbf{x}, \theta_u) \| \ . \tag{9}$$

to be the approximation error for $f_u(\mathbf{x})$ using the FNNs in the integrated models. We can derive the conditions for "yes" answer to the first question in Table 2.

In general, due to $f_u(\mathbf{x})$ is simpler than $f(\mathbf{x})$ as a nonlinearity expression, one is mostly possible to achieve the condition of $\| e_U \| < \| e_C \|$ by using the same number of free parameters. Comparing with the superposition case, one can see that the multiplication case requires one more condition on $\| \mathbf{R}(\mathbf{x}) \| \leq 1$.

The second question is related to the parameter identification when $\mathbf{R}(\mathbf{x})$ is partially known. It is also of importance that one can recover the full function $\mathbf{R}(\mathbf{x})$, in which the parameter set θ_u usually provides a physical meaning to the problem investigated. However, due to the coupling effect, one may fail to obtain the reasonable estimation for the real parameter(s). We present the following definition and theorem.

Definition 1. *Suppose $\theta_j \in \theta_R$ and $\theta_i \in \theta_u$ are two parameter sets for the nonlinear function $z = g(\mathbf{x}, \theta_R, \theta_u), \mathbf{x} \in \Re^n, z \in \Re$. If one has the following relation:*

$$z_1 = g(\mathbf{x}, \theta_i^1, \theta_j^1) \neq z_2 = g(\mathbf{x}, \theta_i^2, \theta_j^2) \ . \tag{10-a}$$

and $\theta_j^1 \in \theta_R, \theta_j^2 \in \theta_R, \theta_i^1 \in \theta_u, \theta_i^2 \in \theta_u, (\theta_j^1 \neq \theta_j^2)$ *or* $(\theta_i^1 \neq \theta_i^2)$. (10-b)

the two parameter sets are defined to be independent. If one has $z_1 = z_2$ while the relation (10-b) remains, θ_j and θ_i are then to be dependent parameters.

Theorem 1. For the integrated model, which approximates a nonlinear function in a form of $z = g(\mathbf{x}, \theta_R, \theta_u)$ by using two parameter sets, $\theta_j \in \theta_R$, and $\theta_i \in \theta_u$, where θ_R is the parameter set for the partially known relationship, if θ_j and θ_i are dependent, the real parameter θ_j will be unable to be estimated from the integrated model.

Proof. For simplicity, we only derive the multiplication case of the integrated model. The same conclusions can be extended to the other cases. Since θ_j and θ_i are dependent, one can obtain the following relation:

$$z = \mathbf{R}(\mathbf{x}, \theta_i^1) * g_u(\mathbf{x}, \theta_j^1) = \mathbf{R}(\mathbf{x}, \theta_i^2) * g_u(\mathbf{x}, \theta_j^2) \ . \tag{11-a}$$

and $\quad \theta_j^1 \in \theta_R, \theta_j^2 \in \theta_R, \theta_i^1 \in \theta_u, \theta_i^2 \in \theta_u, (\theta_j^1 \neq \theta_j^2) \quad \text{or} \quad (\theta_i^1 \neq \theta_i^2) \ . \tag{11-b}$

Since this dependency generate more than one group of θ_j and θ_i for the same output data (see eq. 11-a) even for the relation (eq. 11-b), one is unable to receive a unique solution of the estimation to the real parameter of the partially known relationship from the integrated model. □

4 Numerical Studies

In this section, numerical examples are demonstrated. The objective of the numerical studies is to understand the proposed integrated models in comparing with the conventional FNNs. The FNNs in the integrated models in Fig. 2 are the Gaussian-type RBF approximators, in the following expression:

$$g_u(x, \theta_i) = \sum_{i=1}^{n} w_i e^{\frac{-(x-c_i)^2}{\delta_i^2}} + d \ . \tag{12}$$

where $\theta_u (= \{\mathbf{w}, \mathbf{c}, \delta, d\})$ is a free parameter set for the unknown function. Since the superposition case is similar to multiplication case, only the multiplication case is investigated in this work.

Example 1. About estimation of real parameters.

In this example, the nonlinear function investigated is:

$$y(t) = 1 + A\sin(\omega t - \phi)e^{-(t/5)}, \qquad A = 4, \omega = 5, \phi = 1 \ . \tag{13}$$

For this function, we suppose that only a sinusoidal function is known and is associated to the whole function in a "multiplication" format. An integrated model shown in Fig. 2-b is employed, where $\theta_R (= \{A, \omega, \phi\})$ is the free parameter set for the partially known function $R(t) = A\sin(\omega t - \phi)$; and two RBFs are used in eq. (12). Running on the data ($t \in [0, 4], \Delta t = 0.1$), we obtain the following results:

$$\text{MSE} = \| e_M \| / N_{data} = 0.0222, \quad \theta_R = \{0.976, 5.05, 4.20\}.$$

Table 3. Comparison data between the conventional RBF model and the integrated model ($R(x)=e^{-\alpha x}$) for the regression (Training: $x \in [0,10]$, Testing: $x \in [10,40]$) ($N_{population}=400$, $N_{generation}=200$, $N_{smaple}=10$)

	Conventional RBF model (10 Free Parameters)		Integrated model (8 Free Parameters)	
	MSE (Training)	MSE(Testing)	MSE(Training)	MSE(Testing)
Best	0.00264	0.00916	0.00413	0.000694
Mean	0.00524	0.498	0.00734	0.0813
Std.	0.00207	0.719	0.00225	0.217

A good approximation is obtained from the integrated model. However, some parameters, A and ϕ, are failed to be estimated correctly. Examining $g_u(t)$ and $R(t)$, we can find out that A and ϕ are dependent on **w** and **c**, respectively. They exhibit two effects from the dependency, namely, *"amplitude"* and *"shifting"*. On the other hand, no dependency effect exists to the parameter ω. Therefore, one is able to find a reasonably good estimation of this parameter.

Example 2. About generalization capability.

The example investigated below is for examining the generalization capability of the integrated models.

$$f(x) = (1 - x + 2x^2 - 0.4x^3)e^{-0.5x}, \quad x \in [0, 40], \triangle x = 0.2 \ . \tag{14}$$

Suppose that only $R(x) = e^{-\alpha x}$ is known. An FNN with two RBFs are tested to approximate the polynomial part. For a comparison, a convectional model, using three RBFs, is constructed. Table 3 shows the simulation results of the two models. Since the genetic algorithm tool applied produces nearly global optimization results, statistical data are used for reaching fair conclusions. This example shows that the integrated model produces a better generalization than the conventional model to the testing data. However, we failed to obtain the real estimation of parameter α. Sometimes, it is not straightforward to examine the dependency between the parameter sets.

Example 3. About singularity processing in the integrated model.

In this example, we will demonstrate that the integrated model may suffer from a singularity problem in its training or testing process, while its corresponding neural network in the conventional approach does not exhibit such phenomenon. A two-input-one-output *"sinc"* function is chosen to be a target system for examining the singularity problem in the new modeling strategy. The function is given in the following form:

$$f(x,y) = \frac{\sin(\sqrt{x^2 + y^2})}{\sqrt{x^2 + y^2}} \ . \tag{15}$$

For the given example, we assume that the numerator in the function is fully known, *i.e.*,

$$R(x,y) = \sin(\sqrt{x^2 + y^2}) \ . \tag{16}$$

Therefore, the neural network modular within the integrated model will approximate an unknown function for

$$g_u(x, y, \theta_u) \approx \frac{1}{\sqrt{x^2 + y^2}} \ . \qquad (17)$$

The training data are selected evenly within $x, y \in [-10, 10]$, $\triangle x = \triangle y = 1.0$. The total number of the training samples is 441. Fig. 3a and 3b depict the surface of the exact "*sinc*" function and the reconstructed surface using the integrated model, respectively. It is observed that the significant discrepancies exist at the origin point, *i.e.*, $z(0, 0) = 0$. However, this phenomenon does not appear to the conventional neural networks (Fig. 3c). Examining eq. (15), one can obtain a *removable singularity* when $x = y = 0$. At this singular point, the function fails to be analytic. Because of the existence of its limit, this singularity is removable. The conventional neural networks can handle removable singularity problems without any difficulty. However, the integrated model may suffer from a difficulty of singularity for a reasonable estimation at singular point(s). Considering the present example in eq. (15), when $x = y = 0$, one obtains $R = 1.0$ at the singular point. The sampling data at this point can be properly learnt by the conventional FNNs. On the contrary, the integrated model may present calculations of the function limits, like $0/0$, ∞/∞ or $0 * \infty$, at the singular point in the training or testing process. When this case occurs, either the model fails for processing directly or an error is obtained. In the example, the model shows a big error when $x = y = 0$ since g_u in eq. (17) produces a non-zero value, but zero for R in eq. (16). The output of the integrated model at the singular point is quite distinguishable from the present example. First, it outputs a zero value for z. Second, it exhibits a sudden change to disturb the smoothness of the function.

Usually, one has no information about the singularity property to the function investigated. For overcoming the unknown singularity problem, we propose the following procedures to the application of the integrated model:

Step 1. *Searching for the existence of removable singularity.*

If an integrated model produces an infinitive output at a specific input point, or exhibits a sudden change on the output, one can hypothesize the corresponding point to be a singular point (or called hypothesis point).

Step 2. *Removing or biasing the hypothesis point in the second-running simulation.*

Either in training or testing process, one can re-run the simulation of input data without including the hypothesis point. If possible, it is better to bias the hypothesis point and add the new observation point into the data set.

Step 3. *Confirming the singular point and estimating the output at the singular point.*

After the second-running simulation, one obtains a scalar error (MSE) which is used for the hypothesis testing. If this error is smaller than that of the first-running simulation, one will confirm the hypoth-

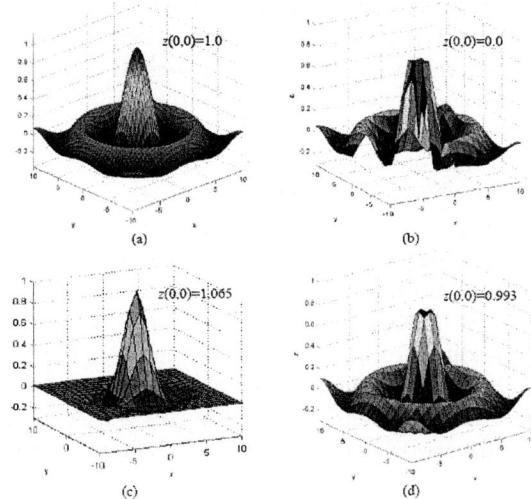

Fig. 3. Surface plots of "*sinc*" function. (a) Exact surface plot. (b) Reconstructed surface from the integrated model (including the singular point for training, MSE=0.00239). (c) Reconstructed surface from the conventional RBFs (MSE=0.0197). (d) Reconstructed surface from the integrated model (removing the singular point for training, MSE=0.000226).

esis is true. The output at the singular point can be estimated by an interpolation approach from the data around the singular point. Otherwise, it is a false hypothesis.

In the present example, we confirm the existence of one removable singularity by applying the procedures above. Fig. 3d shows reconstructed surface of the integrated model, in which we remove the observation data corresponding to the singular point ($x = y = 0$). The scalar error for the data set with 440 samples is MSE=0.000226, which is significantly lower than the model including the singular point (MSE=0.00239). The estimation output at the singular point by using the parabolic interpolation in Fig. 3d is $z(0,0) = 0.993$. Note that each integrated model in Fig 3 use two RBFs, and nine free parameters, but five RBFs for the conventional neural networks.

It seems that the difficulty of singularity introduced from the integrated model is the weakness of the model. However, we consider it as an added value for neural network technique. Singularity is an important property of physical systems. Usually, it is better to acquire such information from numerical models. The conventional neural networks fails to obtain singularity information directly. The proposed integrated model may reveal the existence of removable singularity.

5 Final Remarks

Most scientific and engineering applications require comprehensive power on either analytical models or numerical models. Falling into the category of numerical models, neural networks are also desirable to incorporate and exhibit domain knowledge explicitly. In this work, we explore neural networks by associating partially known relationships. Among several cases of partially known relationships, we investigate two cases of "superposition" and "multiplication". The integrated models are proposed. Two basic questions are discussed. We derive the conditions of guaranteed better approximations of the integrated models over the conventional FNNs, and one theorem about the estimation of real parameters in the partially known functions. Numerical studies demonstrate that while the integrated models present a certain degree of comprehensive power, they do not guarantee to produce better performance.

The natural world is neither *"totally white"* nor *"totally black"*. Therefore, any numerical model should not be built to be a *"completely black box"*. However, the most existing models using neural network technique are falling into this situation. We believe that the issues proposed in this work exhibit a new direction for the study of neural networks, even for other machine learning tools.

References

1. Haykin, S.: Neural Networks, A Comprehensive Foundation, (1999, 2nd ed), Printice Hall, New York
2. Duda, R.O., Hart, P.E. and Stork, D., Pattern Classification (2001, 2nd eds), John Willy, New York
3. Andrews, R. and Geva, S.: On the Effects of Initializing a Neural Network with Prior Knowledge, Proceedings of the International Conference on Neural Information Processing, Perth, Western Australia (August 1999) 251-256
4. Fu, L.M.: Rule Generation from Neural Networks, IEEE Trans. Sys. Man Cyber. 24 (1994) 1114-1124
5. Tickle A., Andrews R., Golea M. and Diederich J.: The Truth will Come to Light: Directions and Hallenges in Extracting the Knowledge Embedded within Trained Artificial Neural Networks, IEEE Trans. Neural Networks. 9(1998) 1057-1068
6. Jang, J.-S. R.: ANFIS: Adaptive-network-based Fuzzy Inference System, IEEE Transactions on Systems, Man and Cybernetics. 23 (1993) 665-685
7. Towell, G. and Shavlik, J.: Knowledge-based Artificial Neural Networks, Artificial Intelligence. 70 (1994) 119-165
8. Yu, T., Jan, T., Debenham, J. and Simoff, S.: Incorporating Prior Domain Knowledge in Machine Learning: A Review, 2004 International Conference on Advances in Intelligent Systems, Luxembourg (November 2004)

SVM Regression and Its Application to Image Compression[1]

Runhai Jiao, Yuancheng Li, Qingyuan Wang, and Bo Li

Digital Media Laboratory, School of Computer Science and Engineering,
Beihang University, Beijing, 100083, China
`runhaijiao@sina.com, dflyc@163.com`
`wangqy2008@sohu.com, boli@buaa.edu.cn`

Abstract. This paper proposes a new image compression algorithm which combines SVM regression with wavelet transform. Compression is achieved by using SVM regression to approximate wavelet coefficients. Based on the characteristic of wavelet decomposition, the coefficient correlation in wavelet domain is analyzed. According to the correlation characteristic at different scales and orientations, three kinds of arranging methods of wavelet coefficients are designed, which make SVM compress the coefficients more efficiently. Moreover, an effective entropy coder based on run-length and arithmetic coding is used to encode the support vectors and weights. Experimental results show that the compression performance of the algorithm achieve much improvement.

1 Introduction

Neural networks have been used in image compression for many years. Namphol [1] employed a multilayer hierarchical neural network using nested training algorithm to obtain image compression, which can exploit the pixel correlation of the whole image but need to define the network topology before training. Some authors [2, 3] applied Kohonen's self-organized feature map to generating the codebook in vector quantization and achieved better performance than that of LBG method. However, the performances of these algorithms are greatly depending on the training images, and compression time may be long if the codebook is large. Recently, Robinson [4] proposes a novel image compression method by combining Support Vector Machine (SVM) with DCT. It avoids defining the topology, which is a common drawback of existing image compression algorithm based on neural network. This method achieves better compression results comparing with JPEG. However the reconstructed image has blockartifact, especially at higher compression ratios. The reason is that the method is based on DCT, and image is compressed block by block. The blockartifact phenomenon of Robinson's method is not easily removed because of the characteristic of its compression procedure.

---

[1] Supported by the National Defense Basic Research Fund and the Program for New Century Excellent Talents in University. The research is made in the State Key Lab of Software Development Environment.

Compared with DCT, wavelet transform has more advantages. First, it compacts most of the image energy into a few wavelet coefficients. Second, it has localization in both space and frequency domains. Based on these facts, we propose a new compression scheme by applying SVM to compress wavelet coefficients. In this scheme, image is first decomposed by wavelet transform, and then wavelet coefficients are quantized and approximated by SVM. Finally, adaptive arithmetic coding is used to encode model parameters of SVM.

The remainder of this paper is organized as follows: Section 2 gives an overview of SVM regression and the basic concept of its application to image compression. Section 3 analyzes the characteristic of wavelet coefficients, and then illustrates the procedure of using SVM to compress wavelet coefficients. In section 4, implementation of the compression algorithm is described in details. Section 5 gives the experimental results. In section 6, conclusion is drawn and direction for future works is also discussed.

2 SVM Regression for Image Compression

SVM, as a new machine learning method, has been widely used in many areas because of its good generalization ability. It is designed to solve pattern recognition problem, which is to find a decision rule with good generalization. Regression is an extension use of classification, which is a non-separable classification that each data point can be thought of being as its own class [5]. It can learn data dependency between the input and the output, which achieves compression.

2.1 SVM Regression

In regression, given a set of training points, the real function is approximated within a predefined error ε by choosing the minimum number of training points. For each training point chosen by SVM, which is termed as Support Vector, there is a corresponding weight. Usually, the number of the support vectors is less than that of the training points, which is the essence that SVM can accomplish data compression. The regression problem can be formulated as follow:

$$f(x,w) = \sum_{i=1}^{N} w_i \phi_i(x). \tag{1}$$

SVM attempts to learn the input-output relationship (or function) from the given training points, $(x_1, y_1), (x_2, y_2), ..., (x_l, y_l)$, where $x_i \in R^n$ and $y_i \in R$. The function can be modeled by formula (1), where N represents the number of support vectors, w_i are the weights to be found, and $\phi_i(x)$ are the kernel functions. In the case of regression, we use Vapnik's linear loss function [4, 6] with ε insensitivity zone as a measure of the error between $f(x)$ and y:

$$error = |f(x,w) - y| = \begin{cases} 0 & if \ |y - f(x,w)| \le \varepsilon \\ |y - f(x,w)| - \varepsilon & if \ |y - f(x,w)| > \varepsilon \end{cases}. \tag{2}$$

Thus, the loss is equal to zero if the difference between the predicted $f(x,w)$ and y is less than ε. Vapnik's insensitivity loss function defines an ε tube such that if the predicted value is within the tube the loss is zero, otherwise the error equals the magnitude of the difference between the predicted value and the radius ε of the tube. From the viewpoint of data compression, the larger the ε is, the smaller number of support vectors is, and then a higher compression ratio can be achieved.

2.2 Discrete Wavelet Transform

Wavelet transform [7] can be applied to any signal with finite energy (i.e. any square-integrable function). Original signal $f(t)$ can be decomposed by using wavelets as the basis function

$$c_{a,b}(f) = \int_{-\infty}^{+\infty} \psi_{a,b}(t) f(t) dt . \qquad (3)$$

Where $\psi_{a,b}(t)$ are the wavelet functions, which are obtained through dilation and translation of the mother wavelet function $\psi(t)$

$$\psi_{a,b}(t) = |a|^{1/2} \psi(at - b) . \qquad (4)$$

Where a is the dilation factor, and b is the translation factor. Generally, we choose a and b as integers, then $\psi_{a,b}(t)$ can be represented as

$$\psi_{j,k}(t) = 2^{j/2} \psi(2^j t - k) \qquad j,k \in Z . \qquad (5)$$

Based on Multi-resolution Analysis (MRA), there must exist a scaling function $\varphi(t)$ corresponding with the wavelet function $\psi(t)$, and they satisfy

$$\begin{cases} \varphi(t) = \sum h_n \varphi(2t - n) \\ \psi(t) = \sum g_n \varphi(2t - n) \end{cases} . \qquad (6)$$

Then $\varphi_{j,k}(t)$ and $\psi_{j,k}(t)$ are basis functions spanning subspace V_j and W_j respectively, and the direct sum of V_j and W_j gives the subspace V_{j-1}, $\{V_j\}$ forms a set of approximation space of $L^2(R)$. Therefore, any $f(t)$ which belong to $L^2(R)$ can be represented by wavelets basis functions. In [8], S. Mallat proposed a fast decomposition algorithm for Discrete Wavelet Transform (DWT) called Mallat algorithm where the wavelet transform can be considered as a low-pass filter and a bandpass filter.

In image processing, 2-D decomposition algorithm can be easily extended from that of 1-D decomposition if the wavelet transform is separable. When performing multilevel wavelet transform on image, it can be considered as a subband filter. After a one-scale wavelet transform we can decompose an image into four finer scale subbands, labeled with LL_1, HL_1, LH_1 and HH_1. Continuous decomposing the lowest frequency subband, LL_k, of each scale, we get four coarser scale subbands, LL_{k+1}, HL_{k+1}, LH_{k+1} and HH_{k+1}. Fig. 1 is the result of a 5-scale transform, which is called tree-decomposition mode. Except LL_5, all other subbands correspond to three orientations, HL, LH and HH, respectively. Thus all HL_k subbands ($1 \leq k \leq 5$) have an identical direction HL, so do all LH_k or HH_k subbands.

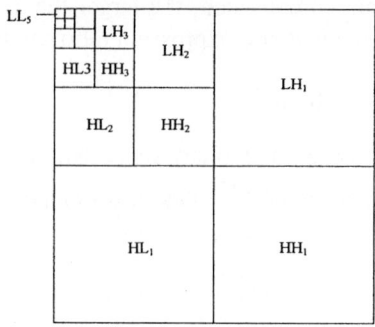

Fig. 1. 5-level discrete wavelet decomposition

The advantages of the wavelet transform over other transforms for image coding stem mainly from the fact that it performs a multiresolution analysis of the image. This multiresolution analysis is very similar to the way in which the human visual system interprets images, and as such is a natural and efficient way to code and store two-dimensional data. Although wavelet transform can remove most redundancy in space domain, there is still redundancy between wavelet coefficients, mainly for two reasons: first, in image compression the widely used wavelet filters are biorthogonal. That means, the scale function $\varphi(t)$ and the wavelet function $\psi(t)$ are not orthogonal, which result in redundancy between low frequency and high frequency; second, wavelet transform has both space and frequency characteristic, adjacent wavelet coefficients in one subband are the convolution results of filters and the adjacent image blocks. Usually, they have similar magnitudes, and are considered to be dependant statistically. In order to improve the compression performance, it is a key issue to remove the redundancy between wavelet coefficients. In this paper, we apply SVM regression to removing these redundancies and achieve compression. In SVM regression model, the inputs to the model are the positions of wavelet coefficients, and the desired outputs are the values of the coefficient. After training, the SVM model parameters are encoded by arithmetic coders.

3 Using SVM to Compress Wavelet Coefficients

After wavelet transform, image is decomposed into a series of subbands from low frequency to high frequency. Each wavelet subband is the representation of the image in different frequency range. Most of the image energy is compacted into the low frequency subbands. According to human visual system model, low frequencies are more sensitive to human eyes, and they play a great important role in image reconstruction. High frequencies describe the image details, which are less sensitive to human eyes. In order to keep the most import information with the given bit rate, we adopt different compression methods for different subbands. In our algorithm, the lowest subband is encoded by DPCM, which is nearly lossless. The finer scale subbands are compressed by SVM, which approximates the wavelet coefficients by using fewer support vectors. Moreover, some of the finer scale subbands are discarded directly because they only contain a little amount of energy and have neglectable effect on the image quality. Below, we will give a detailed description of our compression scheme.

3.1 Wavelet Coefficients Arrangement

From theoretical analysis and statistical experiments, there still exists correlation in wavelet coefficients, which can be utilized by SVM to improve compression efficiency. Generally, there are three kinds of correlation between wavelet coefficients [9, 10]: neighborhood correlation, parent-child correlation and siblinghood correlation. In our algorithm we just utilize the neighborhood correlation because it is the most important one, and the others are not remarkable. In the tree-decomposition mode, the neighborhood correlation has different characteristic at different scales and different orientations. Therefore, the arranging method of wavelet coefficients is a key technique to be solved before SVM regression.

1. Wavelet coefficients are not restricted as nonnegative. They can be defined by both magnitudes and signs. Commonly, it is assumed that no correlation exists between signs. Therefore, the signs and the magnitudes should be separated and compressed using different methods. In our algorithm the signs are directly encoded by entropy coder, and the magnitudes are compressed by SVM.

2. According to the spatial-frequency characteristic of wavelet transform, the magnitudes of the coefficients vary greatly in the whole wavelet domain, which is harmful to regression. However, the magnitudes of adjacent coefficients in one subband are usually similar in values. Based on these facts, wavelet coefficients of each subband are divided into a number of non-overlapping blocks. The coefficients of each block are mapped to produce a one dimension vector, which is compressed by SVM. The block size can be variable and 8×8 is chosen in this paper.

3. As shown in Figure 1, 2-D wavelet decomposition has three orientations, and neighborhood correlation is different at different orientation. In LH, subband gives vertical high frequency of the image, so the correlation between vertical coefficients is stronger than that of horizontal coefficients. That is to say, the vertical adjacent coefficients are more possibly similar in magnitude than that of the horizontal adjacent coefficients. Vice versa, in HL, subband gives horizontal high frequency of the image, so the correlation between horizontal coefficients is stronger than that of vertical coefficients. In HH, the correlation is not remarkable. In SVM regression, we should arrange wavelet coefficients with similar magnitude continuously as possibly as we can. This makes SVM learn data dependency more efficiently, which will cut down the number of the support vectors. According to the neighborhood correlation characteristic, we design three kinds of scan orders when the coefficient block is mapped to produce one dimension vector. They are illustrated in figure 2.

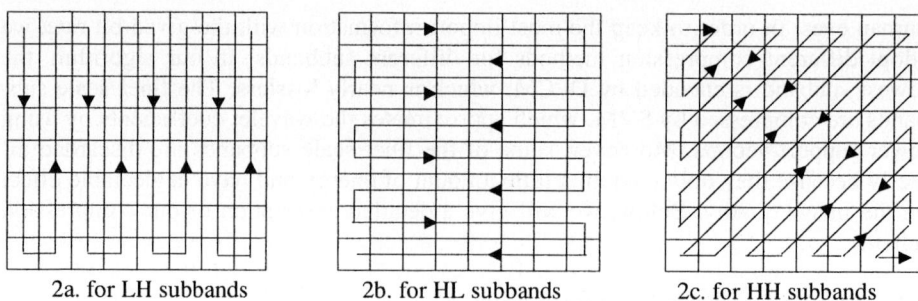

2a. for LH subbands 2b. for HL subbands 2c. for HH subbands

Fig. 2. Three scan orders for mapping coefficient block to produce one dimension vector

3.2 Wavelet Coefficients Normalization

Normalization is an important step in neural network application [11]. In SVM regression, normalizing wavelet coefficients will produce weights that are lower in magnitude and similar in value, which make the weights more compressible. Because the magnitudes of coefficients in the whole wavelet domain vary greatly, it is better to normalize wavelet coefficients of each subband respectively. In this paper, we choose formula (7) for normalization

$$c' = \frac{c - c_{min}}{c_{max} - c_{min}}. \qquad (7)$$

where c_{min} and c_{max} are the minimal and the maximal wavelet coefficients in the subband respectively, c is the coefficient to be normalized and c' is the value after normalization.

3.3 Compression of Wavelet Coefficients

Using the proper scan order, each coefficient block is mapped into a 64-dimension vector, called Y for convenient. The positions of the elements in Y form the vector X, which is also a 64-dimension vector. In SVM regression model, X and Y are the input data and the desired output data respectively. In training process, besides the ε, there are still two parameters in the model that affect the compression efficiency: kernel type and kernel parameter. There are usually three kinds of kernel types can be selected in SVM model: linear, polynomial and Gaussian function. Different kernels suit different types of data. Referring [12], the PDF of the coefficients in one block is approximately considered as Gaussian distribution, so we choose Gaussian function as the regression kernel and experimental results verify its efficiency. Generally, there are no guidelines for setting the value of the kernel parameter, and the optimal value is gotten from experiments in this paper.

3.4 Encoding of the Support Vectors and Weights

After the SVM regression, we get the support vectors and weights. They should be encoded and saved in the coding bit stream. In decoding, the SVM regression model is initialized with the support vectors and weights to predict the original wavelet coefficients.

In SVM regression, the weight is corresponding to the support vector one by one. That is to say, if the input training point is chosen as the support vector, there should be a corresponding weight. In our regression model, support vector has two meanings: first, it represents the input; second, it indicates the position of the input. Therefore, the support vectors and weights can be combined and encoded together [4]. For each 8×8 coefficient block, after SVM regression we combine the support vectors with the weights into a 64 dimensional weight vector. If the input training point is chosen as support vector, the corresponding weight is nonzero; otherwise the weight is set to zero. The nonzero weights are real number and should be quantized before entropy coding. This will induce extra error in the compression procedure. However, the error can be controlled by the quantization step size.

Many elements of the weight vector are zeros. In order to enhance the compression efficiency, a combination of run length coding and arithmetic coding is used to encode the weights. First, the weights are processed by run-length coding, and each group of continuous zeros is formed into a codeword. Next, all the codewords and nonzero weights are encoded by the adaptive arithmetic coder [13].

4 Implementation

We perform 5 level wavelet transform on image, which is usually chosen in image compression algorithm. After decomposition, wavelet coefficients are first quantized using scalar quantizer with deadzone. The optimal deadzone size is 0.6 from the experimental results. Secondly, the coefficients of each subband are processed according to their importance. The LL_5 subband is encoded by DPCM, and the other 15 finer scale subbands are compressed using SVM. For each finer scale subband, the magnitudes and the signs of the coefficients are processed separately. The signs are compressed by arithmetic coder directly, and the magnitudes are compressed by SVM. Thirdly, the coefficients of each subband are normalized and divided into non-overlapped blocks. According to the orientation characteristic of the subband, one of the three scan orders is employed to map the two dimension block into one dimension vector. Finally, after SVM regression, the weights and support vectors are encoded together by arithmetic coder. The main procedure of the encoding algorithm is given briefly as follows:

1. Preprocess image (subtract 128 from each pixel);
2. Perform 5-level wavelet transform;
3. Quantize wavelet coefficients using scalar quantization with deadzone;
4. Encode subband LL_5 using DPCM;
5. For each finer scale subband (from low frequency to high frequency)
    5.1. Normalize wavelet coefficients;
    5.2. Divide wavelet coefficients into coefficient blocks;
    5.3. For each coefficient block
        5.3.1. Map the block into one dimension vector
        5.3.2. Train SVM and combine the weights and support vectors together;

5.3.3. Quantize the weights;
5.3.4. Encode the weights vector using run length and arithmetic coding;
5.4. Encode the signs of the wavelet coefficients ;

It should be noted when a subband is not an integral number of 8×8 blocks, it can be padded with the same value of the boundary wavelet coefficient (i.e., extra pixels are added so that the subband can be divided into an integral number of 8×8 blocks).

The algorithm has been implemented with MATLAB, and we use LibSVM [14] for regression, which is a popular SVM toolkit that can be easily integrated in MATLAB. Daubechies 9/7 wavelet is chosen as the filter bank in our implementation. It has linear phase and good energy compaction ability, which are the most two important properties to be considered in image compression.

5 Experimental Results

In order to demonstrate its coding efficiency, the algorithm is evaluated on two benchmark 512×512 grayscale images: Lena and Goldhill. In Table 1, we present the compression results on Lena in comparison with RK-i algorithm, and compression results on Goldhill are listed in Table 2. From the results, it is clear that our algorithm is competitive against RK-i algorithm in terms of rate-distortion performance, with an improvement of more than 1 dB averagely. The reconstructed images are given in Appendix for comparison in subjective quality. It is obvious that the blockartifact, which is a serious problem in RK-i algorithm, has been well removed, and the image quality get a remarkable improvement.

Table 1. PSNR(dB) results for Lena image compared with RK-i algorithm

Compression Ratio	RK-i	Our Algorithm
25	28.0	29.9
35	/	28.9
44	26.1	27.8

Table 2. PSNR(dB) results for Goldhill image using our algorithm

Compression Ratio	Our Algorithm
20	28.9
30	27.9
42	26.9

6 Conclusion

In this paper, we have proposed a new compression algorithm based on SVM regression and wavelet transform. Experimental results show that it gains better compression performance than that of existing compression algorithm.

The results are only a preliminary investigation of compressing wavelet coefficient using SVM regression, and there is much work that can be done to improve the performance. For example, the arrangement of wavelet coefficients should be more flexible, which makes SVM learn data dependency more efficiently. A better kernel function is another consideration in the future. It can improve the approximation ability of SVM regression, which reduces the number of support vectors and improves the compression ratio.

References

1. Namphol, A., Chin, S.H., Arozullah, M: Image Compression with a Hierarchical Neural Nnetwork. IEEE Transactions on Aerospace and Electronic Systems, Vol. 32, No 1(1996)326–338
2. Chen, O.T.-C., Sheu, B.J., Fang, W.-C.: Image Compression using Self-organization Networks. IEEE Transactions on Circuits and Systems for Video Technology. Vol. 4, No 5(1994)480–498
3. Amerijckx, C., Verleysen, M., Thissen, P., Legat, J.-D.: Image Compression by Self-organized Kohonen Map. IEEE Transactions on Neural Networks, Vol. 9, No 3(1998) 503–507
4. Jonathan Robinson, Vojislav Kecman: Combing Support Vector Machine Learning with the Discrete Cosine Transform in Image Compression. IEEE Transactions on Neural Networks, Vol. 14, No 4(2003)950–958
5. Drucker, H., Burges, C.J.C., Kaufmann, L., Smola, A., and Vapnik, V.: Support Vector Regression Machines. Cambridge,MA: MIT Press(1997)
6. Vapnik ,V. N.: The Nature of Statistical Learning Theory. Springer, 1995
7. Antonini, M., Barlaud, M., Mathieu, P., Daubechies, I.: Image Coding using Wavelet Transform. IEEE Transactions on Image Processing, Vol. 1, No 2(1992)205–220
8. Mallat, S.G.: A Theory for Multiresolution Signal Decomposition: the Wavelet Representation. IEEE Transactions on Pattern Analysis and Machine Intelligence, Vol. 11, No 7(1989)674–693
9. Juan Liu, Pierre Moulin: Information—theoretic Analysis of Interscale and Intrascale Dependencies between Image Wavelet Coefficients. IEEE Transactions on Image Processing, Vol. 10, No 11(2001)1647–1658
10. Li Bo, Wang Hai: Bit Plane Predicting Image Compression Algorithm Based on Wavelet Packet Transform. Chinese J. Computer, Vol. 22, No 7(1999)685–691
11. Jiang, J.: Image Compression with Neural Networks—A survey. Signal Processing: Image Communication, Vol. 14(1999)737–760
12. Wang XL, Han H, and Peng SL: Image Restoration Based on Wavelet-domain Local Guassian Model. Jorunal of Software, Vol. 15, No 3(2004)433–450
13. Witten ,J.H., Neal, R. and Cleary, J. G.: Arithmetic Coding for Data Compression. Comm. ACM, Vol 30(1987)520–540
14. http://www.csie.ntu.edu.tw/~cjlin/libsvm/

Appendix: Subjective Comparison of the Reconstructed Images

Original

RK-i 24:1

RK-i 44:1

Our algorithm 24:1

Our algorithm 44:1

Original

Our algorithm 20:1

Our algorithm 42:1

Reconstruction of Superquadric 3D Models by Parallel Particle Swarm Optimization Algorithm with Island Model

Fang Huang and Xiao-Ping Fan

College of Information Science and Engineering, Central South University,
Changsha 410083, China
hfang@mail.csu.edu.cn

Abstract. In this paper, a new algorithm IPPSO (Parallel Particle Swarm Optimization with Island model) is proposed. It aims at remedying the defect of superquadric parametric fitting problem which is solved with L-M (Levenberg- Marquardt) method in 3D reconstruction and improving the algorithm performance of particle swarm optimization for application to large-scale problems and multi-variable solutions. This paper investigates 3D representation characteristics of superquadrics and makes analysis for the defect of superquadric parametric model fitting by L-M algorithm. It presents the principle and the implementation of superquadric parametric model fitting by using IPPSO. In addition, it describes the design principle and implementation method of IPPSO. In the end, the simulation results are analyzed. The results show the good effectiveness of the proposed approach, especially in the accuracy and discernment of superquadric 3D models reconstruction for objects.

1 Introduction

An optimal representation of whole characteristics for objects is its volume model with space occupancy attribute. Superquadric parametric model is suitable for representing volumetric objects. Thanks to the stable mathematical structure of superquadrics, the method in model reconstruction is also very stable. 3D objects can be represented by superquadrics with highly compressed data, using a few parameters to show various shape. It is usually named parametric model[1]. The 3D modeling is actually the superquadric parametric fitting. Generally, this sort of optimization problem is transferred to a nonlinear least-squares optimization one and is solved by L-M (Levenberg- Marquardt) method[1],[2],[7]. But the L-M method is not effective when the objective function is highly nonlinear or large residual[3]. The simulation results we made on fitting superquadric parametric model by L-M method demonstrate that there is infeasible solution due to local minima and low convergence speed. In this paper, a new superquadric parametric model fitting method is proposed based on parallel particle swarm optimization algorithm.

PSO (Particle Swarm Optimization) is a new evolutionary computation technique based on social behavior. Recently, more and more interests are paid on the research and application of PSO because of its easy implementation, steady performance, excellent effectiveness, and few parameters to adjust[4]. However, similar to genetic algorithms, the computation complexity usually limits the solution efficiency of PSO. Therefore, the study for parallel particle swarm optimization algorithm will bring deep significance for multi-variable solving to large-scale or super large-scale problems[5]. In addition, solution to superquadric parametric model fitting is plagued by multiple local optima and numerical noise. In this case, as same as other global search algorithms, the premature convergence also causes the unsuccessful solution to PSO. The search of PSO is pulled by individual's inertia itself, individual and global guidance factors. If just all populations quickly gather to an individual with high fitness which dominates the whole colony, the premature convergence is ineluctable[6]. In this paper, inspired by genetic algorithms, we propose a new algorithm IPPSO (Parallel Particle Swarm Optimization with Island model). It not only raises solving efficiency but also enhances colony variety and consequently improves the algorithm convergence performance. With using IPPSO to the superquadric parametric model fitting, it is a remedy for the defects of L-M method, and the results show the good effectiveness.

2 3D Representation Characteristics for Superquadrics

2.1 Implicit Function of the Superquadric

A superquadric surface is expressed by the following implicit equation.

$$\left(\left(\frac{x}{a_1}\right)^{\frac{2}{\xi_2}} + \left(\frac{y}{a_2}\right)^{\frac{2}{\xi_2}}\right)^{\frac{\xi_2}{\xi_1}} + \left(\frac{z}{a_3}\right)^{\frac{2}{\xi_1}} = 1. \tag{1}$$

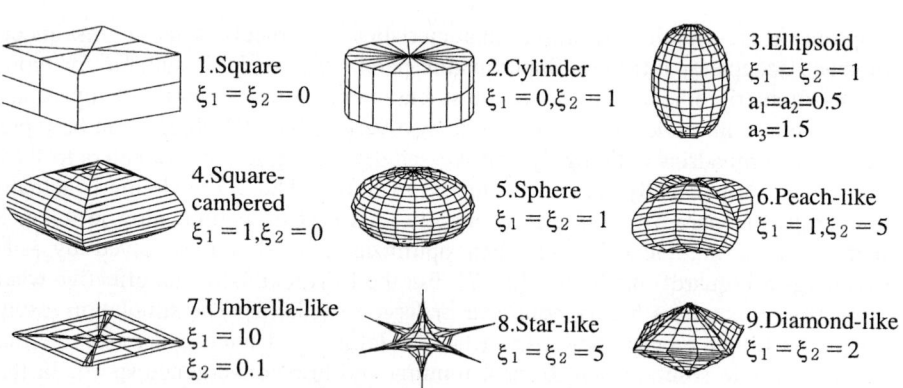

Fig. 1. Typical shapes of superquadric (a1=a2=a3=1, except ellipsoid)

The implicit function defines superquadric surface in an object centered coordinate system. Parameters a_1, a_2, a_3 define the superquadric size in x, y, and z coordinates, respectively. ξ_1, ξ_2 are the deformation parameter. When both ξ_1 and ξ_2 are changed, the superquadric can model a large set of various building blocks, shapes like in Fig.1. Apparently, a plentiful various shapes are represented by only five parameters.

The implicit equation of superquadrics contains location information from 3D points to the superquadric surface. Formula (1) shows that for any point in space we can locate it relative to superquadric surface only with solving the implicit equation's left expression. If the left expression is F(x, y, z), the results are as follows. When F(x, y, z)=1, point(x ,y ,z) is on the surface of the superquadric. When F(x, y, z)>1, the corresponding point lies outside and when F(x, y, z)<1, the corresponding point lies inside the superquadric[1],[7].

2.2 Fitting Superquadric Parametric Model Using 3D Data Points

At first, a set of discrete 3D points on the objects surface of the object are obtained from vision sensors, then the shape of it can be recovered by the superquadric model. A set of optimal parameters is selected in order to accurately represent the shapes. As we know, only the discrete points are more near the surface of the superquadric, more accurate model can be built. Considering the global minimum error, the problem is transferred to a nonlinear least-squares optimization one. The object function is as follows.

$$\min \sum_{i=1}^{N} \left(\sqrt{a_1 a_2 a_3} \left(F^{\xi_1}(x_i, y_i, z_i) - 1 \right) \right)^2 . \quad (2)$$

In practice, input 3D points generally can't form entire envelop, and they are focused on one visual profile. So the superquadric model parameters aren't only one set after fitting, and it results in uncertainty of the model. Therefore, a factor $\sqrt{a_1 a_2 a_3}$ is required in formula (2) for enhancing the global restriction for the model[1],[2]. In addition, an altered expression $\left(F^{\xi_1}(x_i, y_i, z_i) - 1 \right)$ base on formula (1) is brought for numerical value stability [1],[7].

3 Analyzing Defect for Fitting Superquadric Model by L-M

By above analysis, the superquadric parametric model fitting problem is transferred to a nonlinear least-squares optimization one and is solved by L-M method[1],[2],[7]. Because Gauss-Newton method integrates with Trust to L-M, the nonlinear least-squares optimization problem is solved by linearization mode essentially, and it is only suitable for small residual one. So L-M algorithm has the lower convergence speed when the objective function is highly nonlinear or large residual so that some problems cannot be solved[3]. As we know, the superquadric function is a highly nonlinear one. In practice, the range data or image data getting from vision sensors is a set of discrete points without cognitive powers. For the reason the initialization of superquadric parametric models is indeterminate. So the superquadric model fitting by least-squares is also large residual. With the addition of precondition for L-M algorithm solving is that the object function is monotony in the searching scope. The obtained solution is a local optima depending on the initialization[3]. The experiment results are showed in Tab.1. The standard and fitting parameters for various shape models are respectively listed. The standard value of size parameter a1, a1, a3 for models are 1 except ellipsoid (a1=a2=0.5,a3=1.5), and the standard value of deformation parameter ξ_1, ξ_2 are represented the corresponding shapes. The superquadric function is a non-monotony in the scale (0-10) for ξ_1、ξ_2, which have a diversification surface. In this case, the uncertain solutions for model can be found or aren't optimal one by L-M algorithm. For example, when the diversification surface is less smooth such as sphere, ellipsoid and diamond-like, the optimal solution is rapidly found, and the ultimately attained minimal value of the object function approximates zero and the value of fitting parameters are consistent with the standards. When the surface is abrupt such as star-like, peach-like and umbrella-like, the value of fitting parameters depends on the initialization. If the initialization approaches the standard value, the optimal solution can be attained, or else cannot. When the surface of model is orthogonal turning such as square, square-cambered and cylinder, the solution can not be found absolutely due to the lower convergence speed.

Table 1. Results for L-M algorithm

Std. value ($a_1=a_2=a_3=1$)	Initialization			Fitting parameter					Minimum	Solution
	a_1,a_2,a_3	ξ_1	ξ_2	a_1	a_2	a_3	ξ_1	ξ_2		
5. $\xi_1=\xi_2=1$	10,5,0.2	6	8	1	1	1	1	1	1.0396e-20	Yes
	0.2, 4, 6	7	9	1	1	1	1	1	2.4313e-17	Yes
3. $a_1=a_2=0.5$, $a_3=1.5$, $\xi_1=\xi_2=1$	7, 5, 3	10	8	0.5	0.5	1.5	1	1	7.4002e-20	Yes
	2, 4, 6	8	0.1	0.5	0.5	1.5	1	1	4.0924e-20	Yes
9. $\xi_1=\xi_2=2$	10,5,0.1	6	8	1	1	1	2	2	1.7352e-16	Yes
	0.1, 4, 6	7	9	1	1	1	2	2	3.4799e-18	Yes
8. $\xi_1=\xi_2=5$	2, 4, 6	0.1	10	1	1	1	5	5	9.2145e-17	Yes
	9, 5, 0.1	10	0.1	0.7204	0.6895	0.8934	4.5392	0.1256	33.094	No
6. $\xi_1=1,\xi_2=5$	2, 4, 6	0.1	10	1	1	1	1	5	8.368e-20	Yes
	10,5,0.1	10	0.1	0.7803	0.7494	0.9147	0.7401	0.1023	29.783	No
7. $\xi_1=10,\xi_2=0.1$	2, 4, 6	7	10	1	1	1	10	0.1	5.1861e-19	Yes
	0.2, 4, 6	6	9	0.7656	-0.2032	0.7929	7.5327	-0.0368	19.772	No
1. $\xi_1=\xi_2=0$	2, 4, 6	5	9	2	-4	6	5	6.325	23137000	No
	1, 1, 1	1	1	1	-1	1	1	-1.8854	36.312	No
4. $\xi_1=1,\xi_2=0$	7, 5, 3	2	1	7	5	3	2	1	35059	No
	2, 4, 6	0.1	10	2	-4	6	0.0999	7.305	2371000	No
2. $\xi_1=0,\xi_2=1$	10,5,0.1	10	0.1	10.252	0.9595	0.9973	0.0026	1.7008	86.372	No
	2, 4, 6	0.1	10	1.5362	2.7502	-0.2592	0.1049	8.9235	749590	No

The simulation results show that there is a great disadvantage to L-M method for solving the nonlinear least-squares optimization problem with the object function of superquadric. Usually, a method of evolution computation is introduced to solve the nonlinear optimization problems which are difficult to deal with traditional way, such as GA(genetic algorithm). But GA does not satisfy the needs for real-time surroundings modeling[2]. So, a novel approach of evolution computation is proposed that is Particle Swarm Optimization.

4 Fitting Superquadric Parametric Models by PSO

4.1 Particle Swarm Optimization

Particle Swarm Optimization is inspired by the investigation for searching behavior of bird flocks. Every bird can adjust respective flying velocities and position according to the individual's own and accompanier's experience for guider, and the foods are found with the highest efficiency. In PSO, each bird is stated as particle which is defined as a potential solution of a problem. In a D-dimensional space, the ith particle represented as $X_i = (x_{i1}, x_{i2}, \cdots, x_{iD})$. Each particle maintains a memory of its previous best position $P_i = (p_{i1}, p_{i2}, \cdots, p_{iD})$ with the best fitness that is named as the individual best position, and a velocity along each dimension represented as $V_i = (v_{i1}, v_{i2}, \cdots, v_{iD})$. In the whole population, a memory of the global best position P_{gd} also is maintained with the global best fitness. In each generation, the particle can adjust its own a new velocity and position by follow iterative formulae until optimal solution is found[8]. For determining the next flying step, each particle updates own velocity according to three terms which are the current velocity v_{id}^t adjusted by an inertial weight w as a decreasing linear function in t from 0.9 to 0.2, the distance from the particle's current position x_{id}^t to the individual best position influenced by an individual guider factor ψ_1 with 2*rand(1) and the distance from the current position to the global best position impacted by global guider factor ψ_2 also 2*rand(1)[9]. So ψ_1 and ψ_2 produce equal acting force on the searching progresses. It especially depends on an application problem how to set these important parameters.

4.2 Implementing for Fitting Superquadric Parametric Models by PSO

The nonlinear least-squares optimization problem for fitting superquadric parametric model is solved by PSO and its fitness function is defined as formula (2). As well know, fitted parameters(a_1, a_2, a_3, ξ_1, ξ_2) form a 5-dimensional space for particles to search. The deformation parameters ξ_1, ξ_2 are limited in the searching scope from 0 to 10 ($x_{min}=0$, $x_{max}=10$) because of restricting the solution space within super-ellipsoid, and the flying velocity is from $v_{max}=5$ to $v_{min}=-5$. The number of particle is set as 160 and the maximal iterative number is 2000. The algorithm in pseudo-code follows.

```
Begin
   for t = 1 to 2000
     for i =1 to 160
       for d = 1 to 5
         Update velocity:
```
$$v_{id}^{t+1} = w * v_{id}^t + \psi_1 * (p_{id} - x_{id}^t) + \psi_2 * (p_{gd} - x_{id}^t) ;$$

```
Limit velocity: v_{id}^{t+1} = min(v_max, max(v_min, v_{id}^{t+1})) ;
Update position: x_{id}^{t+1} = x_{id}^t + v_{id}^{t+1} ;
Limit position: x_{id}^{t+1} = min(x_max, max(x_min, x_{id}^{t+1})) ;
end - for - d;
```

Compute fitness of X_i^{t+1}: $\sum_{i=1}^{N}\left(\sqrt{a_1 a_2 a_3}\left(F^{\xi_1}(x_i, y_i, z_i) - 1\right)\right)^2$;

```
Update historical information regarding p_i and p_g;
end - for - i;
Terminate if p_g meets problem requirements;
end - for - t;
end.
```

The results for fitting superquadric parametric models by PSO show in Tab. 2. Comparing Tab.1 with Tab.2, the optimal solution can be attained by PSO in these models which cause the unsuccessful solution to L-M algorithm. In addition, the initialization of position and velocity for particles are randomization duo to the global searching of PSO. Whether the optimal solution can be attained, it does not depend on the initialization differing from L-M algorithm. So the cognitive powers for models are enhanced and the method is more worth in application.

Table 2. Results for PSO algorithm (10 runs)

Std. value ($a_1=a_2=a_3=1$)	The average of fitting parameter					Average minimum	Average iterations	Successes
	a_1	a_2	a_3	ξ_1	ξ_2			
5. $\xi_1=\xi_2=1$	1.0000	1.0000	1.0000	1.0000	1.0000	3.2683e-9	306	10
3. $a_1=a_2=0.5$, $a_3=1.5$, $\xi_1=\xi_2=1$	0.5000	0.5000	1.5000	1.0000	1.0000	6.63208e-8	306	10
9. $\xi_1=\xi_2=2$	1.0000	1.0000	1.0000	2.0000	2.0000	5.02731e-9	315	10
8. $\xi_1=\xi_2=5$	1.0000	1.0000	1.0000	5.0000	5.0000	3.72186e-8	313	10
6. $\xi_1=1, \xi_2=5$	1.0000	1.0000	1.0000	1.0000	5.0000	1.78176e-8	314	10
7. $\xi_1=10, \xi_2=0.1$	1.0000	1.0000	1.0000	10.0000	0.1000	2.83339e-8	284	10
1. $\xi_1=\xi_2=0$	1.0000	1.8708	0.9982	3.0000	0.0367	6.28055	306	7
4. $\xi_1=1, \xi_2=0$	1.0009	1.2821	1.0009	1.0020	0.0170	0.0201079	291	9
2. $\xi_1=0, \xi_2=1$	1.0000	1.0000	0.9977	3.7620	0.0000	7.87678	344	6

In Tab.2, the results show that PSO has the advantage of L-M algorithm in superquadric modeling evidently. But also when the surface of model is orthogonal turning such as square, cambered-square and cylinder, the algorithm progresses often falls into the local optima because of PSO's the premature convergence so that the error of these model parameters is caused and even the solution can not be found sometimes. So a parallel particle swarm optimization algorithm with island model is proposed for restraining the premature convergence in order to improve the performance of PSO.

5 Parallel Particle Swarm Optimization Algorithm with Island Model

5.1 Design for IPPSO

Island model is classified as a parallel mode of decomposing[6]. The whole colony is divided into sub-populations, and each sub-population carries out its own evolution in respective processor which communicates information each other in an appropriate interval. Island model is also known as coarse grain model in which there is more than one particle in the sub-population of each processor. Because of a simple implementation of the coarse grain model, it can be simulated implementing on network or stand-alone machine without parallel computer. Currently, the coarse grain model is very popular in parallel genetic algorithm. Therefore, the parallel scheme base on island model for PSO is brought out with migration strategy of centralized topology. As Fig.2, the colony of particles is divided into many sub-populations to form islands. Each sub-population performs global PSO for oneself including computation and evaluation for every particle's fitness only inside the island. Then a region best particle is produced. The evolution in island is executed by a stand-alone sub-process in order to reduce coupling. Each sub-process periodically sends the region best particle to main-process to form main-population by migration strategy of centralized topology. The main-population chooses a global best particle of the colony from them and broadcasts it back, guiding the sub-population evolving to the global optima.

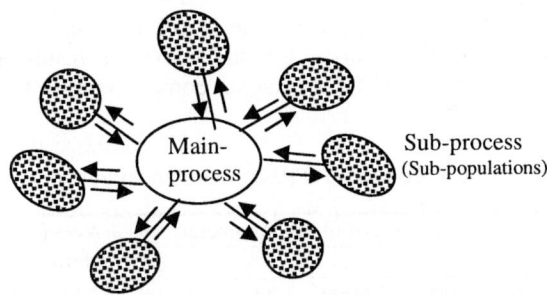

Fig. 2. Centralized topology for IPPSO

5.2 Improving Convergence Capability for IPPSO

PSO is a parallel evolutionary technique. The algorithm performance is greatly influenced by its adjustable parameters. Both of these tuning parameters are as individual and global guidance factors, contributing to the exploration and exploitation tradeoff in searching process. Exploration is the ability to test various regions in the problem space in order to locate a good optimum, hopefully the global one. Exploitation is the ability to concentrate the search around a promising candidate solution in order to locate the optimum precisely[10]. The composition force of the exploration and exploitation tradeoff is produced by the tuning parameters, which can be used to solve given objective function. A larger value of individual guidance factor can lead to more iterative number and cost more time, otherwise a larger value of global guidance factor can lead to premature convergence for local optima. If the premature convergence was restrained by increasing the individual guidance factor exceedingly, the algorithm would degenerate into stochastic search one. The reference [10] contains a guidance way of choosing parameters for PSO after analyzing the dynamic behavior of a particle by the theory of linear discrete-time dynamic system. But the effect of object function has not been taken into account. Since nonlinear objective function often includes some local optima, the tuning parameters can not be suitable for both of the higher convergence speed and the accurate global optimum simultaneously. With the addition of division of sub-populations in IPPSO, the framework of PSO is recomposed. The independent evolutionary processes of PSO are worked in these islands respectively, which constructs a searching process of global optimum for the region. It is equivalent to add the traction of the region between individual guidance factor and global guidance factor to balance the absolute extremes, and to attain its goal of improving the algorithm performance.

5.3 Analysis for Experimental Results

The value of adjustable parameters is set in IPPSO as much as in PSO for comparable results. As formula (3), the inertial weight w is decreased from 0.9 to 0.2 in the progresses. The individual guider factor is $\psi_1 = 2*rand(1)$, the global guider factor is $\psi_2 = 2*rand(1)$ and the number of particles is 160. In IPPSO, the colony of particles is divided into eight sub-populations with 20 particles in each to form islands. The independent sub-population evolutionary is carried on a thread-process in the island because of adopting the Multithreading technique in IPPSO. The migration period between the islands is set as 10 iterations. For the numerical results all optimizations were performed on Pentium4 2.4C GHz personal computers with the Windows operating system.

Table 3. Results for IPPSO algorithm (10 runs)

Std. value ($a_1=a_2=a_3=1$)	The average of fitting parameter					Average minimum	Average iterations	Successes
	a_1	a_2	a_3	ξ_1	ξ_2			
5. $\xi_1=\xi_2=1$	1.0000	1.0000	1.0000	1.0000	1.0000	3.24771e-9	284	10
3. $a_1=a_2=0.5$, $a_3=1.5$, $\xi_1=\xi_2=1$	0.5000	0.5000	1.5000	1.0000	1.0000	6.6255e-8	289	10
9. $\xi_1=\xi_2=2$	1.0000	1.0000	1.0000	2.0000	2.0000	5.02431e-9	292	10
8. $\xi_1=\xi_2=5$	1.0000	1.0000	1.0000	5.0000	5.0000	3.70835e-08	305	10
6. $\xi_1=1,\xi_2=5$	1.0000	1.0000	1.0000	1.0000	5.0000	1.77656e-8	290	10
7. $\xi_1=10,\xi_2=0.1$	1.0000	1.0000	1.0000	10.0000	0.1000	2.82949e-8	250	10
1. $\xi_1=\xi_2=0$	1.0000	1.0060	1.0000	0.0000	0.0000	4.57446e-18	247	10
4. $\xi_1=1,\xi_2=0$	1.0000	1.0247	1.0000	1.0000	0.0051	1.50555e-9	257	10
2. $\xi_1=0,\xi_2=1$	1.0000	1.0000	1.0000	0.0000	1.0000	1.27262e-10	283	10

As Tab.3, the experimental results are improved by IPPSO, obviously. When fitting parametric models for square, cambered-square and cylinder, the more precise solutions are also successfully attained in repetitiously running. In addition, the average iterative numbers of the algorithm are decreased 10% contrasting Tab.2. In Fig.3, the graphs show the convergent progresses fitting square and cylinder, using PSO and IPPSO respectively. As we well know from the progress graphs, the convergence of parallel particle swarm optimization algorithm with island model is greatly enhanced, the iterative numbers are decreased and the fitness value is better. Consequentially, the performance of the algorithm is greatly improved.

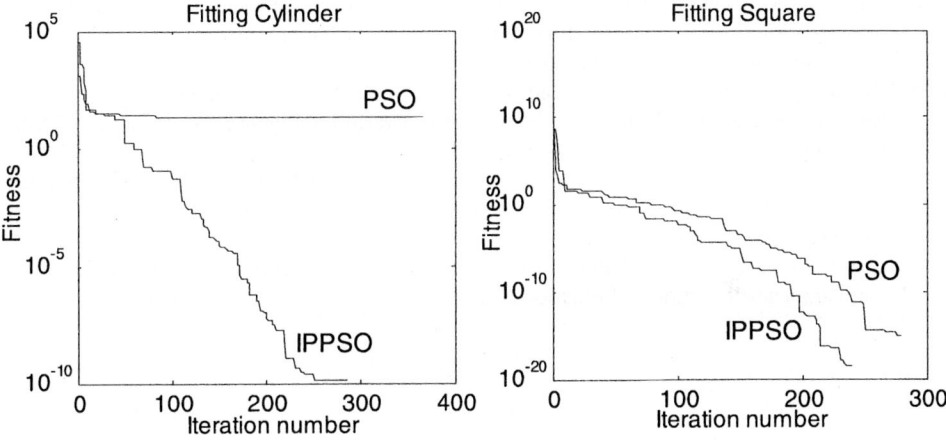

Fig. 3. Progress graphs

6 Conclusion

In this paper, a new superquadric parametric model fitting method is proposed based on parallel particle swarm optimization algorithm with island model. The results show the good effectiveness of the proposed approach, especially in the accuracy and discernment of superquadric 3D models reconstruction for objects. Reconstruction of superquadric 3D models by novel IPPSO is a successful application and practice in solving non-linear and large-scale problems. Furthermore, in results shown the high modeling efficiency and stability of the method would be a foundation for real-time modeling of a scene. We also plan to explore the possibility to use the proposed method for real-time recovery of complicated geometric models.

References

1. Solina, F., and Bajcsy, R.: Recovery of Parametric Models from Range Image: The Case for Superquadrics with Global Deformations. IEEE Trans. Pattern Analysis and Machine Intelligence, vol.12, no.2, pp.131-147, 1990
2. Chevalier, L., Jaillet F., and Baskurt, A.: Segmentation and Superquadric Modeling of 3D Objects. WSCG2003, vol.11, no.1, Plzen (CZ), February 3-7 2003

3. Yuan Ya Xiang, Sun Wen Yu: Optimization Theory and Method. Beijing: Science Press, 1997,312-350
4. Eberhart,R.C., and Shi,Y.: Particle Swarm Optimization: Developments, Applications and Resources. Proceedings of the IEEE Congress on Evolutionary Computation, pp.81-86, Seoul, Korea. 2001
5. Schutte, J.F., Reinbolt, J.A., Fregly B.J., et al.: Parallel Global Optimization with the Particle Swarm Algorithm. Int. J. Numer. Meth. Engng 2004, 61: 2296-2315
6. Yu-lan Hu, Fu-chen Pan, et al.: Parallel Genetic Algorithm Based on Population Size Mutable Coarse-grained. Mini-Micro System, 2003, 24(3):534-536
7. Dimitrios Katsoulas and Ales Jaklic: Fast Recovery of Piled Deformable Objects using Superquadrics. In: Van Gool, Luc, Eds. Proceedings Pattern recognition: 24th DAGM symposium, pages 174-181, Zurich Switzerland, 2002
8. Peram,T., Veeramachaneni, K., and Mohan, C. K.: Fitness-distance-ratio Based Particle Swarm Optimization. Proceedings of the IEEE Swarm Intelligence Symposium 2003, Indianapolis, Indiana, USA. pp. 174-181, 2003
9. Brian Birge: PSOt-a Particle Swarm Optimization Toolbox for Use with Matlab. Proceedings of the IEEE Swarm Intelligence Symposium 2003, Indianapolis, Indiana, USA. pp. 182-186, 2003
10. Ioan Cristian Trelea: The Particle Swarm Optimization Algorithm: Convergence Analysis and Parameter Selection. Information Processing Letters 85 (2003) 317–325

Precision Control of Magnetostrictive Actuator Using Dynamic Recurrent Neural Network with Hysteron

Shuying Cao, Jiaju Zheng, Wenmei Huang, Ling Weng,
Bowen Wang, and Qingxin Yang

Province-Ministry Joint Key Laboratory of Electromagnetic Field and Electrical Apparatus
Reliability, Hebei University of Technology
Box 359, Hebei University of Technology, Tianjin, 300130, China
csy236@eyou.com

Abstract. A control strategy for precision position tracking of the magnetostrictive actuator (MA) with dominant hysteresis is proposed. In this strategy, a dynamic recurrent neural network with hysteron (DRNNH) is adopted as a feedforward controller for on-line learning the inverse model of the MA to remove the effect of the hysteresis of the MA. A proportional-plus-derivative (PD) feedback controller is used to reduce the position tracking error. Simulation results validate the excellent performances of the control strategy.

1 Introduction

The magnetostrictive actuator (MA), with large strain, high force and nanometer solution, has a wide range of applications in precision position tracking technology. However, the MA exhibits dominant hysteretic nonlinearity, which usually exists in many physical systems, such as electronic relay circuits, piezoelectric actuator and shape memory alloy actuator. Such hysteretic nonlinerity can cause position error in the open-loop system, and cause system instability in the closed-loop system. In order to remedy these problems, researchers have proposed various control methods to cancel out the hysteretic effect. Jung et al. [1] use feedforward control to reduce scanning errors. Due to hysteresis modeled as "deterministic hysteretic paths", the controller depends on large amounts of experimental data. Cruz-Hernández and Hayward [2] introduce a variable phase, an operator that shifts its periodic input signal by a phase angle that depends on the magnitude of the input signal. However, this study must redesign for different actuators. Researchers [3,4,5,6] have studied the Preisach-based inverse control technique. In the control technique, a Preisach model is used to capture the hysteresis characteristic, and then the inversion of the Preisach model is constructed to cascade with the hysteretic system so that the system became a linear structure. The inverse control technique can reduce largely the hysteretic nonlinearity, which promotes the control and applications of the hysteretic systems. However, the inverse control technique cannot provide the capability for adapting to changes in operating conditions because the Preisach model and its inverse model are off-line identified using a large number of first-order reversal curves. Hwang et al. [7] propose a feedforward neural network compensator to reduce the hysteretic effect. As the feedforward neural network is only a static mapping system, it is not suitable to describe a hysteresis, which is, in fact, a nonlinear dynamic system. It is apparent from [7] that the problem is far from being solved and general methods for controller design are still not available.

In this paper, a dynamic recurrent neural network with hysteron (DRNNH) is constructed for the description of the inverse model of the hysteresis, and a control strategy combining the DRNNH controller and a proportional-plus-derivative (PD) controller is applied for precision position tracking of the MA with dominant hysteresis. Simulation results show the validity of the control strategy.

2 Model and Reversibility of Magnetostrictive Actuator

2.1 Hysteresis Model of Magnetostrictive Actuator

Calkins et al. [8] combine the Jiles–Atherton model with the magnetostriction model to establish a quasi-static hysteresis model for the MA. In [9], we extend the model [8] to establish a dynamic hysteresis model for the MA by considering the actuator's structural dynamic principle. In the model, the anhysteretic magnetization M_{an}, the irreversible magnetization M_{irr}, the total magnetization M, the magnetostriction λ, the displacement y are quantified through the expressions

$$M_{an} = M_s \left[\coth\left(\frac{H + \tilde{a}M}{a}\right) - \frac{a}{H + \tilde{a}M} \right]. \tag{1}$$

$$\frac{dM_{irr}}{dH} = \frac{M_{an} - M_{irr}}{\delta_1 k - \tilde{a}(M_{an} - M_{irr})}. \tag{2}$$

$$M = (1-c)M_{irr} + cM_{an}. \tag{3}$$

$$\lambda = \frac{3}{2}\frac{\lambda_s}{M_s^2}M^2. \tag{4}$$

$$y = \frac{1}{M_z s^2 + C_z s + K_z}(A_r E \lambda). \tag{5}$$

where $H=nI$ is the magnetic field generated by a solenoid having n turns per unit length under the input current I, δ_1 takes the value 1 or -1 based on the sign of dH/dt. Equations (1)-(5) can be combined to yield displacement values of the actuator in response to the input current I. In the combined model, M_s, a, $\tilde{a}$, k, c, λ_s, A_r, E is respectively, the saturation magnetization, shape parameter, magnetic-stress interactions, average energy of pinning sites, reversibility coefficient, saturation magnetostriction, cross-sectional area and Young modulus of the Terfenol-D rod; M_z, C_z, K_z is respectively, the mass, damping and stiffness of the actuator. The parameters of the hysteresis model for an actual MA are listed in Table 1. Using the above parameter values and the combined model, we give the characteristic of the hysteresis for the

MA as plotted in Fig.1. As shown in [9], the combined model can accurately describe the relation of the input current and the output displacement of the MA, thus meet control applications.

Table 1. Parameters of the hysteresis model for an actual MA

M_s=7.65×10^{5}A/m	λ_s =1005×10^{-6}	a =7012A/m	k=3283A/m
$\tilde{a}$=-0.01	c=0.18	E=3×10^{10}N/m^2	A_r=1.2668×10^{-4}m^2
K_z=3.3613×10^{7}N/m	C_z=4304Ns/m	M_z=0.5kg	n=10435/m

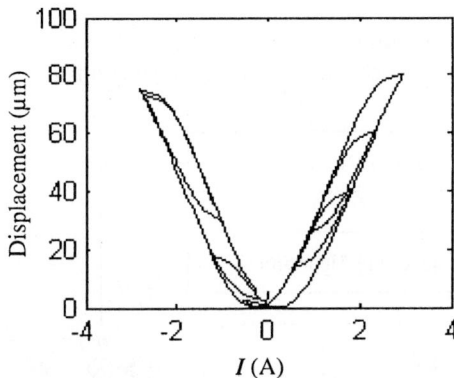

Fig. 1. Characteristic of the hysteresis for the MA

2.2 Reversibility of Magnetostrictive Actuator

Fig.1 shows that the relation between the input current and output displacement for the MA exhibits non-differential, multivalued and butterfly-shaped hysteresis loops. The hysteresis branches are not monotonic and, strictly speaking, they do not admit any inverse. The conditions that guarantee the existence of the hysteresis inverse model of MA can be verified only if input current is limited to the interval [0,+∞] (or[−∞,0]). When input current is limited to the interval [0,+∞] (or[−∞,0]), assume the hysteresis dynamic model for the MA can be described through a Nonlinear Auto Regressive Moving Average model (NARMA)

$$y(k) = \Gamma[y(k-1),\cdots,y(k-n),I(k),\cdots,I(k-m)] . \tag{6}$$

where $y(\cdot)$ and $I(\cdot)$ are discrete sequences of the output displacement and input current of the MA, n, m are the corresponding lags in the output and input($m \leq n$), Γ is a hysteresis function and k is discretized time. Thus, for two arbitrary currents $I_1(k) \neq I_2(k)$, the inequation

$$\Gamma[y(k-1),\cdots,y(k-n),I_1(k),I(k-1),\cdots,I(k-m)] \\ \neq \Gamma[y(k-1),\cdots,y(k-n),I_2(k),I(k-1),\cdots,I(k-m)] . \tag{7}$$

comes into existence, so the hysteresis inverse model of the MA exists at the $[y(k-1),\cdots,y(k-n),I(k-1),\cdots,I(k-m)]^T$. From (6), the inverse dynamic model of the MA is

$$I(k) = \Gamma^{-1}[y(k), y(k-1), \cdots, y(k-n), I(k-1), \cdots, I(k-m)] \ . \tag{8}$$

3 Precision Control of Magnetostrictive Actuator Using DRNNH

3.1 Precision Control System Design for Magnetostrictive Actuator

The precision position tracking control system for MA consists of a fixed gain PD controller and a neural network feedforward inverse controller as shown in Fig.2.

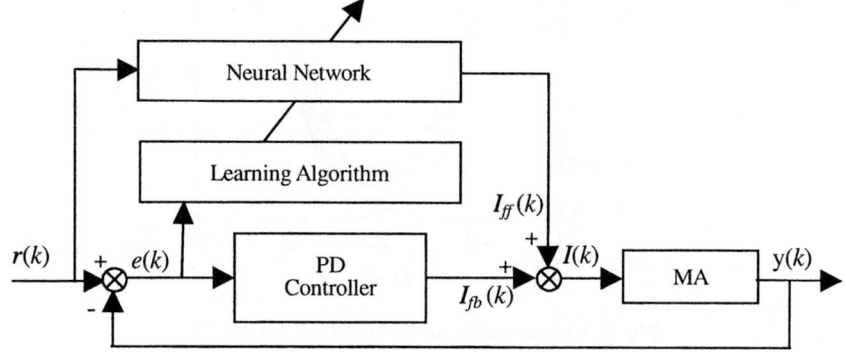

Fig. 2. Precision position tracking control system for MA

In Fig.2, k is discrete time, $r(k)$ is the reference input, $y(k)$ is the MA output, and the input of the MA is $I(k)=I_{ff}(k)+I_{fb}(k)$. Here $I_{ff}(k)$ is the output of the neural network controller, and $I_{fb}(k)$ is the output of the PD controller , i.e

$$I_{fb}(k) = k_p e(k) + k_d [e(k) - e(k-1)] \ . \tag{9}$$

where $e(k)=r(k)-y(k)$ is the tracking error, k_p and k_d is respectively, the proportional gain and differential gain. In this control system, the tracking error $e(k)$ is used as the training signal to modify the parameters of the neural network, which is called as feedback-error learning scheme [10,11]. The learning scheme has the advantage: the learning and control are performed simultaneously in sharp contrast to the conventional "learn-then-control' approach. The PD controller ensures adequate performance prior to convergence of the neural network weights, and reduces the steady-state output errors [10, 11]. When the tracking error is zero, the PD controller produces no output, hence the neural network learning is completed, and the neural network represents the true inverse dynamic model of the MA. As the MA is a dynamic nonlinear device, it will be suitable to use dynamic recurrent neural networks instead of using static neural networks to model the inverse dynamic model of the MA [11, 12]. This paper uses a DRNNH as a feedforward inverse controller.

3.2 Dynamic Recurrent Neural Network with Hysteron

According to the characteristics of the MA as shown in Fig.1, a DRNNH with P input layer nodes, Q hidden layer nodes and one output layer node is constructed. The DRNNH differs from the dynamic neural network [12] in the two ways: 1) a hysteretic neuron model [13] is introduced into the hidden layer activation function. The two parameters of the activation function are tuned in order to maximize its performance. So, it seems that the function provides us with much more flexibility than usual sigmoid activation function, in which there are no parameters to tune; 2) To ensure that the input current is limited to the interval $[0,+\infty]$, the output layer activation function adopts a special smooth function. The special function is constructed by the algebraic addition of a sigmoid function and a linear function with an adjustable parameter. The expressions and the learning algorithm of the DRNNH are as follows.

In the tracking control system, the MA output $y(k)$ is desired to track the reference input $r(k)$, i.e., $y(k)=r(k)$. From (8), the inverse dynamic model of the MA becomes

$$I(k) = \Gamma^{-1}[r(k), y(k-1), \cdots, y(k-n), I(k-1), \cdots, I(k-m)] . \tag{10}$$

Thus, the input and output of the input layer for the DRNNH is defined as

$$Z(k) = [r(k), y(k-1), \cdots, y(k-n), I(k-1), \cdots, I(k-m)]^T$$
$$O_i(k) = I_i(k) \quad i = 1, 2, \cdots, P \tag{11}$$

where $P=n+m+1$, $I_i(k)$ is the ith element of $Z(k)$. The input and output of the hidden layer are given by

$$I_j(k) = w_j O_j(k-1) + \sum_{i=1}^{P} w_{ij} O_i(k) \quad j = 1, 2, \cdots, Q \tag{12}$$
$$O_j(k) = f[I_j(k)] = 0.5 \tanh[r_j(I_j(k) - q_j)]$$

where w_{ij} are the weights connecting the input nodes to the hidden nodes, w_j are recurrent weights of the hidden nodes, $f[I_j(k)]$ is the hidden layer activation function [13], parameter r_j controls the slope of $f[I_j(k)]$ and q_j is the offset of $f[I_j(k)]$. The input and output of the output layer are given by

$$I_l(k) = \sum_{j=1}^{Q} w_{jl} O_j(k) \quad l = 1 \tag{13}$$
$$I_{ff}(k) = g[I_l(k)] = \frac{1}{1+e^{-I_l(k)}} + \beta_l I_l(k)$$

where w_{jl} are the weights connecting the output of the hidden nodes to the output nodes, $g[I_l(k)]$ is the output layer activation function, β_l is an adjustable parameter of $g[I_l(k)]$. Suppose the cost function for training of the DRNNH is defined as

$$J = \frac{1}{2}[r(k) - y(k)]^2 = \frac{1}{2}e^2(k) . \tag{14}$$

The parameters $\theta=[w_{jl}, \beta_l, w_j, w_{ij}, r_j, q_j]$ of the DRNNH are on-line modified by the gradient descent learning law

$$\theta(k) = \theta(k-1) - \eta \frac{\partial J}{\partial \theta}. \tag{15}$$

where η is the learning rate. The gradient $\partial J / \partial \theta$ is computed by using the back-propagation method. Its computations are summarized in the following lemma.

Lemma 1. Given the DRNNH as described by (11)-(13), the gradient $\partial J / \partial \theta$ is

$$\begin{aligned}
\frac{\partial J}{\partial w_{jl}} &= -e(k)\,\text{sgn}\!\left[\frac{\partial y(k)}{\partial I(k)}\right] g'[I_l(k)] O_j(k) \\
\frac{\partial J}{\partial \beta_l} &= e(k) I_l(k)\,\text{sgn}\!\left[\frac{\partial y(k)}{\partial I(k)}\right] \\
\frac{\partial J}{\partial w_j} &= -\delta_j P_j(k) \\
\frac{\partial J}{\partial w_{ij}} &= -\delta_j Q_{ij}(k) \\
\frac{\partial J}{\partial r_j} &= -\delta_j D_j(k) \\
\frac{\partial J}{\partial q_j} &= -\delta_j E_j(k)
\end{aligned} \tag{16}$$

where

$$\delta_j = -e(k)\,\text{sgn}\!\left[\frac{\partial y(k)}{\partial I(k)}\right] g'[I_l(k)] w_{jl} f'[I_j(k)]. \tag{17}$$

and

$$\begin{aligned}
P_j(k) &= O_j(k-1) + w_j P_j(k-1),\; P_j(0) = 0 \\
Q_{ij}(k) &= I_i(k) + w_j Q_{ij}(k-1),\; Q_{ij}(0) = 0 \\
D_j(k) &= (1/r_j)(I_j(k) - q_j + r_j w_j D_j(k-1)),\; D_j(0) = 0 \\
E_j(k) &= w_j E_j(k-1) - 1,\; E_j(0) = 0
\end{aligned} \tag{18}$$

Therefore, the parameters θ of the DRNNH can be modified recursively by (15)-(18). When the tracking error is zero is zero, the learning modification of θ is completed and the DRNNH represents the true inverse model of the MA.

3 Simulation Results and Analysis

In the simulation study, the parameters of the hysteresis model are shown in Table 1; the parameters for the DRNN are $P=n+m+1=7(n=3, m=3)$, $Q=14$, $\eta=0.1$, and all θ are initially 0; the parameters for PD are $k_p=0.1$ and $k_d=0.001$; and sample period T is 0.1ms. We select a square wave and two multisine signals as the reference inputs. The square wave signal is aimed at checking the performance of the proposed control system with respect to a sequence of step inputs. The two multisine signals are chosen in order to evaluate the tracking capability in case of continuous reference and the compensating capability of hysteretic nonlinearity. Numerical simulation results are shown in Figs.3-6. Noted that Fig.3 is the simulation result of the open-loop control system with DRNNH feedforward controller and without PD feedback controller, and Figs.4-6 are the simulation results of the proposed control system with DRNNH feedforward and PD feedback controllers.

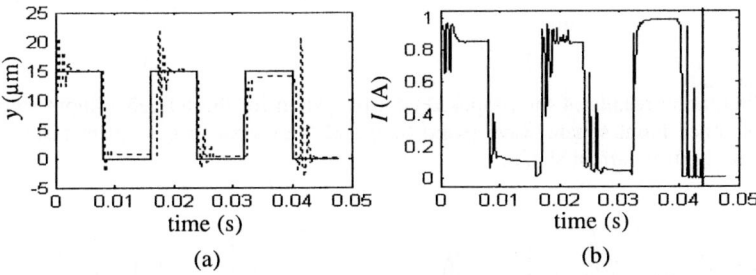

Fig. 3. Simulation results of the open-loop control system for the reference input square wave. (a) the reference input square wave (solid line) and the output displacement response(dashed line); (b) the control current signal.

The simulation results for the reference input square wave with period 0.016s and amplitude 15μm are shown in Fig.3 and Fig.4. From Fig.3 and Fig.4, some conclusions can be drawn: *1)* The open-loop control system only with DRNNH controller produces a limited steady-state error (the maximum steady-state error is only about 0.99μm). The steady-state error is reduced to 0.04μm by the proposed control system; *2)* The open-loop control system only with DRNNH controller can fast respond to the step inputs, but the dynamic control performances, such as overshoot and transient vibration, are big. These performances are improved by the proposed control system; *3)* In short, the inverse model obtained by the DRNNH can rapidly approach the true inverse model of the MA, and the PD feedback controller can compensate the mapping error of the DRNNH and improve the dynamic control performances.

From the analysis of Fig.5 and Fig.6, some conclusions can be drawn: *1)* For the reference input $r=30+11\sin(6\pi t)+12\sin(2\pi t-\pi/2)+9\sin(12\pi t+\pi)$μm, Fig.5 shows that after 3ms time, the control system can track the reference input in the steady-state tracking error [-0.5μm, 0.5μm]. Fig.5 (d), Fig.5 (e) and Fig.5 (f) show the inverse control process of the control system. Here the hysteresis loop in Fig.5(d) is very approximate inverse description of the hysteresis loop in Fig.5 (e), which is verified by Fig.5(c) and Fig.5(f); *2)* For the reference input $r=43+15\sin(10\pi t)+25\sin(40\pi t)\mu m$ with bigger amplitude and higher frequency, Fig.6 shows that after 1ms time, the control system can track the reference input in the steady-state tracking error[-2μm, 6μm]; *3)* Comparison of the inverse control process in Fig.5 and Fig.6 show the hysteresis loops of the MA are very different under the different reference inputs. It is obvious that the proposed control strategy can on-line obtain different inverse model of the MA for the different reference inputs in very short time, thus eliminate the impact of hysteresis and lead to very small tracking error.

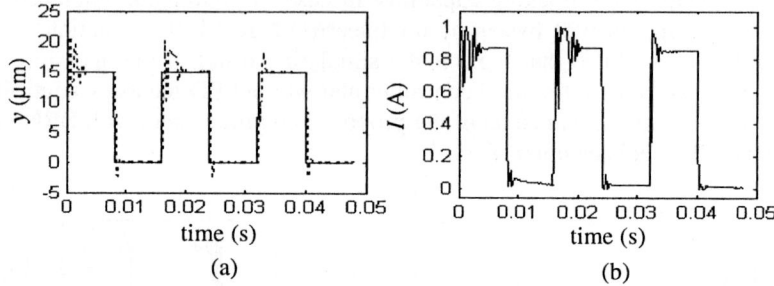

Fig. 4. Simulation results of the proposed control system for the reference input square wave. (a) the reference input square wave (solid line) and the output displacement response(dashed line); (b) the control current signal.

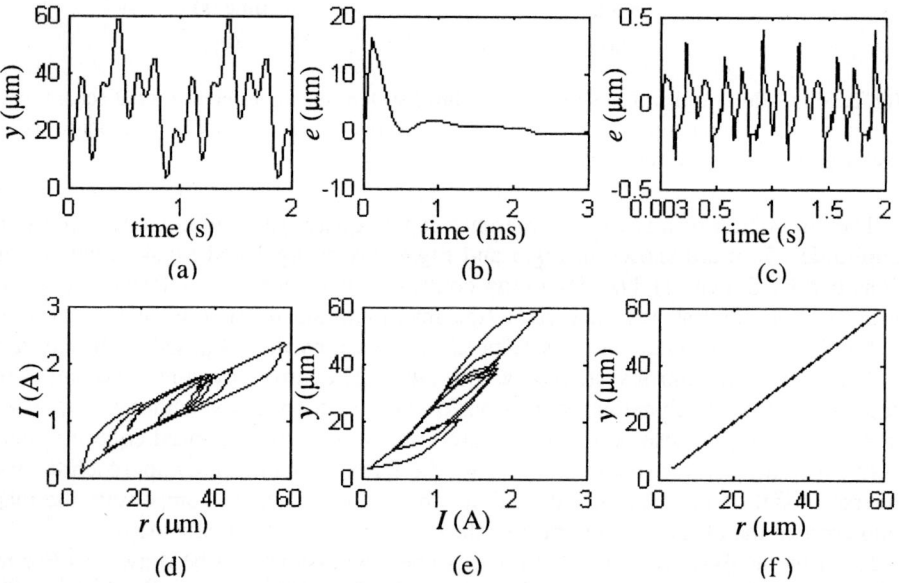

Fig. 5. Simulation results of the proposed control system for the reference input $r=30+11\sin(6\pi t)+12\sin(2\pi t -\pi/2)+9\sin(12\pi t +\pi)\mu m$. (a) the output displacement response; (b) and (c) the tracking error response((b): 0-3ms; (c):3ms-2s); (d) the control signal *I* versus the input *r* from 3ms to 2s; (e) the output displacement *y* versus the control signal *I* from 3ms to 2s; (f) the output displacement *y* versus the input *r* from 3ms to 2s.

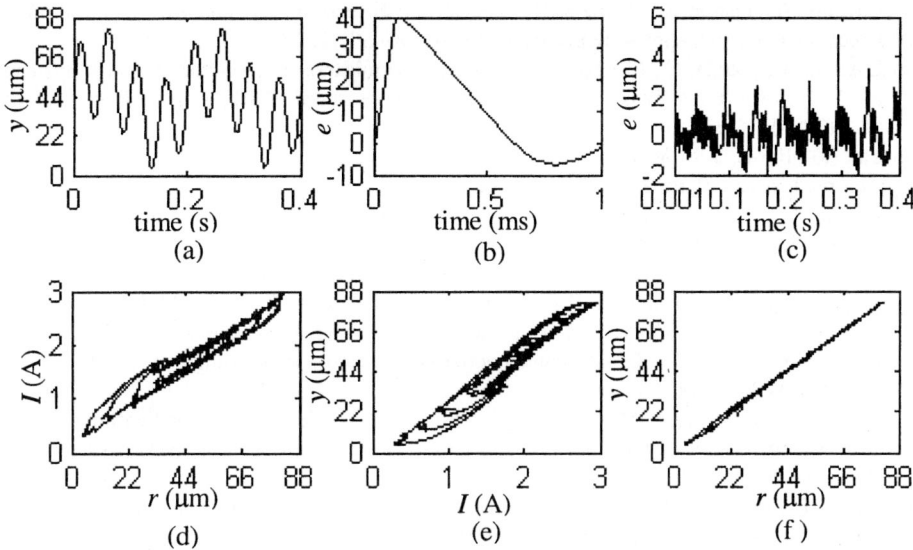

Fig. 6. Simulation results of the proposed control system for the reference input $r=43+15\sin(10\pi t)+25\sin(40\pi t)$μm. (a) the output displacement response; (b) and (c) the tracking error response((b): 0-1ms; (c):1ms-0.4s); (d) the control signal I versus the input r from 1ms to 0.4s; (e) the output displacement y versus the control signal I from 1ms to 0.4s; (f) the output displacement y versus the input r from 1ms to 0.4s.

4 Conclusions

According to the hysteretic nonlinearity of the MA, a DRNNH is constructed and a control strategy combining the DRNNH and PD controllers is proposed. Compared with the previous control methods, the proposed control strategy has the following merits: *1)* Not need to know the mathematical model of the MA; *2)* Learning and controlling are performed simultaneously; *3)* The control strategy can on-line obtain the inverse model of the MA in very short time, thus eliminate the impact of hysteresis and lead to very small tracking error. Thus the control strategy has strong robustness, and adaptability, and is also suitable for the piezoelectric and shape memory alloy actuators with hysteresis.

References

1. Jung, S., Kim, S.: Improvement of Scanning Accuracy of PZT Piezoelectric Actuators by Feedforward Model-Reference Control. Precision Engineering, vol.16, no.1(1994)40-55
2. Cruz-Hernández, J.M., Hayward,V.: Phase Control Approach to Hysteresis Reduction. IEEE Trans.Contr.Syst.Technol., vol.9,no.1(2001)17-26
3. Natale,C., Velardi,F., Visone,C.: Identification and Compensation of Preisach Hysteresis Models for Magnetostrictive Actuators. Physica B 306(2001) 161-165

4. Cavallo, A., Natale, C., Pirozzi, S., Visone, C.: Effects of Hysteresis Compensation in Feedback Control Systems. IEEE Trans. Magn., vol 39, no.3 (2003)1389-1392
5. Cavallo, A., Natale, C., Pirozzi, S., Visone, C., Formisano, A.: Feedback Control Systems for Micropositioning Tasks with Hysteresis Compensation. IEEE Trans. Magn., vol.40, no.2 (2004)876-879
6. Tan, X., Baras, J.S.: Modeling and Control of Hysteresis in Magnetostrictive Actuators. Automatica, vol.40, no.9 (2004) 1469-1480
7. Hwang, C.L., Jan, C., Chen, Y. H.: Piezomechanics using Intelligent Variable Structure Control. IEEE Trans. Ind. Electron., vol. 48,no.1(2001)47–59
8. Calkins, F.T., Smith, R. C., Flatau, A. B.: Energy-Based Hysteresis Model for Magnetostrictive Transducers. IEEE Trans. Magn., vol.36, no.2 (2000)429-439
9. Cao, S.Y., Wang, B.W., Yan, R.G., Huang, W.M., Weng, L.: Dynamic Model with Hysteretic Nonlinearity for Giant Magnetostrictive Actuator. Proceedings of CSEE, vol.23, no.11 (2003)145-149
10. Miyamoto, H., Kawato, M., Setoyama,T., Suzukim, R.: Feedback-Error-Learning Neural Network for Trajectory Control of a Robotic Manipulator. Neural Networks, vol.1 (1988)251-265
11. Rao, D.H., Gupta, M.M.: Dynamic Neural Adaptive Control Schemes. Proc. Of American Control Conference, San Francisco, (1993)1450-1454
12. Ku, C.C., Lee, K.Y.: Diagonal Recurrent Neural Networks for Dynamic systems Control. IEEE Trans. Neural. Networks., vol.6,no.1(1995)144-156
13. Bharitkar, S., Mendel, J. M.: The Hysteretic Hopfield Neural Network. IEEE Trans. Neural Networks, vol.11, no.4(2000)879-888

Orthogonal Forward Selection for Constructing the Radial Basis Function Network with Tunable Nodes

Sheng Chen[1], Xia Hong[2], and Chris J. Harris[1]

[1] School of Electronics and Computer Science,
University of Southampton, Southampton SO17 1BJ, UK
{sqc, cjh}@ecs.soton.ac.uk
[2] Department of Cybernetics, University of Reading,
Reading RG6 6AY, UK
x.hong@reading.ac.uk

Abstract. An orthogonal forward selection (OFS) algorithm based on the leave-one-out (LOO) criterion is proposed for the construction of radial basis function (RBF) networks with tunable nodes. This OFS-LOO algorithm is computationally efficient and is capable of identifying parsimonious RBF networks that generalise well. Moreover, the proposed algorithm is fully automatic and the user does not need to specify a termination criterion for the construction process.

1 Introduction

The radial basis function (RBF) network is a popular artificial neural network structure that has found wide applications in machine learning and engineering. The parameters of the RBF network include its centre vectors and variances of the basis functions as well as the weights that connect the RBF nodes to its output node. The parameters of a RBF network can be learned via nonlinear optimisation using the gradient based algorithm [1], the evolutionary algorithm [2] or the E-M algorithm [3]. Such a nonlinear learning approach is computationally expensive and may encounter the local minima problem. Additionally, the network structure or the number of RBF nodes has to be determined via other means. Alternatively, clustering algorithms can be applied to find the RBF centre vectors as well as the associated basis function variances [4]-[6]. This leaves the RBF weights to be determined by the usual linear least squares solution. Again, the number of the clusters has to be determined via other means, such as cross validation.

A popular approach for constructing RBF networks is to formulate the problem as a linear learning one by considering the training input data points as candidate RBF centres and employing a common variance for every RBF node. A parsimonious RBF network is then identified using the efficient orthogonal least squares (OLS) algorithm [7]-[10]. Similarly, the support vector machine (SVM) and other sparse kernel modelling methods [11]-[17] also fit the kernel centres to the training input data points and adopt a common variance for every kernels. A sparse kernel representation is then sought. Since the common variance is not provided by the learning algorithm, it has to be determined via cross validation. In a recent work [10], a locally regularised OLS

(LROLS) algorithm based on the leave-one-out (LOO) mean square error criterion has been proposed, which compares favourably with other existing state-of-the-art sparse kernel modelling methods, in terms of model sparisty and generalisation performance.

This paper proposes an efficient construction algorithm for the RBF network with tunable nodes. In this approach, each RBF node has a tunable centre vector and a tunable diagonal covariance matrix, and an orthogonal forward selection (OFS) procedure is adopted to append the RBF nodes one by one by incrementally minimising the LOO criterion. Because the RBF centres are not restricted to the training input points and each node has an individually adjusted covariance matrix, the proposed OFS-LOO algorithm can produce sparser representations with excellent generalisation capability, in comparison with the existing sparse RBF or kernel modelling methods. Efficiency of the proposed algorithm is ensured because of the orthogonalisation procedure. Furthermore, the construction process is fully automatic and there is no need for the user to specify any additional termination criterion.

2 Construction of the RBF Network with Tunable Nodes

Consider the regression modelling problem of approximating the N pairs of training data, $\{(\mathbf{x}_k, y_k)\}_{k=1}^N$, with the RBF network defined in

$$y_k = \hat{y}_k + e_k = \sum_{i=1}^{M} w_i g_i(\mathbf{x}_k) + e_k = \mathbf{g}^T(k)\mathbf{w} + e_k \tag{1}$$

where $\mathbf{x}_k \in \mathcal{R}^m$, $\hat{y}_k$ denotes the RBF model output, $e_k = y_k - \hat{y}_k$ is the modelling error, M is the number of RBF nodes, $\mathbf{w} = [w_1\ w_2 \cdots w_M]^T$ is the RBF weight vector, $g_i(\bullet)$ for $1 \le i \le M$ denote the RBF regressors, and $\mathbf{g}(k) = [g_1(\mathbf{x}_k)\ g_2(\mathbf{x}_k) \cdots g_M(\mathbf{x}_k)]^T$. We will consider the general RBF regressor of the form

$$g_i(\mathbf{x}) = K\left(\sqrt{(\mathbf{x} - \boldsymbol{\mu}_i)^T \boldsymbol{\Sigma}_i^{-1} (\mathbf{x} - \boldsymbol{\mu}_i)}\right) \tag{2}$$

where $\boldsymbol{\mu}_i$ is the centre vector of the ith RBF unit, the diagonal covariance matrix has the form $\boldsymbol{\Sigma}_i = \text{diag}\{\sigma_{i,1}^2, \cdots, \sigma_{i,m}^2\}$, and $K(\bullet)$ is the chosen RBF or kernel function. By defining $\mathbf{y} = [y_1\ y_2 \cdots y_N]^T$, $\mathbf{e} = [e_1\ e_2 \cdots e_N]^T$, and $\mathbf{G} = [\mathbf{g}_1\ \mathbf{g}_2 \cdots \mathbf{g}_M]$ with

$$\mathbf{g}_k = [g_k(\mathbf{x}_1)\ g_k(\mathbf{x}_2) \cdots g_k(\mathbf{x}_N)]^T,\ 1 \le k \le M \tag{3}$$

the regression model (1) over the training data set can be written in the matrix form

$$\mathbf{y} = \mathbf{Gw} + \mathbf{e} \tag{4}$$

Note that $\mathbf{g}_k$ denotes the kth column of $\mathbf{G}$ while $\mathbf{g}^T(k)$ is the kth row of $\mathbf{G}$.

Let an orthogonal decomposition of the regression matrix $\mathbf{G}$ be $\mathbf{G} = \mathbf{PA}$, where $\mathbf{A}$ is the upper triangular matrix with unity diagonal elements and $\mathbf{P} = [\mathbf{p}_1\ \mathbf{p}_2 \cdots \mathbf{p}_M]$ with the orthogonal columns that satisfy $\mathbf{p}_i^T \mathbf{p}_j = 0$, if $i \ne j$. The regression model (4) can alternatively be expressed as

$$\mathbf{y} = \mathbf{P}\boldsymbol{\theta} + \mathbf{e} \tag{5}$$

where the weight vector $\boldsymbol{\theta} = [\theta_1\ \theta_2 \cdots \theta_M]^T$ in the orthogonal model space satisfies the triangular system $\mathbf{A}\mathbf{w} = \boldsymbol{\theta}$. Since the space spanned by the original model bases $g_i(\bullet)$, $1 \leq i \leq M$, is identical to the space spanned by the orthogonal model bases, the RBF model output is equivalently expressed by

$$\hat{y}_k = \mathbf{p}^T(k)\boldsymbol{\theta} \tag{6}$$

where $\mathbf{p}^T(k) = [p_1(k)\ p_2(k) \cdots p_M(k)]$ is the kth row of $\mathbf{P}$.

2.1 Orthogonal Forward Selection Based on the Leave-One-Out Criterion

The LOO mean square error is a measure of the model generalisation capability [10]. For the n-term RBF model, the LOO criterion is defined as

$$J_n = \frac{1}{N}\sum_{i=1}^{N}\left(e_i^{(n,-i)}\right)^2 = \frac{1}{N}\sum_{i=1}^{N}\left(\frac{e_i^{(n)}}{\eta_i^{(n)}}\right)^2 \tag{7}$$

where $e_i^{(n,-i)}$ denotes the LOO modelling error of the n-term model, $e_i^{(n)}$ the usual n-term modelling error, and $\eta_i^{(n)}$ the LOO modelling error weighting. Note that $e_k^{(n)}$ and $\eta_k^{(n)}$ can be computed recursively using

$$e_k^{(n)} = y_k - \sum_{i=1}^{n}\theta_i p_i(k) = e_k^{(n-1)} - \theta_n p_n(k) \tag{8}$$

and

$$\eta_k^{(n)} = 1 - \sum_{i=1}^{n}\frac{p_i^2(k)}{\mathbf{p}_i^T\mathbf{p}_i + \lambda} = \eta_k^{(n-1)} - \frac{p_n^2(k)}{\mathbf{p}_n^T\mathbf{p}_n + \lambda} \tag{9}$$

respectively, where $\lambda \geq 0$ is a small regularisation parameter. Therefore, the computation of the LOO criterion J_n is very efficient.

The proposed OFS-LOO algorithm appends the RBF nodes one by one by incrementally minimising the LOO criterion J_n. Specifically, at the nth stage of the construction procedure, the nth RBF node is determined by minimising J_n with respect to the node's centre vector $\boldsymbol{\mu}_n$ and diagonal covariance matrix $\boldsymbol{\Sigma}_n$

$$\min_{\boldsymbol{\mu}_n,\boldsymbol{\Sigma}_n} J_n(\boldsymbol{\mu}_n, \boldsymbol{\Sigma}_n) \tag{10}$$

The construction procedure is automatically terminated if $J_M \leq J_{M+1}$, yielding an M-term RBF network. Note that the LOO criterion J_n is at least locally convex and such an M exists [10]. After the OFS-LOO model construction, the LROLS-LOO algorithm of [10] can be applied to further reduce the model size and to automatically update regularisation parameters. Note that the refinement involving the LROLS-LOO requires a minimal computation, as the selected model size M is typically very small.

2.2 Positioning and Shaping a RBF Node

The task at the nth stage of the model construction is to position and shape the nth RBF node by solving the optimisation problem (10). Since this optimisation problem is non-convex, a gradient-based algorithm may become trapped at a local minimum. We adopt a global search algorithm called the repeated weighted boosting search (RWBS) [18] to determine $\boldsymbol{\mu}_n$ and $\boldsymbol{\Sigma}_n$. The algorithm is summarised as follows. Let $\mathbf{u}$ be the vector that contains $\boldsymbol{\mu}_n$ and $\boldsymbol{\Sigma}_n$. Give the following initial conditions:

$$e_k^{(0)} = y_k \text{ and } \eta_k^{(0)} = 1, \ 1 \leq k \leq N, \text{ and } J_0 = \frac{1}{N}\mathbf{y}^T\mathbf{y} = \frac{1}{N}\sum_{k=1}^{N} y_k^2 \quad (11)$$

Specify the following algorithmic parameters: P_S – population size, N_G – number of generations in the repeated search, and ξ_B – accuracy for terminating the weighted boosting search.

Outer loop: generations For $l = 1 : N_G$

Generation initialisation: Initialise the population by setting $\mathbf{u}_1^{[l]} = \mathbf{u}_{\text{best}}^{[l-1]}$ and randomly generating rest of the population members $\mathbf{u}_i^{[l]}$, $2 \leq i \leq P_S$, where $\mathbf{u}_{\text{best}}^{[l-1]}$ denotes the solution found in the previous generation. If $l = 1$, $\mathbf{u}_1^{[l]}$ is also randomly chosen.

Weighted boosting search initialisation: Assign the initial distribution weightings $\delta_i(0) = \frac{1}{P_S}$, $1 \leq i \leq P_S$, for the population. Then

1. For $1 \leq i \leq P_S$, generate $\mathbf{g}_n^{i)}$ from $\mathbf{u}_i^{[l]}$, the candidates for the nth model column, and orthogonalise them:

$$\alpha_{j,n}^{i)} = \frac{\mathbf{p}_j^T \mathbf{g}_n^{i)}}{\mathbf{p}_j^T \mathbf{p}_j}, \ 1 \leq j < n \quad (12)$$

$$\mathbf{p}_n^{i)} = \mathbf{g}_n^{i)} - \sum_{j=1}^{n-1} \alpha_{j,n}^{i)} \mathbf{p}_j \quad (13)$$

$$\theta_n^{i)} = \frac{\left(\mathbf{p}_n^{i)}\right)^T \mathbf{y}}{\left(\mathbf{p}_n^{i)}\right)^T \mathbf{p}_n^{i)} + \lambda} \quad (14)$$

2. For $1 \leq i \leq P_S$, calculate the LOO cost function value of each $\mathbf{u}_i^{[l]}$:

$$e_k^{(n)}(i) = e_k^{(n-1)} - p_n^{i)}(k)\theta_n^{i)}, \ 1 \leq k \leq N \quad (15)$$

$$\eta_k^{(n)}(i) = \eta_k^{(n-1)} - \frac{\left(p_n^{i)}(k)\right)^2}{\left(\mathbf{p}_n^{i)}\right)^T \mathbf{p}_n^{i)} + \lambda}, \ 1 \leq k \leq N \quad (16)$$

$$J_n^{i)} = \frac{1}{N}\sum_{k=1}^{N} \left(\frac{e_k^{(n)}(i)}{\eta_k^{(n)}(i)}\right)^2 \quad (17)$$

where $p_n^{i)}(k)$ is the kth element of $\mathbf{p}_n^{i)}$.

Inner loop: weighted boosting search $t = 0; t = t + 1$
Step 1: Boosting
1. Find
$$i_{\text{best}} = \arg\min_{1 \leq i \leq P_S} J_n^{i)} \text{ and } i_{\text{worst}} = \arg\max_{1 \leq i \leq P_S} J_n^{i)}$$
Denote $\mathbf{u}_{\text{best}}^{[l]} = \mathbf{u}_{i_{\text{best}}}^{[l]}$ and $\mathbf{u}_{\text{worst}}^{[l]} = \mathbf{u}_{i_{\text{worst}}}^{[l]}$.
2. Normalise the cost function values
$$\bar{J}_n^{i)} = \frac{J_n^{i)}}{\sum_{m=1}^{P_S} J_n^{m)}}, \quad 1 \leq i \leq P_S$$
3. Compute a weighting factor β_t according to
$$\xi_t = \sum_{i=1}^{P_S} \delta_i(t-1) \bar{J}_n^{i)}, \quad \beta_t = \frac{\xi_t}{1 - \xi_t}$$
4. Update the distribution weightings for $1 \leq i \leq P_S$
$$\delta_i(t) = \begin{cases} \delta_i(t-1) \beta_t^{\bar{J}_n^{i)}}, & \text{for } \beta_t \leq 1 \\ \delta_i(t-1) \beta_t^{1 - \bar{J}_n^{i)}}, & \text{for } \beta_t > 1 \end{cases}$$
and normalise them
$$\delta_i(t) = \frac{\delta_i(t)}{\sum_{m=1}^{P_S} \delta_m(t)}, \quad 1 \leq i \leq P_S$$

Step 2: Parameter updating
1. Construct the $(P_S + 1)$th point using the formula
$$\mathbf{u}_{P_S+1} = \sum_{i=1}^{P_S} \delta_i(t) \mathbf{u}_i^{[l]}$$
2. Construct the $(P_S + 2)$th point using the formula
$$\mathbf{u}_{P_S+2} = \mathbf{u}_{\text{best}}^{[l]} + \left(\mathbf{u}_{\text{best}}^{[l]} - \mathbf{u}_{P_S+1} \right)$$
3. Calculate $\mathbf{g}_n^{P_S+1)}$ and $\mathbf{g}_n^{P_S+2)}$ from $\mathbf{u}_{P_S+1}$ and $\mathbf{u}_{P_S+2}$, orthogonalise these two candidate model columns (as in (12) to (14)), and compute their corresponding LOO cost function values $J_n^{i)}$, $i = P_S + 1, P_S + 2$ (as in (15) to (17)). Then find
$$i_* = \arg\min_{i=P_S+1, P_S+2} J_n^{i)}$$
The pair $(\mathbf{u}_{i_*}, J_n^{i_*)})$ then replaces $(\mathbf{u}_{\text{worst}}^{[l]}, J_n^{i_{\text{worst}})})$ in the population
If $\|\mathbf{u}_{P_S+1} - \mathbf{u}_{P_S+2}\| < \xi_B$, exit **inner loop**.
End of inner loop
The solution found in the lth generation is $\mathbf{u} = \mathbf{u}_{\text{best}}^{[l]}$.

End of outer loop

This yields the solution $\mathbf{u} = \mathbf{u}_{\text{best}}^{[N_G]}$, i.e. $\boldsymbol{\mu}_n$ and $\boldsymbol{\Sigma}_n$ of the nth RBF node, the nth model column $\mathbf{g}_n$, the orthogonalisation coefficients $\alpha_{j,n}$, $1 \leq j < n$, the corresponding orthogonal model column $\mathbf{p}_n$, and the weight θ_n, as well as the n-term modelling errors $e_k^{(n)}$ and associated LOO modelling error weightings $\eta_k^{(n)}$ for $1 \leq k \leq N$.

3 Modelling Examples

Example 1. The engine data set [19] was used to demonstrate the effectiveness of the proposed OFS-LOO algorithm. The data were collected from a Leyland TL11 turbocharged, direct injection diesel engine operated at low engine speed, where the input $u(t)$ was the fuel rack position and the output $y(t)$ was the engine speed. The input-output data set, depicted in Fig. 1, contained 410 samples. The first 210 data points were used in training and the last 200 points in model validation. The previous study

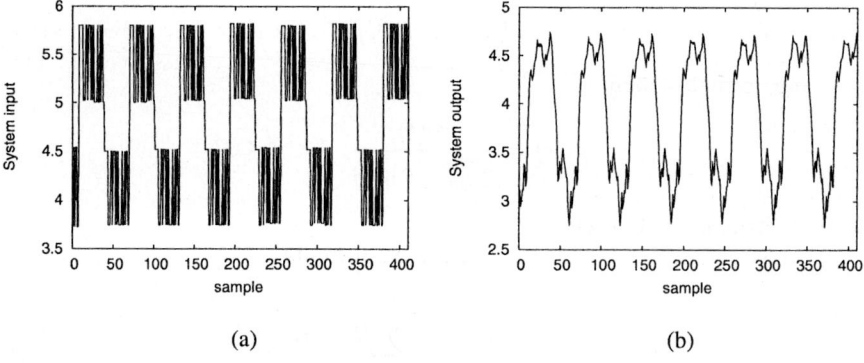

Fig. 1. The engine data set: (a) input $u(t)$ and (b) output $y(t)$

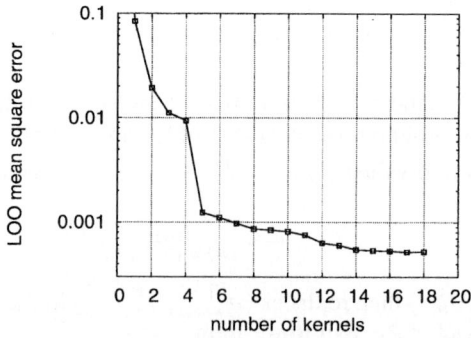

Fig. 2. The LOO mean square error as a function of the model size for the engine data set

Table 1. Comparison of the three models obtained by the SVM, LROLS-LOO and OFS-LOO algorithms for the engine data set

algorithm	RBF type	model size	MSE over training set	MSE over test set
SVM	fixed Gaussian	92	0.000447	0.000498
LROLS-LOO	fixed Gaussian	22	0.000453	0.000490
OFS-LOO	tunable Gaussian	15	0.000466	0.000480

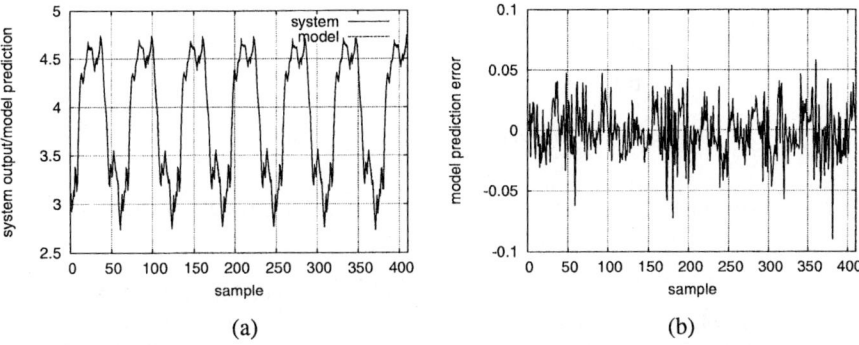

Fig. 3. Modelling performance for the engine data set by the 15-node RBF network constructed by the OFS-LOO algorithm: (a) the model output $\hat{y}_k$ superimposed on the system output y_k, and (b) the modelling error $e_k = y_k - \hat{y}_k$

[9],[10] has shown that this data set can be modelled adequately as $y_i = f_s(\mathbf{x}_i) + e_i$ with $y_i = y(i)$, $\mathbf{x}_i = [y(i-1)\ u(i-1)\ u(i-2)]^T$, where $f_s(\bullet)$ describes the unknown underlying system to be identified and e_i denotes the system noise.

In the work [10], various state-of-the-art RBF and kernel modelling techniques were applied to construct Gaussian RBF network models for this data set, and the LROLS-LOO algorithm produced the best result. We applied the proposed OFS-LOO technique to this data set. Fig. 2 depicts the LOO mean square error (MSE) as a function of the model size during the modelling process using the OFS-LOO. It can be seen that the algorithm automatically constructed a 17-term RBF model, since $J_{18} > J_{17}$. The LROLS-LOO algorithm was then employed to further simplify this obtained model, yielding a final 15-term RBF network. This 15-term model is compared with the model quoted from [10], which was obtained purely by the LROLS-LOO method, in Table 1. As a comparison, the model obtained by the SVM algorithm is also listed in Table 1. Fig. 3 illustrates the modelling performance of the 15-node RBF network constructed by the OFS-LOO algorithm.

Example 2. This example constructed a model for the gas furnace data set (Series J in [20]). The data set, depicted in Fig. 4, contained 296 pairs of input-output points. The input u_k was the coded input gas feed rate and the output y_k represented CO_2 concentration from the gas furnace. All the 296 data points were used in training, and the input vector was defined as $\mathbf{x}_k = [y_{k-1}\ y_{k-2}\ y_{k-3}\ u_{k-1}\ u_{k-2}\ u_{k-3}]^T$. In the

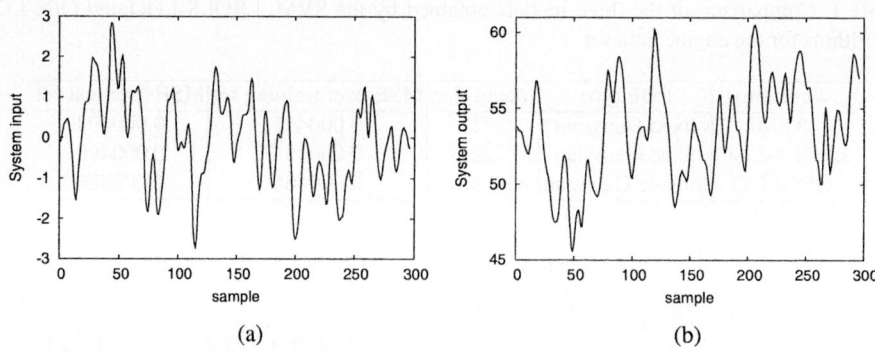

Fig. 4. The gas furnace data set: (a) input $u(t)$ and (b) output $y(t)$

Table 2. Comparison of the three models obtained by the SVM, LROLS-LOO and OFS-LOO algorithms for the gas furnace data set

algorithm	RBF type	model size	training MSE	LOO MSE
SVM	fixed Gaussian	62	0.052416	0.054376
LROLS-LOO	fixed thin-plate-spline	28	0.053306	0.053685
OFS-LOO	tunable Gaussian	15	0.054306	0.054306

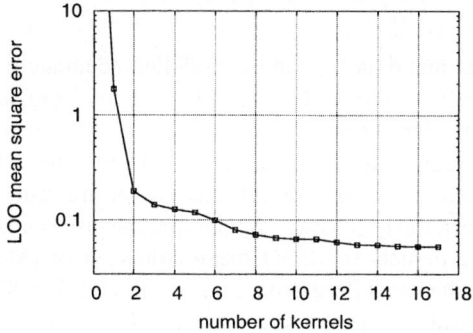

Fig. 5. The LOO mean square error as a function of the model size for the gas furnace data set

study [10], several existing RBF modelling techniques were applied to this data set using the thin-plate-spline basis functions defined by

$$K(\|\mathbf{x} - \mathbf{x}_i\|) = \|\mathbf{x} - \mathbf{x}_i\|^2 \log(\|\mathbf{x} - \mathbf{x}_i\|), \ 1 \leq i \leq N, \tag{18}$$

and the best result was obtained by the LROLS-LOO algorithm. The RBF network constructed by the LROLS-LOO algorithm is given in Table 2, where the LOO MSE was used to indicate the model generalization performance since there was no test data set. We also applied the SVM algorithm to fit a RBF network with the Gaussian basis function to this data set and the resulting model is also listed in Table 2.

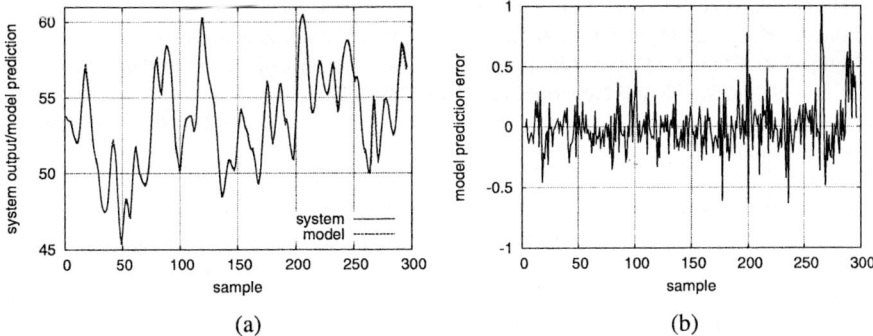

Fig. 6. Modelling performance for the gas furnace data set by the 15-node RBF network constructed by the OFS-LOO algorithm: (a) the model output $\hat{y}_k$ superimposed on the system output y_k, and (b) the modelling error $e_k = y_k - \hat{y}_k$

We applied the proposed OFS-LOO technique to this data set. Fig. 5 depicts the LOO MSE as a function of the model size during the modelling process using the OFS-LOO. It can be seen that the algorithm automatically constructed a 16-term RBF model, since $J_{17} \geq J_{16}$. The LROLS-LOO algorithm was then employed to further simplify this obtained model, yielding a final 15-term RBF network. This 15-term model is compared with the two models obtained by the SVM and LROLS-LOO algorithms, in Table 2. Fig. 6 illustrates the modelling performance of this 15-node RBF network constructed by the OFS-LOO algorithm.

4 Conclusions

A novel construction algorithm has been proposed for RBF networks with tunable nodes. Unlike most of the sparse RBF or kernel modelling methods, the RBF centres are not restricted to the training input data points and each node has an individually adjusted diagonal covariance matrix. The proposed OFS-LOO method appends the RBF nodes one by one by incrementally minimising the LOO mean square error. This construction process is computationally efficient due to the orthogonalisation procedure employed. Moreover, the model construction is fully automatic and the user does not need to specify a termination criterion.

Acknowledgements

S. Chen wish to thank the support of the United Kingdom Royal Academy of Engineering.

References

1. An, P.E., Brown, M., Chen, S., Harris, C.J.: Comparative Aspects of Neural Network Algorithms for On-Line Modelling of Dynamic Processes. Proc. I. MECH. E., Pt.I, J. Systems and Control Eng. **207** (1993) 223–241

2. Whitehead, B.A., Choate, T.D.: Evolving Space-Filling Curves to Distribute Radial Basis Functions Over an Input Space. IEEE Trans. Neural Networks **5** (1994) 15–23
3. Yang, Z.R., Chen, S.: Robust Maximum Likelihood Training of Heteroscedastic Probabilistic Neural Networks. Neural Networks **11** (1998) 739–747
4. Moody, J., Darken, C.J.: Fast Learning in Networks of Locally-Tuned Processing Units. Neural Computation **1** (1989) 281–294
5. Chen, S., Billings, S.A., Grant, P.M.: Recursive Hybrid Algorithm for Non-linear System Identification Using Radial Basis Function Networks. Int. J. Control **55** (1992) 1051–1070
6. Chen, S.: Nonlinear Time Series Modelling and Prediction Using Gaussian RBF Networks with Enhanced Clustering and RLS Learning. Electronics Letters **31** (1995) 117–118
7. Chen, S., Cowan, C.F.N., Grant, P.M.: Orthogonal Least Squares Learning Algorithm for Radial Basis Function Networks. IEEE Trans. Neural Networks **2** (1991) 302–309
8. Chen, S., Wu, Y., Luk, B.L.: Combined Genetic Algorithm Optimisation and Regularised Orthogonal Least Squares Learning for Radial Basis Function Networks. IEEE Trans. Neural Networks **10** (1999) 1239–1243
9. Chen, S., Hong, X., Harris, C.J.: Sparse Kernel Regression Modelling Using Combined Locally Regularized Orthogonal Least Squares and D-Optimality Experimental Design. IEEE Trans. Automatic Control **48** (2003) 1029–1036
10. Chen, S., Hong, X., Harris, C.J., Sharkey, P.M.: Sparse Modelling Using Orthogonal Forward Regression with PRESS Statistic and Regularization. IEEE Trans. Systems, Man and Cybernetics, Part B **34** (2004) 898–911
11. Vapnik, V.: The Nature of Statistical Learning Theory. Springer-Verlag, New York (1995)
12. Gunn, S.: Support Vector Machines for Classification and Regression. Technical Report, ISIS Research Group, Department of Electronics and Computer Science, University of Southampton, UK (1998)
13. Chen, S.S., Donoho, D.L., Saunders, M.A.: Atomic Decomposition by Basis Pursuit. SIAM Review **43** (2001) 129–159
14. Tipping, M.E.: Sparse Bayesian Learning and the Relevance Vector Machine. J. Machine Learning Research **1** (2001) 211–244
15. Schölkopf, B., Smola, A.J.: Learning with Kernels: Support Vector Machines, Regularization, Optimization, and Beyond. MIT Press, Cambridge, MA (2002)
16. Vincent, P., Bengio, Y.: Kernel Matching Pursuit. Machine Learning **48** (2002) 165–187
17. Lanckriet, G.R.G., Cristianini, N., Bartlett, P., Ghaoui, L.E., Jordan, M.I.: Learning the Kernel Matrix with Semidefinite Programming. J. Machine Learning Research **5** (2004) 27–72
18. Chen, S., Wang, X.X., Harris, C.J.,: Experiments with Repeating Weighted Boosting Search for Optimization in Signal Processing Applications. IEEE Trans. Systems, Man and Cybernetics, Part B **35** (2005) 682–693
19. Billings, S.A., Chen, S., Backhouse, R.J.: The Identification of Linear and Non-linear Models of a Turbocharged Automotive Diesel Engine. Mechanical Systems and Signal Processing **3** (1989) 123–142
20. Box, G.E.P., Jenkins, G.M.: Time Series Analysis, Forecasting and Control. Holden Day Inc. (1976)

A Recurrent Neural Network for Extreme Eigenvalue Problem

Fuye Feng[1], Quanju Zhang[2], and Hailin Liu[3]

[1] Missile College,
Engineering University of Air Force,
SanYuan, Shanxi, China
ffye2004@yahoo.com.cn
[2] Department of Mathematics,
Hong Kong Baptist University,
Kowloon Tong, Hong Kong
qjzhang@math.hkbu.edu.hk
[3] Faculty of Applied Mathematics,
Guangdong University of Technology,
Guangzhou, Guangdong, China
lhl@scnu.edu.cn

Abstract. This paper presents a novel recurrent time continuous neural network model for solving eigenvalue and eigenvector problem. The network is proved to be globally convergent to an exact eigenvector of a matrix A with respect to the problem's feasible region. This convergence is called quasi-convergence in the sense of the starting point to be in the feasible set. It also demonstrates that the network is primal in the sense that the network's neural trajectories will never escape from the feasible region when starting at it. By using an energy function, the network's stable point set is guaranteed to be the eigenvector set of the involved matrix. Compared with the existing neural network models for eigenvalue problem, the new model's performance is more effective and more reliable. Moreover, simulation results are given to illustrate further the global convergence and the fundamental validity of the proposed neural network for eigenvalue problem.

1 Introduction

Computing the eigenvalues and corresponding eigenvectors of a matrix $A \in R^{n \times n}$ is necessary in many scientific and engineering problems, e.g. in signal processing, control theory, geophysics, etc. It is an important subject in numerical algebra to find new methods for eigenvalue computation. Traditional methods to solve this problem are included in Golub's book [1] and more references can be found therein. It is known that conventional methods in scientific computation is time-consuming for large-scale problems and hence neural network methods are encouraging in scientific computing from the seminal work of Hopfield and Tank [2]. Neural network method, unlike classical algorithms, behaves two novel characters of which, one is the computation can perform in real-time on line and the other is the hardware implementation designed by application-specific integrated circuits. It has been investigated extensively to use neural network for solving various optimization problems, see [3-5], in the past twenty years since the method being employed for unconstrained optimization problem first [2]. As to the problems of finding roots of polynomials, Huang [6-9] gave an excellent research work by neural network method which opens another field of neural network's application area. An overview of solving optimization problems by neural network method can be found in monograph [10].

Neural networks proposed for solving eigenvalue and eigenvector problems are much less than the existing models for optimization problems. There exist several models for solving this problem by penalty functions which may generate infeasible solutions and hence fail to find true solutions involved [10-11]. Furthermore, any stability results of the existing models can not be guaranteed there either. This drawback may lead to unreliable and inefficient operation in finding the involved problem's solutions. In this paper, as a challenge for the existing ones, a new neural network model is proposed which can perform well in solving the eigenvalue problems. Unlike the current ones, the new model is proved to be always feasible and globally quasi-convergent to an exact eigenvector of a matrix. This new model is amenable to execute computation procedure in true parallel and distributed way which, of course, is superior to any of the conventional methods.

The remaining part of this paper is organized as follows. Section II formulates the extreme eigenvalue problem as a optimization problem and reveals briefly the idea to construct a neural network for solving it. In section III, a multi-layered recurrent neural network is proposed for solving this problem and a detailed description of the new model's architecture is presented. Meanwhile, a global convergence theorem is proved in this section. Simulation results are remained in section IV which is illustrated by numerical experiments that the model is reliable and efficient for solving eigenvalue problem. Finally, in section V, we summarize main results in this paper and make a concluding remark.

2 Problem Formulation

It is well known [1] that the computation of an eigenvalue λ and its corresponding eigenvector $v = [v_1, \cdots, v_n]^T \neq 0 \in R^n$ of a real matrix $A \in R^{n \times n}$ leads to solve the following algebraic system of equations

$$(A - \lambda I)v = 0, \tag{1}$$

where I is the identity matrix of $n \times n$. Clearly, if v is a eigenvector, so is any multiple of v with a nonzero multiplying factor α because $(A - \lambda I)\alpha v = 0$ when $(A - \lambda I)v = 0$. So, in order to eliminate multiplicity of eigenvectors, normalization to unit length is usually employed in computation, i.e., the constraint

$$v^T v = 1, \tag{2}$$

is required.

Based on penalty function method [10-11], several multi-layered artificial neural network models are proposed for solving problem (1-2). It is known [12] that by using the penalty method to construct neural network models or to make classical algorithms has following three explicit defects. 1) There is a penalty parameter to tune and no rule available can be used to guarantee a good choice for the parameter; 2) It is usually occurring in the penalty method to find infeasible points as optimal solutions instead of true optimal solutions; 3) In order to construct neural network with a penalty function, stability result usually can not be guaranteed in most cases, see [4], [10-11]. So, penalty function method is little employed in practical computation due to the existence of these shortcomings in classical optimization algorithms. We claim here that it is not encouraging in constructing neural network models with penalty function either. Therefore, some new neural network models are needed to solve the problem in a more efficient way. To meet this requirement, we are going to propose a new neural network model which can overcome all the existing defects mentioned above owned by the old ones.

In practical scientific computing problems, only the eigenvalues in a given interval $[a, b]$ are in demand and in most applications, e.g. in signal processing, only the external (i.e. minimal or maximal) real eigenvalues and the corresponding eigenvectors are required for a real symmetric matrix A. Throughout the paper, we will focus our attention on the minimal real eigenvalue and its corresponding eigenvector for a symmetric matrix. This problem can be reformulated, Golub [1], by minimizing the Rayleigh quotient $R(x)$ as follows

$$R(x) = \frac{<Ax, x>}{<x, x>} = \frac{x^T A x}{x^T x}, \qquad (3)$$

where $v \in R^n$ is assumed to be never identically zero. In considering the normalized eigenvectors, we can reformulate the minimal eigenvalue problem as following optimization problem

$$\text{minimize} \quad F(x) = x^T A x \qquad (4)$$
$$\text{s.t.} \quad x^T x = 1. \qquad (5)$$

Since $F(x)$ in (4) is continuous in the nonempty compact feasible set, (4-5) has at least one solution x^1 which should satisfy the KKT conditions. This means that there exists a λ_1 such that

$$2Ax^1 + \lambda_1 2x^1 = 0, \qquad (6)$$

then $Ax^1 = -\lambda_1 x^1$, and so x^1 is an eigenvector of A of unit length. By adding an orthogonal requirement to the first eigenvector x^1 as constraint in (4-5) to construct another optimization programming and using the KKT condition again, we get the existence of another eigenvector x^2 which is orthogonal to the first one x^1. Consequently, it could be demonstrated inductively that a real symmetric matrix A has n mutually orthogonal eigenvectors of unit length.

It is easy to obtain the following equation by differentiating the function $R(x)$ defined in equation (3)

$$\nabla R(x) = \frac{2(x^T x)Ax - 2(x^T Ax)x}{x^T x}. \tag{7}$$

Setting the result to be zero, we get

$$(x^T x)Ax - (x^T Ax)x = 0. \tag{8}$$

Equation (8) characters the equation that the minimal eigenvalue's corresponding eigenvector should be satisfied. Accordingly, it suggests us for the construction of the neural network model to solve our problem. That is, to set the equilibrium points of the neural network to be points of zero values of $\nabla R(x)$. Accordingly, we can construct a neural network model described in the following section.

3 The Neural Network and the Global Convergence

Consider the following multi-layered recurrent neural network with the state variables governed by a system of differential equations

$$\frac{dx}{dt} = -(x^T x)Ax + (x^T Ax)x, \tag{9}$$

here $x \in R^n$, and $A \in R^{n \times n}$ is a symmetric real matrix. Such neural networks governed by a system of autonomous differential equations are implementable by analog circuits [10]. One node, e.g. the i-th node of the network's architecture is depicted in **Fig. 1**. From its construction, it can be seen that the network is composed of fundamental units such as integrators, summers and analog multipliers described in [10] and hence, this network could be well implementable with systems of circuits.

First of all, it should be noted that any nonzero equilibrium point x of the neural systems (9) is an eigenvector of A with eigenvalue $\lambda = (x^T Ax)/(x^T x)$. Conversely, if a nonzero x is an eigenvector of A with the corresponding eigenvalue λ, then, it follows from $Ax = \lambda x$ and $\lambda = (x^T Ax)/(x^T x)$ that this x should be an equilibrium point of system (9). This means that to identify the eigenvectors implies to identify the nontrivial equilibrium points of the proposed neural network model. As it is known, stability of system (9) concerns the system's equilibrium state at a longtime elapsing, this going to be discussed here in detail.

First, we give some mathematical definitions of stability for a general autonomous system $\dot{x} = F(x)$.

Definition 1. Let $x(t)$ be a solution of system $\dot{x} = F(x)$. The system is said to be globally convergent to a set X with respect to set W if every solution $x(t)$ starting at W satisfies

$$\rho(x(t), X) \to 0, \quad \text{as} \quad t \to \infty, \tag{10}$$

here $\rho(x(t), X) = \inf_{y \in X} \|x - y\|$ and $x(0) = x_0 \in W$.

A Recurrent Neural Network for Extreme Eigenvalue Problem

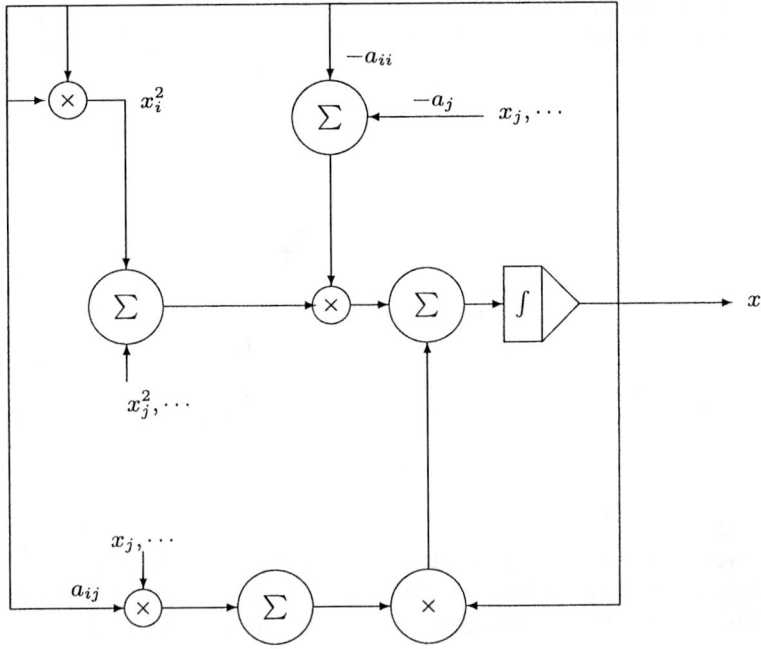

Fig. 1. One node of neural network model (9)

Definition 2. Let $x(t)$ be a solution of system $\dot{x} = F(x)$. The system is said to be primal to W if every solution $x(t)$ starting at W will never escape from this W. If the system is convergent to a set with respect to a primal feasible set W of a mathematical programming, we say that the system is globally quasiconvergent with respect to the feasible set W.

We can now state our main result on the network model's globally convergence as follows.

Theorem 1. Network (9) is primal with respect to the feasible set $W = \{x | x^T x = 1\}$ and the eigenvector set of matix A is globally convergent with respect to this W and hence, network (9) is a globally quasiconvergent neural network system.

Proof: Let $x(t)$ be any trajectory starting at $x_0 \in W$. It follows from (9) that

$$\frac{dx^T x}{dt} = 2x^T \frac{dx}{dt} \tag{11}$$
$$= 2[-(x^T x)x^T Ax + (x^T Ax)x^T x] \tag{12}$$
$$= 0, \tag{13}$$

that is $x(t)$ has to be constant along trajectories of (9), so, $\|x(t)\| = \|x_0\| = 1$, the first part of theorem 1 is proved.

Define an energy function $V(x) = \frac{1}{2}x^T A x$ and compute its total derivative along any neural network trajectory $x(t)$ starting at $x_0 \in W$, we get

$$\frac{dV}{dt} = x^T A \frac{dx}{dt}. \tag{14}$$

It follows from (9) and (14) that

$$\frac{dV}{dt} = -\|x\|^2 x^T A A x + (x^T A x)^2 \tag{15}$$
$$= -\|x\|^2 \|Ax\|^2 + (x^T A x)^2 \tag{16}$$
$$\leq -\|x\|^2 \|Ax\|^2 + \|x\|^2 \|Ax\|^2, \tag{17}$$

where the last inequality comes from Cauchy-Schwartz inequality, and so,

$$\frac{dV}{dt} \leq 0, \tag{18}$$

which means the energy of $V(x)$ is decreasing along any trajectory of (9).

This and the boundedness of $x(t)$ imply that $V(x)$ is a Liapunov function to system (9). So, by LaSalle's invariant principle [13], we know that all these trajectories of (9) will converge to the largest invariant set Σ of set E like

$$\Sigma \subseteq E = \{x \mid \frac{dV}{dt} = 0\}. \tag{19}$$

However, we know that equality holds in Cauchy-Schwartz inequality only if there exists a λ such that $Ax = \lambda x$. Noting that $x(t)$ is primal with respect to set W, that is $x^T x = 1$, we guarantee that $x(t)$ will approach the eigenvector set of matrix A. Theorem 1 is proved to be true then. So, it is demonstrated that the proposed neural network (9) is a promising neural network model both in implementable construction sense and in theoretic convergence sense for solving eigenvalue problem of a symmetric matrix A. In practice, it is also important to simulate a network's effectiveness by numerical experiments to test its performance. Next section will focus on handling some experiment examples to reach this goal.

4 Illustrative Examples

The following two examples are employed as numerical experiments to show the network's good performance.

Example 1. Let A be a matrix of

$$A = \begin{pmatrix} 1 & -2 \\ -2 & 4 \end{pmatrix}. \tag{20}$$

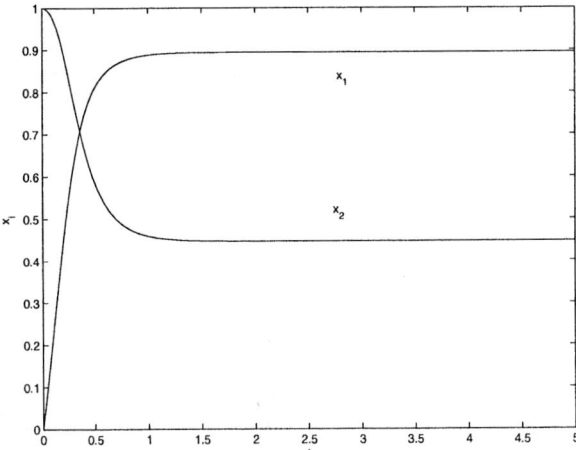

Fig. 2. Transient behaviors of neural trajectories x_1, x_2

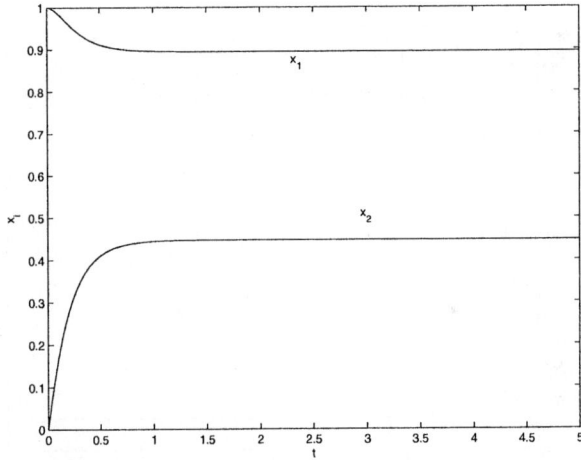

Fig. 3. Neural trajectories x_1, x_2 with starting point $(1, 0)$

It has two eigenvalue $\lambda_1 = 0, \lambda_2 = 5$ with corresponding eigenvectors $x_1 = [0.8944, 0.4472]^T$ and $x_2 = [-0.4472, 0.8944]^T$. Neural network model presented in (9) for this problem can be formulated as

$$\frac{dx}{dt} = \begin{pmatrix} -2x_1^2 + 3x_1x_2^2 + 2x_2^3 \\ 2x_1^3 - 3x_1^2x_2 - 2x_1x_2^2 \end{pmatrix}. \tag{21}$$

ODE23 is employed to solve this system on MATLAB 7.0. The neural trajectories x_1, x_2 at starting point $(0, 1)$ are shown in **Fig. 2**. It can be seen visibly from the figure that the proposed neural network converges to our solutions very soon.

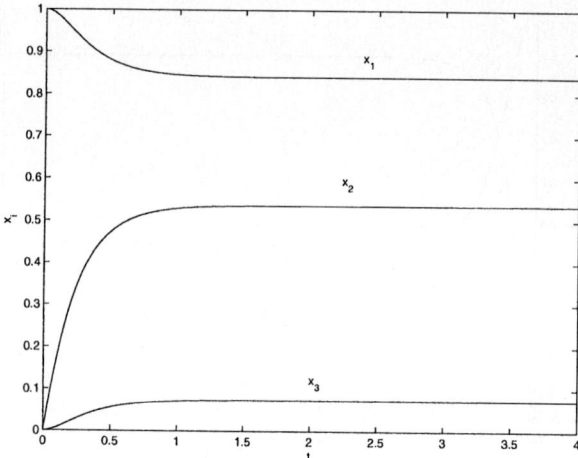

Fig. 4. Transient behaviors of x_1, x_2, x_3 with starting point $(1, 0, 0)$

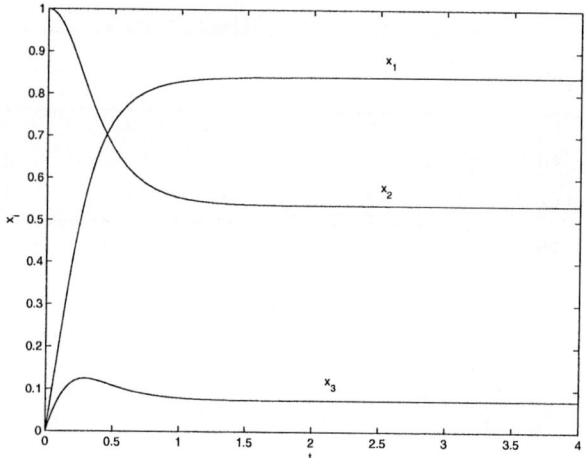

Fig. 5. Neural trajectories x_1, x_2, x_3 from $(0, 1, 0)$

For other initial points, **Fig. 3** presents how the solutions of this problem are located by neural trajectories from a different starting point $(1, 0)$ in a more clearly visible way.

Example 2. Consider A to be the matrix of

$$A = \begin{pmatrix} 2 & -2 & 0 \\ -2 & 4 & -1 \\ 0 & -1 & 8 \end{pmatrix}. \tag{22}$$

The minimal eigenvalue of A is $\lambda_1 = 0.7251$ with corresponding eigenvector $x_1 = [0.8410, 0.5361, 0.0737]^T$. The proposed neural network model for this example is

$$\frac{dx}{dt} = \begin{pmatrix} 2x_1x_2^2 + 6x_1x_3^2 - 2x_1^2x_2 + 2x_2^3 + 2x_2x_3^2 - 2x_1x_2x_3 \\ 2x_1^3 - 2x_1^2x_2 - 2x_1x_2^2 - x_2^2x_3 + 4x_2x_3^2 + 2x_1x_3^2 + x_1^2x_3 + x_3^3 \\ -6x_1^2x_3 - 4x_2^2x_3 - x_2x_3^2 + x_1^2x_2 + x_2^3 - 4x_1x_2x_3 \end{pmatrix}. \quad (23)$$

Conducted on MATLAB 7.0 by ODE23 with initial point $(1, 0, 0)$, the required solutions are guaranteed along the neural trajectories and the transient behaviors of these trajectories are shown in **Fig. 4**.

Also, we choose another initial point $(0, 1, 0)$ as starting point and the corresponding neural trajectories are presented in **Fig. 5** which obviously reveals how the solutions to be located by the proposed neural network model from different initial point.

5 Conclusion

In this paper, we have proposed a neural-network model for solving algebraic eigenvalue and eigenvector problem. It is shown that stability of the proposed neural network governed by a system of differential equations behaves global quasiconvergence with respect to the problem's feasible set. This model has overcome the stability defects and the unfair cases of infeasible solutions may found by the existing models with penalty function. Certainly, the network presented here can perform well on computation in real time online which is also superior to the classical algorithms. Finally, numerical simulation results demonstrate further that the new model can act both effectively and reliably on our purpose of locating solutions to the problems involved.

References

1. Golub, G. H., Van loan, C. F.: Matrix Computations. The Johns Hopkins University Press, Baltimore (1989)
2. Hopfield, J. J., Tank, D. W.: Neural Computation of Decisions in Optimization Problems. Biolog. Cybernetics, Vol. 52, No. 1, (1985) 141-152
3. Bouzerdorm, A., Pattison, T. R.: Neural Network for Quadratic Optimization with Bound Constraints. IEEE Trans. Neural Networks, Vol. 4, No. 2, (1993) 293-304
4. Kennedy, M. P., Chua, L. O.: Neural Networks for Nonlinear Programming. IEEE Trans. on Circuits and Systems, Vol. 35, No. 6, (1988) 554-562
5. Xia, Y. S., Wang, J.: A General Methodology for Designing Globally Convergent Optimization Neural Networks. IEEE Trans. on Neural Networks, Vol. 9, No. 6, (1998) 1311-1343
6. Huang, D.S., Ip, Horace H.S., Chi, Zheru.: A Neural Root Finder of Polynomials Based on Root Moments. Neural Computation, Vol. 16, No. 8, (2004) 1721-1762
7. Huang, D.S.: A Constructive Approach for Finding Arbitrary Roots of Polynomials by Neural Networks. IEEE Transactions on Neural NetworksVol. 15, No. 2, (2004) 477-491

8. Huang, D.S., Ip, Horace H.S., Chi, Zheru., Wong, H.S.: Dilation Method for Finding Close Roots of Polynomials Based on Constrained Learning Neural Networks. Physics Letters A, Vol. 309, No. 5-6, (2003) 443-451
9. Huang, D.S., Ip, Horace H.S., Chi, Zheru., Wong, H.S.: A New Partitioning Neural Network Model for Recursively Finding Arbitrary Roots of Higher Order Arbitrary Polynomials. Applied Mathematics and Computation, Vol. 162, No. 3, (2005) 1183-1200
10. Cichocki, A., Unbehauen, R.: Neural Networks for Optimization and Signal Processing. John Wiley & Sons, (1993)
11. Cichocki, A., Unbehauen, R.: Neural Networks for Computing Eigenvalues and Eigenvectors. Biolog. Cybernetics, Vol. 68, No. 1, (1992) 155-164
12. Bazaraa, M. S., Shetty, C. M.: Nonlinear Programming, Theory and Algorithms. John Wiley and Sons, New York, NY, (1979)
13. LaSalle, J.: The Stability Theory for Ordinary Differential Equations. J. Differential Equations, Vol 4, No. 1, (1983) 57-65

Chaos Synchronization for a 4-Scroll Chaotic System via Nonlinear Control*

Haigeng Luo, Jigui Jian, and Xiaoxin Liao

Department of Control Science and Engineering, Huazhong University,
of Science and Technology, Wuhan, Hubei, 430074, China
moe1968@sina.com

Abstract. This paper investigates the chaos synchronization problem of a new 4-scroll chaotic system. Three nonlinear control approaches via state variables are studied, namely nonlinear feedback control, adaptive control and adaptive sliding mode type of control. Based on Lyapunov stability theory, control laws are derived such that the two identical 4-scroll systems are to be synchronized. Some sufficient conditions for the synchronization are obtained analytically in three cases. Numerical simulation results are given to show the effectiveness of the proposed methods.

1 Introduction

The idea of synchronizing two identical chaotic systems from different initial values was introduced by Pecora and Carroll in 1990 [1]. Synchronization in coupled chaotic systems has been extensively investigated in the last few years [2,3,4,5,6,7,8,9,10,11,12,13,14,15,16,17]. Several types of synchronization features have been proposed: complete synchronization, lag synchronization, generalized synchronization, phase and imperfect phase synchronization and so on [2]. Many possible applications have been discussed in physiology, ecological systems, chemical reaction, nonlinear optics, power systems, fluid dynamics and secure communications by computer simulation and realized in laboratory condition.

Recently, Liu and Chen [18,19,20] found a novel 2- or 4-scroll chaotic attractor(we denote this system by LC system in this paper)which is three-dimensional quadratical continuous autonomous dynamical system with three cross-product nonlinear terms. This system is dissipative and symmetrical, and similar to some known chaotic systems, such as Lorenz system, Chen system and Rössler system, but is not topologically equivalent with them. It displays rich and typical bifurcation and chaotic phenomena for some values of the control parameters.

The aim of this paper is to investigate the synchronization problem of two LC chaotic systems via nonlinear control technique under three cases: When system's parameters are all known, a simple nonlinear feedback controller can guide the response system synchronizing with the drive system. If parameters

---

* Supported by National Natural Science Foundation of China No. 60274007 and No. 60474011.

are all unknown, synchronization can be achieved when an adaptive controller is employed. Furthermore, in order to increase the robustness of the closed loop system, the chaotic system with bounded and time-varied parameters is considered. By designing an adaptive sliding mode type of controller, following the idea of Li and Shi [4], two chaotic systems can mutually robustly synchronize under this case. Lyapunov direct method is used to prove the asymptotic behaviors of solutions for the controlled systems. Numerical simulation results are given to show the effectiveness of the proposed methods.

First we need to recall some concepts and terms of synchronization theory.

Consider the system of differentiable function

$$\dot{x} = f(x), \tag{1}$$

$$\dot{y} = g(y,x), \tag{2}$$

where $x \in R^n, y \in R^n, f, g : R^n \to R^n$ are assumed to be analytic functions.

Let $x(t, x_0)$ and $y(t, y_0)$ be solutions of Eqs. (1) and (2), respectively. The solutions $x(t, x_0)$ and $y(t, y_0)$ are said to be synchronized if

$$\lim_{t \to \infty} \|x(t, x_0) - y(t, y_0)\| = 0. \tag{3}$$

We first reformulate the LC system equation in the following form:

$$\begin{cases} \dot{x} = ax + yz, \\ \dot{y} = -by - xz, \\ \dot{z} = -cz - xy, \end{cases} \tag{4}$$

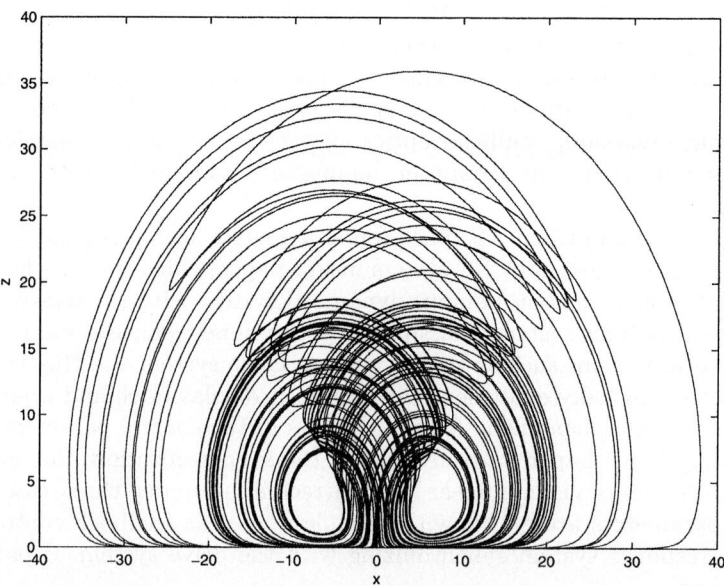

Fig. 1. The 2-scroll chaotic attractor of system (4) with $a = 1, b = 10, c = 4$

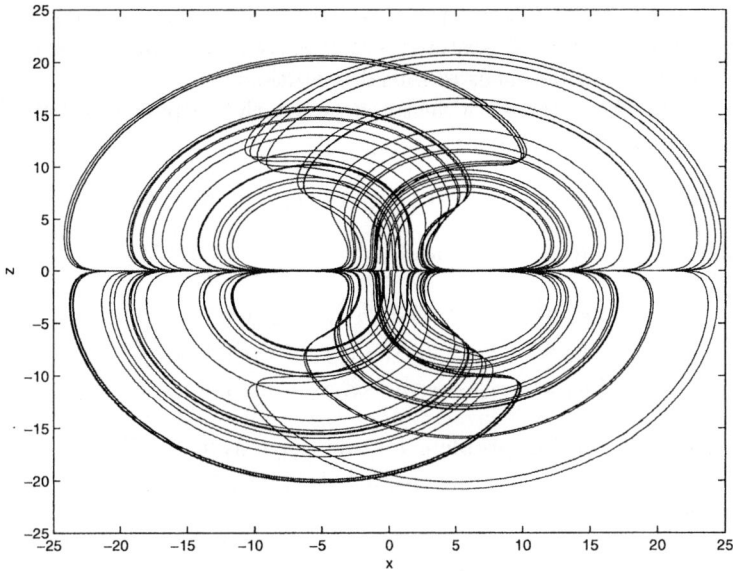

Fig. 2. The 4-scroll chaotic attractor of system (4) with $a = 0.4, b = 10, c = 4$

where a, b, c are positive constants, and satisfy: $a < b+c$. If we select $b = 10, c = 4$, when $a = 1$, or $a = 0.4$, it displays a two-scroll or four-scroll chaotic attractor respectively. Their phase portraits of projection on the $x - z$ plane are shown in Fig. 1 and Fig. 2.

2 Nonlinear State Feedback Synchronization

In order to observe the synchronization behavior in LC system, we have two LC systems where the drive system with three state variables denoted by the subscript 1 drives the response system structurally identical equations denoted by the subscript 2. However, the initial condition on the drive system is different from that of the response system, therefore two LC systems are described, respectively, by the following equations:

$$\begin{cases} \dot{x}_1 = ax_1 + y_1 z_1, \\ \dot{y}_1 = -by_1 - x_1 z_1, \\ \dot{z}_1 = -cz_1 - x_1 y_1, \end{cases} \quad (5)$$

and

$$\begin{cases} \dot{x}_2 = ax_2 + y_2 z_2 + u_1, \\ \dot{y}_2 = -by_2 - x_2 z_2 + u_2, \\ \dot{z}_2 = -cz_2 - x_2 y_2 + u_3, \end{cases} \quad (6)$$

where control input vector $u = [u_1, u_2, u_3]^T$ is to be determined for the purpose of synchronizing the two LC systems with the same and known parameters a, b and c in spite of the differences in initial conditions.

Let us define the state error vector between the response system that is to be controlled and the drive system as

$$e = [e_1, e_2, e_3]^T := [x_2 - x_1, y_2 - y_1, z_2 - z_1]^T. \tag{7}$$

Subtracting Eq. (5) from Eq. (6) and using the notation (7) yields

$$\begin{cases} \dot{e}_1 = ae_1 + z_1e_2 + y_1e_3 + e_2e_3 + u_1, \\ \dot{e}_2 = -be_2 - z_1e_1 - x_1e_3 - e_1e_3 + u_2, \\ \dot{e}_3 = -ce_3 - e_1y_1 - e_2x_1 - e_1e_2 + u_3. \end{cases} \tag{8}$$

Then, the synchronization problem for the drive system (5) and the response system (6) is turned into investigating the stability of zero solution for the error dynamical system (8).

In ideal situation, assuming that the parameters of drive and response system are all known, we design following nonlinear state feedback controller

$$u_1 = -ke_1 - \frac{1}{2}z_1e_2 - \frac{1}{2}y_1e_3, \quad u_2 = x_1e_3, \quad u_3 = x_1e_2. \tag{9}$$

Constructing following Lyapunov function

$$V = \frac{1}{2}(2e_1^2 + e_2^2 + e_3^2).$$

Computing the time derivative of V along the solutions of equation (8), we get

$$\dot{V} = \left.\frac{dV}{dt}\right|_{(8)} = 2e_1\dot{e}_1 + e_2\dot{e}_2 + e_3\dot{e}_3$$

$$= 2ae_1^2 + 2e_1e_2e_3 + 2y_1e_1e_3 - 2e_1(ke_1 + \frac{1}{2}z_1e_2 + \frac{1}{2}y_1e_3)$$
$$-be_2^2 - e_1e_2e_3 - z_1e_1e_2 - x_1e_1e_3 + x_1e_2e_3 - ce_3^2$$
$$-e_1e_2e_3 - y_1e_1e_3 - 2x_1e_2e_3 + x_1e_2e_3$$
$$= 2(a - k)e_1^2 - be_2^2 - ce_3^2.$$

Obviously, $\dot{V}$ is negative definite if $k > a$. Therefore, the trivial solution of error dynamical system (8) is asymptotically stable. This means the drive system (5) completely synchronizes with the response system (6).

3 Adaptive Synchronization

The feedback synchronization method derived in previous section requires that the system parameters must be known a priori. However, in many real applications it can be difficult to determine exactly the values of the system parameters.

Consequently, the feedback gain k cannot be appropriately chosen to guarantee the implement and stability of synchronization. For overestimating k an expensive and too conservative control effort is needed. To overcome these drawbacks, an adaptive synchronization approach with state variable feedback control is derived.

To do this, we choose following state feedback controller:

$$u_1 = -k_1 e_1 - \frac{1}{2} z_1 e_2 - \frac{1}{2} y_1 e_3, \quad u_2 = -k_2 e_2 + x_1 e_3, \quad u_3 = -k_3 e_3 + x_1 e_2, \quad (10)$$

where feedback gain parameters k_1, k_2, k_3 are varied, which are updated according to the following adaption algorithm:

$$\dot{k}_1 = \gamma_1 e_1^2, \quad \dot{k}_2 = \gamma_2 e_2^2, \quad \dot{k}_3 = \gamma_3 e_2^2, \quad (11)$$

where positive constants γ_1, γ_2 and γ_3 are adaption gains. Then the resulting error dynamical system can be expressed by

$$\begin{cases} \dot{e}_1 = a e_1 + \frac{1}{2} z_1 e_2 + \frac{1}{2} y_1 e_3 + e_2 e_3 - k_1 e_1, \\ \dot{e}_2 = -b e_2 - z_1 e_1 - e_1 e_3 - k_2 e_2, \\ \dot{e}_3 = -c e_3 - e_1 y_1 - e_1 e_2 + u_3 - k_3 e_3, \\ \dot{k}_1 = \gamma_1 e_1^2, \quad \dot{k}_2 = \gamma_2 e_2^2, \quad \dot{k}_3 = \gamma_3 e_2^2. \end{cases} \quad (12)$$

Let $\bar{k}_1 = a + \frac{\varepsilon}{2}, \bar{k}_2 = -b + \varepsilon$, and $\bar{k}_3 = -c + \varepsilon$, where ε is arbitrary positive constant. We consider following Lyapunov function

$$V = \frac{1}{2} \left[2 e_1^2 + e_2^2 + e_3^2 + \frac{2}{\gamma_1} \left(k_1 - \bar{k}_1 \right)^2 + \frac{1}{\gamma_2} \left(k_2 - \bar{k}_2 \right)^2 + \frac{1}{\gamma_3} \left(k_3 - \bar{k}_3 \right)^2 \right].$$

Taking the time derivative of V along the solutions of Eq. (12), we get

$$\begin{aligned} \dot{V} = \left. \frac{dV}{dt} \right|_{(12)} &= 2 e_1 \dot{e}_1 + e_2 \dot{e}_2 + e_3 \dot{e}_3 + \frac{2}{\gamma_1} \left(k_1 - a - \frac{\varepsilon}{2} \right) \dot{k}_1 \\ &\quad + \frac{1}{\gamma_2} \left(k_2 + b - \varepsilon \right) \dot{k}_2 + \frac{1}{\gamma_3} \left(k_3 + c - \varepsilon \right) \dot{k}_3 \\ &= -\varepsilon (e_1^2 + e_2^2 + e_3^2) = -\varepsilon \|e^2\| \leqslant 0. \end{aligned} \quad (13)$$

Since V is a positive and decrescent function and $\dot{V}$ is semidefinite, it follows that the equilibrium point $(e_1 = e_2 = e_3 = 0, k_1 = \bar{k}_1, k_2 = \bar{k}_2, k_3 = \bar{k}_3)$ of system (12) is uniformly stable, i.e., $e_1(t), e_2(t), e_3(t) \in L_\infty$, and $k_1(t), k_2(t), k_3(t) \in L_\infty$. In view of Eq. (12), we can easily show that the square of $e_1(t), e_2(t)$, and $e_3(t)$ are integrable with respect to time t, i.e., $e_1(t), e_2(t), e_3(t) \in L_2$. Next by Barbalat's Lemma, Eq. (14) implies that $\dot{e}_1, \dot{e}_2, \dot{e}_3 \in L_\infty$, which in turn implies $e_1(t), e_2(t), e_3(t) \to 0$ as $t \to \infty$. Thus, in the closed-loop system Eq. (12), $x_2(t) \to x_1(t), y_2(t) \to y_1(t), z_2(t) \to z_1(t)$, when $t \to \infty$. This implies that the two LC systems have been globally asymptotically synchronized under the control law (10) associated with adaption law (11).

4 Robust Synchronization via Adaptive Sliding Mode Type of Control

However, the influence of environment may cause instability; i.e., cause a slow varying in the parameters of the chaotic system. It is reasonable to consider two LC systems with different time-varying unknown parameters. Therefore, we reformulate Eq. (5) and Eq. (6) as follows

$$\begin{cases} \dot{x}_1 = a(t)x_1 + y_1 z_1, \\ \dot{y}_1 = -b(t)y_1 - x_1 z_1, \\ \dot{z}_1 = -c(t)z_1 - x_1 y_1, \end{cases} \quad (14)$$

and

$$\begin{cases} \dot{x}_2 = a'(t)x_2 + y_2 z_2 + u_1, \\ \dot{y}_2 = -b'(t)y_2 - x_2 z_2 + u_2, \\ \dot{z}_2 = -c'(t)z_2 - x_2 y_2 + u_3. \end{cases} \quad (15)$$

To guarantee that Eq. (14) and Eq. (15) are dissipative systems, we suppose that $a(t), b(t), c(t), a'(t), b'(t), c'(t)$ are positive time-varying unknown parameters which vary in bounded interval and satisfy: $a(t) < b(t) + c(t), a'(t) < b'(t) + c'(t)$. Let

$$a'(t) = a(t) + \delta a(t), \ b'(t) = b(t) + \delta b(t), \ c'(t) = c(t) + \delta c(t). \quad (16)$$

The boundedness of parameters can be expressed as

$$a(t), a'(t) \in [\underline{a}, \bar{a}], \quad b(t), b'(t) \in [\underline{b}, \bar{b}], \quad c(t), c'(t) \in [\underline{c}, \bar{c}], \quad (17)$$

where $\bar{a}, \bar{b}, \bar{c}$ and $\underline{a}, \underline{b}, \underline{c}$ are positive constants, which denote the upper and lower bound of $a(t), b(t)$ and $c(t)$ respectively. We can easily obtain relation

$$|\delta a(t)| \leqslant \bar{a} - \underline{a}, \quad |\delta b(t)| \leqslant \bar{b} - \underline{b}, \quad |\delta c(t)| \leqslant \bar{c} - \underline{c}. \quad (18)$$

To investigate synchronization problem of Eq. (14) and Eq. (15), subtracting Eq. (14) from Eq. (15) associated with Eq. (16) yields following error dynamical system

$$\begin{cases} \dot{e}_1 = a(t)e_1 + z_1 e_2 + y_1 e_3 + e_2 e_3 + \delta a(t)x_2 + u_1, \\ \dot{e}_2 = -b(t)e_2 - z_1 e_1 - x_1 e_3 - e_1 e_3 - \delta b(t)y_2 + u_2, \\ \dot{e}_3 = -c(t)e_3 - e_1 y_1 - e_2 x_1 - e_1 e_2 - \delta c(t)z_2 + u_3. \end{cases} \quad (19)$$

Following the idea of document [4], we choose an adaptive sliding mode type of controller and parameters updated law as follows

$$\begin{cases} u_1 = -\frac{1}{2}(\eta e_1 + e_2 z_1 + e_3 y_1) - \hat{\bar{a}} e_1 + (\hat{\underline{a}} - \hat{\bar{a}})|x_2|\mathrm{sgn}e_1, \\ u_2 = -\eta e_2 + e_3 x_1 + \hat{\underline{b}} e_2 + (\hat{\underline{b}} - \hat{\bar{b}})|y_2|\mathrm{sgn}e_2, \\ u_3 = -\eta e_3 + e_2 x_1 + \hat{\underline{c}} e_3 + (\hat{\underline{c}} - \hat{\bar{c}})|z_2|\mathrm{sgn}e_3, \\ \dot{\hat{\bar{a}}} = 2|e_1 x_2| + 2e_1^2, \quad \dot{\hat{\bar{b}}} = |e_2 y_2|, \quad \dot{\hat{\bar{c}}} = |e_3 z_2| \\ \dot{\hat{\underline{a}}} = -2|e_1 x_2|, \quad \dot{\hat{\underline{b}}} = -e_2^2 - |e_2 y_2|, \quad \dot{\hat{\underline{b}}} = -e_3^2 - |e_3 z_2|, \end{cases} \quad (20)$$

where η is arbitrary positive constant, $\hat{\bar{a}}, \hat{\bar{b}}, \hat{\bar{c}}, \hat{\underline{a}}, \hat{\underline{b}}, \hat{\underline{c}}$ are the estimation of unknown parameters $\bar{a}, \bar{b}, \bar{c}, \underline{a}, \underline{b}, \underline{c}$ respectively. Let $\tilde{\bar{a}} = \bar{a} - \hat{\bar{a}}, \tilde{\bar{b}} = \bar{b} - \hat{\bar{b}}, \tilde{\bar{c}} = \bar{c} - \hat{\bar{c}}, \tilde{\underline{a}} = \underline{a} - \hat{\underline{a}}, \tilde{\underline{b}} = \underline{b} - \hat{\underline{b}}, \tilde{\underline{c}} = \underline{c} - \hat{\underline{c}}$. Considering the Lyapunov function

$$V = \frac{1}{2}(2e_1^2 + e_2^2 + e_3^2 + \tilde{\bar{a}}^2 + \tilde{\bar{b}}^2 + \tilde{\bar{c}}^2 + \tilde{\underline{a}}^2 + \tilde{\underline{b}}^2 + \tilde{\underline{c}}^2). \tag{21}$$

Taking the time derivative of V along the solutions of Eq. (19), we have

$$\begin{aligned}
\dot{V} = D^+V\big|_{(19)} &= 2e_1\dot{e}_1 + e_2\dot{e}_2 + e_3\dot{e}_3 + \tilde{\bar{a}}\dot{\tilde{\bar{a}}} + \tilde{\bar{b}}\dot{\tilde{\bar{b}}} + \tilde{\bar{c}}\dot{\tilde{\bar{c}}} + \tilde{\underline{a}}\dot{\tilde{\underline{a}}} + \tilde{\underline{b}}\dot{\tilde{\underline{b}}} + \tilde{\underline{c}}\dot{\tilde{\underline{c}}} \\
&= 2a(t)e_1^2 + 2e_1e_2e_3 + 2z_1e_1e_2 + 2y_1e_1e_3 + 2\delta a(t)x_2e_1 - \eta e_1^2 - z_1e_1e_2 \\
&\quad -y_1e_1e_3 - 2\hat{\bar{a}}e_1^2 + 2(\hat{\underline{a}} - \hat{\bar{a}})|x_2e_1 - b(t)e_2^2 - e_1e_2e_3 - z_1e_1e_2 \\
&\quad -x_1e_2e_3 - \delta b(t)y_2e_2 - \eta e_2^2 + x_1e_2e_3 + \hat{\bar{b}}e_2^2 + (\hat{\underline{b}} - \hat{\bar{b}})|y_2e_2| - c(t)e_3^2 \\
&\quad -e_1e_2e_3 - y_1e_1e_3 - x_1e_1e_2 - \delta c(t)z_2e_3 + x_1e_2e_3 + (\hat{\underline{c}} - \hat{\bar{c}})|z_2e_3| \\
&\quad +\hat{\bar{c}}e_3^2 + (\bar{a} - \hat{\bar{a}})(-2e_1^2 - 2|e_1x_2|) - (\bar{b} - \hat{\bar{b}})|e_2y_2| - (\bar{c} - \hat{\bar{c}})|e_3z_2| \\
&\quad -\eta e_3^2 + 2(\underline{a} - \hat{\underline{a}})|e_1x_2| + (\underline{b} - \hat{\underline{b}})(e_2^2 + |e_2y_2|) - (\underline{c} - \hat{\underline{c}})|e_3z_2| \\
&= -\eta(e_1^2 + e_2^2 + e_3^2) + 2[a(t) - \bar{a}]e_1^2 + [-b(t) + \underline{b}]e_2^2 + [-c(t) + \underline{c}]e_3^2 \\
&\quad +2(\underline{a} - \bar{a})|x_2e_1| + 2\delta a(t)x_2e_1 + (\underline{b} - \bar{b})|y_2e_2| - \delta b(t)y_2e_2 \\
&\quad +(\underline{c} - \bar{c})|z_2e_3| - \delta c(t)z_2e_3 \\
&\leqslant -\eta(e_1^2 + e_2^2 + e_3^2) + 2[a(t) - \bar{a}]e_1^2 + [-b(t) + \underline{b}]e_2^2 + [-c(t) + \underline{c}]e_3^2 \\
&\quad +2(\underline{a} + |\delta a(t)| - \bar{a})|x_2e_1| + (\underline{b} + |\delta b(t)| - \bar{b})|y_2e_2| + (\underline{c} + |\delta c(t)| - \bar{c})|z_2e_3| \\
&\leqslant -\eta(e_1^2 + e_2^2 + e_3^2) = -\eta\|e\|^2. \tag{22}
\end{aligned}$$

Therefore, the solutions of Eq. (19) $e_1(t), e_2(t), e_3(t)$ and the estimate values $\hat{\bar{a}}, \hat{\bar{b}}, \hat{\bar{c}}, \hat{\underline{a}}, \hat{\underline{b}}, \hat{\underline{c}}$ of the unknown constants $\bar{a}, \bar{b}, \bar{c}, \underline{a}, \underline{b}, \underline{c}$ are globally bounded for all $t \geqslant 0$. Furthermore we can conclude that $V(t)$ and $e_1(t), e_2(t), e_3(t)$ are also globally bounded for all $t \geqslant 0$ from Eqs. (19) and (21). Then integrating the inequality (22) we get $\int_0^t \|e\|^2 \leqslant [V(0) - V(t)]/\eta$. Obviously $V(0)$ is bounded, thus $e_1(t), e_2(t)$, and $e_3(t)$ are square integrable with respect to time t, i.e., $e_1(t), e_2(t), e_3(t) \in L_2$. Next by Barbalat's Lemma, Eq. (19) implies that $\dot{e}_1, \dot{e}_2, \dot{e}_3 \in L_\infty$, which in turn implies $e_1(t), e_2(t), e_3(t) \to 0$ as $t \to \infty$. Thus, in the closed-loop system Eq. (19) associated with Eq. (20), $x_2(t) \to x_1(t), y_2(t) \to y_1(t), z_2(t) \to z_1(t)$, when $t \to \infty$. Finally, we can draw a conclusion that the zero solution of Eq. (19) is globally asymptotically stable. Therefore, the system (14) can globally asymptotically synchronizes with the system (15) under the control law and adaptive law (20).

5 Numerical Simulation

We will show numerical experiments to demonstrate the effectiveness of the proposed control scheme. Based on simulink 5.0 and the S-function in Matlab 6.5, variable-step Runge-Kutta method is used to integrate the differential equations. The parameters a, b, and c are chosen as $a = 1.5, b = 5$, and $c = 2$ in all simulations to ensure the existence of chaos in the absence of control. The parameters $k, \gamma_1, \gamma_2, \gamma_3$, and η are chosen as $k = 1.65, \gamma_1 = \gamma_2 = \gamma_3 = 0.2$ and $\eta = 2$. In the third case, the time-varying parameters are chosen as follows: $a(t) = a + \frac{1}{5}\sin t$, $b(t) = b(1 + \frac{1}{5}\cos t)$, $c(t) = -c(1 - \frac{1}{5}\sin t)$, $a'(t) = a + \frac{1}{5}\sin t$, $b'(t) = -b(1 + \frac{1}{5}\sin t)$, $c'(t) = -c(1 - \frac{1}{5}\cos t)$.

Throughout all simulations, we suppose initial condition of the drive and response LC chaotic system is $(0.5, 0.6, 0.9, -1, -1, -1)$ and the unknown parameters have zero initial conditions. Numerical simulations show that synchronization are achieved successfully under three cases. Fig. 3, Fig. 4 and Fig. 5 display the results.

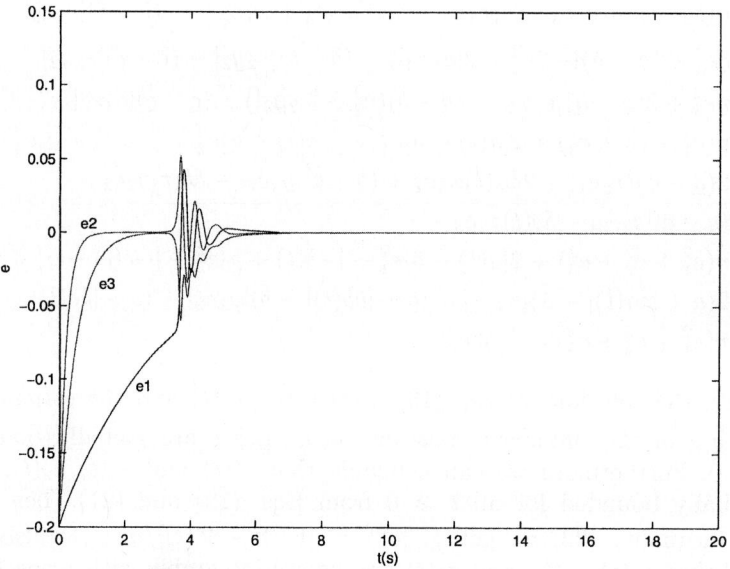

Fig. 3. The time response for the synchronization errors e_1, e_2 and e_3 of system (8) under the nonlinear feedback controller (9)

6 Conclusion

In this paper, the synchronization problem for a new 2- or 4-scroll chaotic system is discussed. According to how much knowledge of system parameters that can be grasped, three types of nonlinear controller are introduced. Among of them, adaptive and adaptive sliding mode type of control can much enhance the stability and robustness of synchronization. Numerical simulations show the validity and feasibility of our proposed approaches.

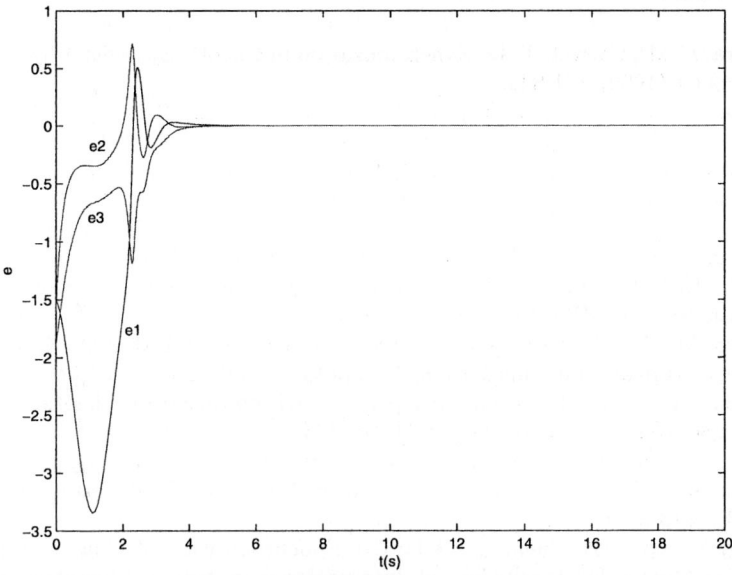

Fig. 4. The time response for the synchronization errors e_1, e_2 and e_3 of system (12) under the adaptive controller (10) and the adaption algorithm (11)

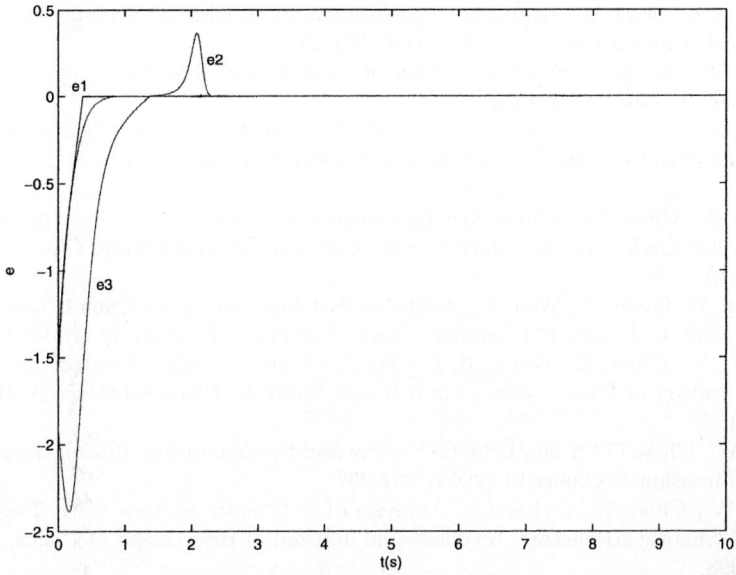

Fig. 5. The time response for the synchronization errors e_1, e_2 and e_3 of system (19) under the adaptive sliding mode type of controller and adaption law (21)

References

1. Pecora, L. M., Carroll, T. L.: Synchronization in Chaotic Systems. Physical Review Letters 64 (1990) 821-824.
2. Boccaletti, S., Kurths, J., Osipov, G., Valladares, D. L., Zhou. C. S.: The Synchronization of Chaotic Systems. Physics Reports, 366 (2002) 1-101.
3. Chen, G., Dong, X.: From Chaos to Order: Methodologies, Perspectives and Applications, World Scientific, Singapore, 1998.
4. Li, Z., Shi, S.: Robust Adaptive Synchronization of Rössler and Chen Chaotic Systems Via Slide Technique. Physics Letters A 311 (2003) 389-395.
5. Li, Z., Han, C. Shi, S.: Modification for Synchronization of Rössler and Chen Chaotic Systems. Physics Letters A 301 (2002) 224-230.
6. Yassen, M. T.: Adaptive Control and Synchronization of A Modified Chua's Circuit System. Applied Mathematics and Computation 135 (2003) 113-128.
7. Yassen, M. T.: Feedback and Adaptive Synchronization of Chaotic Lü System. Chaos,Solitons & Fractals 25 (2005) 379-386.
8. Jiang, G., Chen, G., Tang, K. S.: A New Criterion for Chaos Synchronization Using Linear State Feedback Control. International Journal of Bifurcation & Chaos 13 (2003) 2343-2351.
9. Hong, Y., Qin, H., Chen, G.: Adaptive Synchronization of Chaotic Systems via State or Output Feedback Control. International Journal of Bifurcation and Chaos 11 (2001) 1149-1158.
10. Kocarev, L., Parlitz, U.: General Approach for Chaotic Synchronization with Applications to Communication. Physical Review Letters 74 (1995) 5028-5031.
11. Agiza, H. N., Yassen, M. T.: Synchronization of Rössler and Chen Chaotic Dynamical Systems Using Active Control. Physics Letters A 278 (2001) 191-197.
12. Huang, L., Feng, R., Wang, M.: Synchronization of Chaotic Systems via Nonlinear Control. Physics Letters A 320 (2004) 271-275.
13. Liao, T.: Adaptive Synchronization of Two Lorenz Systems. Chaos, Solitons & Fractals 9 (1998) 1555-1561.
14. Liao, X., Chen, G., Wang, H. O.: On Global Synchronization of Chaotic Systems. Dynamics of Continuous, Discrete and Impulsive Systems, Series B, 10 (2003) 865-872.
15. Liao, X., Chen, G.: Chaos Synchronization of General Lurie Systems via Time-delay Feedback Control. International Journal of Bifurcation and Chaos, 13 (2003) 207-213.
16. Wang, Y., Guan, Z., Wen, X.: Adaptive Synchronization for Chen Chaotic System with Fully Unknown Parameters. Chaos, Solitons & Fractals 19 (2004) 899-903.
17. Wang, Y., Guan, Z., Wang, H. O.: Feedback and Adaptive Control for The Synchronization of Chen System via a Single Variable. Physics Letters A 312 (2003) 34-40.
18. Liu, W., Chen, G.: A new Chaotic System and Its Generation. International Journal of Bifurcation & Chaos 13 (2003) 261-267.
19. Liu, W., Chen, G.: Dynamical Analysis of A Chaotic System With Two Double-scroll Chaotic Attractors. International Journal of Bifurcation & Chaos 14 (2004) 971-998.
20. Lü, J., Chen, G., Chen, D.: A New Chaotic System and Beyond: The Generalized Lorenz-like System. International Journal of Bifurcation & Chaos 14 (2004) 1507-1537.

Global Exponential Stability of a Class of Generalized Neural Networks with Variable Coefficients and Distributed Delays

Huaguang Zhang[1,2] and Gang Wang[1,2]

[1] Key Laboratory of Process Industry Automation of Ministry of Education,
Northeastern University, China
hgzhang@ieee.org, erelong@sohu.com
[2] College of Information Science and Engineering, Northeastern University,
Shenyang, Liaoning, 110004, P.R. China

Abstract. In this paper, the requirement of Lipschitz condition on the activation functions is essentially dropped. By using Lyapunov functional and Young inequality, some new criteria concerning global exponential stability are obtained for generalized neural networks with variable coefficients and distributed delays. Since these new criteria do not require the activation functions to be differentiable, bounded or monotone nondecreasing and the connection weight matrices to be symmetric, they are mild and more general than previously known criteria.

1 Introduction

In recent years, the stability problem of time-delayed neural networks has been deeply investigated for the sake of theoretical interest as well as applications. This is because, in the implementation of artificial neural networks, time delays are unavoidably encountered. In fact, due to the finite switching speed of amplifies, time delays are likely to be present in models of electronic networks. Time delays in the response of neurons can influence the stability of a network, some works have proclaimed that time delays in response of neurons can influence a network creating oscillatory and unstable characteristics. Thus a delay parameter must be introduced into the system model.

In practice, although the use of constant fixed delays in the models of delayed feedback systems serves as a good approximation in simple circuits consisting of a small number of neurons [1,2,3,4,5] [9,10], neural networks usually have a spatial extent due to the presence of an amount of parallel pathway with a variety of axon sizes and lengths. Thus they may result in the distribution of propagation delays. Under these circumstances, the signal propagation is not instantaneous and cannot be modeled with discrete delays. A better way is to incorporate continuous distribute delays. For application of neural networks with continuously distribute delays, interested readers may refer to [6,7,8].

In [1,2,3] [6,8,9,10], the connection weights a_{ij} and b_{ij} involved in system (1)(see Sect. 2) are usually constant. However, if we assume that the connection weights a_{ij} and b_{ij} depend on the neuronal states, this system can be applied in the fields of neurobiological modeling and analogue computing [4] [5]. The presence of the term involving a_{ij} assumes, in addition to the delayed propagation of signals, a set of local interactions in the network whose propagation time is instantaneous.

On the other hand, all of results above assume that the activation functions satisfy the Lipschitz condition. However, in many evolutionary processes as well as optimal control models and flying object motion, there are many bounded monotone-nondecreasing signal functions which do not satisfy the Lipschitz condition. For instance, in the simplest case of the pulse-coded signal function which has received much attention in many fields of applied sciences and engineering, an exponentially weighted time average of sampled pulses is often used which does not satisfy the Lipschitz condition. Therefore, it is very important and, in fact, necessary to study the issue of the stability of such a dynamical neural systems whose activation functions do not satisfy the Lipschitz condition.

In this paper, we essentially drop the requirement of Lipschitz condition on the activation functions, which allows us to include a variety of nonlinearities. Some new criteria concerning the global exponential stability of generalized neural networks with variable coefficients and distributed delays are obtained. Several previous results are improved and generalized.

2 Preliminaries

We consider the following generalized neural networks model by incorporating distributed time delays in different communication channels,

$$\frac{dx_i(t)}{dt} = -d_i(x_i(t)) + \sum_{j=1}^{n} a_{ij}(x_1(t), x_2(t), \cdots, x_n(t)) f_j(x_j(t))$$

$$+ \sum_{j=1}^{n} b_{ij}(x_1(t), x_2(t), \cdots, x_n(t)) \int_{-\infty}^{t} K_{ij}(t-s) g_j(x_j(s)) ds,$$

$$i = 1, 2, \cdots, n, \tag{1}$$

where n denotes the number of neurons in the neural network, $x_i(t)$ is the state of the ith neuron at time t, $f_j(\cdot)$ and $g_j(\cdot)$ denote the activation functions of the jth neuron with $f_j(0) = g_j(0) = 0$, $a_{ij}(\cdot)$ and $b_{ij}(\cdot)$ denote the connection weight and delayed connection weight of the jth neuron on the ith neuron, respectively, $d_i(\cdot)$ represents the rate with which the ith neuron will reset its potential to the resting state when disconnected from the network and external inputs with $d_i(0) = 0$, the delayed kernels $K_{ij}(\cdot)$ satisfy the following property:

(A1) Assume that the delayed kernels $K_{ij}(\cdot)$, $i, j = 1, 2, \cdots, n$, are real-valued nonnegative continuous functions defined on $[0, \infty)$ and

$$\int_0^\infty K_{ij}(s) ds = 1, \quad \int_0^\infty K_{ij}(s) e^{\varepsilon s}(s) ds < +\infty, \tag{2}$$

where $\varepsilon > 0$ is a constant.

System (1) satisfies the initial conditions

$$x_i(s) = \psi_i(s), \quad s \in (-\infty, 0],$$

where $\psi_i(\cdot)$ denotes real-valued continuous function defined on $(-\infty, 0]$.

The following assumptions on system (1) are made throughout this paper:

(A2) There exist positive constants $\overline{D}_i$ and $\underline{D}_i$, $i = 1, 2, \cdots, n$, such that

$$0 < \underline{D}_i \leq \frac{d_i(x_i) - d_i(y_i)}{x_i - y_i} \leq \overline{D}_i, \quad for\ all\ x_i \neq y_i, i = 1, 2, \cdots, n.$$

(A3) the activation function $f_j, g_j (j = 1, 2, \cdots, n)$ satisfy $x_j f_j(x_j) > 0 (x_j \neq 0); x_j g_j(x_j) > 0 (x_j \neq 0)$ and that there exist real numbers m_j and n_j such that

$$m_j = \sup_{x_j \neq 0} \frac{f_j(x_j)}{x_j}, \quad n_j = \sup_{x_j \neq 0} \frac{g_j(x_j)}{x_j}, \quad for\ x_j \neq 0, j = 1, 2, \cdots, n.$$

Remark 1. Note that the hypothesis (A3) is weaker than the Lipschitz condition that is mostly used in literature. Further, if f_j, g_j for each $j = 1, 2, \cdots, n$ are Lipschitz functions, then m_j and n_j for each $j = 1, 2, \cdots, n$ can be replaced by the respective Lipschitz constants.

(A4) the following conditions hold:

(i) $\int_0^\infty K_{ij}(s)|b_{ij}(x_1(z+s), \cdots, x_n(z+s))|^{l_{ij}} ds < +\infty.$

(ii) $\int_0^\infty K_{ij}(s) \int_{-s}^0 |b_{ij}(x_1(z+s), \cdots, x_n(z+s))|^{l_{ij}} e^{\varepsilon z} dz ds < +\infty.$

Remark 2. Hypothesis (A1) and (A4) will ensure that (5), (16), (18), (19) and (21) (see Sect. 3) exist upper boundaries.

We denote

$$\|\psi - \varphi^*\|_r^r = \sup_{-\infty \leq s \leq 0} [\sum_{i=1}^n |\psi_i(s) - \varphi_i^*|^r], \quad r > 1, \tag{3}$$

where $\varphi^* = (\varphi_1^*, \varphi_2^*, \cdots, \varphi_n^*)$.

Definition 1. *The equilibrium* $x^* = (x_1^*, x_2^*, \cdots, x_n^*)$ *of* (1) *is said to be globally exponentially stable, if there exist* $\varepsilon > 0$ *and* $M(\varepsilon) \geq 1$, *such that* $\sum_{i=1}^n |x_i(t) - x_i^*|^r \leq M(\varepsilon)\|\psi - \varphi^*\|_r^r e^{-\varepsilon t}$ *for all* $t > 0$, *and* ε *is called the degree of exponential stability.*

Definition 2. *Let* $f(t): R \to R$ *be a continuous function. The upper right dini derivative* $D^+ f(t)$ *is defined as*

$$D^+ f(t) = \overline{\lim_{h \to 0^+}} \frac{1}{h}(f(t+h) - f(t)).$$

Lemma 1. *Assume that $a \geq 0$, $b \geq 0$, $p > 1$, $q > 1$ with $\frac{1}{p} + \frac{1}{q} = 1$, then the following inequality:*

$$ab \leq \frac{1}{p}a^p + \frac{1}{q}b^q \tag{4}$$

holds(the inequality is called as Young inequality).

3 Main Results

It is obvious that $(0, \cdots, 0)$ is the solution of system (1). So, $x^* = (0, \cdots, 0)$ is one of the equilibrium points of system (1).

Theorem 1. *Consider generalized neural networks with variable coefficients and distributed delays described by system (1) and assume that the hypothesis (A1)-(A4) are satisfied. If there are constants $\varepsilon, h_{ij}, l_{ij}, p_{ij}, q_{ij}, h_{ji}, l_{ji}, p_{ji}, q_{ji} \in R$, $w_i > 0$ $(i, j = 1, 2, ..., n)$ and $r > 1$ such that*

$$\alpha = \max_{1 \leq i \leq n} \sup \left\{ \frac{1}{w_i(r\underline{D}_i - \varepsilon)} \left[w_i(r-1) \sum_{j=1}^{n} |a_{ij}(x_1(t), \cdots, x_n(t))|^{\frac{r-h_{ij}}{r-1}} m_j^{\frac{r-q_{ij}}{r-1}} \right. \right.$$

$$+ \sum_{j=1}^{n} w_j |a_{ji}(x_1(t), \cdots, x_n(t))|^{h_{ji}} m_i^{q_{ji}}$$

$$+ w_i(r-1) \sum_{j=1}^{n} |b_{ij}(x_1(t), \cdots, x_n(t))|^{\frac{r-l_{ij}}{r-1}} n_j^{\frac{r-p_{ij}}{r-1}} \int_0^\infty K_{ij}(s) e^{\frac{\varepsilon s}{r-1}} ds$$

$$\left. \left. + \sum_{j=1}^{n} w_j n_i^{p_{ji}} \int_0^\infty K_{ji}(s) |b_{ji}(x_1(t+s), \cdots, x_n(t+s))|^{l_{ji}} ds \right] \right\} < 1, \tag{5}$$

and

$$\beta = \max_{1 \leq i \leq n} (r\underline{D}_i - \varepsilon) > 0, \tag{6}$$

for all $t > 0$ and $i = 1, 2, \cdots, n$, the equilibrium point $x^* = (0, \cdots, 0)$ of system (1) is globally exponentially stable.

Proof. We choose a Lyapunov functional for system (1) as

$$V(x_1, \cdots, x_n)(t) = \sum_{i=1}^{n} w_i \left[|x_i(t)|^r e^{\varepsilon t} + \sum_{j=1}^{n} n_j^{p_{ij}} \int_0^\infty K_{ij}(s) \right.$$

$$\left. \cdot \int_{t-s}^{t} |b_{ij}(x_1(z+s), \cdots, x_n(z+s))|^{l_{ij}} |x_j(z)|^r e^{\varepsilon z} dz ds \right]. \tag{7}$$

Calculating the upper right derivative D^+V along the solutions of system (1), one can derive that

$$D^+V(x_1,\cdots,x_n)(t) = \sum_{i=1}^n w_i \Bigg[\varepsilon|x_i(t)|^r e^{\varepsilon t} + r|x_i(t)|^{r-1}\mathrm{sgn}(x_i(t))e^{\varepsilon t}$$

$$\left(-d_i(x_i(t)) + \sum_{j=1}^n a_{ij}(x_1(t),\cdots,x_n(t))f_j(x_j(t))\right.$$

$$+ \sum_{j=1}^n b_{ij}(x_1(t),\cdots,x_n(t))\int_{-\infty}^t K_{ij}(t-s)g_j(x_j(s))ds\bigg)$$

$$+ \sum_{j=1}^n n_j^{p_{ij}}\int_0^\infty K_{ij}(s)|b_{ij}(x_1(t+s),\cdots,x_n(t+s))|^{l_{ij}}|x_j(t)|^r e^{\varepsilon t}ds$$

$$- \sum_{j=1}^n n_j^{p_{ij}}\int_0^\infty K_{ij}(s)|b_{ij}(x_1(t),\cdots,x_n(t))|^{l_{ij}}|x_j(t-s)|^r e^{\varepsilon(t-s)}ds\Bigg]. \tag{8}$$

According to the hypothesis (A2) and (A3), we have

$$D^+V(x_1,\cdots,x_n)(t)$$

$$\leq \sum_{i=1}^n w_i e^{\varepsilon t}\Bigg[(\varepsilon - r\underline{D}_i)|x_i(t)|^r + r\sum_{j=1}^n m_j|a_{ij}(x_1(t),\cdots,x_n(t))||x_i(t)|^{r-1}|x_j(t)|$$

$$+r\sum_{j=1}^n n_j|b_{ij}(x_1(t),\cdots,x_n(t))|\int_0^\infty K_{ij}(s)|x_i(t)|^{r-1}|x_j(t-s)|ds$$

$$+ \sum_{j=1}^n n_j^{p_{ij}}\int_0^\infty K_{ij}(s)|b_{ij}(x_1(t+s),\cdots,x_n(t+s))|^{l_{ij}}|x_j(t)|^r ds$$

$$- \sum_{j=1}^n n_j^{p_{ij}}\int_0^\infty K_{ij}(s)|b_{ij}(x_1(t),\cdots,x_n(t))|^{l_{ij}}|x_j(t-s)|^r e^{-\varepsilon s}ds\Bigg]. \tag{9}$$

Estimating the right of (9) by using Yang inequality, i.e., $ab \leq \frac{1}{p}a^p + \frac{1}{q}b^q$, where $p > 1, q > 1$ with $\frac{1}{p}+\frac{1}{q}=1$, we have

$$r\sum_{j=1}^n m_j|a_{ij}(x_1(t),\cdots,x_n(t))||x_i(t)|^{r-1}|x_j(t)|$$

$$= r\sum_{j=1}^n \left[|a_{ij}(x_1(t),\cdots,x_n(t))|^{\frac{r-h_{ij}}{r-1}} m_j^{\frac{r-q_{ij}}{r-1}}|x_i(t)|^r\right]^{\frac{r-1}{r}}$$

$$\cdot \left[|a_{ij}(x_1(t),\cdots,x_n(t))|^{h_{ij}} m_j^{q_{ij}}|x_j(t)|^r\right]^{\frac{1}{r}}$$

$$\leq r \sum_{j=1}^{n} \left[\frac{r-1}{r} |a_{ij}(x_1(t),\cdots,x_n(t))|^{\frac{r-h_{ij}}{r-1}} m_j^{\frac{r-q_{ij}}{r-1}} |x_i(t)|^r \right.$$
$$\left. + \frac{1}{r} |a_{ij}(x_1(t),\cdots,x_n(t))|^{h_{ij}} m_j^{q_{ij}} |x_j(t)|^r \right]. \tag{10}$$

$$r \sum_{j=1}^{n} n_j |b_{ij}(x_1(t),\cdots,x_n(t))| \int_0^\infty K_{ij}(s) |x_i(t)|^{r-1} |x_j(t-s)| ds$$
$$= r \sum_{j=1}^{n} \int_0^\infty K_{ij}(s) \left[|b_{ij}(x_1(t),\cdots,x_n(t))|^{\frac{r-l_{ij}}{r-1}} n_j^{\frac{r-p_{ij}}{r-1}} |x_i(t)|^r e^{\frac{\varepsilon s}{r-1}} \right]^{\frac{r-1}{r}}$$
$$\cdot \left[|b_{ij}(x_1(t),\cdots,x_n(t))|^{l_{ij}} n_j^{p_{ij}} |x_j(t-s)|^r e^{-\varepsilon s} \right]^{\frac{1}{r}} ds$$
$$\leq r \sum_{j=1}^{n} \int_0^\infty K_{ij}(s) \left[\frac{r-1}{r} |b_{ij}(x_1(t),\cdots,x_n(t))|^{\frac{r-l_{ij}}{r-1}} n_j^{\frac{r-p_{ij}}{r-1}} |x_i(t)|^r e^{\frac{\varepsilon s}{r-1}} \right.$$
$$\left. + \frac{1}{r} |b_{ij}(x_1(t),\cdots,x_n(t))|^{l_{ij}} n_j^{p_{ij}} |x_j(t-s)|^r e^{-\varepsilon s} \right] ds. \tag{11}$$

According to (9),(10) and (11), then

$$D^+ V(x_1,\cdots,x_n)(t)$$
$$\leq \sum_{i=1}^{n} w_i e^{\varepsilon t} \left[(\varepsilon - r\underline{D}_i) |x_i(t)|^r \right.$$
$$+ (r-1) \sum_{j=1}^{n} |a_{ij}(x_1(t),\cdots,x_n(t))|^{\frac{r-h_{ij}}{r-1}} m_j^{\frac{r-q_{ij}}{r-1}} |x_i(t)|^r$$
$$+ \sum_{j=1}^{n} |a_{ij}(x_1(t),\cdots,x_n(t))|^{h_{ij}} m_j^{q_{ij}} |x_j(t)|^r$$
$$+ (r-1) \sum_{j=1}^{n} \int_0^\infty K_{ij}(s) |b_{ij}(x_1(t),\cdots,x_n(t))|^{\frac{r-l_{ij}}{r-1}} n_j^{\frac{r-p_{ij}}{r-1}} |x_i(t)|^r e^{\frac{\varepsilon s}{r-1}} ds$$
$$\left. + \sum_{j=1}^{n} n_j^{p_{ij}} \int_0^\infty K_{ij}(s) |b_{ij}(x_1(t+s),\cdots,x_n(t+s))|^{l_{ij}} |x_j(t)|^r ds \right]$$
$$\leq \sum_{i=1}^{n} e^{\varepsilon t} w_i (r\underline{D}_i - \varepsilon) \Big\{ -1$$
$$+ \frac{1}{w_i(r\underline{D}_i - \varepsilon)} \left[w_i(r-1) \sum_{j=1}^{n} |a_{ij}(x_1(t),\cdots,x_n(t))|^{\frac{r-h_{ij}}{r-1}} m_j^{\frac{r-q_{ij}}{r-1}} \right.$$
$$+ \sum_{j=1}^{n} w_j |a_{ji}(x_1(t),\cdots,x_n(t))|^{h_{ji}} m_i^{q_{ji}}$$

$$+w_i(r-1)\sum_{j=1}^{n}|b_{ij}(x_1(t),\cdots,x_n(t))|^{\frac{r-l_{ij}}{r-1}}n_j^{\frac{r-p_{ij}}{r-1}}\int_0^{\infty}K_{ij}(s)e^{\frac{\varepsilon s}{r-1}}ds$$

$$+\sum_{j=1}^{n}w_j n_i^{p_{ji}}\int_0^{\infty}K_{ji}(s)|b_{ji}(x_1(t+s),\cdots,x_n(t+s))|^{l_{ji}}ds\Big]\Big\}|x_i(t)|^r$$

$$\leq \max_{1\leq i\leq n}\{r\underline{D}_i-\varepsilon\}(-1+\alpha)\sum_{i=1}^{n}w_i|x_i(t)|^r e^{\varepsilon t}$$

$$= -\beta(1-\alpha)\sum_{i=1}^{n}w_i|x_i(t)|^r e^{\varepsilon t}. \tag{12}$$

According to (5) and (6), we have $D^+V(x_1,\cdots,x_n)(t)\leq 0$, i.e.,

$$V(x_1,\cdots,x_n)(t)\leq V(x_1,\cdots,x_n)(0),\text{ for all }t\geq 0. \tag{13}$$

Note that

$$V(x_1,\cdots,x_n)(t)\geq \sum_{i=1}^{n}w_i|x_i(t)|^r e^{\varepsilon t}\geq \min_{1\leq i\leq n}\{w_i\}\sum_{i=1}^{n}|x_i(t)-x^*|^r e^{\varepsilon t}, t\geq 0 \tag{14}$$

$$V(x_1,\cdots,x_n)(0) = \sum_{i=1}^{n}w_i\Big[|x_i(0)|^r$$

$$+\sum_{j=1}^{n}n_j^{p_{ij}}\int_0^{\infty}K_{ij}(s)\int_{-s}^{0}|b_{ij}(x_1(z+s),\cdots,x_n(z+s))|^{l_{ij}}|x_j(z)|^r e^{\varepsilon z}dzds\Big]$$

$$\leq \max_{1\leq i\leq n}\{w_i\}\Big[\|\psi-\varphi^*\|_r^r$$

$$+\sum_{j=1}^{n}n_j^{p_{ij}}\int_0^{\infty}K_{ij}(s)\int_{-s}^{0}|b_{ij}(x_1(z+s),\cdots,x_n(z+s))|^{l_{ij}}e^{\varepsilon z}dzds\|\psi-\varphi^*\|_r^r\Big]$$

$$= \max_{1\leq i\leq n}\{w_i\}Q(\varepsilon)\|\psi-\varphi^*\|_r^r, \tag{15}$$

where

$$Q(\varepsilon) = 1+\sum_{j=1}^{n}n_j^{p_{ij}}\int_0^{\infty}K_{ij}(s)\int_{-s}^{0}|b_{ij}(x_1(z+s),\cdots,x_n(z+s))|^{l_{ij}}e^{\varepsilon z}dzds\geq 1. \tag{16}$$

From (13), (14) and (15), we have

$$\sum_{i=1}^{n}|x_i(t)-x^*|^r\leq M(\varepsilon)\|\psi-\varphi^*\|_r^r e^{-\varepsilon t}, \tag{17}$$

where
$$M(\varepsilon) = Q(\varepsilon)\frac{\max\limits_{0\leq i\leq n}\{w_i\}}{\min\limits_{0\leq i\leq n}\{w_i\}} \geq 1.$$

So, for all $t > 0$, the equilibrium point $x^* = (0, \cdots, 0)$ of system (1) is globally exponentially stable.

The proof is complete.

Corollary 1. *Assume that the hypothesis (A1)-(A4) are satisfied. If there are constants $h_{ij}, l_{ij}, p_{ij}, q_{ij}, h_{ji}, l_{ji}, p_{ji}, q_{ji} \in R$, $w_i > 0$ ($i, j = 1, 2, ..., n$) and $r > 1$ such that*

$$\alpha' = \max_{1\leq i\leq n}\sup\left\{\frac{1}{w_i r \underline{D}_i}\left[w_i(r-1)\sum_{j=1}^n |a_{ij}(x_1(t),\cdots,x_n(t))|^{\frac{r-h_{ij}}{r-1}} m_j^{\frac{r-q_{ij}}{r-1}}\right.\right.$$

$$+ \sum_{j=1}^n w_j |a_{ji}(x_1(t),\cdots,x_n(t))|^{h_{ji}} m_i^{q_{ji}}$$

$$+ w_i(r-1)\sum_{j=1}^n |b_{ij}(x_1(t),\cdots,x_n(t))|^{\frac{r-l_{ij}}{r-1}} n_j^{\frac{r-p_{ij}}{r-1}}$$

$$\left.\left.+ \sum_{j=1}^n w_j n_i^{p_{ji}} \int_0^\infty K_{ji}(s)|b_{ji}(x_1(t+s),\cdots,x_n(t+s))|^{l_{ji}}ds\right]\right\} < 1, \quad (18)$$

for all $t > 0$ and $i = 1, 2, \cdots, n$, the equilibrium point $x^* = (0, \cdots, 0)$ of system (1) is globally exponentially stable.

Corollary 2. *Assume that the hypothesis (A1)-(A4) are satisfied. When $h_{ij} = h_{ji} = l_{ij} = l_{ji} = 1$, $p_{ij} = 2 - 2r_2$, $p_{ji} = 2 - 2r_2$, $q_{ij} = 2 - 2r_1$, $q_{ji} = 2 - 2r_1$ ($i, j = 1, 2, \cdots, n$) and $r=2$, such that*

$$\alpha'' = \max_{1\leq i\leq n}\sup\left\{\frac{1}{w_i(2\underline{D}_i - \varepsilon)}\left[w_i\left(\sum_{j=1}^n |a_{ij}(x_1(t),\cdots,x_n(t))|m_j^{2r_1}\right.\right.\right.$$

$$+ \sum_{j=1}^n |b_{ij}(x_1(t),\cdots,x_n(t))|n_j^{2r_2}\int_0^\infty K_{ij}(s)e^{\varepsilon s}ds\bigg)$$

$$+ \sum_{j=1}^n w_j\bigg(|a_{ji}(x_1(t),\cdots,x_n(t))|m_i^{2-2r_1}$$

$$\left.\left.+ \sum_{j=1}^n n_i^{2-2r_2}\int_0^\infty K_{ji}(s)|b_{ji}(x_1(t+s),\cdots,x_n(t+s))|ds\bigg)\right]\right\} < 1, \quad (19)$$

for all $t > 0$ and $i = 1, 2, \cdots, n$, the equilibrium point $x^* = (0, \cdots, 0)$ of system (1) is globally exponentially stable.

Remark 3. By Corollary 2, we can easily obtain the result reported by Liao and Wong ([6], Theorem 2). Therefore, [6] is included in this paper as a special case. However, their method required the activation functions need to satisfy the Lipschize condition, this limitation is relaxed in our analysis. Our activation functions need to satisfy hypothesis (A3) which is weaker than the Lipschize condition.

If we derive $d_i(x_i) = d_i x_i, d_i > 0, a_{ij}(x_1, \cdots, x_n) = a_{ij}(constant), b_{ij}(x_1, \cdots, x_n) = b_{ij}(constant)$, system (1) can be described as follows

$$\frac{dx_i(t)}{dt} = -d_i x_i(t) + \sum_{j=1}^{n} a_{ij} f_j(x_j(t))$$
$$+ \sum_{j=1}^{n} b_{ij} \int_{-\infty}^{t} K_{ij}(t-s) g_j(x_j(s)) ds, \quad i = 1, 2, \cdots, n. \quad (20)$$

Then we have the following results.

Corollary 3. *Assume that the hypothesis (A1) and (A3) are satisfied. If there are constants $\varepsilon, h_{ij}, l_{ij}, p_{ij}, q_{ij}, h_{ji}, l_{ji}, p_{ji}, q_{ji} \in R, w_i > 0$ ($i, j = 1, 2, ..., n$) and $r > 1$ such that*

$$\alpha_{NN} = \max_{1 \le i \le n} \sup \left\{ \frac{1}{w_i(rd_i - \varepsilon)} \left[w_i(r-1) \sum_{j=1}^{n} |a_{ij}|^{\frac{r-h_{ij}}{r-1}} m_j^{\frac{r-q_{ij}}{r-1}} \right.\right.$$
$$+ \sum_{j=1}^{n} w_j |a_{ji}|^{h_{ji}} m_i^{q_{ji}} + w_i(r-1) \sum_{j=1}^{n} |b_{ij}|^{\frac{r-l_{ij}}{r-1}} n_j^{\frac{r-p_{ij}}{r-1}} \int_0^{\infty} K_{ij}(s) e^{\frac{\varepsilon s}{r-1}} ds$$
$$\left.\left. + \sum_{j=1}^{n} w_j |b_{ji}|^{l_{ji}} n_i^{p_{ji}} \right] \right\} < 1, \quad (21)$$

and

$$\beta_{NN} = \max_{1 \le i \le n} (rd_i - \varepsilon) > 0, \quad (22)$$

for all $t > 0$ and $i = 1, 2, \cdots, n$, the equilibrium point $x^* = (0, \cdots, 0)$ of system (20) is globally exponentially stable.

Corresponding to Corollary 3, we have

Corollary 4. *Assume that the hypothesis (A1) and (A3) are satisfied. If there are constants $h_{ij}, l_{ij}, p_{ij}, q_{ij}, h_{ji}, l_{ji}, p_{ji}, q_{ji} \in R, w_i > 0$ ($i, j = 1, 2, ..., n$) and $r > 1$ such that*

$$\alpha'_{NN} = \max_{1 \le i \le n} \sup \left\{ \frac{1}{w_i r d_i} \left[w_i(r-1) \sum_{j=1}^{n} |a_{ij}|^{\frac{r-h_{ij}}{r-1}} m_j^{\frac{r-q_{ij}}{r-1}} + \sum_{j=1}^{n} w_j |a_{ji}|^{h_{ji}} m_i^{q_{ji}} \right.\right.$$
$$\left.\left. + w_i(r-1) \sum_{j=1}^{n} |b_{ij}|^{\frac{r-l_{ij}}{r-1}} n_j^{\frac{r-p_{ij}}{r-1}} + \sum_{j=1}^{n} w_j |b_{ji}|^{l_{ji}} n_i^{p_{ji}} \right] \right\} < 1, \quad (23)$$

for all $t > 0$ and $i = 1, 2, \cdots, n$, the equilibrium point $x^* = (0, \cdots, 0)$ of system (20) is globally exponentially stable.

Remark 4. By corollary 4, when $h_{ij} = h_{ji} = l_{ij} = l_{ji} = p_{ij} = p_{ji} = q_{ij} = q_{ji} = 1 (i, j = 1, 2, \cdots, n)$, condition (23) is the result in ([8],Theorem 1).

For discrete time-delayed system, we have

Remark 5. When $h_{ij} = h_{ji} = l_{ij} = l_{ji} = 1(i, j = 1, 2, \cdots, n)$, and $r = 1$, condition (23) is the results in ([9], Corollary 1) and ([10], Theorem 1).

Remark 6. When $h_{ij} = h_{ji} = l_{ij} = l_{ji} = p_{ij} = p_{ji} = q_{ij} = q_{ji} = 1(i, j = 1, 2, \cdots, n)$, and r=2, condition (23) is the result in ([10], Theorem 2).

Remark 7. [8,9,10] required the activation functions need to satisfy the Lipschitz condition, this limitation is relaxed in our analysis. Our activation functions need to satisfy hypothesis (A3) which is weaker than the Lipschitz condition.

4 Conclusions

In this paper, we have given a family of new criteria concerning global exponential stability of generalized neural networks with variable coefficients and distributed delays by applying Lyapunov functional and Young inequality, and the conditions possess highly important significance in some applied fields, for instance, they can be applied to design globally exponentially stable neural networks, such as Hopfield neural networks, cellular neural networks, bi-directional associative networks. Previous results [6,8,9,10] are included in this paper as special cases, thus the existing results are improved and generalized.

Acknowledgments

This work was supposed by the National Natural Science Foundation of China under Grant 60325311 and 60274017.

References

1. Arik, S., Tavanoglu, V.: Equilibrium Analysis of Delayed CNNs. IEEE Trans. Circuits syst. I 45 (1998) 168–171
2. Cao, J.D., Wang, J.: Absolute Exponential Stability of Recurrent Neural Networks with Lipschitz-continuous Activation Functions and Time Delays. Neural network 17 (2004) 379–390
3. Chen, A.P., Cao, J.D., Huang, L.H.: Periodic Solution and Global Exponential Stability for Shunting Inhibitory Delayed Cellular Neural Networks. Electronic Journal of Differential Equation 29 (2004) 1–16
4. Liang, J.L., Cao, J.D.: Boundedness and Stability for Recurrent Neural Networks with Variable Coefficients and Time-varying Delays. Physics Letters A 318 (2003) 53–64

5. Guo, S.J., Huang, L.H., Dai, B.X., Zhang, Z.Z.,: Global Existence of Periodic Solutions of BAM Neural Networks with Variable Cofficients. Physics Letters A 317 (2003) 97–106
6. Liao, X.F., Wong, K., Li, C.G.,: Global Exponential Stability for A Class of Generalized Neural Networks with Distributed Delays. Nonliner Analysis: Real World Applications 5 (2004) 527-547
7. Hamori, J., Rosks, T.,: The Use of CNN Models in the Subcortical Visual Pathway. IEEE Trans. on Circuits and Systems 40 (1993) 182-194
8. Liang, J.L., Cao, J.D.,: Global Asymptotic Stability of Bi-directional Associative Memory Networks with Distributed Delays. Applied Mathematics and Computation 152 (2004) 415-424
9. Zeng, Z.G., Wang, J., Liao, X.X.: Global Exponential Stability of A General Class of Recurrent Neural Networks with Time-varying Delays. IEEE Trans. on Circuits and Systems 50 (2003) 1353–1358
10. Cao, J.D., Wang, J.: Global Asymptotic Stability of A General Class of Recurrent Neural Networks with Time-varying Delays. IEEE Trans. on Circuits and Systems 50 (2003) 34–44

Designing the Ontology of XML Documents Semi-automatically

Mi Sug Gu, Jeong Hee Hwang, and Keun Ho Ryu

Database/Bioinfomatics Laboratory Chungbuk National University
{gumisug, jhhwang, khryu}@dblab.chungbuk.ac.kr

Abstract. Recently as XML is becoming the standard of exchanging web documents and public documentations, XML data are increasing in many areas. And the semantic web based on the ontology is appearing for the exact information retrieval. The ontology for not only the text data but also XML data is being required. However, the existing ontology has been constructed manually and it is time and cost consuming. Therefore in this paper, we propose the semi-automatic ontology generation method using the data mining technique, the association rules. Applying the association rules to the XML documents, we intend to find out the conceptual relationships to construct the ontology. Using the conceptual ontology domain level extracted from the XML documents, we construct the ontology by using XML Topic Maps (XTM) automatically.

1 Introduction

At present, the development of internet techniques can make the users meet various kinds of information. However, the current information retrieval system can't give the exact information to the users, therefore the semantic web is appearing. As XML is the standard of exchanging the web documents and the public documentations, XML documents are increasing in many areas.

The existing ontology has been generated by the experts. The technique has been rather time-consuming and expensive because the range of data is so wide spread. Therefore in this paper, we propose a novel technique of semi-automatic ontology generation using the data mining method. This technique helps to provide the users with the efficient information of a particular domain.

Semantic web is based on the ontology, and the ontology provides the basic knowledge system to represent particular domain knowledge and helps to provide the information[1,2,3].

The existing ontology has been constructed by the text documents. However, according as the XML documents are increasing, it is necessary to construct the ontology based on the XML documents.

And XTM specifies the concept about the subject and defines the relationships among them. It is the standard of ISO (International Standard Organization). XML Topic map model(XTM) is composed of topic, occurrence, and association. Topic is a subject, and association is relations among subjects. And occurrence has a role that points out the information resource[4,5].

[6,7] suggested a novel approach to find out the non-taxonomic conceptual relations from the text documents generated through the linguistic process about the text data. It used the generalized association rules not only to find out the conceptual relations, but also to determine the appropriate abstract level which can define the relations. It is important to determine how many and what type of conceptual relations we can model in a particular ontology. Therefore it filters the conceptual relations, applying the generalized association rules to the text data.

This paper comprises as follows. Chapter 2 will explain not only the way how to apply the association rule algorithm to the XML documents, but also the generation process of the ontology domain level. Chapter 3 will describe XTM and the automatic ontology construction using TM4J, Topic Map Engine. Chapter 4 will conclude this paper.

2 Generation of Ontology Domain Level

2.1 Topic Map Ontology Model

The topic map models of the ontology are composed of topic, association, and occurrence[4,5].

First, topic is a real thing such as people, objects, concepts, attributes, meanings, and so on. Topic in a particular document comprises the terms which represent the subjects. Topics in a topic map have a topic type classified by a similar topic. And they also have a hierarchy relation composed of super- and sub-class, and inheritance relation[8]. In this paper, we chose tour information as the ontology domain. To construct the ontology we defined the topics and topic types such as area, tour sites, accommodations, transportations, foods, goods, and so on.

Second, occurrence means the information resource which topics refer to. Each topic is linked to one or more real knowledge items. For example, the site address such as http://dblab.chungbuk.ac.kr/~tour refers to the resource linking to the tour information. Occurrence has many types such as document file, image file, special record in a database, and so on. Topics in the topic map used the technique such as HyTime and Xpointer to point to the resource. Dividing between topic map and real knowledge resource provides the indexing method to classify the knowledge items[9].

Third, association defines not only super- and sub- classes among two or more topics, but also semantic relations. There're some relations in the following sentences; "Yongduam is located in Chejudo.", "Jeju Pacific Hotel is one of the accommodations.". In the first sentence, there's a relation, "be located in" between two topics, "Yongduam" and "Chejudo". In the second one, there's a relation "is one of", between two topics, "Jeju Pacific Hotel" and "accommodations". Topic has topic type and association has association type, super-class, sub-class, and so on. Topic type and association type play important function in the topic map for representing, classifying, and organizing the knowledge and the information.

Figure 1 represents a class structure of a conceptual relation hierarchy among topics to construct the ontology. The hierarchical structure about an area has a root class, tour site. And there are sub-classes of it such as "Jejudo", "Kangwondo", "Chonranamdo",

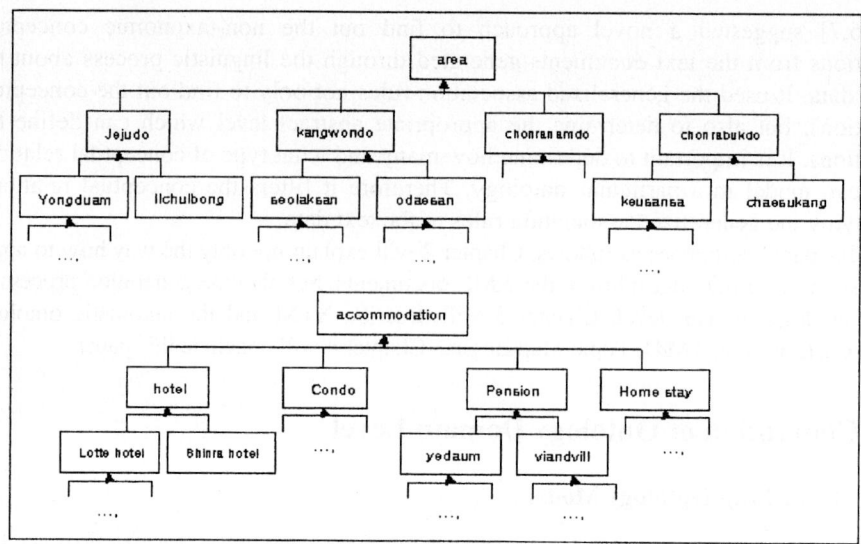

Fig. 1. Conceptual class structure of topics

"Chonrabukdo", and so on. And each of them has sub-classes of the tour sites names. And the class structure of accommodation has sub-classes such as "hotel", "condo", "pension", "home stay", and so on. And there are also sub-classes of them such as the names of the "hotel", "condo", "pension", "home stay", and so on.

2.2 XML Documents for Mining the Ontology Domain Level

In this section, we will explain the association rules used for the automatic ontology generation about the XML documents.

Recently, because XML data are increasing in various areas and describe the meta-data, they can help to provide the users with the exact information. And also the data mining to them is needed. XML documents basically have a hierarchy. Therefore we can utilize the generalized characteristics of the XML tags and the hierarchical structure of XML documents as the taxonomy while preprocessing for the ontology generation.

The ontology generation for the text data used the generalized association rules to construct the taxonomy, because text data don't have a hierarchy. The mining rules traverse the ancestor nodes to find out the frequent conceptual relation pairs[6,7,10].

However, as the XML documents have a hierarchical structure and the generalized characteristics, we used the association rules to find out the frequent patterns in them. For the experiment of XML data, we got the information from the web sites and then we transformed the data into the XML documents by Xgenerator[11]. Not only applying the association rules to, but also joining the XML tags in the XML documents, we find out the frequent patterns of XML tags related to each other.

Then we will explain the preprocessing about the tour information before applying the association rules to it.

First, we got the documents about tour from the web sites, and then transformed them into XML documents by Xgnerator. Second, we parsed the XML documents using Dom Parser. Third, we extracted about 2,500 XML tags from the XML documents. Fourth, we found out the tags with the similar expression which represents the same meaning from the extracted tags in the XML documents by WordNet (http://www.consci.princeton.edu/~wn/wn2.0) [12]. Fourth, after extracting the tags from the XML documents, we made a mapping table numbering each tag to perform the association rules.

After applying the association rules to the preprocessed XML documents, the following steps are necessary to find out the conceptual relations to generate the ontology domain level.

First, we set the transaction. XML documents about the tour information are the transactions. That is, the documents with the tour site, the accommodation, the transportation, and so on are the transactions to apply the association rules. Second, we set the items. Each tag in all the XML documents is the items. Third, we determine the minimum support and the minimum confidence predetermined by the users. Fourth, we find out the relations more than the minimum support and the minimum confidence. Fifth, we prune the items which have no any relations among XML tags. Sixth, we construct the hierarchical structure to generate the ontology.

[6,10] used the generalized association rules to find out the related concepts in the text documents and the ontology abstract level for constructing the ontology. They found out the pairs of the related concepts from the data in the ancestor nodes rather than the leaf nodes. And then they pruned the pairs which have the support and the confidence less than the minimum support and the minimum confidence.

2.3 The Process of Domain Level Generation for the Ontology

To generate the ontology domain level, we constructed a generalized hierarchy using the tags in the XML documents. When constructing the ontology using this hierarchy, it is cost and time consuming due to the huge amount of data. Therefore, we used the association rules to generate the ontology domain level for constructing the ontology. The purpose using the association rules is to determine how many and what types of conceptual relations we will extract. After performing the learning which finds out the relations among the tags in the XML documents, we prune the tags which don't have the relations and then create the hierarchy proper to construct the ontology.

For example, figure 2 shows the tags in the XML documents and figure 3 represents the hierarchy based on it. Each document is part of the XML documents extracted from the web sites. And table 1 is the transaction table based on the XML documents.

In the transactions such as A, B, C, D in the table 1, there are area, address, hotel, domestic area, foreign area, and pension in all the transactions. We perform the association rules which extract the related pairs of the frequent patterns repeatedly

```
<domestic area><area= Jejudo>
        <tour site><location>Jeju city</location>
        <site name>Yongduam</site name></tour site>
    </area>
    <area = Kangwondo >
    <tour site><location>Sokcho</location>
        <site name>Seoraksan national park</site name></tour site>
    </area>
</domestic area>
<foreign area><area= America>
        <tour site><location>Nevada</location>
        <site name>Lasvegas</site name></tour site></area>
    <area = Japan >
        <tour site><location>Shizuoka city</location>
        <site name>Fuji mountain</site name></tour site>
    </area>
</foreign area>
```

Fig. 2. XML document about tour site

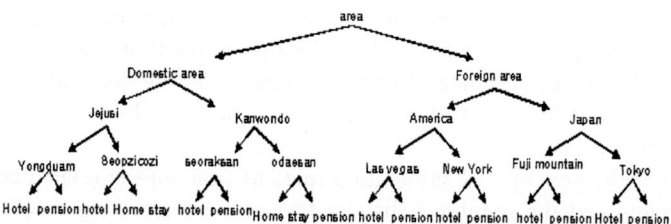

Fig. 3. Hierarchy of XML document about tour site

Table 1. Transaction table generated from XML document

TID	Items
A	domestic area, foreign area, area, tour site, location, hotel, home stay, pension, address, goods
B	accommodation, area, hotel, pension, hotel name, foreign area, room, address, domestic area, room type
C	domestic area, area, tour site, location, food, restaurant, food name, address, phone number
D	domestic area, area, tour site, location, goods, goods name

and then find out the frequent patterns. In the procedure, if the extracted items have the subsumption relations, we choose the more generalized tags and use them to generate the ontology domain level. In case the related pairs of XML tags such as <hotel>, <area> pair and <accommodation>, <area> pair are extracted together as the frequent patterns, we choose the generalized upper tag pairs in the conceptual relations such as <accommodation>, <area>. The reason is because the XML documents have a hierarchy which the upper tags are considered more generalized.

Table 2. Conceptual relation pairs

Tag	Attribute	Tag	Attribute
domestic area	area	hotel	accommodation
foreign area	tour site	restaurant name	food
Kangwondo	area	site name	tour site
goods name	goods	site name	tour site

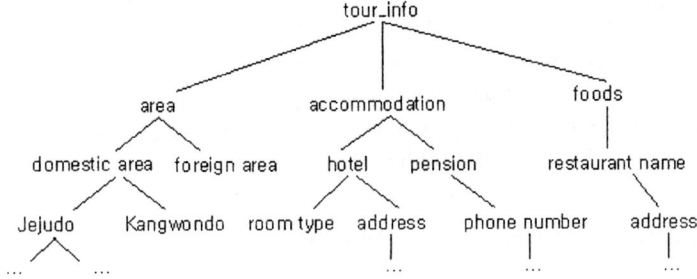

Fig. 4. Generalized hierarchy

Table 3. Discovered relations

discovered relation	confidence	support
(domestic area, hotel)	0.37	0.04
(Jejudo, Jeju Pacific Hotel)	0.11	0.03
(Kangwondo, tour site)	0.39	0.04
(hotel name, hotel)	0.15	0.01
(home stay, address)	0.22	0.02
(goods, Chonranamdo)	0.35	0.03
(tour site, goods)	0.38	0.05

We extract the related terms and attributes in the XML documents. Each document is composed of the transformed XML documents which have the tour information such as "site areas", "accommodations", "domestic areas", "foreign areas", "transportations", and so on. Then we extract 2,500 related XML tags through the linguistic process about the documents. The association rules induce the generalized hierarchy based on the conceptual relations, and then find the proper relations about the concepts to construct the ontology.

Table 2 shows the pairs of the terms and attributes related to the tags in the XML documents. And figure 4 represents the generalized hierarchy of the conceptual relations from the XML documents.

Next, we prune the conceptual relation pairs with low frequency using the support and the confidence, because they are inappropriate to construct the ontology.

Table 3 is the results from performing the association rules. In the process of finding the conceptual relation pairs, the pruned pairs are represented with red lines. (Jejudo, Jeju Pacific Hotel), (hotel name, hotel), (home stay, address) pairs are pruned.

We will explain why the pairs are pruned as follows.

First, the conceptual relation pair (Jejudo, Jeju Pacific Hotel) has the upper concept pair (domestic area, hotel). Because the child concept pairs are included in the ancestor tags which are more generalized, they are pruned. Second, (hotel name, hotel) and (home stay, address) are pruned because they have low support and confidence.

We find out the frequent relation pairs and use them to construct the ontology using the association rules. While constructing the ontology, if the support and the confidence of the upper level are high, the lower level is pruned by subsumption. By the association rules we pruned the inadequate tag relations, and then we determine the most adequate ontology domain levels which are the conceptual relations.

For example, there are the XML documents which include the accommodations (such as hotel, condo, pension, home stay, and so on) related to each tour site. To find out the adequate relations such as {site area}⇒{accommodation} or {site area}⇒{goods} from the documents, we determine the adequate domain level to construct the ontology using the association rules.

3 Automatic Ontology Construction

We construct the taxonomy for the ontology by XTM, using the frequent tag patterns and the conceptual relation pairs in the XML documents, applying the association rules.

3.1 The Process of Topic Map Generation

The ontology about the tour information starts from the root tag, "tour_info", in the XML documents. To classify the extracted conceptual relation pairs about tour site

```
<!-- tour_info -->
<topic id="tour_info">
        <baseName><baseNameString>Tour_Info</baseNameString></baseName>
</topic>
<!-- Tour_Info - Chejudo -->
        <topic id="domestic area">
                <instanceOf><topicRef xlink:href="#tour_info"/></instanceOf>
                <baseName><baseNameString>domestic area</baseNameString></baseName>
        </topic>
        <topic id="Jejudo">
                <instanceOf><topicRef xlink:href="#domestic area"/></instanceOf>
                <baseName><baseNameString>Jejudo</baseNameString</baseName>
        </topic>
        <topic id="tour site">
                <instanceOf><topicRef xlink:href="#Jejudo"/></instanceOf>
                <baseName><baseNameString>tou siter</baseNameString></baseName>
                <occurence><resourceRef xlink:href="http://megalo.wo.to"/></occurrence
        </topic>
        <topic id="Yongduam">
                <instanceOf><topicRef xlink:href="#tour site"/></instanceOf>
                <baseName><baseNameString>Yongduam</baseNameString></baseName>
                <occurence><resourceRef xlink:href="http://megalo.wo.to"/></occurrence>
        </topic>
        <topic id="accommodation">
                <instanceOf><topicRef xlink:href="#Yongduam"/></instanceOf>
                <baseName><baseNameString>accommodation</baseNameString></baseName>
                <occurence><resourceRef xlink:href="http://megalo.wo.to"/></occurrence>
        </topic>
        <topic id="Jeju Pacific hotel">
                <instanceOf><topicRef xlink:href="#accommodation"/></instanceOf>
                <baseName><baseNameString>Jeju Pacific hotel</baseNameString></baseName>
                <occurence><resourceRef xlink:href="http://megalo.wo.to"/></occurrence
        </topic>
```

Fig. 5. XTM document about tour information

and accommodation, we divide them into the domestic area and the foreign area. The XTM document represents the tour site in "Jejudo" among domestic areas and the accommodation related to the areas. Topics in the topic map are "domestic area", "Jejudo", "tour site", "Yongduam", "accommodation", "Jeju Pacific Hotel", and so on. Figure 5 shows part of the XTM ontology source for the "tour_info".

The association in the topic map starts from the root, "tour_info", and the topic "domestic area" is a sub node of "tour_info". Also the topic "Jejudo" is a sub node of "domestic area", the topic "tour site" is a sub node of "Jejudo", and the topic "Yongduam" is a sub node of "Jejudo". And "accommodation" is a sub node of "Youngduam", and "Jeju Pacific Hotel" is a sub node of "Yongduam". Using XTM topic map structure and XML documents, we construct the ontology.

We construct the XTM documents through the previous process and then perform the container, Jakarta-Tomcat. We checked the validity of the documents using the Onmigator, topic map tool, from Ontopia[13]. After checking the validity, we perform parsing to the valid documents through Saxparser and TMparser and then store the ontology into the object-relational ontology database using Hibernate[14]. And then we generate the topic map automatically using TMBuilder in TM4J, Topic Map Engine[13].

3.2 Automatic Ontology Generation Using TM4J

In this paper to generate the ontology automatically, we used TM4J, topic map tool (http://www.ontopia.net). Tm4J is composed as follows[13].

Topic Map Object Wrappers provide the interface to implement the objects. Topic Map Storage Wrappers store each object of topic map into the real storage and then load them..

Topic Map Factory generates each constituent of topic map. It provides the interface through the factory of Topic Map Object Wrapper. Topic Map Provider loads in memory and provides the topic map stored in the Storage. Topic Map Cache Manager retrieves the information about not only a particular topic but also other topics related to it and then stores them in a cache. Therefore if there are topics proper to the retrieval conditions, it doesn't approach to the disk and provide the topic from the cache directly. Topic Map Manager provides the interface to manage the topic map. Topic Map Utils provide the function of topic map import and export.

Figure 6 illustrates the structure of our retrieval engine based on the ontology. HTML, XML, and other documents extracted from the web server by the retrieval engine are transformed to XML documents and then perform parsing to them. After conducting the process of Topic Map Engine, they are stored in the database. Through the user interface the adequate information to the users' queries is provided.

We assume the user query such as "While traveling "Yongduam" in "Jejudo", retrieve the appropriate accommodation.". "Yongduam" is one of "site name" and "hotel" is one of "accommodation". We find out the conceptual relation from each other in the XML documents.

Figure 7 shows the retrieval results about the query using the ontology structure.

Because the left menus on the figure are classified into the area, we can click on the menu with hierarchical structure and then choose "Jejudo". Next we can select "Yongduam" among the tour sites. Through the keyword retrieval on the right upper side, we can enter the information that we want to get such as accommodation information, transportation information, goods, and so on. Then on the right lower side the information is displayed.

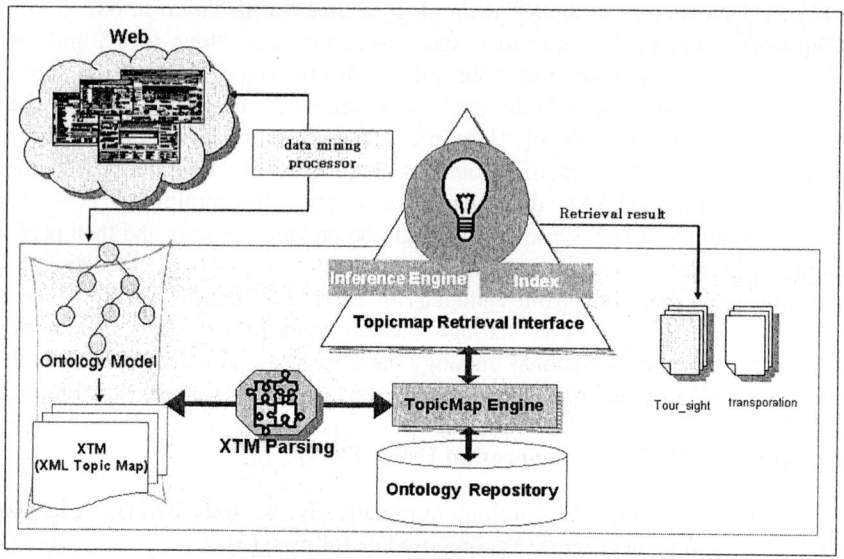

Fig. 6. Ontology based retrieval engine diagram

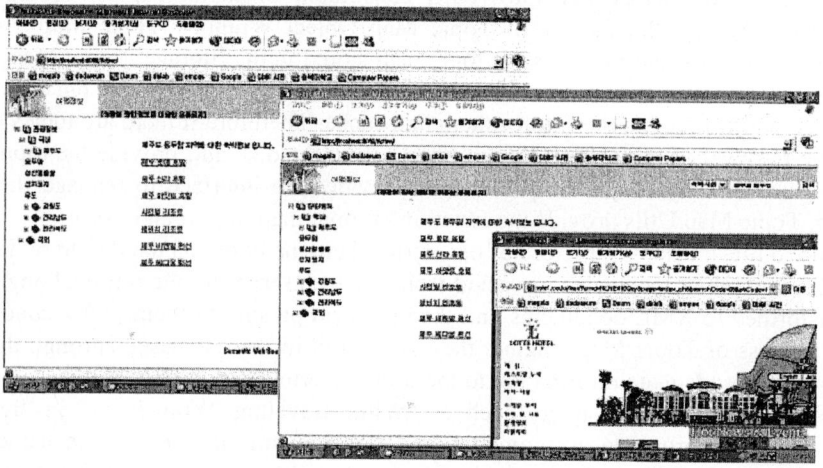

Fig. 7. Ontology based retrieval results

4 Conclusions

According to the development of the internet, the current web has a large amount of data. Therefore, it is difficult for the users to get exact information. To realize the efficient information retrieval system a new paradigm, semantic web is emerging, adding the semantic information to the current web. The goal of the semantic web is to make a program that human and machine can understand the current web.

Recently XML documents are so increasing that the data mining method to them is needed. Constructing the ontology manually is inefficient, because it is cost and time consuming. Therefore, in this paper we proposed to construct the ontology which is the base of the semantic web using the association rules. Also we used XML Topic Map which is one of the ontology languages and TM4J, topic map engine, to construct the ontology and implemented the semantic web retrieval engine. The ontology based semantic web can help human and machine understand the web. Therefore, the system can provide the exact information to the users' requirements. In the future we will consider the XML tags and the contents to construct the ontology.

Acknowledgment

This work was supported by the Regional Research Centers Program of Ministry of Education & Human Resources Development in Korea.

References

1. Gruber, T.R.: Toward Principles for the Design of Ontologies used for Knowledge Sharing. Int. J.Human - Computer Studies 43 (1995) 907-928
2. Staab, S., Schnurr, H.P., Studer, R., Sure, Y.: Ontologies: Principles, Methods and Applications. The Knowledge Engineering Review 11 (2) (1996) 93-136
3. Staab, S., Schnurr, H.P., Studer, R., Sure, Y.: Knowledge Processes and Ontologies. IEEE Intelligent Systems, Special Issue on Knowledge Management (2001)
4. Steve Pepper: The TAO of Topic Maps. XML Conference & Exposition (2000)
5. Pepper, S. Moore, B.: XML Topic Maps(XTM) 1.0. TopicMaps.Org
6. Maedche, A., Staab, S.: Discovering Conceptual Relations from Text. Institute AIFB, Karlsruhe University, Germany (2001)
7. Maedche, A., Staab, S.: Semi-Automatic Engineering of Ontologies from Text. Institute AIFB, Karlsruhe University, Germany (2000)
8. Kim, J.M., Park, C.M., Jung, J.W., Lee, H.J., Jung, H.Y., Min, K.S., Kim, H.J.: K-Box: Ontology Management System based on Topic Maps. KISS Conference, Korea (2004)
9. Jung, H.Y., Kim, J.M., Jung, J.W., Kim, H.J.: Knowledge Map of XTM Base. SIGDB, Korea (2003)
10. Srikant R., Agrawal, R. : Mining Generalized Association Rules. VLDB (1995)
11. http://www.cs.toronto.edu/tox/toxgene/index.html
12. http://www.cogsci.princeton.edu/~wn/wn2.0
13. http://www.ontopia.net
14. http://www.hibernate.org

A Logic Analysis Model About Complex Systems' Stability: Enlightenment from Nature[*]

Naiqin Feng[1,2], Yuhui Qiu[1], Fang Wang[1], Yingshan Zhang[3], and Shiqun Yin[1]

[1] Faculty of Computer & Information Science, Southwest-China Normal University,
Chongqing, 400715, China
fengnq@swnu.edu.cn
[2] Faculty of Computer & Information Science, Henan Normal University,
Xinxiang, 453002, China
fengnq@henannu.edu.cn
[3] Dept of Probability & Statistics, East-China Normal University,
Shanghai, 200062, China

Abstract. A logic model for analyzing complex systems' stability is very useful to many areas of sciences. In the real world, we are enlightened from some natural phenomena such as "biosphere", "food chain", "ecological balance" etc. By research and practice, and taking advantage of the orthogonality and symmetry defined by the theory of multilateral matrices, we put forward a logic analysis model of stability of complex systems with three relations, and prove it by means of mathematics. This logic model is usually successful in analyzing stability of a complex system. The structure of the logic model is not only clear and simple, but also can be easily used to research and solve many stability problems of complex systems. As an application, some examples are given.

1 Introduction

The logic analysis of stability is a concerned problem of many knowledge branches. Why can an atom be a comparatively stable system relatively to the substances world [1]? Why can a cell be the basic unit that makes a living thing? Why can an aggregate model of biological gene form a stable structure? Why can a special structure of gene be the regular cause making a genetic disease? What reason can make a special working procedure of a factory into the stable working procedure [2,3]? Like this, and so on, the problems of intelligent reasoning of stability are usually encountered, but how to define the logic analysis structure of stability, the views of different scholars are different from each other. In the real world, we are enlightened from some concepts and phenomena such as "biosphere", "food chain", "ecological balance" etc. With research and practice, by using the theory of multilateral matrices [4] and analyzing the conditions of symmetry [5] and orthogonality [6-8] what a stable system must satisfy, in particular, with analyzing the basic conditions [9,10] what stable working procedure of good

---

[*] This research is supported by the Science Fund of Henan Province, China (0511012500) and key project of Information and Industry Department of Chongqing City, China (200311014).

product quality must satisfy, we are inspired and find some rules and methods, then present the logic model for analyzing the stability of a complex system. This logic model is usually successful in analyzing stability of a complex system. The structure of logic model is not only clear and simple, but also can be easily used to research and solve many stability problems of complex systems.

This paper is structured as follows. Section 2 defines the concept of logic analysis model with three relations (Definition 3), and proves three basic properties of it. Section 3 builds the logic analysis model of stability (Definition 4), and presents three theorems about the model with proving them. In section 4, five examples are given to illustrate their simple applications about the concept and the model presented above. Finally, section 5 summarizes the paper, and conclusions are given.

2 A Logic Analysis Model with Three Relations

Definition 1. Let set $A \neq \emptyset$, and $\sim$ be a relation on A. Then $\sim$ is called an equivalence relation on A, if and only if $\forall x, y, z \in A$, satisfy:

1. $x \sim x$;
2. If $x \sim y$, then $y \sim x$;
3. If $x \sim y, y \sim z$, then $x \sim z$.

That is, $\sim$ is reflexive, symmetric and conveyable.

Definition 2. Let set $A \neq \emptyset$, and $\rightarrow$ and $\Rightarrow$ are two different relations on A. Then $\rightarrow$ and $\Rightarrow$ is called a neighboring relation and a alternate relation on A respectively, if and only $\forall x, y, z \in A$, satisfy:

1. First triangle reasoning (transition reasoning)
   (1) If $x \rightarrow y, y \rightarrow z$, then $x \Rightarrow z$, i.e. $\rightarrow$ meets $\rightarrow$ with developing transition phenomenon;
   (2) If $x \rightarrow y, x \Rightarrow z$, then $y \rightarrow z$;
   (3) If $x \Rightarrow z, y \rightarrow z$, then $x \rightarrow y$.
2. Second triangle reasoning (atavism reasoning)
   (1) If $x \Rightarrow y, y \Rightarrow z$, then $z \rightarrow x$, i.e. $\Rightarrow$ meets $\Rightarrow$ with developing atavism;
   (2) If $z \rightarrow x, x \Rightarrow y$, then $y \Rightarrow z$;
   (3) If $y \Rightarrow z, z \rightarrow x$, then $x \Rightarrow y$.

The First triangle reasoning (transition reasoning) and the Second triangle reasoning (atavism reasoning) can be represented by the following Fig. 1, where to every triangle, any two sides determine the third side.

3. Equivalence relation $\sim$ meets $\rightarrow$ or $\Rightarrow$, with the following rules of conveying (genetic reasoning):
   (1) If $x \sim y, y \rightarrow z$, then $x \rightarrow z$;
   (2) If $x \sim y, y \Rightarrow z$, then $x \Rightarrow z$;
   (3) If $x \rightarrow y, y \sim z$, then $x \rightarrow z$;
   (4) If $x \Rightarrow y, y \sim z$, then $x \Rightarrow z$.

$$\begin{array}{cc} x & x \\ \Downarrow \searrow & \Downarrow \swarrow \\ z \leftarrow y & y \Rightarrow z \end{array}$$

Fig. 1. First triangle reasoning (left) and Second triangle reasoning (right)

Definition 3. Let V be a set, and there be three relations $\sim$, $\rightarrow$ and $\Rightarrow$ on V. If $\forall x, y \in V$ (x, y can be the same), at least there is one of three relations $\sim$, $\rightarrow$ and $\Rightarrow$ between x and y, and there are not two relations $\rightarrow$ and $\Rightarrow$ between x and y simultaneously, i.e. there are not $x \rightarrow y$ and $x \Rightarrow y$ at the same time, then V is called a logic analysis model.

Obviously, in the model there is not contradiction, because the above two triangle are independent each other, and any two sides of them determine the third side, with coordination in reasoning. There are three basic properties in the model.

Property 1. For $\forall x, y \in V$, only one of five relations $x \sim y, x \rightarrow y, y \rightarrow x, x \Rightarrow y, y \Rightarrow x$ is existent and correct.

Proof. Assume that there are both $x \sim y$ and $x \rightarrow y$ simultaneously. By the symmetry of equivalence relation and genetic reasoning, we can get $x \rightarrow x$. By $x \rightarrow x$, $x \rightarrow y$ and transition reasoning can get $x \Rightarrow y$. But because of definition 3, $x \Rightarrow y$ contradicts $x \rightarrow y$. Therefore, there are not both $x \sim y$ and $x \rightarrow y$ simultaneously.

Assume that there are both $x \sim y$ and $x \Rightarrow y$ i.e. $y \sim x$ and $x \Rightarrow y$ simultaneously. Then we can get $y \Rightarrow y$ by genetic reasoning. Next, we get $y \rightarrow x$ by $y \Rightarrow y$, condition $x \Rightarrow y$ and atavism reasoning. By the above proving, we know that $y \sim x$ contradicts $y \rightarrow x$. So, there are not both $x \sim y$ and $x \Rightarrow y$ on V simultaneously.

Similarly, we can prove that there are not both $x \sim y$ and $y \rightarrow x$, and both $x \sim y$ and $y \Rightarrow x$ simultaneously.

Assume that there are both $y \rightarrow x$ and $x \rightarrow y$ simultaneously. By atavism reasoning, we can obtain $y \Rightarrow y$. In addition, $y \sim y$, this result contradicts the conclusion proved previously.

Assume that there are both $x \rightarrow y$ and $y \Rightarrow x$ simultaneously. By atavism reasoning, we can obtain $x \Rightarrow x$. In addition, $x \sim x$, this result contradicts the conclusion proved previously too.

Similarly, we can prove that there are not both $x \Rightarrow y$ and $y \rightarrow x$, both $x \Rightarrow y$ and $y \Rightarrow x$ simultaneously. The proof is complete.

Property 2. $\forall x, y, z \in V$, if $x \rightarrow y$, $x \rightarrow z$, then $y \sim z$; similarly, if $x \Rightarrow y$ and $x \Rightarrow z$, then $y \sim z$.

Proof. We adopt disproved method. Taking advantage of conditions $x \to y$ and $x \to z$, concerning the relation between y and z. If $y \to z$, then we can obtain $x \Rightarrow z$ by transition reasoning, but $x \Rightarrow z$ contradicts $x \to z$. If $y \Rightarrow z$, then $z \Rightarrow x$ by atavism reasoning, this contradicts $x \to z$. Similarly, we can prove that there are not both $z \Rightarrow y$ and $z \to y$, therefore, $y \sim z$.

It is the similar process to prove another half of property 2.

Property 3. $\forall x, y, z \in V$, if $x \to z$, $y \to z$, then $x \sim y$; similarly, if $x \Rightarrow z$ and $y \Rightarrow z$, then $x \sim y$.

Proof. It is similar to that proof like property 2.

3 Logic Analysis Model of Stability

Definition 4. A logic analysis model is said to be steady, if at least for one of $\to$ and $\Rightarrow$, such as $\to$, there is a cycle chain (or causal circle) like the following form:

$$x_1 \to x_2 \to x_3 \to \cdots \to x_n \to x_1$$

The definition given above, for a relatively stable system, is most essential. If there is not the chain or circle, then there will be some elements without causes or some elements without results in a system. Thus, this system is to be in the state of finding its results or causes, i.e. this system will fall into an unstable state, and there is not any stability to say.

From the stable logic analysis model of complex systems, we can obtain several interesting consequences given below:

Theorem 1. In a stable logic analysis model, there must be the cycle chain that its length is five, and there is not the cycle chain that its length is less than 5.

Proof. The only need is to prove the three cases given below:

1. There are not the cycle chains: their length is 1,2,3 or 4;
2. There is a cycle chain that length is five;
3. For a stable logic analysis model, there must be a cycle chain that length is five.

Three cases given above are proved as follows:

1. Obviously, $x_1 \to x_1, x_1 \to x_2 \to x_1, x_1 \to x_2 \to x_3 \to x_1$ are all impossible. Assume that there is $x_1 \to x_2 \to x_3 \to x_4 \to x_1$, we can obtain $x_1 \Rightarrow x_3 \Rightarrow x_1$ by transition reasoning and $x_1 \to x_1$ by atavism reasoning. This result contradicts Property1.
2. For the cycle chain whose length is 5: $x_1 \to x_2 \to x_3 \to x_4 \to x_5 \to x_1$, can infer that: $x_1 \Rightarrow x_3 \Rightarrow x_5 \Rightarrow x_2 \Rightarrow x_4 \Rightarrow x_1$. There is not any contradiction here.
3. For any one of stable logic analysis models, by Definition 4, there is a cycle chain like this: $x_1 \to x_2 \to x_3 \to \cdots \to x_n \to x_1$. By proving step 1, we know that n≥5.

If n=5, then case 3 has been proved.

If n>5, then $x_1 \to x_2 \to x_3 \to x_4 \to x_5 \to x_6 \cdots$, so we can obtain $x_1 \Rightarrow x_3 \Rightarrow x_5$ by transition reasoning and obtain $x_5 \to x_1$ by atavism reasoning. Therefore, we have $x_1 \to x_2 \to x_3 \to x_4 \to x_5 \to x_1$. The proof is complete.

From proving Theorem 1, we can know that there are two different cycle chains, whose length is five simultaneously. That is, cycle chains $x_1 \to x_2 \to x_3 \to x_4 \to x_5 \to x_1$ and $x_1 \Rightarrow x_3 \Rightarrow x_5 \Rightarrow x_2 \Rightarrow x_4 \Rightarrow x_1$ appear together at the same time.

Theorem 2. To any one of stable logic analysis model V, we can divide all elements of V into 5 categories: V_1, V_2, V_3, V_4, V_5, in which $V_i \cap V_j = \emptyset (i \neq j)$, and $\cup V_i = V$, i=1,2,…,5. Elements in the same category are equivalence each other, and there is the relation $\to$ or $\Rightarrow$ between this category V_i and that category V_j.

Proof. To any V, by Theorem 1, there is a cycle chain as follows:

$$x_1 \to x_2 \to x_3 \to x_4 \to x_5 \to x_1$$

Let $V_i = \{x : x \sim x_i, x \in V\}, i = 1, 2, \cdots, 5$. Firstly, we will prove $V_i \cap V_j = \emptyset (i \neq j)$. Make use of disproving. If $V_i \cap V_j \neq \emptyset$, then $\exists x \in V_i \cap V_j$, make $x \sim x_i, x \sim x_j$, therefore, $x_i \sim x_j$, leading to contradiction.

Secondly, we will prove $\cup_{i=1}^{5} V_i = V$, i.e. $\forall x \in V$, $\exists x_i$, make $x \sim x_i$. We know that there must be one of 5 relations $x \sim x_1, x \to x_1, x_1 \to x, x \Rightarrow x_1, x_1 \Rightarrow x$ between x and x_1. If $x \sim x_1$, then the proof is complete; if $x \to x_1$, in addition, $x_5 \to x_1$, then $x \sim x_5$; if $x \Rightarrow x_1$, in addition, $x_4 \Rightarrow x_1$, then $x \sim x_4$. Similarly, other cases can be proved too.

Theorem 2 indicates that we can research stability of a complex system with 3 relations by researching its 5 equivalence categories.

Theorem 3. To any logic analysis system V with 3 relations $\sim$, $\to$ and $\Rightarrow$, dividing its elements into categories according to equivalence relations, only stable architecture is shown as follows(Fig. 2):

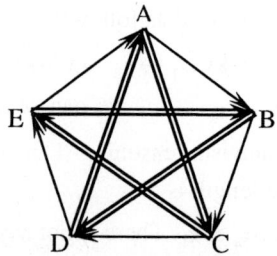

Fig. 2. Only stable architecture

Theorem 3 can be indirectly inferred by theorem 1 and theorem 2.

These theorems have very important significance. Please look at several examples given below.

4 Examples

Example 1. In an on-line control system of product quality, both different relations—working procedure and management are considered generally. For example, let $\rightarrow$ be working procedure, $\Rightarrow$ be managing procedure. The on-line control system given below can be adopted:

Assume that $x_1, x_2, \cdots$, etc. are inspection points of working procedure or managing procedure, then $x_1 \rightarrow x_2 \rightarrow x_3 \rightarrow \cdots$. Where, $x_i \rightarrow x_{i+1}$ is called flow section of adopting ith working procedure. Suppose that, according to design, substandard products rate of each section is q. Assume that the inspector discovers that there may be problems at $x_2 \rightarrow x_3$, then manager wants to inspect that weather working procedure section $x_2 \rightarrow x_3 \rightarrow \cdots \rightarrow x_6$ is to be in a stable producing state. Then he or she can take an inspection of substandard products rate at x_2, recording it as q_1, and take an inspection of substandard products rate at x_4, recording it as q_2, in addition, take another inspection of the rate at x_6, recording as q_3. Assume that:

$$r_1 = 1 - \sqrt{(1-q_2)/(1-q_1)}, r_2 = 1 - \sqrt{(1-q_3)/(1-q_2)}$$

Then, $r_1 > q$ will shows that there may be problems at working procedure section $x_2 \rightarrow x_3 \rightarrow x_4$. But there may be errors in inspection, so we inspect continuously. If finding $r_2 > q$ in the next inspection, then it is reasonable to think that there are some problems in working procedure section $x_2 \rightarrow x_3 \rightarrow \cdots \rightarrow x_6$, and the quality problems may probably be located at section $x_2 \rightarrow x_3 \rightarrow x_4$. Thus, productions or half-productions need to return from x_6 to x_2 to reproduce. From the above analysis, to the above quality inspection management, $\rightarrow$ can be understood as an error of some working procedure inspections, $\Rightarrow$ as an error that found by some above inaspectors, thinking the reasoning below (Fig. 3) reasonable:

The above reasoning rules form the inspection to stability of producing. The rules may be expressed as: "The same mistake can be permitted one or two times but three or four times."

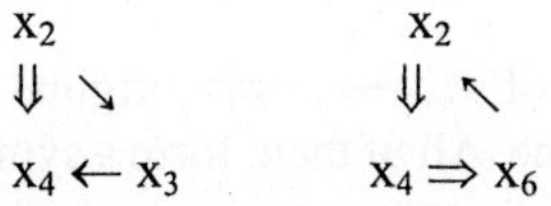

$$\begin{array}{cc} x_2 & x_2 \\ \Downarrow \searrow & \Downarrow \searrow \\ x_4 \leftarrow x_3 & x_4 \Rightarrow x_6 \end{array}$$

Fig. 3. Reasoning of example 1

Example 2. In system design of dependability, people consider a risk function $\lambda(k)$ and scalar variable r:

$$\lambda(k) = \frac{P_k}{\sum_{i=k}^{n} P_i}, r = \frac{P_{k+i}}{P_i} - \frac{P_{k+i+1}}{P_{i+1}}$$

where, $P_1, P_2, \ldots, P_n$ are scattered probability-distribution-density, and n is the product life-span. Analyzing the state of products, we can see that they are to be in flow as follows:

producing stage: $A_1 : \lambda(0) = 0$;
early stage : $B_1 : 0 < \lambda(k) < 1, r < 0$;
accidental stage : $C_1 : 0 < \lambda(k) < 1, r = 0$
breakage stage : $D_1 : 0 < \lambda(k) < 1, r > 0$
life-span stage : $E_1 : \lambda(k) = 1$;
another producing stage : $A_2 : \lambda(0) = 0$;
another early stage : $B_2 : 0 < \lambda(k) < 1,\ r < 0$;
...

Although for designers of a product and for inspectors of product dependability the producing stages is the stage what they pay careful attention to together, the designers may be more concerned with design of accidental stage and life-span stage, and the inspectors may be more concerned with testing to early stage and breakage stage. We can regard the relation between the same kind of stage, for example, between $A_1, A_2 \cdots$, as ~, and the relation between the two continuous stage, for example, between $A_1, B_1 \cdots$, as $\rightarrow$, along with the relation between two alternate stages, such as $B_1, D_1 \cdots$ as $\Rightarrow$. Obviously, the above system forms a stable logic analysis system, and it satisfies Theorem 1 to 3.

Example 3. Assume that F={$(x_1, x_2)'$: $0 \leq x_1 \leq 1, 0 \leq x_2 \leq 2$ } is a plane rectangle, translation is regarded as ~. Define that $\rightarrow$ is to turn F $2\pi/5$ anticlockwise, that is :

$$x \rightarrow y : y = \begin{pmatrix} \cos\frac{2}{5}\pi & -\sin\frac{2}{5}\pi \\ \sin\frac{2}{5}\pi & \cos\frac{2}{5}\pi \end{pmatrix} x, x = (x_1, x_2)'$$

In addition, define that $\Rightarrow$ is to turn F $4\pi/5$ anticlockwise, i.e.

$$x \Rightarrow y : y = \begin{pmatrix} \cos\frac{4}{5}\pi & -\sin\frac{4}{5}\pi \\ \sin\frac{4}{5}\pi & \cos\frac{4}{5}\pi \end{pmatrix} x, x = (x_1, x_2)'$$

Going through the functions of ~, $\rightarrow$, $\Rightarrow$,original plane rectangle will become many plane rectangles in the plane. All of them form a symmetric plane graph. Let A=0, B=2π/5, C=4π/5, D=8π/5, then it is correct to reason as fellows (Fig. 4):

```
    A           A
    ⇓ ↘         ⇓ ↖
  C ← B       C ⇒ D
```

Fig. 4. Reasoning of example 3

The example demonstrates that this system forms a stable logic analysis model too.

Example 4. Ancient Chinese theory "Yin Yang Wu Xing" [11] has been surviving for several thousands of years without dying out, proving it reasonable to some extent. If we regard ~ as the same category, neighboring relation → as consistency and alternate relation ⇒ as conflict, then the above defined logic analysis model of stability is consistent with the logic architecture of reasoning of "Yin Yang Wu Xing". Yin and Yang mean that there are two opposite relations in the world: consistency → and conflict ⇒, as well as general equivalence category ~. There is only one of three relations ~, → and ⇒ between every two objects. Everything makes something, and is made by something; everything restrains something, and is restrained by something; i.e. one thing is overcome by another thing. The ever changing world, following the relations, must be divided into five categories by equivalence relation, being called "Wu Xing": wood, fire, earth, gold, water. The "Wu Xing" is to be "neighbor is friend", and "alternate is foe". We can see, from this, the ancient Chinese theory "Yin Yang Wu Xing" is a reasonable logic analysis system to stability of complex systems.

Example 5. A object is launched, with its elevation α (degree), and its mass m(kg), and momentum G(kg·m/s), then distance that it can arrive there in level is:

$$y = \frac{1}{g}(\frac{G}{m})^2 \sin 2\alpha \equiv f(G, m, \alpha)$$

where m=1.0kg, g=9.8m/s^2(acceleration of gravity). When launching, m, G and α go up and down within ranges of Δ m=0.01m, Δ G=0.02G, $\Delta \alpha$ =0.05α. The value of G and α are unknown. The question is: what are the value of G and α, that can make the launched object most stably approaching 1000 meters (goal) in level direction? So called the most stable means that if it arrives at the distances $y_1, y_2, \ldots, y_n$, with taking G=G_0, $\alpha = \alpha_0$ and testing several times at point (G_0, α_0), then

$$\bar{y} = \frac{1}{n}\sum_{s=1}^{n} y_s = 1.0 km \text{ , while } R_f^2(G, \alpha) = \frac{1}{n}\sum_{s=1}^{n}(y_s - 1000)^2$$

will get the minimum at the point (G_0, α_0).

The above problem is a usually model in control area of missiles, with having important worth in theory and practice. There are many similar problems in many domains such as economic management and prediction, products quality control, stability of working procedure, online automatic control, physics, chemistry and biology, etc. The kind of problems is called the problem of stable center of complex systems. Although there is a lot of that kind of problems, there are few accurate mathematical models to use, and few good methods to find the stable center.

Finding the stable center of a complex system is a important problem on data analysis. In general, people now select the stable center of a complex system by firstly selecting some criteria for judging the stability, then finding the most optimal criteria of stability. However, different school of thought selects different stability criteria, so that different stability criteria bring about different stability center. For improving the above problem, we present the above logic model to help people analyzing stability of complex systems, and proving that its cycle chain length of the stable structure is five. From this, we obtain a novel method analyzing the stable center of a complex system. This method needs only five criteria, and if we can make them the most optimal, then we will find its stable center.

Suppose that the value y of a complex system can be expressed as:
$$y=f(x_1,x_2,\ldots,x_m)+ \mathcal{E}$$
where $f(x_1,x_2,\ldots,x_m)$ is a polybasic function, $x_j \in [a_j, b_j]$, $j=1,2,\ldots,m$, $\mathcal{E}$ is a random variable. Suppose that mb is the goal value.

$\forall x^0 = (x_1^0, x_2^0, \ldots, x_3^0) \in [a_j, b_j]$, set a permitted error ($\Delta x_1, \Delta x_2, \ldots, \Delta x_m$). Suppose that x_{ij} ($i=1,2,\ldots,n$; $j=1,2,\ldots,m$) are some points in $[x_j^0 - \Delta x_j, x_j^0 + \Delta x_j]$, satisfying $|x_{(i+2)j}-x_{(i+1)j}|=|x_{(i+1)j}-x_{ij}|$. We call x^0 as the center point of experiments, and x_{ij} as the sequence points of experiments. If f is known by us, then we can obtain $y_i=f(x_{i1},x_{i2},\ldots,x_{im})$, $i=1,2,\ldots,n$, through putting x_{ij} into formula and computing them. If we don't know f, then we can make some experiments at x_{ij} and get the observation value y_i of $y=f(x_{i1},x_{i2}, \ldots,x_{im})$. Thus, we just obtain the experiment data $y_1, y_2, \ldots, y_n$ nearby the center point x^0 of experiments.

We select five criteria below to carve and paint this system's stability at x^0:

1. χ^2—identification criterion. These observation values $y_1, y_2, \ldots, y_n$ from experiments must sufficiently identify or include those information at point x^0. We have known that if orthogonal design method is adopted, then χ^2 just will reach the most optimal;

2. $\mu = f(x^0)$ — position criterion.

3. $\sigma^2(x^0) = \frac{1}{n-1} E \sum_{i=1}^{n} (\hat{f}(x_{i1}, x_{i2}, \ldots, x_{im}) - \bar{y})^2$ — fluctuation criterion, representing this system's fluctuation at point x^0;

4. $R^2(\hat{f}) = \frac{1}{n} E \sum_{i=1}^{m} (\hat{f}(x_{i1}, x_{i2}, \ldots, x_{im}) - mb)^2$ — risk criterion;

5. $SN = \frac{(Ey)^2}{\sigma^2}$ — Signal Noise ratio criterion.

The five criteria and relations between them form a stable logic analysis system (fig. 5). Its stability at point x^0 can be controlled by using the above 5 criteria.

In fig. 5, → can be understood as positive function, and ⇒ can be understood as negative function. For helping reader comprehension, a example in true world is given in fig. 6. It is too simple and easy to explain more. Where, M and W express a Man and a Woman, respectively, while → and ⇒ express "love" and "kill", respectively.

Our purpose is finding a testing center point x^0 in ranges of $\prod_{j=1}^{m} [a_j, b_j]$, at this point, with satisfying that:

$$\mu = f(x^0) = mb, \qquad R_f^2(x^0) = \min_{x \in \prod_{j=1}^m [a_j,b_j]} R_f^2(x),$$

$$\sigma^2(x^0) = \min_{x \in \prod_{j=1}^m [a_j,b_j]} \sigma^2(x), \qquad SN(x^0) = \max_{x \in \prod_{j=1}^m [a_j,b_j]} SN(x)$$

We call the point x^0 stable center or stable point of this complex data system about target design. Like this, we just give a statistical model of the stable center of a complex data system.

In stability experiments of launched objects and experiment designs of products, the new logic analysis model have been already successfully applied many times, with efficiently reducing testing times and bring us many benefits.

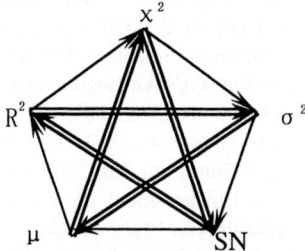

Fig. 5. The five criteria and relations between them form a stable logic analysis system

Fig. 6. Triangle love

5 Conclusions

In this paper, with enlightening from nature, we present a new logic model of intelligent reasoning for analyzing stability of a complex system, and we prove it by means of the mathematics. In the meantime we illustrate the applications of the logic model by using five examples. We think that the logic model is reliable.

The logic model presented by us has been already applied in some areas. For example, in the experiment design and in the analysis of stability of a weapon factory's products, we have used the logic model with reducing the test times, promoting the stability of products, and deriving many economical benefits. Its application practice shows that the logic model is very much effectual to analyzing stability of a complex system. The logic model has very wide uses. Consequently, we can believe that it would bring many benefits for us. Its application algorithm of the logic analysis model will be written in another paper by the authors after.

References

1. Written by Einstein, Translated by Hao Li. Significance of Relativity. Beijing: Science Press (1979)
2. Written by Tiankouxuanyi, Translated by China TQ Institute of Weapon Industry. Quality Engineering in Stage of Developing and Designing. Beijing: Weapon Industry Press of China (1992)
3. Written by Tiankouxuanyi, Translated by China TQ Institute of Weapon Industry. Quality Engineering in Stage of Manufacture. Beijing: Weapon Industry Press of China (1992)
4. Yingshan Zhang. Multilateral Matrix Theory. Beijing: China Statistics Press (1993)
5. Zhang Y.S., Pang S.Q., Jiao Z.M. and Zhao W.Z.: Group Partition and Systems of Orthogonal Idempotents, Linear Algebra and its Applications 278 (1998) 249-262
6. Zhang Y.S., Lu Y.Q. and Pang S.Q.: Orthogonal Arrays obtained by Orthogonal Decomposition of Projection Matrices, Statistica Sinica 9 (1999) 595-604
7. Zhang Y.S., Pang S.Q. and Wang Y.P.: Orthogonal Arrays obtained by Generalized Hadamard Product, Discrete Mathematics 238 (2001) 151-170
8. Zhang Y.S., Duan L., Lu Y.Q. and Zheng Z.G.: Construction of Generalized Hadamard Matrices $D(r^m(r+1), r^m(r+1); p)$, J. Stat. Plan. Inf. 104 (2002) 239-258
9. Zhang Y.S.: Choosing Polybasic Linear Model and Estimated Form in C_p Statistics, Journal of Henan Normal University(Nature) 3 (1993) 31-38
10. Meixia Meng: A Analysis Method of Stable Center of Complex System, Theses of Master Degree of Henan Normal University (2003)
11. Research Center for Chinese and Foreign Celebrities, Developing Center of Chinese Culture Resources, Chinese Philosophy Encyclopedia. Shanghai: Shanghai People Press (1994)

Occluded 3D Object Recognition Using Partial Shape and Octree Model

Young Jae Lee[1] and Young Tae Park[2]

[1] Jeonju University, 1200 Hyo Ja Dong Wansan-Gu Jeonju Jeonbug 560-759, Korea
leeyj@jj.ac.kr
[2] Kyung Hee University, Socheonri Giheungeup Yonginsi Gyeonggido 449-701, Korea
ytpark@khu.ac.kr

Abstract. The octree model, a hierarchical volume description of 3D objects, may be utilized to generate projected images from arbitrary viewing directions, thereby providing an efficient means of the data base for 3D object recognition. The feature points of an occluded object are matched to those of the model object shapes generated automatically from the octree model in viewing directions equally spaced in the 3D coordinates. The best matched viewing direction is calibrated by searching for the 4 pairs of corresponding feature points between the input and the model image projected along the estimated viewing direction. Then the input shape is recognized by matching to the projected shape. Experiment results show good performance of proposed algorithm.

1 Introduction

3D object recognition[1-10] remains as a challenging problem in the area of computer vision. It's not always possible to retrieve the complete shape of objects in the real world, when the objects are occluded or illumination conditions change unexpectedly. Researchers have suggested a wide range of schemes to solve this problem in a framework of partial shape recognition[5-10]. Several techniques using polygon, vertex angle and a string matching may be exemplified[6-7]. For example, in the polygon matching technique[6], the dissimilarity between the input and the model objects is estimated using polygon moments, and the initially matched corner points are used to align other corner correspondence. This method, however, may be applied only when the objects can be approximated by a polygon. The string matching technique[7] using a string for matching but it is difficult with this technique to check which part is matched partially. In general, a large database is required for the 3D object recognition, since objects may have different shapes depending on the viewing direction. This makes the matching process very slow. In this paper we suggest a new and efficient algorithm for recognizing occluded 3D objects.

2 Occluded 3D Object Recognition

In this paper we suggest the method that can recognize the 3D object from the partial image received from a random position and direction. The technique recognizes an object from partial shape and figures out correctly relative distance, volume and rotation, even though the target is occluded or only part of the object is caught with a camera. Our new algorithm first finds local maxima of a smoothed k-cosine function, and approximately identifies feature points for the contour of occluded object. Matching primitives, which have similar magnitude ratios and rotation angles, are detected in order to determine whether there is a model given to the image of an object in the Hough space. The translation vector, that minimizes the mean square error for matching contour segment pairs, is then calculated. Through this process, a 2D input image and the feature points of a projected 3D object can be identified. The viewing direction is fixed with four pairs of feature points, a projected image is produced by octree model according to the viewing direction, and finally the image is projected to the input image to find out whether it matches or not. This is the way the new algorithm works.

2.1   2D Feature Points

A set of contour points abstracted by the Turtle algorithm is defined as $C = \{p_i = (x_i, y_i) | i = 1, \cdots, n\}$. We can calculate the folding ratio of i with i-k and i+k contour points. Using $\boldsymbol{a_{ik}}$ and $\boldsymbol{b_{ik}}$, the vector from p_i to p_{i-k} and p_{i+k}, $\cos\theta_{ik}$ which is the k-cosine value of i contour point, is defined as follows:

$$\cos\theta_{ik} = \frac{\boldsymbol{a_{ik}} \cdot \boldsymbol{b_{ik}}}{|\boldsymbol{a_{ik}}| |\boldsymbol{b_{ik}}|} \tag{1}$$

The k value is a very important parameter to determine the accuracy of the description of the contour. The smaller the value is, the more accurate the description is. The k value is, however, sensitive to noises and minute changes. If the value is large, it becomes harder to detect feature points for the detailed shape of an object. Therefore, the feature points should be detected automatically by smoothing and scale-space filtering in order to determine the k value that best corresponds to the contour information. The scale-space filtering[9] can detect stable feature points while increasing the size of a smoothing kernel. The k-cosine value is put into the smoothing process with the convolution of the Triangular function. The contour feature point that has the local maximum among smoothed k-cosine values is very important to determine the shape of an object. In order to describe the object accurately in relation to the size and complexity of the object, the k value should be selected properly. The scale factor k is increased to choose the k value automatically, and the feature point that doesn't respond to the change of the k value. Let's mark the feature points abstracted by the feature point detection algorithm as S_i $i = 1 \cdots, n$. The contour points are projected on the line constructed with $(S_i, S_{i+\delta})$ vertically to $(S_i, S_{i+\delta})$ and the primitives that compose the shape of an object can be abstracted. δ is the parameter that prescribes the feature of a primitive. The δ value should be determined by the feature point abstraction algorithm. For the accurate recognition that is not influenced by the number and position of feature points, we use two types of the primitives: $\delta = 2$(Type1) and $\delta = 3$(Type2). The primitive recognition is accomplished by matching with the model primitive. First, we calculate the matching score record between the i_{th} primitive L_i of an object and the j_{th} primitive O_i of the model:

$$\eta_{ij} = |L_i - O_i| = \frac{1}{D}\sum_{k=1}^{D}|L_{ik} - O_{jk}| \qquad (2)$$

Here D stands for the number of projection for the primitive, and the low ratio shows that this is an acceptable match. The pair, which has the lowest matching score record between object and model primitives, is selected and saved. Finally the pair whose size ratio and rotation angle matches perfectly is sorted out, and the primitive recognition is completed. The second step uses the Hough technique, in order to search for the primitive set that has the similar magnitude ratio and rotation angle among recognized primitive pairs. First produced is the bin that has a similar magnitude ratio S. Next produced is the bin that has a similar rotation angle θ to the primitives that belong to each bin. The magnitude ration of the bin is assigned ±15 to find out significant similarities. $\eta_{s,\theta}$ shows the number of primitives which belong to the bin that has the similar magnitude ratio S and rotation angle θ. The bin, whose value of the primitives is higher than 2, is checked whether they are continuous or not. The continuous primitives are finally designated as matching primitives. If $\eta_{s,\theta}$ is more than 2 and the primitives are continuous, the model provides vivid information for the present object. The number of the pair of matching primitives in the neighbor is more than 2 and the magnitude ratio and the rotation angle correspond to each other, so that we can say the similarity to the model is significantly high. With this technique, it is even possible to recognize partially damaged objects with their partial object information. The primitives of the model, and the $\eta_{s,\theta}$ matching and neighboring primitive pairs $(q_1, p_1), \cdots, (q_n, p_n)$ the average magnitude ratio S and the average rotation angle θ calculated from those primitive pairs, and their relative position can be shown as follows:

$$S = \frac{1}{\eta_{s,\theta}}\sum_{i=1}^{\eta_{s,\theta}} \frac{length(p_i)}{length(q_i)} \quad , \quad \theta = \frac{1}{\eta_{s,\theta}}\sum_{i=1}^{\eta_{s,\theta}}[length(p_i) - length(q_i)] \qquad (3)$$

$$x' = S \cdot cos(\theta) \cdot x - S \cdot sin(\theta) \cdot y + t_x$$
$$y' = S \cdot sin(\theta) \cdot x + S \cdot cos(\theta) \cdot y + t_y \qquad (4)$$

(x', y') is the coordinate system of the object, and (x, y) is that of the model. The relative position of the object (tx, ty) can be calculated with the primitive coordinates of both the model and the object so that the mean square error can be minimized.

$$t_x = \frac{1}{N}\sum_{(x,y)}[x' - (S \cdot cos(\theta) \cdot x - S \cdot sin(\theta) \cdot y)]$$

$$t_y = \frac{1}{N}\sum_{(x,y)}[y' - (S \cdot sin(\theta) \cdot x + S \cdot cos(\theta) \cdot y)] \qquad (5)$$

In other words, the relative position can be obtained with the difference of average coordinates between the model primitive and the object primitive converted by the magnitude ratio S and the rotation angle θ.

2.2 Octree Model

Octree model has a data structure in which the two-dimensional quadtree extends to a three-dimensional expression. It involves itself in the relation to the three-dimensional space and has its data compressed. This type of expression is the method that enables the searching process to calculate efficiently by utilizing tree searching, and it productively uses the special relation of some basic algorithm of graphics such as translation, scaling, rotation and hidden-surface removal. The octree model is used in a variety of fields to express 3D objects effectively. The examples include computer graphics and its application, animation, CAD, moving object realization by robotics, medical 3D construction with computed tomography, etc[1-4]. The octree structure uses three orthogonal 2D images to produce the 2D image of a 3D model object easily and quickly, which is projected from an arbitrary viewing direction, with the volume intersection algorithm[3] so that it can reduce the size of database impressively. If the surface information of a 3D object is added, the 2D projection image can be expressed with pseudo gray. In short, the octree modelling method can be used to recognize efficiently any 3D objects. Among the surface nodes detected by the process above mentioned, the surface node whose normal vector times viewing direction vector has minus value is the node that can be seen from the viewing direction. The 2D projection image can be produced by parallel-projecting the surface of these nodes to the viewing direction. With the absolute value of the multiplied vectors, the pseudo gray surface information can be expressed on the 2D projection image. Fig. 1. shows the 2D projection images of a airplane and a motor car from an arbitrary viewing direction using octree. Fig. 2. shows the process in which the octree is automatically produced with three orthogonal 2D images projected from three arbitrary viewing directions.

Feature points of a 3D object, which are derived from the 2D feature points detected from top, side and front images, are produced in accordance with the following principle: If the two 2D image feature points P_i and Q_j, which are

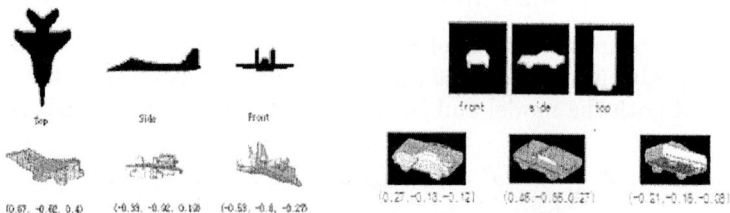

Fig. 1. Example of images projected from octree

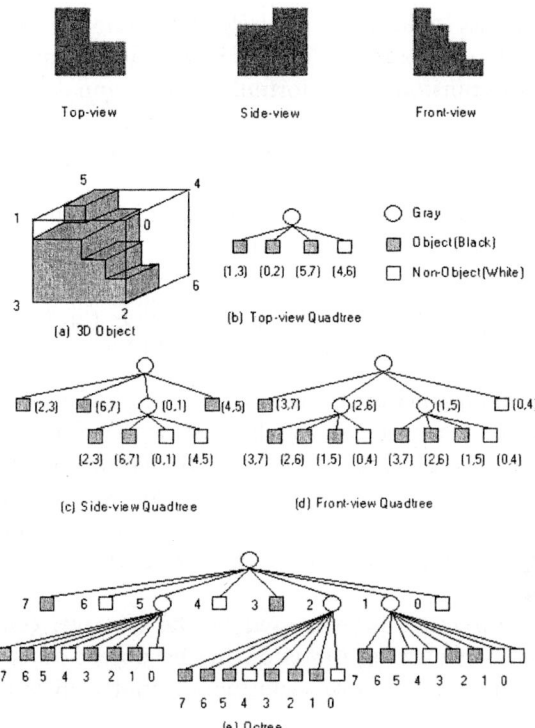

Fig. 2. Octree generated from Top, Side, Front view block images

caught from the two orthogonal viewing directions, are extended to each viewing direction and they intersect, the intersecting point is the feature point of the 3D object. Consequently the 3D feature points can be abstracted from all the feature point pair P_i and Q_j on the 2D image.

2.3 3D Feature Points and Matching

The geometric transformation for the four pairs of feature points is used to find the feature points of the 3D object corresponding to the feature points of the 2D image[8]. If the coordinates of the 2D feature point P_i and the 3D feature point p_j corresponding to P_i are (X_i, Y_i), (x_j, y_j, z_j) and , the projection from p_j to P_i can be expressed as follows:

$$X_i = R_{11}\ x_j + R_{12}\ y_j + R_{13}\ z_j + t_x$$
$$Y_i = R_{21}\ x_j + R_{22}\ y_j + R_{23}\ z_j + t_y \qquad (6)$$

The transform relation between the 2D coordinates and the 3D coordinates is expressed with the rotation matrix factors

$V_1^t = \begin{bmatrix} R_{11} & R_{12} & R_{13} \end{bmatrix}$, and $V_2^t = \begin{bmatrix} R_{21} & R_{22} & R_{23} \end{bmatrix}$, and the translation transform factors t_x and t_y. If four pairs of feature points are given, the transformation can be solved. This transformation formula is orthogonal.

$$V_1 \cdot V_2 = 0, \quad ||V_1|| = c, \quad ||V_2|| = c \tag{7}$$

However taking quantization effect and noises into consideration, the following expression is used to evaluate digital images:

$$|V_1 \cdot V_2| < \delta_1, \quad 1 - \delta_2 < \frac{||V_1||}{||V_2||} < 1 + \delta_2 \tag{8}$$

δ_1 and δ_2 are small constants, and we used 0.3 and 0.2 in the experiments. The four pairs of feature points, which meet with all the conditions already mentioned, are then searched for. If a set of 2D feature points $\{P_1, P_2, \cdots, P_n\}$ and a set of 3D feature points $\{p_1, p_2, \cdots, p_n\}$ are given, the transform relation between the 2D and the 3D coordinates can be obtained by searching for the four pairs of feature points that satisfy the transformation expressions (6) and (8).

If the viewing direction is expressed in the observer-centered coordinates as , the corresponding viewing direction in the object-centered coordinates is transformed this way, where the transformation matrix R is marked:

$$\mathbf{R} = \begin{bmatrix} R_{11} & R_{12} & R_{13} \\ R_{21} & R_{22} & R_{23} \\ R_{31} & R_{32} & R_{33} \end{bmatrix} \tag{9}$$

$R^{-1} = R^t$ (R is an orthogonal matrix)

$$V_{dir} = R^{-1} \begin{bmatrix} 0 \\ 1 \\ 1 \end{bmatrix} = \begin{bmatrix} R_{31} \\ R_{32} \\ R_{33} \end{bmatrix} = \begin{bmatrix} R_{11} \\ R_{12} \\ R_{13} \end{bmatrix} \times \begin{bmatrix} R_{21} \\ R_{22} \\ R_{23} \end{bmatrix} \tag{10}$$

So, the viewing direction(V_{dir}) if four pairs of feature points that satisfy the expression (6) and (8) are determined, the viewing direction can be estimated by the cross product of V1 and V2. If the shape of 3D object is simple, the feature point matching between 2D and 3D is enough to recognize the 3D object. If the shape is rather complex, some false pairs of feature points satisfying (6) and (8) can exist by coincidence. So, the verifying process is necessary to eliminate possible errors. Our algorithm produces the 2D projection image that is composed with the octree in accordance with the calculated viewing direction, and finally confirms if it really matches with the input image. This algorithm, therefore, can find out the original, complete 3D shape from partial shape. Even if another object model is added, it doesn't have to change the proposed recognition algorithm, but simply supplements model database to the primitives of a new object so its recognition system has the merit of being easily extended.

3 Experiments

3.1 Experiment 1

Our algorithm materialized matching in case of two arbitrarily occluded 3D images from a certain viewing direction(Fig. 3(a)and(b)). Feature points were first found with the method above mentioned in order to find two 3D objects in the occluded 3D images of Fig. 3(b), and similar matching points(Fig. 3(c)) and the model were found in the database that was constructed with the octree model. Four distinctive feature points were checked against feature points in the occluded part (red points in Fig. 3(d)). Fig. 3(e) shows it's projection image that found with the octree model and feature points. Fig. 3(f) shows the final matching image after the generated image was projected to the input image in order to check their matching ratio. We found the first object of the two 3D objects in the input image in the process of matching. Similarly Fig. 3(g) shows the input image and its feature points, Fig. 3(h) the partial image of matching points, and Fig. 3(i) the matching model and 4 feature points. Using these feature points, the 2D projection image and feature points are shown in Fig. 3(j). Fig. 3(k) shows the final matching image.

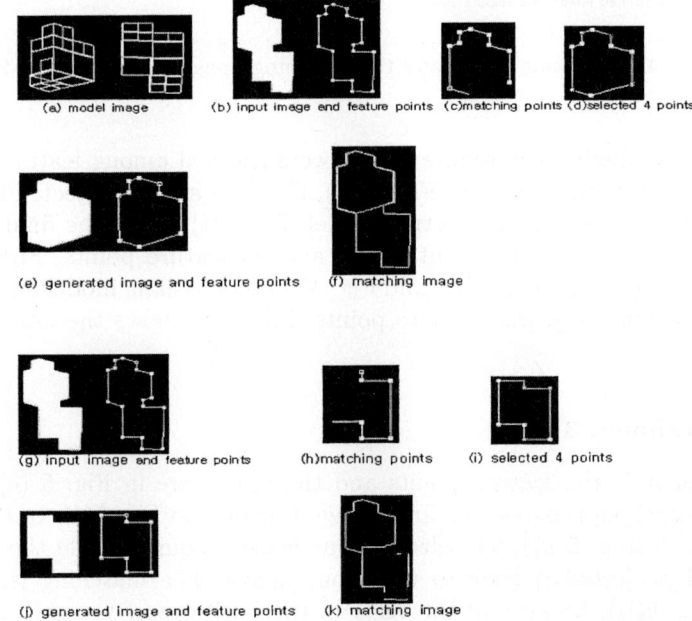

Fig. 3. The input image and the matching image of Experiment 1

3.2 Experiment 2

As we did in Experiment 1, Fig. 4(b) shows the occluded 3D input image and its feature points. The feature points of matching part and the model are shown in

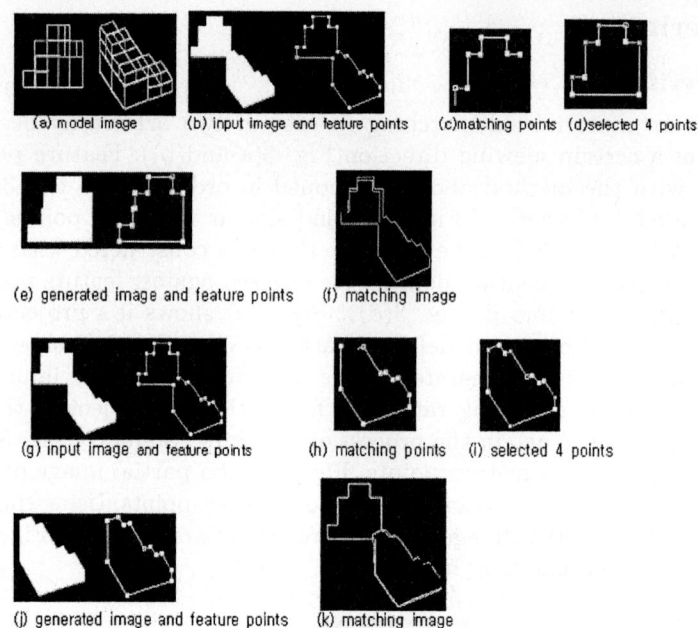

Fig. 4. The input image and the matching image of Experiment 2

Fig. 4(c). Four distinctive feature points were checked among feature points in the occluded part (red points in Fig. 4(d)). Fig. 4(e) shows projected image and feature points found with the octree model. Fig. 4(f) shows the final matching image. Fig. 4(g) shows the input image and its feature points, Fig. 4(h) the partial image of matching points, and Fig. 4(i) the matching model and Fig. 4(h) shows projection image and feature points. Fig. 4(k) shows the final matching image.

3.3 Experiment 3

In Experiment 3, the feature points and the model are in Fig. 5 (c) and (d). After the matching process, the final image is shown in Fig. 5(f). In case of the second object (Fig. 5(g)), we selected four feature points, made the projected image, and projected it back to the input image. The matching is, however, rejected(Fig. 5(j)), because of the error in the positions of feature points. The algorithm could recognize one of the two 3D input images.

3.4 Experiment 4

In Experiment 4, two 3D objects were occluded and the new feature points were created. (arrows in Fig. 6(b)) This made the number of feature points increased to recognize partial shapes so that the matching was rejected.

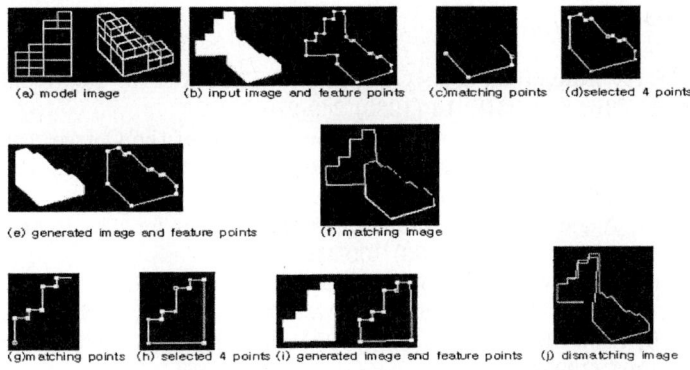

Fig. 5. The input image and the result image of Experiment 3

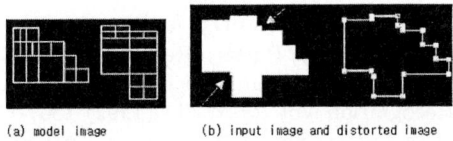

Fig. 6. The input image and the distorted image of Experiment 4

Table 1. Result of occluded 3D object matching

Experiment	object	viewing dir.	Cal. viewing dir.	angular error	pixel error
Exp. 1	object1	-0.47,-0.88,0.01	-0.49,-0.87,0.01	1.43	1.63
	object2	-0.33,0.67,0.67	-0.37,0.68,0.64	3.17	4.75
Exp. 2	object1	-0.65,-0.69,-0.32	-0.67,-0.67,-0.33	2.16	2.51
	object2	-0.71,-0.05,-0.71	-0.66,-0.14,-0.73	5.86	2.51

4 Conclusion

If the 3D object is occluded, some part of the 2D image of 3D models should be used. The 2D image matching was carried on with the primitives and the positions in the Hough space, the viewing direction was estimated with those feature points, and the 3D partial matching was found by projecting the 2D projection image to the input image using the octree model. In cases where the object was excessively occluded so that the number of feature points changed, or many feature points were moved, the matching recognition was rejected. With the exception of these cases, the new algorithm recognized the image of object accurately and calculated the relative positions correctly. It was not influenced by the position, volume or rotation of the object. Even if a new model object is added, the new algorithm does not change its recognition algorithm but simply supplements model database to the primitives of a new object.

References

1. Jackins, C.L., Tanimoto, S.L.: Octrees and Their Use in Representing Three-dimensional Objects. CGIP. 14 (1980) 249-270
2. Noborio, H., Fukuda, S., Arimoto, S.: Construction of the Octree Approximating Three-dimensional Objects by Using Multiple Views. IEEE Trans. PAMI. Vol. 10 No. 6 (1988) 769-782
3. Chien, C.H., Aggarwal, J.K.: Volume/Surface Octrees for The Representation of 3-D Objects. CGIP. Vol. 36 (1986) 100-113
4. Chien, C.H., Aggarwal, J.K.: Model Construction and Shape Recognition from Occluding Contours. IEEE Trans. on PAMI. Vol. 11 No. 4 (1989) 372-389
5. Cho, MS.J., Kim, M.S., Kim, S.D.: Automatic Target Recognition System. Proc. of ACCV 1995 (1995) 100-104
6. Mark W.K., and Kashyap R.L.: Using Polygons to Recognize and Locate Partially Occluded Objects. IEEE Trans. on PAMI. Vol. 9 No. 4 (1987) 483-494
7. Maurice, Maes : Polygonal Shape Recognition Using String Matching. Pattern Recognition Vol. 24 No. 5 (1991) 433-440
8. Silberberg, T.M. , Davis, L., Harwood D. : An Iterative Hough Procedure for Three Dimensional Object Recognition. Pattern Recognition Vol. 17 No. 6 (1984) 621-629
9. Pei, S., Lin, C.: The Detection of Dominant Points on Digital Curves by scale-space filtering. Pattern Recognition Vol. 25 No. 11 (1992) 1307-1314
10. Ansari, N., Delp, E.J.:Partial Shape Recognition : a Landmark-based Approach IEEE Trans. PAMI Vol. 12 No. 5 (1990) 470-483

Possibility Theoretic Clustering

Shitong Wang[1,2], Fu-lai Chung[1], Min Xu[2], Dewen Hu[3], and Lin Qing[2]

[1] Dept. of Computing, Hong Kong Polytechnic University, Hong Kong, China.
[2] School of Information Engineering, Southern Yangtze University, China.
wst_wxs@yahoo.com.cn
[3] School of Automation, National Defense University of Science and Technology, Changsha, China.
dewenhu@nudt.edu.cn

Abstract. Based on the exponential possibility model, the possibility theoretic clustering algorithm is proposed in this paper. The new algorithm is distinctive in determining an appropriate number of clusters for a given dataset while obtaining a quality clustering result. The proposed algorithm can be easily implemented using an alternative minimization iterative procedure and its parameters can be effectively initialized by the Parzon window technique and Yager's probability-possibility transformation. Our experimental results demonstrate its success in artificial datasets.

1 Introduction

Clustering has long been a hot research topic in various disciplines and it has been widely applied in the past decades, e.g., decision making, document retrieval, image segmentation, and pattern classification. Recently, more and more researchers are interested in its application in *large* datasets [1,2]. Clustering algorithms attempt to organize unlabeled input vectors into clusters or "natural groups" such that data points within a cluster are more similar to each other than those belonging to different clusters, i.e., to maximize the intra-cluster similarity while minimizing the inter-class similarity. Based on their way to handle uncertainty in data, the clustering algorithms can be categorized into probabilistic and fuzzy clustering. Fuzzy clustering is our concern here. Among the proposed fuzzy clustering algorithms, the fuzzy c-means (FCM) algorithm [3] is perhaps the most well-known one. Most of the newly proposed fuzzy clustering algorithms originate from it, e.g., possibilistic clustering [4], fuzzy competitive learning, and gravity-based clustering. FCM and its variants realize the clustering task for a dataset by minimizing an objective function subject to some constraints, e.g. the summation of all the membership degrees of every data point to all clusters must be 1. However, this usually makes the objective functions monotonically decrease with the increase of the number of clusters, making the problem of setting an appropriate number of clusters even harder.

Compared with FCM, little attention has been paid to possibilistic clustering [4,5,6]. Based on possibility measure, Krishnapuram and Keller [4] presented a possibilistic clustering algorithm to enhance the robustness to noise and outliers in datasets. They take use of the possibility measures to relax the constraints adhered to the objective functions, or add some penalty terms to the objective functions. The number of clusters must be preset for the objective functions, and their clustering performances heavily depend on a reasonable choice of the number of clusters.

The theory of possibility was introduced by Zadeh [7] in 1978 and it is commonly known as an uncertainty theory devoted to the handling of incomplete information. Recently, Tanaka and his group presented their recent works on possibilistic data analysis for operations research [8] where a new concept called exponential possibility model was introduced. Being inspired by this new achievement in possibility theory, we develop a novel possibilistic clustering algorithm which provides a new perspective on the relationship between possibility theory and clustering. Furthermore, a unified criterion for choosing an appropriate number of clusters and realizing efficient clustering simultaneously is introduced. The new algorithm can be easily implemented by an alternative minimization iterative procedure. In order to avoid the algorithm's sensitivity to initial conditions, Yager's probability-possibility transformation method [9] and Parzon window technique are used to estimate the initial parameter values. The remainder of the paper is organized as follows. Section 2 introduces the fundamentals of exponential possibility model and elaborates this model from a clustering viewpoint. In Section 3, the new clustering algorithm is presented. We demonstrate the effectiveness of the new algorithm in artificial datasets in Section 4. The final section concludes the paper.

2 Exponential Possibility Model and Clustering

Possibility theory, firstly proposed by Zadeh [7], offers a model for effectively representing the compatibility of a crisp value with a fuzzy concept where a fuzzy variable is associated with a possibility distribution in the similar way that a random variable is associated with a probability distribution. It has been further developed for the so-called possibilistic data analysis for which a new concept called exponential possibility model [2,10,11,12] is introduced by Tanaka and his group recently. An exponential possibility distribution is regarded as a representation of evidence, which is represented by an exponential function

$$\Pi_A(x) = \exp\{-(x-a)^t D_A (x-a)\} \tag{1}$$

where the evidence is denoted as A, a is a center vector and D_A is a symmetrical positive definite matrix. It should be pointed out here that there is a big difference between an exponential possibility distribution and an exponential probability distribution.

$$\text{Possibility:} \begin{cases} A = (a, D_A)_e \\ \Pi_A(x) = \exp\{-(x-a)^t D_A (x-a)\} \\ \max_x \Pi_A(x) = 1 \end{cases} \tag{2}$$

Possibility:
$$\begin{cases} x \sim (\mu, \Sigma) \\ p(x) = \dfrac{1}{(2\pi)^{\frac{d}{2}}|\Sigma|^{\frac{1}{2}}} \exp\{-\dfrac{1}{2}(x-\mu)^T \Sigma^{-1}(x-\mu)\} \\ \int p(x)dx = 1 \end{cases} \quad (3)$$

where d denotes the dimensional number, μ denotes the mean and Σ denotes the variance matrix

= = = = = = = = = = = = = = = = = =

Let the joint exponential possibility distribution on the $(n+m)$-dimensional space be

$$\Pi_A(x,y) = \exp\{-(x-a_1, y-a_2)'\begin{bmatrix} D_{11} & D_{12} \\ D_{12}^t & D_{22} \end{bmatrix}(x-a_1, y-a_2)\} \quad (4)$$

where x and y are n- and m-dimensional vectors, respectively; a_1 and a_2 are center vectors in the n- and m-dimensional spaces, i.e. X & Y respectively; and D_{11}, D_{12}, D_{22} are the positive definite matrices. Then, we can define the marginal possibility distributions on X and Y as

$$\Pi_A(x|y) = \exp\{-(x + D_{11}^{-1}D_{12}(y-a_2) - a_1)' D_{11}(x + D_{11}^{-1}D_{12}(y-a_2) - a_1)\} \quad (7)$$

$$\Pi_A(y|x) = \exp\{-(y + D_{12}^{-1}D_{11}(x-a_1) - a_2)' D_{22}(y + D_{12}^{-1}D_{11}(x-a_1) - a_2)\} \quad (8)$$

In [12], Tanaka et al. derived the following theorem:
Theorem 1. If $\Pi_A(x|y)$ & $\Pi_A(y)$ satisfies (7) and (6) respectively, then

$$\Pi_A(x,y) = \Pi_A(x|y)\Pi_A(y) \quad (9)$$

According to this theorem, we immediately have

$$\Pi_A(x|y)\Pi_A(y) = \Pi_A(y|x)\Pi_A(x) \quad (10)$$

which is very important because it provides the solid mathematical foundation of our possibility theoretic clustering algorithm.

For $\Pi_A(x|y)\Pi_A(y) = \Pi_A(y|x)\Pi_A(x)$ in (10), we may view x as the observable pattern in the observable space (i.e. a dataset) and y as its representation (i.e. its cluster) in the representation space. Suppose K is the number of clusters, then the representation space consists of K corresponding clusters. Thus, for the given observable space X, the clustering task attempts to find out all the corresponding clusters in the representation space such that (10) holds true. In other words, if we simplify Π_A as Π, and define

$$\Pi_{C_1}(x,y) = \Pi(y|x)\Pi(x) \quad (11)$$

$$\Pi_{C_2}(x,y) = \Pi(x|y)\Pi(y) \tag{12}$$

then we may explain the clustering task from a new angle. That is, for every datum x in a dataset, clustering tries to obtain its cluster y in the representation space such that $\Pi_{C_1}(x,y) = \Pi_{C_2}(x,y)$. In practice, due to the possible existence of a discrepancy between $\Pi_{C_1}(x,y)$ and $\Pi_{C_2}(x,y)$, we can introduce the so-called *consistent function* F to reach a balance between $\Pi_{C_1}(x,y)$ and $\Pi_{C_2}(x,y)$, i.e. to realize the clustering task by minimizing the following consistent function

$$F(C_1, C_2) = F(\Pi(y|x)\Pi(x), \Pi(x|y)\Pi(y)) \tag{13a}$$

Obviously, F should be a convex function satisfying

$$F(\Pi(y|x)\Pi(x), \Pi(x|y)\Pi(y)) \geq 0 ;$$
$$F(\Pi(y|x)\Pi(x), \Pi(x|y)\Pi(y)) = 0, \quad iff \quad \Pi(y|x)\Pi(x) = \Pi(x|y)\Pi(y) \tag{13b}$$

The minimization of $F(C_1, C_2)$ can be easily implemented by an *alternative minimization iterative procedure*:

Step 1: Fix $C_2 = C_2^{old}$, compute $C_1^{new} = \arg\min_{C_1} F$

Step 2: Fix $C_1 = C_1^{old}$, compute $C_2^{new} = \arg\min_{C_2} F$

which guarantees to reduce F until it converges to a local minimum. The alternative minimization iterative procedure can be implemented by the EM algorithm or using the gradient descent learning rules.

Let us only take the well known *Kullback divergence* here for ease of discussion, i.e. F is defined as

$$KL_1(\Pi(y|x)\Pi(x), \Pi(x|y)\Pi(y)) = \sum_{i=1}^{N_x}\sum_{j=1}^{N_y} \Pi(y_j|x_i)\Pi(x_i) \times \ln\frac{\Pi(y_j|x_i)\Pi(x_i)}{\Pi(x_i|y_j)\Pi(y_j)} \tag{14}$$

$$KL_2(\Pi(y|x)\Pi(x), \Pi(x|y)\Pi(y)) = -\sum_{i=1}^{N_x}\sum_{j=1}^{N_y} \Pi(y_j|x_i)\Pi(x_i) \times \ln \Pi(x_i|y_j)\Pi(y_j) \tag{15}$$

Let

$$F_1(\theta_K, K) = KL_1(\Pi(y|x)\Pi(x), \Pi(x|y)\Pi(y)) = KL_1(\Pi_{C_1}(x,y), \Pi_{C_2}(x,y)) \tag{16}$$

$$F_2(\theta_K, K) = KL_2(\Pi(y|x)\Pi(x), \Pi(x|y)\Pi(y)) = KL_2(\Pi_{C_1}(x,y), \Pi_{C_2}(x,y)) \tag{17}$$

where K is the number of clusters, θ_K is the parameter set, i.e. $\theta_K = \{D_{11}, D_{12}, D_{22}, a_1, a_2\}$. With K fixed, when $F_1(\theta_K, K)$ or $F_2(\theta_K, K)$ achieves its minimum, we can find out the corresponding cluster y to which every datum x in a dataset belongs. In general, K is unknown or we may perhaps have multiple choices. The above alternative minimization procedure can help us to determine an appropriate K by picking the smallest K^* among all possible values of K such that $F_1(\theta_{K^*}, K^*)$ or $F_2(\theta_{K^*}, K^*)$ reaches its minimum for the given dataset. In other words, by minimizing $F_1(\theta_K, K)$ or $F_2(\theta_K, K)$, we can determine the best number of clusters and the corresponding clusters simultaneously, which is a major contribution of the proposed method here.

To save the space, we do not intend to go into the details of how to make $F_1(\theta_K, K)$ or $F_2(\theta_K, K)$ achieve its minimum by using the alternative minimization iterative procedure, since it can be easily done by the EM algorithm or the gradient descent learning rules but the details are tedious.

3 Possibility Theoretic Clustering Algorithm

As we know, the alternative minimization iterative procedure does not guarantee to achieve the global minimum and is very sensitive to the initial parameter values. Therefore, in order to make the new clustering algorithm work well, we should carefully initialize the parameters. In this paper, an effective method based on Yager's probability-possibility transformation [9] is proposed.

In [9], Yager introduced the probability-possibility transformation method as follows. For a given n-dimensional dataset $X = \{x_i = (x_{i1}, x_{i2}, \ldots, x_{in}) \mid i = 1 \cdots N\}$, if the possibility distribution $\pi(x_i)$ of each x_i in the dataset is given, then after ranking all $\pi(x_i)$ in decreasing order, we can use the following transformation equation to obtain the corresponding probability:

$$p(x_1) = \frac{\pi(x_1)}{N}$$
$$p(x_j) = p(x_{j-1}) + \frac{1}{N+1-j}(\pi(x_j) - \pi(x_{j-1})) \quad \text{for } j = 2, \cdots, N$$
(18)

Obviously, if the probability distribution $p(x_i)$ of each x_i is known, we can use (18) to obtain the corresponding possibility distribution. With the well-known *Parzon window technique* [13], for the dataset described above, we may estimate $p(x_i)$ reasonably as:

$$p(x_i) = \frac{1}{N}\sum_{l=1}^{N} K(x_i - x_l) \quad \& \quad K(r) = \frac{1}{h^n}k(\frac{r}{h})$$
(19)

where $k(\bullet)$ denotes a prefixed kernel function. In this paper, for simplicity, we take $k(\frac{r}{h}) = \exp(-\frac{r^2}{h^2})$ for the artificial datasets to be used in the experimental result section. We also estimate $p(x_i)$ using the so-called Epanechnikov kernel function in the large image processing application:

$$p(x_i) = \frac{1}{N}\sum_{j=1}^{N} k(x_{i1} - x_{j1}, \cdots, x_{in} - x_{jn}) \qquad (20)$$

Where

$$k(x_{i1} - x_{j1}, \cdots, x_{in} - x_{jn}) = (\frac{3}{4})^n \frac{1}{B_1 B_2 \cdots B_n} \prod_{1 \le l \le n}(1 - (\frac{x_{xl} - x_{jl}}{B_l})^2) \quad if \left|\frac{x_{il} - x_{jl}}{B_l}\right| \le 1, \forall i$$
$$k(x_{i1} - x_{j1}, \cdots, x_{in} - x_{jn}) = 0 \qquad otherwise \qquad (21)$$

and parameters $B_1, \cdots, B_n$ can be estimated according to Scott's rule [13], i.e. $B_l = \sqrt{5}s_l |N|^{-\frac{1}{n+4}}$, $l=1,2, \ldots, n$, with s_l denoting the standard deviation of the data set along the lth dimension. In [11], it has been shown that the exact shape of the kernel function does not affect the approximation a lot. A polynomial or Gaussian function can work equally well. The standard deviation of the function, i.e. its bandwidth, is much more important. The Epanechnikov kernel function is easy to integrate (the biased sampling procedure [2,14] needs this property for large image processing), and so, for convenience, we adopt this kernel function. With $p(x_i)$ estimated and ranked in decreasing order, in terms of (19-21), we immediately have

$$\pi(x_1) = p(x_1)N$$
$$\pi(x_j) = \pi(x_{j-1}) + (N+1-j)(p(x_j) - p(x_{j-1})) \qquad for\ j = 2, \cdots, N \qquad (22)$$

That is, with Parzen window technique and Yager's probability-possibility transformation method, we can easily obtain the possibility distribution of any dataset. Since exponential possibility distribution is assumed by the proposed clustering algorithm, we have to approximate it appropriately. Such an exponential possibility distribution can be easily found by minimizing the mean-squared-error (MSE) below:

$$MSE = \frac{1}{N}\sum_{i=1}^{N}[\Pi(x_i) - \pi(x_i)]^2 \qquad (23)$$

where $\Pi(x_i) = \max_y \Pi(x_i, y) = \exp\{-(x_i - a_1)'(D_{11} - D_{12}D_{22}^{-1}D_{12}')(x_i - a_1)\}$ according to (5). In this way, we can estimate the initial values of parameters D_{11}, D_{12}, D_{22} & a_1 for our possibility theoretic clustering algorithm. Now, let us state the overall algorithm.

Step 1. For the given dataset, compute its probability distribution using Parzon window technique, and then, obtain its possibility distribution using the probability-possibility transformation method. Set the initial parameter values of $\{D_{11}, D_{12}, D_{22}, a_1\}$ by the results of minimizing (23) using the gradient descent learning rules.

Step 2. Choose a consistent function F_1 or F_2, fix K, and compute
$$\theta_{1,K}^* = \arg\min_{\theta_K} F_1(\theta_K, K) \text{ or } \theta_{2,K}^* = \arg\min_{\theta_K} F_2(\theta_K, K)$$
using the alternative minimization iterative procedure (implemented by the EM algorithm or the gradient descent learning rules).

Step 3. Change K and repeat step 2 until
$$K^* = \arg\min_K F_1(\theta_{1,K}^*, K) \text{ or } K^* = \arg\min_K F_2(\theta_{2,K}^*, K).$$
The final clustering result is K^* and $\theta_{1,K}^*$, or, K^* and $\theta_{2,K}^*$; depending on the adoption of F_1 or F_2.

After obtaining $\theta_{1,K}^*$ or $\theta_{2,K}^*$, we fix all the parameters in (8). For a data point x, we can use (8) to determine its cluster which makes $\prod(y \mid x)$ in (8) achieve the maximum in the representation space.

4 Performance on Artificial Datasets

Despite the introduction of probability-possibility transformation by Yager [9], how to effectively determine the possibility distribution of a dataset still remains a challenging topic. Moreover, our assumption here that a dataset must follow an exponential possibility distribution seems too rigid in practice and the clustering performance might be severely affected. According to our experimental results, this is not a concern, or, at least we can justify that the proposed algorithm is very effective for convex datasets like the artificial ones below.

Fig.1 depicts three artificial datasets of five classes with three different degrees of overlap. With the number of clusters $K=5$, the proposed clustering algorithm is able to converge to the expected cluster centers (the bold points in Fig.1). Fig.2 records the changes in $F_1(K)$ and $F_2(K)$ for $K=3$ to 7. Obviously, $F_1(K)$ and $F_2(K)$ reach the minimum values when $K=5$ for all the three datasets and this is in line with the *ideal* number of clusters. Similar performance can be obtained for the datasets in Fig.3 consisting of seven classes. As demonstrated by Fig.4, the proposed clustering algorithm again is able to identify the ideal number of clusters which is 7. The experimental results here validate the effectiveness of the proposed algorithm in both clustering performance and identifying a reasonable number of clusters.

5 Conclusions and Future Work

In this paper, a new clustering approach based on Tanaka's exponential possibility model [8] is introduced and a possibility theoretic clustering algorithm is proposed. It takes use of the Parzon window technique and Yager's probability-possibility transformation [9] to initialize its parameters and the algorithm can be easily implemented by the alternative minimization iterative procedure. It possesses a distinctive feature that an appropriate number of clusters can be determined automatically during the clustering process. As demonstrated by the simulations with artificial convex datasets, the new algorithm is effective in converging to the clusters' centers and identifying the required number of clusters. Our future works include the applications of this new algorithm to large image processing tasks, data mining problems, pattern recognition, and bioinformatics databases, and the development of its robust version so that a class of robust possibility theoretic clustering algorithms for noisy datasets can be developed.

Acknowledgements

This work is supported by New_century Outstanding Young Teacher Grant of Ministry of Education of China (MOE), The Key Project Grant of MOE, Natural Science Foundation of China (grant No. 60225015), Natural Science Foundation of JiangSu Province (grant No. BK2003017) .

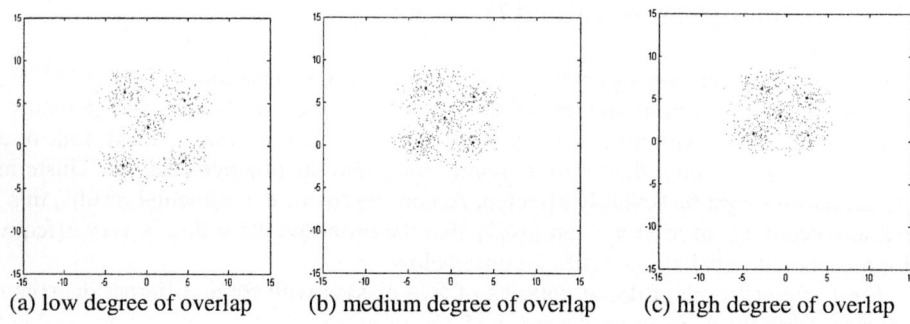

(a) low degree of overlap (b) medium degree of overlap (c) high degree of overlap

Fig. 1. Three artificial datasets of five classes

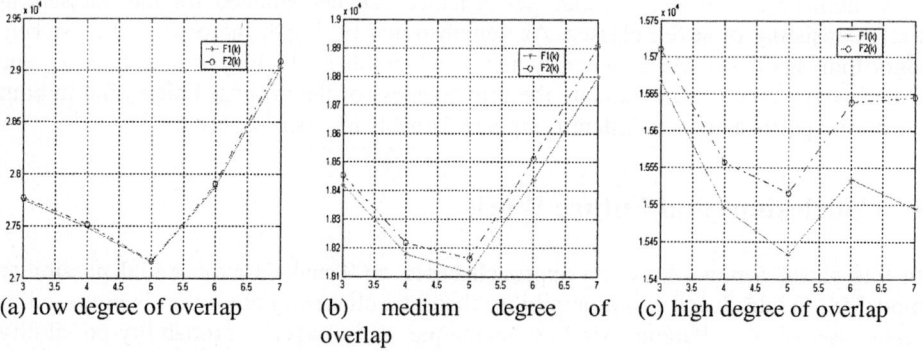

(a) low degree of overlap (b) medium degree of overlap (c) high degree of overlap

Fig. 2. Changes in $F_1(\theta_k,k)$ and $F_2(\theta_k,k)$ for the three datasets in Fig.1 using different number of clusters K

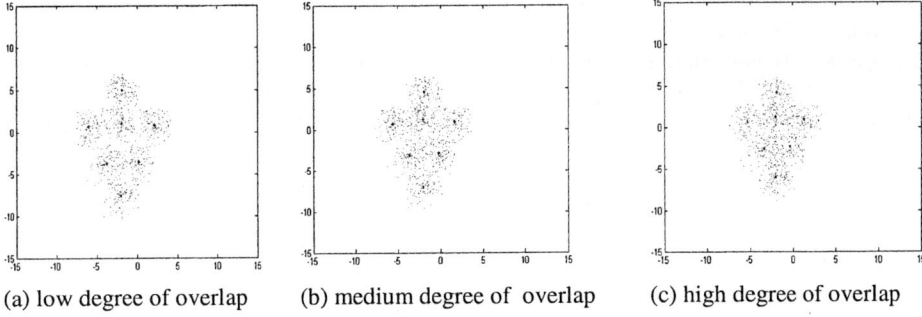

(a) low degree of overlap (b) medium degree of overlap (c) high degree of overlap

Fig. 3. Three artificial datasets of seven classes

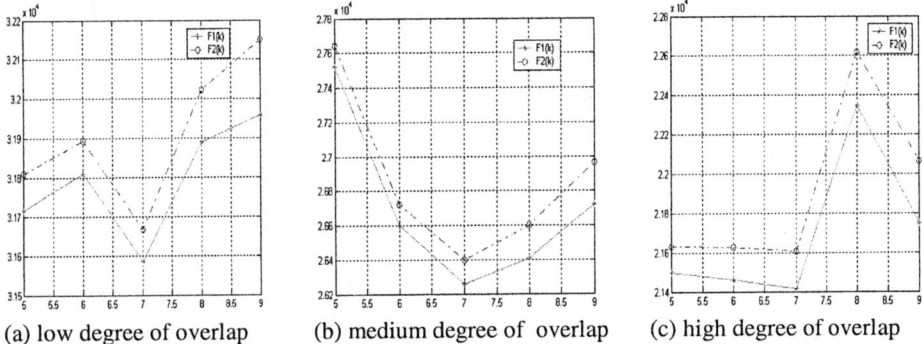

(a) low degree of overlap (b) medium degree of overlap (c) high degree of overlap

Fig. 4. Changes in $F_1(\theta_k,k)$ and $F_2(\theta_k,k)$ for the three datasets in Fig.3 using different number of clusters K

References

1. Pal, N.R., and Bezdek, J.C.: Complexity Reduction for Large Image Processing. IEEE Trans. on Systems, Man and Cybernetics - Part B:Cybernetics, vol.32, no.5 (2002)598-611
2. Kollios, G., Gunopulos, D., Koudas, N., etc.: Efficient Biased Sampling for Approximate Clustering and Outlier Detection in Large Data Sets. IEEE Trans. on Knowledge and Data Engineering, vol.15, no.5 (2003)1170-1187
3. Bezdek, J.C.: Pattern Recognition with Fuzzy Objective Function Algorithm. New York: Plenum (1981)
4. Krishnapuram, R. and Keller, J.: A Possibilistic Approach to Clustering. *IEEE Trans. on Fuzzy Systems*, vol.1, no.2, (1993)98-110
5. Krishnapuram, R., Frigui, H.,and Nasraoui, O.: Fuzzy and Possibilistic Shell Clustering Algorithms and Their Application to Boundary Detection and Surface Approximation - Part I. *IEEE Trans. on Fuzzy Systems*, vol.3, no.1 (1995)29-43
6. Krishnapuram,R.,and Keller, J.: The Possibilistic C-means Algorithm: Insights and Recommendations. *IEEE Trans. on Fuzzy Systems*, vol.4, no.3 (1996)385-393

7. Zadeh, L.A.: Fuzzy Sets as a Basis for a Theory of Possibility. Fuzzy Sets and Systems, vol.1 (1978)3-28
8. Tanaka, H.and Guo, P.: Possibilistic Data Analysis for Operations Research. New York:Physica-Verlag (1999)
9. Yager, R.R.: On the Instantiation of Possibility Distributions. Fuzzy Sets and Systems, vol.128 (2002)261-266
10. Sugihara, K. and Tanaka, H.: Interval Evaluations in the Analytic Hierarchy Process by Possibility Analysis. *Computational Intelligence*, vol.17, no.3, pp.567-579 (2001)
11. Tanaka, H. and Ishibuch, H.: Evidence Theory of Exponential Possibility Distributions. *Int. J. Approximate Reasoning*, vol.8, pp.123-140 (1993)
12. Rosenfeld, A. and Kak, A.C.: Digital Picture Processing. New York:Academic (1982)
13. Scott, D.: Multivariate Density Estimation: Theory, practice and visualization. New York:Wiley (1992)
14. Hong-Qiang Wang and Huang, D.S.: A Novel Clustering Analysis Based on PCA and SOMs or Gene _expression Patterns. Lecture Notes in Computer Science 3174: 476-481, Springer

A New Approach to Predict N, P, K and OM Content in a Loamy Mixed Soil by Using Near Infrared Reflectance Spectroscopy

Yong He[1], Haiyan Song[1], Annia García Pereira[1,2], and Antihus Hernández Gómez[1,2]

[1]College of Biosystems Engineering and Food Science, Zhejiang University,
310029, Hangzhou, China
yhe@zju.edu.cn
[2]Agricultural Mechanization Faculty, Havana Agricultural University, Cuba.
anniagarcia_2000@hotmail.com

Abstract. Near Infrared Reflectance (NIR) spectroscopy technique was used to estimate N, P, K and OM content in a loamy mixed soil of Zhejiang, Hangzhou. A total of 165 soil samples were taken from the field, 135 samples spectra were used during the calibration and cross-validation stage. 30 samples spectra were used to predict N, P, K and OM concentration. NIR spectra and constituents were related using Partial Least Square Regression (PLSR) technique. The r between measured and predicted values of N and OM, were 0.925 and 0.933 respectively, demonstrated that NIR spectroscopy have potential to predict accurately this constituents in this soil, not being this way in the prediction of P and K with r, 0.469 and 0.688 respectively, demonstrated a poorly for P and a less successfully for K prediction. The result also shows that NIR could be a good tool to be combined with precision farming application.

1 Introduction

The actual technological development in positioning, sensing and control system has open a new era, where practices traditionally used in agriculture are staying behind. Precision farming is a term used to describe the management of variability within field, applying agronomic inputs in the right place, at the right time and in the right quantity to improve the economic efficiency and diminish the environmental impact of crop production [1]. The study and understanding of the soil spatial variability is very important due to it is the responsible that crops yield distributed unevenly within the field what forces in some occasions to an excessive use of fertilizers [2],[3], and it is the soil spatial variability the first step to develop a Site Specific Crop Management program (SSCM) that will help to improve soil quality through treatments practical and economically efficient, but to study soil constituents composition, an exhaustive processes have to be dedicated to laboratory analysis where time and economic indicators appear as drawbacks.

On the other hand in testing materials field, Visible-NIR spectroscopy has appeared as a rapid and nondestructive analytical technique that correlates diffusely reflected near-infrared reflection with the chemical and physical properties of materials [4],[5]. It has been used for assessing grain and soil qualities [4], [6], [7], [8], [9], [10] and has been demonstrated this method is rapid, convenient, simple, accurate and able to analyze many constituents at the same time.

The goal of this study is to analyze NIR spectroscopy potential to estimate N, P, K and OM content in a loamy mixed active thermic aeric endoqualfs soil as well as to combine these predicted macronutrients concentration with Geographic Information System (GIS) and statistic using N, P, K and OM spatial variability within the field to obtain it's distribution maps and the correlation among them.

2 Material and Methods

2.1 Soil Sampling

The experimental site is located in Zhejiang, Hangzhou (120°11E, 30°28N), the field is relatively flat. The main monthly temperature fluctuates from 3.5 °C in January to 26.8 °C in July with an annual average temperature of 16.2 °C and rainfall of 279 mm. It has a wet climate, with a long frost-free period and relatively high summer temperatures. The soil is normally not mechanized, penetrating and easy to cultivate and it has a previous crop rotation with corn. The soil type was classified as loamy mixed active thermic aeric endoqualfs according with Zhejiang soil classification [11].

The test field boundary was performed with Trimble AgGPS 132 equipment. The normal grid method was used to sample the soil. Measurements were taken at 30 sample plots; at a depth of 20 cm and grid interval 5 m (two sample plots were taken in the diagonal of the south-east grid). A composite sample was obtained by mixing 5 soil samples (equal volume basis), 1 central plot and the remaining 4, separated 1 m away from each sampling point. Other 135 samples were taken randomized from the field to complete the 165 samples to be analyzed with spectroscopy. The collected soil in each sample was divided in two portions, each portion was placed in a bag properly close to vacuum and was classified in A and B. Group A was taken to the Soil Science laboratory in Zhejiang University and was tested for N, P, K and OM using laboratory analysis. Group B was applied to be analyzed by NIR spectroscopy. All soil samples were air dried and sieved <2 mm.

2.2 Soil Spectral Data Measurement

The soil data collection was performed with a spectrophotometer (ASD FieldSpec Pro FR (350–2500 nm)/ A110070). A total of 165 samples were analyzed. The soil was set in a petri dish and then was flushed the surface. For each sample, reflectance measurements were completed from 350–2500 nm, wavelength increment 2 nm. Three reflectance spectra were taken over the central area of the petri dish rotating the sample in between each reflectance spectra approximately 120 °, for each reflectance spectra the scan number were 20 at exactly the same position, a total scan for each sample were 60, fixed scanning time 0.1 s. Absorbance for the scanned was recorded (log [1/R]). All spectra recoded were checked visually and averaged using ViewSpec pro version 2.14 (from ASD) and exported to multivariate analysis software-

The Unscrambler 8.0 (CAMO ASA, Norway). Multiplicative Scatter Correction (MSC) technique was used as pretreatment to reduce the effect of scattering. Representative mean absorbance soil spectra (log 1/R) of ten samples are shown in fig. 1.

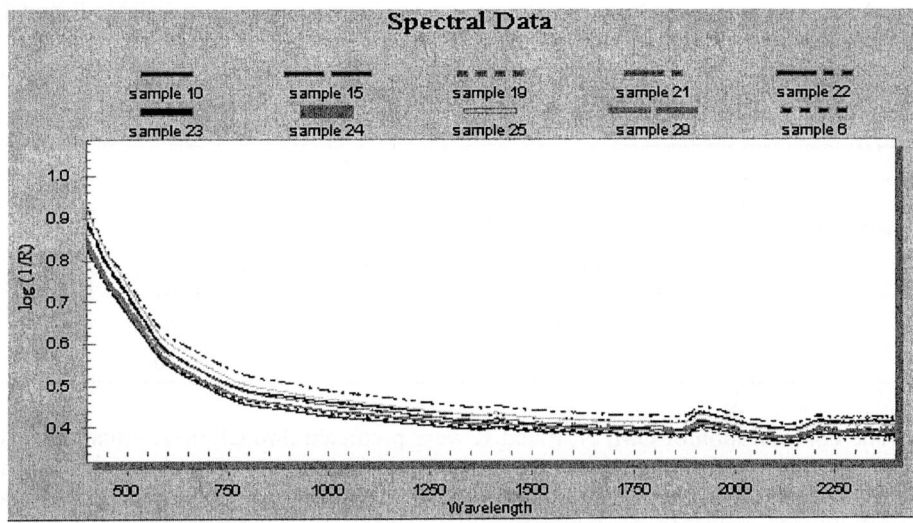

Fig. 1. Representative mean absorbance spectra (log 1/R) of ten samples after apply MSC pretreatment (400-2400 nm)

2.3 Spectral Analysis

The 60 spectra recoded for each sample were averaged to give one spectrum per sample. The reference chemical results for each constituent were added to the NIR spectral file. The total of 165 samples was divided in set (I) and (II). Set (I) with 135 samples was used to develop the calibration model. Calibration equations were developed using Principal Component Analysis (PCA) and Partial Least Square technique (PLS). PLS regression analysis technique was used to relate the reflectance spectra (400-2400 nm) with measured soil properties values. Using set (I), the spectral reflectance data in units of log (1/R) were first reduced by smoothing over several adjacent spectral points (gap sizes) (3, 9, 11, 19, 39) to produce 666, 222, 182, 105 and 51 new data points, respectively. Separated calibration equations were computed for each constituent. Leave-one–out cross–validation was performed on the calibration set to determine the optimum number of factors (F) for the PLS regression calibration model of each soil property. The F giving the smallest RMSECV (Root Mean Square Error of Cross-Validation) between measured and predicted values was chosen for the PLS regression calibration models. To avoid over-fitting, F was always <1/10 of the number of samples in the calibration set.

The calibration was completed when one equation was selected as giving the best results. The best calibrations is the one with lowest Root Mean Square Error of Prediction (RMSEP), Standard Error of Prediction (SEP), the standard deviation (SD), and the highest r (coefficient of correlation). The SEP is considered to be more important than r, as it relates directly to the error expected in the future predictions. Each calibration equation developed from set (I) was used to predict the constituent values for the independent spectra in set (II) (30 samples), the validation set. For each calibration equation, the NIR predicted values for set (II) were correlated with their measured values. Calibration statistic of each selected model for N, OM, P and K content future prediction is shown in table 1.

Table 1. Calibration statistic for N, OM, P and K content

Soil constituent	Gap sizes	F	r	Calibration		Cross-validation		
				SEC	RMSEC	r	SECV	RMSECV
N	7	7	0.967	2.055	2.048	0.938	2.829	2.819
OM	11	5	0.962	0.058	0.058	0.957	0.061	0.062
P	9	6	0.651	26.64	26.31	0.543	29.54	29.43
K	11	6	0.786	49.82	49.64	0.748	53.56	53.36

The concentration of OM, N, P and K were predicted through these equations and following very carefully the same analysis procedure as were the spectra that provided the prediction equation. The r, SEC, and SEP between the predicted and measured values of OM, N, P and K were used to evaluate the prediction ability of NIR spectroscopy technique.

3 Results and Discussions

3.1 Spectral Properties and Prediction

From fig.1, we can see all the spectra have three large absorption peaks in the near infrared region (700-2500 nm) around 1400, 1900 and 2200 nm, that were analogous with the highest absorbance peaks obtained by [4], [12] testing another soil types, and some few small peaks in the region between 1000-1100 nm and 2200-2300 nm.

Due to PLS technique would be helpful to examine how each soil constituent is simply related to individual wavelength so that a better understanding of NIR spectra may be obtained. The correlation coefficient across the entire spectral region between 400-2400 nm was in general very small, nevertheless, for each constituent during the calibration model construction were found some light better coefficients related to individual wavelengths, for example: in N 1902, 2364, 1826 and 2098 nm, only 1826 nm correspond to similar wavelength selected by [9] using Multiple Linear Regression (MLR) technique. The best correlation values for OM were found in 993, 1080, 1951, and 2277 nm; only 1080 correspond with the wavelength selected by [13]. In the previous selected wavelength no one repeats in both constituents. During P and K analysis, in spite of both constituents present a very poor correlation and prediction using NIR, some peaks were found during the correlation analysis that could indicate the presence of these constituents in the spectra, over the NIR wavelength 2380, 2363 and 1906, 2217, 2279 for P and K, respectively.

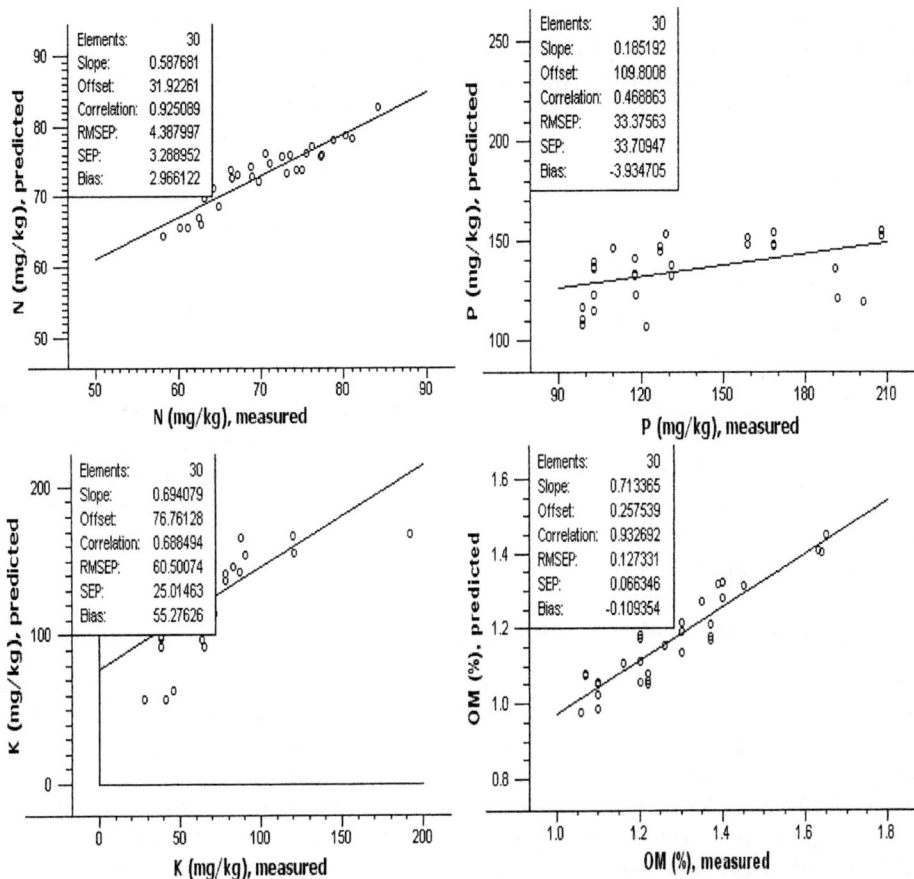

Fig. 2. Correlation between measured and predicted values of N, P, K and OM

The prediction set demonstrated that NIR is able to predict N and OM in soil coinciding with [4], [10], [12], [14], and [15], not being these ways to P and K prediction. Correlation coefficient higher than 0.93 (r>0.93), were obtained during the calibration and cross-validation states for N and OM (table 1). Statistic for predicted N, P, K and OM is shown in fig.2. The r between measured and predicted values of N and OM were 0.925 and 0.933, as well as SEP were 3.289 and 0.066 respectively, demonstrated that NIRS method have potential to predict accurately this constituents in soil. A slight better correlation was found in the OM prediction although the difference is not quite significant. The r=0.933 obtained in the OM prediction is within the range proposed by [15] in their investigation to the same constituent (0.81-0.97) and considered as a good performance. In the case of N, the r=0.925 obtained shows also a good performance, that was lightly higher than those obtained by [4], [9], respectively.

Unfortunately, statistics between measured and predicted P and K with r, 0.469 and 0.688, as well as SEP, 33.376 and 25.015 respectively, demonstrated a poorly for P and a less successfully K prediction in both constituents. One possible explanation for this could be the non-direct relationship among P and K with C-H-O-N bonds, which obtained the same results with [16] by testing Manure-amended Soils. A slight better r coefficient (r=0.688) was obtained for K, but the difference between RMSEP and SEP and the bias value shows not very good accuracy in the prediction and was not found to be repeatable for this reason. Better results in available P and K prediction were obtained by [17] with determination coefficient r^2 =0.42 and r^2 = 0.84 respectively, with an acceptable prediction of K but still a poor prediction of P. Contrary to [17], some researchers found no absorption bands for cations such as Na, K, Ca, and Mg, they also expressed that mineral constituents may also be predictable if they are bound to organics or correlated with organic components, in this way; it may be possible to predict this previous constituents composition [18]. [19] found a variety of bonds containing P, but the wavelengths where these absorb are less well documented than for those between O, H, C, and N, they also obtained that there are numerous known NIR absorbers involving C and N that could explain the predictability of these elements, but it is less clear whether fractions of P are spectrally active, or predictable because of correlation with C and N. Further investigation have to be carried on this field, even including Mid-infrared region where have appeared some bands related with phosphate mineral [20].

3.2 Soil N and OM Spatial Variability

Once obtained N, P, K and OM values, measured (reference parameters) and predicted, all constituents' spatial variability was analyzed. No significant correlation was found among constituents. The tested field can be considered as fairly productive according N and K average composition 80.51 mg/kg and 70.23 mg/kg with a range of values from 57.88 to 93.08 mg/kg and 20.6 to 192 mg/kg, respectively. In the specific case of K, a wide variability across the field can be observed, and the fertility evaluated through this constituent could be classified from very poor to good according to Mallarino considerations [21], whereas it presents good and very good OM and P composition with average of 1.33 % and 139.5 mg/kg with a range of values from 1.06 to 1.79 % and 98.5 to 207.8 mg/kg, respectively. The high P values may correspond with the location of fertilizer bands and as a result of recent or long fertilization history where the reduction or absorption of this constituent was not very good.

Measured and predicted N, P, K and OM spatial distribution was interpolated by ArcView GIS 3.1 software using spline method, the spatial distribution maps were generated and shown in fig.3. The reference maps for the predicted and measured values of N and OM were almost the same, not being this way for P and K due to the non successful prediction of these constituents. The variability of N changes from north to south with a diminution of this nutrient concentration following this direction. This variability must be mainly caused by management practices including tillage and irrigation whereas a quite similar variability is observed in the OM concentration within the field. Through the range of concentration and colors represented in each map, a slight few difference can be found between predicted and reference parameters. In both constituents predicted parameters are illustrated with slight color attenuation what means a few less concentration.

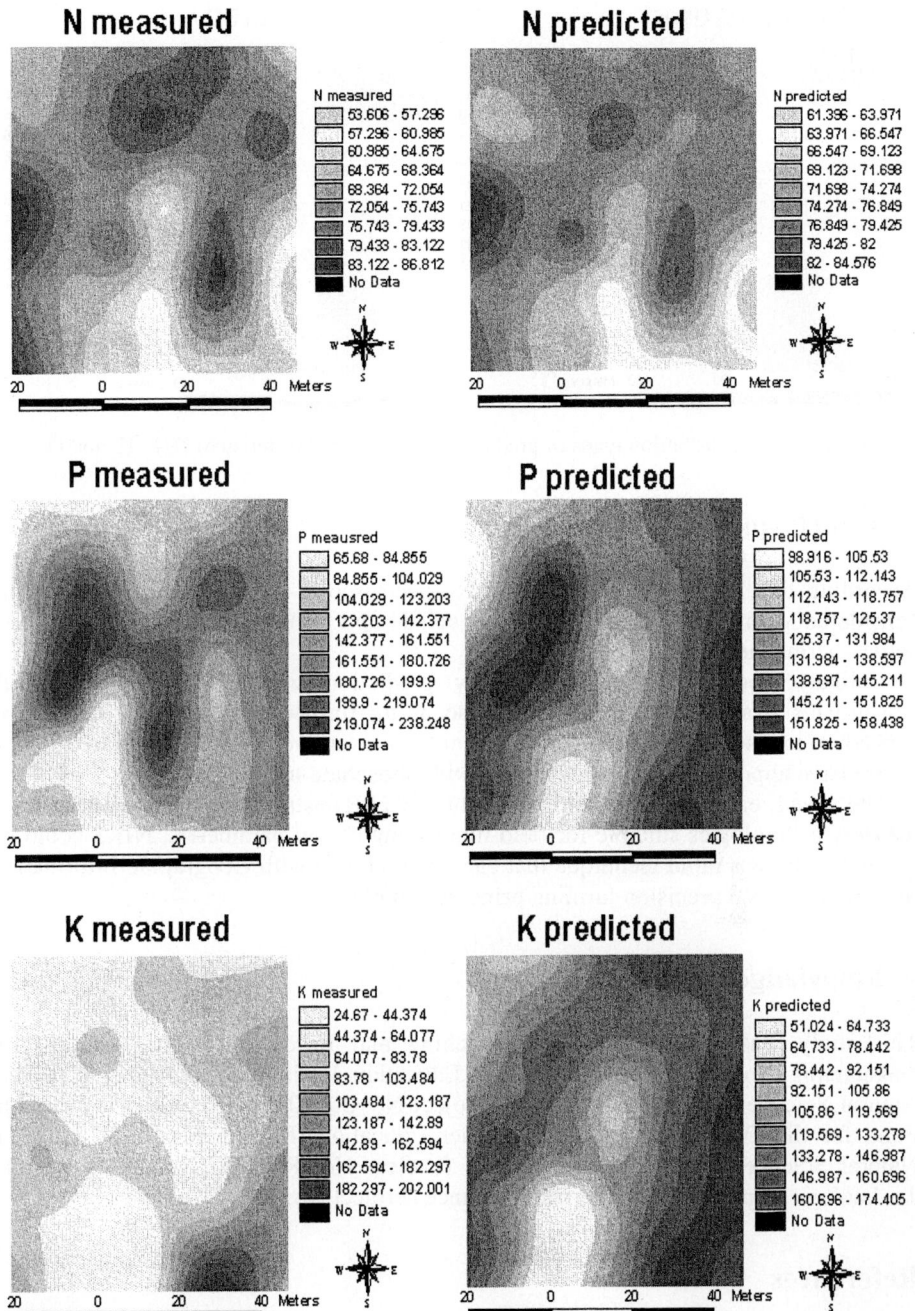

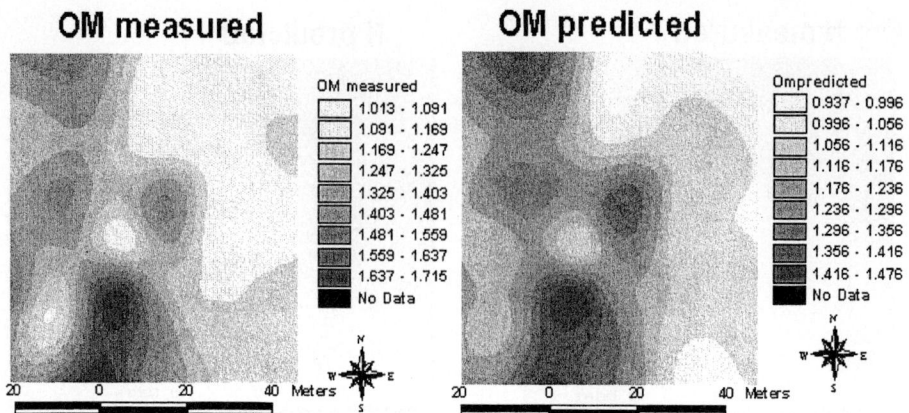

Fig. 3. Spatial distribution maps of predicted and measured (reference) N, P, K, and OM

4 Conclusions

According this investigation results, NIR spectroscopy is a technique that can be considered with good potential to assess soil N and OM content in a loamy mixed active thermic aeric endoqualfs soil.

Unlikely poor and less successfully prediction was obtained testing both constituents P and K in the same soil, so as to, further investigations have to be carried on testing P and K using NIR technology, even including Mid-infrared region where have appeared some bands related with phosphate mineral.

The speed, easy operation and portability of NIR instruments make NIRS one of the only technologies suitable for field monitoring of soil parameters. NIRS could be useful in situ as a rapid technique that can be combined with Geographic Information System (GIS) and precision farming principles application.

Acknowledgements

This study was supported by the Teaching and Research Award Program for Outstanding Young Teachers in Higher Education Institutions of MOE, P. R. C., Natural Science Foundation of China (Project No: 30270773), Specialized Research Fund for the Doctoral Program of Higher Education (Project No: 20040335034), Natural Science Foundation of Zhejiang (Project No: RC02067) and Science and Technology Department of Zhejiang Province (Project No. 2005C21094).

References

1. Earl, R., Thomas, G. and B. S. Blackmore.: The Potential Role of GIS in Autonomous Field Operations. Computers and Electronics in Agriculture. 25 (2000) 107–120
2. Schmidt, M. G., H. Schreier and P. B. Shah.: Factors Affecting the Nutrient Status of Forest Sites in a Mountain Watershed in Nepal. Journal of Soil Science. 44(1993) 417-425

3. Liu, G and J. Kuang.: A Study on Spatial Variability of Soil Nutrient Within Field. ICETS2000-session 6: Technology innovation and sustainable Agriculture (2000)
4. Chang, C. W. and D. A. Laird.: Near-Infrared Reflectance Spectroscopy Analysis of Soil C and N. Soil Science. Vol. 167. No. 2 (2002) 110-116
5. Stark, E., K. Luchter, and M. Margoshes.: Near-infrared Analysis (NIRA): A technology for Quantitative and Qualitative Analysis. Appl. Spectrosc. Rev. 22(1986) 335-399
6. Kawamura, S., M. Natsuga and K. Itoh.: Determination of Undried Rough Rice Constituent Content Using Near-Infrared Transmission Spectroscopy. Transactions of the ASAE. Vol. 42. No. 3(1999) 813-818
7. Delwiche, S. and W. R. Hruschka.: Protein Content of Bulk Wheat from Near- Infrared Reflectance of Individual Kernel. Cereal Chem. Vol.77. No.1 (2000) 86-88
8. Barton, F. E, D. S. Himmelsbach, A. M. McClung and E. T. Champagne.: Rice Quality by Spectroscopic Analysis: Precision of three spectral regions. Cereal Chem. Vol.77. No.5 (2000) 669-672
9. Morra, M. J. M. H. Hall and L. L. Freeborn.: Carbon and Nitrogen Analysis of Soil Fractions Using Near-Infrared Reflectance Spectroscopy. Soil Sci. Soc. Am. J. 55(1991) 288-291
10. Ben-Dor, E. and A. Banin.: Near-Infrared Analysis as a Rapid Method to Simultaneously Evaluate Several Soil Properties. Soil Sci. Soc. Am. J. 59(1995) 364-372
11. Zhang, M. K.: Soil Science to the 21st century. Environment science. Editorial publish. Beijing. China (1999)12-16
12. Fidencio P. H., R. J. Poppi, J. C. De Andrade and H. Cartella.: Determination of Organic Matter in Soil Using Near-infrared Spectroscopy and Partial Least Square Regression. Common. Soil. Sci. Plant Anal. 33(9&10) (2002)1607-1615
13. Krisnan, P., J. D. Alexander, B. J. Butler, and J. W. Hummel.: Reflectance Technique for Predicting Soil Organic Matter. Soil Sci. Am. J. 44(1980) 1282-1285
14. Sudduth, K. A. and J. W. Hummel.: Evaluation of Reflectance Methods for Soil Organic Matter Sensing. Transaction of the ASAE. 34 (1991) 1900-1909
15. Martin P. D and D. F. Malley. : Use of Near-Infrared Spectroscopy for Monitoring and Analysis of Carbon Sequestration in Soil (2003) http://www.iisd.org/pdf/2003/climate_paul_martin.ppt
16. Malley, D. F, L. Yesmin and R. G. Eilers.: Rapid Analysis of Hog Manure and Manure-amended Soils Using Near-infrared Spectroscopy. Division s-8—Nutrient Management & Soil & Plant Analysis. Soil Sci. Soc. Am. J. 66 (2002) 1677-1686
17. Krischenko, V. P., S. G. Samokhvalov, L. G. Fomina and G. A. Novikova (1992). : Use of Infrared Spectroscopy for the Determination of Some Properties in Soil. In *Making Light Work: Advances in near Infrared Spectroscopy*. I. Murray and L. A. Cowe (eds). Developed from the 4th International Conference of Near Infrared Spectroscopy, Aberdeen, Scotland, 19-23 August 1991. VCH, Weinheim, New York, Basel, Cambridge, 239-249
18. Burns, D., and Cziurczak,E.: (ed.) Handbook of Near-infrared Spectroscopy. Marcel Dekker, New York(1992)
19. Williams, P. and K. H. Norris.: Variable Affecting near Infrared Spectroscopic Analysis. Chapter 9. (2001a) 171-185 *in* P. Williams and Norris, (eds.). Near-infrared technology in the griculture and food industries. 2nd ed. The American Association of Cereal Chemists, St. Paul, MN
20. Reeves, J., G. McCarty and T. Mimmo.: The Potential of Diffuse Reflectance Spectroscopy for the Determination of Carbon Inventories in Soil. Environ. Pollut. 116(2002) S277-S284
21. Mallarino A. P and A. M. Blackmer.: Profit-maximizing Critical Values of Soil-test Potassium for Corn. J. Prod. Agric. 7(1994) 261-268

Soft Sensor Modeling Based on DICA-SVR[*]

Ai-jun Chen, Zhi-huan Song, and Ping Li

National Laboratory of Industrial Control Technology,
Institute of Industrial Process Control, Zhejiang University, Hang Zhou, 310027, China
{ajchen, zhsong, pl}@iipc.zju.edu.cn

Abstract. A new feature extraction method, called dynamic independent component analysis (DICA), is proposed in this paper. This method is able to extract the major dynamic features from the process, and to find statistically independent components from auto- and cross-correlated inputs. To deal with the regression estimation, we combine DICA with support vector regression (SVR) to construct multi-layer support vector regression. The first layer is feature extraction that has the advantages of robust performance and reduction of analysis complexity. The second layer is the SVR that makes the regression estimation. This kind of soft-sensor estimator was applied to estimation of process compositions in the simulation benchmark of the Tennessee Eastman (TE) plant. The simulation results clearly showed that the estimator by feature extraction using DICA can perform better than that without feature extraction and with other statistical methods for feature extraction.

1 Introduction

Recently, support vector regression (SVR) has become a popular tool in model forecasting, due to its remarkable generalization performance and elegant statistical learning theory [1]. But designing a soft sensor is usually difficult because its modeling is based on case data. These data have the features of discreteness, nonlinearity, contradiction, and complexity, which could deteriorate the generalization performance of SVR. So developing a SVR, the first important step is feature extraction. In the framework of SVR, several approaches for feature extraction are also available, such as SVR with 1-norm regularized term, the recursive feature elimination method, saliency analysis, genetic algorithm and statistical methods [2,3,4,5].

Independent component analysis (ICA) is a recently developed technique for revealing hidden factors that underlies sets of measurements followed on a non-Gaussian distribution. Its goal is to decompose a set of multivariate data into a base of statistically independent components without a loss of information. However, variables rarely remain at a steady state but rather are driven by random noise and uncontrollable disturbances in chemical processes. These effects make the variables have autocorrelation and the system has dynamic properties. Thus, a feature extraction method taking into account the serial correlations in the data is needed.

---
[*] The project (2003AA412110) was supported by Hi-Tech Research & Development Program of China.

Ku et al. (1995) exhibited this problem first and suggested a modified method, in which PCA is applied to a time lagged data matrix so that it extracts time-dependent linear relations. More recently, several works focused on capturing process dynamics using a state space model have been proposed [7,8,9]. With other developed methods dynamic PCA has been employed for fault detection, state space model, process monitor and multivariate statistical control.

In this paper, applying ICA to the augmenting matrix with time-lagged variables, called *dynamic* ICA (DICA), is suggested for developing dynamic models and improving the feature extracting performance. Compared with ICA, DICA can show more powerful extracting performance in the case of a dynamic process since it can extract source signals that are independent of the auto- and cross-correlation of variables.

The primary object of this investigation is to use DICA extract the essential independent components, which reformulate the SVR estimator by transforming the original inputs into features that are mutually statistically independent.

2 DICA Feature Extraction

Conventional ICA implicitly assumes that the observations at one time are statistically independent of observations at any past time. That is to say, the measured variable at one time has not only serial independence within each variable series at past time, but also statistical inter-independence between the different measured variable series at past time. However, the dynamics of a continuous production process cause the measurements to be time dependent, which means that the data may have both cross-correlation and auto-correlation. To consider temporal correlations of measurements, we should extract the relations between the past and current values of measured variables. So the first step of DICA method is data pretreatment.

2.1 Data Pretreatment

The ICA methods can be extended to take into account the serial correlations, by augmenting each observation vector with the previous l observations and stacking the data matrix in the following manner,

$$x(l) = \begin{bmatrix} x_k^T & x_{k-1}^T & \cdots & x_{k-l}^T \\ x_{k-1}^T & x_{k-2}^T & \cdots & x_{k-l-1}^T \\ \vdots & \vdots & \vdots & \vdots \\ x_{k+l-n}^T & x_{k+l-n-1}^T & \cdots & x_{k-n}^T \end{bmatrix}^T . \quad (1)$$

where x_k is the d-dimensional observation vector in the training set at time k, n + 1 is the number of samples, and l is the number of lagged measurements. By performing ICA on the data matrix in Eq. (1), a multivariate autoregressive (ARX model if the process inputs are included) is extracted directly from the data. With an appropriate order, which is the window length, a data set with this structure can capture the dynamics of a process. $x(l)$ can be defined as a data window in the time dimension, as shown in Fig.1. Ku et al (1995) have suggested a method for automatically determining the number of lags. Experience indicates that a value of $l = 1$ or 2 is usually appropriate.

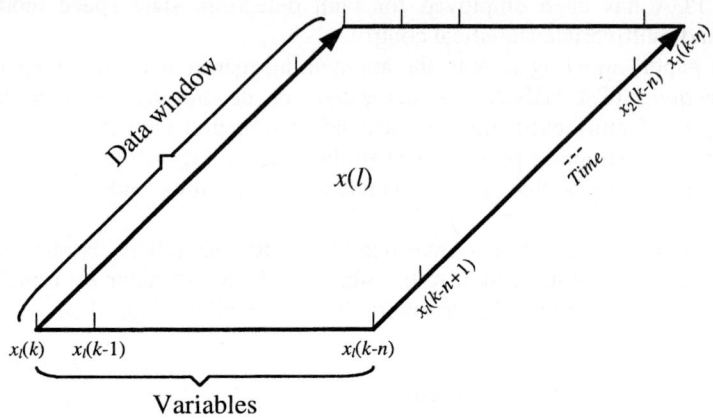

Fig. 1. A data window for a continuous process

Normalize the augmented matrix of modeling data using the mean and standard deviation of each variable. The auto-scaled form is $X \in R^{m \times N}$, where m is the number of augmented variables ($d \times l$) and $N=n-l+1$. Then apply whitening procedure to get the matrix Z, which eliminates the cross-correlation between random variables. After the whitening transformation we have

$$z(k) = Q \cdot x(k). \qquad (2)$$

where x(k) is the kth column vector of X, $Q = \Lambda^{-1/2} \cdot U^T$, Λ is a diagonal matrix with the eigenvalues of the data covariance matrix $Rx = E(x(k) \cdot x^T(k))$ and U is a matrix with the corresponding eigenvectors as its columns.

2.2 ICA Algorithm

In the ICA algorithm, it is assumed that at time k the observed m-dimensional data vector $Z(k) = [z_1(k),\ldots, z_m(k)]^T$ can be expressed as linear combinations of n unknown independent components $S(k) = [s_1(k),\ldots,s_n(k)]^T$, given by the model:

$$Z(k) = A \cdot S(k) + e(k). \qquad (3)$$

where $A \in R^{m \times n}$ is the unknown mixing matrix, e(k) is the residual vector. The objective of ICA can be defined as follows: to find a demixing matrix B^T, so that components of the reconstructed data vector $\hat{s}(k)$ are given as

$$\hat{s}(k) = B^T \cdot z(k) = W \cdot x(k). \qquad (4)$$

where B is an orthogonal matrix as verified by the following relation:

$$E\{Z(k) \cdot Z^T(k)\} = B \cdot E\{S(k) \cdot S^T(k)\} \cdot B^T = B \cdot B^T = I. \tag{5}$$

Then, from Eq. (2), we can estimate $s(k)$ as follows:

$$S(k) = B^T \cdot Z(k) = B^T \cdot Q \cdot X(k). \tag{6}$$

From Eqs. (4) and (6), the relation between W and B can be expressed as

$$W = B^T \cdot Q. \tag{7}$$

After finding B, the demixing matrix W can be obtained from Eq. (7). To calculate B, there are two common measures of non-Gaussianity: kurtosis and negentropy. A detailed description of ICA algorithm can be found in the references [10,11,12]. In this paper we use the Fast ICA algorithm, which is based on the information-theoretic quantity of entropy.

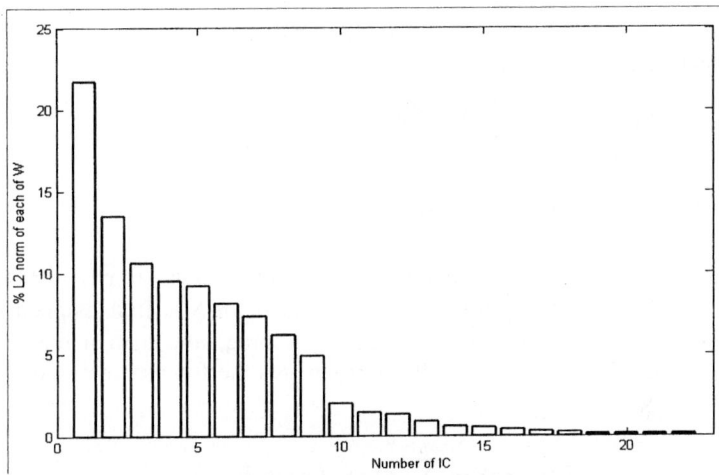

Fig. 2. lot of percent L_2 norm of each row of W against IC number

2.3 Dimension Reduction

A key step in a data dimensionality reduction technique is to determine the order of the reduction. An important drawback of the SVR is that its training phase is sometimes very demanding in computational time, especially in cases of high feature space dimensionality. So reduction of feature dimensionality has two advantages: robust performance and reduction of analysis complexity [13,14]. A number of methods have been suggested to determine the component order example. In the present study we used a Euclidean norm (L2) to sort the rows of the demixing matrix, W. After the ordering of the independent components (ICs), the data dimension can be reduced by selecting a few rows of W based upon the assumption that the rows with the largest sum of squares coefficient have the greatest effect on the variation of

S. Fig. 2 gives a representative plot of the percentage of the L2 norm of the sorted demixing matrix (W) against the IC number. The sorted demixing matrix is obtained from the operating data of the Tennessee Eastman process (Section 4). Note that the L2 norms of last thirteen ICs are much smaller than the rest, indicating a break of some kind between the first nine ICs and the remaining thirteen.

Separating W into the dominant part and the excluded part based on the magnitudes of the norms, then we have

$$W = \begin{bmatrix} W_p \\ W_c \end{bmatrix}. \qquad (8)$$

where W_p is the DICA transform matrix to be calculated.

2.4 On-line Feature Extraction

Once obtain the new observed data vector $x(t)$ at time t. Constitute the augmented vector with time lag l, the data form is $X(t)= [x(t)^T, x(t-1)^T,\ldots, x(t-l)^T]^T$. Then, apply the same scaling as used in modeling and get the scaled data form $X_{new}(t)$. The result of on-line feature extraction $S_{np}(t)$ can be obtained from:

$$S_{np}(t) = W_p X_{new}(t). \qquad (9)$$

3 Support Vector Regression

After feature extraction the training data can be expressed as $\{s_i, y_i\}$, where $s_i \in R^n$, $y_i \in R$, $i=1,\ldots,N$. In support vector regression, the input S is first mapped onto a h-dimensional feature space using some fixed mapping, and then a linear model is constructed in this feature space. SVR approximates the function using the following form:

$$f(s, \omega) = \sum_{j=1}^{h} \omega_j g_j(s) + b. \qquad (10)$$

where $g_j(s)$ denotes a set of nonlinear transformations, and b is the "bias" term. Often the data are assumed to be zero mean, so the bias term in Eq. (10) is dropped.

The quality of estimation is measured by the loss function $L(y, f(s, w))$. Support vector regression uses a new type of loss function called ε-insensitive loss function proposed by Vapnik:

$$L_\varepsilon(y, f(s, \omega)) = \begin{cases} 0 & if \ |y - f(s, \omega)| \leq \varepsilon \\ |y - f(s, \omega)| - \varepsilon & otherwise \end{cases}. \qquad (11)$$

The empirical risk is:

$$R_{emp}(\omega) = \frac{1}{N} \sum_{l=1}^{N} L_\varepsilon(y_i, f(s_i, \omega)). \qquad (12)$$

This loss function provides the advantage of using sparse data points to represent the designed function (10). ε is the tube size of SVR and determined empirically [15].

SVR performs linear regression in the high-dimension feature space using ε-insensitive loss and, at the same time, tries to reduce model complexity by minimizing $\|w\|^2$. This can be described by introducing slack variables ξ_i, ξ_i^* to measure the deviation of training samples outside ε-insensitive zone. Thus SVM regression is formulated as minimization of the following functional:

$$\min \quad \frac{1}{2}\|\omega\|^2 + C\sum_{i=1}^{N}(\xi_i + \xi_i^*)$$

$$s.t \begin{cases} y_i - \omega \cdot g(s_i) - b \leq \varepsilon + \xi_i^* \\ \omega \cdot g(s_i) + b - y_i \leq \varepsilon + \xi_i \\ \xi_i, \xi_i^* \geq 0, i = 1,\ldots,n \end{cases} \quad (13)$$

This optimization problem can be transformed into the dual problem, and its solution is given by

$$f(\mathbf{x}) = \sum_{i=1}^{n_{SV}}(\alpha_i - \alpha_i^*)K(s_i, s) + b \quad (14)$$

$$s.t \quad 0 \leq \alpha_i^* \leq C, \quad 0 \leq \alpha_i \leq C$$

The kernel function $K(s_i, s)$ is a symmetric function satisfying Mercer's conditions. The sample points that appear with non-zero coefficient in Eq. (14) are called support vectors (SVs).

4 Composition Estimator Using DICA-SVR

To illustrate the performance of DICA for feature extraction in the context of SVR, we use the framework of DICA-SVR estimation for process compositions through the Tennessee Eastman process simulation, which is also compared with those of SVR, PCA-SVR, ICA-SVR and DPCA-SVR

4.1 TE Process Introduction

The Tennessee Eastman process simulator has been widely used to compare various control approaches. The process consists of five major unit operations: a reactor, a condenser, a compressor, a separator, and a stripper. The four reactants A, C, D and E and the inert B are fed to the reactor where the products G and H are formed and a byproduct F is also produced. The control structure Listed in Luyben [16] was chosen for this study, which is shown schematically in Fig. 3. The process has 22 continuous process measurements, 19 composition measurements, and 12 manipulated variables. The details on the process description are well explained in [17].

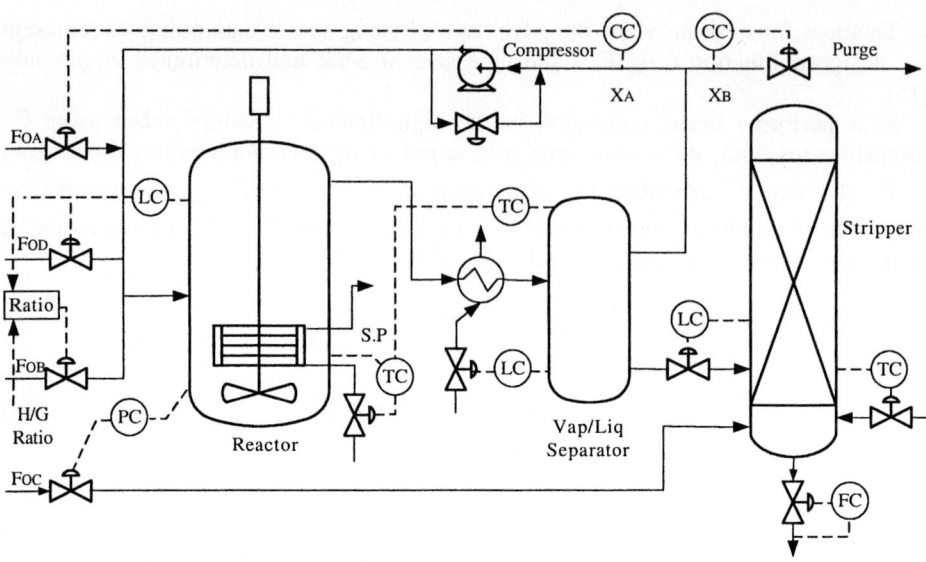

Fig. 3. Control system of the TE process

4.2 Composition Estimator

A DICA-SVR structure shown in Fig. 4 is employed for the composition estimator in this study. In light of the DICA-SVR architecture in Fig. 4, the basic philosophy of this study considers that dynamic process compositions can be predicted from other secondary measurements. It can also be expressed as the following sampled-data equation: $X_t = f(IC1_t, \ldots, ICn_t)$, where X_t is the prediction composition of process output vector at time t, f is a nonlinear function, and $IC1_t, \ldots, ICn_t$ are output vectors of DICA feature extraction to the measured process variables at time t. In order to construct a DICA-SVR estimator, a complete set of plant operation data under Luyben control structure with some fluctuations is collected for DICA-SVR training.

22nd typical collected training data such as temperature, pressure, level, flow rate, and compositions for 200 h are used. For the DPCA and DICA, the lag order l was selected by using a method in [6]. Through this analysis, one lagged variable of each measurement for the DPCA and the DICA were added in the data matrix. It should be noted that $X_{A,pg}$, $X_{B,pg}$ (component A, B in purge) and $X_{G,pd}$ (component G in product stream) are obtained by on-line gas chromatography (GC). The time delay from on-line GCs is then shifted out for each composition variable. The sampling time (or time delay) for $X_{G,pd}$ is 0.25 h, and that for $X_{A,pg}$ (or $X_{B,pg}$) is 0.1 h. Since there are two different sampling periods for measured compositions of the TE process, there have two DICA feature extractors.

The free parameters that produce the smallest sum of the validation errors are used in the experiment. The training parameters and error results for five types of estimator are shown in Table 1. The training results clearly show that the DICA method has the best performance than others. The estimated $X_{B,pg}$ is not as good as others. The reason may be that component B is an inert and those process variables may have less influence on it.

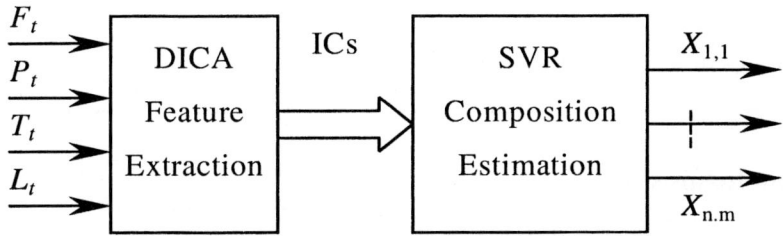

Fig. 4. The architecture of DICA-SVR estimator

Table 1. Training parameters and error results

Methods	(γ, C, ε, δ)	P	Average percent error (%)		
			$X_{A,pg}$	$X_{B,pg}$	$X_{G,pd}$
SVR	(3; 100; 0.05,-)	-	0.89	3.91	1.23
PCA – SVR	(10; 100; 0.01,-)	9	0.69	3.03	0.89
DPCA - SVR	(10; 100; 0.01,-)	14	0.60	2.94	0.75
ICA – SVR	(35; 1000; 0.05; 1.0)	9	0.55	2.75	0.67
DICA - SVR	(10; 100; 0.01; 1.2)	14	0.49	2.04	0.59

γ, C, ε are the parameters of SVR, δ is the parameter of ICA, P is the number of PCs or ICs used.

4.3 Experimental Result

Six sets of testing data under Luyben control structure are chosen for the selection of DICA-SVR topologies in this study. Downs and Vogel originally designed these testing sets for setpoint or disturbance changes in the TE process, and they are labeled as follows:

Test A: product rate change -15%
Test B: product mix change from 50G/50H to 40G/60H
Test C: reactor operating pressure change -60 KPa
Test D: purge gas composition B change +2%
Test E: IDV6
Test F: IDV8

Comparisons of the SVR, PCA-SVR, ICA-SVR, DPCA-SVR and DICA-SVR predictions for $X_{A,pg}$, $X_{B,pg}$ and $X_{G,pd}$ on the testing data (Tests A-F) are illustrated in Table 2. From these tests, it appears that the DICA-SVR estimator can reasonably predict process compositions than other methods. The simulation also shows that SVR by feature extraction using statistical methods can achieve better generalization performance than that without feature extraction.

Table 2. Testing error results

Methods	Average percent error (%)			Maximum percent error (%)		
	$X_{A,pg}$	$X_{B,pg}$	$X_{G,pd}$	$X_{A,pg}$	$X_{B,pg}$	$X_{G,pd}$
SVR	1.29	4.85	1.56	2.37	12.13	2.54
PCA + SVR	0.86	3.91	1.29	1.69	9.03	1.89
DPCA + SVR	0.70	3.62	0.95	1.36	8.75	1.47
ICA + SVR	0.60	3.37	0.86	1.12	8.25	1.37
DICA + SVR	0.54	3.12	0.70	0.87	5.03	1.09

5 Conclusions

In this paper, a new statistical feature extraction method that using ICA to the augmenting matrix with time-lagged variables called DICA is proposed. Since ICA explores higher order information of the original inputs than PCA, ICA can reveal more useful information than PCA. So DICA has the ability to remove the major dynamics from the process and to find statistically independent components from auto- and cross-correlated variables. The proposed method is used for feature extraction, and combined with SVR to construct multi-layer support vector regression.

The superiority of DICA feature extraction method over PCA, DPCA, and ICA ones, is illustrated in the composition estimator on the simulation of TE process. The simulation results clearly show that the regression model and estimator by feature extraction using DICA can perform better than that with DPCA, ICA or PCA for feature extraction. It also demonstrates the fact that SVR by these statistical methods can achieve better generalization performance than that without feature extraction.

References

1. V.N. Vapnik.: The Nature of Statistical Learning Theory, Springer, New York (1995)
2. Bradley,P.S., Mangasarian,O.L.: Feature Selection via Concave Minimization and Support Vector Machines. Proceedings of the 15th International Conference on Machine Learning, Madison, WI, USA (1998) 82–90
3. Cao,L.J., Chua,K.S., Chong,W.K., Lee,H.P., Gu,Q.M.: A Comparative of PCA, KPCA and ICA for Dimensionality Reduction in Support Vector Machine. Neurocomputing, Vol. 55 (2003) 321–336
4. Weston,J., Mukherjee,S., Chapelle,O., Pontil,M., Poggio,T.,Vapnik,V.N.: Feature Selection for SVMs. Adv. Neural Inform. Process. Systems. Vol. 13 (2001) 668–674
5. Tay,F.E.H., Cao,L.J.: Saliency Analysis of Support Vector Machines for Feature Selection. Neural Network World, Vol. 2 (2001) 153–166
6. Ku, W., Storer, R. H., Georgakis, C.: Disturbance Detection and Isolation by Dynamic Principal Component Analysis. Chemometrics and Intelligent Laboratory Systems, Vol. 30 (1995) 179–196
7. Negiz, A., Cinar, A.: Statistical Monitoring of Multivariable Dynamic Processes with State Space Models. A.I.Ch.E. Journal, Vol.43 (1997) 209–221

8. Simoglou, A., Martin, E.B., Morris, A.J.: Statistical Performance Monitoring of Dynamic Multivariate Processes Using State Space Modeling. Computers and Chemical Engineering, Vol. 26 (2002) 909–920
9. Russell, E.L., Chiang, L.H., Braatz, R.D.: Fault Detection in Industrial Processes Using Canonical Variate Analysis and Dynamic Principal Component Analysis. Chemometrics and Intelligent Laboratory Systems, Vol. 51 (2000) 81–93
10. Hyvcarinen,A.: Survey on Independent Component Analysis. Neural Comput. Surveys 2. (1999) 94–128
11. Hyvcarinen,A., Karhunen,J., Oja,E.: Independent Component Analysis. John Wiley & Sons, Inc, New York, USA (2001)
12. Cardoso,J-F., Soulomica,A.: Blind Beam forming for Non- Gaussian Signals. IEEE Proc. F 140 (6) (1993) 362–370
13. Cheung,Y.M., Xu,L.: An Empirical Method to Select Dominant Independent Components in ICA Time Series Analysis. Proc. Int. Joint Conf. Neural Networks (1999) 3883–3887
14. Cheung,Y.M., Xu,L.: Independent Component Ordering in ICA Time Series Analysis. Neurocomputing, Vol. 41 (2001) 145–152
15. Vapnik, V.: The Support Vector Method of Function Estimation. In J. A. K. Suykens & J. Vandewalle, Nonlinear modeling: advanced blackbox techniques.Boston: Kluwer Academic (1998b) 55-85
16. Luyben, W. L.: Simple Regulatory Control of the Eastman Process. Industrial & Engineering Chemistry Research, Vol. 35 (1996) 3280-3289
17. Downs, J. J. Vogel, E. F.: A Plant-wide Industrial Process Control Problem. Computers and Chemical Engineering, Vol.17 (1993)

Borderline-SMOTE: A New Over-Sampling Method in Imbalanced Data Sets Learning

Hui Han[1], Wen-Yuan Wang[1], and Bing-Huan Mao[2]

[1] Department of Automation, Tsinghua University, Beijing 100084, P. R. China
hanh01@mails.tsinghua.edu.cn
wwy-dau@mail.tsinghua.edu.cn
[2] Department of Statistics, Central University of Finance and Economics,
Beijing 100081, P. R. China
maobinghuan@yahoo.com

Abstract. In recent years, mining with imbalanced data sets receives more and more attentions in both theoretical and practical aspects. This paper introduces the importance of imbalanced data sets and their broad application domains in data mining, and then summarizes the evaluation metrics and the existing methods to evaluate and solve the imbalance problem. Synthetic minority over-sampling technique (SMOTE) is one of the over-sampling methods addressing this problem. Based on SMOTE method, this paper presents two new minority over-sampling methods, borderline-SMOTE1 and borderline-SMOTE2, in which only the minority examples near the borderline are over-sampled. For the minority class, experiments show that our approaches achieve better TP rate and F-value than SMOTE and random over-sampling methods.

1 Introduction

There may be two kinds of imbalances in a data set. One is between-class imbalance, in which case some classes have much more examples than others [1]. The other is within-class imbalance, in which case some subsets of one class have much fewer examples than other subsets of the same class [2]. By convention, in imbalanced data sets, we call the classes having more examples the majority classes and the ones having fewer examples the minority classes.

The problem of imbalance has got more and more emphasis in recent years. Imbalanced data sets exists in many real-world domains, such as spotting unreliable telecommunication customers [3], detection of oil spills in satellite radar images [4], learning word pronunciations [5], text classification [6], detection of fraudulent telephone calls [7], information retrieval and filtering tasks [8], and so on. In these domains, what we are really interested in is the minority class other than the majority class. Thus, we need a fairly high prediction for the minority class. However, the traditional data mining algorithms behaves undesirable in the instance of imbalanced data sets, as the distribution of the data sets is not taken into consideration when these algorithms are designed.

The structure of this paper is organized as follows. Section 2 gives a brief introduction to the recent developments in the domains of imbalanced data sets. Section 3 describes our over-sampling methods on resolving the imbalanced problem. Section 4 presents the experiments and compares our methods with other over-sampling methods. Section 5 draws the conclusion.

2. The Recent Developments in Imbalanced Data Sets Learning

2.1 Evaluation Metrics in Imbalanced Domains

Most of the studies in imbalanced domains mainly concentrate on two-class problem as multi-class problem can be simplified to two-class problem. By convention, the class label of the minority class is positive, and the class label of the majority class is negative. Table 1 illustrates a confusion matrix of a two-class problem. The first column of the table is the actual class label of the examples, and the first row presents their predicted class label. TP and TN denote the number of positive and negative examples that are classified correctly, while FN and FP denote the number of misclassified positive and negative examples respectively.

Table 1. Confusion matrix for a two-class problem

	Predicted Positive	Predicted Negative
Positive	TP	FN
Negative	FP	TN

$$\text{Accuracy} = (TP+TN)/(TP+FN+FP+TN) \tag{1}$$

$$\text{FP rate} = FP/(TN+FP) \tag{2}$$

$$\text{TP rate} = \text{Recall} = TP/(TP+FN) \tag{3}$$

$$\text{Precision} = TP/(TP+FP) \tag{4}$$

$$F-value = ((1+\beta^2) \cdot \text{Recall} \cdot \text{Precision})/(\beta^2 \cdot \text{Recall} + \text{Precision}) \tag{5}$$

When used to evaluate the performance of a learner for imbalanced data sets, accuracy is generally apt to predict the majority class better and behaves poorly to the minority class. We can come to this conclusion from its definition (formula (1)): if the dataset is extremely imbalanced, even when the classifier classifies all the majority examples correctly and misclassifies all the minority examples, the accuracy of the learner is still high because there are much more majority examples than minority examples. Under the circumstance, accuracy can not reflect reliable prediction for the minority class. Thus, more reasonable evaluation metrics are needed.

ROC curve [9] is one of the popular metrics to evaluate the learners for imbalanced data sets. It is a two-dimensional graph in which TP rate is plotted on the y-axis and

FP rate is plotted on the x-axis. FP rate (formula (2)) denotes the percentage of the misclassified negative examples, and TP rate (formula (3)) is the percentage of the correctly classified positive examples. The point (0, 1) is the ideal point of the learners. ROC curve depicts relative trade-offs between benefits (TP rate) and costs (FP rate). AUC (Area under ROC) can also be applied to evaluate the imbalanced data sets [9]. Furthermore, F-value (formula (5)) is also a popular evaluation metric for imbalance problem [10]. It is a kind of combination of recall (formula (3)) and precision (formula (4)), which are effective metrics for information retrieval community where the imbalance problem exist. F-value is high when both recall and precision are high, and can be adjusted through changing the value of β, where β corresponds to relative importance of precision vs. recall and it is usually set to 1.

The above evaluation metrics can reasonably evaluate the learner for imbalanced data sets because their formulae are relative to the minority class.

2.2 Methods for Dealing with Imbalanced Data Sets Learning

The solutions to imbalanced data sets can be divided into data and algorithmic levels. categories. The methods at data level change the distribution of the imbalanced data sets, and then the balanced data sets are provided to the learner to improve the detection rate of minority class. The methods at the algorithm level modify the existing data mining algorithms or put forward new algorithms to resolve the imbalance problem.

2.2.1 The Methods at Data Level

At the data level, different forms of re-sampling methods were proposed [1]. The simplest re-sampling methods are random over-sampling and random under-sampling. The former augments the minority class by exactly duplicating the examples of the minority class, while the latter randomly takes away some examples of the majority class. However, random over-sampling may make the decision regions of the learner smaller and more specific, thus cause the learner to over-fit. Random under-sampling can reduce some useful information of the data sets. Many improved re-sampling methods are thus presented, such as heuristic re-sampling methods, combination of over-sampling and under-sampling methods, embedding re-sampling methods into data mining algorithms, and so on. Some of the improved re-sampling methods are as follows.

Kubat et al. presented a heuristic under-sampling method which balanced the data set through eliminating the noise and redundant examples of the majority class [11]. Nitesh et al. over-sampled the minority class through SMOTE (Synthetic Minority Over-sampling Technique) method, which generated new synthetic examples along the line between the minority examples and their selected nearest neighbors [12]. The advantage of SMOTE is that it makes the decision regions larger and less specific. Nitesh et al. integrated SMOTE into a standard boosting procedure, thus improved the prediction of the minority class while not sacrificing the accuracy of the whole testing set [13]. Gustavo et al. combined over-sampling and under-sampling methods to resolve the imbalanced problem [14]. Andrew Estabrooks et al. proposed a multiple re-sampling method which selected the most appropriate re-sampling rate adaptively [15]. Taeho Jo et al. put forward a cluster-based over-sampling method which dealt with between-class imbalance and within-class imbalance simultaneously [16]. Hongyu Guo et al. found out hard examples of the majority and minority classes during the process of boosting, then generated new synthetic examples from hard examples and add them to the data sets [17].

2.2.2 The Methods at Algorithm Level

The methods at algorithm level operate on the algorithms other than the data sets. The standard boosting algorithm, e.g. Adaboost [18], increases the weights of misclassified examples and decreases those correctly classified using the same proportion, without considering the imbalance of the data sets. Thus, traditional boosting algorithms do not perform well on the minority class. Aiming at the disadvantage above, Mahesh V. Joshi et al. proposed an improved boosting algorithm which updated weights of positive prediction (TP and FP) differently from weights of negative prediction (TN and FN). The new algorithm can achieve better prediction for the minority class [19]. When dealing with imbalanced data sets, the class boundary learned by Support Vector Machines (SVMs) is apt to skew toward the minority class, thus increase the misclassified rate of the minority class. Gang Wu et al. proposed class-boundary alignment algorithm which modify the class boundary through changing the kernel function of SVMs [20]. Kaizhu Huang et al. presented Biased Minimax Probability Machine (BMPM) to resolve the imbalance problem. Given the reliable mean and covariance matrices of the majority and minority classes, BMPM can derive the decision hyperplane by adjusting the lower bound of the real accuracy of the testing set [21]. Furthermore, there are other effective methods such as cost-based learning, adjusting the probability of the learners and one-class learning, and so on [22] [23].

3 A New Over-Sampling Method: Borderline-SMOTE

In order to achieve better prediction, most of the classification algorithms attempt to learn the borderline of each class as exactly as possible in the training process. The examples on the borderline and the ones nearby (we call them borderline examples in this paper) are more apt to be misclassified than the ones far from the borderline, and thus more important for classification.

Based on the analysis above, those examples far from the borderline may contribute little to classification. We thus present two new minority over-sampling methods, borderline-SMOTE1 and borderline-SMOTE2, in which only the borderline examples of the minority class are over-sampled. Our methods are different from the existing over-sampling methods in which all the minority examples or a random subset of the minority class are over-sampled [1] [2] [12].

Our methods are based on SMOTE (Synthetic Minority Over-sampling Technique) [12]. SMOTE generates synthetic minority examples to over-sample the minority class. For every minority example, its k (which is set to 5 in SMOTE) nearest neighbors of the same class are calculated, then some examples are randomly selected from them according to the over-sampling rate. After that, new synthetic examples are generated along the line between the minority example and its selected nearest neighbors. Not like the existing over-sampling methods, our methods only over-sample or strengthen the borderline minority examples. First, we find out the borderline minority examples; then, synthetic examples are generated from them and added to the original training set. Suppose that the whole training set is T, the minority class is P and the majority class is N, and

$$P = \{p_1, p_2, ..., p_{pnum}\}, N = \{n_1, n_2, ..., n_{nnum}\}$$

where *pnum* and *nnum* are the number of minority and majority examples. The detailed procedure of borderline-SMOTE1 is as follows.

Step 1. For every $p_i (i = 1,2,..., pnum)$ in the minority class P, we calculate its m nearest neighbors from the whole training set T. The number of majority examples among the m nearest neighbors is denoted by $m'(0 \leq m' \leq m)$.

Step 2. If $m' = m$, i.e. all the m nearest neighbors of p_i are majority examples, p_i is considered to be noise and is not operated in the following steps. If $m/2 \leq m' < m$, namely the number of p_i's majority nearest neighbors is larger than the number of its minority ones, p_i is considered to be easily misclassified and put into a set DANGER. If $0 \leq m' < m/2$, p_i is safe and needs not to participate in the follows steps.

Step 3. The examples in DANGER are the borderline data of the minority class P, and we can see that $DANGER \subseteq P$. We set

$$DANGER = \{p'_1, p'_2, ..., p'_{dnum}\}, \quad 0 \leq dnum \leq pnum$$

For each example in *DANGER*, we calculate its k nearest neighbors from P.

Step 4. In this step, we generate $s \times dnum$ synthetic positive examples from the data in DANGER, where s is an integer between 1 and k. For each p'_i, we randomly select s nearest neighbors from its k nearest neighbors in P. Firstly, we calculate the differences, dif_j ($j = 1,2,...,s$) between p'_i and its s nearest neighbors from P, then multiply dif_j by a random number r_j ($j = 1,2,...,s$) between 0 and 1, finally, s new synthetic minority examples are generated between p'_i and its nearest neighbors:

$$synthetic_j = p'_i + r_j \times dif_j, \quad j = 1,2,...,s$$

We repeat the above procedure for each p'_i in *DANGER* and can attain $s \times dnum$ synthetic examples. This step is similar with SMOTE, for more detail see [12].

In the procedure above, p_i, n_i, p'_i, dif_j and $synthetic_j$ are vectors. We can see that new synthetic data are generated along the line between the minority borderline examples and their nearest neighbors of the same class, thus strengthened the borderline examples.

Borderline-SMOTE2 not only generates synthetic examples from each example in *DANGER* and its positive nearest neighbors in P, but also does that from its nearest negative neighbor in N. The difference between it and its nearest negative neighbor is multiplied a random number between 0 and 0.5, thus the new generated examples are closer to the minority class.

Our methods can be easily understood with the following simulated data set, Circle, which has two classes. Fig. 1 (a) shows the original distribution of the data set, the circle points represent majority examples and the plus signs are minority examples. Firstly, we apply borderline-SMOTE to find out the borderline examples of the minority class, which are denoted by solid squares in Fig. 1 (b). Then, new synthetic examples are generated through those borderline examples of the minority class. The synthetic examples are shown in Fig. 1 (c) with hollow squares. It is easy to find out from the figures that, different from SMOTE, our methods only over-sample or strengthen the borderline and its nearby points of the minority class.

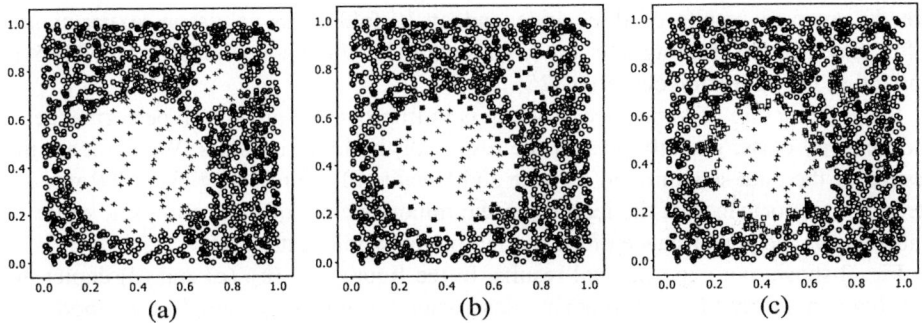

Fig. 1. (a) The original distribution of Circle data set. (b) The borderline minority examples (*solid squares*). (c) The borderline synthetic minority examples (*hollow squares*).

4 Experiments

We use TP rate and F-value for the minority class to evaluate the results of our experiments. TP rate denotes the accuracy of the minority class. And the value of β in F-value is set to 1 in this paper.

The four data sets used in our experiments are shown in Table 2. Among the four data sets, Circle is our simulated data set depicted in Fig. 1, and the others are from UCI [24]. All the attributes in the data sets are quantitative. For Satimage, we choose class label "4" as the minority class and regard the remainders as the majority class, as we only study two-class problem in this paper.

Table 2. The description of the data sets

The name of Data set	number of Examples	number of Attributes	Class label (minority : majority)	Percentage of minority class
Circle(Simulation)	1600	2	1:0	6.25%
Pima(UCI)	768	8	1:0	34.77%
Satimage(UCI)	6435	36	4:remainder	9.73%
Haberman(UCI)	306	3	Die : Survive	26.47%

In our experiments, four over-sampling methods are applied to the data sets: SMOTE, random over-sampling and our methods, borderline-SMOTE1 and borderline-SMOTE2, among which random over-sampling method augments the minority class by exactly duplicating the positive examples partly or completely [1]. Through increasing the number of examples in the minority class, over-sampling methods can balance the distribution of the data sets and improve the detection rate of the minority class.

In order to compare the results conveniently, the value of m in our methods is set in a way that, the number of the minority examples in *DANGER* is about half of the minority class. The value of k is set to 5 like SMOTE. For each method, the TP rates and F-values are attained through 10-fold cross-validation. In order to decrease the randomness in SMOTE and our methods, the TP rates and F-values for these methods are the average results of three independent 10-fold cross-validation experiments. After the original training sets are over-sampled with the methods above, C4.5 is applied as the validation classifier [25].

Since the nature of imbalance problem is to improve the prediction performance of the minority class, we only present the results of the minority class. We compare the results of the data sets through TP rate and F-value of the minority class. TP rate reflects the performance of the learner on the minority class of the testing set, while F-value shows the performance of the learner on the whole testing set.

Fig. 2 shows our experimental results. In the figure, (a), (b), (c) and (d) depict the F-value and TP rate for the minority class when the four over-sampling methods are applied on Circle, Pima, Satimage and Haberman respectively. The x-axis in each figure is the number of the new synthetic examples. The F-value and TP rate of the original data sets with C4.5 are also shown in the figures.

The results illustrated in Fig. 2 reveal the following results. First of all, all the four over-sampling methods improve TP rate of the minority class. For Circle, Pima and Haberman, the TP rates of our methods are better than SMOTE and random over-sampling. Comparing with the original data sets, the best improvements of TP rate for borderline-SMOTE1 and borderline-SMOTE2 on Circle are 20 and 22 per cent, 21.3 and 20.5 per cent on Pima, 10.1 and 10.0 per cent on Satimage, and both 45.2 per cent on Haberman. For Satimage, the TP rates of our methods are lager than that of random over-sampling, and are comparable with SMOTE. Secondly, the F-value of borderline-SMOTE1 is generally better than SMOTE and random over-sampling, and the F-value of borderline-SMOTE2 is also comparable with others. Comparing with the original data sets, the best improvements of F-value for borderline-SMOTE1 and borderline-SMOTE2 on Circle are 12.1 and 10.3 per cent, 2.3 and 1.3 per cent on Pima, 2.3 and 1.4 per cent on Satimage, and 24.7 and 23.0 per cent on Haberman.

As a whole, border-SMOTE1 behaves excellent on both TP rate and F-value, and borderline-SMOTE2 behaves super on TP rate because it generates synthetic examples from both the minority borderline examples and their nearest neighbors of the majority class, however, the procedure causes overlap between the two classes, thus decreases its F-value to some extent.

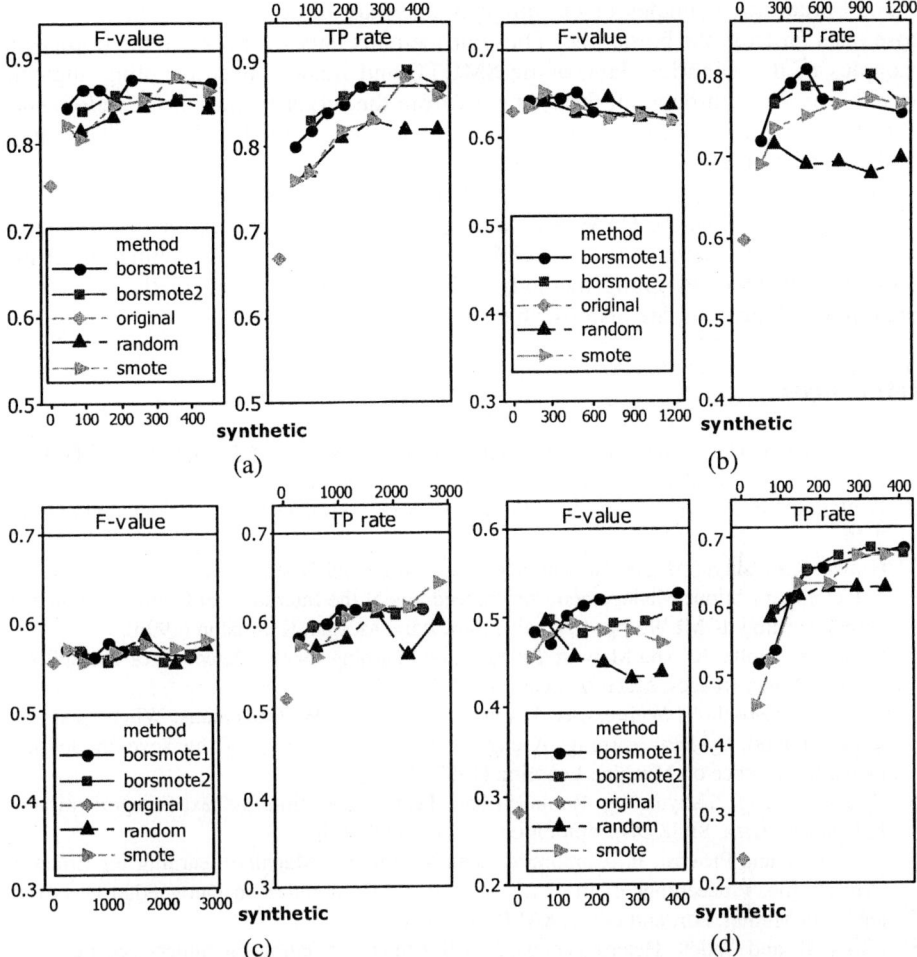

Fig. 2. (a), (b), (c) and (d) illustrate the F-value and TP rate for minority class when proposed over-sampling methods are applied on Circle, Pima, Satimge and Haberman respectively with C4.5. "borsmote1" and "borsmote2" denote borderline-SMOTE1 and borderline-SMOTE2, "random" denotes random over-sampling, and "original" denotes the values of the original data sets. The x-axis is the number of synthetic examples

5 Conclusion

In recent years, learning with imbalanced data sets receives more and more attentions in both theoretical and practical aspects. However, traditional data mining methods are not satisfactory. Aiming to solve the problem, two new synthetic minority over-sampling methods, borderline-SMOTE1 and borderline-SMOTE2 are presented in this paper. We compared the TP rate and F-value of our methods with SMOTE, random over-sampling and the original C4.5 for four data sets.

The borderline examples of the minority class are more easily misclassified than those ones far from the borderline. Thus our methods only over-sample the borderline examples of the minority class, while SMOTE and random over-sampling augment the minority class through all the examples from the minority class or a random subset of the minority class. Experiments indicate that our methods behave better, which validates the efficiency of our methods.

There are several topics left to be considered further in this line of research. Different strategies to define the *DANGER* examples, and automated adaptive determination of the number of examples in *DANGER* would be valuable. The combination of our methods with under-sampling methods and the integration of our methods to some data mining algorithms, are also worth trying.

References

1. Nitesh V.Chawla, Nathalie Japkowicz and Aleksander Kolcz.: Editorial: Special Issue on Learning from Imbalanced Data Sets. SIGKDD Explorations 6 (1) (2004) 1-6
2. G. Weiss: Mining with rarity: A unifying framework. SIGKDD Explorations 6 (1) (2004) 7-19
3. Ezawa, K.J., Singh, M. and Norton, S.W.: Learning Goal Oriented Bayesian Networks for Telecommunications Management. In Proceedings of the International Conference on Machine Learning, ICML'96(pp. 139-147), Bari, Italy, Morgan Kaufmann (1996)
4. Kubat, m., Holte, R., and Matwin, S.: Machine Learning for the Detection of Oil Spills in Satellite Radar Images. Machine Learning 30 195-215
5. A. van den Bosch, T. Weijters, H. J. van den Herik, and W. Daelemans: When small disjuncts abound, try lazy learning: A case study. In Proceedings of the Seventh Belgian-Dutch Conference on Machine Learning (1997) 109-118
6. Zhaohui Zheng, Xiaoyun Wu, Rohini Srihari: Feature Selection for Text Categorization on Imbalanced Data. SIGKDD Explorations 6 (1) (2004) 80-89
7. Fawcett, T.and Provost, F.: Combining Data Mining and Machine Learning for Effective User Profile. Proceedings of the 2nd International Conference on Knowledge Discovery and Data Mining, Portland OR, AAAI Press (1996) 8-13
8. Lewis, D. and Catlett, Heterogeneous, J.: Uncertainty Sampling for Supervized Learning. Proceedings of the 11th International Conference on Machine Learning, ICML'94 (1994) 148-156
9. Bradley A.: The use of the area under the ROC curve in the evaluation of machine learning algorithms. Pattern Recognition 30 (7) (1997) 1145-1159
10. Rijsbergen, C. J. van: Information Retrieval, Butterworths, London (1979)
11. Kubat, M., and Matwin, S. Addressing the Course of Imbalanced Training Sets: One-sided Selection. In ICML'97 (1997) 179-186
12. Chawla, N.V., Bowyer,K.W., Hall, L.O., Kegelmeyer W.P.: SMOTE: Synthetic Minority Over-Sampling Technique. Journal of Artificial Intelligence Research 16 (2002) 321-357
13. Chawla, N.V., Lazarevic, A., Hall, L.O. and Bowyer, K.: SMOTEBoost: Improving prediction of the Minority Class in Boosting. 7th European Conference on Principles and Practice of Knowledge Discovery in Databases, Cavtat Dubrovnik, Croatia (2003) 107-119
14. Gustavo, E.A., Batista, P.A., Ronaldo, C., Prati, Maria Carolina Monard: A Study of the Behavior of Several Methods for Balancing Machine Learning Training Data. SIGKDD Explorations 6 (1) (2004) 20-29

15. Andrew Estabrooks, Taeho Jo and Nathalie Japkowicz: A Multiple Resampling Method for Learning from Imbalanced Data Sets. Comprtational Intelligence 20 (1) (2004) 18-36
16. Taeho Jo, Nathalie Japkowicz: Class Imbalances versus Small Disjuncts. Sigkdd Explorations 6 (1) (2004) 40-49
17. Hongyu Guo, Herna L Viktor: Learning from Imbalanced Data Sets with Boosting and Data Generation: The DataBoost-IM Approach. Sigkdd Explorations 6 (1) (2004) 30-39
18. Yoav Freund, Robert Schapire: A decision-theoretic generalization of on-line learning and an application to boosting. Journal of Computer and System Sciences 55 (1) (1997) 119-139
19. Joshi, M., Kumar,V., Agarwal, R.: Evaluating Boosting Algorithms to Classify Rare Classes: Comparison and Improvements. First IEEE International Conference on Data Mining, San Jose, CA (2001)
20. Gang Wu, Edward Y.Chang. Class-Boundary Alignment for Imbalanced Dataset Learning. Workshop on Learning from Imbalanced Datasets II, ICML, Washington DC (2003)
21. Kaizhu Huang, Haiqin Yang, Irwin King, Michael R. Lyu. Learning Classifiers from Imbalanced Data Based on Biased Minimax Probability Machine. Proceedings of the IEEE Computer Society Conference on Computer Vision and Pattern Recognition (2004)
22. Dietterich, T., Margineantu, D., Provost, F. and P. Turney, edited, Proceedings of the ICML'2000 Workshop on Cost-sensitive Learning (2000)
23. Manevitz, L.M. and Yousef, M.: One-class SVMs for document classification. Journal of Machine Learning Research 2 (2001) 139-154
24. Blake, C., & Merz, C. (1998). UCI Repository of Machine Learning Databases http://www.ics.uci.edu/~mlearn/~MLRepository.html. Department of Information and Computer Sciences, University of California, Irvine
25. Quinlan, J. C4.5: Programs for Machine Learning. Morgan Kaufmann, San Mateo, CA (1992)

Real-Time Gesture Recognition Using 3D Motion History Model

Ho-Kuen Shin, Sang-Woong Lee, and Seong-Whan Lee

Department of Computer Science and Engineering, Korea University
Anam-dong, Seongbuk-gu, Seoul 136-713, Korea
{hkshin, sangwlee, swlee}@image.korea.ac.kr

Abstract. In this paper, we present a novel method for real time gesture recognition with 3D Motion History Model (MHM). There are two difficult problems in gesture recognition: the camera view and the duration of gesture. First, we solved the camera view problem which is very difficult in the environment of single directional camera (e.g., monocular or stereo camera). Utilizing 3D-MHM with the disparity information, not only this problem is solved but also the reliability of recognition and the scalability of system are improved. Second, we proposed the dynamic history buffering (DHB) to solve the duration problem that comes from the variation of gesture velocity at every performing time. DHB improves the problem using magnitude of motion. We implemented a real-time system and performed gesture recognition experiments. The system using 3D-MHM achieves better results of recognition than using only 2D motion information.

1 Introduction

As the day the robot assists in human life is visible on the results of active humanoid robot research, convenient human robot interfaces as well as the appearance of robot become very important components. Hence gesture recognition based on vision sensor is one of the advanced interfaces. The gesture recognition is generally considered as the analysis of human hands and performing given command. However, the research described in this paper focus on analysis about whole body motion in the daily living, and its final goal is analyzing current movement and tendency of human movements. The research of gesture recognition can be divided into two main categories. One is trajectory-based recognition after analyzing components of the human body from the input images. Typically, hidden Markov model is used for recognizing gesture [7]. In this method, it is important to fit 3D human model to a silhouette image and extract articulation information using inverse kinematics [3] or analyze components of the human body from raw data directly [5]. The reliability of this method, however, is very poor in the monocular environments, and also the process is very complex, difficult, and computationally expensive.

The other one is motion-based recognition. This method is fast and easy, so a real-time implementation is possible. Morrison and McKenna [9] presented an experimental comparison above two methods. A representative example of motion-based method is Motion History Image (MHI) [1, 2, 4]. MHI is a simple algorithm and has an advantage that the blurred sequence affects performance less. It has, however, viewpoint problem, false alarms and a limitation of the scalability because only 2D motion is analyzed. Particularly, viewpoint problem without multi-view cameras is one of the most difficult problems in computer vision. Multiple cameras at various directions can be a solution, but that will be a considerable restriction in real-world environments.

In this paper, we propose 3D-MHM using disparity information from stereo input sequence which can discriminate each gesture in the various viewpoints with only one model. The different method utilizing disparity information is shown in [6, 8].

The remainder of this paper is structured as follows. In Section 2, the definition, procedure, and advantages of 3D-MHM are demonstrated. Overview of a real time system, normalization, duration, and recognition method are presented in detail in Section 3. Experimental environments and results, comparing 2D with 3D motion information is described in Section 4. Finally, we conclude the results in Section 5.

2 Proposed 3D Motion History Model

2.1 Definition of 3D-MHM

3D Motion History Model is defined by a virtual 3D model using stereo input sequences which contain motion history information in 3D space. This model is proposed for overcoming a limitation of 2D motion analysis such as the viewpoint problem and scalability.

The procedure of constructing 3D-MHM follows 5 steps:

Step 1. Calculate disparity maps from the sequence of stereo input images.
Step 2. Reconstruct body model in 3D space using the silhouette image and the disparity information.
Step 3. Calculate the difference between two consecutive reconstructed body models.
Step 4. Insert the result of Step 3 into history buffer (DHB is described in Section. 3.2) and then merge all objects of buffer into one 3D space. Each object has a different intensity value over time, and this is the 3D motion history model.
Step 5. Calculate global gradient orientation of 3D-MHM using 3D Sobel operator and then, rotate 3D-MHM about Z axis until it tends toward constant direction. Finally generate projection image of it.

In step 2, the Reconstructed Body Model (RBM) is generated using intensity of disparity map. The instant of RBM at time t is calculated by Equation (1).

$$RBM_t(x,y,z) = \begin{cases} 1 & \text{if } S_t(x,y) = 1 \text{ and } C_d \cdot Dis_t(x,y) = z \\ 0 & \text{otherwise} \end{cases} \quad (1)$$

where S is a silhouette image, C_d is a constant for normalization and Dis is disparity map.

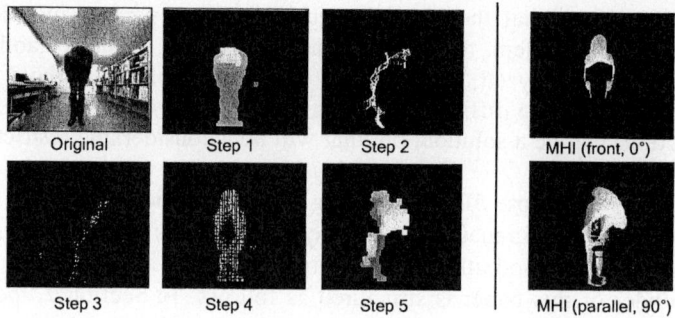

Fig. 1. The procedure of constructing 3D-MHM

The magnitude of motion is used for solving the duration problem (Section 3.2 describe in detail). It is calculated by Equation (2).

$$M_t = \iiint D_t(x,y,z)dxdydz \quad (2)$$

where M is the magnitude of motion and D is the difference of consecutive two RBMs.

The 3D-MHM is constructed from the difference of the consecutive RBMs. That is calculated by Equation (3).

$$MHM_t(x,y,z) = \begin{cases} i_{max} & \text{if } D_t(x,y,z)=1 \\ \max(0, MHM_{t-1}(x,y,z) - C_a \log(M_t)) & \text{otherwise} \end{cases} \quad (3)$$

where i_{max} is maximum intensity (e.g., 255 in 8bit grayscale) and C_a is an attenuating constant.

Fig. 1 shows the procedure of 3D-MHM. Although gesture is bowing in the 0° view (front), 3D-MHM describes motion information well as 90° view (parallel).

2.2 Comparing 3D-MHM with MHI

As mentioned earlier, the MHI is an easy and fast algorithm but it has a limitation in some case because of using only 2D motion information. Fig. 2 (b) and (e) show the case where false alarms occur in the MHI based gesture recognition system. Fig. 2 (b) is the MHI of walking forward, but the motion image does not represent the gesture well. If someone shakes a body a little in the left-right or turns around, the almost same MHI can be generated. Similarly, in the bowing gesture case, when someone shakes a head in the left-right direction, the similar problem occurs in MHI.

The 3D-MHM overcomes these problems using disparity information. In the above examples are the motion in the forward-backward direction is more important than in the left-right direction. Furthermore, the gesture recognition system using the 3D-MHM can recognize asking handshake, putting someone's foot forward and more various gestures. 3D-MHM has the advantages in the scalability and the reliability.

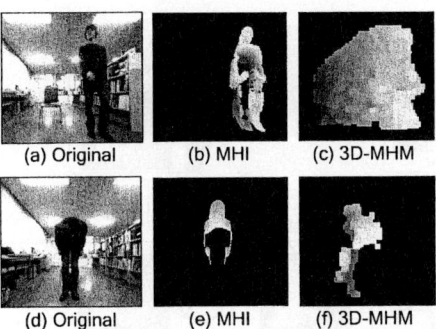

Fig. 2. Comparing 3D-MHM with MHI

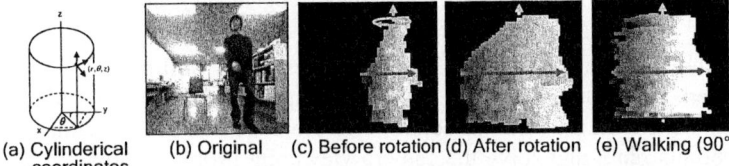

Fig. 3. The virtual viewpoints

2.3 The Viewpoint for Matching 3D-MHM

The 3D-MHM has a function of virtual viewpoint, so we can see the motion model at any other views virtually. In 2D MHI approaches, many templates are necessary for gesture recognition in the various viewpoints. For example, if we want to recognize a gesture in the 0°, 90° view, then we need two different templates of each gesture for training. Therefore, the more viewpoints, and the more training data set should be required and consequently computational cost is increased. Moreover, the performance of the discrimination is deteriorated.

The 3D-MHM has the 3D motion information in one model. The 3D-MHM provides virtual viewpoints through rotating about Z axis (Cylindrical coordinates) as changing a viewpoint. To solve the viewpoint problem, we use directional information of gradient. Equation (4) shows gradient of function f.

$$\nabla f = \left(\frac{\partial f}{\partial x}, \frac{\partial f}{\partial y}, \frac{\partial f}{\partial z} \right) \quad (4)$$

The Sobel 3D operator with 3x3x3 kernel is applied to calculation of 3D-MHM's gradients as shown in Equation (5).

$$\theta(x,y,z) = \arctan \frac{Sobel_z(x,y,z)}{Sobel_x(x,y,z)}, \quad \phi(x,y,z) = \arctan \frac{Sobel_y(x,y,z)}{Sobel_x(x,y,z)} \quad (5)$$

An average of the gradients is global gradient orientation and used as standard direction for matching between input and training data. The global gradient orientation can be presented by $(r, \theta, r\tan\phi)$ in cylindrical coordinates. The 3D-MHM is rotated about Z axis until θ is 90°.

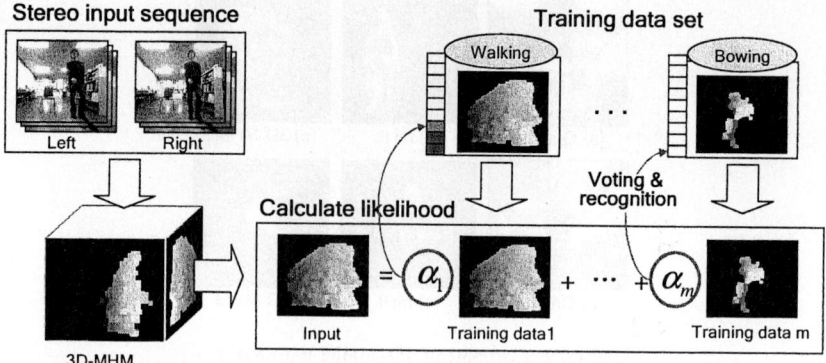

Fig. 4. Overview of gesture recognition system

In 90°, the projection images of 3D-MHMs described well motion information can be generated. This rotation is also used on comparing the input 3D-MHM with the trained 3D-MHMs in the same direction. Therefore we can get the same 3D-MHM in any viewpoints input and solve the camera viewpoint problem shown in Fig. 3. Also, it can find the best projection image which describes the motion well

3 Gesture Recognition Using 3D-MHM

The entire gesture recognition system using 3D-MHM is shown in Fig. 4.

3.1 Spatial Normalization

In [10], 7 Hu-moments are translation and rotation invariant therefore they are useful for normalizing data. However, we could realize it is difficult to classify two motion data using 7 Hu-moments, because the difference between 7 Hu-moments of walking forward MHI and sitting MHI is too small.

Therefore we used height of dominant region and center of gravity which is calculated by the Equation (6) for spatial normalization. Fig. 5 shows the results after spatial normalization.

$$m_{pqr} = \int_{-\infty}^{\infty}\int_{-\infty}^{\infty}\int_{-\infty}^{\infty} x^p y^q z^r \rho(x,y,z) dx dy dz \tag{6}$$

$$\bar{x} = m_{100}/m_{000},$$
$$\bar{y} = m_{010}/m_{000},$$
$$\bar{z} = m_{001}/m_{000},$$

where m is moment, ρ is density function of the RBM and $(\bar{x},\bar{y},\bar{z})$ is center position.

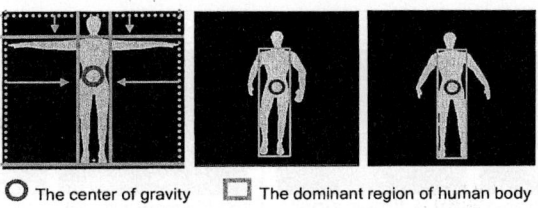

Fig. 5. Examples of the spatial normalization

3.2 Dynamic History Buffering

The variation of the motion velocity at every performing time has a bad influence upon recognition performance. Because of duration problem, MHIs appear different shapes in some cases. In Fig. 6 (a) and Fig. 6 (b), both images are MHIs of arm-up gesture, but Fig. 6 (a) is performing with normal speed and Fig. 6 (b) is performing 4 times slowly. Although some person performs the same gesture but different velocity, recognition is failed due to the different duration. In the prior work [1], Equation (7) is used to solve this problem.

$$H_{\tau-\Delta\tau}(x,y,t) = \begin{cases} (H_\tau(x,y,t) - \Delta\tau) & \text{if } H_\tau(x,y,t) > \Delta\tau \\ 0 & \text{otherwise} \end{cases} \quad (7)$$

where $\Delta\tau = (\tau_{max} - \tau_{min})/(n-1)$ and n is the number of temporal integration windows.

They determined the maximum and the minimum duration and generated all MHIs on the range between the maximum and the minimum. The disadvantage of this method is that the more a number of gesture increase or the range is bigger, the more data are necessary. Also computational cost increase and discriminability is lowered, because it must require each template of the input data in each view. The solution of problem by different capture rate using tMHI is shown in [4]. Current timestamp as silhouette value is used. However, the variation of person's velocity is not covered.

We solved this problem by controlling the history buffer dynamically. Original structure of generating MHI is as follows: Compute the difference image between two consecutive silhouette images. Merge them to one image with different intensity over time. The previous motion information disappears over time although motion is not occurred. This point is the main reason why we had different MHIs according to duration. To solve this problem, we propose the dynamic history buffering (DHB). The DHB has the buffer of difference given image for preventing loss of prior motion information. The difference image is added to buffer, only when larger motion is occurred than the threshold. The operation of DHB is determined by Equation (8).

$$Buffer(n) = \begin{cases} D_\tau(x,y,z) & \text{if } M > Th \\ 0 & \text{otherwise} \end{cases} \quad (8)$$

where D is the difference of consecutive two silhouettes or RBMs, M is the magnitude of motion, x, y, z are position values, n is the contents number in buffer and Th is threshold.

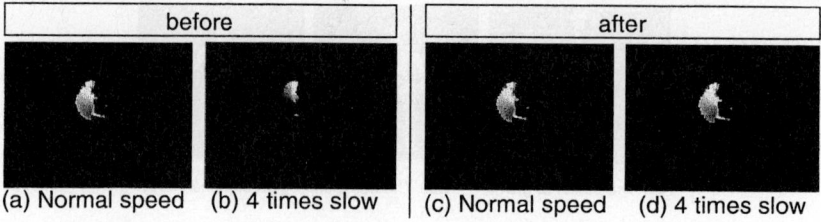

Fig. 6. The results of using the dynamic history buffering (Arm-up)

In this buffer control, only when motion magnitude is higher than the threshold, the current motion is only meaningful, otherwise the current information is dropped and the prior motion data in the buffer is preserved. Therefore when movement is not occurred or its magnitude is small such as a slow motion, buffer's contents are not changed. Eventually, DHB has a characteristic of velocity invariant and can generate consistently the 2D and 3D motion history information as shown in Fig. 6 (c) and (d).

3.3 Continuous Gesture Recognition

3.3.1 Least Square Method

To calculate likelihood, least square method (LSM) is used. The P is a matrix that consists of training data set such as 3D-MHMs, projected image of 3D-MHM, or MHI for each gesture. The α is weight coefficient matrix. Input data, $\tilde{P}$ can be presented by matrix P and α as shown in Equation (9).

$$P\alpha \approx \tilde{P}, \tag{9}$$

where

$$P = \begin{pmatrix} p_1(x_1) & \cdots & p_m(x_1) \\ \vdots & \ddots & \vdots \\ p_1(x_n) & \cdots & p_m(x_n) \end{pmatrix}, \quad \alpha = (\alpha_{11}, \alpha_{12}, \cdots, \alpha_{gm})^T, \tilde{P} = (\tilde{p}(x_1), \cdots, \tilde{p}(x_n))^T \tag{10}$$

So, we defined error function as Equation (11).

$$\tag{11}$$

The optimal coefficient α^* to minimize error function is likelihood and it can be solved by Equation (12).

$$\alpha^* = \arg\min_{\alpha} E(\alpha), \quad \alpha^* = (P^T P)^{-1} P^T \tilde{P} = P^+ \tilde{P} \tag{12}$$

where P^+ is pseudo-inverse of P, g is the number of gesture, each row of P is a variable, and each column of P is an observation.

3.3.2 Voting Algorithm

To use context in time-sequential input data, the voting algorithm is applied as a last part of recognition system. The coefficients of each gesture from least square method which is described in Section 3.3.1 are accumulated by voting. The voting coefficients are calculated by Equation (14). When the number of vote is higher than threshold, finally system presents result of gesture recognition.

$$vote_g(t) = vote_g(t-1) + \sum_{i=1}^{N} \alpha_{gi}^*(t) - C_v \qquad (13)$$

where g is a gesture number, α^* is coefficient from LSM, C_v is an attenuating constant and N is the number of training data for a gesture.

Fig. 7 shows example of using the context information in video sequence through the voting. Sitting gesture and bowing gesture are very similar in the start part. Although the real gesture is sitting, the coefficient of bowing is rather higher than that of sitting at the beginning. But the system correctly recognizes as sitting gesture by voting algorithm after time is passed. Fig. 7 (a) shows current α^* of each gesture in bar graph. Fig. 7 (b) shows voting coefficients of each gesture in bar graph. Fig. 7 (c) shows the history of α^*.

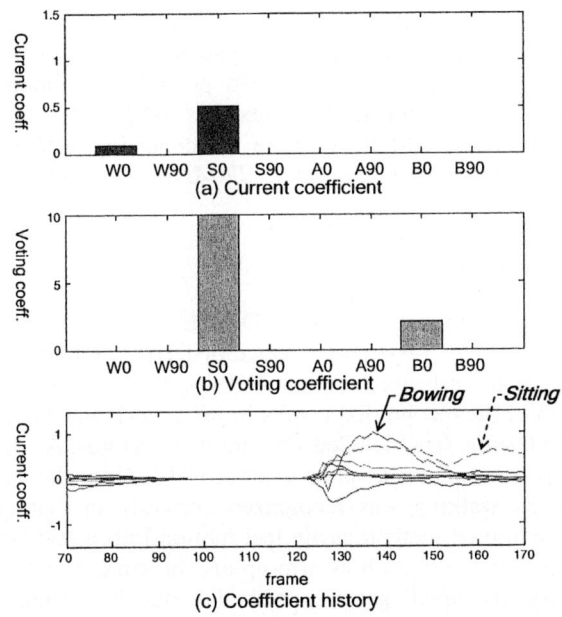

Fig. 7. Result of voting algorithm of 0°, 90° views: Walking, Sitting, Arm-up, Bowing

4 Experimental Results and Analysis

4.1 Experimental Environments

We used a calibrated stereo camera which is manufactured by VIDERE DESIGN MEGA-D. The focal length is 4.8mm. The resolution of input sequence was 320x240

8bit grayscale and the frame rate was 24 frames per second or higher. The experiment of gesture recognition system was implemented on Pentium IV 1.7GHz.

(a) Walking (0 °, 90° view) (b) Sitting (0 °, 90° view)

(c) Arm-up (0 °, 90° view) (d) Bowing (0 °, 90° view)

Fig. 8. Examples of the training data

Four gestures such as walking, sitting, arm-up and bowing were used for experiment. Training data set was acquired with 0°, 90° camera views at the chroma-key background. The training data set in 90° view was only used for training of 2D motion. The training process required three persons. Fig. 8 shows examples of the training data. Testing data set consists of two person's continuous gesture video sequence. We resized original image into 100x100x100 (width x height x depth) for 3D-MHM and 100x100 for MHI, because of denoising effect and decreasing computational cost. The projection images of 3D-MHM after rotation are used for 3D experiments.

4.2 Experimental Results

To measure the efficacy of the proposed 3D-MHM, we performed four kinds of experimental tests. In order to test performance of DHB, the experiments of 2D and 3D motion analysis were performed without DHB and with DHB respectively. Table 1 shows comparison of the recognition results in 3D-MHM, and MHI before and after applying DHB. The testing data included slow motion gestures. As described in Table 1, the recognition performance was poor without DHB. Only the gesture which had large motion such as walking was recognized correctly in that case. This result occurred because traditional method easily lost motion information over time in case of the slow and small gestures such as arm-up and bowing. The DHB improves the variation of velocity and small gestures with dynamic buffering. The recognition results with DHB show that false alarms decrease in 3D-MHM. False alarms occurred

Table 1. Gesture recognition results

Method	Traditional method		Applying DHB	
	2D	3D	2D	3D
Recognition rate	12.5%	12.5%	87.5%	93.7%
Number of false alarm	1	0	7	1

in case of involuntary actions such as turn-around, stand-up, and swing. Those MHIs were very similar to walking's. Utilizing stereo vision, better recognition rate can be obtained. Discrimination between walking forward and turning around was possible by using disparity map. In that case, forward-backward motion was more important, but its presentation is impossible in 2D.

5 Conclusions

In this paper, we proposed a novel method of the motion based gesture recognition with 3D Motion History Model. The 3D-MHM improved false alarms problem in 2D motion analysis using disparity information. Also, the problem of view-based method was solved by using 3D global gradient orientation. 3D-MHM provided virtual viewpoint and could extract the projection images of 3D-MHM which well described each gesture. Concurrently recognizing short-long, slow-rapid and small-large gestures are too difficult at the state of art in gesture recognition. Using Dynamic History Buffering, the duration problem due to the variation of the gesture velocity every performing time was improved remarkably.

As one of future works, expanding the number of gestures for experiments is important. Now constructing 2D and 3D gesture database is in progress.

Acknowledgments

This research was supported by the Intelligent Robotics Development Program, one of the 21st Century Frontier R&D Programs funded by the Ministry of Commerce, Industry and Energy of Korea.

References

[1] Bobick, A.F. and Davis, J.W.: The Recognition of Human Movement Using Temporal Templates. IEEE Trans. on Pattern Analysis and Machine Intelligence 23 (7) (March 2001) 257-267
[2] Shan, C., Wei, Y., Qiu, X. and Tan, T.: Gesture Recognition Using Temporal Template Based Trajectories. Proc. IEEE International Conference on Pattern Recognition, Cambridge, United Kingdom, (2004) 954-957
[3] Sminchisescu, C. and Telea, A.: Human Pose Estimation from Silhouettes a Consistent Approach Using Distance Level Sets. Proc. of International Conference on Computer Graphics, Visualization and Computer Vision, Plzen–Bory, Czech Republic, (2002)
[4] Bradski, G.R. and Davis, J.W.: Motion Segmentation and Pose Recognition with Motion History Gradients. Machine Vision and Applications, Nara, Japan, (2002)
[5] Mori, G., Ren, X., Efros, A. and Malik, J.: Recovering Human Body Configurations: Combining Segmentation and Recognition. IEEE Computer Society Conference on Computer Vision and Pattern Recognition (2004)
[6] Ye, G., Corso, J. and Hager, G.: Gesture Recognition using 3d Appearance and Motion Features. IEEE Conference on Computer Vision and Pattern Recognition Workshops (2004)

[7] Lee,H. and Kim, J.: An HMM-Based Threshold Model Approach for Gesture Recognition. IEEE Trans. on Pattern Analysis and Machine Intelligence 21 (10) (1999)
[8] Bae, K. Koo, J. and Kim, E.: A New Stereo Object Tracking System Using Disparity Motion Vector. Optics Communications (2003)
[9] Morrison, K. and McKenna, S.: An Experimental Comparison of Trajectory-Based and History-Based Representation for Gesture Recognition. Proc. of 5th International Gesture Workshop (2004) 152-163
[10] Hu, M.: Visual Pattern Recognition by Moment Invariants. IRE Trans. Information Theory 8 (2) (1962) 179-187

Effective Directory Services and Classification Systems for Korean Language Education Internet Portal Sites*

Su-Jin Cho[1] and Seongsoo Lee[2]

[1] Department of Korean Language Education, Seoul National University, 151-742, Korea
chosoojin@chollian.net
[2] School of Electronics Engineering, Soongsil University, 156-743, Korea
sslee@ssu.ac.kr

Abstract. Recently, the progress of information and communication technologies leads to web-based education. However, one of these problems is that it is very difficult to find out proper educational materials over the billions of unclassified and unrelated web materials. Well-designed directory services and classification systems of educational materials in the Internet are indispensable for effective web-based education. In this paper, we propose novel directory services and classification systems for effective Korean language learning. We analyzed the elementary components of Korean language learning, and exploit them to develop effective directory services and classification systems. We also propose a guideline to develop them. We also consider peer-to-peer networking service as searching and exchanging educational material.

1 Introduction

Recently, the progress of information and communication technologies (ICT) leads to web-based education (WBE), where computer and information technology make new paradigms in conventional education. However, disappointingly, computer and information technology cannot come beyond the role of automated teacher, and they fail to achieve active and organic education.

One of these problems is that it is very difficult to find out proper educational materials over the billions of unclassified and unrelated web materials. Most of web materials are unprocessed and they are not suitable for real-world class education. Teachers and learners have a hard time in searching proper educational materials fitting their educational purpose. Is it really helpful for the teachers and learners to go on their classes easily and effectively just if there are large libraries and museums beside them?

Consequently, well-designed directory services and classification systems of educational materials in the Internet are indispensable for effective web-based education. In this paper, we propose novel directory services and classification

---

* This work was supported by the Soongsil University Research Fund.

systems for effective Korean language learning. We analyzed the elementary components of Korean language learning, and exploit them to develop effective directory services and classification systems. We also propose a guideline to develop them, and show a simple example. We also consider peer-to-peer networking (P2P) service as searching and exchanging educational material and provide effective directory services and classification systems for P2P service.

2 Directory Services of Internet Portal Sites

In 1967, Stanly Milgram made an experiment to measure the "distance" between two randomly-chosen persons in the United States. In the experiment, he chose some persons and gave them a mission to deliver a letter to an unfamiliar destination person. Each person in the experiment was asked to deliver the letter to his acquaintance that seems to know the destination person, and the receiver also deliver it to his acquaintance in a similar way. In the experiment, Milgram found his famous idea of "six degree of separation," i.e. there are six intermediators between two arbitrary persons. [1]

In a similar way, in 1999, Albert Barabasi found that there are only 19 "links" between two arbitrary web materials. [2]. It means that we can access any arbitrary web materials we want to search on the average in 19 clicks over Internet. In his research, he found that the Internet portal sites are the "hub" of the link, and their roles are very important when we want to find out proper material. Therefore, we have to carefully analyze the directory service and classification system of Internet portal sites to develop effective Korean language education Internet portal sites.

When we look for some web materials in a brain-storming level, i.e. "I'd like to get some multimedia educational material for daily life conversation.", it is difficult to find out exact proper search keyword. Instead, it is better to use directory service and navigate it with screening detailed information unneeded.

Table 1 shows the directory system of several famous Internet portal sites in Korea. As shown in Table 1, directory service of each portal service is quite different. Especiallly, Daum [8] provides quite different directory service, since this Internet portal site is mainly community-based, and the users exchange information and knowledge, and make off-line meeting for collaborative learning through the communities. It shows that we have to consider the communities specialized for given educational purposes.

In Korean language learning, the directory services should reflect the characteristics of language learning. Table 2 shows the directory services of Web Korean [9], a Korean culture and language Internet portal site. Different from other Internet portal sites, it has educational categories such as Korean language education and Korea-related institutional categories such as cultural academies and embassies in its top-level categories. Although the directory services of Web Korean cannot be directly applied, it quite largely reflects them, and it can be a good reference for Korean language learning Internet portal site.

Table 1. Directory services of several famous Internet portal sites in Korea

Yahoo! [3]	Naver [4]	Empas [5]	Google [6]	Paran [7]	Daum [8]
News, Media	Education, Chinese	News, Media	Home, Life	Health, Medical	News
Entertainment	World, Travel	Enterprise, Shopping	Game	Science, Academy	Phone-world
Business, Economy	Computer, Internet	Economy	Science	Education	Finance
Health, Medical	Life Style	Health, Medical	Education	News, Media	Education
Education, Academy	Culture, Art	Education, School	News, Media	Business, Economy	Life
Social Science	Celebrity, People	Academy	Business, Economy	Social, Culture	Kids-world
Government	News, Media	Korea, World	Social	Life, Home	Meeting
Recreation	Game	Entertainment	Shopping	Entertainment, Art	Search
Computer, Internet	Leisure, Sports	Game	Sports	Travel, Sports	Dictionary
Local	Economy	Computer, Internet	Kids, Teens	Local	Entertainment
Social, Culture	Enterprise, Shopping	Travel, Sports	Entertainment	Jobs	Local
Arts	Social, Politics	Culture, Art, Religion	Leisure, Hobby	Computer, Internet	Leisure
Natural Science	Entertainment	Life, Hobby	Internet, WWW		Ms-Net
References	Kids	Social	Local, Countries Computer		MY

Table 2. Directory service of Web Korean

Education	Korean Language, English Education, Japanese Education, Chinese Education
Culture	Traditional Culture, Religion, Kimchi, Taekwondo, History, Celebrity, Koreanology, Korea in the World, Modern Culture
Government	Governmental Institution, Embassy, Culture Center
Economy	Statistics, Finance, Industry, IT
Life	Local Information, Transportation, Map, Travel, Weather, Telephone Number, Survival Korean Language

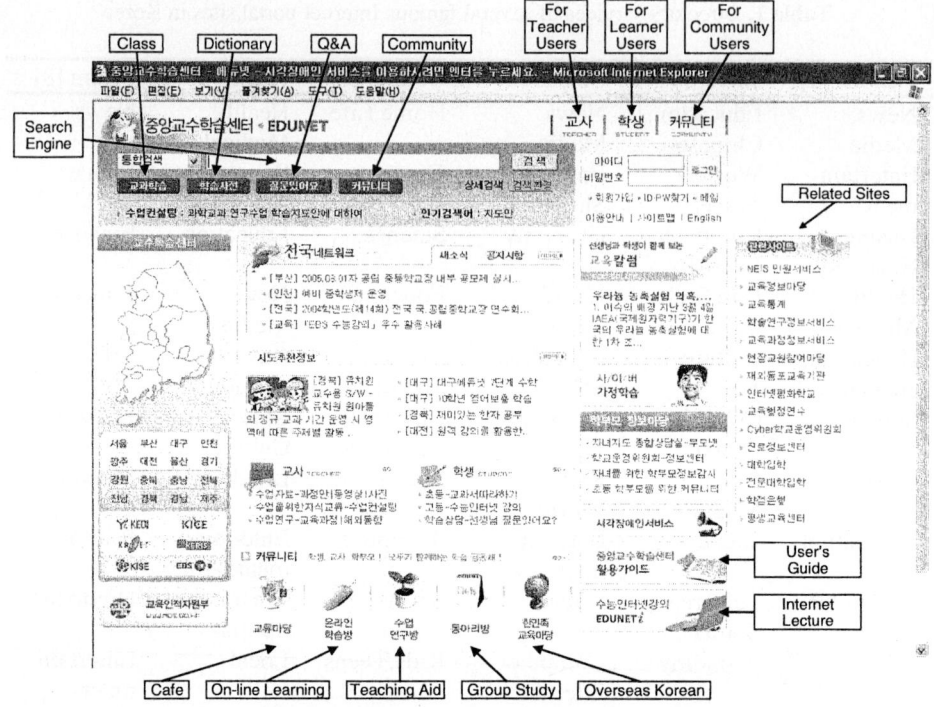

Fig. 1. Top menu of Edunet

Fig. 1 shows the top menu of Edunet [10]. This is a Korean governmental educational web site, and it is regarded to have very similar users and very similar contents with Korean language education Internet portal site. Edunet has three user modes, i.e. teachers, learners, and community users. When the user logs in as one of three user modes, Edunet displays corresponding submenu to fit the user's purpose. This suggests an important aspect, i.e. the directory services and the classification system should be specialized and differentiated according to the user's role.

In Korean language education Internet portal sites, it is desirable to classify the user groups into teachers, learners, and researchers. Fig. 2 shows the proposed prototype Korean language education Internet portal site. It consists of three parts. First, it provides community services between teachers, learners, and researchers, and the directory services and classification systems are differentiated for each user group. Second, it provides various information on Korean language and its instruction and educational materials. Third, various additional information, news, and links on the Korean culture, travel, and life are provided. Third part aims that the learners have more chances to get continuous information in their daily life even if they connect to the site without special purpose. It provides six categories related to Korean language and culture in its directory service – social, education, life, travel, Korean language test, instruction-learning resources, and research resources. Its detailed directory service is shown in Table 3.

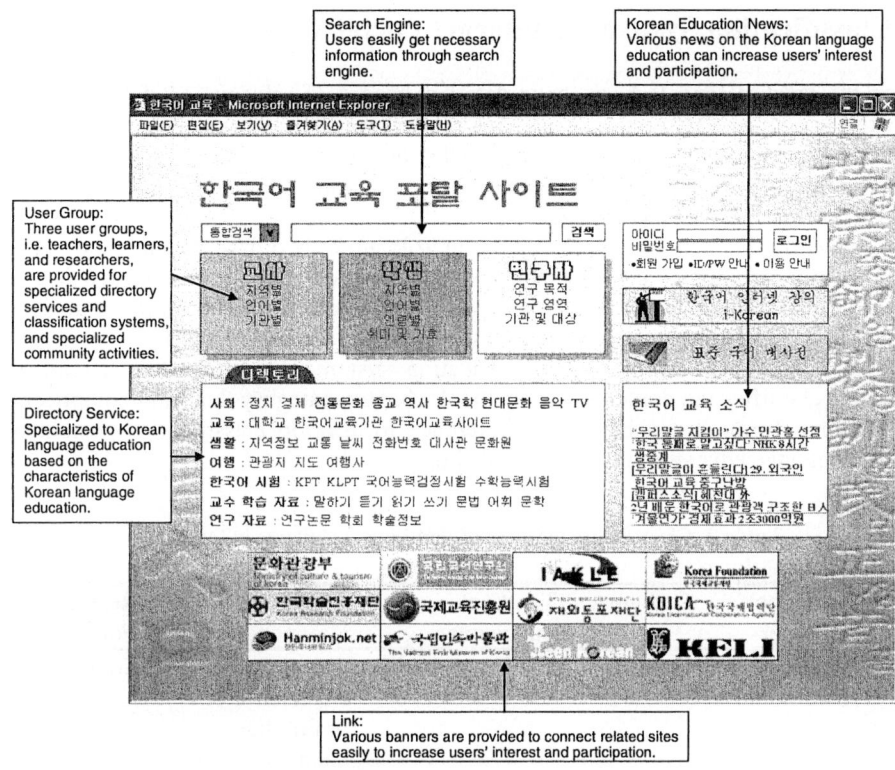

Fig. 2. Top menu of the proposed prototype Korean language education Internet portal site

Table 3. Directory service of the proposed prototype Korean language education Internet portal site

Social	Politics, Economy, Traditional Culture, Religion, History, Koreanology, Modern Culture, Music, TV
Education	University, Korean Language Institute, Korean Language Educational Web Site
Life	Regional Information, Transportation, Weather, Telephone Directory, Embassy, Culture Center
Travel	Sightseeing, Map, Travel Agency
Korean Language Test	Korean Proficiency Test, Korean Language Proficiency Test, Korean Language Ability Certification Test, Collage Scholastic Ability Test
Instruction-Learning Resources	Speaking, Listening, Reading, Writing, Grammar, Vocabulary, Literature
Research Resources	Publication, Conference, Academic Information

3 Classification Systems of Internet Portal Sites

When the teachers and learners have detailed images of information they look for, keyword search is very useful. Some Internet portal sites provide natural language search engines, but they cannot provide satisfactory search performance. In general, it is more effective to use simple keywords and their combination with search operators such as (and/&), (or/+), and (not/-/!).

Classification system plays an important role in keyword search methods. The search engine sometimes directly matches the input keyword in the text, but it also refers the classification system in the database. Therefore, it is desired to develop effective classification system for Korean language educational materials.

Table 3 show an example of classification system of multimedia contents in Edunet. It classifies the contents into textbook chapter, category, learning component, and resource form. It can be a good reference to develop classification system of Korean language education Internet portal site.

Table 3. Classification system of multimedia contents in Edunet

No.	Textbook Chapter	Category	Textbook Subchapter	Learning Component	Resource Form
4	1. Along with mind	Reading	1. With poem	Telling experience of unfamiliar sound	Sound
5	1. Along with mind	Reading	1. With poem	Reading opem with thinking writer's mind	Graphic
12	1. Along with mind	Reading	1. With poem	Discussing after listening the story	Animation
15	1. Along with mind	Reading	2. Further comprehension	Reading poem with creatively understanding	Image
22	1. Along with mind	Reading	1. With poem	How to creatively read poem	Module
27	1. Along with mind	Listening Writing	0. Introduction	Exchanging each other's feeling	Video

In general, most teachers develop their own educational material in person. They usually store them in their personal computers and seldom exchange these materials. Even in the same school, many teachers develop same or similar materials for same educational purpose. This wastes a lot of time and money. It is desired to provide effective way to search and exchange educational material with each other. Recently,

Effective Directory Services and Classification Systems   905

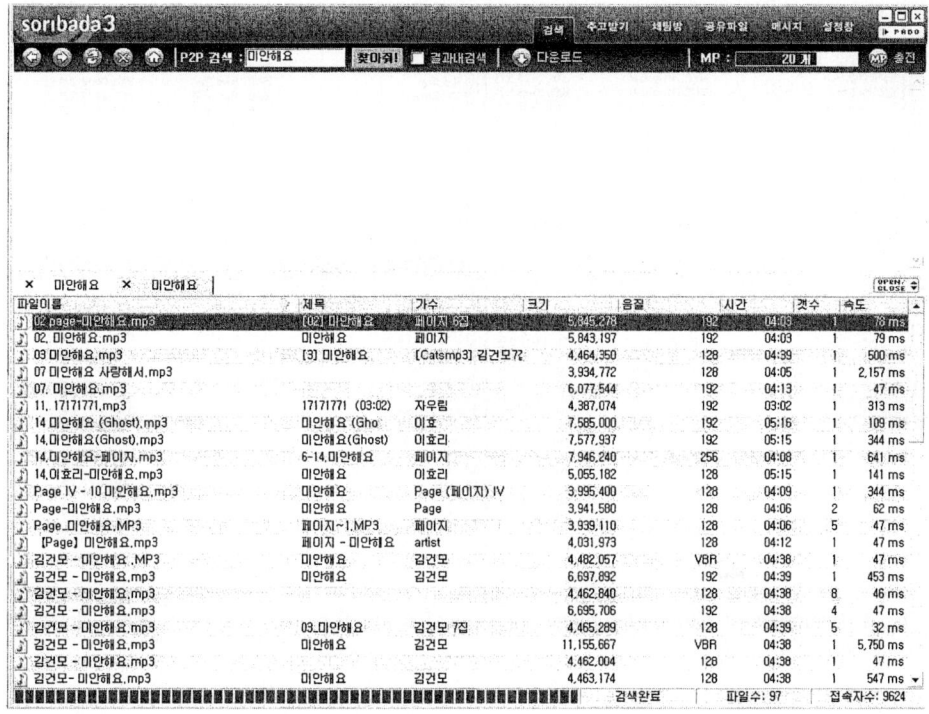

Fig. 3. Search results of Soribada P2P service

peer-to-peer networking (P2P) services such as E-donkey [11] and Soribada [12] enable us to search and exchange many files of other users' personal computers.

Fig. 3 shows Soribada P2P service, when the user looks for the song "I'm sorry!". The files listed in Fig. 3 are stored in other users' personal computer, and the user can search and download them easily. This approach can be adopted in Korean language education, and searching and exchanging Korean language educational materials using P2P service will be very useful for teachers and learners. Most teachers and learners suffer from lack of good educational materials, and P2P can be a breakthrough against this problem.

However, some ID tags are necessary in the educational materials when the user searches the proper materials he/she wants. Soribada P2P service is widely used in searching MP3 files of popular songs, and it requires five ID tags – file name, title of song, singer, name of album, and track number in disk. Therefore, the classification system and the detailed descriptor of Korean language educational materials should provide some ID tags for P2P service.

To determine ID tags of Korean language educational materials, we have to investigate how the educational materials are stored in other users' personal computers. Fig. 4 shows an example how the Korean language teachers store the educational materials. As shown in the Fig. 4, only the names and types of the files are known, and it is difficult to find out proper educational materials from above

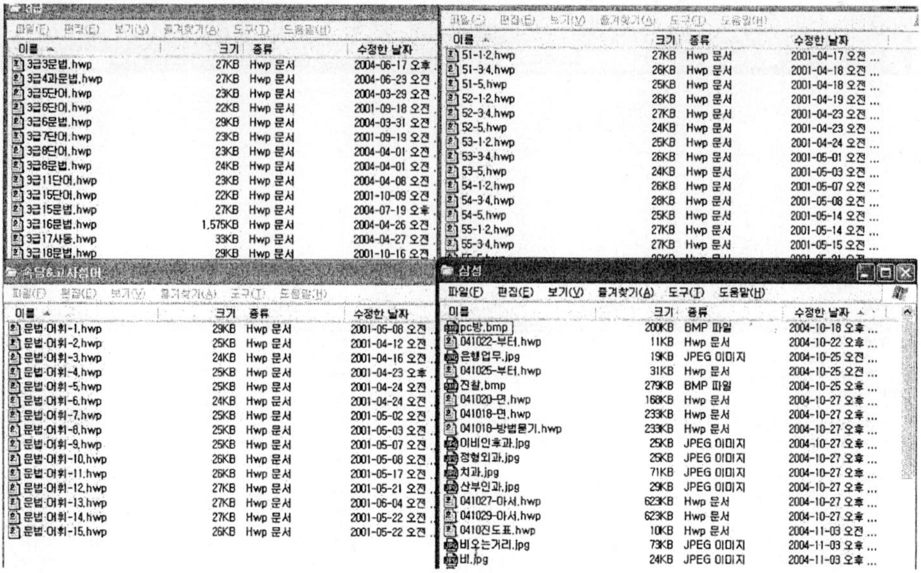

Fig. 4. Examples of Korean language educational material stored in several users' personal computers

information. Consequently, ID tags should be carefully determined for effective searching and exchanging.

For effective search and classification of Korean language educational materials, all instruction-educational materials should have the following ID tags.

(1) Level: Korean Proficiency Test (KPT)-based level 1 to level 6
(2) Textbook: Name of textbook and its publisher
(3) Theme: Name of the educational material
(4) Description: Description of the educational material
(5) Grammar: Grammar to teach in the educational material
(6) Vocabulary: Vocabularies to teach in the educational material
(7) Category: Speaking, listening, writing, reading, grammar, and vocabulary
(8) File Type: Text, sound, image, animation, and video
(9) Related Material: Location of related educational materials
(10) Evaluation: Evaluation score, hit count, and download count
(11) Registrant: Registrant of the educational material
(12) File Information: Information of included files

Fig. 5 is an example of Korean language educational material, and its ID tags are shown in Table 4.

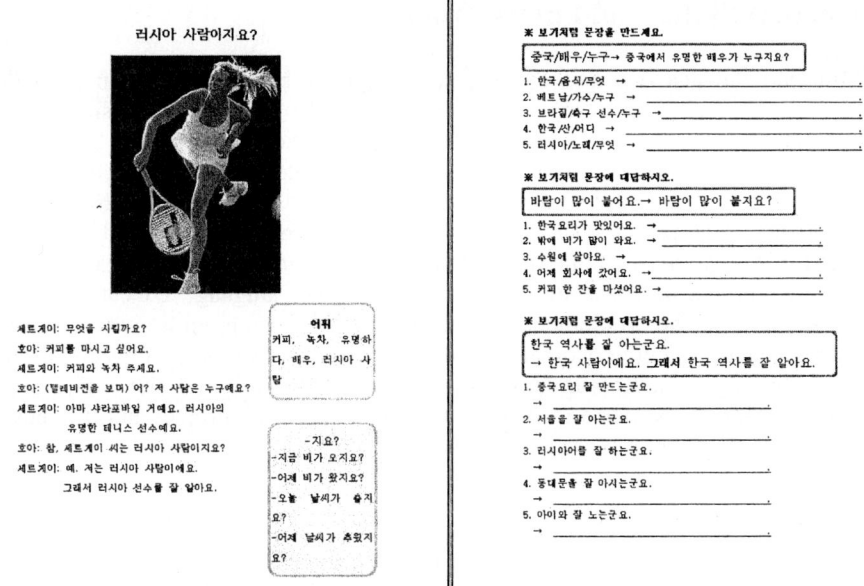

Fig. 5. Example of Korean language educational material

Table 4. ID tags of Korean language educational material in Fig. 5

Level	KPT Level 2	
Textbook	Arisu Korean 2/Lesson 30, Arisu Media Co. Ltd.	
Theme	Are you a Russian?	
Description	Conversation in a restaurant watching television	
Grammar	Are you ~ ?, I wish ~., Therefore ~.	
Vocabulary	Coffee, Green tea, Famous, Actor, Russian	
Category	Grammar, Speaking	
File Type	Text, Image	
Related Material	C:/Documents/admin/sujin/Korean/041218.hwp	
Evaluation	Score	★★★★☆
	Hit Count	1,042
	Download Count	969
Registrant	Sujin Cho	
File Information	Text: 041220.hwp(47kb)	
	Image: Sharapova.jpg(254kb)	
	A_A62053079A.wmv / 3885 kb	

4 Current Status and Future Work

We finished the analysis and experiments to gather basic information to build a Korean language education Internet portal site. We also finished the basic structures and contents of the directory services and the classification systems. A prototype Korean language education Internet portal site was implemented in Apache, php, and MySQL. 7 main directories and 36 sub-directories with about 200 links are implemented in the prototype Korean language education Internet portal site. As for ID tags for P2P services, about 100 contents were transformed into ID tags and they were stored in the database of the prototype Korean language education Internet portal site. Now we are implementing prototype P2P service program with ID tags for searching and exchanging Korean language education materials between different users. To build perfect Korean language education Internet portal site with effective P2P services should be performed in a nation-wide research project and it is beyond the scope of our current research.

References

1. Milgram, S.: Behavioral study of obedience, Journal of Abnormal and Social Psychology 67 (1963) 371-378
2. Albert, R., Jeong, H., Barabasi, A.: The Diameter of the World Wide Web, Nature 401 (1999) 130-131
3. http://kr.yahoo.com
4. http://www.naver.com
5. http://www.empas.com
6. http://www.google.co.kr
7. http://www.paran.com
8. http://www.daum.net
9. http://www.webkorean.org
10. http://www.edunet4u.net
11. http://www.edonkey2000.com
12. http://www.soribada.com

Information-Theoretic Selection of Classifiers for Building Multiple Classifier Systems*

Hee-Joong Kang[1] and MoonWon Choo[2]

[1] Department of Computer Engineering, Hansung University,
389 Samsun-dong 3-ga, Sungbuk-gu, Seoul, Korea
hjkang@hansung.ac.kr
[2] Division of Multimedia, Sungkyul University,
147-2, Anyang-8 Dong, Manan-Gu, Anyang-City, Kyunggi Province, Korea
mchoo@sungkyul.edu

Abstract. Only a few studies have investigated on how to select component classifiers from a classifier pool. But, the performance of multiple classifier systems depends on the component classifiers as well as the combination methods. A couple of information-theoretic methods selecting the component classifiers by considering the relationship among classifiers are proposed in this paper. These methods are applied to the classifier pool and examine the possible classifier sets for building the multiple classifier systems. A classifier set is selected as a candidate and evaluated with the other classifier sets on the recognition of unconstrained handwritten numerals.

1 Introduction

Improved performance by combining multiple classifiers has been shown in a multiple classifier system for more than a decade [1,2]. The performance of a multiple classifier system depends on the component classifiers as well as the combination methods. But, only a few studies have investigated on how to select the component classifiers from a classifier pool [3]. Thus, the selection of component classifiers, how to select them, or how many to select remain important research issues. For example, Woods et al. [2] showed the reason why a strategy should be devised when selecting the mix of classifiers, because they observed that in some cases, fewer classifiers provided superior results to more classifiers. Kang and Lee reported some strategies for selecting the multiple classifiers [3].

In this paper, four information-theoretic methods are reviewed and proposed for building a multiple classifier system. It is assumed that the number of selected component classifiers is fixed in advance, in order to alleviate the selection problem of classifiers. Simple criteria are to select the component classifiers according to the ranks of their forced recognition rate or reliability rate up to the fixed number. Information-theoretic criteria are based on the measure of closeness in [4,5] or the conditional entropy in [6,7] which considers the relationship

---

* This research was financially supported by Hansung University in the year of 2005.

among classifiers, or the minimization of mutual information (mMI) or the maximization of mutual information (MMI) between classifiers. The mMI criterion is proposed to select the component classifiers as complementary to each other as possible. The MMI criterion is proposed to select the component classifiers as highly correlated to each other as possible.

These four information-theoretic criteria for selecting the component classifiers are evaluated together with two simple selection criteria on the recognition of unconstrained handwritten numerals from the Concordia University [8] and University of California, Irvine (UCI) [9] repositories. The selection criteria are applied to the classifier pool and we examine the possible classifier sets, and select one of the classifier sets as the candidate for building a multiple classifier system (MCS). The MCS candidates are evaluated by using the combination methods in [4,6] together with the other classifier sets in the experiments.

The remainder of this paper is organized as follows. Section 2 explains the selection criteria of classifiers. Experimental results for evaluating the selection criteria are provided in Section 3 and a discussion is given in Section 4.

2 Information-Theoretic Selection Criteria

Two simple selection criteria are briefly introduced at first. One is the forced recognition rate (FRR) criterion and the other is the reliability rate (RR) criterion. The FRR criterion evaluates the classifier forcing a decision for every input, and not allowing rejections. The RR criterion considers the accuracy of all non-rejected decisions. And four information-theoretic criteria are explained by considering the first- and second-order dependencies among classifiers. These dependencies enable us to optimally approximate the high order probability distributions with the product of low distributions for Bayesian decision combination methods as in [4,6]. The first is a measure of closeness (MC) criterion [5], and the second is a conditional entropy (CE) criterion, based on the conditional entropy minimization of upper bound of Bayes error rate [7]. The third is a minimization of mutual information (mMI) criterion proposed to select the component classifiers as complementary to each other as possible. The fourth is a maximization of mutual information (MMI) criterion proposed to select the component classifiers as highly correlated to each other as possible.

2.1 Measure of Closeness

The measure of closeness (MC) is used for obtaining the optimal approximations by minimizing the difference between a real distribution $P(C)$ and an approximate distribution $P_a(C)$ where a vector variable C represents both a label class and K classifiers' decisions where K is the number of classifiers. The measure of closeness, $I(P(C), P_a(C))$, is defined in the following expression:

$$I(P(C), P_a(C)) = \sum_c P(c) \log \frac{P(c)}{P_a(c)}. \tag{1}$$

When the dth-order dependency in the $(K+1)$st-order probability distribution of C is considered for the application of the measure of closeness, an approximate formula is defined by the following expression:

$$P_\mathrm{a}(C_1,\cdots,C_{K+1}) = \prod_{j=1}^{K+1} P(C_{n_j}|C_{n_{id(j)}},\cdots,C_{n_{i1(j)}}), \qquad (2)$$

$$(0 \le id(j),\cdots,i1(j) < j),$$

such that C_{n_j} is conditioned on all d terms from $C_{n_{i1(j)}}$ to $C_{n_{id(j)}}$, and where $(n_1,\cdots,n_K,n_{K+1})$ is an unknown permutation of integers $(1,\cdots,K,K+1)$ and where $P(C_{n_j}|C_0,C_{n_{i\cdot(j)}})$ is defined as $P(C_{n_j},C_{n_{i\cdot(j)}})$.

Given the order of dependency d and K classifiers, the optimal product approximation for each classifier set is found by the application of the approximate formula P_a of Eq. (2) to Eq. (1), as in the following expressions by dropping the subscript n of C_{n_j}:

$$I(P(C),P_\mathrm{a}(C)) = \sum_c P(c)\log\frac{P(c)}{P_\mathrm{a}(c)}$$

$$= \sum_c P(c)\log P(c) - \sum_{j=1}^{K+1}\sum_c P(c)\log P(C_j|C_{id(j)},\cdots,C_{i1(j)})$$

$$= -\sum_{j=1}^{K+1} M(C_j;C_{id(j)},\cdots,C_{i1(j)}) + \sum_{j=1}^{K+1} H(C_j) - H(C) \qquad (3)$$

$$H(C) = -\sum_c P(c)\log P(c)$$

$$M(C_j;C_{id(j)},\cdots,C_{i1(j)}) = \sum_c P(c)\log\frac{P(C_j|C_{id(j)},\cdots,C_{i1(j)})}{P(C_j)} \qquad (4)$$

From Eq. (3), minimizing $I(P(C),P_\mathrm{a}(C))$ is equivalent to maximizing $\sum_{j=1}^{K+1} M(C_j;C_{id(j)},\cdots,C_{i1(j)})$ which is the total sum of dth-order mutual information. It is assumed that the larger the total sum of the dth-order mutual information is, the better its associated classifier set. Thus, the MC criterion finds an optimal product approximation relevant to each classifier set by maximizing the total sum of mutual information and then selects as a MCS candidate one classifier set having the largest total sum of mutual information.

2.2 Conditional Entropy

The conditional entropy (CE) relevant to the Bayes error rate is also applied to obtaining the optimal approximations by minimizing the conditional entropy $H(M|E)$ composed of a label class M and a vector variable E of K classifiers' decisions. The Bayes error rate P_e is defined in the following expression by introducing the C-D(Class-Decisions) mutual information $U(M;E)$ as in [6]:

$$P_e \le \frac{1}{2}H(M|E) = \frac{1}{2}(H(M) - U(M;E)) \qquad (5)$$

$$U(M;E) = \sum_m \sum_e P(m,e) \log \frac{P(m,e)}{P(m)P(e)}. \tag{6}$$

When dth-order dependency in the probability distribution of M and E is considered for the application of the minimization of conditional entropy, two approximate formulae are defined by the following expressions, as we consider dependencies among classifiers:

$$P_a(E_1, \cdots, E_K, M) = \prod_{j=1}^K P(E_{n_j}|E_{n_{id(j)}}, \cdots, E_{n_{i1(j)}}, M), \tag{7}$$

$$P_a(E_1, \cdots, E_K) = \prod_{j=1}^K P(E_{n_j}|E_{n_{id(j)}}, \cdots, E_{n_{i1(j)}}), \tag{8}$$

$$(0 \leq id(j), \cdots, i1(j) < j),$$

such that E_{n_j} is conditioned on all d terms from $E_{n_{i1(j)}}$ to $E_{n_{id(j)}}$, and where $(n_1, \cdots, n_K)$ is an unknown permutation of integers $(1, \cdots, K)$. $P(E_{n_j}|E_0, E_0, M)$ is $P(E_{n_j}, M)$, $P(E_{n_j}|E_0, E_{n_{i \cdot (j)}}, M)$ is $P(E_{n_j}|E_{n_{i \cdot (j)}}, M)$, and $P(E_{n_j}|E_0, E_{n_{i \cdot (j)}})$ is $P(E_{n_j}, E_{n_{i \cdot (j)}})$, by definition.

Given the order of dependency d and K classifiers, the optimal product approximation for each classifier set is found by the application of the approximate formulae P_a of Eqs. (7) and (8) to the C-D mutual information, as in the following expressions by dropping the subscript n of E_{n_j}:

$$U(M;E) = \sum_e \sum_m P(e,m) \log \frac{P(e|m)}{P(e)}$$

$$= \sum_{e,m} P(e,m) \log[\frac{1}{P(m)} \prod_{j=1}^K P(E_j|E_{id(j)}, \cdots, E_{i1(j)}, m)]$$

$$- \sum_e P(e) \log \prod_{j=1}^K P(E_j|E_{id(j)}, \cdots, E_{i1(j)})$$

$$= H(M) + \sum_{j=1}^K [D(E_j; E_{id(j)}, \cdots, E_{i1(j)}, m) - D(E_j; E_{id(j)}, \cdots, E_{i1(j)})] \tag{9}$$

$$D(E_j; E_{id(j)}, \cdots, E_{i1(j)}, m) = \sum_{e,m} P(e,m) \log \frac{P(E_j|E_{id(j)}, \cdots, m)}{P(E_j)}$$

$$D(E_j; E_{id(j)}, \cdots, E_{i1(j)}) = \sum_e P(e) \log \frac{P(E_j|E_{id(j)}, \cdots, E_{i1(j)})}{P(E_j)}$$

$$\Delta D(E_j; E_{id(j)}, \cdots, E_{i1(j)}) =$$
$$D(E_j; E_{id(j)}, \cdots, E_{i1(j)}, m) - D(E_j; E_{id(j)}, \cdots, E_{i1(j)}) \tag{10}$$

From Eq. (9), maximizing $U(M;E)$ is equivalent to maximizing $\sum_{j=1}^K \Delta D(E_j; E_{id(j)}, \cdots, E_{i1(j)})$ which is the total sum of Δ dth-order mutual information, since the remaining term is constant. It is assumed that the larger the total sum of Δ dth-order mutual information is, the better its associated classifier set. Thus, the CE criterion finds an optimal product approximation relevant to each classifier set by maximizing the total sum of Δ mutual information and then selects as a MCS candidate one classifier set having the largest total sum of Δ mutual information.

2.3 Minimization of Mutual Information

The minimization of mutual information (mMI) criterion is proposed to select the component classifiers in a pool as complementary to each other as possible. The mutual information is used to relatively measure such complementarity between classifiers. It is assumed that the higher the mutual information is, the higher the complementarity. The mMI criterion is to select classifiers in the pool and is to put them into the classifier set of multiple classifier system up to the number of classifiers. Initially, a classifier set S is empty, and the mutual information between every classifier and a label class set, and the mutual information between classifiers are computed respectively. A procedure to find the classifier set as a MCS candidate is as follows:

1. For each computed mutual information, find a classifier having the maximum mutual information in a pool to the label class set and then put the classifier into the classifier set.
2. In order to find a classifier in a pool as complementary to classifiers in the classifier set as possible, and find a classifier having minimum mutual information in a pool with respect to a classifier in the classifier set, and then put the classifier into the classifier set.
3. Until the number of classifier in the classifier set meet the fixed number of classifiers, repeat the step 2 and then final classifier set will be found.

2.4 Maximization of Mutual Information

The maximization of mutual information (MMI) criterion is proposed to select the component classifiers in a pool as highly correlated to each other as possible. The mutual information is also used to relatively measure such correlation between classifiers. It is assumed that the higher the mutual information is, the higher the correlation. The MMI criterion is to select classifiers in the pool and is to put them into the classifier set of multiple classifier system up to the number of classifiers. Initially, a classifier set S is empty, and the mutual information between every classifier and a label class set, and the mutual information between classifiers are computed respectively. A procedure to find the classifier set as a MCS candidate is as follows:

1. For each computed mutual information, find a classifier having the maximum mutual information in a pool to the label class set and then put the classifier into the classifier set.

2. In order to find a classifier in a pool as highly correlated to classifiers in the classifier set as possible, and find a classifier having maximum mutual information in a pool with respect to a classifier in the classifier set, and then put the classifier into the classifier set.
3. Until the number of classifier in the classifier set meet the fixed number of classifiers, repeat the step 2 and then final classifier set will be found.

3 Experimental Results

A number of multiple classifier systems (MCSs) built from the pool of six classifiers, $E1$, $E2$, $E3$, $E4$, $E5$, $E6$, will be evaluated in this section. These classifiers are developed by using the features or structural knowledge of numerals such as bounding box, centroid, and the width of horizontal runs, at KAIST and Chonbuk National Universities. The used handwritten numeral database is a fairly representative collection of digits. The UCI data sets in [9] are used for optical recognition of handwritten digits and consist of three training data sets tra, cv, $wdep$ and one test data set $windep$. The Concordia data sets in [8] consist of two training data sets A, B and one test data set T.

The performance of individual classifiers on test data sets is shown in terms of recognition and reliability rates in Figure 1. We note that classifiers $E4$ and $E5$ were built by using the structural knowledge obtained from the numerals of the Concordia University source, they are not as good on the numerals from UCI. The *reject* results of a classifier were used in the MC criterion.

Table 1. Overview of individual classifiers

	architecture	classifier	distance function	reference
$E1$	singular	neural network	pixel distance function	[10]
$E2$	modular	neural network	directional distance distribution	[10]
$E3$	singular	neural network	mesh feature	[10]
$E4$	modular	rule-based	modified structural knowledge	[11]
$E5$	modular	rule-based	structural knowledge	[11]
$E6$	singular	neural network	contour feature	[12]

Each neural network based classifier was trained with the training data sets A and tra. The optimal product sets were found by using the two data sets A, B and the three data sets tra, cv, $wdep$. The selection criteria were applied to the possible classifier sets and then we selected the most successful classifier set among them for a fixed number of classifiers. To denote the information-theoretic criteria according to the order of dependency, we use the abbreviations as shown in Table 2. All the MCSs were evaluated by the following combination methods on the test data sets: voting, Borda count, Bayesian combination methods abbreviated as in Table 3. The Bayesian methods are described in [1,4,6].

From the possible 20 MCSs consisting of three classifiers for each data set, the classifier sets by the selection criteria are shown in Table 4. Figures 2 and

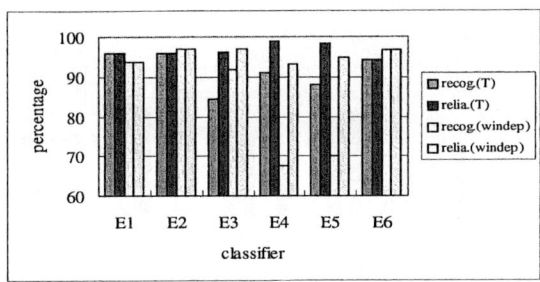

Fig. 1. Results of individual classifiers on test data sets: T, *windep*

Table 2. Selection criteria

notation	criterion	dependency
$MC1$	MC	first-order dependency
$CMC1$	MC	conditional first-order dependency
$MC2$	MC	second-order dependency
$CE1$	CE	first-order dependency
$CE2$	CE	second-order dependency

Table 3. Bayesian combination methods

method	meaning
$CIAB$	Conditional Independence Assumption based Bayesian
$ODB1$	first-Order Dependency based Bayesian
$CODB1$	Conditional first-Order Dependency based Bayesian
$ODB2$	second-Order Dependency based Bayesian
$DODB1$	Δ first-Order Dependency based Bayesian
$DODB2$	Δ second-Order Dependency based Bayesian

Table 4. MCS of three classifiers

data set	selection criterion	classifiers
Concordia	FRR	$E1,E2,E6$
	RR	$E3,E4,E6$
	$MC1,CMC1,MC2$	$E1,E4,E6$
	$CE1,CE2$	$E2,E4,E6$
	mMI	$E1,E3,E5$
	MMI	$E1,E4,E5$
UCI	$FRR,RR,MC1,CMC1,MC2,MMI$	$E2,E3,E6$
	$CE1,CE2$	$E2,E4,E6$
	mMI	$E1,E2,E4$

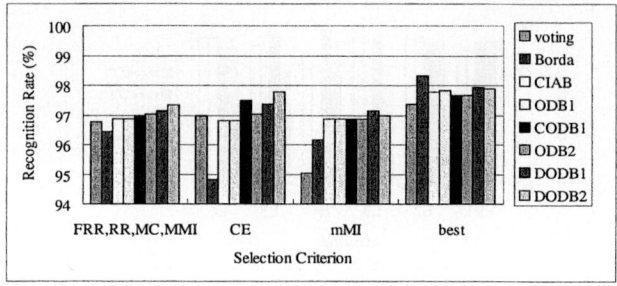

Fig. 2. Results of three classifier MCS on Concordia data set: T

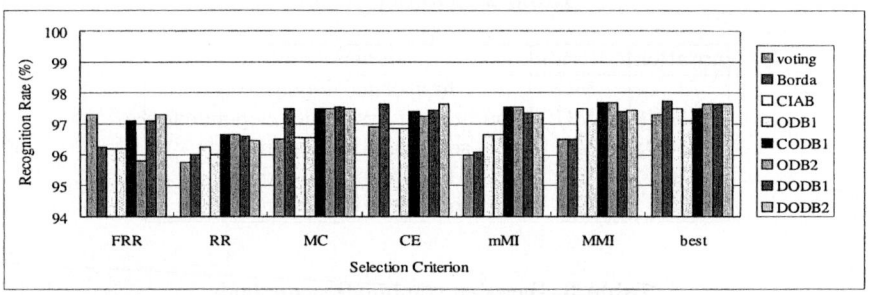

Fig. 3. Results of three classifier MCS on UCI data set: *windep*

Table 5. MCS of four classifiers

data set	selection criterion	classifiers
Concordia	RR	$E1,E3,E4,E5$
	$MC1,CMC1,MC2,MMI$	$E1,E4,E5,E6$
	$FRR,CE1,CE2$	$E1,E2,E4,E6$
	mMI	$E1,E2,E3,E5$
UCI	$FRR,RR,MC1,CMC1,MC2,MMI$	$E1,E2,E3,E6$
	$CE1$	$E2,E3,E4,E6$
	$CE2$	$E3,E4,E5,E6$
	mMI	$E1,E2,E3,E4$

3 show the results of the selected classifier sets in terms of recognition rates together with the highest rates for the combination methods shown with the item *best*. In this case, the CE criterion was better than the other criteria in most combinations, except the Borda count method using the numerals of UCI.

As for four classifiers, 15 MCSs for each data set were examined, and the selected classifier sets were evaluated in terms of recognition rates, as shown in Table 5 and Figures 4 and 5. Although the MCSs by the CE criterion showed worse results than those by the other criteria when they used ODB1, CODB1,

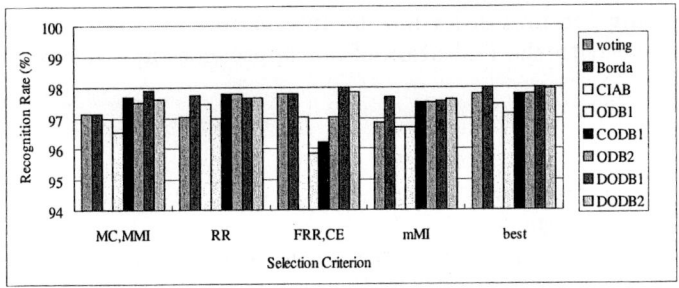

Fig. 4. Results of four classifier MCS on Concordia data set: T

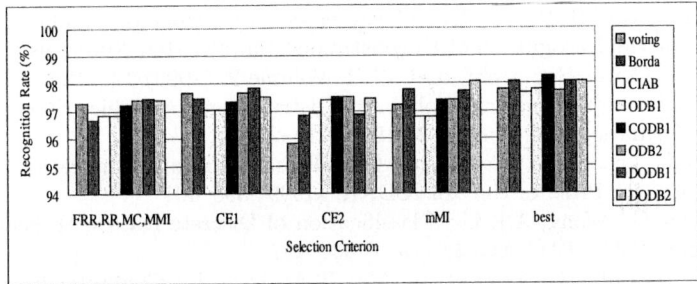

Fig. 5. Results of four classifier MCS on UCI data set: *windep*

or ODB2 method for the numerals of Concordia, the MCSs by the *CE1* criterion showed better results than those by the other criteria in most combinations. Particularly, the mMI criterion showed the best result using DODB2 method as for the numerals of UCI.

From the results, the CE criterion were useful in selecting the most promising classifier sets from the pool of classifiers for building a MCS, although the MCS candidates by the CE criterion did not necessarily coincide with the best. The mMI criterion proposed for the complementarity was not good at this time. The MMI criterion works similarly to the MC criterion. The selection criteria based on information theory would be one of the promising clues when Bayesian combination methods are considered.

4 Discussion

Although the selection criteria based on information theory showed positive evidence and their utility was supported through the recognition experiments, further studies should be needed in that the MCS candidates do not guarantee the best recognition, and the limit lies with the fixed number of classifiers except the mMI and MMI criteria. As for the mMI and MMI criteria, there is a still room to deal with higher order dependency among classifiers, because current version uses only the first-order mutual information between classifiers. It will be useful to deal with the limitation of our approaches as a future work.

References

1. Kittler, J., Hatef, M., Duin, R.P.W., Matas, J.: On Combining Classifiers. IEEE TPAMI **20** (1998) 226–239
2. Woods, K., Kegelmeyer Jr., W.P., Bowyer, K.: Combinition of Multiple Classifiers Using Local Accuracy Estimates. IEEE TPAMI **19** (1997) 405–410
3. Kang, H.J., Lee, S.W.: Experimental Results on the Construction of Multiple Classifiers Recognizing Handwritten Numerals. In: Proc. of the 6th ICDAR. (2001) 1026–1030
4. Kang, H.J., Lee, S.W.: A Dependency-based Framework of Combining Multiple Experts for the Recognition of Unconstrained Handwritten Numerals. In: Proc. of 1999 IEEE Comp. Soc. Conf. on CVPR. Volume 2. (1999) 124–129
5. Lewis, P.M.: Approximating Probability Distributions to Reduce Storage Requirement. Information and Control **2** (1959) 214–225
6. Kang, H.J., Lee, S.W.: Combining Classifiers based on Minimization of a Bayes Error Rate. In: Proc. of the 5th ICDAR. (1999) 398–401
7. Wang, D.C.C., Wong, A.K.C.: Classification of Discrete Data with Feature Space Transform. IEEE TAC **AC-24** (1979) 434–437
8. Suen, C.Y., Nadal, C., Legault, R., Mai, T.A., Lam, L.: Computer Recognition of Unconstrained Handwritten Numerals. Proc. of IEEE (1992) 1162–1180
9. Blake, C., Merz, C.: UCI repository of machine learning databases [http://www.ics.uci.edu/~mlearn/mlrepository.html]. Irvine, CA, Dept. of Infor. and Comp. Sciences (1998)
10. Oh, I.S., Suen, C.Y.: Distance features for neural network-based recognition of handwritten characters. IJDAR **1** (1998) 73–88
11. Oh, I.S., Lee, J.S., Hong, K.C., Choi, S.M.: Class-expert approach to unconstrained handwritten numeral recognition. In: Proc. of the 5th IWFHR. (1996) 35–40
12. Matsui, T., Tsutsumida, T., Srihari, S.N.: Combination of Stroke/Background Structure and Contour-direction Features in Handprinted Alphanumeric Recognition. In: Proc. of the 4th IWFHR. (1994) 87–96

Minimal RBF Networks by Gaussian Mixture Model

Sung Mahn Ahn[1,*] and Sung Baik[2]

[1] Kookmin University, Seoul 136-702, Korea
sahn@kookmin.ac.kr
[2] Sejong University, Seoul 143-747, Korea
sbaik@sejong.ac.kr

Abstract. Radial basis function (RBF) networks have been successfully applied to function interpolation and classification problems among others. In this paper, we propose a basis function optimization method using a mixture density model. We generalize the Gaussian radial basis functions to arbitrary covariance matrices, in order to fully utilize the Gaussian probability density function. We also try to achieve a parsimonious network topology by using a systematic procedure. According to experimental results, the proposed method achieved fairly comparable performance with smaller number of hidden layer nodes to the conventional approach in terms of correct classification rates.

1 Introduction

Radial basis function (RBF) networks have been successfully applied to function interpolation and classification problems among others. Girosi and Poggio [6] have shown that RBF networks possess the property of best approximation and Lowe [8] illustrated the usage of such networks for classification problems. According to Bishop [1], the typical topology of the network is such that it has one hidden layer consisting of a number of units, and the network output can be written as

$$y_k(x) = \sum_{j=1}^{M} w_{kj}\phi_j(x) + w_{k0}, \qquad (1)$$

where $\phi_j(x)$ is the basis function and w_{k0} is the bias. Usually $\phi_j(x)$ is the Gaussian with hyperspherical form of covariance matrix such as

$$\phi_j(x) = \exp\left(-\frac{\|x-\mu_j\|^2}{2\sigma_j^2}\right). \qquad (2)$$

---

[*] The author was supported by the research program of Kookmin University in 2004.

The widely accepted procedure for training RBF networks is a two-stage training, where, in the first stage, the basis function parameters are chosen by an unsupervised learning and then the basis functions are kept fixed while the second-layer weights are found in the second phase of training. There are several approaches to the first stage of training. Among them are K-nearest-neighbor [7], orthogonal least squares [2,3], and K-means clustering algorithm by Moody and Darken [9]. A more principled approach, however, is to think of the basis functions as the components of a mixture density model, where the parameters can be found by re-estimation procedures based on the EM algorithm [4]. And once the mixture model has been optimized, the components can be used as the basis functions by dropping the mixing coefficients of the model [1].

There also is an issue about having a parsimonious network topology. For instance, Musavi et al. [10] proposed an algorithm, which begins with a large number of nodes that are merged if possible, with the associated widths and locations of these nodes changed accordingly. Another approach by Yingwei et al. [14] achieves a minimal topology by pruning redundant hidden nodes.

In this paper, we propose a basis function optimization method using a mixture density model. We generalize the Gaussian radial basis functions to arbitrary covariance matrices, Σ_j, in order to fully utilize the Gaussian probability density function.

We also try to achieve a parsimonious network topology by using a more systematic procedure as opposed to heuristic approaches by Musavi el al. [10] and Yingwei et al. [14].

Our presentation is organized as this. In the following sections, we briefly describe the mixture density model and then explain the main idea of this paper, which is the basis function optimization method. Then are described some experimental results and finally some discussions.

2 Mixture Model and a Parsimonious RBF Network

In finite mixture models, we assume that the true but unknown density is of the form in equation (3), that g is known, and that the nonnegative mixing coefficients, π_j, sum to unity.

$$f(x;\pi,\mu,\Sigma) = \sum_{j=1}^{g} \pi_j N(x;\mu_j,\Sigma_j) \qquad (3)$$

Traditional maximum likelihood estimation leads to a set of normal equations which can only be solved using some type of iterative procedure. A convenient iterative method is the EM algorithm [4, 15], where it is assumed that each observation $x_i, i=1,2,\cdots,N$, is associated with an unobserved state $z_i, i=1,2,\cdots,N$, and z_i is the indicator vector of length g, $z_i = (z_{i1}, z_{i2}, \cdots, z_{ig})'$, and z_{ij} is 1 if and only if x_i is generated by density j and 0 otherwise. The joint distribution of x_i and z_i under Gaussian mixture assumption is [13]

$$f(x_i, z_i|\Theta) = \prod_{j=1}^{g} [\pi_j N(x_i; \mu_j, \Sigma_j)]^{z_{ij}}. \tag{4}$$

Therefore, the log likelihood for the complete-data is

$$\ell_c(\Theta) = \sum_{i=1}^{N} \sum_{j=1}^{g} z_{ij} \log[\pi_j N(x_i; \mu_j, \Sigma_j)] \tag{5}$$

Since z_{ij} is unknown, the complete-data log likelihood cannot be used directly. Thus we instead work with its expectation, that is, we apply the EM algorithm. Once the mixture model has been optimized, the components can be used as the basis functions of an RBF network by dropping the mixing coefficients of the model. There, however, is one more issue we need to address as was mentioned earlier. That is, we would like to have a network having a minimum number of components or basis functions. In the finite mixture model, the number of components is fixed. But in practice, it is not realistic to assume knowledge of the number of components ahead of time. Thus it is necessary to find the minimum number of components with given data. This is so called the problem of model selection, and the method of regularization [1, 6] or the penalized likelihood method [12] can give a solution to that. According to the regularization method we can define a new likelihood function of the form,

$$\tilde{\ell} = \ell + \lambda \Omega,$$

where ℓ is the original likelihood and Ω is a regularization term. The choice of Ω determines the resulting form of our model. When regularization method is applied to function interpolation, the most common form of a regularization term involves the assumption that the input-output mapping function is smooth. In our case, however, we would like to use it to control the number of components of the finite mixture model. A basic assumption we use here is that we begin with an overfitted model in terms of the number of components. That is, the model we have is already fitted to given data by a maximum likelihood method, but some of the components in the model are redundant and, thus, can be removed with the current likelihood level maintained. So Ω is chosen such that the solution resulting from the maximization of the new likelihood function, $\tilde{\ell}$, favors less complex models over the one that we begin with. The goal is achieved by adjusting the mixing coefficients, π, and the resulting form of regularization function is as follows.

$$\sum_{i=1}^{N} \sum_{j=1}^{g} z_{ij} \log[\pi_j N(x_i; \mu_j, \Sigma_j)] + \lambda N \sum_{j=1}^{g} (\alpha - 1) \log \pi_j \tag{6}$$

An intuitive explanation about how the regularization term in (6) works is this. The value of Ω increases as the values of π decrease. Thus $\pi = 0$ maximizes the regularization term. Since, however, we have to satisfy the constraint that the nonnegative mixing coefficients, π, sum to unity as well as to maximize the original likelihood, we expect some (not all) of π's drop to zero. It turns out that the proposed method removes unnecessary components under proper choices of λ and α.

We can find the optimal parameter values that maximize (6). Those are

$$\hat{\pi}_j = \frac{\frac{1}{N}\sum_{i=1}^{N} z_{ij} + \frac{1}{N}(\alpha_j - 1)}{1 + \frac{1}{N}\left(\sum_{j=1}^{g}\alpha_j - g\right)} \quad (7)$$

and

$$\hat{\mu}_j = \frac{\sum_{i=1}^{N} z_{ij} x_i}{\sum_{i=1}^{N} z_{ij}}$$

$$\hat{\Sigma}_j = \frac{\sum_{i=1}^{N} z_{ij}(x_i - \mu_j)(x_i - \mu_j)'}{\sum_{i=1}^{N} z_{ij}} \quad (8)$$

The component reduction algorithm is based on the modified MAP estimate. The algorithm is basically the EM algorithm and the expectation of z_{ij} is estimated using the formula in [15]. Difference from the usual EM algorithm comes when we check if the values of some π's are too small. We check it after each interation step, and if some π's are too small, the correspoding components are eliminated from the model. The current value of (6) can drop down temporarily when we eliminate those insignificant components, in which case we reset the current likelihood value so that the procedure continues with a model having smaller number of components.

3 Experimental Results

In this section, we are going to show a couple of applications of the proposed method to pattern classification problems. The first one is a toy problem just for demonstrating how the proposed method works. And the second example is an application to an aerial image segmentation problem.

3.1 A Two-Class Classification Problem

This example consists of two classes. Data is generated using 18 2-dimensional Gaussian components having equal covariances (2-by-2 identity matrix) and 17 of them have equal mixing coefficients (1/20), and the remaining one has a relatively large mixing coefficient (3/20) (The left half of Figure. 1). 4000 sample points were drawn from it, and the left half of Figure 1 shows the structure of the mixture distribution. The data generated from the 17 small Gaussians constitute one class, which accounts for about 85 percent of the sample and the rest constitutes the other class. An RBF network to solve this classification problem should have 2 input nodes, 19 hidden Gaussian units including a bias node, and one output node. The optimal weights connecting the hidden units and the output node can be determined by linear matrix inversion techniques [1]. The target output values being 0 for one class and 1 for the other, the network output values range between 0 and 1. By selecting proper threshold value, say between .2 and .3, the correct classification rate of the network can be above 97%.

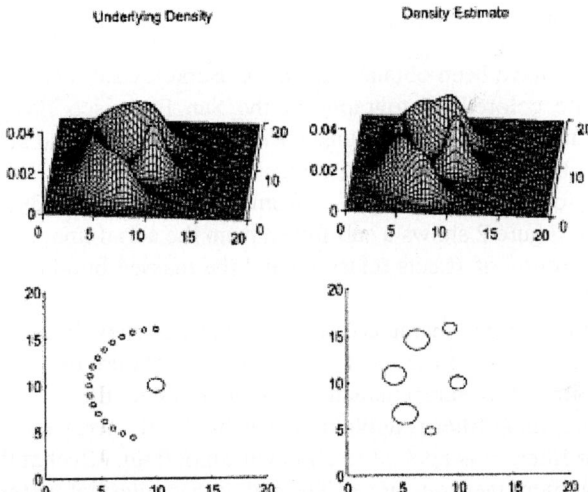

Fig. 1. True data distribution (left half) and estimated data distribution (right half)

Now let us take a look at the right half of Figure. 1 that shows the result of the proposed method explained in the previous section. The resulting mixture turned out to have 6 components, which we number 1 through 6 clockwise starting from the rightmost component. The estimated parameters are as follows.

1. $\pi = 0.15, \mu = \begin{bmatrix} 10.01 \\ 9.98 \end{bmatrix}, \Sigma = \begin{bmatrix} 0.96 & -0.03 \\ -0.03 & 0.83 \end{bmatrix}$

2. $\pi = 0.16, \mu = \begin{bmatrix} 6.88 \\ 4.93 \end{bmatrix}, \Sigma = \begin{bmatrix} 1.43 & -0.44 \\ -0.44 & 1.04 \end{bmatrix}$

3. $\pi = 0.14, \mu = \begin{bmatrix} 7.82 \\ 7.17 \end{bmatrix}, \Sigma = \begin{bmatrix} 0.78 & -0.07 \\ -0.07 & 1.52 \end{bmatrix}$

4. $\pi = 0.16, \mu = \begin{bmatrix} 8.87 \\ 15.69 \end{bmatrix}, \Sigma = \begin{bmatrix} 1.88 & 0.44 \\ 0.44 & 1.18 \end{bmatrix}$

5. $\pi = 0.22, \mu = \begin{bmatrix} 5.40 \\ 13.80 \end{bmatrix}, \Sigma = \begin{bmatrix} 1.87 & 1.25 \\ 1.25 & 2.36 \end{bmatrix}$

6. $\pi = 0.16, \mu = \begin{bmatrix} 4.10 \\ 10.17 \end{bmatrix}, \Sigma = \begin{bmatrix} 1.02 & 0.42 \\ 0.42 & 1.78 \end{bmatrix}$

The RBF network based on this mixture model should have 2 input nodes, 7 hidden Gaussian units including a bias node, and one output node. And the correct classification rate of the resulting network turned out to be as good as the previous network having 19 hidden units. It is, therefore, obvious to say that the second RBF network is better, since it performs as good as the first one and is parsimonious as well.

3.2 An Aerial Image Segmentation Problem

Experimental data have been obtained from UC Berkeley Library Web [16]. They are aerial black/white colored photographs of the San Francisco Bay area, California, where there are natural resource features such as forests, river and sea, and man-made features such as buildings, roads, and bridges. Some sub-images (280 by 280 pixels) can be selected for targets classification from an aerial image (1308 by 1536 pixels) of a certain area. Figure 2 shows a sub-image from the aerial images for the classification of the intersection of streets (class A) and the massed buildings (class B), which look like textures.

For extraction of target characteristics, the image is convoluted by Gabor spectral filtering [5] which is one of the most popular texture feature extraction methods. For efficient extraction of texture characteristics, additional filtering steps are required [11]. The 7x7 averaging filter is applied to estimate local energy response of the filter. Next, non-linear filtering is applied to eliminate smoothing effect at the borderline between distinctive homogeneous areas. The non-linear filter computes standard deviation over five small windows spread around a given pixel. The mean for the lowest deviation window is returned as the output. Values of each texture feature are subject to a normalization process to eliminate negative imbalances in feature distribution.

Gabor filters are useful to deal with the texture characterized by local frequency and orientation information. Gabor filters are obtained through a systematic mathematical approach. A Gabor function consists of a sinusoidal plane of particular frequency and orientation modulated by a two-dimensional Gaussian envelope. A two-dimensional Gabor filter is given by (9).

$$G(x, y) = \exp\left[\frac{1}{2}\left(\frac{x}{\sigma_x^2} + \frac{x}{\sigma_y^2}\right)\right] \cos\left(\frac{2\pi x}{n_0} + \alpha\right) \qquad (9)$$

By orienting the sinusoid at an angle and changing the frequency n_0, many Gabor filtering sets can be obtained. An example of a set of eight Gabor filters is decided with different parameter values (n_0 = 2.82 and 5.66 pixels/cycle and orientations α = 0°, 45°, 90°, and 135°). In this experiment, 4 filters are selected for target classification. Figure. 2 and 3 represent a sample sub-image and four featured images.

Fig. 2. A sub-image used in this experiment. Each white box represents each class.

For the target classification using RBF network, we generated 2000 4-dimensional training data points, 1000 for each class. By applying the proposed method, the resulting mixture model turned out to have 2 components, the parameters of which are as follows.

$$\pi = 0.58, \mu = \begin{bmatrix} 119.33 \\ 32.78 \\ 106.72 \\ 45.80 \end{bmatrix}, \Sigma = \begin{bmatrix} 395.1 & -548.14 & 451.8 & -355.47 \\ -548.14 & 1023.1 & -740.46 & 599.96 \\ 451.8 & -740.46 & 707.52 & -453.99 \\ -355.47 & 599.96 & -453.99 & 478.92 \end{bmatrix}$$

$$\pi = 0.42, \mu = \begin{bmatrix} 47.21 \\ 120.37 \\ 48.77 \\ 119.73 \end{bmatrix}, \Sigma = \begin{bmatrix} 304.71 & -27.23 & 84.01 & -21.14 \\ -27.23 & 135.51 & 27.80 & 142.72 \\ 84.01 & 27.80 & 165.51 & 36.63 \\ -21.14 & 142.72 & 36.63 & 164.59 \end{bmatrix}$$

As a result, the corresponding RBF network model has 4 input nodes, 3 hidden Gaussian units including a bias node, and the one output node. And the correct classi-

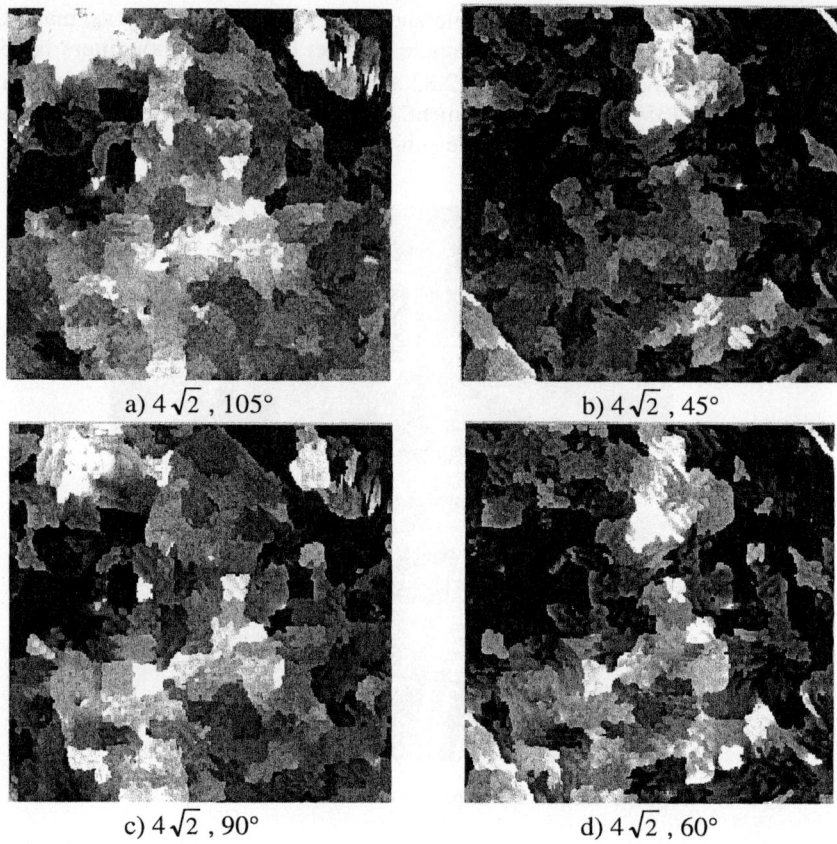

Fig. 3. Featured Images obtained by changing two parameters of Gabor filtering

fication rate of the network turned out to be 92.55%. We also implemented several RBF networks having different numbers of hidden Gaussian units, ranging from 20 to 30 and the resulting correct classification rates have ranged between 92% and 94%.

4 Conclusion

In this paper, we proposed a basis function optimization method using a mixture density model. The parameter estimation method using formulae such as (6) is sometimes called regularization method [1,6]. Or in another context, it is called maximum penalized likelihood method [12].

The proposed method generalizes the Gaussian radial basis functions to arbitrary covariance matrices, Σ_j, in order to fully utilize the Gaussian probability density function. The proposed method also achieves a parsimonious network topology by using a more systematic procedure as opposed to heuristic approaches by Musavi et al. [10] and Yingwei et al. [14]. The proposed method was tested in two problems. The first one is a toy problem just for demonstrating how the proposed method works. And the second example is an application to an aerial image segmentation problem.

According to experimental results, the proposed method achieved fairly comparable performance with smaller number of hidden layer nodes to the conventional approaches in terms of correct classification rates.

References

1. Bishop, C. M.: Neural Networks for Pattern Recognition. Oxford, (1995)
2. Chen, S., Billings, S. A., and Luo, W.: Orthogonal least squares methods and their application to non-linear system identification. International Journal of Control, 50(5), (1989) 1873-1896
3. Chen, S., Cowan, C.F.N., and Grant, P.M.: Orthogonal least squares learning algorithm for radial basis function networks. IEEE transactions on Neural Networks, 2(2) (1991) 302-309
4. Dempster, A. P., Laird, N. M., and Rubin, D. B.: Maximum Likelihood from Incomplete Data via the EM Algorithm. Journal of Royal Statistical Society(B), 39 (1977) 1-38
5. Farrokhnia, M. and Jain, A.: A multi-channel filtering approach to texture segmentation. Proceedings of IEEE Computer Vision and Pattern Recognition Conference, (1990) 346-370
6. Girosi, F. and Poggio, T.: Networks and the best approximation property. Biological Cybernetics, 63 (1990) 169-176
7. Kraaijveld, M. And Duin, R.: Generalization capabilities of minimal kernel-based networks. In proceedings of the International Joint Conference on Neural Networks, vol 1 (1991) 843-848
8. Lowe, D.: Radial basis function networks. In M. A. Arbib (Ed.), The Handbook of Brain Theory and Neural Networks. Cambridge, MA: MIT Press (1995)
9. Moody, J. and Darken., C.J.: Fast learning in networks of locally-tuned processing units. Neural Computation, 1(2) (1989) 281-294
10. Musavi, M., Ahmed, W., Chan, K., Faris, K., and Hummels, D.: On training of radial basis function classifiers. Neural Networks, 5 (1992) 595-603
11. Pachowicz, P.W.: A learning-based Semi-autonomous Evolution of Object Models for Adaptive Object Recognition. IEEE Trans. on Systems, Man, and Cybernetics, 24-8 (1994) 1191-1207
12. Silverman, B., W.: On the Estimation of a Probability Density Function by the Maximum Penalized Likelihood Method. The Annals of Statistics, 10 (1982) 795-810
13. Titterington, D. M., Smith, A. F. M., and Makov, U. E.: Statistical Analysis of Finite Mixture Distributions, Wiley (1985)
14. Yingwei, L., Sundararajan, N., and Saratchandran, P.: A sequential learning scheme for function approximation using minimal radial basis function neural networks. Neural Computation, 9 (1997) 461-478
15. Xu, L. and Jordan, M. I.: On Convergence Properties of the EM Algorithm for Gaussian Mixtures. Neural Computation, 8 (1996) 129-151
16. http://sunsite.berkeley.edu/AerialPhotos/vbzj.html#index.

Learning the Bias of a Classifier in a GA-Based Inductive Learning Environment*

Yeongjoon Kim and Chuleui Hong

Software School, Sangmyung University,
Seoul, Korea
{yjkim, hongch}@smu.ac.kr

Abstract. We have explored a meta-learning approach to improve the prediction accuracy of a classification system. In the meta-learning approach, a meta-classifier that learns the bias of a classifier is obtained so that it can evaluate the prediction made by the classifier for a given example and thereby improve the overall performance of a classification system. The paper discusses our meta-learning approach in details and presents some empirical results that show the improvement we can achieve with the meta-learning approach in a GA-based inductive learning environment.

1 Introduction

The concept of combining multiple classifiers into one classification system has become very popular [1,2,3,4,5]. The main purpose of creating a complex multi-classifier system is to obtain better classification performance than the performance offered by its components – individual base classifiers. Among those works of integrating multiple learned models, Doan et. al.[1] have explored a multistrategy learning approach that applies multiple learner modules to a given problem, then combines the predictions of modules using a meta-learner. Similarly, Fan et. al. [6] also have explored a meta-learning approach to combine several classifiers, each of which is computed by different algorithms, to improve the overall prediction accuracy. In their approach, several classifiers are learned first from the same training data set. Then, the next level classifier, a meta-classifier, is obtained from the predictions made by the first level classifiers on the examples in the training data set. When an instance is being classified, the first level classifiers make their predictions. The predictions are then presented to the meta-classifier, which makes a final prediction. Empirical results reveal that their meta-learning approach can improve the overall prediction accuracy of a classification system.

Inspired by their works, we have explored a meta-learning approach to enhance the classification performance of a classifier. However, our approach is different from

---

* This work was supported by Ministry of Education and Human Resources Development through Embedded Software Open Education Resource Center (ESC) at Sangmyung University.

theirs in the sense that we use a meta-learning approach to obtain a meta-classifier that can decide whether the prediction made by a classifier is correct or not. In our approach, the learning system learns the classification behavior of a classifier, yielding a meta-classifier. Then, in the classification process, the meta-classifier is used to decide whether the prediction made by the classifier for a given example is correct or not. If the meta-classifier considers the prediction made by the classifier as incorrect one, the example is classified again by another classifier which is trained with examples that are classified incorrectly by the first classifier.

The paper is organized as follows. Section 2 explains the meta-learning approach in details. Section 3 discusses briefly a GA-based inductive learning environment in which our meta-learning approach has been explored. Section 4 presents some empirical results obtained with the meta-learning approach. Finally, Section 5 concludesthe paper.

2 Learning the Bias of a Classifier

In the meta-learning approach, a classification system consists of three classifiers CF_1, MC, and CF_2. The classifier CF_1 is a classifier that learns concept descriptions from given training data set to perform regular classification task. The classifier MC is a meta-classifier. The meta-classifier MC learns the classification behavior (i.e., the bias) of CF_1 so that it can decide whether the prediction made by CF_1 is correct or not. The third classifier CF_2 is trained with the set of examples that are classified by MC as incorrectly classified examples by CF_1. The function of CF_2 is to classify examples again that are possibly misclassified by CF_1, giving a second chance to them.

In the meta-learning approach, the training process consists of three phases. In the first training phase, the classifier CF_1 is learned from given training data set. The training data set for the classifiers CF_1, denoted by TCF_1, is a regular training data set, in which each example is represented with a list of attribute values and the classification of the example (see Figure 1-(a)). In the next training phase, the meta-classifier MC is trained for the training data set, denoted by TMC, in which each element is represented with the attribute values of an example in TCF_1, the prediction made by CF_1 for the example, and the classification of the prediction of CF_1 which is either 0 or 1, where 0 represents incorrect decision and 1 represents correct decision (see Figure1-(b)). The training data set for the meta-classifier is obtained as follows:

Step 1. Obtain the classifier CF_1 from given training data set TCF_1.
Step 2. Classify the training examples in TCF_1 with CF_1. For the prediction of CF_1 for each training example, classify the prediction into one of two class, 0 or 1, depending on the correctness of the prediction of CF_1. 0 represents an incorrect decision and 1 represents a correct decision.
Step 3. Using training examples in TCF_1, construct the training data set TMC for the meta-learner as follows:
Element in TMC = (attribute values of an example in TCF_1, P_{CF1}, C_{PCF1})
where P_{CF1} is the prediction of CF_1 and C_{PCF1} is the classification of P_{CF1}.

(a) Example of Training Data Set TCF_1

$(A_1, A_2, A_3, A_4,$ Class $)$
Example 1 : (0.3, 0.0, 1.0, 0.67, 0)
Example 2 : (0.4, 0.1, 0.8, 0.77, 0)
Example 3 : (0.5, 0.3, 0.6, 0.47, 1)
Example 4 : (0.3, 0.2, 0.9, 0.67, 1)
Example 5 : (0.7, 0.1, 0.3, 0.47, 2)
Example 6: (0.9, 0.0, 0.0, 0.37, 2)

* In TCF_1, each example is represented with values for attribute A_1, A_2, A_3, A_4, and the class to which the example belongs

(b) Example of Training Data Set TMC

$(A_1, A_2, A_3, A_4,$ CF_1's ,   Correctness of)
                                Prediction CF_1's Prediction
Example 1 : (0.3, 0.0, 1.0, 0.67,   0,   1   )
Example 2 : (0.4, 0.1, 0.8, 0.77,   0,   1   )
Example 3 : (0.5, 0.3, 0.6, 0.47,   1,   1   )
Example 4 : (0.3, 0.2, 0.9, 0.67,   0,   0   )
Example 5 : (0.7, 0.1, 0.3, 0.47,   1,   0   )
Example 6: (0.9, 0.0, 0.0, 0.37,   2,   1   )

* CF_1 classifies Example 1, 2, 3, 6 correctly and Example 4, 5 incorrectly

(c) Example of Training Data Set TCF_2

$(A_1, A_2, A_3, A_4,$ CF_1's ,   Class)
                                Prediction
Example 4 : (0.3, 0.2, 0.9, 0.67,   0,   1   )
Example 5 : (0.7, 0.1, 0.3, 0.47,   1,   2   )

Fig. 1. Training data set example

Finally, in the third training phase, the classifier CF_2 is trained for the examples that MC classifies as incorrectly classified examples by CF_1. In the training data set for CF_2, denoted by TCF_2, each element is represented with the attribute values of an example in TCF_1, the prediction of CF_1, and the classification of the example (see Figure 1-(c)). The training data set for CF_2 is obtained as follows:

Step 1. Perform the classification task with classifier MC for the training data TMC.
Step 2. For the training examples classified by MC as incorrect one, construct the training data set TCF_2 for the classifier CF_2 as follows:

Element in TCF_2 = (attribute values of an example in TCF_1, P_{CF1}, C) where C is the classification of the example.

Figure 1 depicts the examples of training data set for each classifier and Figure 2 depicts the training phase of the meta-learning approach.

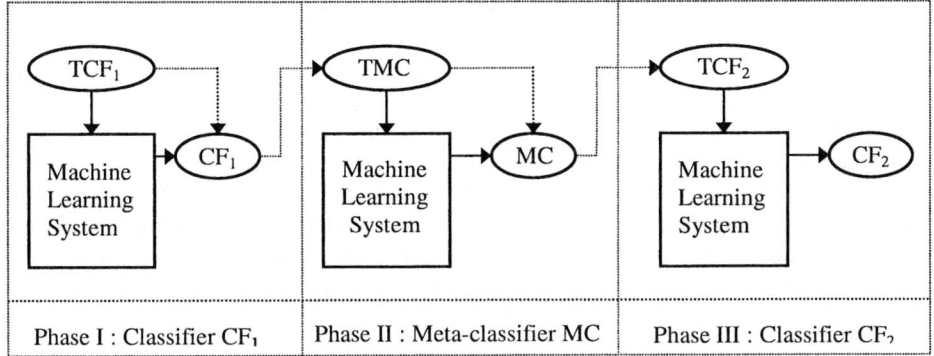

Fig. 2. Training phase of the meta-learning approach

In the meta-learning approach, the system classifies a given unknown example t_e as follows :

Step 1. CF_1 classifies the given example t_e which is represented with a list of attribute values

Step 2. MC decides whether the decision made by CF_1 is correct or not with input data for MC, which consists of attribute values of t_e and the prediction of CF_1.

Step 3. If MC classifies the prediction of CF_1 as correct, the system returns the prediction of CF_1 as its final prediction result. Otherwise attribute values of t_e alongwith the prediction of CF_1 is turned into CF_2 as input data

Step4. CF_2 classifies t_e again using given input data and the system selects the decision of CF_2 as its final decision.

To see how the system classifies a given example, let's assume that t_e is given by t_e=z (0.3, 0.1, 0.9, 0.77, 0) and that CF_1 classifies t_e as class 1. Then MC evaluates the decision made by CF_1 with input data (0.3, 0.1, 0.9, 0.77, 1), where the last attribute value is the prediction of CF_1. If MC classifies the decision of CF_1 as correct one, then the final classification result is class 1. However, if MC classifies the decision of CF_1 as incorrect one, t_e is turned into CF_2 along with the decision of CF_1 for further classification (i.e., (0.3, 0.1, 0.9, 0.77, 1) is turned into CF_2). In this case, the decision of CF_2 is considered as the final decision of the system.

3   Learning Bayesian Classification Rules with Genetic Algorithm

The objective of this section is to give a brief description of a GA-based inductive learning environment in which we have explored our meta-learning approach. A more detailed discussion of it has been given in [7].

From give examples, the system learns PROSPECTOR-style rules [8] that have the form:

If E then H with S =s; N = n

where S and N are odds-multipliers, measuring the *sufficiency* and *necessity* of E for H. In general, PROSPECTOR rules work with odds instead of probabilities, using the following conversion from probabilities to odds: O(H) = P(H)/(1-P(H)). That is, a probability of 0.75 is converted to odds of 3 (=0.75 / 0.25). If we have a rule, "If E then H with S=2.0;N=0.1", it expresses that the presence of E (P(E')=1; P(E') denotes the posterior probability of E) increases the prior odds of H, O(H), by a factor 2, whereas the absence of E (P(E')=0) decreases the prior odds of H by a factor of 0.1. In the case that P(E') is neither 0.0 (yielding an odds-multiplier of N) nor 1.0 (resulting in an odds-multiplier of S), it becomes necessary to interpolate. For example, if P(E') = 0.3, then an odds-multiplier between 0.1 and 2 has to be obtained; using a simple linear interpolation function as an example, it would be interpolated 0.7*N + 0.3*S = 0.7*0.1 + 0.3*2, yielding an odds-multiplier of 0.67 for the rule.

The system learns two kinds of rules from given examples: *is-high* rules and *is-close-to* rules. The syntax of *is-high* rule is

If is-high (A) then D with S=3; N=0.1

For a given example, the *is-high* rule produces an odds-multiplier between 3 and 0.1 based on the relative highness of the value for the attribute A of the given example to other examples. An example of *is-close-to* rule is

If is-close-to (A, a) then D with S = 4; N = 0.2

and it produces an odds-multiplier between 4 and 0.2 based on the closeness of the value for the attribute A to a certain constant *a*.

The posterior odds for a decision D, O(D'), is computed as follows:

If the following rules provide evidence for the decision D

(r_1) If E_1 then D with S=s_1;N=n_1

...

(r_m) If E_m then D with S=s_m ;N=n_m

then the posterior odds of D, O(D'), is computed as follows:

$$O(D')=O(D|E_1' \wedge ... \wedge E_m') = O(D) * \Pi^m_{i=1} \lambda_i$$

where λ_i = O(D|E_i') / O(D) is the odds-multiplier of the rule r_i.

Finally, we have to discuss how decisions are chosen by a rule-set that consists of learned rules. Assuming that we have decision candidates DC={D_1,...,D_n} with prior odds O(D_1),...,O(D_n), these odds are updated by firing rules of the rule-set, yielding posterior odds O(D_1'),...,O(D_n'); finally, the decision with the highest posterior odds is selected. If the decision chosen by the rule-set is the same as the classification of a given example, then we say that the rule-set classified the example correctly; otherwise, we say that it classified the example incorrectly.

A genetic algorithm approach is used to learn appropriate *is-high* rules and *is-close-to* rules from given examples. In the genetic algorithm approach, a population consists of a fixed number of rule-sets and rule-sets themselves are represented in chromosomal representation as ordered sequences of rules $r_1,...,r_n$. At the beginning of rule learning process, the first generation is generated randomly. In the following evolution process, next generation is obtained by applying genetic operators to the current population. 1-point crossover operators and mutation operators are used as genetic operators. The 1-point crossover operator creates two offspring by exchanging some rules of selected parents. The mutation operator selects a rule r from a rule-set and replaces it with a newly generated rule r'. Parents are selected based on their fitness, using the popular roulette wheel method. Fitness of a rule-set is evaluated by the percentage of examples the rule-set classifies correctly. During the evolution process, the system monitors average fitness and maximum fitness of the current generation. If none of them has been increased by a certain amount in a certain number of generations, the system terminates its learning process.

4 Empirical Results

We performed some experiments to evaluate the performance of the meta-learning approach in the GA-based learning environment discussed in Section 3. To evaluate the performance of the meta-learning approach, we used two data collections: glass data collection (GL) and soybean diseases data collection (SBD). The characteristics of each data collection are as follows:

- Glass data collection (GL): the data set contains 214 instances obtained from 6 different kinds (building-windows-float processed, building-windows-non-float processed, containers,...) of glass, in which each instance is represented with 9 numeric-valued attributes (Na, Fe, K, ...).
- Soybean diseases data collection (SBD): each of fifteen diseases that affect soybean plants is described by 35 attributes. The data set contains 290 instances.

For each data set, training / testing data set were generated by equally dividing the data set into two subsets. Then, for each training / testing data set, the classifier CF_1 was learned first. In the following steps, the meta-classifier MC and the classifier CF_2 were learned and the performance of the basic classifier (CF_1) and that of the meta-learning approach were evaluated. This whole process was repeated ten times.

The results of experiments are depicted in Table 1 and Table 2. In the tables, "Basic" denotes the performance of the basic classifier and "Meta" denotes the performance of the meta-learning approach. For the glass data collection, the meta-learning approach improves classification performance by more than 12.3% for training data set and by 1.2% for testing data set. For the soybean diseases data collection, the classification performance is increased quite significantly for both training and testing data set, improving the performance of the system by more than 24.7% and 18.1%, respectively.

Table 1. Comparison of Performance: Basic vs. Meta-learning

Data Set	Basic		Meta	
	Train	Test	Train	Test
GL	70.3	60.0	82.6	61.2
SBD	61.9	52.7	86.6	70.8

To find a proper explan ation why the meta-learning approach could only slightly improve the testing performance for the glass data collection, the performance of MC was evaluated for the two data collections. Table 2 summarizes the evaluation results. As we can see, the performance of MC for the glass data collection is much worse than that of MC for the soybean diseases data collection especially for the testing data set. Hence, the bad performance of the meta-learning approach in testing for the glass data collection can be explained by the bad performance of MC, which might be caused by bad performance of CF_1.

Table 2. Performance of the meta-classifier MC

Data Set	Train	Test
GL	81.5	63.7
SBD	89.4	79.5

Bagging and boosting are well-known ensemble learning methods [9,10,11,12,13]. They combine multiple learned base models with the aim of improving generalization performance. Similarly, we have explored an approach, called *multi-classifier learning approach*, to combine multiple classifiers in our GA-based learning environment. In the multi-classifier learning approach, the search process consists of two steps. In the first step, a certain number of different individual classifiers are learned for the same training data set. Then, in the second step, a search algorithm is applied to the classifiers obtained in the first step to find the subset of classifiers that provides the best performance. After finding a set of classifiers, a classification system is constructed using the classifiers in the set and decisions are made using classifiers relying on a decision-making scheme such as voting.

To see the effect of the meta-learning approach in the multi-classifier learning environment, we applied our multi-classifier learning approach to the basic system described in Section 3 and the meta-learning approach discussed in Section 2, and performance of classification systems were evaluated. Table 3 summarizes the results of experiments. In the table, "Ind" denotes the performance of individual classifier and "Mult" denotes the performance of a multi-classifier system. In the multi-classifier learning environment, the meta-learning approach improves the performance of the system further, increasing the performance of the system by 15.3% for the soybean diseases data collection.

Table 3. Comparison of Performance in Multi-classifier Learning Environment

Data Set	Basic				Meta			
	Ind		Mult		Ind		Mult	
	Train	Test	Train	Test	Train	Test	Train	Test
GL	70.3	60.0	80.4	65.8	82.6	61.2	92.7	66.7
SBD	61.9	52.7	76.1	68.2	86.6	70.8	94.5	83.5

5 Conclusions

A meta-learning approach has been explored. In the meta-learning approach, a meta-classifier learns the bias of a classifier so that it can evaluate the prediction made by the classifier for a given example. In the meta-learning approach, the prediction of the classifier for an unknown example is evaluated by the meta-classifier. If the meta-classifier classifies the prediction as correct one, the prediction of the classifier is accepted as the final decision of the classification system. Otherwise, the example is classified again by the second classifier which is trained with examples that are misclassified by the first classifier.

Experiments reveal that the meta-learning approach improves the performance of the learning system quite significantly. Even though, we have discussed our meta-learning approach in a GA-based inductive learning environment, we can easily apply our meta-learning approach to other inductive learning environments such as neural networks or decision trees learning environments.

In the meta-learning approach, the performance of the system depends on the performance of the meta-classifier. As a future work, it might be interesting to develop a system that can bias the first classifier in a certain direction so that the meta-classifier can achieve best performance. Moreover, further study should be performed to find a better way to construct the training data set for MC and CF_2.

References

1. Doan, A., Domingos, P., Halevy, A. : Learning to Match the Schemas of Data Sources : A Multistrategy Approach. Machine Learning, 50(3) (2003) 279-301
2. Major, R. L., Ragsdale, C. T.: An Aggregation Approach to the Classification Problem Using Multiple Prediction Experts. Information Processing and Management. 36 (2000) 683-696
3. Ishibuchi, H., Nakashima, T., Morisawa, T.: Voting in Fuzzy Rule-based Systems for Pattern Classification Problems. Fuzzy Sets and Systems (1999)
4. Giraud-Carrier, C., Vilalta, R., Brazdil, P.: Introduction to the Special Issue on Meta-Learning. Machine Learning, 54(3) (2004) 187-193
5. Esposito, F., Semerano, G., Fanizzi, N., Ferilli, S. : Multistrategy Theory Revision: Induction and Abduction in INTHELEX. Machine Learning, 38 (2000) 133-156
6. Fan, D. W., Chan, P. K., Stolfo, S. J. : A Comparative Evaluation of Combiner and Stacked Generalization. IMLM-96.(1996) 40-46

7. Eick, C. F., Kim, Y-J., Secomandi, N., Toto, E. : DELVAUX - An Environment that Learns Bayesian Rule-sets with Genetic Algorithms. The Third World Congress on Expert Systems (1996)
8. Duda, R., Hart, P., Nilsson, J.:Subjective Bayesian methods for rule-based inference systems. in Proc. National Computer Conference (1976) 1075-1082
9. Meir, R., Rätsch, G. : An Introduction to Boosting and Leveraging. Advanced Lectures on Machine Learning, LNCS (2003) 119-184
10. Rätsch, G., Warmuth, M. K. :. Maximizing the Margin with Boosting. Annual Conference on Computational Learning Theory, LNAI 2375 (2002) 334-350
11. Dietterich, T. G. : An Experimental Comparison of Three Methods for Constructing Ensembles of Decision Trees : Bagging, boosting, and randomization. Machine Learning, 40(2) (1999) 139-157
12. Lugosi, G., Vayatis, N. :A Consistent Strategy for Boosting Algorithms. Annual Conference on Computational Learning Theory, LNAI 2375 (2002) 303-318
13. El-Yaniv R., Derbeko, P., Meir, R. : Variance Optimized Bagging. 13th European Conference on Machine Learning (2002)

GA-Based Resource-Constrained Project Scheduling with the Objective of Minimizing Activities' Cost

Zhenyuan Liu and Hongwei Wang

Institute of Systems Engineering, Huazhong University of Science and Technology,
Wuhan, Hubei, 430074, China

Abstract. Resource-Constrained Project Scheduling Problem(RCPSP) with the Objective of Minimizing Activities' Cost is one of the critical sub-problems in partner selection of construction supply chain management. In this paper, this type of RCPSP is mathematically modelled firstly. The analysis on the characteristic of the problem shows that the objective function is non-regular and the problem is NP-complete. The basic idea for the solution of the problem is clarified. A genetic algorithm is developed and the parameters of the algorithm are analyzed based on the tests of an example. The proposed GA is demonstrated to be effective based on the results of a computational study on the updated PSPLIB.

1 Introduction

Partner selection is an important problem in supply chain management. When we design a construction supply chain in which general contractor is the kernel entity, the project scheduling general contractor's will be constrained by the capacities of the renewable resources supplied by the partners such as subcontractor, ready-mix concrete vendors. We should consider how to get the least activities' cost in the project with the constraints of due date and resource capacities of partners. Thereinto, activities' cost consists of static cost and completed activity holding cost that is dynamic and varies with the changing of the finish time of the activity and project makespan. In other words, a type of RCPSP with the objective of minimizing activities' cost that is a critical sub-problem in partner selection of construction supply chain management should be solved.

RCPSP is an important problem in the research of project management and has got much attention from academics. The performance of RCPSPs measures include time objectives (i.e. minimizing makespan, mean tardiness, mean completion time) and cost objectives (i.e. maximizing net present value, minimizing total costs including the costs of resource consumption, overhead and tardiness penalties), and single objective is often considered [1]. The constraints of the problem usually comprise precedence relations between activities, renewable resource-constraints, non-renewable resource-constraints and doubly resource-constraints[1]. According to the theory of computing complexity, RCPSPs belongs to the class of NP-hard problems[2]. In the related literatures, the approaches to solve the RCPSPs can be classified to two categories which are known as optimal procedures and heuristic approaches. Optimal procedures are integer programming, implicit emulation with branch and bound as well as dynamic programming[1]. These procedures are feasible to solve the project scheduling problems with a small scale. When facing to the large-scale project scheduling, we generally fall back on the heuristic approaches in which meta-heuristics such as Genetic Algorithm, Simulated Annealing have been used[3-5]. We can find that there has been an abundant research on RCPSPs, but people gave little attention to the activities' cost in project. Smith and Dodin have ever considered the integration of materials ordering charge, materials holding cost and completed activities holding cost[6,7], but resource constraint has not been included in their work. In this paper a genetic algorithm is proposed to solve RCPSPs with the objective of minimizing activities' cost where every activity is executed in single mode and the project has renewable resource-constraints as a result of the limited capacity of the partners. Some examples will be given to illustrate the efficiency of the algorithm.

2 Problem Description

The classical RCPSPs can be stated as follows. We consider a single project which consists of $j = 1, 2, \cdots, J$ activities with a non-preemptive duration of d_j periods, respectively. Due to technological requirements, precedence relations between some of the activities enforce that an activity $j = 2, \cdots, J$ may not be started before all its immediate predecessors $i \in P_j$ (P_j is the set of immediate predecessors of activity j) have been finished. In the description of activity-on-node method, the structure of the project is a network where the nodes represent the activities and the arcs the precedence relations. The network is acyclic and numerically labelled, where an activity has always a higher label than all its predecessors. Without loss of generalization, we can assume that 1 is the only start activity and activity J is the only finish activity. K-types of renewable resources supplied by the partners will be consumed during the project. It is assumed that the project needs r_{jk} units of resource k to process activity j during every period of its duration. The static cost of activity j is v_j. Let F_j be the finish time of activity j and A_t the set of activities being in progress in period t.

We set h is the cost per period of holding completed activities stated as a percent of activity fixed cost. The capacity of resource k supplied by some partner candidate is noted by R_k and the unit price C_k. The due date of the project is H. With a given H, we can get the earliest finish time e_j and the latest finish time l_j of activity j by using CPM. The time parameters in the problem are all integer valued. V_t is the holding cost for completed activities of the project through period t. We use binary decision variables $X_{jt}, j = 1, \cdots, J, t = e_j, \cdots, l_j$.

$$X_{jt} = \begin{cases} 1, \text{if activity } j \text{ is completed at the end of period } t \\ 0, \text{otherwise} \end{cases}$$

The model of RCPSPs with the objective of minimizing activities' cost can be presented as follows:

$$\min \quad TV = \sum_{t=1}^{F_J} V_t + \sum_{j=1}^{J} v_j . \tag{1}$$

$$\text{s.t.} \quad \sum_{t=e_i}^{l_i} tX_{it} \leq \sum_{t=e_j}^{l_j} [(t-d_j)X_{jt}], \forall i \in P_j, j=1,\cdots,J . \tag{2}$$

$$\sum_{t=e_j}^{l_j} X_{jt} = 1, j=1,\cdots,J . \tag{3}$$

$$F_J = \sum_{t=e_J}^{l_J} tX_{Jt} \leq H . \tag{4}$$

$$\sum_{j \in A_t} r_{jk} \leq R_k, k=1,\cdots,K, t=1,\cdots,F_J . \tag{5}$$

$$v_j = v'_j + \sum_{k=1}^{K} C_k r_{jk} d_j, j=1,\cdots,J . \tag{6}$$

$$V_t = V_{t-1} + h \sum_{j=1}^{J} v_j X_{jt}, t=1,\cdots,F_J . \tag{7}$$

$$X_{jt} \in \{0,1\}, j=1,\cdots,J . \tag{8}$$

LAs formula (1), the objective is to minimize TV, the activities' total cost of the project which includes two items. The first item is the sum of holding cost of the completed activities from the start time of the project to its finish time, i.e., the dynamic cost related with the activities in the project. The second is the amount of the activities' static cost. Constraints (2) ensure the precedence relations can be satisfied. (3) indicates that the activity must be executed within the interval (e_j, l_j) and (4) means that the project can not be finished after the due date. (5) is the resource-constraints which guarantees that the per-period availabilities of the renewable resources are not violated. (6) gives the composition of the activities' static cost which consists of two parts, the first part, v'_j, is the part that is not related with the selection of partners, while the later is the cost for resources needed to complete the activity j. Finally, (7) illustrates how to get the holding cost for the completed activities through period t where $V_0 = 0$.

With a candidate for partner, the capacities and unit prices of its resources are decided, so v_j in (6) can be seen as a constant. Therefore, the later item in objective (1) is fixed no matter how to schedule the project. As a result, the focus will be placed on the dynamic cost of the activities. Moreover, (7) indicates that there is holding cost for each activity from the finish time of the activity to the finish time of the project. Based on these, the objective function can be changed to the following where only the dynamic cost is considered:

$$\min \quad TV' = h \sum_{j=1}^{J} v_j (\sum_{t=e_J}^{l_J} tX_{Jt} - \sum_{t=e_j}^{l_j} tX_{jt}) . \qquad (9)$$

The objective function (9) is non-regular According to the definition proposed by Sprecher[8]. In addition, the feasible solutions of the problem is the same as that of traditional RCPSPs with makespan objective so that this type of RCPSPs is also NP-hard. To seek the heuristics, we firstly consider the case where the resources are unconstrained. From the objective function (9), the later the finish time of the activity is, the less the holding cost after its completion. This results in a lower of the total activities' cost of the project. The best scheduling is that every activity starts at its latest start time. Correspondingly, when resource constraints are considered, we also hope each activity is completed as late as possible with the presupposition that the capacity of each resource is not violated.

3 Design of Genetic Algorithm

3.1 Encoding and Decoding

In the research of RCPSPs with makespan objective, activities chain with precedence relationship as chromosomes is usually used[3]. For example, Fig.1 is an activity network represented by activity-on-node. (1,2,3,4,5,6,7,8, 9,10) and (1,2,5,4,3,7,6,9,8,10) are two activities chains with precedence relationship. According to the model, we also use this type of activities chain to represent a solution. The decoding procedure is a serial scheduling schema as follows:

For $i = J$ to 1
Begin
$\pi R_k := R_k - \sum_{j \in A_t} r_{jk}, k = 1, \cdots, K, t = 1, \cdots, H$
$j^* := s(i)$
$LS_{j^*} := \min\{ST_j | j \in S_{j^*}\} - d_{j^*}$
$ST_{j^*} := \max\{t | ES_{j^*} \leq t \leq LS_{j^*}, r_{j^* k} \leq \pi R_{kr}, k = 1, \cdots, K, r = t, t+1, \cdots, t + d_{j^*} - 1\}$
$FT_{j^*} := ST_{j^*} + d_{j^*}$
End;

where s is an activities chain with precedence relationship, $s(j)$ is the activity in the position j of s, S_{j^*} is the immediate successors of j^* and $ST_{j^*}, FT_{j^*}, ES_{j^*}, LS_{j^*}$ are respectively the factual start time, the factual finish time, the earliest start time and the latest start time. πR_{kt} is the quantity of resource k unoccupied in period t. The procedure includes J stages and the time complexity can be measured as $O(J^2 KH + JH^2 K)$.

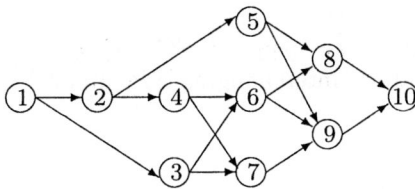

Fig. 1. An Example Activity Network

3.2 Initial Population

According to the schema of encoding, an initial chromosome can be created at random by using the following algorithm where E_n is the feasible activities whose successors are all listed in the chain. The procedure is called repeatedly to reach the population size.

$n := J, s(n) := J$
While $n>1$ Do Stage n
Begin
Computing $E_n := \{j | S_j \subseteq \{s(n), s(n+1), \cdots, s(J)\}, j = 1, 2, \cdots, J\}$
Choose at random: $j^* \in E_n$
$s(n-1) := j^*; n := n - 1$
End;

3.3 Fitness Function and Selection Scheme

The objective is to minimize the activities' cost so that a fitness measure is proposed according to the general principle of genetic algorithm $f(s) = TV_m - TV'(s)$ where TV_m is a very large constant, and $TV'(s)$ is the value of TV' after chromosome s has been decoded. Selection scheme is represented by stochastic tournament model with elitist preservation.

3.4 Crossover Operators

The operation of crossover must enable the created chromosomes also preserve precedence relationship. The following One-point Crossover Operator and Two-point Crossover Operator are designed in accord with the encoding, where SP_1 and SP_2 are the parent chains, SC_1 and SC_2 are the children.

1. One-Point Crossover
Generate an integer at random: $r \in (1, J)$.
for $j = J$ to r $SC_1(j) := SP_1(j)$;
for $j = J$ to 1 {if $SP_2(j) \notin SC_1$ then $r := r - 1; SC_1(r) := SP_2(j);$}
2. Two-Point Crossover
Generate two integers at random: r_1, r_2, s.t. $1 \leq r_1 < r_2 \leq J$.
for $j \in [1, r_1]$ and $j \in [r_2, J]$ $SC_1(j) := SP_1(j)$;

for $j = J$ to 1
{if $SP_2(j) \notin SC_1$ then $\quad r_1 := r_1 - 1; SC_1(r_1) := SP_2(j);$}
In the above two operators, interchanging SP_1 and SP_2 will produces another child.

3.5 Mutation Operators

Let s be an activities chain with precedence relationship where the activity in position j is selected to be mutated and $p(k)$ is the position of activity k in s. Therefore, we can get two parameters: LP, the largest position in s of the immediate predecessors of $s(j)$, and SS, the smallest position in s of the immediate successors of $s(j)$. The mutation procedure is as follows:

Choose at random: $j \in (1, J)$;
Defining: $LP := \max\{p(i)|i \in P_{s(j)}\}, SS := \min\{p(i)|i \in S_{s(j)}\}$
Choose at random: $k \in (LP, SS)$ and $k \neq j$;
$temp := s(j)$;
if $k < j$ {for $i = j - 1$ to $k \quad s(i+1) := s(i);$ }
else {for $i = j$ to $k - 1 \quad s(i) := s(i-1);$ }
$s(k) := temp$

4 Computational Study

4.1 A Simple Example

The project is described in Fig.1 where only one type of renewable resource is considered. The parameters of the activities are listed in Table 1 and the available amount of the resource is $R = 30$. Giving $h = 0.01$ and $H = 30$.

The computation is executed under different configurations of the parameters of the algorithm where we set total chromosome is 6000, the population size $PopSize = 150, 200, 300, 500$, the crossover probability $p_c = 0.4, 0.5, 0.6, 0.7$, the mutation probability $p_m = 0.05, 0.1, 0.15, 0.2$, the preservation scale $PreScale = 2, 3, 4$ and the tournament scale $TourScale = 2, 3, 4$. The results show that convergence is fine. The best result shown in Fig.2 is $TV' = 41.24$, which is consistent with the result by using AMPL and CPLEX. Selection of Crossover Operators, $PopSize, p_c$ and $PreScale$ has little impact on the efficiency of the algorithm. In contrast, the selection of p_m and $TourScale$ has more impact. The more p_m

Table 1. The Parameters of the Activities

Activity	1	2	3	4	5	6	7	8	9	10
d_j	0	1	5	3	2	6	5	3	5	0
r_{j1}	0	10	20	14	16	16	12	16	18	0
v_j	0	10	100	42	32	96	60	48	90	0

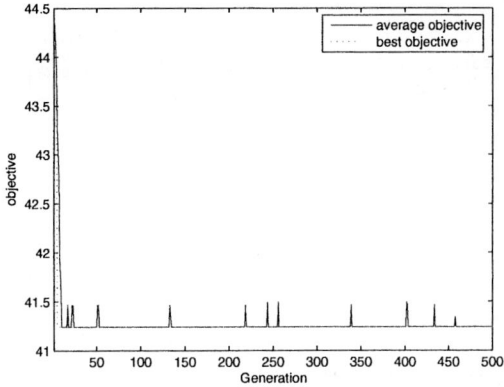

Fig. 2. The average objective and best objective in generations with $PopSize = 12$, $Generation = 500$, One-point Crossover, $p_c = 0.7$, $p_m = 0.1$, $PreScale = 2$, $TourScale = 2$

Table 2. The results of different heuristics

Heuristics	PAM	SSS	PSS	GA
Average Error	2.96%	2.32%	1.41%	0.33%
maximal Error	25.6%	29.1%	21.4%	7.1%

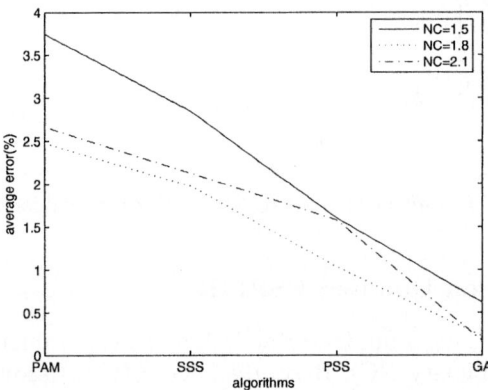

Fig. 3. The effect of network complexity on the algorithms

or $TourScale$ is, the better the effect of convergence. However, a small p_m is a disadvantage to the diversity of population and more time will be required in computing under a larger $TourScale$.

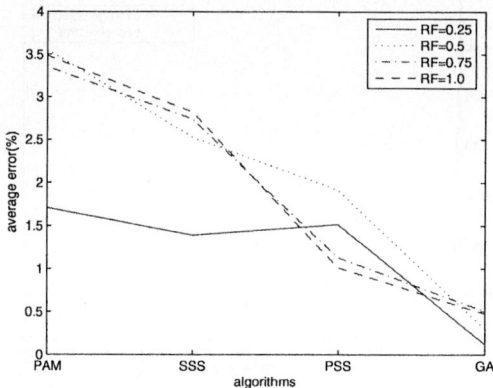

Fig. 4. The effect of resource factor on the algorithms

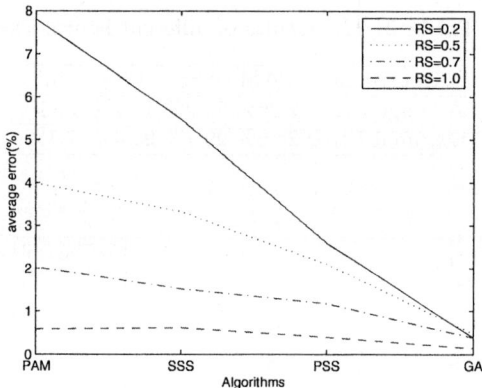

Fig. 5. The effect of resource strength on the algorithms

4.2 Tests Based on Updated PSPLIB

PSPLIB created by using a full factorial design of the parameters of activity network(Network Complexity(NC), Resource Factor(RF), Resource Strength(RS)) is a general instance library widely used for testing the algorithms of RCPSPs with makespan objective[9], where $NC = 1.5, 1.8, 2.0, RF = 0.25, 0.5, 0.75, 1.0$, $RS = 0.2, 0.5, 0.7, 1.0$. In the computation, the instance set with 30 activities that consists of 480 activity networks is updated. The static cost of each activity and the due date of each project are given. The parameters of the genetic algorithm are set as $PopSize = 30, Generation = 40, p_c = 0.7, p_m = 0.1, PreScale = 3$, $TourScale = 2$, One-point Crossover. The procedure shows fine convergence. In

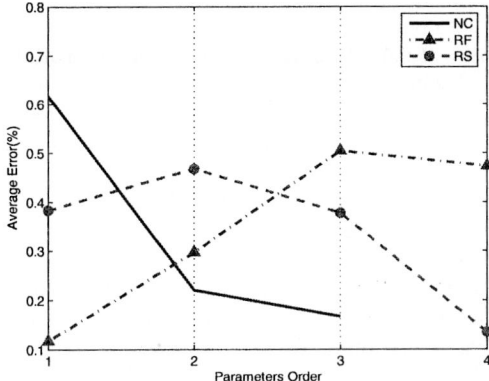

Fig. 6. The effect of the parameters of activity network on GA

addition, the instances are also solved with the other three heuristics we have proposed[10]: Preposing Activity Matrix-based Scheduling(PAM), Priority-rule-based Serial Scheduling Schema(SSS) and Parallel Scheduling Schema(PSS) under the principle of deterministic multi-pass scheduling. The best solution is decided based on the results by using the above four approaches. The average error and the maximal error listed in Table 2 and the effect of the parameters of activity network on the algorithms in Fig.3-5 show that the GA proposed in this paper is the most effective. Fig.6 show that the result of the GA will be better when NC is increasing, but RF has a nearly contrary impact. In addition, the more RS is, the less the average error of the GA.

5 Conclusion

The RCPSP with the objective of minimizing the activities' cost is a critical subproblem of the partner selection in construction supply chain management. In the literatures on RCPSPs, there is few to consider this type of objectives with resource constraints. Its mathematic model is given in this paper and the characteristic of the model is analyzed. To solve the problem, a genetic algorithm is designed which is demonstrated to be effective based on the results of Computational study.

Acknowledgments

The work was supported by the National Science Foundation of China(7017015) and The Teaching and Research Award Fund for Outstanding Young Teachers in Higher Education Institutions of MOE, PR China.

References

1. Kolisch, R., Padman, R.: An Integrated Survey of Deterministic Project Scheduling. Omega,**3** (2001) 249-272
2. Blazewicz, J., Lenstra, J.K., Rinnooy, Kan-A.H.G.: Scheduling Subject to Resource Constraints: Classification and Complexity. Discrete Applied Mathematics, **1** (1983) 11-24
3. Kolisch, R., Hartmann, S.: Heuristic Algorithms for the Resource-Constrained Project Scheduling Problem: Classification and Computational Analysis. WWeglarz, J. ed.: Project Scheduling-Recent Models, Algorithms and Applications. Boston: Kluwer Academic Publishers, (1999) 147-178
4. Hindi, K. S., Yang, H., Fleszar, K.: An Evolutionary Algorithm for Resource-Constrained Project Scheduling. IEEE Transactions on Evolutionary Computation,**5** (2002) 512-518
5. Lee, J. K., Kim, Y. D.: Search Heuristics for Resource-Constrained Project Scheduling. Journal of the Operational Research Society,**3**(1996) 678-689
6. Smith, D. D., Smith, D. V.: Optimal Project Scheduling with Materials Ordering. IIE Transactions,**4** (1987) 122-129
7. Dodin, B., Elimam, A. A.: Integrated Project Scheduling and Material Planning with Variable Activity Duration and Rewards. IIE Transactions,**11**(2001) 1005-1018
8. Sprecher, A., Kolisch, R., Drexl, A.: Semi-active, Active, and Non-delay Schedules for the Resource-Constrained Project Scheduling Problem. European Journal of Operational Research,**1**(1995) 94-102
9. Kolisch, R., Sprecher, A.: PSPLIB-A Project Scheduling Problem Library. European Journal of Operational Research,**1**(1996) 205-216
10. Liu, Zh. Y.: Resource-Constrained Project Scheduling Problem and Its Application in Designing Construction Supply Chain. Ph.D. Dissertation, Huazhong University of Science and Technology(2005)

Single-Machine Partial Rescheduling with Bi-criterion Based on Genetic Algorithm*

Bing Wang[1], Xiaoping Lai[1], and Lifeng Xi[2]

[1] School of Information Engineering, Shandong University at Weihai,
Weihai 264209, China
[2] School of Mechanical Engineering, Shanghai Jiaotong University,
Shanghai 200030, China
wangbing@sdu.edu.cn

Abstract. A partial rescheduling (PR) heuristic is presented for single machine with unforeseen breakdowns. Unlike a full rescheduling strategy where all unfinished jobs are considered, a PR strategy reschedules partial unfinished jobs which form a PR problem, and shifts the rest jobs to the right according to the solution of the PR problem. The rescheduling problem considers a bi-criterion that optimizes both shop efficiency (i.e. makespan performance of the schedule) and stability (i.e. deviation from the original schedule). A genetic algorithm is developed to solve the PR problem. Extensive computational testing was conducted. The computational results show that the PR heuristic with bi-criterion can significantly improve schedule stability with little sacrifice in efficiency, and provide a reasonable trade-off between solution quality and computational efforts.

1 Introduction

Medium to long term schedules are required a priori in order to effectively plan shop activities and sufficiently make use of shop resources. However, myriad unforeseen disturbances, such as rush orders, excess processing time, and machine breakdown etc, will arise during execution. The disturbances that were not accounted for in the original schedule would impossibly make the executed schedule completely consistent to the original schedule. The problem of shop rescheduling on occurrence of an unforeseen disruption is addressed here. The practical solution of this problem requires satisfaction of two often conflicting goals. Bi-criterion heuristics are developed which attempt to (1) reschedule the shop to retain schedule efficiency, and (2) simultaneously minimize the cost impact of the schedule change.

The typical rescheduling schemes in the existing literatures are reactive rescheduling. Reactive rescheduling [1] can be performed timely driven by failure events or periodically, in which original schedule is generated without considering any disruptions and then revised on occurrence of disruptions. Three types

---

* Partly Supported by National Natural Science Foundation (60274013) and Natural Science Foundation for Youths of Shandong University (11010053187075).

of reactive rescheduling strategy with different computational costs were used in the existing literatures. Full rescheduling (FR), where all unfinished jobs are rescheduled to satisfy certain objective, can be merely applied in small or medium size problems with though it can result in an optimal solution. PR can provide a trade-off between solution quality and computational cost through only considering partial unfinished jobs. Right-shift rescheduling (RSR), where all unfinished jobs are just slid to the right as far as necessary to accommodate (absorbing idle time) the disruption, can not guarantee the solution quality though it just requires the least computational efforts.

Wu et al. [2] addressed bi-criterion problems by use of several local search algorithms, which could solve only small or medium rescheduling problems due to their computational complexity. In order to deal with large-scale scheduling problems, Wang et al. [3,4] used rolling horizon scheduling procedures, in which only partial jobs were considered at one iteration. When Sabuncuoglu et al. [1] used a PR strategy to deal with the large-scale rescheduling problems, schedule stability is often poor because of only considering schedule efficiency. In this paper, we extend PR strategy to deal with large-scale rescheduling problems for single machine with schedule efficiency and schedule stability. Two types of objective function are defined for PR problems based on a procedural horizon and a terminal horizon respectively.

2 Single-Machine Rescheduling with Efficiency and Stability

Consider a single-machine problem with release time to minimize the makespan. There are n jobs to be scheduled. A job denoted as i has a release time r_i, a processing time p_i, and a tail q_i. These three parameters of each job are known a priori. For a schedule S of this problem, the makespan which is denoted as $M(S)$ is defined as

$$M(S) = \max_{i \in S} (b_i + p_i + q_i) \quad (1)$$

where b_i is the beginning time of job i in S. This problem is NP-hard [5].

A minimal makespan original schedule S^0 can be generated without considering any disruptions. However, after a disruption, e.g. a machine breakdown, occurs, at the moment u, when the machine returns to service, the unfinished jobs should be rescheduled. The release time of each unfinished job is updated as follows:

$$r_i' = \max(u, r_i) \quad (2)$$

In order to deal with the rescheduling problem with efficiency and stability, we should give the measure for schedule deviation. In this paper, stability of schedule is measured by the schedule deviation based on the original schedule, which is denoted as $D(S)$

$$D(S) = \sum_{i \in S} |b_i - b_i^0| \qquad (3)$$

where b_i^0 is the beginning time of job i in S^0.

The rescheduling problems with bi-criterion are to minimize (1) and (3). An optimization problem with such dual objectives can be converted into a single overall objective problem. Let the criteria be

$$\min_S J(S) = w_D D(S) + w_M M(S) \qquad (4)$$

where w_D is the weight of schedule deviation and w_M is the weight of schedule efficiency. The amount of weights represents importance degree of the corresponding objective. If a pair of weights is specified, for the optimal solution S^* of the overall problem, $(D(S^*), M(S^*))$ is a pair of effective solution of the corresponding bi-criterion problem. If the weights vary, other effective solutions can be obtained. This single objective problem is also NP-hard [2].

3 PR Strategy for Single-Machine Rescheduling

Non-preemptive jobs are assumed so that a job being processed at the time of disruption must be restarted. Let the RSR solution for a rescheduling problem be S^R, where the jobs are ordered according to the original sequence and, the beginning times of jobs are decided as follows:

$$b_1^R = \max(u, r_1') \quad \text{(the interrupted job)}$$

$$b_j^R = \max(b_j, r_j', b_{j-1}^R + p_{j-1}) \quad (j > 1)$$

where job 1 is the first unfinished job, i.e. the interrupted job. b_j^R denotes the beginning time of job j in S^R. Obviously, for a specified original schedule, the larger the right-shift interval Δt between the beginning time b_1^R and the beginning time b_1^0 in S^0 is, the larger the schedule deviation is. In order to distinguish RSR solutions with different right-shift interval, we give the the following definition.

Definition 1. *Let the original schedule of a single-machine scheduling problem be S^0, a RSR solution S^R is called Δt-RSR solution if the first unfinished job is shifted to the right by Δt.*

Although the job sequence change of a RSR solution is the smallest, no optimality of objectives is performed while just naturally absorbing idle time in the original schedule to adapt to breakdown disruptions.

Let N be the set of all jobs. If $\hat{N}$ denotes the set of finished jobs on occurrence of a breakdown, $S(\hat{N})$ represents the finished schedule at that time. The duration of machine breakdown is D. At u, all unfinished jobs form the job set N'. We identify a rescheduling horizon from the first unfinished job on, where partial unfinished jobs is included into a PR problem following the original schedule sequence and form the job set N_p. The job number of N_p, which is used to formulate the horizon size for PR, can be specified as a fixed number or a fixed proportion of all jobs. The job set $\tilde{N}$ consists of other unfinished jobs except for jobs of N_p. The job number of $\tilde{N}$ is denoted as $|\tilde{N}|$. Then we have $N' = N_p \bigcup \tilde{N}$ and $N = \hat{N} \bigcup N'$.

Definition 2. *The partial unfinished job set N_p of on occurrence of a disruption, where fully rescheduling with certain criteria will be performed, is called a PR-horizon. A PR-horizon size refers to the job number of N_p, denoted as k.*

Match up rescheduling (MUR) [6, 7, 8] was proposed by Bean and Birge. Instead of minimizing the total job deviation, MUR adapts the original schedule during a transient period following a disruption and the transient schedule is designed to match-up with the original schedule in a finite amount of time. The new schedule is completely consistent to the original schedule from match-up point on. For certain duration of machine breakdown D, the imposition f a match-up time may cause delay in match-up point if no enough idle time exists in the transient period of the original schedule.

The following, we give the definition of "non-delay" match-up schedule different from that in Wu et al. [2].

Definition 3. *A match-up schedule is said to be non-delay if the schedule has a completion time identical to the completion time of the original schedule.*

Let $S^0(N_p)$ be the partial schedule for N_p in S^0 and C_p^0 be the completion time of $S^0(N_p)$. The PR solution for N_p is denoted as $S^p(N_p)$ and the completion time of $S^p(N_p)$ is denoted as C_p^p. We let C_p^0 be the moment of match-up point and the processing time of $S^0(N_p)$ be the match-up time. Thus if the new schedule can't match-up the original schedule within the match-up time, the delay of match-up point can be formulated as $\Delta C_p = C_p^p - C_p^0$.

Let $S^0(\hat{N})$ be the partial schedule for job set $\hat{N}$. Let $S^0(\tilde{N})$ be the partial schedule for $\tilde{N}$ in S^0. If we let $S^{PR}(\tilde{N})$ be the ΔC_p-RSR solution for $S^0(\tilde{N})$, the new schedule $S^p(N')$ for N' consists of $S^p(N_p)$ and $S^{PR}(\tilde{N})$. The new global schedule S^p consists of $S^0(\hat{N})$ and $S^p(N')$.

Two types of criteria for PR problem are defined as follows based on PR-horizon N_p locating at the course or the terminal of the original schedule respectively :

$$\min_{S(N_p)} J_p = \{w_D \sum_{i \in N_p} |b_i^p - b_i^0| + w_D|\tilde{N}|(C_p^p - C_p^0)\} \quad |\tilde{N}| > 0 \quad (5)$$

$$\min_{S(N_p)} J_p = \{w_D \sum_{i \in N_p} |b_i^p - b_i^0| + w_M M(S^P(N_p))\} \quad |\tilde{N}| = 0 \quad (6)$$

(5) is designed for N_p locating at the processing course of the original schedule. The objective of PR is to minimize both the delay ΔC_p and the schedule

deviation. Since the consideration of match-up delay would make the new schedule inserted by less idle time in case more idle time greatly puts off later jobs, it is reasonable to use the number of later jobs as the weight for match-up delay. In such a manner, the schedule deviation of the latter job set $\tilde{N}$ is considered in the PR problem. When N_p locates the terminal of the original schedule, the job set $\tilde{N}$ is empty and we directly consider the makespan in the PR problem (6).

Definition 4. *This type of PR with objectives formulated as (5) and (6) is termed PRDO for short.*

The algorithm of PRDO is given roughly as follows except for the detailed algorithm of PR problem:

Step 1. Minimizing the makespan of a problem to generate the original schedule S^0 without considering any disruption;

Step 2. Implement S^0 until a breakdown occurs, note the finished partial schedule $S^0(\hat{N})$ at this moment;

Step 3. For a given breakdown duration D, compute the moment u for the machine returning to service, the release times of jobs in N' are updated according to (2);

Step 4. The first k jobs from the beginning of $S^0(N')$ are included into the PR-horizon N_p, note the completion time C_p^0 of $S^0(N_p)$, compute $|\tilde{N}| = n - (|\hat{N}| + k)$; (Here k is the specified PR-horizon size)

Step 5. If $|\tilde{N}| > 0$, PR is performed according to (5), and the solution $S^p(N_p)$ as well as the completion time C_p^p of $S^p(N_p)$ and the delay of match-up point ΔC_p can be obtained. The new partial schedule $S^{PR}(\tilde{N})$ is the ΔC_p-RSR solution of the partial original schedule $S^0(\tilde{N})$. Thus the new global schedule is $S^p = S^0(\hat{N}) + S^p(N_p) + S^{PR}(\tilde{N})$; If $|\tilde{N}| = 0$, PR is performed according to (6) and the solution $S^p(N_p)$ can be obtained. The new global schedule is $S^p = S^0(\hat{N}) + S^p(N_p)$;

Step 6. Compute the global schedule makespan $M(S^p)$ and the schedule deviation $D(S^p)$. For a pair of specified weights (w_d, w_M), the overall objective $J(S^p)$ defined as (4) can be obtained;

The PR problem will be solved by the local search procedure based on genetic algorithm and a detailed description will be given in the next section.

4 Genetic Algorithm for PR Problems

In this paper, since the size of PR problem is limited, genetic algorithm can be applied to solve the problem (5) or (6). We can encode the schedules and generate the initial population in the similar manners to Wu et al. [2]. However, since our objectives differ from those of Wu et al. [2], the genetic algorithm should be modified.

The success of a genetic algorithm relies on an effective chromosomal encoding of solutions. That is, they require an encoding which allows the genetic operators to exploit the best solutions currently available, and at the same time robustly explore the solution space. For scheduling problems, encoding of the solutions is most difficult. The obvious method is to encode the solutions by their sequences. However, with such encodings, a standard crossover operator which simply crosses the parent sequences will result in infeasible schedules. Past approaches to this problem entail modifying the crossover operator to guarantee schedule feasibility. This often results in the failure of child solutions to inherit important parental features, and often involves unacceptable computational burden.

For the bi-criterion scheduling problem, Wu et al. [2] resolved the difficulty in encoding by converting the sequence into a string of "artificial tails". An artificial tail is defined as follows:

Let $j = 1, \ldots, n$ index jobs in N' ordered according to the sequence to be encoded. Then define

$$q_n' = 0, \qquad q_{j-1}' = q_j' + p_j$$

Thus the earlier a job appears in the sequence, the longer its artificial tail will be. Specifically, a list, $\{q_1', \ldots, q_n'\}$, ordered by the job number is maintained for each solution as its "chromosome". This encoding can be manipulated by a typical crossover operator, and can be evaluated by applying Schrage's heuristic, which guarantees schedule feasibility.

To start the genetic algorithm, an initial solution population must be generated. Two desirable properties of the initial population are (1) high quality solutions (corresponding to fitness), and (2) diverse solutions. A heuristic method developed by Wu et al. is used to generate the initial population. The method for initial population generation uses the solutions produced by a heuristic termed "α-ε grid search", which allow varying emphasis to be placed on the two criteria.

Once the initial population is generated, each solution in the population is encoded by the string of artificial tails, $Q_y = \{q_1', \ldots, q_n'\}$, used to generate the schedule. From the current population, P, a number of "more fit" solutions will be selected for reproduction based on the fitness measure.

A critical element of a genetic algorithm is the measure of solution fitness. The fitness measure represents the selection probability of a solution during the search. For equation (5), the fitness measure for a given PR solution, $S^y(N_p)$, is then defined as follows:

$$f_y = (J_{pmax} - J_{py})^l / \sum_{j \in P} (J_{pmax} - J_{pj})^l$$

where $J_{pmax} = \sum_{i \in S^c(N_p)} |b_i^c - b_i^0| + |\tilde{N}|(C_p^m - C_p^0)$

$$J_{py} = \sum_{i \in S^y(N_p)} |b_i^y - b_i^0| + |\tilde{N}|(C_p^y - C_p^0)$$

$S^c(N_p)$ is the PR solution by Carlier's algorithm [9], b_i^c and b_i^y are the beginning times of job i in $S^c(N_p)$ and $S^y(N_p)$ respectively. C_p^m is the completion time of the minimum deviation PR solution, and C_p^y is the completion time of

$S^y(N_p)$. l is a constant used for tuning purposes. As l increases, the genetic algorithm becomes more selective when generating off-spring solutions. With a large l-value, only the "most fit" solution will survive and the algorithm converges prematurely in one step. For equation (6), the fitness measure of a solution is defined by a similar format.

The genetic algorithm can be summarized as follows:

Step 1. Initialization: start with an initial set of tuning parameters [i.e. l, population size (ps), mutation probability (mp), and number of generations (ng)], generate an initial population by " α-ε grid search", compute the fitness $f_y, \forall S^y(N_p) \in P$, and save the set of artificial tails, Q_y, corresponding to each solution $S^y(N_p)$.

Step 2. $g = 1$, start iterations.

Step 3. Copy the set of mb best solutions from the current population to the next generation.

Step 4. Perform crossover ($ps - mb$)/2 times as follows: select a distinct pair of solutions ($S^y(N_p), S^j(N_p)$) randomly from the current population based on probabilities: f_y and f_j; (b) generate two crossover sites x_1 and $x_2(x_1 < x_2)$ from a discrete uniform distribution between 1 and $|N'|$; (c) swap the sublist $\{q'_{y1}, \ldots, q'_{yn}\}$ of Q_y with the corresponding sublist in $Q_j = \{q'_{j1}, \ldots, q'_{jn}\}\}$, apply Schrage's heuristic to obtain the new schedule, compute the criteria J_{py}, J_{pj}; (d) copy both off-spring solution to the next generation, compute f_y, f_j, and save Q_y, Q_j.

Step 5. For each solution $S^y(N_p)$, apply mutation with probability, mp. Mutation is performed by setting $\alpha = uniform[0, 1]$ then recomputed all the artificial tails in Q_y by $q_i'' = (1 - \alpha)q_i' + \alpha q_i$.

Step 6. Set the generation as the current population, $g = g + 1$, if $g < ng$, go to step 3, else stop.

5 Computational Testings and Results

In the following testings, we assigned the same weights for dual objectives, i.e. $w_D = w_M = 1$. The original schedule was created using Schrage's algorithm [9]. All procedures were coded in C language and ran on the Microsoft Visual c++ 6.0 under the Windows XP operating environment. All tests ran on a computer with Pentium 4-M CPU 1.80GHz. Only one machine breakdown disruption was generated. The duration of the disruptions range from five percent to fifteen percent of the processing time of the original schedule. We assumed that the disruption would not occur among the last twenty jobs because the number of the rescheduled jobs is too small to make rescheduling trivial in those cases.

Problems were randomly generated using a format similar to that used by Chand et al. [10].

In genetic algorithm for PR problems, we used $l = 2$, population size $ps = 100$, mutation probability $mp = 0.05$, and the number of generations $ng = 200$. When the α-ε procedure was used to generate initial population, $\Delta\alpha = 0.01$ and $\Delta\varepsilon = 0.5$.

Since FR under genetic algorithm can only deal with rescheduling problems with small or medium size, it is impossible to compare PRDO with FR in large-size problems. However, we can use a RSR solution as a baseline where our approach is compared due to its low computational burden. Testing was conducted to compare the PR solution with the RSR solution for each problem.

When ρ value is 0.20, jobs arrive rather rapidly and largely overlap each other so that almost no idle time exists in the original schedule. The situation when ρ value is 2.00 is the reverse. Therefore, the problems with three ρ values actually represent three situations where different amount of idle time exists in the original schedule. Experiment under each situation was conducted in order to investigate the correlation between the solution quality and the computational efforts. For problems with each ρ value, PR-horizon in PRDO was specified as 10, 20, 30, and 40 jobs respectively. For each PR-horizon size, 20 problems were generated. The improvement percentage of PRDO over RSR is calculated as $(RSR - PRDO)/PRDO$. The computational results for 200-job problems with three ρ values show in Fig. 1, Fig. 2 and Fig. 3 respectively.

In Fig. 1, the continuous lines represent the improvement percentage in stability and the broken lines represent that in efficiency. Fig. 1 demonstrates that PRDO greatly improves the schedule stability over RSR with little sacrifice in efficiency. As the PR-horizon size increases, the improvements consistently get larger for all ρ values. The improvements of PRDO over RSR increase more rapidly for the problems with smaller ρ values, where less idle time exists in the original schedules. For the problems with 2.00 ρ value, where much more idle time exists in the original schedule, the improvements are less.

Fig. 2 presents the improvements of the overall objective of PRDO over RSR for problems with three ρ values. Though the efficiency of PRDO improves little in some situations, the improvements of the overall objectives are consistently getting larger for all problems as the PR-horizon size increases.

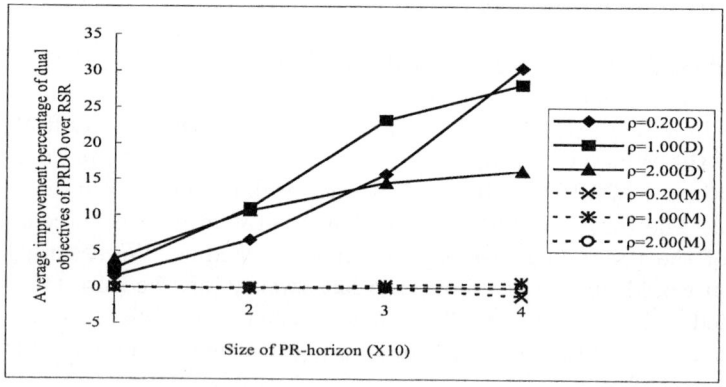

Fig. 1. Improvements of dual objectives of PRDO over RSR

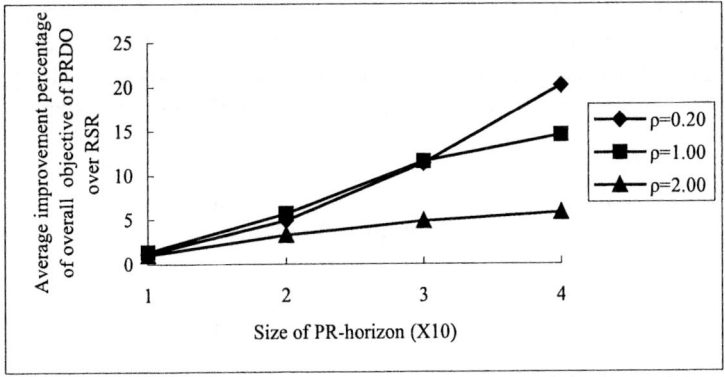

Fig. 2. Improvements of overall objective of PRDO over RSR

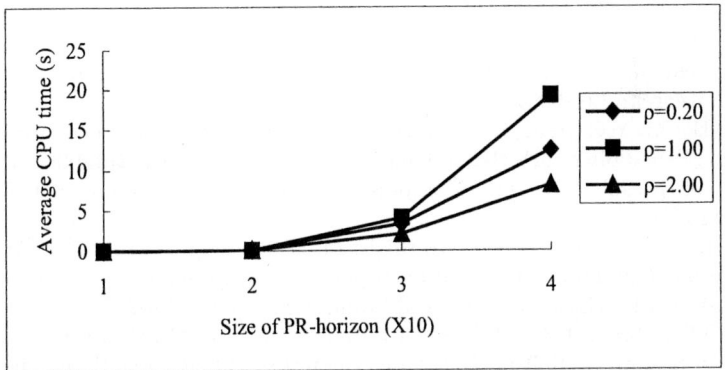

Fig. 3. Average CPU time of PRDO

In Fig. 3, the average CPU time of PRDO is presented when PR-horizon varying from 10 to 40. It is shown that as the PR-horizon size increases, the average CPU time increases. The computational results demonstrate that genetic algorithm is effective for PR problems and we must pay more computational cost for larger improvements of PRDO. Among different ρ value problems, those with $\rho = 1.00$, which refer to the hardest scheduling problems in Hariri et al. [11], expend the highest computational cost. From the computational results, we estimate that the almost best trade-off between the solution quality and the computational cost may be achieved at the 30-job PR-horizon.

6 Conclusions

PR heuristic with bi-criterion for single-machine with a breakdown is developed in this paper. The rescheduling problem considers both efficiency and stability. PR strategy addresses the computational complexity of large-size rescheduling problems. Two types of objective function are designed for PR problems. The global objectives are partly reflected in the criterion of PR problems. Genetic algorithm is used to solve the bi-criterion PR problem. Extensive computational experiments are performed to show the impact of PR-horizon size on solution quality and solution time. The computational results show that schedule stability is largely improved over that of RSR with little sacrifice in efficiency. This kind of heuristic can achieve a trade-off between solution quality and solution time and is effective in large-scale rescheduling problems.

References

1. Sabuncuoglu, I., Bayiz, M.: Analysis of Reactive Scheduling Problems in a Job Shop Environment. European Journal of Operational Research. **126**(2000) 567–586
2. Wu, S.D., Storer, R.H., Chang, P.C.: One-machine Rescheduling Heuristics with Efficiency and Stability as Criteria. Computers in Operations Research. **20**(1993) 1–14
3. Wang, B., Xi, Y.G., Gu, H.Y.: Terminal Penalty Rolling Scheduling Based on an Initial Schedule for Single-machine Scheduling Problem. Computers and Operations Research. **32**(2005) 3059–3072
4. Wang, B., Xi, Y.G., Gu, H.Y.: An Improved Rolling Horizon Procedure for Single-machine Scheduling with Release Times. Kongzhi yu Juece. **20**(2005) 257–260
5. Garey, M.R., Johnson, D.S.: Computers Intractability. Freeman, San Francisico, Calif. (1979)
6. Bean, J.C., Birge, J.R.: Match-up Real-time Scheduling.In Proceeding of the Symposium on Real Time Optimization in Automated Manufacturing Facilities, NBS publication 724, National Bureau of Standards. (1985) 197–212
7. Bean, J.C., Birge, J.R.: Mittenehal J and Noon C E Match-up Scheduling with Multiple Resources, Release Dates and Disruption. Operations Research. **39**(1991) 470–483
8. Akturk, M.S., Gorgulu, E.: Match-up Scheduling under a Machine Breakdown. European Journal of Operational Research. **12**(1999) 81–97
9. Carlier, J.: The One-machine Sequencing Problem. European Journal of Operational Research. **11**(1982) 42–47
10. Chand, S., Traub, R., Uzsoy, R.: Rolling Horizon Procedures for the Single Machine Deterministic Total Completion Time Scheduling Problem with Release Dates. Annals of Operations Research. **11**(1997) 115–125
11. Hariri, A.M.A., Potts, C.N.: An Algorithm for Single Machine Sequencing with Release Dates to Minimize Total Weighted Completion Time. Discrete Applied Mathematics. **11**(1983) 99–109

An Entropy-Based Multi-population Genetic Algorithm and Its Application

Chun-lian Li[1], Yu sun[2], Yan-shen Guo[3], Feng-ming Chu[3], and Zong-ru Guo[3]

[1] School of Computer Science, Changchun University, Changchun 130022, China
[2] Institute of Special Education, Changchun University, Changchun 130022, China
[3] Institute of Material Medical Chinese Academy of Medical Sciences &
Peking Union Medical College, Beijing 1000050, China
zrguo@imm.ac.cn

Abstract. An improved genetic algorithm based on information entropy is presented in this paper. A new iteration scheme in conjunction with multi-population genetic strategy, entropy-based searching technique with narrowing down space and the quasi-exact penalty function is developed to solve nonlinear programming problems with equality and inequality constraints. A specific strategy of reserving the most fitness member with evolutionary historic information is effectively used to approximate the solution of the nonlinear programming problems to the global optimization. Numerical examples and an application in molecular docking demonstrate its accuracy and efficiency.

1 Introduction

Genetic Algorithm (GA) [1] is a time-consuming method and this has greatly hindered it from applied to some optimization problems. A new method built on multi-population technique developed in prior work [2] is proposed in this paper, in which an entropy-based searching technique is developed to ensure rapid and steady convergence, and uses a specific strategy of reserving the most fitness member with evolutionary historic information to obtain the global solution. As application, a drug molecular docking experiment is given.

2 Implementation of Entropy-Based GA

2.1 Transformation of Optimization Models

The general constraint non-linear programming problem can be stated as follows:

$$\begin{aligned} &\min \quad f(\mathbf{d}) \\ &s.t. \quad g_j(\mathbf{d}) \le 0, \quad j = 1,2,\cdots q \end{aligned} \quad (1)$$

where $f(\mathbf{d})$ is the objective function, $\mathbf{d}=\{d_1,d_2,\ldots,d_n\}^T$ is a vector of n design variables, $g_j(\mathbf{d})(j=1,2,\ldots,q)$ are the constraint functions. Problem (1) is the general constrained optimization model, and has many and widespread applications. Engineering design applications usually involve non-linear functions f and/or g and almost invariably have active constraints at the optimum. In order to solve problem (1) efficiently, first, define a Parametric Constraint Evaluation (PCE) Function.

Definition 1. If ψ is a positive real variable, and $G=\{g_j(d)\}$, $(j=1,2,...,q)$, is a set of constraint functions, then

$$E(G) = (1/\psi) \ln \sum_{j=1}^{q} \exp(\psi g_j(\mathbf{d})) \qquad (2)$$

is a parametric constraint evaluation (PCE) function. Problem (1) is transformed into the following model by means of PCE function:

$$\min \quad f(\mathbf{d})$$
$$\text{s.t.} \quad g_\psi(\mathbf{d}) = (1/\psi) \ln \sum_{i=1}^{q} \exp(\psi g_i(\mathbf{d})) \leq 0 \qquad (3)$$

In discussing the relationship between problems (1) and (3) the following definitions and lemma are introduced.

Definition 2. If, for any $F(\mathbf{d})=\{f_1(\mathbf{d}), f_2(\mathbf{d}), ..., f_q(\mathbf{d})\}$, and $\overline{F}(\mathbf{d}) = \{\overline{f}_1(\mathbf{d}), \overline{f}_2(\mathbf{d}), ..., \overline{f}_q(\mathbf{d})\}$, with $f_j(\mathbf{d}) \leq \overline{f}_j(\mathbf{d})$, $j=1,2,\cdots,q$, and there exists at least one j_0, $(1 \leq j_0 \leq q)$, such that $f_{j_0}(\mathbf{d}) < \overline{f}_{j_0}(\mathbf{d})$, then $F(\mathbf{d}) \leq \overline{F}(\mathbf{d})$ or, simply $F \leq \overline{F}$.

Definition 3. If, for any, $F, \overline{F} \in E^q$, with $F \leq \overline{F}$, $E(F) < E(\overline{F})$, then $E(F)$ is a strictly monotone increasing function of F.

Lemma 1. The PCE function $E(G)$ is a strictly monotone increasing function of G, and if $\psi \to \infty$ then

$$(1/\psi) \ln \sum_{j=1}^{q} \exp(\psi g_j(\mathbf{d})) = \max \ g_j(\mathbf{d}), \quad j=1,2,\cdots,q. \qquad (4)$$

Proof. Let

$$G = \{g_j(\mathbf{d})\} \leq \overline{G} = \{\overline{g}_j(\mathbf{d})\} \quad j=1,2,\cdots,q \qquad (5)$$

According to the above three definitions and lemma 1, the proof will be easily completed and the concrete proving process is abbreviated.

The PCE function plays an important role in the proposed method. The following theorem aids understanding of its properties.

Theorem 1. If $\psi \to \infty$, then the optimization problems (1) and (3) have the same Kuhn-Tucker (K-T) points.

Proof. The Lagrange augmented function Problem (3) is

$$L(\mathbf{d}, \mu) = f(\mathbf{d}) + (\mu/\psi) \ln \sum_{j=1}^{q} \exp(\psi g_j(\mathbf{d})) \qquad (6)$$

where $\mu > 0$ is the Lagrange multiplier of corresponding constraint. The K-T condition for problem (3) is given as

$$\partial f(\mathbf{d})/\partial d_i + (\mu/\psi)\left\{\sum_{j=1}^{q}\exp(\psi g_j(\mathbf{d}))\cdot\partial g_j(\mathbf{d})/\partial d_i\right\}/\sum_{j=1}^{q}\exp[\psi g_j(\mathbf{d})] = 0 \qquad (7)$$

$$(1/\psi)\ln\sum_{j=1}^{q}\exp(\psi g_j(\mathbf{d})) \leq 0 \qquad (8)$$

$$(\mu/\psi)\ln\sum_{j=1}^{q}\exp(\psi g_j(\mathbf{d})) = 0, \mu \geq 0. \qquad (9)$$

By means of Lemma 1 and equation (8), if $\psi \to \infty$, then

$$(1/\psi)\ln\sum_{j=1}^{q}\exp(\psi g_j(\mathbf{d})) = \text{Max } g_j(\mathbf{d}) \leq 0, j = 1,2,\cdots,q. \qquad (10)$$

i.e.

$$g_j(\mathbf{d}) \leq 0 \qquad (11)$$

Substituting

$$\lambda_j = \{\mu\cdot\exp(\psi g_j(\mathbf{d}))\}/\left\{\psi\cdot\sum_{j=1}^{q}\exp(\psi g_j(\mathbf{d}))\right\} \qquad (12)$$

into equations (7) and (9) gives

$$\partial f(\mathbf{d})/\partial d_i + \sum_{j=1}^{q}\lambda_j g_j(\mathbf{d}) = 0 \qquad (13)$$

Combining equations (9), (11) and (12), if $\psi \to \infty$, then

$$\lambda_j g_j(\mathbf{d}) = 0 \qquad (14)$$

i.e.

$$\begin{cases} g_j(\mathbf{d}) = o & \text{if } \lambda_j > 0 \\ g_j(\mathbf{d}) < 0 & \text{if } \lambda_j = 0 \end{cases} \qquad (15)$$

Equations (11), (13) and (15) are identical the Kuhn-Tucker condition of the problem (1). Hence the problems (3) and (1) have the same Kuhn-Tucker points and vice versa. The theorem is proved.

Kuhn-Tucker points are obtained by solving the Kuhn-Tucker conditions, which are necessary condition for the optimum solution of non-linear programming with equality and inequality constraints. Then solving problem (1) is substituted by problem (3). Unlike some optimality criteria methods, there is no need to find active constraints. The λ_j in equation (12) can give the active level of the constraints

2.2 Elements of Genetic Search

In this paper, the five basic GA operators are binary coding, integer-decimal selection, two point crossover, uniform mutation. As for the fitness function, it is solved by using quasi-exact penalty function [3], so problem (3) can be transformed to:

$$\varphi_\psi(\mathbf{d}) = f(\mathbf{d}) + (\alpha/\psi)\ln\left\{1 + \sum_{i=1}^{q} \exp(\psi g_i(\mathbf{d}))\right\} \quad (16)$$

the parameter ψ can be chosen in the range $10^3 - 10^5$ and $\alpha > 0$ is penalty factor. Fitness function of GA by means of equation (16) may be written as:

$$\max \quad F(\mathbf{d}) = C - \varphi_\psi(\mathbf{d}) \quad (17)$$

where C is a large positive number to ensure $F > 0$.

Besides, elitist maintaining is employed to ensure the global convergence.

2.3 Information-Entropy-Based Searching Technique

As mentioned above, the GA begins from generating arbitrarily m populations with the initial design space. If $F_j(\mathbf{d})$ ($j=1,2,\ldots,m$) represent the best value of the fitness function occurs in the jth population, then we need to maximum $F_j(\mathbf{d})$ ($j=1,2,\ldots,m$) by means of a genetic search, i.e., to solve the following optimum problem:

$$\min \quad -F_j(\mathbf{d}), \quad j = 1, 2, \cdots, m \quad (18)$$

It is difficult to solve problem (18) completely, We need only to get efficient narrowing coefficients for the searched space. By information entropy principle [4], an entropy based-optimization model can be constructed as follows:

$$\begin{cases} \min \quad -\sum_{j=1}^{m} p_j F_j(\mathbf{d}) \\ \min \quad H = -\sum_{j=1}^{m} p_j \ln(p_j) \\ \text{s.t.} \quad \sum_{j=1}^{m} p_j = 1, \ p_j \in [0\ 1] \end{cases} \quad (19)$$

where H is the information entropy, p_j is here defined as a probability that the optimal solution of the problem (18) occurs in the population j. In discussing the relationship between problems (18) and (19) the following theorem is introduced.

Theorem 2. The problem (18) and (19) have the same optimal solution.

Proof. Suppose that $\mathbf{d}^*$ and $p^* = \{p^*_1, p^*_2, \ldots, p^*_m\}$ are the optimal solution of problem (19), so that

$$\min \quad H^* = -\sum_{j=1}^{m} p^*_j \ln(p^*_j) = p^*_l \ln(1) = 0 \quad (20)$$

where $p^*_{l}=1$, $p^*_{i}=0$, for $i \neq l$. Hence

$$\min \ - \sum_{j=1}^{m} p_j F_j(\mathbf{d}^*) = \min \ - F_l(\mathbf{d}^*) \tag{21}$$

Obviously, $\mathbf{d}^*$ and p^* are also the optimal solution of problem (18). It can be similarly proved that the optimal solution of problem (18) is also the optimal solution of problem (19), and the proof is completed.

By means of the weighted coefficient method for solving multi-objective optimization, problem (19) can be transformed into the single objective optimization problem:

$$\min \ -(1-\beta)\sum_{j=1}^{m} p_j F_j(\mathbf{d}) - \beta \sum_{j=1}^{m} (p_j \ln p_j) \tag{22}$$

$$s.t. \ \sum_{j=1}^{m} p_j = 1, \ p_j \geq 0, \ j = 1,2,\cdots,m$$

where $\beta \geq 0$ is weight coefficient. The Lagrange augmented function of (22) is

$$L_H(\mathbf{d},\mathbf{p},\beta,\eta) = -(1-\beta)\sum_{j=1}^{m} p_j F_j(\mathbf{d}) - \beta \sum_{j=1}^{m}(p_j \ln p_j) + \eta\left(\sum_{j=1}^{m} p_j - 1\right) \tag{23}$$

where η is Lagrange multiplier. The stationary conditions of L_H with respect to $\mathbf{p}$, η and $\mathbf{d}$ give

$$\ln p_j = [(\beta-1)/\beta]F_j(\mathbf{d}) + \eta/\beta - 1, \ \sum_{j=1}^{m} p_j = 1 \tag{24}$$

$$(1-\beta)\sum_{j=1}^{m} p_j \cdot [\partial F_j(\mathbf{d})/\partial d_i] = 0, \ i = 1,2,\cdots,N \tag{25}$$

The solution of equation (24) is

$$p^*_j = \exp(\gamma F_j(\mathbf{d})) / \sum_{j=1}^{m} \exp(\gamma F_j(\mathbf{d})) \tag{26}$$

in which

$$\gamma = (\beta-1)/\beta \tag{27}$$

is called as quasi-weight coefficient. The $(1-p_j)$ can be used as the coefficients of narrowing searching space in the modified GA. When the optimal solution occurs in the lth population, then $(1-p^*_l)=0$, and its searching space is not narrowing.

Design space is defined as initial searching space $D(0)$. m populations with N members are generated in the given space. After a few generations are independently evolved in each population (only two generations in this paper), Searching space of each population is narrowed according to the following equation:

$$D(K) = (1 - p_j)D(K-1) \tag{28}$$

$$\underline{d}_i(K) = \max\left\{\left[d_i^*(K) - 0.5(1-p_j)D(K)\right], \underline{d}_i(0)\right\}$$
$$\overline{d}_i(K) = \min\left\{\left[d_i^*(K) - 0.5(1-p_j)D(K)\right], \overline{d}_i(0)\right\}$$

where $\underline{d}_i(K)$ and $\overline{d}_i(K)$ are the modified lower and upper limits of ith design variable at Kth generation respectively. $d_i^*(K)$ is the value of design variable i of the best member in the population j.

2.4 Algorithm Organization

The implementation of the proposed GA consists of the following steps:

Step 1. Give the initial design space $D(0) = [\underline{d}(0), \overline{d}(0)]$ and parameters. Generate m populations with given space $D(0)$

Step 2. Perform GA with elitist maintaining processes in section 3 a few generations (only one generation in this paper)

Step 3. Perform the information-entropy-based searching process with narrowing down space, see equation (28)

Step 4. Check convergence: if searching space of in the best population has been reduced to a very small area (a given tolerance), then stop; else go to step 2

3 Numerical Tests

The following three examples are used to clarify the efficiency and accuracy of GA.

Table 1. Examples: Comparison of various methods

E1 [5]	max s.t.	$f(x) = -x_1^2 - x_2^2 - 2x_3^2 - x_4^2 + 5x_1 + 5x_2 + 21x_3 - 7x_4$ $x_1^2 + x_2^2 + x_3^2 + x_4^2 + x_1 - x_2 + x_3 - x_4 \leq 8$ $x_1^2 + 2x_2^2 + x_3^2 + 2x_4^2 - x_1 - x_4 \leq 10$ $2x_1^2 + x_2^2 + x_3^2 + 2x_1 - x_2 - x_4 = 5$
E2 [6]	max s.t	$f(x) = -(x_1 - 3)^2 - (x_2 - 2)^2$ $x_1^2 + x_2^2 \geq 5$ $x_1 + 2x_2 = 4$ $x_1, \quad x_2 \geq 0$
E3 [7]	min s.t.	$f(x,y) = (4 - 2.1x^2 + x^4/3)x^2 + xy + (-4 + 4y^2)y^2$ $-3 \leq x \leq 3$ $-2 \leq y \leq 2$

The parameters used are: number of populations $m=4$, population size $N=50$, crossover probability $p_c=0.65$, mutation probability $p_m=0.1$; and control parameters: $\psi = 10^3, \alpha = 1.2 \times 10^5$. Figure 1 shows the evolution history of genetic design. Table 2 gives the comparison of various methods.

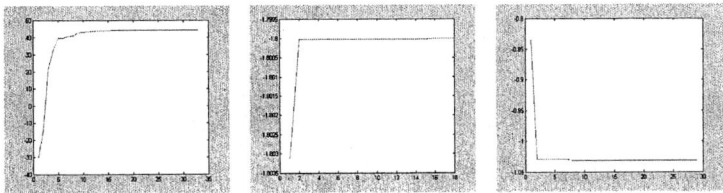

Fig. 1. The evolution history of the three examples: vertical coordinate represents the objective value and the horizontal represents the generations

Table 2. Examples: Comparison of various methods

Test problem		Theoretical solution	This paper	Other literatures [2,8]
P1	Design variables	(0.000, 1.000, 2.000, -1.000)	.008362, 1.043144, 1.993470, -.997657)	(-0.03032, 0.9752, 2.0277, -0.9702) [2]
	Objective function	44.0000	43.988990	43.98178[2]
	Generations		33	50[2]
P2	Design variables	(2.400, 0.800)	(2.398946, 0.800527)	(2.4150, 0.7925) [2]
	Objective function	-1.8000	-1.800001	-1.80028[2]
	Generations		18	100[2]
P4	Design variables	(-0.0898, 0.7126)	(-0.089215, 0.713274)	(-0.0864, 0.7119) [8]
	Objective function	-1.031628	-1.031623	-1.0316[8]
	Generations		29	100[8]

4 Application

The molecular docking design is generally cast as a problem of finding the low-energy binding modes of a small molecule or ligand based on the "lock and key mechanism"[9] within the active site of a macromolecule, or receptor, whose structure is known. However, protein-ligand docking (PLD) for drug molecular docking design is an ideal approach to virtual screening, i.e., to search large sets of compounds for putative new lead structure. Because of the computationally expensive nature of searching problem, drug molecular docking design still needs more efficient computation methods algorithms. In this paper, a PLD process is mimicked using the proposed GA to search the optimal conformation of the flexible ligand.

4.1 Protein-Ligand Docking Design Using GA

In this paper, we assume that the ligand is flexible and the receptor is rigid. The optimization problem of drug molecular docking design can be written as follows

$$\begin{aligned}
\text{Min} \quad & E = f(T_x, T_y, T_z, R_x, R_y, R_z, T_{b1}, \cdots, T_{bn}) \\
\text{s.t.} \quad & \underline{X} \le T_x \le \overline{X} \\
& \underline{Y} \le T_y \le \overline{Y} \\
& \underline{Z} \le T_z \le \overline{Z} \\
& -\pi \le angle \le \pi, \quad angle = R_x, R_y, R_z, T_{b1}, \cdots, T_{bn}
\end{aligned} \quad (29)$$

where design variables $T_{b1}, \cdots, T_{bn}$ are the torsion angles of the rotatable bonds for flexible ligand, $T_x, T_y, T_z, R_x, R_y, R_z$, are the position coordinates and rotational angles of the rigid substructure (segment that not include rotatable bonds) in ligand for the matching-based orientation search, and objective function E is intermolecular interaction energy

$$E = \sum_{i=1}^{lig} \sum_{j=1}^{rec} \left(\frac{A_{ij}}{r_{ij}^a} - \frac{B_{ij}}{r_{ij}^b} + 332.0 \frac{q_i q_j}{Dr_{ij}} \right) \quad (30)$$

where each term is a double sum over ligand atoms i and receptor atoms j, r_{ij} is distance between atom i in ligand and atom j in receptor, A_{ij}, B_{ij} are Van der waals repulsion and attraction parameters, a, b are Van der waals repulsion and attraction exponents, q_i, q_j are point charges on atoms i and j, D is dielectric function, and 332.0 is factor that converts the electrostatic energy into kilocalories per mole.

In the PLD process, the binding free energy should been transferred to relate to the ligand atoms' Cartesian coordinates only. The Cartesian coordinates of all the ligand atoms can be determined by solving optimization problem (49). It means that the optimal conformation of flexible ligand is formed by the translation (T_x, T_y, T_z), rotation (R_x, R_y, R_z) and the torsion motions (T_{bi}, i=1,2, …, n, n is the number of torsion bonds). The former six variables are the six degrees of freedom for rigid segment, it can also be seemed as the orientation of the ligand. T_{bi} is the angle of the ith flexible bond. For GA, each chromosome consists of the above three design variables that represent a ligand in a particular conformation and orientation. The design space of (T_x, T_y, T_z) must be limited in the spheres (get by MS package) space of the receptor. A Circum-cuboid of the sphere space is here used to confine the rigid body's coordinates, which can greatly avoid the computational complexity of resolving the actual boundary. The rest variables are allowed to vary between $-\pi$ and π rad.

4.2 Drug Molecular Docking Example

As an example, the PLD process between sc-558 and its receptor in known cyclooxygenase-2 (COX-2) inhibitors is investigated on a distributed memory MIMD parallel computer Tianchao Dawning 3000. Parameters used by GA are: population numbers

6, population size 20, crossover rate 0.65, mutation rate 0.1 and length 19 of each chromosome. The docking energy score of the best conformation obtained is -3.27 kcal/mol kcal/mol lower than -1.65 kcal/mol kcal/mol of DOCK5.0 [10] program. Fig.2 shows the comparisons of conformations corresponding to the model molecule. The CPU times needed are 476.64 seconds and 15.26 seconds respectively

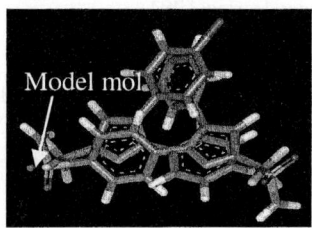 

Fig. 2. Conformations corresponding to the model molecule: The left one is resulting conformation profile of this paper and the right is the resulting conformation profile of DOCK5.0

5 Conclusions

An entropy-based multi-population Genetic Algorithm for nonlinear programming problems has been presented. The method is based on entropy-based searching technique with narrowing down space of multi-population. A new iteration scheme in conjunction with multi-population genetic strategy, entropy-based searching technique with narrowing down space and the quasi-exact penalty function is developed to ensure rapid and steady convergence. Numerical examples and an application in molecular docking show that the proposed method results in considerable savings in computer time and approximate to global solution in design optimization. Although an application for drug molecular docking design is here given only, it is also suitable for other engineering fields.

Acknowledgments

The authors gratefully acknowledge financial support for this work from the National Natural Science Foundation （10272030） and the Subsidized by the Special Funds for Major State Basic Research Project (G1999032805) of China. Additionally, appreciate the Kuntz group (UCSF) for supplying DOCK program.

References

1. D. E. Goldberg: Genetic Algorithms in Search, Optimization & Machine Learning, Reading. Addison Wesley, (1989)
2. Wu, Jin-ying, Wang, Xi-cheng: A Parellel Genetic Design Method with Coarse Grain. Chinese Journal of Computational Mechanics, 19(2)148-153(2002), (in Chinese)
3. Li, Xing-si: A Quasi-exact Penalty Function Method for Nonlinear Program [J]. Chinese Science Bulletin, 36,1451-1453 (1991)

4. Elements of Information Theory, New York: Wiley, (c1991)
5. Charalambous C.: Nonlinear Least Pth Optimization and Nonlinear Programming. Mathematics Programming, 12, 195-225 (1977)
6. Bazaraa MS, Shetty LM: Non-linear Programming: Theory and Algorithms. New York: Wiley (1993)
7. Floudas C A, Pardalos P M.: A Collection of Test Problems for Constrained Global Optimization Algorithms. LNCS, Vol. 455, Springer-Verlag(1987)
8. He, Xiong-jun, Sun, Guo-zhen, Liu,Gang: Random Perturbation Method of Genetic Algorithms. J. Wuhan Univ. (Nat. Sci. Ed.), 47, 285-288 (2001) (In Chinese)
9. Kubinyi H,Burger's Medicinal Chemistry and Drug Discovery, Fifth Edition, Vol.(1): Principles and Practice (ed. Wolff M E), John Wiley & Sons, Inc., New York, 497-571 (1995)
10. DOCK5.0.0. Demetri Moustakas. Kuntz Laboratory, UCSF, 4,15 (2002)

Reinforcement Learning Based on Multi-agent in RoboCup[*]

Wei Zhang, Jiangeng Li[1], and Xiaogang Ruan

School of Electronic Information & Control Engineering
Beijing University of Technology, Beijing, 100022, P.R. China
dorm204@hotmail.com
lijg@bjut.edu.cn, adrxg@bjut.edu.cn

Abstract. Multi-agent systems form a particular type of distributed artificial intelligence systems. As an important character of players in game, autonomous agents' learning has become the main direction of researchers. In this paper, based on basic reinforcement learning, multi-agent reinforcement learning with specific context is proposed. The method is applied to RoboCup to learn coordination among agents. In the learning, the game field is divided into different areas, and the action choice is made dependent on the area in which the ball is currently located. This makes the state space and the action space decrease. After learning the optimal joint policy is determined. Comparison experiment between stochastic policy and this optimal policy shows the effectiveness of our approach.

1 Introduction

Multi-agent systems form a particular type of distributed artificial intelligence systems. In fact, many real-world problems such as engineering design, intelligent research, medical diagnosis, robotics, etc require multiple agents. From an AI perspective, we can think of a multi-agent system as a collection of agents that coexist in an environment, interact (explicitly or implicitly) with each other, and try to optimize a performance measure [1].

RoboCup (Robot World Cup) offers an integrated research task which covers many areas of AI and robotics [2], [3]. As a testing platform for the designing and research of the multi-agent system, RoboCup mainly studies the advanced functions of Robot football team, which include design principles of autonomous agents, multi-agent collaboration, strategy acquisition, real-time reasoning, reactive behavior, real-time sensor fusion, learning, vision, motor control, intelligent robot control, and any more. As presented in detail by [2], simulation RoboCup soccer is a fully distributed, multi-agent domain with both teammates and adversaries. There are hidden states, meaning that each agent has only a partial world view at any given moment. The agents also have noisy sensors and actuators, meaning that they do not perceive the world exactly as it is, nor can they affect the world exactly as intended. In addition,

---

[*] This paper is supported by the National Science Foundation of China (60375017).
[1] Corresponding author.

the perception and action cycles are asynchronous, prohibiting the traditional AI paradigm of using perceptual input to trigger actions. Communication opportunities are limited, and the agents must make their decisions in real-time. These italicized domain characteristics combine to make simulated robot soccer a realistic and challenging domain. As an important character of players in game, autonomous agents' learning has become the main direction of researchers, because the robot soccer has the features such as dynamic, real-time, competitive, not-full information, etc.

Multi-agent systems have been successfully utilized for reinforcement learning, which is a learning technique that requires almost nothing about the dynamics of the environment to learn about. Autonomous agents learn from their environment by receiving reinforcement signals after interacting with the environment. Learning from an environment is robust because agents are directly affected from the dynamics of the environment.

This paper presents a reinforcement learning method with specific context based on multi-agent for players' coordination in simulation RoboCup soccer. In the learning, the game field is divided into different areas, and the action choice is made dependent on the area in which the ball is currently located. This makes the state space and the action space decrease. After learning the optimal joint policy is determined.

The rest of the paper is organized as follows. The concept and algorithm of basic reinforcement learning are introduced in section 2. Reinforcement learning algorithm with specific context based on multi-agent is presented in section 3. The application of the multi-agent group reinforcement learning in RoboCup is given in section 4. Conclusions and future work are included in section 5.

2 Reinforcement Learning

2.1 Reinforcement Learning Model and Algorithm

Reinforcement learning requires learning from interactions in an environment in order to achieve certain goals [4]. The method finds an optimal policy through trail-and-error, different from supervised learning that informs the agent which action is taken by positive examples and counter examples. The entity interacting with its environment by actions is called agent. At each time step, an agent observes its environment and selects the next actions based on that observation. In the next time step, the agent obtains the new observation that may reflect the effects of its previous action and a payoff value indicating the quality of its previous action. Shown in Fig. 1 is the generic Reinforcement Learning framework, which consists of an environment and an agent.

An agent is composed of three main components, namely sensors, actuators and the control system. An agent observes the environment through its sensors and presents an indication of that (denoted by s) to the control system. The control system in turn decides on the action to be taken (denoted by a) and reports that action to its actuators. This action changes the environment. Upon this change, the environment reports a reward or punishment value (denoted by r) back to the agent. Based on this cycle, the control system tries to learn optimal action selection (optimal policy) in the

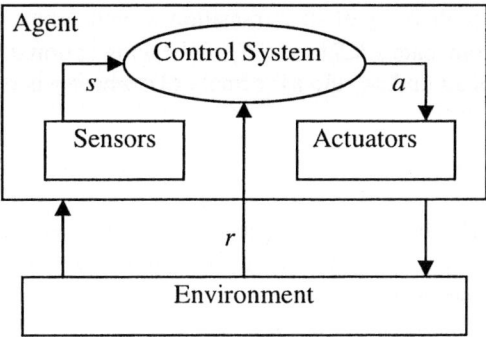

Fig. 1. Reinforcement learning framework

agent's environment. The dynamic of the environment is called its model. The objective function (value function) is set to $V^\pi(s)=r_0+\gamma r_1+\gamma^2 r_2+...$, where γ is discount factor, $0<\gamma<1$. The agent's task is to learn the control policy $\pi: S \to A$, which can maximize the expectation of the reward sum, that is, in the changing series of environment, select actions

$$S_0 \xrightarrow{a_0 \cdot \gamma_0} S_1 \xrightarrow{a_1 \cdot \gamma_1} S_2 \xrightarrow{a_2 \cdot \gamma_2} \cdots$$

to maximize the objective function. So we obtain the optimal control policy through the formula (1):

$$\pi^* = \arg\max_\pi V^\pi(s), \quad \forall s \in S . \tag{1}$$

Reinforcement learning can solve a large dynamic programming with any a prior. A general form of reinforcement learning is [5]:

(1) Initial the inner state S of a learner as S_0,
(2) Loop:
    Observe the current state s;
    Select an action a by value function V;
    Perform action a;
    Let r be the immediate reward from action a at state s;
    Update inner state by updating function, s→s'.

2.2 Q-Learning

Q-learning is a kind of reinforcement learning independent of model proposed by Watkins. In the iteration Q-learning introduces the sum of reward value $Q^*(s, a)$ of state-action pairs as evaluation function, the target is the map from state-action to its corresponding Q value. The optimal value Q^* is defined as the maximization of the sum of discounted reward value which an agent gets when action a is taken in state s, as followed by formula (2),

$$Q^*(s,a) = r(s,a) + \gamma \sum_{s \in S} T(s,a,s') \max_{a'} Q^*(s',a') . \tag{2}$$

where $T(s, a, s')$ is probability of state transition, which denotes the probability of the agent transferring from state $s \in S$ to $s' \in S$ after taking action a.

And Q value has an update rule as formula (3), where γ is discount factor, $0<\gamma<1$; α is learning rate, $0<\alpha<1$.

$$Q(s_t, a_t) \leftarrow Q(s_t, a_t) + \alpha[r_t + \gamma \max_a Q(s_{t+1}, a) - Q(s_t, a_t)] \ . \tag{3}$$

First initialize $Q(s, a)$ arbitrarily, after each action a is taken, the value $Q(s, a)$ is updated by the formula (3), and the value Q will gradually converge to Q^* through repeatedly searching state space.

3 Reinforcement Learning Based on Multi-agent

Reinforcement learning is based on single agent, and what we concern is how multi-agent can learn to cooperate and coordinate in a game so as to fulfill a common goal. In Q-learning, the research is mainly on the interaction between a single agent and the environment, and other agents are only taken as a part of the environment. But in practice, an agent is of initiative and is different from the environment, and its action has intention and adaptation, so there needs some extension of the existing Q-learning so as to make it fit for the environment.

The most direct approach to extend Q-learning to a multi-agent environment is to regard the complete system as one large single agent in which the joint action is regarded as a single action [6]. Although this approach leads to the optimal solution, it is infeasible to solve problems with many agents by this approach, since the joint state-action space, which is exponential in the number of agents, becomes intractable.

To solve the problem mentioned above, a concept of chief agent is introduced [7]. In RoboCup, the player controlling the ball is of course the chief agent. By taking the chief agent as the learner and taking the other agents as assistant agents, who cooperate with each other in specific state, and through posts shift between chief agent and assistant agent, the learning of the whole group is realized.

So on the basis of the learning algorithm presented in [8], [9], we propose one method of reinforcement learning based on multi-agent under the specific state, as shown in Fig. 2. Here $S_{\_end}$ represents the state which the ball is out of the game field or is kicked into one of the goals. The characters of this solving method are:

(1) Multiple agents have a common long-time goal, so they can form an agent group to compete against other agent group.
(2) At some specific time, with specific context the policy of agent group only is embodied in the action selection of the chief agent, because only the chief agent can perform learning. Other assistant agents can only serve the chief agent.
(3) The action performed by chief agent may make other assistant agents the chief agent, and it self the assistant agent, that is, chief and assistant agents may transform their posts mutually.

> Initialize $Q(s, a)$ arbitrarily, assign chief agent 0 and assistant agents 1~n according to the specific state, and let agent policy be $\pi = \{\pi_1, \pi_2, ..., \pi_n\}$
> Repeat (for each episode):
>   Initialize S (including the beginning state S_0 and end state S_{end})
>   For agent i=1~n
>   Observe state s, take action a, get reward r, s'
>   Compute Q_i, learn optimal policy π
>   Transfer action a to chief agent 0
>   For all the agents, update the Q value
> $Q(s, a) \leftarrow (1-\alpha)Q(s, a) + \alpha[r + \gamma \max_{a' \in A} Q(s', a')]$
> $S \leftarrow S'; a \leftarrow a'$;
> Until $S = S\_{end}$

Fig. 2. Algorithm of reinforcement learning based on multi-agent

4 Experiments

We applied our method to the RoboCup to realize the actions' coordination of players. The RoboCup soccer server provides a fully distributed dynamic multi-robot domain with both teammates and adversaries. It models many real-world complexities such as noise in object movement, noisy sensors and actuators, limited physical ability and restricted communication.

In RoboCup, a player is regarded as an autonomous agent. Autonomous agents' learning has become the main direction of researchers, because the robot soccer has the features such as dynamic, real-time, competitive, not-full information, etc. Agents have a common goal, which is to win the game. But every agent also has itself temporary goal according to its current state. In order to satisfy his goals, an autonomous agent must select at each moment in time the most appropriate action among all possible actions that he can execute. In Simulation RoboCup Soccer, the team should possess basic ability of counterwork to ensure agents having the chance of learning. In this paper, we assume that every player has mastered basic actions such as kick, turn, dash, etc.

Since soccer server provides a highly dynamic environment it is difficult to create a fixed attacking plan or any other form of premeditated action sequence that works in every situation. For decreasing state-action space we apply multi-agent reinforcement learning with specific context to realize action coordination among several players in this section.

4.1 State-Action Selection

In order to reduce state-action space, we have divided the field into different areas (see Fig. 3) and made the action choice dependent on the area in which the ball is currently located [10]. For example, an agent who has the ball will consider different actions when he is on his own half from those when he is close to the opponent's goal. Assuming that an agent has the ball on the area 3a, we analyze the player's state-action space so as to study actions' coordination between the player who controls the ball and other players.

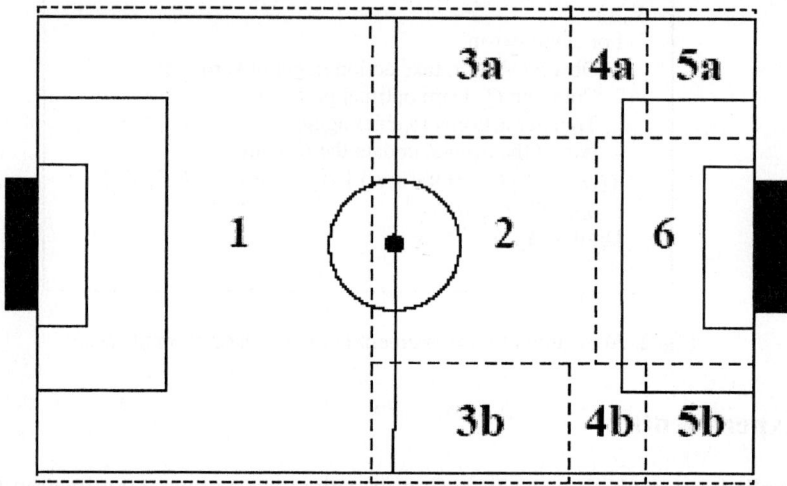

Fig. 3. Areas on the field which are used for action selection when the ball is kickable

In the experiment, let the player who controls the ball be chief agent, and other teammates within chief agent's range of vision be assistant agents (where *range of vision* refers to the range in which the chief agent can distinguish both of team and number). Let coordination be limited among about 3 players. Here we only consider states in close relation with action choice. The components of a state include position of the ball p, angle of chief agent relative to the ball θ_1, distance between chief agent and the closest teammate d_1, distance between chief agent and the closest opponent d_2, angle of chief agent relative the closest teammate θ_2, angle of chief agent relative to the closest opponent θ_3. The actions ball-controller can take include *dribble* (), *passTo* (k), and *turn* (). Here *passTo* (k) denotes that the ball-controller kick the ball directly towards teammate k.

At the beginning of learning in action selecting, because lack of experience about the state, the Q value not always represent the reinforcement value correctly. And usually because selecting an action of a highest Q, agent will seek along the same random factors when agent selecting action. So the selected actions according to the current Q value are not the optimal. In order to explore unknown space effectively, agent should select action stochastically. The usually used method is Boltzmann distribution which is denoted as formula (4):

$$prob(a_i) = \frac{e^{Q(s,a_i)/T}}{\sum_{a_k \in A} e^{Q(s,a_k)/T}}, \quad a_i \in A \ . \tag{4}$$

where *prob(a_i)* represents the probability of performing action a_i, A is the action set from which the chief agent select one action a_i, T is the temperature, which represents the size of randomicity. As T gets larger, then randomicity gets larger too. At the beginning of learning, set T a larger value, because learning experience is less then, and needs to increasing the searching ability after, as learning carries on, lower T gradually, lest the obtained learning effect be destroyed.

4.2 Algorithm Implementation

At first we define value of the reinforcement. The goal of RoboCup is to kick the ball in the opponent's goal. But because score is fewer in the match, if regarding it as the only factor of rewards and punishments signal to learn, the result will be certainly very bad. Hence we regard $r = \Delta d/5$ as reinforcement signal, where Δd denotes change of distance between the ball and opponent's goal. Δd is positive when the ball going forward and negative otherwise. Additional rewards will be given when our team scores.

We applied the algorithm which was introduced in section 3 to RoboCup when the ball is located in area 3a. For example, one scene appeared possibly in the game is shown in Fig. 4. In the figure, the ball-controller is agent 0, while other teammates in his vision range are agent 1 and agent 2. Every state value can be calculated. And actions the ball-controller can execute possibly are *dribble* (), and *passTo* (*i*) (here *i* = 1, 2). Through learning the ball-controller can take correct action. And cooperation in small scope would be realized. The specific implementation of the algorithm in RoboCup should be as follows in this paper.

Firstly we designated state sets and action sets including *passTo* (*k*), *dribble* (), and *turn* (). Then we initialized Q value randomly, appointed the ball-controller as chief agent and its teammates within its range of vision as assistant agents (usually including 2~3 agents). The chief agent chose an action to execute according to the probability determined by formula (4) and got rewards according to the result of implementation, where $T = 0.1$. And then Q value was updated through formula (3) mentioned by section 2.2, where discount factor $\gamma = 0.9$, learning rate $\alpha = 0.5$. Here Q value was computed by artificial neural network of three layers [7]. The network's input was every state value, and output was Q value obtained when one action was taken. And the hidden layer is set to 12 neurons. Finally the actions leading to the maximal Q value would be chosen to execute.

In the experiment we selected the robot soccer team of Beijing Institute of Technology, Everest, as training opponent, while our team adopted stochastic policy or policy yielded by our approach. Because score is fewer in the real match, here we only analyzed three parameters which have important influence on the result of the match such as success ratio of passing ball, dribbling ball, and time of controlling the ball, the results shown as Table 1. Table 1 shows the average values of parameters after 100 games. In the table, we can find the success ratio of passing ball changes from 48.8% to 55.3% after our team took policies yielded by our method. In the same time the success ratio of dribbling ball changes from 66.2% to 71.6% while time of controlling ball increase to 32.8% from 23.5%. It indicates that the ability of coordination and controlling ball of the team improved obviously. The policy yielded by our method is effective in the experiments.

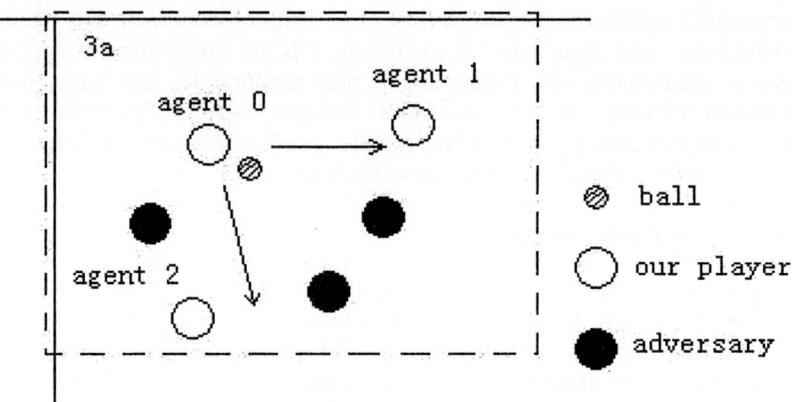

Fig. 4. One scene when the ball is located in Area 3a

Table 1. Contrast of parameters under different policies

Average (after 100 games)	Success ratio of passing ball (%)	Success ratio of dribbling ball (%)	Time of controlling ball (%)
Stochastic policy	48.8	66.2	23.5
Policy yielded by our approach	55.3	71.6	32.8

5 Conclusions and Future Work

Reinforcement learning is a very effective online machine learning method. In the typical multi-agent system of RoboCup, we used the algorithm of reinforcement learning based on multi-agent in specific context to realize coordination among 2~3 agents, and then realize common goal, which is to win the game. In the experiment, we considered only players' cooperation with specific state, so the state-action space was decreased. And then the optimal policies would be obtained after many matches. Experiment shows that the approach plays a good role to the improvement of team's integral level.

From the experiment we can conclude that the number of agents (players) using the learning approach must be fixed carefully. As to a group of 11 players, the test result is not ideal, but to a group of 2~3 players, the corresponding small scope cooperation can get a much better result. So only allowing a few players in a certain scope around the ball-controller into the learning can deal with the cooperation problem in a small scope. And because the convergence velocity of learning is slow, we can get the better policy only after many matches. So there is one problem of computational cost in practical.

In this paper, we only discussed cooperation under special state. As future work, we would like to adjust our approach to realize the cooperation in a bigger scope including more players. Another interesting direction is to find some better efficient algorithms to save computational cost.

References

1. Wooldridge, M.: An Introduction to Multiagent Systems. John Wiley and Sons Ltd. (Feb. 2002)
2. Kitano, H., Tambe, M., Stone, P., et al: The RoboCup Synthetic Agent Challenge 97[A]. RoboCup-97: Robot Soccer World Cup I[C]. Berlin. Springer Verlag. (1998) 62 - 73
3. Kitano, H., Asada, M., Noda, I., Matsubara, H.: RoboCup: Robot World Cup. Robotics & Automation Magazine. IEEE. Vol. 5. Issue 3. (Sept. 1998) 30 – 36
4. Sutton, R.S., Barto, A.G.: Reinforcement Learning: An Introduction. MIT Press. Cambridge. MA. (1998)
5. Mitchell, T.M.: Machine Learning [M]. McGraw-Hill Companies. Inc. (1997)
6. Abul, O., Polat, F., Alhajj, R.: Multiagent Reinforcement Learning Using Function Approximation. IEEE Transactions on Systems, Man, and Cybernetics-Partc. Application and Reviews. Vol. 30. (Nov. 2000)
7. Wei Meng, Bingrong Hong, Xuedong Han: Application of Reinforcement Learning to Robot Soccer [J]. Application Research of Computers. (2002) 79 - 81
8. Xianyi Cheng, Xiaohua Yuan, Linghan Pan, Deshen Xia: Reinforcement Learning in Simulation RoboCup Soccer. Proceedings of the Third International Conference on Machine Learning and Cybernetics. Shanghai. Vol. 1. (Aug. 26-29. 2004) 244 - 248
9. GuoQuan Wang, Haibin Yu: Multi-Agent Reinforcement Learning: An Approach Based on Agents' Cooperation for A Common Goal. The 8th International Conference on Computer Supported Cooperative Work in Design Proceedings. Vol. 1. (May 26-28. 2004) 336 - 339
10. R. de Boer, Kok, J.R.: The Incremental Development of A Synthetic Multi-Agent System. The UvA Trilearn 2001 Robotic Soccer Simulation Team. Master's thesis. University of Amsterdam. The Netherlands. (Feb. 2002)

Comparison of Stochastic and Approximation Algorithms for One-Dimensional Cutting Problems

Zafer Bingul and Cuneyt Oysu

Department of Mechatronics Engineering,
Kocaeli University, 41040, Kocaeli, Turkey
{zaferb, coysu}@kou.edu.tr

Abstract. The paper deals with the new algorithm development and comparison of three one-dimensional stock cutting algorithms regarding trim loss. Three possible types of problems used in this study are identified as easy, medium and hard. Approximate method is developed which enables a comparison of solutions of all three types of problems and of the other two stochastic methods. The other two algorithms employed here are Genetic Algorithms (GA) with Improved Bottom-Left (BL) and Simulated Annealing (SA) with Improved BL. Two examples of method implementation for comparison of three algorithms are presented. The approximate method produced the best solutions for easy and medium cutting problems. However, GA works very well in hard problems because of its global search ability.

1 Introduction

One dimensional cutting problem is to cut a given set of items of different sizes into a minimum number of equal-sized bins. The basic goal of the one dimensional cutting problem is to determine optimal patterns as well as how many times each pattern should be used. It is encountered in many industrial processes such as wood, glass, steel and paper industries. One-dimensional stock cutting problems are known to be Non-deterministic polynomial complete. No procedure can solve these problem instances in polynomial time. In the one-dimensional cutting stock problem, (CSP) smaller lengths are generally to be cut from a minimum number of identical stock pieces. The solution to this problem can be found by many ways like approximate methods, evolutionary methods and heuristics.

The cutting problems may be categorized into dimensionality, kind of assignment, assortment of large objects and assortment of small items. Their solutions may be categorized into item-oriented and pattern oriented approach [1]. Traditional objectives in CSPs are to minimize the trim loss and stock usage.

In this paper, three different CSP solution methods are compared based on both trim loss and stock usage. These methods are the approximate method, genetic algorithm (GA) and simulated annealing (SA) approach methods. The first method is based on column generation technique with local search algorithm. Last two methods however find overall solution to the problem using bottom left algorithm. Thus stochastic algorithms outperform the LP based solution for CSPs with different stock lengths. Especially, GA has been applied to cutting and packing problems by many researchers [2].

The method's performances are tested using 6 benchmark problems selected from [3]. These benchmark data sets are classified through the number of items, the bin capacity and the length of all items. The benchmark problems are grouped into three different data sets. Two problems from each data set were selected to be used in this study.

2 Placement Algorithms

Solution approach of the cutting problems can be divided into two sections: determining of permutation for pieces layout sequences and applying placement algorithm for placing all pieces. There are many placement algorithms based on sliding technique such as difference process (DP) algorithm [4], bottom left (BL) algorithm [5], improved bottom left (I-BL) algorithm [6] and bottom left fill (BLF) algorithm [7]. The placement algorithm using the permutation is executed to place rectangular pieces into the main object.

The improved bottom left algorithm is mentioned below. The aim of this algorithm is that a piece cannot be moved any further to the left and downwards without collision.

A layout pattern can be represented by a permutation π as shown below:

$\pi=(i_1...i_n)$ – Permutation   i: index of item (L_i)

Allocation of some of pieces is illustrated in Fig. 1. As can be seen in Fig. 1, piece L_3 is placed first to the bottom and then to the left as far as possible. Pieces L_2 and L_1 are placed in the same manner. The arrows in the figure depict the movement of the piece L_8 to its optimal location.

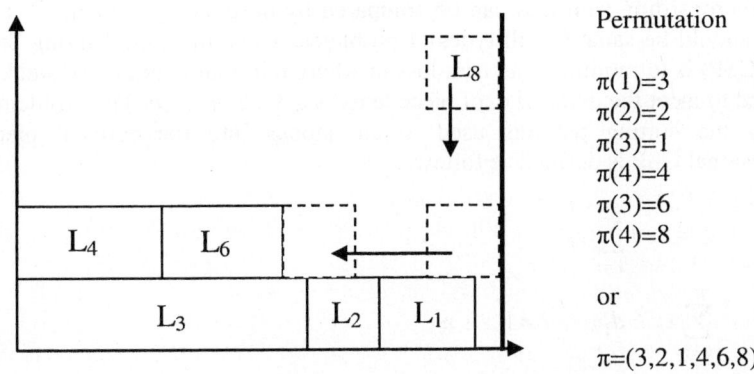

Fig. 1. Illustration of the improved bottom left algorithm

The number of possible layout patterns allocated by BL algorithm is $2^n \cdot n!$ for given the number of n pieces [5]. Fig. 2 shows the number of possible layout patterns versus number of pieces. As can be seen in Fig. 2, a number of possible solutions to the cutting problems are getting very high as the number of pieces increases. Therefore, they require very large search space and make the cutting problems very hard to solve.

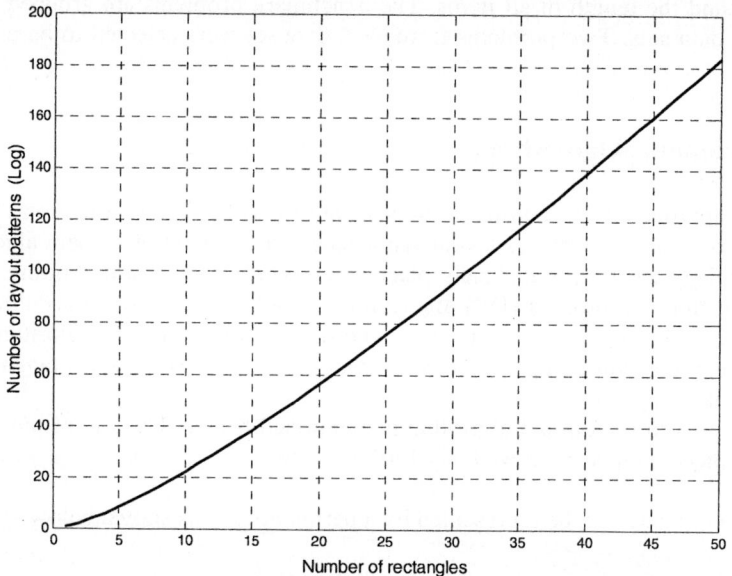

Fig. 2. Number of layout patterns according to number of items

3 The Cutting Stock Problem

The trim loss is based on different cutting stock problems and approaches. Solution methods regarding trim loss can be compared by introducing the trim loss definition which should be same for all types of problems. The aim of the Cutting Stock Problems (CSP) is to minimize the trim loss in where minimum number of stock lengths L are used to meet the demand d_i of piece lengths a_i for $i=1, …,n$. The problem solutions specify the cutting patterns used to cut stocks into the demand pieces. One-Dimensional CSP is defined as follows:

$$z = \min \sum_{k=1}^{K} y_k$$

$$\sum_{k=1}^{K} x_k \geq d_i \quad i=1,……,n$$

$$\sum_{i=1}^{n} a_i x_{ik} \leq L y_k \quad k=1,……K \tag{1}$$

$$y_k \in \{0,1\} \quad k=1,……K$$

$$x_{ik} \geq 0 \quad and \quad i=1,……,n; k=1,……,M$$

where $y_k=1$ if the piece k is used or 0 otherwise, x_{ik} is the frequency of item i used in the piece k and K is the number of stock lengths expected to be used. The total length of the items for a piece is L. Trim loss is the difference between the stock length and L.

3.1 The Approximate Method

The approximate method can be defined as follow. First of all, the following algorithm searches for combinations with minimum trim losses.

$$\begin{aligned}
&\text{For } i_1=1 \text{ to } n \\
&\text{For } i_2=1 \text{ to } n \\
&\quad . \\
&\text{For } i_p=1 \text{ to } n \\
&\text{Max}(d(i_1) + d(i_2)+....d(i_p)) \leq L \\
&\text{End for } i_p \\
&\quad .. \\
&\text{End for } i_2 \\
&\text{End for } i_1
\end{aligned} \quad (2)$$

where $p = \text{int}[L / avg(\sum_{i=1}^{n} d_i)]$ is permutation level.

Then the cutting list is extracted from the main demand list and new search starts. The fundamental idea of the search algorithm is to take permutation at a predefined level p. The permutation level however is found from the mean of the demand list. Thus, the search is localized to a certain number of combination members. This decreases the computation time considerably. If various levels of permutations are to be searched for minimum trim loss, the computation time increases significantly for a solution. Since the demand list is sorted in descending order, the biggest piece is the first item in the list. Taking the first member initially in every search makes it sure that worst piece is removed from the list. Experience shows that when smaller pieces are to be combined, the trim losses are less significant. In the search algorithm the computation time is reduced by using some extra stop criteria. The search is stopped when the total length of the cutting pattern is equal to the stock length during the search. The flow chart in Fig. 3 illustrates the approximate solution algorithm.

3.2 Genetic Algorithm Approach

GA is a search algorithm inspired by the natural reproduction and evolution of the living creatures. GA, firstly developed by Holland [8], is applied effectively to solve various combinatorial optimization problems. Before applying the GA, a representation scheme to encode problem solutions into placement algorithm must be defined.

In the GA solution approach applied here, an order-based GA is combined with improved BL algorithm to solve the one dimensional rectangular cutting problems. This solution approach is known as hybrid GA. In the first part of GA solution

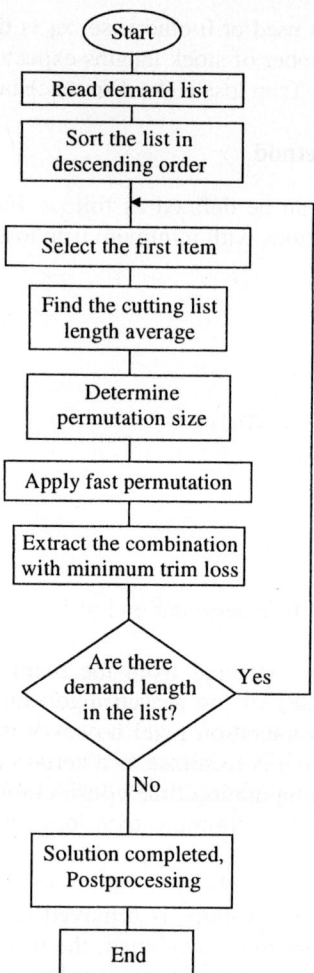

Fig. 3. Flow chart of the approximate solution for CSP

approach, the influences of different population sizes and mutation rates were examined to find the best GA parameters for the cutting problems. The hybrid GA was employed to solve different test problems consisting of different pieces. For each run, the number of iterations was set to 1000. The GA parameters, population sizes and mutation rates, were changed between 20-100 and 0.01-1 respectively. During this part of work, Stefan Jakobs crossover technique was used. The best GA parameters were taken as population size of 80 and mutation rate of 0.7. In the second part of GA solution approach, the influences of different crossover techniques (OBX, CX, OX, PMX, UX, SJX) on the solutions of the cutting problem were studied using the best GA parameters obtained from the first part of GA solution approach. In order to see clearly the influences of different crossover techniques on the solution of the cutting problem, much more difficult cutting problems were chosen. Six crossover techniques; order based crossover (OBX), cycle crossover (CX), order crossover (OX), partially matched crossover (PMX), uniform crossover (UX), Stefan Jakobs Crossover (SJX) were used for illustrating effects of crossover on the solution. It was seen that OBX crossover technique produced better results than the other technique. The individuals obtained by using OBX technique are changed extensively as compared to other crossover techniques. Thus there exists enough diversity in population when OBX crossover technique is used. All results related with GA solution approach were given below.

GA parameters used in this study are summarized below:

Representation   : Permutation
Population size   : 20, 40, 60, 80, 100
Selection   : Roulette Wheel
Mutation rate   : 0.01, 0.1, 0.3, 0.6, 0.7, 0.9, 1.
Mutation technique : Swap mutation.
Crossover techniques : OBX, CX, OX, PMX, UX, SJX.
Number of generation to termination: 100-1000

3.3 Simulated Annealing Approach

SA is based on the analogy between the process of finding a possible best solution of a combinatorial optimization problem and the annealing process of a solid to its minimum energy state in statistical physics. SA, firstly developed by Kirkpatrick [9], is a local search algorithm.

In our SA solution approach, SA and improved BL algorithm (known as hybrid SA) are combined to solve the same cutting problems. In the first part of SA solution approach, it was tried to find the best SA parameters for cutting problems. Therefore, many different SA parameters were used to study influences of the parameters on the solutions. The hybrid SA was applied to solve the different test problems. The influences of different cooling schedules (proportional decrement schedule and Lundy and Mees schedule), neighborhood moves (swapping move and shifting move) and values for equilibrium condition (3, 5, 10) on the solution of the cutting problems were examined for the different temperature values (initial temperatures were changed between 0.1 and 0.8). In the second part of SA solution approach, the solution for test problems was studied using the best SA parameters obtained from previous work. It was seen that swapping move worked better than shifting move in the different temperature values. The swapping move and two different cooling schedules (proportional decrement schedule and Lundy and Mees schedule) were used to compare the cooling schedules. Possible number of neighborhood moves is changed between 3 and 10. It was seen that Lundy and Mees schedule yielded better than proportional decrement schedule and the best results were obtained with using 3 for the possible number of neighborhood moves. All results related with SA solution approach were given below.

SA parameters used in this study are summarized below:

Representation : Permutation
Initial Temperatures : 0.1, 0.2, 0.3, 0.4, 0.5, 0.6, 0.7, 0.8
Final Temperature : 0.01
Neighborhood move : Swapping and Shifting.
Cooling schedule : Proportional Decrement Schedule and Lundy and Mees Schedule.
Termination criteria : Maximum number of iteration (100-1000)
Possible number of neighborhood moves (pnm): If there is no improvement in the foregoing (3, 5, 10) moves.

3 Results and Discussions

In order to examine and evaluate the performance of the three algorithms introduced above, computational experiments were conducted. Six data sets have been generated for testing the algorithms. These data sets are available via the internet (http://www.wiwi.uni-jena.de/Entscheidung/binpp/) which is designed as a benchmark problem sets by Scholl and Klein [3]. Data sets are constructed as in Table 1.

Table 1. Details for data sets

	Pr.1	Pr.2	Pr.3	Pr.4	Pr.5	Pr.6
# of items	50	50	100	100	200	200
Stock length	1000	1000	1000	1000	100000	100000

All tests were performed on a Pentium 4 (2.4 GHz) computer. There are three categories of problems sets easy, medium and hard. Two problems were selected from each category. The results obtained from Problems 4 and 6 is shown in Fig. 4. All results are summarized in Table 2. As can be seen in table easy and medium problems can be solved with minimum waste using approximate method. When the problems are getting harder GA outperforms the other methods. The reason for this is the size of search space. As the search space increases GA performance for searching becomes more efficient compared to other methods. The approximate method gives very accurate results for easy and medium cases in a very short computational time. GA gives better results compared to SA when the global search is the nature of the problem. However, SA is more efficient in local search problems. Because of stochastic nature of GA and SA algorithms, all problems were run several times.

Table 2. Percentage of trim loss for each problem

	Pr.1	Pr.2	Pr.3	Pr.4	Pr.5	Pr.6
Approximate meth.	0.15	0.12	0	4.05	5.02	7.48
SA	1.44	12.59	3.76	10.41	4.80	7.12
GA	1.31	10.45	3	9.4	4.44	6.82

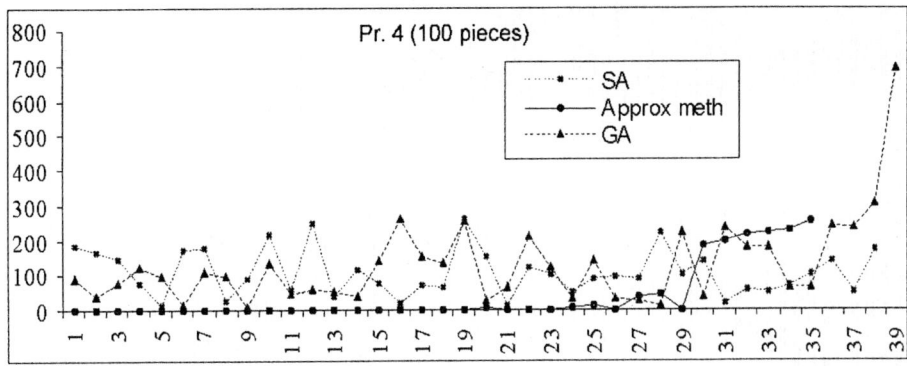

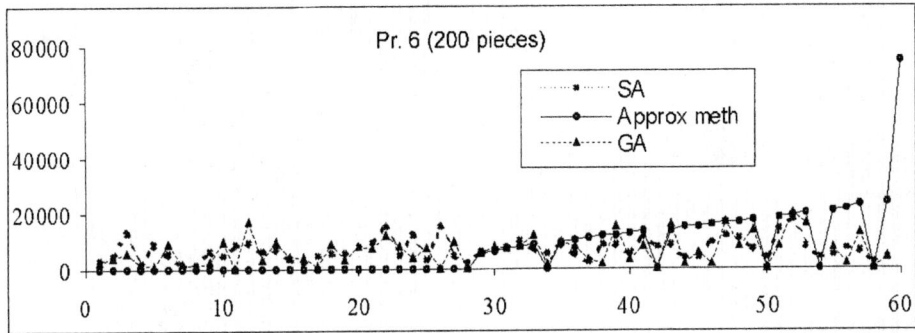

Fig. 4. Comparison of three algorithm's solutions for Pr.4 and Pr.6

Fig 5. illustrates the results obtained from approximate method in the form of column generation. As can be seen in Fig 5., very best combinations extracted from the cutting stock at the first half of the computation. However, in SA and GA approach trim losses are distributed homogenously through the computation.

4 Conclusions

The article examines the three algorithms (GA, SA and Approximate method) for one-dimensional stock cutting problems. The traditional optimization techniques (i.e., linear programming and integer programming) suffer some drawbacks when they are used to solve the one-dimensional CSP. There are many algorithms and methods for one-dimensional stock cutting with different factors that are taken into account. But the most important common factor is the trim loss. The three algorithms were compared based on trim loss for one-dimensional stock cutting problems. Conclusions based on the results obtained in this study, the best search algorithm for easy and medium problems is the approximate method. On the contrary, GA works very well in hard problems because of its global search ability.

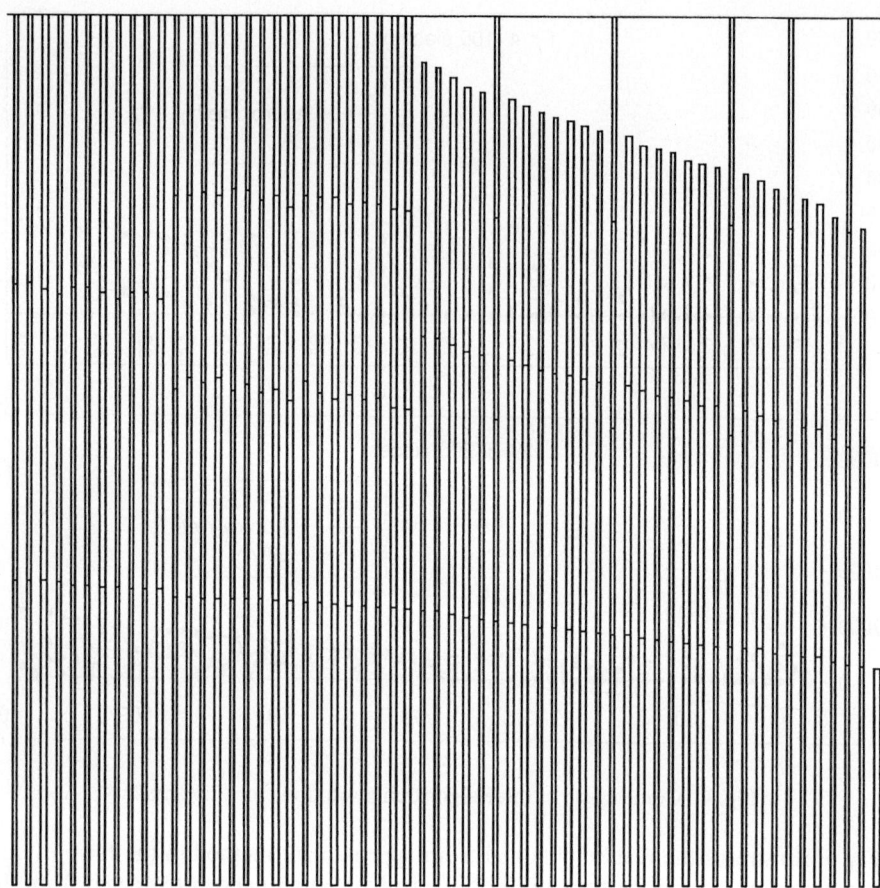

Fig. 5. Results generated from the approximate algorithm for Pr.5

References

1. Dyckhoff H.: A typology of cutting and packing problems, European Journal of Operational Research 44 (1990) 145-159
2. Liang K., Yao X., Newton C. and Hoffman D.: A new evolutionary approach to cutting stock problems with and without contiguity, Computers & Operations Research, 29 (2002) 1641-1659
3. Scholl A., Klein R., and Jürgens C.: BISON: A fast hybrid procedure for exactly solving the one-dimensional bin packing problem, Computers & Operations Research 24 (1997) 627-645
4. Leung T. W., Yung C. H., Chan C .K.: Applications of Genetic Algorithm and Simulated Annealing to The 2-Dimensional Non-Guillotine Cutting Stock Problem, IFORS '99, (1999) Beijing China
5. Jakobs, S.: On genetic algorithms for the packing of polygons, European Journal of Operational Research, Vol. 88, (1996) 165-181

6. Liu, D. and Teng, H.: An Improved BL Algorithm for Genetic Algorithm of the Orthogonal Packing of Rectangles, European Journal of Operational Research, Vol. 112. (1999). 413-420
7. Hopper, E., Turton, B.C.H.: A genetic algorithms for a 2D industrial packing problem, Computers in Engineering, Vol 37, (1999) 375-378
8. Holland, J.H.: Adaptation in Natural and Artificial Systems, University of Michigan Pres, Ann Arbor, MI, (1975)
9. Kirkpatrick, S., Gelatt, C.D. and Vecchi, M.P.: Optimization by Simulated Annealing. Science, New Series, Vol. 220, (1983) 671-680
10. Goulimis C.: Optimal solutions for the cutting stock problem, European Journal of Operational Research 44 (1990) 197-208
11. Gradišar M., Kljajić M., Resinovič G.: A hybrid approach for optimization of one dimensional cutting, European Journal of Operational Research 119 (1999) 165-174
12. Stadtler H.: A one-dimensional cutting stock problem in the aluminum industry and its solution, European Journal of Operational Research 44 (1990) 209-223

Search Space Filling and Shrinking Based to Solve Constraint Optimization Problems

Yi Hong[1], Qingsheng Ren[1], Jin Zeng[2], and Ying Zhang[1]

[1] Department of Computer Science and Engineering,
Shanghai Jiaotong University, Shanghai 200030, P.R. China
{goodji,zhangying}@sjtu.edu.cn
ren-qs@cs.sjtu.edu.cn
[2] Department of Mathematics,
Shanghai Jiaotong University, Shanghai 200030, P.R. China
ZengJin@sjtu.edu.cn

Abstract. Genetic algorithm (GA) is an effective method to tackle combinatorial optimization problems. Since the limitation of encoding method, the search space of GA should be regular. Unfortunately, for constraint optimizations, this precondition is unsatisfied. To obtain a regular search space, a commonly used method is penalty functions. But the setting of a good penalty function is difficult. In this paper, a novel algorithm, called search space filling and shrinking algorithm (SSFSA), is proposed. SSFSA first seeks a smaller search space which covers all the feasible domains, then fills the unfeasible search space to acquire a regular search space. Search space shrinking diminishes the search space, so shortens the searching time. Search space filling repairs the irregular search space, and makes GA execute effectively. Experimental results show that SSFSA outperforms penalty methods'.

1 Introduction

Genetic algorithm (GA) [1] is a method of evolutionary computation which draws inspiration from the principle of Darwinian evolution. The main idea of Darwinian evolution is survival of the fittest. According to this principle, GA employs such operators as selection, crossover and mutation to generate new populations. GA is simple to execute, and can search the global optimal point in the whole search space not depending on the gradient of target functions. Consequently, it has been applied to a very wide range of problems.

GA is natural and effective to solve unconstraint optimization problems, but it has to be adapted a lot for the constraint cases. The searched optimal point should be a high fitness point, and also a feasible point. In real-world applications, including in hardware design, automatic control, flow-shop, scheduling and so on, a majority of problems have some constraints. A constraint optimization problem can be defined as:

$$\text{Minimize: } f(x) \quad x \in R^n$$
$$s.t \quad g_j(x) \geq 0 \quad j = 1,....,m \quad (1)$$

Where m is the number of constraints.

To solve constraint optimization problems with GA, many measures have been proposed [2] [3], such as unfeasible individual repaired [4], special representation [5], special genetic operators [6] [7], hybrid search [8] and so on. The most commonly used method to handle constraints is penalty function. Penalty function was originally proposed by Courant in 1940s and later expanded by Carroll. Its main idea is adding a penalty for all individuals locating in the unfeasible domain. Some penalty methods have been adopted [9] [10], including static penalty, dynamic penalty, co-evolutionary penalty and death penalty. But all of them exist the clumsiness: if the penalty is too small, GA can't make sure to converge to a feasible point; if it is small, when the global optima locates at the constraint boundary, the search speed of GA is very slow [11] [12].

The remainder of this paper is arranged as follows. Section 2 studies the search space filling method. Section 3 presents the search space shrinking method. Search space filling and shrinking algorithm (SSFSA) is proposed in section 4. The comparison between SSFSA and penalty function method in section 5.

2 Search Space Filling

For a constraint optimization problem:

$$\text{Minimize: } f(x) \quad x \in R^n$$
$$s.t \quad g_j(x) \geq 0 \quad j = 1,....,m \tag{2}$$

If $g_j(x) = x - a$ or $g_j(x) = b - x$ $a \in R^n, b \in R^n, j = 1,....,m$, we assume that the search space is regular, and the optimization problem is unconstrained. For other kinds of $G_j(x)$, the search space is irregular, and the optimization problem is constrained. The most commonly used method to handle constraint optimization problem is switching it into an unconstrained one [13]. That is the process of search space regularization. Penalty function is a method of search space regularization. Here we give a novel method called search space filling to realize it. The target function after filled is:

$$\text{Minimize: } F(x) = \begin{cases} f(x) & x \in \text{feasible domain} \\ h(x) & \text{otherwise} \end{cases} \tag{3}$$

Where
$h(x)$ is the filling function.
$h(x)$ should satisfy the following conditions:

1) $\min\{f(x), x \in \text{feasible domain}\} \leq \min\{h(x), x \notin \text{feasible domain}\}$;
2) $f(x) \equiv h(x)$ $x \in S$, S is the boundary of constraints;
3) $h(x)$ should be as simple as possible.

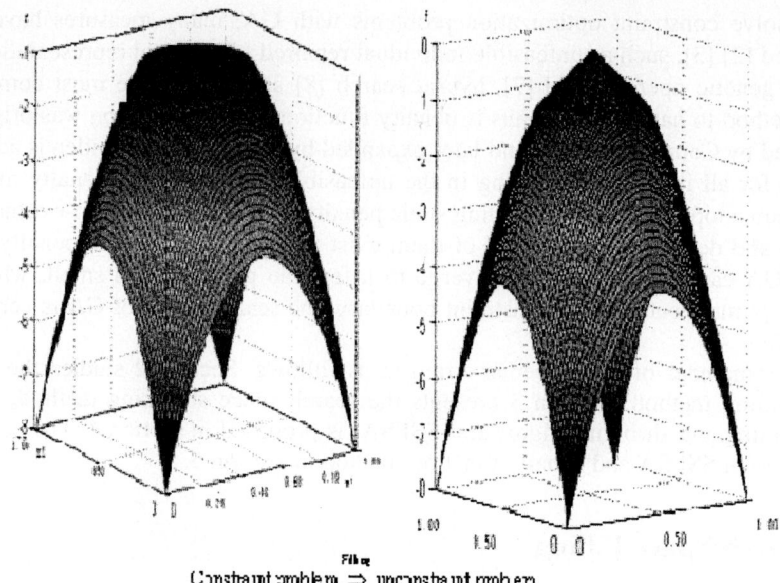

Fig. 1. Transforming from constraint problem to unconstraint case

Example 1:

$$\text{Minimize } f(x) = -x_1^2 - x_2^2$$
$$s.t : x_1^2 + x_2^2 \geq 1 \quad -2 \leq x_i \leq 2$$

To solve this problem with GA, according to this analysis above, we transform it into an unconstraint case as:

$$h(x) = -\sqrt{x_1^2 + x_2^2} \quad x_1^2 + x_2^2 \leq 1$$

$$F(x) = \begin{cases} -x_1^2 - x_2^2 & x_1^2 + x_2^2 \geq 1 \\ -\sqrt{x_1^2 + x_2^2} & others \end{cases}$$

Figure 1 is the transforming process. From figure 1, we can see that the global minima haven't been changed. The new target function is an unconstrained optimization problem, which is simple to be tackled. But not all problems are as simple as the example 1 given above. In fact, for most of problems, we can't set $h(x)$ directly. But that is not to say filling is useless. We can use $f(x)$ to approximate $h(x)$.

In each generation, individuals are classified into two subsets F and NF:

$$\begin{cases} X \in F & if \ X \text{ satisfies all constraints} \\ X \in NF & others \end{cases} \quad (4)$$

If $X \in F$, we can calculate its fitness directly; while $X \in NF$, its fitness is estimated by the nearest feasible individual Y:

$$fitness(X) = fitness(Y) \times \exp(-\lambda \times dist(X,Y)) \quad (5)$$

Where:
> Y: The nearest individual to X who satisfies all the constraints;
> $\lambda: \lambda > 0$, adjusting parameter;
> $dist(X,Y)$: The distance between individual X and Y.

$$dist(X,Y) = \sqrt{\sum_{k=1}^{n}(X_k - Y_k)^2}$$

This measure is simple and effective.

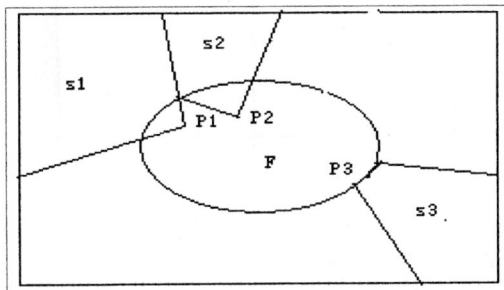

Fig. 2. Feasible points and their nearest areas

Let's analysis the figure 2, if point k is located at s_1, then

$$fitness(k) = fitness(P_1) \times \exp(-\lambda \times dist(P_1, K)) \quad (6)$$

Since $dist(p1,k)$ is continuous in s1, so $fitness(k)$ is continuous in s1; and if the global minimal is at the constrained boundary such as P3, with the evolution of population, the individual will gathered around P3, whether the individual is feasible point or unfeasible point, just because when a unfeasible point P approaching P3, the distance between P3 and P will approximate 0.

$$fitness(p) \approx fitness(p3) \quad (7)$$

From these analysis, we know that this method can approximately realize requires given above.

3 Search Space Shrinking

Search space filling aims at transforming constraint optimization problem to unconstrained one; while search space shrinking aims at diminishing the searching space and shortening the searching time.

Example 2:

$$\text{Minimize } x_1^2 + x_2^2$$
$$s.t. \quad -1 \leq x_1 + x_2 \leq 1$$
$$-1 \leq x_1 - x_2 \leq 1$$
$$-2 \leq x_i \leq 2 \quad i = 1,2$$

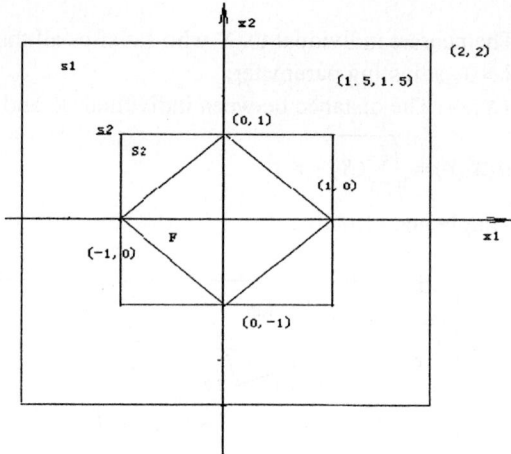

Fig. 3. The Filled search space, the feasible space and the shrunk search space. S1: The regular space; F: The feasible space; S2: The shrank space

From figure 3, we can see that the regular search space (s1) is sometimes larger than the shrunk regular space (s2) containing all feasible individuals. So if we adopt the shrunk regular space as our searching space, the born probability of unfeasible individual will be small. At the same time, the time consumption can also be cut down. The most important fact is that it makes the novel algorithms presented below viable: the precondition of SSFSA is there exist a certain proportional feasible individuals in the population.

The shrunk search space should satisfy:

1) It should cover all the feasible individuals;
2) In the boundary of constraints, it should have some unfeasible space, which will favor the search when the global optimal is at the boundary;

Here an algorithm to calculate the shrunk search space is presented. We give each constraint a value, and define a target function $GF(X)$:

$$GF(X) = \sum_{k=1}^{m} T(g_j(X)) \quad m \text{ is the number of constraints}.$$

$$T(g_j(X)) = \begin{cases} 1 & X \text{ satisfies } g_j(X) \\ \exp(\alpha \times g_i(X)) & X \text{ dissatisfies } g_j(X) \end{cases} \quad (8)$$

$GF(X)$ reflects the degree that individual X satisfies constraints. In experiments, $\alpha = 0.01$

The frame to calculate the shrunk search space is as follows:

a) Initialize the population and the shrunk search space O,
$O = ((\min x_1, \max x_1), ..., (\min x_k, \max x_k), ..., (\min x_n, \max x_n)) = ((0,0), ..., (0,0), ..., (0,0))$;
b) Decode;

c) For an individual X in the population, if it satisfies all the constraints, then modifying the shrunk parameter space O:

$$\max x_i = \max\{\max x_i, x_i(X)\} \text{ and } \min x_i = \min\{\min x_i, x_i(X)\}$$

d) Throw off all individuals who satisfies all the constraints, and compensate the population with random new individuals;
e) Selection;
f) Crossover;
g) Mutation;
h) If the finish condition satisfies, goto i), else go to b;
i) Expand the shrunk search space O:

$$\max x_i = \max x_i + \Delta x_i \quad \Delta x_i > 0 \text{ and } \min x_i = \min x_i - \Delta x_i \quad \Delta x_i > 0$$

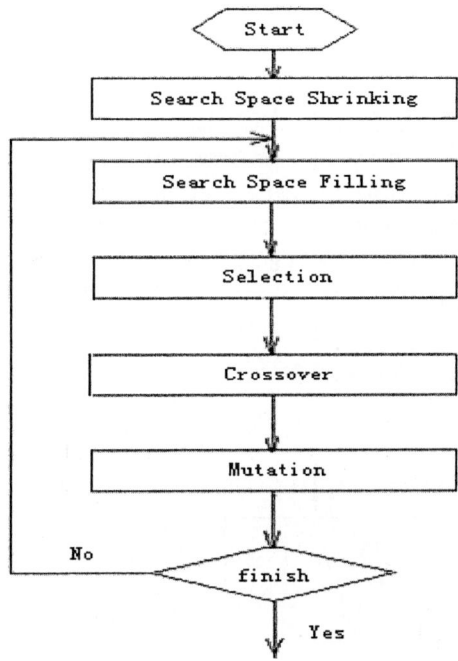

Fig. 4. The flow of SSFSA

4 Search Space Filling and Shrinking Algorithm

Combining search space shrinking method and search space filling method, we propose a novel algorithm to tackle constraint optimizations, called search space filling and shrinking algorithm (SSFSA). The main framework of SSFSA is first seeking a smaller search space which containing all the feasible area, and later filling the irregular search. The finishing condition is the limited evolutionary generation. The filling function $h(x)$ is approximated by its neighbour feasible point. $h(x)$ is dynamical with the evolution of population. Search space shrinking is achieved at the beginning. The flow of SSFSA is given as figure 4.

5 Numerical Experiments

Problem1:

$$\text{Minimize } G_1(x) = (x_1-10)^2 + 5(x_2-12)^2 + x_3^4 + 3(x_4-11)^2 + 10x_5^6 \\ + 7x_6^2 + x_7^4 - 4x_6x_7 - 10x_6 - 8x_7$$

$$\text{S.t.} \begin{cases} 127 - 2x_1^2 - 3x_2^4 - x_3 - 4x_4^2 - 5x_5 \geq 0 \\ 282 - 7x_1 - 3x_2 - 10x_3^2 - x_4 + x_5 \geq 0 \\ 196 - 23x_1 - x_2^2 - 6x_6^2 + 8x_7 \geq 0 \\ -4x_1^2 - x_2^2 + 3x_1x_2 - 2x_3^2 - 5x_6 + 11x_7 \geq 0 \\ -10.0 \leq x_i \leq 10.0, i = 1, \ldots, 7 \end{cases}$$

Problem2 :

$$\text{Minimize } G_2(x) = \frac{3x_1 + x_2 - 2x_3 + 0.8}{2x_1 - x_2 + x_3} - \frac{4x_1 - 2x_2 + x_3}{7x_1 + 3x_2 - x_3}$$

$$\text{S.t.} \begin{cases} x_1 + x_2 - x_3 \leq 1 \\ -x_1 + x_2 - x_3 \leq -1 \\ 12x_1 + 5x_2 + 12x_3 \leq 34.8 \\ 12x_1 + 12x_2 + 7x_3 \leq 29.1 \\ -6x_1 + x_2 + x_3 \leq -4.1 \\ 0 \leq x_i \leq 10 \quad i = 1, 2, 3 \end{cases}$$

Problem3:

$$\text{Maximize } G_3(x) = \left| \frac{\sum_{i=1}^{20} \cos^4(x_i) - 2\prod_{i=1}^{20} \cos^2(x_i)}{\sqrt{\sum_{i=1}^{20} ix_i^2}} \right|$$

$$\text{s.t.} \begin{cases} \sum_{i=1}^{20} x_i \leq 7.5 \times 20 \\ \prod_{i=1}^{20} x_i \geq 0.75 \\ 0 \leq x_i \leq 10 \quad i = 1, 2, \ldots, 20 \end{cases}$$

Parameter setting:
    Size of population: 100
    Length of each parameter: 20
    Rate of Crossover   0.90
    Rate of mutation   0.001
    λ 0.02

Table 1. Average result of problems solved by SSFSA, Dynamic penalty and Static penalty after 30 executions

Method	Problem 1	Problem 2	Problem 3
SSFSA	680.7413	-2.4630	0.7982
Dynamic penalty	680.7930	-2.4360	0.7600
Static penalty	681.1132	-2.4262	0.7139

Table 2. Result of problem 3 solved with different adjusting parameter λ by SSFSA

λ	0.001	0.01	0.02	0.04	0.10
Best value	0.7553	0.7880	0.7982	0.7996	0.7901

Table 3. Shrunk search space for all problems

	Problem 1	Problem 2	Problem 3
Regular search space	$-10 \leq x_i \leq 10, i=1...7$	$0 \leq x_i \leq 10, i=1...3$	$0 \leq x_i \leq 10, i=1...20$
Shrunk search space	$-5.8310 \leq x_1 \leq 4.5606$	$0 \leq x_1 \leq 1.5973$	$0 \leq x_i \leq 10, i=1...20$
	$-2.4907 \leq x_2 \leq 2.3403$	$0 \leq x_2 \leq 0.4572$	
	$-5.0358 \leq x_3 \leq 5.3264$	$0 \leq x_3 \leq 1.5030$	
	$-6.1458 \leq x_4 \leq 6.3769$		
	$-10.0000 \leq x_5 \leq 10.0000$		
	$-7.3986 \leq x_6 \leq 6.3736$		
	$-0.4515 \leq x_7 \leq 10.0000$		

SSFSA outperforms static penalty method and dynamic penalty method which is shown in table1. For all the three problems, the result of SSFSA is the best. The performance of SSFSA is affected by the adjusting parameter λ. If λ is too large, it will go against the search of boundary optima; if λ is too small, the filling search space will became very flat which will delay the convergent time. Table 2 shows that when 0.04λ=, GA can get a satisfactory result. Table 3 is the search space after shrunk. For problem 1 and problem 2, the shrunk search space is much smaller than the original regular search space.

6 Conclusions

Constrained optimization is a promising area. Many real-world problems can be transformed into it. Commonly used penalty methods exist some deficiencies. This study attempts to tackle constraint optimization with search space filling and shrinking method. Since in the unfeasible area, SSFSA have neither global optimal point nor local optimal point, it is impossible for SSFSA converges to an unfeasible point. Our future work will aim at finding a more effective filling method.

Acknowledgement

This project was supported by the National Natural Science Foundation of China (Nos. 60271033). The first author thanks his friend Ying Guo.

References

1. Holland J H.: Adaptation in Nature and Artificial Systems. The University of Michigan Press, 1975, MIT Press, 1992
2. Hajalo p, Yoo J.: constraint handling in genetic search - a comparative study. Collection of Technical Papers AIAA/ASME/ASCE/AHS/ASC Structures, Structural dynamic & Materials Conference, n 4, 1995, 2176-218
3. Deb Kalyanmoy.: Efficient constraint handling method for genetic algorithms. Computer Methods in Applied Mechanics and Engineering, v 186, n 2, Jun, 2000, 311-338
4. Hyun Myung, Jong-Hwan Kim.: Hybrid evolutionary programming for heavily constrained problems. BioSystems, 1996,38: 29~43
5. Wu Wen-Hong, Lin Chyi-Yeu.: Hybrid-coded crossover for binary-coded genetic algorithms in constraint optimization. Engineering Optimization,2004, 101-122
6. Mu Sheng-Jin, Su Hong-Ye, Chu Jian, Wang Yue-Xuan.: An infeasibility degree selection based genetic algorithms for constraint optimization problems. Proceedings of the IEEE International Conference on Systems, Man and Cybernetics, v 2, 2003, 1950-1954
7. Coelle Carlos A, Montes Efren Mezure.: Constraint-handling in genetic algorithms through the use of dominance-based tournament selection. Advanced Engineering Informatics, v 16, n 3, July, 2002, 193-203
8. Barnier Nicolas, Brisset Pascal.: Optimization by hybridization of a genetic algorithm with constraint satisfaction techniques. Proceedings of the IEEE Conference on Evolutionary Computation, ICEC, 1998, 645-649
9. Handa Hisashi, Katai Osuma, Babo Norio, Sawarage Tetsuo.: Solving constraint satisfaction problems by using coevolutionary genetic algorithms. Proceedings of the IEEE Conference on Evolutionary Computation, ICEC, 1998, 21-26
10. Hajalo p, Yoo J.: constraint handling in genetic search - a comparative study. Collection of Technical Papers - AIAA/ASME/ASCE/AHS/ASC Structures, Structural dynamic & Materials Conference, n 4, 1995, 2176-2186
11. Miettinen Kaisa, Makela Markom M, Toivanen Jari.: Numerical comparison of some penalty-based constraint handling techniques in genetic algorithms. Journal of Global Optimization v 27, n 4, December, 2003, 427-446
12. Yeniay Ozgur.: Penalty function methods for constrained optimization with genetic algorithms. Mathematical and Computational Applications, v 10, n 1, April, 2005, 45-56
13. Homaifar, Abdollah, Charlene H. Qi, Steven H. Lai.: Constraint optimization via genetic algorithms. Simulation, v 62, n 4, Apr, 1994, 242-254

A Reinforcement Learning Approach for Host-Based Intrusion Detection Using Sequences of System Calls*

Xin Xu[1,2] and Tao Xie[1]

[1] School of Computer, National University of Defense Technology,
410073, Changsha, P. R. China
xuxin_mail@263.net
[2] Institute of Automation, National University of Defense Technology,
410073, Changsha, P. R. China

Abstract. Intrusion detection has emerged as an important technique for network security. Due to the complex and dynamic properties of intrusion behaviors, machine learning and data mining methods have been widely employed to optimize the performance of intrusion detection systems (IDSs). However, the results of existing work still need to be improved both in accuracy and in computational efficiency. In this paper, a novel reinforcement learning approach is presented for host-based intrusion detection using sequences of system calls. A Markov reward process model is introduced for modeling the behaviors of system call sequences and the intrusion detection problem is converted to predicting the value functions of the Markov reward process. A temporal different learning algorithm using linear basis functions is used for value function prediction so that abnormal temporal behaviors of host processes can be predicted accurately and efficiently. The proposed method has advantages over previous algorithms in that the temporal property of system call data is well captured in a natural and simple way and better intrusion detection performance can be achieved. Experimental results on the MIT system call data illustrate that compared with previous work, the proposed method has better detection accuracy with low training costs.

1 Introduction

With the rapid development of computer networks and related applications, computer security has become a critical problem not only in industry but also in our whole society. To defend various cyber attacks and computer viruses, lots of computer security techniques have been studied, which include cryptography, firewalls and intrusion detection, etc. Among these techniques, intrusion detection is considered to be more promising for defending complex intrusion behaviors since different behavior models or patterns can be developed to detect intrusions. Thus, one of the

---

* Supported by the National Natural Science Foundation of China Under Grants 60303012, 60225015, Specialized Research Fund for the Doctoral Program of Higher Education under Grant 20049998027, Chinese Post-Doctor Science Foundation under Grant 200403500202, and A Project Supported by Scientific Research Fund of Hunan Provincial Education Department

central problems for intrusion detection is to build effective behavior models or patterns to distinguish normal behaviors from abnormal behaviors by observing certain kind of audit data. According to the different types of audit data, IDSs can be divided into two categories, i.e., network-based IDS and host-based IDS. A network-based IDS monitors the contents as well as the formats of network data which are usually irrelevant to the operating systems of host computers. Host-based IDS detects possible attacks or viruses into host computers by collecting information specific to the operating systems of the target computers.

The earliest intrusion detection model was proposed by Denning [1] and many research works have been devoted to the construction of effective intrusion detection models. Until now, there are two general approaches to building intrusion detection models, i.e., misuse detection and anomaly detection. Misuse detection extracts feature rules or behavior patterns based on the audit data of known attacks so that novel attacks may not be detected by misuse detection systems. Anomaly detection establishes normal behavior models to detect suspicious attack behaviors. Although new attacks can be detected by anomaly detection systems, high rates of false alarms usually occur.

In addition to the above deficiencies of misuse and anomaly detection systems, another challenge for intrusion detection systems (IDSs) is the increasing amounts of attack types and audit data so that conventional manually constructed intrusion detection models can not be adaptive to dynamic and complex intrusion behaviors. In recent years, the applications of machine learning and data mining techniques in intrusion detection systems have attracted lots of research interests [2]. By making use of data mining or machine learning algorithms, adaptive intrusion detection models can be automatically constructed based on labeled or unlabeled audit data.

Until now, although many adaptive intrusion detection methods, such as neural networks, support vector machines, decision trees [3][4][5], etc., have been proposed in the literature, there are still much work to do to improve the performance of IDSs. One of the main reasons for the performance problem of IDS is that many complex intrusions usually involve temporal sequences of dynamic behaviors while many conventional supervised learning methods only deal with temporally isolated labeled samples. Recently, dynamic behavior models for intrusion detection have been studied in the literature [6][7], where Markov chain models and Hidden Markovian models (HMMs) were proposed for constructing temporal dynamic models of intrusion or normal behaviors. Nevertheless, the performance of HMM-based IDSs still needs to be improved and the training algorithms for HMMs usually have high computational costs.

Unlike supervised learning and unsupervised learning, reinforcement learning (RL) [8][9] is another class of machine learning methods. A RL system learns an optimal sequential decision policy by interacting with the environment and only receiving delayed rewards. RL is different from supervised learning in that no explicit teacher signal is required and sequential optimal policies can be automatically constructed for complex stochastic tasks. In this paper, by introducing a novel Markov reward process model for host-based intrusion detection, the intrusion detection problem is transformed to a value function prediction task of Markov reward processes and a reinforcement learning algorithm is proposed for constructing intrusion detection models automatically. Compared to previous adaptive intrusion-detection approaches, the proposed method is more suitable to build models for detecting complex intrusion behaviors. Experimental results on the MIT lpr system call data illustrate that the proposed method not only has higher detection accuracy but also requires lower computational costs.

This paper is organized as follows. In Section 2, the Markov reward model for the host-based intrusion detection problem is given. In Section 3, a temporal-difference learning algorithm is proposed for learning prediction of the Markov process based on system call traces. Experimental results on the MIT lpr data are provided in Section 4 to illustrate the effectiveness of the proposed method. Some conclusions are given in Section 5.

2 Markov Reward Model for Host-Based Intrusion Detection

2.1 Host-Based Intrusion Detection Using Sequences of System Calls

As discussed in [7], host-based intrusion detection can be realized by observing sequences of system calls, which are related to the operating systems in the host computer. The system calls are recorded by monitoring the execution of different processes in the host computer. The execution trajectories of different processes form different traces of system calls. Here, each trace is defined as the list of system calls issued by a single process from the beginning of its execution to the end. This is a simple definition, but the meaning of a process, or trace, varies from program to program. For some programs, a process corresponds to a single task; for example, in lpr, a SunOS program, each print job generates a separate trace. In other programs, multiple processes are required to complete a task. As discussed in [10], a simple case of a process trace consisting 7 system calls is shown as follows.

open, read, mmap, mmap, open, read, mmap

In the construction of host-based intrusion detection model using sequences of system calls, a certain amount of normal traces as well as attack traces are collected and labeled by human experts. To detect abnormal behavior or attacks based on the system call traces, state transition models are commonly used to distinguish normal traces from abnormal traces, where the states can be defined as short sequences of system calls in a single trace. For example, if we select a sequence of 4 system calls as one state and the sliding length between sequences is 1, the state transitions corresponding to the above simple trace are

State 1: open, read, mmap, mmap
State 2: read, mmap, mmap, open
State 3: mmap, mmap, open, read
State 4: mmap, open, read, mmap

In many cases, the sliding length between sequences can also be greater than 1. Since there are uncertainties in the modeling of the state transition model, Markov chains or Hidden Markov processes (HMMs) are very promising to realize accurate estimation of the underlying state transitions. However, previous works on Markov chain modeling of system call traces [6][7] only employ traditional probability estimation techniques, which are computational expensive, and their detection performance still needs to be improved.

2.2 Markov Reward Process for Host-Based Intrusion Detection

In this sub-section, we will give a brief introduction to Markov reward process as well as some basic notions of reinforcement learning. A Markov process is a stochastic process whose past has no influence on the future if its present state is specified and a Markov chain is a Markov process having a countable number of states. In previous work on Markov chains model for intrusion detection [6][7], only a probability structure of state transitions is considered, which may be inefficient in estimating the temporal patterns of normal and abnormal behaviors. In this paper, we will present a novel Markov reward model for intrusion detection, where rewards are defined for state transitions to improve temporal pattern prediction of intrusion detection. In the following, by introducing some reward structure into a Markov chain, a Markov reward process model can be formulated, which has been widely applied in many areas including fault diagnosis, et al.

Consider a Markov chain whose states lie in a finite or countable infinite space S. The states of the Markov chain can be indexed as $\{1,2,\ldots,n\}$, where n is possibly infinite. Let the state trajectory generated by the Markov chain be denoted by $\{x_t | t= 0,1,2,\ldots; x_t \in S\}$. The dynamics of the Markov chain is described by a transition probability matrix P whose (i,j)-th entry, denoted by p_{ij}, is the transition probability for $x_{t+1} = j$ given that $x_t = i$.

For each state transition from x_t to x_{t+1}, a scalar reward r_t is defined, i.e., the reward function has the following form

$$r_t = r(x_t, x_{t+1}) \tag{1}$$

To realize the prediction of future rewards starting from a state, value functions are commonly used to facilitate computation. The value function of each state is defined as follows:

$$V(i) = E\{\sum_{t=0}^{\infty} \gamma^t r_t | x_0 = i\} \tag{2}$$

where $0 < \gamma \leq 1$ is a discount factor.

As discussed in subsection 2.1, a state transition model can be introduced for host-based intrusion detection using sequences of system calls. By selecting short sequences of system calls as states, a single trace can be regarded as a trajectory of an absorbing Markov chain. Since traces can be labeled as normal or abnormal, we design a reward function for each trace, where normal traces have a terminal reward of -1 and abnormal traces have a terminal reward of $+1$ and the rewards for every intermediate state-transitions are 0. By the definition in (2), the value function of a state will give a predicting probability of the underlying trace to be normal or abnormal. If we get accurate value function estimations, we can determine a trace as normal or abnormal without any difficulties. Thus, the intrusion detection problem can be transformed to a value function prediction problem of Markov reward processes.

In operations research, lots of work has been done to compute the value functions of Markov reward processes. However, when the models of Markov reward processes are unknown, which is usually the truth in many applications including intrusion detection, conventional algorithms in operational research do not work. Based on the pioneering work of Sutton [9], reinforcement-learning algorithms have been widely studied to solve value function prediction problems of model-free Markov chains. In the following, we will present the reinforcement-learning algorithm for host-based IDS by solving the value function prediction problem of the underlying Markov reward processes.

3 The Reinforcement Learning Algorithm for Host-Based IDS

As has been discussed in Section 2, host-based intrusion detection can be tackled by value function prediction methods in reinforcement learning, where the temporal difference learning algorithms called TD(λ) [9][12] are the most popular ones.

In the TD(λ) algorithm, there are two basic mechanisms which are the temporal difference and the eligibility trace, respectively. Temporal differences are defined as the differences between two successive estimations and have the following form.

$$\delta_t = r_t + \gamma \tilde{V}_t(x_{t+1}) - \tilde{V}_t(x_t) \tag{3}$$

where x_{t+1} is the successive state of x_t, $\tilde{V}(x)$ denotes the estimate of the value function $V(x)$ and r_t is the reward received after the state transition from x_t to x_{t+1}.

The eligibility traces can be viewed as an algebraic trick to improve learning efficiency without recording all the data of a multi-step prediction process. This trick is based on the idea of using the truncated return of a Markov chain. In temporal-difference learning with eligibility traces, an n-step truncated return is defined as

$$R_t^n = r_t + \gamma r_{t+1} + \ldots + \gamma^{n-1} r_{t+n-1} + \gamma^n \tilde{V}_t(s_{t+n}) \tag{4}$$

For an absorbing Markov chain whose length is T, the weighted average of truncated returns is

$$R_t^\lambda = (1-\lambda) \sum_{n=1}^{T-t-1} \lambda^{n-1} R_t^n + \lambda^{T-t-1} R_T \tag{5}$$

where $0 \leq \lambda \leq 1$ is a decaying factor and $R_T = r_t + \gamma r_{t+1} + \ldots + \gamma^T r_T$ is the Monte-Carlo return at the terminal state. In each step of the TD(λ) algorithm, the update rule of the value function estimation is determined by the weighted average of truncated returns defined above. The corresponding update equation is

$$\Delta \tilde{V}_t(s_i) = \alpha_t (R_t^\lambda - \tilde{V}_t(s_i)) \tag{6}$$

where α_t is a learning factor.

The update equation (6) can be used only after the whole trajectory of the Markov chain is observed. To realize incremental or online learning, eligibility traces are defined for each state as follows:

$$z_{t+1}(s_i) = \begin{cases} \gamma \lambda z_t(s_i) + 1, & \text{if } s_i = s_t \\ \gamma \lambda z_t(s_i), & \text{if } s_i \neq s_t \end{cases} \tag{7}$$

The online TD(λ) update rule with eligibility traces is

$$\tilde{V}_{t+1}(s_i) = \tilde{V}_t(s_i) + \alpha_t \delta_t z_{t+1}(s_i) \tag{8}$$

where δ_t is the temporal difference at time step t, which is defined in (3) and $z_0(s)=0$ for all s.

Since the state space of a Markov chain is usually large or infinite in practice, function approximators are commonly used to approximate the value function. TD(λ) algorithms with linear function approximators are the most popular and well-studied ones[12]. In our implementation of TD(λ) algorithms for host-based intrusion detection, a linear basis function is chosen as follows

$$\phi(x) = (\phi_1(x), \phi_2(x), \ldots, \phi_n(x))^T \tag{9}$$

The estimated value function can be denoted as

$$\tilde{V}_t(x) = \phi^T(x) W_t \tag{10}$$

where $W_t = (w_1, w_2, \ldots, w_n)^T$ is the weight vector.

The corresponding incremental weight update rule is

$$W_{t+1} = W_t + \alpha_t (r_t + \gamma \phi^T(x_{t+1}) W_t - \phi^T(x_t) W_t) \vec{z}_{t+1} \tag{11}$$

where the eligibility trace vector $\vec{z}_t(s) = (z_{1t}(s), z_{2t}(s), \ldots, z_{nt}(s))^T$ is defined as

$$\vec{z}_{t+1} = \gamma \lambda \vec{z}_t + \phi(x_t) \tag{12}$$

For more detailed discussion on temporal difference learning algorithms, please refer to [9] and [12].

In the following, we will present the complete framework for RL-based intrusion detection as well as a formal description of the temporal different algorithm for learning prediction. Firstly, the proposed RL-based intrusion detection framework is shown in Fig. 1. In the framework, there are two separate processes, which are the RL prediction model training process and the online detection process, respectively. During the model training process, the audit data from a host computer are divided into two classes of traces, i.e., the normal traces and the attack traces. Then, a reward function discussed above as well as some kind of state feature extraction method is introduced so that the trace data are transformed to the sample data of the underlying Markov chains. The linear TD learning algorithm is employed to perform learning prediction of the Markov chains, where the value functions of the Markov chains are predicted. When the learning prediction is completed, a value function prediction model is constructed, which can be used to realize online detection.

During the online detection process, state features are extracted from the input trace data and the vale function prediction model is used to compute value functions of the states. Then the normal or abnormal properties of the trace can be determined by the state value function and a pre-selected or optimized threshold V_0.

The formal description of the TD learning algorithm for the RL-based intrusion detection scheme is given as follows.

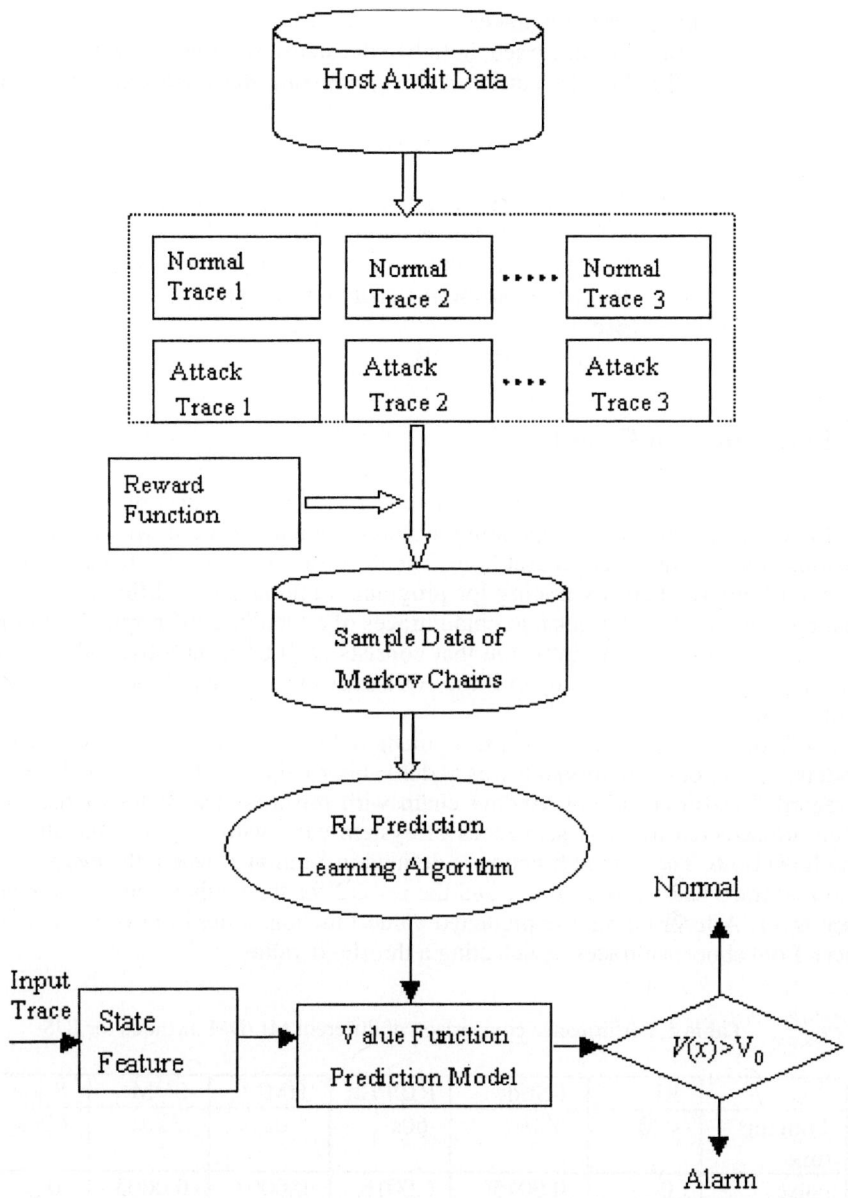

Fig. 1. Framework of the RL-based intrusion detection scheme

(*Algorithm* 1: Linear TD learning algorithm for host-based IDS)
Given: Training sample data of Markov chains generated from host audit data, a stop criterion for training, and linear basis functions for state features.

    (1)   Initialize the weights for value function prediction.
    (2)   While the stop criterion is not satisfied,

Loop for every trace,
- (a) For one trace, initialize the states, set time step $t=0$.
- (b) For the current state s_t, compute the predicted value function $V(s_t)$.
- (c) Observe the state transition from s_t to the next state s_{t+1}, and get the current reward.
- (d) Compute the predicted value function $V(s_{t+1})$ and the temporal difference δ_t using equation (3).
- (e) Update the weights using equation (11) and (12).
- (f) If s_{t+1} is an absorbing state, return to (a),
  Else
  $t=t+1$, return to (b).

4 Experimental Results

The proposed reinforcement learning method for intrusion detection is evaluated on the lpr trace data in SunOS operating systems, which can be downloaded at http://www.cs.unm.edu/~immsec/dataset.html. The data for lpr were collected at the MIT AI laboratory environment by tracing lpr programs running on 77 different hosts, each running SunOS, for two weeks, to obtain traces of a total of 2766 normal print jobs. A single lprcp symbolic link intrusion that consists of 1001 print jobs is also obtained. Detection of an anomaly in any of these 1001 traces is considered successful detection of the intrusion.

To employ the proposed RL-based method for constructing intrusion detection models, we use only 10 normal traces and 20 abnormal traces for training. Every trace is regarded as an absorbing Markov chain with rewards. The states of the Markov chain are selected as short sequences of system calls with length 6 and the sliding length is also 6. The reward function is defined in Section 3, where the ending state of a normal trace has a reward of −1 and the reward for the ending state of an abnormal trace is +1. After training, the predicted value function is used to distinguish normal traces from abnormal traces by selecting a threshold value.

Table 1. Performance comparison of different ML/DM methods for IDS

	RL	t-Stide	RIPPER	HMM	SVM	Stide
Training time	<20s	600s	60s	5 days	350s	600s
False alarm rate	0	0.0075	0.0016	0.0003	0.0003	0

In [10], a support vector machine (SVM) method is proposed for host-based IDS using sequences of system calls and the performance of SVM-based IDS is compared with other machine learning or data mining (ML/DM) methods used for host IDS. The performance is evaluated by the training time of each method as well as the false alarm rate under 100% detection rate. Table 1 shows the performance comparison of the proposed RL method with other learning or data mining algorithms for the same lpr data. It is illustrated that the proposed RL-based intrusion detection approach not only has high detection precision (zero false alarm rate with 100% detection rate) but also has low computational costs.

5 Conclusion and Future Work

In this paper, a novel reinforcement learning approach for host-based intrusion detection is proposed. Different from previous work on Markov chain models of host-based intrusion detection, the proposed method establishes a Markov reward process model and transforms the intrusion detection problem to a value function prediction problem. Thus, temporal difference algorithms in reinforcement learning can be used to realize model-free value function prediction so that accurate intrusion detection can be realized in a computationally efficient way. Experimental results show that the proposed method can achieve better performance than previous methods with lower computational costs. Future work may include a more comprehensive study both in theory and in experiments to make the method be applied in real intrusion detection systems.

References

1. Denning D.: An Intrusion-Detection Model. IEEE Transactions on Software Engineering, Vol.13, No 2 (1987)
2. Lee W.K., Stolfo S.J.: A Data Mining Framework for Building Intrusion Detection Model. In: Gong L., Reiter M.K. (eds.): Proceedings of the IEEE Symposium on Security and Privacy. Oakland, CA: IEEE Computer Society Press (1999) 120~132
3. Mukkamala S., Janoski G., Sung A. H.: Intrusion Detection Using Neural Networks and Support Vector Machines. In: Proceedings of IEEE International Joint Conference on Neural Networks (2002) 1702-1707
4. Ryan J., Lin M-J., Miikkulainen R.: Intrusion Detection with Neural Networks. In: Advances in Neural Information Processing Systems 10, Cambridge, MA, MIT Press (1998)
5. Lane T., Brodley C.: Temporal Sequence Learning and Data Reduction for Anomaly Detection. ACM Transactions on Information and System Security, 2(3) (1999) 295–331
6. Jha S., Tan K., Maxion R.: Markov Chains, Classifiers, and Intrusion Detection. In: Proceddings of the Computer Security Foundations Workshop (CSFW), June (2001)
7. Warrender C., Forresr S., Pearlmutter B.: Detecting Intrusions using System Calls: Alternative Data Models. In: Gong L., Reiter M.K. (eds.): Proceedings of the 1999 IEEE Symposium on Security and Privacy. Oakland, CA: IEEE Computer Society Press (1999) 133~145
8. Kaelbling L.P., Littman M.L., Moore A.W.: Reinforcement Learning: a Survey. Journal of Artificial Intelligence Research, 4 (1996) 237-285
9. Sutton R.: Learning to Predict by the Method of Temporal Differences. Machine Learning, 3(1), (1988) 9-44
10. Hofmeyr, S. A., Forrest, S., Somayaji, A.: Intrusion Detection Using Sequences of System Calls. Journal of Computer Security, 6(3) (1998) 151-180
11. Rao X., Dong C.X., Yang S.Q.: An Intrusion Detection System based on Support Vector Machine. Journal of Software, 14(4) (2003) 798~803
12. Xu X., He H.G., Hu D.W.: Efficient Reinforcement Learning Using Recursive Least-Squares Methods. Journal of Artificial Intelligence Research, 16, (2002) 259-292

Evolving Agent Societies Through Imitation Controlled by Artificial Emotions

Willi Richert, Bernd Kleinjohann, and Lisa Kleinjohann

University of Paderborn / C-Lab, Germany
{richert, bernd, lisa}@c-lab.de

Abstract. An architecture is proposed that combines a simple learning method with one of the most natural evaluation systems: imitation controlled by emotions. Using this architecture agents develop behavioral clusters and form a society that improves its ability to reach a given goal over time. Imitation works by observing and applying behavior sequences (episodes). This leads to new and diverse episodes, because the observation introduces small errors. On the other hand, bad episodes are forgotten if they don't help the agents to satisfy their emotional system that plays the role of an inherent performance measurement. After a while, the agents can be grouped by their typical behavioral patterns. Since these imitated sequences can be seen as "memes" similar to genes in the biological world, this paper explores imitation from the view of memetic proliferation.

We show by simulation that using imitation combined with emotions as evaluation measure tasks can be performed by an agent society without having to specify them in detail. The society's performance is quantified using an entropy measure to qualitatively evaluate the emerging behavioral clusters.

1 Introduction

Intelligent agents are entering more and more dynamic application fields — fields in which it is no longer feasible to provide instructions for every possible situation the agent might run into. Instead a means for automatic adaptation is needed. A straightforward approach in this case could be to choose a learning method, create the needed adaptive system (e.g. devise a non-deterministic automaton in case of Reinforcement Learning [1]) and let the software learn using the chosen parameters in order to maximize a predefined fitness function. However, in many situations that won't work, because not all possible states the system might run into can be known in advance or the fitness function is not exactly known. Here, we need a new approach both for learning and for evaluating the effort. In addition to this, if there is not only one agent but instead a whole society of agents trying to perform a predefined goal, it is most likely that an agent has already conquered a sub-problem another agent will be faced with later on. So, why not letting the agents benefit from each other's experience and spread the newly gathered experience to other agents in that society? In this paper, we propose an architecture that provides such possibility by combining a simple learning method with one of the most natural evaluation systems: imitation controlled by emotions. Imitation has been studied so far mainly in terms of learning by demonstration, where a simple action of a demonstrator is imitated by an imitator. E.g., Gatsoulis et al. show how foraging can be trained via learning through imitation [2]. In their experiments they point out, that the imitator can generalize beyond its training data. Billard looks at the imitation learning problem from the biological point of view, when presenting a model for motor skills imitation [3]. Demiris and Hayes distinguish active and passive imitation [4] in order to handle the situation, in which the demonstrated action is already known to the observer and that one, in which the observer has to imitate completely new actions, differently. These two cases are combined in their *dual-route architecture*. Borenstein and Ruppin developed a framework called *imitation enhanced evolution* (IEE) in order to explore, how imitation can be utilized in evolutionary processes of agent populations [5]. However, in many application fields the basic behaviors in question are already known, and instead the proper sequence of these behaviors has to be learned, which is the focus of this paper. When one subject copies an information unit from another one, the object that is being transferred underlies several rules. Dawkins was the first one to state that information units that can be transferred evolve according to similar rules that govern biological evolution [6]. The Oxford English Dictionary describes this information unit, called *meme* according to the biological counterpart *gene*, as an "element of a culture or system of behaviour that may be considered to be passed from one individual to another by non-genetic means, esp. imitation". For Dawkins, examples of memes are "tunes, ideas, catch-phrases, clothes fashions, ways of making pots or of building arches". In a similar manner, an episode can also be seen as a meme that is transferred from one agent to another at every observation. After a while, groups of agents performing different behavior sequences should emerge, because different agents will have observed different behavioral patterns at different times.

In this paper we present an emotion based control architecture that makes of these so-called memes by its imitation system. The architecture is evaluated with a simulation application, in which the performance improvement is measured by means of the agents' emotion systems. This is a heuristic composed of emotions and drives, which are the driving force behind the agent's behavior. Our emphasis is on adapting this heuristic to solve technical problems and not on the discussion about the real emotional model of human beings. Every agent strives for feeling good emotions and avoiding bad ones. Since their emotions are the consequence of external stimulations [7] the agents have to modify the cause for these stimulations — either by neutralizing the stimulus in case it aroused bad feelings or by supporting it in the other case. To accomplish this they have to find out and select the proper sequence of actions or behaviors in their behavior repertoire that can be applied by the behavior system. In our approach this selection is learned by the imitation system of an agent which strives to imitate sequences that have been observed with other agents in the past and have been considered to be successful, meaning that the agent felt better afterwards, as registered by its emotion system. In the long run, cultural clusters emerge consisting of similar behaving agents.

2 Architecture

Our imitation learning approach adapts the emotional architecture from the robot head MEXI [8] and extends it with an imitation system that enables agents to imitate each other. To achieve this it provides an interface by which agents can read from other agents their executed action and emotional state. The overall architecture has to perform the following three tasks, which are combined in the proposed architecture as shown in Fig. 1.

- Imitation System: Observe agents and apply previously observed episodes appropriate for improving the current emotional situation.
- Emotion System: Choose a behavior that should be applied if the imitation system has no better episode to offer. The calculation of the behavior is based only on the current emotional state.
- Behavior System: Map the chosen behavior to more detailed action instructions (e. g. Behavior $X \rightarrow$ "turn 10, move forward 5mm, open gripper").

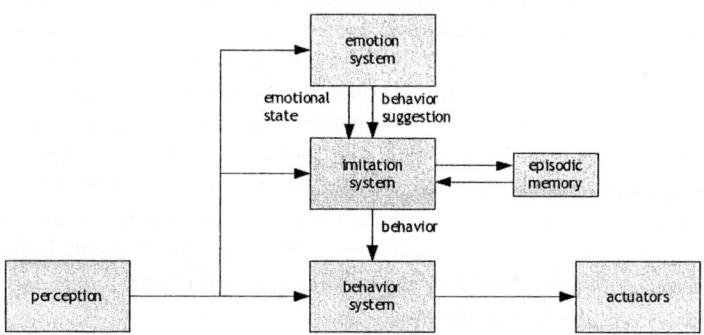

Fig. 1. The agent's architecture

Using the perception data the emotion system consisting of emotions and drives, which will be described in detail later on, calculates which action should be executed so that every emotion will be satisfied as an effect. Together with its action choice it delivers the current emotional state to the imitation system. The imitation system evaluates the emotional state and switches into one of three modes, which are represented by according behaviors in the behavior system. OBSERVE: Look at the nearest agent and record its emotional state together with the executed action for a similar situation in the future, where it might be applied. APPLY: Find in the episodic memory an episode which matches the current situation and execute it. Otherwise, perform some random action like e. g. foraging. In the OBSERVE and APPLY mode the action selection by the emotion system has no influence. The behavior system is entirely controlled by the imitation system. Otherwise, the emotion system takes over the control. If the imitation system has decided which action to execute it forwards this choice to the behavior system which has the task to transform it into a set of detailed action commands that can be passed to the actuators. Arkin showed how this can be straightforwardly accomplished using Motor Schemes [9].

The behavior, emotion, and imitation system will now be described in more detail using an example application to illustrate and evaluate the concepts. For this purpose, a two-dimensional simulation environment has been set up, in which ten agents are simulated for 10,000 time steps per simulation run. Their aim is to feel good emotions and avoid bad ones. To achieve this an agent has to collect flags which are distributed over the whole simulation field using its gripper and carry them to one of four bins which are located in the corners of the field. The agents feel the positive emotion joy, the negative emotion anger, and are equipped with the drive imitation.

2.1 The Behavior System

The behavior system translates the abstract behaviors into more detailed atomic actions that can be executed by its actuators. To the emotion and imitation system it offers a repertoire of ten complex behaviors. The behavior repertoire can be grouped into two different behavior classes: normal and imitation behaviors, which are triggered only on behalf of the imitation system. All behaviors are modeled using the Motor Schemas architecture enhanced by application specific functions like usage of the gripper or using the imitation interface. The normal behaviors are STOP, WANDER, ACQUIRE, DELIVER1 to DELIVER4, and DROP and have the following function:

STOP Stand still and do nothing with the gripper.
WANDER Wander around in random direction in order to find some flags. Gripper is opened.
ACQUIRE Approach the nearest flag with gripper in trigger modus, i.e. that means that it closes its gripper if the flag is in the gripper.
DELIVER(1-4) Move toward bin 1, 2, 3, or 4. Only applicable if the agent holds a flag in its gripper.
DROP Lay down the flag in the gripper and move away from the flag's position.

The imitation system's functionality has been implemented by the two behaviors OBSERVE and APPLY, which have the possibility to access the agent's episodic memory and to read the observed agent's emotion system's state directly.

OBSERVE An agent can observe any agent within a predefined radius. This behavior takes care of keeping close distance to the observed agent so that the observation is not interrupted. While observing, the executed behavior and emotional state change of the observed agent is recorded at every time step, and it has to make a decision on the following three possibilities: 1) Observe further since the collected data is useful so far. 2) Stop observation, because the recorded data contains no valuable information. 3) Stop observation and prepare the recorded data for storing into the episodic memory. Its result is based on the emotional change of the observed agent, described in section 2.2. If the quality (cf. Eq. 1) has changed significantly ($\Delta G \geq 0.2$) in the considered time frame the episode will be extracted and saved in the episodic memory. The episode's start is set to the behavior immediately preceding the first rise of quality. Its end is set to the behavior immediately succeeding the last rise of quality. Then, the episode is compressed, meaning that only behaviors that at least increase the quality ($\Delta G \geq 0.09$) are left in the episode. Both thresholds for the change of the quality ΔG have been empirically determined.
APPLY In this behavior the agent has to choose which of the previously collected episodes is the best one to apply in the current situation. To solve this problem

the agent looks at the G values of every episode and at the individual emotion's changes. But, since these values have been observed only at another agent, it does not mean this episode will result in similar emotional changes if the observing agent is applying the episode's behavior by itself. Hence, the emotional differences are adjusted after every application. This way, the agent can filter out observed episodes that do not work for him.

2.2 The Emotion System

The driving force behind the propagation of memes in this work is the agent's emotion system: a meme will only be copied from one agent to another if it serves the well being of the imitating agent, i. e. that the execution of its behavior sequence proved as being advantageous. How emotions can be modeled and expressed using an algorithmic approach MEXI [8] and Kismet [10] have successfully shown. In this work, MEXI's architecture has been adopted, offering a triple-tower architecture with its middle tower enhanced by an emotion system. This is responsible for setting the action selection bias towards actions that will result in a better emotional feeling. Differently from MEXI, the emotion system in this work does not directly affect the behavior system in action selection. Instead, the output of the emotion system works only as an advice for the imitation system which has been inserted between the behavior system and the emotion system (cf. Fig 1). Ultimately, the imitation system chooses which actions are executed. At the beginning when no episodes have been observed yet, the emotion system's choices for behavior are accepted by the imitation system and forwarded directly to the behavior system. In the long run, when the imitation system has gathered many favorable episodes these episodes are preferred to the emotion system's choices.

The emotion system controls a set of emotions and drives. Emotions can be positive or negative ones and are real values in the range $[0, 1]$ having a threshold value. In the evaluation application we implemented the positive emotion "joy" and the negative emotion "anger". If the positive emotion's threshold is exceeded or the value for the negative emotion falls below the threshold those emotions are said to be satisfied and the agent feels good. Otherwise, the emotion system tries to get back to the satisfied area. We used the drive "imitation" to switch between the recurrent phases OBSERVE and APPLY. The drive's value ranges in the interval $[-1, 1]$ and has a positive and a negative threshold. The area between those thresholds is called the homeostatic area, meaning that the drive is satisfied. The imitation drive oscillates with a sinus between -1 and 1 if no stimuli are present in order to realize a cyclical behavior. It is connected to the OBSERVE and APPLY phases with their corresponding behaviors. If the value for imitation exceeds the positive threshold it is an indicator that the emotion system has the bias to observe other agents. In the other case, if the value crossed the negative threshold, the imitation system would try to apply some previously observed episode. In both cases the emotion system is not directly switching to the corresponding behavior. Instead, it only modifies its inclination toward the behavior by means of its configuration, as with MEXI. It is important to note that the choice of behaviors is no discrete action. Instead, as modeled in the MEXI architecture, emotions configure the individual behavior weights. In the end, the final configuration is calculated by the configuration suggestions of each emotion [8].

Impact of Stimuli on the Emotions Joy and Anger Emotions and rives in real life have interesting time dependent properties. E. g. the emotion anger decays over time if the initial anger evoking stimulus has ceased. This behavior is modeled with the help of excitation functions. For the evaluation example the following perceptual predicates that can be extracted by the agent's perception system have been chosen: FlagVisible is true, if a flag outside a bin is visible. AgentVisible is true, if another agent is visible. FlagInGripper is true, if the agent holds a flag. StayingInBin is true, if the agent stays in one of the four bins. These predicates' impact on the agent's emotion system is listed in Table 1.

Table 1. *Impact of perceptions on emotions and emotions/drive on the choice of behaviors.* "↑"/"↓": increase/decrease of the preference of the behavior, ">": exceeding its threshold, "<": falling below the threshold. "—": the homeostatic area

Perception	true	false
FlagVisible	Joy ↑↑ Anger ↓↓	Joy ↓ Anger ↑
AgentVisible	—	—
FlagInGripper	Joy ↑↑ Anger ↓	Joy ↓ Anger ↑
StayingInBin	Joy ↑↑	—

Emo/Drive	Thresh.	Impact
Joy	>	OBSERVE ↓
Anger	>	—
Imitation	>	OBSERVE ↑
	—	OBSERVE ↓
		APPLY ↓
	<	APPLY ↑

The Emotional Quality The internal state of an agent is completely described by its emotions. To ease the comparison between two emotional states, which is necessary in the OBSERVE phase, the quality G describing the desirability of an emotional state is introduced. Let P and N be the number of positive and negative emotions, respectively. Further, let $E_{k,p}$ denote the positive and $E_{l,n}$ the negative emotions with $k \in [1, \ldots, P]$, $l \in [1, \ldots, N]$, and w_i the individual emotion weights. Then the quality G is calculated as defined in Formula (1).

$$G = \sum_{k=1}^{P} w_{k,p} \cdot E_{k,p} + \sum_{l=1}^{N} w_{l,n} \cdot (1 - E_{l,n}) \qquad (1)$$

Now, an episode Q lasting from time t_1 to t_2 can be easily classified depending on the quality value before t_1 and after t_2: Episode Q is favorable iff $G_{t_2} > G_{t_1}$.

2.3 The Imitation System

The act of imitation is made up of the following three processes [11]:

1. **Recognition** of successful action sequences.
2. **Transformation** of these sequences from the perspective of the demonstrator into the perspective of the imitator.
3. **Generation** of the according action sequence.

Recognition and transformation is the task of the OBSERVE behavior. Generation is the domain of APPLY. The question, when to observe other agents and when to apply the observed episodes is not in the scope of the imitation system. We implemented the drive imitation in the emotion system, and used its cyclic property of oscillating between the according two phases.

Because the agents have no possibility to see some kind of gesture or facial expression of the observed agent they need another means by which they can judge the outcome of the recently executed episode. Because the agent society is run in a simulation environment a direct interface to the observed agents mind was implemented into the architecture, enabling reading of the recently executed action and emotional state. Using it, after an observation the observer has three data entities at his disposal: the observed emotional state prior to an action, the action itself, and the observed emotion state immediately after the behavior execution. To quantify an emotional state in order to make it comparable the quality G that is computed out of the emotional state's components as described above is used. The more favorable an emotional state is the higher is the value of G. Taken an episode that lasts from time step t_1 till t_2 an episode is said to be favorable if $G_{t_2} > G_{t_1}$. In this case the agent has to process the episode's data and save it for the according emotional state. While executing the OBSERVE behavior the quality of the observed agent's emotions is monitored at every time step. In case that it is not increasing over a predefined period of time the OBSERVE behavior is canceled. It does not suffice to record only the executed behaviors. If, e. g., an agent only records the action sequence (A_1, A_2, A_3) it does not know when the state changes $A_1 \rightarrow A_2$ and $A_2 \rightarrow A_3$ have occurred. In addition, the events that triggered the behaviors have to be saved. Here, the changes of the emotions are treated as event triggers. If an emotion has increased or decreased more than a predefined value this change is considered as being a trigger for the following executed behavior of the observed agent. If the observer is executing this episode in the future, it will not switch to the next episode step or behavior until it is registering at least the same emotional changes.

The episodic memory Z consists of up to z_{max} memory entries $Z_i = (E, O, T)$, which stands for episode, originator, and time-stamp, respectively. The triggers that must be met before the next behavior in that episode can be executed are stored together with the behavior itself. Thus, an episode entry is a tuple (C, b) with C being the emotional triggers for the behavior b and b the number of the corresponding behavior. Every time the agent has observed an apparently advantageous episode and saves it in its episodic memory it also updates its mapping from emotional states to episodes. This way it always knows which episode to execute in the current emotional state.

Impact of the Emotions and the Drive on the Choice of Behaviors The values of the emotions and drives affect the bias toward their preferred behaviors. If an emotion or drive is not satisfied it strengthens its bias toward an behavior that could satisfy it. Their impacts are listed in Table 1.

3 Evaluation

An agent society consisting of ten agents was simulated to analyze whether behavioral clusters emerge and how the imitation capability affects the overall agent society performance. We used the TeamBots package which provides a full range of simulation supporting software modules. The agent society's learning progress can be investigated in two ways: 1) By simply measuring the performance increase, and 2) by analyzing the agent society's diversity, i.e. how the agents subdivide to groups using similar behavioral patterns. Each simulation (also called a simulation round) consists of 10,000 steps. To achieve meaningful results, the simulation has been consecutively executed 100 times. After the first round each simulation used the learned episodes from the previous round, but randomized the agents' and flags' positions. This simulation of 100 rounds was repeated 80 times from scratch (called a simulation run). Afterward, an average run is calculated, in which every average nth round is calculated of the average of all the nth rounds in the 80 simulation runs. To monitor the agent's individual learning progress the Wellness-Test has been devised. It calculates the performance measure w for every agent as follows: For every simulation step increase w by one for every satisfied emotion, i.e. if the positive emotion joy is above its threshold or the negative emotion anger is below its threshold. Since every simulation round consists of 10,000 simulation steps, an agent may collect up to 20,000 Wellness points. This value is reachable only in theory, though, because even the best agent will start its observed action sequences only if one of the emotions is dissatisfied, and this means that it will not get maximum Wellness points for the according simulation step. The average of all ten agents which accounts for the performance of the total agent society is displayed in Fig. 2. The Wellness interval [0, 20000] is transferred to [0, 1]. Starting with 0.178 after step 0 the average performance amounts to 0.397 after step 99, which means an increase by approximately 120%. The maximum value of 1.0 is not achievable for practical reasons: The observation/application process is only invoked if the emotional system is not satisfied, which leads to a suboptimal result in the same simulation step. Furthermore, the longer the simulation has run the fewer flags are available

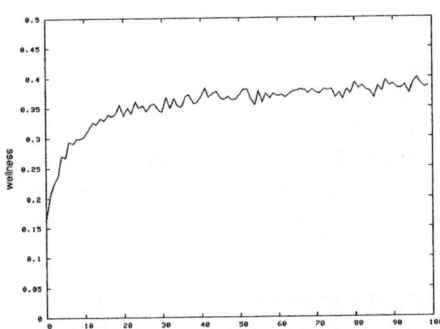

Fig. 2. The average performance in the Wellness-Test over 100 rounds. Every round is the average over 80 runs. The performance increases $\approx 120\%$

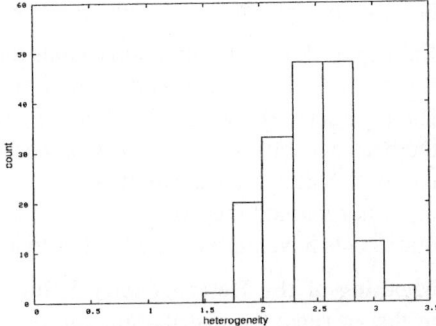

Fig. 3. Histogram of the heterogeneity. In the heterogeneity interval [0, 3.322] the average of all runs is $H_{avg} = 2.57$

4 Results

The overall diversity of the agent society with ten agents in this work can be computed using Shannon's information entropy [12] (cf. [13]) with $H(\mathbb{A}) = -\sum_{i=1}^{c} P_i \log_2(P_i)$. Here a society $\mathbb{A}$ is divided into n clusters $C_1 \ldots C_n$. The proportion of cluster C_i is denoted with $P_i = \frac{|C_i|}{\sum |C_i|}$. Hence, with ten agents the value for $H(\mathbb{A})$ lies within the interval $[H_{min}, H_{max}]$, where H_{min} denotes the diversity for a totally homogeneous society (all agents in one cluster) and H_{max} is the diversity value if the society is totally diverse, i.e. ten clusters each containing one agent: $H_{min} = 0$, $H_{max} = -10 \cdot (0.1 \cdot \log_2(0.1)) \approx 3.322$ The clusters are calculated on the basis of the resemblance of the episodic memory of the two agents A_i and A_j, denoted as $D(A_i, A_j)$ in Formula (2).

$$D(A_i, A_j) = \frac{1}{S} \sum_k |\pi_i(k) - \pi_j(k)| \qquad (2)$$

The number of different states is S. $\pi_a(k)$ stands for the episode that agent a is selecting in state k. But how can Episodes be subtracted? The expression $|\pi_i(k) - \pi_j(k)|$ is set to 1, if $E_i \neq E_j$ and 0 otherwise, with E_k being the episodic memory of agent A_k. Two episodic memories are considered different, if there is at least one state for which both agents select different episodes to execute. The state is a two-bit number, with joy corresponding to the higher bit and anger to the lower bit. "1" means that the emotion is satisfied, "0" otherwise. The state $(joy = 1, anger = 1)$ is omitted because in that case the agent is totally content with the actual situation and no episode has to be executed. Having three different states $D(A_i, A_j)$ can have the values $1/3$, $2/3$ and 1.0. The grouping into clusters can now be performed for a given resemblance threshold ϵ. Two agents are considered to be sufficiently equal for belonging to the same cluster if $D(A_i, A_j) < \epsilon$. In the following the diversity and clustering of the agent society is applied to an example, which is arbitrarily taken from the 80 simulation runs. Diversity and clusters are calculated dependent on the value for ϵ. This has to be set somewhere between two possible consecutive values of $D(A_i, A_j)$: in this case 0.4, 0.7 and 1.0.

Example After 100 simulation rounds the difference $D(A_i, A_j)$ of the learned episodes is calculated. E.g. agent A_1 differs from agents A_0 to A_9 in $[1/3, 0, 1, 1, 1, 1, 1, 1, 1/3, 2/3]$. Having ϵ set to 0.4 we get the clustering $C = \{\{0, 1, 8\}, \{2\}, \{3, 6, 7\}, \{4, 5\}, \{9\}\}$ and the heterogeneity $H \approx 2.171$. Comparing this value for H with the heterogeneity interval $[0, 3.322]$ we can say that neither all agents learned the same (then H would be zero) nor learned they very different episodes. Learning has taken place which shows that agents have imitated each other to a certain degree.

Meaning of the Heterogeneity Values Relating the heterogeneity of all last rounds in the 80 runs, the distribution can be seen in the histogram of Fig. 3. It divides the histogram interval in seven non-zero areas. The height of each box stands for the number of runs in which the heterogeneity of the agent society after the last round lied in the area's interval. The average heterogeneity is $H_{avg} = 2.57$. Hence, in this work for the heterogeneity interval $[0, 3.322]$ the average agent society is approximately $\frac{3.322 - 2.57}{3.322} \approx 0.23$, that means 23% homogeneous. This means that although some measurable amount of the agent society adopted behaviors of other agents, there has been left room for the development of new behaviors.

5 Discussion and Outlook

An agent architecture has been described that enables the emergence of different cultural clusters in an agent society based on the agents' experience. Using imitation of behavioral sequences (episodes) constrained by an emotion system this leads to behavioral diversity and has the advantage that the agents try out different promising ways to reach that goal. On the one hand, the more promising an episode seems to be, the more likely it will be imitated by another agent. By this, errors will be introduced which lead to new episodes ("memes"). On the other hand, an episode that does not help to satisfy the agent's emotion system will be quickly forgotten. It could be shown by simulation that using this kind of architecture the average learning performance of the agent society can substantially be improved, i.e. cultural clusters of agents are emerging, where the individual agent's performance increases with more and more imitated episodes. Future work should investigate how memes, i. e. the episodes, change in detail when being transferred from one agent to another, if more realistic randomization effects could be deployed.

References

1. Sutton, R.S., Barto, A.G.: Reinforcement Learning: An Introduction. MIT Press, Cambridge (1998)
2. Gatsoulis, Y., Maistros, G., Marom, Y., Hayes., G.: Learning to forage through imitation. In: Proceedings of the Second IASTED International Conference on Artificial Intelligence and Applications (AIA2002). (2002) 485–491
3. Billard, A.: Learning motor skills by imitation: a biologically inspired robotic model (2000)
4. Demiris, J., Hayes, G.: Imitation as a dual-route process featuring predictive and learning components: a biologically-plausible computational model. In K. Dautenhahn, C. Nehaniv, eds.: Imitation in animals and artifacts, Cambridge, MA, USA, MIT Press (2002) 327–361
5. Borenstein, E., Ruppin, E.: Enhancing autonomous agents evolution with learning by imitation. In: Second International Symposium on Imitation in Animals and Artifacts. (2003)
6. Dawkins, R.: The Selfish Gene. Oxford University Press, Oxford (1976)
7. Plutchik, R.: The Emotions. University Press of America (1991)
8. Esau, N., Kleinjohann , B., Kleinjohann, L., Stichling, D.: MEXI - machine with emotionally extended intelligence: A software architecture for behavior based handling of emotions and drives. In: Proceedings of the 3rd International Conference on Hybrid and Intelligent Systems (HIS'03), IEEE Systems, Man and Cybernetics Society, Melbourne, Australia (2003)
9. Arkin, R. C.: Behaviour-Based Robotics. MIT Press (1998)
10. Breazeal, C., Scassellati, B.: How to build robots that make friends and influence people (1999)
11. Blackmore, S.: The Meme Machine. Oxford University Press (1999)
12. Shannon, C.E.: A mathematical theory of communication. Bell System Technical Journal **27** (1948) 379–423 and 623–656
13. Balch, T.: Behavioral Diversity in Learning Robot Teams. PhD thesis, Georgia Institute of Technology (1998)

Study of Improved Hierarchy Genetic Algorithm Based on Adaptive Niches

Qiao-Ling Ji[1], Wei-Min Qi[1], Wei-You Cai[1], Yuan-Chu Cheng[1], and Feng Pan[2]

[1] College of Power and Mechanical engineering, Wuhan University, Wuhan, 430072, China
{Qiao-ling, Wei-Min, Wei-You, Yuan-Chu, qwmin}@126.com
[2] Three Gorges Hydropower Plant, Yichang, 443133, China
{Feng, Feng_Pan}@126.com

Abstract. Canonical genetic algorithms have the defects of pre-maturity and stagnation when applied in optimizing problems. In order to avoid the shortcomings, an adaptive niche hierarchy genetic algorithm (ANHGA) is proposed. The algorithm is based on the adaptive mutation operator and crossover operator to adjust the crossover rate and probability of mutation of each individual, whose mutation values are decided using individual gradient. This approach is applied in Percy and Shubert function optimization. Comparisons of niche genetic algorithm (NGA), hierarchy genetic algorithm (HGA) and ANHGA have been done by establishing a simulation model and the results of mathematics model and actual industrial model show that ANHGA is feasible and efficient in the design of multi-extremum.

1 Introduction

Since Genetic Algorithms (GAs) were firstly put forward by J.H.Holland in 1970s [1], it has been widely used in optimizing complex functions, identifying parameters, optimizing neural networks and so on. GAs are stochastic optimization methods based on the mechanics of natural evolution and natural genetics [2,3]. They have been successfully applied to finding a global optimum of a single objective problem [4]. In the optimization of multimodal functions, however, the standard GA converges to only one peak since it cannot maintain controlled competition among the competing operation corresponding to different peaks.

In recent years much work has been done with the aim of extending genetic algorithms to make it possible to find more than one local optimum of a function. One of the techniques developed for this purpose is known as a niching method [5]. In natural ecosystems, a niche can be viewed as an organism's task, which permits species to survive in their environment.

Though niche GA has strong searching ability and is easy to find many global optimums, its local searching ability should be improved. In this paper, a modified niche GA is proposed in order to improve its performance.

In Section 2, the niche genetic algorithm (NGA) is reviewed. In order to improve the local searching ability of NGA, we combined niche with hierarchy technology and introduced adaptive mutation probability in Section 3, then the adaptive niche hierarchy genetic algorithm (ANHGA) is proposed. In Section 4, the simulation result shows that the ANHGA has strong searching ability and is easy to find many global optimums. An industry model is used to test the reliability of ANHGA in Section 5. Finally, we give some comments in Section 6.

2 Niche Genetic Algorithm

Niching methods have been developed to minimize the effect of genetic drift resulting from the selection operator in the traditional GA in order to allow the parallel investigation of many solutions in the population. The niche technology mainly adjusts the fitness of individuals and replacement strategy when generating the new generation. This makes the individuals evolve in special environment, ensures diversity of evolution population and gets many global optimums at the same time [6]. Representative niche methods are preselecting, crowding and sharing technology.

Table.1 describes the procedure of NGA in simple programs.

Table 1. Genetic algorithm with niching method

Generate initial population: parents		
Iterate		
	Choose: =random-value	
	Case choose	
	generation	mutation	crossover
	Find smallest HD (child, parents)	
	of those find parent with worst fitness	
	If better fitness: exchange (child, parent)	
Show best designs		

As the niche technology is an effective measure to maintain diversity when GAs are applied to optimize functions with many apices or tasks with many targets, it is mainly used to improve GA operators and doesn't change encoding structure. The research found the niche technology and hierarchy GA are mutual complementary in mechanism. The advantages of their combination will be better than those of single method. Based on this thought, we put forward adaptive niche hierarchy genetic algorithm (ANHGA). ANHGA changes in the following aspects: hierarchy structure is used in encoding method, niche technology is used in individuals operation and mutation probability is changed adaptively.

3 The Modified Niche Hierarchy Genetic Algorithm

Using niche technology, ANHGA adopts speedup strategies during the process of encoding, selection and replacement to maintain reasonable population diversity and make GA not only converge but also discover many apices. ANHGA uses hierarchy structure to encode. Before selection, it adjusts individual fitness based on sharing strategy to increase selecting probability of small-scale species. During replacement, it selects individuals as the new generation ones based on density and fitness [7,8].

3.1 Hierarchy Encoding Structure

In hierarchy encoding method, each chromosome is composed of two parts: control gene and constitution gene. Control gene determines whether constitution genes are active. Chromosome includes dominant genes and recessive genes. The active constitution genes are dominant and effective. The inactive constitution genes are recessive and ineffective. The two kinds of genes are inherited to the next generation at the same time. The corresponding control genes determine whether they are transformed [9].

The control genes in hierarchy encoding are often binary encoding. The constitution genes are float encoding or binary encoding in allusion to practical problems. The number of constitution genes controlled by each control gene is variable with specific problems. The structure of hierarchy encoding is shown as Fig.1.

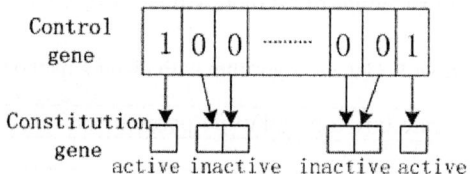

Fig. 1. Structure of hierarchy encoding

In Fig.1, the upper-layer is control genes. The below-layer is constitution genes. In control genes layer, "1" indicates the active and dominant corresponding constitution genes and "0" expresses that the corresponding constitution genes are inactive and recessive. Decoding the dominant genes gets the solutions of the given problems. The recessive genes are inherited to the next generation with the dominant genes and may be activated during the process of evolution. The effective gene segments are adjusted continually until getting the satisfied solutions.

3.2 Individual Density

Similarity of individual gene codes embodies close degree among individuals. We define sharing function to express the density of individuals. So we introduce the concepts of sharing function and individual density.
In order to describe the problem conveniently, some definition should be made as follows:

$\vec{X}(t)$: Population of t th generation;
$X_i(t)$: Constitution genes of i th individual;
N : The size of population;
$d(i, j)$: The distance between i th individual and j th individual.

Then sharing function is defined in Eq.(1):

$$sh(d(i,j)) = \begin{cases} 1 - \left(\dfrac{d(i,j)}{\Delta(t)}\right)^{\alpha} & d(i,j) \leq \Delta(t) \\ 0 & else \end{cases} \quad (1)$$

Where $\Delta(t)$ is a variable, which describes the close degree between i th and j th constitution genes. Its value is determined according to practical problems. α is a parameter that controls the shape of sharing function. Usually α is set to be 1.

It can be seen from Eq.(1) that when individuals are similar, the value of sharing function is bigger. Whereas when individuals are different, the value of sharing function is small.

Equation 2 set $sh(d(i,j))$ as sharing function of i th individual and j th individual. The density of i th individual is defined as:

$$C_i(t) = \frac{1}{N}\sum_{j=1}^{N} sh(d(i,j)) \,. \quad (2)$$

Obviously individual density can be used to appraise population diversity. The larger $C_i(t)$ is, the more number of individuals whose constitution genes are similar to those of i th individual is. In such case, the population may concentrate and lose diversity.

3.3 Fitness Sharing

In fact, for individuals in certain range $\Delta(t)$ can be regarded as a species [10,11]. Large individual density means the corresponding species are large too. Whereas, if some species' density are too large, the fitness of all individuals in this species should be reduced and their selecting probabilities be decreased to maintain the population diversity, create niche evolution environments and encourage small number species to multiply. As for population of t generation, the fitness of $X_i(t)$ after sharing can be defined as:

$$f_i^{'}(t) = f_i(t)/C_i(t) \,. \quad (3)$$

Where, $f_i(t)$ is individual fitness of $X_i(t)$ before sharing. GA carries out selection according to the Eq.(3).

Supposing that the fitness of all individuals in i th species is f_i, the number of individuals in this species is N_i, k is the number of species. Then the stable state of fitness sharing can be expressed as

$$f_i/N_i = f_j/N_j \,. \quad (4)$$

where $i \neq j$ and $\sum_{i=1}^{k} N_i = N$.

3.4 Adaptive Mutation

In fact, for individuals in certain range $\Delta(t)$ can be regarded as a species. Large individual density means the corresponding species are large too. Whereas, if some species' density are too large, the fitness of all individuals in this species should be reduced and their selecting probabilities be decreased to maintain the population diversity, create niche evolution environments and encourage small number species to multiply. As for population of t generation, the fitness of $X_i(t)$ after sharing can be defined as: when individual density $C_i(t)$ is bigger, larger probability should be applied to mutation operation. Considering both the evolution time and individual density as well as ensuring algorithm to converge, the mutation probability should be limited to (0,0.5). Based on these considerations, we put forward the following adaptive mutation probability.

$$P_m(t) = 1 - \frac{1}{1+\exp(-t/C_i(t))} \tag{5}$$

3.5 Individual Replacement Based on Crowding Strategy

If we define parent population as $\vec{X}(t)$ and the population after mutation as $\vec{X'}(t)$, then mix population $\vec{X}(t)$ and $\vec{X'}(t)$, adjust individual fitness according to the following equation:

$$fit_j'(t) = \beta \frac{f_j(t)}{\sum_{k=1}^{N} f_k(t)} + (1-\beta) \frac{\frac{1}{C_j(t)}}{\sum_{k=1}^{N} \frac{1}{C_j(t)}}, \quad j=1,2,\ldots,2N. \tag{6}$$

Where β is a weight coefficient. The adjustment of fitness balances between individual fitness and density. Rank $fit_j'(t)$ in descent order and select the first N individuals of parent generation to compose the next generation. It is obvious that the individual replacement strategy based on crowding method can make uniform distribution and maintain population diversity preferably.

4 Simulations

In order to compare and test the searching ability of two function optimization problems named as Percy and Shubert function, the simulations of GA, NGA, HGA and ANHGA are performed respectively.

Percy function is defined as:

$$\max f_1(x_x, x_2) = \frac{x_1^2 + x_2^2}{2} + \cos(20\pi x_1)\cos(20\pi x_2) + 2 \ , \ -10 \le x_1, x_2 \le 10 \ . \quad (7)$$

The figure of Percy function is shown in Fig.2. It includes 4 global optimal solutions and the Apex's height is 103.0.

Shubert function is defined as:

$$\min f_2(x_1, x_2) = \left\{\sum_{i=1}^{5} i \times \cos[(i+1)x_1 + i]\right\} \times \left\{\sum_{i=1}^{5} i \times \cos[(i+1)x_2 + i]\right\}, \quad (8)$$
$$-10 \le x_1, x_2 \le 10$$

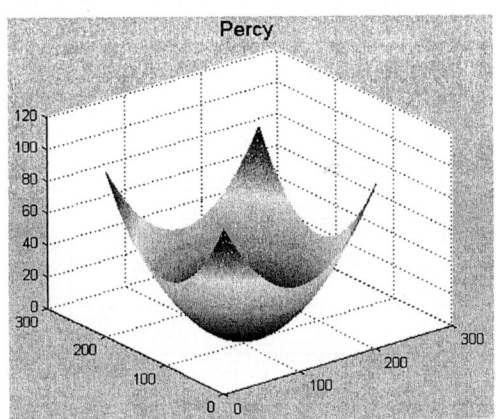

Fig. 2. Percy function

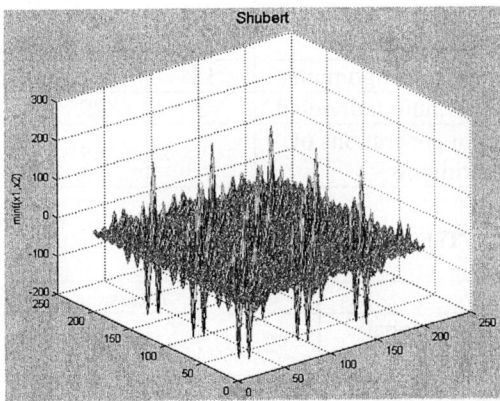

Fig. 3. Shubert function

As shown in the Fig.3, it's known that Shubert function has 760 local optimal solutions. Among which there are 18 optimums and the value is -186.7310. The objective function is converted to fitness value using the following equation:

$$F(x_1, x_2) = \begin{cases} 1 - 0.05 f_2(x_1, x_2), & \text{if } f_2(x_1, x_2) < 0 \\ 1, & \text{if } f_2(x_1, x_2) \geq 0 \end{cases} \quad (9)$$

Simple hierarchy genetic algorithm (HGA) can improve the diversity of population by hierarchy encoding [12], but the simple genetic operators adopted causes limitations on premature convergence avoidance and efficiency improvement. In following simulations, all four genetic algorithms (GA, NGA, HGA, ANHGA) utilize the same control parameters listed in Table 2. All four algorithms adopt proportional selection, single point crossover, $P_c = 0.8$. Each algorithm began with the same initial population and was simulated 20 cycles for each optimized function respectively. The simulation results are listed in Table 3 and 4.

Table 2. Control parameters

	Population size	Chromosome length	Termination time	Weight Coefficient	$\Delta(t)$
Percy	30	20	200	0.75	0.5/t
Shubert	50	20	300	0.75	0.5

Table 3. Percy function simulation results

	GA	NGA	HGA	ANHGA
Number of average global optimums	1.2	3.8	2.3	3.94
Number of global optimum individual	3	29	10	31
Average evolutional generations of global optimums		57	30	42

Table 4. Shubert function simulation results

	GA	NGA	HGA	ANHGA
Average numbers of global optimums	2.1	5.5	3.9	12
Average individual numbers of global optimums	3	18	11	30

Comparing the tables above, it can be seen from simulation results that the space searching ability of GA and simple HGA is unsatisfactory and its optimizing efficiency is inferior to NGA and ANHGA apparently. Under the same condition the solution quality of NGA and ANHGA performs better, which demonstrates that the fitness sharing and the crowding strategy are effective in optimizing functions with many apices. Shubert function is a complicated optimization problem. It is very difficult to find 18 global optimums at the same time. Although GA is easy to carry out, there lie some problems such as premature and convergence speed. HGA can found 2~5 global optimums and the number of global optimum individual is 11. However ANHGA can find a dozen of optimums and sometimes even get 18 global optimums, which proved that hierarchy encoding with niche methods can maintain population diversity effectively and converge rapidly.

5 Example of Application

A reliability optimum design theory and method of a jaw clutch has been introduced [13]. The design goal is to seek a set of design parameters (outside diameter of clutch: D; number of gear: Z; structure coefficient K_1; structure coefficient K_2) under the condition of probability restriction. The relation of S_b and design parameters are defined as:

$$S_b = \frac{15.6T_n}{\left(\frac{Z}{2}+0.5\right)\left[\frac{\pi D(1-K_2)}{4Z}\right]^2 (1+K_1)D} \tag{10}$$

In Eq.(10), T_n is a constant. Obviously, S_b is the nonlinear function of the design parameters. The paper presents the values of (S_b, D, Z, K_1, K_2) using the restraint random directional methods. Under the completely same design condition, we employ ANHGA and NGA and HGA to genetic optimization calculation. After 400 iterative cycle the calculation is shown in Table 5.

Table 5. Optimization result of a jaw clutch

	Z	K_1	K_2	D(mm)	S_b (MPa)
NGA	25	0.74	0.34	45	2099.264
HGA	25	0.74	0.341	42	2203.24
ANHGA	25	0.7442	0.3348	64.777	692.934

From Table 5, it is obvious that ANHGA is more efficient than NGA and SHGA in optimum design of jaw clutch.

6 Conclusion

In this paper, niche technology and hierarchy genetic algorithm were combined and adaptive niche hierarchy GA based on sharing and crowding was put forward. The adaptive niche hierarchy genetic algorithm improves GA from not only encoding but also operators. These measures also increase searching ability of the GA effectively, ensuring population diversity and finding many solutions of complicated problems. The application of the modified genetic algorithm shows it is great effective to the multimodal optimization and appropriate for the design of multi-extremum complex system.

References

1. Holland, J.H.: Adaptation in Nature and Artificial Systems. Ann Arber, MI, University of Michigan Press (1975)
2. Sareni, B., Krahenbuhl, L., Nicolas, A.: Niching Genetic Algorithms for Optimization in Electromagnetics. In Proc. 11th COMPUMAG'97, Rio de Janeiro (1997) 563-564
3. Goldberg, D. E.: Genetic Algorithms in Search, Optimization and Machine Learning. Reading, MA, Addison Wesely (1989)
4. Rudolph, G.: Convergence Analysis of Canonical Genetic Algorithms. IEEE Trans. Neural networks, Special Issue on Evolution Computing, Vol.5, (1994) 96-101
5. Mahfoud, S.W.: Niching Methods for Genetic Algorithms, Ph.D. dissertation, Univ. Illinois at Urbana-Champaign, Illinois Genetic Algorithm Lab., Urbana, IL (1995)
6. Lee, C. G., Cho, D. H., Jung, H. K.: Niching Genetic Algorithm with Restricted Competition Selection for Multimodal Function Optimization. IEEE Trans. Magn., Vol.34, No.1, (1999) 1722-1755
7. Zhou, Beiyue., Deng, Bin., Guo, Guanqi.: Research of A Class of Improved Genetic Algorithm Based on Niches[J]. Journal of Mechanical Strength, Vol.24, No.1, (2002) 13-16
8. Yu, Shouyi., Guo, Guanqi.: A Class of Niche Used in Genetic Algorithms for Improving Efficiency of Searching Global Optimum. Information and Control, Vol.30, No.6, (2001) 326-331
9. Gong, Dunwei., Pan, Fengping., Xu, Shifan.: Adaptive Niche Hierarchy Genetic Algorithms. Proc. Of the 2002 IEEE Region 10 Conf. On Computers, Communicatona, Control and Power Engineering, Beijing: Posts & Telecom Press (2002) 39-42
10. Yu, Xin-jie., Zan-ji Wang.: Fitness Sharing Crowding Genetic Algorithm, Control and Decision, Vol.16, No.6, (2001) 926-929
11. Liu, Zhiyong., Liu, Mandan., Qian, Feng.: The Application of One Improved Niche Genetic Algorithm for Elman Recurrent Neural Networks. Proceedings of the 5th World Congress on Intelligent Control and Automation, Hangzhou, (2004)1978-1981
12. Gong, Dunwei., Sun, Xiaoyan., Guo, Xijin., Zhou, yong.: Adaptive Hierarchy Genetic Algorithm. Proceedings of IEEE TENCON'02, V ol.1, (2002)81-84
13. Liu, Weishan.: Optimization of Reliability Design of Machine Components. Beijing: China Science and Technology Press (1993)

Associativity, Auto-reversibility and Question-Answering on Q'tron Neural Networks

Tai-Wen Yue and Mei-Ching Chen

Dept. of Computer Science and Engineering,
Tatung University, Taipei, Taiwan, R.O.C.
twyu@mail.cse.ttu.edu.tw, mcchen@ttu.edu.tw

Abstract. Associativity, auto-reversibility and question-answering are the three intrinsic functions to be investigated for the proposed Q'tron Neural Network (NN) model. A Q'tron NN possesses these functions due to its property of local-minima free if it is built as a *known-energy system* which is equipped with the proposed *persistent noise-injection mechanism*. The so-built Q'tron NN, as a result, will settle down if and only if it 'feels' feasible, i.e., the energy of its state has been low enough truly. With such a nature, the NN is able to accommodate itself 'everywhere' to reach a feasible state autonomously. Three examples, i.e., an associative adder, an N-queen solver, and a pattern recognizer are demonstrated in this paper to highlight the concept.

1 Introduction

The monotonically decreasing nature of the energy of the Hopfield NN model [5,6] attracts many researchers to apply the model for constraint satisfaction [4,11] and combinatorial optimization [10]. However, the local-minima problem makes the NN frustrated [8,9]. Several stochastic approaches based on *simulated annealing* and other techniques have been proposed to resolve the local-minima problem [12]. Such attempts have resulted in several alternative neural-network models including Boltzmann [1,2], Cauchy [7] and Gaussian Machines [3]. These approaches, in fact, tackle the problem in an 'unknown-energy' fashion. Roughly speaking, these machines do not know how low their energies must be in order to get a good enough and/or a feasible solution.

This paper introduces the concept of the *known-energy systems* based on the *Q'tron (quantum neuron) NN model* [13], which is a significantly extended version of the Hopfield NN model [5,6]. A known-energy system considered in the paper is a device that consists of a set of entangled processing elements (PEs) with discrete outputs. The entanglement among these PEs will lead the system to move macroscopically toward lower-energy states. Microscopically, however, the system can become stable if and only if it has reached a state whose energy value approaches that of global minimum within a prespecified precision, i.e., the energy is low enough certainly. This implies that the energy value of global minimum must be 'known' by the system a priori in some possible sense. The known-energy property will allow us to determine the *bounded noise spectra* for Q'trons of the NN systematically. Persistently injecting such pieces of random noise into Q'trons will enable the NN to settle down if and only if it has reached a state whose energy is low enough surely [13].

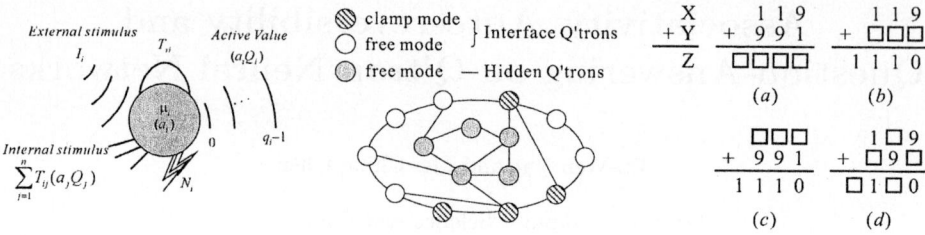

Fig. 1. The Q'tron model

Fig. 2. The application model of a known-energy system built using Q'tron NN model

Fig. 3. The functions of an associative adder

A known-energy system discussed in the paper can be considered as an associative memory that stores a set of patterns whose corresponding energies are all lying within a low-and-known energy range. To retrieve a particular pattern, we can feed a key, i.e., partially available information, to the system by 'keying' it at some Q'trons. The operation, if applicable, will trigger the NN to roam to the next stable state corresponding to a stored pattern associated with that key. Considering the key to be a question, we are then to interact with the system in a question-answering mode. If the system is also 'keyable everywhere', the computation performed by the system is hence reversible automatically.

The organization of this paper is as follows: Section 2 gives a brief overview on the Q'tron NN model. Section 3 defines the known-energy systems and their content-addressabilities. Section 4 to 6 provide three examples to manifest the general guidelines to building known-energy systems that are fully content-addressable using the Q'tron NN model. They are the associative adder, the N-queen solver, and the pattern recognizer, respectively. Finally, a conclusion is drawn in Section 7.

2 The Q'tron NN Model

The basic processing element of a Q'tron NN is called a Q'tron, a shorthand of a *quantum neuron*, schematically shown in Fig. 1. Let μ_i represent the ith Q'tron in a Q'tron NN. The output-level of μ_i, denoted as $Q_i \in \{0, 1, ..., q_i - 1\}$ with q_i (≥ 2). The actual output of μ_i, called *active value*, is $a_i Q_i$, where $a_i > 0$ is the unit excitation strength, called *active weight*. In a Q'tron NN, for a pair of connected Q'trons μ_i and μ_j, there is only one connection strength, i.e. $T_{ij} = T_{ji}$, and $T_{ii} < 0$ usually. The *noise-injected stimulus* $\hat{\mathcal{H}}_i$ for the Q'tron μ_i is defined as:

$$\hat{\mathcal{H}}_i = \mathcal{H}_i + \mathcal{N}_i = \sum_{j=1}^{n} T_{ij}(a_j Q_j) + I_i + \mathcal{N}_i, \qquad (1)$$

where $\mathcal{H}_i$ denotes the *noise-free net stimulus* of μ_i, which apparently is equal to the sum of *internal stimuli*, namely, $\sum_{j=1}^{n} T_{ij}(a_j Q_j)$, and *external stimulus* I_i. The term $\mathcal{N}_i$ denotes the piece of random noise fed into μ_i, and n denotes the number of Q'trons in the NN. In case that $P(\mathcal{N}_i = 0) = 1$, $i = 1, \ldots, n$, the Q'tron NN is said to run in *simple mode*; otherwise, it is said to run in *full mode*.

At each time step only one *free* Q'tron is selected for level transition subject to the following rule:

$$Q_i(t+1) = Q_i(t) + \Delta Q_i(t) \text{ with } \Delta Q_i(t) = \begin{cases} +1 & \hat{\mathcal{H}}_i(t) > \frac{1}{2}|T_{ii}a_i| \text{ and } Q_i(t) < q_i - 1; \\ -1 & \hat{\mathcal{H}}_i(t) < -\frac{1}{2}|T_{ii}a_i| \text{ and } Q_i(t) > 0; \\ 0 & \text{otherwise,} \end{cases} \quad (2)$$

where assuming that the ith Q'tron is selected at time $t+1$.

2.1 System Energy — Stability

The system energy embedded in a Q'tron NN, say, $\mathcal{S}$ is defined as

$$\mathcal{E}_\mathcal{S}(Q) = -\frac{1}{2}\sum_{i=1}^{n}\sum_{j=1}^{n}(a_iQ_i)T_{ij}(a_jQ_j) - \sum_{i=1}^{n}I_i(a_iQ_i) + K; \quad (3)$$

where $Q = \{(Q_1, \ldots, Q_n) : Q_i \in \{0, \ldots, q_i - 1\}\}$ is a state of $\mathcal{S}$, n is total number of Q'trons in the NN, and K can be any suitable constant. It was shown that, to run a Q'tron NN in simple mode, the energy $\mathcal{E}_\mathcal{S}$ defined above will monotonically decrease with time [13]. This implies that a Q'tron NN running in simple mode performs a greedy search. In case the majority of local-minima of the NN corresponds to poor or, even worse, infeasible solutions of the underlying problem, then the NN will be useless almost.

2.2 Bounded Noise Spectra

To enable the NN to escape 'only' from local-minima representing unsatisfactory solutions, each Q'tron is allowed to have a bounded noise spectrum only, i.e., $\mathcal{N}_i \in [\mathcal{N}_{i-}, \mathcal{N}_{i+}]$. As will be seen in the following sections, the values of $\mathcal{N}_{i-}$ and $\mathcal{N}_{i+}$ for each Q'tron can be systematically determined if the Q'tron NN is constructed with known-energy concept. One convenient way for computer simulation is to generate only *bang-bang noise*. In this way, the distribution of noise is specified by the so-called *noise ratio specification* (NRS) of a Q'tron. It is defined as:

$$\text{NRS}_i = P(\mathcal{N}_i = \mathcal{N}_i^-) : P(\mathcal{N}_i = 0) : P(\mathcal{N}_i = \mathcal{N}_i^+) \quad (4)$$

with $P(\mathcal{N}_i = \mathcal{N}_i^-) + P(\mathcal{N}_i = 0) + P(\mathcal{N}_i = \mathcal{N}_i^+) = 1$, where NRS_i represents the NRS of μ_i, and $P(\mathcal{N}_i = x)$ represents the probability that $\mathcal{N}_i = x$ is generated. Clearly, if a Q'tron NN runs in *simple mode*, then $\text{NRS}_i = 0 : 1 : 0$ for all i. In the following, we will assume that $P(\mathcal{N}_i = \mathcal{N}_i^-) \neq 0$ and $P(\mathcal{N}_i = \mathcal{N}_i^+) \neq 0$ for all i if a Q'tron NN runs in full mode.

2.3 The Application Model

To make a Q'tron NN versatilely accessible, each Q'tron can either be operated in *clamp mode*, i.e., its output-level is clamped fixed at a particular level, or in *free mode*, i.e., its output-level is allowed to be updated according to the level transition rule specified in Eq. (2). Furthermore, the Q'trons in an NN are categorized into two types: *interface Q'trons* and *hidden Q'trons*, see Fig. 2. The former provides an interface for user's interaction, whereas the latter is functionally necessary to make the NN to a known-energy one. We will assume that hidden Q'trons are always free unless otherwise specified. Interface Q'trons operated in clamp-mode are used to feed the available or affirmative information (a question) into the NN. The other free-mode interface Q'trons, on the other hand, are used to perform association to 'fill in' the missing or uncertain information (an answer).

3 Known-Energy Systems

In this section, we define the concept of known-energy systems and investigate their content addressabilities.

3.1 Definition of Known-Energy Systems

Given a Q'tron NN, say $\mathcal{S}$, we will use $\mu^I = \{i_1, \ldots, i_n\}$ and $\mu^H = \{h_1, \ldots, h_m\}$ to denote the sets of all its interface Q'trons and hidden Q'trons, respectively, use $Q^I = \{(Q_{i_1}, \ldots, Q_{i_n}) : Q_{i_k} \in \{0, \ldots, q_{i_k} - 1\}\}$ and $Q^H = \{(Q_{h_1}, \ldots, Q_{h_m}) : Q_{h_k} \in \{0, \ldots, q_{h_k} - 1\}\}$ to denote their corresponding state spaces, respectively, and $Q^{IH} = Q^I \times Q^H$. Furthermore, $Q^i \in Q_I$, $Q^h \in Q_H$, and $Q^{ih} \in Q^{IH}$ are called a *surface state*, a *hidden state*, and a *system state* of $\mathcal{S}$, respectively. Now, we have the following useful definitions:

Definition 1. itLet $\mathcal{E}_\mathcal{S}^*$ be a lower-bound of $\mathcal{E}_\mathcal{S}$, i.e., $\mathcal{E}_\mathcal{S}^* \leq \min\{\mathcal{E}_\mathcal{S}(Q) : Q \in Q^{IH}\}$, whose value is assumed known explicitly, and let $L_{\mathcal{S},\delta} = [\mathcal{E}_\mathcal{S}^*, \mathcal{E}_\mathcal{S}^* + \delta]$, where $\delta \geq 0$, be an energy range. Then, corresponding to δ, the α-floor set, denoted as $\mathcal{F}_\alpha(\mathcal{S}, \delta)$, and the β-floor set, denoted as $\mathcal{F}_\beta(\mathcal{S}, \delta)$, of $\mathcal{S}$ are defined as:

$$\mathcal{F}_\alpha(\mathcal{S}, \delta) = \{Q^{ih} : Q^{ih} \in Q^{IH} \text{ and } \mathcal{E}_\mathcal{S}(Q^{ih}) \in L_{\mathcal{S},\delta}\}$$
$$\mathcal{F}_\beta(\mathcal{S}, \delta) = \{Q^i : Q^i \in Q^I \text{ and } \exists Q^{ih} = (Q^i, Q^h) \in Q^{IH} \text{ such that } Q^{ih} \in \mathcal{F}_\alpha(\mathcal{S}, \delta)\}$$

Clearly, if δ is small, $L_{\mathcal{S},\delta}$ in fact specifies a low-and-known energy range. Hence, $\mathcal{F}_\alpha(\mathcal{S}, \delta)$ will represent a set of low-energy wells, i.e., the states corresponding to the cost effective solutions of a problem provided that the problem has been properly mapped. In applications, however, we are only interested in the surface state of an NN. Therefore, $\mathcal{F}_\beta(\mathcal{S}, \delta)$ is really the one of interest.

Definition 2. itAssume that $\mathcal{F}_\alpha(\mathcal{S}, \delta) \neq \emptyset$. Then, $\mathcal{S}$ is said to be a known-energy system with depth δ if all Q'trons in μ^I are free, it can 'ever' reach a system state, say, Q^{ih} such that $Q^{ih} \in \mathcal{F}_\alpha(\mathcal{S}, \delta)$ with probability one. Furthermore, suppose that $\mathcal{S}$ reaches $Q^{ih} = (Q^i, Q^h)$ at time t_1. Then, $Q^i(t_2) \in \mathcal{F}_\beta(\mathcal{S}, \delta)$ for all $t_2 \geq t_1$.

The above definition tells us that, initializing a known-energy system at any state, the system finally will be trapped into an energy well in its α-floor set if it is nonempty, and, from then on, the surface state of the system will become stable (Its hidden state, however, might keep changing). Therefore, the solution reported by the system must be good enough as its surface state becomes stable.

3.2 Content Addressabilities

To interact with a known-energy system, we can let certain interface Q'trons be clamped at a particular state, which represents a key or a question, see Fig. 2. Suppose that the system is able to enter its α-floor set in this sense, whenever possible. Then, the system, in fact, serves as a content addressable memory, i.e., the state of the free interface Q'trons will give a recall or an answer corresponding to that key or question when they settle down. In the sequel, the interface Q'trons used to clamp a key or a question will be called *key-nodes*, and the free interface Q'trons are called *recall-nodes*. For simplicity, in the following, we assume that the depth, i.e., δ, of known-energy system $\mathcal{S}$ is implicitly specified, and its α-floor set and β-floor set will be abbreviated as $\mathcal{F}_\alpha(\mathcal{S})$ and $\mathcal{F}_\beta(\mathcal{S})$, respectively.

Definition 3. itLet $\mathcal{S}$ be a known-energy system, and let $\mu^I = \{i_1, \ldots, i_n\}$ be the set of all interface Q'trons of $\mathcal{S}$. Without loss of generality, let $\mu^K = \{i_1, \ldots, i_{n'}\} \subseteq \mu^I$ denote a set key nodes, and define

$$K_\mathcal{S} = \{Q^k = (Q_{i_1}, \ldots, Q_{i_{n'}}) : \exists (Q^k, Q^r) \in \mathcal{F}_\beta(\mathcal{S})\}$$

as a set of legal keys. Then, $\mathcal{S}$ is said to be content addressable by μ^K if, Q'trons in μ^K are clamped with a legal key, say, $Q^k \in K_\mathcal{S}$, it can ever enter its α-floor set with probability one. Furthermore, $\mathcal{S}$ is said to be fully content-addressable if it is content-addressable by any $\mu^K \subseteq \mu^I$.

The above definition ensures that a content addressable system will always give a meaningful recall when a legal key is clamped at key-nodes. However, if an illegal key is clamped, the answer may be unpredictable.

In the remainder of the paper, we will provide some examples as guidelines to solve problems based on the known-energy concept. The so-built known-energy systems are all fully content-addressable intrinsically and, hence, auto-reversible.

4 The Associative Adder

This simple example demonstrates the local-minima escaping capability of the Q'tron NN model, and highlights that the full content-addressability of a problem solver is easily achievable if it is built as a known-energy system.

4.1 The Addition — The Forward Problem

Given two n-digit decimal numbers, say $X = x_{n-1} \cdots x_1 x_0$ and $Y = y_{n-1} \cdots y_1 y_0$ where $x_i, y_i \in \{0, \ldots, 9\}, i = 0, \ldots, n-1$, the goal of an adder is to report an $(n+1)$-digit decimal number, say $Z = z_n \cdots z_1 z_0$ where $z_i \in \{0, \ldots, 9\}, i = 0, \ldots, n-1$ and $z_n \in \{0, 1\}$, such that $Z = X + Y$, i.e., to answer the question of Fig. 3(a).

To construct a Q'tron NN adder, say $\mathcal{A}_n$ (or $\mathcal{A}$; for short), we prepare three Q'tron sets $\mu^X = \{\mu_0^X, \ldots, \mu_{n-1}^X\}$, $\mu^Y = \{\mu_0^Y, \ldots, \mu_{n-1}^Y\}$ and $\mu^Z = \{\mu_0^Z, \ldots, \mu_n^Z\}$, to deal with the summand X, the addend Y, and the sum Z, respectively. Clearly, Q'trons in $\mu^X \cup \mu^Y$ are for feeding questions, and Q'trons in μ^Z are for reporting answers. They are all interface Q'trons. To represent digits of decimal numbers, we let $q_i^X = q_i^Y = q_i^Z = 10$, i.e., $Q_i^X, Q_i^Y, Q_i^Z \in \{0, \ldots, 9\}, i = 0, \ldots, n-1$, and $q_n^Z = 2$, i.e., $Q_n^Z \in \{0, 1\}$. Moreover, corresponding to the position of each digit, the active weight of each Q'tron can be determined naturally, namely $a_i = a_i^X = a_i^Y = 10^i, i = 0, \ldots, n-1$, and $a_j = a_j^Z = 10^j$, $j = 0, \ldots, n$.

With these Q'trons, the goal of addition is, then, to minimize the energy function, say, $\mathcal{E}_\mathcal{A}$ defined by:

$$\mathcal{E}_\mathcal{A} = \frac{1}{2} \left(\sum_{i=0}^{n-1} a_i Q_i^X + \sum_{i=0}^{n-1} a_i Q_i^Y - \sum_{i=0}^{n} a_i Q_i^Z \right)^2. \quad (5)$$

It will be seen in the next section that the above energy function is in *integer-programming type*. Referring to the form of Eq. (3), the other parameters, i.e., connection strengths and external stimuli, of $\mathcal{A}$ can then be determined by mapping $\mathcal{E}_\mathcal{A}$ to it. So,

$$T_{gh} = \begin{cases} -1 & \text{for } g, h \in \mu^X \cup \mu^Y \text{ or } g, h \in \mu^Z \\ 1 & \text{for } g \in \mu^X \cup \mu^Y \text{ and } h \in \mu^Z; \text{ or } g \in \mu^Z \text{ and } h \in \mu^X \cup \mu^Y \end{cases} \quad (6)$$

$$I_g = 0 \text{ for all } g \in \mu^X \cup \mu^Y \cup \mu^Z \quad (7)$$

4.2 The Known-Energy System $\mathcal{A}$ and Its Content-Addressability

Although the structure of $\mathcal{A}$ is simple, running $\mathcal{A}$ in simple mode, it usually functions poorly. For example, consider $\mathcal{A}_3$. Suppose now that Q'trons in μ^X and μ^Y are clamped with values $Q_2^X Q_1^X Q_0^X = 250$ and $Q_2^Y Q_1^Y Q_0^Y = 251$, respectively, and Q'trons in μ^Z are free and at state $Q_3^Z Q_2^Z Q_1^Z Q_0^Z = 1000$. Applying the Q'tron's dynamics, i.e., Eq. (2), onto this state, one will see that this state is a local minimum and, hence, $\mathcal{A}_3$ gets stuck. Even in case that all Q'trons are set free, the NN also cannot be liberated from local minima completely. For example, $Q_2^X Q_1^X Q_0^X = 009$, $Q_2^Y Q_1^Y Q_0^Y = 009$, and $Q_3^Z Q_2^Z Q_1^Z Q_0^Z = 0020$ is a local minimum in simple mode.

It is clear that when $\mathcal{A}$ reaches a true answer state, we must have

$$\left| \sum_{i=0}^{n-1} a_i Q_i^X + \sum_{i=0}^{n-1} a_i Q_i^Y - \sum_{i=0}^{n} a_i Q_i^Z \right| \leq \frac{\Delta}{2}, \quad (8)$$

where $0 \leq \frac{\Delta}{2} < 1$. $\frac{\Delta}{2}$ is the so-called *solution qualifier* in [13], which can be used to tune the solution performance of a known-energy system. For example, setting $\frac{\Delta}{2} \geq 1$ implies that some degree of error will be tolerable for the adder. Substituting Eq. (8) into Eq. (5), then the feasible energy range corresponding to a correct answer can thus be 'known', i.e.,

$$0 \leq \mathcal{E}_{add} \leq \frac{1}{2} \left(\frac{\Delta}{2} \right)^2 \quad (9)$$

With such a piece of knowledge, the bounded noise-spectra for Q'trons in $\mathcal{A}$ can then be determined to fulfill the goal of Eq. (8) or (9) using the method proposed in [13]. Specifically, the bounded noise-spectrum $[\mathcal{N}_i^-, \mathcal{N}_i^+]$ for Q'tron $\mu_i \in \mu^{\mathcal{A}}$ satisfies

$$-\mathcal{N}_i^- = \mathcal{N}_i^+ = \max\left(0, \frac{1}{2}|T_{ii} a_i| - \frac{\Delta}{2}\right) \quad (10)$$

Theorem 1. *Q'tron NN $\mathcal{A}$ is a known-energy system with depth $\frac{1}{2}\left(\frac{\Delta}{2}\right)^2$ for any $\frac{\Delta}{2} \geq 0$ if it runs in full mode with Q'trons' noise-spectra specified by Eq. (10). Furthermore, $\mathcal{A}$ is fully content-addressable.*

The theorem is a straightforward conclusion of the *Completeness Theorem* described in [13]. The facts described in the Theorem 1 indicates that $\mathcal{A}$ can be operated in an associative manner as depicted in Fig. 3.

5 The n-Queen Solvers

An n-queen problem is to ingeniously set n queens on an $n \times n$ chessboard so that all of them can live in peace with each other. Only one queen can be in each row, each column, while at most one queen can be in each diagonal.

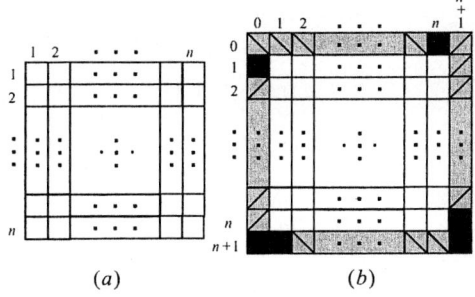
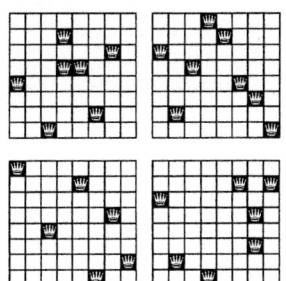

Fig. 4. (a) The $n \times n$ interface Q'tron-plane of the n-queen solver; (b) The $(n+2) \times (n+2)$ Q'tron plane of the n-queen solver; the surrounding Q'trons are hidden Q'trons

Fig. 5. Four sample local-minima obtained by running Q_s in simple-mode

5.1 The n-Queen Program as an Integer Program

Although a known-energy system without any hidden Q'tron is applicable to solve n-queen problem [14], we here adapt another approach instead to highlight the role of hidden Q'trons.

We will use an $n \times n$ Q'tron-plane to represent the chessboard of n-queen problem, as was shown in Fig. 4(a). Let $Q_{ij} \in \{0,1\}$, where $i, j \in \{1, \ldots, n\}$, denote the number of queen in cell (i, j). Then, the n-queen problem can be formulated as the following integer program

$$\sum_{j=1}^{n} Q_{ij} = 1, \quad i = 1, \ldots, n \tag{11}$$

$$\sum_{i=1}^{n} Q_{ij} = 1, \quad j = 1, \ldots, n \tag{12}$$

$$\sum_{\substack{i,j \in \{1,\ldots,n\} \\ i+j = m}} Q_{ij} \leq 1, \quad m = 2, \ldots, 2n \tag{13}$$

$$\sum_{\substack{i,j \in \{1,\ldots,n\} \\ i-j = m}} Q_{ij} \leq 1, \quad m = -n+1, \ldots, n-1 \tag{14}$$

$$Q_{ij} \in \{0,1\}, \quad i, j = 1, \ldots, n. \tag{15}$$

To achieve known energy, Eq. (13) and (14) are converted to equalities by introducing some *slack* Q'trons. For conceptually easier, the chessboard is extended to $(n+2) \times (n+2)$, as shown in Fig. 4(b). The Q'trons that marked black are useless. However, keeping them there is harmless for problem formulation. The surrounding Q'trons which are not marked black are slack Q'trons. They serve as hidden Q'trons, see Fig. 2. Accordingly, Eq. (13) to (15) are then replaced as

$$\sum_{\substack{i,j \in \{1,\ldots,n\} \\ i+j = m}} Q_{ij} + Q_{i_m, j_m} = 1, \quad m = 2, \ldots, 2n \tag{16}$$

$$\sum_{\substack{i,j \in \{1,\ldots,n\} \\ i-j=m}} Q_{ij} + Q_{i'_m,j'_m} = 1, \quad m = -n+1,\ldots,n-1 \qquad (17)$$

$$Q_{ij} \in \{0,1\}, \qquad i,j = 0,\ldots,n+1. \qquad (18)$$

where i_m, j_m, i'_m and j'_m are functions for aligning a slack Q'tron to the corresponding diagonal-constraint term, i.e.,

$$(i_m, j_m) = \begin{cases} (m, 0), & 2 \le m \le n \\ (m-n-1, n+1), & n+1 \le m \le 2n \end{cases}$$

$$(i'_m, j'_m) = \begin{cases} (0, -m), & -n+1 \le m \le 0 \\ (n+1, n+1-m), & 1 \le m \le n-1 \end{cases}$$

The integer program to solve the n-queen problem now is in a *standard form*.

5.2 The Known-Energy System $\mathcal{Q}$

Denote the n-queen solver to be constructed as $\mathcal{Q}_n$ (or $\mathcal{Q}$; for short). Referring to the integer program defined above, then the goal of $\mathcal{Q}$ is to minimize the energy function, say, $\mathcal{E}_\mathcal{Q}$ defined by

$$\mathcal{E}_\mathcal{Q} = \mathcal{E}_- + \mathcal{E}_| + \mathcal{E}_\backslash + \mathcal{E}_/, \qquad (19)$$

where

$$\mathcal{E}_- = \frac{1}{2}\sum_{i=1}^{n}\left(\sum_{j=1}^{n} Q_{ij} - 1\right)^2, \quad \mathcal{E}_| = \frac{1}{2}\sum_{j=1}^{n}\left(\sum_{i=1}^{n} Q_{ij} - 1\right)^2,$$

$$\mathcal{E}_\backslash = \frac{1}{2}\sum_{m=2}^{2n}\left(\sum_{\substack{i,j \in \{1,\ldots,n\} \\ i+j=m}} Q_{ij} + Q_{i_m,j_m} - 1\right)^2,$$

$$\mathcal{E}_/ = \frac{1}{2}\sum_{m=-n+1}^{n+1}\left(\sum_{\substack{i,j \in \{1,\ldots,n\} \\ i-j=m}} Q_{ij} + Q_{i'_m,j'_m} - 1\right)^2.$$

An energy function derived from a standard integer program in the above form is called in an integer-programming-type in [13,14]. It is not hard to see that a feasible n-queen configuration is reached if and only if

$$0 \le \mathcal{E}_\mathcal{Q} \le \frac{1}{2}\left(\frac{\Delta}{2}\right)^2 \qquad (20)$$

provided $\frac{\Delta}{2}$ is set sufficiently small, e.g., $0 \le \frac{\Delta}{2} < 1$. Hence, the known-energy property of $\mathcal{E}_\mathcal{Q}$ is confirmed.

It is clear that each Q'tron in $\mathcal{Q}$ has two output-levels. So, $q_{ij} = 2$ for all $0 \le i,j \le n+1$. Furthermore, since each Q'tron in $\mathcal{Q}$ plays the same role, simply set $a_{ij} = 1$ for all $0 \le i,j \le n+1$. Referring to the form of Eq. (3), the other parameters, i.e., connection strengths and external stimuli, of $\mathcal{Q}$ can then be determined by mapping $\mathcal{E}_\mathcal{Q}$ to it. To save space, they are not detailed here.

With the piece of knowledge specified in Eq. (20), we have the following theorem.

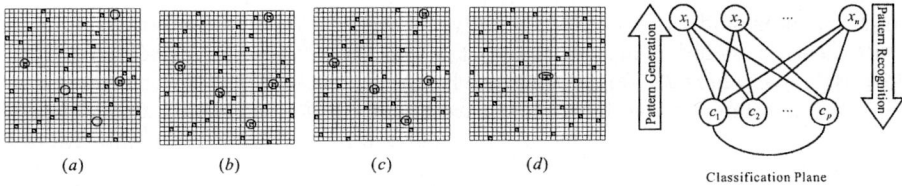

Fig. 6. (a) A legal n-queen configuration obtained by setting all Q'trons free; the circle cells will be clamped to one so as to retrieve an interesting answer; (b) the state right after the circled cells of (a) are clamped; this state is unstable;(c) a legal n-queen configuration corresponding to the setting of (b); (d) the NN in this setting will never settle down if its setting is kept unchanged

Fig. 7. The known-energy system to perform pattern recognition classification, pattern generation and pattern association

Theorem 2. $\mathcal{Q}$ is a known-energy system with depth $\frac{1}{2}\left(\frac{\Delta}{2}\right)^2$ for any $\frac{\Delta}{2} \geq 0$ if it runs in full mode with Q'trons' noise-spectra $[-\mathcal{N}_{ij}^-, \mathcal{N}_{ij}^+]$ for all $\mu_{ij} \in \mu^{\mathcal{Q}}$ satisfying

$$-\mathcal{N}_{ij}^- = \mathcal{N}_{ij}^+ = \max\left(0, \frac{1}{2}|T_{ij,ij} a_{ij}| - \frac{\Delta}{2}\right). \tag{21}$$

Furthermore, $\mathcal{Q}$ is fully content-addressable.

The theorem is true due to $\mathcal{Q}$ is constructed from an integer-programming-type energy function, see [13].

5.3 Discussion

Without noise injection, $\mathcal{Q}$ usually malfunctions. For example, Fig. 5 shows several wrong answers reported by running $\mathcal{Q}_8$ in simple-mode. Since $\mathcal{Q}$ is fully content-addressable when it runs in full-mode, we can demand it to only offer answer(s) that we are interested in. For example, Fig. 6 shows an operating scenario that interact with $\mathcal{Q}_{30}$ in a question-answering mode.

6 The Pattern Recognizer/Generator

The two example known-energy systems described above are both constructed based on the integer-programming-type energy functions. However, some problems are not intrinsically in that nature. Yet, by proper reformulation, their energy functions, though not in that type, can also be endowed with the known-energy property. Accordingly, one can build the corresponding known-energy systems with the desirable content-addressability by carefully designing the noise spectra for Q'trons in such systems.

The known-energy system to be described is a pattern recognizer. Since the system is also fully content-addressable, it is also a pattern generator by viewing it oppositely, see Fig. 7. The system is originally constructed for pattern recognition. However, its inverse function, i.e., pattern generation, is automatically functionable.

6.1 The Pattern Recognition — The Forward Problem

Here, we consider the pattern-recognition task to be the one that classifies an input binary pattern into a satisfactory class corresponding to one of stored patterns. Specifically, let $W = \{W^j | W^j = (x_1^j, \ldots, x_n^j) \in \{0,1\}^n, 1 \leq j \leq p\}$ be a set of different stored patterns, and let $Q_x = (Q_{x_1}, \ldots, Q_{x_n}) \in \{0,1\}^n$ denote an arbitrary input pattern (a feature vector). Then, to classify Q_x into the kth class is considered proper if $d(Q_x, W^k) \leq \epsilon$, where $d(\cdot, \cdot)$ represents the Hamming distance between the two patterns and $\epsilon \geq 0$ is a nonnegative integer. Apparently, if ϵ is set sufficiently small, then the process corresponds to classifying an input pattern into the best representative class. Fig. 7 shows the Q'tron NN that performs the pattern recognition task. All Q'trons in the NN are binary, Q'trons in $\mu^X = \{x_1, \ldots, x_n\}$ are for holding input pattern, and Q'trons in $\mu^C = \{c_1, \ldots, c_p\}$ are for reporting classification result. Given an input pattern, if a proper classification is possible, we want the NN to report a proper class, say, c_k by setting $Q_{c_k} = 1$ and $Q_{c_j} = 0$ for all $j \neq k$ when it settles down; otherwise, the NN either settles down with $Q_{c_j} = 0$ for all j or never settles down.

The following function can be used for the Hamming distance measurement

$$d(Q_x, W^k) = \sum_{i=1}^{n}[x_i^k(1-Q_{x_i}) + (1-x_i^k)Q_{x_i}] = \sum_{i=1}^{k} x_i^k - \sum_{i=1}^{n}(2x_i^k - 1)Q_{x_i} \quad (22)$$

Accordingly, the energy function to fulfill the pattern-recognition task now is defined as follows:
$$\mathcal{E}_\mathcal{P} = \mathcal{E}_{proper} + \mathcal{E}_{inh} + K \quad (23)$$
where K can be any suitable constant, and

$$\mathcal{E}_{proper} = \sum_{j=1}^{p} Q_{c_j} \left[\sum_{i=1}^{k} x_i^k - \sum_{i=1}^{n}(2x_i^k - 1)Q_{x_i} - b \right], \quad \mathcal{E}_{inh} = \frac{\lambda}{2} \sum_{j=1}^{p} \sum_{\substack{k=1 \\ k \neq j}}^{p} Q_{c_j} Q_{c_k},$$

where b is a positive integer which satisfies $b > \epsilon$, and λ is a positive weighting factor whose value is set sufficiently large, e.g., $\lambda \geq b + \epsilon$. With a careful investigation, one then see that $\mathcal{E}_{proper}$ measures the properness of a classification, and $\mathcal{E}_{inh}$ is for mutual inhibition.

The following lemma describes the known-energy property of $\mathcal{E}_\mathcal{P}$.

Lemma 1. *Let $b > \epsilon, \lambda \geq b + \epsilon$ and $K = b$. Then, $\mathcal{E}_\mathcal{P} \in [0, \epsilon]$ if and only if a proper classification for the pattern-recognition task mentioned above is made.*

6.2 The Known-Energy System $\mathcal{P}$

Denote the Q'tron NN to be constructed as $\mathcal{P}$. Following the similar procedure described above, its parameters can be obtained. They are summarized as follows:

1. active weights: $a_{x_i} = a_{c_j} = 1$ for all $1 \leq i \leq n$ and $1 \leq j \leq p$;
2. number of output-level: $q_{x_i} = q_{c_j} = 2$ for all $1 \leq i \leq n$ and $1 \leq j \leq p$;
3. connection strengths:

$$\begin{array}{ll} T_{c_i c_j} = (\delta_{ij} - 1)\lambda & \text{for all } 1 \leq i,j \leq p, \\ T_{x_i c_j} = T_{c_j x_i} = 2x_i^j - 1 & \text{for all } 1 \leq i \leq n \text{ and } 1 \leq j \leq p, \\ T_{x_i x_j} = 0 & \text{for all } 1 \leq i,j \leq n, \end{array}$$

where δ_{ij} is the Kronecker delta function, i.e., $\delta_{ij} = \begin{cases} 1 & i = j \\ 0 & i \neq j \end{cases};$

4. external stimuli:
$$I_{c_j} = b - \sum_{i=1}^{n} x_i^j \quad \text{for all } 1 \leq j \leq p,$$
$$I_{x_i} = 0 \quad \text{for all } 1 \leq i \leq n;$$

Since $\mathcal{E}_\mathcal{P}$ is not in integer-programming type, we cannot use the aforementioned method to determine Q'trons' noise-spectra. However, by considering the range of tolerable noise strengths for each type of Q'trons so as to fulfill our goal, we obtain the following noise-spectra for Q'trons:

5. noise spectra:
$$\mathcal{N}_{c_j} \in [-b + \epsilon + 0.5, 0.5] \quad \text{for all } 1 \leq j \leq p,$$
$$\mathcal{N}_{x_i} \in [-0.5, 0.5] \quad \text{for all } 1 \leq i \leq n;$$

The following theorems describe the patterns stored in $\mathcal{P}$ and investigate its content-addressability. Their proofs are skipped here.

Theorem 3. $\mathcal{P}$ *is a known-energy system with depth* $\epsilon \geq 0$ *if it runs in full mode with noise-spectra defined above, and its α-floor set $\mathcal{F}_\alpha(\mathcal{P})$ and β-floor set $\mathcal{F}_\beta(\mathcal{P})$ are*

$$\mathcal{F}_\alpha(\mathcal{P}) = \mathcal{F}_\beta(\mathcal{P}) = \{(Q_x, Q_c^k) : d(Q_x, W^k) \leq \epsilon, 1 \leq k \leq p\},$$

where $Q_x = (Q_{x_1}, \ldots, Q_{x_n}) \in \{0,1\}^n$, *and* $Q_c^k = (Q_{c_1}, \ldots, Q_{c_p}) \in \{0,1\}^p$ *such that* $Q_{c_k} = 1$ *and* $Q_{c_j} = 0$ *for all* $j \neq k$.

The above theorem tells us that a pattern-class pair is stored in $\mathcal{P}$ if and only if it corresponds to a proper classification.

Theorem 4. $\mathcal{P}$ *is fully content-addressable.*

The theorem indicates that $\mathcal{P}$ automatically functions as a pattern generator if we operate it oppositely. This shows the auto-reversibility of known-energy systems.

7 Conclusions

This paper introduces the concept of known-energy systems based on the Q'tron NN model. A known-energy system always targets at keeping its energy low enough, and is able to settle down only if its energy does so already. To this end, Q'trons in the system are persistently noise-injected according to their bounded noise-spectra, which can be determined systematically at design time, so that the system can react to goal violation immediately when its environment changes. Hence, known-energy systems can be considered as continuously thinking machines. It is also shown that the known-energy systems are usually fully content-addressable. This allows users to access a known-energy system through many different ways, e.g., to perform memory association, to perform inverse function, or to do question-answering. Three simple examples, i.e., the associative adders, the n-queen solvers, and the pattern recognizers, are demonstrated to confirm these. More information is available at http://www.cse.ttu.edu.tw/twyu/qtron/ks/.

References

1. Aarts, E., Korst, J.: Simulated Annealing and Boltzmann Machines. Wiley and Sons, Chichester, England (1989)
2. Ackley, D. H., Hinton, G. E. and Sejnowski, T. J.: A Learning Algorithm for Boltzmann Machine. Cognitive Science **9** (1985) 147-169
3. Akiyama, Y., Yamashita, A., Kajiura, M. and Aiso, H.: Combinatorial Optimization with Gaussian Machines. Proceedings of IEEE International Joint Conference on Neural Networks **1** (1989) 533–540
4. Bartak, R.: On-line Guide to Constraint Programming. http://kti.mff.cuni.cz/bartak/constraints/, Prague (1998)
5. Hopfield, J. J.: Neural Networks and Physical Systems with Emergent Collective Computational Abilities. Proceedings of Natl. Acad. Sci. USA **79** (1982) 2554-2558
6. Hopfield, J. J., Tank, D. W.: Neural Computation of Decisions in Optimization Problems. Biological Cybernetics **52** (1985) 141-152
7. Jeong, H., Park, J. H.: Lower Bounds of Annealing Schedule for Boltzmann and Cauchy Machines. Proceeding of IEEE International Joint Conference on Neural Networks **1** (1989) 581-586
8. Lee, B. W., Sheu, B. J.: Modified Hopfield Neural Networks for Retrieving the Optimal Solution. IEEE Transactions on Neural Networks **6** (1991) 137-142
9. Lu, M., Zhan, Y., Mu, G.: Bipolar Optical Neural Network with Adaptive Threshold. Optik **91** (1992) 178-182
10. Papadimitriou, C. H., Steiglitz, K.: Combinatorial Optimization: Algorithms and Complexity. Prentice-Hall (1998)
11. Tsang, E.: Foundations of Constraint Satisfaction. Academic Press, London and San Diego (1993)
12. Wong, E: Stochastic Neural Networks. Algorithmica (1991) 466-478
13. Yue, T. W., Chiang, S. C.: Quench, Goal-Matching and Converge The Three-Phase Reasoning Of a Q'tron Neural Network. Proceedings of the IASTED International Conference on Artificial and Computational Intelligence (2002) 54-59
14. Yue, T. W., Chen, M. C.: Q'tron Neural Networks for Constraint Satisfaction. Proceedings of Fourth International Conference on Hybrid Intelligent Systems (HIS'04) (2004) 398-403

Associative Classification in Text Categorization

Jian Chen, Jian Yin, Jun Zhang, and Jin Huang*

Department of Computer Science,
Zhongshan University, Guangzhou, China
ellachen@163.com

Abstract. Text categorization has become one of the key techniques for handling and organizing text data. This model is used to classify new article to its most relevant category. In this paper, we propose a novel associative classification algorithm ACTC for text categorization. ACTC aims at extracting the k-best strong correlated positive and negative association rules directly from training set for classification, avoiding to appoint complex support and confidence threshold. ACTC integrates the advantages of the previously proposed effective strategies as well as the new strategies presented in this paper. An extensive performance study reveals that the improvement of ACTC outperform other rule-based classification approaches on accuracy.

1 Introduction

Text categorization refers to the task of automatically assigning documents into one or more predefined classes or categories. In recent years, there has been an increasing number of statistical and machine learning techniques that automatically generate text categorization knowledge based on training examples. Such techniques including decision trees, k-nearest-neighbor system, rule induction, gradient descent neural networks [1], regression models, Linear Least Square Fit, and support vector machines assume the availability of a large pre-labeled or tagged training corpus. The goal of text categorization is to classify documents into a certain number of predefined categories. Each document can be in multiple, exactly one or no category at all. Using machine learning, the objective is to learn classifiers from examples which perform the category assignments automatically.

In recent years, a new classification technique, called **associative classification**, is proposed to combine the advantages of association rule mining and classification [2]. In general, this model has three phases as follows: (1) Generates all the class association rules (CARs) satisfying certain user-specified

---

* This work is supported by the National Natural Science Foundation of China (60205007), Natural Science Foundation of Guangdong Province (031558, 04300462), Research Foundation of National Science and Technology Plan Project (2004BA721A02), Research Foundation of Science and Technology Plan Project in Guangdong Province (2003C50118) and Research Foundation of Science and Technology Plan Project in Guangzhou City (2002Z3-E0017).

minimum support threshold as candidate rules by association rule mining algorithm, such as Apriori or FPgrowth. These discovered rules have the form of (*attributes* ⇒ *class_label*); (2) Evaluates the qualities of all CARs discovered in the previous phase and pruning the redundant and low effective rules. Usually the *minimum confidence* is taken as a criterion to evaluate the qualities of the rules. Just "useful" rules with higher confidence than a certain threshold are selected to form a classifier; (3) Assigns a class label for a new data object. The classifier scores all the rules consistent with new data object and select some or all suitable rules to make a prediction.

The recent studies [3] show that this classification model achieves higher accuracy than traditional rule-based classification approaches such as C4.5 [4] and Ripper [5]. However, this model still suffers from accuracy and efficiency because of its well-known support-confidence framework. In this paper, we propose a novel associative classification algorithm based on correlation analysis, which aims at extracting the k-best strong correlated positive and negative association rules directly from training set for classification, avoiding to appoint complex support and confidence threshold. The remainder of the paper is organized as follows: Section 2 introduces the basic definitions and notations of associative classification in text categorization. In Section 3, we introduce our approach in detailed on literal selection, correlation analysis and how to use these correlated rules for text categorization. Experimental results are described in Section 4 along with the performance of our algorithm in compared with other previous rule-based classification approaches. Finally, we summarize our research work and draw conclusions in Section 5.

2 Basic Concepts and Terminology

Let $A = \{A_1, A_2, \ldots, A_k\}$ is a set of k attributes. $V[A] = \{v_1, v_2, \ldots, v_j\}$ is the domain of attribute A. Let $C = \{c_1, c_2, \ldots, c_m\}$ is a set of predefined class label for text categories. Let $\mathcal{D}$ is a domain of documents, where each d in $\mathcal{D}$ follows the scheme $\{v_1, v_2, \ldots, v_k\}(v_i \in V[A], 1 \leq v_i \leq n)$.

Definition 1. *(Literal) A literal l is an attribute-value pair of the form $A_i = v$, $A_i \in A, v \in V[A]$. An example $d = \{v_1, v_2, \ldots, v_k\}$ satisfies literal l iff $v_i = v$, where v_i is the value of the ith attribute of d.*

Definition 2. *(Association Rule) An **association rule** r, which takes the form of $a_r \Rightarrow c_r$ with a_r of the form $l_1 \wedge l_2 \wedge \cdots \wedge l_i$ called the **antecedent** of rule and c_r of the form $l'_1 \wedge l'_2 \wedge \cdots \wedge l'_j$ called the **consequence** of rule, $a_r \cap c_r = \emptyset$. We say an example d satisfies rule's antecedent iff it satisfies every literal in a_r and it satisfies the whole rule iff it satisfies c_r as well.*

Definition 3. *(Support and Confidence) For a given rule r is of the form $a_r \Rightarrow c_r$,*

$$sup(r) = |\{d | d \in \mathcal{D}, d \text{ satisfies } r\}|$$

is called the **support** of r on T,

$$conf(r) = \frac{sup(r)}{|\{d|d \in \mathcal{D}, d\ satisfies\ a_r\}|}$$

is the **confidence** of rule r.

The main task of associative classification is to discover a set of rules from the training set with the attributes in the antecedents and the class label in the consequence, and use them to build a classifier that is used later in the classification process. So only association rules relating the rule antecedent to a certain class label are of interest. In the literature on associative classification the term class association rule has been introduced to distinguish such rules from regular association rules whose consequence may consist of an arbitrary conjunction of literals.

Definition 4. *(Class Association Rule) An class association rule r, which takes the form of $a_r \Rightarrow c_r$ with a_r of the form $l_1 \wedge l_2 \wedge \cdots \wedge l_i$ called the **antecedent** of rule and c_r of the form c_i called the **consequence** of rule, $c_i \in C$, where C is the set of class label.*

The traditional associative classification model based on well-known support-confidence framework suffers from accuracy and efficiency due to the following facts [6]:

- Deciding on a good value of support is not easy. Most previous work usually set the threshold value to 1%. But low support threshold may cause the generation of a very huge rules set, in part of which are useless and redundant. And it is challenging to store, retrieve, prune and sort such a large number of rules efficiently for classification. However, it is bad if the support is assigned too high, some highly predictive rules with low support but high confidence will probably be missed.
- The minimum confidence, which is used as the criterion to evaluate CARs, has been reported that may cause underfitting of rules.
- Typical association rules in classification are those referred to positive association rules of the form $(a_r \Rightarrow c)$. In fact, negative association $(\overline{a_r} \Rightarrow c_r)$ can provide valuable information as same as positive association rules. There are few algorithm use negative association rules to do classification.

3 Associative Classification in Text Categorization

Text categorization is the task of approximating the unknown target function $\Phi : \mathcal{D} \times \mathcal{C} \rightarrow \{P, N\}$ by means of a function $\hat{\Phi} : \mathcal{D} \times \mathcal{C} \rightarrow \{P, N\}$ called the **classifier**. If $\Phi(d_j, c_i) = P$, then d_j is called a **positive example** of c_i, while if $\Phi(d_j, c_i) = N$ it is called a **negative example** of c_i. Depending on different applications, text categorization may be either **single-label** (i.e. exactly one $c_i \in \mathcal{C}$ must be assigned to each $d_j \in \mathcal{D}$), or **multi-label** (i.e. any number

$0 \leq n_j \leq m$ of categories may be assigned to each $d_j \in \mathcal{D}$). A classifier for c_i is then a function $\hat{\Phi} : \mathcal{D} \times \mathcal{C} \rightarrow \{P, N\}$ that approximates the unknown target function $\Phi_i : \mathcal{D} \times \mathcal{C} \rightarrow \{P, N\}$.

3.1 Literal Selection

The first step in text categorization is to transform documents, which typically are strings of characters, into a representation suitable for the learning algorithm and the classification task. Information Retrieval research suggests that word stems work well as representation units and that their ordering in a document is of minor importance for many tasks. This leads to an attribute value representation of text. Each distinct word w_i corresponds to a feature. To avoid unnecessarily large feature vectors, words are considered as features only if they are not "stop-words" (like "and", "the", etc.). This representation scheme makes the antecedent of rule be a set of words. After this, some techniques such as clustering methods [7] can reduce feature spaces by joining similar words into clusters. Then we can use these clusters as features for text categorization in real applications.

Instead of generating candidate rules, we will use Foil Gain [8] to evaluate the goodness of the current rule r, and generate a small set of predictive rules directly from the dataset. This gain values are calculated by using the total weights in the positive and negative example sets instead of simply counting up the number of records in the training sets.

Definition 5. *(Foil Gain) Given a literal p and the current rule $r : a_r \Rightarrow c_r$, $p \notin a_r$, the Foil Gain of p is defined as follows:*

$$FoilGain(p) = |P(r+p)| \left(\log \frac{|P(r+p)|}{|P(r+p)| + |N(r+p)|} - \log \frac{|P|}{|P| + |N|} \right) \quad (1)$$

where there are $P(r)$ positive examples and $N(r)$ negative examples satisfying r in database $\mathcal{D}$. And after appending p to r, there will be $P(r+p)$ positive and $N(r+p)$ negative examples satisfying the new rule $r' : a_r \wedge p \Rightarrow c_r$.

Essentially, $FoilGain(p)$ represents the total number of bits saved in representing all the positive examples by adding p to r.

3.2 Correlation Measure

Piatetsky-Shapiro [9] argued that a rule $X \Rightarrow Y$ is *not interesting* if $sup(X \cup Y) \approx sup(X)sup(Y)$. One interpretation of this proposition is that the task of association rule mining is to explore the "relationship" among attributes in rule. So a rule is called *not interesting* if its antecedent and consequence are approximately independent. For a given class association rule $r : a_r \Rightarrow c_r$, we focus on two types of relationship between antecedent a_r and class label c_r. A **positive correlation** is evidence of a general tendency that when the example has attribute a_r it is very likely to be classed as c_r. A **negative correlation** occurs when we discover the example does belong to c_r while it hasn't the attribute a_r. The mostly used correlated rules in previous classification approaches are those referred to positive association rules like $(a_r \Rightarrow c_r)$. In fact, the negative association $(\overline{a_r} \Rightarrow c_r)$ can also provide valuable information as same as positive association rules.

Based on Piatetsky-Shapiro's argument, for a given class association rule $r : a_r \Rightarrow c_r, c_r \in C$, where C is the set of class label, we can write the interestingness of an association between a_r and c_r in the form of their statistical dependence as equation (2) [10]:

$$Dependence(a_r, c_r) = \frac{sup(a_r \cup c_r)}{sup(a_r)sup(c_r)} = \frac{conf(r)}{sup(c_r)} \qquad (2)$$

There are the following three possible cases:

1. $Dependence(a_r, c_r) = 1 \Rightarrow conf(r) = sup(c_r)$, a_r and c_r are independent.
2. $Dependence(a_r, c_r) > 1 \Rightarrow conf(r) > sup(c_r)$, a_r and c_r are positively correlated, and the following holds:

$$0 < conf(r) - sup(c_r) \leq 1 - sup(c_r)$$

In particular, we have:

$$0 < C_m = \frac{conf(r) - sup(c_r)}{1 - sup(c_r)} \leq 1$$

The nearer the ratio C_m is close to 1, the higher the positive correlation between antecedent a_r and class label c_r.

3. $Dependence(a_r, c_r) < 1 \Rightarrow conf(r) < sup(c_r)$, a_r and c_r are negatively correlated, and the following holds:

$$-sup(c_r) \leq conf(r) - sup(c_r) < 0$$

In particular, we have:

$$-1 \leq C_m = \frac{conf(r) - sup(c_r)}{sup(c_r)} < 0$$

The nearer the ratio C_m is close to -1, the higher the negative correlation between antecedent a_r and class label c_r.

To differentiate the subtlety of negative correlations, we develop a new measure, C_m(Equation (3)), as the positive and negative correlation measure:

$$C_m(a_r, c_r) == \begin{cases} \frac{conf(r) - sup(c_r)}{1 - sup(c_r)} & \text{(if } conf(r) \geq sup(c_r) \wedge sup(c_r) \neq 1) \\ \frac{conf(r) - sup(c_r)}{sup(c_r)} & \text{(if } conf(r) < sup(c_r) \wedge sup(c_r) \neq 0) \end{cases} \qquad (3)$$

To evaluate and measure both positive and negative association rules, we can take C_m as the correlation of the class association rule between antecedent a_r and class label c_r.

3.3 More General Rule

To make the classification more accurate and effective, ACTC prunes rules whose information can be expressed by other simpler but more essential rules.

Definition 6. *(More General Rule)* Given two rules $r_1 : a_{r1} \Rightarrow c_{r1}$ and $r_2 : a_{r2} \Rightarrow c_{r2}$, we says r_1 is **more general than** r_2 iff (1) $Accuracy(r_1) \geq Accuracy(r_2)$; (2) $a_{r1} \subset a_{r2}$ and (3) $c_{r1} = c_{r2}$.

If r_1 is more general than r_2, that means r_1 does more contribution to classification but occupies smaller memory space. ACTC just keeps these rules have higher rank and fewer attributes in its antecedent and other more specific rules with low rank should be pruned.

3.4 Our Algorithm

In this paper, we develop a novel associative classification algorithm based on correlation analysis, ACTC, i.e., *Associative Classification in Text Categorization*, which integrates the advantages of the previously proposed effective strategies as well as some ones newly developed here. ACTC outperforms other associative classification methods by the following advantages:

1. extracts a smaller set of rules with higher quality and lower redundancy directly from the training dataset, avoiding repeated database scan in rules generation;
2. prunes the weak correlation rules directly in the process of rules generation. The advantage is that this is not only more efficient (no post-pruning is necessary) but also more elegant in that it is a direct approach;
3. uses both positive and negative association rules of interest for classification, which take into account the *presence* as well as *absence* of attributes as a basis of rules.

ACTC performs a depth-first-search rule generation process that builds rules one by one. At each step, every possible literal is evaluated by FoilGain and the best one is appended to the current rule. Moreover, instead of selecting only the best one, ACTC also keeps all close-to-the-best literals in rules generation process so that it will not miss some important rules. For the current rule r, ACTC calculates its correlation $C_m(r)$. If $C_m(r)$ is greater than a predefined threshold $\phi_{\min}$, then $(a_r \Rightarrow c_r)$ will be added into the positive rules set PR. But if $C_m(r)$ is lower than $-\phi_{\min}$, then $(\overline{a_r} \Rightarrow c_r)$ will be added into the negative rules set NR. After building each rule, all positive target tuples satisfying that rule are removed. ACTC repeatedly searches for the current best rule and removes all the positive examples covered by the rule until all the positive examples in the data set are covered.

Once the classifier $R = (PR \cup NR)$ has been established in the form of a ordered list of rules, ACTC just selects a small set of strong correlated association rules to make prediction by the following procedure:

Step 1: For those rules satisfying the new data in antecedent, selects the first k-best rules for each class according to rules' consequences;
Step 2: Calculates the rank of each group by summing up the C_m of relevant rules. We think the *average* accuracy is not advisable because many trivial rules with low accuracy will weaken the effect of the whole group;
Step 3: The class label of new data will be assigned by that of the highest rank group.

Algorithm 1. Classify a new document d

```
Input: Class labels set C, Classifier R, a new document d
Output: Class label attached to d
begin
    foreach c_i ∈ C do
        S_i ← ∅;
    end
    foreach r ∈ R do
        if a_r satisfies d then
            if c_r = c_i then
                //categorize r according its class label
                S_i ← S_i ∧ r;
                if |S_i| = k then
                    S_i.rank = ∑_{i=1}^{k} C_m(r_i);
                    break;
                end
            end
        end
    end
    //put the new document in the class that has the highest rank
    C(d) ← c_i with S_i.rank = max(S_1.rank, ..., S_m.rank)
end
```

---

4 Experimental Results and Performance Study

All the following experiments are performed on a 2.4GHz Pentium-4 PC with 512MB main memory. In the rule generation algorithm, $\phi_{\min}$ is set to 0.2 and the best 3 rules are used in prediction.

4.1 Reuters-21578 Database

We firstly conduct an extensive performance study on Reuters-21578 Database. The Reuters-21578 corpus contains 21578 articles taken from the Reuters newswire. The documents from the top 5 categories of the Reuters- 21578 document collection based on the ModApte split are used in our experiments. Each article is designated to zero or more semantic categories such as earn, acq, money-fx, trade or crude. In both the training and test sets we preprocessed each article so that any additional information except for the title and the body was removed. Each document is represented as feature vector, in which words are considered as features only if they occur in the training data at least n times. The table 1 gives the relationship between n and the total number of distinct features in these 5 categories.

As can be seen from figure 1, ACTC achieves better prediction accuracy than C4.5 on different datasets of Reuters-21578 Database.

Table 1. Number of words with different frequency

	Frequency of words (n)	Number of words
Dataset1	2	6819
Dataset2	3	4189
Dataset3	4	2722
Dataset4	5	1869

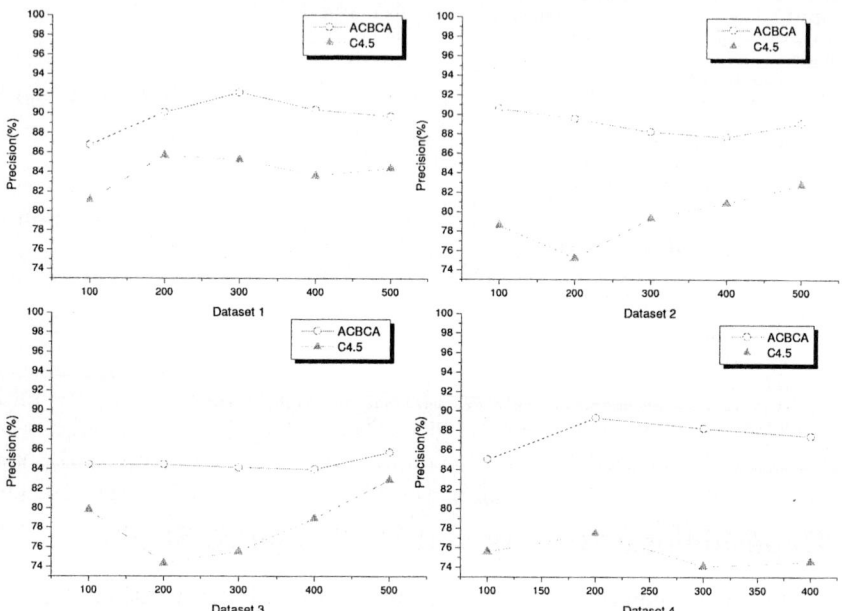

Fig. 1. Comparison of Accuracies on Reuters-21578 Database

4.2 Chinese Biological Abstracts Database

The Chinese biological abstracts database is created by by Shanghai Information Center of Life Sciences, Chinese Academy of Sciences. The database contains documents on the biological research in China, including that on general biology, cytology, genetics, biophysics, molecular biology, etc. All abstract have been categorized by experts and researchers of relevant fields.

Firstly, five categories which contain more than 10000 abstracts respectively are selected randomly from the database. They are *R282* (including 63866 abstracts), *R289* (including 46593 abstracts), *R256* (including 24664 abstracts) and *R241* (including 24338 abstracts). Each abstract can be in one or more categories.

Comparing the results on figure 2, we observe that ACTC can achieve a better accuracy than CBA on all 15 experiments with different number of the abstracts.

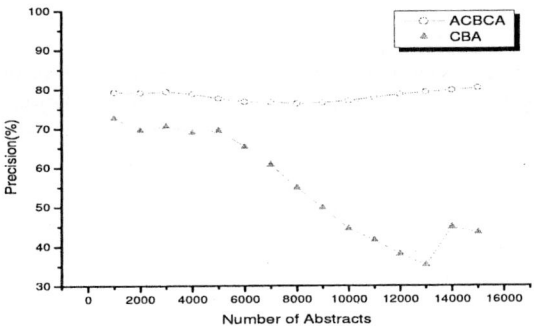

Fig. 2. Comparison of Accuracies on Chinese Biological Abstracts Database

The accuracy of CBA heavily depends on the number of instants and decreases dramatically when the number of the abstracts increasing. As can be seen from figure 2, the effect of CBA is not stable, too.

5 Conclusions

We proposed a new associative classification algorithm based on correlation analysis, ACTC, for text categorization. ACTC differs from existing algorithms for this task insofar, that it does not employ the support-confidence framework and take both positive and negative rules under consideration. It uses an exhaustive and greedy algorithm based Foil Gain to extract CARs directly from the training set. During the rule building process, instead of selecting only the best literals, ACTC inherits the basic idea of CPAR, keeping all close-to-the-best literals in rules generation process so that it will not miss the some important rule.

Since the class distribution in the dataset and the correlated relationship among attributes should be taken into account because it is conforming to the usual or ordinary course of nature in the real world, we present a new rule scoring schema named C_m to evaluate the correlation of the CARs. The experimental results of the ACTC show that a much smaller set of positive and negative association rules can efficient enhancing the prediction accuracy of algorithms.

Actually, we just use the association rules of the type $(a_r \Rightarrow c_r)$ and $(\overline{a_r} \Rightarrow c_r)$ for classification. These two type rules have an direct association with the class label, so they can be considered together. However, if there are more than two classes in the dataset, $(a_r \Rightarrow \overline{c_r})$ and $(\overline{a_r} \Rightarrow \overline{c_r})$ just provide information that "not-belong-to", instead of giving the association between data attributes and class label directly. We are currently investigating reasonable and effective methods to use these two kinds of rules.

References

1. Huang, De-Shuang.: Systematic Theory of Neural Networks for Pattern Recognition. Publishing House of Electronic Industry of China, Beijing (1996)
2. Liu, B., Hsu, W., Ma, Y.: Integrating Classification and Association Rule Mining. In Proceedings of International Conference on Knowledge Discovery and Data Mining (1998), New York, AAAI. 80-86.
3. Li, W., Han, J., Pei, J.: CMAR: Accurate and Efficient Classification Based on Multiple Class-Association Rules. In: Proceedings of the 2001 IEEE International Conference on Data Mining, San José, California, USA, IEEE Computer Society (2001)
4. John Ross Quinlan.: C4.5: Programs for Machine Learning. Morgan Kaufmann, San Mateo, CA(1993)
5. Cohen, W. W.: Fast Effective Rule Induction. In Proceedings of 12th International Conference on Machine Learning, Tahoe City, California, USA, Morgan Kaufmann (1995) 115-123
6. Yin, X., Han, J.: CPAR: Classification Based on Predictive Association Rules. In Proceedings of SIAM International Conference on Data Mining (SDM'03), San Fransisco, CA, May (2003)
7. Zhang, Jun., Henry, S.H., Chung, W. L. LO: Adaptive Genetic Algorithms using Clustering Technique. IEEE Trans. Evolutionary Computation (Accepted with revision)
8. Quinlan, J. R., Cameron-Jones, R. M.: FOIL: A midterm report. In Proceedings of European Conference. Machine Learning, Vienna, Austria, (1993) 3-20
9. Piatetsky-Shapiro, G.: Discovery, Analysis, and Presentation of Strong Rules. Knowledge Discovery in Databases, G. Piatetsky-Shapiro and WJ Frawley (Eds.), AAAI/MIT Press, (1991) 229-238.
10. Wu, Xindong., Zhang, Chengqi., Zhang, Shichao.: Efficient Mining of Both Positive and Negative Association Rules. ACM Transactions on Information Systems, 22(3).(2004) 381-405

A Fast Input Selection Algorithm for Neural Modeling of Nonlinear Dynamic Systems

Kang Li and Jian Xun Peng

School of Electrical & Electronic Engineering,
Queen's University Belfast,
Ashby Building, Stranmillis Road, Belfast BT9 5AH, UK
{K.Li, J.Peng}@qub.ac.uk

Abstract. In neural modeling of non-linear dynamic systems, the neural inputs can include any system variable with time delays. To obtain the optimal subset of inputs regarding a performance measure is a combinational problem, and the selection process can be very time-consuming. In this paper, neural input selection is transformed into a model selection problem and a new fast input selection method is used. This method is then applied to the neural modeling of a continuous stirring tank reactor (CSTR) to confirm its effectiveness.

1 Introduction

In modeling non-linear complex systems using neural networks, the selection of inputs for the network is an important stage [5], [9]. The neural inputs can include any system variable of interest with various time delays for time-series or dynamic systems, and the number of candidate neural inputs can be extremely large. For example for a simple multi-input-single-output dynamic system with 5 possible system inputs, the maximal time delays for both inputs and output are supposed to be 5, then the total number of candidate neural inputs is 30. Now suppose that 5 significant neural inputs are to be selected in neural network construction over a network selection criterion, then more than 140000 explicit searches are required. Obviously this is computationally very expensive. Moreover, if the input selection is considered together with the neural structure selection (i.e. the selection of hidden layers and hidden neurons), the total selection process is extremely complex even for neural modeling of simple nonlinear dynamic systems.

To alleviate the computational complexity in selecting the neural inputs, suboptimal neural input selection methods have been proposed. For example in [9], a number of operating points were chosen for the nonlinear dynamic system under study, and the system was linearized around all operating points. Then a systematic approach was used to select the neural inputs iteratively for these linearized models. Generally speaking, there is a natural linkage between input selection in neural modeling and the variable selection in nonlinear system modeling, and the former can be regarded as a special case of the later problem. Various approaches exist in the literature. For example, in [7], the variable selection procedure is proposed from a set of linearized models, and cluster analysis was used for division of sub-regions. Then genetic algorithm was used to address the combinatory optimization problem, and each variable selection process intended to explicitly solve the least squares problem of a linear system once. For these two approaches, the total sub-regions are a combination of sub-ranges of all variables, which is again a combinational problem. Therefore they can be difficult to implement in cases when very limited number of samples are available for each sub-region or the system has large number of candidate variables. Other researches used statistical procedures to guide the selection procedures [1], such as hypothesis tests, information criteria and cross validation, etc, however the combinational problem and the computational complexity were not addressed.

In [5], modeling is regarded as an alternative function approximation approach therefore functional approximation of a complex system using other base functions can achieve equivalent performance, e.g. Volterra series or simply the power series, given that certain condition is met [8]. One of the differences between functional approximation using power series or Volterra series and using neural networks are that a neural model is 'nonlinear-in-the-parameter' in general and a polynomial regression model using Volterra series or power series is linear-in-the-parameter. Therefore if a polynomial regression model is viewed as an equivalent counterpart of the neural model, then the variable selection problem can be transformed into a model selection problem, and standard algorithms to solve the least square problem, such as orthogonal least squares algorithm [2] or the fast non-orthogonal least square method [6], can be used. In this paper a fast input selection algorithm is proposed and the computational complexity of the proposed algorithms is given. This algorithm is used to select inputs in modeling a continuous stirring tank reactor (CSTR) using neural networks to confirm the effectiveness.

This paper is organized as follows. Section 2 gives the problem formulation and section 3 gives the fast new neural input selection method. The complexity analysis of the proposed algorithm is also compared with the previously proposed method [5] in section 3. Section 4 presents the application study. Section 5 is the conclusion.

2 Problem Formulation

Consider a MISO nonlinear dynamic system of the following form,

$$y(t) = f(y^{t-1}, u_1^{t-d_1}, \cdots, u_m^{t-d_m}, \varepsilon^t) + \varepsilon(t) \tag{1}$$

where t is the time index, $y(t)$, $u_1(t),...,u_m(t)$ and $\varepsilon(t)$ are the system output, system inputs and white noise series respectively, $f(\bullet)$ is some unknown non-linear function, d_i's are time delays, $y^{t-1} = [y(t-1),...,y(t-n_y)]^T$, $\varepsilon^{t-1} = [\varepsilon(t-1),...,\varepsilon(t-n_\varepsilon)]^T$, and $u_i^{t-d_i} = [u_i(t-d_i-1),...,u_i(t-d_i-n_{ui})]^T$ for $i=1,\cdots,m$, where n_y, n_{ui} and n_ε are the maximal time lags for output variable, system input variables and the white noise series, respectively.

A neural network such as multi-layer percetron (MLP) with one hidden layer or radial basis function neural network can be used to approximate the above system in the following form [3][4],

$$y_{nn} = \Theta \bullet \begin{bmatrix} h \\ 1 \end{bmatrix}, h_i = \psi\left(W_{in}^{(i)} \bullet \begin{bmatrix} u \\ 1 \end{bmatrix}\right), i=1,2,...,p \tag{2}$$

where $\mathbf{h}^T = [h_1, \cdots, h_p]$ are outputs of hidden nodes, $\mathbf{u}^T = [u_1, \cdots, u_p]$ are inputs to the neural network, $\mathbf{\Theta}^T \in \Re^{p+1}$ and $(\mathbf{W}_{in}^{(i)})^T \in \Re^{p+1}$, $i = 1, \cdots, p+1$, are weights/bias for output node and hidden nodes, ψ is the activation function for hidden nodes.

The input selection problem for neural network model in (2) is expressed as to select an appropriate subset of system variables with appropriate delays from y^{t-1} and $u_i^{t-d_i}$ in (1). However, (2) is nonlinear-in-the-parameter, and each variable selection requires to train the parameters in (2) once, which is computationally expensive.

Now if (1) is approximated by a polynomial regression model, then the variable selection can be converted into a model selection problem, which is to solve a least square problem of the following model,

$$y(t) = \sum_{i=0}^{P} \theta_i \varphi_i(t) + \varepsilon(t) \quad (3)$$

where, $\varphi_0(t) = 1$, $\varphi_i(t) = \prod_{j=n_{y1}}^{n_{yi}} y(t-j) \prod_{k=1}^{m} (\prod_{j=n_{uk1}}^{n_{uki}} u_k(t - d_k - j))$, $i = 1, \cdots, P$,

$0 \le n_{y1} \le \cdots \le n_{yi} \le n_y$, $0 \le n_{uk1} \le \cdots \le n_{uki} \le n_{uk}$, and in term $\varphi_i(t)$, if $n_{yi} = 0$, then $\varphi_i(t)$ does not include $y(\bullet)$; if $n_{uki} = 0$, then $\varphi_i(t)$ does not include $u_k(\bullet)$. Obviously, $\varphi_i(t)$ may have many forms, and in this paper we only use the following types for $\varphi_i(t)$,

$$\varphi_i(t) = \begin{cases} (u_i(t-k))^j, & i=1,\ldots,m,\ j=1,\ldots,n_{oi},\ k=1,\ldots,n_{ui} \\ (y(t-k))^j, & j=1,\ldots,n_{oy};\ k=1,\ldots,n_y \end{cases} \quad (4)$$

where n_{oi} is the maximal order of power for u_i, n_{oy} is the maximal power order for y, and the multiplication terms of two or more variables are not considered in this paper for simplicity.

The number of possible terms in form of (4) for a saturated model is $\sum_{i=1}^{m} n_{oi} n_{ui} + n_{oy} n_y$. The variable selection proposed in this paper is to select most significant terms from the term pool and then identify the neural inputs from the selected significant terms.

3 The Input Selection Algorithm

Before introducing the new proposed selection algorithm, the non-orthogonal forward regression model selection algorithm [5] is discussed. Consider the nonlinear polynomial model (3) that is used to represent a non-linear dynamic system with terms defined in (4). Now if N samples are used for parameter estimation, (3) becomes:

$$Y = \Phi \Theta + \Xi \quad (5)$$

where $Y^T = [y(1), \cdots, y(N)]$, $\Phi = [\varphi_0, \varphi_1, \ldots, \varphi_p]$, $\varphi_i = [\varphi_i(1), \ldots, \varphi_i(N)]^T$, $i = 0,1, \cdots, p$, $\Xi^T = [\varepsilon(1), \cdots, \varepsilon(N)]$.

The loss function is defined as:

$$E(\Theta) = \Xi^T \Xi \tag{6}$$

Paper [6] suggests that if all vectors in the information matrix in (4) are preprocessed such that. $\varphi_i^T \varphi_i = 1$, $i = 1, \cdots, p$, and $Y^T Y = 1$, then the cost function can be computed in interactively as each new term is added into the system model sequentially:

$$\begin{cases} E_{k+1} - E_k = -(Y^T \varphi_{k+1}^{(k)})^2 /((\varphi_{k+1}^{(k)})^T \varphi_{k+1}^{(k)}), E_0 = 1 \\ \varphi_i^{(k+1)} = \varphi_i^{(k)} - (((\varphi_i)^T \varphi_{k+1}^{(k)}) \varphi_{k+1}^{(k)}) /((\varphi_{k+1}^{(k)})^T \varphi_{k+1}^{(k)}) \\ \varphi_i^{(0)} = \varphi_i, i = 1, \cdots, p, k = 1, \cdots, p-1 \end{cases} \tag{7}$$

where E_k is the cost function for the first k terms, $E_{k+1} - E_k$ is the net contribution to the cost function when a new term φ_{k+1} is added into the model. $\delta E_{k+1} = (Y^T \varphi_{k+1}^{(k)})^2 /((\varphi_{k+1}^{(k)})^T \varphi_{k+1}^{(k)})$ is the difference of the cost function when a new term φ_{k+1} is added. The geometric interpretation of δE_{k+1} is that it is the ratio of the squared projection of the transformed observation vector $\varphi_{k+1}^{(k)}$ on the output vector Y over the length of $\varphi_{k+1}^{(k)}$. δE_{k+1} is acquired explicitly by (7) without solving the least square problem and no matrix decomposition is used. This algorithm is therefore computationally efficient. In the following, the algorithm proposed above can be further simplified.

3.1 The New Fast Algorithm

Firstly, define a matrix series $M_k \in \Re^{k \times k}$, $k = 1, \cdots, p$, which takes the form of

$$M_k \equiv \Phi_k^T \Phi_k \tag{8}$$

where $\Phi_k \in \Re^{N \times k}$, $k = 1, \cdots, p$ contains the first k columns of the full regression matrix Φ in (5). It follows that

$$\left. \begin{array}{l} \hat{\Theta}_k = M_k^{-1} \Phi_k^T y \\ E_k = y^T y - \hat{\Theta}_k^T \Phi_k^T y \end{array} \right\}, \quad k = 1, \cdots, p \tag{9}$$

where $\hat{\Theta}_k \in \Re^k$, $k = 1, \cdots, p$ is the estimate of the first k parameters in Θ in (5), and E_k is the cost function when the first k columns in Φ are selected.

Before introducing the new fast algorithm, two propositions are introduced.

Proposition 1. Define anther matrix series $R_k \in \Re^{N \times N}$ as follows

$$R_k \equiv \begin{cases} I - \Phi_k M_k^{-1} \Phi_k^T, & 0 < k \le p \\ I, & k = 0 \end{cases} \tag{10}$$

where Φ_k, $k = 1, \cdots, p$, is of full column rank. Then following holds true

$$R_{k+1} = R_k - \frac{R_k \varphi_{k+1} \varphi_{k+1}^T R_k^T}{\varphi_{k+1}^T R_k \varphi_{k+1}}, \quad k = 0,1,\cdots,p-1 \tag{11}$$

Proposition 2. Suppose $\varphi_i, i=1,\cdots,p$ in Φ are mutually linear independent, then the matrices $R_k \in \Re^{N \times N}$, $k=1,\cdots,p$ defined in (10) have the following properties:

(1) $$R_k^T = R_k, \quad (R_k)^2 = R_k \tag{12}$$

(2) $$R_k R_j = R_j R_k = R_k \quad \text{for all } k \geq j \tag{13}$$

(3) $$R_k \varphi_i = 0, \quad \forall i \in \{1,\cdots,k\} \tag{14}$$

Having established these two propositions, the new fast forward model selection algorithm can now be derived.

According to (9) and the definition of R_k in (9), it is obvious that

$$E_k = y^T y - \hat{\Theta}_k^T \Phi_k^T y = y^T R_k y \tag{15}$$

Then according to (11), it follows that

$$E_{k+1} = y^T R_{k+1} y = E_k - \frac{y^T R_k \varphi_{k+1} \varphi_{k+1}^T R_k^T y}{\varphi_{k+1}^T R_k \varphi_{k+1}} \tag{16}$$

Furthermore, define

$$\varphi_i^{(k)} \equiv R_k \varphi_i, \quad k,i = 1,\cdots,p \tag{17}$$

and noting proposition 2, then (17) becomes

$$\delta E_{k+1} = E_{k+1} - E_k = -(y^T \varphi_k^{(k)})^2 / (\varphi_{k+1}^{(k)})^T \varphi_{k+1}^{(k)} \tag{18}$$

Equation (18) computes the net contribution of term φ_{k+1} to the cost function when it is included in the model. To further simplify the computational complexity, define

$$\left.\begin{array}{l} a_{k,i} \equiv (\varphi_k^{(k-1)})^T \varphi_i^{(k-1)}, i = k,\cdots,n \\ a_{k,y} \equiv (\varphi_k^{(k-1)})^T y \end{array}\right\} k = 1,2,\cdots,n \tag{19}$$

Using proposition 2 for R_k and the definition of $\varphi_i^{(k)}$, then produce the following

$$a_{k,i} = (\varphi_k^{(k-1)})^T \varphi_i^{(k-1)} = (\varphi_k^{(k-2)})^T \varphi_i^{(k-2)} - \frac{a_{k-1,k} a_{k-1,i}}{a_{k-1,k-1}} \tag{20}$$

This derivation continue until

$$a_{k,i} = \varphi_k^T \varphi_i - \sum_{j=1}^{k-1} \frac{a_{j,k} a_{j,i}}{a_{j,j}}, \quad k=1,\cdots,n, i=k,\cdots,n \tag{21}$$

Similarly,

$$a_{k,y} = \varphi_k^T \mathbf{y} - \sum_{j=1}^{k-1} \frac{a_{j,k} a_{j,y}}{a_{j,j}}, k = 1, \cdots, n \qquad (22)$$

Accordingly, we have

$$\mathbf{y}^T \varphi_i^{(k)} = \mathbf{y}^T \varphi_i - \sum_{j=1}^{k} \frac{a_{j,y} a_{j,i}}{a_{j,j}} \qquad (23)$$

and

$$(\varphi_i^{(k)})^T \varphi_i^{(k)} = (\varphi_i)^T \varphi_i - \sum_{j=1}^{k} a_{j,i}^2 / a_{j,j} \qquad (24)$$

Finally, the net contribution of φ_{k+1} to the cost function is explicitly and more efficiently computed as follows

$$\delta E_{k+1} = -\frac{\left(\mathbf{y}^T \varphi_{k+1} - \sum_{j=1}^{k+1} \frac{a_{j,y} a_{j,k+1}}{a_{j,j}}\right)^2}{(\varphi_{k+1})^T \varphi_{k+1} - \sum_{j=1}^{k} \frac{a_{j,k+1}^2}{a_{j,j}}} \qquad (25)$$

3.2 Input Selection for Neural Modeling

Step 1. Build a saturated regression model with all possible terms being included, and denote T_{sa} as the terms pool for the saturated full model, T_{fi} as the terms pool that consists of all selected terms.

Step 2. Forward model selection, which selects the best terms φ_{k+1} from T_{sa} into T_{fi}, satisfying the following criterion:

$$\max_{\varphi_i} \left\{ \frac{\left(\mathbf{y}^T \varphi_{k+1} - a_{k+1,y} \sum_{j=1}^{k+1} a_{j,k+1} / a_{j,j}\right)^2}{(\varphi_{k+1})^T \varphi_{k+1} - \sum_{j=1}^{k} a_{j,k+1}^2 / a_{j,j}}, \varphi_i \in T_{sa} \right\} \qquad (26)$$

Step 3. Check phase. Check whether $index(\delta E_{k+1})$, which is some selection criterion based on the error information, such as AIC, FPE, etc., satisfies the desired target. In this paper final prediction error (FPE) is used, which is defined as $FPE = E_k(1 + \lambda k / N)$, where E_k is the cost function, k is the model complexity, N is the number of training samples, λ is a constant. If the criterion is satisfied, go to step 5, otherwise continue.

Step 4. Update. Add φ_{k+1} into term pool T_{fi} and delete φ_{k+1} from T_{sa}, go to step 2.

Step 5. Group all terms in T_{fi} from which identify all system variables and their corresponding time delays. All system variables with their time delays appearing in the term pool T_{fi} will be used as the neural network inputs.

3.3 Computational Complexity Analysis

The basic arithmetic operations involved in algorithm implementation are addition/subtraction, and multiplication/division. For the first algorithm given in (7), the total number of addition/subtraction operations for select k terms out of p terms with N training samples in model identification is

$$C_+ = 2kN(2p-k) - 2kp + k^2 \tag{27}$$

and the total number of multiplication-division operations is

$$C_\times = 2kN(2p-k) + 3kp - n(3n-1)/2 \tag{28}$$

The total computational complexity for algorithm given in (7) is then given by:

$$C_1 = C_+ + C_\times = 4kN(2p-k) + kp - k(k-1)/2 \tag{29}$$

Similarly for the proposed algorithm (25) is given by:

$$C_2 = [(2p-k)(k+1) + 2p]N + kp(k+4) - 6p - \frac{2k(k+5)(k-2)}{3} - 2 \tag{30}$$

According to (29) and (30), the new proposed algorithm is significantly more efficient than the previously proposed algorithm.

4 Application

Fig. 3 shows a schematic representation of a chemical system common to many chemical processing plants, known as a Continuously Stirred Tank Reactor (CSTR). Within a CSTR two chemicals are mixed and react to produce a product compound at a concentration $C_a(t)$, with the temperature of the mixture being $T(t)$. The differential equations representing the CSTR reaction are shown as follows.

$$\dot{C}_a(t) = \frac{q}{v}(C_{ao} - C_a(t)) - k_o C_a(t) \exp\left(-\frac{E}{R \cdot T(t)}\right) \tag{31}$$

$$\dot{T}(t) = \frac{q}{v} \cdot (T_o - T(t)) + k_1 \cdot C_a(t) \cdot \exp\left(-\frac{E}{R \cdot T(t)}\right)$$
$$+ k_2 \cdot q_c(t) \cdot \left(1 - \exp\left(-\frac{k_3}{q_c(t)}\right)\right) \cdot (T_{co} - T(t)) \tag{32}$$

where C_a is product concentration, C_{ao} is the inlet feed concentration, q the process flow-rate, T_o and T_{co} the inlet feed and coolant temperatures respectively. k_o, E/R, v, k_1, k_2 and k_3, are thermodynamic and chemical constants relating to this particular problem. The physical plant was simulated using the nominal values given in Table 1.

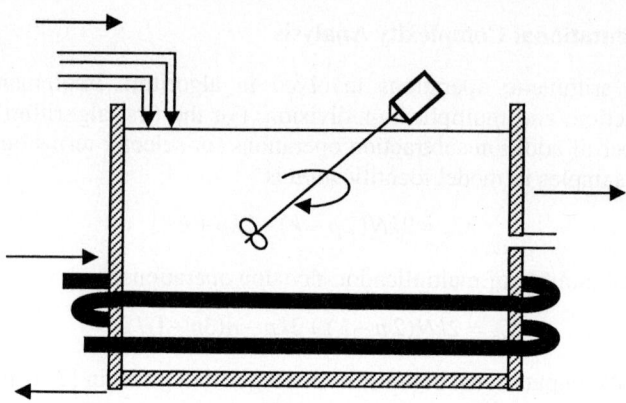

Fig. 1. Schematic representation of CSTR

Table 1. Nominal values for CSTR parameters

Parameter	Nominal Value
Process flowrate q	100 l/min
Reactor volume v	100 l
Reaction rate constant k_0	7.8×10^{10} min^{-1}
Activation energy E/R	1×10^4 K
Feed temperature T_0	350 K
Inlet coolant temperature T_{co}	350 K
Heat of reaction ΔH	-2×10^5 cal/mol
Specific heats C_p, C_{pc}	1 cal/g/K
Liquid densities ρ, ρ_c	1×10^3 g/l
Heat transfer coefficient h_a	7×10^5 cal/min/K
Inlet feed concentration C_{ao}	1.0 mol/l

Generation of modeling data: For a steady-state output concentration of $C_a(t) = 0.1$ mol/l, equations (40), (41) yield values of $T(t) = 438.54$ K and $q_c(t) = 103.41$ l/min. The system was simulated with a sampling interval of 0.2 seconds, and subjected to input consisting of uniformly distributed random perturbations in the input $q_c(t)$ over the range [-10, 10] l/min from the operating point with a zero-order-hold of 2 seconds. A normally distributed random signal was added to the output to simulate measurement noise. The statistics of the data as compared to their nominal values are shown in Table 2.

Table 2. Statistics of CSTR data

Data	Nominal	Mean	Variance
$q_c(t)$ (l/min)	103.41	103.42	28.84
$C_a(t)$ (mol/l)	0.1000	0.1052	0.00093

The candidate neural inputs are $u=q_c(t)$ and $y=C_a(t)$ with maximal time delay of order 6, and the maximal degree of nonlinearity for each variable is 3. There are totally 36 terms in the saturated model with 12 possible network inputs.

A training data set of 1500 samples is used. The proposed algorithm is used to rank these 36 terms and Table 3 lists the first few significant terms.

Based on Table 3, PFE starts to increase instead of decreasing when the 7th term $y(t-1)^3$ is added into the model. Therefore the first 6 terms shows that the following 6 variables will be used as the neural inputs to build an MLP model, $y(t-1)$, $y(t-2)$, $y(t-3)$, $y(t-4)$, $u(t-1)$ and $u(t-2)$, thus reducing another 6 possible neural inputs.

Table 3. Significant terms in order

Term	ΔE	FPE
$y(t-1)$	0.85061	0.14938
$y(t-2)$	0.03600	0.11870
$(u(t-1))^3$	0.00538	0.11813
$(y(t-3))^3$	0.01128	0.11032
$y(t-4)$	0.00558	0.10824
$(u(t-2))^3$	0.00443	0.10704
$y(t-1)^3$	0.00284	0.10747
$y(t-5)$	0.00172	0.10912
$(u(t-3))^3$	0.00090	0.11173
$y(t-4)^2$	0.00162	0.11323

Table 4. Comparison of two MLP models using different inputs

Data Sets	ANN_1	ANN_2
Training set	0.01824	0.02077
Validation	0.02655	0.02773

Table 4 compares the performance of MLP model (ANN_1) using the selected 6 neural inputs and the MLP model (ANN_2) using all 12 variables. These two neural networks trained over half of the data samples and validated over the unseen second half of data. The values in Table 4 are sum of squared errors over the training data and validation data sets.

Table 4 shows that both networks can produce similar modeling performance with the second neural network consisting of only half of the inputs in the first network.

5 Conclusion

In this paper, input selection for neural modelling of nonlinear systems is converted to a model selection problem, and a fast algorithm has been proposed. The computational complexity analysis shows that the proposed new method can achieve significant reduction in computational load. The application case study shows that the proposed method can produce a neural model of equivalent model performance with significantly reduced number of neural inputs.

Acknowledgement

Dr K. Li wishes to acknowledge the financial support of the UK Engineering and Physical Sciences Research Council (EPSRC Grant GR/S85191/01).

References

1. Anders, U., Korn, O.: Model selection in neural networks. Neural Networks, vol. 12 (1999) 309-323
2. Billings, S. A., Chen, S., Korenberg, M. J.: Identification of MIMO nonlinear systems using a forward-regression orthogonal estimator. Int. J. Control, vol. 49 (1989) 2157-2189
3. Huang, D.S., Zhao, W. B.: Determining the centers of radial basis probabilities neural networks by recursive orthogonal least square algorithms. Applied Mathematics and Computation, vol 162 (2005) 461-473.
4. Huang, D.S.: The local minima free condition of feedforward neural networks for outer-supervised learning. IEEE Trans on Systems, Man and Cybernetics, Vol.28 (1998) 477-480.
5. Li, K.: A New Input Selection Method for Neural Modeling of Nonlinear Complex Systems. Proceedings of the 5th World Congress on Intelligent Control and Automation (WCICA 2004), Hangzhou, China, June 15-19, 2018-2021
6. Li, K., Thompson, S., Peng, J.: Modelling and prediction of NOx emission in a coal-fired power generation plant. Control Engineering Practice, vol. 12 (2004) 707-723.
7. Mao, K. Z., Billings, S. A.: Variable selection in nonlinear systems modelling. Mechanical Systems and Signal Processing, vol. 13 (1999) 240-255
8. Yasida, K.: Functional analysis. Springer-Verlag, NY, 1978.
9. Yu, D. L., Gomm, J. B., Williams, D.: Neural model input selection fro a MIMO chemical process. Engineering Application of Artificial Intelligence, vol. 13 (2000) 15-12

Delay-Dependent Stability Analysis for a Class of Delayed Neural Networks*

Ru-Liang Wang[1] and Yong-Qing Liu[2]

[1] Department of Information Technology, Guangxi Teachers' University,
Nanning, 530001, P. R. China
awangruliang@tom.com
[2] College of Automation Science and Engineering, South China Univ. of Tech.,
Guangzhou, 510640, P. R. China

Abstract. In this paper, we consider a class of time-delay artificial neural networks and obtain practical criteria to test asymptotic stability of the equilibrium of the time-delay artificial neural networks, with or without perturbations. These criteria require verification of the definiteness of a certain matrix, or verification of a certain inequality. Furthermore, we discuss the exponential stability and estimate the exponential convergence rate for time-delay artificial neural networks. The applicability of our results is demonstrated by means of two specific examples.

1 Introduction

Time delay is commonly encountered in biological and artificial neural networks (see [1]- [6] and so on), and its existence is frequently a source of oscillation and instability (see [2], [7]). Therefore, the stability of time-delay neural networks has long been a focused topic of theoretical as well as practical importance. This stability issue has also gained increasing attention for its essential role in signal and image processing, artificial intelligence, industrial automation, and so on.

Over the years, one class of neural networks which has received a great deal of attention (see [8],[9]) is described by equations of the form:

$$\dot{x}(t) = -Cx(t) + FS(x(t)) + b \tag{1}$$

where x is a real n-vector, b is a real n-vector, C is a real $n \times n$ diagonal matrix, F is a real $n \times n$ matrix (representing neuron interconnections), and $S(x)$ is a real n-vector valued function. For the neural networks with time delays, they have received attention (see [10], [11]) are described by delay equations of the form:

$$\dot{x}(t) = -Cx(t) + FS(x(t-h)) + b \tag{2}$$

---

* The research is supported by the National Natural Science Foundation of China under Grant 69934030.

where $C, F, S(\cdot)$ and b are defined as in (1) and $h > 0$ denotes transportation delay. In (2) the delays associated with the various neurons are of the same size h. The effects of time delays on Hopfield neural networks described by (2) with F assumed to be symmetric have also been extensively investigated (see [12]). However, because of parameter inaccuracies and measurement errors introduced during the implementation process, it is in general not possible to realize artificial neural networks with interconnections that are precisely symmetric. In such cases, the analysis of the local qualitative properties of the network's equilibria assumes great importance.

In this paper, we do not assume that the interconnecting structure F in (1) or (2) is symmetric. We establish two types of sufficient conditions for the asymptotic stability of the equilibrium of the time-delay artificial neural networks (2), with or without perturbations. These criteria require verification of the definiteness of a certain matrix, or verification of a certain inequality. Furthermore, we discuss the exponential stability and estimate the exponential convergence rate by the method modified Lyapunov function for time-delay artificial neural networks. Two examples are given to illustrate the application of the obtained results.

2 Notations

Let $x^T = (x_1, \cdots, x_n)$ denotes the transpose of x. Let $\lambda_{M(m)}(W)$ denote the maximum (minimum) eigenvalues of a matrix W. Let $W > 0$ ($W < 0$) denote a symmetric positive-(negative-) definite matrix W. For $x \in R^n$, let $\|x\|$ denote the Euclidean vector norm, $\|x\| = (x^T x)^{1/2}$, and for $A \in R^{n \times n}$, let $\|A\|$ denote the norm of A induced by the Euclidean vector norm. Let $h > 0$, $x \in C[[-h, +\infty), R^n]$, and $t > 0$. We use x_t to represent a segment of $x(\theta)$ on $[t-h, t]$, with $\|x_t\| = \sup_{t-h \leq \theta \leq t} \|x(\theta)\|$.

3 Main Results

3.1 Stability Analysis of Time-Delay Neural Networks

In the present section we consider a family of artificial neural networks with delays:

$$\dot{x}(t) = -Cx(t) + FS(x(t-h)) + b \tag{3}$$

where $x \in \Omega \subset R^n$, $C = diag\{c_1, \cdots, c_n\}$ with $c_i > 0, 1 \leq i \leq n$, $F = [F_{ij}]_{n \times n} \in R^{n \times n}$, $b = (b_1, \cdots, b_n)^T \in R^n$ is a constant vector, and $S(x) = [s_1(x_1), \cdots, s_1(x_n)]^T$ with $s_i \in C^1[R, (-1,1)]$ where s_i is monotonically increasing, for $i = 1, \cdots, n$. A special case of (3), when $h = 0$ and F is symmetric, has been studied widely (see, e.g.,[8], [9]) and is called the Hopfield neural network. In this paper, we consider the asymptotic stability of the equilibrium x_e of (3). Without loss of generality, we can assume that $x_e = 0, s(0) = 0$, and $b = 0$. We have $s_i(0) = 0$ for $i = 1, \cdots, n$. We assume that $s_i(x_i)$ satisfies the sector conditions

$$0 \leq \sigma_i^m \leq \frac{s_i(x_i)}{x_i} \leq \sigma_i^M \tag{4}$$

$1 \leq i \leq n$. The new form of the neural network is now give by

$$\dot{x}(t) = -Cx(t) + FS(x(t-h)) \tag{5}$$

where C, F, and S are given in (3), with the sector condition (4) satisfied. The time-delay neural network (5) has been studied widely [see [10] and [11]]. In both [10] and [11], it is assumed that C is a diagonal matrix with identical positive elements, i.e., associated with each neural is the same capacitance and input resistance. The results in [10] rely on the linearization of (5) and yield local results. We now apply the method of the Lyapunov function to discuss the time-delay artificial neural networks (5). We have the following Theorem1:

Theorem 1: System (5) is asymptotically stable for any arbitrarily bounded delay h, if there exists a positive matrix P, and a matrix $\Gamma = diag\{\lambda_1, \cdots, \lambda_n\}, \lambda_j > 0$, $j = 1, \cdots, n$, such that

$$U^M \stackrel{\Delta}{=} -(C^T P + PC) + \Gamma + PF\Sigma^M \Gamma^{-1} \Sigma^M F^T P \tag{6}$$

is negative definite, where $\Sigma^M = diag\{\sigma_1^M, \cdots, \sigma_n^M\}$, and σ_i^M are defined by (4) for $i = 1, \cdots, n$.

Proof: We can rewrite (5) as

$$\dot{x}(t) = -Cx(t) + F\Sigma(x(t-h))x(t-h) \tag{7}$$

where $\Sigma(x) = diag\{\sigma_1(x_1), \cdots, \sigma_n(x_n)\}$, and $\sigma_i(x_i) = \frac{s_i(x_i)}{x_i}, i = 1, \cdots, n$.

Then $\sigma_i(x_i) \in [\sigma_i^m, \sigma_i^M]$, for $i = 1, \cdots, n$. Using the Lyapunov functional as the form

$$V(x(t)) = x^T(t)Px(t) + \int_{t-h}^{t} x^T(\theta)\Gamma x(\theta)d\theta$$

we obtain along the solutions of (7)

$$\begin{aligned}
\dot{V}(x(t)) &= \dot{x}^T(t)Px(t) + x^T(t)P\dot{x}(t) + x^T(t)\Gamma x(t) - x^T(t-h)\Gamma x(t-h) \\
&= -x^T(t)(C^T P + PC)x(t) + x^T(t)\Gamma x(t) - x^T(t-h)\Gamma x(t-h) \\
&\quad + x^T(t-h)\Sigma^T(x(t-h))F^T Px(t) + x^T(t)PF\Sigma(x(t-h))x(t-h) \\
&\leq -x^T(t)(C^T P + P^T C)x(t) + x^T(t)\Gamma x(t) \\
&\quad + x^T(t)P^T F\Sigma(x(t-h))\Gamma^{-1}\Sigma^T(x(t-h))F^T Px(t)
\end{aligned} \tag{8}$$

Let $y^T(t) = (y_1(t), \cdots, y_n(t)) = x^T(t)PF$, then the last term of (8) assumes the form

$$y^T(t)\Sigma\Gamma^{-1}\Sigma^T y(t) = \sum_{i=1}^{n} y_i^2(t)\lambda_i^{-1}\sigma_i^2(x_i(t-h))$$

$$\leq \sum_{i=1}^{n} y_i^2(t)\lambda_i^{-1}(\sigma_i^M)^2 \tag{9}$$

$$= x^T(t)PF\Sigma^M\Gamma^{-1}\Sigma^M F^T Px(t)$$

Then,

$$\dot{V}(x(t)) \leq x^T(t)[-(C^TP+PC)+\Gamma+PF\Sigma^M\Gamma^{-1}\Sigma^M F^TP]x(t)$$
$$= x^T(t)U^M x(t).$$

Therefore, system (5) is asymptotically stable if U^M is negative definite.

Corollary 1: The equilibrium $x = 0$ of (5) is asymptotically stable if

$$\|P_0 F\| < \frac{1}{\overline{\sigma}^M} \tag{10}$$

where $P_0 = diag\{\frac{1}{c_1}, \cdots, \frac{1}{c_n}\}$, and $\overline{\sigma}^M = \max\{\sigma_i^M : 1 \leq i \leq n\}$.

Proof: Suppose that there exists $P_0 = diag\{\frac{1}{c_1}, \cdots, \frac{1}{c_n}\}$, $\overline{\sigma}^M = \max\{\sigma_i^M : 1 \leq i \leq n\}$ such that (10) hold. From condition (4), it is easy to see that U^M is negative definite. By Theorem 1, we obtain that the equilibrium $x = 0$ of (5) is asymptotically stable.

Remark 1: If $\Omega = R^n$, then the conditions of Theorem 1 or Corollary 1 imply that $x = 0$ is the only equilibrium of neural network (5), and this equilibrium is asymptotically stable in the large.

In [9], the neural network

$$\dot{x} = -x + FS(x) + b \tag{11}$$

is considered, where $F = [F_{ij}]_{n \times n}, b = (b_1, \cdots, b_n), S(x) = (s_1(x_1), \cdots s_n(x_n))^T$ and $0 < s_i'(x_i) \leq 1, i = 1, \cdots, n$, where $s_i'(x_i) = \frac{ds_i}{dx_i}(x_i)$. It is shown in [9] that the equilibrium $x = 0$ of (11) is asymptotically stable if $\|F\| < 1$. This condition can easily be obtained by applying Corollary 1 (letting $P_0 = I$ the identity matrix, and

$\overline{\sigma}^M = 1$). Furthermore Corollary 1 shows that the condition $\|F\| < 1$ is so strong that it guarantees not only the asymptotic stability of the equilibrium $x = 0$ of the nondelayed neural network (11), but also the asymptotic stability of the equilibrium $x = 0$ of system (11) with an additional arbitrarily bounded delay, given by $\dot{x}(t) = -x(t) + TS(x(t-h)) + b$.

3.2 Robust Stability Analysis of Time-Delay Neural Networks

In this section we consider a perturbation model of (5) given by the equation:

$$\dot{x}(t) = -(C + \Delta C)x(t) + (F + \Delta F)S(x(t-h)) \quad (12)$$

where $C = diag\{c_1, \cdots, c_n\}$ with $c_i > 0, i = 1, \cdots, n, \Delta C = diag\{\Delta c_1, \cdots, \Delta c_n\}$, $F, \Delta F \in R^{n \times n}$ and $S(x) = [s_1(x_1), \cdots, s_n(x_n)]^T$, with $0 \leq \sigma_i^m \leq \dfrac{s_i(x_i)}{x_i} \leq \sigma_i^M$ for $i = 1, \cdots, n$.

Theorem 2: The equilibrium $x = 0$ of neural network (12) is asymptotically stable for any arbitrarily bounded delay if

$$2\|P_0\| \cdot \|\Delta C\| + \|P_0 F\|^2 (\overline{\sigma}^M)^2 + \|P_0\|^2 \cdot \|\Delta F\|^2 (\overline{\sigma}^M)^2 + 2\|P_0 F\| \cdot \|\Delta F\| \overline{\sigma}^M < 1 \quad (13)$$

where $P_0 = diag\{\dfrac{1}{c_1}, \cdots, \dfrac{1}{c_n}\}$, and $\overline{\sigma}^M = \max\{\sigma_i^M : 1 \leq i \leq n\}$.

Remark 2: Theorem 2 is a consequence of Theorem 1. We omit the details of the proof.

3.3 Exponential Stability Analysis of Time-Delay Neural Networks

We first transform system (5) into the following form:

$$\dot{x}(t) = -Cx(t) + FS(x(t-h)), \quad (14)$$
$$x(t) = \phi(t), \quad t \in [-h, 0],$$

where $x = [x_1, x_2, \cdots x_n]^T$ is the state vector of the transformed system, C, F and S are given in (3), with the sector condition (4) satisfied, h is a positive number for delay time and $\phi(t)$ is an initial value. Before stating the results, we first need the following definitions and lemmas.

Definition 1. If there exist $k > 0$ and $\gamma(k) > 0$ such that

$$\|x(t)\| \leq \gamma(k) e^{-kt} \sup_{-h \leq \theta \leq 0} \|x(\theta)\|, \quad \forall t > 0, \quad (15)$$

Then system (14) is said to be exponentially stable, where k is called the degree of exponential stability.

Lemma 1. (Sanchez et al., 1999) Given any real matrices $\Sigma_1, \Sigma_2, \Sigma_3$ of appropriate dimensions and a scalar $\varepsilon > 0$ such that $0 < \Sigma_3 = \Sigma_3^T$. Then, the inequality holds:

$$\Sigma_1^T \Sigma_2 + \Sigma_2^T \Sigma_1 \leq \varepsilon \Sigma_1^T \Sigma_3 \Sigma_1 + \varepsilon^{-1} \Sigma_2^T \Sigma_3^{-1} \Sigma_2. \tag{16}$$

Proof: Factorizing $\Sigma_3 = (\Sigma_3^{1/2})^T (\Sigma_3^{1/2})$ and defining the matrix $\Sigma \overset{\Delta}{=} \Sigma_3^{1/2} (\varepsilon^{1/2} \Sigma_1 - \varepsilon^{-1/2} \Sigma_3^{-1} \Sigma_2)$, we have

$$\begin{aligned}\Sigma^T \Sigma &= (\varepsilon^{1/2} \Sigma_1^T - \varepsilon^{-1/2} \Sigma_2^T \Sigma_3^{-1}) \Sigma_3 (\varepsilon^{1/2} \Sigma_1 - \varepsilon^{-1/2} \Sigma_3^{-1} \Sigma_2) \\ &= \varepsilon \Sigma_1^T \Sigma_3 \Sigma_1 - \Sigma_2^T \Sigma_1 - \Sigma_1^T \Sigma_2 + \varepsilon^{-1} \Sigma_2^T \Sigma_3^{-1} \Sigma_2 \\ &\geq 0\end{aligned} \tag{17}$$

By simple rearrangement of terms in Equation (17), we immediately obtain Equation (16).

Lemma 2. (Schur complement) (Boyd et al., 1994) The following matrix inequality:

$$\begin{bmatrix} Q(x) & S(x) \\ S(x)^T & R(x) \end{bmatrix} > 0, \tag{18}$$

where $Q(x) = Q(x)^T, R(x) = R(x)^T$, and $S(x)$ depend affinely on x, is equivalent to

$$R(x) > 0, \quad Q(x) - S(x) R(x)^{-1} S(x)^T > 0. \tag{19}$$

Theorem 3: suppose that there exists a positive matrix P, a matrix $\Gamma = diag\{\lambda_1, \cdots, \lambda_n\}, \lambda_j > 0, j = 1, \cdots, n$, and a scalar $k > 0$, such that

$$\Theta^M \overset{\Delta}{=} -(C^T P + PC - 2kP) + \Gamma + PF\Sigma^M \Gamma^{-1} \Sigma^M F^T P \tag{20}$$

is negative definite, where $\Sigma^M = diag\{\sigma_1^M, \cdots, \sigma_n^M\}$, and σ_i^M are defined by (4) for $i = 1, \cdots, n$. Then system (14) is exponentially stable. Moreover

$$\|x(t)\| \leq \sqrt{\frac{\lambda_M(P) + \lambda_M(\Gamma) \frac{1 - e^{-2kh}}{2k}}{\lambda_m(P)}} \|\phi\| e^{-kt} \tag{21}$$

Proof: We can rewrite (14) as

$$\begin{aligned}\dot{x}(t) &= -Cx(t) + F\Sigma(x(t-h))x(t-h), \\ x(t) &= \phi(t), \quad t \in [-h, 0],\end{aligned} \tag{22}$$

where $\Sigma(x) = diag\{\sigma_1(x_1), \cdots, \sigma_n(x_n)\}$, and $\sigma_i(x_i) = \dfrac{s_i(x_i)}{x_i}, i = 1, \cdots, n$.

Then $\sigma_i(x_i) \in [\sigma_i^m, \sigma_i^M]$, for $i = 1, \cdots, n$. In order to study the exponentially stability of system (14), the Lyapunov function used in Theorem 1 is modified as follows:

$$V(x(t)) = e^{2kt}x^T(t)Px(t) + \int_{t-h}^{t} e^{2k\theta}x^T(\theta)\Gamma x(\theta)d\theta \tag{23}$$

The time derivative of the Lyapunov function along the trajectories of system (22) is

$$\dot{V}(x(t)) = 2ke^{2kt}x^T(t)Px(t) + e^{2kt}\dot{x}^T(t)Px(t) + e^{2kt}x^T(t)P\dot{x}(t) + e^{2kt}x^T(t)\Gamma x(t)$$
$$- e^{2k(t-h)}x^T(t-h)\Gamma x(t-h) \tag{24}$$
$$= e^{2kt}[x^T(t)(2kP - C^TP - PC)x(t) + x^T(t-h)\Sigma^T(x(t-h))F^T Px(t)$$
$$+ x^T(t)PF\Sigma(x(t-h))x(t-h)] + e^{2kt}x^T(t)\Gamma x(t)$$
$$- e^{2k(t-h)}x^T(t-h)\Gamma x(t-h)$$

Let $\Sigma_1 = x(t-h), \Sigma_2 = \Sigma^T(x(t-h))F^T Px(t)$. Then, by Lemma 1, we obtain the following inequality:

$$x^T(t-h)\Sigma^T(x(t-h))F^T Px(t) + x^T(t)PF\Sigma(x(t-h))x(t-h)$$
$$- x^T(t-h)\Gamma x(t-h) \tag{25}$$
$$\leq x^T(t)PF\Sigma(x(t-h))\Gamma^{-1}\Sigma^T(x(t-h))F^T Px(t)$$

Substituting Eqs.(25) into Eqs.(24) gives

$$\dot{V}(x(t)) \leq e^{2kt}[x^T(t)(2kP - C^TP - PC)x(t)] + e^{2kt}x^T(t)\Gamma x(t)$$
$$+ e^{2kt}x^T(t)PF\Sigma(x(t-h))\Gamma^{-1}\Sigma^T(x(t-h))F^T Px(t) \tag{26}$$
$$- e^{-2kh}x^T(t-h)\Gamma x(t-h)$$
$$\leq e^{2kt}[x^T(t)(2kP - C^TP - PC)x(t)x(t) + x^T(t)\Gamma x(t)]$$
$$+ e^{2kt}x^T(t)PF\Sigma(x(t-h))\Gamma^{-1}\Sigma^T(x(t-h))F^T Px(t)$$

Let $y^T(t) = (y_1(t), \cdots, y_n(t)) = x^T(t)PF$, then the last term of (26) assumes the form

$$y^T(t)\Sigma\Gamma^{-1}\Sigma^T y(t) = \sum_{i=1}^{n} y_i^2(t)\lambda_i^{-1}\sigma_i^2(x_i(t-h))$$
$$\leq \sum_{i=1}^{n} y_i^2(t)\lambda_i^{-1}(\sigma_i^M)^2$$
$$= x^T(t)PF\Sigma^M\Gamma^{-1}\Sigma^M F^T Px(t)$$

Then,

$$\dot{V}(x(t)) \leq e^{2kt}[x^T(t)(2kP - C^TP - PC)x(t)x(t) + x^T(t)\Gamma x(t)]$$
$$+ e^{2kt}x^T(t)PF\Sigma^M\Gamma^{-1}\Sigma^M F^T Px(t)$$
$$= e^{2kt}x^T(t)\Theta^M x(t)$$

where $\Theta^M = -(C^T P + PC - 2kP) + \Gamma + PF\Sigma^M \Gamma^{-1}\Sigma^M F^T P$. Since $\Theta^M < 0$, we have $\dot{V}(x(t)) < 0$. Therefore, we obtain $V(x(t)) \leq V(x(0))$. However

$$V(x(0)) = x^T(0)Px(0) + \int_{-h}^{0} e^{2k\theta} x^T(\theta)\Gamma x(\theta) d\theta$$

$$\leq \lambda_M(P)\|\phi\|^2 + \lambda_M(\Gamma)\|\phi\|^2 \int_{-h}^{0} e^{2k\theta} d\theta$$

$$= [\lambda_M(P) + \lambda_M(\Gamma)\frac{1-e^{-2kh}}{2k}]\|\phi\|^2$$

and $V(x(t)) \geq e^{2kt}\lambda_m(P)\|x(t)\|^2$. Therefore $\|x(t)\| \leq \sqrt{\dfrac{\lambda_M(P) + \lambda_M(\Gamma)\dfrac{1-e^{-2kh}}{2k}}{\lambda_m(P)}}\|\phi\|e^{-kt}$.

The proof of Theorem 3 is thus completed.

4 Examples

To demonstrate the applicability of some of the proceeding results, we now consider specific example.

Example 1: Consider a class of neural work represented by

$$\dot{x}(t) = -Cx(t) + FS(x(t-h)) \tag{27}$$

where $x \in R^{10}$

$$C = diag\{3.5, 1.8, 3.6, 3.6, 1.49, 1.95, 1.74, 1.55, 2.89, 3.62\} \tag{28}$$

$$S(x) = (s_1(x_1), \cdots, s_n(x_n))^T \text{ with } s_i(x_i) = \frac{2}{\pi}\tan^{-1}(x_i) \tag{29}$$

for $i = 1, \cdots, n$, and $T = \dfrac{T_0}{2}$, where

$$F_0 = \begin{bmatrix} 0.1 & 0.5 & 0.5 & 0.5 & 0.5 & 0.5 & -0.5 & 0.5 & 0.5 & 0.5 \\ 0.5 & -3.6 & 0.5 & -0.5 & 0.5 & 0.5 & 0.5 & 0.5 & 0.5 & 0.5 \\ 0.5 & 0.5 & 0.8 & 0.5 & 0.5 & -0.5 & 0.5 & 0.5 & 0.5 & -0.5 \\ 0.5 & 0.5 & 0.5 & 0.2 & 0.5 & 0.5 & 0.5 & 0.5 & -0.5 & 0.5 \\ -0.5 & 0.5 & -0.5 & 0.5 & -2.65 & 0.5 & 0.5 & 0.5 & -0.5 & 0.5 \\ 0.5 & 0.5 & 0.5 & 0.5 & -0.5 & -2.45 & 0.5 & 0.5 & 0.5 & -0.5 \\ -0.5 & 0.5 & -0.5 & 0.5 & 0.5 & 0.5 & -2.35 & 0.5 & 0.5 & -0.5 \\ 0.5 & 0.5 & 0.5 & -0.5 & 0.5 & 0.5 & 0.5 & -2.9 & 0.5 & 0.5 \\ -0.5 & 0.5 & 0.5 & 0.5 & 0.5 & -0.5 & 0.5 & 0.5 & -3.35 & 0.5 \\ 0.5 & 0.5 & 0.5 & 0.5 & 0.5 & 0.5 & -0.5 & 0.5 & 0.5 & -2.5 \end{bmatrix} \tag{30}$$

This system satisfies the condition $\|P_0 F\| < \dfrac{1}{\overline{\sigma}^M}$, since

$P_0 = C^{-1}$, $\|P_0 F\| = 1.19$, $\overline{\sigma}^M = s_i'(x_i) = 0.6366$, and $\dfrac{1}{\overline{\sigma}^M} = 1.5708$. In view of Corollary 1, the equilibrium $x = 0$ of system (27) and it is asymptotically stable for any arbitrarily bounded delay h.

Example 2: We consider the robustness problem of neural networks represented by the equation

$$\dot{x}(t) = -(C + \Delta C)x(t) + (F + \Delta F)S(x(t-h)) \qquad (31)$$

where C and S are given by (28) and (29), $F = F_0$ is given by (30), and $h = h(t) \leq 0.05$. We assume that it is known that $\|\Delta C\| \leq 0.1$ and $\|\Delta F\| \leq 0.1$. Using Theorem 2, we obtain that the equilibrium $x = 0$ of system (31) is asymptotically stable.

5 Conclusions

In this paper, we have established the sufficient criteria of symptotic stability for the time-delay artificial neural networks. These criteria require verification of the definiteness of a certain matrix, or verification of a certain inequality. It is obvious that these criteria are easy test and convenient for the application in the practice. Furthermore, we discuss the exponential stability and estimate the exponential convergence rate by the method modified Lyapunov function for time-delay artificial neural networks. Two examples are given to illustrate the application of the obtained results.

References

1. Zeng, Z.G., Huang, D.S., Wang, Z.F.: Attractability and location of equilibrium point of cellular neural networks with time-varying delays. International Journal of Neural Systems 5 (2004), 337-345
2. Baldi, P., Atiya, A. F.: How delays affect neural dynamics and learning. IEEE Trans. on Neural Networks 5 (1994), 612-621
3. Zeng, Z.G., Huang, D.S., Wang, Z.F.: Pattern recognition based on stability of discrete-time cellular neural networks. Lecture Notes in Computer Science, Vol 3173 (2004), 1008-1014
4. Pang, S.L., Wang,Y., Bai, Y.: Credit scoring model based on neural network.. The International Conference on Machine Learning and Cybernetics, Beijing, China, 4(2002), 1742-1746
5. Zeng, Z.G., Huang, D.S., Wang, Z.F.: Stability analysis of discrete-time cellular neural networks. Lecture Notes in Computer Science, Vol 3173 (2004), 114-119
6. Zeng, Z.G., Huang, D.S., Wang, Z.F.: Practical stability criteria for cellular neural Networks described by a template. WCICA'04 (2004), 160-162
7. Liao, X. F., Wong, K. W., Yu, J. B.: Novel stability conditions for cellular neural networks with time delay. International Journal of Bifurcation and Chaosism, 4 (2001), 1853-1864

8. Matsuoka, K.: Stability conditions for nonlinear continuous neural networks with asymmetric connection weights. Neural networks 5 (1992), 495-500
9. Kelly D. G.: Stability in contractive nonlinear neural networks. IEEE Trans. Biomed.Eng. 37 (1990), 231-242
10. Marcus, C. M., Westervelt, R. M.: Stability of analog neural networks with delay. Physical Rev. A 39 (1989), 347-359
11. Burton, T. A.: Averaged neural networks. Neural networks 6 (1993), 677-680
12. Civalleri, P.P., Gilli, M., Pandolfi, L.: On stability of cellular neural networks with delay. IEEE Trans. Circuits Syst. I, Vol. 40 (1993), 157-165
13. Sanchez, E.N., Perez, J.P.: Input-to-state stability (ISS) analysis for dynamic NN. IEEE Trans. on Circuits Systems--I 46 (1999), 1395-1398
14. Boyd, S., Ghaoui, L.EI, Feron, E., Balakrishnan, V.: Linear matrix inequalities in system and control theory. Philadephia:SIAM.(1994)

The Dynamic Cache Algorithm of Proxy for Streaming Media

Zhiwen Xu, Xiaoxin Guo, Zhengxuan Wang, and Yunjie Pang

Faculty of Computer Science and Technology, Jilin University,
Changchun City, 130012, Jilin Province, China
xuzhiwen@public.cc.jl.cn

Abstract. The transmission streaming media becomes a challenging study problem for the Web application. The proxy cache for streaming media is an efficient method to solve this problem. In proxy cache, prefix cache is to cache the initial part of the streaming media in the proxy cache, so that there is no startup delay for the inquest of the clients. Segmentation based cache is to cache the length of streaming media, according to the inquest frequency of the clients so as to save the web resources; The proxy cache and the segmentation based cache are both pre-drawing cache methods. In this paper, we proposed the algorithm of high efficient dynamic cache method, based on prefix cache and segmentation based cache strategy. The algorithm carries out real time dynamic cache in the proxy cache and makes the dynamic cache with batch and patching algorithm, transmitted and proxy cached by the server, deal with the requests of more than one clients within a relatively short period of time. Therefore, the web resources used by patching channel and regular channel and release the network burden of the server. Event-driven simulation, introduced to evaluate this algorithm, is very efficient.

1 Introduction

At present, web proxy cache has broad application, since the proxy cache technology for caching text and image objects does not apply to streaming media. The reasons are follows: First of all, video files require large memory volume. A single file may demands 10 M to 10 G memory volume, which is determined by the quality and length of the video. Most content cached should be stored in the hard disc, and hard disc for proxy cache and memory cache must be organized with great care. Secondly, the fact that real-time media transmission requires obviously large disc volume, bandwidth of network and support within a long period of time, which demands that effective cache algorithm should be used in proxy cache so as to avoid using too much disc volume for caching new content. The characteristic of the streaming media determines that streaming media is its caching objects instead of web objects. The research of cache technology for streaming media is a challenging subject.

2 Related Work

2.1 Prefix Cache

In order to solve the problem of startup delay and realize smooth data transmission, Sen.S and other experts propose the method of prefix Cache[1, 2]. When transmitting the media, they divide the media into two parts. The smaller preceding part is called prefix cache. When the users make an application, what we store in the prefix media will be played first. Meanwhile, the part stored in suffix cache is transmitted from the content server to the proxy cache. When media stored in prefix cache is over, the part stored in the suffix media starts to play. In this way, the problem of startup delay is solved effectively. The literatures[3, 4, 5]give a introduction to the research in such areas as the management and organization of proxy cache based on prefix cache, the connecting schemes of the server, the schedule of batch and patch in the proxy cache.

2.2 The Segmented Cache of Streaming Media

From the point of view of proxy cache, the initial portion of a media stream in the proxy cache is more important than the latter portion. According to the great importance of the initial part of most media objects and the investigation that most media objects should be cached partially, Wu Kun-Lung [6] and his team to develop a segmented approach to high-speed cache media objects. Blocks of a media object received by the proxy server are grouped into variable-sized, distance-sensitive segments. In fact, the segmented size increases exponentially from the beginning segment. For simplicity, the video i is 2i-1 blocks and contains media blocks $2^{i-1}, 2^{i-1}+1, \cdots, 2^i - 1.(i \in (1,2,\cdots,M))$. The literatures[7, 8, 9]give a introduction to the research in such areas as the dynamic management and organization of proxy cache based on segmented cache.

3 The Dynamic Cache Algorithm of Proxy Cache

The segmentation-based and prefix cache are both pre-fetching cache methods. This paper puts forward a high efficient dynamic cache algorithm, which carries out real time dynamic cache in the proxy cache and makes the dynamic cache transmitted and proxy cached by the server, deal with the requests of more than one clients within a relatively short period of time. Therefore, the web resources used by patching channel and regular channel and release the network burden of the server.

3.1 Usage Scheme of Dynamic Cache Algorithm

In the proxy cache of streaming media based on segmentation, the size of the media cached varies with the changes of the requested frequency of a certain media. In general, only a part of the media object is saved in the proxy cache, whereas the other part is not saved. When a user requests the latter media object, the part not saved should be released from the content server to proxy cache and then can be transmitted to the client. When the media has request by more than one client at the same time, it should be transmitted from the content server to the proxy cache for several times. We put forward a dynamic efficient organizing algorithm of cache, utilizing the characteristic of proxy cache for segmented streaming media. Within the certain duration of time, if two or more users request the same video material, we may cache a segmentation of this video dynamically to meet those users' requirements. It is unnecessary for the proxy cache to apply for the video material frequently. Once is enough to satisfy all the users. The length of this duration of time should be shorter than that of the video material and the permitted over cache threshold.

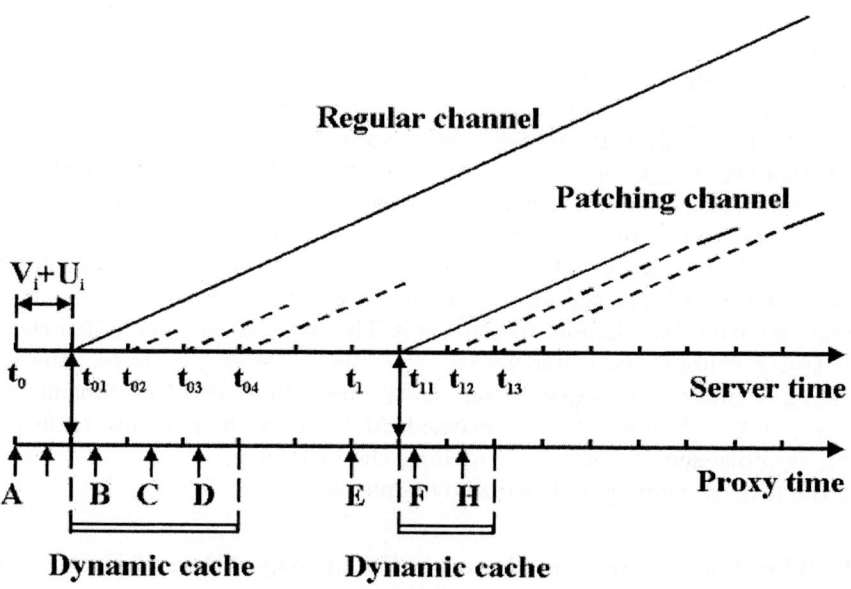

Fig. 1. The process of dynamic cache

Fig.1 shows the dynamic cache algorithm deal with the process of streaming media for proxy cache. V_i, U_i represent the time length of prefix cache and segmented cache. Thus in proxy cache, the time length of the streaming media cached is $V_i + U_i$. At time t_0, client A first request the streaming media, and the server of this media starts to transmit it through regular channel at time $V_i + U_i$. The length of the streaming media transmitted is the value that we get after subtracting the length in the proxy cache from the length applied by client A. clients requests put forward within time interval $V_i + U_i$, after time t_0, can all be processed by using patching algorithm and multicast, which is not a method of dynamic cache. There are two conditions as to the disposal process of dynamic cache algorithm.

Case 1: the time intervals between $t_0, t_{01}, t_{02}, t_{03}, t_{04}$ are equal. Client B put forward his request within time interval between t_{01} and t_{02}, client C between t_{02} and t_{03}, Client D between t_{03} and t_{04}; when not using dynamic cache, the request of client B is processed by server at t_{02} through patching channel (dashed in Fig.1 indicate this process); the request of client C is processed by server at t_{03} through patching channel; the request of client D is processed by server at t_{04} through patching channel. The requests of client B, C, D need to occupy the server of the patching channel and network resources. In the dynamic cache algorithm, cache lasts from t_{01} to t_{04}, then the time length of dynamic cache is $t_{04}-t_{01}$, occupying cache space $(t_{04} - t_{01})^2$, saving the resources patching channel occupies at t_{02}, t_{03}, t_{04}.

Case 2: the time intervals between $t_1, t_{11}, t_{12}, t_{13}$ are the same. Client E put forward his request at t_1; Client F request between t_{11} and t_{12}; client H between t_{12} and t_{13}. When not using dynamic cache, the request of client E is processed through patching channel at t_{11} (the real line of Fig.1 indicate this process); the request of client F is processed through patching channel at t_{12} (the real line of Fig.1 indicate this process); the request of client H is processed through patch channel at t_{13}. The requests of clients E, F, H need to occupy the server of patching channels and net work resources. In dynamic cache algorithm, cache lasts from t_{11} to t_{13}, then the time's length of dynamic cache is $(t_{13} - t_{11})(t_{13} - t_{01})$, saving the resources that patching channels occupies at t_{11}, t_{12}. The patching channel at time t_{11}, t_{12}, t_{13} adopt dynamic cache, whose length is $t_{13} - t_{11}$ and the longest patching channel to process it The disposal process is like the part of patching channel, indicated by the real line in the Fig.1. At t_{11}, the whole patching channel is in transmission. At t_{12}, only the part, not transmittedthe part shown by real lineat t_{11} is processed. At t_{13}, only the part, not transmitted at t_{12} is processed, so that the patching channels at t_{11}, t_{12}, t_{13} are combined into the longest channel and realize dynamic cache.

3.2 The Analysis of The Dynamic Cache Algorithm of Proxy Cache

For simplicity, we ignore the transmission delay of network. In order to give our scheme a qualitative analysis, we first make the following summary. The streaming media is a video, whose length is L_i and with the request rate λ_m. The length of the prefix cache in the proxy is V_i, while the length of segmentation-based cache in the proxy is U_i, so the total cache length cached in the proxy is $V_i + U_i$. Supposing the request rate of media object m is represented by a Poison process with the parameter λ_m, namely $P = e^{-\lambda_m(U_i+V_i)}$ is a vacant request rate, with a duration $V_i + U_i$. Supposing that x_{ij-1} represents the requests, arrived within the (j-1)th batch interval. Supposing $x_{i0}, x_{i1}, \cdots, x_{iN}$ is an independent random variable sequence with popularized probability distribution. Whether the proxy need to obtain patching data at t_i through extra patching channel depends on the value of $x_{i0}, x_{i1}, \cdots, x_{iN}$ and x_i. If $x_{i0} = x_{i1} = \cdots = x_{iN} = 0$ or $x_{i0} = x_{i1} = \cdots = x_{iN} \neq 0$, then the proxy does not need to apply to the server for any patching data. If the proxy applies to the server for the patching data, whose size is η at the start window, it is obvious that η is probably any value among $0,(V_i+U_i),2(V_i+U_i),\cdots,(N-1)(V_i+U_i)$. Similarly, we may use P_i to represent the probability of $\eta = k(V_i+U_i)$, namely, $P_k(\eta = k(V_i+U_i)) = P_k, k = 0, 1, 2, \cdots, N-1$. According to Poison distribution, we may make the conclusion: $P_0 = 1, P_1 = e^{-\lambda_m(U_i+V_i)}, P_2 = e^{-2\lambda_m(U_i+V_i)}, \cdots, P_{N-1} = e^{-(N-1)\lambda_m(U_i+V_i)}$.

According to these assumptions, we can get the request patching channel number μ of N-1.

$$\mu = \sum_{k=1}^{N-1} P_k \qquad (1)$$

The patching channel resources saved is R.

$$R = (U_i + V_i) \sum_{k=1}^{N-1} k P_k \qquad (2)$$

We process the patching channels through dynamic cache of proxy. Under case 1 of dynamic cache algorithm, there is no need to process the patching; under case 2, the transmission of patching channel is needed and the transmission length equals the longest patching channel and all the other patching procession is substituted by dynamic cache. Then we will analyze the performance of dynamic cache algorithm from the point of view of value needed by the system to transmit streaming media. L_i is the length of streaming media. C_s and C_p represent respectively the value modulus transmitted from the server to the proxy cache and from the proxy cache to the client. V_i is the length of prefix cache. The media segmentation $[V_i, L_i]$ needs to be made from the original server. $C_i(V_i)$ is the average value of video transmitted by using prefix cache.

$$C_i(V_i) = (C_s \frac{L_i - V_i}{1 + \lambda_i V_i} + C_p L_i)\lambda_i B_i \qquad (3)$$

V_i is prefix cache length, and U_i is segmented cache length. Media segmentation $[V_i + U_i, L_i]$ needs to be drawn from the original sever and $C_i(V_i, U_i)$ is the average value of video transmitted by using segmented cache.

$$C_i(V_i, U_i) = (C_s \frac{L_i - V_i - U_i}{1 + \lambda_i(V_i + U_i)} + C_p L_i)\lambda_i B_i \qquad (4)$$

When using dynamic cache, under case 1, the average request addition is $1+\mu$, then the transmitted value $C_{id}(V_i, U_i)$ is.

$$C_{id}(V_i, U_i) = (C_s \frac{L_i - V_i - U_i}{1 + \lambda_i(V_i + U_i)} \frac{1}{1 + \mu} + C_p(L_i + \mu(V_i + U_i)))\lambda_i B_i \qquad (5)$$

When using dynamic cache, under case 2, the average request addition is $1+\mu$, W is the average length of patching channel, then the transmitted value $C_{id}(V_i, U_i)$ is.

$$C_{id}(V_i, U_i) = (C_s \frac{W}{1+\lambda_i(V_i+U_i)} \frac{1}{1+\mu} + C_p(L_i + \mu(V_i+U_i)))\lambda_i B_i \quad (6)$$

The former item represent the transmission value between content server and proxy server, and the later item is the transmission value between proxy server and the client. Our major objective is to reduce the resources of backbone network, and that is the smaller the first item, the better. We considered if two or more applicants apply for the same video within time $V_i + U_i$, we may save the resource of the backbone network for μ times on average, and improve the byte hit ratio of proxy cache, by making those latter applicants not take up network resources of the backbone.

This strategy considers the significance of the initial part of a requested media, and ensures higher efficiency of the proxy cache; however, because the segmented strategy doesn't consider adequately the effect imposed on it by the same media stored in the proxy. We design the algorithm of dynamic cache with the main intention to consider adequately the case when many clients apply for the same media within the duration of time and on the basis of segmented cache, dynamically cache the segmentation of the streaming media on demand by multi-user so as to ensure that it is necessary for only the very first applicant to take out the part of media not cached in the proxy cache from the server and cache this segmentation in the proxy cache. This length of time is regarded as the length of the media cached. We adopt FIFO replacement policy to make this part of cached media meet the need of multi-user within the duration of time. The efficiency of proxy cache will be influenced by the determination of this time length. If we determine this time length by calculating users' behavior, then the proxy cache allocates the cache length according to this time length.

4 Performance Evaluation

4.1 Methodology

We utilize an event-driven simulator to stimulate the proxy cache service and furthermore to evaluate the algorithm of the dynamic cache based on variable-size. Let's suppose that the media objects are videos and the size of these videos are uniformly distributed between 0.5 B and 1.5 B blocks, where B represents video size. The default value of B is 2,000. The playing time for a block is assumed to be 1.8 seconds. In other words, the playing time for a video is between 30 minutes and 90 minutes. The size of cache is expressed on the basis of the quantitative description of media blocks. The default cache size is 400,000 blocks. The inter-arrival time distributes with the exponent λ. The default value of λ is 60.0 seconds. The requested video titles are selected from a total of the distinct video titles. The popularity of each video title M accords to the Zipf-like distribution. The Zipf-like distribution brings two parameters, x and M. the former has something to do with the degree of skew. The distribution is given by $p_i = C/i^{1-x}$ for each i$\in (1, \cdots, M)$, where $C = 1/\sum_{i=1}^{M} 1/i^{1-x}$ is a normalized constant. Suppose x = 0 corresponds to a pure Zipf distribution, which is highly skew. On the other hand, suppose x = 1 corresponds to a uniform distribution with no skew. The default value for x is 0.2 and that for M is 2,000. The popularity of each video title changes with time. It is very likely that a group of users may visit different video titles at different periods of time and the users' interest may be different. In our simulations, the distribution of the popularity changes every request R. The correlation between two Zipf-like distributions is modeled by using a single parameter k which can be any integer value between 1 and M. First, the most popular video in the first Zipf-like distribution finds its counterpart, the r_1-th most popular video in Zipf-like distribution 1, where r_1 is chosen randomly between 1 and k. Then, the most popular video in the second Zipf-like distribution finds its counterpart, the r_2-th most popular video. r_2 is chosen randomly between 1 and min (M, k+10), except r_1. The rest may be deduced by analog. When k represents the maximum position in popularity, a video title may shift from one distribution to the next. k = 1 expresses perfect conformity, and k = M expresses the random case or unconformity.

We compared the dynamic cache algorithm with the full video approach, the variable-sized segmented approach, and the prefix schemes in terms of the impact they imposed on byte hit ratio and startup delay from the following aspects: the cache size, the skew of the video popularity, users' viewing behavior and other related system parameters.

4.2 Impact of Cache Size

We study the impacts imposed by the cache size on the byte hit ratio and startup delay. For a fairly wide range of cache size, the dynamic cache method has the highest byte hit ratio and the lowest fraction of requests with startup delay, whose byte hit ratio is higher than the variable-sized segmented approach and the prefix schemes with the same startup delay. Fig.2 shows the impact cache size imposes on the byte hit ratio. Fig.3 presents the impact imposed by cache size on the fraction of requests with startup delay. The full video approach and the prefix have comparable byte hit ratio, with the full video approach having a slight advantage over the prefix scheme. For a smaller cache size, the advantage of byte hit ratio managed by the variable-sized segmented approach is quite evident. The dynamic cache method proves to have the highest byte hit ratio. Even though the full video and the prefix approaches perform almost equally in byte hit ratio, they differ dramatically in the fraction of requests with startup delay. The full video approach has a significantly higher fraction of requests with startup delay (Fig.3). For example, for a cache size of 400,000 blocks, 0.615 of the requests cannot start immediately using the full video approach. However, only 0.162 of applicants encounter startup delay using dynamic cache, variable-size segmentation and prefix approaches. Within the whole range of cache size, the effect of the dynamic cache approach, variable-size segmented method and the prefix strategy are basically the same.They all effectively solve the problem of startup delay.

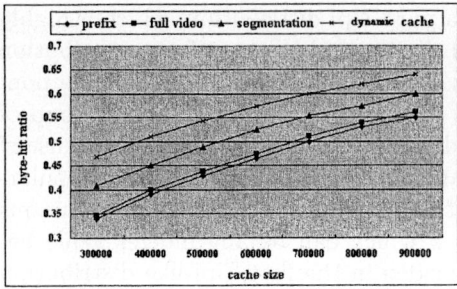
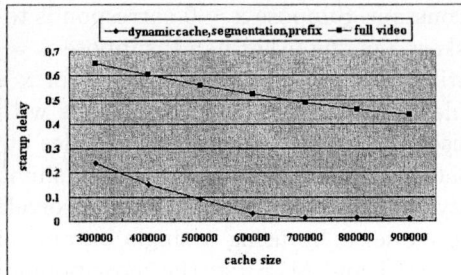

Fig. 2. Impact of byte-hit ratio

Fig. 3. Impact of startup delay

4.3 Impact of Video Popularity

Let us examine the impact that the video popularity imposes on the byte hit ratio and startup delay. The dynamic cache method has the highest byte hit ratio when the video popularity makes changes of wide scope. The dynamic cache approach, the variable-sized segmentation and the prefix schemes all have the same fewest request time with startup delay, which is superior to the whole video. Fig.4 shows the impact of skew in video popularity on byte hit ratio, while Fig.5 shows its impact on the startup delay. In addition to the parameter of Zipf, x, we also studied the changes of the popularity distribution and the impact of the maximum video shifting position k. The request R of the video shift was set to be 200. Fig.6 shows the impact of the maximum shifting position of a video. When the maximum shifting distance increases, the byte hit ratio of the dynamic cache, the variable-sized segmentation and the prefix approaches will fall, but only very slightly. The dynamic cache method is always better than the variable-sized segmentation and the prefix approach, which is closely related with the popularity distributions of the video titles and with the range of k, which is from 5 to 40.

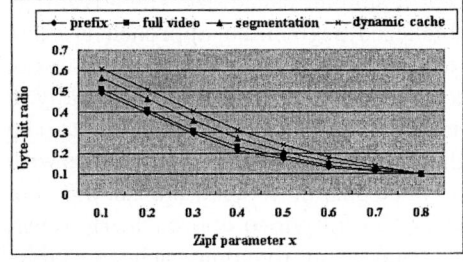
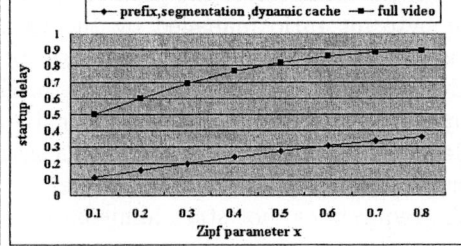

Fig. 4. Impact of byte-hit radio

Fig. 5. impact of startup delay

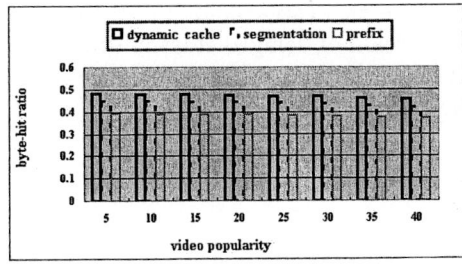
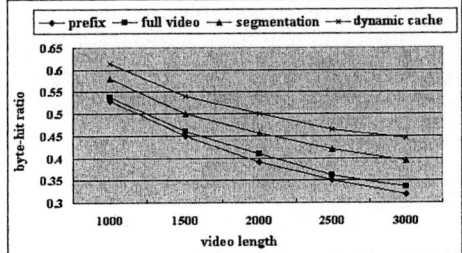

Fig. 6. Impact of popularity

Fig. 7. impact of video length

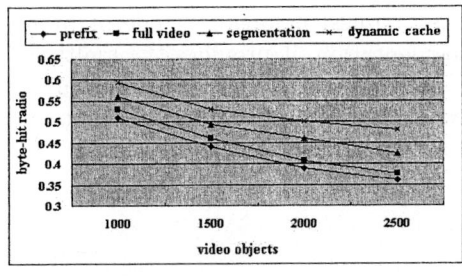
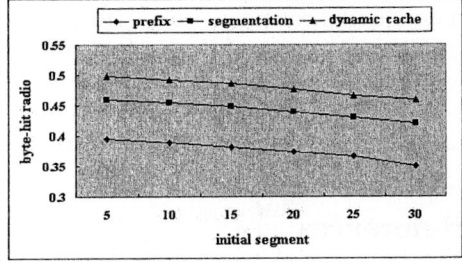

Fig. 8. Impact of video objects

Fig. 9. Impact of initial segmentation

4.4 Impact of Other System Parameters

Fig.7 shows the impact of video length imposes on the byte hit ratio. In general, as the size of the media file increases, the byte hit ratio will fall, this is true for all the four approaches. When the size of a media file is very large, the dynamic cache algorithm can ensure higher byte hit ratio than the segmentation and other two approaches. As to a video with the length of 3000 blocks, the byte hit ratios of dynamic cache strategy and variable-sized segmented strategy are respectively 0.331 and 0.284. If the length falls to 1000 blocks, the byte hit ratios may reach 0.623 and 0.594 respectively. No matter which approach we use, dynamic cache strategy or variable-size segmented approach, caching large media will cause the byte hit ratio in proxy cache to fall. However, dynamic cache strategy is better than variable-sized segmented strategy.

Fig.8 shows the cases of applicants for distinct media objects from the angle of quantity. Once again, the dynamic cache strategy and the variable-sized segmented approach have much more advantage over the other two approaches, even when the conditions for caching are less favorable for both of them. Comparatively speaking, the advantage of the cache strategy is more outstanding. Fig.9 examines the percentage of cache dedication for storing the initial segmentation. Because the cache for the suffixes is reduced, the byte hit ratio falls with the increase in using initial segmentation. This slight decrease in byte hit ratio can be offset by increasing benefits substantially by the means of reducing start delay. For example, let us compare these two cases,0.052 and 0.154. The byte hit ratio is barely decreased, but the fraction of delayed startup drops substantially. However, no more benefits can be derived once the percentage of the initial segmentation cached increases beyond 0.20.

5 Conclusion

In this paper, we put forward the algorithm for dynamic cache, based on the variable-sized segmented approach. The segmented approach groups media blocks into variable-sized segmentations. This method differs from the way we handle a web object, which is usually handled as a whole. The algorithm of dynamic cache considers adequately the users' request behavior. While maintaining the advantage of the variable-sized segmentation, it provides the multi-user within a period of time with the same media they request by using dynamic cache. The algorithm of cache greatly saves traffics resource on the backbone of network, and enhance the byte hit ratio and efficiency of the proxy cache.

References

1. Sen, S., Rexford, J., Towsley, D.: Proxy Prefix Caching for Multimedia Steams. in Proc. of IEEE Inforcom'99, New York, USA 3 (1999) 1310-1319
2. White, P.P., Crowcroft, J.: Optimized Batch Patching with Classes of Service. ACM Communications Review, Vol. 30, No. 4 (2000)
3. Verscheure, O., Verkatramani, C., Froassard, P., Amini, L.: Joint Server Scheduling and Proxy Caching for Video Delivery. Computer Communications, Vol. 25, No. 4 (2002) 413-423
4. Wang, B., Sen, S., Adler, M., Towsley, D.: Optimal Proxy Cache Allocation for Efficient Streaming Media Distribution. IEEE Trans. on Multimedia, Vol. 6, No. 2 (2004)
5. Pascal Frossard, Olivier Verscheure: Batched Patch Caching for Streaming Media. IEEE Communications Letters, Vol. 6, No. 4 (2002)
6. Wu, K.L., Yu, P.S.: Segment-Based Proxy Caching of Multimedia Streams. In Proc. Of IEEE INFOCOM (2001)
7. Xu, Z.W., Guo, X.X., Wang, Z.X., Pang, Y.J.: The Dynamic Cache-Multicast Method for Streaming Media. In proceeding of 3rd European Conference on Universal Multiservice Networks (2004) 278-290
8. Xu, Z.W., Guo, X.X., Pang, Y.J., Wang, Z.X.: The Strategy of Batch Using Dynamic Cache for Streaming Media. in proceeding of IFIP International Conference on Network and Parallel Computing (2004) 508-513
9. Xu, Z.W., Guo, X.X., Pang, Y.J., Wang, Z.X.: The Transmitted Strategy of Proxy Cache Based on Segmented Video. in proceeding of IFIP International Conference on Network and Parallel Computing (2004) 502-508

Fusion of the Textural Feature and Palm-Lines for Palmprint Authentication

Xiangqian Wu[1], Fengmiao Zhang[1], Kuanquan Wang[1], and David Zhang[2]

[1] School of Computer Science and Technology,
Harbin Institute of Technology (HIT), Harbin 150001, China
{xqwu, zhangfm, wangkq}@hit.edu.cn
[2] Biometric Research Centre, Department of Computing,
Hong Kong Polytechnic University, Kowloon, Hong Kong
csdzhang@comp.polyu.edu.hk

Abstract. There are many features on a palm and different features reflect the different characteristic of a palmprint. Fusion of multiple palmprint features may enhance the performance of palmprint authentication system. In this paper, we investigate the fusion of the textural feature (PalmCode) and the palm-lines. Several fusion strategies have been compared. The experimental results show that the original PalmCode scheme is optimal for the very high security systems, while the fusion of the PalmCode and palm-lines using the Weighted Sum Strategy is the best choice for other systems.

1 Introduction

Computer-aided personal recognition is becoming increasingly important in our information society. Biometrics is one of the most important and reliable methods in this field [1]. The most widely used biometric feature is the fingerprint and the most reliable feature is the iris. However, it is very difficult to extract small unique features (known as minutiae) from unclear fingerprints and the iris input devices are very expensive . Other biometric features, such as the face and voice, are less accurate and they can be mimicked easily. The palmprint, as a relatively new biometric feature, has several advantages compared with other currently available features [1]: palmprints contain more information than fingerprint, so they are more distinctive; palmprint capture devices are much cheaper than iris devices; palmprints also contain additional distinctive features such as principal lines and wrinkles, which can be extracted from low-resolution images; a highly accurate biometrics system can be built by combining all features of palms, such as palm geometry, ridge and valley features, and principal lines and wrinkles, etc. It is for these reasons that palmprint recognition has recently attracted an increasing amount of attention from researchers [2, 3, 4, 5, 6, 7].

A palmprint contains following basic elements: principal lines, wrinkles, delta points and minutiae, etc. [8]. And these basic elements can constitute various palmprint features. For example, principal lines and wrinkles constitute line features (called palm-lines) and all of these basic elements constitute textural feature, etc. Different palmprint features reflect the different characteristic of a palmprint. Fusion of multiple palmprint features may enhance the performance of palmprint authentication system. The palm-lines and textural features (such as PalmCode) have been widely used for palmprint recognition. In this paper, we will investigate the fusion of the palm-lines and PalmCode for palmprint authentication.

When palmprints are captured, the position, direction and amount of stretching of a palm may vary so that even palmprints from the same palm may have a little rotation and translation. Furthermore, palms differ in size. Hence palmprint images should be orientated and normalized before feature extraction and matching. The palmprints used in this paper are captured by a CCD based palmprint capture device [5]. In this device, there are some pegs between fingers to limit the palm's stretching, translation and rotation. These pegs separate the fingers, forming holes between the forefinger and the middle finger, and between the ring finger and the little finger. In this paper, we use the preprocessing technique described in [5] to align the palmprints. In this technique, the tangent of these two holes are computed and used to align the palmprint. The central part of the image, which is 128 × 128, is then cropped to represent the whole palmprint. Such preprocessing greatly reduces the translation and rotation of the palmprints captured from the same palms. Figure 1 shows a palmprint and its cropped image.

The rest of this paper is organized as follows. Section 2 reviews feature extraction and matching. Section 3 presents several fusion strategies. Section 4 contains some experimental results. And in Section 5, we provide a conclusion.

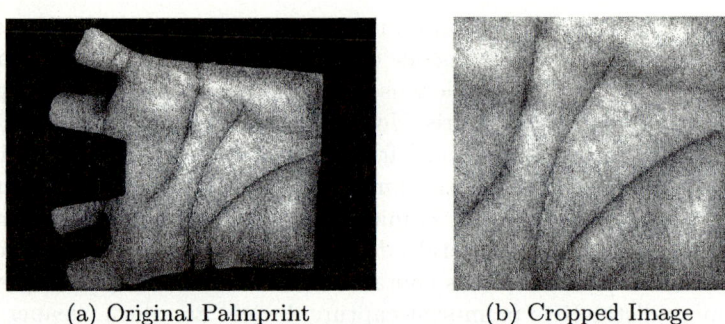

(a) Original Palmprint   (b) Cropped Image

Fig. 1. An example of the palmprint and its cropped image

2   Feature Extraction and Matching

In this section, we will review the PalmCode and palm-lines extraction and matching.

2.1 PalmCode Extraction

Let I denote a palmprint image, its PalmCode can be extracted as below [5, 9]. Suppose G_α ($\alpha = 0°, 45°, 90°$ and $135°$) be a circular Gabor filter with the orientation α [10] and use it to filter I as following:

$$I_{G_\alpha} = I * G_\alpha \tag{1}$$

where "*" represents the convolution operation and G_j

The magnitude of the filtered image I_{G_j} is defined as:

$$M_\alpha(x,y) = \sqrt{I_{G_\alpha}(x,y) \times \bar{I}_{G_\alpha}(x,y)} \tag{2}$$

where "-" represents the complex conjugate.

And the orientation of the point (x, y) is decided by the following equation:

$$O(x,y) = \arg\max_\alpha(M_j(x,y)) \tag{3}$$

According to Eq. (1) and Eq. (3), the PalmCode can be computed as following [9]:

$$P_R(x,y) = \begin{cases} 1, & \text{if } \mathbf{Re}[I_{G_{O(x,y)}}(x,y)] \geq 0; \\ 0, & \text{otherwise.} \end{cases} \tag{4}$$

$$P_I(x,y) = \begin{cases} 1, & \text{if } \mathbf{Im}[I_{G_{O(x,y)}}(x,y)] \geq 0; \\ 0, & \text{otherwise.} \end{cases} \tag{5}$$

where P_R and P_I are called the real part and imaginary part of the PalmCode.

For each palmprint, only the pixels at $(4i, 4j)(i = 0, \ldots, 31, j = 0, \ldots, 31)$ are used to extract the PalmCode [5, 9].

Figure 2 shows some palmprints and their PalmCodes.

2.2 PalmCode Matching

Suppose P_R, Q_R, P_I and Q_I be the real part and imaginary part of two Palm-Codes. The normalized hamming distance can be used to measure the similarity of these two PalmCodes [9].:

$$D = \frac{\sum_{i=1}^{N}\sum_{j=1}^{N} P_M(i,j) \cap Q_M(i,j) \cap ((P_R(i,j) \otimes Q_R(i,j) + P_I(i,j) \otimes Q_I(i,j)))}{2\sum_{i=1}^{N}\sum_{j=1}^{N} P_M(i,j) \cap Q_M(i,j)} \tag{6}$$

where P_M and Q_M are the mask of these two PalmCodes respectively. These masks are used for denoting the non-palmprint pixels [5].

In this paper, we translate the distance to a matching score by following equation:

$$S = 1 - D \tag{7}$$

Obviously, $0 \leq S \leq 1$ and the larger the matching score, the greater the similarity between these two PalmCodes.

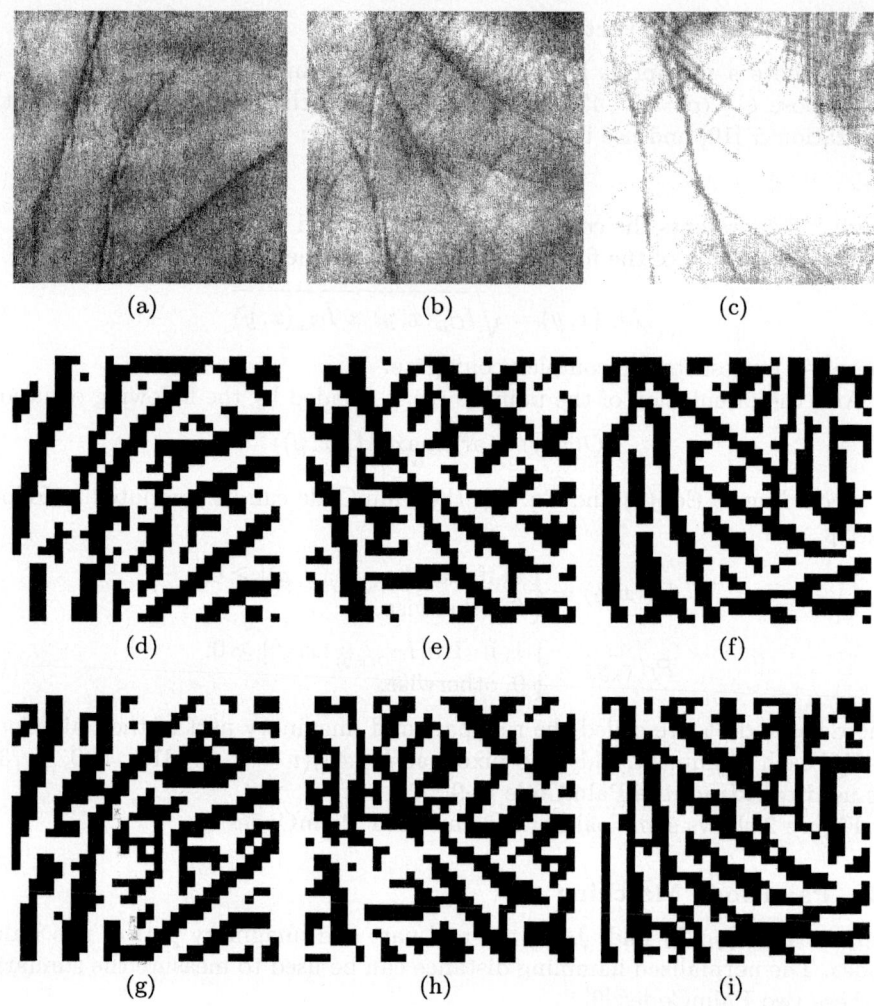

Fig. 2. Some examples of the PalmCodes: (a)–(c) are original palmprints; (d)–(f) are the real parts of the PalmCodes; (g)–(i) are the imaginary parts of the PalmCodes

2.3 Palm-Line Extraction

Palm-lines can be extracted using the morphological operations described in [11]. In the gray-scale morphology theory, two basic operations, namely *dilation* and *erosion* for image f are defined as follows:
Dilation:

$$(f \oplus b)(s,t) = \max\{f(s-x, t-y) + b(x,y)| \\ (s-x, t-y) \in D_f \text{ and } (x,y) \in D_b\} \quad (8)$$

Erosion:

$$(f \ominus b)(s,t) = \min\{f(s-x, t-y) - b(x,y) | \\ (s-x, t-y) \in D_f \text{ and } (x,y) \in D_b\} \quad (9)$$

where D_f and D_b represent the domains of image f and structuring element b. Furthermore, two additional operations *opening* and *closing* are defined by combining the dilation and erosion operations:

opening:

$$f \circ b = (f \ominus b) \oplus b \quad (10)$$

closing:

$$f \bullet b = (f \oplus b) \ominus b \quad (11)$$

And using the closing operation, *bothat* operation is defined as below:

bothat:

$$h = (f \bullet b) - f \quad (12)$$

The bothat operation can be used to detect the valley in an image. Because all palm-lines are valley in a palmprint, the bothat operation is suitable for palm-line extraction. The shape of the structuring element heavily affects the result of line extraction. Since the directions of the palm-lines are very irregular, we should extract these lines in different directions. The directional structuring element $b_{0°}$ used to extract the palm-lines in $0°$ direction is shown in Figure 3(a) and the directional structuring element b_θ used to extract the palm-lines in direction θ can be obtained by rotating $b_{0°}$ with degree θ. $b_{45°}$, $b_{90°}$ and $b_{135°}$ are also shown in Figure 3.

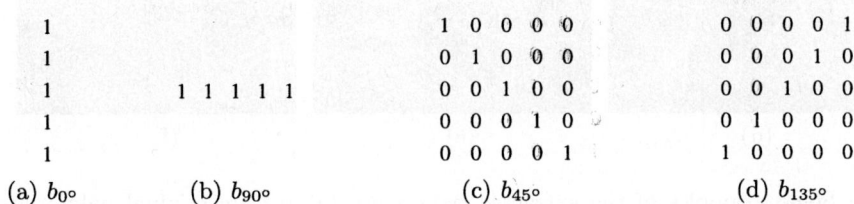

(a) $b_{0°}$ (b) $b_{90°}$ (c) $b_{45°}$ (d) $b_{135°}$

Fig. 3. The directional structuring elements used for palm-line extraction

The palm-lines in θ direction can be extracted by the following process:

1. Smoothing the original image I by convolving the original image with $b_{\theta+90°}$;
2. Processing the smoothed image by using bothat operation with structuring element b_θ and get the θ-directional magnitude M_θ;
3. Looking for the local maximum points along direction $\theta + 90°$ in M_θ;
4. Thresholding the maximum magnitude image.

The palm-lines are first extracted in several directions and the binary image L containing the final palm-lines can be obtained as follows:

$$L = \bigvee_{\text{all } \theta} L_\theta, \tag{13}$$

where L_θ is the binary image containing the extracted palm-lines in θ direction and "$\bigvee$" is the logical "OR" operation.

After conducting the closing and thinning operations, we obtain the resultant palm-line image.

In this paper, we extract the palm-lines in four directions: $0°$, $45°$, $90°$ and $135°$. Figure 4 shows some palmprints and their extracted palm-lines.

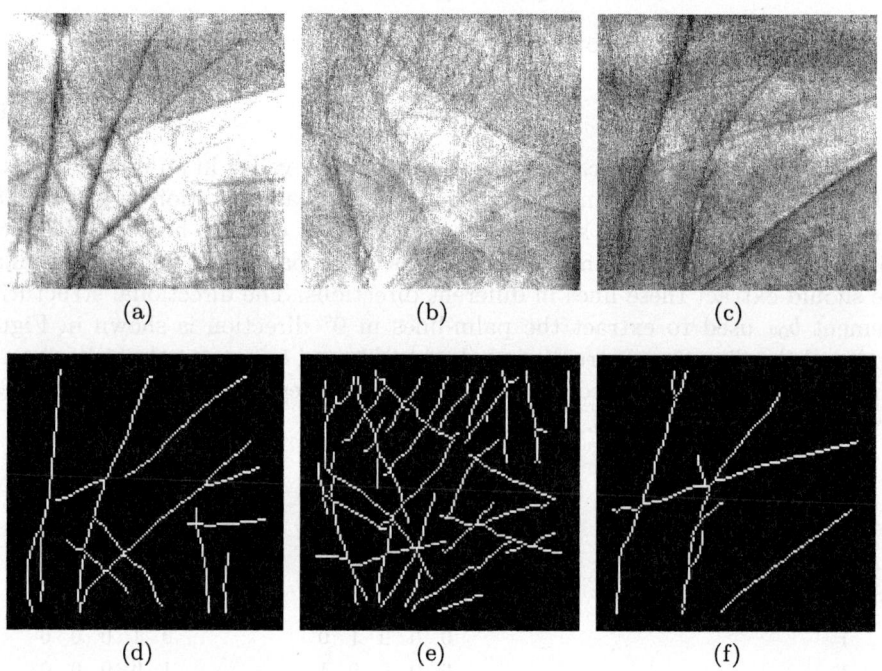

Fig. 4. Some examples of the extracted palm-lines: (a)–(c) are original palmprints; (d)–(f) are the extracted palm-lines

2.4 Palm-Line Matching

Let I_1 and I_2 denote two palmprints, and L_1 and L_2 denote the binary images containing their palm-lines. The most natural way to match L_1 against L_2 is to compute the proportion of the line points that are in the same position in L_1 and L_2 in the palm-line image. However, because of the existence of noise, the line points of the same lines may be not superposed on the palmprints captured from the same palm at a different time. Fortunately, the point shift that results from the noise should be small. Therefore we can first dilate L_1 to get L_D and then count the overlapping points between L_2 and L_D. These overlapping points are regarded as the matched points of L_1 and L_2. The matching score between L_1 and L_2 is defined as the proportion of the matched points to the total line points in L_1 and L_2:

$$S(L_1, L_2) = \frac{2}{M_{L_1} + M_{L_2}} \times \sum_{i=1}^{M} \sum_{j=1}^{N} [L_2(i,j) \wedge L_D(i,j)] \qquad (14)$$

where "$\wedge$" is logical "AND" operator; $M \times N$ is the size of the image; M_{L_1} and M_{L_2} are the number of the points on the palm-lines in L_1 and L_2, respectively. L_D is the dilating result of L_1.

Obviously, $0 \leq S \leq 1$, and the larger the matching score, the greater the similarity between these two palmprints.

3 Fusion Strategies

Denote x_1 and x_2 as the matching scores of the PalmCode and palm-lines, respectively. We fuse these two scores by following strategies to obtain the final matching score x.

S_1: **Maximum Strategy:**

$$x = \max(x_1, x_2) \qquad (15)$$

S_2: **Product Strategy:**

$$x = \sqrt{x_1 x_2} \qquad (16)$$

S_3: **Sum Strategy:**

$$x = \frac{x_1 + x_2}{2} \qquad (17)$$

S_4: **Weighted Sum Strategy:**

$$x = a x_1 + b x_2, \qquad a + b = 1; \qquad (18)$$

Obviously, when $a = b = 0.5$, the weighted sum strategy is same as the sum strategy. We find that when $a = 0.8$ and $b = 0.2$, this strategy can get the best performance.

4 Experimental Results

We test the proposed approach on a database containing 3240 palmprints collected from 324 different palms with the CCD-based device [5]. Each palm provided 10 samples. These palmprints were taken from the people of different ages and both sexes in the Hongkong Polytechnic University. The size of the images in the database is 384×284. Using the preprocessing technique described in [5],

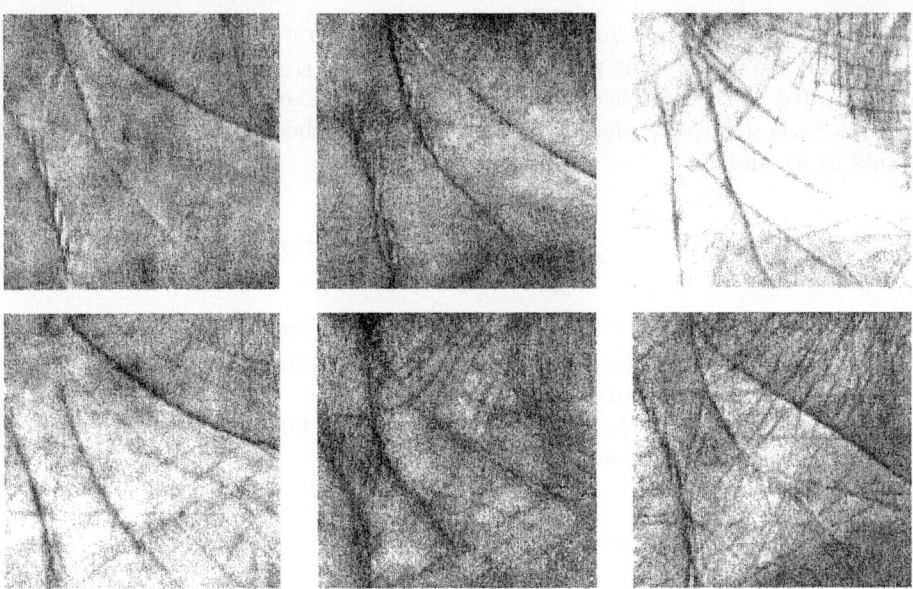

Fig. 5. Typical Samples in the database

the central 128 × 128 part of the image was cropped to represent the whole palmprint. Some typical samples in the database are shown in Figure 5.

All of the described fusion strategies were tested. To test the performance of these fusion strategies, each sample is matched against the other palmprints in the database. If the matching score exceeds a given threshold, the input palmprint is accepted. If not, it is rejected. The performance of a biometric method is often measured by the false accept rate (FAR) and false reject rate (FRR). While it is ideal that these two rates should be as low as possible, they cannot be lowered at the same time. So, depending on the application, it is necessary to make a trade-off: for high security systems, such as some military systems, where security is the primary criterion, we should reduce FAR, while for low security systems, such as some civil systems, where ease-of-use is also important, we should reduce FRR. To test the performance with respect to the FAR and FRR trade-off, we usually plot the so-called Receiver Operating Characteristic (ROC) curve, which plots the pairs (FAR, FRR) with different thresholds [12]. And the ROC curve of each fusion strategy is plotted in Figure 6. And their equal error rate (EER) are listed in Table 1.

Table 1. EERs of the original PalmCode (PC), the line matching (LM), and the different fusion strategies (S_1, S_2, S_3 and S_4).

Scheme	PC	LM	S_1	S_2	S_3	S_4
EER (%)	0.23	0.37	0.21	0.22	0.21	0.17

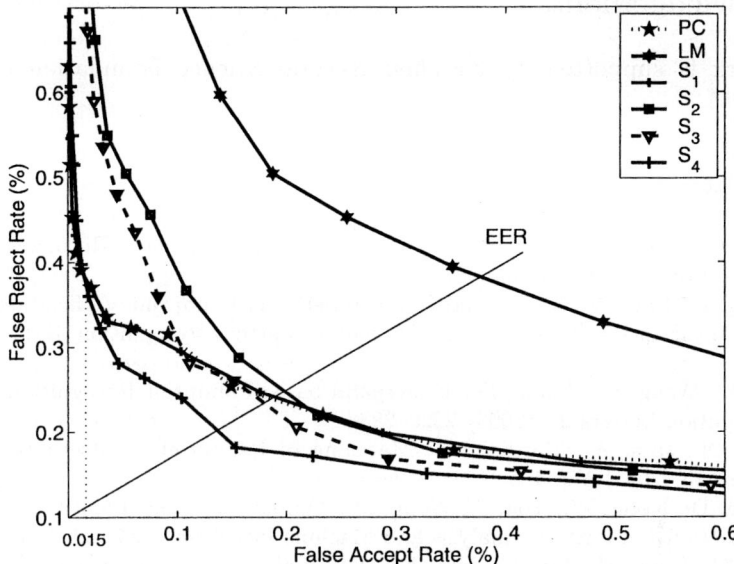

Fig. 6. The ROC Curves of the original PalmCode (PC), the line matching (LM), and the different fusion strategies (S_1, S_2, S_3 and S_4)

According to the figure, when $FAR < 0.015\%$, the original PalmCode scheme perform better than others. In other cases, the weighted sum fusion strategy outperform other strategies. That is, in all of these strategies, the original Palm-Code is the best for the very high security systems, such as some military systems, in which FAR should be very low, while for other systems, the fusion of the PalmCode and palm-lines using the weighted sum strategy is the best choice.

5 Conclusion

Palmprint is an important complement of the available biometric features. There are so many features on a palm and the different type features reflect the different characteristic of the palm. Among these features, the textural features and the line features have been extensive investigated for palmprint authentication. This paper examined the fusion of these two features. Several fusion strategies have been implemented on a database containing 3240 palmprints and the experimental results show that the original PalmCode scheme is optimal for the very high security systems, while the fusion of the PalmCode and palm-lines using the weighted sum strategy is the best choice for other systems. The EER of the weighted sum fusion strategy is about 0.17, which is comparable with the other palmprint recognition approaches.

Acknowledgements

This work is supported by National Natural Science Foundation of China (60441005).

References

1. Jain, A., Ross, A., Prabhakar, S.: An Introduction to Biometric Recognition. IEEE Transactions on Circuits and Systems for Video Technology **14** (2004) 4–20
2. Zhang, D., Shu, W.: Two Novel Characteristics in Palmprint Verification: Datum Point Invariance and Line Feature Matching. Pattern Recognition **32** (1999) 691–702
3. Wu, X., Wang, K., Zhang, D.: Fisherpalm based Palmprint Recognition. Pattern Recognition Letters **24** (2003) 2829–2838
4. Duta, N., Jain, A., Mardia, K.: Matching of Palmprint. Pattern Recognition Letters **23** (2001) 477–485
5. Zhang, D., Kong, W., You, J., Wong, M.: Online Palmprint Identification. IEEE Transactions on Pattern Analysis and Machine Intelligence **25** (2003) 1041–1050
6. Han, C., Chen, H., Lin, C., Fan, K.: Personal Authentication using Palm-print Features. Pattern Recognition **36** (2003) 371–381
7. Wu, X., Wang, K., Zhang, D.: Wavelet Energy Feature Extraction and Matching for Palmprint Recognition. Journal of Computer Science and Technology **20** (2005) 411–418
8. Zhang, D.: Automated Biometrics–Technologies and Systems. Kluwer Academic Publishers (2000)
9. Kong, W., Zhang, D.: Feature-level Fusion for Effective Palmprint Authentication. Internatrional Conference on Biometric Authentication, LNCS **3072** (2004) 761–767
10. Daugman, J.: High Confidence Visual Recognition of Persons by a test of Statistical Independence. IEEE Transactions on Pattern Analysis and Machine Intelligence **15** (1993) 1148–1161
11. Wu, X., Wang, K., Zhang, D.: A Novel Approach of Palm-line Extraction. In: Proceedings of the Third International Conference on Image and Graphics, Hong Kong, P.R. China, IEEE Computer Society (2004) 230–233
12. Maio, D., Maltoni, D., Cappelli, R., Wayman, J.L., Jain, A.: FVC2000: Fingerprint Verification Competition. IEEE Transactions on Pattern Analysis and Machine Intelligence **24** (2002) 402–412

Nonlinear Prediction by Reinforcement Learning

Takashi Kuremoto, Masanao Obayashi, and Kunikazu Kobayashi

Dept. of Computer Science and Systems Eng., Eng. Fac., Yamaguchi Univ.,
Tokiwadai 2-16-1, Ube, Yamaguchi 755-8611, Japan
{wu, m.obayas, koba}@yamaguchi-u.ac.jp

Abstract. Artificial neural networks have presented their powerful ability and efficiency in nonlinear control, chaotic time series prediction, and many other fields. Reinforcement learning, which is the last learning algorithm by awarding the learner for correct actions, and punishing wrong actions, however, is few reported to nonlinear prediction.

In this paper, we construct a multi-layer neural network and using reinforcement learning, in particular, a learning algorithm called Stochastic Gradient Ascent (SGA) to predict nonlinear time series. The proposed system includes 4 layers: input layer, hidden layer, stochastic parameter layer and output layer. Using stochastic policy, the system optimizes its weights of connections and output value to obtain its prediction ability of nonlinear dynamics. In simulation, we used the Lorenz system, and compared short-term prediction accuracy of our proposed method with classical learning method.

1 Introduction

Artificial neural network models, as a kind of soft-computing methods, have been considered as effective nonlinear predictors [1,2,3,4] in last decades. Casdagli employed the radial basis function network (RBFN) in chaotic time series prediction in early time [1]. Leung and Wang analyzed the structure of hidden-layer in RBFN, and proposed a technique called the cross-validated subspace method to estimate the optimum number of hidden units, and applied the method to prediction of noisy chaotic time series [3]. Oliveira ,Vannucci and Silva suggested a two-layered feed-forward neural network, where the hyperbolic tangent activation function was chosen for all hidden units, the linear function for the final output unit, and obtained good results for the Lorenz system, Henon and Logistic maps [2]. Such of neural network models are not only developed on fundamental studies of chaos, but also applied in many nonlinear predictions, e.g., oceanic radar signals [3], financial time series [4], etc. Kodogiannis and Lolis compared the performance of some neural networks, i.e., Multi-layer perceptron (MLP), RBFN, Autoregressive recurrent neural network (ARNN), etc., and fuzzy systems, used for prediction of currency exchange rates [4].

Meanwhile, reinforcement learning, a kind of goal-directed learning, is of great use for a learner (agent) adapting unknown environments [5,6]. When the environment belongs to Markov decision process (MDP), or Partially observable Markov decision process (POMDP), an learner acts some trial-and-error searches according to certain policies, and receives reward or punishment.

Through the interactions between the environment and learner, both exploration and exploitation are carried out, the learner adapts to environment gradually. Though reinforcement learning has been showing more contributions on artificial intelligence, optimal control theory and other fields, however, this algorithm of machine learning is hardly applied in nonlinear prediction [7].

We have proposed a self-organized fuzzy neural network, which using a reinforcement learning algorithm called Stochastic Gradient Ascent (SGA) [6] to predict chaotic time series [7], and obtained a high precision result in simulation. However, the prediction system used multiple softcomputing techniques including fuzzy system, self-organization function, stochastic policy and so on, so it was complained too complex to use. In this paper, we intend to use a simple multi-layer neural network but apply SGA on it, to predict nonlinear time series. This system includes 4 layers: input layer, hidden layer, stochastic parameter layer and output layer. Using stochastic policy, the system optimizes its weights of connections and output value to obtain its prediction ability of nonlinear dynamics. In simulation, we used the Lorenz system [8], and compared short-term prediction accuracy of our proposed method with classical learning method, i.e. error back propagation (BP) [9].

2 Prediction Systems

2.1 Conventional Prediction System

Traditionally, multiple-layer feedforward neural networks serve as a good preditor of nonlinear time series [1,2,4]. Fig. 1 gives an example diagram of the networks. Units in each layer are linear functions, or monotonous functions i.e. sigmoid function, generally. Output of units are transfer by weighted connection to the units in next layer, and by adjusting the weights, network output approach to a teacher signal, e.g. training data of time series here. The optimal structure for chaotic time series prediction of this kind of networks are researched detailly in Ref. [2]. In convenient, multiple nodes in hidden layer accept input with weights w_{kn}, and their output is given by:

$$H_k(t) = \frac{1}{1 + e^{-\beta_H \sum x_n(t) w_{kn}}} \quad (1)$$

where β_H is a constant.

Similarly, output of unit in output layer of system can be described as:

$$\hat{Y}(t+1) = \frac{1}{1 + e^{-\beta_Y \sum H_k(t) w_{yk}}} \quad (2)$$

where β_Y is a constant.

2.2 Proposal Prediction System

To deal with nonlinear dynamcs, we could not neglect stochastic methods, which are more effective on resolving problems in real world. We propose a multiple-

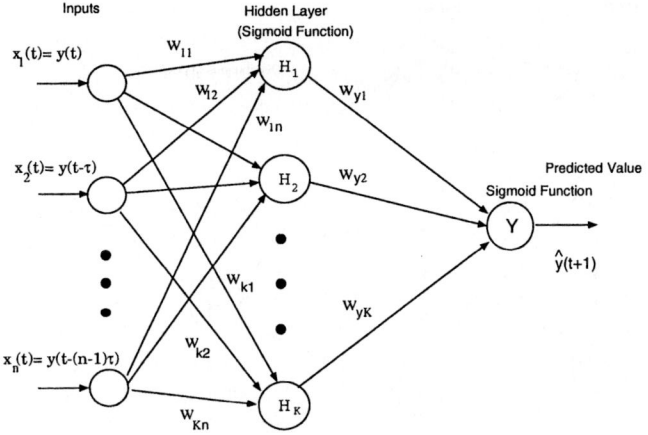

Fig. 1. Architecture of a prediction system using error back propagation learning algorithm (conventional system)

layer neural network here, as a nonlinnear predictior, using reinforcement learning algorithm which has a stochastic policy (Fig. 2).

This hierarchical network is composed by 4 layers:

1) Input layer which receiving information of environment, i.e., reconstructed data of time series;
2) Hidden layer which is constructed by multiple nodes of sigmoid function;
3) Stochastic layer (Distribution of prediction layer in Figure 2) which are parameters of probability function, and the nodes fire according to sigmoid function too;
4) Output layer which is a probability function, we use Gaussian function here. Stochastic gradient ascent (SGA) [6], which respects to continuous action, is naturally served into the learning of our predictor. The prediction system and its learning method will be described in detail in this section.

Reconstructed Inputs. According to the Takens embedding theorem [10], the inputs of prediction system on time t, can be constructed as a n dimensions vector space $X(t)$, which includes n observed points with same intervals on time series $y(t)$.

$$X(t) = (x_1(t), x_2(t), \cdots, x_n(t)) \qquad (3)$$
$$= (y(t), y(t-\tau), \cdots, y(t-(n-1)\tau)) \qquad (4)$$

where τ is time delay (interval of sampling), n is the embedding dimension.

If we set up a suitable time delay and embedding dimension, then a track which shows the dynamics of time series will be observed in the reconstructed state space $X(t)$ when time step t increases.

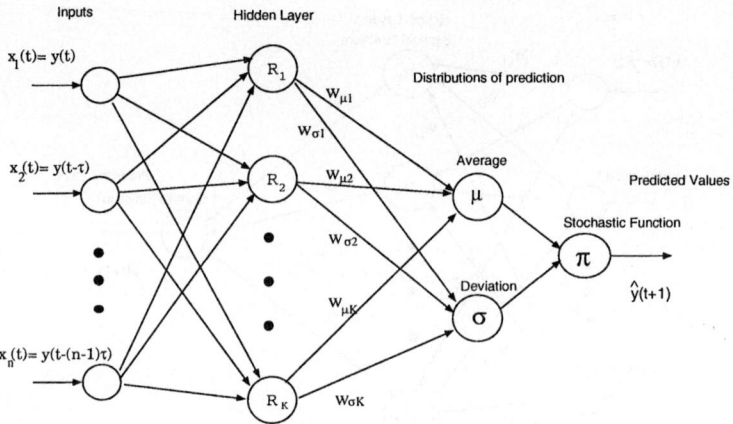

Fig. 2. Architecture of a prediction system using reinforcement learning algorithm (proposal system)

Hidden Layer. Multiple nodes accept input with weights w_{ij}, and their output is given by:

$$R_j(t) = \frac{1}{1 + e^{-\beta_R \sum x_i(t) w_{ij}}} \quad (5)$$

where β_R is a constant.

Stochastic Layer. To each hidden node $R_j(t)$ in hidden layer, parameters of distribution function are connected in weight $w_{j\mu}$ and weight $w_{j\sigma}$ when we consider the output is according to Gaussian distribution. Nodes in stochastic layer give their output μ, σ as:

$$\mu(R_j(t), w_{j\mu}) = \frac{1}{1 + e^{-\beta_\mu \sum R_j(t) w_{j\mu}}} \quad (6)$$

$$\sigma(R_j(t), w_{j\sigma}) = \frac{1}{1 + e^{-\beta_\sigma \sum R_j(t) w_{j\sigma}}} \quad (7)$$

where β_μ, β_σ is constant, respectively.

Output Layer. The node in output layer means a stochastic policy in reinforcement learning. Here we use a 1-dimension Gaussian function $\pi(\hat{y}(t+1), W, X(t))$ simply, to predict time series data:

$$\pi(\hat{y}(t+1), W, X(t)) = \frac{1}{\sqrt{2\pi}\sigma} e^{-\frac{(\hat{y}(t+1) - \mu)^2}{2\sigma^2}} \quad (8)$$

where $\hat{y}(t+1)$ is the value of one-step ahead prediction, produce by regular random numbers. W means weights w_{ij}, $w_{j\mu}$ and $w_{j\sigma}$. This function causes learner's action so it is called stochastic policy in reinforcement learning.

Reinforcement Learning: SGA Algorithm. Kimura and Kobayashi suggested a reinforcement learning algorithm called stochastic gradient ascent (SGA), to respect to continuous action[6]. Using this stochastic approximation method, we train the proposed multiple-layer neural network to be nonlinear predictor. The SGA algorithm is given under.

1. Accept an observation $X(t)$ from environment.
2. Predict a future data $\hat{y}(t+1)$ under a probability $\pi(\hat{y}(t+1), W, X(t))$.
3. Collate training samples of times series, take the error as reward r_i.
4. Calculate the degree of adaption $e_i(t)$, and its history for all elements ω_i of internal variable W. where γ is a discount($0 \leq \gamma < 1$).

$$e_i(t) = \frac{\partial}{\partial \omega_i} ln(\pi(\hat{y}(t+1), W, X(t))) \qquad (9)$$

$$D_i(t) = e_i(t) + \gamma D_i(t-1) \qquad (10)$$

5. Calculate $\Delta \omega_i(t)$ by under equation.

$$\Delta \omega_i(t) = (r_i - b) D_i(t) \qquad (11)$$

where b is a constant.

6. Improvement of policy: renew W by under equation.

$$\Delta W(t) = (\Delta \omega_1(t), \Delta \omega_2(t), \cdots, \Delta \omega_i(t), \cdots) \qquad (12)$$

$$W \leftarrow W + \alpha(1-\gamma)\Delta W(t) \qquad (13)$$

where α is a learning constant, non-negative.

7. Advance time step t to $t+1$, return to (1).

3 Simulation

Using time series data of the Lorenz system [8], we examine efficiency of proposed prediction system and compare with error back propagation (BP) method, a classical learning of hierarchical neural network. Both prediction system act in same procedure: observe the Lorenz time series till 1,500 steps, use the beginning 1,000 steps to be training samples, then perform learning loops till prediction errors going to a convergence. After the architecture of system becomes stable, it is employed to predict data from 1,001 step to 1,500 step.

3.1 The Lorenz System

The Lorenz system, which is leaded from convection analysis, is composed with ordinary differential equations of 3 variables $o(t), p(t), q(t)$. Here, we use their discrete difference equations (Equ. 14 – 16), and predicts the variable $o(t)$(Equ. 16).

$$o(t+1) = o(t) + \Delta t \cdot \sigma \cdot (p(t) - o(t)) \qquad (14)$$
$$p(t+1) = p(t) - \Delta t(o(t) \cdot q(t) - r \cdot o(t) + p(t)) \qquad (15)$$
$$q(t+1) = q(t) + \Delta t(o(t) \cdot p(t) - b \cdot q(t)) \qquad (16)$$

here, we set $\Delta t = 0.005, \sigma = 16.0, \gamma = 45.92, b = 4.0$.

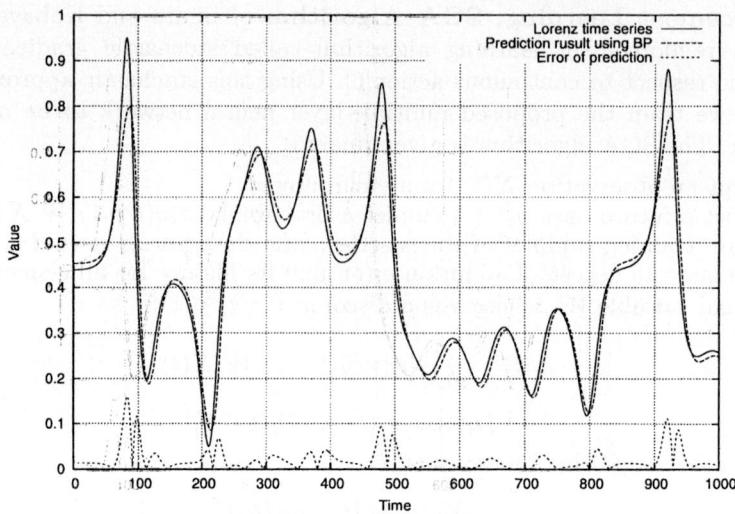

Fig. 3. Error back propagation (BP): learning result (2,000 iterations)

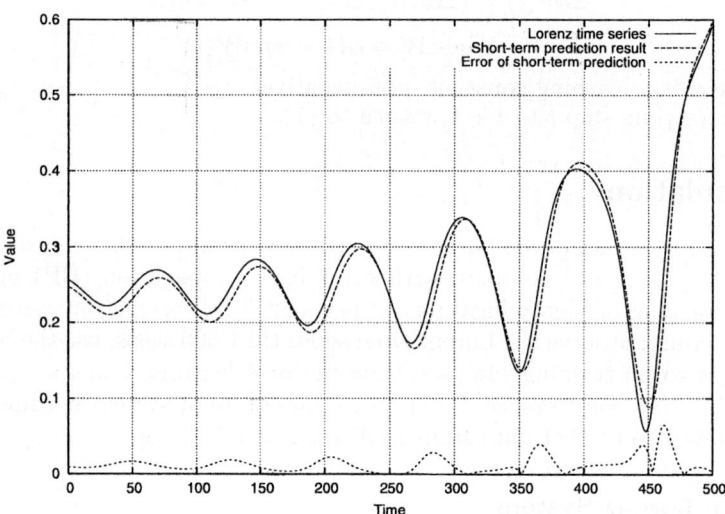

Fig. 4. Error back propagation (BP): Short-term (1-step ahead) prediction result

3.2 Parameters of Prediction System

Parameters in every part of prediction system are reported here.

1. Reconstruction of input space by embedding(Equ.(1),(2)): Embedding dimension $n : 3$, Time delay $\tau : 1$, (i.e.,in the case of input to be data of step 1,2,3, then the data of step 4 will be predicted).

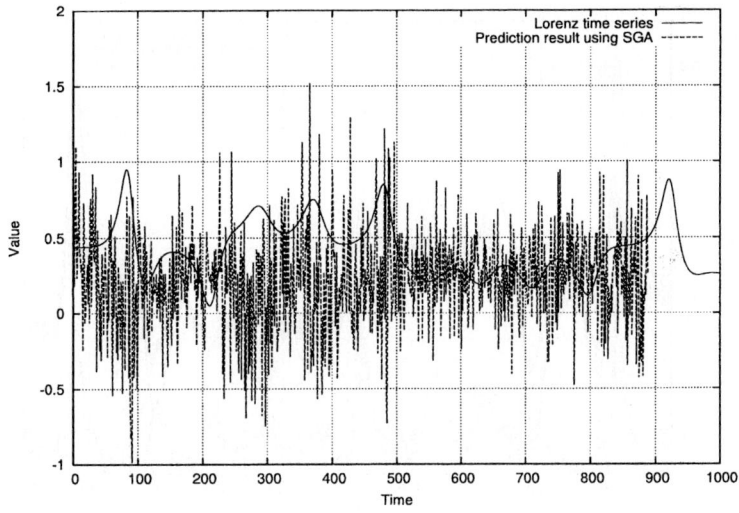

Fig. 5. Stochastic Gradient Ascent (SGA): Before learning (0 iteration)

2. Multiple neural network using BP:
   Number of hidden layer: 1, Number of hidden layer's nodes: 6, Constant β_H of units in hidden layer: 1.0, Constant β_Y of unit in output layer: 1.0, Learning rate: 0.01, Maximum value of error: 0.1.
3. Proposed neural network:
   Number of nodes R_j: 60, Constant β_R of units in hidden layer: 10.0, Constant β_μ of unit μ in stochastic layer: 8.0, Constant β_σ of unit σ in hidden layer: 18.0, Learning constant: For weight w_{ij}, α_{ij}: 2.0E-6, for weight $w_{j\mu}$, $\alpha_{j\mu}$: 2.0E-5, for weight $w_{j\sigma}$, $\alpha_{j\sigma}$: 2.0E-6, Reinforcement learning of SGA: Reward from prediction error r_t is

$$r_t = \begin{cases} 4.0E-4 & if |\hat{y}(t+1) - y(t+1)| \le \varepsilon \\ -4.0E-4 & if |\hat{y}(t+1) - y(t+1)| > \varepsilon \end{cases}$$

Limitation of errorsε: 0.1, Discountγ: 0.9.

3.3 Simulation Result

For conventional learning algorithm (BP), Fig. 3 shows its learning result after 2,000 times iteration, and one-head prediction result is shown in Fig. 4. The average value of prediction error in 500 steps short-term prediction is 0.0129 (values of time series data are regularized into (0, 1)).

For proposed system using reinforcement learning, Fig. 5 and Fig. 6 show its learning aspects, Fig. 7 shows its learning result after 30,000 times iteration, and one-head prediction result is shown in Fig. 8. The average value of pre-

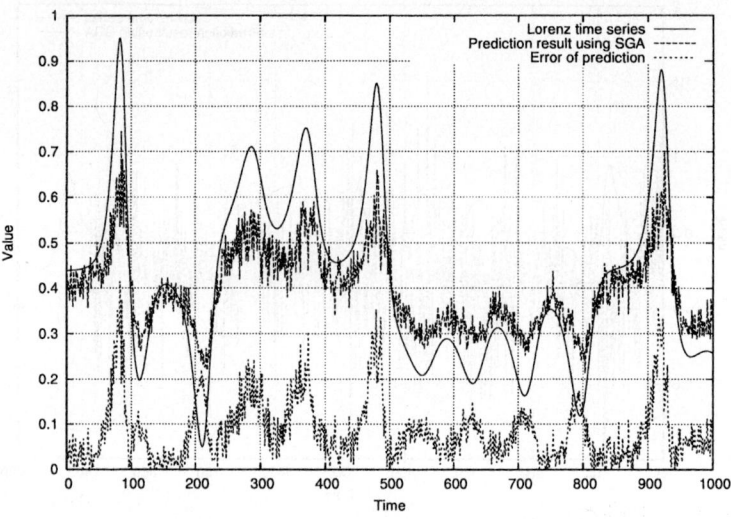

Fig. 6. Stochastic Gradient Ascent (SGA): Learning result (5,000 iterations)

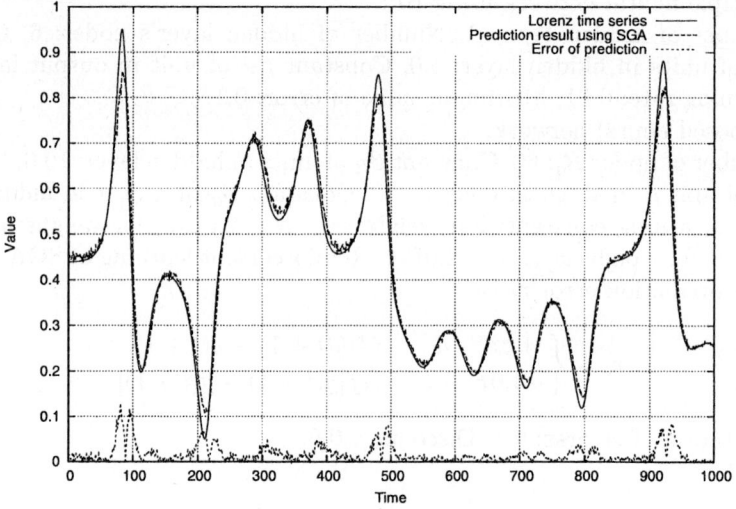

Fig. 7. SGA: Learning result (30,000 iterations)

diction error in 500 steps short-term prediction is 0.0112 (values of time series data are regularized into (0, 1)), i.e., prediction precision is raised 13.2% than conventional algorithm.

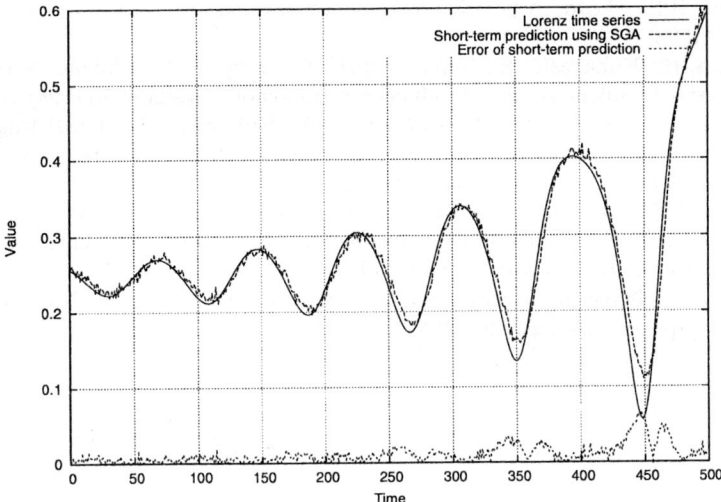

Fig. 8. SGA: Short-term (1-step ahead) prediction result

4 Conclusions

An algorithm of reinforcement learning with stochastic policy, SGA, was applied to a multi-layer neural network to predict nonlinear time series in this paper. Using Lorenz chaotic time series, prediction simulation demonstrated the proposed system provided successful learning results and prediction results comparing with conventional learning algorithm. For its stochastic output, the proposed system is expected to be applied on the noise contained complex dynamics, or particially observable Markov decision process in real world.

Acknowledgments

A part of this work was supported by MEXT KAKENHI (No. 15700161).

References

1. Casdagli, M.: Nonlinear Prediction of Chaotic Time Series. Physica D: Nonlinear Phenomena 35 (1989) 335-356
2. de Oliveira, K. A., Vannucci, A., da Silva, E. C.: Using Artificial Neural Networks to Forecast Chaotic Time Series. Physica A 284 (1996) 393-404
3. Leung, H., Lo, T., Wang, S.: Prediction of Noisy Chaotic Time Series Using an Optimal Radial Basis Function. IEEE Trans. on Neural Networks 12 (2001) 1163-1172
4. Kodogiannis, V., Lolis, A.: Forecasting Financial Time Series Using Neural Network and Fuzzy System-based Techniques. Neural Computing & Applications 11 (2002) 90-102

5. Sutton, R.S., Barto, A.G.: Reinforcement Learning: An Introduction. The MIT Press (1998)
6. Kimura, H., Kobayashi, S.: Reinforcement Learning for Continuous Action Using Stochastic Gradient Ascent. Intelligent Autonomous Systems 5 (1998) 288-295
7. Kuremoto, T., Obayashi, M., Yamamoto, A., Kobayashi, K.: Predicting Chaotic Time Series by Reinforcement Learning. Proc. of The 2nd Intern. Conf. on Computational Intelligence, Robotics and Autonomous Systems (CIRAS 2003)
8. Lorenz, E. N.: Deterministic Nonperiodic Flow. J. atomos. Sci. 20 (1963) 130-141
9. Rumelhart, D. E., Hinton, G.E., Williams, R. J.: Learning Representation by Back-propagating Errors. Nature 232 (9) (1986) 533-536
10. Takens, F.: Detecting Strange Attractor in Turbulence. Lecture Notes in Mathematics (Springer-Verlag) 898 (1981) 366-381

Author Index

Aghaeinia, Hasan II-998
Aguilar, Jose I-514
Ahn, Hyun Soo I-155
Ahn, Seongjin II-781
Ahn, Sung Mahn I-919
Alirezaee, Shahpour II-998
Amin, Hesham H. I-621

Bai, Huixian I-485
Baik, Sung I-919
Bengtsson, Ewert II-851
Bi, Ran II-11
Bi, Yan-Zhong II-588
Bingul, Zafer I-976
Byun, Kyung Jin I-184

Cai, Liang I-97
Cai, Wei-You I-1014
Cai, Yuanli II-218
Cao, Gexiang I-707
Cao, Guang-yi II-179
Cao, Shuying I-767
Cha, Jae Sang II-695, II-704, II-713
Chang, Faliang I-77
Chang, Zhiguo II-139
Chen, Ai-jun I-868
Chen, Chun I-591
Chen, Dechang I-543
Chen, Enhong I-475
Chen, ErKui I-581
Chen, Gen-Cai I-591
Chen, Guanrong II-149
Chen, Guimin I-533
Chen, Jian I-1035
Chen, Jianhua II-346
Chen, Ling I-591
Chen, Mei-Ching I-1023
Chen, Panjun II-880
Chen, Qiumei I-505
Chen, Sheng I-777
Chen, Tierui II-771
Chen, Xiyuan I-447
Chen, Xuhui I-380
Chen, Zonghai II-502

Cheng, Shiduan II-968
Cheng, Xueqi II-771
Cheng, Yiping II-616
Cheng, Yuan-Chu I-1014
Chi, Zheru II-492
Chiang, Ziping II-257
Cho, Kyoung Rok II-675, II-695
Cho, Sang-Hyun II-870
Cho, Seong Chul II-695
Cho, Su-Jin I-899
Cho, Sung-Bae II-228
Choi, Jongmoo I-223
Choi, Won-Chul II-675
Choo, MoonWon I-909
Chu, Feng-ming I-957
Chu, Tianguang II-645
Chun, Kwang-Ho II-811
Chung, Chin Hyun II-414
Chung, Fu-lai I-849
Conilione, Paul C. II-61
Cui, Jianfeng II-405
Cui, Jianzhong II-90
Cui, Peiling II-405

Dai, Honghua I-465
Dai, Huaping II-21, II-109
Dai, Ruwei I-40
Dai, Wenzhan I-505
Dang, Haifeng I-485
Deng, Shengchun I-400
Djemame, Karim II-266
Do, Yongtae I-301
Dong, Jinxiang II-99
Du, Ji-Xiang I-87, I-253, I-282
Duan, Hai-Xin I-429
Duan, Jianyong II-548

Eo, Ik Soo I-184

Faez, Karim II-998
Fan, Xiao-Ping I-757
Fang, Bin I-272, I-688
Fang, Bing I-631
Fang, Lei I-263
Fang, Ping I-631

Fang, Yi   I-263
Fang, Yong   II-880
Fard, Alireza Shayesteh   II-998
Fei, Yan-qiong   II-179
Feng, Chen   I-650
Feng, Chunbo   I-697
Feng, Ding-zhong   II-900
Feng, Fuye   I-787
Feng, Naiqin   I-828
Feng, Yueping   I-350
Feng, Zhilin   II-99
Fujii, Robert H.   I-621

Gan, Liangzhi   II-159
Gao, Lin   II-80
Ge, Cheng   I-679
Ge, YunJian   I-631
Giusto, D.D.   II-751
Gökmen, Muhittin   II-860
Gómez, Antihus Hernández   I-859
Gong, Dun-Wei   I-571
Gu, Mi Sug   I-818
Gu, Wen-jin   II-326
Gu, Xiao   I-253, I-282
Gu, Yajun   I-707
Gulez, Kayhan   I-243
Guo, Feng   II-841
Guo, Jun   I-456
Guo, Lei   I-697
Guo, Xiaoxin   I-30, I-661, I-1065
Guo, Yan-shen   I-957
Guo, Zhenyu   II-346
Guo, Zong-ru   I-957

Ha, Jong-Shik   II-396
Hahn, Minsoo   I-184
Han, Chong-Zhao   I-146
Han, Fei   II-189
Han, Hui   I-878
Han, Jiuqiang   II-218
Han, Jun-Hua   I-165
Han, Qi   I-533
Harris, Chris J.   I-777
He, Hai-Tao   II-433
He, Na   II-890
He, Xing-Jian   I-165
He, Yong   I-859
He, Zengyou   I-400
He, Zhenyu   I-272, I-688
Heng, Pheng-Ann   II-958

Heng, Xing-Chen   II-209
Heo, Joon   II-821
Hong, Chuleui   I-928
Hong, Hyun-Ki   I-641
Hong, Xia   I-777
Hong, Yi   I-986
Hong, Yan   II-41
Hou, Zeng-Guang   II-443
Hu, Bao-Gang   I-737
Hu, Dewen   I-849
Hu, Jun   II-51
Hu, Laizhao   I-707
Hu, Yan-Jun   II-978
Hu, Yi   II-548
Hu, Yunhua   I-485
Hu, ZeLin   II-169
Huang, Fang   I-757
Huang, Guang-Bin   I-717, II-189
Huang, Houkuan   I-495
Huang, Jian   I-688
Huang, Jin   I-1035
Huang, Tingzhu   I-370
Huang, Wenmei   I-767
Huang, Yanxin   II-119
Hwang, Bon-Woo   I-611
Hwang, Jeong Hee   I-818
Hwang, Yong-Ho   I-641

Iketani, Takashi   II-880

Jalili-Kharaajoo, Mahdi   II-286
Jang, Euee S.   I-155, I-194
Jang, Jun Yeong   II-414
Jeon, Hyoung-Goo   II-675
Jeong, He Bum   I-184
Ji, Chunguang   II-910, II-929
Ji, Guangrong   I-650
Ji, Hong-Wei   I-49, I-671
Ji, Qiao-Ling   I-1014
Jia, Jianyuan   I-533
Jia, Xiaojun   I-126
Jia, Yunde   I-679
Jian, Jigui   I-797
Jiang, Aiguo   II-316
Jiang, Chunhong   I-310
Jiang, Minghui   II-238
Jiang, Mingyan   II-482
Jiang, Peng   II-578
Jiang, Weijin   I-20
Jiao, Li-cheng   I-59

Jiao, Runhai I-747
Jin, Shiyao I-291
Jin, Weidong I-707
Jin, Zhong II-958
Jung, Young Jin II-296
Jwo, Jung-Sing II-99

Kang, Hang-Bong II-870
Kang, HeauJo II-665, II-685
Kang, Hee-Joong I-909
Kang, Hwan Il II-530
Khajepour, Mohammad II-286
Khan, Muhammad Khurram II-723
Kim, Dong Hwa II-366
Kim, Dong Phil II-521
Kim, Gi-Hong II-821
Kim, Hak-Man II-704
Kim, Hong-jin II-801
Kim, Hyuncheol II-781
Kim, Jin Ok II-414
Kim, Jin Up II-695
Kim, Jinmook II-831
Kim, Myoung-Jun II-811
Kim, Sang-Tae II-396
Kim, Sang Wook II-521
Kim, Seongcheol II-598
Kim, Sungmin I-611
Kim, Tai-hoon II-665
Kim, Wook Joong I-155
Kim, Yeongjoon I-928
Kimachi, Masatoshi II-880
Kleinjohann, Bernd I-1004
Kleinjohann, Lisa I-1004
Kobayashi, Kunikazu I-1085
Koh, Seok Joo II-396, II-521
Kong, Qingqing II-21
Kook, Hyung Joon II-512
Kruger, Uwe I-727
Kuremoto, Takashi I-1085
Kwak, Hoon-Sung I-194

Labrador, Miguel I-514
Lai, Xiaoping I-947
Lee, Hyun II-675
Lee, Jong-Joo II-704, II-713
Lee, Keun-Soo II-801
Lee, Keun-Wang II-801
Lee, Kyoung-Mi II-733
Lee, Mal Rey II-665, II-685
Lee, Sangkeon II-791

Lee, Sangwoon II-791
Lee, Seongsoo I-899
Lee, Seong-Whan I-213, I-233,
  I-601, I-611, I-888, II-1
Lee, Sang-Woong I-213, I-888
Lee, Sukhan II-423
Lee, Vincent C.S. I-410
Lee, Won-Joo I-438
Lee, YangSun II-665
Lee, Yang-Weon I-552
Lee, Young Jae I-839
Li, Bo I-747
Li, Chun-lian I-957
Li, Gang I-465
Li, Hong-Nan II-139
Li, Houqiang I-117
Li, Jiangeng I-967
Li, Jun II-139
Li, Kan II-742
Li, Kang I-390, I-1045
Li, Miao II-169
Li, Ming I-360, II-502
Li, Ping I-868
Li, Pingkang I-727
Li, Shanping II-306
Li, Shao II-31
Li, Shipeng II-771
Li, Shiyong II-910, II-929
Li, Shouju II-276
Li, Xin II-71
Li, Xing I-429
Li, Xu-Qin II-189
Li, Yanjun II-568
Li, Yuan II-538
Li, Yuancheng I-747
Li, Yuejun I-340
Li, Zhi II-751
Li, Zushu II-920, II-929
Lian, Yong II-751
Liang, Wei I-679, II-558
Liao, Xiaoxin I-524, I-797
Lim, Soo-Yeon I-438
Lim, Sungsoo II-228
Lin, Ruizhong II-568, II-578
Lin, Xin II-306
Lin, Xinggang I-203
Lin, Xu-Mei II-463
Lin, Yue-Song II-607
Liu, Bao-Lu II-433
Liu, Chunmei I-40

Liu, Guangwu II-90
Liu, Hailin I-787
Liu, Heng II-920, II-929
Liu, Hui II-548
Liu, Jiming II-988
Liu, Ju I-320
Liu, Li I-263
Liu, Liping II-453
Liu, Qing I-1
Liu, Ruizhen II-761
Liu, Tao I-456
Liu, Wenbin II-80
Liu, Wu I-429
Liu, Xiaoming II-99
Liu, Yingxi II-276
Liu, Yong-Qing I-1055
Liu, Zhanhui I-30, I-661
Liu, Zhengkai I-117
Liu, Zhenqiu I-543
Liu, Zhenyuan I-937
Lok, Tat-Ming I-717, II-189
Lu, Ruzhan II-548
Lu, Yinan I-30, I-661
Luo, Haigeng I-797
Luo, Qi II-472
Luo, Xiao-Nan II-433
Lyu, Michael R. I-717, II-189

Ma, Janjun I-1
Ma, Jianfeng II-376
Ma, Jun I-340
Ma, Li I-77
Mao, Bing-Huan I-878
Mao, Xin II-880
Mao, Xuerong II-238
Megson, Graham I-136
Mei, Li I-97
Mei, Tao II-463
Min, Seung-Hyun II-811
Molina, Martin II-199
Moon, Song-Hyang I-213
Moon, Young-Jun II-791
Moradi, Hadi II-423

Nian, Rui I-650

Obayashi, Masanao I-1085
Oh, Hyun-Seo II-675
Onar, Omer Caglar I-243
Ono, Osamu I-562
Oysu, Cuneyt I-976

Pan, Chen I-263
Pan, Feng I-1014
Pan, Quan II-405
Pang, Yanwei I-117
Pang, Yunjie I-30, I-350, I-661, I-1065
Park, Chang-Beom II-1
Park, Jae-Pyo II-801
Park, Jin Ill II-366
Park, Sang-Cheol I-233
Park, Sung-Kee I-601
Park, Tae-Yoon I-194
Park, Young Tae I-839
Pei, Jihong I-68
Peng, Fuyuan I-10
Peng, Jian Xun I-1045
Peng, Zhenrui II-929
Pereira, Annia García I-859
Premaratne, Prashan I-107

Qi, Feihu II-880
Qi, Wei-Min I-1014
Qian, Gang II-841
Qiao, Jianping I-320
Qiao, Yizheng I-77
Qin, Ting II-502
Qin, Zheng II-209
Qing, Lin I-849
Qiu, Yuhui I-828
Qu, Han-Bing I-737
Quan, Zhong-Hua I-49, I-671

Ren, Hai Peng II-149
Ren, Qingsheng I-986
Richert, Willi I-1004
Roudsari, Farzad Habibipour II-286
Ruan, Xiaogang I-967, II-939
Ruensuk, Songpol II-248
Ryou, Hwangbin II-831
Ryu, Keun Ho I-818, II-296

Safaei, Farzad I-107
Seok, Bo-Ra I-641
Shang, Yanlei II-968
Shen, Fei I-631
Shen, Xingfa II-453, II-578
Shen, Yi I-524, II-238
Shi, Wei II-306
Shi, Xiaolong II-71
Shi, Xinling II-346
Shin, Ho-Kuen I-888
Shin, Myong-Chul II-704, II-713

Sohn, Hong-Gyoo II-821
Song, Haiyan I-859
Song, Jiatao II-492
Song, Rui I-475
Song, Zhi-huan I-868
Su, Guangda I-203, I-310
Sun, Guoxia I-320
Sun, Jun I-420
Sun, Li-Min II-588
Sun, Qibin II-751
Sun, Quan-Sen II-958
Sun, Suqin II-356
Sun, Xia I-485
Sun, Youxian II-21, II-159, II-453, II-568, II-578
Sun, Yu I-957
Sun, Zonghai II-159

Tan, Min II-443
Tang, Chunming II-851
Tang, Daquan II-326
Tang, Fang II-636
Tang, Yuanyan I-272, I-688
Tao, Jian I-126
Thammano, Arit II-248
Tian, Fengzhan I-495
Tian, Guang II-880
Tian, Tian I-340
Tian, Yan I-10, II-548
Treasure, Richard J. I-727
Tsuboi, Yusei I-562
Tu, Yiqing I-465

Uzunoglu, Mehmet I-243

Vural, Bulent I-243

Wang, Bing I-947
Wang, Bowen I-767
Wang, Chao II-376
Wang, Chuangcun II-41
Wang, Chunheng I-40
Wang, Dianhui II-61
Wang, Fang I-828
Wang, Gang I-807
Wang, Guizeng II-405
Wang, Hongwei I-937
Wang, Huanbao II-538
Wang, Hui-Jing II-463
Wang, Junyan I-203

Wang, Kuanquan I-1075
Wang, Ling II-636
Wang, Linze II-929
Wang, Long II-645
Wang, Qingyuan I-747
Wang, Ruifeng II-218
Wang, Ru-Liang I-1055
Wang, Sheng II-316
Wang, Shitong I-849
Wang, Wei I-1, II-492
Wang, Weiqiang II-11
Wang, Wen-Yuan I-878
Wang, Xi-li I-59
Wang, Xiao-Feng I-87, I-253, I-282
Wang, Xue II-316
Wang, Ya-hui II-179
Wang, Yan II-119, II-761
Wang, Yong I-737
Wang, Zhengxuan I-1065
Wang, Zhengyou II-492
Wang, Zhi II-453, II-568, II-578
Wang, Zhihai I-495
Wei, Xiao-yong I-174
Wei, ZhiGuo II-169
Wen, Shitao II-910, II-929
Weng, Ling I-767
Won, Jong Woo I-155
Woo, Seon-Kyung II-1
Wu, Hao II-636
Wu, Jian II-890
Wu, Jian-Ping I-429
Wu, Jin II-266
Wu, Lijiang II-31
Wu, Mingqiao I-291
Wu, Qiu-xuan II-179
Wu, Shuanhu II-41
Wu, Xiangqian I-1075
Wu, Yue II-880
Wu, Yuehua II-988
Wu, Yunsong I-136
Wu, ZhongCheng I-631

Xi, Hongsheng II-129
Xi, Lifeng I-947
Xia, De-Shen II-958
Xia, Feng II-453
Xia, Guoqing II-336
Xia, Xiao-Lei I-717
Xiang, Wei II-502
Xiao, Mei I-146

Xie, Bo   I-591
Xie, Guangming   II-645
Xie, Tao   I-995
Xie, Weixin   I-68
Xu, Dan   I-174
Xu, Dian-guo   II-890
Xu, Guandong   II-80
Xu, Jian   II-306
Xu, Jin   II-71
Xu, Min   I-849
Xu, Qingjiu   II-326
Xu, Wenbo   I-420
Xu, Xiaofei   I-400
Xu, Xin   I-995
Xu, Yao-Hua   II-978
Xu, Yuhui   I-20
Xu, Yunpeng   II-949
Xu, Yusheng   I-20
Xu, Zhiwen   I-30, I-661, I-1065
Xue, An-ke   II-607
Xue, Fei   II-655
Xue, Jianru   I-330

Yalçýn, Ýlhan Kubilay   II-860
Yan, Jingqi   II-920, II-929
Yan, Lanfeng   I-1
Yan, Shaoze   II-386
Yan, Ting-Xin   II-588
Yang, Beibei   II-939
Yang, Benkun   II-336
Yang, Hee-Deok   I-601
Yang, Jie   II-988
Yang, Jing   II-90, II-949
Yang, Qingxin   I-767
Yang, Xingjian   I-410
Yang, Xuan   I-68
Yang, Yahong   I-380
Yang, Yongqing   I-581
Yao, Yan-Sheng   II-463
Yi, Juneho   I-223
Yi, YingNan   I-340
Yin, Baoqun   II-129
Yin, Jian   I-1035
Yin, Jianwei   II-99
Yin, Shiqun   I-828
Yin, Zhixiang   II-90
You, Kang-Soo   I-194
You, Xinge   I-272, I-688
Yu, Dongmei   I-380
Yu, Haibin   II-558

Yu, Jian   I-495
Yu, Jinyong   II-326
Yu, Nenghai   I-117
Yu, Qingcang   I-126
Yu, Zhenhua   II-218
Yu, Zhezhou   II-119
Yuan, Dongfeng   II-482
Yuan, Guo-wu   I-174
Yuan, Yubo   I-370
Yue, Tai-Wen   I-1023

Zeng, Jin   I-986
Zeng, Peng   II-558
Zhan, Daqi   II-356
Zhang, Caifang   I-10
Zhang, Dali   II-129
Zhang, David   I-1075, II-920
Zhang, Duan   II-21
Zhang, Fengmiao   I-1075
Zhang, Gexiang   I-707
Zhang, Guo-Jun   I-87
Zhang, Haitao   II-502
Zhang, Huaguang   I-807
Zhang, Jian   II-169
Zhang, Jian-Hua   I-571
Zhang, Jiashu   II-723
Zhang, Jing   II-51
Zhang, Jing-Ru   I-390
Zhang, Jun   I-390, I-1035
Zhang, Lei   I-117, I-146
Zhang, Li-bin   II-900
Zhang, Lin   II-386
Zhang, Qiang   II-80
Zhang, Qiuyu   I-380
Zhang, Quanju   I-787
Zhang, Ru   II-910, II-929
Zhang, Shiwu   II-988
Zhang, Wei   I-967
Zhang, Xiao-Ping   II-742
Zhang, Ying   I-986
Zhang, Yingshan   I-828
Zhang, Yong   I-571
Zhang, Yousheng   II-538
Zhang, Yu-Tian   II-472
Zhang, Yuan-Yuan   II-978
Zhang, Yue   I-165
Zhang, Yufeng   II-346
Zhang, Yumin   I-697
Zhang, Zheng   II-71
Zhao, Fuqing   I-380

Zhao, Guoying   II-238
Zhao, Min   I-475
Zhao, Qianchuan   II-626
Zhao, Qiang   II-386
Zhao, Wencang   I-650
Zhao, Wenli   II-929
Zhao, Yun   I-126
Zheng, Da-Zhong   II-636, II-655
Zheng, Jiaju   I-767
Zheng, Kai   II-386
Zheng, Nanning   I-330
Zheng, Qinghua   I-485
Zheng, Sheng   I-10
Zhong, Huixiang   I-350
Zhong, Xiaopin   I-330

Zhou, Chunguang   II-119
Zhou, Qiuyong   I-1
Zhou, Wengang   II-119
Zhou, Yanhong   II-11
Zhu, Bin B.   II-771
Zhu, Daqi   I-581
Zhu, Hong-Song   II-588
Zhu, Jianming   II-376
Zhu, Xiangou   II-80
Zhu, Zhongliang   I-291
Zong, Xiaojun   I-524
Zou, An-Min   II-443
Zou, Hongxing   II-949
Zuo, Guoyu   II-939
Zuwairie, Ibrahim   I-562

Lecture Notes in Computer Science

For information about Vols. 1–3541

please contact your bookseller or Springer

Vol. 3659: J.R. Rao, B. Sunar (Eds.), Cryptographic Hardware and Embedded Systems – CHES 2005. XIV, 458 pages. 2005.

Vol. 3654: S. Jajodia, D. Wijesekera (Eds.), Data and Applications Security XIX. X, 353 pages. 2005.

Vol. 3653: M. Abadi, L.d. Alfaro (Eds.), CONCUR 2005 – Concurrency Theory. XIV, 578 pages. 2005.

Vol. 3649: W.M.P. van der Aalst, B. Benatallah, F. Casati, F. Curbera (Eds.), Business Process Management. XII, 472 pages. 2005.

Vol. 3645: D.-S. Huang, X.-P. Zhang, G.-B. Huang (Eds.), Advances in Intelligent Computing, Part II. XXVIII, 1013 pages. 2005.

Vol. 3644: D.-S. Huang, X.-P. Zhang, G.-B. Huang (Eds.), Advances in Intelligent Computing, Part I. XXVII, 1101 pages. 2005.

Vol. 3639: P. Godefroid (Ed.), Model Checking Software. XI, 289 pages. 2005.

Vol. 3638: A. Butz, B. Fisher, A. Krüger, P. Olivier (Eds.), Smart Graphics. XI, 269 pages. 2005.

Vol. 3636: M.J. Blesa, C. Blum, A. Roli, M. Sampels (Eds.), Hybrid Metaheuristics. XII, 155 pages. 2005.

Vol. 3634: L. Ong (Ed.), Computer Science Logic. XI, 567 pages. 2005.

Vol. 3633: C. Bauzer Medeiros, M. Egenhofer, E. Bertino (Eds.), Advances in Spatial and Temporal Databases. XIII, 433 pages. 2005.

Vol. 3632: R. Nieuwenhuis (Ed.), Automated Deduction – CADE-20. XIII, 459 pages. 2005. (Subseries LNAI).

Vol. 3627: C. Jacob, M.L. Pilat, P.J. Bentley, J. Timmis (Eds.), Artificial Immune Systems. XII, 500 pages. 2005.

Vol. 3626: B. Ganter, G. Stumme, R. Wille (Eds.), Formal Concept Analysis. X, 349 pages. 2005. (Subseries LNAI).

Vol. 3625: S. Kramer, B. Pfahringer (Eds.), Inductive Logic Programming. XIII, 427 pages. 2005. (Subseries LNAI).

Vol. 3624: C. Chekuri, K. Jansen, J.D.P. Rolim, L. Trevisan (Eds.), Approximation, Randomization and Combinatorial Optimization. XI, 495 pages. 2005.

Vol. 3623: M. Liśkiewicz, R. Reischuk (Eds.), Fundamentals of Computation Theory. XV, 576 pages. 2005.

Vol. 3621: V. Shoup (Ed.), Advances in Cryptology – CRYPTO 2005. XI, 568 pages. 2005.

Vol. 3620: H. Muñoz-Avila, F. Ricci (Eds.), Case-Based Reasoning Research and Development. XV, 654 pages. 2005. (Subseries LNAI).

Vol. 3619: X. Lu, W. Zhao (Eds.), Networking and Mobile Computing. XXIV, 1299 pages. 2005.

Vol. 3615: B. Ludäscher, L. Raschid (Eds.), Data Integration in the Life Sciences. XII, 344 pages. 2005. (Subseries LNBI).

Vol. 3614: L. Wang, Y. Jin (Eds.), Fuzzy Systems and Knowledge Discovery, Part II. XLI, 1314 pages. 2005. (Subseries LNAI).

Vol. 3613: L. Wang, Y. Jin (Eds.), Fuzzy Systems and Knowledge Discovery, Part I. XLI, 1334 pages. 2005. (Subseries LNAI).

Vol. 3612: L. Wang, K. Chen, Y. S. Ong (Eds.), Advances in Natural Computation, Part III. LXI, 1326 pages. 2005.

Vol. 3611: L. Wang, K. Chen, Y. S. Ong (Eds.), Advances in Natural Computation, Part II. LXI, 1292 pages. 2005.

Vol. 3610: L. Wang, K. Chen, Y. S. Ong (Eds.), Advances in Natural Computation, Part I. LXI, 1302 pages. 2005.

Vol. 3608: F. Dehne, A. López-Ortiz, J.-R. Sack (Eds.), Algorithms and Data Structures. XIV, 446 pages. 2005.

Vol. 3607: J.-D. Zucker, L. Saitta (Eds.), Abstraction, Reformulation and Approximation. XII, 376 pages. 2005. (Subseries LNAI).

Vol. 3606: V. Malyshkin (Ed.), Parallel Computing Technologies. XII, 470 pages. 2005.

Vol. 3603: J. Hurd, T. Melham (Eds.), Theorem Proving in Higher Order Logics. IX, 409 pages. 2005.

Vol. 3602: R. Eigenmann, Z. Li, S.P. Midkiff (Eds.), Languages and Compilers for High Performance Computing. IX, 486 pages. 2005.

Vol. 3599: U. Aßmann, M. Aksit, A. Rensink (Eds.), Model Driven Architecture. X, 235 pages. 2005.

Vol. 3598: H. Murakami, H. Nakashima, H. Tokuda, M. Yasumura, Ubiquitous Computing Systems. XIII, 275 pages. 2005.

Vol. 3597: S. Shimojo, S. Ichii, T.W. Ling, K.-H. Song (Eds.), Web and Communication Technologies and Internet-Related Social Issues - HSI 2005. XIX, 368 pages. 2005.

Vol. 3596: F. Dau, M.-L. Mugnier, G. Stumme (Eds.), Conceptual Structures: Common Semantics for Sharing Knowledge. XI, 467 pages. 2005. (Subseries LNAI).

Vol. 3595: L. Wang (Ed.), Computing and Combinatorics. XVI, 995 pages. 2005.

Vol. 3594: J.C. Setubal, S. Verjovski-Almeida (Eds.), Advances in Bioinformatics and Computational Biology. XIV, 258 pages. 2005. (Subseries LNBI).

Vol. 3593: V. Mařík, R. W. Brennan, M. Pěchouček (Eds.), Holonic and Multi-Agent Systems for Manufacturing. XI, 269 pages. 2005. (Subseries LNAI).

Vol. 3592: S. Katsikas, J. Lopez, G. Pernul (Eds.), Trust, Privacy and Security in Digital Business. XII, 332 pages. 2005.

Vol. 3591: M.A. Wimmer, R. Traunmüller, Å. Grönlund, K.V. Andersen (Eds.), Electronic Government. XIII, 317 pages. 2005.

Vol. 3590: K. Bauknecht, B. Pröll, H. Werthner (Eds.), E-Commerce and Web Technologies. XIV, 380 pages. 2005.

Vol. 3588: K.V. Andersen, J. Debenham, R. Wagner (Eds.), Database and Expert Systems Applications. XX, 955 pages. 2005.

Vol. 3587: P. Perner, A. Imiya (Eds.), Machine Learning and Data Mining in Pattern Recognition. XVII, 695 pages. 2005. (Subseries LNAI).

Vol. 3586: A.P. Black (Ed.), ECOOP 2005 - Object-Oriented Programming. XVII, 631 pages. 2005.

Vol. 3584: X. Li, S. Wang, Z.Y. Dong (Eds.), Advanced Data Mining and Applications. XIX, 835 pages. 2005. (Subseries LNAI).

Vol. 3583: R.W. H. Lau, Q. Li, R. Cheung, W. Liu (Eds.), Advances in Web-Based Learning – ICWL 2005. XIV, 420 pages. 2005.

Vol. 3582: J. Fitzgerald, I.J. Hayes, A. Tarlecki (Eds.), FM 2005: Formal Methods. XIV, 558 pages. 2005.

Vol. 3581: S. Miksch, J. Hunter, E. Keravnou (Eds.), Artificial Intelligence in Medicine. XVII, 547 pages. 2005. (Subseries LNAI).

Vol. 3580: L. Caires, G.F. Italiano, L. Monteiro, C. Palamidessi, M. Yung (Eds.), Automata, Languages and Programming. XXV, 1477 pages. 2005.

Vol. 3579: D. Lowe, M. Gaedke (Eds.), Web Engineering. XXII, 633 pages. 2005.

Vol. 3578: M. Gallagher, J. Hogan, F. Maire (Eds.), Intelligent Data Engineering and Automated Learning - IDEAL 2005. XVI, 599 pages. 2005.

Vol. 3577: R. Falcone, S. Barber, J. Sabater-Mir, M.P. Singh (Eds.), Trusting Agents for Trusting Electronic Societies. VIII, 235 pages. 2005. (Subseries LNAI).

Vol. 3576: K. Etessami, S.K. Rajamani (Eds.), Computer Aided Verification. XV, 564 pages. 2005.

Vol. 3575: S. Wermter, G. Palm, M. Elshaw (Eds.), Biomimetic Neural Learning for Intelligent Robots. IX, 383 pages. 2005. (Subseries LNAI).

Vol. 3574: C. Boyd, J.M. González Nieto (Eds.), Information Security and Privacy. XIII, 586 pages. 2005.

Vol. 3573: S. Etalle (Ed.), Logic Based Program Synthesis and Transformation. VIII, 279 pages. 2005.

Vol. 3572: C. De Felice, A. Restivo (Eds.), Developments in Language Theory. XI, 409 pages. 2005.

Vol. 3571: L. Godo (Ed.), Symbolic and Quantitative Approaches to Reasoning with Uncertainty. XVI, 1028 pages. 2005. (Subseries LNAI).

Vol. 3570: A. S. Patrick, M. Yung (Eds.), Financial Cryptography and Data Security. XII, 376 pages. 2005.

Vol. 3569: F. Bacchus, T. Walsh (Eds.), Theory and Applications of Satisfiability Testing. XII, 492 pages. 2005.

Vol. 3568: W.-K. Leow, M.S. Lew, T.-S. Chua, W.-Y. Ma, L. Chaisorn, E.M. Bakker (Eds.), Image and Video Retrieval. XVII, 672 pages. 2005.

Vol. 3567: M. Jackson, D. Nelson, S. Stirk (Eds.), Database: Enterprise, Skills and Innovation. XII, 185 pages. 2005.

Vol. 3566: J.-P. Banâtre, P. Fradet, J.-L. Giavitto, O. Michel (Eds.), Unconventional Programming Paradigms. XI, 367 pages. 2005.

Vol. 3565: G.E. Christensen, M. Sonka (Eds.), Information Processing in Medical Imaging. XXI, 777 pages. 2005.

Vol. 3564: N. Eisinger, J. Małuszyński (Eds.), Reasoning Web. IX, 319 pages. 2005.

Vol. 3562: J. Mira, J.R. Álvarez (Eds.), Artificial Intelligence and Knowledge Engineering Applications: A Bioinspired Approach, Part II. XXIV, 636 pages. 2005.

Vol. 3561: J. Mira, J.R. Álvarez (Eds.), Mechanisms, Symbols, and Models Underlying Cognition, Part I. XXIV, 532 pages. 2005.

Vol. 3560: V.K. Prasanna, S. Iyengar, P.G. Spirakis, M. Welsh (Eds.), Distributed Computing in Sensor Systems. XV, 423 pages. 2005.

Vol. 3559: P. Auer, R. Meir (Eds.), Learning Theory. XI, 692 pages. 2005. (Subseries LNAI).

Vol. 3558: V. Torra, Y. Narukawa, S. Miyamoto (Eds.), Modeling Decisions for Artificial Intelligence. XII, 470 pages. 2005. (Subseries LNAI).

Vol. 3557: H. Gilbert, H. Handschuh (Eds.), Fast Software Encryption. XI, 443 pages. 2005.

Vol. 3556: H. Baumeister, M. Marchesi, M. Holcombe (Eds.), Extreme Programming and Agile Processes in Software Engineering. XIV, 332 pages. 2005.

Vol. 3555: T. Vardanega, A.J. Wellings (Eds.), Reliable Software Technology – Ada-Europe 2005. XV, 273 pages. 2005.

Vol. 3554: A. Dey, B. Kokinov, D. Leake, R. Turner (Eds.), Modeling and Using Context. XIV, 572 pages. 2005. (Subseries LNAI).

Vol. 3553: T.D. Hämäläinen, A.D. Pimentel, J. Takala, S. Vassiliadis (Eds.), Embedded Computer Systems: Architectures, Modeling, and Simulation. XV, 476 pages. 2005.

Vol. 3552: H. de Meer, N. Bhatti (Eds.), Quality of Service – IWQoS 2005. XVIII, 400 pages. 2005.

Vol. 3551: T. Härder, W. Lehner (Eds.), Data Management in a Connected World. XIX, 371 pages. 2005.

Vol. 3548: K. Julisch, C. Kruegel (Eds.), Intrusion and Malware Detection and Vulnerability Assessment. X, 241 pages. 2005.

Vol. 3547: F. Bomarius, S. Komi-Sirviö (Eds.), Product Focused Software Process Improvement. XIII, 588 pages. 2005.

Vol. 3546: T. Kanade, A. Jain, N.K. Ratha (Eds.), Audio- and Video-Based Biometric Person Authentication. XX, 1134 pages. 2005.

Vol. 3544: T. Higashino (Ed.), Principles of Distributed Systems. XII, 460 pages. 2005.

Vol. 3543: L. Kutvonen, N. Alonistioti (Eds.), Distributed Applications and Interoperable Systems. XI, 235 pages. 2005.

Vol. 3542: H.H. Hoos, D.G. Mitchell (Eds.), Theory and Applications of Satisfiability Testing. XIII, 393 pages. 2005.